High Entropy Alloys

High Entropy Alloys

Innovations, Advances, and Applications

Edited by
T.S. Srivatsan and Manoj Gupta

CRC Press is an imprint of the
Taylor & Francis Group, an **informa** business

First edition published 2020
by CRC Press
6000 Broken Sound Parkway NW, Suite 300, Boca Raton, FL 33487-2742

and by CRC Press
2 Park Square, Milton Park, Abingdon, Oxon, OX14 4RN

Library of Congress Cataloging-in-Publication Data

Names: Srivatsan, T. S., editor. | Gupta, M. (Manoj), editor.
Title: High entropy alloys : innovations, advances, and applications / edited by T. S. Srivatsan and Manoj Gupta.
Description: Boca Raton : CRC Press, 2020. | Includes bibliographical references and index. | Summary: "The book gives a cohesive overview of innovations, advances in processing and characterization, and applications for high entropy alloys (HEAs) in performance-critical and non-performance-critical sectors. It covers manufacturing and processing, advanced characterization and analysis techniques and evaluation of mechanical and physical properties"--Provided by publisher.
Identifiers: LCCN 2019054139 (print) | LCCN 2019054140 (ebook) | ISBN 9780367356330 (hardback : acid-free paper) | ISBN 9780367374426 (ebook)
Subjects: LCSH: Alloys. | Order-disorder in alloys.
Classification: LCC TN690 .H575 2020 (print) | LCC TN690 (ebook) | DDC 669/.9--dc23
LC record available at https://lccn.loc.gov/2019054139
LC ebook record available at https://lccn.loc.gov/2019054140

ISBN: 978-0-367-35633-0 (hbk)
ISBN: 978-0-367-37442-6 (ebk)

Typeset in Times
by Deanta Global Publishing Services, Chennai, India

Contents

Preface

In the domain enveloping alloy development, consolidation and selection for the purpose of use in a spectrum of products and/or applications has traditionally relied on the existence of a solvent element to which various solute atoms are added for the purpose of either achieving or merely improving specified properties of the resultant alloy. This gradually led to the development and emergence of alloys that were either named or identified based on the major element in the alloy i.e., aluminum (Al)-base alloys, copper (Cu)-base alloys, iron (Fe)-base alloys, magnesium (Mg)-base alloys, and nickel (Ni)-base alloys. Sustained research and development efforts, particularly in the domain enveloping innovation induced Professor Brian Cantor and Professor Jien-Weh Yeh to not only investigate, but to also study, multicomponent solid solutions i.e., alloys having equal, or near-equal, molar ratios, during their investigation way back in 1981 and 1995. These two research scholars eventually published the outcome and findings of their research study in scientific journals in 2004. According to these two researchers, their "unique" alloys, in direct contrast to the traditional alloys, based on one or two principal elements, had one striking characteristic i.e., an unusually high entropy of mixing. These two researchers opted to name their new class of alloys "high entropy alloys" (HEAs). Through the years since first being found and their eventual emergence, these alloys have attracted considerable interest and attention among academicians, researchers and technologists spread through the domains of academic institutions, national research laboratories, and industries. Interestingly, much prior to the publication of the papers of these two researchers in 2004, Professor S. Ranganathan (associated with the Indian Institute of Science [Bangalore, India]) felt the importance of this new class of alloys and wrote about them in his classic paper "Alloyed Pleasures: Multimetallic Cocktails" in 2003. This paper has through the years been cited a few hundred times as the first publication on this class of alloys i.e., high entropy alloys. Professor Jien-Weh Yeh was the first to refer to these alloys as "high entropy alloys" (HEAs) since the configurational entropy of these alloys was expected to be high for a random solution state. Such a high entropy was expected to be conducive for the formation of simple solid solutions (crystalline or amorphous) rather than complex microstructures having many compounds. The emergence of these alloys did get interest, attention, and contributions from many researchers, as is evident from the fact that the last decade alone has seen several hundred papers being published on HEAs achieved through various elemental combinations. The significant outcomes from the research studies conducted to date are the following:

(i) The alloys do form simple solid solutions in most cases, and
(ii) The number of phases observed in these alloys is noticeably less than the maximum predicted from the use of Gibbs phase rule.

This bound volume, or book, makes an attempt at providing a cohesively complete and convincing overview of the following:

(i) A few recent innovations in the domain specific to high entropy alloys (HEAs).
(ii) Concurrent advances on aspects specific to processing and characterization and the resultant emerging developments.
(iii) Potential viable applications for these alloys for the purpose of selection and use in both performance-critical and non-performance-critical applications.

This can be considered to be both useful and essential at the prevailing moment in an attempt to not only record, but also present, the noticeable studies being made in the domain specific to the development of high entropy alloys (HEAs) and resultant technology-related applications.

In the introductory chapter, the contributing authors provide a careful, complete, and comprehensive overview of the numerous intricacies and facets specific to the high entropy alloys while concurrently providing a neat summary of the different approaches that could be taken with specific reference to alloy design. The authors also present and discuss key aspects of microstructural development for the single-phase alloy, multiphase alloy, and eutectic microstructure. In the process, the authors allude to the influence of materials and processes on the development of properties. They also list the challenges and potential for these alloys. They conclude their comprehensive write-up by listing all of the possible strategies for the further advancement of these alloys.

Overall, this book contains three sections. Each section, i.e., **Section A, Section B,** and **Section C** contains a few well laid-out technical chapters. To make every effort to meet with the needs and requirements of the different readers, each chapter has been written and presented by one or more authors to ensure that it offers a clean, clear, complete, and, importantly, a convincing presentation and discussion of the intricacies specific to their analysis, observations, and interpretations of the subject matter in a cohesively convincing and compelling manner.

In the first section of the book (**Section A**), the focus is on "**Innovations**" specific to the family of high entropy alloys (HEAs). This section has three chapters. Chapter 1 introduces the interested reader to aspects both pertinent and relevant to interstitial alloy structuring specific to the family of high entropy alloys. By summarizing the relationship between microstructure and properties, the authors predict compositional opportunities for designing new alloys having a unique combination of properties to offer. Chapter 2 provides an in-depth analysis, in a lucid and convincing manner, of the key aspects specific to the challenges in the domain of processing and resultant properties, which are almost always far superior to the conventionally developed alloys. The authors of this chapter focus their presentation on engineering the development of lightweight high entropy alloys (LWHEAs) and clearly present and briefly discuss a few of the developments in the domain specific to new alloy compositions and the numerous challenges they present, especially with reference to processing. The following chapter, Chapter 3, is devoted to a discussion of the challenges specific to processing of these alloys in the bulk form and the viable remedial solutions to safely overcome any and all of these challenges. The authors present and adequately discuss the relevant aspects that are key to both the conventional solidification route and powder metallurgy (P/M) route. Through the length of their chapter, the authors attempt to provide plausible remedies to mitigate the problems that arise in the use of different processing routes for bulk high entropy alloys (HEAs).

The **second section** of this book (**Section B**) the focus is on "**Advances**" and it consists of eight exhaustive and well laid out chapters. The first chapter in this section (Chapter 4) attempts to provide the interested reader with key highlights specific to the effect and/or influence of the addition of carbon on microstructural development and the resultant properties of the CoCuFeMnNi high entropy alloy. This chapter can safely be categorized as being a healthy refresher for the knowledgeable reader and learned engineer, while concurrently providing the novice and inquisitive learner with an introductory highlight into the specific role of element addition to an existing high entropy alloy. In the following chapter (Chapter 5), key aspects specific to the conjoint influence of processing and addition of the element silicon to the CoCrCuFeNiSi high entropy alloy are well presented and neatly discussed from both a scientific and engineering perspective. In the next chapter (Chapter 6), the theme for presentation and discussion is the corrosion and thermal behavior of the $AlCr_{1.5}CuNi_2FeCo_x$ high entropy alloys. The specific role of cobalt content in influencing the corrosion behavior and thermal response characteristics of the chosen high entropy alloy was well studied and the findings neatly presented and sufficiently discussed. In Chapter 7, the author presents his views, resulting from a comprehensive study, analysis, and the resultant interpretations to come up with a good understanding of the fracture resistance of the family of high entropy alloys (HEAs). In his exhaustive and comprehensive paper, this single author attempts to summarize well over a decade of relevant research investigations specific to studying and understanding the fracture behavior of the family of high entropy alloys (HEAs). Aspects specific to fracture characterization of these alloys, through the synergism of fractographic characteristics and fracture mechanisms at both the microscopic level and macroscopic level are well presented and neatly discussed. In the following chapter (Chapter 8), the contributing authors clearly present and thoroughly discuss key aspects specific to the thermodynamics of high entropy alloys (HEAs). In their chapter, the authors also attempt to provide a discussion on the application of thermodynamic principles to engineering high entropy alloy compositions. The required guidelines in synergism with phase diagram inspection and Gibbs energy composition plots are also well presented and discussed. In the following chapter (Chapter 9), key aspects specific to electrodeposition of high entropy alloy coatings are well presented and discussed with specific reference to their microstructure through the thickness of the coating and the role and/or influence of the coating on corrosion response or corrosion properties. The contributing authors clearly present and thoroughly discuss electrodeposition of the CrFeCoNiCu, AlCrFeCoNiCu, and MnCrFeCoNi high entropy alloy coating systems. In Chapter 10, the theme for presentation in concurrence with discussion of all the intricacies and aspects that must be considered in studying disordered high entropy alloys using X-ray diffraction is discussed using a truly scientific perspective. The formation of multicomponent disordered solid solutions from a careful study of the X-ray diffraction pattern is also well presented so as to convince both the "experienced" researcher and the new entrant or "learner" to this field of emerging alloys and materials. This chapter, based both on content and description, can safely be considered to be both educative and enriching from a standpoint of the analysis and rationalization of the findings. In the following chapter (Chapter 11), key aspects specific to microstructural development and any and all cracking-related noise that could arise during the development of high entropy alloys is well presented and aided with focused discussion.

The third section of this book (**Section C**) is devoted to aspects both related to and relevant to "**Applications**." In the opening chapter of this section (Chapter 12), the contributing authors elegantly present and discuss the numerous benefits that can and does arise in the selection and use of high entropy alloys (HEAs) for high-temperature thermoelectric applications. The authors present and discuss the basic principles behind the thermoelectric materials and the overall properties that are necessary to be controlled to achieve efficient thermoelectric materials. In the following chapter (Chapter 13), the contributing authors make a comprehensive review of microstructural effects on the fatigue behavior of various high entropy alloys as a follow-on to the fatigue properties of multiple high entropy alloys that have been investigated and reported in the published literature. The authors attempt to focus their review on studying and rationalizing the influence of intrinsic microstructural features and/or microstructural effects on fatigue response of these alloys. In the following chapter (Chapter 14), the contributing authors provide a neat and convincing review on the development of high entropy alloys (HEAs) having specific functional properties to offer. The authors note these functional properties to be the following: magnetic, magnetocaloric, hydrogen storage, corrosion resistance, thermoelectric, superconducting, irradiation, catalytic, and biomedical. A comparison of the functional properties of the high entropy alloys (HEAs) with the conventional alloys is also presented. In the following chapter (Chapter 15), the contributing authors provide a lucid and well-written overview of the few to many challenges of seeing in reality the true commercialization of the family of high entropy alloys (HEAs). They present and briefly discuss the many attributes the newly developed alloys must possess so as to be able to replace the existing alloys for the purpose of selection and use both in existing and emerging applications. They also try to list and discuss the key considerations the newly developed alloys must possess. They also include in their chapter a brief discussion of the future of these alloys from the standpoint of commercialization. In the following chapter (Chapter 16), the author provides a comprehensive review of the fatigue and fracture behavior of high entropy alloys. Besides providing a brief outline of the concepts that were developed and put to use for the purpose of evaluating and analyzing the fatigue and fracture behavior of these alloys, this author also spends a brief time presenting and discussing the deformation mechanisms, which are both required and operative for achieving an improvement in the fatigue and fracture behavior of these advanced alloys. The following chapter (Chapter 17) presented and equally well discussed key aspects specific to the welding of high entropy alloys (HEAs). Since welding does have the tendency to deteriorate the properties of the base material due to synergism of "local" changes in both composition and microstructure, there is a need to study the ability of these alloys to be receptive to welding. The contributing authors provide a review of the published works on the welding of these alloys. In the following chapter (Chapter 18), the contributing authors try to summarize the key results for the magnetic high entropy alloys. They repeatedly highlight the fact that by careful variation of the alloying elements, the magnetic, electrical, and mechanical properties of these alloys can be modified. In the next chapter (Chapter 19), the contributing author presents and discusses the role and significance of high entropy alloy fibers and specifically, their high mechanical strength and ductility. They emphasize the need for high strength fibers for use as heavy-duty fasteners in industrial applications. Due to manufacturing processes and size effect, the microstructure and resultant mechanical properties of the fibers are different from the bulk ingots. In Chapter 20, the contributing authors provide a neat overview of the key aspects available in the published literature specific to high entropy alloy films, and these include the following: (i) preparation methods, (ii) composition design, (iii) phase structures, and (iv) performance. They also mention the prospects of high entropy films for high-throughput experiments, while concurrently discussing the technologies that can be used for high-throughput preparation of the high entropy alloys or materials.

Overall, this archival monograph on aspects related to the family of high entropy alloys does provide a background that should enable an interested reader to not only understand but to also comprehend the immediate past, the prevailing present, and the possible future trends and/or approaches in the domain specific to the development and emergence of these alloys with an emphasis on innovations coupled with an applicability for its selection and use in a wide range of applications. Thus, based entirely on the contents included in this bound volume, it can very well serve as a single reference book or even as a textbook for the following:

(a) Students spanning seniors in the undergraduate program of study in the fields of materials science and engineering, mechanical engineering and chemical/manufacturing engineering.
(b) Fresh graduate students pursuing graduate degrees in materials science and engineering, mechanical engineering, and chemical/manufacturing engineering.
(c) Researchers spread through both research laboratories and industries striving to specialize and excel on aspects related to research on materials science and engineering and resultant product development.
(d) Engineers striving and seeking for novel and technically viable materials for selection and use in both performance-critical and non-performance-critical applications.

We certainly anticipate this bound volume to be of much interest to scientists, engineers, technologists, and even entrepreneurs.

Acknowledgments

The editors graciously and generously acknowledge the understanding and valued support by way of co-operation they received from the contributing authors of the different chapters contained in this bound volume. Efforts made by the contributing authors to both present and discuss the different topics included in this bound volume have certainly enabled enhancing both the scientific and technological content and this is greatly appreciated. The useful comments, corrections, recommendations, and suggestions made by the referees on the different chapters have certainly helped in enhancing the technical content and merit of the final version of each chapter included in this bound volume.

Our publisher, CRC Press, has been very supportive and patient through the entire process initiating from the moment of conception of this intellectual project. We extend an abundance of thanks, valued appreciation, and recurring gratitude to the editorial staff at CRC Press. Specifically, we mention of **Ms Allison Shatkin** (Senior Acquisitions Editor, Books (Materials, Chemical, & Petroleum Engineering), Taylor & Francis, Inc. [Florida]) and **Ms Camilla Michael** (Editorial Assistant – Engineering at CRC Press/Taylor & Francis, Inc.) for their sustained interest, involvement, attention, and almost always spontaneous assistance arising from an understanding of the situation coupled with their enthusiastic disposition toward extending help to both the editors and the contributing authors. This certainly helped to a large extent in ensuring the timely execution of the far too many to detail intricacies related to smooth completion of this bound volume from the moment of its approval up until compilation and publication. At moments of need, we the editors have found the presence of **Ms Camilla Michael** at CRC Press coupled with her energetic willingness and voluntary participation to be a strong pillar of support, almost always favorably disposed to extending help. Due acknowledgment both by way of recurring appreciation and resonating applause is reserved and extended to Mr Bryan Moloney (Project Manager) for his diligent, dedicated, and devoted services shadowed with salient dynamism on all aspects related and relevant to the compilation and presentation of the bound volume through the services of Deanta Global.

Most importantly and worthy of recording is that the timely compilation and publication of this bound volume would not have been possible without: (i) the understanding, cooperation, assistance, and importantly, the patience extended by several of the contributing authors, and (ii) the positive contributions of the peer-reviewers. One of the editors (**Dr T.S. Srivatsan**) would like to record in print his ceaseless, endless, and unbounded gratitude to **Dr K. Manigandan** (Associate Professor in the Department of Mechanical Engineering at **The University of Akron**) for his sustained willingness and enthusiastically driven efforts, stemming from a desire to extend timely and highly valued assistance by way of identifying, locating, and narrowing down the list of potential researchers and scholars to whom invitations were extended to for the purpose of submitting their precious contribution for inclusion in this bound volume.

EDITORS

Dr. T.S. Srivatsan, **F ASM, F AAAS, F ASME**
Professor (Emeritus)
Department of Mechanical Engineering
Auburn Science and Engineering Center
The University of Akron
Akron, Ohio 44325, USA
E-Mail: tsrivatsan@uakron.edu

Dr. Manoj Gupta
Associate Professor
Department of Mechanical Engineering
National University of Singapore [NUS]
EA-07-08, No. 9, Engineering Drive 1
Singapore 117576, Singapore
E-Mail: mpegm@nus.edu.sg

Editors

Dr. T.S. Srivatsan, Professor (Emeritus) of the Department of Mechanical Engineering at the University of Akron. He received his Bachelor of Engineering in Mechanical Engineering (BEng, 1980) from **Bangalore University** (Bangalore, India) and, subsequently, graduate degree(s) [**Master of Science** in Aerospace Engineering (MS 1981) and **Doctor of Philosophy** in Mechanical Engineering (PhD 1984)] from **Georgia Institute of Technology**. Dr. Srivatsan joined the faculty in the Department of Mechanical Engineering at the University of Akron in August 1987. Since joining, he has instructed undergraduate and graduate courses in the areas of (a) Advanced Materials and Manufacturing Processes, (b) Mechanical Behavior of Materials, (c) Fatigue of Engineering Materials and Structures, (d) Fracture Mechanics, (e) Introduction to Materials Science and Engineering, (f) Mechanical Measurements, (g) Design of Mechanical Systems, and (h) Mechanical Engineering Laboratory. His research areas currently span the fatigue and fracture behavior of advanced materials to include monolithic(s), intermetallic, nano-materials and metal-matrix composites; processing techniques for advanced materials and nanostructure materials; inter-relationship between processing and mechanical behavior; electron microscopy; failure analysis; and mechanical design. He has authored/edited/co-edited a total of **66 books** (Bound volumes: **61**; Research monographs: **5**) in areas cross-pollinating mechanical design; processing and fabrication of advanced materials; deformation, fatigue and fracture of ordered intermetallic materials; machining of composites; failure analysis; and technology of rapid solidification processing of materials. He serves as co-editor of *International Journal on Materials and Manufacturing Processes* and on the editorial advisory board of five other journals in the domain of Materials Science and Engineering. His research has enabled him to deliver over 240 technical presentations in national and international meetings and symposia; technical/professional societies; and research and educational institutions. He has authored and co-authored over **700-plus** technical publications in archival international journals (**365**), chapters in books (10), proceedings of national and international conferences (240), scholarly reviews of technical books (79), and technical/scientific reports (75). His RG score is **45** with a h-index of **52** and **Google Scholar citations of 9000**; ranking him among the **top 2 percent** of researchers in the world. In recognition of his efforts, contributions, and thir impact on furthering science, technology, and education he has been elected

(a) Fellow of of American Society for Materials, International (ASM Int.)
(b) Fellow of American Society of Mechanical Engineers (ASME)
(c) Fellow of American Association for the Advancement of Science (AAAS)

He has also been recognized as: (a) Outstanding Young Alumnus of Georgia Institute of Technology, (b) Outstanding Research Faculty, of the College of Engineering (University of Akron) and (c) Outstanding Research Faculty at the University of Akron.

He offers his knowledge in research services to the U.S. Government (U.S. Air Force and U.S. Navy), National Research Laboratories, and industries related to aerospace, automotive, power-generation, leisure-related products, and applied medical sciences.

Dr. Manoj Gupta, was a former Head of Materials Division of the Mechanical Engineering Department and Director-designate of Materials Science and Engineering Initiative at NUS, Singapore. He did his PhD in the University of California, Irvine (1992), and postdoctoral research at the University of Alberta, Canada (1992). In August 2017, he was highlighted among the Top 1% of Scientists in the World by the Universal Scientific Education and Research Network and among 2.5% among scientists as per ResearchGate. To his credit are: (i) the disintegrated melt deposition technique and (ii) the hybrid microwave sintering technique, an energy-efficient solid-state processing method to synthesize alloys/micro/nanocomposites. He has published over 535 peer reviewed journal papers and owns two US patents and one Trade Secret. His current *h*-index is 61, RG index is >47, and citations are greater than 14,000. He has also co-authored six books, published by John Wiley, Springer, and MRF – USA. He is Editor-in-Chief/Editor of 12 international peer-reviewed journals. In 2018, he was announced as World Academy Championship Winner in the area of Biomedical Sciences by the International Agency for Standards and Ratings. A multiple award winner, he actively collaborates with others in Japan, France, Saudi Arabia, Qatar, China, the United States, and India and spends time in those countries as a visiting researcher, professor, and chair professor.

Contributors

Ahmed Aliyu
Department of Materials Engineering
Indian Institute of Science, Bangalore
Bangalore, India

A.A. Bagdasaryan
Sumy State University
Sumy, Ukraine

Srinivasa Rao Bakshi
Department of Metallurgical and Materials Engineering
Indian Institute of Technology, Madras
Chennai, India

Kantesh Balani
Department of Materials Science and Engineering
Indian Institute of Technology, Kanpur
Kanpur, India

Gaurav Kumar Bansal
Materials Engineering Division
CSIR National Metallurgical Laboratory
Jamshedpur, India

I.K. Bhat
Manav Rachna University
Faridabad, India

Krishanu Biswas
Department of Materials Science and Engineering
Indian Institute of Technology, Kanpur
Kanpur, India

P. Chakravarthy
Indian Institute of Space Science and Technology
Valiamala, Trivandrum, India

Avanish Kumar Chandan
Materials Engineering Division
CSIR National Metallurgical Laboratory
Jamshedpur, India

Manoj Chopkar
Department of Metallurgical and Materials Engineering
National Institute of Technology, Raipur
Chhattisgarh, India

Sheng Guo
Chalmers University of Technology
Gothenburg, Sweden

Anshul Gupta
Department of Materials Science and Engineering
Indian Institute of Technology, Kanpur
Kanpur, India

Manoj Gupta
Department of Mechanical Engineering
National University of Singapore (NUS)
Singapore

N.P. Gurao
Department of Materials Science and Engineering
Indian Institute of Technology, Kanpur
Kanpur, India

K. Guruvidyathri
School of Engineering Sciences and Technology
University of Hyderabad
Hyderabad, India

Anirudha Karati
Department of Chemistry
Indian Institute of Technology, Madras
Chennai, India

Vikas Kukshal
Mechanical Engineering Department
National Institute of Technology
Uttarakhand, India

Anil Kumar
Department of Mechanical Engineering
Bhilai Institute of Technology
Chhattisgarh, India

Dong Yue Li
SKL for Advanced Metals and Materials
USTB
Beijing, China

Rui Xuan Li
SKL for Advanced Metals and Materials
USTB
Beijing, China

Weidong Li
Akron Innovation Center
The Goodyear Tire & Rubber Company
Akron, Ohio

and

Department of Materials Science and Engineering
University of Tennessee
Knoxville, Tennessee

K. Liu
Center for Friction Stir Processing
Department of Materials Science and Engineering
University of North Texas
Denton, Texas

Gopi Kishor Mandal
Materials Engineering Division
CSIR National Metallurgical Laboratory
Jamshedpur, India

K.V. Mani Krishna
Materials Science Division
Bhabha Atomic Research Center (BARC)
Mumbai, India

Joydev Manna
Department of Metallurgical and Materials Engineering
Indian Institute of Technology, Madras
Chennai, India

Sukriti Mantri
School of Materials Science and Engineering
University of New South Wales
Sydney, Australia

Rajesh K. Mishra
Department of Physics
School of Physical and Chemical Sciences
Central University of South Bihar
Gaya, India

R.S. Mishra
Center for Friction Stir Processing
Department of Materials Science and Engineering
Advanced Materials and Manufacturing Processes Institute
University of North Texas
Denton, Texas

Soumyaranajan Mishra
Department of Metallurgical and Materials Engineering
Indian Institute of Technology, Madras
Chennai, India

B.S. Murty
Indian Institute of Technology, Hyderabad
Hyderabad, Andhra Pradesh, India

V. Muthupandi
Advanced Materials Processing Laboratory
Department of Metallurgical and Materials Engineering
National Institute of Technology
Tiruchirappalli, Tamil Nadu, India

Rameshwari Naorem
Department of Physics
Indian Institute of Technology, Kanpur
Kanpur, India

S.V.S. Narayana Murthy
Material Characterization Division
Materials and Metallurgy Group
Vikram Sarabhai Space Centre
Trivandrum, India

P. Neelima
Indian Institute of Space Science and Technology
Valiamala, Trivandrum, India

S.S. Nene
Center for Friction Stir Processing
Department of Materials Science and Engineering
University of North Texas
Denton, Texas

Amar Patnaik
Mechanical Engineering Department
Malaviya National Institute of Technology
Jaipur, India

A.D. Pogrebnjak
Sumy State University
Sumy, Ukraine

Rahul Ravi
Department of Metallurgical and Materials Engineering
Indian Institute of Technology, Madras
Chennai, India

Kalyan Kumar Ray (Retired)
Department of Metallurgical and Materials Engineering
Indian Institute of Technology, Kharagpur
Khargapur, India

M.Y. Rekha
Department of Materials Engineering
Indian Institute of Science, Bangalore
Bangalore, India

Samrand Shafeie
Chalmers University of Technology
Gothenburg, Sweden

Rohit R. Shahi
Department of Physics
School of Physical and Chemical Sciences
Central University of South Bihar
Gaya, India

and

Department of Physics
MNNIT
Allahabad, India

Shivakant Shukla
Center for Friction Stir Processing
Department of Materials Science and Engineering
University of North Texas
Denton, Texas

K. Sivaprasad
Advanced Materials Processing Laboratory
Department of Metallurgical and Materials Engineering
National Institute of Technology
Tiruchirappalli, Tamil Nadu, India

Reshma Sonkusare
Department of Materials Science and Engineering
Indian Institute of Technology, Kanpur
Kanpur, India

R. Sokkalingam
Advanced Materials Processing Laboratory
Department of Metallurgical and Materials Engineering
National Institute of Technology
Tiruchirappalli, Tamil Nadu, India

Chandan Srivastava
Department of Materials Engineering
Indian Institute of Science, Bangalore
Bangalore, India

Vikas Chandra Srivastava
Materials Engineering Division
CSIR National Metallurgical Laboratory
Jamshedpur, India

T.S. Srivatsan
Department of Mechanical Engineering
Auburn Science and Engineering Center
The University of Akron
Akron, Ohio

Anandh Subramaniam
Department of Materials Science and Engineering
Indian Institute of Technology, Kanpur
Kanpur, India

Khin Sandar Tun
Department of Mechanical Engineering
National University of Singapore [NUS]
Singapore

Surekha Yadav
Department of Materials Science and Engineering
Indian Institute of Technology, Kanpur
Kanpur, India

and

Indian Institute of Technology, Madras
Chennai, India

Xue Hui Yan
SKL for Advanced Metals and Materials
USTB
Beijing, China

Yong Zhang
SKL for Advanced Metals and Materials
USTB
Beijing, China

High Entropy Alloys
An Overview on Current Developments

Gaurav Kumar Bansal, Avanish Kumar Chandan, Gopi Kishor Mandal, and Vikas Chandra Srivastava

CONTENTS

I.1 INTRODUCTION

High entropy alloys are one of the recent and promising discoveries in materials development, a pursuit that has revolutionized the thought process researchers were accustomed to when dealing with alloy development for specialized applications requiring unprecedented properties. The pioneering work of Cantor and co-workers [1] and Yeh and co-workers [4] on such alloys in the 2004 led to increasing interest in these alloys and has culminated in worldwide exploration efforts for new alloy systems with unseen properties. This entails finding new applications or replacement of existing materials. This helped the research community in identification of the feasible combination of elements which can yield suitable microstructures. The synthesis of simple microstructures from such complex concentrated alloys was like a challenge and was a matter of fascination among the researchers until it was realized that the feasibility of high entropy alloys would not be justifiable unless such high-cost alloys exhibited some extraordinary properties. Since high entropy alloy systems were new to the scientific community, the effectiveness of various alloying elements on the development of microstructures needed to be assessed carefully. In this context, several studies [1–16] were devoted to decipher the role of the constituent alloying elements on the microstructural evolution. This was done by varying one individual element at a time in the alloy, followed by a microstructural investigation. The study of variation in the alloying content individually helped in understanding the compositions which should be avoided or included for the accomplishment of a particular set of microstructures and subsequent improvement in properties.

The premise on which these alloys have been conceptualized is high configurational entropy, due to possible random mixing of constituent alloying elements in these multi-principal alloy systems, which reaches a maximum when the alloy is equiatomic [4, 17]. Therefore, high entropy alloys are defined as the alloys having at least five elements with configurational entropy of more than 1.5 R (R being the gas constant) [18, 19]. High mixing entropy of any random solid solution enhances its stability, particularly at high temperatures. However, later, it was recognized that many other thermodynamic and topological parameters conclusively affect the solid solution formation ability e.g., atomic radius, mixing enthalpy, valence electron concentration, electronegativity, etc. [20–26]. This led to a frantic search for new alloy compositions suitable for the formation of single-phase solid solutions. Using these parameters, a few new alloy systems were found and were experimentally explored. Recently, it was also advocated that it could be deceptive to consider only the entropy as a determining factor. Therefore, these alloys are also christened as multi-principal multicomponent alloys or compositional complex alloys, eliminating the concept of entropy, though keeping the contribution of entropy intact [13]. Considering the large number of elements available for alloying and the multitude of compositional combinations possible, the compositional space to be explored became extremely vast rendering conventional techniques of alloy design inadequate [27–29]. This led to the employment of many new approaches to efficiently explore the compositional space and find the most promising alloy systems for experimental trials. Each high entropy alloy engenders different microstructures and properties, varying with process conditions, whether structural or functional. Therefore, alloy phase diagrams and their predictions came to the rescue, and different tools, in combination with the phase diagram calculations, led to a paradigm shift in alloy design approaches.

Despite these theoretical alloy exploration activities, a large amount of experimental data on microstructures, mechanical properties, wear characteristics, oxidation, and corrosion behavior, etc. have been continuously generated during the last 15 years [10, 11, 16, 30–41]. However, most of these studies were concentrated on alloys prepared by vacuum-arc melting, which mostly led to solidification in non-equilibrium conditions. These experiments and the ensuing properties of materials made it possible to categorize these alloy systems in several high entropy alloy groups, e.g., structural, lightweight, high temperature, and precious metal, etc. [13]. The microstructural features containing body-centered cubic, face-centered cubic, a combination of body-centered cubic and face-entered cubic, eutectics, and precipitation in metastable phases have been recognized and explored. The face-centered cubic solid-solution phases have invariably shown high ductility but low strength, whereas body-centered cubic solid solutions phases have been characterized with high strength but low ductility. The face-centered cubic phases, or a combination of face-centered cubic and body-centered cubic phases have been shown to also engender superplastic behavior. However, only a few alloy systems have been characterized in terms of mechanical response under tensile loading. The eutectic alloy systems are one of the new entrants in high entropy alloys, with great promise for a superior combination of strength and ductility [5, 11, 42]. However, only a few such systems have been explored as yet. It has also been proposed in many of the review articles that attempts must be made to explore the possibility of precipitation strengthening in high entropy alloys so as to make it possible to achieve a desired blend of properties by tailoring the content, size, and distribution of precipitates using appropriate process conditions. In addition to structural properties, high entropy alloys have been observed to possess exemplary magnetic and electrical properties [43], which have not been explored extensively due to the fact that the current knowledge on the interaction of various alloying elements in near-equiatomic compositions is limited.

It is obvious from the above that the area of high entropy or multicomponent multi-principal or compositional complex alloys has been under intense investigation by different research groups addressing thermodynamic aspects of high entropy alloys, accelerated exploration of alloys, generating experimental data, and predicting possible microstructures and properties of alloys. However, all these areas suffer from intrinsic limitations, which make it difficult to concretize and single out a tool to find optimized alloy systems for a given application. The thermodynamic basis of exploration suffers from non-availability of reliable experimental databases, which leads to inaccuracy or uncertainty in the predictions of phase constitution and microstructures. The tools available for predictions of properties are still in their nascent stages and require far-reaching efforts to incorporate different microstructural features. The experimental exploration of various alloys has been restricted by the cost and time as well as the absence of concrete alloy design strategies. In view of the above, this chapter is an attempt to revisit the developments in the sphere of high entropy alloys e.g., the thermodynamics involved, alloy design approaches being employed, microstructure and mechanical properties obtained for various alloy systems, and the challenges and potential of these alloys. The chapter brings out the latest developments in high entropy alloys and the possible strategies for the further advancement of these alloys.

1.2 THERMODYNAMIC BASIS

A search within the vast composition space in high entropy alloys results in a varied range of phase constitution. One of the main focus areas of high entropy alloys studied earlier was to find a single-phase solid solution by controlling the configurational entropy. A single-phase solid solution alloy generally provides good strength retaining the ductility. Apart from a single-phase solid solution, multi-principal element alloys may form other phases, including intermetallic compounds, precipitates, and amorphous phases. The presence of these phases, particularly intermetallics, in the alloys may deteriorate properties by embrittlement. However, recent works suggest that an excellent balance of structural properties of some of these alloy systems can be obtained by tailoring the microstructure and controlling the volume fraction, size, distribution, and morphology of intermetallics and precipitates [13]. It is, therefore, necessary to have prior knowledge of various phases, including solid solutions as well as intermetallics and precipitates, which are expected to be present in these alloy systems. An accurate thermodynamic prediction of various stable phases present in an alloy system is a prerequisite for further experimental exploration. However, accurate prediction of phase diagrams for these alloy systems is difficult due to the lack of experimental databases. The proposed phase formation rules based on limited studies needs to be revised thoroughly for precise prediction of the phase diagrams of high entropy alloys [44]. In this section, the current status pertaining to the thermodynamic basis of phase formation in high entropy alloys has been discussed.

From the fundamental of phase transformation, a negative change in Gibbs free energy ($\Delta G = G_{Product} - G_{parent} =$ driving force) of parent and product phases results in the formation of thermodynamically stable structures, which grow in the existing phase until the whole of the parent phase is transformed. This driving force, i.e., negative change in Gibbs free energy, is necessary for the spontaneous phase transformation. However, it is not a sufficient requirement as the progress of transformation is generally decided by the kinetic factors also. The Gibbs free energy change of parent and product phases includes the contribution from change in enthalpy as well as entropy. In multicomponent high entropy alloys, an accurate estimation of ΔG is extremely complicated due to the involvement of many variables, even at a fixed temperature and composition. For multicomponent systems, Gibbs free energy of mixing (ΔG_{mix}) can be expressed as:

$$\Delta G_{mix} = \Delta H_{mix} - T\Delta S_{mix} \quad \text{(I.1)}$$

where, ΔH_{mix} is the enthalpy change, ΔS_{mix} is the entropy change, and T is the temperature.

The stability of random solid-solution phase in high entropy alloys is mainly attributed to their high configurational entropy [4, 45]. In the early studies on high entropy alloys, it was assumed that these alloys behave like an ideal solid solution, in which mixing enthalpies and non-configurational entropies of mixing are expected to be negligible. Under these conditions, the change in Gibbs free energy due to mixing is mainly contributed by the configurational entropy [45, 46], as

$$\Delta G_{mix} = -T\Delta S_{mix} = RT\sum c_i \ln(c_i) \quad \text{(I.2)}$$

where R is the ideal gas constant and c_i is the concentration of ith element in the mixture.

Equation (I.2), however, is applicable for $T\Delta S_{mix} >> \Delta H_{mix}$ [4, 47]. In general, a solid solution at high temperature or solution in the liquid state has sufficiently high thermal energy to form a randomized solution. Thus, the expression for the change in configurational entropy, based on the equation $\Delta S_{mix} = R\ln(n)$ (where n is the number of equiatomic elements in the mixture), is expected to overestimate the real mixing entropy of an alloy that can be valid only at high temperature [46].

Based on the above discussion, it can be concluded that multi-principal multicomponent alloys having single random solid-solution phase, due to high configurational entropy of mixing, can be defined as high entropy alloys [46]. Figure I.1 shows the variation of estimated mixing configurational entropy of alloys having an ideal random solid solution as a function of atomic percentage aluminum with an increase in the number of constituent elements. The figure clearly illustrates that an increase in the number of elements in the alloy results in an increase in configurational entropy and the magnitude of configurational entropy becomes maximum for equiatomic concentrations. The figure also reveals that the estimated configurational entropies, based on an ideal random solid solution of equiatomic binary and five element alloys, are 5.76 and 13.37 J mol^{-1} K^{-1}, respectively. Yeh and co-workers [4] stated that the configurational entropy of mixing in the alloy containing a minimum of five constituent elements, with atomic concentrations in the range of 5–35%, is generally sufficient to stabilize a single-phase solid solution by minimizing the Gibbs free energy via counterbalance of the enthalpy of mixing [46]. However, there may be a significant difference between the configurational entropy of mixing in the real alloy system and the ideal entropy of mixing, particularly at low temperatures [48]. In real alloys, the formation of an ideal random solid solution is generally not possible due to interatomic interactions, which result in chemical ordering. In as-cast high

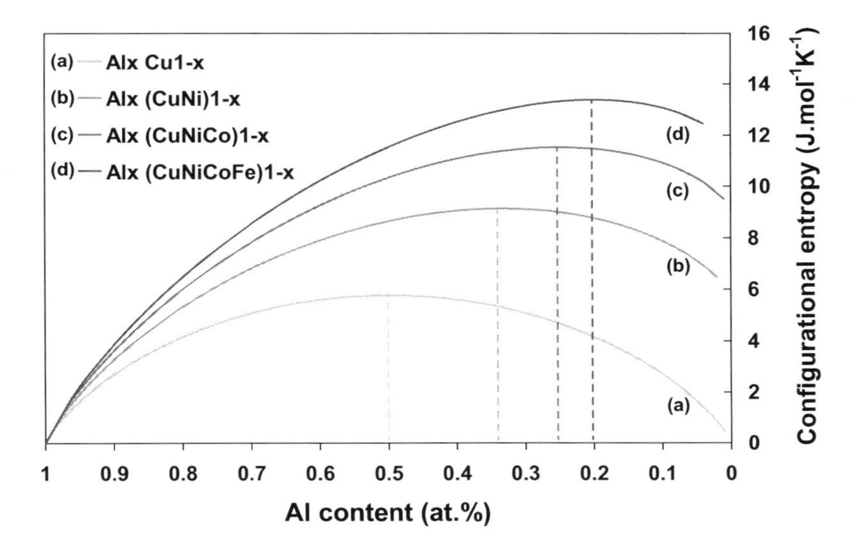

FIGURE I.1 Variation of estimated configurational entropy of mixing of alloys, having an ideal random solid solution, as a function of atomic percent aluminum.

entropy alloys, apart from solid solutions, the presence of intermetallic compounds and precipitates are also observed at room temperature. The presence of a number of phases in the as-cast high entropy alloys indicates that the approach based on only ΔS_{mix} parameters is unsuccessful in predicting single-phase solid solution-forming ability in high entropy alloys. Therefore, attempts were made to assess the phase stability based on semi-empirical approaches considering the synergistic effect of many variables including enthalpy of mixing (ΔH_{mix}), atomic size difference (δ), valence electron concentration, electronegativity, etc. These approaches mostly follow the Hume–Rothery rules for the formation of solid solution [49]. The value of ΔH_{mix} can be calculated using mixing enthalpy of equiatomic binary alloys, as below [20, 25, 47, 50]:

$$\Delta H_{mix} = \sum 4\Delta H_{ij}^{mix} c_i c_j \tag{I.3}$$

where ΔH_{ij}^{mix} is the enthalpy of mixing of the equiatomic liquid binary alloy between the ith and j^{th} elements.

The thermodynamic consideration of phase selection among different phases is determined based on the competition between ΔH_{mix} and $T\Delta S_{mix}$ [25]. Yang and Zhang [44] proposed the dimensionless parameter Ω, by considering the contribution both from ΔH_{mix} and ΔS_{mix} for the design of high entropy alloys, as:

$$\Omega = T_m \Delta S_{mix} / |\Delta H_{mix}| \tag{I.4}$$

where T_m is the weighted average melting point of the alloy system.

Based on Equation (I.4), it can be stated that the probability of solid solution-forming ability of high entropy alloys will be more for the larger value of the dimensionless parameter Ω. However, the values of Ω for different phases in high entropy alloys witness significant overlap [47]. It is to be noted that atomic size difference has not been incorporated for the estimation of the parameter Ω [44]. Although the dimensionless parameter Ω enhances the understanding of the formation of a single-phase random solid solution, this parameter was also unsuccessful as the sole descriptor of the solid solution-forming ability in high entropy alloys, similar to configurational entropy of mixing. Therefore, for the phase selection in high entropy alloys, this parameter (Ω) can be used along with other relevant variables [47].

The geometry effect is identified as another important variable determining the formation of a random solid solution-phase in high entropy alloys due to the atomic size mismatch among the constituent elements [25]. The atomic size difference (δ) can be expressed as:

$$\delta = \sqrt{\sum c_i \left(1 - \frac{r_i}{\sum c_j r_j}\right)^2} \tag{I.5}$$

where, r_i and r_j are the atomic radius for the ith and j^{th} element, respectively.

The combined effect of the two parameters (namely, ΔH_{mix} and δ), proposed by Zhang and co-workers [20], provided a better explanation for the solid solution-forming ability in high entropy alloys. It is observed that the tendency for the formation of a single-phase solid solution increases with low values of enthalpy of mixing (generally in the range of −15 to +5 kJ/mol) and small atomic size difference ($0 < \delta < 5$) [47, 51]. With an increase in mixing enthalpy and atomic size mismatch, the presence of different phases in the high entropy alloy microstructure is obtained. The alloys with a favorable negative value of mixing enthalpy are likely to form intermetallic compounds. The combined parameters ΔH_{mix} and δ provide satisfactory requirements for the formation of a single-phase solid solution; however, these two parameters are still insufficient for true predictions.

The inconsistency between experimental and theoretical investigations, based on the above parametric approach, is also due to the prediction of configurational entropy of mixing (ΔS_{mix}) based on a random ideal solution of alloying elements having an identical atomic size [47]. However, in real solutions, constituent elements may have significantly different atomic sizes and an increased packing density with a decrease in temperature [24, 47, 52]. The mixing entropy of an alloy depends not only on chemical composition but also on other parameters, including atomic configuration, atomic mismatch, etc. [13, 23, 47, 51]. Therefore, total configurational entropy (S_T) of a real alloy system is considerably lower than the configurational entropy of mixing (ΔS_{mix}) of a random ideal solution, particularly at low temperatures. The total configurational entropy (S_T) for a real solution can be represented as follows [53]:

$$S_T(c_i, r_i, \xi) = S_C(c_i) - \left|S_E(c_i, r_i, \xi)\right| \tag{I.6}$$

where S_C is the configurational entropy of mixing of a random ideal solution.

The functional form of excessive mixing entropy (S_E) is not straightforward, and it is generally a function of many parameters including composition (c_i), atomic size (r_i), atomic packing density (ξ), etc. [47, 54]. In order to predict the solid solution-forming ability in new high entropy alloys, another parameter (ϕ) was proposed by Ye and co-workers [23, 24] as per Equation (I.7), which considers the excessive mixing entropy (S_E). When the value of ϕ is greater than a critical value, high entropy alloys form solid solutions. However, empirically estimated critical value of ϕ depends on as-cast and heat treatment conditions of high entropy alloys as well.

$$\phi = (S_C - \left|\Delta H_{mix}\right|/T_m)/\left|S_E\right| \tag{I.7}$$

It is to be noted that enthalpy of mixing (ΔH_{mix}) of the multicomponent alloys is calculated as the sum of the formation enthalpies of the respective equiatomic liquid binary alloys, based on the Miedema model [55], where the Gibbs energies of the formation of other phases are not considered. One of the major shortcomings in these approaches is that the summation of formation enthalpies of the respective binary alloys may cancel each other out, and that may result in the estimated formation enthalpy of the multicomponent alloy in an optimal range for the formation of solid solution. For example, one set of binary alloys may have a large negative enthalpy of formation, which encourages precipitation of intermetallic compounds, whereas another binary set may have a large positive value of enthalpy, which encourages phase separation [49]. The summation of these two values of enthalpy will cancel each other out, which results in small values of enthalpy of mixing for the alloy indicative of the formation of a solid solution. Although these approaches are useful to get an idea about the single solid-solution phase-formation ability in multicomponent alloys, more reliable predictive approaches are required, as these parametric approaches have several limitations.

In order to study the stability of high entropy alloys, the Gibbs energy function of all the relevant phases needs to be established. In a multi-principal alloy system, the phases can be random and/or ordered solutions as well as intermetallic compounds. In a real alloy system, several major parameters including formation enthalpy and configurational entropy of all the relevant phases play a significant role in phase selection. In general, it is assumed that intermetallic phases have site occupancy of constituent elements, which are fixed and deterministic. However, intermetallic phases present in high entropy alloys can have various quasi-identical atoms of different elements in each sublattice. In each sublattice, atoms of different elements can also have random or real mixing. Therefore, the configurational entropy of ordered intermetallic phases in high entropy alloys can have a significant contribution, particularly if the number of elements is more than the number of sublattices in the intermetallic compound [13]. Therefore, it is always a great challenge to develop a simplified model with a specific focus on selected parameters for phase selections in high entropy alloys.

The CALPHAD (computer calculation of phase diagrams) approach is being used as an alternative for the prediction of the amounts and compositions of equilibrium phase/s in the high entropy alloys based on the minimization of total Gibbs energy of the system with respect to independent variables subjected to material balance, electroneutrality, and phase rule constraints. The sum of the Gibbs energies of the relevant phases including pure components, and random and ordered solutions, as well as intermetallic compounds is defined as the total Gibbs energy of the entire system. In the CALPHAD approach, the Gibbs free energy function of each phase (stable and metastable phases) is modeled with respect to composition and temperature at a given pressure. By neglecting other (physical, magnetic, etc.) contributions, the Gibbs energy of a relevant phase (ϕ) can be expressed as follows [13, 47, 56]:

$$G^{\phi} = \sum \left(G^{ref} + G^{ideal} + G^{E} \right) \tag{I.8}$$

where G^{ref} is Gibbs energy contribution from the pure elemental phase constituents, G^{ideal} is the ideal Gibbs energy of mixing, and G^{E} is the excess Gibbs energy term.

Neglecting the enthalpy of mixing in ideal solution, G^{ideal} in solution phase can be expressed based on the ideal configuration entropy of mixing. For a multicomponent system, excess energy (G^{E}) for solution phases can be calculated from the excess properties of binary and ternary alloys as follows [56]:

$$G^{E} = \sum_{i} \sum_{j>i} c_i c_j L_{ij} + \sum_{i} \sum_{j>i} \sum_{k>j} c_i c_j c_k L_{ijk} \tag{I.9}$$

where L_{ij} and L_{ijk} are the binary and ternary interaction parameters, which can be determined from the experimental data of respective binary and ternary systems.

The presence of ordered solution, intermetallic, and Laves phases in high entropy alloys are modeled based on sublattice formulation by considering the site occupancy of constituent elements [56]. The expression for Gibbs energy [57, 58] for the coexistence of ordered and disordered phases in the high entropy alloys is established based on the partitioning model. The non-disorder partitioning model is also used to describe the Gibbs energy for sigma (σ) and mu (μ) phases in high entropy alloys. A detailed formulation of these Gibbs energy terms based on the CALPHAD approach can be found elsewhere [56, 57].

The minimization of total Gibbs energy, in a multicomponent alloy system, results in the formation of a single phase, or a number of phases, which are thermodynamically stable. In the CALPHAD approach, databases for conventional alloys are generally developed by evaluating the binary and ternary interaction parameters based on vast experimental data. In principle, all the binary and ternary systems pertaining to the multicomponent alloy system need to be assessed for the creation of the high entropy alloy database. However, there is an exponential increase in the number of possible ternary alloy systems with an increase in the number of elements, which imposes a great challenge in database creation [59]. Moreover, only a limited number of elements are considered in the high entropy alloy database. The estimated amount and composition of phases in high entropy alloys, based on the CALPHAD approach, are generally well in agreement with the available experimental data for a wide temperature range. However, some of the CALPHAD predictions are not in agreement with the experimentally obtained as-cast microstructure of high entropy alloys. The inconsistency in the validation may be due to database limitation or metastable as-cast microstructure resulting from sluggish diffusion [60]. Apart from thermodynamics, several kinetic issues related to phase selection in the high entropy alloys need to be addressed. Most of the reported high entropy alloys are metastable in as-cast condition at room temperature [48]. The experimental evidence reveals the higher activation barrier for the diffusion of elements in high entropy alloys in comparison to conventional alloys [49, 61]. As a consequence, nucleation and growth for the formation of new phases will be slower. Therefore, sufficient measures need to be adopted to obtain the equilibrium structure experimentally in order to compare the CALPHAD results with the experiments.

1.3 APPROACHES FOR ALLOY DESIGN

As discussed above, the major emphasis on high entropy alloys has mostly been concentrated upon achieving single-phase alloy systems i.e., disordered solid solution. This favors their applications in low load-bearing structural parts and also circumvents additional complications related to strict microstructural control e.g., grain size and volume fraction, size, and distribution of second phase [60]. Looking at key concepts of alloy design of high entropy alloys, it seems that high configuration entropy above 1.5 R may lead to the formation of a single-phase solid solution. However, many of the alloy systems studied by researchers do not follow this rule perfectly, and large deviations are witnessed. The multiphase nature of these alloys indicates that high configurational entropy is not the only defining criterion for single-phase random solid solution formation [60, 62]. There are mainly two single-phase structures that have been shown to appear in a range of high entropy alloys i.e., face-centered cubic and body-centered cubic structures. The face-centered cubic structures generally give rise to low strength, but high ductility, whereas body-centered cubic structures are characterized by high strength, but ductility is highly compromised [63]. A single-phase high entropy alloy solid solution need to be understood in terms of deformation mechanisms, which are governed by the alloy composition, making it possible to predict their mechanical properties. Strengthening of single-phase high entropy alloy solid solutions has already been theorized by Varenne and co-workers [82], where experimental and predicted yield strengths matched successfully for the CoCrFeMnNi face-centered cubic solid solution. This theory maintained that the strength is not directly related to the number of elements in high entropy alloy and is not maximized for equiatomic compositions. In addition, it was concluded in this study that local structural disorder generates an additional contribution to the strength. However, Pickering and Jones [65] found that the most studied CoCrFeMnNi alloy system forms precipitates in a face-centered cubic matrix, after a long duration of annealing at intermediate temperatures. This is a clear indication of a metastable state at room temperature and small-scale phase separation, if any, goes undetected by conventional characterization techniques [13]. As most of the studies on high entropy alloys employed the vacuum-arc melting process for melting and solidification, the fast cooling of the alloys may lead to non-equilibrium solidification and thus does not show the possible phase separation at room temperature.

Despite these limitations and fundamental questions on high entropy alloys, the structural and functional properties of several alloys, including single- as well as multiphase alloys, have been found to be promising when compared with conventional alloys systems [62, 66–69]. Today, conscious efforts are being made to exploit the metastability of high entropy alloys and formation of second phases or nano-precipitates, in a high-strength single-phase matrix, to tailor the physical and mechanical properties of materials [67, 70–72]. For example, incorporation of aluminum and titanium in CoCrFeNi leads to precipitation of nano-sized coherent phase in a face-centered cubic CoCrFeNi matrix and thermomechanical processing of this alloy engenders yield strength of 1 GPa with 17% elongation [179]. These multi-principal alloy systems can serve different applications requiring high strength and ductility, wear resistance, high-temperature applications, corrosion resistance, superconductivity, cryogenic applications, and so on [18, 68]. Therefore, it is of paramount importance to explore application-driven compositions and process routes for high entropy alloys and understand the effect of different elements and their content on the microstructural features, phase evolution, second-phase content, and properties. However, the vast expanse of possible compositions of high entropy alloys poses difficulties in arriving at suitable and optimum alloy compositions for a given application or intended microstructural content. The experimental exploration of these alloys can be a time-consuming and a costly affair. A consideration of the structural alloys being commercially used today, for example, different steel grades, superalloys, aluminum alloys, and so on, suggests that the average time span in the development of an altogether different alloy system requires around 20 years. Despite the well-established concept of high entropy alloy, only seven new alloy families have been proposed in the last 15 years [69]. The enormous possibilities available with high entropy alloys can reduce the developmental time-span, provided a coordinated effort is made to screen the alloys for their properties and shorten the window for experimental validation. Therefore, researchers have made attempts to adopt different alloy design approaches for arriving at suitable compositions of alloys, particularly single-phase structures at room temperature or multiphase microstructures for given applications requiring specific properties [27, 67]. These approaches include high-throughput computational materials design [74], accelerated screening of alloys through mathematical modeling and phase diagram calculations [29], parametric estimation [25], combinatorial synthesis [191], and machine learning [76]. In the following sections, various approaches for alloy screening and selection have been discussed in detail.

I.3.1 Parametric Approach

With the advent of high entropy alloys, the parameters affecting the solid solution formation ability were first considered and attempts were made to find out a correlation among the factors, such as mixing enthalpy (ΔH_{mix}), mixing entropy (ΔS_{mix}), and atomic size mismatch (δ) with the solid solution-forming ability of an alloy system [20, 22], as already discussed in Section I.2. These parameters were calculated for a range of alloy systems reported in the literature, and the same was tested for identification of the range of parameters that can lead to a single-phase solid solution high entropy alloy. These parametric studies mainly led to a very basic alloy screening strategy reducing the alloy composition space window for possible experimental design [21, 23, 24]. Guo and Liu [77] found that the formation of a solid solution requires: $-22 \leq \Delta H_{mix} \leq 7$ kJ/mol, $0 \leq \delta \leq 8.5$, and $11 \leq \Delta S_{mix} \leq 19.5$ J/(K·mol). The atomic size mismatch is identified as one of the most critical parameters for the formation of a disordered solid solution [77]. However, the crystal structure of the solid-solution phase is difficult to predict based on the above-mentioned parameters [25, 47], although a study carried out by Tripathy and co-workers [78] showed that $\delta < 5$ favored the formation of a face-centered cubic phase, whereas $\delta > 5.4$ generally led to the stabilization of body-centered cubic phase. In order to obtain better prediction capability, additional parameters such as electronegativity ($\Delta\chi$), valence electron concentration, melting temperature (T_m), e/a ratio, and entropy of fusion [20, 25, 77–80] were also included in these parametric approaches. It has been demonstrated that valence electron concentration is a critical parameter to predict the crystal structure of high entropy alloys [47, 78, 81]. A higher value of valence electron concentration (≥ 8) tends to stabilize the face-centered cubic phase, whereas lower values (<6.87) engender the body-centered cubic phase. Tian and co-workers [80] also studied the valence electron concentration effect and found that an average value of valence electron concentration in the range of 4.33–7.55 forms body-centered cubic and 7.80–9.50 form face-centered cubic type solid solutions. It is also reported that electronegativity difference is a useful parameter to get an idea about elemental segregation in the alloy [25, 47]. These attempts at prediction through parametric and topological properties did enhance the understanding of the effect of these parameters on solid solution formation tendency. However, this did not lead to widespread acceptance, as the prediction capabilities of these techniques were limited by their parametric values only.

1.3.2 Computational Accelerated Alloy Screening

The realization of the limitations of parametric estimation led researchers to employ phase diagram calculation tools to predict the various equilibrium phases in these multi-principal component high entropy alloys and adopt rapid screening of possible alloy systems for single-phase solid solution. Senkov and co-workers [29] investigated a palette of 26 elements for structural high entropy alloys using the CALPHAD approach. They considered equiatomic alloys with combinations of 3, 4, 5, and 6 elements and demonstrated that the number of single-phase solid solution alloys at 600°C decreased from thirty-one for equiatomic ternary alloy systems to zero for equiatomic alloys with six elements, whereas the number of alloy systems consisting of combinations of solid solution and intermetallic compounds increased with an increase in the number of elements in the alloy. This study indicated that a search for single-phase high entropy alloy solid solutions in a vast world of possible high entropy alloy compositions will not accrue many benefits. This approach is required to be combined with the tools for the prediction of solid-solution strengthening [64]. The ability to predict solid solution formation along with their properties will be an advantage in designing alloys for a given application. Varvenne and co-workers [82] and Coury and co-workers [83] have already demonstrated a good correlation between experimental and theoretical mechanical properties of a high entropy alloy consisting of a face-centered cubic solid solution. Therefore, research should not only be limited to finding smaller composition window of possible alloys, but also to screen the composition in the given composition space for the best properties. Raghavan and co-workers [21] concluded in their study that the CALPHAD framework predicts body-centered cubic phases compositions better than that of face-centered cubic phases, which they attributed to the fact that the more open body-centered cubic structure possesses the greater presence of kinetic effects. They also observed that the formation tendency of the body-centered cubic phase is favored when the atomic size difference is large. Similar observations were made by Kube and co-workers [84] when they scanned 2,478 alloys using combinatorial synthesis route of co-sputtering. However, the reliable prediction of phases using CALPHAD requires thermodynamic databases for high entropy alloys which are far from complete [13, 85, 86]. This may lead to erroneous predictions of phase stability, which will affect the predicted property data also.

In this context, *ab initio* calculations have also been attempted to achieve single-phase solid solution compositions, for meaningful predictions from the accelerated alloy screening. Troparevsky and co-workers [49] proposed to predict the formation of a single solid solution using the high-throughput computation [74, 87–89], using density functional theory, of the formation enthalpies of binary alloys. They constructed an enthalpy matrix of all possible combinations of the constituent binary elements by considering the lowest energy level of each compound. Based on the lowest formation enthalpy criteria, the most likely combination of alloying elements was found out for a single-phase solid solution formation. However, this approach necessitates finding out the lowest enthalpies of formation for all the possible combinations of constituent elements including binary, ternary, quaternary, etc. Moreover, the first principles-based approaches are difficult to handle, particularly for multicomponent systems, as these approaches are computationally intensive and not suitable for the accelerated screening of alloys for the design of new high entropy alloys.

1.3.3 High-Throughput Alloy Design

As discussed above, attempts are being made to screen the vast composition space of high entropy alloys to arrive at alloy compositions that give rise to single-phase solid solutions. Similarly, approaches have been proposed to successfully predict the solid solution hardening in high entropy alloys [64, 83]. These attempts were mainly concentrated on equiatomic alloy systems. However, there are ample possibilities of better compositions with single-phase stability in non-equiatomic compositions also. In addition, the observations that multiphase high entropy alloys can give a tremendous possibility to tailor the material properties have gained considerable attention from researchers. If non-equiatomic compositions are also considered for the design of alloy systems, the number of possible alloy compositions will increase severalfold. Keeping this vast space of compositions in view, it is a difficult proposition to decide upon the composition for a given application. Grosse and co-workers [67] and Miracle and co-workers [90] have categorized this compositional space for structural applications considering the effect of principal elements on properties such as Young's modulus, tensile yield strengths, density, and temperature. These categories are the 3d transition metal, refractory metal, light metal alloys, brasses, and bronzes. Despite this classification, the compositional space for a given category remains large for any Edisonian optimization of composition. Therefore, both experimental and computational frameworks need to be in place to explore the compositional space in an effective manner, not only to screen the range of alloy compositions for single or multiphase stability, but also to suggest alloys that can exhibit a combination of properties [83, 90]. The high-throughput alloy design is defined as an automatic flow from ideas to results, where data being dealt with is far too large to produce or analyze. Miracle and co-workers [90] proposed a three-stage strategy for high-throughput materials development. The stages are as follows:

(i) High-throughput computations to narrow down compositional space for desired microstructure e.g., single-phase or multiphase. The computational alloy screening within the CALPHAD framework is already available, though the required databases are still incomplete restricting reliable exploration of the full range of compositional space, as has been discussed in Senkov et al. [29].
(ii) This stage uses output alloy systems from stage-1 and high-throughput experimental evaluation of the alloy system is carried out e.g., microstructure insensitive properties like modulus, density, thermal conductivity, thermal expansion coefficient, etc. using various characterization techniques.
(iii) Tensile strength, ductility, fracture toughness, etc. are evaluated in this stage.

The rapid evaluation is proposed to be accomplished by fixed composition and controlled microstructural gradients. The evaluation with microstructural gradient is necessary due to the fact that the properties depend not only on the composition but also the processes; and the microstructural content thereof e.g., grain size, precipitate size and volume fraction, orientation relationships, etc. The microstructural gradients in a fixed composition can be obtained by different methods proposed by Miracle et al. [90] e.g., annealing a rod of fixed composition in a furnace with a temperature gradient. The materials library generated using these three stages, along with the integration of other computational tools in the materials discovery loop, can be of immense help in designing novel materials with promising properties or with yet undiscovered features. In addition to the above, it is also important to have some basic materials property library with a compositional gradient. The experimental evaluation of phase equilibria and/or property changes (e.g., nano-hardness and Young's modulus) with a compositional gradient will also augment the materials data in the integrated materials discovery loop. There are different routes available for the preparation of compositionally graded high entropy alloys e.g., laser additive manufacturing, diffusion couples or multiples and magnetron sputtering [83, 191]. These routes are the variants of combinatorial materials synthesis, which will be discussed in the next section.

I.3.4 Combinatorial Synthesis

Combinatorial materials synthesis allows simultaneous synthesis and screening of large window of different material compositions. The technique was pioneered and extensively used by pharmaceutical industries for drug discovery. This method is now being considered for accelerating the discovery and optimization of high entropy alloys. As against the computational high-throughput materials design, combinatorial synthesis employs experimental arrangements to screen the properties of several compositions simultaneously in a short time, giving an idea of the phase stability and initial material properties [91, 92, 191]. In this way, making and designing diffusion multiple of different elements is adopted. A schematic of four and five component diffusion multiples, used for combinatorial high entropy alloy research, is shown in Figure I.2. In the four-component systems, four pure metal blocks can be assembled and hot-pressed at high temperature and the high entropy alloy treated to get a quaternary mixing zone in the center along with a large compositional gradient (Figure I.2a). Similarly, five-component diffusion couples can be made by initially bonding two binary alloys (AB and CD) followed by making a diffusion multiple with the fifth element (E) (Figure I.2b). After sectioning and polishing, localized property evaluation is attempted at the microscale to reach a composition-structure-property correlation of various solid-solution and/or intermetallic compounds. Wilson and co-workers [93] used equiatomic Fe-Mn, Ni-Co, and Cr multiple diffusion couples and heated the same to 1,200°C for 8 h under a load of 8 kN in a hot press to map hardness and composition. This study revealed that parametric estimation criteria did not support the phase stability criteria and hardness values did not have a direct correspondence with compositional complexity and atomic size mismatch. Coury and co-workers [83] used similar diffusion couples and tested for nano-hardness

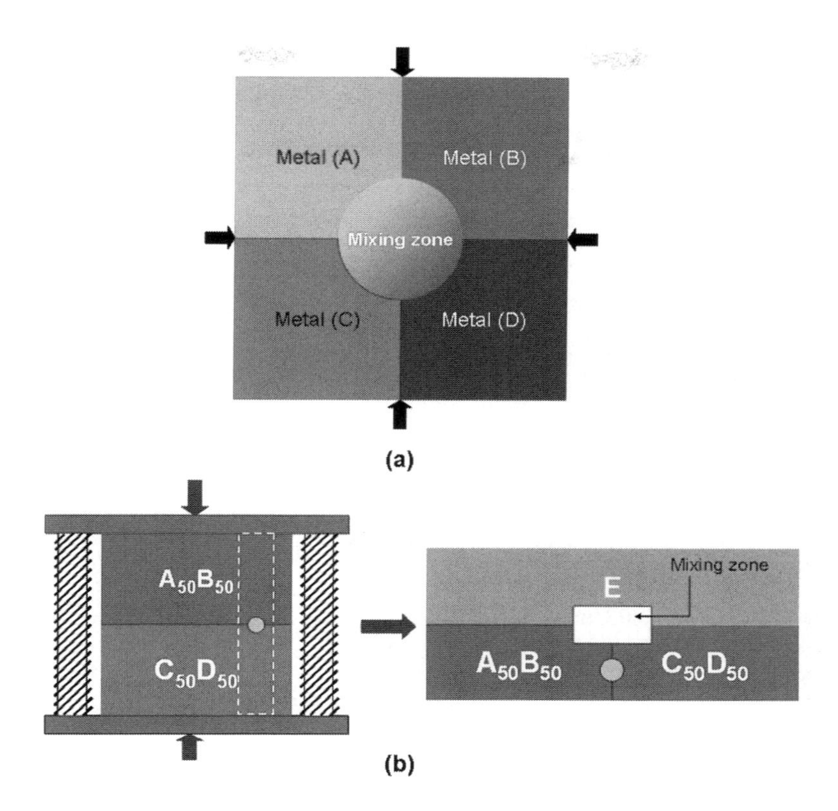

FIGURE I.2 A schematic showing diffusion for combinational high entropy alloy research: (a) four-component alloy systems and (b) five-component systems.

and composition. It was concluded in their study that nanoindentation can be used to estimate yield strength of single-phase face-centered cubic alloy with about 10% error.

Laser additive manufacturing is also one of the techniques that can be employed to synthesize different compositions as well as diffusion multiples [94]. The directed energy deposition technique of additive manufacturing can be used for this purpose [95]. As shown in Figure I.3, different elemental powders can be simultaneously deposited, where the composition is controlled by the powder flow rate, to achieve different alloy compositions. The gradient created between the alloy layers can be characterized at the microscale for structure–property correlations. Similarly, a large number of coupons can be made with varying compositions. This process of mass production of different alloys may lead to faster alloy screening. Borkar and co-workers [94] used a laser additive manufacturing system to make a compositionally graded rod (10 mm diameter and 25 mm length) of $Al_xCrCuFeNi_2$ (x = 0 to 1.5), with aluminum content varying along the length, for rapid assessment of composition–microstructure–property relationships. They observed that microstructure and phase constitution change accompanied the variation in hardness with aluminum content. These techniques have been widely used for studying the compositional space of a

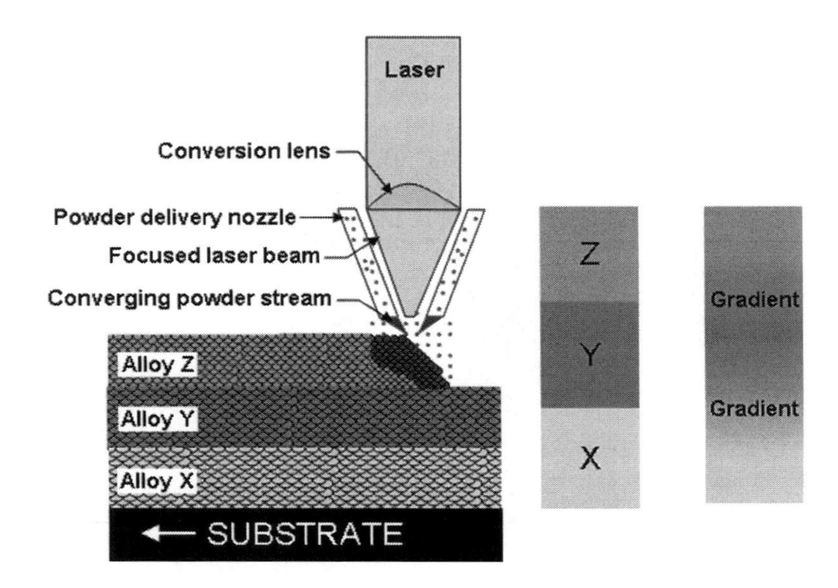

FIGURE 1.3 A schematic showing the additive manufacturing process producing multilayered bulk alloys. (Premixed or elemental powders can be fed into the interaction zone of the laser beam. A large number of compositions can be obtained by controlling different powder flow rates. The final build can be sectioned to obtain compositionally gradient materials for analysis).

given set of principal elements and the related properties. These experimental evaluation strategies by combinatorial methods can be fruitfully employed for materials database creation.

1.3.5 Machine Learning

Machine learning is the ability of machines to automatically learn from observations or data, experience, and instructions. It uses algorithms that can access data and learn from it to find a pattern, which helps in making better decisions or predictions. This tool has been employed by material scientists for many applications such as materials informatics [96], glass-forming ability [97], structural flow defects in disordered solids [98], high entropy alloys [99–101], etc. Huang and co-workers [101] employed this tool and used 401 numbers of experimental datasets of high entropy alloys on three common machine-learning algorithms to predict solid solution, intermetallic phases, and a combination of the two. They could predict the phases with an average accuracy of up to 94% using a multilayer feed-forward neural network. They also evaluated the relative importance of ΔH_{mix}, ΔS_{mix}, δ, $\Delta\chi$, and valence electron concentration in testing accuracy, and found that δ and valence electron concentration are more crucial for the phase selection. Wen and co-workers [76] also employed a machine-learning algorithm to predict alloy compositions with desired properties e.g., high hardness. A dataset of 155 samples with measured hardness was used for $Al_xCo_yCr_zCu_uFe_vNi_w$ ($15 < x < 47$; $5 < y < 22$; $6 < z < 34$; $5 < u < 16$;

$5 < v < 31$; $5 < w < 22$ at.%). Implicit mapping of composition and properties were established through a machine-learning algorithm. New candidate alloys for experimentation were found by combining the predictions and utility functions for maximizing the expected utility parameter. They could find 17 newly synthesized alloys having more than 10% high hardness compared to maximum hardness in the training dataset. Some of these new alloys with hardness in the range of 859–883 HV are $Al_{43-47}Co_{14-25}Cr_{18-23}Cu_{0-5}Fe_{5-12}Ni_{5-7}$ (at.%). These strategies can be employed well for optimization of other properties in high entropy alloys and extended to superalloys and bulk metallic glasses which suffer similar limitations as high entropy alloys.

1.4 MICROSTRUCTURAL DEVELOPMENT IN HIGH ENTROPY ALLOYS

The microstructure is of prime interest to the metallurgist while contemplating the development of an alloy. Its importance can be understood from the fact that the properties and suitability of an alloy for a given application can be modeled and predicted based on the microstructure. Therefore, modern alloy development efforts emphasize the forecast of a microstructure in order to understand the behavior of the alloy. Before the high entropy alloy concept was floated in 2004, researchers mainly focused on the development of simple alloys, based on a single principal element acting as a solvent, yielding a non-complex microstructure, i.e., solid solution (as either substitutional or interstitial elements) and/or intermetallic compound. In fact, metallurgists were reluctant to explore alloys beyond a single principal element concept. It was a general belief, consistent with the Gibbs phase rule, that the greater the number of principal elements, the higher will be the chance of formation of intermetallics and compounds along with a number of other phases, posing difficulty in fabrication and handling. Moreover, the lack of available databases on the interaction between elements, beyond the binary combination, added further to the apprehension in dealing with such systems. However, the formation of simple solid-solution structures like face-centered cubic and body-centered cubic in compositionally complex systems ($CuCoNiCrAl_xFe$, CoCrFeNiMn, etc.) indicated the possibility of the presence of certain factors which tend to stabilize the simple solid-solution phases in these alloys [2, 4]. Yeh and co-workers [2, 4] proposed that the tendency of formation of intermetallic phases or compounds in the as-cast structure would be suppressed by the high configurational entropy of mixing in these alloys. This proposition indicated the formation of simple solid-solution phases at higher temperatures. Further, these microstructures might undergo phase transformation during the slow cooling, resulting in the formation of complex structures such as precipitation during cooling, spinodal decomposition, intermetallic, compound formation, etc. However, phase decomposition would be significantly retarded by the sluggish nature of diffusion kinetics and consequently, no long-range diffusion in such complex concentrated alloys. It is due to this fact that the alloy conceivably retained a simple structure in the as-cast condition.

This particular section aims to bring out the scenario of microstructural development in high entropy alloys. As per Miracle and co-workers [13], alloys can be classified based on the constituent elements in the alloy as 3d transition metal high entropy alloys, refractory metal high entropy alloys, and others which includes metals like aluminum, beryllium, lithium, magnesium, scandium, silicon, tin, titanium, and zinc. The present study considers only the alloys with their constituent elements being mostly the 3d transition elements, as this alloy class is most studied. For the present study, the high entropy alloys have been classified in terms of the complexity of the microstructure e.g., single-phase, multiphase, eutectic, and metastable. In the following, independent sections have been devoted to all these types of microstructural features. Additionally, an exhaustive review of microstructural development in various high entropy alloys consisting of transition elements (iron, manganese, cobalt, chromium, copper, and nickel) along with some other elements such as aluminum, gold, molybdenum, silver, carbon, tin, tungsten, vanadium, titanium, niobium, and boron, produced either through vacuum induction melting or vacuum-arc melting have been presented in Table I.1. This table methodically enlists the high entropy alloys systems developed since the inception of high entropy alloys in the year 2004 up to the latest developments in 2019. As a first step, categorization of various high entropy alloy systems has been done on the basis of several phases formed in the as-cast condition. Second, the alloy compositions have been arranged in alphabetical order of the constituent elements which allows for effortless screening of the alloys containing a particular phase. The effect of alloying elements in the solidification microstructure evolution in different high entropy alloy systems can be accessed from Table I.1.

I.4.1 Single-Phase Microstructure

The definition of a high entropy alloy points toward the favorable condition for stabilization of simplest microstructure, i.e., a single-phase in the alloy system. However, only a few multi-principal alloy systems possess a complete single-phase microstructure, defying the high entropy contribution in stabilizing simple microstructure. In the last 15 years, a few important single-phase high entropy alloys have been discovered, which formed a single-phase face-centered cubic structure [1]. This particular class of high entropy alloy consisting of all or some of the transition elements like iron, manganese, chromium, cobalt, and nickel in various combinations has been thoroughly investigated over a broad range of composition. This huge focus is driven by the excellent mechanical properties shown by such alloys, as compared to the established engineering materials. Gludovatz and co-workers [68] reported that the equiatomic CrMnFeCoNi alloy displayed extraordinary damage tolerance with ultimate tensile strengths (UTS) of ~1 GPa, fracture toughness of ~200 MPa.$m^{1/2}$ at crack initiation, and ~300 MPa.$m^{1/2}$ for stable crack growth at −196°C. Several compositional variants of the equiatomic CrMnFeCoNi alloy were also investigated, which demonstrated encouraging mechanical properties [14, 15, 142–146]. In this regard, the equiatomic CrCoNi alloy showed superior tensile properties and fracture toughness than the equiatomic CrMnFeCoNi alloy both at room temperature and sub-zero temperatures [147, 148]. Wung and co-workers reported a non-equiatomic $Fe_{60}Co_{15}Ni_{15}Cr_{10}$ (at.%) alloy which displayed an extraordinary combination of tensile properties with the ultimate tensile strength of ~1.5 GPa and total elongation of ~87% at −196°C [15]. The fine-grain equimolar FeCrCoNi alloy wires displayed tensile yield strength of ~1.2 GPa at −50°C [149].

TABLE I.1
List of Phases Stabilized in Different High Entropy Alloy Compositions

Phase	Alloy system	References
Face-centered cubic	$Al_{0.3}CoCrCuFeNi$	[16]
	$Al_{0.5}CoCrCuFeNi$	[10, 16, 40]
	$Al_{0.5}CoCrCu_{0.5}FeNi$	[102]
	$Al_{0.5}CoCrCuFeNi^*$	[30, 35]
	$Al_8Co_{17}Cr_{17}Cu_8Fe_{17}Ni_{33}{}^*$	[34]
	$Al_{0.5}CoCrCuFeNiTi_x$ (x=0, 0.2)	[103]
	$Al_{0.5}CoCrCuFeNiV_x$ (x=0, 0.2)	[104]
	$Al_{0.3}CoCrFeMn_xNi^*$ (x=0.1, 0.3)	[105]
	$Al_x(CoCrFeMnNi)_{100-x}$ (x=2, 3, 4, 7)	[36]
	$Al_{0.3}CoCrFe\ Mo_{0.1}Ni$	[106]
	$Al_xCoCrFeNi$ (x=0.1, 0.4)	[3]
	$Al_xCoCrFeNi$ (x=0.25, 0.375)	[107]
	$Al_{0.3}CoCrFeNi$	[3, 106, 108–109]
	$Al_{0.3}CoCrFeNi^*$	[105, 110]
	$Al_{0.5}CoCrFeNi$	[111]
	$Al_{2.4}Co_{24.4}Cr_{24.4}Fe_{24.4}Ni_{24.4}$	[3]
	$Al_{7.2}Co_{23.2}Cr_{23.2}Fe_{23.2}Ni_{23.2}$	
	$Al_{9.1}Co_{22.7}Cr_{22.7}Fe_{22.7}Ni_{22.7}$	
	$Al_{0.3}\ CoCrFeNiTi_{0.1}$	[106]
	$Al_xCrCuFeNi_2$ (x=0.2, 0.4, 0.7)	[33]
	$Al_{0.5}CrCuFeNi_2$	[33, 112]
	$Al_5\ Cr_{12}Fe_{35}Mn_{28}Ni_{20}$	[113]
	$Al_{10}Cr_{12}Fe_{35}Mn_{23}Ni_{20}$	
	$Al_{7.5}Cr_6Fe_{40.4}Mn_{34.8}Ni_{11.3}$	[114]
	$Al_{7.5}Cr_6Fe_{40.4}Mn_{34.8}Ni_{11.3}$ + 1.1 at.% C	
	$Al_{0.6}CrFeNi_{2.4}$	[115]
	$CoCrCuFeMnNi^*$	[1]
	CoCrCuFeNi	[16, 38, 40, 116, 117]
	$CoCrCu_{0.5}FeNi$	[117]
	$(CoCrCuFeNi)_{96}Nb_4$	[38]
	CoCrFeMnNi	[22, 36, 118, 119]
	$CoCrFeMnNi^*$	[1]
	$CoCrFeMo_{0.3}Ni$	[71, 108]
	$CoCrFeMnNiNb^*$	[1]
	CoCrFeMnNiNbC	[119]
	CoCrFeNi	[3, 22, 71, 107, 109, 117, 120]
	$CoCrFeNiTi_{0.3}$	[121]
	$CrCuFeMn_2Ni_2$	[122]
	$CrCu_2Fe_2MnNi_2$	
	$Cr_2CuFe_2Mn_2Ni_2$	
	CrCuFeMoNi	[116]

(Continued)

TABLE I.1 (CONTINUED)
List of Phases Stabilized in Different High Entropy Alloy Compositions

Phase	Alloy system	References
	CoFeMnNi	[137]
	$CoFeMnNiTi_{0.25}$	
Body-centered cubic	$AlCoCrCu_{0.5}FeNi$	[10, 102]
	$Al_{3.0}CoCrCuFeNi$	[16]
	$Al_{23}Co_{15}Cr_{23}Cu_8Fe_{15}Ni_{16}^*$	[34]
	$Al_x(CoCrFeMnNi)_{100-x}$ (x = 16, 20)	[36]
	$Al_{2.0}CoCrFeMo_{0.5}Ni$	[123]
	AlCoCrFeNi	[3, 7, 107, 116, 123, 124]
	$Al_xCoCrFeNi$ (x = 0.875, 1.25, 1.50, 2.0)	[107]
	$Al_xCoCrFeNi$ (x = 0.9, 1.2, 1.5, 1.8, 2.0)	[3]
	$Al_xCoCrFeNi$ (x = 1.5, 2.0, 2.5, 3.0)	[7, 116]
	$AlCoCrFeNiNb_{0.1}$	[124]
	$Al_{2.5}CoCuFeNiSn_x$ (x = 0,0.03)	[125]
	$Al_xCrCuFeNi_2$ (x = 1.8, 2.0, 2.2, 2.5)	[33]
	AlCoFeNi	[129]
	$Al_x(Cr_{20}Fe_{40}Mn_{25}Ni_{15})_{100-x}$ (x = 10, 14)	[126]
	$Al_{0.5}CrFe_{1.5}MnNi_{0.5}$	[127]
	$AlCrFeMo_xNi$ (x = 0, 0.2, 0.5)	[32]
	AlCrFeNi	[22]
	AlNbTiV	[128]
Face-centered cubic + Body-centered cubic	$AlB_xCoFeNi$ (x = 0.05, 0.10, 0.15, 0.20)	[129]
	AlCoCrCuFeNi	[5, 10, 16, 40, 130]
	$Al_xCoCrCuFeNi$ (x = 1.5, 2.0)	[16, 40]
	$Al_xCoCrCuFeNi$ (x = 0.8, 1.3, 1.8, 2.3, 2.5, 2.8)	[16]
	$AlCo_{0.5}CrCuFeNi$	[10]
	$AlCoCr_{0.5}CuFeNi$	
	$AlCoCrCuFe_{0.5}Ni$	
	$AlCoCrCuFeNi_{0.5}$	
	$AlCoCrCuFeNi^*$	[37]
	$Al_{1.5}CoCrCu_{0.5}FeNi$	[102]
	$Al_{0.5}CoCrCuFeNiV_x$ (x = 0.4,1.2,1.4,1.6,1.8,2.0)	[104]
	$Al_{0.5}CoCrCuFeNiTi_x$ (x = 0.4, 0.6)	[103]
	AlCoCrCuFeNiTi	[5]
	AlCoCrCuFeNiV	
	AlCoCrCuNi	[130]
	$Al_x(CoCrFeMnNi)_{100-x}$ (x = 8, 9, 10, 11, 12, 13, 14, 15)	[36]
	$Al_xCoCrFeNi$ (x = 0.5, 0.7, 0.8)	[3]
	$Al_xCoCrFeNi$ (x = 0.6, 0.9)	[109]
	$Al_xCoCrFeNi$ (x = 0.5, 0.75)	[107]
	$AlCoCrFeNi_{2.1}^*$	[11]

(Continued)

TABLE I.1 (CONTINUED)
List of Phases Stabilized in Different High Entropy Alloy Compositions

Phase	Alloy system	References
	$(AlCoCrFeNi)_{100-x}C_x$ (x=2, 4, 6, 8)	[131]
	$Al_xCoCrFeNiTi_{(1-x)}$ (x=0.5, 0.8)	[132]
	AlCrCuFeNi	[116]
	$Al_xCrCuFeNi_2$ (x=0.8, 0.9, 1.2, 1.5)	[33]
	$Al_{19}Co_{20}Fe_{20}Ni_{41}$	[133]
	$Al_x(Cr_{20}Fe_{40}Mn_{25}Ni_{15})_{100-x}$ (x=2, 6)	[126]
	$Al_xCr_{0.75}FeMnNi$ (x=0.25, 0.50, 0.75)	[75]
	$Al_{0.3}CrFe_{1.5}MnNi_{0.5}$	[127]
	$(Al_{21.7}Cr_{15.8}Fe_{28.6}Ni_{33.9})_x(Al_{9.4}Cr_{19.7}Fe_{41.4}Ni_{29.5})_{100-x}$ (x=10, 30, 50, 70, 90)	[134]
	$Al_xCrFeNi_{3-x}$ (x=0.7, 0.8, 0.9, 1.0)	[115]
	CoCrFeMnNiTi	[1]
	CoCrFeMnNiV	
	CrCuFeMnNi	[116, 122]
	Cr_2CuFe_2MnNi	[122]
	$CrCu_2Fe_2Mn_2Ni$	
	$Cr_2Cu_2FeMn_2Ni$	
	$Cr_2Cu_2FeMnNi_2$	
	$Cr_{20}Fe_{40}Mn_{25}Ni_{15}$	[126]
Face-centered cubic + Others[a]	AlAuCoCrCuNi	[130]
	$Al_{0.5}B_xCoCrCuNi$* (x=0.2, 0.6, 1.0)	[30]
	$Al_8Co_{17}Cr_{17}Cu_8Fe_{17}Ni_{33}$*	[54]
	$Al_{0.5}CoCrFeMo_{0.5}Ni$	[123]
	$(Al_{7.5}Cr_6Fe_{40.4}Mn_{34.8}Ni_{11.3})$ + 0.5 and 1.0 at.% B	[114]
	$C_{1.84}(CoCrFeMnNi)_{98.16}$	[118]
	$(CoCrCuFeNi)_{100-x}Nb_x$ (x=8, 12, 16)	[38]
	CoCrFe $Mo_{0.5}Ni$	[71, 123]
	$CoCrFeMo_{0.85}Ni$	[71, 135]
	$CoCrFeNiNb_x$ (x=0.45, 0.50, 0.65)	[136]
	$CoCrFeNiTi_x$ (x=0.3, 0.5)	[120]
	$Co_xCrFeNiTi_{0.3}$ (x=0.6, 0.8)	[121]
	$CoFeMnNiTi_x$ (x=0.50, 0.75)	[137]
	$Cr_xFe_{50-x}Mo_{25}Ni_{25}$ (x=5, 10, 15, 20, 25)	[138]
	$Cr_{25}Fe_{50-x}Mo_xNi_{25}$ (x=5, 10, 15, 20)	
	$Cr_xFe_{75-2x}Mo_xNi_{25}$ (x=5, 10, 15, 20)	
	$Co_2Mo_xNi_2VW_x$ (x=0.5, 0.6, 0.8)	[139]
Body-centered cubic + Others[b]	$AlCoCrFeMo_{0.5}Ni$	[8, 123]
	$Al_xCoCrFeMo_{0.5}Ni$ (x=1.0, 1.5)	[123]
	$AlCo_xCrFeMo_{0.5}Ni$ (x=0.5, 1.0, 1.5)	
	$AlCoCr_xFeMo_{0.5}Ni$ (x=0.5, 1.5, 2.0)	[8, 123]

(*Continued*)

TABLE I.1 (CONTINUED)
List of Phases Stabilized in Different High Entropy Alloy Compositions

Phase	Alloy system	References
	$AlCoCrFe_xMo_{0.5}Ni$ (x=0.6, 1.0, 1.5, 2.0)	[123]
	$AlCoCrFeMo_xNi$ (x=0.5,0.9)	
	$AlCoCrFeMoNi_x$ (x=0,0.5,1.0)	
	$AlCoCrFeNiNb_x$ (x=0.25, 0.50, 0.75)	[124]
	$Al_{0.9}CoCrFeNiTi_{0.5}$	[109]
	$Al_{2.5}CoCuFeNiSn_x$ (x=0.05, 0.07, 0.10)	[125]
	AlCrCuFeNiTi	[140]
	$AlCoFeMo_{0.5}Ni$	[8, 123]
	$AlCrFeMo_{0.5}Ni$	[123]
	$AlCrFeMo_xNi$ (x=0.8, 1.0)	[32]
Face-centered cubic + Body-centered cubic + Others[c]	AlAgCoCrCuNi	[130]
	AlCoCrCuFeMnNi	[5]
	$Al_{0.5}CoCrCuFeNiTi_x$ (x=0.8,1.0,1.2,1.4,1.6,1.8,2.0)	[103]
	$Al_{0.5}CoCrCuFeNiV_x$ (x=0.6, 0.8, 1.0)	[104]
	$AlCoCrFeMo_{0.5}Ni_x$ (x=1.5, 2.0)	[123]
	$AlCo_{2.0}CrFeMo_{0.5}Ni$	
	$Al_{0.5}CoCrNiTi_{0.5}$	[141]
	$Co_2Mo_xNi_2VW_x$ (x=1.0, 1.5, 1.75)	[139]

[a] Hexagonal closed pack, Laves, σ, μ, η, R, γ', M7C3, M23C6, AuCu, (Cr,Fe)B.
[b] Laves, Ni17Sn3, σ, Fe2Ti, compounds.
[c] σ, μ, Ti2Ni, CoCr, Ag – either one or in combination.
*Vacuum induction melting (all others without * mark are from vacuum-arc melting).

Usually, the as-cast, well-homogenized microstructure of such alloys consists of single-phase face-centered cubic structure. The equiatomic CrMnFeCoNi alloy showed a face-centered cubic structure produced via various routes and thereafter processed through different processing techniques. The alloy synthesized by vacuum-arc melting [17, 65, 143, 150] or vacuum induction melting [151], processed through hot rolling [143], cold rolling [17, 150] or cryo-rolling [157], exhibited single-phase face-centered cubic structure. Even the severely deformed alloy, up to the shear strain of more than 500% by high-pressure torsion, is reported to retain the face-centered cubic structure [153]. Another compositionally deviated alloy from the equiatomic CrMnFeCoNi alloy, the $Fe_{40}Mn_{27}Ni_{26}Co_5Cr_2$ alloy exhibited a stable disordered face-centered cubic solid solution both in as-cast and homogenized state, despite having a considerably lower configurational entropy of mixing [154, 155]. Further, in the quest to obtain superior mechanical properties, many compositional variants of the Fe-Mn-Co-Cr-Ni system were produced. In this regard, the stacking fault energy (SFE) of the equiatomic CrMnFeCoNi alloy can be further reduced by an alloying element adjustment. The reduced stacking fault energy of the alloy may trigger deformation twins even at room temperature and the mechanical properties may be improved further. Stacking fault energy (Γ^{SFE}) of the face-centered cubic system can be related with the Gibbs free energy ($\Delta G^{FCC \to HCP}$) change required to create a platelet of ε hexagonal closed pack of a thickness of two atomic layers as [156]:

$$\Gamma^{SFE} = 2\rho \Delta G^{FCC \to HCP} + 2\sigma^{FCC \to HCP} \quad (I.10)$$

where $\sigma^{FCC \to HCP}$ is the energy required for the creation of new surfaces between the face-centered cubic and the newly forming hexagonal closed pack phase.

As per Equation (I.10), the stacking fault energy of the alloy can be lowered if the chemical free energy change $\Delta G^{FCC \to HCP}$ is lowered by adjusting the chemical composition of the alloy. To this end, alloys with even lower stacking fault energy can be synthesized by eliminating nickel and adjusting the other elements from the equiatomic CrMnFeCoNi alloy. Li and co-workers [14] showed the effect of nickel on the $\Delta G^{FCC \to HCP}$ of $Co_{20}C_{20}Mn_{20}Fe_{40-x}Ni_x$ alloy and found that the $\Delta G^{FCC \to HCP}$ of the alloy decreased with reducing the nickel content in the alloy and the lowest $\Delta G^{FCC \to HCP}$ was observed in the alloy without any nickel addition (Figure I.4a). Therefore, it will be logical to produce a nickel-free alloy for obtaining the lowest stacking fault energy. Further, the effect of manganese on the $\Delta G^{FCC \to HCP}$ in a nickel-eliminated type of alloy, $Fe_{80-x}Mn_xCo_{10}Cr_{10}$, was also shown as depicted in Figure I.4b [14].

It is evident from Figure I.4b that an alloy, $Fe_{50}Mn_{30}Cr_{10}Co_{10}$ (at.%), can be produced with the lowest stacking fault energy. However, the particular alloy along with the $Fe_{45}Mn_{35}Cr_{10}Co_{10}$ alloy formed a two-phase structure [12]. In contrast to the alloy system, the low-cost alloy, $Fe_{40}Mn_{40}Cr_{10}Co_{10}$, with nickel-eliminated and reduced cobalt content compared to the equiatomic CrMnFeCoNi alloy showed a stable single face-centered cubic phase structure, elucidating the importance of manganese in a face-centered cubic phase stabilization [14, 142]. The manganese dependence of the microstructure evolution in $Fe_{80-x}Mn_xCo_{10}Cr_{10}$ type alloy has been presented in Figure I.5 [12]. It can be observed that the volume fraction of the face-centered cubic phase decreased with the decreasing manganese content, showing that manganese is a face-centered cubic phase stabilizer.

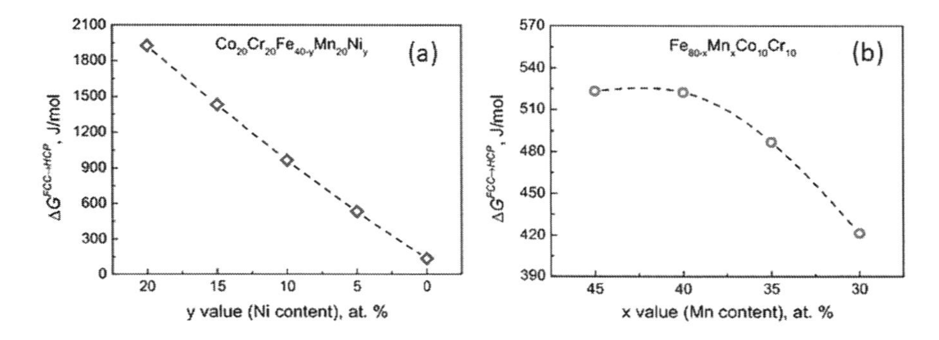

FIGURE I.4 Variation in Gibbs free energy change for face-centered cubic to hexagonal closed pack transformation in (a) $Co_{20}C_{20}Mn_{20}Fe_{40-x}Ni_x$ alloy with different nickel content and (b) $Fe_{80-x}Mn_xCo_{10}Cr_{10}$ alloy with varying manganese content [14].

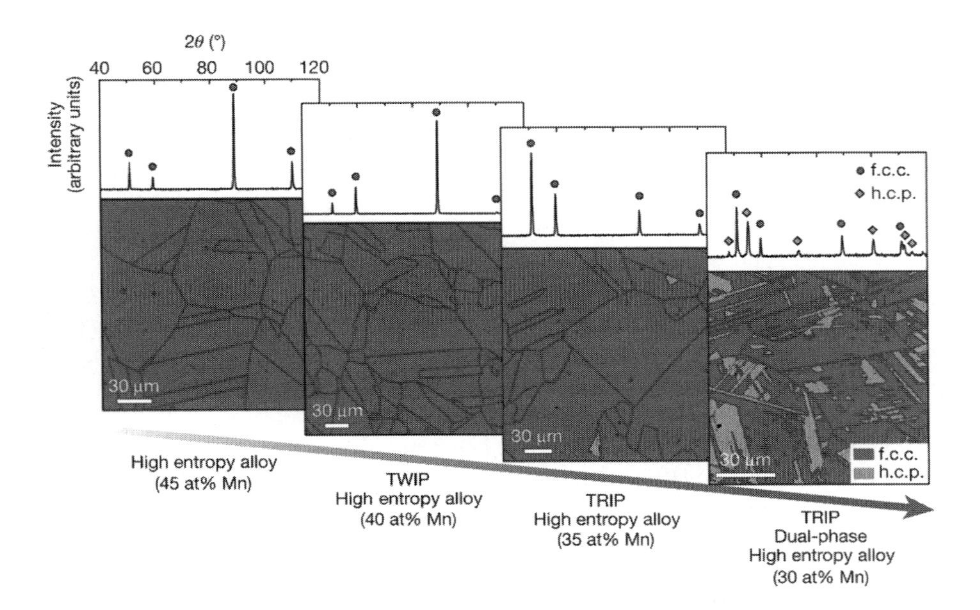

FIGURE I.5 X-ray diffraction patterns and corresponding electron back-scattered diffraction maps of $Fe_{80-x}Mn_xCo_{10}Cr_{10}$ alloy with x=45, 40, 35, and 30 at.% [12].

A very similar alloy, $Fe_{37}Mn_{45}Co_9Cr_9$ (at.%), without any nickel content, also resulted in a stable face-centered cubic solid solution in the as-cast and homogenized state [154]. Stepanov and co-workers [152] presented a few cobalt-free high entropy alloys such as $Fe_{40}Mn_{28}Ni_{28}Cr_4$, $Fe_{40}Mn_{28}Ni_{20}Cr_{12}$, and $Fe_{40}Mn_{28}Ni_{14}Cr_{18}$ displaying single-phase face-centered cubic structure in a cold rolled and annealed state. An equiatomic manganese-free FeCoCrNi alloy also showed a stable disordered face-centered cubic solid solution [149]. Wung and co-workers [15] harnessed the metastability-engineering strategy to produce two manganese-free high entropy alloys, $Fe_x(CoNi)_{90-x}Cr_{10}$ (with x=55 and 57.5 at.%), which showed single-phase face-centered cubic structure at room temperature. The corresponding face-centered cubic phase in these two alloys was unstable and progressively transformed into the body-centered cubic phase with an increase in strain level during the deformation process, thus providing a high work hardening [15]. An iron-free high entropy alloy has also been studied by the present authors to produce non-equiatomic CoCrMnNi alloys. The alloy was prepared by vacuum induction melting (2 kg), followed by hot rolling of the as-cast structure at 1,200°C. Interestingly, the alloy in the as-cast condition was indicated as a single-phase face-centered cubic structure by the X-ray diffraction pattern (Figure I.6a), instead of the fact that the microstructure was composed of dendrite (DR) and interdendrite (IDR) regions, which could be compositionally different (Figure I.6b). Further, the microstructure in the homogenized and hot rolled state was also single-phase face-centered cubic structure (as is evident from the X-ray diffraction pattern in Figure I.6a) with polygonal grains. The compositional variation between the dendrite and interdendrite region in the as-cast alloy was revealed by performing the elemental mapping using wavelength dispersive

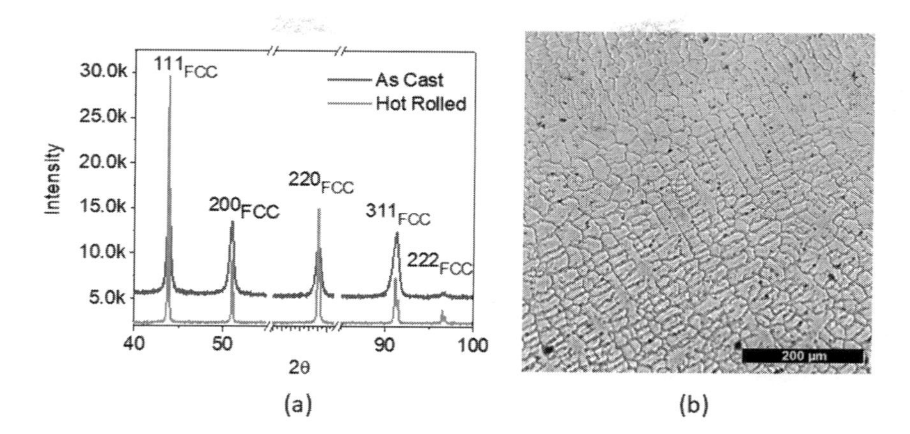

FIGURE I.6 (a) X-ray diffraction pattern and (b) optical micrograph of CoCrMnNi alloy in the as-cast condition.

spectroscopy technique in an electron probe micro-analyzer. Figure I.7 presents the elemental mapping of the alloy, clearly indicating the composition variation in the dendrite and interdendrite regions. It is evident that manganese and nickel preferentially segregate at the interdendrite region leaving behind cobalt and chromium. This could be explained on the basis of the highest negative enthalpy of formation of the manganese–nickel (–8 kJ/mol) pair among all possible binaries in the system. Therefore, the affinity of manganese and nickel is greater when together in the form of some intermetallic compound. However, in the present case, no separate compound formation is observed.

Apart from the system consisting of only 3d transition elements (other than copper), a large number of studies have been devoted to the one containing copper and aluminum. The alloy system AlCoCrCuFeNi is one of the alloys which was introduced as an example to demonstrate the so-called high entropy effect on the microstructure [4]. It was shown that such a compositionally complex alloy system retained a simple microstructure consisting of face-centered cubic and body-centered cubic phases. This particular alloy system was studied with compositional variations. The Al_xCoCrCuFeNi alloy has been reported to be single-phase face-centered cubic structure for the aluminum content up to $x=0.5$ (x is in molar ratio), after which the body-centered cubic phase appeared in the microstructure. The face-centered cubic + body-centered cubic microstructure further transitioned into a full body-centered cubic structure, with the subsequent addition of aluminum ($x>2.8$) [4, 10, 16]. Tung and co-workers [10] investigated the AlCoCrCuFeNi system with the variation of each element at two levels i.e., 0.5 and 1 molar ratios. It was observed that the alloy exhibited a full face-centered cubic structure with a reduction in the aluminum content to 0.5 molar ratios. However, the structure changed to the full body-centered cubic structure when copper was reduced to 0.5 molar ratios. This showed that aluminum and copper have an opposite tendency in terms of stabilizing different microstructures i.e., aluminum is a body-centered cubic stabilizer, whereas copper is a face-centered cubic stabilizer. Singh and co-workers [9] carried out an important

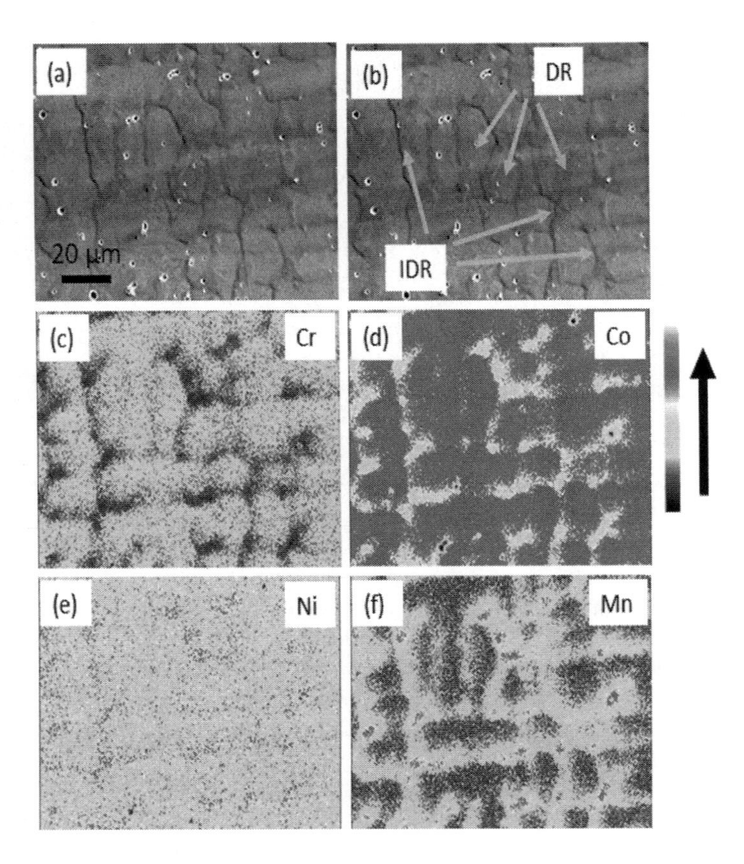

FIGURE I.7 Elemental mapping of the CoCrMnNi alloys in the as-cast condition showing segregation of alloying elements in the dendrite (DR) and interdendrites (IDR) regions.

investigation on the development of solidification microstructure with varying cooling rate. The equiatomic alloy AlCoCrCuFeNi was prepared by the conventional melting and casting route with the achievable cooling rate of 10–20 K/s and also by the splat quenching technique (cooling rate: 10^6–10^7 K/s). The splat quenched specimen developed a microstructure composed of a single body-centered cubic phase. It was claimed that: "The successful preparation of a body-centered cubic phase AlCoCrCuFeNi high entropy alloy by rapid cooling (10^6–10^7 K s^{-1}) is reported in the present work for the first time."

The presence of copper in the AlCoCrCuFeNi alloy tends to segregate in the interdendrite regions, which could lead to the deterioration in the mechanical properties [9, 10, 16]. Therefore, the six-component AlCoCrCuFeNi alloy system was modified by eliminating copper in order to yield one of the most studied high entropy alloy systems AlCoCrFeNi. Kao and co-workers [107] published an important report revealing the microstructure of $Al_xCoCrFeNi$ in as-cast, as-homogenized, and as-cold rolled condition. The microstructure of the as-cast alloy was observed to be a full face-centered cubic structure (up to $x = 0.375$) and a complete body-centered cubic structure for $x > 0.875$. Li and co-workers [7] also studied the effect of aluminum on the microstructure of the $Al_xCoCrFeNi$ alloys with the value of x varied as 1, 1.5, 2, 2.5, and 3 (molar ratio). This study confirmed that all the alloy compositions were single phase ordered body-centered cubic structure. Aluminum being the largest atomic size element in the alloy, the lattice constant was reported to increase with the increase in the aluminum content. The as-cast microstructure was found to be dendritic in all the cases. However, the morphology of the microstructure varied differently in alloys with different aluminum content.

I.4.2 Multiphase Microstructure

Single-phase forming high entropy alloys were found to be limited in number, thus raising serious concern about the notion of entropic contribution toward phase stabilization. The majority of the high entropy alloys showed the formation of multiple numbers of phases including precipitates, intermetallics, and compounds. From the application point of view, the presence of appropriately controlled multiphase microstructures can be helpful in the accomplishment of the desired set of mechanical properties. Understanding the various process parameters which control the microstructural evolution is important and the proper knowledge of same could be helpful in obtaining suitable microstructures. The present section reviews various kinds of multiphase high entropy alloy systems containing a wide variety of microstructures. The effect of alloying additions on the microstructural evolution of several high entropy alloy systems has been presented.

As discussed in the previous section, the AlCoCrCuFeNi alloy system showed the face-centered cubic phase up to aluminum content of 0.5 molar ratios. A body-centered cubic phase appeared in the microstructure with a further increase in the aluminum content. Together with this, spinodal decomposition occurs, leading to modulated structure formation consisting of ordered (B2) phase and disordered (A2) phase [4]. The sluggish diffusion kinetics were attributed to the formation of nanostructured crystallites. Here, it is important to consider the contribution of slow kinetics in the retention of simple microstructure. The sluggish diffusion due to massive alloying can effectively retard any phase transformation, which can be aided by the decrease in temperature during solidification. Since the high-temperature phases are usually a simple solid solution, their retention is supported by the insufficient diffusion of alloying elements. In this regard, it has been shown that high entropy alloys retain those solid-solution phases which actually form first upon cooling [158]. The preferred formation of solid solutions over intermetallics at higher temperatures is due to the large entropy aid (thermal + configurational). However, fine precipitates and intermetallic compounds are usually seen at low temperatures. The nanometer-sized precipitate formation is dictated by the sluggish nature of the diffusion process in such concentrated systems. Therefore, it can be conceived that the contribution of entropy in favoring simple solid-solution phases has been overestimated. Tong and co-workers [16] investigated the $Al_xCoCrCuFeNi$ alloy in detail and has shown the microstructural development with increasing aluminum content from 0 to 3 molar ratios. They constructed the first phase diagram of such kind of systems based on the microstructure and differential thermal analyses, as shown in Figure I.8. The occurrence of

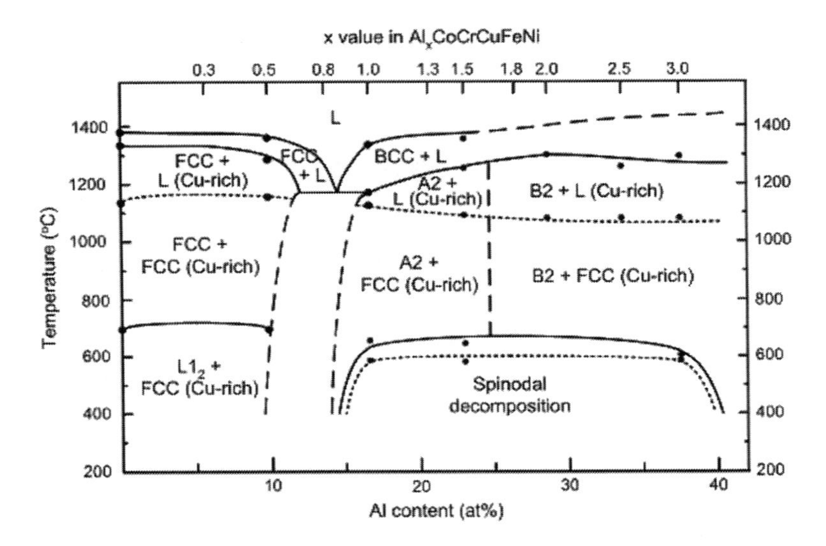

FIGURE I.8 Phase diagram of the $Al_xCoCrCuFeNi$ alloy system [16].

nano-precipitates was observed. The alloys with low aluminum content were composed of ordered face-centered cubic nano-precipitates dispersed in a face-centered cubic matrix, while fine nano-sized precipitate in spinodally decomposed phase was observed in the alloys with high aluminum content. The decreased elemental solubility of the face-centered cubic and body-centered cubic phases at low temperatures might have resulted in precipitation. The occurrence of nanoscale precipitates, rather than coarse precipitates during the solidification, was ascribed to the restricted diffusion kinetics in such alloys.

As elucidated in the previous section, rapid cooling (achieved by splat quenching) of the equiatomic AlCoCrCuFeNi alloy resulted in the metastable single-phase body-centered cubic structure, unlike the one produced from the conventional casting route which developed a series of precipitates [9]. The cast microstructure was found to possess all kinds of possible precipitates and compositionally segregated phases. Figure I.9 shows the classification of microstructure development based on the cooling rate for AlCoCrCuFeNi alloy [9]. The figure clearly depicts that the microstructure of the slowly cooled alloy exhibits copper-rich precipitates in dendrite regions and copper segregation in the interdendrite regions. The copper segregation in the microstructure could impair the mechanical properties of the alloy system. Therefore, proper measures must be adopted to take care of the segregation issue. Further, several copper-based precipitates also formed during the slow rate of solidification.

Srivastava and co-workers [159] studied $AlCoCu_{0.50}Cr_{0.75}FeNi$ alloy (molar ratio) both by casting as well as spray forming. Both the as-cast and spray formed materials showed a two-phase structure, consisting of body-centered cubic (aluminum–nickel-rich) and face-centered cubic (chromium-iron-rich) phases. However, the copper segregation at the body-centered cubic/face-centered cubic interface was reduced due to a comparatively high cooling rate achieved during spray forming. The overspray

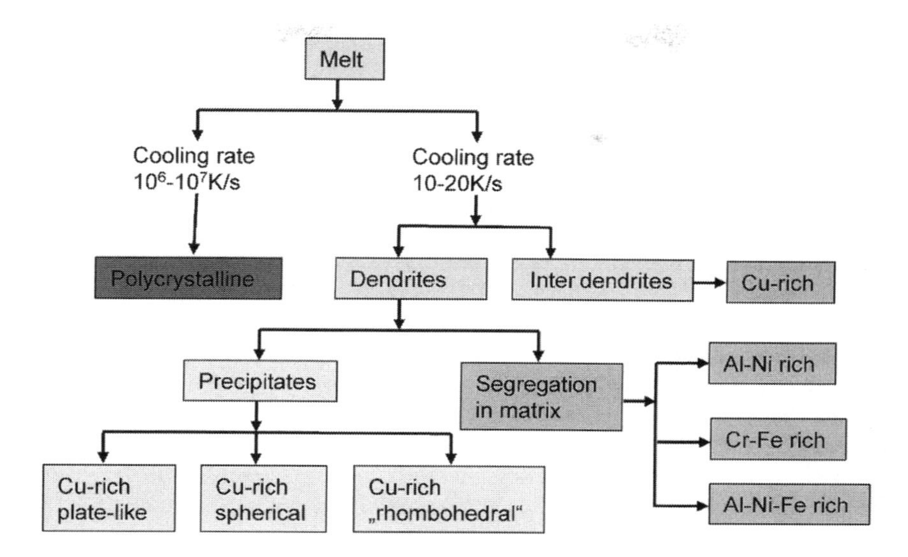

FIGURE I.9 AlCoCrCuFeNi alloy system: schematic of the phase evolution during solidification at different cooling rates [9].

powders, produced as a by-product of spray forming, invariably showed only single-phase body-centered cubic structures due to their high cooling rate, as was seen by other investigators also. The spray forming of the high entropy alloy also led to drastic grain refinement (25–30 μm) compared to the as-cast grain structure (600–1,000 μm), as shown in Figure I.10. As these two-phase materials are difficult to process by any deformation process, the near-net-shape technique of spray forming may be promising for achieving dense and refined microstructural features.

As mentioned earlier, in order to evade the problem of copper segregation in microstructures, the copper-free AlCoCrFeNi alloy system was introduced. Wang and co-workers [160] showed the microstructure of the AlCrFeCoNi alloy depicting polygonal grains with segregation in the form of dendrites. A micro-analysis of these microstructures revealed aluminum and nickel-rich body-centered cubic dendrites with chromium and iron-enriched interdendritic regions. The distribution of cobalt

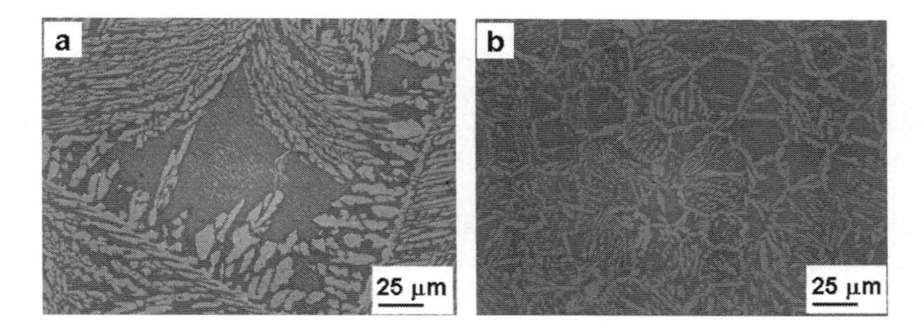

FIGURE I.10 Optical micrographs of $AlCoCu_{0.50}Cr_{0.75}FeNi$ alloy in the (a) as-cast and (b) spray-formed conditions, showing a drastic grain-size refinement.

was almost uniform in the matrix. This has been clearly explained by the fact that the enthalpy of mixing of the aluminum–nickel combination is –30 kJ/mol, which is the highest of any of the combinations in the alloy [55]. Apart from the major body-centered cubic phase, (111) ordered reflection corresponding to the $L1_2$ face-centered cubic phase was also observed in the as-cast alloy. Kao and co-workers [107] found that the $Al_xCoCrFeNi$ alloy exhibited face-centered cubic + body-centered cubic duplex microstructures in the composition range $0.5 \leq x \leq 0.875$ (molar ratio). Wang and co-workers [3] presented a dedicated report on the composition-microstructure correlation in the $Al_xCoCrFeNi$ alloy system, with aluminum variation from 0 to 2 molar ratios. The alloy without aluminum was composed of full face-centered cubic structure, whereas the aluminum addition led to the appearance of a body-centered cubic structure with an increase in the aluminum content of the alloy. Further, the face-centered cubic phase was fully eliminated in the $x = 0.9$ alloy. Interestingly, lattice parameters increased with the aluminum addition for both the phases [3]. However, the lattice parameters of both the phases were constant in the face-centered cubic + body-centered cubic two-phase region. This is due to the fact that the composition of both the phases was basically constant, with the increasing aluminum content owing to the changing phase fraction (the increasing amount of body-centered cubic phase and decreasing amount of face-centered cubic phase) and subsequent partitioning of aluminum to the body-centered cubic phase. A detailed compositional analysis revealed that a maximum of 11 at.% of aluminum can be accommodated into the face-centered cubic phase, after which the aluminum is rejected to the interdendritic regions to form the body-centered cubic phase with a spinodal structure. Further, at least 18.4 at.% aluminum is required to form a complete body-centered cubic structure. In this regard, the body-centered cubic stabilizing tendency of aluminum and its preferential partitioning to the body-centered cubic structure can be explained on the basis of a large atomic radius of aluminum and the body-centered cubic being a relatively open structure. The lattice strain could significantly increase with the increase in the aluminum content, as aluminum is an element with a large atomic radius [3, 161]. Therefore, a face-centered cubic to body-centered cubic transformation may be triggered by the aluminum addition in order to accommodate the lattice distortion induced by the introduction of aluminum in the system. A cobalt-free AlCrFeNi alloy investigated by Dong and co-workers [32] showed a mixture of two body-centered cubic structures with simple separation of aluminum–nickel and iron–chromium binary combinations.

The influence of the titanium addition was investigated in $Al_{0.5}CoCrCuFeNi$ alloy [103]. Two body-centered cubic phases appeared in the $Al_{0.5}CoCrCuFeNiTi_x$ alloy, with titanium up to 0.4 molar ratios. Further increase in the titanium content led to the formation of Ti_2Ni and CoCr-like structures. Chen and co-workers [103] claimed that, unlike aluminum, titanium does not promote ordering of the body-centered cubic structure. Zhou and co-workers [162] studied the effect of addition of titanium to the copper-free $AlCoCrFeNiTi_x$ alloy. Titanium addition in the alloy up to 1.5 molar ratios developed a Fe_2Ti-type Laves phase, apart from the body-centered cubic solid solution. The presence of titanium in the solid solution, however, provided massive solid-solution strengthening, yielding compressive fracture strength as high as 3.14 GPa, along with decent compressive ductility of 23.3%. Similarly, a nanoscale cellular structure developed through spinodal decomposition with the addition of silicon (up to $x=0.4$) in the $AlCoCrFeNiSi_x$ alloy [6]. Excessive addition of silicon ($x \geq 0.6$) led to precipitation of detrimental chromium and silicon-rich δ phase at the grain boundaries. Ma and Zhang [124] studied the effect of niobium addition on the microstructure and properties of the $AlCoCrFeNiNb_x$ ($x=0$ to 0.75) alloy. A body-centered cubic phase was found, along with a (CoCr)Nb-type Laves phase in the alloys with $x>0.1$ molar ratios. Hsu and co-workers studied the effect of the presence of chromium in the alloy $AlCoCr_xFeMo_{0.5}Ni$ ($x=0$ to 2) [8]. Only B2 and the chromium-rich σ phase were present in the as-cast microstructure despite the fact that the alloy was composed of six principal elements. The matrix phase of the dendrite changed from the B2 phase to the σ phase with the increase in chromium content of the alloy. The volume fraction of the σ phase increased with the increase in chromium content.

Many authors reported about the morphological transition with the addition of different alloying elements in high entropy alloys systems. It is a well-known fact that morphology of phases can significantly affect the mechanical properties of the alloys. In this context, Li and co-workers investigated the effect of manganese, titanium, and vanadium additions, separately, in modifying the microstructure of the AlCrFeCoNiCu alloy [5]. The manganese addition did not affect the microstructure except the presence of a long strip-shaped chromium-enriched phase. Similar to AlCrFeCoNiCu alloy, the AlCrFeCoNiCuV alloy also possessed a dendritic structure, but the morphology of the dendrites changed from modulated plates to ellipsoidal particles. However, the titanium addition changed the microstructure from dendritic to eutectic cells. The typical as-cast dendritic structure of the AlCoCrFeNi alloy changed to a eutectic structure on the addition of molybdenum of more than 0.1 molar ratios in $AlCoCrFeNiMo_x$ alloy [31]. Dong and co-workers [32] investigated the effect of molybdenum on the microstructure of a cobalt-free $AlCrFeNiMo_x$ with x varying from 0 to 1 molar ratio. With the addition of molybdenum, the eutectic structure of AlCrFeNi alloy witnessed a transition from a eutectic to a hypo-eutectic to a hyper-eutectic structure. Hsu and co-workers [163] studied the effect of cobalt addition on the microstructure of a newly developed high entropy alloy $AlCo_xCrFeMo_{0.5}Ni$ ($x=0.5$ to 2.0). The morphology of the as-cast structure was reported to change from dendritic to polygonal with the increase in the cobalt content. Typically, the alloy showed simple phases (body-centered cubic, face-centered cubic, and σ) in the microstructure. However, the suppression of the σ phase was observed with the increase in the cobalt content.

1.4.3 Eutectic Microstructure

As discussed earlier, the initial fascination with high entropy alloys was to produce single or multiphase structures containing simple phases such as face-centered cubic, body-centered cubic, etc. These aims were further modified from the microstructure toward their deliverables i.e., the properties that ultimately determine suitability for an application. Single-phase high entropy alloys with face-centered cubic structure usually exhibit high tensile ductility, but are lacking in strength. For example, equiatomic CrMnFeCoNi alloy showed elongation ~50% and YS and UTS of ~410 MPa and ~750 MPa, respectively [68]. Similarly, the alloy $Fe_{40}Mn_{40}Co_{10}Cr_{10}$ possessed a tensile elongation of ~58%, but with limited YS and UTS of ~240 MPa and ~520 MPa, respectively [142]. On the other hand, high entropy alloys with body-centered cubic structure invariably show high strength, but low ductility e.g., single crystals of equiatomic NbMoTaW showed high strength levels ~4–4.5 GPa, but lower ductility (less than 0.2%) [164]. Therefore, the single-phase high entropy alloys either featured high strength or high ductility. A superior combination of strength and ductility is a major limitation of most of these alloys. Any attempt with microstructural engineering to optimize the strength led to a loss in the ductility and vice versa, an effect commonly known as the strength–ductility trade-off. As discussed in the previous section, to obtain a balanced combination of strength and ductility, it would be sensible to combine the individual properties of face-centered cubic and body-centered cubic phases in the microstructure. Several works on the mechanical properties in high entropy alloys with face-centered cubic + body-centered cubic microstructure revealed that just an alloy with some combination of face-centered cubic and body-centered cubic phases will not be able to engender the desired superior strength–ductility combination [165]. Instead, novel alloy design strategies will be required to obtain microstructures that could be helpful to overcome the strength–ductility trade-off. In this regard, Lu and co-workers [11] proposed the idea of combining the high entropy concept with eutectic alloy compositions, referred to as eutectic high entropy alloy. The eutectic mixture consisting of high entropy soft face-centered cubic and hard body-centered cubic phases was expected to overcome the strength–ductility trade-off. In addition to the superior strength–ductility combination, a few major characteristics of the eutectic high entropy alloys were also highlighted which could make it a possible candidate for high-temperature applications [11]. In eutectic high entropy alloys, the equilibrium microstructure could be stable and the low energy phase boundaries could be less vulnerable to any major transformation at higher temperatures.

In the process of realizing the eutectic high entropy alloy concept, Lu and co-workers [11] produced the first such alloy of its kind i.e., $AlCoCrFeNi_{2.1}$ (molar fraction). The as-cast alloy (at a scale of ~2.5 Kg) showed the eutectic microstructure with fine lamellae. The liquid phase transformed to face-centered cubic (cobalt-chromium-iron-rich) and body-centered cubic (aluminum–nickel-rich) phase during isothermal eutectic reaction. The room temperature fracture strength was found up to be 944 MPa with ductility of 25.6%. It is noteworthy that the alloy showed extraordinary properties at elevated temperatures, retaining the strength and ductility of 806 MPa and 33.7%, respectively, at 600°C. At 700°C, however, the alloy did exhibit a marginal decrease in strength and ductility i.e., 538 MPa and 22.9%, respectively.

Before moving further, it would be worthwhile to mention that various reports on several high entropy alloys also showed the eutectic microstructure, which were purely accidental [5, 31, 32, 42, 124, 166]. Li and co-workers [5], while studying the effect of various alloying additions on the microstructure of AlCrFeCoNiCu alloy, encountered the transition of dendritic microstructure to the eutectic cells with the addition of titanium. In another report, Ma and co-workers [124] reported the formation of eutectic microstructure with body-centered cubic and ordered Laves phase

(CoCr)Nb-type with niobium addition in an AlCoCrFeNi high entropy alloy. Gou and co-workers [42] found the formation of a near-eutectic $Al_2CrCuFeNi_2$ alloy in which a microstructure with a sunflower-like pattern was observed (Figure I.11a).

After the successful illustration of a eutectic high entropy alloy in terms of microstructural stability and the mechanical properties by Lu and co-workers [11], many other investigations were also undertaken for the development of such alloys. He and co-workers [52] designed the $CoCrFeNiNb_x$ eutectic high entropy alloy and found the microstructural transition from divorced eutectic to hypo-eutectic and hyper-eutectic structures with the niobium addition from 0.1 to 0.8 molar ratios. In their very next report, He and co-workers [73] presented the stability of lamellar structures in the previously designed $CoCrFeNiNb_x$ alloy. It was revealed that the alloy exhibited microstructural stability up to 750°C, after which rapid coarsening of the lamellar structure occurred. Jiang and co-workers [39] claimed to discover few new eutectic

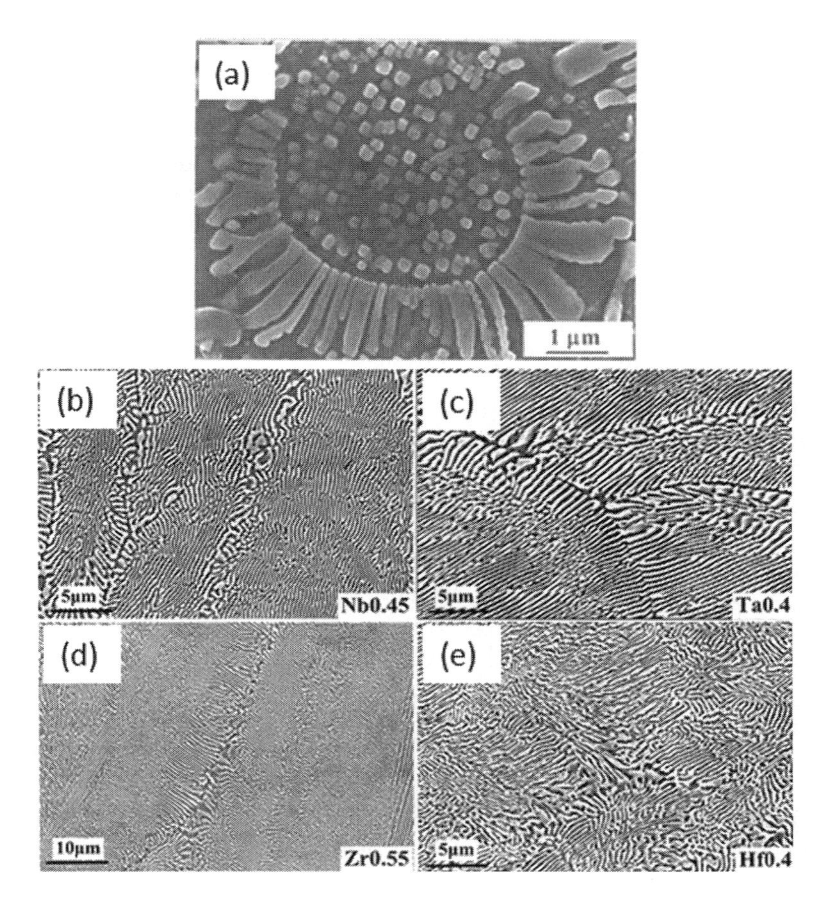

FIGURE I.11 (a) Sunflower-like microstructure in the $Al_2CrCuFeNi_2$ eutectic high entropy alloy [42] and (b–e) eutectic high entropy alloy composed on face-centered cubic and Laves phases for compositions $CoCrFeNiNb_{0.45}$, $CoCrFeNiTa_{0.4}$, $CoCrFeNiZr_{0.55}$, and $CoCrFeNiHf_{0.4}$, respectively [39].

high entropy alloys, namely, $CoCrFeNiNb_{0.45}$, $CoCrFeNiTa_{0.4}$, $CoCrFeNiZr_{0.55}$, and $CoCrFeNiHf_{0.4}$ (Figure I.11b–e) using their design strategy of combining the enthalpy of mixing and constituent binary eutectic composition. This new design strategy for eutectic high entropy alloys can be useful for locating eutectic compositions. Recently, a cost-effective cobalt-free alloy with composition $CrFeNi_{(3-x)}Al_x$ was reported, in which the variant with x = 0.8 was observed to possess face-centered cubic + ordered body-centered cubic (B2 phase) eutectic structure [115]. The full eutectic microstructure for x = 0.8 for the alloy $Co_2Mo_xNi_2VW_x$ possessed highest compressive strength of 2,364 MPa among all the other variants with x ranging from 0.5 to 1.75 molar ratio [139]. In another important report on the most successful eutectic high entropy alloy, $AlCoCrFeNi_{2.1}$, Gao and co-workers [167] investigated the deformation behavior of individual phases in the alloy, which resulted in a superior combination of strength and ductility. It was found that the face-centered cubic ($L1_2$) phase with low stacking fault energy deformed due to a planar dislocation slip along with the formation of massive stacking faults, which led to the large strain hardening in the alloy. In the second eutectic phase, the body-centered cubic (B2), a high density of dislocations were found to get pinned by chromium-rich nano-precipitates resulting in high strength owing to the restricted dislocation motion. Zhu and co-workers [31] studied the influence of molybdenum on the microstructure development of the as-cast $AlCoCrFeNiMo_x$ alloy (x = 0 to 0.5). The typical dendritic structure of the AlCoCrFeNi alloy changed to the eutectic structure when $x > 0.1$.

Although eutectic high entropy alloys can potentially overcome the so-called strength–ductility trade-off and are an important candidate for the high-temperature application, this class of high entropy alloy did not get ample attention from the research community. Therefore, eutectic high entropy alloys should be studied extensively in order to harness the known advantages of the eutectic microstructure with high-temperature stability of microstructure and low energy interfaces. These alloys show relatively better castability too [11].

I.4.4 Metastability

The stability of the microstructure in compositionally concentrated high entropy alloys has been a debatable topic since its inception. From the metallurgist point of view, it is intriguing to obtain a non-complex microstructure consisting of two–three numbers of simple solid-solution phases in an equiatomic five–six element system. Aided by various illustrations, the initial belief of massive entropic contribution toward the solid-solution phase stabilization gradually shifted to the presence of metastability in the microstructure of such alloys [12, 168, 169]. Research in the field of metastability in high entropy alloys was provoked under the prevailing condition of the dismal performance of high entropy alloys with simple microstructures. Moreover, an outstanding combination of strength and ductility achieved in certain high entropy alloys was attributed to various forms of microstructural metastability in various studies [12, 14]. Metastability in a system can be classified into three forms: (i) compositional metastability; (ii) structural metastability; and (iii) morphological metastability [170]. Various types of metastability, if present in high entropy alloys, can offer an astronomical number of possibilities of obtainable microstructure. This can create a pool of microstructure and the corresponding processing condition to choose from and achieve the desired properties. In the following text, compositional and structural metastability pertaining to high entropy alloys has been discussed.

It has been proven in a number of instances that the solid-solution phases in high entropy alloys are the ones which solidify first from the liquid state. The particular phase may get retained if it is thermodynamically stable in the entire solidification temperature range. However, if that particular phase is metastable, it may form other phases, including precipitates, depending on the transformation kinetics. The presence of metastable phases in an alloy system is related to the time and temperature-dependent kinetics of elemental diffusion. To this end, the equiatomic FeMnCrCoNi alloy has been found to possess a highly stable face-centered cubic structure at room temperature. Stability of the face-centered cubic structure can be assessed from the fact that the alloys fabricated by various routes such as induction melting [151] or arc melting [45, 143], further processed by hot rolling [143], cold rolling [17], cryogenic rolling [157], or deformed severely by high pressure torsion (HPT) [153] etc., are all composed of face-centered cubic structures. However, in recent years, a few studies have revealed that the particular alloy can decompose into precipitates and other phases by heat treatment at intermediate temperatures. In this context, Schuh and co-workers [153] revealed that highly deformed nano-grained alloy developed a mixture of a chromium-rich phase and a (nickel + manganese)-rich phase after annealing it at 450°C for a short duration of five minutes. What followed next was the development of a new (iron + cobalt)-rich phase after a long annealing of 15 h. In another study, Pickering and co-workers [65] found that the alloy decomposed into a chromium-rich σ phase during annealing at 700°C for at least 500 h and its volume fraction increased when the annealing duration increased. $M_{23}C_6$ type carbides were also observed at relatively shorter annealing duration of 125 hours. These studies clearly showed that the face-centered cubic phase remains stable above 800°C [168, 171], and even the precipitates disappeared at a temperature above 800°C. This implied that the face-centered cubic phase was a high-temperature phase. From these observations, it can be inferred that the strain-free equiatomic FeMnCrCoNi system is fairly stable and can decompose at intermediate temperatures only when kept for longer durations. This depicted the sluggishness in the diffusion of constituent elements at lower temperatures. Insufficient diffusion kinetics thus helped in retention of the high-temperature phase, face-centered cubic, which of course is a metastable phase. The other high entropy alloy class, AlCoCrFeNi alloys, is also reported to be compositionally metastable. Several works in the literature have shown the decomposition of the parent structure when annealed at some intercritical temperature. To this end, the as-cast microstructure of a AlCoCrFeNi alloy consisting of a face-centered cubic (dendrite region) and body-centered cubic (interdendrite region) showed precipitation of the brittle σ phase when annealed at 850°C [172]. Annealing at 975°C caused the transformation of the σ phase back to a body-centered cubic phase, depicting that the body-centered cubic phase in the as-cast structure was the high-temperature phase which got retained at room temperature during solidification. The initial face-centered cubic structure in the cold rolled and homogenized $Al_{0.3}CrCuFeNi_2$ alloy formed ordered $L1_2$ nano-precipitates after annealing at 550°C [169]. Several other variants of FeMnCrCoNi and AlCoCrFeNi class of high entropy alloys containing a small fraction of titanium, molybdenum, niobium, etc. showed alteration in metastability in terms of precipitation and second phase formation [66, 106, 135, 141, 73].

In one of the studies on the equiatomic AlCuCrFeNi system by the present authors, the as-cast structure consisted of chromium-iron-rich dendritic regions, with fine precipitation of aluminum–nickel-rich precipitates, and interdendritic region showed an aluminum–nickel-rich phase with a fine chromium-iron-rich phase (Figure I.12). Copper segregation was not observed in the as-cast condition. However, a heat treatment at 900°C for 1 h led to precipitation of copper in interdendritic regions. This shows the metastable state of the as-cast alloy prepared by vacuum-arc melting and casting. It has also been observed that slow cooling in the furnace engendered a larger size of the primary phase and the precipitates.

In another study to assess the solidification behavior and evolution of metastability of $AlCoCu_{0.50}Cr_{0.75}FeNi$ alloy, the spray formed alloy (as shown in Figure I.10b) was reheated to 1,300°C for 1 h in a two-phase region (liquid + body-centered cubic; body-centered cubic dissolution temperature 1,330°C as obtained by Thermo-Calc™ using the HEA1 database) and allowed to cool in the furnace. The structure evolved was similar to the as-cast structure (Figure I.13a). However, an equiaxed, seemingly single-phase, microstructure evolved when the alloy was water quenched from the same temperature (Figure I.13b). The composition of different grains was almost the same. When the spray formed alloy was kept at 1,200°C for 3 h, followed by water quenching, a two-phase structure was observed. The matrix grains (gray contrast) were rich in aluminum–nickel, whereas the grain boundary phase, as well as small precipitates in the matrix grains (white contrast), were rich in chromium-iron (Figure I.13c). However, the compositional richness of these phases was not as prominent as observed in the as-cast condition. The alloy quenched from 1,300°C (Figure I.13b) was further subjected to heat treatment at 700°C (Figure I.13d), 850°C (Figure I.13e), and 1,000°C (Figure I.13f) for 2 h each. With an increase in the heat treatment temperature, the precipitate size has increased. Grain boundary precipitates were also observed, which increased in thickness with temperature and enveloped the grain boundaries at 1,000°C. These precipitates, both grain boundaries as well as grain interior, were chromium–iron-rich. These microstructural features indicate that various compositional metastabilities can be generated by

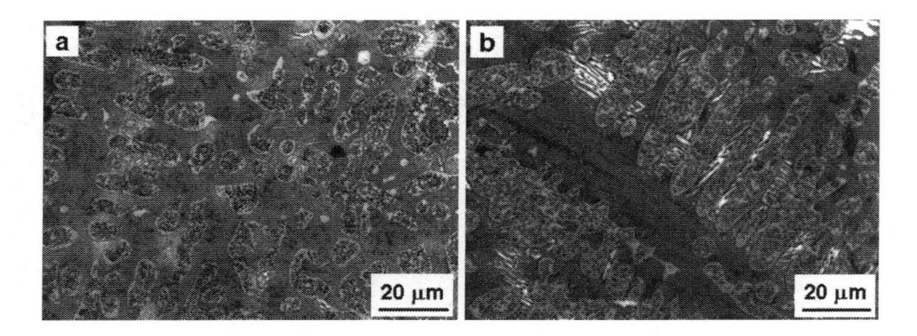

FIGURE I.12 Optical micrographs showing AlCuCrFeNi alloy in the (a) as-cast condition and (b) as-cast alloy heat treated at 900°C for 1 h.

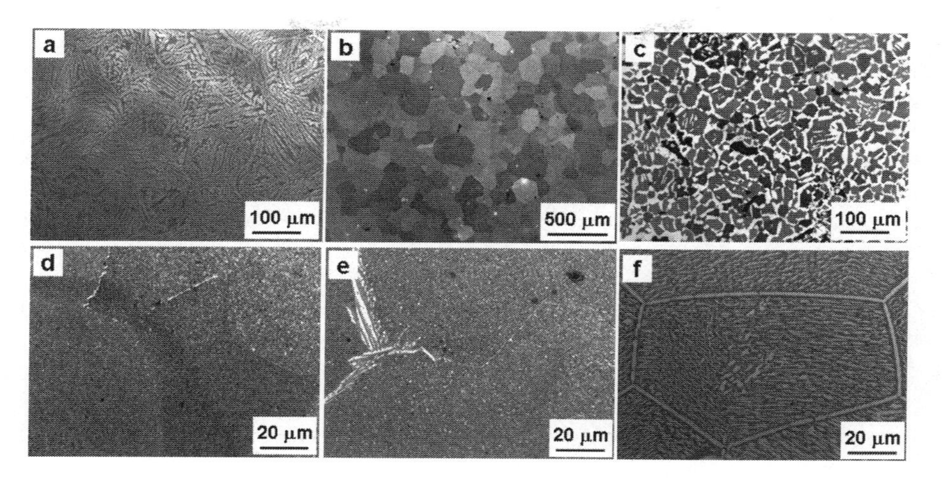

FIGURE I.13 Optical micrographs obtained for spray formed $AlCoCu_{0.50}Cr_{0.75}FeNi$ alloy in different conditions (a) 1,300°C for 1 h + furnace cooled; (b) 1,300°C for 1 h + water quenched; (c) 1,200°C for 3 h + water quenched; (d–f) 1,300°C for 1 h + water quenched followed by heat treatment for 2 h at 700°C, 850°C, and 1,000 °C and air cooling, respectively.

different solidification states and further subjected to controlled heat treatments to obtain desired second-phase precipitation. This compositional metastability in high entropy alloys can open up a large research domain in terms of microstructural engineering. The precipitation hardening in high entropy alloys can be utilized judiciously along with massive solid solution in order to obtain excellent mechanical properties.

The other kind of metastability, structural metastability, is a phenomenon harnessed very often in tuning the mechanical response of the most important structural material, steel. In this context, the stacking fault energy of the metastable face-centered cubic phase plays an important role. The face-centered cubic phase with low stacking fault energy can transform into martensite during deformation, thus providing work hardening via the transformation-induced plasticity (TRIP) mechanism [156]. Similarly, structural metastability engineering in high entropy alloys can be a potent way to overcome the strength–ductility trade-off. In a pioneering work by Li and co-workers [12], the first of its kind, dual-phase transformation-induced plasticity high entropy alloy was introduced. The nickel-free $Fe_{50}Mn_{30}Co_{10}Cr_{10}$ system showed an unprecedented combination of strength and ductility with tensile strength of ~880 MPa and fracture elongation of ~75% which were way ahead of that obtained in the equiatomic FeMnCoCrNi system [17]. The accomplishment of superior room temperature mechanical properties has been primarily attributed to the destabilized face-centered cubic phase, which during deformation transformed into a hexagonal closed pack phase. This provided the required work hardening via the so-called transformation-induced plasticity mechanism. Recently, Wung and co-workers [15] unraveled a series of manganese-free high entropy alloy performing outstandingly at −196 °C. The alloys showed strength levels as high as ~1,500 MPa and elongation to fracture as high as ~130%. In particular, the $Fe_{60}Co_{15}Ni_{15}Cr_{10}$ alloy displayed cryogenic tensile strength of ~1,500 MPa along with superior ductility of ~87%. Apart from the dislocation activities, the metastability of the face-centered cubic phase triggered the face-centered cubic to body-centered cubic transformation during deformation, playing a major role in providing work hardening. Despite the above appealing investigations, work in the field of structural metastability is limited and requires dedicated attention. In light of the exceptional mechanical properties obtained in structurally metastable high entropy alloys, it would be worthwhile to venture into the subject further in order to discover new such systems with even more exciting properties.

I.5 MATERIALS, PROCESSES, AND PROPERTY RELATIONSHIPS

The importance of various microstructures and their possible effect on mechanical properties has been discussed in the preceding Section I.4. In this section, a comprehensive and leading-edge review on the structure–property relationship has been provided for different high entropy alloys. As high entropy alloys have received widespread attention, there have been a large number of publications in this area. Hence, the high entropy alloys mainly consisting of the transition elements, along with some other elements, such as molybdenum, aluminum, carbon, vanadium, titanium, and boron, produced either through vacuum induction melting or vacuum-arc melting, have been reviewed. The data collected are up to date to the best of author's knowledge and preference has been given to recent research contributions with novel results. The dataset includes alloy compositions, types of microstructures and mechanical properties under tensile loading. The reported results of tests under compressive loading have not been considered in the present analysis. The materials with different synthesis and processing conditions and tested at different temperatures, such as room temperature, sub-zero temperatures and high temperatures, have been included. This dataset is expected to help researchers to identify the most attractive alloys among a wide range of high entropy alloys and provide a guide for future design, which is of vital importance to accelerate its development mainly in the structural applications.

The microstructures and mechanical properties, namely the yield strength, ultimate tensile strength, total elongation, and the product of ultimate tensile strength and total elongation achieved at different temperatures are given in Table I.2. The alloy compositions corresponding to a particular phase or a combination of phases have been arranged in alphabetical order for better readability and easiness to find the desired alloy composition. In contrast to Table I.1, where only the as-cast condition was considered, here we provide the microstructural constitution and mechanical properties for different high entropy alloys under a wide range of processing conditions, i.e., as-cast, homogenized, hot rolled, cold rolled, cryo-rolled, aged, and annealed (either any one or in combination). As can be seen in Table I.2, there are only a few high entropy alloys which have been produced through vacuum induction melting (marked by an asterisk at the end of composition) in comparison to vacuum-arc melting.

All the ultimate tensile strength and total elongation combinations achieved in high entropy alloys have been plotted in Figure I.14a. The data has been categorized

TABLE I.2
Mechanical Properties Achieved in High Entropy Alloys at Room, Sub-Zero, and High Temperatures for Different Alloy Compositions and Processing (YS – Yield Strength, UTS – Ultimate Tensile Strength, TE – Total Elongation)

Alloy composition	Processing history	Phases	Temperature (°C)	YS (MPa)	UTS (MPa)	TE (%)	UTS×TE (GPa%)	References
$Al_4(CoCrFeMnNi)_{96}$	As cast			219	501	58.7	29.41	[36]
$Al_7(CoCrFeMnNi)_{93}$				243	528	47.2	24.92	
$Al_{0.3}CoCrFeMn_{0.1}Ni^*$				149	336	49	16.46	[105]
$Al_{0.3}CoCrFeMn_{0.3}Ni^*$	As cast			158	371	46	17.07	
$Al_{0.4}CoCrFeMnNi^*$	Cast+Hom.+CR+Ann. (1,100°C – 3 h)			–	620	40.6	25.17	[176]
$Al_{0.25}CoCrFe_{1.25}Ni_{1.25}$	As cast	Face-centered cubic	Room temperature (20/25°C)	–	430	46	19.78	[177]
	Cast+CR (13.3%)			–	450	25.3	11.38	
	Cast+CR(26.7%)			–	507	5.9	29.91	
	Cast+CR(40%)			–	596	5.5	32.78	
	Cast+CR(60%)			–	642	3.3	21.19	
	Cast+CR(80%)			–	703	2.7	18.98	
$Al_{0.3}CoCrFeNi^*$	As cast			119	295	59	17.40	[105]
$Al_{0.3}CoCrFeNi$	Cast+Hom.+CR+Ann. (1,150°C)			–	400	65	26.00	[178]
$Al_4(CoCrFeNi)_{94}Ti_2$	Cast+Hom. (1,200°C –4 h)			–	503	67.5	33.95	[179]
$Al_{0.5}CrCuFeNi_2$	As cast			361	500	16	80.00	[112]
	Cast+CR+Ann. (1,100°C – 1 day)			355	633	3.3	20.89	
$Al_5Cr_{12}Fe_{35}Mn_{28}Ni_{20}$	As cast			280	556	48.2	26.80	[113]
$Al_{10}Cr_{12}Fe_{35}Mn_{23}Ni_{20}$				320	625	42.45	26.53	
$Al_5Cr_{12}Fe_{35}Mn_{28}Ni_{20}$	Cast+CR (90%)			1253	1392	8.2	11.41	
$Al_{10}Cr_{12}Fe_{35}Mn_{23}Ni_{20}$				1400	1500	7.85	11.77	

(*Continued*)

TABLE I.2 (CONTINUED)
Mechanical Properties Achieved in High Entropy Alloys at Room, Sub-Zero, and High Temperatures for Different Alloy Compositions and Processing (YS – Yield Strength, UTS – Ultimate Tensile Strength, TE – Total Elongation)

Alloy composition	Processing history	Phases	Temperature (°C)	YS (MPa)	UTS (MPa)	TE (%)	UTS×TE (GPa%)	References
$Al_{7.5}Cr_6Fe_{40.4}Mn_{34.8}Ni_{11.3}$	As cast			–	390	48.4	18.88	[114]
$Al_{7.5}Cr_6Fe_{40.4}Mn_{34.8}Ni_{11.3}$+1.1 at.%C				–	750	57.7	43.27	
$Al_{0.6}CrFeNi_{2.4}$				441	757	45	34.06	[115]
$Co_{10}Cr_{10}Fe_{50}Mn_{30}$	Cast+Hom.+HR+Ann.+HR (60% at 800°C)			–	974	64	62.34	[180]
	Cast+Hom.+HR+Ann.+HR (60% at 1,000°C)			–	775	57	44.17	
	Cast+Hom.+ HR+Ann. + HR (60% at 1,250°C)			–	780	39	30.42	
CoCrFeMnNi	As cast	Face-centered cubic	Room temperature (20/25 °C)	210	495	61.8	30.59	[36]
CoCrFeMnNi*	Cast+HR+Hom.+CR+Ann. (900°C – 3 min)			–	545	68	37.06	[181]
	Cast+HR+Hom.+CR+Ann. (900°C – 3 min)+H-charge			–	571	71	40.54	
CoCrFeMnNi	Cast+Hom. (1,100°C –12 h)			196.7	489.8	69.3	33.94	[182]
	Cast+Hom.+CR+Ann. (800°C – 1 h)			378.7	700.1	54	37.80	
	Cast+Hom.+CR+Ann. (800°C – 1 h)			–	660	60	39.60	[17]

(*Continued*)

TABLE I.2 (CONTINUED)
Mechanical Properties Achieved in High Entropy Alloys at Room, Sub-Zero, and High Temperatures for Different Alloy Compositions and Processing (YS – Yield Strength, UTS – Ultimate Tensile Strength, TE – Total Elongation)

Alloy composition	Processing history	Phases	Temperature (°C)	YS (MPa)	UTS (MPa)	TE (%)	UTS×TE (GPa%)	References
CoCrFeMnNi	Cast+Hom.+CR+Ann. (1,150°C – 1 h)			–	535	78	41.73	[17]
CoCrFeMnNi*	Cast+Hom.+CR+Ann. (800°C – 10 min)			387	690	66.3	45.75	[183]
	Cast+Hom.+CR+Ann. (1,100°C – 3 h)			–	530	52	27.56	[176]
$Co_{10}Cr_{25}Fe_{20}Mn_5Ni_{40}$*	Cast+Hom.+CR+Ann. (900°C – 10 min)			477	745	55.5	41.35	[183]
	Cast+Hom.+CR+Ann. (950°C – 10 min)			466	730	59.6	43.51	
$Co_{20}Cr_{25}Fe_{10}Mn_5Ni_{40}$*	Cast+Hom.+CR+Ann. (900°C – 10 min)			481	790	65.8	51.98	
CoCrFeMnNi+0.5 at.% C	Cast+Hom. (1,100°C –12 h)			260	570	50	28.50	[182]
CoCrFeMnNi+1.0 at.% C				406.1	748.7	32.2	24.11	
CoCuFeMnNi	Cast+Hom.+CR+Ann. (1,000°C – 2 h)			499	823	32.03	26.36	[184]
CoCrFeNi	Cast+Hom. (1,200°C – 4 h)			–	453	69	31.26	[179]
$(CoNi)_{35}Cr_{10}Fe_{55}$*	Cast+Hom.+CR+Ann. (800°C – 10 min)			–	550	68	37.40	[15]
$(CoNi)_{32.5}Cr_{10}Fe_{57.5}$*				–	565	86	48.59	
$CoCrFeNiTi_{0.2}$	Cast+Hom.+CR+Ann.+Aged (800°C – 1 h)			702	1240	36	44.64	[185]
CoCrNi	Cast+Hom.+CR (70%)			–	1300	15	19.50	[186]
	Cast+Hom.+CR+Ann. (600°C – 1 h)		Room temperature (20/25 °C)	1112	1260	23	28.98	
	Cast+Hom.+CR+Ann. (600°C – 4 h)			797	1100	34	37.40	
	Cast+Hom.+CR+Ann. (700°C – 1 h)			597	1000	52.6	52.60	
	Cast+Hom.+CR+Ann. (900°C – 1 h)			346	850	76	64.60	
	Cast+Hom.+CR+Ann. (925°C – 1 h)			–	850	66.3	56.35	[144]
	Cast+Hom.+CR+Ann. (800°C – 1 h)			440	880	68	59.84	[148]

(Continued)

TABLE I.2 (CONTINUED)
Mechanical Properties Achieved in High Entropy Alloys at Room, Sub-Zero, and High Temperatures for Different Alloy Compositions and Processing (YS – Yield Strength, UTS – Ultimate Tensile Strength, TE – Total Elongation)

Alloy composition	Processing history	Phases	Temperature (°C)	YS (MPa)	UTS (MPa)	TE (%)	UTS×TE (GPa%)	References
CoFeMnNi	Cast+Hom.+CR+Ann. (1,000°C – 2 h)			215	545	57.09	31.11	[184]
$CoFeMo_{0.2}Ni_2V_{0.5}$*	As cast			252	547	82.1	44.91	[174]
FeMn	Cast+Hom.+CR+Ann. (1,000°C – 2 h)			191	459	36.64	16.82	[184]
FeMnNi				357	680	57.42	39.04	
$Al_8(CoCrFeMnNi)_{92}$	As cast	Face-centered cubic + Body-centered cubic		283	645	36.5	23.54	[36]
$Al_9(CoCrFeMnNi)_{91}$				333	728	30.4	22.13	
$Al_{10}(CoCrFeMnNi)_{90}$				528	995	19	18.90	
$Al_{11}(CoCrFeMnNi)_{89}$				832	1175	7.7	90.47	
$Al_{0.6}CoCrFeMnNi$*	Cast+Hom.+CR+Ann. (1,100°C – 3 h)			–	800	29.4	23.52	[176]
$AlCoCrFeNi_{2.1}$*	As cast			–	944	25.6	24.17	[11]
	Cast+CR (8%)			–	1145	10.4	11.91	
$AlCoCrFeNi_{2.1}$	Cast+CR (90%)			–	1800	6.8	12.24	[187]
	Cast+CR (90%)+Ann. (800°C – 1 h)			1100	1175	11	12.92	
	Cast+CrR (90%)			–	1780	5.3	9.43	
	Cast+CrR (90%)+Ann. (800°C – 1 h)			1437	1562	14	21.87	
$Al_{0.3}CoCrFeNi$	Cast+Hom.+CR+Ann. (550°C – 24 h)			–	1850	5	9.25	[178]
	Cast+Hom.+CR+Ann. (550°C – 50 h)			–	1650	7	11.55	
$Al_{0.45}CoCrFeNi$	Cast+Hom. (1,250°C – 5 h)			–	580	30.3	17.57	[41]
	Cast+Hom.+CR+Ann. (700°C – 1 h)			–	1445	2.5	3.61	
	Cast+Hom.+CR+Ann. (800°C – 1 h)			–	1270	6	7.62	

(Continued)

TABLE I.2 (CONTINUED)
Mechanical Properties Achieved in High Entropy Alloys at Room, Sub-Zero, and High Temperatures for Different Alloy Compositions and Processing (YS – Yield Strength, UTS – Ultimate Tensile Strength, TE – Total Elongation)

Alloy composition	Processing history	Phases	Temperature (°C)	YS (MPa)	UTS (MPa)	TE (%)	UTS×TE (GPa%)	References
	Cast+Hom.+CR+Ann. (800°C – 2 h)			–	1160	14.8	17.17	
	Cast+Hom.+CR+Ann. (900°C – 1 h)			–	960	24.5	23.52	
	Cast+Hom.+CR+Ann. (1,000°C – 1 h)			–	920	30.2	27.78	
	Cast+Hom.+CR+Ann. (1,100°C – 1 h)			–	820	31.6	25.91	
$Al_{0.5}CrCuFeNi_2$	Cast+CR+Ann. (700°– 1 day)			630	921	4.1	37.76	[112]
	Cast+CR+Ann. (900°C –1 day)			706	1089	5.5	59.89	
$Al_{19}Co_{20}Fe_{20}Ni_{41}$	As cast			577	1103	18.7	20.67	[133]
$Al_{0.7}CrFeNi_{2.3}$				461	835	30	25.05	[115]
$Al_{0.8}CrFeNi_{2.2}$				479	956	12.7	12.14	
$Al_{0.9}CrFeNi_{2.1}$				610	1173	9.1	10.67	
$Al_{1.0}CrFeNi_{2.0}$				774	1357	6.4	8.68	
$(CoNi)_{30}Cr_{10}Fe_{60}{}^*$	Cast+Hom.+CR+Ann. (800°C – 10 min)			–	575	92	52.90	[15]
$Co_{10}Cr_{10}Fe_{50}Mn_{30}{}^*$	Cast+HR+Hom. (1,200°C – 2 h)	Face-centered cubic+Hexagonal closed pack		–	870	74	64.38	[12]
	Cast+HR+Hom.+CR+Ann. (900°C – 3 min)			–	730	50.2	36.65	
$Al_4(CoCrFeNi)_{94}Ti_2$	Cast+Hom.+CR+Ann.+Aged (800°C – 18 h)	Face-centered cubic+γ'+Ni_2AlTi		645	1094	39	42.67	[179]
	Cast+Hom.+CR+Aged (650°C – 4 h)			1005	1273	17	21.64	

(*Continued*)

TABLE I.2 (CONTINUED)
Mechanical Properties Achieved in High Entropy Alloys at Room, Sub-Zero, and High Temperatures for Different Alloy Compositions and Processing (YS – Yield Strength, UTS – Ultimate Tensile Strength, TE – Total Elongation)

Alloy composition	Processing history	Phases	Temperature (°C)	YS (MPa)	UTS (MPa)	TE (%)	UTS×TE (GPa%)	References
$Al_{0.3}CoCrFeNi$	Cast+Hom.+CR+Ann.+Ann. (620°C – 50 h)	Face-centered cubic + L12		–	850	45	38.25	[178]
$Al_{7.5}Cr_6Fe_{40.4}Mn_{34.8}Ni_{11.3}$+0.5 at.% B	As cast	Face-centered cubic+Precipitates		–	444	41.5	18.43	[114]
$Al_{7.5}Cr_6Fe_{40.4}Mn_{34.8}Ni_{11.3}$+1.0 at.%B				–	500	35.9	17.95	
CoCrFeMnNi+0.5 at.% C	Cast+Hom.+CR+Ann. (800°C – 1 h)	Face-centered cubic+M23C6		520.9	791.4	40.1	31.73	[182]
CoCrFeMnNi+1.0 at.% C				643.8	887.6	31.6	28.05	
$Al_{0.3}CoCrFeNi$	Cast+Hom.+CR+Ann. (620°C – 50 h)	Face-centered cubic+body-centered cubic+σ		–	1050	35	36.75	[178]
CoCrFeMnNi	Cast+Hom.+CR+Ann. (800°C – 1 h)	Face-centered cubic	−196	–	1115	91	101.46	[17]
	Cast+Hom.+CR+Ann. (1150°C – 1 h)			–	915	106.8	97.72	
$(CoNi)_{35}Cr_{10}Fe_{55}$*	Cast+Hom.+CR+Ann. (800°C – 10 min)			–	1000	125	125.00	[15]
$(CoNi)_{32.5}Cr_{10}Fe_{57.5}$*				–	1160	118	136.88	
$CoCrFeNiTi_{0.2}$	Cast+Hom.+CR+Ann.+Aged (800°C – 1 h)			860	1580	46	72.68	[185]
CoCrNi	Cast+Hom.+CR+Ann. (925°C – 1 h)			–	1250	76.5	95.62	[144]
	Cast+Hom.+CR+Ann. (800°C – 1 h)			660	1310	90.3	118.29	[148]
			−75	550	1050	73	76.65	
$CoFeMo_{0.2}Ni_2V_{0.5}$*	As cast		−196	419	712	34.6	24.63	[174]
			−70	350	591	28.8	17.02	

(Continued)

TABLE I.2 (CONTINUED)
Mechanical Properties Achieved in High Entropy Alloys at Room, Sub-Zero, and High Temperatures for Different Alloy Compositions and Processing (YS – Yield Strength, UTS – Ultimate Tensile Strength, TE – Total Elongation)

Alloy composition	Processing history	Phases	Temperature (°C)	YS (MPa)	UTS (MPa)	TE (%)	UTS×TE (GPa%)	References
$Fe_{60}(CoNi)_{30}Cr_{10}^{*}$	Cast+Hom.+CR+Ann. (800°C – 10 min)	Face-centered cubic+body-centered cubic	−196	–	1510	87	131.37	[15]
CoCrFeMnNi	Cast+Hom.+CR+Ann. (800°C – 1 h)	Face-centered cubic	200	–	540	43.5	23.49	[17]
CoCrFeMnNi	Cast+Hom.+CR+Ann. (800°C – 1 h)	Face-centered cubic	400	–	490	37.5	18.37	[17]
			600	–	420	51	21.42	
			800	–	135	64.7	8.73	
	Cast+Hom.+CR+Ann. (1,150°C – 1 h)		200	–	425	61	25.92	
			400	–	395	65	25.67	
			600	–	325	56	18.20	
			800	–	181	26.5	4.80	
$CoFeMo_{0.2}Ni_2V_{0.5}^{*}$	As cast		600	214	455	189.2	86.09	[174]
			700	212	397	154.3	61.26	
			800	187	317	119.5	37.88	

*Vacuum induction melting (all others without * mark are from vacuum-arc melting).
HR – Hot Rolled; CR – Cold Rolled; CrR – Cryo-Rolled; Hom. – Homogenized; Ann. – Annealed

with respect to different microstructural content as well as to the tensile test temperatures. It reveals that most of the face-centered cubic high entropy alloys (depicted by circles) can achieve a very high ductility while maintaining strength levels in the range of ~400 and 1,300 MPa. For example, the maximum value of total elongation has been reported to be ~85% at room temperature, 120% at sub-zero temperatures, and 190% at high temperatures. On the contrary, in the high entropy alloys with mixed microstructures, containing a combination of face-centered cubic and body-centered cubic phases (depicted by a triangle), the room temperature ductility has been compromised i.e., total elongation values are <40% in most of the cases. However, the room temperature strength level has been found to reach a maximum value of ~1800 MPa, in contrast to 1,300 MPa for single-phase face-centered cubic high entropy alloys. These variations in the dataset i.e., high strength for body-centered cubic and high ductility for face-centered cubic, corroborate the generally known trend for these types of phases in structural grades of steel. There are also some findings on high entropy alloys containing a small number of other phases, such as hexagonal close packed, γ', σ phase, $M_{23}C_6$ carbides, Ni_2AlTi intermetallics, etc. in a face-centered cubic matrix. The ultimate tensile strength and total elongation combinations for such type of high entropy alloys have been found to be in between those with single-phase face-centered cubic and dual-phase face-centered cubic + body-centered cubic alloys.

The data on ultimate tensile strength and total elongation achieved in high entropy alloys have also been assessed in terms of the product of these two values, which denotes the energy absorption capability of any material during the course of tensile deformation. As shown in Figure I.14a, each dotted isopleth represents a constant value (theoretical) of energy absorption capability in GPa%, which is mentioned at the end of the curve. The single phase face-centered cubic high entropy alloys show energy absorption capability value below ~70 GPa% at room temperature, whereas these values are in the range of ~70–140 GPa% at sub-zero temperatures and ~10–90 GPa% at high temperatures. The room temperature energy absorption capability of mixed microstructures, containing other phases in addition to the face-centered cubic phase, has been observed to reduce significantly in comparison to the single phase face-centered cubic high entropy alloys. For example, the high entropy alloys containing a mixture of face-centered cubic and body-centered cubic phases have been reported to reach a maximum energy absorption capability of ~30 GPa%. This value was improved i.e., in the range of 10–50 GPa%, when the high entropy alloys contain other phases (hexagonal closed pack, γ', σ, $M_{23}C_6$, Ni_2AlTi, L12, precipitates) in a face-centered cubic matrix.

The major deviation from the cluster of the dataset, for a particular class of high entropy alloys, is marked by A and B in Figure I.14a. The dataset in "A" shows exceptionally high ductility at high temperatures for $CoFeMo_{0.2}Ni_{2.0}V_{0.5}$ alloy, even in the as-cast condition, which was reported by Jiang and co-workers [174] very recently. On the contrary, the other dataset at high temperatures was reported by Otto and co-workers [17] after cold rolling and annealing of the as-cast and homogenized CoCrFeMnNi high entropy alloy. As both of the above alloys contain single-phase face-centered cubic structure, these findings highlight the remarkable role of alloy composition in enhancing the ductility of high entropy alloys at high temperatures. This exceptionally high ductility is reported to be the result of unusual

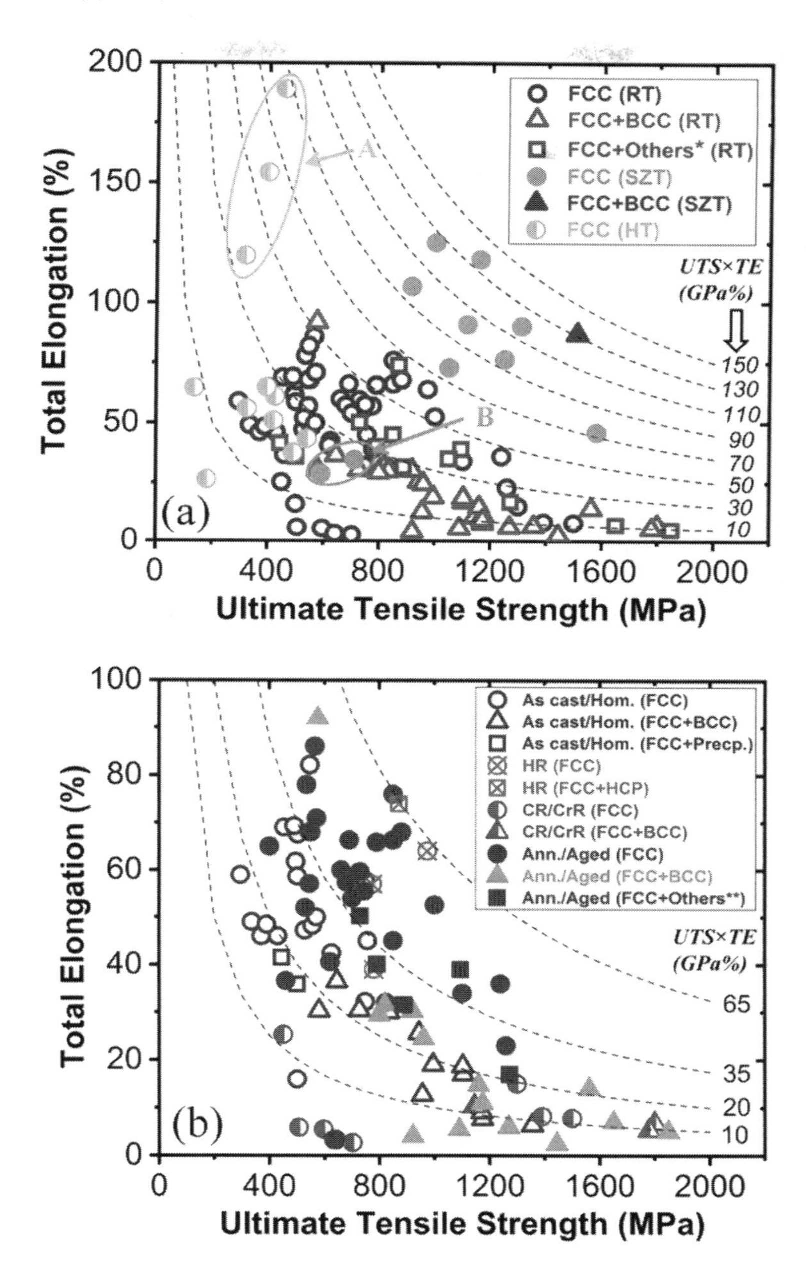

FIGURE I.14 Ultimate tensile strength and total elongation combinations for different phases in high entropy alloys (a) tested at different temperatures and (b) tested at room temperature after different processing conditions (each isopleths represent a constant value (theoretical) of energy absorption capability, as indicated in the plot; * – hexagonal closed pack, γ', σ, $M_{23}C_6$, Ni_2AlTi, L12, precipitates; ** – hexagonal closed packed, γ', $M_{23}C_6$, Ni_2AlTi; RT – room temperature; HT – high temperature; SZT – sub-zero temperature; dataset marked as A and B highlights the major deviation in properties).

strain hardening ability of the studied alloy system. This high strain hardening is the consequence of the coexistence of planar and cross-slip leading to a simultaneous increase in both the strength and the ductility. Interestingly, the $CoFeMo_{0.2}Ni_{2.0}V_{0.5}$ high entropy alloy, which showed exceptional ductility at high temperatures, showed poor combinations of ultimate tensile strength and total elongation at sub-zero temperatures (marked by "B" in Figure I.14a). For example, the energy absorption capability of most of the high entropy alloys at sub-zero temperatures was found to vary in the range of 70–140 GPa%, which was reduced to below ~30 GPa% for the $CoFeMo_{0.2}Ni_{2.0}V_{0.5}$ high entropy alloy, despite having similar single-phase face-centered cubic structure. As can be seen in Table I.2, all the high entropy alloys tested at sub-zero temperatures were in the annealed state after cold rolling, whereas the $CoFeMo_{0.2}Ni_{2.0}V_{0.5}$ high entropy alloy was in the as-cast state. Hence, this reduction in energy absorption capability is expected to arise from the microstructural differences that ensued after various processing conditions prior to tensile testing.

The effect of processing conditions on ultimate tensile strength and total elongation values of high entropy alloys tested at room temperature is shown in Figure I.14(b). The whole gamut of ultimate tensile strength and total elongation combinations has been categorized into four different types of processing conditions i.e., as-cast (open symbols), hot rolled (open symbols with cross mark), cold/cryo-rolled (half-filled symbols), and annealed/aged (filled symbols). It is clearly evident from the figure that annealing or aging treatment on high entropy alloys, without any body-centered cubic phase, leads to a significant improvement in the ultimate tensile strength and total elongation combinations, in comparison to the as-cast or cold/cryo-rolled alloys. As discussed earlier, this enhancement of ultimate tensile strength and total elongation values after the annealing treatment, in comparison to the as-cast condition ("B" in Figure I.14a), is also applicable to sub-zero temperature data. The majority of the as-cast high entropy alloys have been shown to achieve a maximum energy absorption capability value of ~35 GPa%. Among these, single-phase face-centered cubic alloys have achieved higher ductility and lower strength. The opposite is true for the alloys with face-centered cubic + body-centered cubic phases. After the cold rolling of the as-cast structure, although the ultimate tensile strength increased for some of the alloys, the total elongation values have been observed to drop drastically to below 10%. Interestingly, the annealed/aged high entropy alloys, with dual-phase microstructures (body-centered cubic + face-centered cubic), also did not lead to any improvement in the energy absorption capability in comparison to the as-cast condition. However, the annealed/aged high entropy alloys, containing single-phase face-centered cubic and face-centered cubic with hexagonal closed pack phase or precipitates, show a significant improvement in the energy absorption capability i.e., in the range of ~20–65 GPa%. Most of the single-phase face-centered cubic alloys have been reported to be ductile while maintaining a similar strength level compared to that of face-centered cubic high entropy alloys with hexagonal closed pack phase or precipitates. Similar to these alloys, the hot rolled high entropy alloys have also been shown to achieve excellent combinations of the ultimate tensile strength (~1,000 MPa) and total elongation (~75%).

Although the high entropy alloys have been shown to achieve better energy absorption capability, it is important to assess these values with respect to the other

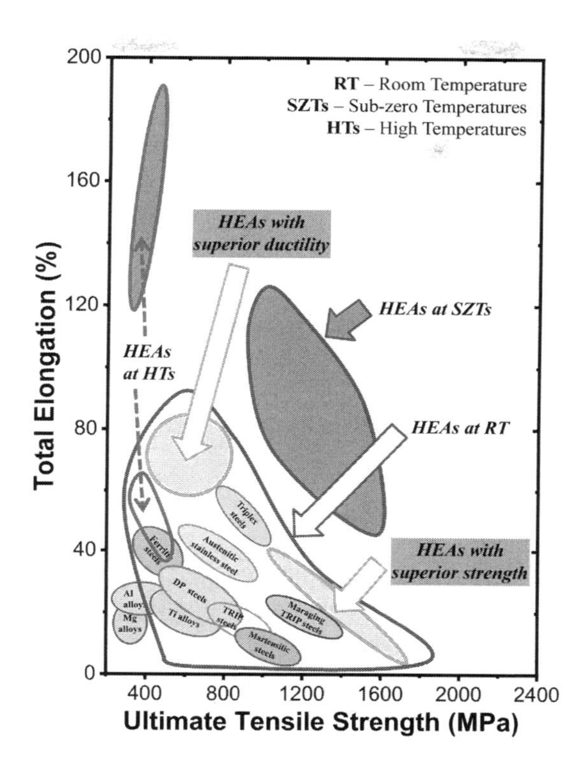

FIGURE I.15 Comparison of ultimate tensile strength and total elongation combinations achieved in high entropy alloys (HEAs) with the structural alloys. The data for ferrous and non-ferrous materials have been taken from [175] and has been superimposed with the high entropy alloy dataset from Table I.2.

structural alloys. Figure I.15 shows a banana diagram (ultimate tensile strength vs total elongation plot) for structural materials, including the high entropy alloy datasets. The ultimate tensile strength and total elongation combinations achieved in high entropy alloys have been incorporated in the existing diagram. The data on high entropy alloys have been categorized according to tensile test temperatures e.g., room temperature, sub-zero temperatures, and high temperatures for better comparison. The eminence of the high entropy alloys at sub-zero temperatures is clearly evident. This makes high entropy alloys a potential candidate for applications, which requires better ultimate tensile strength and total elongation combinations at sub-zero temperatures. At high temperatures, some of the high entropy alloys can achieve an exceptionally high ductility with reasonable strength. The high entropy alloys at room temperature show a wide range of ultimate tensile strength and total elongation values. This highlights the importance of alloy composition and processing conditions to achieve high energy absorption capability. It is observed that some of the ultimate tensile strength and total elongation combinations attained in high entropy alloys coincide with the high-grade steels such as austenitic stainless steel and triplex steels. In this context, the cost and arduous processing of high entropy alloys should be taken into account while looking for a replacement for such high-grade steels with high entropy alloys. Despite the similar properties of some of the high entropy alloys and steels, there is a vast range of ultimate tensile strength and total elongation combinations which are far superior to the presently known structural materials. Such regions with either superior ductility or superior strength have been marked in Figure I.15. Hence, this wide range of high entropy alloys can be considered for any specific application. It seems from the present discussion that composition, microstructural homogeneity achieved after thermomechanical processing, the precipitation of second phases, and microstructural constitution play a determining role in the evolution of mechanical response of these materials.

I.6 CHALLENGES AND PERSPECTIVES ON FUTURE DEVELOPMENTS

Since the first report on high entropy alloys in 2004, researchers have adopted a multipronged strategy to identify promising alloy systems from the vast composition space available [13]. In the beginning, the design principles were based on parametric judgment of whether a certain composition will give rise to a single-phase solid solution [20, 25, 77, 79, 80]. In addition, several studies were conducted based on the trial-and-error method where a small tweaking of composition or addition/deletion of an element was adopted. This definitely led to a significant experimental database on microstructure, phase constitution, and properties. Later, phase diagram calculation was found to be the most effective tool to address the limitations of alloy design and this led to rapid screening of alloys. In addition, several new approaches of materials design, as discussed in Section I.3, were borrowed from other materials classes and applied to high entropy alloys. All these alloy design strategies have culminated in a better understanding of high entropy alloys. However, we have witnessed only the tip of the iceberg, as there is a vast possibility which needs to be uncovered, from both a fundamentals, as well as an applications viewpoint. Moreover, the efforts of the researchers are restricted by the unavailability of reliable thermodynamic and mobility databases required for effective phase predictions. The difficulty is mainly in extrapolating the existing data into metastable regions and higher-order systems. The exploration of data for ternary systems is needed, which is currently limited and inconsistent [59]. Both the parametric and theoretical methods have their own limitations; the former being an oversimplification to find an empirical relationship and the latter being expensive and with inadequate accuracy. In addition, many of the studies on high entropy alloys have been done by making small samples using vacuum-arc melting where the cooling conditions are too far from equilibrium. Even if these materials show a single phase in as-cast condition, mostly they are supersaturated solid solutions which is manifested during long-duration annealing treatment at intermediate temperatures [188, 189]. Therefore, it is difficult to believe the single-phase formation phenomenon in many of the studies on as-cast high entropy alloys, particularly the small-scale castings that experience relatively high cooling rates. This might lead to inaccuracies in empirical correlations observed. Moreover, in the majority of the studies on these small-scale castings, strength, and ductility have been estimated using compressive tests. This makes it difficult to compare these properties with the conventional tensile test data of other high entropy alloys or any other structural materials.

In addition, the search for a single-phase solid solution using the computational high-throughput strategy does not ensure good properties of the material, according to a given application e.g., structural. Furthermore, solid solution hardening is a major gap in scientific understanding, particularly for multi-principal multicomponent alloys, and requires focused effort to develop physics-based models to ensure better predictability for high entropy alloy compositions. Thereafter, solid-solution strengthening fundamentals/tools need to be effectively integrated with the alloy screening framework to achieve desired results [13]. These can be further augmented by experimental high-throughput synthesis of high entropy alloys, with compositional gradient under the combinatorial search framework, and their microstructure and property evaluation. Laser-engineered net-shaping (LENS) with directed energy deposition (DED) can be of immense promise to produce compositionally gradient materials for evaluation and making a materials' library. Despite the principles and definition of high entropy alloys i.e., more than five elements, high entropy, single-phase, solid solution, current interest is growing in the direction of multiphase precipitation-hardened high entropy alloys, wherein solid-solution strengthening along with nanoscale precipitation is expected to surpass the present level of properties achieved [34, 66, 186]. Also, precipitation hardening is one of the promising mechanisms for high-temperature strengthening. Therefore, one of the strategies should be to design alloys which can give a certain amount of supersaturation during non-equilibrium solidification and predicting the amount and composition of precipitates that will form at a given temperature. The role of controlled volume fraction and distribution of intermetallics in a high strength solid solution can be an important direction for new alloy design and development [190]. This can be achieved within the CALPHAD framework. In addition, the properties of high entropy alloys also depend upon the process route, thermomechanical processing, and heat treatments that determine the recrystallization phenomenon, grain size, phase constitution, precipitate volume fraction, precipitate size and distribution, and dislocation density within the materials. These require the creation of a detailed database of different alloy systems. Rapid alloy prototyping of bulk high entropy alloys can be one of the best experimental combinatorial alloy screening processes to explore the optimized alloy composition, within the computationally screened alloys, for structural materials [191].

Though the fundamental scientific understanding of high entropy alloys has reached a significant level, and newer alloys for application are imminent, the question that still remains is the cost. As high entropy alloys are multi-principal component alloys, the inclusion of expensive materials to achieve given properties e.g., specific strength, can offset the advantages derived over conventional alloys [190]. According to the calculations of Senkov and co-workers [29], only 10% of three-component and 2.4% of six-component equiatomic alloys, out of 4983 unique alloys, cost <10$/kg. Therefore, in all the alloy design attempts, cost reduction on raw materials as well as the processing route must be a paramount concern to be addressed.

I.7 POTENTIAL APPLICATIONS OF HIGH ENTROPY ALLOYS

The field of high entropy alloy has witnessed tremendous interest in the last decade and this is mainly due to the promising structural and functional properties of these materials. Concentrated efforts are being made by research groups all over the world to search for potential materials' compositions, processes, and properties for given applications. The potential of a new material system for an application, functional or structural, is seen based on their capability to replace existing material due to performance efficiency, economy, availability, environmental effect/resistance, multifunctionality, and material saving, or altogether new and unprecedented properties. As discussed in the previous sections, high entropy alloys have shown a vast possibility of new material compositions, microstructures, and properties despite the fact that only a miniscule portion of the wide composition space has been explored. Although a comparison of properties obtained and the properties required suggests the suitability of high entropy alloys for a given application, the perception of future development is mostly limited by the knowledge available today. However, as of today, several high entropy alloy families have been identified keeping in view the prospective application areas, e.g., 3d transition metal structural materials, refractory metals for high-temperature applications, low-density structural metals for aerospace and transportation, brasses and bronzes for higher strength levels, and precious metals alloys for catalysis applications [13]. In addition, high entropy alloy carbides, nitrides, and borides have also been considered as one of the promising materials to be employed for thin-film and bulk coatings for protection from wear damages and for diffusion barriers [43]. These high entropy alloys have attracted much of the attention due to their many properties, such as high strength, high-temperature stability, high corrosion and wear resistance, resistance to irradiation, superplasticity, magnetic, electrical, and thermal properties etc. However, it is impossible to achieve all these properties in a single high entropy alloy composition. Therefore, alloy design to achieve a combination of properties for a specified application requires thorough exploration of the existing possibilities and future database generation. Although a large number of publications are devoted to high entropy alloys, the work related to the functional properties evaluation of high entropy alloys is not being undertaken widely. The alloy design approaches being employed to find out suitable compositions have already been discussed in Section I.3.

The high entropy alloy application areas have already been discussed by several researchers [13, 18, 43, 51, 192, 193]. The conventional materials based on one or two principal elements have already reached their limit, particularly with reference to their performance. However, the technological advancement is restricted due to the availability of high-performance materials in the application areas, such as engine materials requiring high-temperature strength, better oxidation, and improved corrosion and creep resistance; nuclear materials with low irradiation damages; hardfacing and cutting tool materials with high wear and oxidation resistance, high elevated temperature strength and toughness, and low friction coefficient; materials with high corrosion and wear resistance for marine structures and chemical plants; coating materials with high hot hardness, toughness, and corrosion and wear resistance; light materials for transportation industries with high specific strength and toughness, formability, and fatigue strength. These application areas are not exhaustive and many new and demanding applications exist [18, 192]. It is expected that more possible applications of high entropy alloys will come up as the alloy designs and property evaluation move toward maturity. Yeh and Lin [18] pointed out some possible breakthrough applications of high entropy alloys, such as turbine blades for higher turbine efficiency, thermal spray deposition for a protection layer with high corrosion and wear resistance [161], molds and dies for high-temperature extrusion of steels, sintered carbide bonded with different amounts of high entropy alloys for cutting tool applications, and hard high entropy nitride coating of cutting tools so as to increase the hardness and toughness. Gludovatz and co-workers [68] showed exceptional damage tolerance of CrMnFeCoNi single-phase face-centered cubic solid solution with fracture toughness exceeding 200 $MPa.m^{1/2}$ at –196°C. The mechanical properties of the alloy improved at the sub-zero temperature of −196°C, particularly due to mechanical nano-twinning with decreasing temperature, indicating the possibility of their applications at cryogenic temperatures. Cost-effective production of complex shaped high entropy alloys components would require good formability along with strength. Shaysultanov and co-workers [194] showed high-temperature superplastic behavior of multiphase AlCoCrCuFeNi alloy at 1,000°C. The superplastic behavior, with 1,240% total elongation was achieved after drastic grain refinement, with an average grain size of 2.1 μm, using a multi-directional hot isothermal forging operation at 950°C. Similarly, Shahmir and co-workers [195, 196] showed superplasticity in CoCrFeNiMn after high-pressure torsion processing to obtain an average grain size of ~10 nm. They could achieve an elongation of >400% at temperatures between 600–700°C. Reddy and co-workers [197] achieved superplastic elongation of 300% for CoCrFeMnNi at 750°C. The high wear resistance of high entropy alloys has been reported by several researchers [30, 104, 198, 199]. In many of the studies, it has been found that a change in composition alters the wear resistance drastically. These changes are particularly attributed to the microstructural content achieved for the given compositions and the phase constitution. The increase in the fraction of the body-centered cubic phase with high hardness generally led to high wear resistance. The high-temperature technological applications in the aerospace, energy, oil, and gas sectors are mostly served by superalloyed materials. High-temperature superalloys have also been produced with better high-temperature mechanical properties compared to many superalloys currently in use. The $Al_xCo_{1.5}CrFeNi_{1.5}Ti_{(0.5-x)}$ with face-centered cubic matrix and uniform distribution of precipitate exhibited hardness values better than IN718 in the range of room temperature to 1,000°C [43, 51] These examples show the potential of high entropy alloys to surpass the properties obtained for conventional alloys.

The recent emphasis on additive manufacturing technology has also made its way into the high entropy alloy arena mainly due to its capability of producing components in a single step [200–202]. Fujieda and co-workers [200] were the first to demonstrate the successful selective electron beam melting (SEBM) method to process AlCoCrFeNi alloy. They showed drastic microstructural refinement as well as improved properties (~1,700–1,800 MPa compressive strength and >30% strain) compared to its as-cast counterpart. The atomized powders and as-cast materials showed the presence of only a body-centered cubic phase, whereas the selective electron beam melting sample exhibited a combination of face-centered cubic and body-centered cubic phases. Ocelik and co-workers [201] used the re-melting technique on AlCoCrFeNi alloy, using a high power laser at speeds of 5 and 10 mm/s and laser power of 300 and 450 W, respectively, and concluded that fast solidification led preferentially to body-centered cubic phase formation with increased hardness. At the same time, they could not see any heat affected zone due to the high thermal stability of the alloy system chosen. The additive manufacturing method can be of promise to produce complex-shaped components, both for structural and functional applications, particularly due to the formability issues in many of the high entropy alloys. Although the load-bearing structural material application has not been very successful with additive manufacturing as yet, non-load-bearing applications can be an option currently. The maturity of additive manufacturing in reducing defects and other metallurgical issues might be possible in the future. Chen and co-workers [202] reviewed the additive manufacturing of high entropy alloys. It is pointed out that most of the studies on additive manufacturing of high entropy alloys are on $Al_xCoCrFeNi$ or CoCrFeMnNi systems generally leading to single-phase microstructure. The additive manufacturing products of high entropy alloy showed better mechanical properties, which is attributed mainly to the refined microstructures obtained due to the rapid solidification effect and cooling in the fabrication process. Figure I.16 shows the yield strength vs elongation obtained after high entropy alloy processing using different additive manufacturing processes. The post-treatment of additively manufactured materials led to further improvement in properties due to the elimination of structural defects and reduction of residual stresses. The process of additive manufacturing can be an added advantage in the search for high entropy alloy composition, particularly considering the processing of these alloys. However, it is also a challenge to produce additive manufacturing quality alloys powders/wires at a commercial scale. A blend of elemental powders definitely can be an option. The economic viability of the additive manufacturing process has not yet been

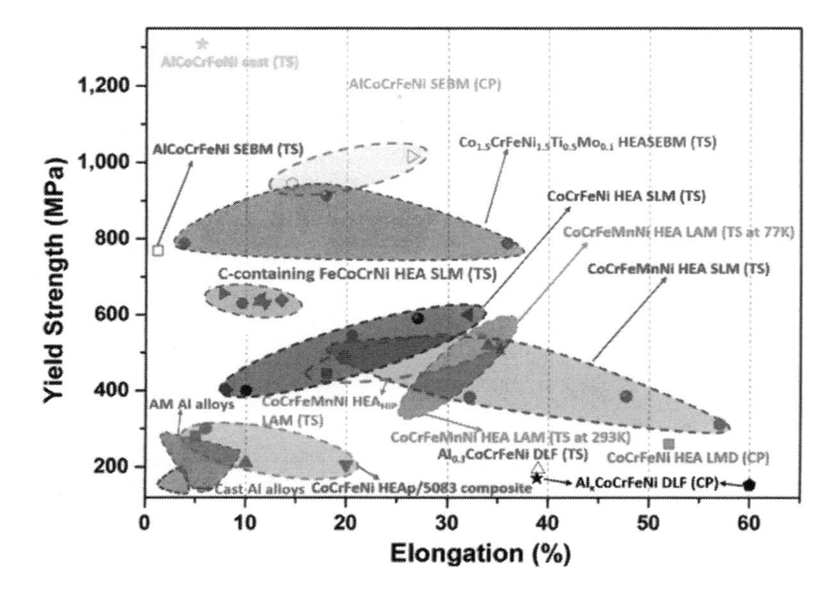

FIGURE I.16 Yield strength vs elongation using various additive manufacturing methods (TS – Tensile, CP – Compression) [202].

clearly understood, and it would be possible as processing with additive manufacturing gets wide impregnation in the market.

1.8 SUMMARY

The advent of high entropy alloys is considered to be one of the most promising developments of the recent past. It has led to a paradigm shift in the thought process hitherto employed for materials design. These developments in the area have witnessed the enormous possibilities for high entropy alloys creating avenues for new applications and/or replacing existing materials for better performance. An astronomical extent of compositional space complemented by the vast scope of microstructural development has been a strong source of motivation toward the aggravated exploration of high entropy alloy systems since their inception. Given that such quantum of scope exists, the chance of inventing new alloys with suitable microstructure and with extraordinary properties, is tremendous. However, based on the research to date, it is understandable that not all combinations of constituent alloying elements and the microstructure will be a winner. There are plenty of outliers, and only a handful of outstandingly performing high entropy alloy systems. To this end, it is important to understand that the development of an alloy system requires a strategic approach right from the alloy design to the processing treatment for tailoring the microstructure for a given application. Keeping this in mind, the present chapter aimed to bring out a perspective on the current developments in high entropy alloys in terms of thermodynamics, alloy design criteria, microstructure, and the ensuing mechanical properties and future prospects. Separate dedicated sections have been devoted to each of the above topics, covering the pertinent developments in high entropy alloy. The transition from the simplistic thermodynamic belief of the major role of just the "configurational entropy in the microstructural stabilization" to other factors such as enthalpy and non-configurational entropy has been clearly brought out. It has been shown that the simple approach of configurational entropy does have several caveats and thus was due for improvement, which was reflected in various recent studies, exploring more relevant approaches for accurate prediction of various phases. A comprehensive description of various approaches currently being adopted for alloy design in high entropy alloys has been given. These included parametric approaches, accelerated alloy screening, and computational, as well as experimental, high-throughput techniques. These alloy design approaches indicated that strategically organized efforts are being made for alloy selection in order to zero in on the best elemental combinations. Apart from these endeavors on alloy design, a large amount of experimental results is already available in the public domain. These results on microstructural development and the ensuing mechanical properties in high entropy alloys have been thoroughly analyzed based on casting route, thermomechanical processes, and phase constitution. The major focus is given to elements such as aluminum, iron, manganese, cobalt, chromium, and nickel, along with several other elements. It has been observed that the analysis of the effect of alloying elements on microstructural development of various high entropy alloy systems could help in understanding maneuvering the alloying additions in order to achieve a suitable high entropy alloy microstructure e.g., single-phase, multiphase, eutectic or metastable microstructures. A properly designed composition engendering the desired microstructure can lead to cutting-edge properties in comparison to conventional alloys. A methodical analysis of the recent literature data pertaining to the effect of process routes, composition, and microstructure on tensile behavior, at different temperatures, revealed extraordinary properties of high entropy alloys, particularly under certain process conditions such as: after cold/hot deformation and/or in annealed/aged condition. It has been observed that single-phase materials with precipitation strengthening can be a promising proposition for a good combination of strength and ductility. The sub-zero temperature tensile strength and ductility combinations surpass many of the currently known structural steel properties. The study has clearly elucidated the strong dependence of the energy absorption capability in high entropy alloys on microstructure and processing conditions.

Finally, a true approach for realizing the most from the high entropy alloys is to simultaneously explore new compositions using advanced computing with reliable thermodynamics databases; use available experimental datasets on microstructure, composition, and properties to reach optimum compositions of an alloy system; generate extensive materials data using the most versatile techniques available for high-throughput experiments; and develop a strong computing platform for accurate prediction of properties of single-phase solid solutions, as well as multiphase materials. An integration of such databases and computational tools will have a far-reaching effect on the efforts and time required and the cost to be incurred for the development of new high entropy alloys.

REFERENCES

1. Cantor, B, I T H Chang, P Knight, and A J B Vincent. 2004. "Microstructural Development in Equiatomic Multicomponent Alloys." *Materials Science and Engineering: Part A* 375–377: 213–18. doi:10.1016/j.msea.2003.10.257.
2. Yeh, Jien-Wei, Swe-kai Chen, Jon-yiew Gan, Su-jien Lin, Tsung-shune Chin, Tao-Tsung Shun, Chung-Huei Tsau, and Shou-Yi Chang. 2004. "Formation of Simple Crystal Structures in Cu-Co-Ni-Cr-Al-Fe-Ti-V Alloys with Multiprincipal Metallic Elements." *Metallurgical and Materials Transactions A* 35: 2533–36. doi.org/10.1007/s11661-006-0234-4.
3. Wang, Woei-ren, Wei-lin Wang, Shang-chih Wang, Yi-chia Tsai, Chun-hui Lai, and Jien-wei Yeh. 2012. "Effects of Al Addition on the Microstructure and Mechanical Property of $Al_xCoCrFeNi$ High-Entropy Alloys." *Intermetallics* 26: 44–51. doi:10.1016/j.intermet.2012.03.005.
4. Yeh, Jien-Wei, Swe Kai Chen, Su Jien Lin, Jon Yiew Gan, Tsung Shune Chin, Tao Tsung Shun, Chun Huei Tsau, and Shou Yi Chang. 2004. "Nanostructured High-Entropy Alloys with Multiple Principal Elements: Novel Alloy Design Concepts and Outcomes." *Advanced Engineering Materials* 6(5): 299–303. doi:10.1002/adem.200300567.
5. Li, B S, Y P Wang, M X Ren, C Yang, and H Z Fu. 2008. "Effects of Mn , Ti and V on the Microstructure and Properties of AlCrFeCoNiCu High Entropy Alloy." *Materials Science and Engineering: Part A* 498(1–2): 482–86. doi:10.1016/j.msea.2008.08.025.
6. Zhu, J M, H M Fu, H F Zhang, A M Wang, H Li, and Z Q Hu. 2010. "Synthesis and Properties of Multiprincipal Component $AlCoCrFeNiSi_x$ Alloys." *Materials Science and Engineering: Part A* 527(27–28): 7210–14. doi:10.1016/j.msea.2010.07.049.
7. Li, C, J C Li, M Zhao, and Q Jiang. 2010. "Effect of Aluminum Contents on Microstructure and Properties of $Al_xCoCrFeNi$ Alloys." *Journal of Alloys and Compounds* 504: 515–18. doi:10.1016/j.jallcom.2010.03.111.

8. Hsu, Chin-you, Chien-chang Juan, Woei-ren Wang, Tsing-shien Sheu, Jien-wei Yeh, and Swe-kai Chen. 2011. "On the Superior Hot Hardness and Softening Resistance of AlCoCrxFeMo0.5 Ni High-Entropy Alloys." *Materials Science and Engineering: Part A* 528(10–11): 3581–88. doi:10.1016/j.msea.2011.01.072.
9. Singh, S, N Wanderka, B S Murty, U Glatzel, and J Banhart. 2011. "Decomposition in Multi-Component AlCoCrCuFeNi High-Entropy Alloy." *Acta Materialia* 59(1): 182–90. doi:10.1016/j.actamat.2010.09.023.
10. Tung, Chung-chin, Jien-wei Yeh, Tao-tsung Shun, Swe-kai Chen, Yuan-sheng Huang, and Hung-cheng Chen. 2007. "On the Elemental Effect of AlCoCrCuFeNi High-Entropy Alloy System." *Materials Letters* 61(1): 1–5. doi:10.1016/j.matlet.2006.03.140.
11. Lu, Y, Yong Dong, Sheng Guo, Li Jiang, Huijun Kang, Tongmin Wang, Bin Wen, and co-workers. 2014. "A Promising New Class of High-Temperature Alloys: Eutectic High-Entropy Alloys." *Scientific Reports* 4: 1–5. doi:10.1038/srep06200.
12. Li, Z, Konda Gokuldoss Pradeep, Yun Deng, Dierk Raabe, and Cemal Cem Tasan. 2016. "Metastable High-Entropy Dual-Phase Alloys Overcome the Strength-Ductility Trade-Off." *Nature* 534(7606): 227–30. doi:10.1038/nature17981.
13. Miracle, D B, and O N Senkov. 2017. "A Critical Review of High Entropy Alloys and Related Concepts." *Acta Materialia* 122: 448–511. doi:10.1016/j.actamat.2016.08.081.
14. Li, Z, and D Raabe. 2017. "Strong and Ductile Non-Equiatomic High-Entropy Alloys: Design , Processing , Microstructure, and Mechanical Properties." *Jom* 69(11): 2099–106. doi:10.1007/s11837-017-2540-2.
15. Wung, J, Jae Bok, Jongun Moon, Seok Su, Min Ji, Ho Yong, Byeong-joo Lee, and Hyoung Seop. 2018. "Exceptional Phase-Transformation Strengthening of Ferrous Medium- Entropy Alloys at Cryogenic Temperatures." *Acta Materialia* 161: 388–99. doi:10.1016/j.actamat.2018.09.057.
16. Tong, Chung-jin, Yu-liang Chen, Swe-kai Chen, Jien-wei Yeh, Tao-tsung Shun, Chun-huei Tsau, Su-jien Lin, and Shou-yi Chang. 2005. "Microstructure Characterization of AlxCoCrCuFeNi High-Entropy Alloy System with Multiprincipal Elements." *Metallurgical and Materials Transactions: Part A* 36A: 881–93.
17. Otto, F, A Dlouchy, Ch Somsen, H Bei, G Eggeler, and E P George. 2013. "The Influences of Temperature and Microstructure on the Tensile Properties of a CoCrFeMnNi High-Entropy Alloy." *Acta Materialia* 61(15): 5743–55. doi:10.1016/j.actamat.2013.06.018.
18. Murty, B S, J W Yeh, and S Ranganathan. 2014. *High-Entropy Alloys*. 1st ed. Elsevier, Butterworth-Heinemann, London.
19. Yeh, Jien-Wei. 2015. "Physical Metallurgy of High-Entropy Alloys." *Jom* 67(10): 2254–61. doi:10.1007/s11837-015-1583-5.
20. Zhang, By Yong, Yun Jun Zhou, Jun Pin Lin, Guo Liang Chen, and Peter K Liaw. 2008. "Solid-Solution Phase Formation Rules for Multi-Component Alloys" *Advanced Engineering Materials* 10(6) 534–38. doi:10.1002/adem.200700240.
21. Raghavan, R, K C Hari Kumar, and B S Murty. 2012. "Analysis of Phase Formation in Multi-Component Alloys." *Journal of Alloys and Compounds* 544: 152–58. doi:10.1016/j.jallcom.2012.07.105.
22. Ren, Ming-xing, Bang-sheng Li, and Heng-zhi Fu. 2013. "Formation Condition of Solid Solution Type High-Entropy Alloy." *Transactions of Nonferrous Metals Society of China* 23(4): 991–95. doi:10.1016/S1003-6326(13)62557-1.
23. Ye, Y F, Q Wang, J Lu, C T Liu, and Y Yang. 2015. "Design of High Entropy Alloys: A Single-Parameter Thermodynamic Rule." *Scripta Materialia* 104: 53–5. doi:10.1016/j.scriptamat.2015.03.023.
24. Ye, Y F, Q Wang, J Lu, C T Liu, and Y Yang. 2015. "The Generalized Thermodynamic Rule for Phase Selection in Multicomponent Alloys." *Intermetallics* 59: 75–80. doi:10.1016/j.intermet.2014.12.011.
25. Guo, S. 2015. "Phase Selection Rules for Cast High Entropy Alloys: An Overview." *Materials Science and Technology* 31(10): 1223–30. doi:10.1179/1743284715y.0000000018.

26. Cao, Z, Hui Jiang, Sheng Guo, Yong Dong, Tingju Li, Xiaoxia Gao, Tongmin Wang, and co-workers. 2018. "Preparing Bulk Ultrafine-Microstructure High-Entropy Alloys via Direct Solidification." *Nanoscale* 10(4): 1912–19. doi:10.1039/c7nr07281c.
27. Yeh, Jien-Wei. 2013. "Alloy Design Strategies and Future Trends in High-Entropy Alloys." *Jom* 65(12): 1759–71. doi:10.1007/s11837-013-0761-6.
28. Pickering, E J, and N G Jones. 2016. "High-Entropy Alloys: A Critical Assessment of Their Founding Principles and Future Prospects." *International Materials Reviews* 61(3): 183–202. doi:10.1080/09506608.2016.1180020.
29. Senkov, O N, J D Miller, D B Miracle, and C Woodward. 2015. "Accelerated Exploration of Multi-Principal Element Alloys with Solid Solution Phases." *Nature Communications* 6: 1–10. doi:10.1038/ncomms7529.
30. Hsu, Chin-you, Jien-wei Yeh, Swe-kai Chen, and Tao-tsung Shun. 2004. "Wear Resistance and High-Temperature Compression Strength of Face Centered Cubic $CuCoNiCrAl_{0.5}Fe$ Alloy with Boron Addition." *Metallurgical and Materials Transactions: Part A* 35(5): 1465–69.
31. Zhu, J M, H M Fu, H F Zhang, A M Wang, H Li, and Z Q Hu. 2010. "Microstructures and Compressive Properties of Multicomponent $AlCoCrFeNiMo_x$ Alloys." *Materials Science and Engineering: Part A* 527(26): 6975–79. doi:10.1016/j.msea.2010.07.028.
32. Dong, Y, Yiping Lu, Jiaorun Kong, Junjia Zhang, and Tingju Li. 2013. "Microstructure and Mechanical Properties of Multi-Component $AlCrFeNiMo_x$ High-Entropy Alloys." *Journal of Alloys and Compounds* 573: 96–101. doi:10.1016/j.jallcom.2013.03.253.
33. Guo, S, Chun Ng, and C T Liu. 2013. "Anomalous Solidification Microstructures in Co-Free $Al_xCrCuFeNi_2$ High-Entropy Alloys." *Journal of Alloys and Compounds* 557: 77–81. doi:10.1016/j.jallcom.2013.01.007.
34. Manzoni, A, H Daoud, S Mondal, S Van Smaalen, R Völkl, U Glatzel, and N Wanderka. 2013. "Investigation of Phases in $Al_{23}Co_{15}Cr_{23}Cu_8Fe_{15}Ni_{16}$ and $Al_8Co_{17}Cr_{17}Cu_8Fe_{17}Ni_{33}$ High Entropy Alloys and Comparison with Equilibrium Phases Predicted by Thermo-Calc." *Journal of Alloys and Compounds* 552: 430–36. doi:10.1016/j.jallcom.2012.11.074.
35. Sheng, H F, M Gong, and L M Peng. 2013. "Microstructural Characterization and Mechanical Properties of an $Al_{0.5}CoCrFeCuNi$ High-Entropy Alloy in as-Cast and High Entropy alloyt-Treated/Quenched Conditions." *Materials Science and Engineering: Part A* 567: 14–20. doi:10.1016/j.msea.2013.01.006.
36. He, J Y, W H Liu, H Wang, Y Wu, X J Liu, T G Nieh, and Z P Lu. 2014. "Effects of Al Addition on Structural Evolution and Tensile Properties of the FeCoNiCrMn High-Entropy Alloy System." *Acta Materialia* 62: 105–13. doi:10.1016/j.actamat.2013.09.037.
37. Nayan, N, Gaurav Singh, S V S N Murty, Abhay K Jha, Bhanu Pant, Koshy M George, and Upadrasta Ramamurty. 2014. "Hot Deformation Behaviour and Microstructure Control in AlCrCuNiFeCo High Entropy Alloy." *Intermetallics* 55: 145–53. doi:10.1016/j.intermet.2014.07.019.
38. Qin, G, Shu Wang, Ruirun Chen, Xue Gong, Liang Wang, Yanqing Su, Jingjie Guo, and H Fu. 2017. "Microstructures and Mechanical Properties of Nb-Alloyed CoCrCuFeNi High-Entropy Alloys." *Journal of Materials Science and Technology* 34(2): 365–69. doi:10.1016/j.jmst.2017.11.007.
39. Jiang, H, Kaiming Han, Xiaoxia Gao, Yiping Lu, Zhiqiang Cao, Michael C Gao, Jeffrey A Hawk, and T Li. 2018. "A New Strategy to Design Eutectic High-Entropy Alloys Using Simple Mixture Method." *Materials and Design* 142: 101–5. doi:10.1016/j.matdes.2018.01.025.
40. Liu, Y Y, Z Chen, Y Z Chen, J C Shi, Z Y Wang, and J Y Zhang. 2019. "The Effect of Al Content on Microstructures and Comprehensive Properties in $Al_xCoCrCuFeNi$ High Entropy Alloys." *Vacuum* 161: 143–49. doi:10.1016/j.vacuum.2018.12.009.
41. Hou, J, Xiaohui Shi, Junwei Qiao, Yong Zhang, Peter K Liaw, and Yucheng Wu. 2019. "Ultrafine-Grained Dual Phase $Al_{0.45}CoCrFeNi$ High-Entropy Alloys." *Materials and Design* 180(15). doi:10.1016/j.matdes.2019.107910.

42. Guo, S, Chun Ng, and C T Liu. 2013. "Sunflower-Like Solidification Microstructure in a Near- Eutectic High-Entropy Alloy." *Materials Research Letters* 1(4): 228–32. doi:10.1080/21663831.2013.844737.
43. Gao, Michael C, Jien-wei Yeh, Peter K Liaw, and Yong Zhang. 2016. *High-Entropy Alloys: Fundamentals and Applications*. 1st ed. Springer.
44. Yang, X, and Y Zhang. 2012. "Prediction of High-Entropy Stabilized Solid-Solution in Multi-Component Alloys." *Materials Chemistry and Physics* 132(2–3): 233–38. doi:10.1016/j.matchemphys.2011.11.021.
45. Otto, F, Y Yang, H Bei, and E P George. 2013. "Relative Effects of Enthalpy and Entropy on the Phase Stability of Equiatomic High-Entropy Alloys." *Acta Materialia* 61(7): 2628–38. doi:10.1016/j.actamat.2013.01.042.
46. Yeh, Jien-Wei. 2006. "Recent Progress in High-Entropy Alloys." *European Journal of Control*: 1–19. doi:10.3166/acsm.31.633-648.
47. Ye, Y F, Q Wang, J Lu, C T Liu, and Y Yang. 2016. "High-Entropy Alloy: Challenges and Prospects." *Materials Today* 19(6): 349–62. doi:10.1016/j.mattod.2015.11.026.
48. He, Q F, Z Y Ding, Y F Ye, and Y Yang. 2017. "Design of High-Entropy Alloy: A Perspective from Nonideal Mixing." *Jom* 69(11): 2092–98. doi:10.1007/s11837-017-2452-1.
49. Troparevsky, M Claudia, James R Morris, Markus Daene, Yang Wang, Andrew R Lupini, and G Malcolm Stocks. 2015. "Beyond Atomic Sizes and Hume-Rothery Rules: Understanding and Predicting High-Entropy Alloys." *Jom* 67(10): 2350–63. doi:10.1007/s11837-015-1594-2.
50. Takeuchi, A, and A Inoue. 2000. "Calculations of Mixing Enthalpy and Mismatch Entropy for Ternary Amorphous Alloys." *Materials Transactions, JIM* 41(11): 1372–78.
51. Zhang, Y, Ting Ting, Zhi Tang, Michael C Gao, Karin A Dahmen, Peter K Liaw, and Zhao Ping. 2014. "Microstructures and Properties of High-Entropy Alloys." *Progress in Materials Science* 61: 1–93. doi:10.1016/j.pmatsci.2013.10.001.
52. He, F, Zhijun Wang, Peng Cheng, Qiang Wang, Junjie Li, Yingying Dang, Jincheng Wang, and C T Liu. 2016. "Designing Eutectic High Entropy Alloys of $CoCrFeNiNb_x$." *Journal of Alloys and Compounds* 656: 284–89. doi:10.1016/j.jallcom.2015.09.153.
53. Mansoori, G A, N F Carnahan, K E Starling, and T W Leland Jr. 1971. "Equilibrium Thermodynamic Properties of the Mixture of Hard Spheres." *The Journal of Chemical Physics* 54(4).
54. Daoud, H M, A Manzoni, R Volkl, N Wanderka, and U Glatzel. 2013. "Microstructure and Tensile Behavior of $Al_8Co_{17}Cr_{17}Cu_8Fe_{17}$Ni33 (at.%) High-Entropy Alloy." *Jom* 65(12): 1805–14. doi:10.1007/s11837-013-0756-3.
55. Boer, F R de, R Boom, W C M Mattens, A R Miedema, and A K Niessen. 1988. *Cohesion in Metals: Transition Metal Alloys (Cohesion and Structure)*. North Holland Publishing Company.
56. Mao, H, Hai-lin Chen, and Qing Chen. 2017. "TCHEA1: A Thermodynamic Database Not Limited for "High Entropy Alloys" *Journal of Phase Equilibria and Diffusion* 38(4): 353–68. doi:10.1007/s11669-017-0570-7.
57. Ansara, I, N Dupin, H L Lukas, and B Sundman. 1997. "Thermodynamic Assessment of the Al-Ni System." *Journal of Alloys and Compounds* 247(1–2): 20–30.
58. Duplin, N, I Ansara, and B Sundman. 2001. "Thermodynamic Re-Assessment of the Ternary System Al-Cr-Ni." *Calphad* 25(2): 279–98.
59. Chen, Hai-lin, Huahai Mao, and Qing Chen. 2018. "Database Development and Calphad Calculations for High Entropy Alloys: Challenges, Strategies, and Tips." *Materials Chemistry and Physics* 210(1): 279–90. doi:10.1016/j.matchemphys.2017.07.082.
60. Miracle, D B, Jonathan D Miller, Oleg N Senkov, Christopher Woodward, Michael D Uchic, and Jaimie Tiley. 2014. "Exploration and Development of High Entropy Alloys for Structural Applications." *Entropy* 16(1): 494–525. doi:10.3390/e16010494.

61. Tsai, K Y, M H Tsai, and J W Yeh. 2013. "Sluggish Diffusion in Co–Cr–Fe–Mn–Ni High-Entropy Alloys." *Acta Materialia* 61(13): 4887–97.
62. MacDonald, B E, Z Fu, B Zheng, W Chen, Y Lin, F Chen, L Zhang, and co-workers. 2017. "Recent Progress in High Entropy Alloy Research." *Jom* 69(10): 2024–31. doi:10.1007/s11837-017-2484-6.
63. Senkov, O N, G B Wilks, D B Miracle, C P Chuang, and P K Liaw. 2010. "Refractory High-Entropy Alloys." *Intermetallics* 18(9): 1758–65.
64. Varvenne, C, G P M Leyson, M Ghazisaeidi, and W A Curtin. 2016. "Solute Strengthening in Random Alloys." *Acta Materialia* 124: 660–83.
65. Pickering, E J, H J Stone, and N G Jones. 2016. "Precipitation in the Equiatomic High-Entropy Alloy CrMnFeCoNi." *Scripta Materialia* 113: 106–9. doi:10.1016/j.scriptamat.2015.10.025.
66. Wei, S, Feng He, and Cemal Cem Tasan. 2018. "Metastability in High-Entropy Alloys: A Review." *Journal of Materials Research* 33(19): 2924–37. doi:10.1557/jmr.2018.306.
67. Gorsse, S, Jean-philippe Couzinié, and Daniel B Miracle. 2018. "From High-Entropy Alloys to Complex Concentrated Alloys." *Comptes Rendus Physique* 19(8): 721–36. doi:10.1016/j.crhy.2018.09.004.
68. Gludovatz, B, A Hohenwarter, Dhiraj Catoor, Edwin H Chang, Easo P George, and Robert O Ritchie. 2014. "A Fracture-Resistant High-Entropy Alloy for Cryogenic Applications." *Science* 345(6201): 1153–58. doi:10.1126/science.1254581.
69. Niu, C, Carlyn R Larosa, Jiashi Miao, Michael J Mills, and Maryam Ghazisaeidi. 2018. "Magnetically-Driven Phase Transformation Strengthening in High Entropy Alloys." *Nature Communications* 9(1363): 1–9. doi:10.1038/s41467-018-03846-0.
70. Huang, Yuan-sheng. 2009. "Recent Patents on High-Entropy Alloy." *Recent Patents on Materials Science* 2(2): 154–57.
71. Shun, Tao-tsung, Liang-yi Chang, and Ming-hua Shiu. 2012. "Microstructure and Mechanical Properties of Multiprincipal Component $CoCrFeNiMo_x$ Alloys." *Materials Characterization* 70: 63–7. doi:10.1016/j.matchar.2012.05.005.
72. Rao, J C, V O Diao, D Vainchtein, C Zhang, C Kuo, Z Tang, W Guo, and co-workers. 2017. "Secondary Phases in $Al_xCoCrFeNi$ High-Entropy Alloys: An In Situ TEM High Entropy Alloy Study and Thermodynamic Appraisal." *Acta Materialia* 131: 206.
73. He, F, Zhijun Wang, Xuliang Shang, Chao Leng, Junjie Li, and J Wang. 2016. "Stability of Lamellar Structures in $CoCrFeNiNb_x$ Eutectic High Entropy Alloys at Elevated Temperatures." *Materials and Design* 104: 259–64. doi:10.1016/j.matdes.2016.05.044.
74. Curtarolo, S, Gus L W Hart, Marco Buongiorno Nardelli, Natalio Mingo, Stefano Sanvito, and Ohad Levy. 2013. "The High-Throughput Highway to Computational Materials Design." *Nature Materials* 12(3): 191–201. doi:10.1038/nmat3568.
75. Li, R, Weiwei Zhang, Yong Zhang, and Peter K Liaw. 2018. "The Effects of Phase Transformation on the Microstructure and Mechanical Behavior of $FeNiMnCr_{0.75}Al_x$ High-Entropy Alloys." *Materials Science and Engineering: Part A* 725: 138–47. doi:10.1016/j.msea.2018.04.007.
76. Wen, C, Yan Zhang, Changxin Wang, Dezhen Xue, Yang Bai, S Antonov, L Dai, T Lookman, and Y Su. 2019. "Machine Learning Assisted Design of High Entropy Alloys with Desired Property." *Acta Materialia* 170: 109–17. doi:10.1016/j.actamat.2019.03.010.
77. Guo, S, and C T Liu. 2011. "Phase Stability in High Entropy Alloys: Formation of Solid-Solution Phase or Amorphous Phase." *Progress in Natural Science: Materials International* 21(6): 433–46. doi:10.1016/S1002-0071(12)60080-X.
78. Tripathy, S, Gaurav Gupta, and Sandip Ghosh Chowdhury. 2018. "High Entropy Alloys: Criteria for Stable Structure." *Metallurgical and Materials Transactions: Part A* 49(1): 7–17. doi:10.1007/s11661-017-4388-z.
79. Guo, S, Qiang Hu, Chun Ng, and C T Liu. 2013. "More than Entropy in High-Entropy Alloys: Forming Solid Solutions or Amorphous Phase." *Intermetallics* 41: 96–103. doi:10.1016/j.intermet.2013.05.002.

80. Tian, F, Lajos K Varga, Nanxian Chen, Jiang Shen, and Levente Vitos. 2015. "Empirical Design of Single Phase High-Entropy Alloys with High Hardness." *Intermetallics* 58: 1–6. doi:10.1016/j.intermet.2014.10.010.
81. Guo, S, Chun Ng, Jian Lu, and C T Liu. 2011. "Effect of Valence Electron Concentration on Stability of Face Centered Cubic or Body Centered Cubic Phase in High Entropy Alloys." *Journal of Applied Physics* 109(103505): 1–5. doi:10.1063/1.3587228.
82. Varvenne, C, Aitor Luque, and William A Curtin. 2016. "Theory of Strengthening in Face Centered Cubic High Entropy Alloys." *Acta Materialia* 118: 164–76. doi:10.1016/j.actamat.2016.07.040.
83. Coury, F G, Paul Wilson, Kester D Clarke, Michael J Kaufman, and Amy J Clarke. 2019. "High-Throughput Solid Solution Strengthening Characterization in High Entropy Alloys." *Acta Materialia* 167: 1–11. doi:10.1016/j.actamat.2019.01.029.
84. Kube, S A, Sungwoo Sohn, David Uhl, Amit Datye, Apurva Mehta, and Jan Schroers. 2019. "Phase Selection Motifs in High Entropy Alloys Revealed through Combinatorial Methods: Large Atomic Size Difference Favors BODY CENTERED CUBIC over FACE CENTERED CUBIC." *Acta Materialia* 166: 677–86. doi:10.1016/j.actamat.2019.01.023.
85. Tsai, M-H. 2016. "Three Strategies for the Design of Advanced High-Entropy Alloys." *Entropy* 18(252): 1–14. doi:10.3390/e18070252.
86. Walle, Axel Van De, and Mark Asta. 2019. "High-Throughput Calculations in the Context of Alloy Design." *MRS Bulletin* 44(4): 252–56. doi:10.1557/mrs.2019.71.
87. Curtarolo, S, D Morgan, K Persson, J Rodgers, and G Ceder. 2003. "Predicting Crystal Structures with Data Mining of Quantum Calculations." *Physical Review Letters* 91(13).
88. Hart, G L W, S Curtarolo, T B Massalski, and O Levy. 2013. "Comprehensive Search for New Phases and Compounds in Binary Alloy Systems Based on Platinum-Group Metals, Using a Computational First-Principles Approach." *Physical Review X*. http://www.ncbi.nlm.nih.gov/pubmed/041035.
89. Morgan, D, G Ceder, and S Curtarolo. 2004. "Computational Crystal Structure Prediction with High-Through-Put Ab Initio and Data Mining Methods." *Jom* 56(70).
90. Miracle, D B, Bhaskar Majumdar, Katelun Wertz, and Stéphane Gorsse. 2017. "New Strategies and Tests to Accelerate Discovery and Development of Multi-Principal Element Structural Alloys." *Scripta Materialia* 127: 195–200. doi:10.1016/j.scriptamat.2016.08.001.
91. Takeuchi, I, Jochen Lauterbach, and Michael J Fasolka. 2005. "Combinatorial Materials Synthesis." *Materials Today* 8(10): 18–26. doi:10.1016/S1369-7021(05)71121-4.
92. Zhao, Ji-cheng. 2006. "Combinatorial Approaches as Effective Tools in the Study of Phase Diagrams and Composition–Structure–Property Relationships." *Progress in Materials Science* 51(5): 557–631. doi:10.1016/j.pmatsci.2005.10.001.
93. Wilson, P, Robert Field, and Michael Kaufman. 2016. "The Use of Diffusion Multiples to Examine the Compositional Dependence of Phase Stability and Hardness of the Co-Cr-Fe-Mn-Ni High Entropy Alloy System." *Intermetallics* 75: 15–24. doi:10.1016/j.intermet.2016.04.007.
94. Borkar, T, B Gwalani, D Choudhuri, C V Mikler, C J Yannetta, X Chen, R V Ramanujan, M J Styles, M A Gibson, and R Banerjee. 2016. "A Combinatorial Assessment of $AlxCrCuFeNi_2$ ($0 < x < 1.5$) Complex Concentrated Alloys: Microstructure, Microhardness, and Magnetic Properties." *Acta Materialia* 116: 63–76. doi:10.1016/j.actamat.2016.06.025.
95. Debroy, T, H L Wei, J S Zuback, T Mukherjee, J W Elmer, J O Milewski, A M Beese, A Wilson-heid, A De, and W Zhang. 2018. "Additive Manufacturing of Metallic Components – Process , Structure and Properties." *Progress in Materials Science* 92: 112–224. doi:10.1016/j.pmatsci.2017.10.001.

96. Ramprasad, R, R Batra, G Pilania, A Mannodikanakkithodi, and C Kim. 2017. “Machine Learning in Materials Informatics: Recent Applications and Prospects.” *NPJ Computational Materials* 3: 1–13.
97. Sun, Y T, H Y Bai, M Z Li, and W H Wang. 2017. “Machine Learning Approach for the Prediction and Understanding of Glass Forming Ability.” *The Journal of Physical Chemistry Letters* 8(14): 3434–39.
98. Cubuk, E D, S S Schoenholz, J M Rieser, D Malone, J Rottler, D J Durian, E Kaxiaras, and A J Liu. 2015. “Identifying Structural Flow Defects in Disordered Solids Using Machine-Learning Methods.” *Physical Review Letters* 114(10). http://www.ncbi.nlm.nih.gov/pubmed/108001.
99. Tancret, F, I Toda-Caraballo, E Menou, and P E J R Diaz-Del-Castillo. 2017. “Designing High Entropy Alloys Employing Thermodynamics and Gaussian Process Statistical Analysis.” *Materials and Design* 115(486e497).
100. Abu-Odeh, A, E Galvan, T Kirk, H Mao, Q Chen, P Mason, R Malak, and R Arróyave. 2018. “Efficient Exploration of the High Entropy Alloy Composition-Phase Space.” *Acta Materialia* 152: 41–57. doi:10.1016/j.actamat.2018.04.012.
101. Huang, W, Pedro Martin, and Houlong L Zhuang. 2019. “Machine-Learning Phase Prediction of High-Entropy Alloys.” *Acta Materialia* 169: 225–36. doi:10.1016/j.actamat.2019.03.012.
102. Li, Bao-yu, Kun Peng, Ai-ping Hu, Ling-ping Zhou, Jia-jun Zhu, and De-yi Li. 2013. “Structure and Properties of $FeCoNiCrCu_{0.5}Al_x$ High-Entropy Alloy.” *Transactions of Nonferrous Metals Society of China* 23(3): 735–41. doi:10.1016/S1003-6326(13)62523-6.
103. Chen, Min-rui, Su-jien Lin, Jien-wei Yeh, Swe-kai Chen, Yuan-sheng Huang, and Chin-pang Tu. 2006. “Microstructure and Properties of $Al_{0.5}CoCrCuFeNiTi_x$ (x=0–2.0) High-Entropy Alloys.” *Materials Transactions* 47(5): 1395–401. doi:10.2320/matertrans.47.1395.
104. Chen, Min-rui, Su-jien Lin, Jien-wei Yeh, Swe-kai Chen, Y-S Huang, and M-H Chuang. 2006. “Effect of Vanadium Addition on the Microstructure , Hardness , and Wear Resistance of $Al_{0.5}CoCrCuFeNi$ High-Entropy Alloy.” *Metallurgical and Materials Transactions: Part A* 37(5): 1363–69.
105. Wong, Sze-kwan, Tao-tsung Shun, Chieh-hsiang Chang, and Che-fu Lee. 2018. “Microstructures and Properties of $Al_{0.3}CoCrFeNiMn_x$ High-Entropy Alloys.” *Materials Chemistry and Physics* 210: 146–51. doi:10.1016/j.matchemphys.2017.07.085.
106. Shun, Tao-tsung, Cheng-hsin Hung, and Che-fu Lee. 2010. “The Effects of Secondary Elemental Mo or Ti Addition in $Al_{0.3}CoCrFeNi$ High-Entropy Alloy on Age Hardening at 700oC.” *Journal of Alloys and Compounds* 495(1): 55–8. doi:10.1016/j.jallcom.2010.02.032.
107. Kao, Yih-farn, Ting-jie Chen, Swe-kai Chen, and Jien-wei Yeh. 2009. “Microstructure and Mechanical Property of As-Cast, -Homogenized, and -Deformed AlxCoCrFeNi ($0 \leq x \leq 2$) High-Entropy Alloys.” *Journal of Alloys and Compounds* 488(1): 57–64. doi:10.1016/j.jallcom.2009.08.090.
108. Shun, Tao-tsung, Cheng-hsin Hung, and Che-fu Lee. 2010. “Formation of Ordered/Disordered Nanoparticles in FACE CENTERED CUBIC High Entropy Alloys.” *Journal of Alloys and Compounds* 493: 105–9. doi:10.1016/j.jallcom.2009.12.071.
109. Qiu, Y, S Thomas, D Fabijanic, A J Barlow, H L Fraser, and N Birbilis. 2019. “Microstructural Evolution, Electrochemical and Corrosion Properties of $Al_xCoCrFeNiTi_y$ High Entropy Alloys.” *Materials and Design* 170: 1–15. doi:10.1016/j.matdes.2019.107698.
110. Shun, Tao-tsung, and Yu-chin Du. 2009. “Microstructure and Tensile Behaviors of FACE CENTERED CUBIC $Al_{0.3}CoCrFeNi$ High Entropy Alloy.” *Journal of Alloys and Compounds* 479(1–2): 157–60. doi:10.1016/j.jallcom.2008.12.088.
111. Lin, Chun-ming, and Hsien-lung Tsai. 2011. “Evolution of Microstructure, Hardness, and Corrosion Properties of High-Entropy.” *Intermetallics* 19(3): 288–94. doi:10.1016/j.intermet.2010.10.008.

112. Ng, C, Sheng Guo, Junhua Luan, Qing Wang, Jian Lu, Sanqiang Shi, and C T Liu. 2014. "Phase Stability and Tensile Properties of Co-Free $Al_{0.5}CrCuFeNi_2$ High-Entropy Alloys." *Journal of Alloys and Compounds* 584: 530–37. doi:10.1016/j.jallcom.2013.09.105.
113. Elkatatny, S, Mohamed A H Gepreel, Atef Hamada, Koichi Nakamura, Kenta Yamanaka, and Akihiko Chiba. 2019. "Effect of Al Content and Cold Rolling on the Microstructure and Mechanical Properties of $Al_5Cr_{12}Fe_{35}Mn_{28}Ni_{20}$ High-Entropy Alloy." *Materials Science and Engineering: Part A* 759: 380–90. doi:10.1016/j.msea.2019.05.056.
114. Wang, Z, and Ian Baker. 2016. "Interstitial Strengthening of a f.c.c. FeNiMnAlCr High Entropy Alloy." *Materials Letters* 180: 153–56. doi:10.1016/j.matlet.2016.05.122.
115. Jin, X, Juan Bi, Lu Zhang, Yang Zhou, Xingyu Du, Yuxin Liang, and Bangsheng Li. 2019. "A New $CrFeNi_2Al$ Eutectic High Entropy Alloy System with Excellent Mechanical Properties." *Journal of Alloys and Compounds* 770: 655–61. doi:10.1016/j.jallcom.2018.08.176.
116. Li, C, J C Li, M Zhao, and Q Jiang. 2009. "Effect of Alloying Elements on Microstructure and Properties of Multiprincipal Elements High-Entropy Alloys." *Journal of Alloys and Compounds* 475(1–2): 752–57. doi:10.1016/j.jallcom.2008.07.124.
117. Hsu, Yu-jui, Wen-chi Chiang, and Jiann-kuo Wu. 2005. "Corrosion Behavior of $FeCoNiCrCu_x$ High-Entropy Alloys in 3.5% Sodium Chloride Solution." *Materials Chemistry and Physics* 92(1): 112–17. doi:10.1016/j.matchemphys.2005.01.001.
118. Ko, Jun Yeong, and Sun Ig Hong. 2018. "Microstructural Evolution and Mechanical Performance of Carbon-Containing CoCrFeMnNi-C High Entropy Alloys." *Journal of Alloys and Compounds* 743: 115–25. doi:10.1016/j.jallcom.2018.01.348.
119. Abbasi, E, and Kamran Dehghani. 2019. "Effect of Nb-C Addition on the Microstructure and Mechanical Properties of CoCrFeMnNi High Entropy Alloys during Homogenisation." *Materials Science and Engineering: Part A* 753: 224–31. doi:10.1016/j.msea.2019.03.057.
120. Shun, Tao-tsung, Liang-yi Chang, and Ming-hua Shiu. 2012. "Microstructures and Mechanical Properties of Multiprincipal Component CoCrFeNiTix Alloys." *Materials Science and Engineering: Part A* 556: 170–74. doi:10.1016/j.msea.2012.06.075.
121. Hung, Wei-jhe, Tao-tsung Shun, and Cheng-ju Chiang. 2018. "Effects of Reducing Co Content on Microstructure and Mechanical Properties of $Co_xCrFeNiTi_{0.3}$ High-Entropy Alloys." *Materials Chemistry and Physics* 210(1): 170–75. doi:10.1016/j.matchemphys.2017.07.024.
122. Ren, B, Z X Liu, D M Li, L Shi, B Cai, and M X Wang. 2010. "Effect of Elemental Interaction on Microstructure of CuCrFeNiMn High Entropy Alloy System." *Journal of Alloys and Compounds* 493(1–2): 148–53. doi:10.1016/j.jallcom.2009.12.183.
123. Hsu, Chin-you, Chien-chang Juan, Shin-tsung Chen, Tsing-shien Sheu, Jien-wei Yeh, and Swe-kai Chen. 2013. "Phase Diagrams of High-Entropy Alloy System Al-Co-Cr-Fe-Mo-Ni." *Jom* 65(12): 1829–39. doi:10.1007/s11837-013-0773-2.
124. Ma, S G, and Y Zhang. 2012. "Effect of Nb Addition on the Microstructure and Properties of AlCoCrFeNi High-Entropy Alloy." *Materials Science and Engineering: Part A* 532: 480–86. doi:10.1016/j.msea.2011.10.110.
125. Liu, L, L J He, J G Qi, B Wang, Z F Zhao, J Shang, and Y Zhang. 2016. "Effects of Sn Element on Microstructure and Properties of $Sn_xAl_{2.5}FeCoNiCu$ Multi-Component Alloys." *Journal of Alloys and Compounds* 654: 327–32. doi:10.1016/j.jallcom.2015.09.093.
126. Stepanov, N D, D G Shaysultanov, R S Chernichenko, M A Tikhonovsky, and S V Zherebtsov. 2019. "Effect of Al on Structure and Mechanical Properties of Fe-Mn-Cr-Ni-Al Non-Equiatomic High Entropy Alloys with High Fe Content." *Journal of Alloys and Compounds* 770: 194–203. doi:10.1016/j.jallcom.2018.08.093.

127. Chen, S T, Wei Yeh Tang, Yen Fu Kuo, Sheng Yao Chen, Chun Huei Tsau, Tao Tsung Shun, and Jien Wei Yeh. 2010. "Microstructure and Properties of Age-Hardenable $Al_xCrFe_{1.5}MnNi_{0.5}$ Alloys." *Materials Science and Engineering: Part A* 527(21–22): 5818–25. doi:10.1016/j.msea.2010.05.052.
128. Stepanov, N D, D G Shaysultanov, GA Salishchev, and M A Tikhonovsky. 2015. "Structure and Mechanical Properties of a Light-Weight AlNbTiV High Entropy Alloy." *Materials Letters* 142: 153–55. doi:10.1016/j.matlet.2014.11.162.
129. Hou, L, Jiatao Hui, Yuhong Yao, Jian Chen, and Jiangnan Liu. 2019. "Effects of Boron Content on Microstructure and Mechanical Properties of $AlFeCoNiB_x$ High Entropy Alloy Prepared by Vacuum Arc Melting." *Vacuum* 164(December 2018): 212–18. doi:10.1016/j.vacuum.2019.03.019.
130. Hsu, U S, U D Hung, J W Yeh, S K Chen, Y S Huang, and C C Yang. 2007. "Alloying Behavior of Iron , Gold and Silver in AlCoCrCuNi-Based Equimolar High-Entropy Alloys." *Materials Science and Engineering: Part A* 461: 403–8. doi:10.1016/j.msea.2007.01.122.
131. Qin, G, Wentian Xue, Ruirun Chen, Huiting Zheng, Liang Wang, Yanqing Su, Hongsheng Ding, Jingjie Guo, and Hengzhi Fu. 2019. "Grain Refinement and FCC Phase Formation in AlCoCrFeNi High Entropy Alloys by the Addition of Carbon." *Materialia* 6: 100259. doi:10.1016/j.mtla.2019.100259.
132. Jiang, S, Zhifeng Lin, Hongming Xu, and Yongxing Sun. 2018. "Studies on the Microstructure and Properties of $Al_xCoCrFeNiTi_{1-x}$ High Entropy Alloys." *Journal of Alloys and Compounds* 741: 826–33. doi:10.1016/j.jallcom.2018.01.247.
133. Jin, X, Yang Zhou, Lu Zhang, Xingyu Du, and Bangsheng Li. 2018. "A Novel $Fe_{20}Co_{20}Ni_{41}Al_{19}$ Eutectic High Entropy Alloy with Excellent Tensile Properties." *Materials Letters* 216(1): 144–46. doi:10.1016/j.matlet.2018.01.017.
134. Yu, W, Yingdong Qu, Chengze Li, Zhe Li, Yufeng Zhang, Yaozu Guo, Junhua You, and R Su. 2019. "Phase Selection and Mechanical Properties of $(Al_{21.7}Cr_{15.8}Fe_{28.6}Ni_{33.9})_x(Al_{9.4}Cr_{19.7}Fe_{41.4}Ni_{29.5})_{100\text{-}x}$ High Entropy Alloys." *Materials Science and Engineering: Part A* 751: 154–59. doi:10.1016/j.msea.2019.02.067.
135. Shun, Tao-tsung, Liang-yi Chang, and Ming-hua Shiu. 2013. "Age-Hardening of the $CoCrFeNiMo_{0.85}$ High-Entropy Alloy." *Materials Characterization* 81: 92–6. doi:10.1016/j.matchar.2013.04.012.
136. Chanda, B, and Jayanta Das. 2019. "An Assessment on the Stability of the Eutectic Phases in High Entropy Alloys." *Journal of Alloys and Compounds* 798: 167–73. doi:10.1016/j.jallcom.2019.05.241.
137. Cui, P, Yimo Ma, Lijun Zhang, Mengdi Zhang, Jiantao Fan, Wanqing Dong, Pengfei Yu, G Li, and R Liu. 2018. "Effect of Ti on Microstructures and Mechanical Properties of High Entropy Alloys Based on CoFeMnNi System." *Materials Science and Engineering: Part A* 737: 198–204. doi:10.1016/j.msea.2018.09.050.
138. Yin, Y, Jingqi Zhang, Qiyang Tan, Wyman Zhuang, Ning Mo, Michael Bermingham, and Ming-xing Zhang. 2019. "Novel Cost-Effective Fe-Based High Entropy Alloys with Balanced Strength and Ductility." *Materials and Design* 162: 24–33. doi:10.1016/j.matdes.2018.11.033.
139. Jiang, H, Huanzhi Zhang, Tiandang Huang, Yiping Lu, Tongmin Wang, and Tingju Li. 2016. "Microstructures and Mechanical Properties of $CO_2Mo_xNi_2VW_x$ Eutectic High Entropy Alloys." *Materials and Design* 109: 539–46. doi:10.1016/j.matdes.2016.07.113.
140. Pi, Jin-hong, Ye Pan, Lu Zhang, and Hui Zhang. 2011. "Microstructure and Property of AlTiCrFeNiCu High-Entropy Alloy." *Journal of Alloys and Compounds* 509(18): 5641–45. doi:10.1016/j.jallcom.2011.02.108.
141. Lee, Che-fu, and Tao-tsung Shun. 2014. "Age Hardening of the $Al_{0.5}CoCrNiTi_{0.5}$ High-Entropy Alloy." *Metallurgical and Materials Transactions: Part A* 45(January): 191–95. doi:10.1007/s11661-013-1931-4.
142. Deng, Y, C C Tasan, K G Pradeep, H Springer, A Kostka, and D Raabe. 2015. "Design of a Twinning-Induced Plasticity High Entropy Alloy." *Acta Materialia* 94: 124–33.

143. Gali, A, and E P George. 2013. "Tensile Properties of High- and Medium-Entropy Alloys." *Intermetallics* 39: 74–8. doi:10.1016/j.intermet.2013.03.018.
144. Miao, J, C E Slone, T M Smith, C Niu, H Bei, M Ghazisaeidi, G M Pharr, and M J Mills. 2017. "The Evolution of the Deformation Substructure in a Ni-Co-Cr Equiatomic Solid Solution Alloy." *Acta Materialia* 132: 35–48. doi:10.1016/j.actamat.2017.04.033.
145. Slone, C E, S Chakraborty, J Miao, E P George, M J Mills, and S R Niezgoda. 2018. "Influence of Deformation Induced Nanoscale Twinning and FCC-HCP Transformation on Hardening and Texture Development in Medium- Entropy CrCoNi Alloy." *Acta Materialia* 158: 38–52. doi:10.1016/j.actamat.2018.07.028.
146. Wang, Y, Bin Liu, Kun Yan, Minshi Wang, Saurabh Kabra, Yu-lung Chiu, David Dye, Peter D Lee, Yong Liu, and Biao Cai. 2018. "Probing Deformation Mechanisms of a FeCoCrNi High-Entropy Alloy at 293 and 77 K Using In Situ Neutron Diffraction." *Acta Materialia* 154: 79–89. doi:10.1016/j.actamat.2018.05.013.
147. Laplanche, G, A Kostka, C Reinhart, J Hunfeld, G Eggeler, and E P George. 2017. "Reasons for the Superior Mechanical Properties of Medium-Entropy CrCoNi Compared to High-Entropy CrMnFeCoNi." *Acta Materialia* 128: 292–303. doi:10.1016/j.actamat.2017.02.036.
148. Gludovatz, B, Anton Hohenwarter, Keli V S Thurston, Hongbin Bei, Zhenggang Wu, Easo P George, and Robert O Ritchie. 2016. "Exceptional Damage-Tolerance of a Medium- Entropy Alloy CrCoNi at Cryogenic Temperatures." *Nature Communications* 7: 1–8. doi:10.1038/ncomms10602.
149. Huo, W, Feng Fang, Hui Zhou, Zonghan Xie, Jianku Shang, and Jianqing Jiang. 2017. "Remarkable Strength of CoCrFeNi High-Entropy Alloy Wires at Cryogenic and Elevated Temperatures." *Scripta Materialia* 141: 125–28. doi:10.1016/j.scriptamat.2017.08.006.
150. Otto, F, N L Hanold, and E P George. 2014. "Microstructural Evolution after Thermomechanical Processing in an Equiatomic, Single-Phase CoCrFeMnNi High-Entropy Alloy with Special Focus on Twin Boundaries." *Intermetallics* 54: 39–48.
151. Laplanche, G, P Gadaud, O Horst, F Otto, G Eggeler, and E P George. 2015. "Temperature Dependencies of the Elastic Moduli and Thermal Expansion Coefficient of an Equiatomic , Single-Phase CoCrFeMnNi High-Entropy Alloy." *Journal of Alloys and Compounds* 623: 348–53. doi:10.1016/j.jallcom.2014.11.061.
152. Stepanov, N D, D G Shaysultanov, M A Tikhonovsky, and G A Salishchev. 2015. "Tensile Properties of the Cr –Fe–Ni–Mn Non-Equiatomic Multicomponent Alloys with Different Cr Contents." *Materials and Design* 87: 60–5. doi:10.1016/j.matdes.2015.08.007.
153. Schuh, B, F Mendez-martin, B Völker, E P George, H Clemens, R Pippan, and A Hohenwarter. 2015. "Mechanical Properties, Microstructure and Thermal Stability of a Nanocrystalline CoCrFeMnNi High-Entropy Alloy after Severe Plastic Deformation." *Acta Materialia* 96: 258–68. doi:10.1016/j.actamat.2015.06.025.
154. Tasan, C C, Y Deng, K G Pradeep, M J Yao, H Springer, and D Raabe. 2014. "Composition Dependence of Phase Stability, Deformation Mechanisms, and Mechanical Properties of the CoCrFeMnNi High-Entropy Alloy System." *Jom* 66(10): 1993–2001. doi:10.1007/s11837-014-1133-6.
155. Yao, M J, K G Pradeep, C C Tasan, and D Raabe. 2014. "A Novel, Single Phase, Non-Equiatomic FeMnNiCoCr High-Entropy Alloy with Exceptional Phase Stability and Tensile Ductility." *Scripta Materialia* 72–73: 5–8. doi:10.1016/j.scriptamat.2013.09.030.
156. Allain, S, J Chateau, O Bouaziz, S Migot, and N Guelton. 2004. "Correlations between the Calculated Stacking Fault Energy and the Plasticity Mechanisms in Fe – Mn – C Alloys." *Materials Science and Engineering: Part A* 387–389: 158–62.
157. Stepanov, N D, M Tikhonovsky, N Yurchenko, D Zyabkin, M Klimova, S Zherebtsov, A Efimov, and G Salishchev. 2015. "Effect of Cryo-Deformation on Structure and Properties of CoCrFeNiMn High-Entropy Alloy." *Intermetallics* 59: 8–17. doi:10.1016/j.intermet.2014.12.004.

158. Ng, C, S Guo, J Luan, S Shi, and C T Liu. 2012. "Entropy-Driven Phase Stability and Slow Diffusion Kinetics in an $Al_{0.5}$CoCrCuFeNi High Entropy Alloy." *Intermetallics* 31: 165–72.
159. Srivastava, V C, G K Mandal, N Ciftci, V Uhlenwinkel, and L Mädler. 2017. "Processing of High-Entropy $AlCoCr_{0.75}Cu_{0.5}$FeNi Alloy by Spray Forming." *Journal of Materials Engineering and Performance* 26(12): 5906–20. doi:10.1007/s11665-017-3071-2.
160. Wang, Y P, B S Li, M X Ren, C Yang, and H Z Fu. 2008. "Microstructure and Compressive Properties of AlCrFeCoNi High Entropy Alloy." *Materials Science and Engineering: Part A* 491(1–2): 154–58. doi:10.1016/j.msea.2008.01.064.
161. Huang, By Ping-kang, Jien-wei Yeh, Tao-tsung Shun, and Swe-kai Chen. 2004. "Multi-Principal-Element Alloys with Improved Oxidation and Wear Resistance for Thermal Spray Coating." *Advanced Engineering Materials* 6(1–2): 74–8. doi:10.1002/adem.200300507.
162. Zhou, Y J, Y Zhang, Y L Wang, and G L Chen. 2007. "Solid Solution Alloys of $AlCoCrFeNiTi_x$ with Excellent Room-Temperature Mechanical Properties." *Applied Physics Letters* 90(18): 1–3. doi:10.1063/1.2734517.
163. Hsu, By Chin-you, Woei-ren Wang, Wei-yeh Tang, Swe-kai Chen, and Jien-wei Yeh. 2010. "Microstructure and Mechanical Properties of New AlCoxCrFeMo0.5Ni High-Entropy Alloys." *Advanced Engineering Materials* 12(1–2): 44–9. doi:10.1002/adem.200900171.
164. Zou, Y, Soumyadipta Maiti, Walter Steurer, and Ralph Spolenak. 2014. "Size-Dependent Plasticity in an $Nb_{25}Mo_{25}Ta_{25}W_{25}$ Refractory High-Entropy Alloy." *Acta Materialia* 65: 85–97. doi:10.1016/j.actamat.2013.11.049.
165. Kuznetsov, A V, D G Shaysultanov, N D Stepanov, G.A Salishchev, and O N Senkov. 2012. "Tensile Properties of an AlCrCuNiFeCo High-Entropy Alloy in as-Cast and Wrought Conditions." *Materials Science and Engineering: Part A* 533: 107–18. doi:10.1016/j.msea.2011.11.045.
166. Mishra, A K, Sumanta Samal, and K Biswas. 2012. "Solidification Behaviour of Ti–Cu–Fe–Co–Ni High Entropy Alloys." *Transaction of Indian Institute Od Metals* 65(6): 725–30. doi:10.1007/s12666-012-0206-x.
167. Gao, X, Yiping Lu, Bo Zhang, Ningning Liang, Guanzhong Wu, Gang Sha, Jizi Liu, and Yonghao Zhao. 2017. "Microstructural Origins of High Strength and High Ductility in an $AlCoCrFeNi_{2.1}$ Eutectic High-Entropy Alloy." *Acta Materialia* 141: 59–66. doi:10.1016/j.actamat.2017.07.041.
168. Otto, F, A Dlouhý, K G Pradeep, M Kubenova, D Raabe, G Eggeler, and E P George. 2016. "Decomposition of the Single-Phase High-Entropy Alloy CrMnFeCoNi after Prolonged Anneals at Intermediate Temperatures." *Acta Materialia* 112: 40–52. doi:10.1016/j.actamat.2016.04.005.
169. Gwalani, B, V Soni, D Choudhuri, M Lee, J Y Hwang, S J Nam, H Ryu, S H Hong, and R Banerjee. 2016. "Stability of Ordered L12 and B2 Precipitates in Face Centered Cubic Based High Entropy Alloys – $Al_{0.3}$CoFeCrNi and $Al_{0.3}CuFeCrNi_2$." *Scripta Materialia* 123: 130–34. doi:10.1016/j.scriptamat.2016.06.019.
170. Turnbull, D. 1981. "Metastable Structures in Metallurgy." *Metallurgical Transactions: Part B* 12(2): 217.
171. Laplanche, G, S Berglund, C Reinhart, A Kostka, F Fox, and E P George. 2018. "Phase Stability and Kinetics of s -Phase Precipitation in CrMnFeCoNi High-Entropy Alloys." *Acta Materialia* 161: 338–51. doi:10.1016/j.actamat.2018.09.040.
172. Munitz, A, S Salhov, S Hayun, and N Frage. 2016. "High Entropy Alloyt Treatment Impacts the Micro-Structure and Mechanical Properties of AlCoCrFeNi High Entropy Alloy." *Journal of Alloys and Compounds* 683: 221–30. doi:10.1016/j.jallcom.2016.05.034.
173. He, F, Zhijun Wang, Jing Wang, Qingfeng Wu, Da Chen, Bin Han, Junjie Li, Jincheng Wang, and J J Kai. 2018. "Abnormal Γ - ε Phase Transformation in the CoCrFeNiNb0.25 High Entropy Alloy." *Scripta Materialia* 146: 281–85. doi:10.1016/j.scriptamat.2017.12.009.

174. Jiang, L, Y P Lu, M Song, C Lu, K Sun, Z Q Cao, T M Wang, F Gao, and L M Wang. 2019. "A Promising $CoFeNi_2V_{0.5}Mo_{0.2}$ High Entropy Alloy with Exceptional Ductility." *Scripta Materialia* 165: 128–33. doi:10.1016/j.scriptamat.2019.02.038.
175. Li, Z, Cemal Cem Tasan, Hauke Springer, Baptiste Gault, and Dierk Raabe. 2017. "Interstitial Atoms Enable Joint Twinning and Transformation Induced Plasticity in Strong and Ductile High-Entropy Alloys." *Scientific Reports* 7(40704): 1–7. doi:10.1038/srep40704.
176. Cao, C M, W Tong, S H Bukhari, J Xu, Y X Hao, P Gu, H Hao, and L M Peng. 2019. "Dynamic Tensile Deformation and Microstructural Evolution of $Al_xCrMnFeCoNi$ High-Entropy Alloys." *Materials Science and Engineering: Part A* 759: 648–54. doi:10.1016/j.msea.2019.05.095.
177. Wang, Z, M C Gao, S G Ma, H J Yang, Z H Wang, M Ziomek-Moroz, and J W Qiao. 2015. "Effect of Cold Rolling on the Microstructure and Mechanical Properties of $Al_{0.25}CoCrFe_{1.25}Ni_{1.25}$ High-Entropy Alloy." *Materials Science and Engineering: Part A* 645: 163–69. doi:10.1016/j.msea.2015.07.088.
178. Gwalani, B, Stephane Gorsse, Deep Choudhuri, Yufeng Zheng, R S Mishra, and R Banerjee. 2019. "Tensile Yield Strength of a Single Bulk $Al_{0.3}CoCrFeNi$ High Entropy Alloy Can Be Tuned from 160 MPa to 1800 MPa." *Scripta Materialia* 162: 18–23. doi:10.1016/j.scriptamat.2018.10.023.
179. He, J Y, H Wang, H L Huang, X D Xu, M W Chen, Y Wu, X J Liu, T G Nieh, K An, and Z P Lu. 2016. "A Precipitation-Hardened High-Entropy Alloy with Outstanding Tensile Properties." *Acta Materialia* 102: 187–96. doi:10.1016/j.actamat.2015.08.076.
180. Hassanpour-esfahani, M, A Zarei-hanzaki, H R Abedi, H S Kim, and D Yim. 2019. "The Enhancement of Transformation Induced Plasticity e Ff Ect through Preferentially Oriented Substructure Development in a High Entropy Alloy." *Intermetallics* 109(December 2018): 145–56. doi:10.1016/j.intermet.2019.03.013.
181. Luo, H, Zhiming Li, and Dierk Raabe. 2017. "Hydrogen Enhances Strength and Ductility of an Equiatomic High- Entropy Alloy." *Scientific Reports*(August): 1–7. doi:10.1038/s41598-017-10774-4.
182. Guo, L, Xiaoqin Ou, Song Ni, Yong Liu, and Min Song. 2019. "Effects of Carbon on the Microstructures and Mechanical Properties of FeCoCrNiMn High Entropy Alloys." *Materials Science and Engineering: Part A* 746: 356–62. doi:10.1016/j.msea.2019.01.050.
183. Choi, Won-mi, Yong Hee Jo, Seok Su Sohn, Sunghak Lee, and Byeong-joo Lee. 2017. "Understanding the Physical Metallurgy of the CoCrFeMnNi High-Entropy Alloy: An Atomistic Simulation Study." *Computational Materials*: 1–9. doi:10.1038/s41524-017-0060-9.
184. Agarwal, R, Reshma Sonkusare, Saumya R Jha, N P Gurao, Krishanu Biswas, and Niraj Nayan. 2018. "Understanding the Deformation Behavior of CoCuFeMnNi High Entropy Alloy by Investigating Mechanical Properties of Binary Ternary and Quaternary Alloy Subsets." *Materials and Design* 157(5): 539–50. doi:10.1016/j.matdes.2018.07.046.
185. Tong, Y, D Chen, B Han, J Wang, R Feng, T Yang, C Zhao, and co-workers. 2019. "Outstanding Tensile Properties of a Precipitation-Strengthened $FeCoNiCrTi_{0.2}$ High-Entropy Alloy at Room and Cryogenic Temperatures." *Acta Materialia* 165: 228–40. doi:10.1016/j.actamat.2018.11.049.
186. Slone, C E, J Miao, E P George, and M J Mills. 2019. "Achieving Ultra-High Strength and Ductility in Equiatomic CrCoNi with Partially Recrystallized Microstructures." *Acta Materialia* 165: 496–507. doi:10.1016/j.actamat.2018.12.015.
187. Bhattacharjee, T, I S Wani, S Sheikh, I T Clark, T Okawa, S Guo, P P Bhattacharjee, and N Tsuji. 2018. "Simultaneous Strength-Ductility Enhancement of a Nano-Lamellar Entropy Alloy by Cryo-Rolling and Annealing." *Scientific Reports*(January): 1–8. doi:10.1038/s41598-018-21385-y.

188. Pickering, E J, H J Stone, and N G Jones. 2015. "Fine-Scale Precipitation in the High-Entropy Alloy $Al_{0.5}$CrFeCoNiCu." *Materials Science and Engineering: Part A* 645: 65–71. doi:10.1016/j.msea.2015.08.010.
189. Jones, N G, K A Christofidou, and H J Stone. 2015. "Rapid Precipitation in an $Al_{0.5}$CrFeCoNiCu High Entropy Alloy." *Materials Science and Technology* 31(10): 1171–77. doi:10.1179/1743284715y.0000000004.
190. Lu, Z P, H Wang, M W Chen, I Baker, J W Yeh, C T Liu, and T G Nieh. 2015. "An Assessment on the Future Development of High-Entropy Alloys: Summary from a Recent Workshop." *Intermetallics* 66: 67–76. doi:10.1016/j.intermet.2015.06.021.
191. Li, Z, A Ludwig, A Savan, H Springer, and D Raabe. 2018. "Combinatorial Metallurgical Synthesis and Processing of High-Entropy Alloys." *Journal of Materials Research* 33(19): 3156–69. doi:10.1557/jmr.2018.214.
192. Gao, Michael C, and Junwei Qiao. 2018. "High-Entropy Alloys (HEAs)." *Metals* 8(108): 1–3. doi:10.3390/met8020108.
193. Yeh, Jien-Wei, and Su-jien Lin. 2018. "Breakthrough Applications of High-Entropy Materials." *Journal of Materials Research* 33(19): 3129–37. doi:10.1557/jmr.2018.283.
194. Shaysultanov, D G, N D Stepanov, A V Kuznetsov, G A Salishchev, and O N Senkov. 2013. "Phase Composition and Superplastic Behavior of a Wrought AlCoCrCuFeNi High-Entropy Alloy." *Jom* 65(12): 1815–28. doi:10.1007/s11837-013-0754-5.
195. Shahmir, H, Junyang He, Zhaoping Lu, Megumi Kawasaki, and Terence G Langdon. 2017. "Evidence for Superplasticity in a CoCrFeNiMn High-Entropy Alloy Processed by High-Pressure Torsion." *Materials Science and Engineering: Part A* 685: 342–48. doi:10.1016/j.msea.2017.01.016.
196. Shahmir, H, Mahmoud Nili-ahmadabadi, Ahad Shafiee, and Terence G Langdon. 2018. "Effect of a Minor Titanium Addition on the Superplastic Properties of a CoCrFeNiMn High-Entropy Alloy Processed by High-Pressure Torsion." *Materials Science and Engineering: Part A* 718: 468–76. doi:10.1016/j.msea.2018.02.002.
197. Reddy, S R, S Bapari, P P Bhattacharjee, and A H Chokshi. 2017. "Superplastic-Like Flow in a Fine-Grained Equiatomic CoCrFeMnNi High-Entropy Alloy." *Materials Research Letters* 5(6): 408–14. doi:10.1080/21663831.2017.1305460.
198. Wu, Jien-min, Su-jien Lin, Jien-wei Yeh, Swe-kai Chen, Yuan-sheng Huang, and Hung-cheng Chen. 2006. "Adhesive Wear Behavior of Al_xCoCrCuFeNi High-Entropy Alloys as a Function of Aluminum Content." *Wear* 261(5–6): 513–19. doi:10.1016/j.wear.2005.12.008.
199. Chuang, Ming-hao, Ming-hung Tsai, Woei-ren Wang, Su-jien Lin, and Jien-wei Yeh. 2011. "Microstructure and Wear Behavior of AlxCo1.5CrFeNi1.5Tiy High-Entropy Alloys." *Acta Materialia* 59(16): 6308–17. doi:10.1016/j.actamat.2011.06.041.
200. Fujieda, T, Hiroshi Shiratori, Kosuke Kuwabara, Takahiko Kato, K Yamanaka, Y Koizumi, and A Chiba. 2015. "First Demonstration of Promising Selective Electron Beam Melting Method for Utilizing High-Entropy Alloys as Engineering Materials." *Materials Letters* 159: 12–5. doi:10.1016/j.matlet.2015.06.046.
201. Ocelík, V, N Janssen, S N Smith, and J Th M De Hosson. 2016. "Additive Manufacturing of High-Entropy Alloys by Laser Processing." *Jom* 68(7): 1810–18. doi:10.1007/s11837-016-1888-z.
202. Chen, S, Yang Tong, and Peter K Liaw. 2018. "Additive Manufacturing of High-Entropy Alloys: A Review." *Entropy* 20(12): 1–18. doi:10.3390/e20120937.

Section A

Innovations

1 Interstitial Alloy Structuring of High Entropy Alloys

A.D. Pogrebnjak and A.A. Bagdasaryan

CONTENTS

1.1 INTRODUCTION

The ever-increasing demands from the aerospace, aviation, atomic, machine-building, and other industries for high-quality, functionally flexible, and environmentally friendly materials necessitate giving special attention to the choice of constituent elements and the means of control or modification of the structural-phase state [1–5]. Modern material science has deviated from the development of simple single-element products toward the combination of a large number of constituent elements in order to give them a set of excellent performance characteristics. This evolution is associated with the development of the new technology industries, which require highly reliable and cost-effective products.

According to the conventional alloy design paradigm (one-component or one-element alloys), modern engineering coatings have been developed and intensively investigated in the last decades. Among them, nitride titanium (TiN) and chromium nitride (CrN) were the most investigated and widely used protective coatings due to their excellent mechanical characteristics [6–8]. The next stage of evolution of protective materials was the tuning of their composition space. For example, the addition of Al to TiN significantly increases the oxidation resistance from 500°C to 800°C; the addition of Al to CrN gives better abrasion wear resistance

[8, 9]. The most impressive example of the second generation coatings was Ti-Si-N with its complex intrinsic architecture: grain size of few nanometers (n-TiN) and a soft amorphous phase a-Si_3N_4. It was shown that such structure allows obtaining materials which can be considered as a "prototype" of superhard coatings (hardness > 40 GPa) [5, 10, 11]. Through similar thinking, scientists developed and investigated a lot of new protective materials, like Me1Me2Me3N, where Me1, Me2 and Me3 in general are transition metals [11]. However, this design route did not provide materials that would meet the requirements for harsh environments, since it was believed that the use of a large number of constituent elements in large concentrations would lead to the formation of intermetallic compounds, which are brittle in nature.

Remarkable progress has been achieved in comparison to one-layer materials in the development of multilayer protective coatings. Such intrinsic architecture is a superposition of layers of different elements with specific thicknesses on each other, like: TiN/MoN [12], CrAlSiN/TiVN [13], ZrN/TiAIN [14], and others [15, 16]. The obtaining of such materials is primarily due to the robust development of sputtering methods, like magnetron sputtering and vacuum-arc deposition. As we will see further, the application of different techniques of deposition allow tuning of composition-microstructure properties space in a wide range. Many scientists report that using a multilayer structure significantly improves wear and oxidation resistance, mechanical characteristics, and electrical and optical properties. Therefore, such materials caused high industrial interest because of the possibility of using them as protective coatings.

Nowadays, extensive research work continues aiming to improve the mechanical and physical properties and thermal stability of protective coatings. Also, it is still an ongoing challenge to develop new materials with new structure and specific or multiple properties. In last decade, increasing attention is being paid to the development of a new class of materials, high entropy alloys (HEA) [17–19]. Every year, an enormous number of articles are published in scientific journals devoted to the studies of their structure at the nano-level, their strength, and their electronic properties. According to the thermodynamic conception, high entropy alloys are disordered solid-solution phase alloys, which consist of at least five constituent elements with concentrations of between 5 and 35%. Many works have shown that HEAs possess many superior properties compare with traditional alloys. Special attention is paid to research materials in combination with some interstitial elements, like O, C, B, and especially N [20]. These systems can combine superior mechanical properties with excellent thermal stability and enhanced strength with good ductility compared with pure metals and alloys. For example, it was found that formation of nitrides based on HEAs allows obtaining materials with superior mechanical properties (e.g., hardness > 40 GPa).

This chapter reveals the closest look to date at the state of composition-microstructure properties relating to nitride coatings based on high entropy alloys, the so-called high entropy alloy nitrides (HEANs) [21–76]. Also, we discuss the new directions of modification of the high entropy nitrides.

1.2 COMPOSITION SPACE OF HIGH ENTROPY NITRIDES

To analyze the composition space of high entropy nitrides as elemental families, we investigated 33 original HEANs systems presented in scientific works. The multi-principal alloys occupy the central region of phase diagrams, which provide a new way to design alloys with specific properties. In the case of HEANs, composition space is more narrowed, since it is necessary to take into account the possibility of the constituent elements forming stable bonds with nitrogen.

It is well known that nitrogen occupies the octahedral interstitial sites of a close-packed lattice and have two metal (Me) atoms in opposite directions. In general, nitrides form a simple crystal structure, like face-centered cubic (FCC) and hexagonal close-packed (HCP) structures, depending on elemental composition and methods of deposition. For example, in general, aluminum reacts with nitrogen to form a hexagonal wurtzite structure, B4. However, the application of high pressure and temperatures can promote the formation of a rock salt structure, B1.

It should be noted that nitrogen can accept electrons from element and form an s^2p^6 configuration and give electrons with the formation of a stable configuration sp^3. In the first case, the compounds possess a clear ionic type of bonding; in the second, metallic and covalent types of bonding [77]. Thus, nitrides of non-metals likely to form covalent types of bonds and transition metals—mixed ionic-metal-covalent bond types and rare-earth nitrides—have the predominant ionic character of bonding.

The family of transition metals are most commonly used with Nitrogen for obtaining materials with high mechanical properties, corrosion, and oxidation resistance. The nitrides of the metals of group IV (titanium, zirconium, and hafnium) and group V (vanadium, niobium, and tantalum) are the refractory transition-metal nitrides (TMNs). Since the TMNs can form different type of bonds, there are several forms of nitrides, like: MeN, Me_2N, Me_3N_5, and others. Nitrides of the metals of group VI (chromium, molybdenum, and tungsten) have a lower acceptor capacity and their strong Me-Me interactions break at elevated temperatures. This circumstance is corroborated by consideration of enthalpy of formation of nitrides (see Table 1.1).

TABLE 1.1
Enthalpy of Formation of Binary Nitrides (kj/mol)

Strong nitride formers		Non-strong nitride formers	
ΔH_{mix} (Ti-N)	−190	ΔH_{mix} (Cr-N)	−107
ΔH_{mix} (Zr-N)	−230	ΔH_{mix} (Mo-N)	−115
ΔH_{mix} (Hf-N)	−218	ΔH_{mix} (W-N)	−103
ΔH_{mix} (V-N)	−143	ΔH_{mix} (Al-N)	−92
ΔH_{mix} (Nb-N)	−174	ΔH_{mix} (Si-N)	−81
ΔH_{mix} (Ta-N)	−173	ΔH_{mix} (Fe-N)	−32
		ΔH_{mix} (Mn-N)	−64

More negative enthalpy formation (more negative free energy) indicates the stability of nitrides. The stability of nitrides decreases in a lower group and for the transition metal in VI–VIII groups. For example, F_2N and Mn_2N nitrides are very unstable because their free energy is positive. Depending on the type of bonding, different TMNs can be good superconductors, have high mechanical and corrosion properties, high melting point, etc.

As we stated above, 33 original HEANs are considered in this chapter. In general, these systems contain transition metals: Cr, Ti, Zr, Nb, V, Ta, Mo, Ni, Fe, Cu, Hf, Co, Mn, Y, Ru, and specific elements, like: Al, Si, B, and C. The most widely used elements are: Cr (79%), Ti (79%), Al (78%), Zr (64%), Nb (42%), and V (36%). 15–30% of the HEANs contain Ta (33%), Si (30%), Mo (27%), Ni (24%), Fe (15%), Cu (12%), Hf (12%), and Co (12%). A few elements, like: Mn (8%), Y (6%), and Ru (3%) appear only in one or two nitride systems. The interstitial elements carbon and boron appear only in two systems (AlCrTaTiZr)NC and (AlCrMnMoNiZrB)N. Almost all systems contain non-strong nitride elements; only three HEANs consist of refractory elements. The list of all HEANs are presented in Table 1.2.

1.3 METHODS OF DEPOSITION AND CRYSTAL STRUCTURE

In the last decade the most anticipated methods of obtaining protective coatings are PVD methods, among them magnetron sputtering and vacuum-arc deposition, which are widely used for the deposition of high entropy nitrides. The deposition temperature (400–450°C) of PVD methods is significantly higher than room temperature, which creates the thermal mismatch between substrate and coating. As a result, compressive stresses develop in coatings, which favorably affects mechanical characteristics. Due to the high deposition temperature and cooling rates (close to 10^{10} K/s) the incident particles have a high probability of maintaining a thermodynamically stable position resulting the formation of dense coatings with a simple crystal structure (in most cases, FCC). Unlike vacuum-arc deposition, magnetron sputtering due to the low degree of ionization of flow particles leads to the formation of amorphous structure or/and ordered phases in the resultant coatings. In contrast, a high degree of ionization of vacuum-arc plasma and the possibility of adjusting the parameters of deposition (working pressure, substrate bias, and temperature deposition) allow the provision of widespread tuning of microstructure properties of the materials. One of the biggest disadvantages of vacuum-arc deposition is the presence of macroscopic inclusions due to the formation of irregularities in materials with a low melting point during evaporation. Such particulate defects on the coatings surface significantly reduce their tribological characteristics. Magnetron sputtering and vacuum-arc deposition with their pros and cons are the key methods for preparation of high-quality HEANs with a wide range of microstructures and properties.

In general, binary transition metal nitrides form solid solutions with simple FCC crystal structures. In the case of HEANs, high entropy of mixing also promotes the formation of thermodynamically stable solid solutions and prevents the appearance of ordered phases (intermetallic compounds). However, as we will see below, there are some HEANs which undergo the phase transition from amorphous structure to amorphous structure+FCC (Figure 1.1).

TABLE 1.2
Method of Deposition, Structure, and Mechanical Properties of HEANs

HEAN	Method of deposition	Structure	Hardness (H) and Young's modulus (E), GPa	References
(AlCrNiSiTi)N	Magnetron sputtering	Amorphous state	H=15,1 E=156	[44]
(AlCoCrCuFeNi)N		Amorphous state	H=12,4 E=131	[44]
(AlCrNbSiTiV)N		FCC	H=41 E=360	[32]
(AlCrFeNiTi)N		Amorphous state – FCC+Amorphous state – FCC	H=21,28 E=253,8	[26]
(AlCrSiTiZr)N		Amorphous state+Low crystallinity	H=19,6 E=231,5	[60]
(AlCrMoTaTiSi)N		FCC	H=35,5 E=243	[37]
(AlCrMoTaTiZr)N		FCC	H=40,2 E=420	[42]
(AlCrMoSiTi)N		FCC	H=34 E=375	[59]
(AlCrNbSiTi)N		FCC	H=36 E=335	[39]
(AlCrTaTiZr)N		FCC	H=35 E=350	[43]

(Continued)

TABLE 1.2 (CONTINUED)
Method of Deposition, Structure, and Mechanical Properties of HEANs

HEAN	Method of deposition	Structure	Hardness (H) and Young's modulus (E), GPa	References
AlCrMoNbZr/ (AlCrMoNbZr)N		Amorphous + FCC	are absent	[24]
(AlCrMnMoNiZrB)N		FCC	H = 10 E = 180	[58]
(AlCrMnMoNiZr)N		Amorphous state – FCC + Amorphous state – FCC	H = 11,9 E = 202	[31]
(AlMoNbSiTaTiVZr)N		FCC	are absent	[55]
(AlCrTaTiZr)NC		FCC	H = 32 E = 280	[57]
(CrTaTiVZr)N		FCC	H = 36,4 E = 273,5	[38]
(FeCoNiCrCuAl)N		Amorphous state	H = 10,4	[61]
(FeCoNiCuVZrAl)N		Amorphous state	H = 12 E = 166	[34]
(FeCoNiCrCuAlMn)N		Amorphous state	H = 11,8	[61]
(TiZrNbHfTa)N		FCC	H = 32,9 E = 179	[40]
(TiVCrZrHf)N		FCC	H = 48 E = 316	[56]
(TiVCrZrNbMoHfTaWAlSi)N		Amorphous state – FCC + Amorphous state – FCC	H = 34.8 E = 276.5	[28]

(Continued)

TABLE 1.2 (CONTINUED)
Method of Deposition, Structure, and Mechanical Properties of HEANs

HEAN	Method of deposition	Structure	Hardness (H) and Young's modulus (E), GPa	References
(TiAlCrSiV)N		Amorphous state – FCC+Amorphous state – FCC	H=31,2 E=305	[33]
(TiTaCrZrAlRu)N		FCC	are absent	[78]
(TiVCrZrY)N		FCC	H=17,5 E=160	[50]
(ZrTaNbTiW)N		BCC+FCC	H=13,5 E=179	[23]
(AlTiVNbCr)N	Vacuum-arc deposition	FCC	are absent	[63]
(TiN-Cu)/(AlNbTiMoVCr)N		FCC	H=24,5	[66]
(TiZrNbAlYCr)N		FCC+BCC	H=47	[64]
(TiHfZrVNbTa)N		FCC	H=42,2	[67]
(TiHfZrVNb)N		FCC	H=44,3	[71]
(TiAlCrZrNb)N		FCC	H=36,6 E=849	[72]
(ZrTiNbSiCr)N		FCC	H=29	[69]

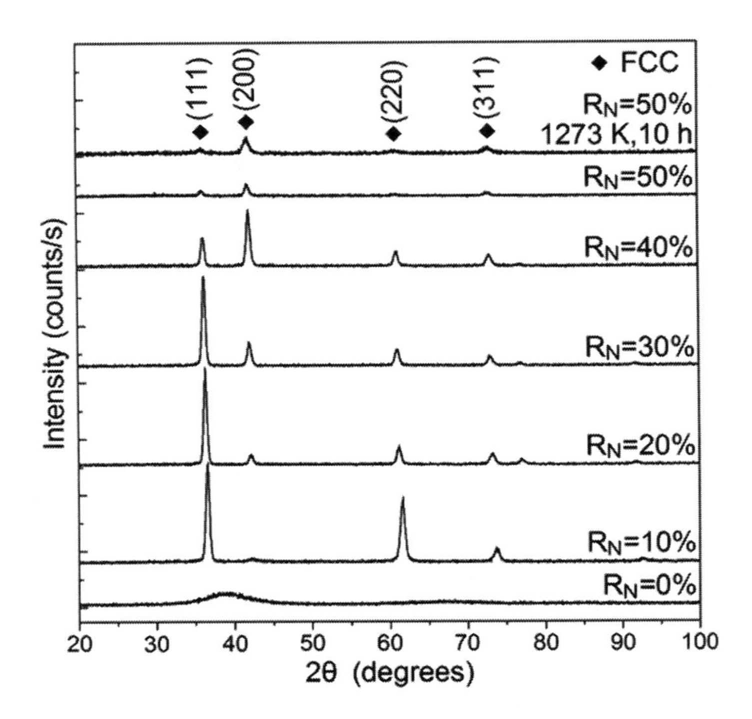

FIGURE 1.1 XRD patterns of the (AlCrMoTaTiZr)N HEANs deposited on Si (100) substrates under different N_2 flow ratios (R_N) [42].

For N-free high entropy films obtained by magnetron sputtering, the inherent situation is the formation of the amorphous phase. According to the authors [42], this is attributed to the sluggish long-range diffusion, large atomic size difference, and high mixing entropy effect. We also assume that the low ionization energy does not allow the development of crystallization processes. With the addition of N (R_N ~ 10–30%) into the processing chamber, (AlCrMoTaTiZr)N films exhibit a simple NaCl-type FCC crystal structure with high-intense (111), (200), and (220) diffraction peaks. It should be noted that grain size doesn't depend on different N_2 flow ratios. The presented results are in concurrence with similar research on some HEANs, presented in Table 1.1. The opposite situation was reported in investigation of (AlCoCrCuFeNi)N coatings [44]. It was found that deposition in N-free atmosphere leads to formation of simple FCC and mixed BCC+FCC solid solution. The increase of R_N leads to transition from crystalline to amorphous structure (see Figure 1.2).

In the case of NEAHs which contain Al, Ta, Si elements, a two-phase structure can be formed. In the work of Chang [28], it was found that the coexistence of FCC and HCP (in small amounts) phases in (TiVCrZrNbMoHfTaWAlSi)N coatings, when R_N increases to 20% and 30%. According to the author, there are several factors of the phase separation during deposition: AlN, TaN, and Si_3N_4 exhibit an HCP phase; a large lattice mismatch among binary nitrides. However, it should be noted that the FCC still remains the dominant phase. Tsai et al. [50] deposited (TiVCrZrY)N coatings by a DC magnetron sputtering and found the formation of HCP phase

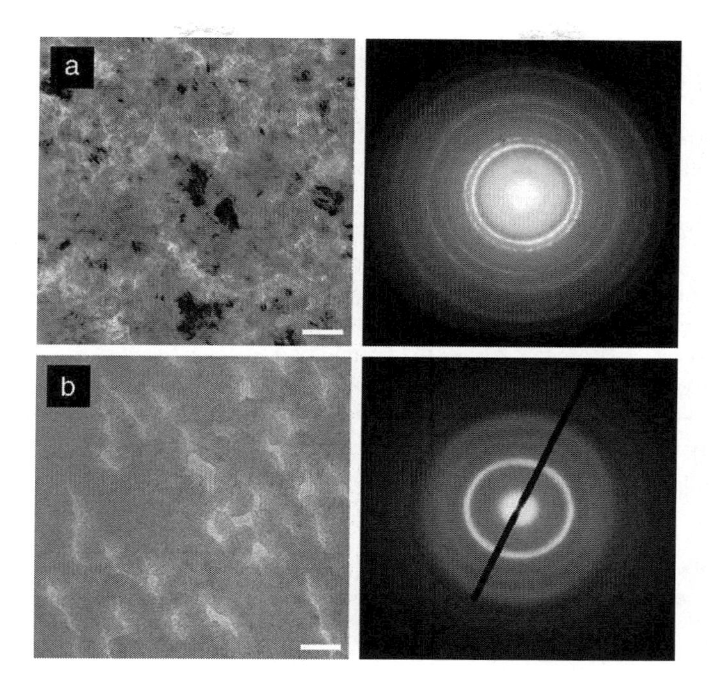

FIGURE 1.2 TEM images and their corresponding SAD patterns of the $Al_{0.5}CoCrCuFeNi$ HEAN, deposited at N_2/Ar fraction: (a) 0.1; (b) 0.5. Scale bars, 50 nm [44].

at R_N = 0%. The (TiVCrZrY)N coatings have a mixed FCC structure and a near-amorphous phase.

Besides the formation of FCC phase, Feng et al. [23] reported the formation of a BCC+FCC solid solution in (ZrTaNbTiW)N films, deposited by magnetron sputtering deposition and nitrogen plasma-based ion implantation. According to the authors, a two-phase structure is formed as a result of a lack of implanted nitrogen. Moreover, the addition of C content into $(AlCrTaTiZr)N_xC_y$ system did not lead to any structural phase change [57]. The authors claimed that the carbides and nitrides of constituent elements were randomly mixed to compose only an FCC solid-solution structure with higher density.

Few efforts have been made to realize the nanocomposite structure in HEANs. Such a structure is characterized by the formation of an amorphous (Si_3N_4)/crystalline (TiN) phase. Despite the low solubility of Si in TMNs, for formation of amorphous (Si_3N_4) phase, silicon concentration from 5 to 12% is required [5]. Unfortunately, the synthesis of (AlCrNbSiTiV)N [51], (TiAlCrSiV)N [33], (AlCrNiSiTi)N [44], (AlCrSiTiZr)N [60], $Si_x(AlCrMoTaTi)_{1-x}N$ [37], (AlCrMoSiTi)N [59], $(Al_{1.5}CrNb_{0.5}Si_{0.5}Ti)N_x$ [39], (AlMoNbSiTaTiVZr)N [55], and (ZrTiNbSiCr)N [69] by magnetron sputtering and vacuum-arc deposition did not bring about the desired results. Moreover, high solubility of Si in HEANs was found: 8% in (AlCrNbSiTiV)N [51] and 5.6% in $(Al_{1.5}CrNb_{0.5}Si_{0.5}Ti)N_x$ [39] coatings. However, Niu et al. [21] reported the formation of a fractional nanocomposite structure in

(AlCrTiZrV)-Si_x-N, deposited by a magnetron sputtering system. Through the high-magnification HRTEM observation, the authors detected the equiaxed regions with an ordered lattice structure and a large number of amorphous interfacial phases (areas between two yellow dotted lines: Figure 1.3d).

No evidence of formation of amorphous structure in HEANs obtained by vacuum-arc deposition had been detected. Pogrebnjak et al. [71] synthesized the (TiHfZrVNb)N coatings under a wide range of deposition parameters: $P_N = 0.08 \div 0.7$ Pa, $U_{sb} = -40 \div -230$ V. As we can see from Figure 1.4, XRD lines belong to 111, 200, 220, and 311 reflections from FCC lattice (NaCl-type). For N-free coating, the XRD pattern can be indexed as a BCC crystal structure, with 110 the preferred orientation of crystallites. During the cathode arc discharge, the largest number of ionized atoms of the target is generated, near 50÷100%, for magnetron sputtering—it does not exceed 5%. In this regard, it should be noted that the coatings obtained by vacuum-arc deposition demonstrate a crystal structure in an N-free working atmosphere, comparing with magneton sputtering.

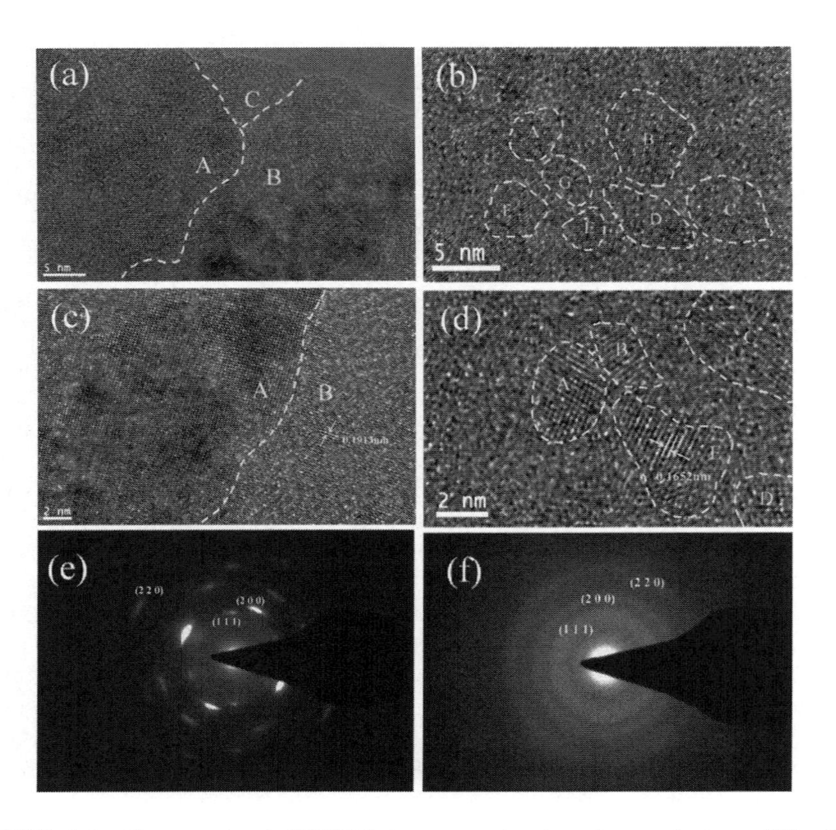

FIGURE 1.3 Cross-sectional HRTEM images and selected-area electron diffraction (SAED) patterns of the (a, c, e) (AlCrTiZrV)N and (b, d, f) (AlCrTiZrV)-$Si_{0.08}$-N films: (a, b) low-magnification HRTEM images; (c, d) high-magnification HRTEM images; (e, f) SAED patterns [21].

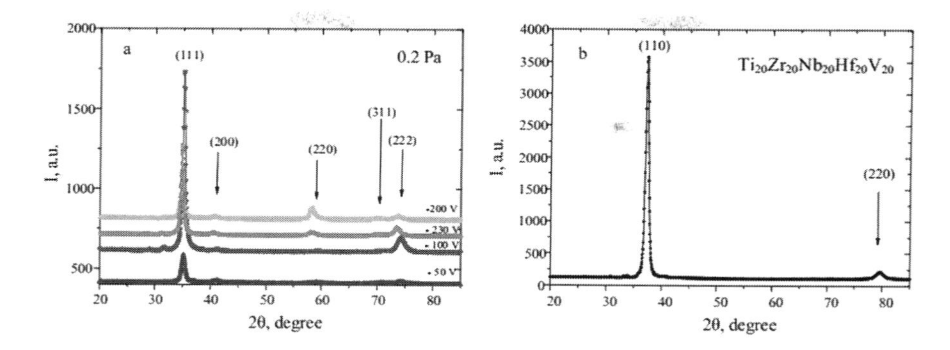

FIGURE 1.4 The results of X-ray diffraction analysis of (TiHfZrVNb)N, obtained by vacuum-arc deposition under: $P_N = 0.2$ Pa; b) N-free sample [71].

The formation of a two-phase crystal structure in (TiZrHfVNbTa)N coatings was found in the work of Pogrebnjak et al. [67]. It was established that increasing nitrogen in a vacuum chamber led to alignment of the FCC and BCC ratio, as 1 to 1. Additional implantation of low-energy negative Au^- ions with a 1×10^{17} cm^{-2} dose led to the formation of a thin surface layer (~34 nm) with mixed amorphous-nanocrystalline structure.

In general, most of the high entropy alloy nitrides with randomly distributed atoms tend to form disordered solid solutions with an FCC crystal structure. In most cases, the binary nitrides of the constituent elements (Ti, Zr, V, Hf, Nb, Mo, W) have a cubic sodium chloride phase, except AlN, TaN, and Si_3N_4. The high mixing entropy effect promotes the solubility of metal elements and can stabilize the formation of simple structures. However, use of non-strong nitride elements like Fe, Mn, Cu, Co, Ni, Cr leads to decreasing of crystallinity and the formation of an amorphous phase, even when the nitrogen concentration reaches 40%. One can conclude it is important to select constituent elements with similar crystal structures, lattice parameters, and high melting temperatures to minimize the risk of crack initiation, layer delamination, and provide improved mechanical characteristics.

1.4 INFLUENCE OF SUBSTRATE BIAS ON STRUCTURE OF HEANs

For PVD methods it is possible to change the conditions of condensation during the deposition process by changing the energy of the deposited ions by applying a negative potential on the substrate. It should be noted that for each coating there is a boundary value of the energy of ionization (depending on U_{sb}), at which the chemical bonds between the metal atoms and the nitrogen are formed. In general, the application of substrate bias leads to the following effects: activation of chemical reaction on the surface, incorporation of defects, densification of structure, and grain refinement which are favorable effect on mechanical properties. However, care is needed in applying high substrate bias due to the beginning of intense processes of re-sputtering.

Chang et al. [38] investigated the influence of substrate bias on the structure and properties of (CrTaTiVZr)N coatings deposited by reactive radio-frequency magnetron sputtering. As we can see from Figure 1.5, no evidence of changes in the crystal structure had been detected. However, a shift of XRD peaks to lower angles and change of the preferential orientation is observed. The increasing of energy of the bombarding ions leads to intense generation of Frenkel pairs and anti-Schottky defects induced by the ion peening effect. This is supported by the shift of XRD peaks toward lower angles and increasing of lattice parameter. Their experimental results also indicated that increase of substrate bias results in refinement of grain size and a transition from slightly tensile (0.52 GPa) to compressive stresses (−4.61 GPa). Such effects were also reported for (TiVCrZrHf)N [41], (TiHfZrVNb)N [71], and $(Al_{1.5}CrNb_{0.5}Si_{0.5}Ti)N_x$ [39].

Special attention is paid to researching evolution of the preferential orientation under different substrate biases. A few investigations have discussed the development of (200), (220) at a higher substrate bias from the point of thermodynamics, where the minimal total energy consists of surface, stopping, and strain energy that are responsible for evolution of the (200), (220), and (111) planes, respectively (overall energy minimization model) [79, 80]. Several authors [81–83] have proposed kinetic driving forces (anisotropy in surface diffusivities, collisional cascade effects, and adatom mobilities) as an explanation for the mechanism of growth process. With a certain confidence, we can claim that the textural evolution of HEANs under different substrate biases can be described through a complex interplay between thermodynamic and kinetic forces. However, the understanding of textural evolution of multicomponent nitrides still remains elusive, so further intensive research is needed in this direction.

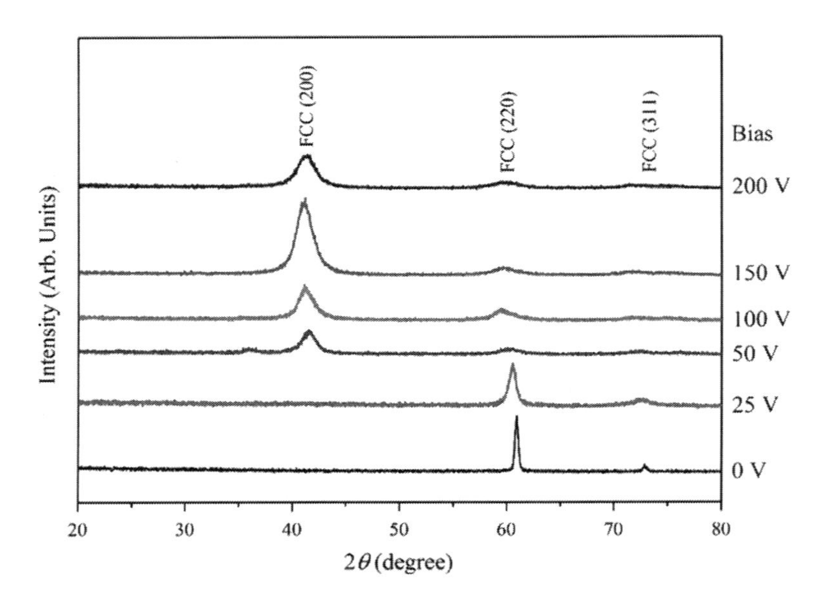

FIGURE 1.5 X-ray diffraction pattern of the (CrTaTiVZr)N coatings deposited at different substrate biases [38].

It is well known that the microstructure of most coatings obtained by magnetron sputtering consists of columnar grains oriented approximately parallel to the direction of growth. At higher ionization, flow energies' atomic mobility is increased, which leads to the activation of surface diffusion processes. As a result, coatings have a dense equiaxial grain structure. Such microstructural evolution is in agreement with the structure zone model proposed by Messier [84].

Shen et al. [39] synthesized the single-phased $(Al_{1.5}CrNb_{0.5}Si_{0.5}Ti)_{50}N_{50}$ HEAN by direct current magnetron sputtering under different substrate bias. According to the research conducted by the authors, coating, deposited without substrate bias, has a typical V-shaped columnar structure. As is clearly visible in Figure 1.6, that application of $U_{sb}=-50$ V leads to significant reduction of the number of microvoids

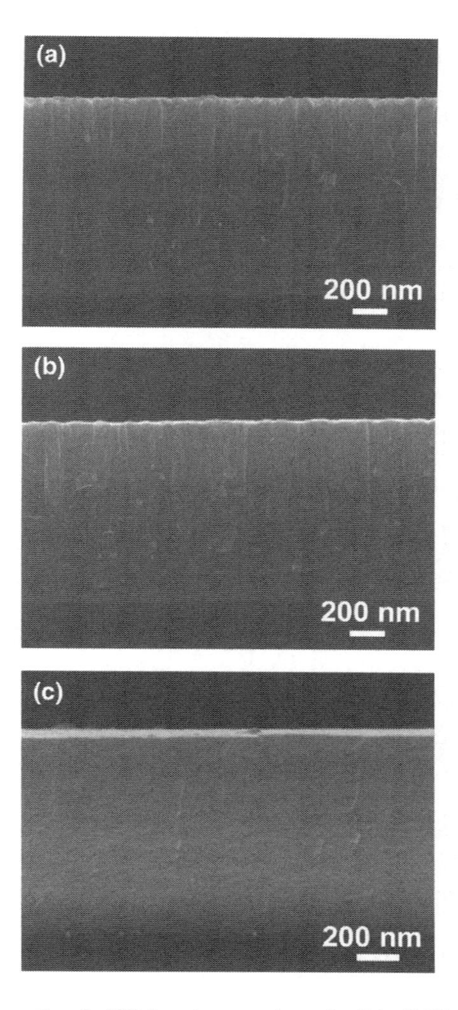

FIGURE 1.6 Cross-sectional SEM micrographs of $(Al_{1.5}CrNb_{0.5}Si_{0.5}Ti)_{50}N_{50}$ HEAN obtained at substrate bias of (a) 0 V, (b) –50 V, and (c) –150 V [39].

between columns. Further increase of substrate bias contributes to the evolution of a fine equiaxed grain structure with dense boundaries.

1.5 THERMAL STABILITY OF HEANs

The key to long-term stability of properties of HEANs is the reduction of the content of impurity atoms and optimization of thermal conditions of deposition. The durability and stability of the coatings structure during high-temperature annealing make it possible to predict their properties at high temperatures, as well as determine their service life. Tsai et al. [29] found an excellent thermal stability of (TiVCrZrHf)N HEANs in a vacuum even at 1,073 K, deposited by magnetron sputtering. However, with the increase of the temperature up to 1,173 K, $CrSi_3$ precipitates with a diamond-like structure are formed. It was found that a slight increase in the grain size—from 6.4 to 9.3 nm—apparently is associated with the mechanism of inhibition of the coarsening of grain, which is described in more detail in the work of Huang and Yeh [54]. According to this mechanism, slight changes in grain size are caused by low energy grain boundaries and non-intensive intensity diffusion. Due to the high density, the effective diffusion distances decrease, which leads to the inhibition of grain growth.

Tsai et al. [45] also investigated the thermal stability of (TiVCrZrHf)N HEAN in air. It was shown that when the temperature reaches 600–700°C, the coating is fully oxidized with the formation of hexagonal $ZrTiO_4$ oxide with subsequent phase transformation in rutile TiO_2 and monoclinic ZrO_2 phases (thickness is 3,160 nm). The formation of such type of oxides is due to the fact that the TiN, ZrN, and HfN have a relatively strong tendency for oxidation (high enthalpy of formation).

Tsai et al. [35] studied the oxidation resistance of (AlCrMoTaTi)-Si_x-N coating deposited via magnetron sputtering depending on concentration of Si and temperature of annealing. The FCC crystal structure of the Si-free coating (AlCrMoTaTi)N was significantly destroyed with formation rutile TiO_2 oxide with a thickness of 379 nm after annealing at 1,073 K in air for 2 h. The nitride phase totally disappeared at 1,173 K. According to the authors' observations, the addition of Si (near 7.51%) leads to significant improving of oxidation resistance of (AlCrMoTaTi)-Si_x-N coatings. After annealing at 1,073 K, no evident oxide peaks were detected. Further increasing of temperature of annealing to 1,173 K leads to the appearance of peaks, which corresponds to the rutile TiO_2 phase (thickness near 202 nm).

The superior oxidation resistance of $(Al_{23.1}Cr_{30.8}Nb_{7.7}Si_{7.7}Ti_{30.7})_{50}N_{50}$ (1) and $(Al_{29.1}Cr_{30.8}Nb_{11.2}Si_{7.7}Ti_{21.2})_{50}N_{50}$ (2) films deposited at −100 and −150 V, respectively, after annealing at 900°C in air for two hours was found in the work of Hsieh et al. [46]. The thickness of the oxide layer was 100 nm for (1) coatings and 80 nm for (2). The application of a higher substrate bias promotes elimination of microvoids and the formation of more dense structure.

Special attention should be paid to works on the development of HEANs, obtained by reactive radio-frequency magnetron sputtering, for thermally stable diffusion barriers [22, 47, 48, 55]. Such materials are widely used in integrated circuits to eliminate resistance-capacitance (RC) delay. However, the binary nitrides can't provide effective protection from inter-diffusion of Cu and Si into the intermediate layer.

That's why the high entropy concept can be used as an alternative method for preparation of new diffusion barrier materials.

Chang et al. [48] reported the thermal stability of (AlCrTaTiZr)N layer with a nanocomposite structure at the extremely high temperature of 900°C at a vacuum of 2×10^{-5} Torr for 30 mins. No evidence of the formation of silicides due to the interdiffusion of Cu and Si had been detected. Li et al. [22] added Mo into (AlCrTaTiZr)N layer and examined the thermal stability at 600°C for 7 h. The layer an remains amorphous structure without any grain boundaries and maintained excellent interface adhesion with the Cu and Si. The excellent thermal and structural stabilities for $(AlMoNbSiTaTiVZr)_{50}N_{50}$ layer were shown at 850°C for 30 mins [55]. However, at 900°C, a peak of $CuSi_3$ was detected, which indicates the destruction of the barrier.

1.6 MECHANICAL PROPERTIES OF HEANs

In many cases, HEANs are used as protective coatings for providing resistance to heat, corrosion and wear, restoring of worn machining parts to original dimensions, and other functions. In general, such materials possess high hardness and a Young's modulus, low coefficient of friction, good adhesion, and thermal stability. The mechanical properties of coatings depend on their structural-phase state (grain size, phase composition, internal stresses, etc.) and methods of preparation. However, many works [19, 20] maintain that the elemental composition of the HEANs is responsible for the hardness and Young's modulus in the first approximation. As we can see from data in Table 1.2, the HEANs with non-strong nitride elements (see Table 1.1) possess low values of mechanical characteristics (H: 10÷15.1 GPa, E: 131÷202 GPa): (FeCoNiCuVZrAl)N, (FeCoNiCrCuAlMn)N, (FeCoNiCrCuAl)N, (AlCoCrCuFeNi)N, (AlCrNiSiTi)N, (AlCrMnMoNiZrB)N, (AlCrMnMoNiZr)N. The low affinity to formation of nitride phase promotes film amorphization, which adversely affects mechanical properties.

Liang et al. [56] fabricated (TiVCrZrHf)N coatings by magnetron sputtering with superior mechanical properties (H: 48 GPa, E: 316 GPa) (see Figure 1.7).

According to the authors [56], with increasing of substrate temperature, the processes of filling micropores in the coatings are more intensive due to the higher mobility of atoms on the surface. Also, the contribution to hardness is provided by the mechanism of grain boundary hardening due to the low dislocation activity.

Investigations of substrate bias effect on the HEANs hardness were provided in several works [38, 41, 71]. Chang et al. [38] showed that the hardness and elastic modulus of the as-deposited (CrTaTiVZr)N films increased from 11.3 GPa to 36.4 GPa and from 200.3 GPa to 273.8 GPa, respectively, as the substrate bias was increased from 0 to −100 V. Further increasing of U_{sb} leads to slight decreasing of hardness decrement of compressive stresses and the evolution of the (200) preferential orientation. Similar results were obtained by Shen et al. [39] during the investigation of the influence of U_{sb} on the mechanical properties of $(Al_{1.5}CrNb_{0.5}Si_{0.5}Ti)N_x$ coatings. Higher values of substrate bias (up to −200 V) lead to the significant softening of coatings due to the inverse Hall–Petch effect.

The effect of Si on the mechanical properties of $(AlCrTiZrV)$-Si_x-N films with nanocomposite structure were studied in the work of Niu et al. [21]. It was shown

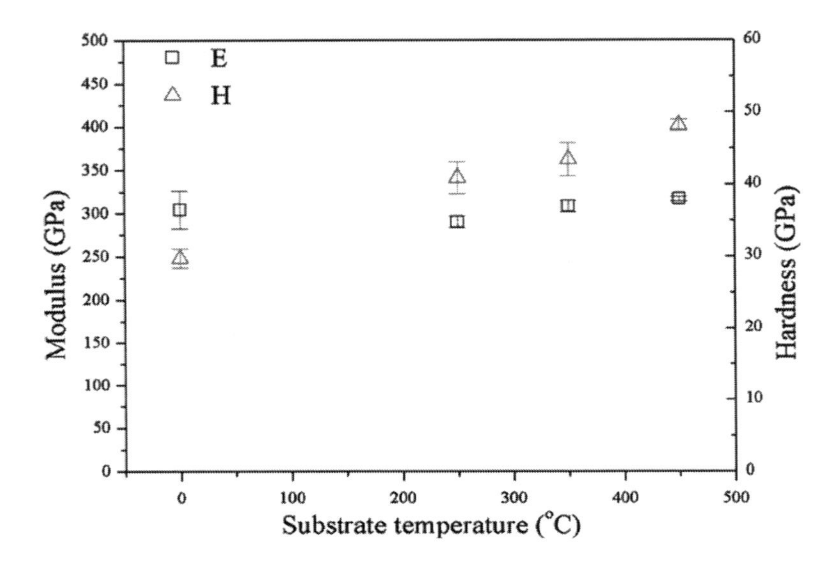

FIGURE 1.7 Mechanical characteristics of the (TiVCrZrHf)N coatings deposited at various substrate temperatures [56].

that the hardness and elastic modulus reaches maximum values 34.3 and 301.5 GPa, respectively, under concentration of Si (8%). Further increasing of Si content (up to 16%) caused a decrease of mechanical properties due to the increasing thickness of amorphous interfacial phases. The authors claimed that in this case, the hardness is mainly governed by the properties of the amorphous interfacial phase. This trend is corroborated by the work of Tsai et al. [37], which showed that the maximum hardness of 35.5 GPa of $Si_x(AlCrMoTaTi)_{1-x}N$ is obtained when Si concentration is 7.51 at.%.

The addition of carbon into $(AlCrTaTiZr)N_xC_y$ HEANs caused an increase of coating hardness from 20 GPa to 32 GPa. According to the authors of this work [57], enhanced hardness can be attributed to the formation of covalent-like carbon bonds, grain refinement, and a denser columnar structure.

An unusually high value of elastic modulus was obtained in the work [72] on the study of (TiAlCrZrNb)N coatings, obtained by vacuum-arc deposition. With the application of substrate bias 120 V and arc current on the Zr-Nb cathode 135 A, the hardness and elastic modulus reach 36.6 and 849 GPa, respectively.

Especially interesting are the works in which the structure and properties of coatings with multilayer architecture are investigated: AlCrMoNbZr/(AlCrMoNbZr)N [24], (TiZrNbTaHf)N/MoN [85], and (TiZrNbHfTa)N/WN [86]. Bagdasaryan et al. [85] prepared TiZrNbTaHf)N/MoN and (TiZrNbHfTa)N/WN, two types of protective coatings, by vacuum-arc deposition. The formation of (TiZrNbTaHf)N and Mo_2N phases with FCC crystal structures was shown in all coatings regardless of substrate bias. The maximum hardness of approximately 29 GPa was obtained at a bias voltage of −200 V and the thinnest modulation period of bilayer (20 nm). The enhancement of mechanical properties was attributed to Hall–Petch strengthening, solid-solution strengthening, formation of strong MeN chemical bonds, and a low modulation period. In the case of (TiZrNbHfTa)N/WN coatings [86], the application of a high substrate caused the increasing of the peak intensities of (TiZrNbTaHf)N (111) and $W_2N(111)$ and precipitation of the BCC metallic phase. The hardness and Young's modulus of the (TiZrNbHfTa)N/WN coating is 34 and 325 GPa at a substrate bias of –90 V, and 31 and 337 GPa at –280 V.

High hardness and Young's modulus are not the only characteristics crucial for protective coatings; tribological parameters, like low coefficient of friction and wear rate, are important as well. Lai et al. [49] investigated tribological behavior of (AlCrTaTiZr)N coatings prepared by reactive RF magnetron sputtering onto differently biased substrates. The friction coefficient remains unchanged under different substrate bias and reaches the steady state with a value 0.76. However, the wear rate decreases with the application of high U_{sb}: 3.66×10^{-6} mm^3/N·m for $U_{sb}=-150$ V; 6.5 $\times 10^{-6}$ mm^3/N·m for $U_{sb}=0$ V. Similar results have been reported for (AlCrMoTaTiZr)N coatings, deposited by reactive RF magnetron sputtering [42]. The steady value of friction coefficients and wear rate for the nitride films of $R_N = 40\%$ and $R_N=50\%$ are 0.74 and 0.80, and 2.8×10^{-6} mm^3/N·m and 2.9×10^{-6} mm^3/N·m, respectively. In our opinion, the better wear resistance of the presented coatings is obviously due to the presence of chemical elements (Cr, Ta, Mo) which are more resistant to abrasion.

Ren et al. [31] examined the tribological behavior of $(AlCrMnMoNiZr)N_x$ coatings, fabricated by reactive DC magnetron sputtering. It was shown that the coatings deposited at $R_N=0$ and 0.2 have low friction coefficients of 0.14 and 0.16. As R_N further increases, the friction coefficient decreases due to the high surface roughness. Pogrebnjak et al. [71] deposited (TiHfZrVNb)N nitride coating by vacuum-arc deposition. They found that the coatings deposited at $U_{sb}=230$ B and the pressure of working atmosphere $P_N=0{,}2$ Pa possess worse values of tribological characteristics: steady value of coefficient of friction: 1.19; wear rate: 0.039×10^{-5} mm^3/N·mm). According to the authors, such tribological behavior is due to the presence of droplet fractions on the surface of the coatings.

1.7 FUTURE DIRECTIONS AND NEW IDEAS

1. **Nanocomposite structure is still a challenge.** The basic idea of the design HEANs with nanocomposite structures is the formation of multiphase solid materials with the properties which will depend on the interaction between two distinct regions: soft components and hard grains. More efforts should be provided for obtaining such materials.
2. **HEANs as a part of materials with adaptive behavior.** The development and improvement of so-called "smart" materials and systems have constituted one of the most promising areas of tribological materials research. One possible way of achieving coatings with excellent tribotechnical characteristics is to combine materials with different properties into a single composite. In such complex structures, each structural component performs its function, like a hard matrix based on the nitride of high entropy alloys, which consists of transition metals, and provides hardness and wear resistivity. Another component can be dichalcogenides of transition metals,

which have proven themselves as solid lubricants, for reducing the coefficient of friction at low and medium temperatures. In spite of the complexity of realization of such materials, new types of HEANs should be designed.

3. **Modification by ion implantation.** Such methods cause the activation of the processes of radiation-stimulated diffusion, the formation of a large number of radiation defects, which lead to the hardening of the surface layer, so it is essential to examine the influence of ion implantation on the phase state, structure, and properties of the surface layer of HEANs.
4. **More properties, more applications.** TTMN materials are well-known for their widespread application as electrocatalysts in hydrogen storage, supercapacitors, superconductors, biocompatible materials, electrical contacts, and others. It will be interesting to examine the superconductivity of HEANs with different compositions.
5. **HEANs with multilayer structures.** The multilayer architecture of protective coatings allow them to achieve excellent functional properties, especially high hardness and low coefficient of friction due to the action of additional mechanisms of hardening: hindering of dislocation across the layer interfaces; coherence stress field at the interfaces as obstacles for dislocation movements; Orowan strengthening. That's why more efforts for realizing multilayer structure in HEANs should be made.

1.8 CONCLUSION

This paper presents and describes the mechanisms for the formation of the structural-phase state of nitride coatings based on high entropy alloys. In most cases, HEANs form a disordered solid solution with a simple FCC crystal structure. However, this tendency decreases with the incorporation of metal group numbers of the periodic table for the coatings, which consist of non-strong nitride elements inherent in the formation of FCC + BCC structures or amorphous phase.

The application of high energy conditions induces high compressive stresses and makes the structure denser, which are favorable influences on mechanical properties. In general, the mechanical properties of these types of interstitial compounds are due to the metal/non-metal covalent bonding, solid solution hardening, and severe lattice distortion. With a certain confidence, we can conclude that a high-entropy approach allows the achievement of coatings tailored to service conditions in a wide range.

REFERENCES

1. Pogrebnjak, A.D., Bagdasaryan, A.A., Pshyk, A., Dyadyura, K. 2017. Adaptive multi-component nanocomposite coatings in surface engineering. *Phys. Usp.* 60(6):586–607.
2. Cavaleiro, A., De Hosson, J.T.M. 2006. *Nanostructured Coatings*. New-York: Springer-Verlag.
3. Sundgren, J.H., Hentzell, H.T.G. 1986. A review of the present state of art in hard coatings grown from the vapor phase. *J. Vac. Sci. Technol. A* 4(5):2259–79.
4. Tjong, S.C., Chen, H. 2004. Nanocrystalline materials and coatings. *Mater. Sci. Eng. R* 45(1–2):1–88.

5. Veprek, S. 2013. Recent search for new superhard materials: Go nano! *J. Vac. Sci. Technol. A* 31(5):050822-1-050822-33.
6. Łępicka, M., Grądzka-Dahlke, M., Pieniak, D., Pasierbiewicz, K., Kryńska, K., Niewczas, A.A. 2019. Tribological performance of titanium nitride coatings: A comparative study on TiN-coated stainless steel and titanium alloy. *Wear* 422–423:68–80.
7. Rasaki, S.A., Zhang, B., Anbalgam, K., Thomas, T., Yang, M. 2018. Synthesis and application of nano-structured metal nitrides and carbides: A review. *Prog. Sol. State Chem.* 50:1–15.
8. Aouadi, S.M., Wong, K.C., Mitchell, K.A.R., Namavar, F., Tobin, E., Mihut, D.M., Rohde, S.L. 2004. Characterization of titanium chromium nitride nanocomposite protective coatings. *Appl. Surf. Sci.* 229(1–4):387–94.
9. Koller, C.M., Hollerweger, R., Sabitzer, C., Rachbauer, R., Kolozsvári, S., Paulitsch, J., Mayrhofer, P.H. 2014. Thermal stability and oxidation resistance of arc evaporated TiAlN, TaAlN, TiAlTaN, and TiAlN/TaAlN coatings. *Surf. Coat. Technol.* 259:599–607.
10. Martina, P.J., Bendavid, A., Cairney, J.M., Hoffman, M. 2005. Ti-Si-N, Zr-Si-N, Ti-Al-Si-N, Ti-Al-V-Si-N thin film coatings deposited by vacuum arc deposition. *Surf. Coat. Technol.* 200(7):2228–35.
11. Tareen, A.K., Priyanga, G.S., Behara, S., Thomas, T., Yang, M. 2019. Mixed ternary transition metal nitrides: A comprehensive review of synthesis, electronic structure, and properties of engineering relevance. *Prog. Sol. State Chem.* 53:1–26.
12. Pogrebnjak, A.D., Eyidi, D., Abadias, G., Bondar, O.V., Beresnev, V.M., Sobol, O.V. 2015. Structure and properties of arc evaporated nanoscale TiN/MoN multilayered systems. *Int. J. Refract. Met. Hard Mater.* 48:222–28.
13. Chang, Y.Y., Chiu, W.T., Hung, J.P. 2016. Mechanical properties and high temperature oxidation of CrAlSiN/TiVN hard coatings synthesized by cathodic arc evaporation. *Surf. Coat. Technol.* 303:18–24.
14. Vladescu, A., Kiss, A., Popescu, A., Braic, M., Balaceanu, M., Braic, V., Tudor, I., Logofatu, C., Negrila, C.C., Rapeanu, R. 2008. Influence of bilayer period on the characteristics of nanometer scale ZrN/TiAlN multilayers. *J. Nanosci. Nanotechnol.* 8(2):717–21.
15. Sodenberg, H., Oden, M. 2005. Nanostructure formation during deposition of TiN/SiNx nanomultilayer films by reactive dual magnetron sputtering. *J. Appl. Phys.* 97. http://www.ncbi.nlm.nih.gov/pubmed/114327-1-114327-8.
16. Li, Y., Liu, Z., Luo, J., Zhanga, S., Qiu, J., He, Y. 2019. Microstructure, mechanical and adhesive properties of CrN/CrTiAlSiN/WCrTiAlN multilayer coatings deposited on nitrided AISI 4140 steel. *Mater. Charac.* 147:353–64.
17. Yeh, J.W., Chen, Y.L., Lin, S.J., Chen, S.K. 2007. High-entropy alloys – A new era of exploitation. *Mater. Sci. For.* 560:1–9.
18. Miracle, D.B., Senkov, O.N. 2017. A critical review of high entropy alloys and related concepts. *Acta Mater.* 122:448–511.
19. Pogrebnjak, A.D., Bagdasaryan, A.A., Yakushchenko, I.V., Beresnev, V.M. 2014. The structure and properties of high-entropy alloys and nitride coatings based on them. *Russ. Chem. Rev.* 83(11):1027–61.
20. Li, W., Liu, P., Liaw, P.K. 2018. Microstructures and properties of high-entropy alloy films and coatings: A review. *Mater. Res. Lett.* 6(4):199–299.
21. Niu, J., Li, W., Liu, P., Zhang, K., Ma, F., Chen, X., Feng, R., Liaw, P.K. 2019. Effects of silicon content on the microstructures and mechanical properties of (AlCrTiZrV)-Six-N high-entropy alloy films. *Entropy* 21(1):1–11.
22. Li, R., Li, M., Jiang, C., Qiao, B., Zhang, W., Xe, J. 2019. Thermal stability of AlCrTaTiZrMo-nitride high entropy film as a diffusion barrier for Cu metallization. *J. Alloy Compd.* 773:482–89.

23. Feng, X., Tang, G., Ma, X., Sun, M., Wang, L. 2013. Characteristics of multi-element (ZrTaNbTiW)N films prepared by magnetron sputtering and plasma based ion implantation. *Nucl. Instr. Meth. Phys. Res.* 301:29–35.
24. Zhang, W., Tang, R., Yang, Z.B., Liu, C.H., Chang, H., Yang, J.J., Liao, J.L., Yang, Y.Y., Liu, N. 2018. Preparation, structure, and properties of high-entropy alloy multilayer coatings for nuclear fuel cladding: A case study of AlCrMoNbZr/(AlCrMoNbZr)N. *J. Nucl. Mater.* 512:15–24.
25. Hsieh, T.H., Hsu, C.H., Wu, C.Y., Kao, J.Y., Hsu, C.Y. 2008. Effects of deposition parameters on the structure and mechanical properties of high-entropy alloy nitride films. *Curr. Appl. Phys.* 18(5):512–18.
26. Zhang, Y., Yan, X.H., Liao, W.B., Zhao, K. 2018. Effects of nitrogen content on the structure and mechanical properties of (Al0.5CrFeNiTi0.25)Nx high-entropy films by reactive sputtering. *Entropy* 20(9):1–12.
27. Chang, K.S., Chen, K.T., Hsu, C.Y., Hong, P.D. 2018. Growth (AlCrNbSiTiV)N thin films on the interrupted turning and properties using DCMS and HIPIMS system. *Appl. Surf. Sci.* 440:1–7.
28. Chang, Z.C. 2018. Structure and properties of duodenary (TiVCrZrNbMoHfTaWAlSi) N coatings by reactive magnetron sputtering. *Mater. Chem. Phys.* 220:98–110.
29. Tsai, D.C., Chang, Z.C., Kuo, B.H., Lin, T.N., Shiao, M.H., Shieu, F.S. 2014. Interfacial reactions and characterization of (TiVCrZrHf)N thin films during thermal treatment. *Surf. Coat. Technol.* 240:160–66.
30. Lin, C.H., Duh, J.G. 2008. Corrosion behavior of (Ti-Al-Cr-Si-V)xNy coatings on mild steels derived from RF magnetron sputtering. *Surf. Coat. Technol.* 203(5–7):558–61.
31. Ren, B., Shen, Z., Liu, Z. 2013. Structure and mechanical properties of multi-element (AlCrMnMoNiZr)Nx coatings by reactive magnetron sputtering. *J. Alloy Compd.* 560:171–76.
32. Huang, P.K., Yeh, J.W. 2009. Effects of substrate temperature and post-annealing on microstructure and properties of (AlCrNbSiTiV)N coatings. *Thin Solid Films* 518(1):180–84.
33. Lin, C.H., Duh, J.G., Yeh, J.W. 2007. Multi-component nitride coatings derived from Ti-Al-Cr-Si-V target in RF magnetron sputter. *Surf. Coat. Technol.* 201(14):6304–08.
34. Liu, L., Zhu, J.B., Hou, C., Li, J.C., Jiang, Q. 2013. Dense and smooth amorphous films of multicomponent FeCoNiCuVZrAl high-entropy alloy deposited by direct current magnetron sputtering. *Mater. Des.* 46:675–79.
35. Tsai, D.C., Deng, M.J., Chang, Z.C., Kuo, B.H., Chen, E.C., Chang, S.Y., Shieu, F.S. 2015. Oxidation resistance and characterization of (AlCrMoTaTi)-Six-N coating deposited via magnetron sputtering. *J. Alloy Compd.* 647:179–88.
36. Shen, W.J., Tsai, M.H., Yeh, J.W. 2015. Machining performance of sputter-deposited (Al0.34Cr0.22Nb0.11Si0.11Ti0.22)50N50 high-entropy nitride coatings. *Coatings* 5(3):312–25.
37. Tsai, D.C., Chang, Z.C., Kuo, B.H., Chang, S.Y., Shieu, F.S. 2014. Effects of silicon content on the structure and properties of (AlCrMoTaTi)N coatings by reactive magnetron sputtering. *J. Alloy Compd.* 616:646–51.
38. Chang, Z.C., Tsai, D.C., Chen, E.C. 2015. Structure and characteristics of reactive magnetron sputtered (CrTaTiVZr)N coatings. *Mater. Sci. Semincond Proc.* 39:30–9.
39. Shen, W.J., Tsai, M.H., Chang, Y.S., Yeh, J.W. 2012. Effects of substrate bias on the structure and mechanical properties of (Al1.5CrNb0.5Si0.5Ti)Nx coatings. *Thin Sold Films* 520(19):6183–88.
40. Braic, V., Vladescu, A., Balaceanu, M., Luculescu, C.R., Braic, M. 2012. Nanostructured multi-element (TiZrNbHfTa)N and (TiZrNbHfTa)C hard coatings. *Surf. Coat. Technol.* 211:117–21.

41. Tsai, D.C., Liang, S.C., Chang, Z.C., Lin, T.N., Shiao, M.H., Shieu, F.S. 2012. Effects of substrate bias on structure and mechanical properties of (TiVCrZrHf)N coatings. *Surf. Coat. Technol.* 207:293–99.
42. Cheng, K.H., Lai, C.H., Lin, S.J., Yeh, J.W. 2011. Structural and mechanical properties of multi-element (AlCrMoTaTiZr)Nx coatings by reactive magnetron sputtering. *Thin Solid Films* 519(10):3185–90.
43. Lai, C.H., Tsai, M.H., Lin, S.J., Yeh, J.W. 2007. Influence of substrate temperature on structure and mechanical, properties of multi-element (AlCrTaTiZr)N coatings. *Surf. Coat. Technol.* 201(16–17):6993–98.
44. Chen, T.K., Wong, M.S., Shun, T.T., Yeh, J.W. 2005. Nanostructured nitride films of multi-element high-entropy alloys by reactive DC sputtering. *Surf. Coat. Technol.* 200(5–6):1361–65.
45. Tsai, D.C., Chang, Z.C., Kuo, L.Y., Lin, T.J., Lin, T.N., Shiao, M.H., Shieu, F.S. 2013. Oxidation resistance and structural evolution of (TiVCrZrHf)N coatings. *Thin Solid Films* 544:580–87.
46. Hsieh, M.H., Tsai, M.H., Shen, W.J., Yeh, J.W. 2013. Structure and properties of two Al-Cr-Nb-Si-Ti high-entropy nitride coatings. *Surf. Coat. Technol.* 221:118–23.
47. Liang, S.C., Tsai, D.C., Chang, Z.C., Lin, T.N., Shiao, M.H., Shieu, F.S. 2012. Thermally stable TiVCrZrHf nitride films as diffusion barriers in copper metallization. *Electrochem. Solid State Lett.* 15(1):H5–H8.
48. Chang, S.Y., Chen, M.K., Chen, D.S. 2009. Multiprincipal-element AlCrTaTiZr-nitride nanocomposite film of extremely high thermal stability as diffusion barrier for Cu metallization. *J. Electrochem. Soc.* 156(5):G37–G42.
49. Lai, C.H., Cheng, K.H., Lin, S.J., Yeh, J.W. 2008. Mechanical and tribological properties of multi-element (AlCrTaTiZr)N coatings. *Surf. Coat. Technol.* 202(15):3732–38.
50. Tsai, D.C., Huang, Y.L., Lin, S.R., Liang, S.C., Shi, F.S. 2010. Effect of nitrogen flow ratios on the structure and mechanical properties of (TiVCrZrY)N coatings prepared by reactive magnetron sputtering. *Appl. Surf. Sci.* 257(4):1361–67.
51. Huang, P.K., Yeh, J.W. 2009. Effects of nitrogen content on structure and mechanical properties of multi-element (AlCrNbSiTiV)N coating. *Surf. Coat. Technol.* 203(13):1891–96.
52. Chang, S.Y., Lin, S.Y., Huang, Y.C., Wu, C.L. 2010. Mechanical properties, deformation behaviors and interface adhesion of (AlCrTaTiZr)Nx multi-component coatings. *Surf. Coat. Technol.* 204(20):3307–14.
53. Tsai, D.C., Chang, Z.C., Kuo, L.Y., Lin, T.J., Lin, T.N., Shieu, F.S. 2013. Solid solution coating of (TiVCrZrHf)N with unusual structural evolution. *Surf. Coat. Technol.* 217:84–7.
54. Huang, P.K., Yeh, J.W. 2010. Inhibition of grain coarsening up to 1000 °C in (AlCrNbSiTiV)N superhard coatings. *Scipr Mater.* 62(2):105–8.
55. Tsai, M.H., Wang, C.W., Lai, C.H., Yeh, J.W., Gan, J.Y. 2008. Thermally stable amorphous (AlMoNbSiTaTiVZr)50N50 nitride film as diffusion barrier in copper metallization. *Appl. Phys. Lett.* 92(5). http://www.ncbi.nlm.nih.gov/pubmed/052190-1-052109-4.
56. Liang, S.C., Chang, Z.C., Tsai, D.C., Lin, Y.C., Sung, H.S., Deng, M.J., Shi, F.S. 2011. Effects of substrate temperature on the structure and mechanical properties of (TiVCrZrHf)N coatings. *Appl. Surf. Sci.* 257(17):7709–13.
57. Chang, S.Y., Lin, S.Y., Huang, Y.C. 2011. Microstructures and mechanical properties of multi-component (AlCrTaTiZr)NxCy nanocomposite coatings. *Thin Solid Films* 519(15):4865–69.
58. Ren, B., Liu, Z.X., Shi, L., Cai, B., Wang, M.X. 2011. Structure and properties of (AlCrMnMoNiZrB0.1)Nx coatings prepared by reactive DC sputtering. *Appl. Surf. Sci.* 257(16):7172–78.

59. Chang, H.W., Huang, P.K., Yeh, J.W., Davison, A., Tsau, C.H., Yang, C.C. 2008. Influence of substrate bias, deposition temperature and post-deposition annealing on the structure and properties of multi-principal-component (AlCrMoSiTi)N coatings. *Surf. Coat. Technol.* 202(14):3360–66.
60. Hsueh, H.T., Shen, W.J., Tsai, M.H., Yeh, J.W. 2012. Effect of nitrogen content and substrate bias on mechanical and corrosion properties of high-entropy films (AlCrSiTiZr)100−xNx. *Surf. Coat. Technol.* 206(19–20):4106–12.
61. Chen, T.K., Shun, T.T., Yeh, J.W., Wong, M.S. 2004. Nanostructured nitride films of multi-element high-entropy alloys by reactive DC sputtering. *Surf. Coat. Technol.* 188–189:193–200.
62. Chen, W., Yan, A., Meng, X., Wu, D., Yao, D., Zhang, D. 2018. Microstructural change and phase transformation in each individual layer of a nano-multilayered AlCrTiSiN high-entropy alloy nitride coating upon annealing. *Appl. Surf. Sci.* 462:1017–28.
63. Yalamahchili, K., Wang, F., Schramm, I.C., Andersson, J.M., Johansson Joesaar, M.P., Tasnadi, F., Mucklich, F., Ghafoor, N., Oden, M. 2017. Exploring the high entropy alloy concept in (AlTiVNbCr)N. *Thin Solid Films* 636:346–52.
64. Pogrebnjak, A.D., Beresnev, V.M., Smyrnova, K.V., Kravchenko, Ya.O., Zukowski, P.V., Bondarenko, G.G. 2018. The influence of nitrogen pressure on the fabrication of the two-phase superhard nanocomposite (TiZrNbAlYCr)N coatings. *Mater. Lett.* 211:316–18.
65. Nyemchenko, U.S., Beresnev, V.M., Gorban, V.F., Novikov, V.Ju., Yaremenko, O.V. 2015. Comparing the Tribological Properties of the Coatings (Ti-Hf-Zr-V-Nb-Ta)N and (Ti-Hf-Zr-V-Nb-Ta)N + DLC. *JNEP* 7:1–4.
66. Beresnev, V.M., Sobol, O.V., Lytovchenko, S.V., Nyemchenko, U.S., Stolbovoy, V.A., Kolesnikov, D.A., Meylehov, A.A., Postelnyk, A.A., Turbin, P.V., Malikov, L.V. 2016. Effect of high-entropy components of nitride layers on nitrogen content and hardness of (TiN-Cu)/(AlNbTiMoVCr)N vacuum-arc multilayer coatings. *JNEP* 8:1–4.
67. Pogrebnjak, A.D., Yakushchenko, I.V., Bondar, O.V., Beresnev, V.M., Oyoshi, K., Ivasishin, O.M., Amekura, H., Takeda, Y., Opielak, M., Kozak, C. 2016. Irradiation resistance, microstructure and mechanical properties of nanostructured (TiZrHfVNbTa)N coatings. *J. Alloy Compd.* 679:155–63.
68. Pogrebnjak, A.D., Bondar, O.V., Borba, S.O., Abadias, G., Konarski, P., Plotnikov, S.V., Beresnev, V.M., Kassenova, L.G., Drodziel, P. 2016. Nanostructured multielement (TiHfZrNbVTa)N coatings before and after implantation of N+ ions (1018 cm−2): Their structure and mechanical properties. *Nucl. Instr. Meth. Phys. Res. B* 385:74–83.
69. Pogrebnjak, A.D., Bagdasaryan, A.A., Beresnev, V.M., Nyemchenko, U.S., Ivashchenko, V.I., Kravchenko, Ya.O., Shaimardanov, Z.H.K., Plotnikov, S.V., Maksakova, O. 2017. The effects of Cr and Si additions and deposition conditions on the structure and properties of the (Zr-Ti-Nb)N coatings. *Ceram. Int.* 13(1):771–82.
70. Gorban, V.F., Zakiev, I.M., Sarzhan, G.F. 2016. Comparative friction characteristics of high-entropy mononitride coatings. *J. Fric. Wear* 37(3):263–67.
71. Pogrebnjak, A.D., Yakushchenko, I.V., Bagdasaryan, A.A., Bondar, O.V., Krause-Rehberg, R., Abadias, G., Chartier, P., Oyoshi, K., Takeda, Y., Beresnev, V.M., Sobol, O.V. 2014. Microstructure, physical and chemical properties of nanostructured (Ti-Hf-Zr-V-Nb)N coatings under different deposition conditions. *Mat. Chem. Phys.* 147(3):1079–91.
72. Blinkov, I.V., Volkhonskii, A.O., Anikin, V.N., Petrzhik, M.I., Derevtsova, D.E. 2011. Phase composition and properties of wear resistant Ti-Al-Cr-Zr-Nb-N coatings manufactured by the arc-physical deposition method. *Inorg. Mater. Appl. Res.* 2(3):261–67.
73. Yu, R.S., Huang, R.H., Lee, C.M., Shieu, F.S. 2012. Synthesis and characterization of multi-element oxynitride semiconductor film prepared by reactive sputtering deposition. *Appl. Surf. Sci.* 263:58–61.

74. Bagdasaryan, A.A., Smirnova, E., Konarski, P., Miśnik, M., Zawada, A. 2014. The analysis of elemental composition and depth profiles of nitride nanostructured coating based on the TiHfVNbZr high-entropy alloy. *JNEP* 6:1–5.
75. Pogrebnjak, A.D., Borisyuk, V.N., Bagdasaryan, A.A., Maksakova, O.V., Smirnova, E.V. 2014. The multifractal investigation of surface microgeometry of (Ti-Hf-Zr-V-Nb) N nitride coatings. *JNEP* 6:1–4.
76. Pogrebnjak, A.D., Bagdasaryan, A.A., Beresnev, V.M., Kupchishin, A.I., Plotnikov, S.V., Kravchenko, Y.O. 2017. Microstructure and tribological properties of nitride coatings based on Zr, Ti, Cr, Nb, and Si elements. *High Temp. Mat. Proc.* 21(3):267–75.
77. Lengauer, W., Eder, A. 2005. Nitrides: Transition Metal Solid-State Chemistry. In: *Encyclopedia of Inorganic Chemistry.* J.Wiley&Sons.
78. Chang, S.Y., Huang, Y.C., Li, C.E., Hsu, H.F., Yeh, J.W., Lin, S.J. 2013. Improved diffusion-resistant ability of multicomponent nitrides: From unitary TiN to senary high-entropy (TiTaCrZrAlRu)N. *JOM* 65(12):1790–96.
79. Pelleg, J., Zevin, L.Z., Lungo, S., Croitoru, N. 1991. Reactive-sputter-deposited TiN films on glass substrates. *Thin Solid Films* 197(1–2):117–28.
80. Je, J.H., Noh, D.Y., Kim, H.K., Liang, K.S. 1997. Preferred orientation of TiN films studied by a real time synchrotron x-ray scattering. *J. Appl. Phys.* 81(9):6126–33.
81. Gall, D., Kodambaka, S., Wall, M.A., Petrov, I., Greene, J.E. 2003. Pathways of atomistic processes on TiN(001) and (111) surfaces during film growth: An ab initio study. *J. Appl. Phys.* 93(11):9086–94.
82. Patsalas, P., Gravalidis, C., Logothetidis, S. 2004. Surface kinetics and subplantation phenomena affecting the texture, morphology, stress, and growth evolution of titanium nitride films. *J. Appl. Phys.* 96(11):6234–35.
83. Schell, N., Matz, W., Bøttiger, J., Chevallier, J., Kringhøj, P. 2002. Development of texture in TiN films by use of in situ synchrotron x-ray scattering. *J. Appl. Phys.* 91(4):2037–44.
84. Messier, R., Giri, A.P., Roy, R.A. 1984. Revised structure zone model for thin film physical structure. *J. Vac. Sci. Technol. A* 2(2):500–3.
85. Bagdasaryan, A.A., Pshyk, A.V., Coy, L.E., Konarski, P., Misnik, M., Ivashchenko, V.I., Kempiński, M., Mediukh, N.R., Pogrebnjak, A.D., Beresnev, V.M., Jurga, S. 2018. A new type of (TiZrNbTaHf)N/MoN nanocomposite coating: Microstructure and properties depending on energy of incident ions. *Comps. Part B* 146:132–44.
86. Bagdasaryan, A.A., Pshyk, A.V., Coy, L.E., Kempiński, M., Pogrebnjak, A.D., Beresnev, V.M., Jurga, S. 2018. Structural and mechanical characterization of (TiZrNbHfTa)N/WN multilayered nitride coatings. *Mater. Lett.* 229:364–67.

2 Processing Challenges and Properties of Lightweight High Entropy Alloys

Khin Sandar Tun, Manoj Gupta, and T.S. Srivatsan

CONTENTS

2.1 INTRODUCTION

Emission of greenhouse gases (GHGs), such as carbon dioxide (CO_2), is the key to atmospheric pollution. A sustained rise in the amount of greenhouse gases has been a noticeable phenomenon since the 1950s. As of this very moment, we are reaching a tipping point where if the greenhouse gas emissions are not arrested and reduced, the change in climate could become irreversible [1]. As greenhouse gas emissions have both a direct impact and an adverse influence on both climate change and global warming, it is necessary to reduce GHG emissions. Among the different primary greenhouse gases (carbon dioxide, nitrous oxide, and methane), carbon dioxide (CO_2) which originates from fuel burning has been mainly responsible for global warming [2]. Table 2.1 lists the sectors releasing CO_2 and their contributions. One of the sectors contributing to CO_2 emissions was ground and air transportation [1].

TABLE 2.1
Global Emission of Carbon Dioxide (CO_2) by the Different Sectors

Sector	Contribution (%)
Electricity/Heat	40
Transport	24
Industrial	19
Residential	7
Others	10

In the transportation sector, several approaches can be used to reduce the emission of CO_2, including the following: (a) an improvement in powertrain energy efficiency, (b) optimization of vehicle design, (c) use of alternative fuels, and (d) material substitution. From a technical standpoint and an economic aspect, a substitution of the conventionally used materials by alternative lightweight materials is considered to be reasonable, and it does show potential [3]. Better performance, or functionality, such as higher acceleration of the vehicle, is achievable with the use of a lighter vehicle. In terms of sustainability, every 10% reduction in weight of the vehicle can enable savings of 6% to 8% in consumption of fuel [4]. In addition, reduction of CO_2 emissions can be up to 1.25 g/km by reducing weight of vehicle by as much as 10 kg while concurrently lowering the environmental impact [5].

About 80% of the total vehicle weight is made of metallic materials. The commonly used metallic materials in the automobile industry include steel, cast iron, titanium, aluminum, and magnesium. Among these materials, alloys of aluminum and alloys of magnesium are considered to be the most promising lightweight materials for the present and continue to be technically feasible and potentially lightweight material until 2030 [4, 6, 7]. However, there is a concern over the potential rise of GHG emissions related to the production and recycling of aluminum and its alloys [8]. On the other hand, despite exceptional savings in weight, there is minimal utilization of the magnesium alloys, which are less than 0.5% of the average vehicle weight. Challenges associated with (i) manufacturing and processing, (ii) assembly, (iii) in-service performance, and (iv) cost hinder the selection and use of magnesium alloys in automotive applications [9, 10]. An additional drawback is the limited number of commercially available magnesium alloys, and the combination of properties that they have to offer is also limited. Besides, in terms of materials properties, magnesium alloys have two major disadvantages. These are the following: (i) low high-temperature strength, and (ii) a relatively poor corrosion resistance. Hence, the development of new alloys with enough properties is required to fulfil the requirements put forth by the automotive industry [10–12].

Recently, a new alloy system based on the use of multiple principal elements has emerged. These alloys are named "high entropy alloys" (HEAs), as they exhibit significantly higher mixing entropies ($\geq$1.5R, R is gas constant) than those of the conventional alloys ($\leq$1R) [13]. The high entropy alloys are fundamentally different from the traditional alloys because the alloy design is based on the use of multiple principal alloying elements. Typically, it consists of at least five major alloying elements in order to maximize the mixing entropy [13–15]. With the use of five or more elements, at near equiatomic ratios, the high entropy alloys exhibit large configurational entropy or mixing entropy, which does enhance the formation of solid solutions while concurrently making the microstructure much simpler than expected [13]. Many of the high entropy alloys studied to date usually contain the transition elements, such as cobalt (Co), chromium (Cr [77]), iron (Fe), nickel (Ni), and copper (Cu). With the use of heavy metals as principal alloying elements, most of the high entropy alloy compositions exhibit density value greater than 10 g/cm^3 [15]. The high density of such high entropy alloy systems is not desirable for selection and use in lightweight structural applications. In order to fabricate low density HEAs, refractory high entropy alloy systems, which are composed of tantalum (Ta), tungsten (W), niobium (Nb), molybdenum (Mo), vanadium (V), hafnium (Hf), and titanium (Ti) were investigated [16]. In the recent past, limited research efforts have been made to develop lighter HEA systems through the selection and use of light metals, such as magnesium (Mg), aluminum (Al), lithium (Li), beryllium (Be), titanium (Ti), and silicon (Si) [17–25]. A list of the lightweight high entropy alloys (LWHEAs) developed by other researchers is provided in Table 2.2.

Lightweight HEAs (LWHEAs) can be defined as those with an overall density close to 3 g/cm^3; and the reported LWHEAs with a density range of 2.20 g/cm^3 to 3.96 g/cm^3 are listed in Table 2.1. As can be seen from the list, in the past ten years (2009–2019), only a handful of research studies were conducted on this promising lightweight alloy system. The reported hardness and compressive properties of the LWHEAs from the literature are summarized in Table 2.3 [21–26]. The lightweight high entropy alloys (LWHEAs) showed exceptionally high hardness, which can be as much as an order of magnitude higher than the conventional alloys, such as AZ31B. A significantly high compressive yield strength and peak strength with limited fracture strain up to 5% were realized for the LWHEAs. Comparatively higher compressive strengths for the LWHEAs when compared to alloy AZ31B exhibit the

TABLE 2.2
List of Reported Lightweight High Entropy Alloys (LWHEAs) from the Literature

Alloy	Density (g/cm^3)	Processing route	Year of publication
$Al_{20}Be_{20}Fe_{10}Si_{15}Ti_{35}$	3.91	Casting	2009 [21]
$Mg_{33}(MnAlZnCu)_{67}$	3.26	Induction melting	2010 [22]
$Mg_{43}(MnAlZnCu)_{57}$	2.51		
$Mg_{45.6}(MnAlZnCu)_{54.4}$	2.30		
$Mg_{50}(MnAlZnCu)_{50}$	2.20		
$Al_{20}Li_{20}Mg_{10}Sc_{20}Ti_{30}$	2.67	Mechanical alloying	2015 [23]
AlLiMgZnSn	3.88	Induction melting	2014 [24]
$AlLi_{0.5}MgZn_{0.5}Sn_{0.2}$	2.98		
$Al_{35}Cu_5Fe_5Mn_5Si_{30}V_{10}Zr_{10}$	3.96	Vacuum die casting	2018 [25]
$Al_{50}Ca_5Cu_5Ni_{10}Si_{20}Ti_{10}$	3.33		

TABLE 2.3
Hardness and Compressive Properties of Reported LWHEAs

Alloy	Density (g/cm³)	Hardness (HV)	Yield strength (MPa)	Peak strength (MPa)	Fracture strain (%)
$Al_{20}Be_{20}Fe_{10}Si_{15}Ti_{35}$	3.91	911	2976	–	–
$Mg_{33}(MnAlZnCu)_{67}$	3.26	315	437	437	3.41
$Mg_{43}(MnAlZnCu)_{57}$	2.51	255	500	500	3.72
$Mg_{45.6}(MnAlZnCu)_{54.4}$	2.30	225	482	482	4.06
$Mg_{50}(MnAlZnCu)_{50}$	2.20	178	340	400	4.83
$Al_{20}Li_{20}Mg_{10}Sc_{20}Ti_{30}$	2.67	499.6	–	–	–
AlLiMgZnSn	3.88	–	600	615	1.2
$AlLi_{0.5}MgZn_{0.5}Sn_{0.2}$	2.98	–	–	546	0
$Al_{35}Cu_5Fe_5Mn_5Si_{30}V_{10}Zr_{10}$	3.96	751	2455	–	–
$Al_{50}Ca_5Cu_5Ni_{10}Si_{20}Ti_{10}$	3.33	437	1429	–	–
AZ31B	1.77	63	133	444	12.6

potential for this new lightweight alloy system to replace the conventional alloy in strength-related structural applications. This also encourages further development of the LWHEAs as a viable alternative light alloy.

Aiming to develop next generation lightweight structural metal alloys, several equiatomic and non-equiatomic lightweight high entropy alloys (LWHEAs) have been developed in the current study. To achieve a target density close to 3 g/cm^3, light metals, such as magnesium (Mg), aluminum (Al), and lithium (Li) are commonly used with other alloying elements. Different elemental combinations as well as compositional variations were attempted during the designing of these alloys. Some alloy compositions were only theoretically feasible, as they failed during the synthesis stage. Failed compositions are also reported here, as it might be beneficial to colleague researchers for further exploration on these alloy compositions using alternative processing approaches. Successfully processed alloy compositions are presented in this study for purpose of further analyses or an extraction of the different properties by interested researchers. In this technical manuscript, a characterization of the microstructure, phase composition, and mechanical properties was performed with the primary purpose of understanding the potential of these alloys for selection and use in weight-critical applications.

2.2 SYNTHESIS AND CHARACTERIZATION OF LIGHTWEIGHT HIGH ENTROPY ALLOYS (LWHEAs)

Lightweight high entropy alloys are processed using the liquid metallurgy route known as the disintegrated melt deposition (DMD) technique. This technique is a hybrid of conventional casting and spray atomization and deposition. This technique has shown a capability for giving a fine dendritic, or equiaxed, structure in cast ingots of metallic materials. The alloying elements are weighed according to the designated chemical composition of the alloys and placed in a graphite crucible. The materials are heated up to the desired superheated temperature (typically 750°C) using an electrical resistance furnace and in an atmosphere of inert argon gas. The superheated melt is then stirred at 450 rpm for 5 mins and bottom-poured into a steel mold. During the pouring stage the melt is disintegrated using two jets of argon gas and subsequently deposited to get an ingot which measures 40 mm in diameter. The as-deposited materials were used for further characterization.

The density of the alloys was measured using Archimedes' principle. Three samples were randomly selected, and precision weighed in laboratory air and when fully immersed in distilled water. An A&D ER-182A electronic balance, with an accuracy of 0.0001 g, was used for recording the weights. The theoretical density of the sample was calculated using the rule-of-mixture principle.

Characterization of the microstructure was performed using scanning electron microscopes (JEOL JSM-6010) equipped with energy dispersive X-ray analysis (EDX). X-ray diffraction analysis was conducted using an automated Shimadzu LAB-XRD-6000 (Cu Kα: λ = 1.54056 A°) spectrometer with a scan speed of 2 degrees per minute.

Microhardness of the samples was measured using a Shimadzu HMV automatic digital microhardness tester (Kyoto, Japan) with a Vickers indenter. An indenting load of 9.807 N was used for a dwell time of 15 sec. The room temperature (27°C) compressive tests were performed on the samples in conformation with procedures detailed in the Standard ASTM E9-89a and using a fully automated servo-hydraulic test machine [Model: MTS810]. The samples were tested at a strain rate of 5×10^{-3} per minute. Rectangular samples having a width (w) to height (h) ratio of 1 were used for purpose of testing in compression.

2.3 RESULTS AND ANALYSIS

2.3.1 Alloy Development

Development of both equiatomic and non-equiatomic high entropy alloys was attempted in the current research study. A total of 16 attempted alloy compositions are listed in Table 2.4 together with their respective mixing entropy, theoretical density, and processing status. Different combinations of alloying elements as well as compositional variation within the same alloying element combination were attempted during alloy development. Out of the 16 attempts of alloy synthesis, six attempts failed at the processing stage while ten attempts were successfully processed and moved forward to testing and characterization. The characterization results are presented and briefly discussed in the forthcoming sections. The processing remark describing the specific issues encountered for the failed high entropy alloy compositions are briefly stated in Table 2.5. For most of the failed compositions, the primary causes for the failure in processing were related to non-melting and fumes emission, coupled with the formation of the hardened layer at the bottom of the crucible. Hence, other processing methods, such as arc melting or induction melting, can be employed as an alternative approach to the bottom-pouring casting method, such as in disintegrated melt deposition (DMD). Since the fumes and ashes

TABLE 2.4
List of Attempted High Entropy Alloy Compositions, Mixing Entropy, Theoretical Density, and Synthesis Outcome

No.	Composition (atm. %)	Entropy of mixing (ΔS_{mix})	Theoretical density (g/cm³)	Condition
1	AlCuNiSnZn*	13.38 (1.61R)	6.79	Fail
2	MgCuNiSnZn*	13.38 (1.61R)	6.23	Fail
3	AlMgCuNiLiZn*	14.90 (1.79R)	4.11	Fail
4	$Al_{26}Mg_{25}Si_{25}Ti_{10}Zn_{10}Ca_4$	13.79 (1.66R)	3.45	Fail
5	$Al_{35}Li_{20}Si_{20}Zn_{15}Ti_{10}$	12.69 (1.52R)	2.83	Fail
6	$Mg_{35}Si_{15}Zn_{15}Sn_{15}Y_{10}Ca_{10}$	13.98 (1.68R)	3.53	Fail
7	AlMgLiCaCuZn*	14.90 (1.79R)	2.87	Pass
8	$Mg_{35}Al_{33}Li_{15}Zn_7Ca_5Cu_5$	12.50 (1.5R)	2.27	Pass
9	$Mg_{35}Al_{33}Li_{15}Zn_7Ca_5Y_5$	12.50 (1.5R)	2.25	Pass
10	$Al_{35}Mg_{30}Si_{13}Zn_{10}Y_7Ca_5$	12.97 (1.56R)	2.73	Pass
11	$Al_{35}Li_{20}Mg_{15}Zn_{15}Si_{10}Ca_5$	13.62 (1.64R)	2.41	Pass
12	AlLiMgZnSi*	13.38 (1.61R)	2.61	Pass
13	$Al_{35}Li_{20}Zn_{20}Si_{15}Mg_{10}$	12.69 (1.53R)	2.74	Pass
14	$Al_{35}Li_{20}Mg_{20}Si_{15}Zn_{10}$	12.69 (1.53R)	2.27	Pass
15	$Mg_{35}Cu_{15}Zn_{15}Li_{15}Y_{10}Ca_{10}$	13.98 (1.68R)	3.02	Pass
16	$Mg_{35}Cu_{20}Zn_{20}Li_{15}Y_{10}$	12.68 (1.52R)	3.66	Pass

can be related to the evaporation of certain low melting metals, a higher pressure for the inert gas used is recommended.

2.3.2 Designing of Lightweight High Entropy Alloys

The basic principles of designing high entropy alloys are applied in the design of lightweight high entropy alloys (LWHEAs) by selecting light metals as the principle alloying elements, since the final alloy density is dependent on them. Light metals such as magnesium (Mg), aluminum (Al), and lithium (Li) were selected to achieve the preferred density close to 3 g/cm³ in the newly developed alloys. In order to form simple microstructures in the HEAs, Gibbs free energy is reduced by maximizing mixing entropy (ΔS_{mix}) [27, 28]. Based on the concept of mixing entropy, the multicomponent alloys having a high mixing entropy value which is equivalent to or greater than 12.471 J/K mol are regarded as high entropy alloys. The mixing entropy of the alloy is calculated using the following relationship:

$$\Delta S_{mix} = -R\sum_{i=1}^{n} c_i \ln c_i \tag{2.1}$$

where R is the gas constant, c_i the mole fraction of the ith element, and n the total number of alloying elements.

TABLE 2.5
List of Failed High Entropy Alloy Compositions, Mixing Entropy, and Processing Remarks

No.	Composition (atm. %)	Entropy of mixing (ΔS_{mix})	Processing remarks
1	AlCuNiSnZn*	13.38 (1.61R)	• Heated to a maximum temperature of 950°C. Materials were not melted well and there was one hardened layer at the bottom of the crucible.
2	MgCuNiSnZn*	13.38 (1.61R)	• Heated to a maximum temperature of 950°C. Materials were partially melted and stirred. During stirring stage, fumes and white color ashes were coming out. This was possibly due to the evaporation of Sn. • Upon bottom-pouring, only small drops came out, leaving un-melted materials inside the crucible.
3	AlMgCuNiLiZn*	14.90 (1.79R)	• Heated to a maximum temperature of 900°C. Materials were not melted well and there was one hardened layer at the bottom of the crucible.
4	$Al_{26}Mg_{25}Si_{25}Ti_{10}Zn_{10}Ca_4$	13.79 (1.66R)	• Heated to a maximum temperature of 950°C and held for 30 minutes. Materials were not melted. The presence of Ti which forms a peritectic system with Al is the likely cause.
5	$Al_{35}Li_{20}Si_{20}Zn_{15}Ti_{10}$	12.69 (1.52R)	• Heated to a maximum temperature of 950°C and held for 30 minutes. Materials were not melted. The presence of Ti which forms a peritectic system with Al is the likely cause.
6	$Mg_{35}Si_{15}Zn_{15}Sn_{15}Y_{10}Ca_{10}$	13.98 (1.68R)	• Heated to a temperature up to 800°C. Materials melted partially at 800°C but materials seem to evaporate as the white fumes are emitted. Able to do stirring of melted portion, but only a few drops of melts came out upon down-pouring. Processing is feasible but under high gas pressure.

*Denotes equiatomic composition.

However, practical results may not be fully coherent with theoretical principles. Formation of multiple phases can be expected instead of the sole formation of a single-phase solid solution. Research findings have shown that not only solid solutions, but also many other phases, such as intermetallic, were detected in the cast high entropy alloys (Ye, 2016). This indicates that ΔS_{mix} cannot be the only influencing parameter in the design of HEAs and other thermodynamic parameters should be taken into consideration for purpose of the alloy design [29].

Other than entropy of mixing (ΔS_{mix}), the enthalpy of mixing (ΔH_{mix}) is also suggested as a parameter to form a solid solution. With the value of ΔH_{mix} closer to zero, the constituent elements in the high entropy alloy can be better distributed in a random manner, thereby allowing for the formation and presence of stable solid solution phases [27, 30]. The enthalpy of mixing is calculated using the relationship:

$$\Delta H_{mix} = \sum_{i=1,i\neq j}^{N} 4\Delta H_{ij}^{mix} c_i c_j \tag{2.2}$$

where $\Delta H_{ij}^{\mathrm{mix}}$ is the enthalpy of mixing between i-th and j-th elements at equimolar compositions, c_i is the mole fraction of the i-th element, and c_j is the mole fraction of the j-th element. The values of $\Delta H_{ij}^{\mathrm{mix}}$ can be obtained using the Miedema macroscopic model for binary liquid alloys [31].

Moreover, with the relationship between ΔS_{mix} and ΔH_{mix}, a new thermodynamic parameter, Ω, was introduced to predict the formation of a solid solution [24, 32, 33]. When $\Omega > 1.1$, the influence of T_m will be greater than the effect of ΔH_{mix} for the formation of solid solution. Hence, the HEA is predicted to consist of mainly solid solutions [34].

This parameter Ω is expressed as:

$$\Omega = \frac{T_m \Delta S_{mix}}{\left|\Delta H_{mix}\right|} \tag{2.3}$$

$$T_m = \sum_{i=1}^{n} c_i \left(T_m\right)_i \tag{2.4}$$

where T_m is the average melting temperature of the number of elements in the alloy, $(T_m)_i$ is the melting point of the i-th element of the alloy, and c_i is the mole fraction of the i-th element.

Another important factor that does exert an influence on the formation of solid solution will be the atomic size mismatch effect. Since the elemental atoms in the high entropy alloy randomly occupy the lattice sites based on a statistical average probability of occupancy, the large atomic size difference does cause significant lattice distortion in the alloy that tends to reduce the overall stability of the solid solution [27]. The atomic size mismatch that affects lattice distortion can be expressed using the following relationship [27, 35]:

$$\delta = 100\sqrt{\sum_{i=1}^{n} c_i \left(1 - \frac{r_i}{\bar{r}}\right)^2} \tag{2.8}$$

where n is the number of constituent elements in the alloy system, c_i is the mole fraction of the i-th element, r_i is the atomic radius of the i-th element, and $\bar{r}$ is the average atomic radius.

TABLE 2.6
Calculated Thermodynamic Parameters of Successfully Synthesized Lightweight High Entropy Alloys

No.	Material	ΔH_{mix} (kJ/mol)	ΔS_{mix} (J/Kmol)	δ	Ω
	Reference range	$-22 \leq \Delta H_{mix} \leq 7$	$11 \leq \Delta S_{mix} \leq 19.5$	$0 \leq \delta \leq 8.5$	$\Omega \geq 1.1$
1	AlMgLiCaCuZn	−9.03	14.90	14.8	1.5
2	$Mg_{35}Al_{33}Li_{15}Zn_7Ca_5Cu_5$	−4.51	12.50	9.3	2.4
3	$Mg_{35}Al_{33}Li_{15}Zn_7Ca_5Y_5$	−6.92	12.50	9.9	1.6
4	$Al_{35}Mg_{30}Si_{13}Zn_{10}Y_7Ca_5$	−13.05	12.97	10.22	1.1
5	$Al_{35}Li_{20}Mg_{15}Zn_{15}Si_{10}Ca_5$	−5.94	13.62	12.1	2.0
6	AlLiMgZnSi	−6.46	13.38	12.1	1.9
7	$Al_{35}Li_{20}Zn_{20}Si_{15}Mg_{10}$	−5.11	12.69	10.2	2.2
8	$Al_{35}Li_{20}Mg_{20}Si_{15}Zn_{10}$	−5.39	12.69	11.1	2.2
9	$Mg_{35}Cu_{15}Zn_{15}Li_{15}Y_{10}Ca_{10}$	−9.40	13.98	13.1	1.5
10	$Mg_{35}Cu_{20}Zn_{20}Li_{15}Y_{10}$	−9.90	12.68	11.0	1.3

According to Guo and co-workers [36], apart from entropy of mixing (ΔS_{mix}), other thermodynamic parameters, namely (i) enthalpy of mixing (ΔH_{mix}), and (ii) atomic size difference (δ_r), should also be considered for the formation of a solid solution. These researchers suggested that formation of a solid solution is possible if the parameters are in the suitable range: (i) $11 \leq \Delta S_{mix} \leq 19.5$ J/K mol, (ii) $-22 \leq \Delta H_{mix} \leq 7$ kJ/mol, and (iii) $0 \leq \delta \leq 8.5$. Another thermodynamic parameter, Ω, will also be taken into account, as Yang and co-workers [32] suggested that a solid solution microstructure can be expected only when Ω is greater than or equal to 1.1. These parameters were used for the purpose of theoretical calculation and designing of the current lightweight high entropy alloys. Thermodynamic parameters of the current LWHEAs were calculated and summarized in Table 2.6. As can be seen from Table 2.6, the calculated parameters, entropy and enthalpy of mixing and Ω, of the alloys were well within the reference parameter range. However, the value for δ was larger than the reference value for all the developed alloys. Since the choice of the alloying elements is limited due to the constraints imposed by the target density of the LWHEAs, it is difficult to fulfill the reference δ value, which is related to atomic size difference among the chosen alloying elements.

2.3.3 Alloy Density

Based on a disordered solid solution and the assumption of the rule of mixtures, the theoretical density of the alloys was calculated using the relationship [37]:

$$\rho_{\text{theo}} = \frac{\sum_{i=1}^{n} c_i A_i}{\sum_{i=1}^{n} \frac{c_i A_i}{\rho_i}}$$

TABLE 2.7
Results of Theoretical and Experimental Density in the Developed Lightweight High Entropy Alloys

No.	Composition (at.%)	Theoretical density (g/cm³)	Experimental density (g/cm³)
1	AlMgLiCaCuZn	2.87	3.43 ± 0.003
2	$Mg_{35}Al_{33}Li_{15}Zn_7Ca_5Cu_5$	2.27	2.44 ± 0.01
3	$Mg_{35}Al_{33}Li_{15}Zn_7Ca_5Y_5$	2.25	2.27 ± 0.01
4	$Al_{35}Mg_{30}Si_{13}Zn_{10}Y_7Ca_5$	2.73	2.73 ± 0.002
5	$Al_{35}Li_{20}Mg_{15}Zn_{15}Si_{10}Ca_5$	2.41	2.70 ± 0.02
6	AlLiMgZnSi	2.61	2.75 ± 0.05
7	$Al_{35}Li_{20}Zn_{20}Si_{15}Mg_{10}$	2.74	2.81 ± 0.01
8	$Al_{35}Li_{20}Mg_{20}Si_{15}Zn_{10}$	2.27	2.44 ±0.01
9	$Mg_{35}Cu_{15}Zn_{15}Li_{15}Y_{10}Ca_{10}$	3.02	2.94 ± 0.02
10	$Mg_{35}Cu_{20}Zn_{20}Li_{15}Y_{10}$	3.66	3.55 ± 0.03

In this expression, c_i, A_i, and ρ_i are the atomic percentage, atomic weight, and density of ith respective constituent element, and n is the total number of alloying elements. The experimental density of the two alloys is provided in Table 2.7. The experimental density was found to be higher for most of the compositions. This indicates the formation and presence of ordered phases (i.e., intermetallic or second-phase particles) in these alloys. The occurrence of an increased experimental density due to the formation and presence of ordered second-phase particles is commonly encountered in high entropy alloys and other multicomponent alloys as reported by few other researchers [24, 37]. For the $Al_{35}Mg_{30}Si_{13}Zn_{10}Y_7Ca_5$ alloy, the experimental density was found to be the same as the theoretical density, indicating the formation and presence of secondary phases coupled with minimal porosity. For the $Mg_{35}Cu_{15}Zn_{15}Li_{15}Y_{10}Ca_{10}$ and $Mg_{35}Cu_{20}Zn_{20}Li_{15}Y_{10}$ compositions, an abnormal finding of lower experimental density value when compared to theoretical density was observed. This is an indication of the high porosity level present in these alloys. The evidence of a porous structure on the alloys fracture surface is presented and briefly discussed in the following section (Section 2.3.4).

2.3.4 Equiatomic AlMgLiCaCuZn Alloy and Related Non-Equiatomic Alloys

The microstructures of the equiatomic AlMgLiCaCuZn alloy and the non-equiatomic AlMgLiCaCuZn alloy with a composition, $Mg_{35}Al_{33}Li_{15}Zn_7Ca_5Cu_5$ are shown in Figure 2.1. A simple microstructure with two visible phases (black and gray) was observed for both the equiatomic and non-equiatomic alloys. The intermetallic phases formed and present in the alloys were identified based on X-ray diffraction (XRD) (Figure 2.2) analysis. From the phase analysis, two common phases were found to have the same constituent elements but different elemental compositions:

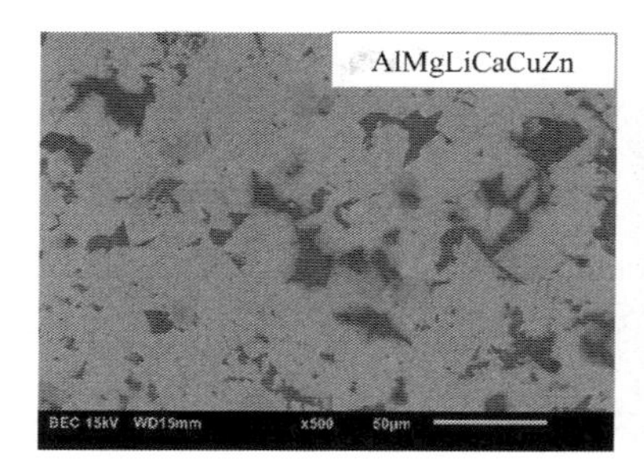

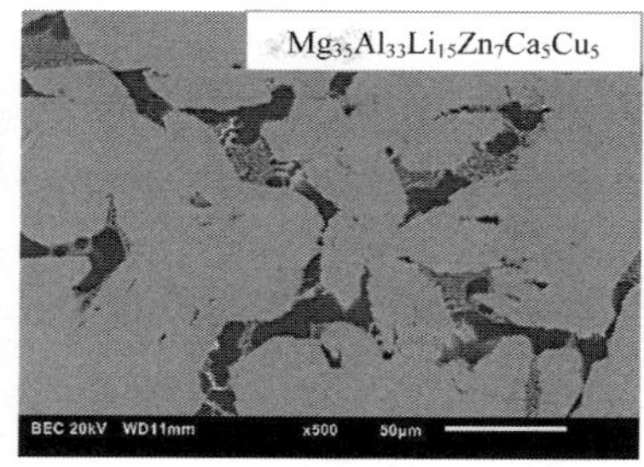

FIGURE 2.1 Scanning electron micrographs of the equiatomic AlMgLiCaCuZn alloy and the non-equiatomic $Mg_{35}Al_{33}Li_{15}Zn_7Ca_5Cu_5$ alloy.

(i) AlCuMg and $CaZn_3$ in the equiatomic alloy, and (ii) Al_2CuMg and CaZn in the non-equiatomic alloy. With the use of an equal atomic percentage of the alloying element, the microstructure is composed of intermetallic phases in the AlMgLiCaCuZn alloy. For the $Mg_{35}Al_{33}Li_{15}Zn_7Ca_5Cu_5$ non-equiatomic alloy, αMg phase was formed coupled with other intermetallic phases due to the presence of a higher amount of magnesium (Mg).

A non-equiatomic alloy, $Mg_{35}Al_{33}Li_{15}Zn_7Ca_5Y_5$, was developed by replacing yttrium (Y) in place of copper (Cu) in the $Mg_{35}Al_{33}Li_{15}Zn_7Ca_5Cu_5$ alloy. A comparison of the microstructure of the two alloys is shown in Figure 2.3, and the corresponding elemental compositions designated as A, B, C, D in the microstructure are listed in Table 2.8. The XRD patterns of two high entropy alloys (HEA) are shown in Figure 2.4. It is to be noted that the unknown phases having very low intensity peaks were not indexed in the XRD pattern.

From the XRD analysis in Figure 2.4a, the diffraction peaks corresponding to αMg, Al_2Y, CaZn, $AlMg_2Zn$, $AlMg_4Zn_{11}$, and $Li_{6.46}Mg$ were detected for the $Mg_{35}Al_{33}Li_{15}Zn_7Ca_5Y_5$ alloy. In the microstructure shown in Figure 2.3a, four phases having different contrasts [Area: A, B, C, and D] were observed through energy dispersive X-ray (EDX) analysis. The dark phase marked as A represents a αMg phase, the white phase marked as B the Al_2Y phase. The light gray phase (area C) and the dark gray phase (area D) were found to be two-layered phases. This implied the combined presence of the phases CaZn, $AlMg_2Zn$, and $AlMg_4Zn_{11}$ appear to be in the area C and area D. Area A in Figure 2.3a can be identified to be the αMg phase (Mg-Zn solid solution) having an atomic composition of 97.59 atomic pct. Mg, and 2.41 atomic pct. Zn, based on quantitative analysis of the phase composition acquired from energy dispersive X-ray (EDX) analysis [38]. For area B in Figure 2.3a, the measured atomic ratio of 61.15 atomic pct. Al, and 32.29 atomic pct. Y, which closely corresponds to the Al_2Y phase. These observations agree with the X-ray diffraction results presented in Figure 2.4a. For the phases CaZn, $AlMg_2Zn$ and $AlMg_4Zn_{11}$, an inaccuracy in chemical compositions of the phases was observed, although matching peaks of these compound/intermetallic phases were detected in the X-ray diffraction pattern (Figure 2.4a). It is logical to assume that the presence of the phases complicates the prediction of composition and associated accuracy.

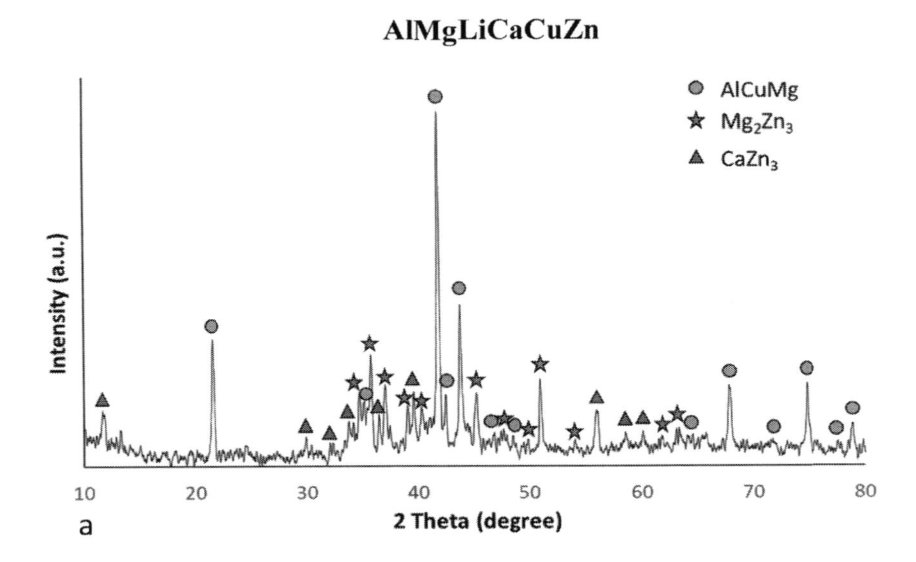

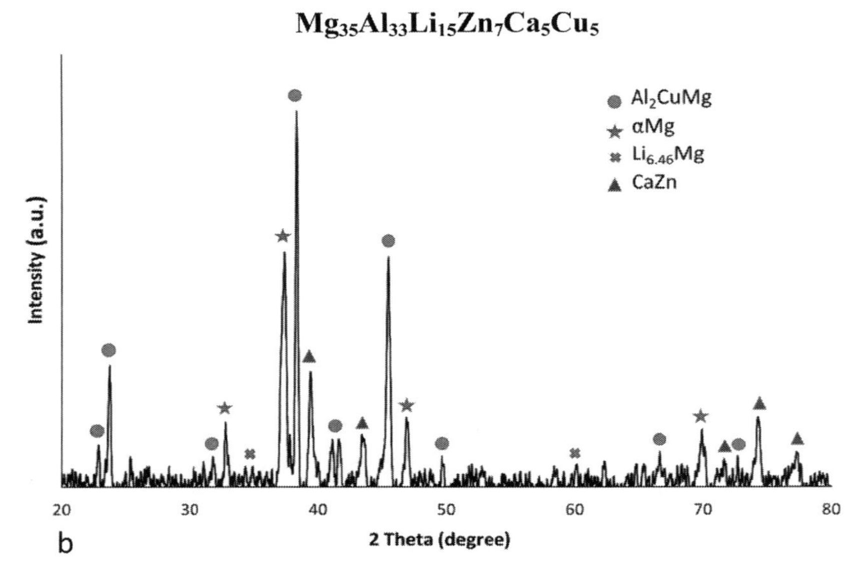

FIGURE 2.2 X-ray diffraction (XRD) analysis of the (a) equiatomic AlMgLiCaCuZn alloy and (b) non-equiatomic $Mg_{35}Al_{33}Li_{15}Zn_7Ca_5Cu_5$ alloy.

In case of the $Mg_{35}Al_{33}Li_{15}Zn_7Ca_5Cu_5$ alloy, the phases αMg, Al_2CuMg, $Li_{6.46}Mg$, and CaZn were found match the X-ray diffraction peaks shown in Figure 2.4b. However, in the microstructure shown in Figure 2.3b, only two major phases are observed, and these are αMg (area A) and Al_2CuMg (area B). From the energy dispersive X-ray (EDX) analysis, the measured chemical composition in area A was found to be 97.91 atomic pct. Mg, and 2.09 atomic pct. Zn, indicating the phase to be αMg. In area B, the chemical compositions of constituents were found to be

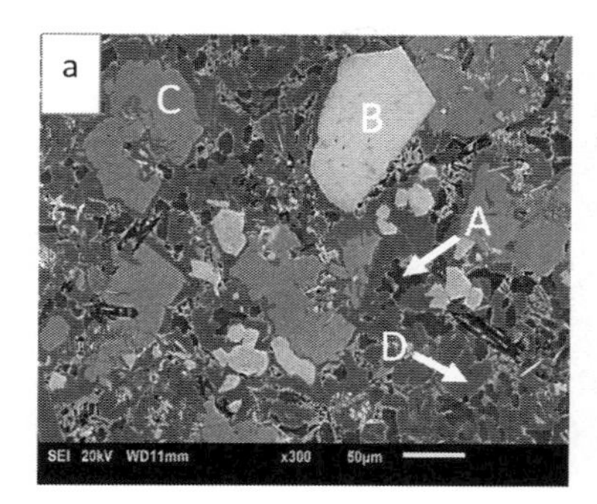

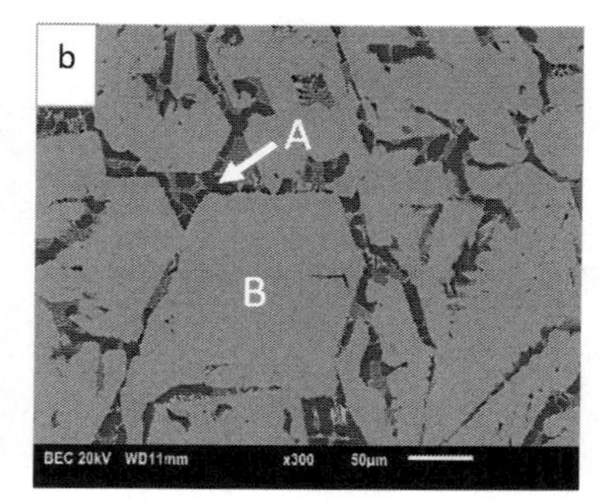

FIGURE 2.3 Scanning electron micrographs of the (a) $Mg_{35}Al_{33}Li_{15}Zn_7Ca_5Y_5$ alloy and (b) $Mg_{35}Al_{33}Li_{15}Zn_7Ca_5Cu_5$ alloy.

TABLE 2.8
Chemical Compositions (Atomic pct.) of Phases in the MgAlLiZnCaY and MgAlLiZnCaCu Alloys

Alloy system	Designated area	Measured chemical compositions (at.%)						
		Mg	Al	Li	Zn	Ca	Y	Cu
MgAlLiZnCaY	A	97.59	–	–	2.41	–	–	
	B	4.39	61.15	–	2.19	–	32.29	
	C	30.18	49.08	–	15.06	5.68	–	
	D	54.56	36.66	–	4.74	4.03	–	
MgAlLiZnCaCu	A	97.91	–	–	2.09	–		–
	B	27.40	49.65	–	6.98	5.89		10.08

Note: Li is not detectable due to the limitation of EDX system used.

27.4 atomic pct. Mg, 49.65 atomic pct. Al, 6.98 atomic pct. Zn, 5.89 atomic pct. Ca, and 10.08 atomic pct. Cu. This suggested the combined presence of the two phases Al_2CuMg and CaZn. However, observing only one contrast (gray color) in the microstructure, the major phase was the Al_2CuMg phase, which matches well with the highest intensity peak in the X-ray diffraction (XRD) pattern (Figure 2.4b). Based on chemical composition from the energy dispersive X-ray (EDX) analysis, the CaZn phase is likely to be present in area B but was not seen in the microstructure, indicating the presence of CaZn to be a minor phase.

For the MgAlLiZnCaCu composition, only two major phases, i.e., (i) Al_2MgCu and (ii) αMg, were observed. When copper was replaced with yttrium, a complex microstructure was observed in the MgAlLiZnCaY alloy when compared to the MgAlLiZnCaCu alloy (Figure 2.3). For the MgAlLiZnCaY alloy, the presence of yttrium together with aluminum caused an observable reduction in the solubility of yttrium in magnesium, forming the Al_2Y phase [39, 40]. The availability of excess magnesium reacted with aluminum and zinc making the microstructure complex.

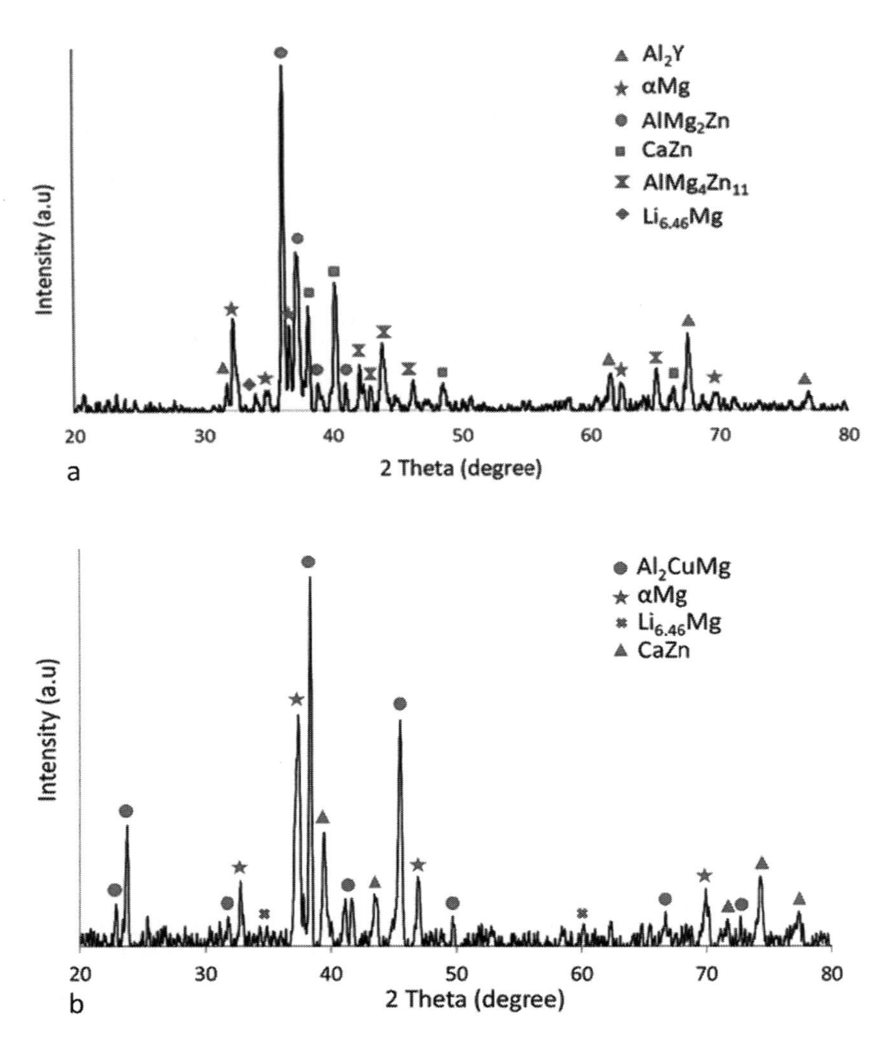

FIGURE 2.4 X-ray diffraction (XRD) pattern of the (a) $Mg_{35}Al_{33}$ $Li_{15}Zn_7Ca_5Y_5$ alloy and (b) $Mg_{35}Al_{33}Li_{15}Zn_7Ca_5Cu_5$ alloy.

With the use of a different alloying element, silicon (having different chemical properties), in place of lithium, together with the same alloying elements—aluminum, magnesium, zinc, yttrium, and calcium—another non-equiatomic alloy with a composition of $Al_{35}Mg_{30}Si_{13}Zn_{10}Y_7Ca_5$ was developed. The effect on microstructural development of using silicon and lithium in the respective alloys is shown in Figure 2.5. The corresponding phase compositions designated as A, B, C, D in the microstructure of $Al_{35}Mg_{30}Si_{13}Zn_{10}Y_7Ca_5$ alloy are summarized in Table 2.9. The X-ray diffraction (XRD) pattern is shown in Figure 2.6.

For the case of the $Al_{35}Mg_{30}Si_{13}Zn_{10}Y_7Ca_5$ alloy, five phases corresponding to (i) Mg_2Si, (ii) Y_5Si_3, (iii) Ca_2Si, (iv) $AlMg_2Zn$, and (v) $AlMg_4Zn_{11}$ were observed in

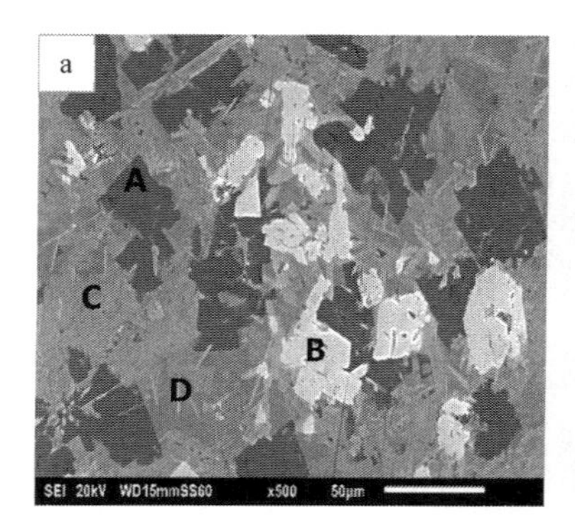

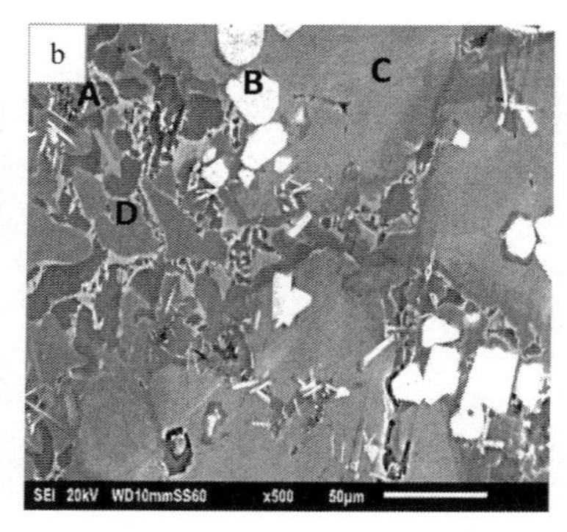

FIGURE 2.5 Scanning electron micrographs of the (a) $Al_{35}Mg_{30}Si_{13}Zn_{10}Y_7Ca_5$ alloy and (b) $Mg_{35}Al_{33}Li_{15}Zn_7Ca_5Y_5$ alloy.

TABLE 2.9
Measured Elemental Composition from Energy Dispersive X-ray Analysis

Alloy system	Designated area	Phase composition (Atomic pct.)					
		Mg	Al	Si	Zn	Y	Ca
AlMgSiZnYCa	**A**	66.51	–	33.16	0.34	–	–
	B	11.67	7.75	38.82	0.98	39.86	
	C	19.35	34.69	16.43	12.75	–	16.77
	D	41.13	47.94	–	10.55	–	0.37

Note: Li is not detectable due to the limitation of EDX system used.

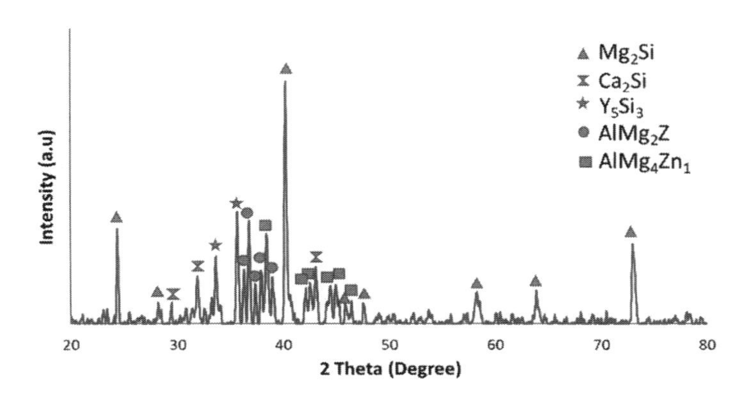

FIGURE 2.6 X-ray diffraction (XRD) pattern of the $Al_{35}Mg_{30}Si_{13}Zn_{10}Y_7Ca_5$ alloy.

the X-ray diffraction pattern. The composition of the phases formed was examined using energy dispersive X-ray analysis. The formation of phases having different contrasts (Area: A, B, C and D) can be seen in the microstructure shown in Figure 2.5a. The area designated as A represents the Mg_2Si phase, while area B represents the Y_5Si_3 phase, and the phases Ca_2Si, $AlMg_2Zn$, and/or $AlMg_4Zn_{11}$ appear to be coexistent in area C. The $AlMg_2Zn$ and $AlMg_4Zn_{11}$ phases appear to coexist in area D. Based on standardless quantitative microanalysis, the observed atomic ratio, 66.51 atomic pct. Mg and 33.16 atomic pct. Si (see Table 2.9) corresponds closely with the Mg_2Si phase (area A in Figure 2.5a) as identified by the XRD pattern shown in Figure 2.6. For rest of the phases, the composition related to each phase (Ca_2Si, $AlMg_2Zn$, and $AlMg_4Zn_{11}$) could be predicted from the EDX analysis as shown in Table 2.9. However, due essentially to the coexistence and complex morphology of the phases, coupled with the uncertainty of standardless analysis of the complex phases, the chemical composition of the phases present in the microstructure could not be closely matched to their formation, as revealed in the XRD analysis.

For the case of $Mg_{35}Al_{33}Li_{15}Zn_7Ca_5Y_5$ alloy, six phases corresponding to (i) αMg, (ii) Al_2Y, (iii) CaZn, (iv) $AlMg_2Zn$, (v) $AlMg_4Zn_{11}$, and (vi) $Li_{6.46}Mg$ were observed in the XRD pattern (Figure 2.4a). The formation of phases having different contrasts (Area: A, B, C and D) can be seen in the microstructure shown in Figure 2.5b. The area designated as A represents the α-Mg phase, area B represents the Al_2Y phase, and the combined presence of the phases CaZn, $AlMg_2Zn$, and $AlMg_4Zn_{11}$ appears in area C and area D. Applying standardless quantitative analysis of the phase composition acquired from EDX analysis, α-Mg (area A in Figure 2.5b) is identified to be the Mg-Zn solid solution phase and area B corresponds to the Al_2Y phase (Table 2.9). For the phases CaZn, $AlMg_2Zn$, and $AlMg_4Zn_{11}$, an inaccuracy in the chemical composition of each phase was observed for the $Al_{35}Mg_{30}Si_{13}Zn_{10}Y_7Ca_5$ alloy. The formation and presence of these phases was confirmed through XRD analysis as shown in Figure 2.4a.

For the silicon-containing alloy, AlMgSiZnYCa, no elemental or disordered solid solution phase was formed. For the lithium-containing alloy, i.e., MgAlLiZnCaY, the formation and presence of the solid solution phase (Mg-Zn) was observed coupled with the presence of ordered phases. Based on a careful observation of the microstructure of the two alloys AlMgSiZnYCa and MgAlLiZnCaY, the alloying element (Y) tends to form blocky and separated intermetallic phases, causing a depletion of yttrium in the phases present in the two alloy systems. The alloying elements, aluminum (Al), magnesium (Mg), zinc (Zn), and calcium (Ca) coexist and form a continuous layer essentially composed of different ordered phases (intermetallic or second-phase particles).

From an observation of the previous alloy compositions, $Mg_{35}Al_{33}Li_{15}Zn_7Ca_5Y_5$ and $Al_{35}Mg_{30}Si_{13}Zn_{10}Y_7Ca_5$, an introduction of the alloying element yttrium (Y) in these alloys led to a complex microstructure when compared to the alloy compositions equiatomic AlMgLiCaCuZn and $Mg_{35}Al_{33}Li_{15}Zn_7Ca_5Cu_5$ without using yttrium. In view of these previous results, another non-equiatomic composition, $Al_{35}Li_{20}Mg_{15}Si_{10}Zn_{15}Ca_5$ was developed without the addition of yttrium (Y) as one of the alloying elements. Unexpectedly, the resultant microstructure was found to be complex, as shown in Figure 2.7. From the microstructure and as verified by both EDX Point and ID analysis

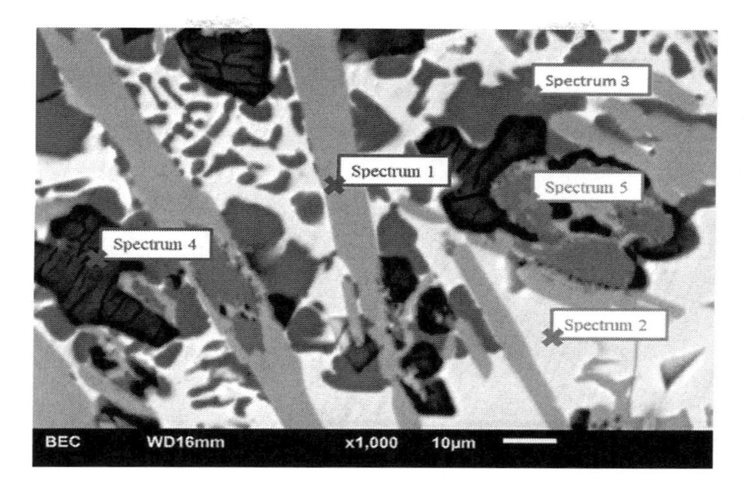

FIGURE 2.7 Scanning electron micrograph of the $Al_{35}Li_{20}Mg_{15}Si_{10}Zn_{15}Ca_5$ alloy.

(Table 2.10), at least five major phases must be present in the alloy. Based on the XRD analysis results from Figure 2.8 and quantitative microanalysis, the phase composition for spectrum 2, spectrum 3, and spectrum 4 can be predicted to be AlLiZn, αAl, and Mg_2Si. However, the composition of spectrum 1 and spectrum 5 from the EDX point and ID results provided in Table 2.10 could not be closely matched with composition of the intermetallic phases found from XRD analysis due to the coexistence and complex morphology of the phases present in the microstructure.

The microhardness and compressive properties of the developed lightweight high entropy alloys presented in this section are summarized in Table 2.11. The alloys showed significantly high hardness, which is approximately 4–6 times higher than the conventional alloys, such as AZ31B (63 HV), following the same nature of superhard lightweight alloy reported in the published literature (Table 2.3). Among the currently developed alloy compositions, the hardness of the alloys, $Al_{35}Mg_{30}Si_{13}Zn_{10}Y_7Ca_5$ and $Al_{35}Li_{20}Mg_{15}Zn_{15}Si_{10}Ca_5$, was found to be higher than the other three alloys. With the

TABLE 2.10
EDX Point and ID Analysis for $Al_{35}Li_{20}Mg_{15}Si_{10}Zn_{15}Ca_5$

	Elemental composition (at.%)				
Spectrum	**Al**	**Mg**	**Si**	**Zn**	**Ca**
1	35.70	3.73	22.83	16.75	20.99
2	46.64	16.24	0.61	36.36	0.15
3	91.23	3.42	0.30	5.05	0
4	8.96	40.46	47.22	2.96	0.40
5	22.23	16.30	40.16	3.44	17.87

Note: Li is not detectable due to the limitation of EDX system used.

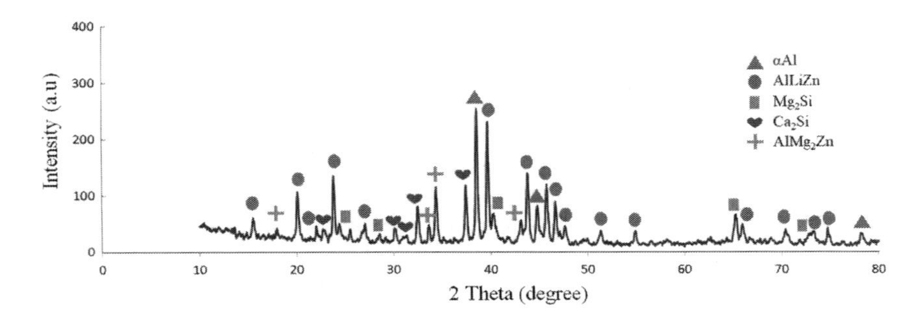

FIGURE 2.8 X-ray diffraction (XRD) results for the $Al_{35}Li_{20}Mg_{15}Si_{10}Zn_{15}Ca_5$ alloy indicating the different phases present.

TABLE 2.11
Microhardness and Compressive Properties of Lightweight High Entropy Alloys

Alloy composition	Microhardness (HV)	Peak strength (MPa)	Fracture strain (%)
AlMgLiCaCuZn	255 ± 14	440 ± 17	3.6 ± 0.4
$Mg_{35}Al_{33}Li_{15}Zn_7Ca_5Cu_5$	267 ± 15	280 ± 23	3.2 ± 0.3
$Mg_{35}Al_{33}Li_{15}Zn_7Ca_5Y_5$	237 ± 10	276 ± 17	3.1 ± 0.1
$Al_{35}Mg_{30}Si_{13}Zn_{10}Y_7Ca_5$	406 ± 15	–	–
$Al_{35}Li_{20}Mg_{15}Zn_{15}Si_{10}Ca_5$	371 ± 12	369 ± 19	3.0 ± 0.3

use of silicon (Si) as one of the alloying elements in these two alloy compositions, the formation and presence of the Mg_2Si intermetallic seems to elevate the hardness of the alloy due to its hard and intrinsically brittle nature [41]. As seen in Table 2.11, the highest hardness was observed for the $Al_{35}Mg_{30}Si_{13}Zn_{10}Y_7Ca_5$ alloy. From the microstructure of the alloy shown in Figure 2.5a (area marked as A) and the XRD)pattern shown in Figure 2.6, Mg_2Si was found to be the dominant phase, which accounts for the resultant high hardness of this alloy. However, the $Al_{35}Mg_{30}Si_{13}Zn_{10}Y_7Ca_5$ alloy was too brittle to perform the compression test. Among other LWHEA compositions, a higher peak strength resulted for the equiatomic alloy, AlMgLiCaCuZn, when compared to the non-equiatomic alloys. Both the equiatomic and non-equiatomic lightweight high entropy alloys failed without exhibiting any evidence of gross plastic deformation. The compressive fracture strain was limited to ~4% for all alloy compositions.

2.3.5 Equiatomic and Non-Equiatomic AlLiMgZnSi Alloys

In the previous section, the developed LWHEAs alloys composed of six alloying elements were presented. In this section, the lightweight high entropy alloys composed of five alloying elements are listed, and the microstructure and mechanical properties of the alloys are presented and briefly discussed.

The scanning electron microscopy images of both the equiatomic and non-equiatomic alloys essentially revealed a complex microstructure with the presence of secondary phases in the as-cast samples. Preliminary inspection of the SEM images of all as-cast alloys revealed the microstructure of the equiatomic $Al_{20}Li_{20}Zn_{20}Si_{20}Mg_{20}$ (Figure 2.9a), to be simpler than the microstructure of the non-equiatomic alloy, having compositions (i) $Al_{35}Li_{20}Zn_{20}Si_{15}Mg_{10}$, and (ii) $Al_{35}Li_{20}Zn_{10}Si_{15}Mg_{20}$. A closer inspection, at 500× magnification, revealed the presence of lamellar structures in both the $Al_{35}Li_{20}Zn_{20}Si_{15}Mg_{10}$ and $Al_{35}Li_{20}Zn_{10}Si_{15}Mg_{20}$ alloys (Figure 2.9b and Figure 2.9c), which was noticeably absent for the equiatomic alloy. The lamellar structures can be seen to be both more prevalent and more well-defined for the $Al_{35}Li_{20}Zn_{10}Si_{15}Mg_{20}$ alloy (Figure 2.9c), although the reasons for its formation, presence, and distribution are unknown due to the overall complexity of the alloy system.

From a combined analysis of both the EDX and XRD (Figure 2.10 and 2.11) results, the phases detected in microstructure of the equiatomic AlLiMgZnSi alloy were similar to Mg_2Si (Area A and C), $AlMg_4Zn_{11}$ (Area B), and Al_4Si (Area D), respectively. Additional diffraction peaks from the XRD pattern were identified to be AlZn and Mg_7Zn_3. Although these phases were not distinctly visible in the microstructure, this clearly indicates that there were three major phases and two minor phases present in the equiatomic AlLiMgZnSi alloy.

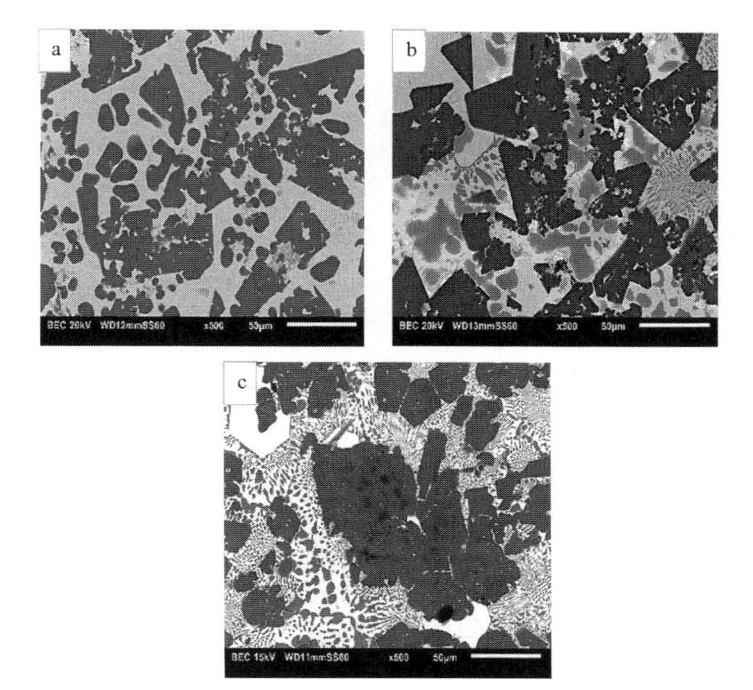

FIGURE 2.9 Scanning electron microscopy images of the (a) equiatomic AlLiMgZnSi, (b) non-equiatomic $Al_{35}Li_{20}Zn_{20}Si_{15}Mg_{10}$, and (c) non-equiatomic $Al_{35}Li_{20}Zn_{10}Si_{15}Mg_{20}$ alloy.

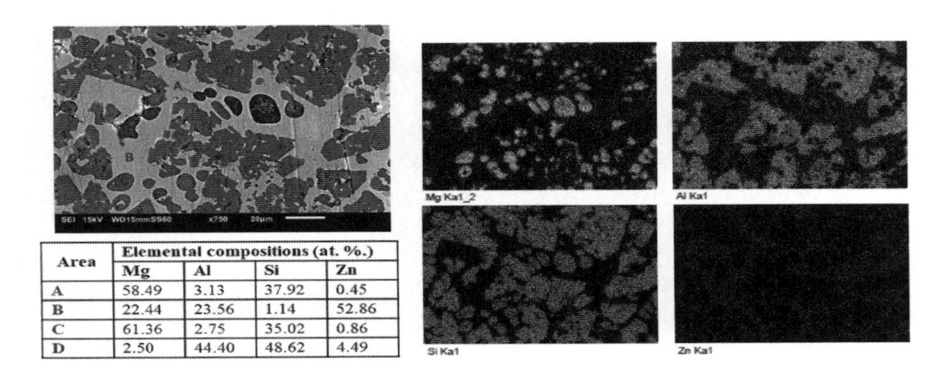

Area	Elemental compositions (at. %.)			
	Mg	Al	Si	Zn
A	58.49	3.13	37.92	0.45
B	22.44	23.56	1.14	52.86
C	61.36	2.75	35.02	0.86
D	2.50	44.40	48.62	4.49

FIGURE 2.10 Microstructure of equiatomic AlLiMgZnSi alloy with corresponding EDX mapping result and chemical compositions of alloying elements from EDX point and ID analysis.

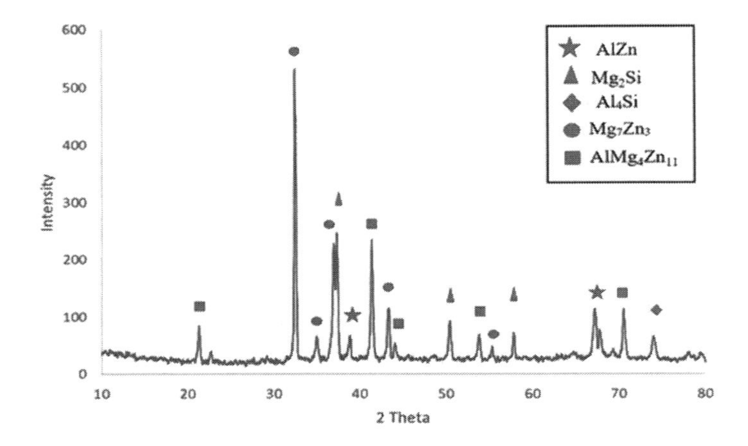

FIGURE 2.11 X-ray diffraction (XRD) analysis results of the equiatomic AlLiMgZnSi alloy.

The presence of phases in the microstructure of the $Al_{35}Li_{20}Zn_{20}Si_{15}Mg_{10}$ alloy designated as areas A, B, C, D, and E were found to be the LiAlSi, AlZn, Mg_4Zn_7, $Mg_{32}(AlZn)_{49}$, and Mg_2Si phases, respectively (Figure 2.12). These phases were identified using both EDX and XRD analyses (Figures 2.12 and 2.13). The lamellar structure seen in the microstructure consists of alternating layers of AlZn and $Mg_{32}(AlZn)_{49}$ (Figure 2.12).

The phases present in the microstructure of the $Al_{35}Li_{20}Zn_{10}Si_{15}Mg_{20}$ alloy are designated as A, B, C, and D and the corresponding elemental composition of each of the phases were measured using EDX point and ID analysis (Figure 2.14). Due to the overall complex nature of the phases, such as overlapping phases and co-presence of the phases, those phases could not be accurately identified using the results provided by the EDX analysis and based on standardless quantitative analysis. Hence, the XRD analysis was performed on the alloy and the diffraction peaks were matched

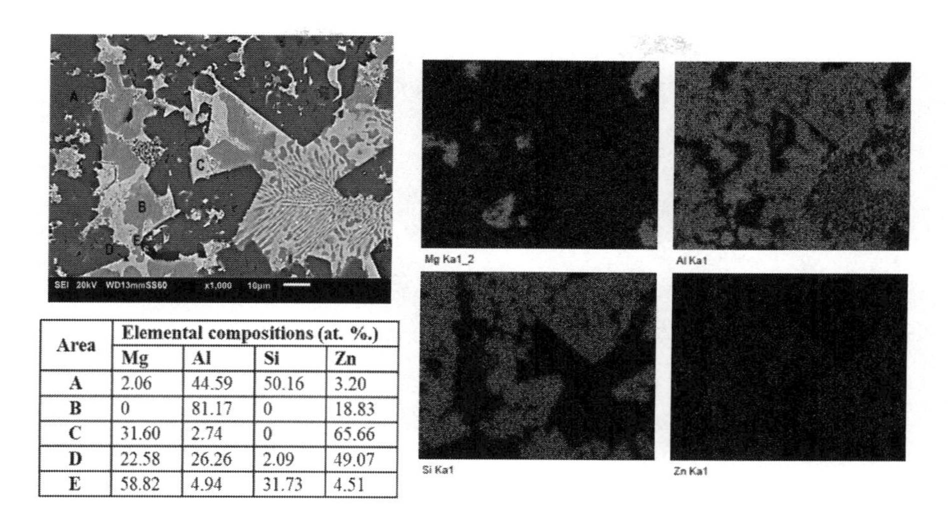

Area	Elemental compositions (at. %.)			
	Mg	Al	Si	Zn
A	2.06	44.59	50.16	3.20
B	0	81.17	0	18.83
C	31.60	2.74	0	65.66
D	22.58	26.26	2.09	49.07
E	58.82	4.94	31.73	4.51

FIGURE 2.12 Microstructure of the $Al_{35}Li_{20}Zn_{20}Si_{15}Mg_{10}$ alloy with corresponding EDX mapping result and chemical compositions of alloying elements from EDX point and ID analysis.

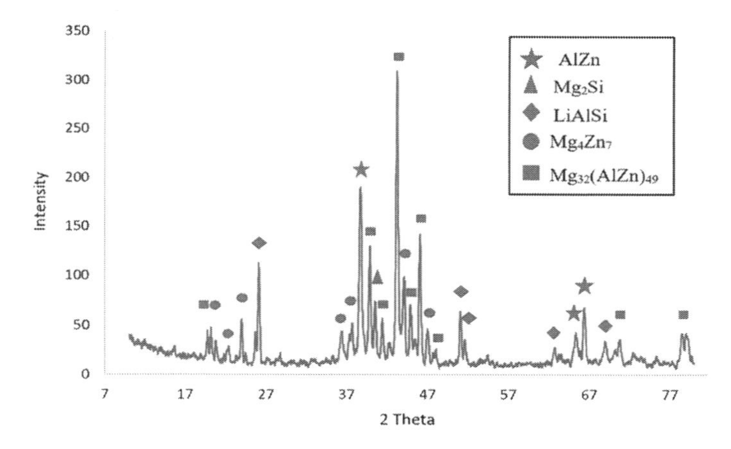

FIGURE 2.13 X-ray diffraction (XRD) analysis result of the $Al_{35}Li_{20}Zn_{20}Si_{15}Mg_{10}$ alloy.

to the phases present based on EDX results. In area A, it appears to have a combined presence of Mg_2Si and AlLiSi phases and a mixture of the dark gray phase and black phase can be seen in the microstructure (Figure 2.14). Area B corresponds to the AlZn phase, area C corresponds to the AlLiSi phase, and area D corresponds to the $Mg_{32}(AlZn)_{49}$ phase. The lamellar structure seen in the microstructure resembles an alternating layer of AlZn and $Mg_{32}(AlZn)_{49}$. Additional diffraction peaks from the XRD pattern were identified to be the Mg_4Zn_7, although this phase was not easily visible in the microstructure.

From the microhardness results provided in Table 2.12, the highest hardness was obtained for the equiatomic AlLiMgZnSi alloy. The resultant high hardness for the

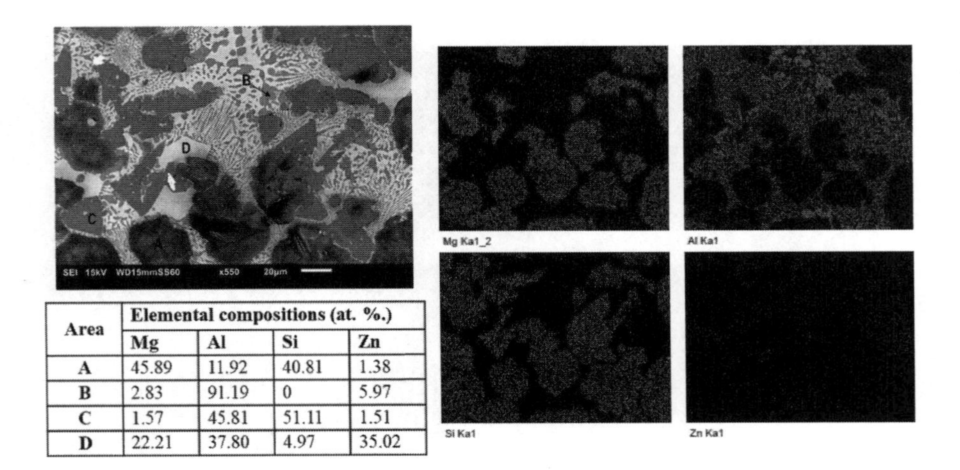

Area	Elemental compositions (at. %.)			
	Mg	Al	Si	Zn
A	45.89	11.92	40.81	1.38
B	2.83	91.19	0	5.97
C	1.57	45.81	51.11	1.51
D	22.21	37.80	4.97	35.02

FIGURE 2.14 Microstructure of $Al_{35}Li_{20}Zn_{10}Si_{15}Mg_{20}$ alloy with corresponding EDX mapping result and chemical compositions of alloying elements from EDX point and ID analysis.

TABLE 2.12
Microhardness and Compressive Properties of Equiatomic and Non-Equiatomic AlLiMgZnSi Alloys

Composition	Microhardness	Peak strength (MPa)	Fracture strain (%)
AlLiMgZnSi	382 ± 16	184 ± 35	2.2 ± 0.4
$Al_{35}Li_{20}Zn_{20}Si_{15}Mg_{10}$	307 ± 24	273 ± 24	3.6 ± 0.1
$Al_{35}Li_{20}Mg_{20}Si_{15}Zn_{10}$	358 ± 18	269 ± 5	3.6 ± 0.3

equiatomic alloy can be partly attributed to solid solution strengthening. Being an equiatomic alloy, equal proportions of elements were added, and it enhances the chance for the formation of a multi-element solid solution, as each element has an equal chance of forming a solid solution with another element. However, the compressive strength and fracture strain of the equiatomic alloy was observed to be inferior to that of the non-equiatomic alloy (Table 2.12). This clearly indicates that the equiatomic alloy was hard and less tough than the non-equiatomic alloys (Figure 2.15).

2.3.6 Magnesium-Based Lightweight High Entropy Alloys

Focusing on the primary goal of developing lightweight high entropy alloys (LWHEAs), magnesium and aluminum were chosen as the strategic alloying elements, together with other alloying elements. With the combined use of magnesium and aluminum, a density of less than 3 g/cm^3 was easily achieved in all the

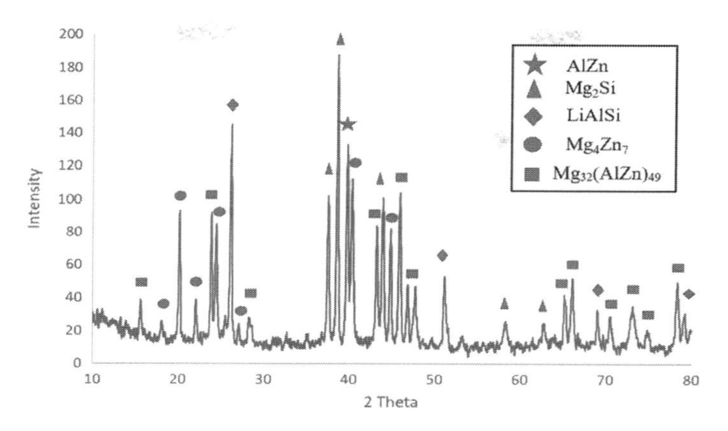

FIGURE 2.15 X-ray diffraction (XRD) analysis result of the $Al_{35}Li_{20}Zn_{10}Si_{15}Mg_{20}$ alloy.

previously developed alloys, validating these alloys as being in the lightweight category. However, a careful observation of the microstructure of the alloys revealed the presence of phases composed of both the magnesium and aluminum elements and this resulted in a complex microstructure for some of the lightweight high entropy alloy compositions. Removing aluminum from the alloy system, two alloy compositions, namely (i) $Mg_{35}Y_{10}Zn_{15}Li_{15}Cu_{15}Ca_{10}$ and (ii) $Mg_{35}Y_{10}Zn_{20}Li_{15}Cu_{20}$, were developed. Yttrium (Y) was used as one of the alloying elements due to its relative low density (4.47 g/cm^3) and relatively high solid solubility (~3.79 atomic pct.) in magnesium [42, 43].

From the microstructure of $Mg_{35}Y_{10}Zn_{15}Li_{15}Cu_{15}Ca_{10}$ alloy (Figure 2.16a), four phases were labeled as A, B, C, and D. There were three observable phases in the $Mg_{35}Y_{10}Zn_{20}Li_{15}Cu_{20}$ alloy (Figure 2.16b) and these were labeled as I, II, and III. These phases were identified using EDX and XRD analysis.

For the $Mg_{35}Y_{10}Zn_{15}Li_{15}Cu_{15}Ca_{10}$ alloy, through the EDX analysis (Table 2.13), the phase corresponding to A is shown to reveal magnesium (Mg), zinc (Zn), and copper (Cu) to be the major elements with possible traces of yttrium (Y). The phase corresponding to B is composed of magnesium (Mg), yttrium (Y), and zinc (Zn).

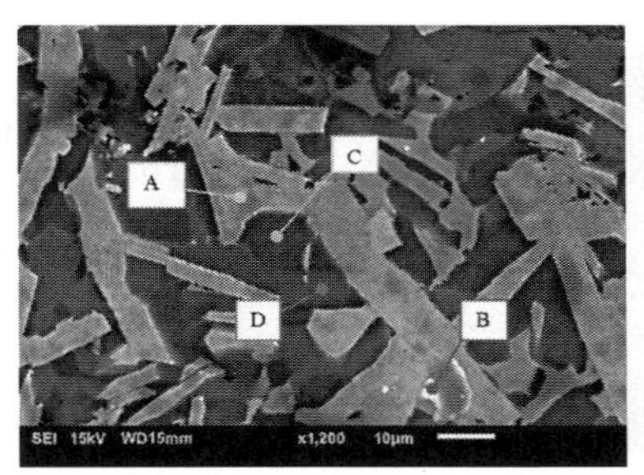

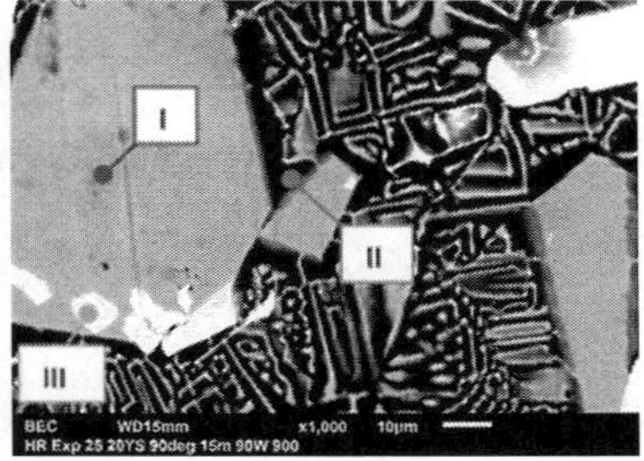

FIGURE 2.16 Scanning electron micrographs of the (a) $Mg_{35}Y_{10}Zn_{15}Li_{15}Cu_{15}Ca_{10}$ alloy and (b) $Mg_{35}Y_{10}Zn_{20}Li_{15}Cu_{20}$ alloy.

TABLE 2.13
Measured Elemental Composition from EDX Analysis

		Elemental composition (at.%)				
Alloy	**Designated area**	**Mg**	**Y**	**Zn**	**Cu**	**Ca**
MgYZnLiCuCa	A	22.1	7.08	36.4	34.2	0.20
	B	41.7	19.32	39.0	–	–
	C	92.8	–	7.23	–	–
	D	75.9	–	3.98	–	20.1
MgYZnLiCu	I	18.7	0.740	44.9	35.3	N.A.
	II	85.3	1.09	11.9	1.72	N.A.
	III	10.0	38.3	46.4	5.26	N.A.

Note: Li is not detectable due to the limitation of EDX system used.

The phases corresponding to C and D show traces of magnesium (Mg), zinc (Zn), and magnesium (Mg), calcium (Ca), and zinc (Zn), respectively. The phases were identified using XRD analysis (Figure 2.17a). Area A corresponds to the CuMgZn phase and the white phase overlaying area A, which is designated as B, is confirmed to be the Mg_3Y_2Zn phase. The area C in the microstructure is confirmed to be the αMg (Mg-Zn solid solution), and area D is identified to be the $Ca_2Mg_5Zn_3$ phase.

Based on phase analysis using XRD (Figure 2.17b), the phases formed and present in the microstructure of the $Mg_{35}Y_{10}Zn_{20}Li_{15}Cu_{20}$ alloy were identified to be the following: (i) CuMgZn (area I), (ii) $Mg_{102.08}Zn_{39.60}$ (area II), and (iii) ZnY (area III). The lamellae structures observed in the microstructure of the $Mg_{35}Y_{10}Zn_{20}Li_{15}Cu_{20}$ alloy is an alternating layer of CuMgZn and MgZn phase. From an observation of the microstructure and phase analysis, the formation of a simple microstructure with fewer intermetallic phases was evident for the $Mg_{35}Y_{10}Zn_{20}Li_{15}Cu_{20}$ alloy when compared to the $Mg_{35}Y_{10}Zn_{15}Li_{15}Cu_{15}Ca_{10}$ alloy.

Mechanical properties of the alloys were assessed in terms of both microhardness and compressive properties. The average microhardness values for the $Mg_{35}Y_{10}Zn_{15}Li_{15}Cu_{15}Ca_{10}$ alloy and the $Mg_{35}Y_{10}Zn_{20}Li_{15}Cu_{20}$ alloy were measured to be 196 ± 11 HV and 131 ± 7 HV, respectively, revealing the $Mg_{35}Y_{10}Zn_{15}Li_{15}Cu_{15}Ca_{10}$ alloy to be the harder alloy. From the compression test results (Figure 2.18), the compressive strength of the $Mg_{35}Y_{10}Zn_{15}Li_{15}Cu_{15}Ca_{10}$ alloy was found to be lower than that of the $Mg_{35}Y_{10}Zn_{20}Li_{15}Cu_{20}$ alloy. In addition, the $Mg_{35}Y_{10}Zn_{15}Li_{15}Cu_{15}Ca_{10}$ alloy failed without revealing any evidence of gross plastic deformation. In contract, the $Mg_{35}Y_{10}Zn_{20}Li_{15}Cu_{20}$ alloy plastically deformed under compressive loading. Hence, it is concluded that the formation and presence of apparently hard, brittle, and elastically deforming intermetallic particles coupled with a complex microstructure for the $Mg_{35}Y_{10}Zn_{15}Li_{15}Cu_{15}Ca_{10}$ alloy led to an overall hard and brittle alloy. It may also be noted that not only the microstructural simplicity but also the type of intermetallic

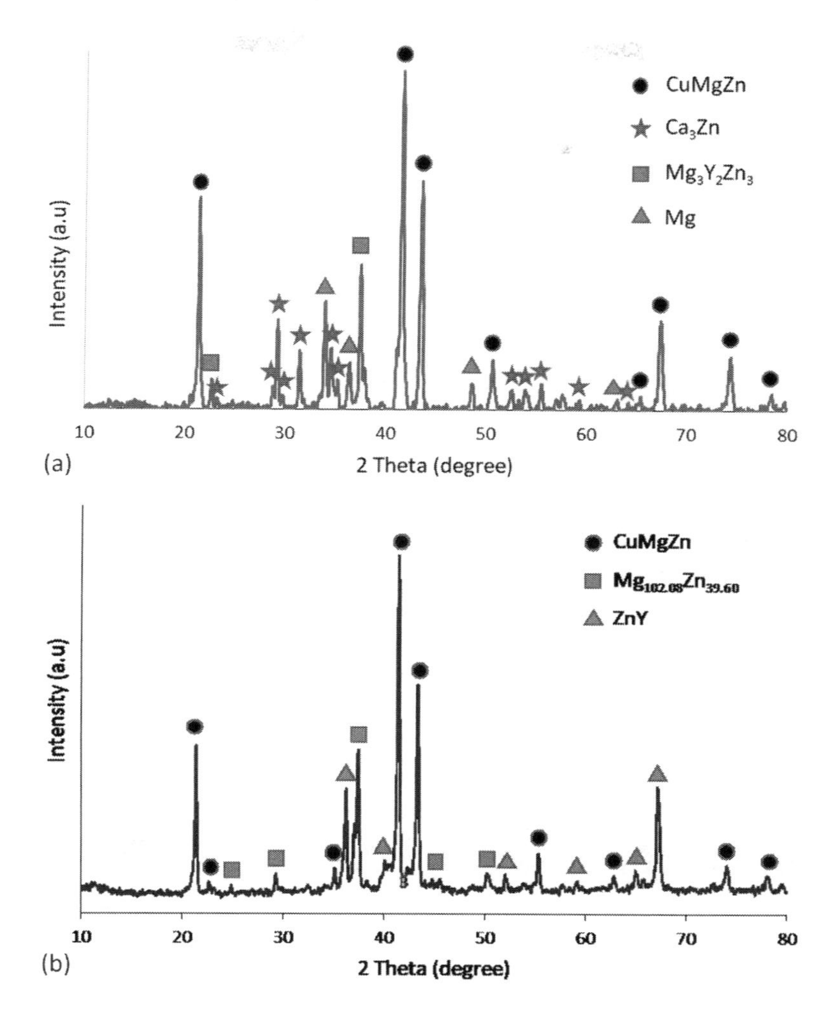

FIGURE 2.17 X-ray diffraction (XRD) analysis results on the (a) $Mg_{35}Y_{10}Zn_{15}Li_{15}Cu_{15}Ca_{10}$ alloy and (b) $Mg_{35}Y_{10}Zn_{20}Li_{15}Cu_{20}$ alloy.

particles formed and present in the alloy does exert an influence on mechanical properties of the alloy.

The fractographs shown in Figure 2.19 essentially reveal a brittle failure for the two chosen alloys. The $Mg_{35}Y_{10}Zn_{20}Li_{15}Cu_{20}$ alloy deformed plastically, but it is not a ductile alloy due essentially to the low failure strain value of 3.7% (Figure 2.18). With evidence of visible pore formation on the fracture surfaces of the alloys (shown by arrows), this could be one of the reasons for the inferior fracture strain, or failure strain, of the two chosen alloys. The presence of relatively large-size pores in the two alloys is also a reason for the resultant reduced experimental density observed for both the $Mg_{35}Y_{10}Zn_{15}Li_{15}Cu_{15}Ca_{10}$ alloy and the $Mg_{35}Y_{10}Zn_{20}Li_{15}Cu_{20}$ alloy (Table 2.7).

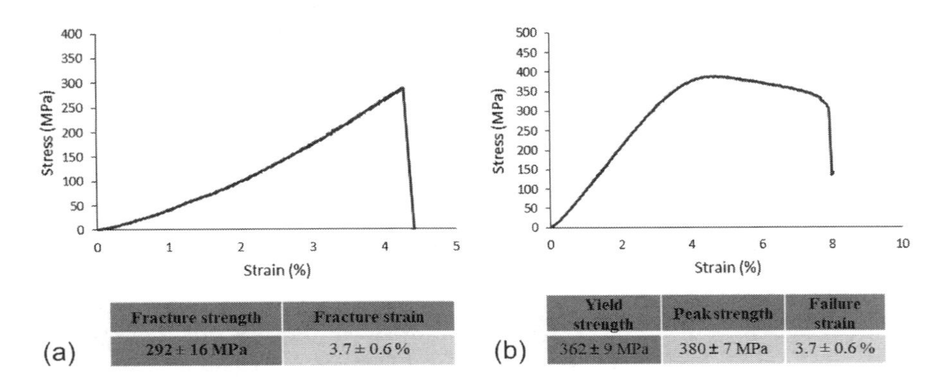

FIGURE 2.18 Compressive stress-strain curve along with corresponding compressive properties for the (a) $Mg_{35}Y_{10}Zn_{15}Li_{15}Cu_{15}Ca_{10}$ alloy and (b) $Mg_{35}Y_{10}Zn_{20}Li_{15}Cu_{20}$ alloy.

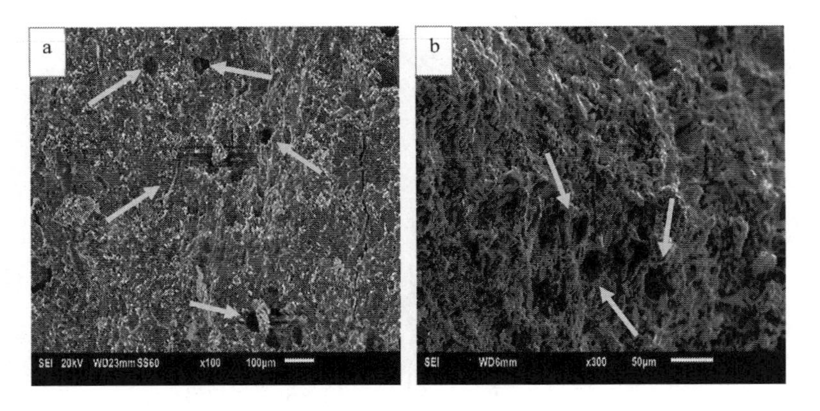

FIGURE 2.19 Scanning electron micrographs of the compressive fracture surfaces of the (a) $Mg_{35}Y_{10}Zn_{15}Li_{15}Cu_{15}Ca_{10}$ alloy and (b) $Mg_{35}Y_{10}Zn_{20}Li_{15}Cu_{20}$ alloy.

2.4 SUMMARY AND CONCLUSIONS

1. High entropy alloys are a new class of metallic alloys that exhibit certain properties that are far superior to conventional alloys.
2. Many of the high entropy alloys (HEAs) possess very high hardness values, wear resistance, and compressive properties.
3. In view of their uniqueness, attempts have been made by researchers to develop these alloys with a lower density to address weight-critical applications targeted to reduce both fuel consumption and carbon dioxide emissions.
4. In recent research studies on lightweight high entropy alloys, light elements, such as magnesium (Mg), aluminum (Al), lithium (Li), and titanium (Ti) were selected as the alloying elements to meet the target density of 3 g/cm^3.
5. The lightweight high entropy alloys (LWHEAs) emerged as a new, unconventional alloy system, which has the possibility of reducing weight of the structural application, such as in the transport sector.

6. To our knowledge, there have only been limited alloy composition and experimental results of such LWHEAs published to date. As such, a series of equiatomic and non-equiatomic LWHEAs were explored in the current research.
7. During alloy development, compositional variations within the same alloy system as well as various alloy systems with different combination of alloying elements were attempted. The effect of these variations on both microstructural development and mechanical properties of the related alloy systems is reported.
8. In this research study, we have demonstrated that LWHEAs can be fabricated using the technique of disintegrated melt deposition (down pouring casting). As some of the LWHEA compositions failed to process using DMD, alternative processing methods, such as (i) arc melting and (ii) induction melting, are recommended for the failed alloy compositions.
9. The microstructural results of the successfully processed alloys revealed the formation and presence of multiple intermetallic phases coupled with an overall complex microstructure for most of the alloy compositions. Between equiatomic and non-equiatomic LWHEAs, a simpler microstructure was observed for the equiatomic alloys because of higher mixing entropy.
10. Study of the thermodynamic parameter revealed microstructural simplicity in some non-equiatomic alloy systems to be easily attainable if the value of δ, the thermodynamic parameter which is related to atomic size differences, can be observably reduced.
11. In terms of microhardness, the hardness variation of the alloys was dependent on both quantity and the nature (hard and tough or hard and brittle) of the intermetallic formed in the microstructure, regardless of the mixing entropy and/or δ value. The harder alloy did not necessarily lead to better compressive properties. The hardness of some of the alloy compositions was significantly high due to the presence of intermetallic particles, which have superior high hardness, or because of the presence of many intermetallic particles.
12. The compressive strengths of the as-fabricated alloys were found to be reduced due to the presence of brittle and superhard intermetallic particles coupled with the formation of an overall complex microstructure due essentially to the presence of multiple intermetallic phases.

We have explored several lightweight high entropy alloy (LWHEA) compositions, and only microstructure, hardness, and mechanical properties have been elaborated. Additional research efforts are very much essential and are required to contain the types and number of secondary phases present in the microstructure through smart compositional control with the prime objective of providing a better understanding of the composition–microstructure–property relationships in LWHEAs to develop this unconventional and potential alloy system having superior mechanical properties.

ACKNOWLEDGMENTS

The authors gratefully acknowledge the Ministry of Education Academic Research Funding (WBS# R-265-000-586-114) for providing the financial support that made this exhaustive and encompassing research study possible.

REFERENCES

1. Quadrelli, R., and Peterson, S. 2007. "The energy–climate challenge: Recent trends in CO2 emissions from fuel combustion." *Energy Policy* 35(11):5938–5952. doi: 10.1016/j.enpol.2007.07.001.
2. 2007. "IPCC, 2007: Summary for policymakers." In: *Climate Change 2007: The Physical Science Basis. Contribution of Working Group I to the Fourth Assessment Report of the Intergovernmental Panel on Climate Change*, edited by S. Solomon, D. Qin, M. Manning, Z. Chen, M. Marquis, K. B. Averyt, M.Tignor and H. L. Miller, Cambridge University Press, Cambridge, United Kingdom and New York, NY, USA.
3. Schubert, E., Klassen, M., Zerner, I., Walz, C., and Sepold, G. 2001. "Light-weight structures produced by laser beam joining for future applications in automobile and aerospace industry." *Journal of Materials Processing Technology* 115(1):2–8. doi: 10.1016/S0924-0136(01)00756-7.
4. Wenlong, S., Xiaokai, C., and Lu, W. 2016. "Analysis of energy saving and emission reduction of vehicles using light weight materials." *Energy Procedia* 88:889–893. doi:10.1016/j.egypro.2016.06.106.
5. Elalem, A., and El-Bourawi, M. S. 2010. "Reduction of automobile carbon dioxide emissions." *International Journal of Material Forming* 3(S1):663–666. doi: 10.1007/s12289-010-0857-2.
6. Hirsch, J. 2014. "Recent development in aluminium for automotive applications." *Transactions of Nonferrous Metals Society of China* 24(7):1995–2002. doi: 10.1016/s1003-6326(14)63305-7.
7. Kulkarni, S., Edwards, D. J., Parn, E. A., Chapman, C., Aigbavboa, C. O., and Cornish, R. 2018. "Evaluation of vehicle lightweighting to reduce greenhouse gas emissions with focus on magnesium substitution." *Journal of Engineering, Design and Technology* 16(6):869–888. doi: 10.1108/jedt-03-2018-0042.
8. Modaresi, R., Pauliuk, S., Lovik, A. N., and Muller, D. B. 2014. "Global carbon benefits of material substitution in passenger cars until 2050 and the impact on the steel and aluminum industries." *Environmental Science and Technology* 48(18):10776–10784. doi: 10.1021/es502930w.
9. Joost, W. J., and Krajewski, P. E. 2017. "Towards magnesium alloys for high-volume automotive applications." *Scripta Materialia* 128:107–112. doi: 10.1016/j.scriptamat.2016.07.035.
10. Blawert, C. H. N., and Kainer, K. U. 2004. "Automotive applications of magnesium and its alloys." *Transactions of the Indian Institute of Metals* 57(4):397–408.
11. Liu, M., Guo, Y., Wang, J., and Yergin, M. 2018. "Corrosion avoidance in lightweight materials for automotive applications." *npj Materials Degradation* 2(1). doi: 10.1038/s41529-018-0045-2.
12. Monteiro, W. A., Buso, S. J., and da, L. V. 2012. "Application of magnesium alloys in transport." In: *New Features on Magnesium Alloys*.
13. Yeh, J.-W., Lin, S.-J., Chin, T.-S., Gan, J.-Y., Chen, S.-K., Shun, T.-T., Tsau, C.-H., and Chou, S.-Y. 2004. "Formation of simple crystal structures in Cu-Co-Ni-Cr-Al-Fe-Ti-V alloys with multiprincipal metallic elements." *Metallurgical and Materials Transactions. Part A* 35(8):2533–2536.

14. Cantor, B., Chang, I., Knight, P., and Vincent, A. 2004. "Microstructural development in equiatomic multicomponent alloys." *Materials Science and Engineering: Part A* 375:213–218.
15. Kumar, A., and Gupta, M. 2016. "An insight into evolution of light weight high entropy alloys: A review." *Metals* 6(9). doi: 10.3390/met6090199.
16. Senkov, O. N., Miracle, D. B., Chaput, K. J., and Couzinie, J.-P. 2018. "Development and exploration of refractory high entropy alloys—A review." *Journal of Materials Research* 33(19):3092–3128. doi: 10.1557/jmr.2018.153.
17. Gupta, M., and Tun, K. S. 2017a. "An insight into the development of light weight high entropy alloys." *Research & Development in Material Science* 2(2):RDMS.000534. doi: 10.31031/RDMS.2017.02.000534.
18. Gupta, M., and Tun, K. S. 2017b. "Light weight high entropy alloys: Processing challenges and properties." *Recent Patents on Materials Science* 10(2):116–121. doi: 10.2174/1874464811666180327121606.
19. Tun, K. S., and Gupta, M. 2018. "Microstructural evolution in MgAlLiZnCaY and MgAlLiZnCaCu multicomponent high entropy alloys." *Materials Science Forum* 928:183–187. doi: 10.4028/www.scientific.net/MSF.928.183.
20. Tun, K. S., Srivatsan, T. S., Kumar, A., and Gupta, M. 2017. "Synthesis of light weight high entropy alloys: Characterization of microstructure and mechanical response." Twenty-sixth International Conference on the Processing and Fabrication of the Advanced Materials (PFAM XXVI), Jeonju, South Korea.
21. Juan, C.-C., Yeh, J., and Chin, T. 2009. "A novel light high-entropy alloy Al 20 Be 20 Fe 10 Si 15 Ti 35." E-MRS Fall Meeting.
22. Li, R., Gao, J. C., and Fan, K. 2010. "Study to microstructure and mechanical properties of Mg containing high entropy alloys." *Materials Science Forum* 650:265–271. doi: 10.4028/www.scientific.net/MSF.650.265.
23. Youssef, K. M., Zaddach, A. J., Niu, C., Irving, D. L., and Koch, C. C. 2015. "A novel low-density, high-hardness, high-entropy alloy with close-packed single-phase nanocrystalline structures." *Materials Research Letters* 3(2):95–99. doi: 10.1080/21663831.2014.985855.
24. Yang, X., Chen, S. Y., Cotton, J. D., and Zhang, Y. 2014. "Phase stability of low-density, multiprincipal component alloys containing aluminum, magnesium, and lithium." *Jom* 66(10):2009–2020. doi: 10.1007/s11837-014-1059-z.
25. Sanchez, J. M., Vicario, I., Albizuri, J., Guraya, T., Koval, N. E., and Garcia, J. C. 2018. "Compound formation and microstructure of As-cast high entropy aluminums." *Metals* 8(3):167.
26. Nguyen, Q., and Gupta, M. 2008. "Enhancing compressive response of AZ31B magnesium alloy using alumina nanoparticulates." *Composites Science and Technology* 68(10–11):2185–2192. doi: 10.1016/j.compscitech.2008.04.020.
27. Murty, B. S., Yeh, J. W., and Ranganathan, S. 2014. *High-Entropy Alloys*. Elsevier Science.
28. Yeh, J. W., Chen, S. K., Lin, S. J., Gan, J. Y., Chin, T. S., Shun, T. T., Tsau, C. H., and Chang, S. Y. 2004. "Nanostructured high-entropy alloys with multiple principal elements: Novel alloy design concepts and outcomes." *Advanced Engineering Materials* 6(5):299–303. doi: 10.1002/adem.200300567.
29. Tasan, C. C., Deng, Y., Pradeep, K. G., Yao, M. J., Springer, H., and Raabe, D. 2014. "Composition dependence of phase stability, deformation mechanisms, and mechanical properties of the CoCrFeMnNi high-entropy alloy system." *Jom* 66(10):1993–2001. doi: 10.1007/s11837-014-1133-6.
30. Miedema, A. R., de Châtel, P. F., and de Boer, F. R. 1980. "Cohesion in alloys — Fundamentals of a semi-empirical model." *Physica Part B+C* 100(1):1–28. doi: 10.1016/0378-4363(80)90054-6.

31. Debski, A., Debski, R., and Gasior, W. 2014. "New Features of Entall Database: Comparison of experimental and model formation enthalpies/ Nowe funkcje bazy danych entall: Porównanie Doświadczalnych I Modelowych entalpii Tworzenia." *Archives of Metallurgy and Materials* 59(4):1337. doi: 10.2478/amm-2014-0228.
32. Yang, X., and Zhang, Y. 2012. "Prediction of high-entropy stabilized solid-solution in multi-component alloys." *Materials Chemistry and Physics* 132(2):233–238.
33. Zhang, Y., Lu, Z. P., Ma, S. G., Liaw, P. K., Tang, Z., Cheng, Y. Q., and Gao, M. C. 2014. "Guidelines in predicting phase formation of high-entropy alloys." *MRS Communications* 4(2):57–62. doi: 10.1557/mrc.2014.11.
34. Zhang, Y., Yang, X., and Liaw, P. K. 2012. "Alloy design and properties optimization of high-entropy alloys." *Jom* 64(7):830–838. doi: 10.1007/s11837-012-0366-5.
35. Zhang, Y., Zhou, Y. J., Lin, J. P., Chen, G. L., and Liaw, P. K. 2008. "Solid-solution phase formation rules for multi-component alloys." *Advanced Engineering Materials* 10(6):534–538. doi: 10.1002/adem.200700240.
36. Guo, S., and Liu, C. T. 2011. "Phase stability in high entropy alloys: Formation of solid-solution phase or amorphous phase." *Progress in Natural Science: Materials International* 21(6):433–446. doi: 10.1016/s1002-0071(12)60080-x.
37. Senkov, O. N., Wilks, G. B., Miracle, D. B., Chuang, C. P., and Liaw, P. K. 2010. "Refractory high-entropy alloys." *Intermetallics* 18(9):1758–1765. doi: 10.1016/j.intermet.2010.05.014.
38. Mendis, C. L., Muddle, B. C., and Nie, J. F. 2006. "Characterization of intermetallic particles in a Mg–8Zn–4Al–0.5Ca (wt%) casting alloy." *Philosophical Magazine Letters* 86(12):755–762. doi: 10.1080/09500830601023779.
39. Rokhlin, L. L. 2003. *Magnesium Alloys Containing Rare Earth Metals: Structure and Properties*. Taylor & Francis.
40. Chang, H. W., Qiu, D., Taylor, J. A., Easton, M. A., and Zhang, M. X. 2013. "The role of Al2Y in grain refinement in Mg–Al–Y alloy system." *Journal of Magnesium and Alloys* 1(2):115–121. doi: 10.1016/j.jma.2013.07.006.
41. Milekhine, V., Onsøien, M. I., Solberg, J. K., and Skaland, T. 2002. "Mechanical properties of FeSi (ε), FeSi2 (ζα) and Mg2Si." *Intermetallics* 10(8):743–750. doi: 10.1016/S0966-9795(02)00046-8.
42. Mezbahul-Islam, M., Mostafa, A. O., and Medraj, M. 2014. "Essential magnesium alloys binary phase diagrams and their thermochemical data." *Journal of Materials* 2014:1–33. doi: 10.1155/2014/704283.
43. Wu, Y., and Hu, W. 2008. "Comparison of the solid solution properties of Mg-RE (Gd, Dy, Y) alloys with atomistic simulation." *Research Letters in Physics* 2008:1–4. doi: 10.1155/2008/476812.

3 High Entropy Alloys in Bulk Form

Processing Challenges and Possible Remedies

Reshma Sonkusare, Surekha Yadav, N.P. Gurao, and Krishanu Biswas

CONTENTS

3.1 INTRODUCTION

Material development has been the cornerstone for any civilization, as the design and development of new materials, especially alloys, has provided the necessary physical infrastructure since the beginning of civilization [1]. Although we still use some materials in their pure form, such as copper, alloys are considered to be the best gift for humans. This is primarily because of the fact that the large number of metals and metalloids in the periodic table provides extensive ways to mix different species in varying proportions to design alloys with superior properties, as demanded for various applications. Hence, the last century saw the development of novel but conventional alloys: stainless steels, superalloys, and titanium alloys, to name a few [2–4]. However, the conventional alloy design strategy uses a single base element with attractive properties and addition of minor alloying elements to improve the balance of properties. Bronze was the first alloy developed by humans by adding small amount of tin to copper, and this strategy has been used since then. Alloy development has become sophisticated with time and even 10–12 elements can be added in a controlled manner, as in the case of superalloys, shown in Figure 3.1 [5]. Even then, all alloy families have a dominant base element. The conventional alloy design strategy has provided us numerous alloys with extravagant amalgamation of properties and has been the foundation of many civilizations. But this approach has now reached its limit and new alloys cannot be unearthed which are required for new technological challenges. This has fueled a new alloy design concept where a minimum of five elements are considered in equiatomic or near-equiatomic concentrations, and this concentrated blend of elements are known as complex concentrated alloys (CCAs), high entropy alloys (HEAs), or multi-principal multicomponent alloys [6–8].

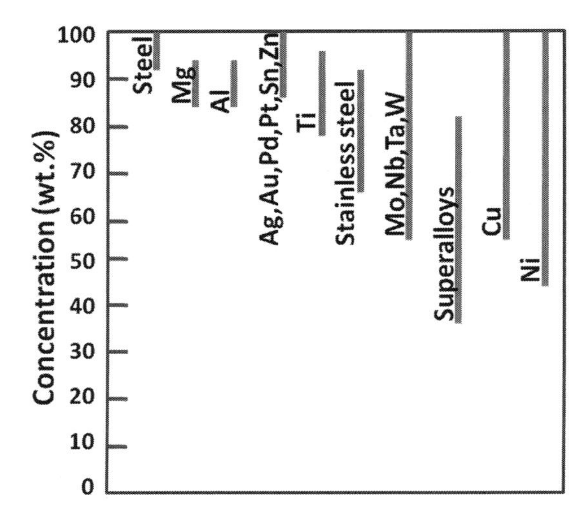

FIGURE 3.1 Concentration of base element in some of the industrially important alloy systems (from [109]).

The idea of HEAs is "to investigate the unexplored central region of the multi-component alloy phase space," which gives a cosmically wide number of unexplored and new alloy bases. The vast number of alloy bases comes from the large number of r principal elements that can be selected from n number of candidate elements [9].

$$C_r^n = \frac{n!}{r!(n-r)!} \tag{3.1}$$

Hence, the new concept provides us novel ways to design and develop "astronomical" numbers of alloys by exploring central portion of the phase diagram [10]. Therefore, HEAs are considered a current area of vigorous research activity in material science and engineering. It has also been realized that it is equally important to develop novel processing routes to prepare HEAs in bulk form. It is well known that the majority of structural applications demand materials in the bulk form. Although vigorous research activities have been carried out in this field, most of the investigations are restricted to lab scale. Therefore, it is important to take a step forward, and industrial scale production of the HEAs is needed to utilize these alloys for future applications. In this regard, the present chapter aims to provide an account of ways and means to synthesize HEAs in bulk form. In particular, three routes, namely, casting, powder metallurgy (P/M), and additive manufacturing (AM) have been discussed, with recently available literature. The problems associated with processing of HEAs in bulk form and plausible remedies have been focused on to provide future directions for research on HEAs. We shall start with the basic concepts of HEAs and then move on to various processing routes, the challenges associated with them, and plausible remedies.

3.2 HIGH ENTROPY ALLOYS: SOME BASIC CONCEPTS

HEAs can be defined on the basis of composition and entropy [11]. According to the composition-based definition, they are defined as alloy systems having more than four principal elements, each with an atomic percentage between 5 and 35. The entropy-based definition defines HEAs as alloy systems having configurational entropy of mixing greater than or equal to 1.5 R (where R is the universal gas constant). According to Boltzmann's hypothesis, the configurational entropy change per mole can be given by the following equation.

$$\Delta S_{\text{config}} = k \ln \omega = -R \sum_{i=1}^{n} X_i \ln X_i \tag{3.2}$$

These alloys are important from an advanced application point of view as well as for developing scientific understanding [12–14]. Many promising properties of the HEAs have been reported up until now, like outstanding high-temperature strength [15, 16], excellent low-temperature fracture toughness, remarkable irradiation resistance [17], good thermal stability [18], exceptional wear resistance [19], high saturation magnetization, and high fatigue resistance, and hence they have attracted the attention of the materials community.

3.2.1 Four Core Effects

HEAs are characterized by their four core effects, which are also responsible for their unique and versatile properties [20]. These effects are more pronounced in HEAs than in conventional alloys. They will now be discussed one by one in detail.

3.2.1.1 High Entropy Effect

The high entropy stabilizes the solid-solution phases (according to the maximum entropy production principle [21]), instead of brittle intermetallic compounds or any other phases, and produce a simpler microstructure. There are three possible competing phases in a solidified alloy, as given in Table 3.1, among which random solid-solution phases tend to have the lowest Gibbs free energy of mixing ($\Delta G_{\text{mix}} = \Delta H_{\text{mix}} - T\Delta S_{\text{mix}}$), especially at high temperatures, and hence they get stabilized. Moreover, it has been found that the number of phases formed in HEAs are much lower than those predicted by Gibbs phase rule ($P = C + 1 - F$), specifying that the high entropy effect increases the mutual solubility between the elements.

Until now, only the configurational entropy was believed to be the main factor for stabilizing the phases in HEAs. But Ma et al. [22] questioned the originally postulated importance of configurational entropy, and the contribution of other entropies (electronic, vibrational, and magnetic) was calculated using a finite temperature *ab initio* method for a prototype five-element HEA, i.e., CoCrFeMnNi. It was found that vibrational entropy increases by up to four times the configurational entropy when the temperature is raised, and electronic and magnetic entropy increases to 50% of configurational entropy on increasing the temperature. This study highlighted the point that all entropy should be given equal noteworthiness while considering phase stabilities in HEAs.

3.2.1.2 Sluggish Diffusion Effect

Phase transformation requires co-operative movement of atoms. But in HEAs, every atom has a different neighborhood, and hence lattice potential energy (LPE) will vary from site to site. Due to large fluctuation in the LPE from one site to another, it is expected that activation energy will be higher and diffusion will be sluggish [23].

TABLE 3.1
Comparison of Thermodynamic Parameters (ΔH_{mix}, ΔS_{mix}, and ΔG_{mix}) for Elemental Phases, Compounds, and Random Solid-Solution Phases [21]

Possible states	Elemental phases	Intermetallic compounds	Random solid solution
ΔH_{mix}	~0	Large negative	Medium negative
ΔS_{mix}	~0	~0	*R ln (n)*
ΔG_{mix}	~0	Large negative	Larger negative

The low-LPE sites acts as deep traps and slow down the diffusion of atoms [24]. Because of the sluggish diffusion effect, one can easily get a supersaturated state, nano-precipitates, decreased grain growth, increased recrystallization temperature, and improved creep properties. The applications which require slow diffusion kinetics (like diffusion barriers) can make use of this sluggish diffusion effect in high entropy materials. In fact, a few studies have shown that HEAs and their nitrides have phenomenal effectiveness as diffusion barriers [25–27]. However, this effect is debatable and there are new studied that question this effect in HEAs [28]. Dabrowa et al. [29] studied the inter-diffusivities in CoCrFeMnNi HEA and no signs of slow diffusion were found when the data was compared with binary and ternary systems. Kottke et al. [30] explained that sluggish diffusion is not properly defined and it should be compared on both the absolute, as well as the homologous temperature scale.

3.2.1.3 Severe Lattice Distortion

HEAs are a whole solute matrix, where every atom is surrounded by different kind of atoms, and hence there is a non-symmetrical neighborhood from site to site. The lattice distortion arises not only from the atomic size differences, but also from the different chemical bonding and crystal structure preferences of elemental species present [31]. The distorted lattice affects the optical, chemical, thermal, mechanical, and electrical properties of the material. For example, there is a decrease in thermal and electrical conductivity (large phonon and electron scattering), but a simultaneous increase in the hardness and yield strength of the material due to large solution hardening. The equation for atomic size difference is as given below.

$$\delta = 100\sqrt{\sum_{i=1}^{n} C_i\left(\frac{1-r_i}{\bar{r}}\right)^2} \tag{3.3}$$

where C_i and r_i are the atomic percentage and atomic radius of ith element, respectively, and $\bar{r}$ is the average radius. The lattice distortion parameter (g) has been calculated for various FCC and BCC HEAs by Wang [32], using atomic structure modeling (Monte Carlo simulations) to generate maximum entropy configurations. In case of FCC FeCoCrNi, FCC CoCrFeMnNi, BCC AlCoCrFeNi, and BCC AlCoCrCuFeNi, g was found to be 0.0085, 0.0070, 0.0210, and 0.0150, respectively, which shows that BCC has more lattice distortions compared to the FCC HEAs.

3.2.1.4 Cocktail Effect

The term "multimetallic cocktail" was first coined by Ranganathan [33], which indicates a synergistic mixture where the end results are interesting and unpredictable by the rule-of-mixture. The composite effect comes not only from the constituent elements but also from their mutual interaction. The cocktail effect in the $Al_xCoCrCuFeNi$ system was studied by Yeh et al. [34], indicating that microstructure and properties can be tailored by varying Al content. Even though Al has FCC crystal structure, its addition to FCC single phase CoCrCuFeNi high entropy alloy results in the transition from an FCC dominated microstructure to a BCC dominated microstructure, with simultaneous increase in hardness and lattice parameter, which can be attributed to the bigger size of the Al atom among all the constituent elements. One more example of the cocktail effect, studied by Zhang et al. [35], showed that $FeCoNi(AlSi)_{0.2}$ has the optimum combination for a soft magnetic material. The ferromagnetic elements (Fe, Ni, Co) form a ductile FCC matrix and non-magnetic elements (Al and Si) increase the lattice distortion, resulting in a positive cocktail effect by achieving high strength, low coercivity, and high magnetization.

3.3 PROCESSING ROUTES FOR HEAS IN BULK FORM

Structural applications of HEAs require fabrication of these alloys in the bulk form. Bulk HEAs can be fabricated using two routes: (a) casting (liquid state route) and (b) powder metallurgy (solid state route). 80% of the HEAs reported in the literature are produced via the liquid state route. In addition, HEAs have recently been fabricated in bulk form using additive manufacturing (AM) technique, mostly utilized for rapid alloy prototyping [36, 37]. The AM approach is in its infancy for HEAs, but a brief overview shall be provided at the end to showcase the potential, present status, and the way forward for all three routes. This shall help us develop a bird's eye view of the bulk processing of HEAs.

3.3.1 CASTING

Casting is the most ancient technique of fabricating metallic components [38], in which molten metal is poured into a mold cavity and subsequently allowed to solidify [39]. The solidified part (known as casting) is then taken out by taking the mold apart (e.g., die casting) or breaking the mold (e.g., sand casting).The former, known as permanent mold casting, has been widely used to fabricate HEAs in the laboratory in the form of suction or drop casting. In the case of drop/suction casting, a water-cooled copper mold is utilized to obtain a cylindrical-shaped rod or plate. There are also other variations of the casting routes, such as Bridgman solidification casting (BST). Interestingly, both single crystals as well as polycrystalline HEAs can be processed via this route. A vacuum-arc melting unit is used for melting and casting metals at lab scale (Figure 3.2). Metals to be melted are placed in a crucible (depression) in the water-cooled copper hearth. A non-consumable tungsten electrode is used as a power source and an electric arc struck between the electrode and metals placed in the crucible is used to melt the metals. First, a vacuum of the order of 10^{-7} mbar is reached using rotary pump (RP) and diffusion pump (DP) and then argon is purged inside the vacuum chamber. Nowadays, the VAM unit comes with a turbomolecular pump (TMP) instead of a diffusion pump, which can reach vacuum level of 10^{-9} mbar in a very short time. Subsequently, the chill casting route in the form of suction casting has been widely utilized to obtain HEAs in bulk form.

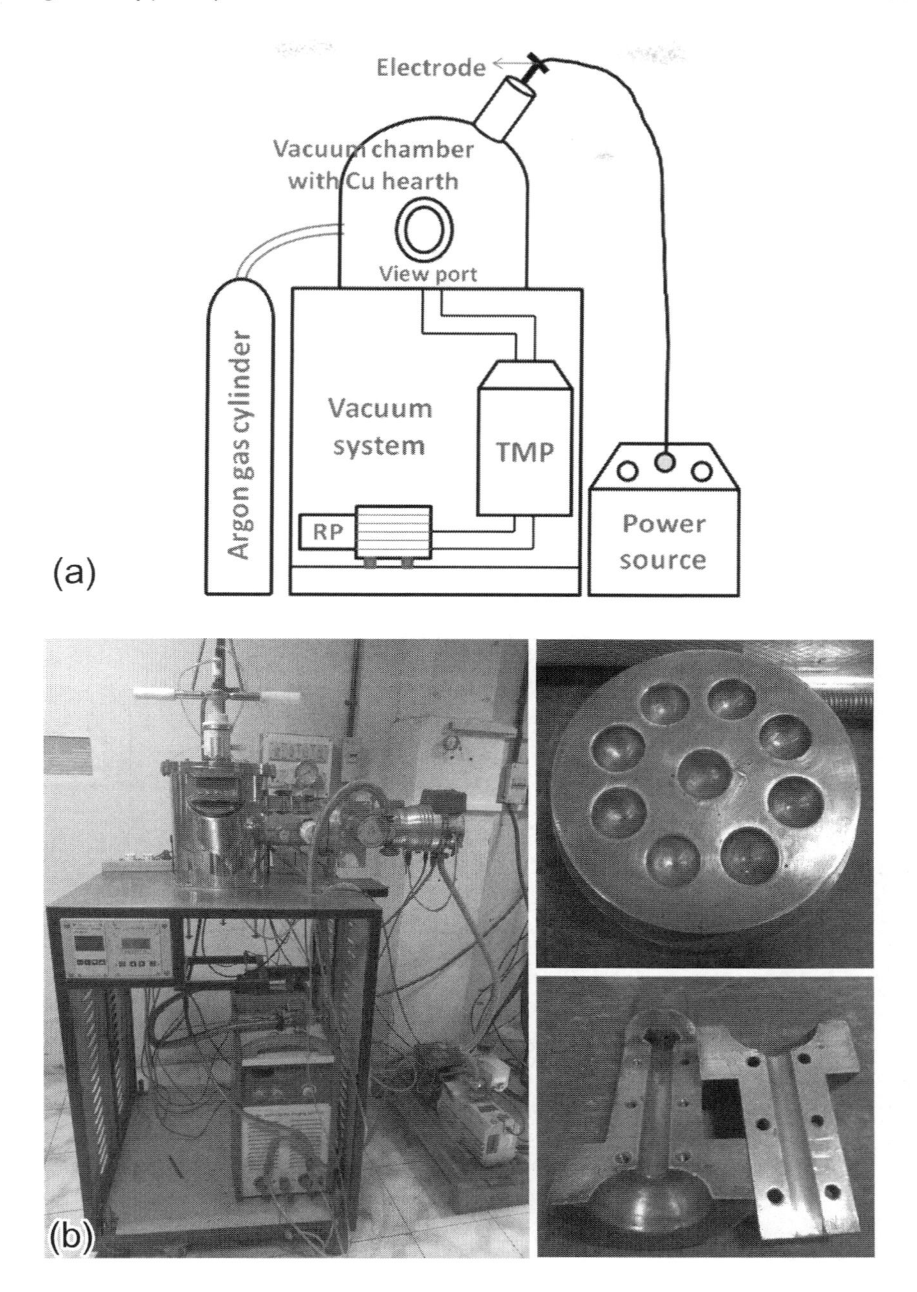

FIGURE 3.2 (a) Schematic of vacuum-arc melting unit and (b) photograph of arc melting-cum-suction casting equipment with copper crucible and two types of copper molds.

3.3.2 Powder Metallurgy (P/M)

Powder metallurgy is a solid-state technique, which uses powder as the raw material [40]. The first step is ball milling (BM) or mechanical alloying (MA), in which not only particle size reduction takes place but alloying between different elements also occurs. MA is a high energy non-equilibrium processing. MA is followed by compaction of the powder in a die cavity to consolidate the loose powder and form a green compact which is then subjected to sintering. Sintering involves heating the green compact in a furnace to a temperature below the melting point of the alloy in an inert atmosphere which causes the particles to fuse together by developing a metallurgical bond. Nowadays, MA is followed by spark plasma sintering (SPS) in which compaction and sintering takes place simultaneously, in a very short time, compared to conventional sintering techniques like hot isostatic pressing (HIP), hot pressing (HP), etc. Figure 3.3 shows schematic and photograph of SPS unit. P/M can be used for high melting point metals, cemented carbides, and metal oxides to produce net-shaped or near net-shaped object.

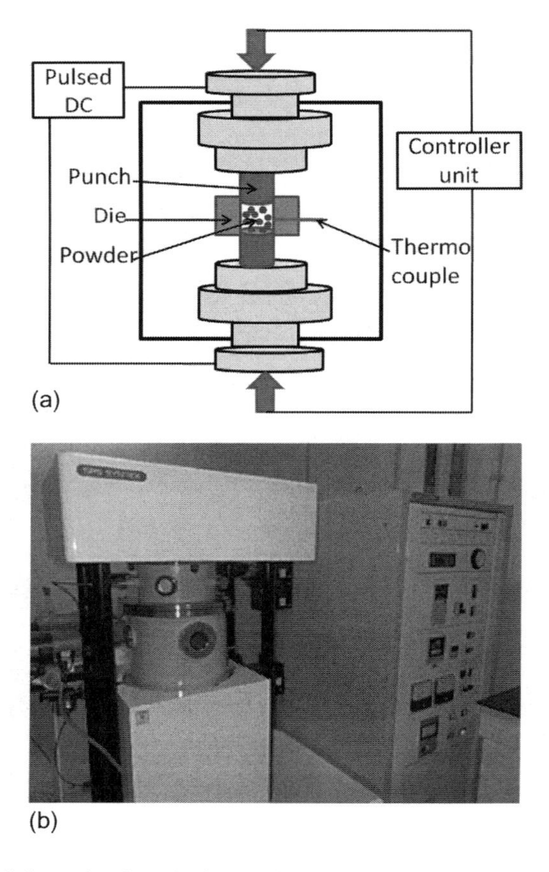

FIGURE 3.3 (a) Schematic of spark plasma sintering (SPS) setup and (b) photograph of SPS facility at IIT Kanpur.

3.3.3 Additive Manufacturing

Additive manufacturing (AM), rapid prototyping, or 3D printing, is a novel fabrication technique in which a computer-aided design (CAD) model can be converted into a 3D object via layer-by-layer deposition of the material, with layer thickness in the range of 20–100 μm [41, 42]. It requires very short time as compared to the conventional fabrication techniques and can also produce complex and intricate shapes. AM samples have fine elongated grains, metastable phases, solute trapping, non-equilibrium microstructures due to directional solidification, and rapid cooling rates [43]. In AM, powder or wire is melted either by laser or electron beam and transformed into a solid part. The widely used AM processes for metals are (a) laser beam melting (LBM) or selective laser melting (SLM), (b) electron beam melting (EBM), and (c) laser metal deposition (LMD), direct metal deposition (DMD), laser-engineered net-shaping (LENS), or laser cladding [44].

LBM/SLM: This is a powder-based technique in which the powder is fed by a hopper and a leveling blade/roller is used for its uniform distribution. The deposited powder is then exposed to a laser beam with spot size between 50–180 μm and power between 20 W–1 kW. During solidification, the new layer gets fused to the layer below it. For the next layer, the platform is lowered and this process is repeated until the object is completely fabricated (Figure 3.4a).

EBM: This is also a powder-based technique. But instead of laser, an electron beam is used as a heat source to melt the powder. An electron gun generates a beam and it is then accelerated with an accelerating voltage of 60 kV. The beam is focused using electromagnetic lenses and is directed on the powder bed using magnetic scan coils. The beam not only melts the powder but also leads to sintering of the powder particles. After solidification, the build plate is lowered and the process is repeated. This process requires a vacuum of less than 10^{-2} Pa.

LMD/DMD/LENS: In this process, a laser is used to melt the surface and then powder is deposited on that surface/substrate (Figure 3.4b) [45]. A multi-jet or coaxial nozzle is used to feed the powder at rates between 4–30 g/min. and a Nd:YAG or CO_2 laser is used as a heat source with spot size and scan speed between 0.3–3 mm and 150 mm/min–1.5 m/min, respectively. This process has a high build rate compared to LBM and EBM and a larger volume object can be fabricated. The nozzle approach has been changed now and it uses wire as the feedstock instead of powder.

3.4 PROCESSING CHALLENGES IN CASTING

We shall now discuss the challenges involved in processing of HEA via each route in detail. Solidification is ubiquitous for synthesis of HEAs via castings, big or small, with a large or smaller cross-section. In fact, this age-old technique is best suited for both single crystals as well as multi-grained castings. However, solidification of alloys involves solute partitioning, which plays a significant role in phase formation and microstructural evolution. The advancing solid–liquid interface is expected to reject solute into the liquid as the solubility of the solute elements in any solid is always smaller than that in the liquid. In a binary dilute alloy, this is straightforward and this behavior is controlled by the slope of the liquidus (m)

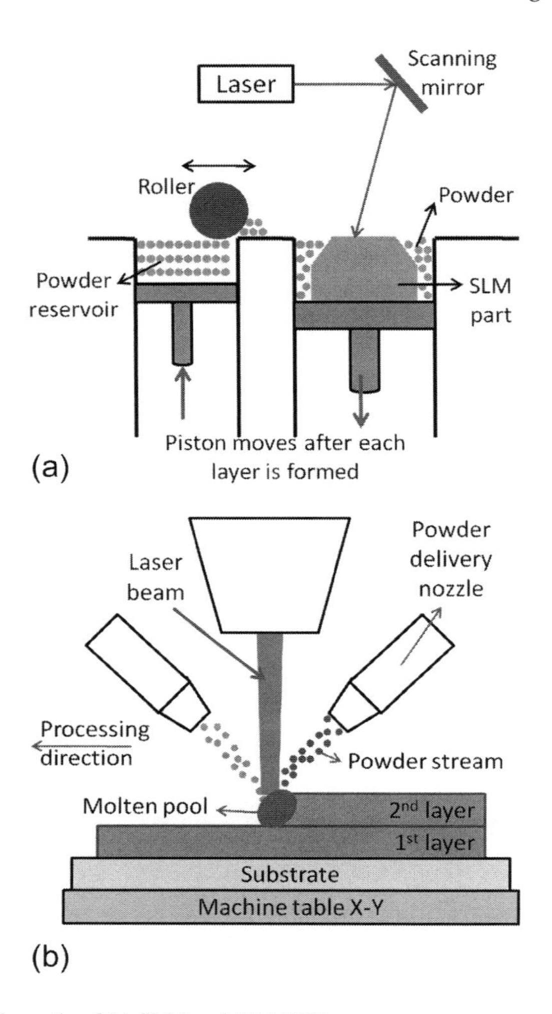

FIGURE 3.4 Schematic of (a) SLM and (b) LENS setup.

and the partition coefficient (k). For negative values of m and k<1, solids will contain lower solutes than liquid. However, when m is positive and k >1, solubility is greater in solid than liquid and hence, the solute will move from liquid to solid, leading to formation of a zone depleted of solute. However, the former is widely observed and reported. The same situation cannot be directly extended to ternary, quaternary, or even quinary alloys. For the sake of brevity, we shall discuss ternary alloy. Let us discuss the case of a ternary alloy containing two solutes, A and B forms single phase during steady-state plane front solidification without any convection.

Figure 3.5 shows the solute distribution in the liquid for both the elements [46]. It is important to realize that the diffusion boundary layer, indicating the characteristic diffusion distance ahead of the liquid–solid front, will not be same for the two solutes because their diffusion coefficients will not be same. The conservation of

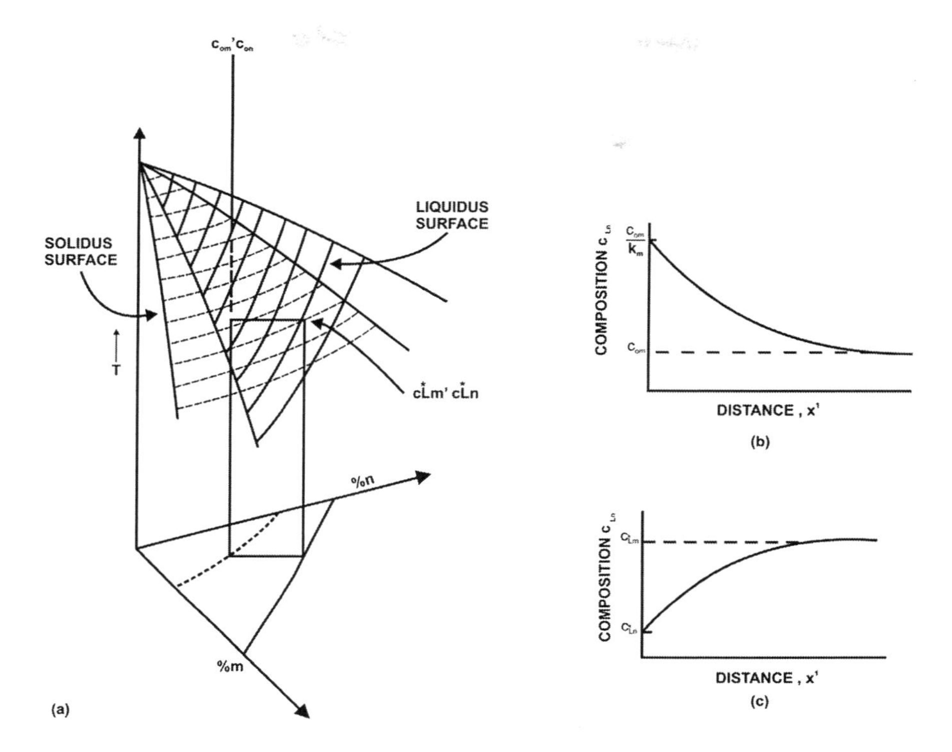

FIGURE 3.5 (a) Part of ternary phase diagram and (b–c) distribution of solute in front of a single-phase ternary alloy solidifying with plane front.

solute can be applied at the solid–liquid interface during steady-state solidification, yielding:

$$D_{AA}\left(\frac{dC_{LA}}{dx}\right)+D_{AB}\left(\frac{dC_{LB}}{dx}\right)=-V(C_{LA}^{*}-C_{0A}) \tag{3.4a}$$

$$D_{BA}\left(\frac{dC_{LA}}{dx}\right)+D_{BB}\left(\frac{dC_{LB}}{dx}\right)=-V(C_{LB}^{*}-C_{0B}) \tag{3.4b}$$

Here, D_{AA} and D_{BB} are the on-diagonal diffusion coefficients of the solutes in liquid and D_{AB} and D_{BA} are off-diagonal coefficients because of interaction terms. Therefore, the solute distribution profiles for both the solutes will significantly depend on the off-diagonal terms. Upon extending this analogy to quinary alloys, the higher order terms will alter the diffusion profile and hence, solute partitioning will be a complex function of the local slope of the liquidus surface, on and off-diagonal diffusion coefficients and other processing conditions. Hence, the complex interplay among elements will significantly affect the solute partitioning and hence, the microstructural evolution of HEAs. Although this analysis can be extended to quinary and higher order systems, the applicability is difficult as determination of diffusion coefficient and partition ratios of different solutes from phase diagram information is tedious and in the most of the cases, it is not available. Solute partitioning during solidification of large castings can cause severe segregation and interdendritic porosity due to hindrance of fluid flow, which are the common defects in casting. In the following section, we shall discuss the challenges involving understanding segregation and possible remedies of the same. First, a brief description of micro-segregation will be provided and this will be followed by macrosegregation.

During alloy solidification, we have two type of segregation in the solidified material: (a) microsegregation and (b) macrosegregation. In the first one, compositional changes occur over the length scale of secondary dendritic arm spacing, and the latter one occurs over large distances of the order of specimen size due to mass flow.

3.4.1 Microsegregation

Let us consider a hypothetical binary phase diagram with straight solidus and liquidus lines, as shown in Figure 3.6a. Here we can define equilibrium partition coefficient k, as follows:

$$k = \frac{X_S}{X_L} \tag{3.5}$$

where, X_S = mole fraction of solute in the solid and X_L = mole fraction of solute in the liquid at a given temperature, in equilibrium condition [47], and k is independent of temperature in this simplified case. Alloy solidification in practice is not so simple and depends on many factors like temperature gradient, cooling rate, and growth rate and therefore a planar solid/liquid interface movement is assumed for simplification. Now we have three limiting cases for the unidirectional alloy solidification: (a) extremely slow (equilibrium) solidification, (b) no diffusion in solid but perfect mixing in liquid, (c) no diffusion in solid and partial mixing (only diffusional) in liquid.

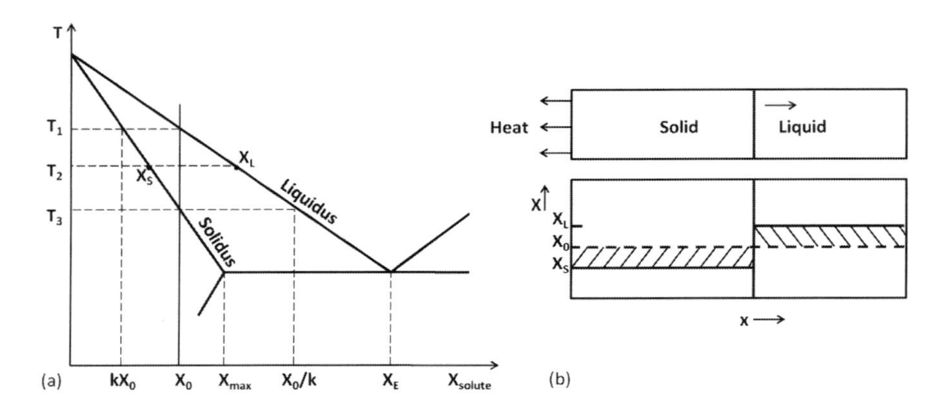

FIGURE 3.6 (a) A hypothetical phase diagram with straight solidus and liquidus lines. (b) Unidirectional solidification of X_0 alloy.

Equilibrium solidification: During equilibrium solidification, there is complete diffusion in solid as well as liquid and fraction of both the phases follow lever rule. In this case, the alloy X_0 will start solidifying at T_1 and the first solid formed will have a composition of kX_0 (Figure 3.6). As the temperature decreases, the solid and liquid composition will follow the solidus and liquidus lines and their relative fraction can be calculated from the lever rule at any temperature. At T_3, the last drop of liquid will have a composition of X_0/k. Since the solidification is very slow, the complete solid will have X_0 composition throughout. The solute balance equation can be given as follows:

$$X_L = \frac{X_0}{1-(1-k)f_s} \tag{3.6}$$

Here, f_s = volume fraction solidified.

No diffusion in solid and complete mixing in liquid: In industries, the cooling rates are often very high during alloy solidification and it can be assumed that there is not sufficient time for diffusion in solid to occur, but liquid has a homogeneous composition due to efficient stirring. In such a case, there will be solute redistribution during solidification and local equilibrium at the solid/liquid interface. For the alloy X_0, the first solid will form at T_1 with kX_0 composition (Figure 3.7). Since $kX_0<X_0$, the solid will reject the solute into the liquid and increase the liquid concentration above X_0. Now, the temperature of the interface has to come below T_1 in order to continue solidification. The next layer of solid formed will have richer solute concentration than the first and this continues as the solidification progresses. Moreover, solidification occurs at a relatively lower temperature than the equilibrium one, and local equilibrium can be assumed at the interface. Since every layer of solid will have different composition due to no diffusion, the average composition of the solid, denoted

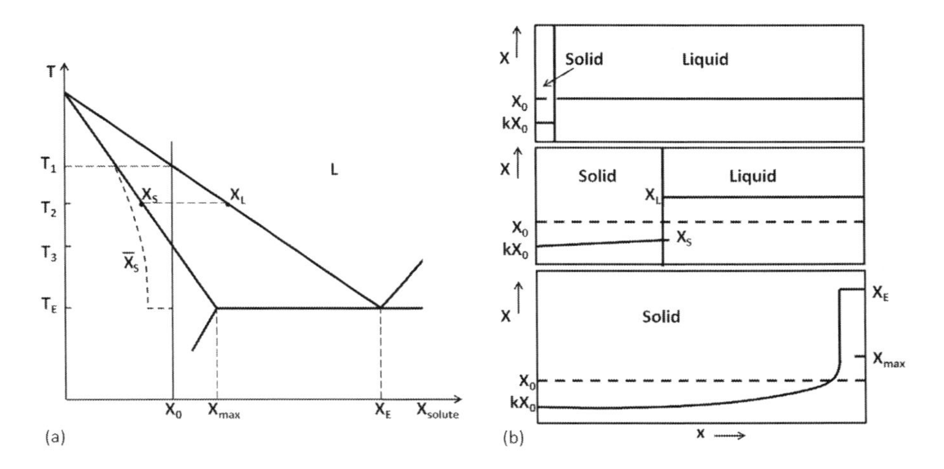

FIGURE 3.7 (a) Phase diagram showing mean composition of solid for the case when there is no diffusion in solid and complete mixing in liquid. (b) Composition profile at T_1, T_2, and below T_E.

by $\overline{X_S}$, will be lower than the composition at the solid/liquid interface, as shown in Figure 3.7. The liquid can become richer in solute (than X_0/k) such that it reaches the eutectic composition X_E. The variation in X_S can be calculated by equating solute rejected into the liquid with increase in solute in the liquid, as given below.

$$(X_L - X_S)df_s = (1 - f_S)dX_L \tag{3.7}$$

Integrating the above equation with boundary condition as $X_S = kX_0$, when $f_s = 0$, we get

$$X_S = kX_0(1 - f_s)^{(k-1)} \text{ and } X_L = X_0 f_L^{(k-1)} \tag{3.8}$$

The above equations are known as Scheil equations or the non-equilibrium lever rule.

No diffusion in solid and no stirring in liquid: In this case, the solute rejected from the solid into the liquid will be transported only by diffusion and therefore there will be a buildup of solute atoms at the solid/liquid interface ahead of the solid and rapid concentration increase in solid, which is known as initial transient. If a constant solidification rate (v) is assumed, a steady state will be reached with the interface temperature as T_3. The solid will form with a bulk composition of X_0 and the liquid in contact with the solid will have a composition of X_0/k. During the steady state, the rate at which diffusion of solute occurs away from the interface should be balanced by rejection of solute in the liquid, given by the following equation:

$$-DC_L' = v(C_L - C_S) \tag{3.9}$$

Here, D = diffusivity in liquid, C_S and C_L are equilibrium solute concentration in the solid and liquid at the interface, and $C_L' = dC_L/dx$ at the interface. The concentration profile for the liquid can be given by the following equation:

$$X_L = X_0\left\{1 + \frac{1-k}{k}\exp\left[-\frac{x}{(D/v)}\right]\right\} \tag{3.10}$$

X_L decreases from X_0/k at $x = 0$ to X_0 at a large distance, with a characteristic width of D/V. In the final transient, the composition rises rapidly and eutectic formation takes place at the end of the bar (Figure 3.8a). During practical alloy solidification, features from all three cases will appear.

The three cases of solidification discussed above are for a binary system where solute and solvent can be clearly defined. But in case of high entropy alloys, there is no solute and solvent, as all the elements have equiatomic or near-equiatomic compositions. Therefore, the solidification route for HEAs will be complex and will depend not only on the partition coefficient, but also on other factors like difference in enthalpy of mixing and difference in melting point. Because of the compositional

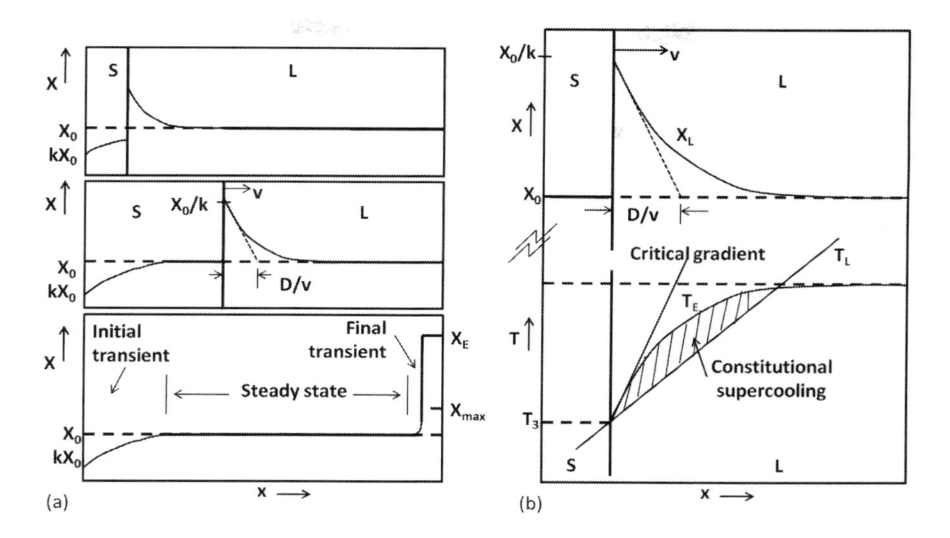

FIGURE 3.8 (a) Composition profile between T_2 and T_3, at T_3, and below T_E. (b) Composition and temperature profile at the solid/liquid interface to show constitutional supercooling.

complexity in HEAs, it is not clear whether the Scheil equation can be directly applied to them.

Figure 3.8b shows the concentration profile during the steady-state solidification and corresponding temperature profile. Here, T_L represents the liquid temperature ahead of the interface and T_E represents equilibrium temperature. The actual liquid temperature can follow any line T_L and at the interface, $T_E = T_L = T_3$. Now, if the liquid temperature gradient is less than the critical gradient, T_L will be below its equilibrium temperature and liquid will be supercooled/undercooled. This is known as constitutional supercooling because it arises from constitutional or compositional effects. Constitutional supercooling is a necessary condition for the formation of stable protrusions on the planar interface. When the temperature gradient in front of the interface is greater than critical gradient, the protrusion will melt back because its tip temperature will be more than the liquidus temperature. When the tip temperature remains below the T_E, the protrusion can grow. The equation for critical gradient under steady-state conditions can be given as follows:

$$\frac{T_1 - T_3}{D / v} \tag{3.11}$$

If $T_L' >$ critical gradient, where $T_L' = dT_L / dx$ at the interface, a stable planar interface will form. Here, $(T_1 - T_3)$ is known as the equilibrium freezing range of the alloy. Microsegregation is not desirable for the alloys and it can be controlled by the proper heat treatment procedure after the alloy solidification. When we go from lab scale to industrial scale, the alloy size increases manyfold and macrosegregation comes into the picture.

3.4.2 Role of Segregation in Multicomponent Alloys

In multicomponent alloys, with varying interactions among different elements, segregation is bound to occur during solidification. The problem associated with these alloys processed via solidification thus can lead to a varying microstructure from the center to the periphery of the surface, with inhomogeneous distribution of as-cast dendrites in morphology, size, and variation of macroscopic properties. Additionally, this can cause large number of inevitable casting defects—elemental segregation, suppression of equilibrium phases, microporosity, microscopic and macroscopic residual stresses, and cracks—hampering mechanical properties. Hence, we need to take measures to reduce or even eliminate these defects during casting of HEAs and the modification of existing routes or development of novel routes are sought after.

3.4.3 Macrosegregation

In this section, the principles of solidification in castings and ingots will be discussed. In industry, alloys are poured into a mold and a casting is obtained. When they are worked later, like in rolling or forging, they are called ingots. The factors affecting macrosegregation are (i) shrinkage, (ii) density difference between solid and liquid, (iii) density difference in the liquid contained in the interdendritic region, and (iv) convection currents due to temperature-induced density difference in the liquid.

Figure 3.9 shows complex segregation patterns due to the combination of the above effects. Due to the large size of the ingot, there will be a high-temperature gradient during solidification and therefore the microstructure near to the mold wall will be different from that of the ingot center. Macrosegregation is harmful for the properties of the alloy, and it can only be minimized by good control of the solidification process and not by the heat treatment process.

3.4.4 Microsegregation in Multicomponent Alloys

In an interesting study, Samal et al. [48] investigated a solidification path for NiTi-based binary $Ni_{60}Ti_{40}$, ternary $Ni_{50}Cu_{10}Ti_{40}$, quaternary $Ni_{48}Cu_{10}Co_{2}Ti_{40}$, and quinary $Ni_{48}Cu_{10}Co_{2}Ti_{38}Ta_{2}$ alloy systems to understand the solidification behavior of quinary HEA. The alloys were prepared via arc melting followed with suction casting in a copper mold. The solidification pathways for all four alloys were determined experimentally as well as by using two analytical solidification models, i.e., the lever rule and the Gulliver–Scheil model in Thermocalc software, to predict equilibrium and non-equilibrium paths, respectively. It was found that the experimentally observed solidification path for binary matched with the Gulliver–Scheil simulations and there was deviation for ternary, quaternary, and quinary alloys. It indicates that the solidification path for binary alloy can be easily predicted, but for higher order systems the same theory does not stand true. This discrepancy can be attributed to some solute diffusion during real solidification which is not considered in the simulations.

In another study, Wu et al. [49] investigated the solidification behavior of $CoCrCu_xFeMoNi$ ($x = 0.1$ to 1) HEA. When $x = 0$, the microstructure showed a

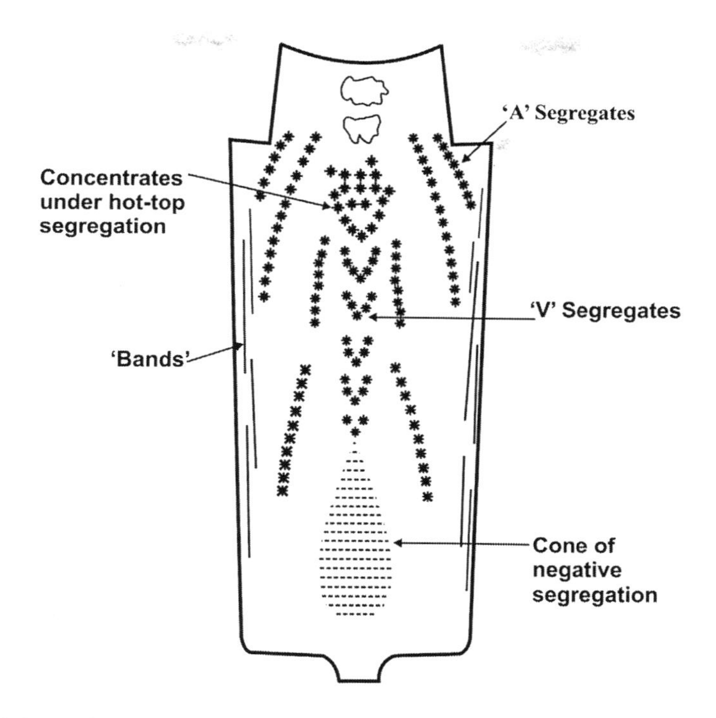

FIGURE 3.9 Positive (+) and negative (–) segregation in a steel ingot.

FeCoNi-rich dendritic region and a MoCr-rich interdendritic region. When x = 0.3, three types of microstructures were observed: the quasi-peritectic zone, the first eutectic zone, and the second eutectic zone. It was concluded that first the Mo-rich phase nucleated from liquid via a peritectic reaction producing a (Co,Fe,Ni)-rich phase and a Mo-rich phase. But this reaction does not advance completely and a eutectic reaction follows to form the first eutectic microstructure. Again, one eutectic reaction occurs in the residual liquid and second eutectic microstructure forms. Further increase in Cu% (x = 0.5, 0.8) led to the formation of spherical structures rich in copper and feather-like structures rich in chromium. In equiatomic CoCrCuFeMoNi, the copper-rich spheres were surrounded by a border rich in Fe, Co, and Ni separating the copper-rich and copper-lean regions. Cr-rich feathers were present in the Cu-lean regions and (Fe,Co,Ni)-rich petal-shaped dendrites were present in Cu-rich regions.

Mishra et al. [50] studied the solidification behavior of arc melted TiCuFeCoNi HEA with different molar ratio of Ti/Cu (x) (x = 1/3, 3/7, 3/5, 9/11, 1, 11/9, 3/2). The microstructure of the alloy with x = 1/3, 3/7, and 3/5 reveals three dendritic phases: a Ti-rich solid solution which forms first from the liquid, and the remaining liquid undergoes a liquid phase separation to form a Cu-rich solid solution and a Co-rich solid solution. The microstructure of the alloy with x = 9/11, 1, 11/9, and 3/2 reveals an ultrafine eutectic between the Laves phase (Ti_2Co type) and a Cu-rich solid solution along with dendrites of Ti-rich and Co-rich solid solution. The pathways for solidification of HEAs are complex and not completely understood to date. It requires the study of microstructural features and phases formed. It has been observed that HEAs form a lower number of phases than predicted using the Gibbs phase rule and it can be because of high configurational entropy. Anomalous solidification microstructures were observed by Guo et al. [51] in $Al_xCrCuFeNi_2$ HEA. When x = 0.9, the alloy showed FeCr-rich dendrites and Cu-rich interdendritic region with 200 nm long rods dispersed in the interdendritic region. When x was increased to 1.2, no dendritic structure was observed and only rods were present. Microstructures in x = 2, 2.2, and 2.5 alloys showed sunflower or petal-like morphology (eutectic NiAl-rich A2/Cr-rich B2 phase) and small precipitates (spinodal decomposed A2 phase) dispersed inside the disk floret (primary NiAl-rich B2 phase). The sunflower-like microstructure is not yet observed in any other HEA systems. Hemphill et al. [52] studied the effect of defects on fatigue behavior of arc melted $Al_{0.5}CoCrCuFeNi$ HEA. The air cooled and last solidification side of the casting had greater segregation, more shrinkage pores, and inclusions which induced microcracks during cold rolling. Moreover, SEM and EDS analysis indicated formation of Al oxide-rich particles during solidification which provided nucleation sites for microcracks. These defects led to variable fatigue life of the samples depending on the position of sample in the mold during solidification.

Now we will discuss some examples of microsegregation during solidification in high entropy alloys. Equiatomic CoCrFeMnNi HEA is a single-phase FCC alloy and is the most widely studied composition among all the HEA compositions [53]. In the as-cast condition, this alloy shows microsegregation of elements. The low melting point Mn segregates in the interdendritic regions and dendritic region is depleted of Mn. Another single-phase FCC alloy which shows segregation is CoCuFeMnNi HEA. Sonkusare et al. [54] have studied the equiatomic CoCuFeMnNi HEA synthesized using a vacuum-arc melting facility. It was found that the as-cast microstructures have Co and Fe in the dendritic regions, Cu and some Mn in the interdendritic regions and Ni is uniformly distributed (Figure 3.10a). Cu has positive enthalpy of mixing with other four elements and hence segregates in the interdendritic regions. To remove this segregation, the alloy was subjected to heat treatment at 1,273 K for ten hours, followed by water quenching. This gives rise to a single phase microstructure with a grain size of the order of 50–100 μm. Apart from elemental segregation, a lot of porosity can also be observed in the BSE image which can be minimized using mechanical processing like rolling, forging, compression, etc. Figures 3.10b and c show the 10 g button and 40 g billet prepared using vacuum-arc melting in the lab, whereas Figure 3.10d shows a 17 kg billet of the same HEA produced in industry. Figure 3.11 shows the difference in microstructure of a small casting and a large casting. Extensive porosity can be observed in the large casting even after homogenization, whereas small casting shows much fewer pores.

In another study, Wang et. al. [56] synthesized AlCrFeCoNi HEA. The as-cast alloy consisted of polygonal grains with intragranular dendritic segregation. The interdendritic region was rich in Co, Cr, Fe and the dendritic region was rich in Al and Ni owing to their most negative enthalpy of mixing (-22 kJ/mol) among the binary pairs of the constituent elements. The addition of copper to AlCrFeCoNi HEA gives rise to an interesting microstructure with even more segregation, studied by Singh et al. [57] where the equiatomic AlCoCrCuFeNi HEA was synthesized via casting route. Figure 3.12 shows bright-field TEM images of the cast alloy showing dendritic and interdendritic regions. The dendritic regions have Cu-rich plate-like

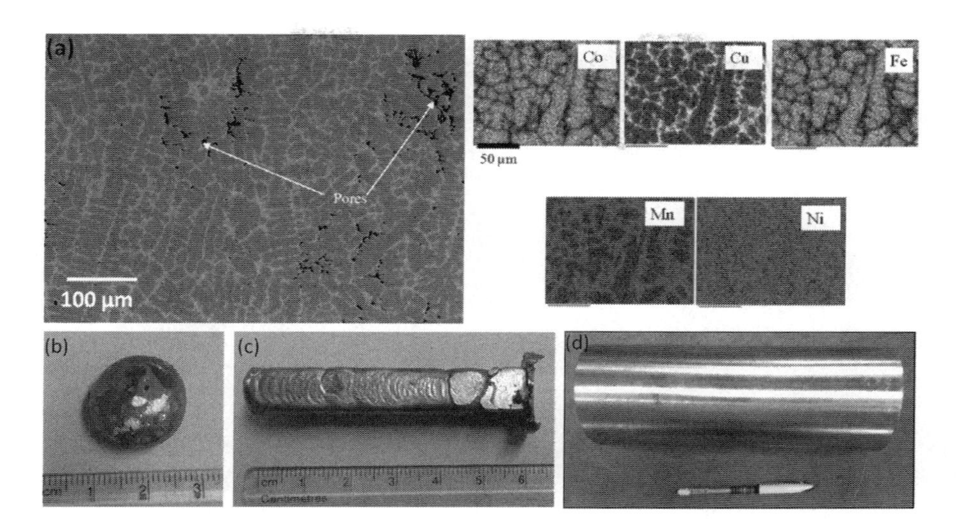

FIGURE 3.10 (a) Back-scattered image and energy dispersive spectroscopy of CoCuFeMnNi HEA, (b) as-cast button (10 g), (c) suction cast billet (40 g), (d) large billet (17 kg) (from [54]).

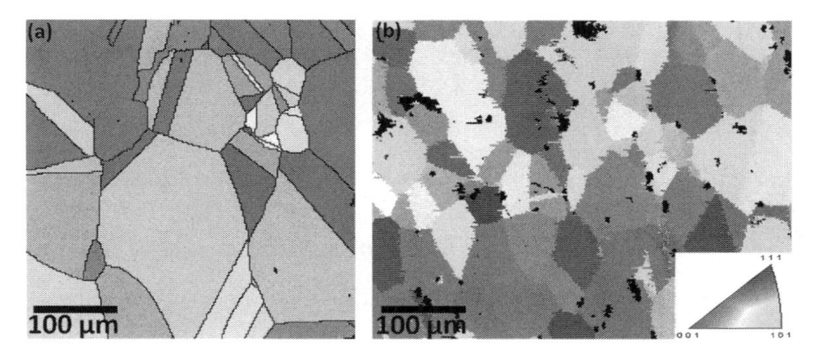

FIGURE 3.11 Inverse pole figure (IPF) maps of annealed sample of CoCuFeMnNi. (a) Small casting (10 g), (b) large casting (17 kg) [55].

precipitates (Figure 3.12b) (200–500 nm long and 50 nm thick) with B2 structure (a = 0.288 nm), oriented along the <110> direction as well as Cu-rich spherical (5–15 nm) and rhombohedron-shaped precipitates (Figure 3.12c). The plate-like precipitates are coherent with the BCC matrix (a = 0.288 nm). The diffraction pattern of rhombohedron precipitates in the inset of Figure 3.12c reveals a weak superlattice reflection, indicating that it has $L1_2$ structure (a = 0.364 nm). The interdendritic region is Cu-rich with $L1_2$ structure (a = 0.358 nm). The regions marked as A, B, and C are rich in Al-Ni, Cr-Fe, and Ni, respectively. Figure 3.12e shows a schematic for the phases formed during different processing routes.

Segregation during casting has also been observed in refractory HEAs. One such study by Senkov et al. [58] on refractory WNbMoTaV HEA shows that W solidifies in the dendritic region, Mo, Nb, V segregates in the interdendritic region, and Ta

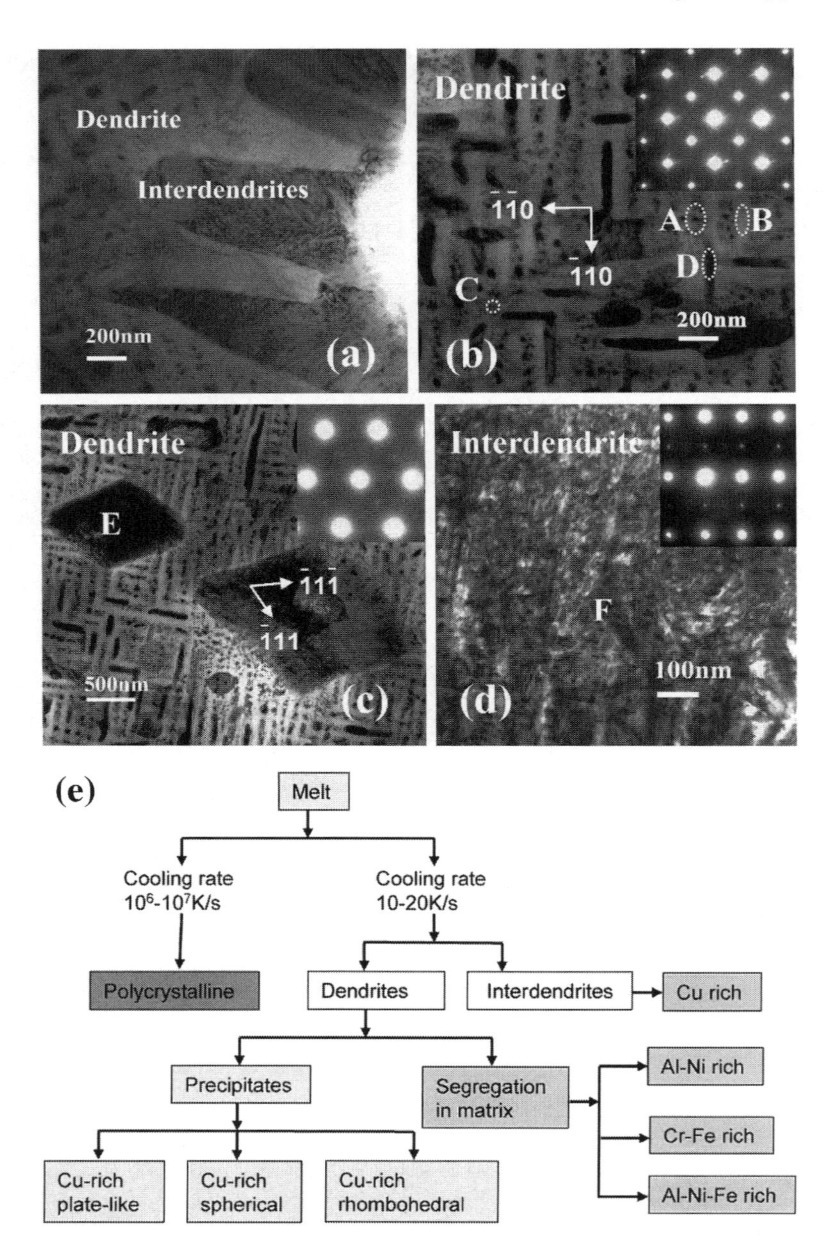

FIGURE 3.12 (a–d) Bright-field TEM micrographs of as-cast equiatomic AlCoCrCuFeNi HEA. (e) Phase segregation observed in AlCoCrCuFeNi HEA (from [57)].

is uniformly distributed. For multiphase HEAs, it is difficult to control the volume fraction of the phases if microsegregation occurs. The possible remedy to remove/minimize microsegregation is proper thermomechanical treatment after the casting of the alloy.

3.4.5 Possible Remedies

In the following section, we shall focus on some of the processing routes, which are expected to provide remedies for the above-mentioned problems associated with conventional casting technique.

Bridgman solidification casting: It is well known in the literature that the Bridgman solidification technique, popularly known as BST, is an effective route to obtain solidified products with microstructural control, leading to property enhancement [59]. In this technique (Figure 3.13), a rod-shaped specimen is obtained by specific control of the direction of heat extraction. This is akin to single crystal growth in which unidirectional heat extraction is carried out. This controlled heat extraction guarantees the microstructural growth direction by controlling temperature gradient and grain direction. The growth direction is controlled by the withdrawal speed whereas temperature gradient is maintained by furnace design and water flow. Figure 3.13 shows a schematic diagram of the process.

The target HEAs are initially cast into rod-shaped samples by the drop or suction casting method. Subsequently, the crushed pieces of the alloy are placed in a long alumina crucible which is then inductively heated to melt the alloy pieces. This is followed by BST with varying withdrawal speed and temperature gradients.

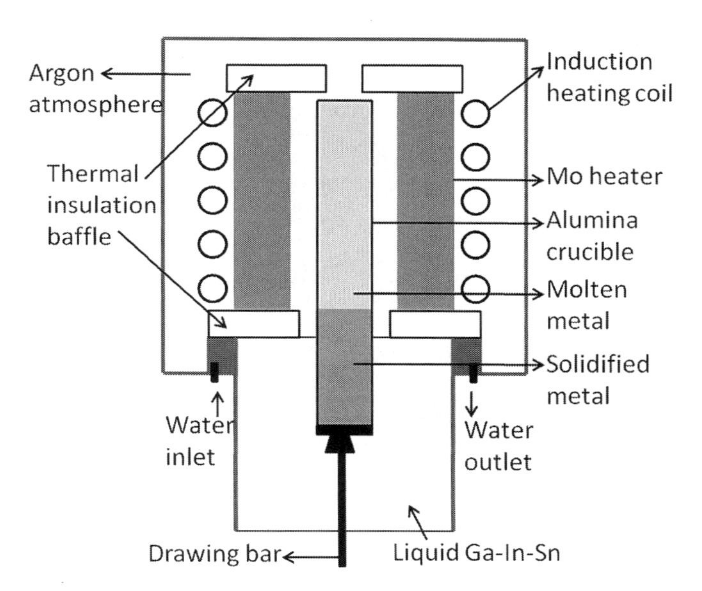

FIGURE 3.13 Schematic of Bridgman solidification technique setup.

Figure 3.14 reveals optical metallographs of quinary AlCoCrFeNi HEA synthesized at varying withdrawal speeds. It is evident that BST processing leads to formation of equiaxed grains of the complex alloy. Figure 3.15 shows SEM micrographs of the same alloy. It shows a honeycomb-like structure revealing nanoscaled precipitates forming due to spinodal-like transformation in solid state. It can be inferred that the BST technique can be utilized to grow single crystals of HEAs [60].

Single crystals of HEAs are considered for potential high-temperature applications, e.g., turbine blades for aircraft engine. The preparation of single crystals, from the perspective of solidification, requires the growth of single nucleus or seed. Fabrication of seed crystal is difficult, as large number of nuclei form during typical copper mold casting. Hence, the competitive growth with preferred crystallographic orientation is needed, which can be realized by BST. Upon selecting the proper ratio of temperature gradient to growth velocity, growth of a single seed into a single crystal can be achieved.

Figure 3.16 reveals the microstructural evolution of quinary $CoCrFeNiAl_{0.3}$ alloy by BST with a pulling speed of 5 μm/s [61]. The schematic diagram reveals growth direction and the SEM microstructures from different regions being shown. It is evident that the HEA undergoes microstructural transition from dendritic to equiaxed to columnar grains and finally to single crystal. The solidification starts with typical dendritic microstructure having multiple crystals (zone A). In zone B, large number of equiaxed grains of the size 50–300 μm is formed. Gradually, the solid/liquid

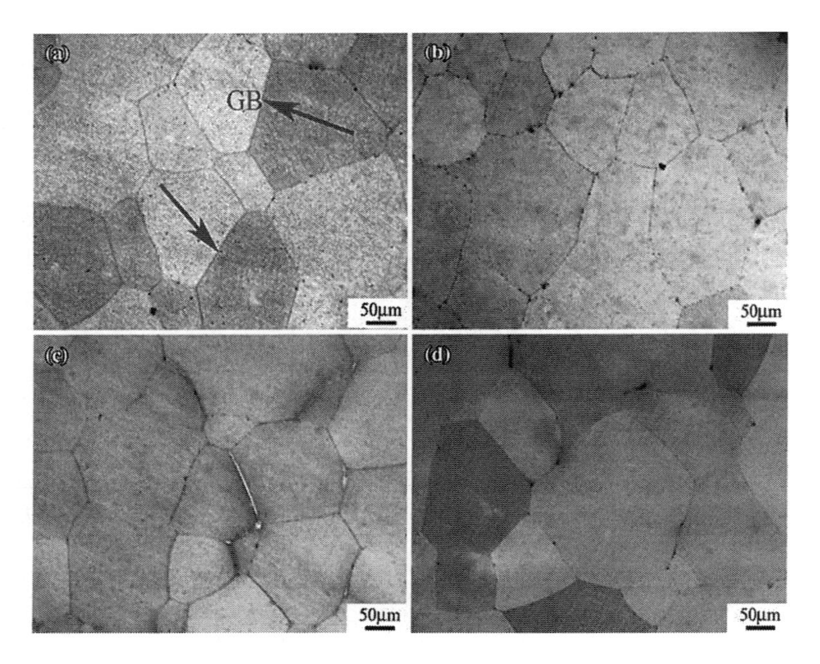

FIGURE 3.14 Optical metallographs of AlCoCrFeNi HEA synthesized using BST at withdrawal speed of (a) 200, (b) 600, (c) 1,000, and (d) 1,800 μm/s (from [60]).

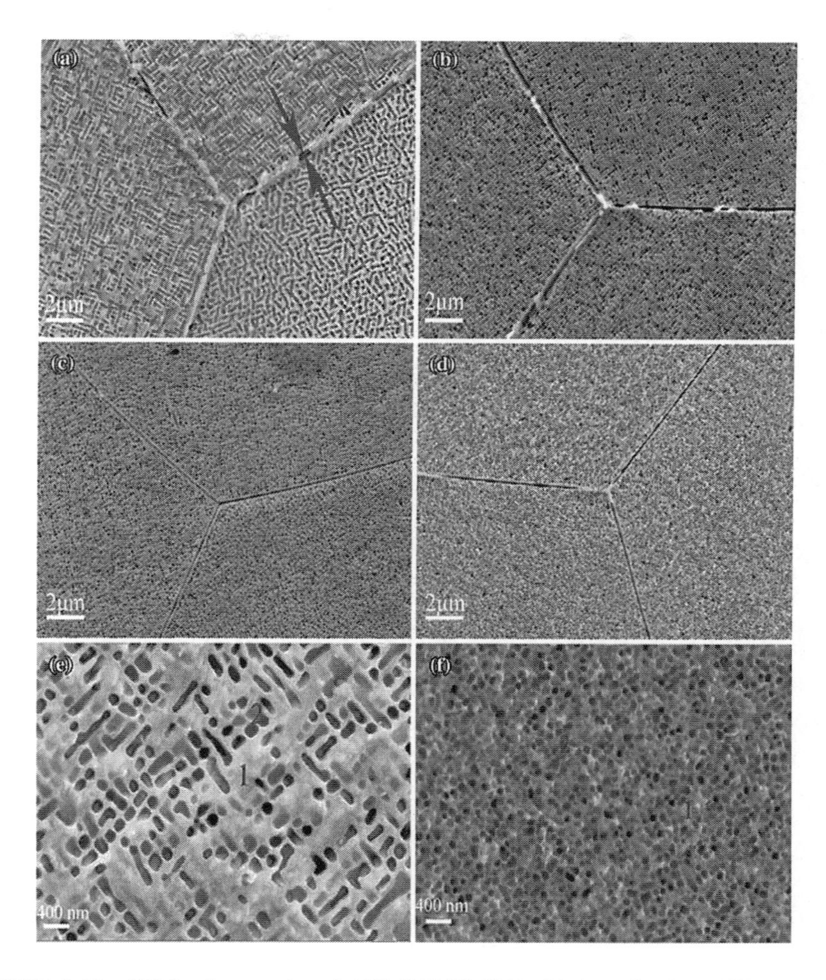

FIGURE 3.15 SEM micrographs of AlCoCrFeNi HEA fabricated using BST at a withdrawal speed of (a) 200, (b) 600, (c) 1,000, and (d) 1,800 μm/s. (e) and (f) are high magnification images of (a) and (d), respectively (from [60]).

interface moves into region C, where large columnar grains (mm size) form, giving rise to preferred growth of a single crystal with preferred crystallographic orientation. In a nutshell, BST can effectively be used for single crystal growth of HEAs.

Figure 3.17a,b shows the crystal orientation or inverse pole figure map from electron back-scatter diffraction (EBSD) for the single crystal $CoCrFeNiAl_{0.3}$ HEA, revealing that the single crystal grows along the <001> direction with narrow and Gaussian distribution of the misorientation [62]. Figure 3.17c illustrates engineering stress-strain plots of both $CoCrFeNiAl_{0.3}$ as-cast and single crystals. The single crystal shows a lower yield and ultimate tensile strength with higher elongation. These features indicate the critical role of grain boundaries in strengthening of HEAs.

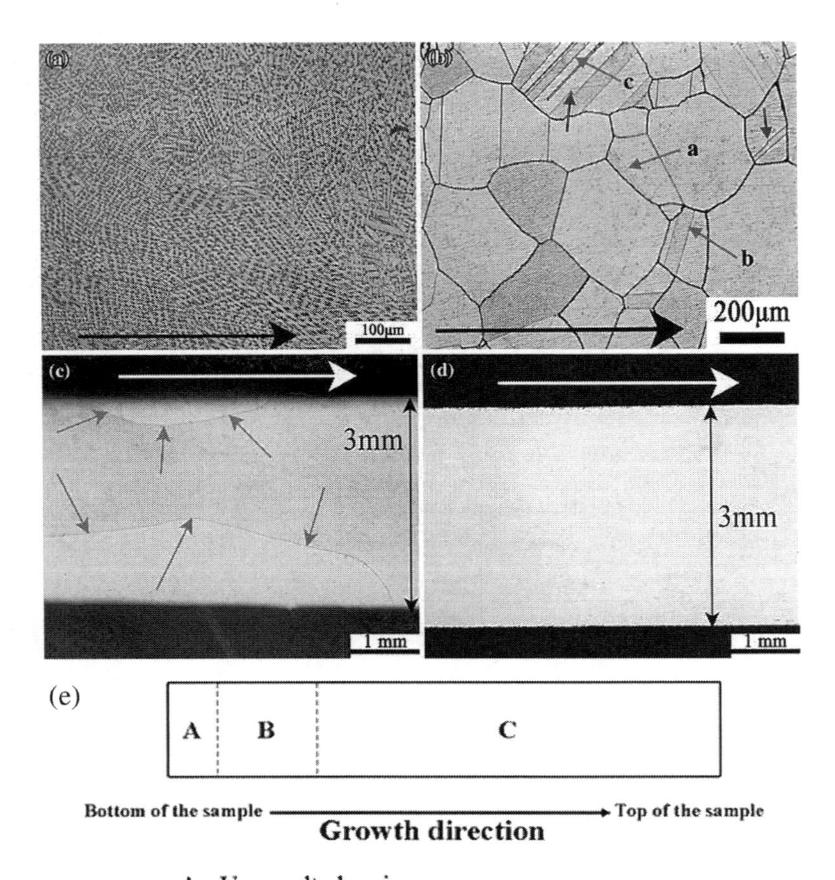

FIGURE 3.16 (a–d) Optical metallographs of $CoCrFeNiAl_{0.3}$ alloy synthesized by BST showing microstructural evolution. (e) Schematic showing growth direction and different zones (from [61]).

3.5 PROCESSING CHALLENGES IN THE POWDER METALLURGY ROUTE

3.5.1 Sintering Methods and Processing Challenges during Sintering

As discussed above, there are certain drawbacks associated with the solidification route, such as casting irregularities (e.g., porosity, segregation), inhomogeneous microstructure, large grain size, etc. Also, synthesizing the alloys having elements with low melting temperature and low vapor pressure is a tedious task (near-impossible) via solidification route. The shortcomings of the solidification and casting route can be overcome using the powder metallurgy route. The powder metallurgy route

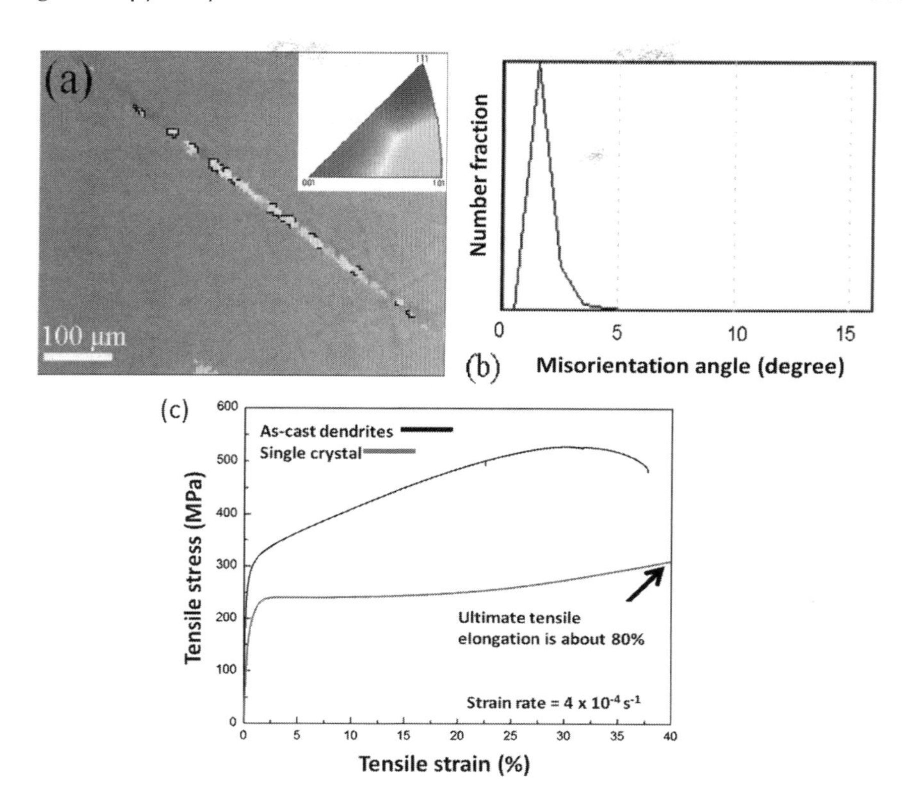

FIGURE 3.17 Engineering stress-strain curve of single crystal and as-cast dendritic $CoCrFeNiAl_{0.3}$ HEA (from [62]).

involves various techniques such as mechanical alloying (MA), atomization method (gas and water atomization). However, there are very few reports available in the literature which uses an atomization method for preparation of HEAs. On the other hand, MA is widely used for the fabrication of HEAs due to its ability to synthesize a wide variety of materials with uniformly distributed homogenous microstructure. MA is a non-equilibrium solid-state powder processing technique which utilizes repeated welding and fracturing of powder particles using high energy ball mills (Figure 3.18) to achieve the alloying of powder particles [63]. MA uses the stored mechanical energy by deformation as a driving force for the activation of powder particles. A variety of high energy mills are available which can be used for MA. Depending upon the type and quantity of powder and desired final constitution, a mill can be chosen. A comparative list of their features is presented in following Table 3.2.

MA is followed by sintering to obtain the final product in the bulk form. For sintering, both the conventional sintering route (such as hot isostatic pressing (HIP), hot pressing (HP), etc.) or the spark plasma sintering route can be used. However, spark plasma sintering (SPS) is preferable to other sintering techniques due to its shorter sintering time, high heating rate, short holding time, limited grain growth, better relative density, and protective environment. SPS uses the pulsed DC current (up to

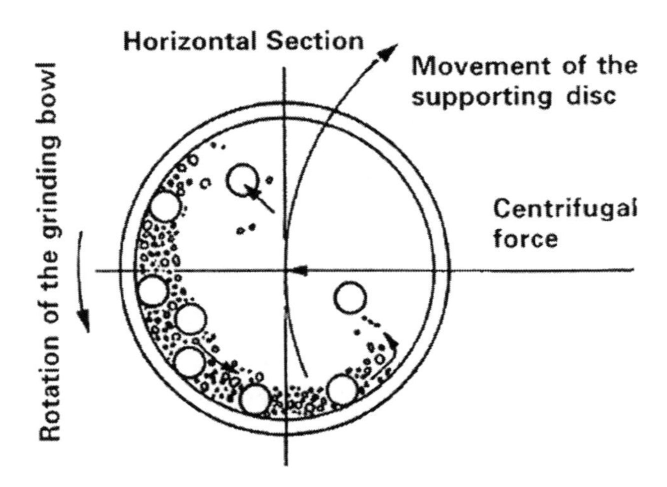

FIGURE 3.18 Schematic of movement of balls (during milling) inside planetary ball mill. (Image courtesy: Gilson Company, Inc., Worthington, OH.) (From [63].)

TABLE 3.2
Comparison of Different Types of Mills

Type of mill	Attributes
SPEX shaker mills	i. High energy mills. ii. Sample size: 10–20 gm at a time. iii. Back and forth shaking motion with combined lateral movement.
Planetary ball mills	i. Planet-like movement, i.e., vials are attached on rotating disk and also rotate on their own axis, opposite to disk movement. ii. Simultaneous application of friction and impact force for milling and alloying. iii. Laboratory scale mill.
Attritor mills	i. Large sample size (0.5 to 40 kg at a time). ii. Uses shear and impact force for milling. iii. Faster than conventional mills.

5,000 A) and uniaxial pressure (30–100 MPa) to achieve the bulk form with stable phases and significant mechanical properties. Spark plasma sintering opens up the possibilities to synthesize the complex microstructure with suitable properties. After solidification and casting, MA followed by SPS is widely used for bulk preparation of HEAs and HEA composites [64].

MA has certain advantages associated with it which makes it interesting to use for HEA preparation. Some of these are listed below:

1. MA is a non-equilibrium technique which involves repeated fracture and welding of powder particles for alloying. MA also leads to activation of powder particles which subsequently helps during sintering.

2. A wide variety of materials such as amorphous, nanocrystalline, intermetallics, ceramics, and metallic alloys can easily be synthesized using MA.
3. It is also possible to synthesize the alloys having elements of different density, low melting temperature, and low vapor pressure using MA, to achieve the homogenous alloy composition and microstructure.

However, there are certain issues associated with MA which includes contamination of alloying powder mix from vial or process control agent (PCA). These issues can be resolved by proper selection of milling media and PCA, optimizing the milling speed and protective inert environment during MA.

For consolidation of MA powder, sintering techniques such as HIP, HP, or SPS can be used, depending upon the required properties. However, conventional sintering methods HIP and HP need a longer time to reach a high sintering temperature which can lead to excessive grain growth and subsequently can affect the mechanical properties of alloy. On the other hand, SPS leads to fully dense alloy with stable phase in short time duration. SPS has certain advantages over other sintering techniques which makes it an interesting choice [4].

Cheng et al. [65] have carried out microstructural and mechanical properties investigation on $FeCoCrNiMnAl_x$ (x = 0–0.5) HEA prepared by MA and hot-pressed sintering. In another similar study, Sun et al. [66] have synthesized non-equiatomic $Fe_{18}Ni_{23}Co_{25}Cr_{21}Mo_8WNb_3C_2$ HEA composite via the MA and HPS route. The MA powder was sintered at three different temperatures of 400°C, 600°C, and 1,050°C during HPS. Phase transition was observed at 600°C temperature where, single austenite phase transforms into austenite A (matrix phase) and M_6C carbide (precipitate phase). Similar behavior was also observed for the 1,050°C temperature. The microstructural investigation revealed the fine grained M_6C particles were distributed on the austenite grain boundary.

In an interesting study, Lv et al. [67] have investigated the microstructure and mechanical behavior of a novel refractory CrMoNbWTi-C HEA. The refractory HEA was fabricated via the MAHPS route. Single-phase BCC solid solution peaks were observed after MA. However, after sintering at 1,450°C, three different phases—the BCC solid-solution phase (volume fraction 57.75%), the fine-grained Lave phase (volume fraction 26.68%), and the high-temperature carbide phase (volume fraction 16.16%)—were observed in the microstructure. The refractory CrMoNbWTi-C HEA also exhibited high fracture strength (3,094 MPa) and hardness (8.26 GPa). The exceptional mechanical properties of HEA were attributed to the precipitation strengthening due to the presence of the intermetallic Lave phase (Cr_2Nb) and carbide phase ((Ti,Nb)C), and grain boundary strengthening.

Colombini et al. [68] in an interesting study have prepared the equimolar FeCoNiCrAl HEA by MA followed by SPS (MASPS) or microwave heating for consolidation. It was observed that samples prepared via SPS had better densification than samples processed via microwave heating. The effect of it also observed on the mechanical properties with SPS processed samples showing better mechanical behavior. Moravcik et al. [69] have carried out a comparative study on the mechanical properties of equiatomic CoCrNi medium entropy alloy prepared via the MASPS route and the casting route. They observed the high tensile strength (1,024 MPa) and elastic modulus (222 GPa) for alloy prepared via MASPS route. In a similar study by Jiang et al. [70], the effect of the synthesis method on the microstructure and mechanical properties of the equiatomic AlCrFeNi medium entropy alloy (MEA) has been investigated. The equiatomic AlCrFeNi MEA was prepared using MASPS, casting and melting, and SPS of elemental powder mix, respectively. Ordered and disordered BCC phases were observed for all three preparation methods, suggesting little influence on the phase formation. However, the microstructure was influenced by synthesis route. A chrysanthemum-like eutectic microstructure was observed in as-cast sample, while SPS samples exhibited a spinodal decomposed microstructure. Sintered MEA exhibited higher yield strength compared to as-cast MEA, which can be attributed to grain boundary strengthening due to the finer grain size of sintered MEA. Mohanty et al. [71] have carried out microstructural investigation on AlCoCrFeNi HEA prepared via MASPS route. Eutectoid transformation (BCC $\rightarrow$ $L1_2$ + σ phase) was observed in mechanically alloyed powders when subjected to spark plasma sintering at the temperature range 973–1,273 K.

MASPS is widely used for synthesis of composites used in tribological application. In an interesting investigation, the first of its kind, Yadav et al. [72–74] from the present research group have successfully designed and synthesized self-lubricating HEA composites with uniformly distributed soft dispersoids in the HEA matrix via the MPSPS route (Figure 3.19). Tribological studies revealed the excellent wear resistance behavior of these HEA composites. Thus, MASPS route can be used to achieve complex microstructures with homogenous distribution of second phases. Studies by different research groups [75–77] on the tribological behavior of HEA composites synthesized via the MASPS route exhibits the potential of the powder metallurgy route.

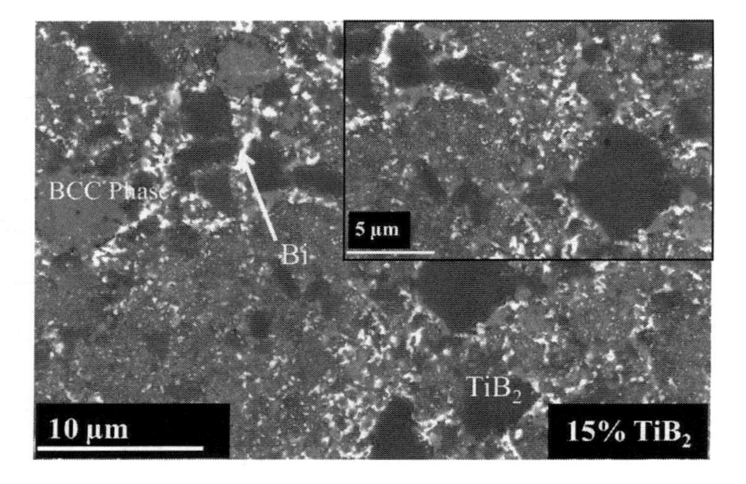

FIGURE 3.19 BSE SEM images of sintered samples of 15 wt-%TiB_2-HEA composite (from [73]).

3.5.2 Choice of Processing Variables

During MA, several variables such as milling time, type of milling media, process control agent, and the milling environment affect the final outcome. There are a few studies available in the literature which have investigated the effect of processing variables on the alloying powder mix. In one such study by Laurent-Brocq et al. [78], the authors investigated the effect of milling condition on the CoCrFeMnNi HEA. The FCC CoCrFeMnNi HEA was milled under planetary and cryomill for different time duration both separately and in combination. The increased milling time resulted in finer crystallite size. It was found that a combination of cryomilling and planetary ball milling followed by SPS leads to nanometric crystallite size, better densification, and microstructural stability. A comparative nanoindentation study shows the increased hardness of sintered HEA compared to casting.

There are very few studies available where PCA has been used during MA [79, 80]. However, contamination due to use of PCA is not observed. PCA (methanol or toluene) was mainly used for prevention of oxidation and cold-welding during milling.

Guo et al. [81] have studied the effect of sintering temperature on the phase formation, microstructural evolution, and mechanical properties of refractory NbTaTiV HEA. The initial powder blend was milled in a protective Ar atmosphere which was then consolidated by sintering at three different temperatures of 1,500°C, 1,600°C, and 1,700°C using spark plasma sintering. The increased sintering temperature does not lead to any phase transformation. A BCC solid-solution phase was observed for all the sintering temperature. However, samples sintered at 1,700°C exhibited better densification and homogenous microstructure. HEA sintered at 1,700°C also exhibited better hardness (510 HV), yield strength (1.37GPa), and compressive fracture strength (2.19 GPa).

In another similar yet separate investigation, Kang et al. [82] and Long et al. [83] have also investigated the effect of varying sintering temperature on the phase stability, microstructure, and mechanical properties of refractory HEAs. In an interesting study, Mane and Panigrahi [84] have studied the effect of alloyed powder treatment on the sintering kinetics. The MA powder was annealed at two different temperatures of 1,100°C and 1,150°C and then subjected to spark plasma sintering. Higher activation energy was observed during sintering for annealed powder compared to milled powder. The higher activation energy was attributed to the decrease in the grain boundaries number and lower defect concentration due to annealing of MA powder. Cheng et al. [85] have investigated the effect of sintering temperature on the phase stability and microstructure of the equiatomic FeCoCrNiMn high entropy alloy prepared via MA and vacuum hot-pressed sintering (VHPS). The X-ray diffraction results of MA powder revealed the presence of major FCC phase (matrix) and minor BCC and amorphous phase. The MA powder was sintered at four different temperatures of 700°C, 800°C, 900°C, and 1,000°C at 50 MPa pressure. During sintering, the BCC phase precipitated out as σ phases and $M_{23}C_6$ carbides. It was observed that increasing sintering temperature led to grain growth in the precipitate (nano to micron). The effect of sintering temperature was also observed on the mechanical properties. With increasing sintering temperature, yield strength decreased (54% decrease), whereas strain to failure ratio increased from 4.4% to 38.2%. The optimum mechanical behavior was observed for FeCoCrNiMn HEA, consolidated at 800°C and 900°C.

3.5.3 Effect of Contamination in Phase Equilibria

Milling media and PCA are primary sources of contamination during MA. The possibility of this can be reduced to a significant level by proper selection of milling media. For example, steel vials can be a possible source of C and Fe contamination if used during the milling of hard materials. This can best be avoided by using vials made of hard materials such as tungsten carbide or zirconia. Wet milling can also lead to carbon-related impurities in the alloy. Most of the HEA reported in the literature (prepared via MA and sintering) does not reveal significant amounts of contamination from the milling source and during sintering. However, there are few reports are available which will be discussed in this section.

Guo et al. [81] have reported the presence of C and O in the sintered NbTaTiV refractory HEA. The presence of impurities was attributed to oxidation of metals during milling and the diffusion of C from the graphite sheet during the sintering. The alloys were consolidated at high sintering temperature of 1,700°C. However, no adverse effect of impurities on the mechanical behavior was observed. Lv et al. [67] have used stainless steel vials for milling of refractory CrMoNbWTi-C HEA. The hard metals resulted in exfoliation of steel vials which then led to contamination of Fe in the matrix. Carbon impurities were also observed which were the result of carbon-based PCA (stearic acid). Thus, use of PCA and milling media resulted in unwanted contamination of Fe and C. However, this contamination was not significant enough to affect the properties of the alloy. Similar studies were also found where the primary source of contamination was PCA or milling media [82, 64, 86, 87].

3.5.4 Microstructure (Micron to Nano-Sized Grains)

Yim et al. [88] have used water atomized powder as a starting material for the mechanical milling of TiC-reinforced CoCrFeMnNi HEA composite. The mechanically milled (MM) powders were then consolidated into bulk shape using SPS. The microstructural investigation and mechanical properties of HEA composite were studied in detail. To study the effect of the TiC addition, water atomized powders were also MM without addition of TiC. It was found out that the TiC addition does not affect the particle size. However, particle morphology is greatly influenced by TiC (from irregular to flaky morphology). The nano-grained TiC also had significance influence on the mechanical properties of the HEA-composite. The addition of TiC leads to increase in yield strength (from 507 MPa to 698 MPa) and fracture strength (from 1,527 MPa to 2,216 MPa) of the HEA composite.

Nam et al. [89] have prepared dual phase (FCC + BCC) $Al_{0.5}CoCrCuFeNiTi_x$ (x = 0, 1, and 2) HEA using powder metallurgy (MA + SPS) route. The presence of Cr-carbide ($Cr_{23}C_6$) and TiC was also observed in sintered samples. High energy milling led to grain refinement at nanoscale. Ti addition also affected the average grain size and a decrease in grain size is observed with increased Ti concentration. The nanocrystalline solid solution and carbide formation helped to achieve good mechanical properties with highest yield strength of 2,877 MPa for $Al_{0.5}CoCrCuFeNiTi_2$ HEA.

In an interesting study, Kang et al. [64] achieved crystallite size of nano level (66.1 nm) after 6 hours of mechanical alloying for equiatomic single-phase BCC WNbMoTaVHEA. The MA HEA was consolidated at 1,500–1,700°C temperature range. The sintering does not lead to any phase transformation; however, a small fraction of oxide inclusions was observed after sintering. These oxide inclusions were the results of oxidation during MA. The mechanical properties of refractory HEA were studied in terms of compressive behavior. The best compressive behavior was observed for a sample sintered at 1,500°C with a yield strength of 2,612 MPa and failure strain of 8.8%. A comparative study with other refractory alloys prepared via casting route also established the superior mechanical behavior of WNbMoTaV HEA prepared via the MASPS route.

Moravcik et al. [90] have carried out an interesting study on the microstructural evaluation and mechanical behavior of $Ni_{1.5}Co_{1.5}CrFeTi_{0.5}$ HEA strengthened by nano-sized Ni_3Ti precipitates and TiO dispersoids. The intermetallic Ni_3Ti was result of precipitation from FCC phase during the sintering. The HEA was subjected to annealing at three different temperatures: 700°C, 900°C, and 1,100°C. The annealing at 700°C and 900°C does not lead to any phase transformation. However, phase fraction, average grain size, and morphology of grains were significantly influenced by annealing. The average grain size of Ni_3Ti precipitates decreased from 40 nm to 33 nm and volume fraction increased from 6.1% to 21.4%, respectively. The increase in size was observed for TiO grains with increased annealing temperature. TiO formed during MA from the initial oxide layer present on the metal surface and remained after sintering due to their high stability. Both the nano-sized intermetallic precipitate and titanium oxide helped the HEA to achieve excellent mechanical properties.

In another yet similar study, Kang et al. [82] have studied the microstructure and mechanical properties of refractory single-phase BCC $Nb_{42}Mo_{20}Ti_{13}Cr_{12}V_{12}Ta_1$HEA prepared via the MASPS route. Microstructural investigation revealed the presence of TiC dispersoids in the BCC matrix, which were formed during the sintering. The primary source of carbon was reported to be a carbon-based process control agent used during milling. The nanometric grain size was attributed to impedance of grain growth due to the presence of carbon dispersoids and the short sintering time. This grain refinement leads to excellent compressive behavior with a yield strength of 2,680 MPa and high Vickers hardness of 741 HV.

Pan et al. [91] have successfully prepared the nano-crystalline single-phase BCC $Nb_{25}Mo_{25}Ta_{25}W_{25}$ and $Ti_8Nb_{23}Mo_{23}Ta_{23}W_{23}$ HEAs via MA. The average crystallite size after MA was 10 nm. The sintering of MA powder resulted in grain size of micron scale. The excellent mechanical behavior of HEA was attributed to fine grain size.

3.5.5 Phase Stability Issues

MA alloying leads to the formation of metastable phases which then transform to stable phases during sintering. Unlike casting, phases formed after sintering are relatively stable and heat treatment is not required to achieve the homogenous microstructure and better properties.

Cao et al. [92] have studied the precipitation behavior of refractory $TiNbTa_{0.5}ZrAl_{0.5}$HEAs. The HEA was synthesized via MA and cold isostatic pressing which was then sintered in a vacuum furnace at 1,300°C. The as-sintered samples were subjected to hot forging at 1,200°C and then subsequently annealed at 600–1,200°C. The X-ray diffraction results show the presence of (Zr, Al)-rich HCP precipitates in the BCC matrix (initial as-sintered phase). Annealing at temperatures greater than 1,000°C leads to coarsening of (Zr, Al)-rich precipitates and decomposition of BCC matrix, which leads to deterioration in compression properties of refractory HEA.

Long et al. [83] studied the phase evolution of refractory NbMoTaWVCr HEA as a function of sintering temperature. MA NbMoTaWVCr HEA was consolidated at temperatures 1,400–1,700°C using spark plasma sintering. MA resulted in the formation of a supersaturated BCC solid-solution phase. However, at temperatures 1,400°C and 1,500°C, the BCC solid solution precipitated $(Cr,V)_2(Ta,Nb)$ (Laves phase) and Ta_2VO_6 particles. Further increases in sintering temperature to 1,600°C and 1,700°C leads to a crystal structure change of Laves phase from C15 to C14, and the volume fraction of it also decreased. An increase in average grain size and BCC volume fraction (76% to 94%) were reported with increased sintering temperature. The mechanical properties of HEA were affected by volume fraction and the size of the Laves phase and oxide particles.

3.6 PROCESSING CHALLENGES IN ADDITIVE MANUFACTURING

AM has substantial superiority over the conventional fabrication techniques but this process has its own deficiencies [93]. The main problem related to AM is heterogeneous microstructure of the alloy leading to anisotropic behavior and inconsistent mechanical properties like poor fatigue life, poor ductility, etc. [94]. Another issue is the print orientation, reported by Entsfellner et al. [95] who found that the change in print orientation altered the stiffness of the printed object. Void formation between the layers, development of residual stresses, pores, hot tears, cracks, swelling, inaccuracies, and defects while transferring CAD into printed objects, are also drawbacks of AM [96]. High energy input at a localized region causes a thermal gradient across the layer and leads to the development of residual stress in the material. Porosity in the AM part could be because of powder, process, or artifact of solidification. Shrinkage porosity (also known as hot tears) forms when sufficient metal does not flow into the melt region [97]. The voids can lead to delamination and affect the mechanical performance of the alloy. The as-deposited alloy consists of columnar grains with elemental segregation at the grain boundaries and this also affects the mechanical properties. Another drawback of AM is poor surface finish of the part which can be attributed to the stair-stepping effect [98]. It can be reduced by minimizing the thickness of layers.

Many HEAs have been fabricated using additive manufacturing (AM) like AlCoCrFeNi [99, 100], Al_xCoCrCuFeNi, CoCrFeNi [101], TiZrNbMoV, ZrTiVCrFeNi with laser-engineered net-shaping (LENS) being the most widely used AM technique for fabricating HEAs [102]. The idea to use AM for HEAs is to avoid the interdendritic segregation in the microstructure that occurs during casting, by increasing the cooling rate [103]. Sun et al. [104] printed equimolar CoCrFeNi HEA using selective laser melting (SLM) and used gas atomized pre-alloyed powder. It was found that intergranular hot cracks were formed during the solidification and the severe residual stresses led to hot tearing of the alloy under tensile loading. Since the APT result showed uniform distribution of all four elements at the crack, the hot tearing was not attributed to elemental segregation. To avoid hot cracking, the fine-grained microstructure can be tailored, because smaller grains provide large grain boundary area and accommodate the residual stresses. Tong et al. [105] prepared equiatomic FeCrCoMnNi HEA using the laser additive manufacturing (LAM) technique, followed by heat treatment under a different laser power. The optical microscope images of the AM sample showed many pores, microvoids and cracks. Increasing the laser power resulted in lesser and smaller pores and microvoids (Figure 3.20).

Joseph et al. [106] studied the effect of hot isostatic pressing (HIP) on the structure and properties of direct laser fabricated (DLF) $Al_xCoCrFeNi$ HEA. The sample with 0.3 at% Al, subjected to HIP revealed no segregation from the as-deposited condition and elimination of second phase which was present at the grain boundaries, leading to improved tensile ductility. Serrated flow behavior, large pores, and grain boundary fracture observed in the as-deposited sample were eradicated because of HIP and true strain at necking increased from 0.38 to 0.6.

In an interesting study, Joseph et al. [107] investigated the difference between arc melted and direct laser fabricated $Al_xCoCrFeNi$ HEA by comparing three different microstructures achieved by varying Al% (x = 0.3, 0.6, and 0.85). The single-phase FCC ($Al_{0.3}$) and BCC ($Al_{0.85}$) alloys showed almost similar properties and microstructure, but the FCC + BCC ($Al_{0.6}$) alloys were drastically different. As shown in Figure 3.21, the arc melted sample shows a dendritic structure, whereas DLF sample shows a Widmanstätten structure which can be attributed to a low-temperature gradient in the melt pool and a low cooling rate (10–100 K/s) during arc melting and high-temperature gradient (10^5–10^7 K/m) and rapid cooling rate (10^3–10^6 K/s) during DLF. The stress-strain curves reveal that DLF single-phase FCC and BCC alloys have better work hardening behavior compared to arc melted alloys, which was attributed to residual thermal stresses associated with AM technique. The DLF alloy with FCC + BCC phases shows lower ductility than arc melted alloy, because of the Widmanstätten structure.

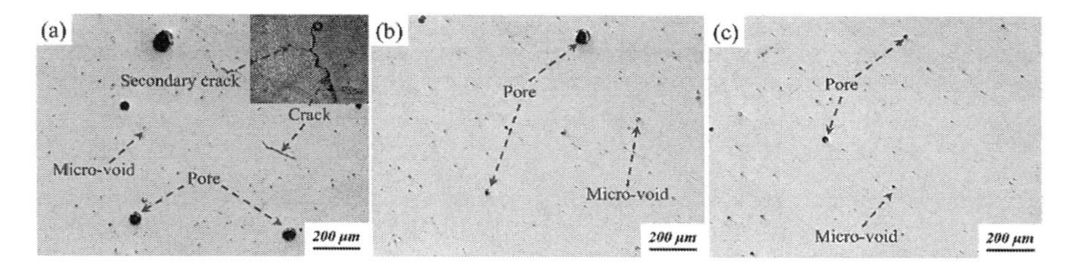

FIGURE 3.20 Optical microscope images of specimens under laser power of (a) 600 W, (b) 800 W, (c) 1,000 W (cross-section is perpendicular to building direction) from [103].

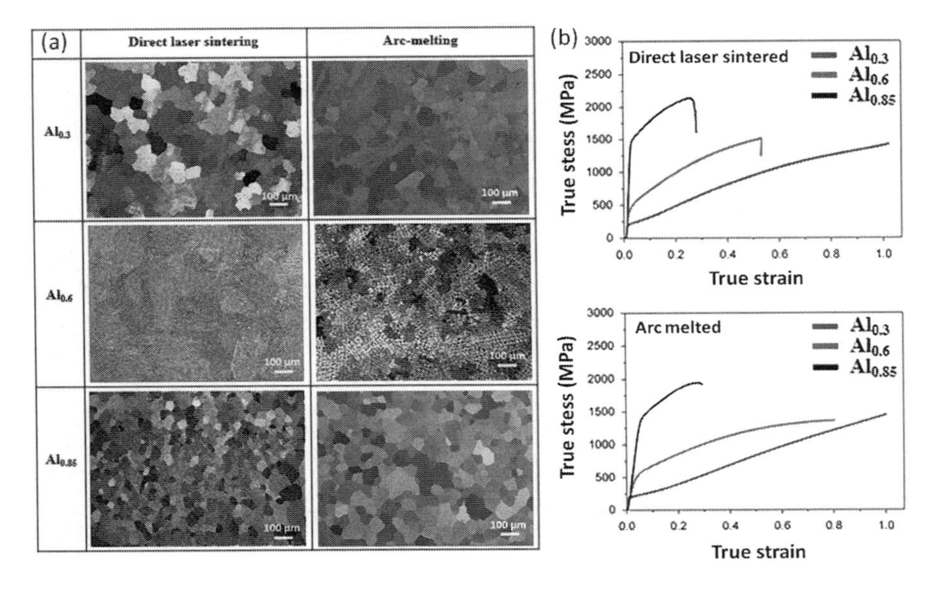

FIGURE 3.21 Comparison between (a) microstructure (normal to cross-section) (b) stress-strain curves of samples fabricated using direct laser and arc melting (from [108]).

Choudhuri et al. [108] also compared the microstructure of CoCrCuFeNiAlTi HEA fabricated using the arc melting and DLF methods and found similar microstructures in both the cases, but finer ones in DLF due to high cooling rate. Borkar et al. [109] investigated the composition–microstructure–microhardness–magnetic property relationship in LENS fabricated compositionally graded $Al_xCrCuFeNi_2$ HEA. A gradual transition from FCC to BCC-based microstructures was observed as a function of Al% with simultaneous increase in hardness. Moreover, it was also observed that the alloy was weakly ferromagnetic at lower Al% and transformed to strongly ferromagnetic at higher Al%.

Li et al. [110] printed the equiatomic CoCrFeMnNi HEA using the selective laser melting (SLM) technique in which lot of porosity was observed. It was found that density of the AM part gradually increased with increasing volumetric energy density (VED) (Figure 3.22). VED can be defined as follows:

$$\text{VED} = \frac{\text{laser power}\,(W)}{\text{scan speed}\left(\frac{\text{mm}}{\text{s}}\right)\times \text{hatching space}\,(\text{mm})\times \text{layer thickness}\,(\text{mm})} \tag{3.12}$$

Increasing VED leads to an increase in temperature of the powder bed. Slow scan speed also increases VED and improves the flowability of the molten pool and thereby enhances layer by layer bonding, finally leading to densification and a good surface finish of the AM part. An increase in VED also led to higher UTS values in the tensile tested sample. The SEM image of the AM part (Figure 3.22) showed

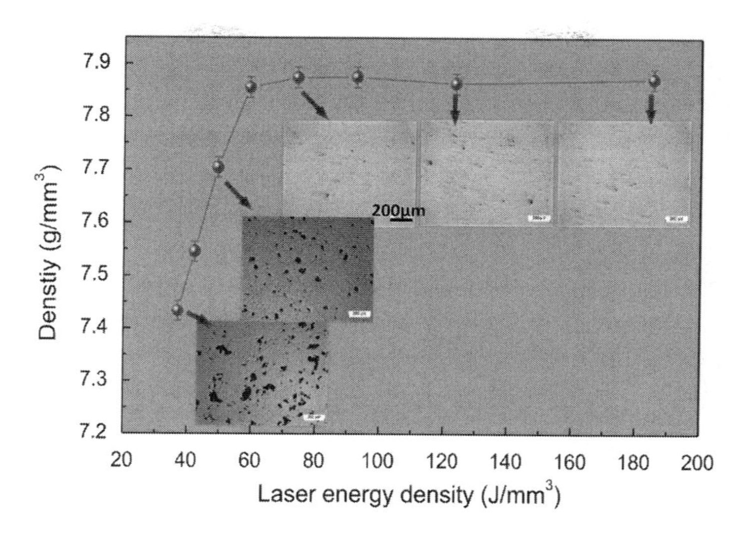

FIGURE 3.22 Effect of laser energy on density and microstructure of AM part (from [110]).

many metallurgical defects like microcracks and micropores. Hot isostatic pressing was carried out on these samples in order to eliminate these defects and increase the density. HIP also led to homogeneous microstructures by removing the Mn segregation at the boundary of the melt pool. The image quality of the Kikuchi pattern was used to analyze residual stress in the AM part. Residual stress could not be minimized using the HIP.

3.7 SUMMARY

The successful synthesis of high entropy alloys (HEAs) in bulk form poses many challenges. These are solid solution-based alloys containing minimum five principal elements, having concentrations varying from 5 to 35 atomic percentage, opening up a vast compositional space to explore for design and development of novel alloys with extraordinary properties. However, most of the research activities are restricted to lab-scale, limiting the prospect of any potential application. Hence, it is important now to move from lab- to industrial-scale production to utilize these alloys for potential applications. In this regard, the present chapter is aimed at providing the processes to synthesize the HEAs in bulk, the challenges involved, and possible remedies. In particular, three processing routes to synthesize bulk HEAs—casting, powder metallurgy, and additive manufacturing—have been discussed in detail.

1. The liquid state route involving casting of these multicomponent alloys is, no doubt, the best possible route to HEAs in bulk form of various sizes. However, the problem of microsegregation and macrosegregation associated with casting can be reduced by proper heat treatment and proper control of solidification process, respectively. The Bridgman solidification technique (BST) can also be used to tackle casting problems by fabrication

of single crystals of HEAs. The control of segregation is considered vital to obtain large casting of solid solution-based multicomponent alloys.

2. The powder metallurgy (P/M) route provides a good alternative to obtain chemically homogeneous HEAs in bulk form with nanostructured microstructures and hence, improvement of properties. Mechanical alloying (MA), followed by spark plasma sintering (SPS) or hot forging, is found to be the most popular route to prepare various HEAs and their composites consisting of multiple phases for various applications (furnace part, tribological, high-temperature components, etc.). However, P/M routes suffer from contamination from the milling media and atmosphere. The consolidation of the MA powder also requires strict process control. Contamination in P/M samples can be reduced by proper selection of the milling media and vial during MA. Proper parameters during SPS results in an alloy with a better microstructure and mechanical property.
3. Additive manufacturing (AM) in various forms (laser beam melting, selective laser melting, laser metal deposition, direct metal deposition, laser engineered net shaping, etc.) has recently been explored to prepare HEAs in bulk form from the powder. Use of a laser beam to melt and solidify HEA powder, widely been utilized for other alloys, can effectively be extended to obtain bulk components of HEAs. However, this route also suffers from microsegregation, voids, and porosity, hampering the properties of AM parts. Voids and porosity in additive manufactured parts can be minimized by increasing the laser energy density. Hot isostatic pressing is also effective in reducing AM defects.

In a nutshell, the successful synthesis of HEAs and their composites in bulk form will form the basis for future prospects of HEAs and hence, research and development activities in this direction will be much sought after in the next five to ten years.

ACKNOWLEDGMENTS

The authors would like to thank the funding agencies, SERB-DST, ISRO, and BRNS for being generous enough to fund the research activities on high entropy alloys at IIT Kanpur.

REFERENCES

1. M.C. Gao, J.W. Yeh, P.K. Liaw, Y. Zhang. *High-Entropy Alloys: Fundamentals and Applications*. Springer (2015).
2. D.B. Miracle, O.N. Senkov. A critical review of high entropy alloys and related concepts. *Acta Materialia* 122 (2017) 448–511.
3. Stephane Gorsse, Daniel B. Miracle, Oleg N. Senkov. Mapping the world of complex concentrated alloys. *Acta Materialia* 135 (2017) 177–187.
4. S. Sharma, S. Yadav, K. Biswas, B. Basu. High-entropy alloys and metallic nanocomposites: Processing challenges, microstructure development and property enhancement. *Materials Science and Engineering: R: Reports* 131 (2018) 1–42.

5. Stéphane Gorsse, Jean-Philippe Couzinié, Daniel B. Miracle. From high-entropy alloys to complex concentrated alloys. *Comptes Rendus Physique* 19(8) (2018) 721–736.
6. Reshma Sonkusare, Aditya Swain, M.R. Rahul, Sumanta Samal, N.P. Gurao, Sudhanshu S. Singh, Krishanu Biswas, N. Nayan. Establishing processing-microstructure-property paradigm in complex concentrated equiatomic CoCuFeMnNi alloy. *Materials Science and Engineering: A* 759 (2019) 415–429.
7. Rani Agarwal, Reshma Sonkusare, Saumya R. Jha, N.P. Gurao, Krishanu Biswas, N. Nayan. Understanding the deformation behavior of CoCuFeMnNi high entropy alloy by investigating mechanical properties of binary ternary and quaternary alloy subsets. *Materials and Design* 157 (2018) 539–550.
8. Reshma Sonkusare, Nimish Khandelwal, Pradipta Ghosh, Krishanu Biswas, N.P. Gurao. A comparative study on the evolution of microstructure and hardness during monotonic and cyclic high pressure torsion of CoCuFeMnNi high entropy alloy. *Journal of Materials Research* 34(5) (2019) 732–743.
9. Prashant K. Sarswat, Sayan Sarkar, Arun Murali, Wenkang Huang, Wenda Tan, Michael L. Free. Additive manufactured new hybrid high entropy alloys derived from the AlCoFeNiSmTiVZr system. *Applied Surface Science* 476 (2019) 242–258.
10. B. Cantor, I.T.H. Chang, P. Knight, A.J.B. Vincent. Microstructural development in equiatomic multicomponent alloys. *Materials Science and Engineering: A* 375–377 (2004) 213–218.
11. J.W. Yeh, S.K. Chen, S.J. Lin, J.Y. Gan, T.S. Chin, T.T. Shun, C.H. Tsau, S.Y. Chang. Nanostructured high-entropy alloys with multiple principal elements: Novel alloy design concepts and outcomes. *Advanced Engineering Materials* 6(5) (2004) 299–303.
12. Zezhou Li, Shiteng Zhao, Robert O. Ritchie, Marc A. Meyers. Mechanical properties of high-entropy alloys with emphasis on face-centered cubic alloys. *Progress in Materials Science* 102 (2019) 296–345.
13. Mageshwari Komarasamy, Tianhao Wang, Kaimiao Liu, Luis Reza-Nieto, Rajiv S. Mishra. Hierarchical multi-phase microstructural architecture for exceptional strength-ductility combination in a complex concentrated alloy via hightemperature severe plastic deformation. *Scripta Materialia* 162 (2019) 38–43.
14. Zhiming Li, K.G. Pradeep, Yun Deng, Dierk Raabe, Cemal CemTasan. Metastable high-entropy dual-phase alloys overcome the strength-ductility trade-off. *Nature* 534(7606) (2016) 227–230.
15. Chin-You Hsu, Chien-Chang Juan, Woei-Ren Wang, Tsing-Shien Sheu, Jien-Wei Yeh, Swe-Kai Chen. *Materials Science and Engineering: A* 528(10–11) (2011) 3581–3588.
16. O.N. Senkov, G.B. Wilks, J.M. Scott, D.B. Miracle. Mechanical properties of Nb25Mo25Ta25W25 and V20Nb20Mo20Ta20W20 refractory high entropy alloys. *Intermetallics* 19(5) (2011) 698–706.
17. N.A.P. Kiran Kumar, C. Li, K.J. Leonard, H. Bei, S.J. Zinkle. Microstructural stability and mechanical behavior of FeNiMnCr high entropy alloy under ion radiation. *Acta Materialia* 113 (2016) 230–244.
18. M.H. Tsai, C.W. Wang, C.W. Tsai, W.J. Shen, J.W. Yeh, J.Y. Gan, W.W. Wu. Thermal stability and performance of NbSiTaTiZr high entropy alloy barrier for copper metallization. *Journal of the Electrochemical Society* 158(11) (2011) 1161–1165.
19. M.H. Chuang, M.H. Tsai, W.R. Wang, S.J. Lin, J.W. Yeh. Microstructure and wear behavior of AlxCo1.5CrFeNi1.5Tiy high entropy alloys. *Acta Materialia* 59(16) (2011) 6308–6317.
20. Ming-Hung Tsai, Jein-Wei Yeh. High-entropy alloys: A critical review. *Materials Research Letters* 2(3) (2014) 107–123.
21. Yong Zhang, Ting Zuo, Zhi Tang, Michael C. Gao, Karin A. Dahmen, Peter K. Liaw, Zhao Ping Lu. Microstructures and properties of high-entropy alloys. *Progress in Materials Science* 61 (2014) 1–93.

22. Duancheng Ma, Blazej Grabowski, Fritz Körmann, Jörg Neugebauer, Dierk Raabe. Ab initio thermodynamics of the CoCrFeMnNi high entropy alloy: Importance of entropy contributions beyond the configurational one. *Acta Materialia* 100 (2015) 90–97.
23. B.S. Murty, J.W. Yeh, S. Ranganathan. *High Entropy Alloys*. Elsevier (2014).
24. K.Y. Tsai, M.H. Tsai, J.W. Yeh. Sluggish diffusion in Co–Cr–Fe–Mn–Ni high-entropy alloys. *Acta Materialia* 61(13) (2013) 4887–4897.
25. Ming-Hung Tsai, Jien-Wei Yeh, Jon-Yiew Gan. Diffusion barrier properties of AlMoNbSiTaTiVZr high-entropy alloy layer between copper and silicon. *Thin Solid Films* 516(16) (2008) 5527–5530.
26. Shou-Yi Chang, Chen-Yuan Wang, Ming-Ku Chen, Chen-En Li. Ru incorporation on marked enhancement of diffusion resistance of multi-component alloy barrier layers. *Journal of Alloys and Compounds* 509(5) (2011) 85–89.
27. Shou-Yi Chang, Dao-Sheng Chen. 10-nm-thick quinary (AlCrTaTiZr)N film as effective diffusion barrier for Cu interconnects at 900C. *Applied Physical Letters* 94 (2009) 231909.
28. M. Vaidya, G. Mohan Muralikrishna, S.V. Divinski, B.S. Murty. Experimental assessment of the thermodynamic factor for diffusion in CoCrFeNi and CoCrFeMnNi high entropy alloys. *Scripta Materialia* 157 (2018) 81–85.
29. Juliusz Dabrowa, Marek Zajusz, Witold Kucza, Grzegorz Cieslak, Katarzyna Berent, Tomasz Czeppe, Tadeusz Kulik, Marek Danielewski. Demystifying the sluggish diffusion effect in high entropy alloys. *Journal of Alloys and Compounds* 783 (2019) 193–207.
30. Josua Kottke, Mathilde Laurent-Brocq, Adnan Fareed, Daniel Gaertner, Loïc Perrière, Łukasz Rogal, Sergiy V. Divinski, Gerhard Wilde. Tracer diffusion in the Ni-CoCrFeMn system: Transition from a dilute solid solution to a high entropy alloy. *Scripta Materialia* 159 (2019) 94–98.
31. Zhipeng Wang, Qihong Fang, Jia Li, Bin Liu, Yong Liu. Effect of lattice distortion on solid solution strengthening of BCC high-entropy alloys. *Journal of Materials Science & Technology* 34(2) (2018) 349–354.
32. S. Wang. Atomic structure modeling of multi-principal-elemental alloys by the principle of maximum entropy. *Entropy* 15(12) (2013) 5536–5548.
33. S. Ranganathan. Alloyed pleasures: Multimetallic cocktails. *Current Science* 85 (2003) 1404–1406.
34. Jien-Wei Yeh. Recent progress in high entropy alloys. *Annales de Chimie – Science des Materiaux* 31(6) (2006) 633648.
35. Y. Zhang, T. Zuo, Y. Cheng, P.K. Liaw. High-entropy alloys with high saturation magnetization, electrical resistivity, and malleability. *Scientific Reports* (2013).
36. Jing Zhang, Yeon-Gil Jung. *Additive Manufacturing*. Elsevier (2018).
37. Andreas Gebhardt. *Understanding Additive Manufacturing*. Copyright © Carl Hanser Verlag (2012).
38. Beenesh Maduriya, N.P. Yadav. Prediction of solidification behaviour of alloy steel ingot casting. *Materials Today: Proceedings* 5(9) (2018) 20380–20390.
39. William D. Callister. *Materials Science and Engineering: An Introduction*, 7th edition. John Wiley & Sons, Inc. (2007).
40. Shagil Akhtar, Mohammad Saad, Mohd Rasikh Misbah, Manish Chandra Sati. Recent advancements in powder metallurgy: A review. *Materials Today: Proceedings* 5(9) (2018) 18649–18655.
41. Ryan T. Shafranek, S. Cem Millik, Patrick T. Smith, Chang-Uk Lee, Andrew J. Boydston, Alshakim Nelson. Stimuli-responsive materials in additive manufacturing. *Progress in Polymer Science* 93 (2019) 36–67.
42. Costanza Culmone, Gerwin Smit, Paul Breedveld. Additive manufacturing of medical instruments: A state-of-the-art review. *Additive Manufacturing* 27 (2019) 461–473.

43. Prashanth Konda Gokuldoss, Sri Kolla, Jürgen Eckert. Additive manufacturing processes: Selective laser melting, electron beam melting and binder jetting—Selection guidelines. *Materials* 10(6) (2017) 672.
44. Dirk Herzog, Vanessa Seyda, Eric Wycisk, Claus Emmelmann. Additive manufacturing of metals. *Acta Materialia* 117 (2016) 371–392.
45. Hooyar Attara, Shima Ehtemam-Haghighi, Damon Kenta, Xinhua Wude, Matthew S. Dargusch. Comparative study of commercially pure titanium produced by laser engineered net shaping, selective laser melting and casting processes. *Materials Science and Engineering: A* 705 (2017) 385–393.
46. Merton C. Flemings. *Solidification Processing*. Copyright © McGraw-Hill (1974).
47. D.A. Porter, K.E. Easterling. *Phase Transformations in Metals and Alloys*, 2nd edition. Chapman and Hall (1992).
48. Sumanta Samal, Krishanu Biswas, Gandham Phanikumar. Solidification behavior in newly designed Ni-rich Ni-Ti-based alloys. *Metallurgical and Materials Transactions A* 47A(12) (2016) 6214–6223.
49. P.H. Wu, N. Liu, W. Yang, Z.X. Zhu, Y.P. Lu, X.J. Wang. Microstructure and solidification behavior of multicomponent CoCrCuxFeMoNi high-entropy alloys. *Materials Science and Engineering: A* 642 (2015) 142–149.
50. Ajit Kumar Mishra, Sumanta Samal, Krishanu Biswas. Solidification behaviour of Ti–Cu–Fe–Co–Ni high entropy alloys. *Transactions of the Indian Institute of Metals* 65(6) (2012) 725–730.
51. Sheng Guo, Chun Ng, C.T. Liu. Anomalous solidification microstructures in Co-free AlxCrCuFeNi2 high-entropy alloys. *Journal of Alloys and Compounds* 557 (2013) 77–81.
52. M.A. Hemphill, T. Yuan, G.Y. Wang, J.W. Yeh, C.W. Tsai, A. Chuang, P.K. Liaw. Fatigue behavior of Al0.5CoCrCuFeNi high entropy alloys. *Acta Materialia* 60(16) (2012) 5723–5734.
53. M. Vaidya, K. Guruvidyathri, B.S. Murty. Phase formation and thermal stability of CoCrFeNi and CoCrFeMnNi equiatomic high entropy alloys. *Journal of Alloys and Compounds* 774 (2019) 856–864.
54. Reshma Sonkusare, P. Divya Janini, N.P. Gurao, S. Sarkar, S. Sen, K.G. Pradeep, Krishanu Biswas. Phase equilibria in equiatomic CoCuFeMnNi high entropy alloy. *Materials Chemistry and Physics* 210 (2018) 269–278.
55. Reshma Sonkusare, Krishanu Biswas, Nowfal Al-Hamdany, H.G. Brokmeier, R. Kalsar, N.P. Gurao. A critical evaluation of heterogeneity in texture and microstructure of high pressure torsion processed CoCuFeMnNi high entropy alloy. Manuscript under preparation.
56. Y.P. Wang, B.S. Li, M.X. Ren, C. Yang, H.Z. Fu. Microstructure and compressive properties of AlCrFeCoNi high entropy alloy. *Materials Science and Engineering: A* 491(1–2) (2008) 154–158.
57. S. Singh, N. Wanderka, B.S. Murty, U. Glatzel, J. Banhart. Decomposition in multi-component AlCoCrCuFeNi high-entropy alloy. *Acta Materialia* 59(1) (2011) 182–190.
58. O.N. Senkov, G.B. Wilks, D.B. Miracle, C.P. Chuang, P.K. Liaw. Refractory high-entropy alloys. *Intermetallics* 18(9) (2010) 1758–1765.
59. X.F. Ding, J.P. Lin, L.Q. Zhang, Y.Q. Su, G.L. Chen. Microstructural control of TiAl–Nb alloys by directional solidification. *Acta Materialia* 60(2) (2012) 498–506.
60. M. Feuerbacher, E. Würtz, A. Kovács, C. Thomas. Single-crystal growth of a FeCoCrMnAl high entropy alloy. *Materials Research Letters* 5(2) (2017) 128–134.
61. S.G. Ma, S.F. Zhang, M.C. Gao, P.K. Liaw, Y. Zhang. A successful synthesis of the CoCrFeNiAl0.3 single-crystal, high-entropy alloy by Bridgman solidification. *Journal of Minerals, Metals and Materials Society* 65(12) (2013) 1751–1758.

62. S.G. Ma, S.F. Zhang, J.W. Qiao, Z.H. Wang, M.C. Gao, Z.M. Jiao, H.J. Yang, Y. Zhang. Superior high tensile elongation of a single-crystal CoCrFeNiAl0.3 high-entropy alloy by Bridgman solidification. *Intermetallics* 54 (2014) 104–109.
63. C. Suryanarayana. Mechanical alloying and milling. *Progress in Materials Science* 46(1–2) (2001) 1–184.
64. C. Velmurugan, V. Senthilkumar, Krishanu Biswas, Surekha Yadav. Densification and microstructural evolution of spark plasma sintered NiTi shape memory alloy. *Advanced Powder Technology* 29(10) (2018) 2456–2462.
65. H. Cheng, X. Liu, Q. Tang, W. Wang, X. Yan, P. Dai. Microstructure and mechanical properties of FeCoCrNiMnAlx high-entropy alloys prepared by mechanical alloying and hot-pressed sintering. *Journal of Alloys and Compounds* 775 (2019) 742–751.
66. C. Sun, P. Li, S. Xi, Y. Zhou, S. Li, X. Yang. A new type of high entropy alloy composite Fe18Ni23Co25Cr21Mo8WNb3C2 prepared by mechanical alloying and hot pressing sintering. *Materials Science and Engineering: A* 728 (2018) 144–150.
67. S. Lv, Y. Zu, G. Chen, X. Fu, W. Zhou. An ultra-high strength CrMoNbWTi-C high entropy alloy co-strengthened by dispersed refractory IM and UHTC phases. *Journal of Alloys and Compounds* 788 (2019) 1256–1264.
68. E.C. Cutiongco, Y.-W. Chung. Prediction of scuffing failure based on competitive kinetics of oxide formation and removal: Application to lubricated sliding of AISI 52100 steel on steel. *Tribology Transactions* 37(3) (1994) 622–628.
69. I. Moravcik, J. Cizek, Z. Kovacova, J. Nejezchlebova, M. Kitzmantel, E. Neubauer, I. Kubena, V. Hornik, I. Dlouhy. Mechanical and microstructural characterization of powder metallurgy CoCrNi medium entropy alloy. *Materials Science and Engineering: A* 701 (2017) 370–380.
70. Z. Jiang, W. Chen, Z. Xia, W. Xiong, Z. Fu. Influence of synthesis method on microstructure and mechanical behavior of Co-free AlCrFeNi medium-entropy alloy. *Intermetallics* 108 (2019) 45–54.
71. S. Mohanty, T.N. Maity, S. Mukhopadhyay, S. Sarkar, N.P. Gurao, S. Bhowmick, Krishanu Biswas. Powder metallurgical processing of equiatomic AlCoCrFeNi high entropy alloy: Microstructure and mechanical properties. *Materials Science and Engineering: A* 679 (2016) 299–313.
72. Surekha Yadav, Arvind Kumar, Krishanu Biswas. Wear behavior of high entropy alloys containing soft dispersoids (Pb, Bi). *Materials Chemistry and Physics* 210 (2018) 222–232.
73. Surekha Yadav, Akash Aggrawal, Arvind Kumar, Krishanu Biswas. Effect of TiB2 addition on wear behavior of (AlCrFeMnV)90Bi10 high entropy alloy composite. *Tribology International* 132 (2019) 62–74.
74. Surekha Yadav, S. Sarkar, Akash Aggarwal, Arvind Kumar, Krishanu Biswas. Wear and mechanical properties of novel (CuCrFeTiZn)100-xPbx high entropy alloy composite via mechanical alloying and spark plasma sintering. *Wear* 410–411 (2018) 93–109.
75. Aijun Zhang, Jiesheng Han, Bo Su, Pengde Li, Junhu Meng. Microstructure, mechanical properties and tribological performance of CoCrFeNi high entropy alloy matrix self-lubricating composite. *Materials & Design* 114 (2016) 253–263.
76. Aijun Zhang, Jiesheng Han,Bo Su,Junhu Meng. A Novel CoCrFeNi high entropy alloy matrix self-lubricating composite. *Journal of Alloys and Compounds* 725 (2017) 700–710.
77. Arun Raphel, S. Kumaran, K. Vinoadh Kumar, Lovin Varghese. Oxidation and corrosion resistance of AlCoCrFeTi high entropy alloy materials today. *Proceedings* 4 (2017) 195–192.

78. M. Laurent-Brocq, P.A. Goujon, J. Monnier, B. Villeroy, L. Perrière, R. Pirès, G. Garcin. Microstructure and mechanical properties of a CoCrFeMnNi high entropy alloy processed by milling and spark plasma sintering. *Journal of Alloys and Compounds* 780 (2019) 856–865.
79. V. Shivam, J. Basu, V.K. Pandey, Y. Shadangi, N. Mukhopadhyay. Alloying behaviour, thermal stability and phase evolution in quinary AlCoCrFeNi high entropy alloy. *Advanced Powder Technology* 29(9) (2018) 2221–2230.
80. A. Raza, H.J. Ryu, S.H. Hong. Strength enhancement and density reduction by the addition of Al in CrFeMoV based high-entropy alloy fabricated through powder metallurgy. *Materials & Design* 157 (2018) 97–104.
81. W. Guo, B. Liu, Y. Liu, T. Li, A. Fu, Q. Fang, Y. Nie. Microstructures and mechanical properties of ductile NbTaTiV refractory high entropy alloy prepared by powder metallurgy. *Journal of Alloys and Compounds* 776 (2019) 428–436.
82. B. Kang, J. Lee, H.J. Ryu, S.H. Hong. Ultra-high strength WNbMoTaV high-entropy alloys with fine grain structure fabricated by powder metallurgical process. *Materials Science and Engineering: A* 712 (2018) 616–624.
83. Y. Long, X. Liang, K. Su, H. Peng, X. Li. A fine-grained NbMoTaWVCr refractory high-entropy alloy with ultra-high strength: Microstructural evolution and mechanical properties. *Journal of Alloys and Compounds* 780 (2019) 607–617.
84. R.B. Mane, B.B. Panigrahi. Comparative study on sintering kinetics of as-milled and annealed CoCrFeNi high entropy alloy powders. *Materials Chemistry and Physics* 210 (2018) 49–56.
85. H. Cheng, Y.C. Xie, Q.H. Tang, C. Rao, P.Q. Dai. Microstructure and mechanical properties of FeCoCrNiMn high-entropy alloy produced by mechanical alloying and vacuum hot pressing sintering. *Transactions of Nonferrous Metals Society of China* 28(7) (2018) 1360–1367.
86. B. Kang, T. Kong, A. Raza, H.J. Ryu, S.H. Hong. Fabrication, microstructure and mechanical property of a novel Nb-rich refractory high-entropy alloy strengthened by in-situ formation of dispersoids. *International Journal of Refractory Metals and Hard Materials* 81 (2019) 15–20.
87. R. Anand Sekhar, S. Samal, N. Nayan, S.R. Bakshi. Microstructure and mechanical properties of Ti-Al-Ni-Co-Fe based high entropy alloys prepared by powder metallurgy route. *Journal of Alloys and Compounds* 787 (2019) 123–132.
88. D. Yim, P. Sathiyamoorthi, S.J. Hong, H.S. Kim. Fabrication and mechanical properties of TiC reinforced CoCrFeMnNi high-entropy alloy composite by water atomization and spark plasma sintering. *Journal of Alloys and Compounds* 781 (2019) 389–396.
89. S. Nam, M.J. Kim, J.Y. Hwang, H. Choi. Strengthening of Al0.15CoCrCuFeNiTix–C (x= 0, 1,2) high-entropy alloys by grain refinement and using nanoscale carbides via powder metallurgical route. *Journal of Alloys and Compounds* 762 (2018) 29–37.
90. I. Moravcik, L. Gouvea, V. Hornik, Z. Kovacova, M. Kitzmantel, E. Neubauer, I. Dlouhy. Synergic strengthening by oxide and coherent precipitate dispersions in high-entropy alloy prepared by powder metallurgy. *Scripta Materialia* 157 (2018) 24–29.
91. J. Pan, T. Dai, T. Lu, X. Ni, J. Dai, M. Li. Microstructure and mechanical properties of Nb25Mo25Ta25W25 and Ti8Nb23Mo23Ta23W23 high entropy alloys prepared by mechanical alloying and spark plasma sintering. *Materials Science and Engineering: A* 738 (2018) 362–366.
92. Y. Cao, Y. Liu, Y. Li, B. Liu, J. Wang, M. Du, R. Liu. Precipitation strengthening in a hot-worked TiNbTa0.5ZrAl0.5 refractory high entropy alloy. *Materials Letters* 246 (2019) 186–189.

93. J.P. Kruth, M.C. Leu, T. Nakagawa. Progress in additive manufacturing and rapid prototyping. *CIRP Annals* 47(2) (1998) 525–540.
94. R. Nowell, B. Shirinzadeh, L. Lai, J. Smith, Y. Zhong. Design of a 3-DOF parallel mechanism for the enhancement of endonasal surgery. *IEEE/ASME International Conference on Advanced Intelligent Mechatronics, AIM* (2017) 749–754.
95. K. Entsfellner, I. Kuru, T. Maier, J.D.J. Gumprecht, T.C. Lueth. First 3D printed medical robot for ENT surgery - Application specific manufacturing of laser sintered disposable manipulators. *IEEE International Conference on Intelligent Robots and Systems* (2014) 4278–4283.
96. Tuan D. Ngo, Alireza Kashani, Gabriele Imbalzano, Kate T.Q. Nguyen, David Hui. Additive manufacturing (3D printing): A review of materials, methods,applications and challenge. *Composites Part B: Engineering* 143 (2018) 172–196.
97. W.J. Sames, F.A. List, S. Pannala, R.R. Dehoff, S.S. Babu. The metallurgy and processing science of metal additive manufacturing. *International Materials Reviews* 61(5) (2016) 315–360.
98. Rasheedat M. Mahamood, Esther T. Akinlabi, Mukul Shukla, Sisa Pityana. Revolutionary additive manufacturing: An overview. *Lasers in Engineering* 27(3) (2014) 161–178.
99. I. Kunce, M. Polanski, K. Karczewski, T. Plocinski, K.J. Kurzydlowski. Microstructural characterisation of high-entropy alloy AlCoCrFeNi fabricated by laser engineered net shaping. *Journal of Alloys and Compounds* 648 (2015) 751–758.
100. Tadashi Fujieda, Hiroshi Shiratori, Kosuke Kuwabara, Takahiko Kato, Kenta Yamanaka, Yuichiro Koizumi, Akihiko Chiba. First demonstration of promising selective electron beam melting method for utilizing high-entropy alloys as engineering materials. *Materials Letters* 159 (2015) 12–15.
101. Yevgeni Brif, Meurig Thomas, Iain Todd. The use of high-entropy alloys in additive manufacturing. *Scripta Materialia* 99 (2015) 93–96.
102. Stéphane Gorsse, Christopher Hutchinson, Mohamed Gouné, Rajarshi Banerjee. Additive manufacturing of metals: A brief review of the characteristic microstructures and properties of steels, Ti-6Al-4V and high-entropy alloys. *Journal of Science and Technology of Advanced Materials* 18(1) (2017) 584–610.
103. Brian A. Welk, Robert E.A. Williams, Gopal B. Viswanathan, Mark A. Gibson, Peter K. Liaw, Hamish L. Fraser. Nature of the interfaces between the constituent phases in the high entropy alloy CoCrCuFeNiAl. *Ultramicroscopy* 134 (2013) 193–199.
104. Z. Sun, X.P. Tan, M. Descoins, D. Mangelinck, S.B. Tor, C.S. Lim. Revealing hot tearing mechanism for an additively manufactured high-entropy alloy via selective laser melting. *Scripta Materialia* 168 (2019) 129–133.
105. Zhaopeng Tong, Xudong Ren, Jiafei Jiao, Wangfan Zhou, Yunpeng Ren, Yunxia Ye, Enoch Asuako Larson, Jiayang Gu. Laser additive manufacturing of FeCrCoMnNi high-entropy alloy: Effect of heat treatment on microstructure, residual stress and mechanical property. *Journal of Alloys and Compounds* 785 (2019) 1144–1159.
106. Jithin Joseph, Peter Hodgson, Tom Jarvis, Xinhua Wu, Nicole Stanford, Daniel Mark Fabijanic. Effect of hot isostatic pressing on the microstructure and mechanical properties of additive manufactured AlxCoCrFeNi high entropy alloys. *Materials Science and Engineering: A* 733 (2018) 59–70.
107. Jithin Joseph, Tom Jarvis, Xinhua Wu, Nicole Stanford, Peter Hodgson, Daniel Mark Fabijanic. Comparative study of the microstructures and mechanical properties of direct laser fabricated and arc-melted AlxCoCrFeNi high entropy alloys. *Materials Science and Engineering: A* 633 (2015) 184–193.
108. D. Choudhuri, T. Alam, T. Borkar, B. Gwalani, A.S. Mantri, S.G. Srinivasan, M.A. Gibson, R. Banerjee. Formation of a Huesler-like L21 phase in a CoCrCuFeNiAlTi high-entropy alloy. *Scripta Materialia* 100 (2015) 36–39.

109. T. Borkar, B. Gwalani, D. Choudhuri, C.V. Mikler, C.J. Yannetta, X. Chen, R.V. Ramanujan, M.J. Styles, M.A. Gibson, R. Banerjee. A combinatorial assessment of AlxCrCuFeNi2 ($0 < x < 1.5$) complex concentrated alloys: Microstructure, microhardness, and magnetic properties. *Acta Materialia* 116 (2016) 63–76.
110. Ruidi Li, Pengda Niu, Tiechui Yuan, Peng Cao, Chao Chen, Kechao Zhou. Selective laser melting of an equiatomic CoCrFeMnNi high-entropy alloy: Processability, non-equilibrium microstructure and mechanical property. *Journal of Alloys and Compounds* 746 (2018) 125–134.
111. Joseph R. Davis. *Metals Handbook Desk Edition*, 2nd edition. Taylor & Francis. (1998).

Section B

Advances

4 Effect of Carbon Addition on the Microstructure and Properties of CoCrFeMnNi High Entropy Alloys

Rahul Ravi and Srinivasa Rao Bakshi

CONTENTS

4.1 INTRODUCTION

The discovery of high entropy alloys has enlarged the horizons of material science toward exotic materials with unique properties [1, 2]. A lot of work has been done to date in these multicomponent systems, predominantly aimed at synthesizing these materials and making them viable for commercial use by tailoring their properties through thermomechanical processing [3–5]. CoCrFeMnNi alloy (also known as Cantor alloy) is one of the primary alloy systems in this class. It has a disordered single-phase FCC structure strengthened by lattice distortion and solid-solution strengthening [1, 6]. The strengthening is largely due to the variations in size and modulus of the constituent atoms. In spite of this, CoCrFeMnNi still has relatively lower yield strength for structural applications [7, 8]. The room temperature yield strength of equiatomic CoCrFeMnNi is found to be in the range of 200–350 MPa for a grain size range of 144–4.4 μm [8]. Hence, research on CoCrFeMnNi has traditionally been oriented toward strengthening the alloy through compositional modification and thermomechanical treatment [8–11]. Strengthening can be further enhanced by the presence of interstitial solute additions of C, N, and B [12–14]. The equivalence of the CoCrFeMnNi crystal structure with that of high manganese steel suggests that carbon should be a promising candidate for interstitial addition to CoCrFeMnNi with the aim of enhancing its mechanical and structural properties with a special emphasis on minimizing the strength–ductility trade-off as much as possible [15–17].

Addition of interstitial solutes, especially carbon, brings about profound changes in CoCrFeMnNi. The primary effect of interstitial atoms on the FCC lattice of CoCrFeMnNi is to introduce huge lattice distortions into the structure. These interstitial solute atoms induce a higher work hardening rate compared to substitutional solute additions. This is because substitutional solute additions give rise to spherically symmetric stress fields which interact only with defects that have a hydrostatic component in their stress field, namely the edge dislocations, whereas interstitial additions result in spherically symmetric stress fields, as well as stress fields which induce a tetragonal distortion in the lattice. In other words, interstitial solute additions can interact with both edge and screw dislocations [18].

This review deals with the effect of interstitial carbon addition at a lattice level, microstructural level, and macro level, as shown schematically in Figure 4.1. At a lattice level, this review describes the effect of carbon on the lattice of CoCrFeMnNi, namely, variation in stacking fault energy, solid-solution strengthening, and the effect of interstitial carbon on dislocation motion. At a microstructural level, formation of carbide precipitates and its effects on the microstructure are described. The effect

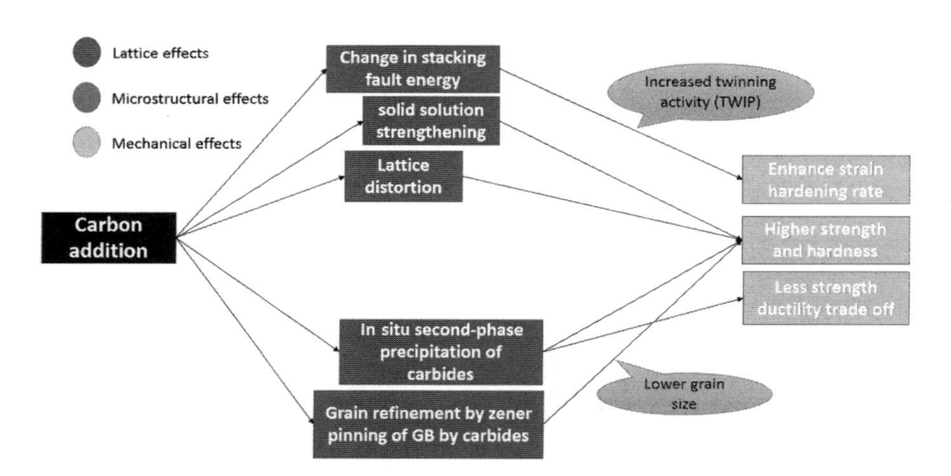

FIGURE 4.1 Flowchart showing the effect of carbon addition on high entropy alloys. The effects are classified into: lattice effects, microstructural effects, and mechanical effects.

of carbon content, annealing temperature, and annealing time on the shape, size, and distribution of carbide precipitates are also discussed. Furthermore, the effect of carbon on the mechanical properties such as strain hardening rate, yield strength, hardness, and tensile strength is also reviewed.

Powder metallurgy as a processing route for carbon addition to high entropy alloys has not been explored to its full potential. Hence, as a first step in this direction, work done on the addition of graphite flakes to mechanically alloyed CoCrFeMnNi has been discussed together with the evaluation of sintered compacts using spark plasma sintering. The effects of graphite flake addition on the powder size, powder morphology, and phase evolution in powders during mechanical alloying and on the microstructure of the compacts after spark plasma sintering have been presented in detail.

4.1.1 Effect of Carbon Addition at a Lattice Level

A primary effect of interstitial solute additions like carbon is to modify the stacking fault energy (SFE) of the lattice which can lead to a change in the deformation mechanism of the alloy. Manipulating stacking fault energy has always been an adopted route to counter the loss in ductility inherent to strengthening [19–21]. The importance of SFE stems from the fact that, when optimized, it is a handle to invoke deformation mechanisms like transformation induced plasticity (TRIP) and twinning induced plasticity (TWIP) [22–25]. This leads to strengthening and a high degree of work hardening of the alloy, especially at cryogenic temperatures [26]. Strengthening happens due to the decrease in the mean free path of moving dislocations due to the incorporation of structural features like nanoscale twin boundaries in case of TWIP effect or a deformation induced displacive transformation from the FCC γ matrix to the HCP ε phase as in a TRIP dual phase HEA [27–30. In such a scenario, there is minimal compromise to ductility as the enhanced work hardening delays the onset of necking.

Generally, solute additions decrease the stacking fault energy of a conventional alloy at a given temperature [31]. In case of CoCrFeMnNi, there are contrasting reports from Wu et al. [26] and Stepanov et al. [32] regarding the effect of carbon addition on the stacking fault energy of CoCrFeMnNi alloy. Wu et al. have reported enhanced yield strength (YS) and tensile strength (TS) when 0.5 at.% C was added to CoCrFeMnNi. The reason was primarily attributed to the formation of nano-twins resulting from the decrease in stacking fault energy by carbon addition. This was also the first instance of carbon addition to CoCrFeMnNi. It should be noted that CoCrFeMnNi is reported to show increased strength, ductility, and fracture toughness at low temperatures [8, 33]. With the addition of carbon, a similar trend was observed wherein TS at 77 K increased from 820 MPa to 1,090 MPa (33% increase) with a 40% elongation to fracture. Stepanov et al. [32] noticed slower twinning kinetics and higher dislocation activity upon 1 at.% C addition to cold rolled-annealed CoCrFeMnNi as shown in Figure 4.2. They reasoned that, since room temperature SFE values of CoCrFeMnNi fall in the range of 18.3–27.3 mJ/m^2 [22, 34], which is below than that of TWIP steels [35] and stainless steels [36], each having a strong compositional and mechanical similarity with CoCrFeMnNi, it can be proposed that carbon addition raises SFE values, thus retarding twin initiation. In contrast,

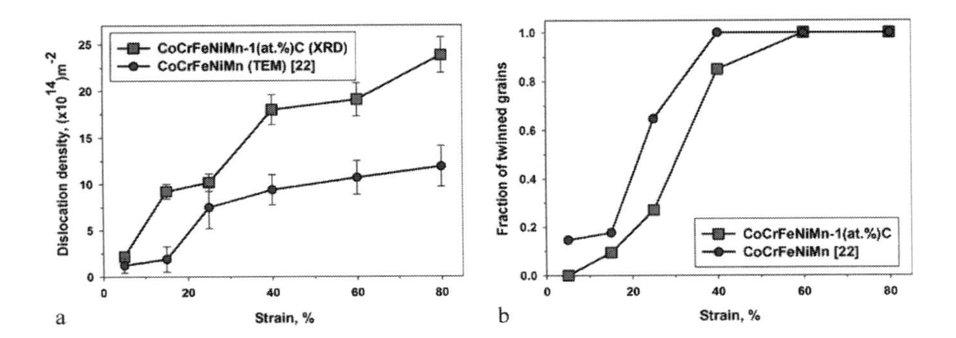

FIGURE 4.2 Higher dislocation densities (a) and slower twinning fraction (b) observed with carbon addition on CoCrFeMnNi [32] with increasing strain. Carbon-containing CoCrFeMnNi is compared with CoCrFeMnNi [30].

Ko et al. [37] have reported a higher twinning tendency with carbon content, as shown in Figure 4.3 where the twin boundary/grain boundary ratio and grain size are plotted as a function of carbon content in the alloy. Some other works showing such contradictory studies on the effect of SFE variation on twinning tendency are also worth noting [38–42].

CoCrFeMnNi deforms by planar dislocation glide during the initial stages of deformation. At later stages of deformation, there is activation of twinning mode under favorable conditions. It has been reported that at room temperature, the twinning mode is activated as late as when the strain reaches fracture stage [27]. At very high temperatures and strains, dislocation motion is aided by microbands [39]. Carbon addition reduces the critical stress for twinning. Hence, twinning gets activated at lower values of strain. Minor carbon addition at room temperature (<0.1 at.%)

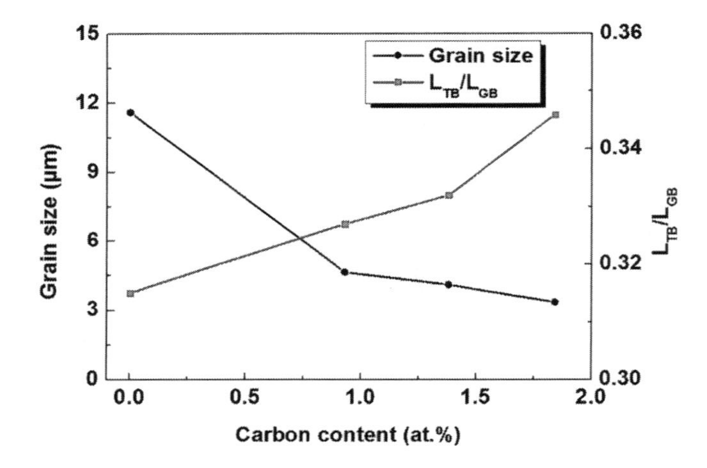

FIGURE 4.3 Changes of twin boundary fraction (L_{TB}/L_{GB}) and grain size with increasing carbon content for annealed CoCrFeMnNi. The grain size decreases while the fraction of twin boundaries increases with carbon content [37].

essentially changes the deformation mode from dislocation glide dominated plasticity to a combination of dislocation glide and twinning/microbands, as reported by Chen et al. [43] and shown in Figure 4.4.

Initially, for the carbon-free CoCrFeMnNi alloy, the TEM image shows dislocations piling up and fragmenting the grain with no evidence of twins (see Figure 4.4a). With a slight carbon addition, deformation twins could be noticed (see Figure 4.4b). This leads to the "dynamic Hall–Petch effect" resulting in increase of both strength and hardness. With carbon addition, planarity of the dislocation increases. This phenomenon is also noticed in Fe-30Mn alloy, as shown in Figure 4.5 [44].

A well-developed dislocation cell structure can be seen in Fe-30Mn (Figure 4.5a), while Fe-22Mn-0.6C shows a large number of twins (Figure 4.5b). Interstitial additions can segregate to twin boundaries and pin them in the process increasing the stability of the twin boundary. Interstitials hinder cross-slip or combination of partial

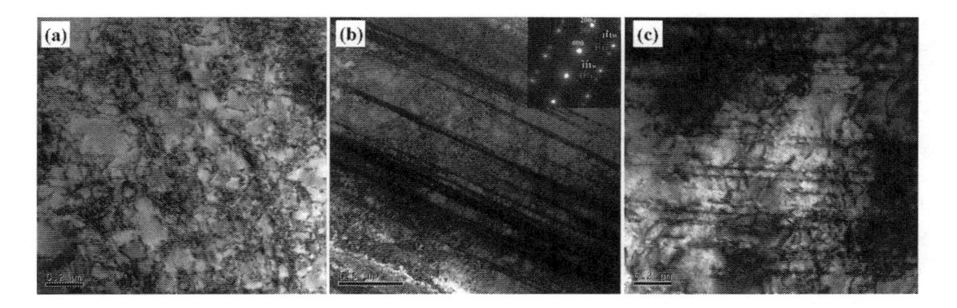

FIGURE 4.4 TEM image for $CoCrFeMnNiC_x$ alloys for carbon contents (a) x=0; (b) x=0.05; (c) x = 0.1 undergoing a tensile strain of 0.25 taken from literature [43]. Inset of SAD pattern is also shown.

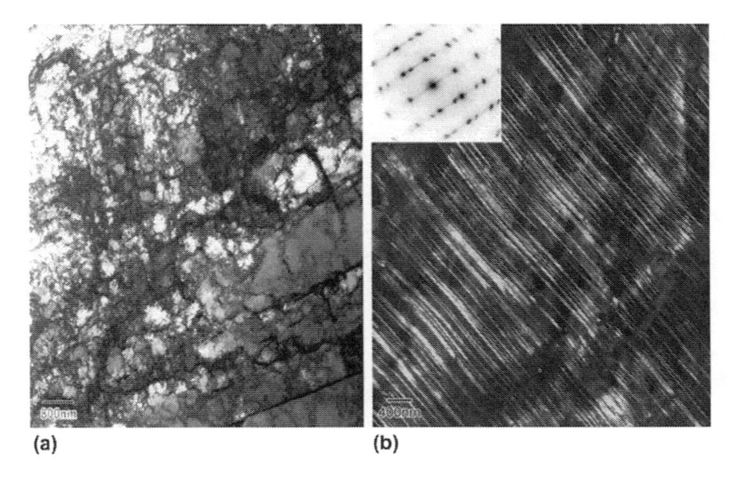

FIGURE 4.5 TEM images of (a) dislocation cells in Fe-30Mn deformed to 20% strain in tension and (b) mechanical twins in Fe-22Mn-0.6C when deformed to 50% taken from the literature [44].

dislocation and hence promotes slip and twinning over the formation of dislocation cell structures which need an excessive cross-slipping tendency [45].

4.1.2 Effect of Carbon Addition at the Microstructural Level

So far, the changes at a lattice level upon carbon addition to CoCrFeMnNi have been discussed. Now, at a microstructural level, carbon addition gives raise to nanoscale carbide precipitates which leads to strengthening and grain refinement. The carbon is usually added in the form of graphite powder or as carbon in alloyed form. The primary fabricating routes employed have been arc melting or casting. Precipitation hardening is always preferred over solid-solution strengthening because of the extent to which a material can be strengthened [46]. Also, it is easier to control the properties of the material by adjusting the shape, size, and distribution of the precipitates through processing routes and/or thermomechanical treatment. Carbon addition to CoCrFeMnNi gives rise to $M_{23}C_6$ and M_7C_3 type carbides, which precipitate from the microstructure when annealed at a temperature of 600°C or above. This is corroborated by the phase diagram of the $FeCoCrNiMnC_{0.1}$ HEA calculated by Thermo-Calc software [32] (Figure 4.6) and also from the XRD pattern of CoCrFeMnNi alloy in cast condition with (Figure 4.7b) and without carbon addition (Figure 4.7a) of 1.84 at.% [37]. Figure 4.6 shows that a mixture of FCC phase and $M_{23}C_6$ carbides is stable at low temperatures and M_7C_3 replaces the $M_{23}C_6$ at higher temperatures.

It has been observed that up to 1 at.% carbon addition, carbide particles form on grain boundaries in the FCC structure of as-cast CoCrFeMnNi and as carbon content increases, carbide particles are seen nucleating in grain interiors too [37]. This is shown in Figure 4.8 where carbon-free alloy has no carbides visible in grain boundaries or grain interiors, whereas as carbon content increases from 0–0.93 at.%

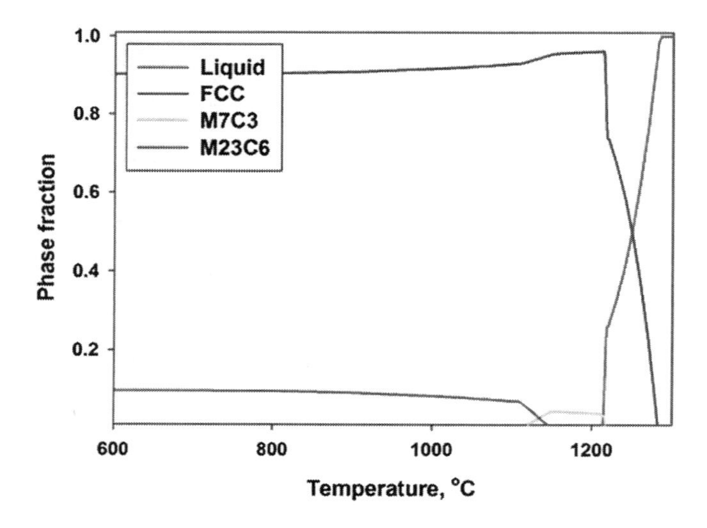

FIGURE 4.6 Phase diagram of the $FeCoCrNiMnC_{0.1}$ alloy calculated by Thermo-Calc software [40].

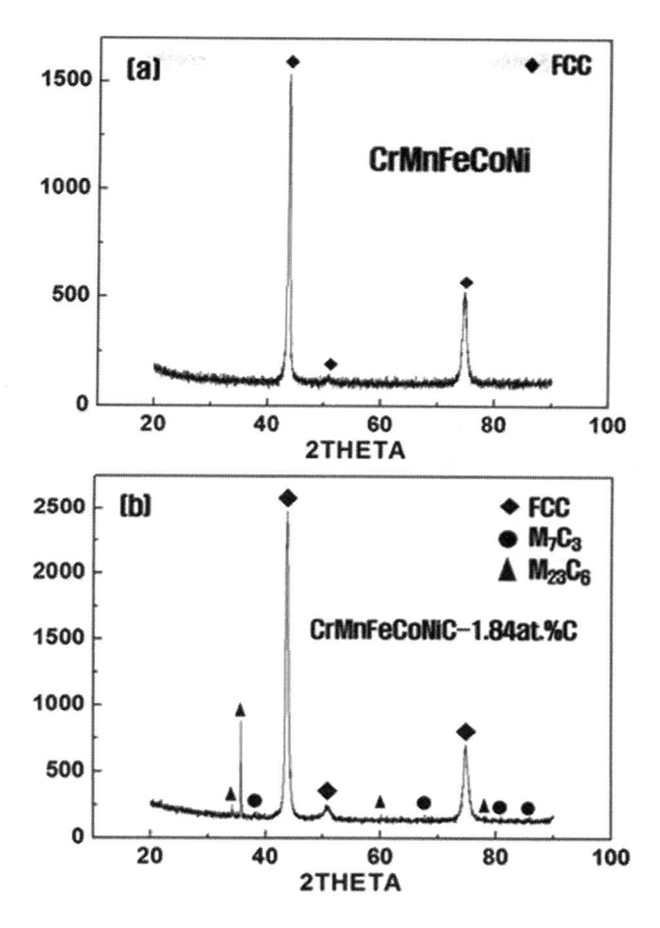

FIGURE 4.7 XRD diffraction pattern from (a) as-cast CoCrFeMnNi alloy, and (b) CoCrFeMnNi alloy with 1.84 at.% carbon [37].

in subsequent images (Figure 4.8c–f), the carbide precipitates can be seen to nucleate preferentiatlly at the grain boundaries. At high carbon content (Figure 4.8g–h) of above 1.38 at.%, carbide precipitates can also be seen in the grain interiors. Cold rolling above 80% strain can disperse the carbides in the matrix.

There is significant increase in strength and hardness upon carbon addition, accompanied by a reduction in ductility. To make the alloy suitable for applications, annealing and thermomechanical treatments have to be carried out. Stepanov et al. [40] have carried out elaborate annealing studies at different temperatures on arc melted CoCrFeMnNi added with varying carbon content. From these studies, the solubility of carbon in CoCrFeMnNi was found to be 1.85–2 at.% at room temperature by XRD lattice parameter measurements. Also, M_7C_3- and $M_{23}C_6$-type carbides were detected. Annealing at 800°C and beyond gave rise to $M_{23}C_6$ type carbides, as shown in Figure 4.9. The $M_{23}C_6$- type carbide precipitates become predominant at higher carbon addition and annealing temperature.

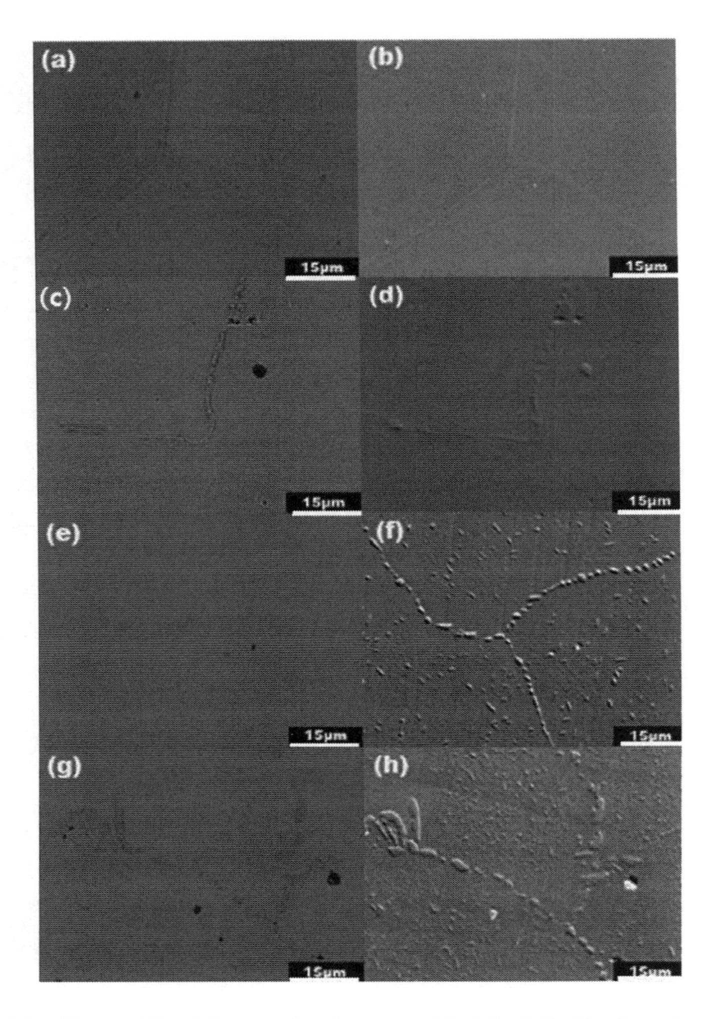

FIGURE 4.8 Compositional (a, c, e, g) and topographical (b, d, f, g) back-scattered electron (BSE) images of carbon-free (a, b) and carbon-containing CoCrFeMnNi alloy with 0.93 at.% (c, d), 1.38 at.% (e, f), and 1.84 at.% (g, h) carbon [37]. Carbide precipitation can be noticed along the grain boundary and grain interiors at higher carbon contents.

The most pronounced change in microhardness and lattice parameter was observed at an annealing temperature of 800°C, as shown in Figure 4.10. This shows that 800–900°C would be an ideal temperature for annealing studies for CoCrFeMnNi. The increase in hardness (Figure 4.10a) and the change in lattice parameter (Figure 4.10b) is due to the precipitation of $M_{23}C_6$ carbides having an average size of 40–80 nm as observed in TEM images. The carbides formed in the as-solidified condition of the alloy were coarse and inhomogeneous, as seen in Figure 4.11. Carbides formed here are M_7C_3 as per the SAED patterns provided in the inset. In Figure 4.12, coarse carbide particles could be seen evenly distributed. These particles are formed due to annealing at 800°C for 14 hours. SAED patterns reveal the carbides to be $M_{23}C_6$ type.

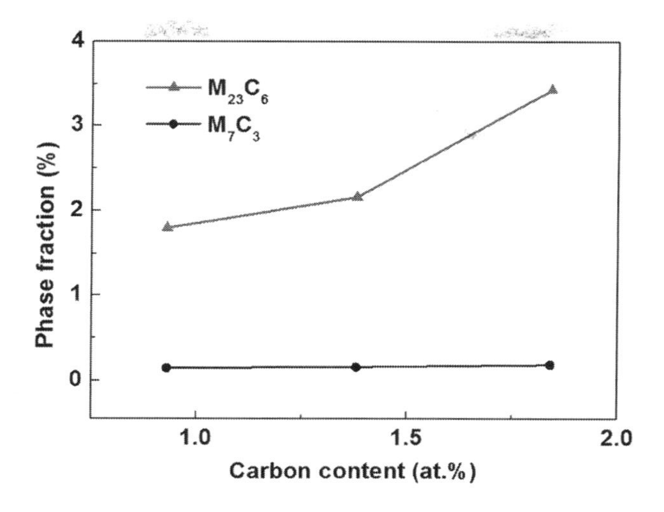

FIGURE 4.9 Variation in phase fraction of M_7C_3 and $M_{23}C_6$ type carbide precipitates in annealed CoCrFeMnNi with carbon content [37]. $M_{23}C_6$ is predominant at higher carbon content.

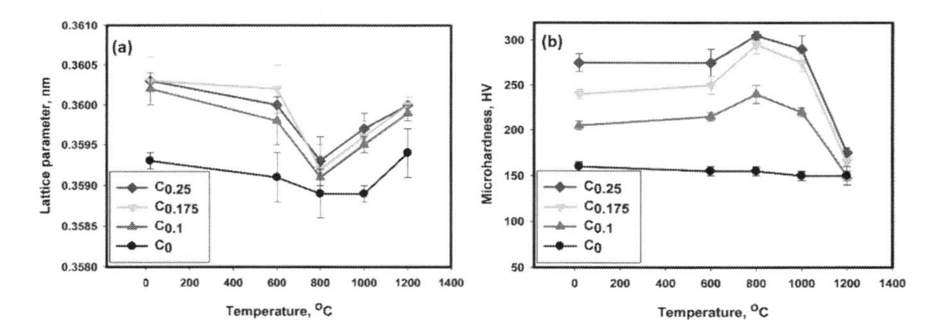

FIGURE 4.10 Variation of (a) lattice parameter and (b) microhardness on annealing temperature at different carbon contents of the CoCrFeNiMnC$_x$ [40]. The minimum value of lattice parameter and maximum value of microhardness coincide at 800°C.

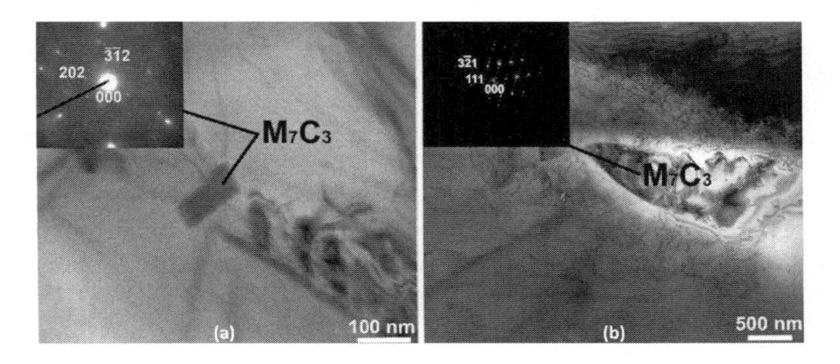

FIGURE 4.11 TEM bright-field images of the (a) CoCrFeNiMnC$_{0.1}$ and (b) CoCrFeMnNiC$_{0.175}$ in the as-solidified condition [40].

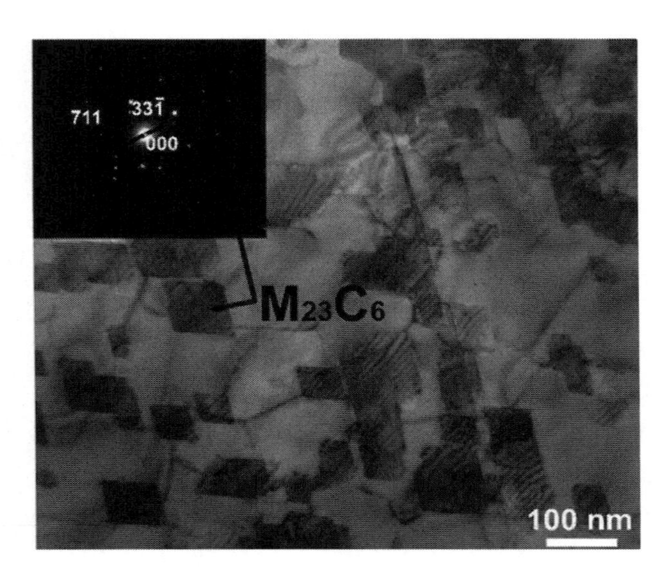

FIGURE 4.12 TEM image of $CoCrFeMnNiC_{0.25}$ annealed to 800°C for 14 h. SAED of area containing second phase is also shown [40].

The carbides formed earlier after annealing treatment of the as-solidified alloy are coarse and the resulting microstructure was quite inhomogeneous. Hence, cold rolling up to 80% strain with annealing can help to resolve this by making the carbide distribution smaller and more uniform, and by lowering the recrystallization temperature. Cold rolling induces Cr rich sigma phase in the annealed-CoCrFeMnNi alloy [47] which otherwise is predominately single-phase FCC even when annealed in the range of 600–1,000°C for 14h [40]. The SEM images of cold rolled (80%) and annealed (800°C for 1 h) CoCrFeMnNi with different carbon content namely, 0 at.%, 0.5 at.%, and 1 at.% are shown in Figure 4.13 [48]. These alloys (0 at.%, 0.5 at.%, 1 at.%) are named C0, C0.5, and C1.0. EBSD analysis on these SEM images was carried out as shown in Figure 4.13. It is seen that with increasing carbon content, the population of $M_{23}C_6$-type carbides increases. While the volume fraction of carbides increases with an increase in carbon content, it is important to note that no significant change in the size of the carbide was observed [48]. This is proved in Figure 4.14 where the particle size distributions of carbide particles are shown at different carbon contents. Also, average grain size decreases with carbon content, as seen in Figure 4.13d–f. This is due to the carbide nano-precipitates pinning the grain boundary during annealing, hence slowing down the grain boundary migration.

Carbide interaction with grain boundary is shown in Figure 4.15. A second phase is shown precipitated along the grain boundary. The SAED patterns confirmed the phase to be $M_{23}C_6$ type carbide. Carbides pinning grain boundaries during annealing have been reported in other works as well [26, 32, 37, 49]. Recrystallized microstructures for carbon containing CoCrFeMnNi were studied by Klimova et al. [50]. The work reported isochronal annealing (same annealing

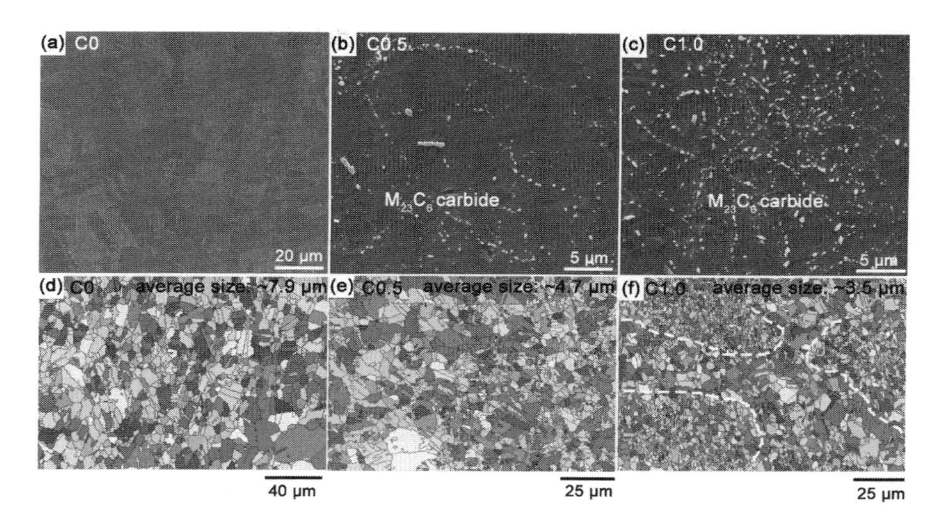

FIGURE 4.13 SEM images of (a) C0, (b) C0.5, and (c) C1.0 of CoCrFeMnNi alloy system and EBSD maps of (d) C0, (e) C0.5, and (f) C1.0 from the same alloy after cold rolling and subsequent annealing at 800°C for 1 h [48]. C0, C0.5, and C1.0 refers to CoCrFeMnNi alloys having carbon content 0 at.%, 0.5 at.%, and 1.0 at.%, respectively.

time, different temperature), as shown in Figure 4.16 and Figure 4.17, as well as isothermal annealing (same temperature, different annealing time) as shown in Figure 4.18 for C containing CoCrFeMnNi. It was observed that annealing starts for CoCrFeMnNi after 973 K and complete recrystallization occurs during annealing in the range 1,173–1,373 K. At lesser temperatures, recrystallization is incomplete and carbides are noticed to align along the elongated grains in a chain-like manner (Figure 4.17b). This phenomenon ceases to exist at annealing temperatures higher than 1,173 K as seen in Figure 4.17d–e. The size of the carbides increases and their volume fraction decreases with an increase in annealing temperature. The large size of carbides and coarser grain size can also be noticed in Figure 4.17d–e as the annealing temperature increases.

Isothermal annealing was also done on carbon containing CoCrFeMnNi at 1,173 K for 2, 5, 10, and 50 h. The microstructures obtained are shown in Figure 4.18. Grain growth was found to be normal and increased with annealing time. Stacking faults and triple junctions can be noticed. Carbide particle size increased with annealing time with no change in their volume fraction observed.

4.1.3 Effect of Carbon Addition at the Macro-Mechanical Level

At a macro level, the discussion on the mechanical properties will be largely focused on the effect of carbon addition on yield strength, hardness, tensile strength, and strain hardening rate of CoCrFeMnNi. Traditionally, in pure FCC metals, yield strength decreases with increase in temperature. The same behavior has been observed by Wu et al. [26] who did tensile tests on CoCrFeMnNi and CoCrFeMnNi-0.5 at.% C at 77 K and 293 K. The results of Wu et al. are shown in Table 4.1. They observed

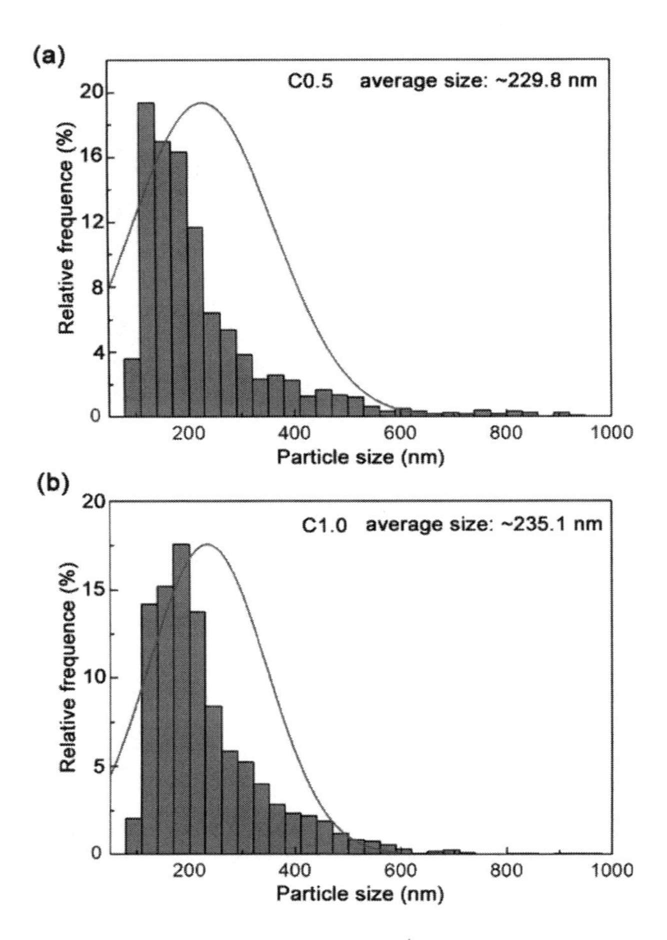

FIGURE 4.14 Particle size distribution of the $M_{23}C_6$ carbides for (a) C0.5 and (b) C1.0 HEAs [48]. Precipitate size doesn't vary much with carbon.

an increase in yield strength with the drop in temperature from 293 K to 77 K. Also, carbon addition increased the yield strength at a particular temperature. Another factor to note is the rise in the extent of work hardening for a particular alloy with decrease in temperature. This phenomenon was attributed to the increased tendency for deformation twins to form in a carbon doped alloy at low temperatures. Even though high-temperature tensile tests are yet to be done on carbon doped CoCrFeMnNi, it is noticed that yield strength and tensile strength increases with carbon addition, as indicated by room temperature tensile tests conducted by Chen et al. [43] (Figure 4.19).

In Figure 4.19, C00 is the carbon-free alloy and C05 is the alloy with 0.05 at.% of carbon. The same nomenclature is adopted for C10, C15, and C20. From Figures 4.19 and 4.20, plastic strain increases for C05 initially and decreases at higher carbon contents. This shows that at a low percentage of carbon, the carbon dissolves completely

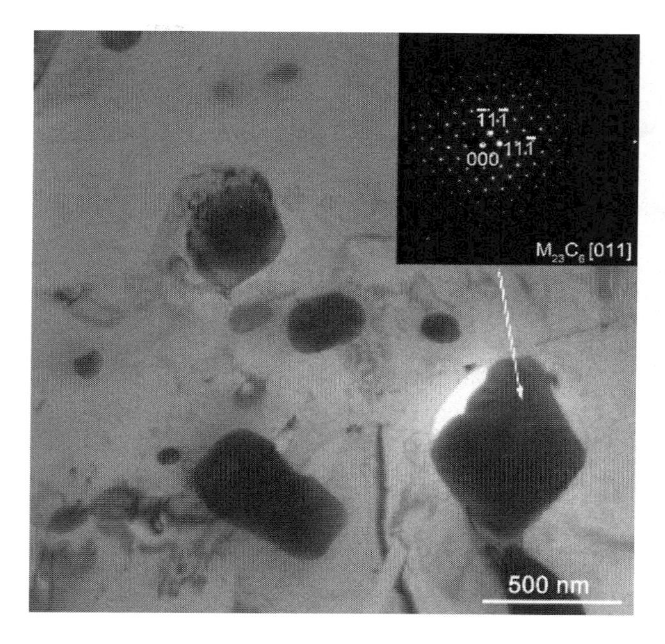

FIGURE 4.15 TEM image of C1.0 HEA [48]. SAED pattern of are also shown. The second phases are seen precipitating along the GB.

FIGURE 4.16 Optical microstructures of the C-containing CoCrFeNiMn-type alloy after annealing at (a) 973 K, (b) 1,073 K, (c) 1,173 K, (d) 1,273 K, (e) 1,373 K for 1 h [50].

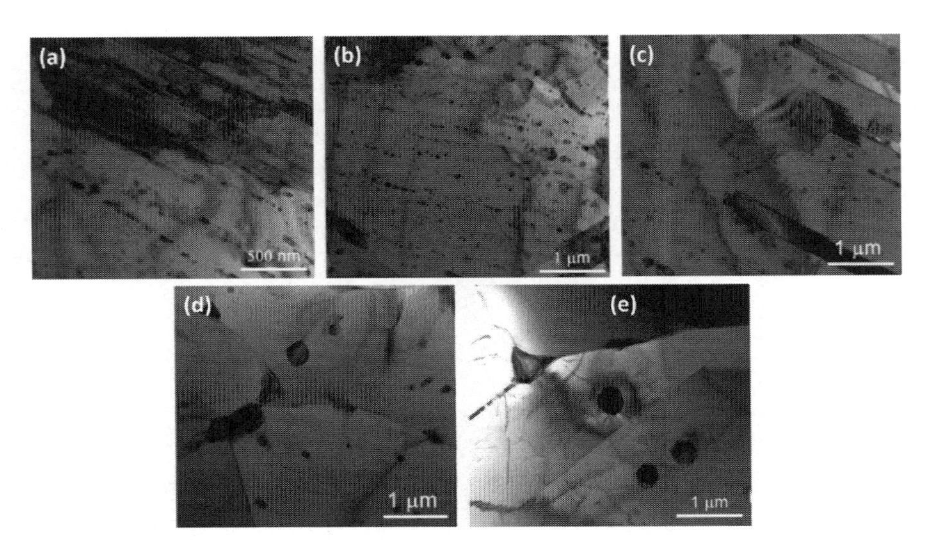

FIGURE 4.17 TEM microstructure of the C-containing CoCrFeNiMn-type alloy after annealing at (a) 973 K, (b) 1,073 K, (c) 1,173 K, (d) 1,273 K, (e) 1,373 K for 1 h [50].

FIGURE 4.18 Optical microstructure of the C-containing CoCrFeNiMn-type alloy alloys after annealing at 1,173 K for: (a) 2 h, (b) 5 h, (c) 10 h, (d) 50 h [50].

TABLE 4.1
0.2% Offset Yield Stress (Σy), the Ultimate Tensile Strength (UTS), the Uniform Elongation to Fracture (Ef), Extent of Work Hardening for the Investigated Alloy (Δσ = UTS – Σy)

	CoCrFeMnNi [8]		CoCrFeMnNi-0.5C [26]	
	77K	293K	77K	293K
σ_y (MPa)	350	165	590	225
UTS (MPa)	820	520	1090	655
e_f (%)	85	65	69	38
$\Delta\sigma$ (MPa)	470	355	580	430

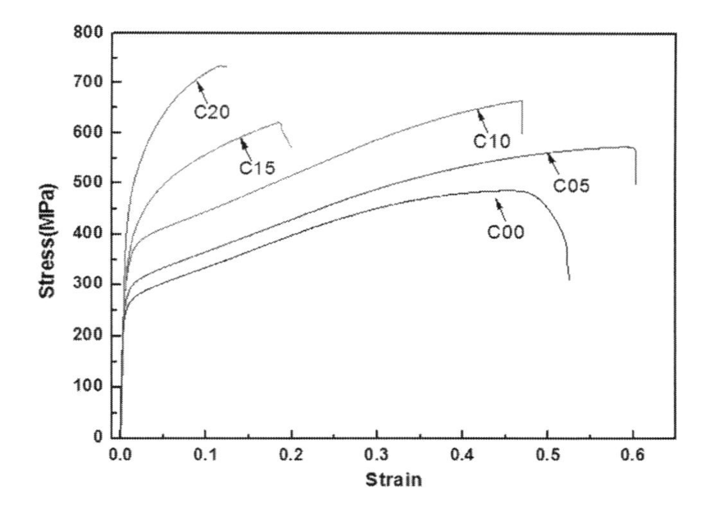

FIGURE 4.19 Engineering stress-strain curves under tension of the C00, C05, C10, C15, and C20 alloys [43]. C00 is the carbon-free alloy and C05 is the alloy containing 0.05 at.% of carbon. The same nomenclature is adopted for C10, C15, and C20.

into the FCC structure of CoCrFeMnNi and there is an increase in strength and ductility. This trend ceases to exist at higher carbon contents where strength increases at the expense of ductility. This problem can be alleviated by employing thermomechanical processes like cold working and/or annealing as seen in Figure 4.21 taken from Cheng et al. [41].

In Figure 4.21, C0 is the as-homogenized carbon-free alloy and C2 is the as-homogenized alloy into which 2 at.% of carbon content is added. The 80% suffix denotes the degree of cold rolling employed with annealing on the as-homogenized alloy. It is observed that the YS and TS of C2 alloy is higher than the C0 alloys and more importantly, it can be noticed that cold rolling with annealing can improve the properties of the carbon containing the as-homogenized alloy. Of all the alloys, the

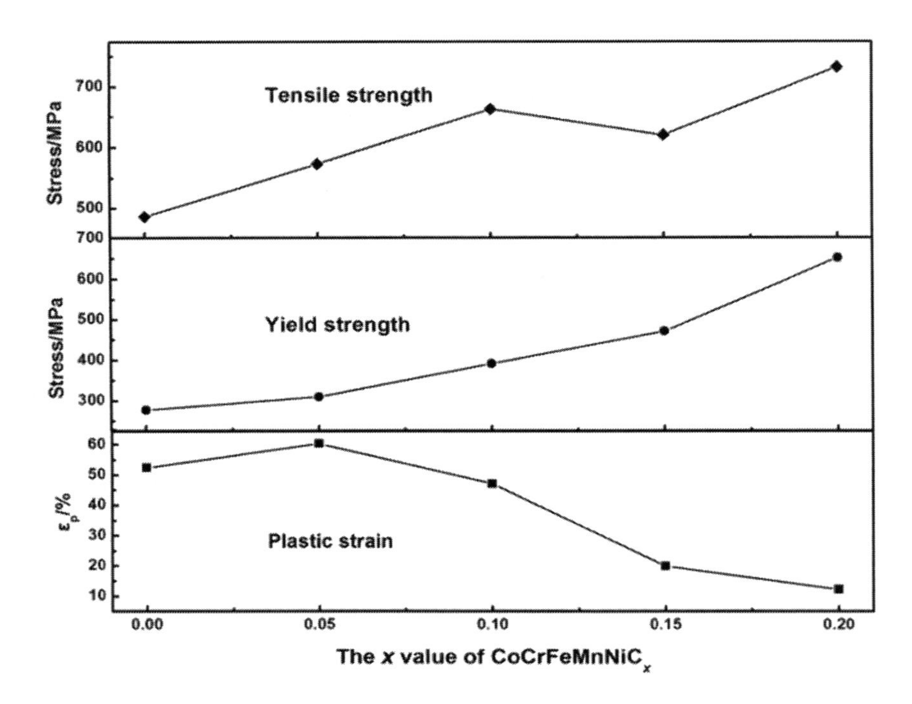

FIGURE 4.20 Effect of C content on the mechanical properties of the CoCrFeNiMnCx alloys [43].

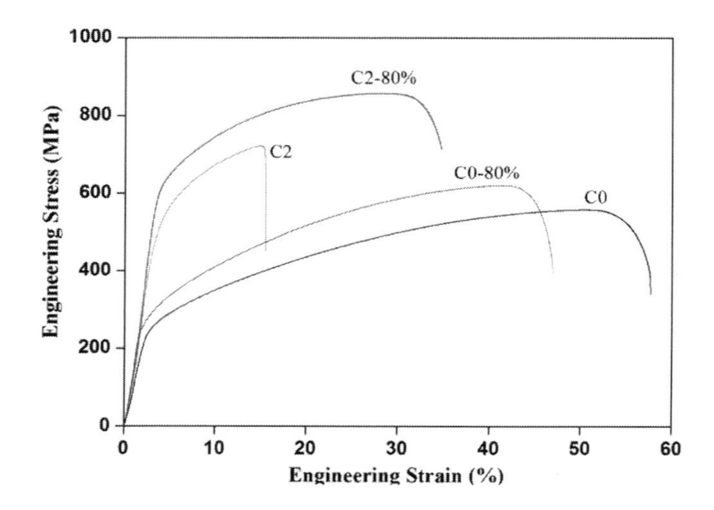

FIGURE 4.21 Engineering tensile stress-strain curves of alloys C0, C0–80%, C2, and C2–80% at room temperature [41].

stress-strain plot of C2–80% has a better strength–ductility combination as seen from Table 4.2. The YS, TS and elongation to fracture for C2–80% were found to be 581 MPa, 857 MPa, and 28%, respectively.

Figure 4.22 shows the variation of micro hardness with and without carbon content (1 at.%) is shown with respect to rolling strain and annealing temperatures, as reported by Stepanov et al. [32]. The microhardness increases with rolling strain for both alloys as in Figure 4.22a and the hardening behavior is the same for carbon-doped and carbon-free alloy. From Figure 4.22b, it can be seen that hardness remains (Table 4.4) when annealed up to 600°C. At higher annealing temperatures, hardness drops due to rapid coarsening of the microstructure for both the alloys.

The strain hardening rate is an important property which imparts stability to a material especially at various temperature regimes and is also a factor in evaluating the formability of a material. Variation of this parameter with strain and carbon contents were studied for cold rolled and annealed CoCrFeMnNi by Ko et al. [37] (Figure 4.23).

TABLE 4.2
Mechanical Properties of Alloys C0, C0–80%, C2, and C2–80% [41]

Alloys	YS (0.2%) (MPa)	UTS (MPa)	Elongation to fracture (%)	Uniform elongation (%)
C0	243	558	50	47
C0–80%	249	621	41	40
C2	446	723	15	12
C2–80%	581	857	28	25

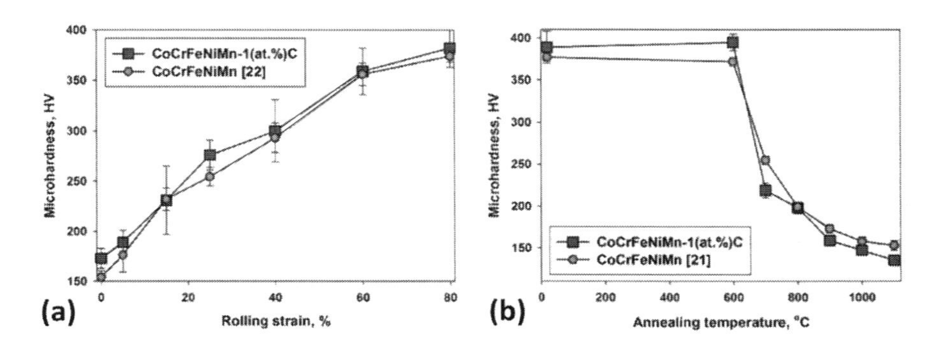

FIGURE 4.22 Variation of Vickers microhardness of the CoCrFeNiMn-1(at. %)C alloy with respect to: (a) rolling strain; (b) subsequent annealing temperature [32]. The comparison for microhardness and annealing temperature plots are taken from literature [22, 30]. The annealing time was 1 h.

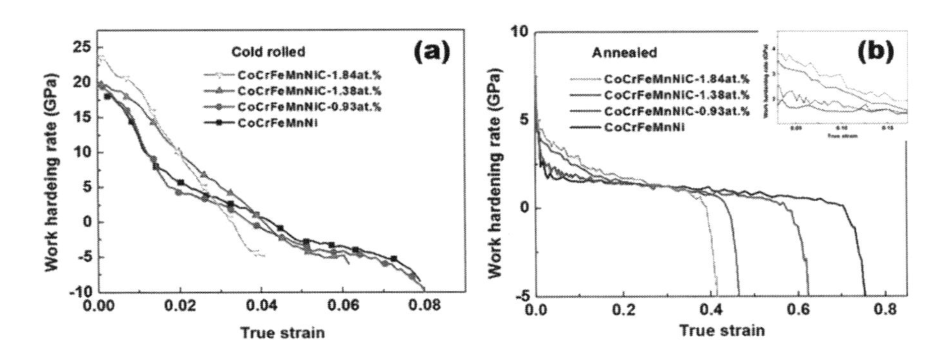

FIGURE 4.23 Hardening rates are plotted against true strain for (a) cold-rolled and (b) annealed CoCrFeMnNi alloy and those with 0.93, 1.38, and 1.84 at.% carbon contents [37].

Cold rolling was done on the homogenized samples for 91% reduction in thickness and annealing of these rolled plates was done for 1 h at a temperature of 900°C. Samples having a carbon content of 0%, 0.93%, 1.38%, and 1.84 at.% C were studied. From the results for the cold rolled and annealed samples, it can be seen that strain hardening rate is the greatest for the sample with the highest carbon content during initial straining. As deformation proceeds, the hardening rate decreases more rapidly with increasing carbon content. Annealing decreases the hardening rate for a particular sample and flattens the curves more, as shown in Figure 4.23b. The hardening rate becomes nearly constant for most parts of the deformation when the sample is annealed. This can be explained by the dislocation sub-structures formed during the initial and final stage of deformation as shown in Figure 4.24 [50]. The microstructure during the initial stage of deformation is inhomogeneous, with dislocation cells and pile-ups (Figure 4.24a), whereas during the later stages of deformation, the microstructure is uniform with enhanced twinning (Figure 4.24b) making the hardening rate more or less constant with strain until a fracture happens.

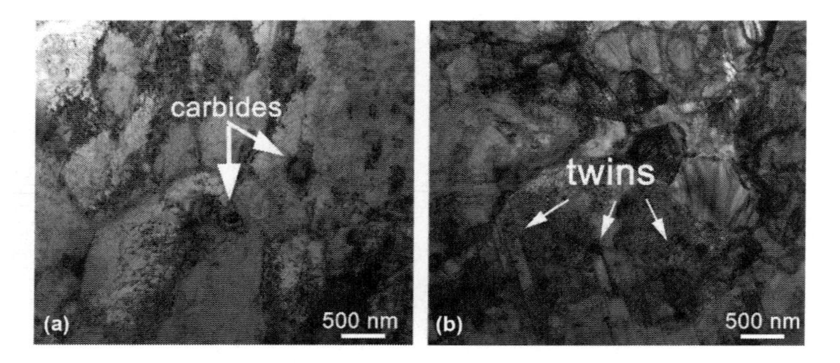

FIGURE 4.24 Dislocation cells, pile-ups, and twins observed in the microstructure of the C-containing CoCrFeNiMn-type alloy after 1 h annealing at 1,173 K and room temperature tension to (a) 5% and (b) 22% [50].

4.2 INTRODUCTION TO CARBON ADDITION ON CoCrFeMnNi THROUGH POWDER METALLURGY FOR WEAR APPLICATIONS

The primary fabricating routes employed for high entropy alloys can be categorized as liquid state processes, such as vacuum-arc melting (VAM), casting, vacuum induction melting (VIM), and solid-state processes, such as mechanical alloying (MA) followed by spark plasma sintering (SPS). While melting and casting processing like VAM/VIM are suitable for the commercial production of high entropy alloys, these processing techniques come with their limitations. They are difficult to control and unsuitable for metals with different densities and melting points. Vapor pressure of elements can create pores in the final product and during cooling and there is also the issue of segregation of elements during solidification. Melting and cooling can also lead to suppression of equilibrium phases. All of these can lead to compositional inhomogeneity in the casting which can deteriorate its properties significantly [51, 52]. In contrast, mechanical alloying is a very versatile process [53]. It involves intimate mixing of its constituent elements in solid state. This extends the solid solubility among the constituent elements of the mixture and also promotes finer microstructures and chemical homogeneity in the matrix. The disadvantages of MA are its unsuitability for large-scale production and the effect of contamination from the balls, the vials, and the medium (process controlling agent).

Even though a large body of literature is available on the study of interstitial HEAs, many specific areas of knowledge deficit can be identified. First, the processing routes employed for studying carbon-added CoCrFeMnNi have been largely melting/casting or VAM/VIM. The benefits offered by the MA route haven't been exploited as required in studying the effects of carbon on CoCrFeMnNi [54]. Second, the primary aim, thus far, has been to alloy the carbon into the FCC matrix of CoCrFeMnNi in order to produce nano-precipitates and nanostructures for enhanced structural properties [26, 48, 55]. Third, wear studies on high entropy alloys were largely confined to refractory high entropy alloys or compositions of Al-containing 3D transition elements [56]. Zhang et al. [57] have reported a high entropy alloy matrix self-lubricating composite exhibiting superior strength and wear properties for the first time using powder metallurgy in 2017. This HEA composite of CoCrFeNi showed excellent tribological properties at room and elevated temperatures [58]. It was established that these properties could be further enhanced by adding solid lubricants like graphite, molybdenum disulphide [57], and sulfur [59].

Due to the metastable nature of mechanical alloying processes, a single-phase FCC may not be the only phase formed in CoCrFeMnNi upon ball milling its constituent elements. Additional phases observed for this alloy are BCC and chromium-rich σ (sigma) phase [60, 61]. As mentioned previously, carbon addition to CoCrFeMnNi through the powder metallurgy route is still largely unreported and hence, the work described below attempts to provide a starting point in this direction i.e., to investigate the effect of carbon addition to the CoCrFeMnNi alloy through mechanical alloying followed by spark plasma sintering.

Following Zhang et al. [57], this work attempts to fabricate a high entropy alloy composite of CoCrFeMnNi matrix added with graphite. The effect of 2 wt.% graphite flake addition (~8.7 at.%) on mechanically alloyed CoCrFeMnNi with special focus on the microstructural evolution was studied. The addition of excess graphite was to fabricate alloys with the aim of having a lower coefficient of friction for wear resistance applications. Incidentally, literature on wear studies on high entropy alloys is sparse [62]. The first study of wear behavior on CoCrFeMnNi was done by Ayyagari et al. [63] in which they compared wear resistance of CoCrFeMnNi and $Al_{0.1}$CoCrFeNi in dry and marine environments and reported that the latter alloy has better wear resistance. Later, a high-temperature wear study on CoCrFeMnNi was also reported [64]. Graphite flake addition on mechanically alloyed CoCrFeMnNi hasn't been reported thus far, although recently, work has started on graphene-reinforced HEA composite of CoCrCuFeNi [65]. Mechanical alloying can result in tungsten carbide contamination due to erosion from the WC balls and vials of the planetary mills. Effect of WC and chromium carbides on the properties and microstructure of CoCrFeMnNi has been of particular interest. For example, in the work by Park et al. [66], the effects of tungsten and chromium carbide from FSW tool on the mechanical properties of friction stir welds for CoCrFeMnNi were studied. Dispersions of fine (W, Cr) carbides were produced to give maximum tensile strength. The effect of WC contamination from mechanical alloying on the mechanical properties (especially wear resistance) of high entropy alloys is still unknown and needs to be investigated.

4.2.1 Effect of Graphite Flake Addition on the Microstructure of Mechanically Alloyed and Spark Plasma Sintered CoCrFeMnNi

Elemental powders of cobalt, chromium, iron, nickel, and manganese were subjected to mechanical alloying by a Fritsch Pulverisette 5 high energy planetary ball mill. All the powders were procured from Alfa Aesar with purity in excess of 99.9% and particle size of around 10 μm. Graphite flakes (7–10 μm particle size, 99% purity) were obtained from Alfa Aesar and 2 wt.% was added to CoCrFeMnNi mixture before ball milling. This was done by dispersing the required amount of graphite flake into the alloy powders using an ultrasonic vibrator. The powders were milled at 300 RPM in a toluene medium with a ball to powder weight ratio of 10:1. The planetary ball mills were cooled for 15 minutes for every hour of operation. Four sets of powders were milled and subjected to spark plasma sintering, as described in Table 4.3.

TABLE 4.3
Nomenclature of the Different Compositions Used in this Work

Si. No	Materials and methods	Nomenclature
1	Co,Cr,Fe,Mn,&Ni mixture ball milled for 10 hours	HEA-10H
2	Co,Cr,Fe,Mn,&Ni +2 wt.% graphite flakes ball milled for 10 hours	HEA-GR-10H
3	Co,Cr,Fe,Mn,&Ni ball milled for 20 hours	HEA-20H
4	Co,Cr,Fe,Mn,&Ni +2 wt.% graphite flakes ball milled for 20 hours	HEA-GR-20H

Powder samples of all the sets were taken at regular intervals during the ball milling to study the phase evolution. All powders were spark plasma sintered at a temperature of 1,273 K at a pressure of 55 MPa. For this, compacts were initially heated at a rate of 100 K/min to reach 1,273 K and subsequently held at that temperature for a dwell time of 5 minutes, and thereafter cooled down to room temperature. The temperature and pressure were selected after giving due consideration to the literature on the sintering of CoCrFeMnNi powders [67, 68]. A Dr Sinter SPS-5000 machine was employed for spark plasma sintering. The sintered samples are named as in Table 4.3, with a prefix of SPS added to the powder names.

4.2.1.1 Powder Characteristics after Milling

Figures 4.25 and 4.26 show the XRD patterns of CoCrFeMnNi with and without graphite addition. Alloying is observed for CoCrFeMnNi powders and an FCC phase is formed after 10 h of milling, along with a BCC phase (Figure 4.25). This was also noticed in the work done by Vaidya et al. [69]. Tungsten carbide from the wear of vials and balls of the planetary ball mill was observed in the powders after prolonged (>10 h) milling times. The CoCrFeMnNi milled powders consist of FCC, BCC, and WC phases after 20 h of ball milling. Milling for more than 20 h results in high WC contamination, as can be seen from the XRD pattern of 30 h milled powder (Figure 4.25).

Figure 4.26 shows the phase evolution during 20 hours of milling of CoCrFeMnNi with 2 wt.% graphite flake addition. The alloying tendency of the powders is inhibited with graphite flake addition, as the BCC Fe peak is maintained throughout the milling time up to 20 h. This is unlike the case of

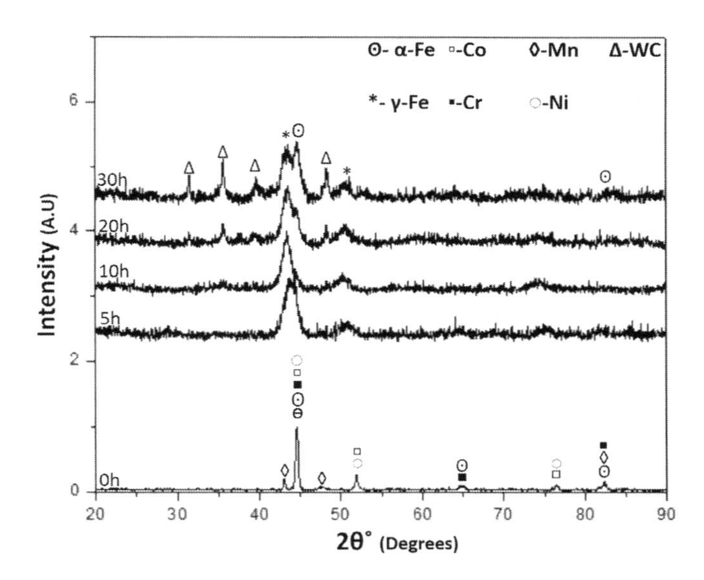

FIGURE 4.25 XRD plot of CoCrFeMnNi showing phases formed when ball milled up to 30 hours. The 30 h ball milled powder has been put there just to show the increased contamination at higher milling duration.

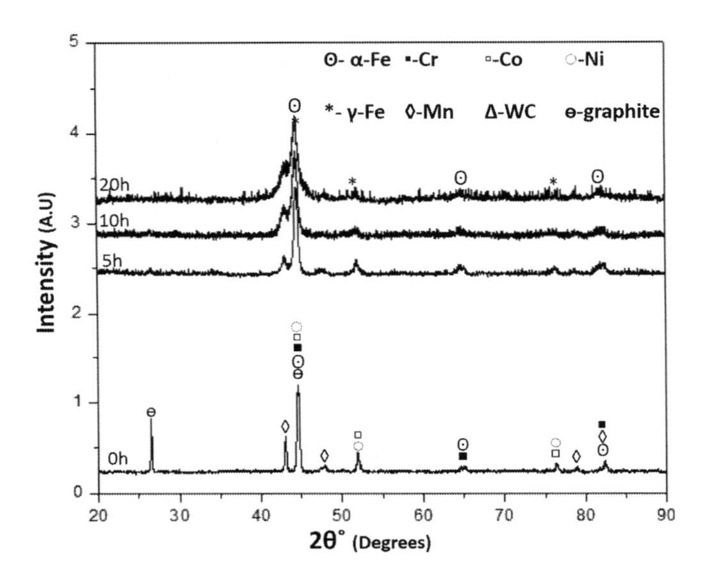

FIGURE 4.26 XRD plot of CoCrFeMnNi+2 wt.% graphite showing phases formed when ball milled up to 20 hours. Notice the decreased WC contamination.

CoCrFeMnNi, where the BCC peak of iron reduces after 10 h of milling and the FCC phase is more predominant. A manganese peak is noticed as a shoulder to the left of the primary peak of BCC Fe in Figure 4.26. This means Mn hasn't dissolved to form the primary FCC phase even after 20 hours of milling. This confirms the inhibition of alloying in presence of graphite. Also, tungsten carbide contamination is not observed upon graphite addition in CoCrFeMnNi. This is due to the fact that graphite flakes act as a lubricant and exfoliate and cover the powder particles, thus inhibiting its alloying.

To observe the effect of graphite flake addition, both the powder morphology as well as the back-scattered electron images of polished powder cross-sections of HEA-10H, HEA-GR-10H, HEA-20H, and HEA-GR-20H have been studied using a scanning electron microscope. Figure 4.27 shows the SEM images of the milled powders. The milled powders of CoCrFeMnNi have irregular shapes. It can be noticed from Figure a,b that powder size decreases upon milling from 10 h to 20 h. It is also observed that the powder size decreases upon graphite flake addition for a given duration of milling. The particle size distributions of the 20 h milled HEA-20H and HEA-GR-20H powders are shown in Figure 4.28. It clearly shows that the HEA-GR-20H powders are smaller compared to HEA-20H. This is due to the presence of graphite/exfoliated graphite layers at the interfaces which prevent cold welding of the powders during impact.

Table 4.4 shows the powder size distribution parameters for the milled powders. It is observed that the D_{10}, D_{50}, and D_{90} values for graphite-added sample are lower than the powders without graphite. From all this, it can be concluded that graphite addition reduces the size of ball milled powders of CoCrFeMnNi and inhibits mechanical alloying.

FIGURE 4.27 Powder morphology of (a) HEA-10H, (b) HEA-GR-10H, (c) HEA-20H, and (d) HEA-GR-20H. Variation in powder size upon graphite addition for similar milling duration can be clearly noticed.

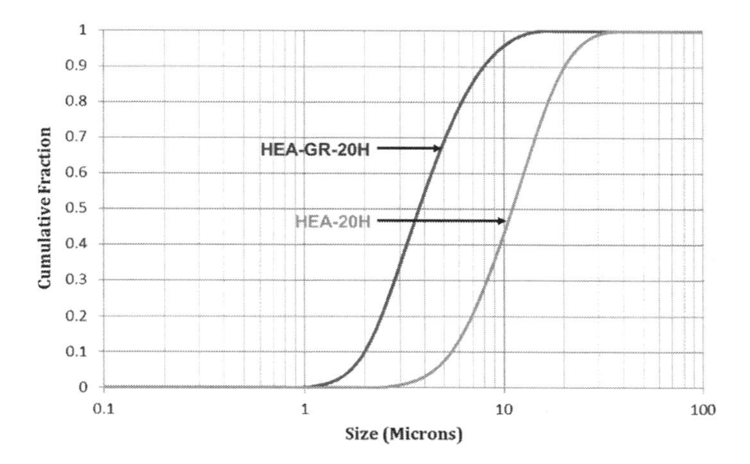

FIGURE 4.28 Powder size distribution of HEA-20H and HEA-GR-20H clearly showing the decreased powder size upon graphite addition.

TABLE 4.4
Powder Size Distribution Data for 20 h Milled Powders

Si. No	Sample	D10 (μm)	D50 (μm)	D90 (μm)
1	HEA-20H	5.4	11	20.1
2	HEA-GR-20H	2	3.7	7.9

Figure 4.29 shows the back-scattered electron images of the polished cross-section of the powders mounted using a resin. Lesser agglomeration is observed at higher milling times, leading to the fine size of the powders. Tungsten carbide is picked up by the powders due to the erosion of the vials and balls of the high energy planetary mills. This erosion increases with the time duration of milling. WC can be seen as white spots and regions in Figure 4.29a, c, and e. It is clearly observed that tungsten carbide contamination is negligible upon graphite addition. Tungsten carbide contamination is very low in Figure 4.29b, e, and f compared to Figure 4.29a, c, and d.

Variation of lattice parameters of the FCC and BCC phases of the powders with ball milling time and post-sintering are produced in Figure 4.30. It is seen that the FCC lattice parameter for CoCrFeMnNi increases with milling time (Figure 4.30a). This indicates alloying of the constituent elements to form a primary FCC phase occurs upon ball milling. The BCC lattice parameter for CoCrFeMnNi is noticed up to 5 hours of ball milling time, after which the peaks are not distinct. In the case of CoCrFeMnNi with graphite addition, as shown in Figure 4.30b, the lattice parameter of FCC as well as BCC doesn't vary with the duration of ball milling.

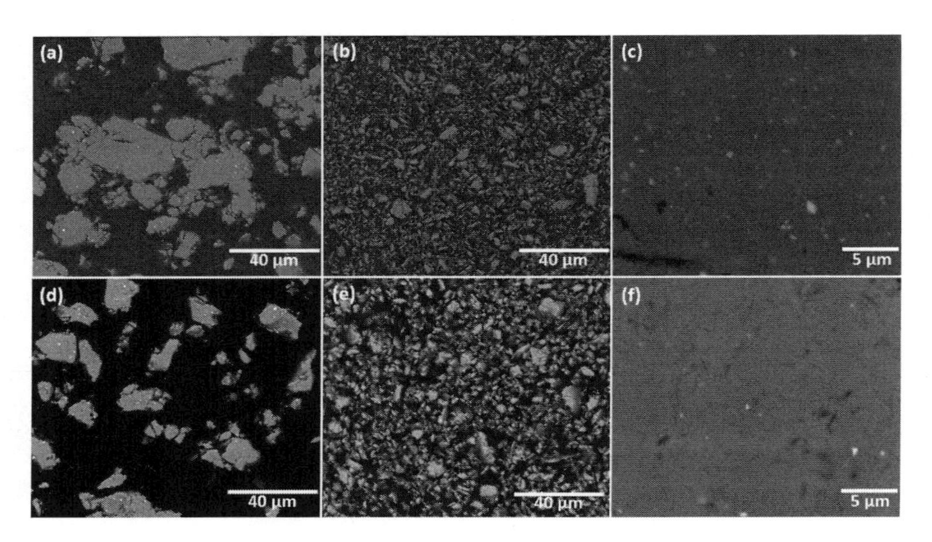

FIGURE 4.29 BSE micrographs of polished powder embedded in resin of (a) HEA-10H, (b) HEA-GR-10H (c) and (d) HEA-20H, (e) and (f) HEA-GR-20H.

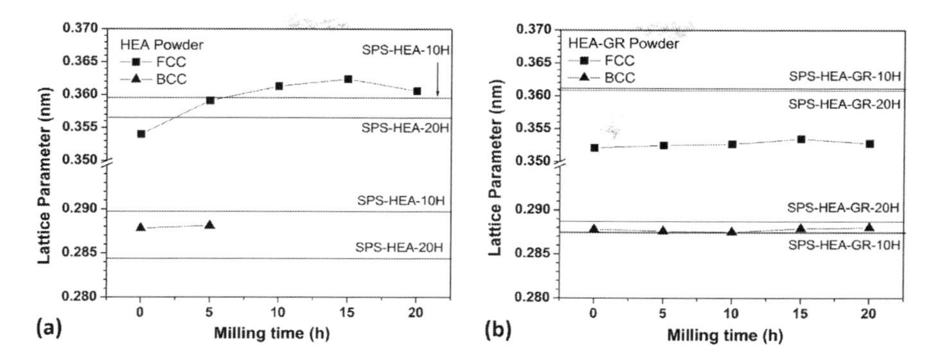

FIGURE 4.30 Variation of lattice parameters of the (a) FCC and (b) BCC phases with ball milling time at various stages of ball milling and SPS for the alloy.

In fact, FCC lattice parameter of CoCrFeMnNi + 2 wt.% graphite is more or less constant with ball milling time and is similar to nickel (0.3524 nm).This indicates the decreased alloying among the constituent elements upon addition of graphite flakes.

4.2.1.2 Characterization of SPS Compacts

Figure 4.31 shows the XRD patterns for the sintered compacts of 10 h and 20 h milled powders.

The XRD pattern of SPS-HEA-10H sample shows a primary FCC phase along with M_7C_3 carbide. This is in agreement with the available literature on mechanical

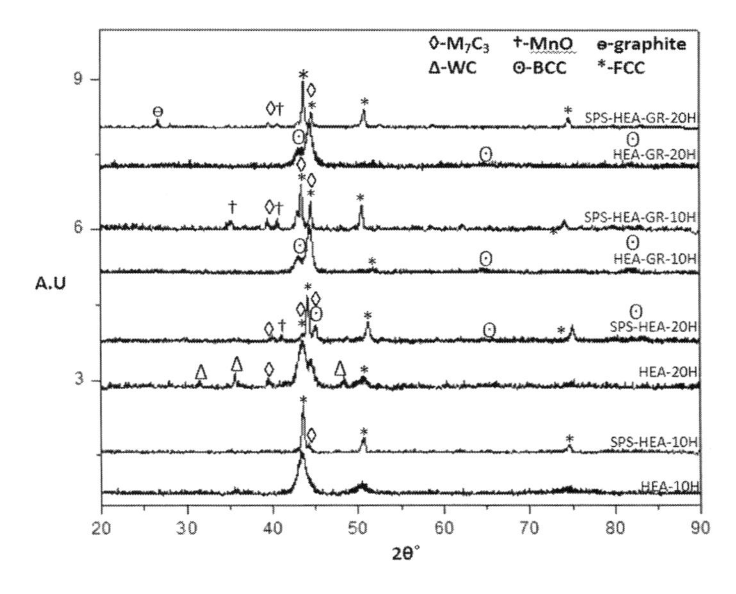

FIGURE 4.31 XRD plots of all sets of powders and their corresponding sintered compacts.

alloying of CoCrFeMnNi [68, 69]. The chromium carbide peak was observed adjacent to the primary FCC peak. The SPS-HEA-20H sample has carbides of tungsten, chromium, and manganese oxide. The presence of MnO is observed in the XRD pattern of the SPS-HEA-20H, SPS-HEA-GR-10H, and SPS-HEA-GR-20H samples. Figure 4.26 showed a peak corresponding to undissolved manganese to the left of the primary BCC peak in HEA-GR-20H powder. This Mn undergoes oxidation upon spark plasma sintering to form MnO.

The lattice parameter of the sintered compacts is shown in Figure 4.30 in the form of horizontal lines on the plot. From Figure 4.30a and b, it can be seen that FCC lattice parameter for the sintered compact is increased when graphite is added. Stepanov et al. [40] have also reported an increase in FCC lattice parameter upon carbon addition. The slight decrease in FCC lattice parameter upon sintering for SPS-HEA-10H and SPS-HEA-20H samples is due to the evolution of carbides. This is not observed in the graphite-added composition (Figure 4.30b). Here, the FCC lattice parameter for the sintered compact is greater than its powder, while the BCC lattice parameter is the same for both powder and the sintered compact. This is due to the inhibited alloying and decreased carbide formation which occurs due to graphite flake addition.

4.2.1.2.1 Microstructure of SPS Compacts

Figure 4.32 shows the BSE images of the SPS compacts. The dark gray and light gray regions in Figure 4.32 correspond to the BCC and FCC phases, respectively. Table 4.5 gives the proportion of FCC and BCC phases in the images obtained using ImageJ software. The FCC phase is predominant in SPS-HEA-10H sample (Figure 4.32a). The round dark spots observed in the microstructure are chromium carbide precipitates. In Figure 4.32b, undissolved manganese in powders of HEA-GR-10H gives rise to formation of MnO in SPS-HEA-GR-10H upon sintering. This is also confirmed by the EDS element distribution maps as shown in Figure 4.33.

The SPS-HEA-GR-10H sample has the least amount of the FCC phase along with oxides and carbides as shown in Figure 4.32b. The FCC phase is formed when every constituent element of CoCrFeMnNi dissolves due to mechanical alloying to form a single phase. Hence, the inhibited alloying among constituent elements brought about by graphite addition leads to lesser amount of FCC phase. For similar reasons, the microstructure is coarser and the phases are larger in size when compared with other compacts, as shown in Figure 4.32b. The white spots correspond to WC and the black dots are chromium carbide in SPS-HEA-20H (Figure 4.32c). Figures 4.34 and 4.35 give the EDS elemental maps for SPS-HEA-20H and SPS-HEA-GR-20H, respectively along with the BSE image of the area mapped. The WC phase can be seen in the EDS map of SPS-HEA-20H as shown in Figure 4.34. EDS maps further show that the FCC phase is rich in nickel, cobalt, and iron and the BCC phase is rich in manganese and chromium. This is also seen in the EDS maps of HEA-GR-20H (Figure 4.35). The elements seem to have segregated in SPS-HEA-GR-20H compared to SPS-HEA-20H, which is more homogenous.

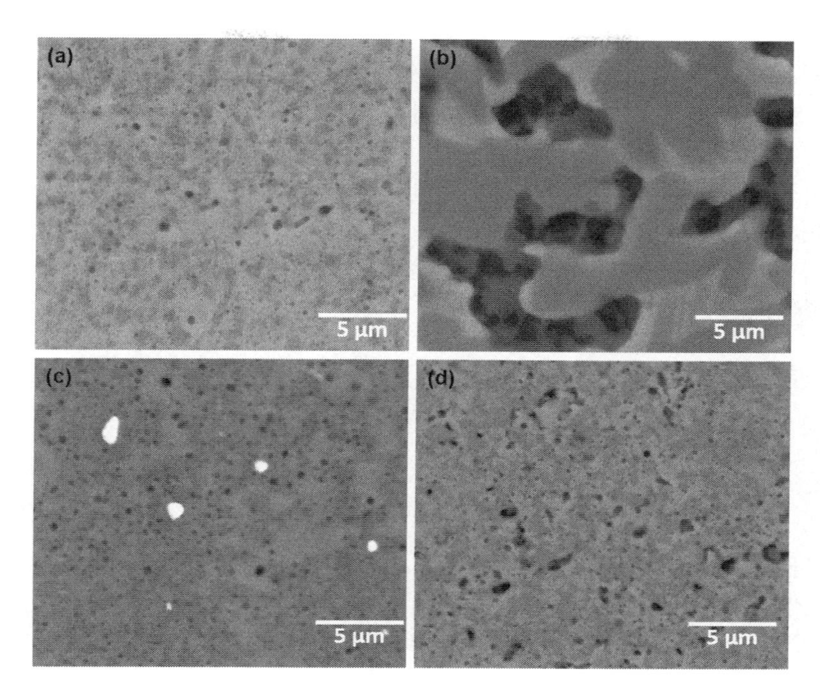

FIGURE 4.32 BSE micrographs of (a) SPS-HEA-10H, (b) SPS-HEA-GR-10H, (c) SPS-HEA-20H, (d) SPS-HEA-GR-20H. WC is observed only in SPS-HEA-20H.

TABLE 4.5
Percentage of BCC and FCC Phases in the Alloys

	BCC phase (%)	FCC phase (%)
SPS-HEA-10H	24	70
SPS-HEA-GR-10H	33	50
SPS-HEA-20H	25	64
SPS-HEA-GR-20H	26	61

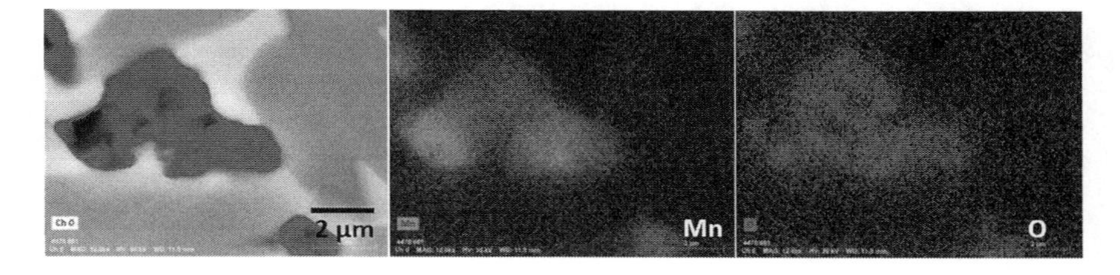

FIGURE 4.33 EDS element map of SPS-HEA-GR-10H showing the presence of MnO.

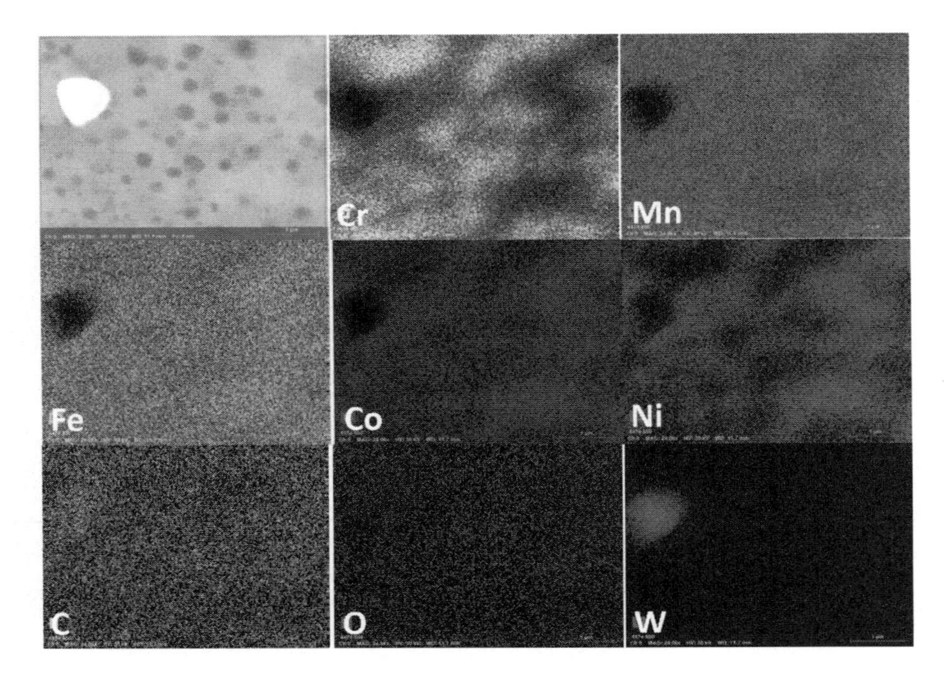

FIGURE 4.34 EDS maps showing the elemental distributions for Co, Mn, Cr, Ni, C, Fe, W for SPS-HEA-20H sample.

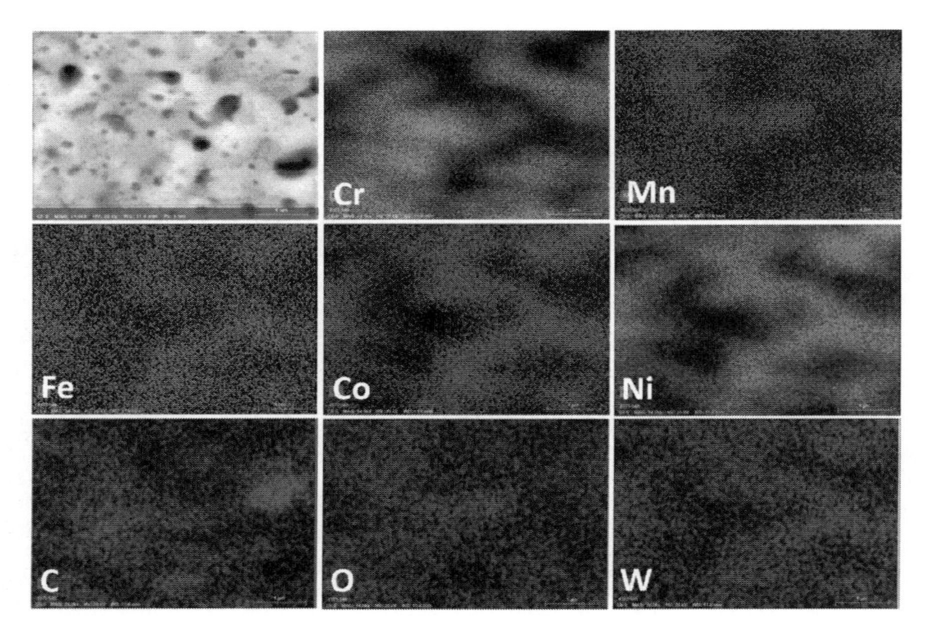

FIGURE 4.35 EDS maps showing the elemental distributions for Co, Mn, Cr, Ni, C, Fe, W for SPS-HEA-GR-20H sample.

Figure 4.36 shows the TEM bright-field images of SPS-HEA-20H and SPS-HEA-GR-20H samples. The images confirm the nano-grained microstructure of the alloys with grain sizes observed as 370 ± 100 nm and 416 ± 100 nm for CoCrFeMnNi and graphite-added CoCrFeMnNi sintered samples, respectively. Chromium carbide can be noticed in the triple junctions of the grains. Visible carbides in the TEM images (Figure 4.36) were measured to be around 170 ± 55 nm in CoCrFeMnNi and 110 ± 35 nm in graphite-added CoCrFeMnNi. Usually, grain refinement happens by carbon addition due to the grain boundary pinning effect of carbides and also during annealing of the alloy after deformation processing when there is a release of stored internal energy resulting in active recrystallization [37, 49, 70]. However, here the grains appear bigger for SPS-HEA-GR-20H as compared to SPS-HEA-20H. This phenomenon of larger grain size due to graphite addition can be attributed to the decreased pick-up of tungsten carbide from the balls and vials as well as lesser evolution of chromium carbides as mentioned before. Also, serrated flow has been reported upon carbon addition on CoCrFeMnNi [49, 71], thus proving the pinning effect of carbides.

4.2.1.3 Hardness of Spark Plasma Sintered CoCrFeMnNi Compacts

Figure 4.37 shows the results of the Vickers hardness test conducted on the SPS compacts. The minimum and maximum hardness is seen for SPS-HEA-10H and SPS-HEA-20H, respectively. The high hardness of SPS-HEA-20H (732 ± 87 VHN) are due to the presence of carbides of chromium and tungsten. This is more than the maximum hardness reported by Kilmametov et al. [72] (6,700 MPa) for nano-crystalline CoCrFeMnNi (grain size of 50 nm) alloy with smaller chromium oxide precipitates (of 7–10 nm) produced by high pressure torsion methods. Upon graphite addition to the powders of Co, Cr, Fe, Mn, and Ni, the hardness value is found to reduce to 550 ± 100 VHN for both SPS-HEA-GR-10H and SPS-HEA-GR-20H. Also, it is observed that SPS-HEA-GR-10H samples have higher hardness compared to SPS-HEA-10H sample. However, the hardness of SPE-HEA-20H samples is more than the SPS-HEA-GR-20H sample. This is due to the presence of WC in the 20 h milled powders, which is very low in the case of graphite-added powders.

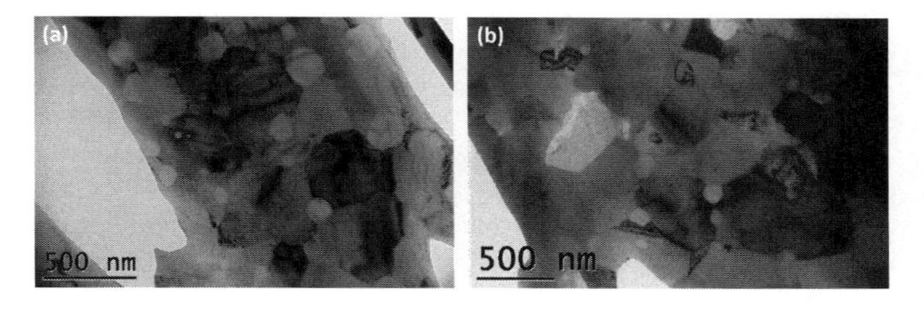

FIGURE 4.36 TEM images of (a) CoCrFeMnNi and (b) CoCrFeMnNi + 2 wt.% graphite alloy. Increased grain size of CoCrFeMnNi upon graphite addition may be noticed.

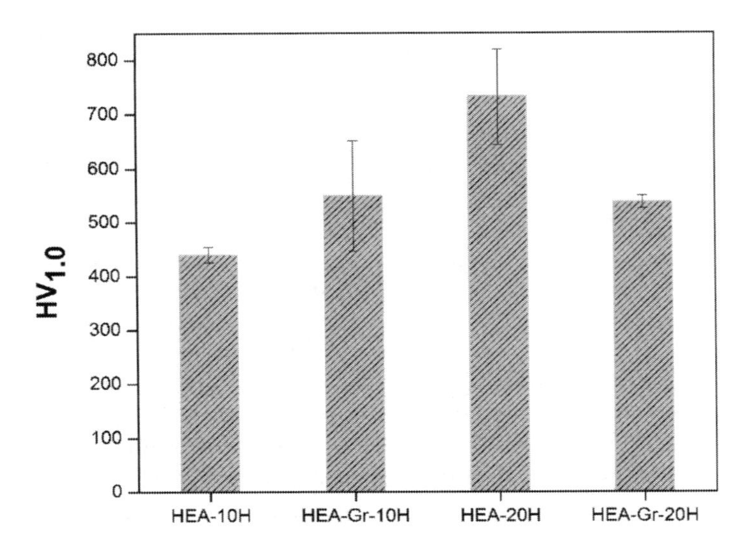

FIGURE 4.37 Vickers hardness measurements of various alloys sintered from the mentioned powders. Hardness decreases for the sintered alloy of HEA-GR-20H powder due to the absence of carbides by graphite addition.

4.3 CONCLUSION

This chapter discusses the effect of carbon addition on CoCrFeMnNi at the lattice level, microstructural level, and macro-mechanical level. The following are some key points from the literature:

1. At the lattice level, carbon addition brings increased distortion to the lattice and changes in stacking fault energy. Changes in SFE activate new deformation mechanisms like TWIP and TRIP in the lattice.
2. Carbon addition to CoCrFeMnNi reduces the critical stress needed for twinning and changes the room temperature deformation mode from dislocation glide dominated plasticity to a combination of dislocation glide and twinning/microbands.
3. Planarity of dislocation increases with carbon addition to CoCrFeMnNi. This is because carbon addition hinders cross-slip and promotes twinning over the formation of dislocation cell structures.
4. At a microstructural level, annealing of carbon containing CoCrFeMnNi alloys gives rise to nanoscale $M_{23}C_6$ and M_7C_3 type carbides precipitates which leads to strengthening and grain refinement. Strengthening happens due to precipitation hardening and grain refinement due to carbides pinning the grain boundaries during annealing. The ideal annealing temperature is in the range 800–900°C for obtaining maximum hardness. The size of the carbides increases and their volume fraction decreases with increase in annealing temperature.

5. At low carbon content, (<1 at.%), carbides precipitate at the grain boundaries. At higher carbon contents, carbides are found to precipitate on grain boundaries as well as grain interior. Cold rolling helps to disperse the precipitates in a uniform manner. The $M_{23}C_6$-type carbides phase fraction is predominant at higher carbon content compared to the M_7C_3 type. Carbide phase fraction increases with carbon addition but no change in the size of the precipitates was noticed with increasing carbon content.
6. Carbon addition up to 1 at.% increases the overall strength as well as ductility. Larger carbon content results in higher strength, but with a drastic reduction in ductility. This can be controlled by employing proper thermomechanical treatments. Carbon addition increases the extent of work hardening. The increase in the extent of work hardening, especially at low temperatures, can be attributed to the higher twinning tendency upon carbon addition.

The scope of powder metallurgy in providing superior carbon-containing-CoCrFeMnNi alloys are explored by studying the effect of 2 wt.% graphite flake addition on the microstructure of CoCrFeMnNi made by mechanical alloying and spark plasma sintering. The salient features of this work are discussed below:

1. Graphite addition inhibits the mechanical alloying of the constituent elements in CoCrFeMnNi to form single phase FCC. An FCC phase with WC contamination is noticed in 20 h milled powder of CoCrFeMnNi. A BCC phase rich in chromium is predominant in 20 milled powder of CoCrFeMnNi + 2 wt.% graphite along with the FCC phase.
2. Graphite addition eliminates the tungsten carbide contamination from the wear of balls and vials of the high energy planetary mills.
3. Graphite addition reduces the size of the milled powders.
4. Upon sintering, CoCrFeMnNi powders forms an FCC and BCC phase along with M_7 C_3 carbides and WC phases. MnO formation was also observed.
5. The 10 h milled graphite-containing powder developed a coarse microstructure due to inhibition of alloying. For the 20 h milled and sintered compacts, grain size was observed to be higher for graphite added CoCrFeMnNi due to decreased chromium and tungsten carbide precipitates which can pin the grain boundary. The 20 h milled powders resulted in ultrafine-grained compacts after SPS.
6. Hardness of 732 VHN (the maximum reported so far) was observed for CoCrFeMnNi due to fine-grained microstructure and presence of WC and Cr_7C_3. Graphite-added CoCrFeMnNi showed a constant hardness of around 550 VHN for both milling times of 10 h and 20 h.

ACKNOWLEDGMENTS

Dr Srinivasa Rao Bakshi would like to acknowledge funding from IIT Madras under the Institute Research and Development Award (Grant No. MET1617839RFIRSRRB).

REFERENCES

1. Yeh, J. W.; Chen, S. K.; Lin, S. J.; Gan, J. Y.; Chin, T. S.; Shun, T. T.; Tsau, C. H.; Chang, S. Y. Nanostructured High-Entropy Alloys with Multiple Principal Elements: Novel Alloy Design Concepts and Outcomes. *Adv. Eng. Mater.* 2004, *6*(5), 299–303+274. doi:10.1002/adem.200300567.
2. Ye, Y. F.; Wang, Q.; Lu, J.; Liu, C. T.; Yang, Y. High-Entropy Alloy: Challenges and Prospects. *Mater. Today* 2015. doi:10.1016/j.mattod.2015.11.026.
3. Miracle, D. B.; Senkov, O. N. A Critical Review of High Entropy Alloys and Related Concepts. *Acta Mater.* 2017, *122*, 448–511. doi:10.1016/j.actamat.2016.08.081.
4. Diao, H. Y.; Feng, R.; Dahmen, K. A.; Liaw, P. K. Fundamental Deformation Behavior in High-Entropy Alloys: An Overview. *Curr. Opin. Solid State Mater. Sci.* 2017, *21*(5), 252–266. doi:10.1016/j.cossms.2017.08.003.
5. Zhang, Y.; Zuo, T. T.; Tang, Z.; Gao, M. C.; Dahmen, K. A.; Liaw, P. K.; Lu, Z. P. Microstructures and Properties of High-Entropy Alloys. *Prog. Mater. Sci.* 2014, *61*(October 2013), 1–93. doi:10.1016/j.pmatsci.2013.10.001.
6. Cantor, B.; Chang, I. T. H.; Knight, P.; Vincent, A. J. B. Microstructural Development in Equiatomic Multicomponent Alloys. *Mater. Sci. Eng. A* 2004, *375–377*(1-2 SPEC. ISS.), 213–218. doi:10.1016/j.msea.2003.10.257.
7. Gali, A.; George, E. P. Tensile Properties of High- and Medium-Entropy Alloys. *Intermetallics* 2013, *39*, 74–78. doi:10.1016/j.intermet.2013.03.018.
8. Otto, F.; Dlouhý, A.; Somsen, C.; Bei, H.; Eggeler, G.; George, E. P. The Influences of Temperature and Microstructure on the Tensile Properties of a CoCrFeMnNi High-Entropy Alloy. *Acta Mater.* 2013, *61*(15), 5743–5755. doi:10.1016/j.actamat.2013.06.018.
9. Yao, M. J.; Pradeep, K. G.; Tasan, C. C.; Raabe, D. A Novel, Single Phase, Non-Equiatomic FeMnNiCoCr High-Entropy Alloy with Exceptional Phase Stability and Tensile Ductility. *Scr. Mater.* 2014, *72–73*, 5–8. doi:10.1016/j.scriptamat.2013.09.030.
10. Otto, F.; Hanold, N. L.; George, E. P. Microstructural Evolution after Thermomechanical Processing in an Equiatomic, Single-Phase CoCrFeMnNi High-Entropy Alloy with Special Focus on Twin Boundaries. *Intermetallics* 2014. doi:10.1016/j.intermet.2014.05.014.
11. Ma, D.; Yao, M.; Pradeep, K. G.; Tasan, C. C.; Springer, H.; Raabe, D. Phase Stability of Non-Equiatomic CoCrFeMnNi High Entropy Alloys. *Acta Mater.* 2015, *98*, 288–296. doi:10.1016/j.actamat.2015.07.030.
12. Cheng, H.; Chen, W.; Liu, X.; Tang, Q.; Xie, Y.; Dai, P. Effect of Ti and C Additions on the Microstructure and Mechanical Properties of the FeCoCrNiMn High-Entropy Alloy. *Mater. Sci. Eng. A* 2018, *719*(January), 192–198. doi:10.1016/j.msea.2018.02.040.
13. Xie, Y.; Cheng, H.; Tang, Q.; Chen, W.; Chen, W.; Dai, P. Effects of N Addition on Microstructure and Mechanical Properties of CoCrFeNiMn High Entropy Alloy Produced by Mechanical Alloying and Vacuum Hot Pressing Sintering. *Intermetallics* 2018, *93*(September 2017), 228–234. doi:10.1016/j.intermet.2017.09.013.
14. Seol, J. B.; Bae, J. W.; Li, Z.; Chan Han, J.; Kim, J. G.; Raabe, D.; Kim, H. S. Boron Doped Ultrastrong and Ductile High-Entropy Alloys. *Acta Mater.* 2018, *151*, 366–376. doi:10.1016/j.actamat.2018.04.004.
15. Ritchie, R. O. The Conflicts Between Strength and Toughness. *Nat. Mater.* 2011, *10*(11), 817–822. doi:10.1038/nmat3115.
16. Li, Z.; Pradeep, K. G.; Deng, Y.; Raabe, D.; Tasan, C. C. Metastable High-Entropy Dual-Phase Alloys Overcome the Strength-Ductility Trade-Off. *Nature* 2016, *534*(7606), 227–230. doi:10.1038/nature17981.

17. Bouaziz, O.; Allain, S.; Scott, C. P.; Cugy, P.; Barbier, D. High Manganese Austenitic Twinning Induced Plasticity Steels: A Review of the Microstructure Properties Relationships. *Curr. Opin. Solid State Mater. Sci.* 2011, *15*(4), 141–168. doi:10.1016/j.cossms.2011.04.002.
18. Meyers, M.; Chawla, K. *Mechanical Behavior of Materials*. Cambridge university press, 2009; pp. 559–562.
19. Zhao, Y. H.; Zhu, Y. T.; Liao, X. Z.; Horita, Z.; Langdon, T. G. Tailoring Stacking Fault Energy for High Ductility and High Strength in Ultrafine Grained Cu and Its Alloy. *Appl. Phys. Lett.* 2006, *89*(12), 2004–2007. doi:10.1063/1.2356310.
20. Gong, Y. L.; Wen, C. E.; Li, Y. C.; Wu, X. X.; Cheng, L. P.; Han, X. C.; Zhu, X. K. Simultaneously Enhanced Strength and Ductility of Cu-XGe Alloys through Manipulating the Stacking Fault Energy (SFE). *Mater. Sci. Eng. A* 2013, *569*, 144–149. doi:10.1016/j.msea.2013.01.022.
21. Sun, P. L.; Zhao, Y. H.; Cooley, J. C.; Kassner, M. E.; Horita, Z.; Langdon, T. G.; Lavernia, E. J.; Zhu, Y. T. Effect of Stacking Fault Energy on Strength and Ductility of Nanostructured Alloys: An Evaluation with Minimum Solution Hardening. *Mater. Sci. Eng. A* 2009, *525*(1–2), 83–86. doi:10.1016/j.msea.2009.06.030.
22. Zaddach, A. J.; Niu, C.; Koch, C. C.; Irving, D. L. Mechanical Properties and Stacking Fault Energies of NiFeCrCoMn High-Entropy Alloy. *JOM* 2013, *65*(12), 1780–1789. doi:10.1007/s11837-013-0771-4.
23. Su, J.; Raabe, D.; Li, Z. Hierarchical Microstructure Design to Tune the Mechanical Behavior of an Interstitial TRIP-TWIP High-Entropy Alloy. *Acta Mater.* 2019, *163*, 40–54. doi:10.1016/j.actamat.2018.10.017.
24. Wang, M.; Li, Z.; Raabe, D. In-Situ SEM Observation of Phase Transformation and Twinning Mechanisms in an Interstitial High-Entropy Alloy. *Acta Mater.* 2018, *147*, 236–246. doi:10.1016/j.actamat.2018.01.036.
25. Liu, S. F.; Wu, Y.; Wang, H. T.; Lin, W. T.; Shang, Y. Y.; Liu, J. B.; An, K.; Liu, X. J.; Wang, H.; Lu, Z. P. Transformation-Reinforced High-Entropy Alloys with Superior Mechanical Properties via Tailoring Stacking Fault Energy. *J. Alloys Compd.* 2019, *792*, 444–455. doi:10.1016/j.jallcom.2019.04.035.
26. Wu, Z.; Parish, C. M.; Bei, H. Nano-Twin Mediated Plasticity in Carbon-Containing FeNiCoCrMn High Entropy Alloys. *J. Alloys Compd.* 2015, *647*, 815–822. doi:10.1016/j.jallcom.2015.05.224.
27. Laplanche, G.; Kostka, A.; Horst, O. M.; Eggeler, G.; George, E. P. Microstructure Evolution and Critical Stress for Twinning in the CrMnFeCoNi High-Entropy Alloy. *Acta Mater.* 2016, *118*, 152–163. doi:10.1016/j.actamat.2016.07.038.
28. Lu, K.; Lu, L.; Suresh, S. Strengthening Materials by Boundaries at the Nanoscale. *Science* 2009, *349*(April), 349–353. doi:10.1126/science.1159610.
29. Choi, W. M.; Jo, Y. H.; Sohn, S. S.; Lee, S.; Lee, B. J. Understanding the Physical Metallurgy of the CoCrFeMnNi High-Entropy Alloy: An Atomistic Simulation Study. *npj Comput. Mater.* 2018, *4*(1), 1–9. doi:10.1038/s41524-017-0060-9.
30. Stepanov, N.; Tikhonovsky, M.; Yurchenko, N.; Zyabkin, D.; Klimova, M.; Zherebtsov, S.; Efimov, A.; Salishchev, G. Effect of Cryo-Deformation on Structure and Properties of CoCrFeNiMn High-Entropy Alloy. *Intermetallics* 2015, *59*, 8–17. doi:10.1016/j.intermet.2014.12.004.
31. Smallman, R. E. Modern Physical Metallurgy and Materials Engineering. *Mod. Phys. Metall. Mater. Eng.* 2016. doi:10.1016/b978-0-7506-4564-5.x5000-9.
32. Stepanov, N. D.; Shaysultanov, D. G.; Chernichenko, R. S.; Yurchenko, N. Y.; Zherebtsov, S. V.; Tikhonovsky, M. A.; Salishchev, G. A. Effect of Thermomechanical Processing on Microstructure and Mechanical Properties of the Carbon-Containing CoCrFeNiMn High Entropy Alloy. *J. Alloys Compd.* 2016, *693*, 394–405. doi:10.1016/j.jallcom.2016.09.208.

33. Gludovatz, B.; Hohenwarter, A.; Catoor, D.; Chang, E. H.; George, E. P.; Ritchie, R. O. ChemInform Abstract: A Fracture-Resistant High-Entropy Alloy for Cryogenic Applications. *ChemInform* 2014, *45*(47), no-no. doi:10.1002/chin.201447007.
34. Huang, S.; Li, W.; Lu, S.; Tian, F.; Shen, J.; Holmström, E.; Vitos, L. Temperature Dependent Stacking Fault Energy of FeCrCoNiMn High Entropy Alloy. *Scr. Mater.* 2015, *108*, 44–47. doi:10.1016/j.scriptamat.2015.05.041.
35. Saeed-Akbari, A.; Imlau, J.; Prahl, U.; Bleck, W. Derivation and Variation in Composition-Dependent Stacking Fault Energy Maps Based on Subregular Solution Model in High-Manganese Steels. *Metall. Mater. Trans. A Phys. Metall. Mater. Sci.* 2009, *40*(13), 3076–3090. doi:10.1007/s11661-009-0050-8.
36. Schramm, R. E.; Reed, R. P. Stacking Fault Energies of Seven Commercial Austenitic Stainless Steels. *Metall. Trans. A* 1975. doi:10.1007/BF02641927.
37. Ko, J. Y.; Hong, S. I. Microstructural Evolution and Mechanical Performance of Carbon-Containing CoCrFeMnNi-C High Entropy Alloys. *J. Alloys Compd.* 2018, *743*, 115–125. doi:10.1016/j.jallcom.2018.01.348.
38. Li, Z.; Tasan, C. C.; Springer, H.; Gault, B.; Raabe, D. Interstitial Atoms Enable Joint Twinning and Transformation Induced Plasticity in Strong and Ductile High-Entropy Alloys. *Sci. Rep.* 2017, *7*(January), 1–7. doi:10.1038/srep40704.
39. Liu, T.; Gao, Y.; Bei, H.; An, K. In Situ Neutron Diffraction Study on Tensile Deformation Behavior of Carbon-Strengthened CoCrFeMnNi High-Entropy Alloys at Room and Elevated Temperatures. *J. Mater. Res.* 2018, *33*(19), 3192–3203. doi:10.1557/jmr.2018.180.
40. Stepanov, N. D.; Yurchenko, N. Y.; Tikhonovsky, M. A.; Salishchev, G. A. Effect of Carbon Content and Annealing on Structure and Hardness of the CoCrFeNiMn-Based High Entropy Alloys. *J. Alloys Compd.* 2016, *687*, 59–71. doi:10.1016/j.jallcom.2016.06.103.
41. Cheng, H.; Wang, H. Y.; Xie, Y. C.; Tang, Q. H.; Dai, P. Q. Controllable Fabrication of a Carbide-Containing FeCoCrNiMn High-Entropy Alloy: Microstructure and Mechanical Properties. *Mater. Sci. Technol. (United Kingdom)* 2017, *33*(17), 2032–2039. doi:10.1080/02670836.2017.1342367.
42. Klimova, M.; Stepanov, N.; Shaysultanov, D.; Chernichenko, R.; Yurchenko, N.; Sanin, V.; Zherebtsov, S. Microstructure and Mechanical Properties Evolution of the Al, C-Containing CoCrFeNiMn-Type High-Entropy Alloy during Cold Rolling. *Materials (Basel)* 2017, *11*(1), 1–14. doi:10.3390/ma11010053.
43. Chen, J.; Yao, Z.; Wang, X.; Lu, Y.; Wang, X.; Liu, Y.; Fan, X. Effect of C Content on Microstructure and Tensile Properties of as-Cast CoCrFeMnNi High Entropy Alloy. *Mater. Chem. Phys.* 2018, *210*, 136–145. doi:10.1016/j.matchemphys.2017.08.011.
44. Bouaziz, O.; Zurob, H.; Chehab, B.; Embury, J. D.; Allain, S.; Huang, M. Effect of Chemical Composition on Work Hardening of Fe—Mn—C TWIP Steels. *Mater. Sci. Technol.* 2011, *27*(3), 707–709. doi:10.1179/026708309x12535382371852.
45. Hong, S. I.; Laird, C. Mechanisms of Slip Mode Modification in F. C. C. Solid Solutions. *Acta Metall. Mater.* 1990, *38*(8), 1581–1594. doi:10.1016/0956-7151(90)90126-2.
46. He, J. Y.; Zhu, C.; Zhou, D. Q.; Liu, W. H.; Nieh, T. G.; Lu, Z. P. Steady State Flow of the FeCoNiCrMn High Entropy Alloy at Elevated Temperatures. *Intermetallics* 2014, *55*, 9–14. doi:10.1016/j.intermet.2014.06.015.
47. Li, J.; Gao, B.; Wang, Y.; Chen, X.; Xin, Y.; Tang, S.; Liu, B.; Liu, Y.; Song, M. Microstructures and Mechanical Properties of Nano Carbides Reinforced CoCrFeMnNi High Entropy Alloys. *J. Alloys Compd.* 2019, *792*, 170–179. doi:10.1016/j.jallcom.2019.03.403.
48. Guo, L.; Ou, X.; Ni, S.; Liu, Y.; Song, M. Effects of Carbon on the Microstructures and Mechanical Properties of FeCoCrNiMn High Entropy Alloys. *Mater. Sci. Eng. A* 2019, *746*(January), 356–362. doi:10.1016/j.msea.2019.01.050.

49. Li, J.; Gao, B.; Tang, S.; Liu, B.; Liu, Y.; Wang, Y. T.; Wang, J. High Temperature Deformation Behavior of Carbon-Containing FeCoCrNiMn High Entropy Alloy. *J. Alloys Compd.* 2018, *747*, 571–579. doi:10.1016/j.jallcom.2018.02.332.
50. Klimova, M. V.; Shaysultanov, D. G.; Chernichenko, R. S.; Sanin, V. N.; Stepanov, N. D.; Zherebtsov, S. V.; Belyakov, A. N. Recrystallized Microstructures and Mechanical Properties of a C-Containing CoCrFeNiMn-Type High-Entropy Alloy. *Mater. Sci. Eng. A* 2019, *740–741*(September 2018), 201–210. doi:10.1016/j.msea.2018.09.113.
51. Li, Z. Interstitial Equiatomic CoCrFeMnNi High-Entropy Alloys: Carbon Content, Microstructure, and Compositional Homogeneity Effects on Deformation Behavior. *Acta Mater.* 2019, *164*, 400–412. doi:10.1016/j.actamat.2018.10.050.
52. Li, Z.; Raabe, D. Influence of Compositional Inhomogeneity on Mechanical Behavior of an Interstitial Dual-Phase High-Entropy Alloy. *Mater. Chem. Phys.* 2018, *210*, 29–36. doi:10.1016/j.matchemphys.2017.04.050.
53. Suryanarayana, C.; Ivanov, E.; Boldyrev, V. V. The Science and Technology of Mechanical Alloying. *Mater. Sci. Eng. A* 2001. doi:10.1016/S0921-5093(00)01465-9.
54. Torralba, J. M.; Alvaredo, P.; García-Junceda, A. High-Entropy Alloys Fabricated via Powder Metallurgy. A Critical Review. *Powder Metall.* 2019, *62*(2), 84–114. doi:10.1080/00325899.2019.1584454.
55. Gao, N.; Lu, D. H.; Zhao, Y. Y.; Liu, X. W.; Liu, G. H.; Wu, Y.; Liu, G.; Fan, Z. T.; Lu, Z. P.; George, E. P. Strengthening of a CrMnFeCoNi High-Entropy Alloy by Carbide Precipitation. *J. Alloys Compd.* 2019, *792*, 1028–1035. doi:10.1016/j.jallcom.2019.04.121.
56. Sharma, A. S.; Yadav, S.; Biswas, K.; Basu, B. Materials Science & Engineering R High-Entropy Alloys and Metallic Nanocomposites: Processing Challenges , Microstructure Development and Property Enhancement. *Mater. Sci. Eng. R* 2018, *131*(June), 1–42. doi:10.1016/j.mser.2018.04.003.
57. Zhang, A.; Han, J.; Su, B.; Meng, J. A Novel CoCrFeNi High Entropy Alloy Matrix Self-Lubricating Composite. *J. Alloys Compd.* 2017, *725*, 700–710. doi:10.1016/j.jallcom.2017.07.197.
58. Zhang, A.; Han, J.; Su, B.; Li, P.; Meng, J. Microstructure, Mechanical Properties and Tribological Performance of CoCrFeNi High Entropy Alloy Matrix Self-Lubricating Composite. *Mater. Des.* 2017, *114*, 253–263. doi:10.1016/j.matdes.2016.11.072.
59. Zhang, A.; Han, J.; Su, B.; Meng, J. A Promising New High Temperature Self-Lubricating Material: CoCrFeNiS0.5 High Entropy Alloy. *Mater. Sci. Eng. A* 2018, *731*(February), 36–43. doi:10.1016/j.msea.2018.06.030.
60. Ji, W.; Wang, W.; Wang, H.; Zhang, J.; Wang, Y.; Zhang, F.; Fu, Z. Alloying Behavior and Novel Properties of CoCrFeNiMn High-Entropy Alloy Fabricated by Mechanical Alloying and Spark Plasma Sintering. *Intermetallics* 2014, *56*, 24–27. doi:10.1016/j.intermet.2014.08.008.
61. Cheng, H.; Xie, Y. Chong; Tang, Q. Hua; Rao, C.; Dai, P. Qiang. Microstructure and Mechanical Properties of FeCoCrNiMn High-Entropy Alloy Produced by Mechanical Alloying and Vacuum Hot Pressing Sintering. *Trans. Nonferrous Met. Soc. China (English. Ed.)* 2018, *28*(7), 1360–1367. doi:10.1016/S1003-6326(18)64774-0.
62. Ayyagari, A.; Hasannaeimi, V.; Grewal, H.; Arora, H.; Mukherjee, S. Corrosion, Erosion and Wear Behavior of Complex Concentrated Alloys: A Review. Metals 2018, *8*(8). doi:10.3390/met8080603.
63. Ayyagari, A.; Barthelemy, C.; Gwalani, B.; Banerjee, R.; Scharf, T. W.; Mukherjee, S. Reciprocating Sliding Wear Behavior of High Entropy Alloys in Dry and Marine Environments. *Mater. Chem. Phys.* 2018, *210*, 162–169. doi:10.1016/j.matchemphys.2017.07.031.

64. Joseph, J.; Haghdadi, N.; Shamlaye, K.; Hodgson, P.; Barnett, M.; Fabijanic, D. The Sliding Wear Behaviour of CoCrFeMnNi and AlxCoCrFeNi High Entropy Alloys at Elevated Temperatures. *Wear* 2019, *428–429*(March), 32–44. doi:10.1016/j.wear.2019.03.002.
65. Rekha, M. Y.; Mallik, N.; Srivastava, C. First Report on High Entropy Alloy Nanoparticle Decorated Graphene. *Sci. Rep.* 2018, *8*(1), 1–10. doi:10.1038/s41598-018-27096-8.
66. Park, S.; Park, C.; Na, Y.; Kim, H. S.; Kang, N. Effects of (W, Cr) Carbide on Grain Refinement and Mechanical Properties for CoCrFeMnNi High Entropy Alloys. *J. Alloys Compd.* 2019, *770*, 222–228. doi:10.1016/j.jallcom.2018.08.115.
67. Eißmann, N.; Klöden, B.; Weißgärber, T.; Kieback, B. High-Entropy Alloy CoCrFeMnNi Produced by Powder Metallurgy. *Powder Metall.* 2017, *60*(3), 184–197. doi:10.1080/00325899.2017.1318480.
68. Joo, S. H.; Kato, H.; Jang, M. J.; Moon, J.; Kim, E. B.; Hong, S. J.; Kim, H. S. Structure and Properties of Ultrafine-Grained CoCrFeMnNi High-Entropy Alloys Produced by Mechanical Alloying and Spark Plasma Sintering. *J. Alloys Compd.* 2017, *698*, 591–604. doi:10.1016/j.jallcom.2016.12.010.
69. Vaidya, M.; Karati, A.; Marshal, A.; Pradeep, K. G.; Murty, B. S. Phase Evolution and Stability of Nanocrystalline CoCrFeNi and CoCrFeMnNi High Entropy Alloys. *J. Alloys Compd.* 2019, *770*, 1004–1015. doi:10.1016/j.jallcom.2018.08.200.
70. Ko, J. Y.; Song, J. S.; Hong, S. I. Effect of Carbon Addition and Recrystallization on the Microstructure and Mechanical Properties of CoCrFeMnNi High Entropy Alloys. *J. Korean Inst. Met. Mater.* 2018. doi:10.3365/KJMM.2018.56.1.26.
71. Ko, J. Y.; Hong, S. I. Effect of Carbon Addition on the Cast and Rolled Microstructures of FeCoCrNiMn High Entropy Alloys. *Key Eng. Mater.* 2017. doi:10.4028/www.scientific.net/kem.737.16.
72. Kilmametov, A.; Kulagin, R.; Mazilkin, A.; Seils, S.; Boll, T.; Heilmaier, M.; Hahn, H. High-Pressure Torsion Driven Mechanical Alloying of CoCrFeMnNi High Entropy Alloy. *Scr. Mater.* 2019, *158*, 29–33. doi:10.1016/j.scriptamat.2018.08.031.

5 Role of Processing and Silicon Addition to $CoCrCuFeNiSi_x$ High Entropy Alloys

Anil Kumar and Manoj Chopkar

CONTENTS

5.1 INTRODUCTION

The concept of high entropy alloys was introduced by Cantor and co-workers in 1981. They discovered a new alloy having a single-phase face-centered cubic solid solution consisting of iron, chromium, manganese, nickel, and cobalt elements in equal proportions. Further, in the mid-1900s, Yeh and his colleagues explored the world of multicomponent alloys [1]. The unique ability of high entropy alloys to form an often single solid solution with excellent properties such as wear resistance, high hardness, softening resistance at the higher temperature, corrosion, and oxidation resistance has created immense interest among researchers [2–8].

High entropy alloys can be obtained in different forms ranging from ingots or slabs to powders, and from bulk to thin film. However, high entropy alloys are commonly prepared by two routes viz., powder metallurgy and vacuum-arc melting [9–26]. The CoCrCuFeNi high entropy alloy has been prepared by the powder metallurgy route by Praveen and co-workers [15]. They reported the formation of two face-centered cubic (FCC) phases and a small amount of sigma (σ) phase. Tong and co-workers [27] prepared the same alloy through the casting route and observed the presence of a single face-centered cubic phase. Further, Ji and co-workers investigated the phase evolution on AlCoCrFeNi alloy prepared by mechanical alloying (MA) and subsequently consolidated by spark plasma sintering (SPS) [28]. Ji and co-workers observed the evolution of mixed phases of face-centered and body-centered cubic (BCC) after spark plasma sintering, whereas a single body-centered cubic phase evolved when the same composition of the alloys was prepared through the arc melting route by Wang and co-workers [29]. Similarly, Qui and co-workers [30] prepared the AlCrFeNiCoCu high entropy alloy by the powder metallurgy route and observed a mixture of body-centered cubic and face-centered cubic phases after sintering the hydraulically pressed alloy. This alloy was also synthesized by Shaysultanov and co-workers [31] through the induction melting technique and they observed three phases which include body-centered cubic, face-centered cubic, and B2 phases. Therefore, it can be approximated that phase formation is strongly dependent on the processing route for a particular high entropy alloy and this effect is worthy of investigation. Furthermore, different processing routes can offer a variety of transformations to obtain phases in high entropy alloys from the multicomponent elements. Consequently, the processing route is certain to affect the phase evolution in high entropy alloys in a severe way.

Furthermore, the majority of works reported in the literature were on the atomic percentage variation of any one transition element of the constituent composition in high entropy alloys [32–44]. It is important to note that, in conventional materials like steel, the addition of silicon is always favorable for structural as well as mechanical properties. In conventional materials such as steels, a silicon addition is always constructive to structure and also for improvement of the mechanical properties. Kumar and co-workers [45] and Liu and co-workers [46] observed that the silicon addition promotes the body-centered cubic phase evolution compared to face-centered cubic on $AlCoCrCuFeNiSi_x$ and $Al_{0.5}CoCrCuFeNiSi_x$ alloy systems, respectively. It is important to note that there are plenty of previous reports available to investigate the effect of metallic elements for phase evolution and mechanical property improvements of high entropy alloys. However, there are very few studies that have been done to explore the effect of non-metallic elements such as silicon. Hence, in the present work, an attempt has been made to investigate the effect of silicon on phase evolution and mechanical properties of CuCrCoFeNi alloy and also to explore the effect of the processing route on the phase evolution and physical properties of high entropy alloys prepared by vacuum-arc melting, spark plasma sintering, and mechanical alloying.

5.2 SYNTHESIS THROUGH SPARK PLASMA SINTERING AND ARC MELTING

Bulk samples were obtained by two different processing routes. In the first route, the mixed powders were consolidated by spark plasma sintering (Dr. Sinter SPS-625, Fuji Electronic Industrial Co. Ltd, Japan) in a graphite die with an inner diameter of

10 mm at 1,000°C for 5 min under uniform pressure of 60 MPa. The initial heating rate of 100°C/min was used between 570°C to 800°C and reduced to 50°C/min up to 1,000°C. In the second route, the green compacts were melted in a water-cooled copper mold under an argon (Ar) atmosphere. The samples were remelted at least six times to improve chemical homogeneity, and finally cast. The processed cast rods were subjected to metallographic sample preparation methods to attain the mirror polish surface. The structures of the samples were characterized by X-ray diffractometer (Bruker D8 Advance) using Cu $K\alpha$ (λ = 0.1541 nm) radiation (with 0.02° step size and 10°/min scan rate). The microstructure of the bulk samples and compositional analysis was studied by a scanning electron microscope (Zeiss EVO 18, Germany) equipped with Oxford Inca energy dispersive X-ray spectroscopy.

5.2.1 X-Ray Diffraction Analysis

The X-ray diffraction pattern of $CoCrCuFeNiSi_x$ (x = 0, 0.3, 0.6, and 0.9) alloys prepared by spark plasma sintering indicates the formation of two face-centered cubic phases (F1 and F2) along with sigma (σ) phase (Figure 5.1a). As copper has high positive mixing enthalpy with other constituent elements in the alloy, resulting in low miscibility between copper and other elements [47], copper thus tends to segregate at the grain boundaries. Although a positive enthalpy of mixing of copper with other elements, the strong affinity of copper and nickel inhibits the complete segregation of copper into the face-centered cubic phase and leads to the formation of the F2 phase after spark plasma sintering. It is well-known that nickel is a face-centered cubic austenite stabilizer in nickel-containing steels. Hence, the F1 phase could be the nickel-iron-cobalt face-centered cubic phase with some amount of chromium and copper. The similar phase formation behavior of the CoCrCuFeNi high entropy alloy after spark plasma sintering was also observed in the work of Praveen and co-workers and Wang and co-workers [15, 39]. The sigma phase can be associated with the compound of cobalt-chromium or iron-chromium, which is also evident from their phase diagrams [15, 48]. The sigma phase has also been reported in cobalt, chromium, and iron-containing alloys viz., CoCrFeMnNi [49] and $AlCoCr_xFeMo_{0.5}Ni$ [50]. This

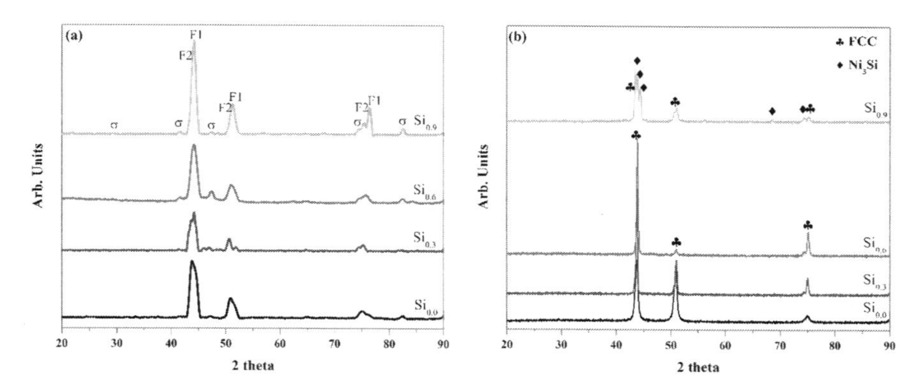

FIGURE 5.1 X-ray diffraction pattern of bulk samples prepared by (a) spark plasma sintering and (b) vacuum-arc melting [16].

sigma phase was observed in previous studies and mainly attributed to the cobalt-chromium/iron-chromium-related tetragonal phase [51]. Fu and co-workers [14] discussed the existence of a sigma phase in the $Co_{0.5}FeNiCrTi_{0.5}$ high entropy alloy prepared by mechanical alloying and spark plasma sintering. They observed that densification at 1,273 K for 8 min has resulted in a stable σ phase. Therefore, in the present study, the sigma phase evolved after spark plasma sintering might also be a stable phase at 1,273 K. Table 5.1 reports the phase fraction (PF) and lattice parameter (LP) of $CoCrCuFeNiSi_x$ alloys. It is observed that the increase in the amount of silicon favors the formation of sigma phase.

The pseudo-Voigt function was used for fitting the X-ray diffraction peak profile and silicon was used as the standard sample to deduct the instrumental contribution. Phase fraction was calculated by Equation (5.1):

$$X_{pi} = I_{pi}^{(h_i k_i l_i)} / \sum I_{pi}^{(h_i k_i l_i)} \tag{5.1}$$

where X_{pi} is the phase fraction of ith phase and $(h_i k_i l_i)$ represents the integrated intensity of the $(h_i k_i l_i)$ peak [10].

To show the splitting between F1 and F2 phases, the most intense peak of $CoCrCuFeNiSi_{0.9}$ high entropy alloy is selected for de-convolution. Figure 5.2 clearly indicates the difference in the peaks of F1 and F2 phases of the $CoCrCuFeNiSi_{0.9}$ high entropy alloy.

The X-ray diffraction patterns for alloys prepared by the casting route are illustrated in Figure 5.1b. The CoCrCuFeNi high entropy alloy with Si_0, $Si_{0.3}$, and $Si_{0.6}$ shows a single face-centered cubic phase solid solution. In cast samples, up to

TABLE 5.1
Phase Fractions (PF) and Lattice Parameters (LP) of $CoCrCuFeNiSi_x$ (x = 0, 0.3. 0.6, and 0.9) High Entropy Alloys after Spark Plasma Sintering and Casting [16]

		Spark plasma sintering		As cast	
Alloys	**Phases**	**PF (%)**	**LP (Å)**	**PF (%)**	**LP (Å)**
$CoCrCuFeNiSi_{0.0}$	F1	49	3.5024	FCC – 100	3.5764
	F2	47	3.5999		
	σ	4	2.9014		
$CoCrCuFeNiSi_{0.3}$	F1	46	3.4985	FCC – 100	3.5733
	F2	41	3.5632		
	σ	13	2.8841		
$CoCrCuFeNiSi_{0.6}$	F1	44	3.4524	FCC – 100	3.5632
	F2	37	3.5365		
	σ	19	2.8147		
$CoCrCuFeNiSi_{0.9}$	F1	41	3.4102	FCC – 91	3.5282
	F2	34	3.5342	Ni_3Si – 9	2.5174
	σ	25	2.7842		

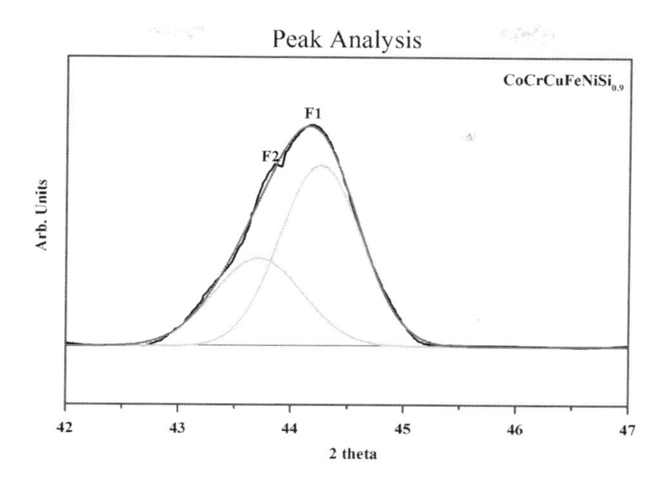

FIGURE 5.2 De-convoluted X-ray diffraction pattern of $CoCrCuFeNiSi_{0.9}$ high entropy alloy [16].

0.9 at.% of silicon addition, there is no other phase formed except the face-centered cubic phase, which confirms that the face-centered cubic phase is a stable phase in these alloys at elevated temperatures. The face-centered cubic phase remains intact after cooling due to sluggish diffusion in multicomponent alloys [52], whereas in the $CoCrCuFeNiSi_{0.9}$ alloy, the presence of the intermetallic compound was detected in X-ray diffraction and identified as a Ni_3Si compound. The formation of the intermetallic compound may be attributed to high negative mixing enthalpy between silicon and nickel elements, which has attractive interaction (Table 5.2). Similarly, Zuo and co-workers [53] also concluded that the silicon addition in $CoFeNiSi_x$ ($x = 0$, 0.25, 0.5, and 0.75) alloy leads to the formation of Ni_3Si if silicon content is equal or above to 0.5.

TABLE 5.2
Mixing Enthalpy (kJ/mol) of Binary Pairs of Constituent Elements Calculated by Meidema's Model [54]

Element	Si	Cr	Fe	Co	Ni	Cu
Melting point (°C)	1414	1890	1538	1492	1453	1083
Atomic radius (pm)	115	128	126	125	124	128
Electronegativity (Pauling)	1.9	1.6	1.8	1.8	1.8	1.9
Si	Si	−37	−35	−38	−40	−19
Cr		Cr	−1	−4	−7	12
Fe			Fe	−1	−2	13
Co				Co	0	6
Ni					Ni	4
Cu						Cu

The lattice parameters (LP) of the alloys decreased as the silicon content increased, as shown in Table 5.1. This is attributed to the small atomic size of the silicon atom, which can easily replace other atoms in the lattice. Therefore, having a mismatch atom in the lattice encourages the increase in the lattice distortion energy considerably, which has a positive impact on solid-solution strengthening. Furthermore, the improvement of solid-solution strengthening provides better wear and hardness properties with higher silicon content in high entropy alloys.

5.2.2 Microstructure Characterization

Scanning electron micrographs in the back-scattered mode of $CoCrCuFeNiSi_x$ alloys prepared by spark plasma sintering are illustrated in Figure 5.3 along with the compositional details by energy dispersive X-ray spectroscopy analysis. The gray and black regions in the scanning electron micrographs indicate the formation of two phases. The gray contrast region has higher nickel, copper, cobalt, and iron content, whereas the black contrast region is chromium-rich. A comparative analysis of X-ray diffraction results (Figure 5.1a) and scanning electron micrographs (Figure 5.3) points out that the gray region comprises two face-centered cubic phases. The outer layers are rich in copper, whereas the inner parts are nickel, cobalt, and iron-rich. Hence, it is conceivable that one of the face-centered cubic phases is iron (nickel, cobalt), i.e., F1 and the other one would be copper (nickel), i.e., F2. The black region is the chromium-rich sigma phase. As the atomic numbers of these components are close to each other, the contrast difference is very low between the two face-centered cubic (F1 and F2) phases in the back-scattered electron mode images. The aforementioned

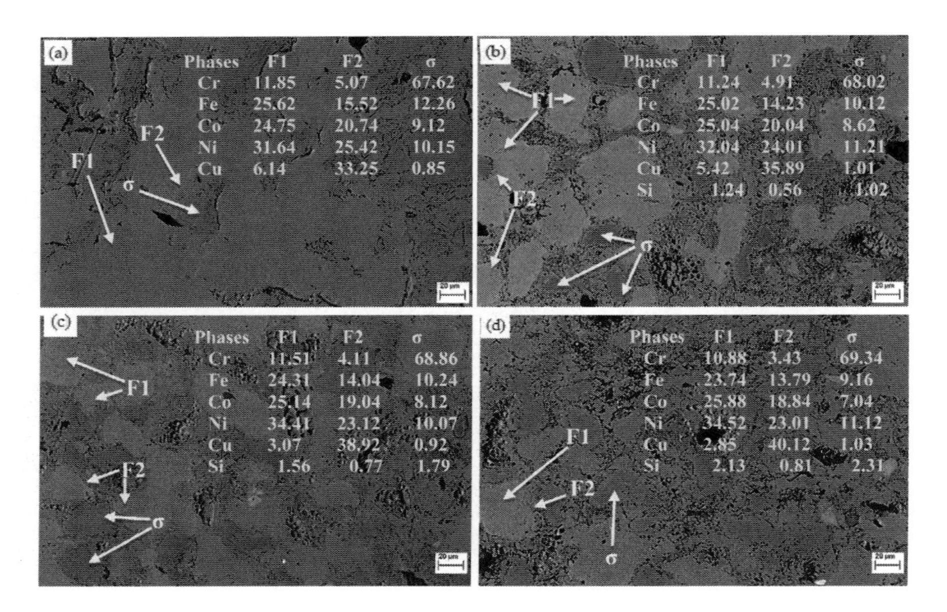

FIGURE 5.3 The scanning electron micrographs of (a) CuCrCoFeNi, (b) $CuCrCoFeNiSi_{0.3}$, (c) $CuCrCoFeNiSi_{0.6}$, and (d) $CuCrCoFeNiSi_{0.9}$ high entropy alloys prepared by the spark plasma sintering process [16].

discussion about the different regions is also in accord with the elemental mapping shown in Figure 5.4. A large amount of copper is segregated at the grain boundaries, and chromium is accumulated inside the matrix. These observations may be explained by the positive mixing enthalpy of chromium with the copper, and its negative mixing enthalpies with the other constituent elements. Thus, chromium is well adopted inside the matrix, while copper is repelled by the majority of the elements due to positive mixing enthalpy of copper with other elements. This results in the separation of copper to form a copper-rich face-centered cubic (F2) phase.

Microstructures of the alloys prepared by casting route are shown in Figure 5.5. The presence of dendritic structure, which is a typical casting characteristic is present in all the alloys. The constituents get divided between the dendritic (DR) and interdendritic (ID) regions based upon their chemical nature and the type of interaction between them. The changes in the microstructure of the $CoCrCuFeNiSi_x$ alloys are different and get intricate with the higher silicon content. For a lower amount of silicon addition (x = 0, 0.3, and 0.6), no significant variations in microstructures were observed (Figure 5.5a,b,c). Whereas, as the silicon content reaches 0.9, a significant variation was observed (Figure 5.5d). A careful analysis of the X-ray diffraction results (Figure 5.1b) along with the scanning electron micrographs (Figure 5.5), indicates that, up to 0.6 of silicon addition, only the face-centered cubic phase is observed, whereas if a further addition of silicon equals 0.9, the microstructure comprises two phases (face-centered cubic and Ni_3Si). The segregation of copper in the interdendritic region evolved as face-centered cubic phase, while silicon gets accumulated in the dendritic region to form a Ni_3Si phase. No significant segregation has been observed for the rest of the elements (iron, nickel, chromium, cobalt) in any region. The formation of a copper-rich phase is attributed to a positive enthalpy of mixing with all other constituent elements. The enthalpy of mixing of all constituents with each other is listed in Table 5.2. The enthalpy of mixing indicates that copper does not have any attractive tendency toward other elements, whereas the

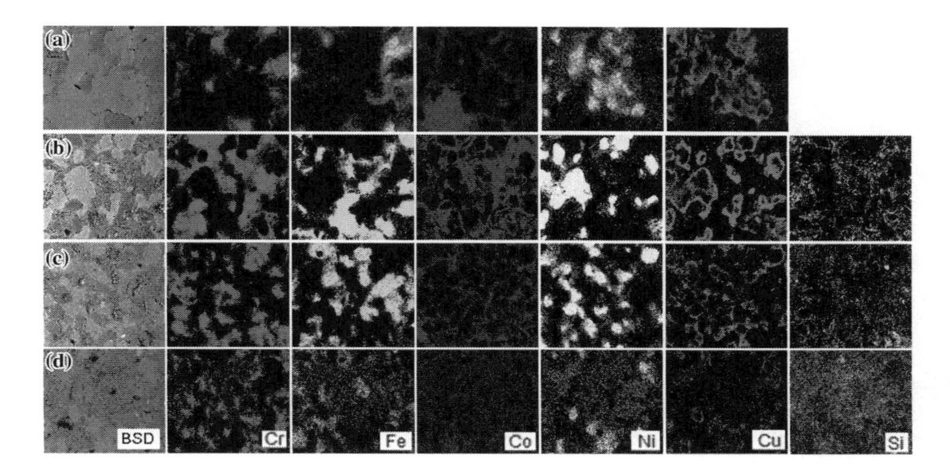

FIGURE 5.4 Elemental mapping of (a) CuCrCoFeNi, (b) $CuCrCoFeNiSi_{0.3}$, (c) $CuCrCoFeNiSi_{0.6}$, and (d) $CuCrCoFeNiSi_{0.9}$ high entropy alloys prepared by the spark plasma sintering process [16].

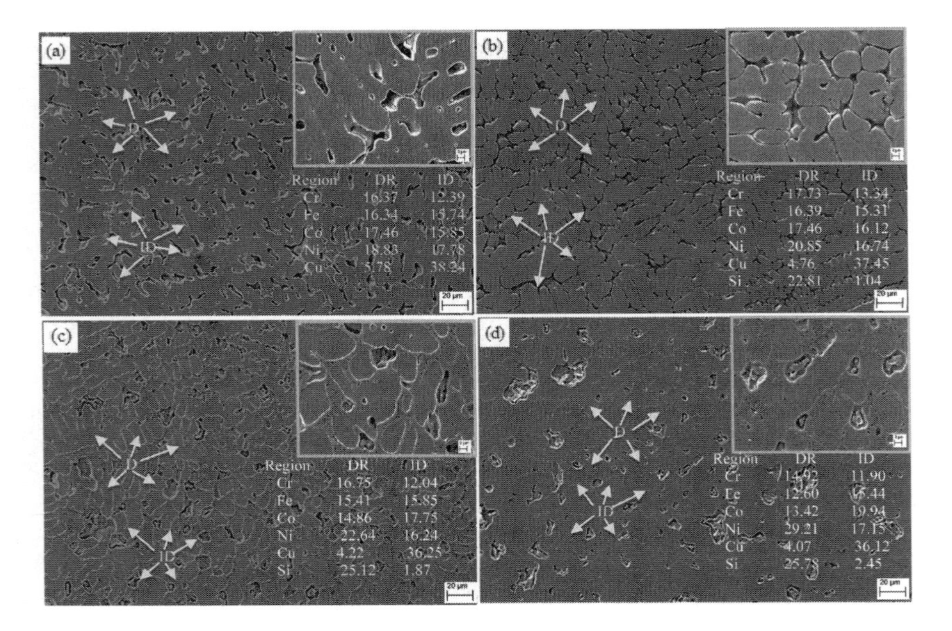

FIGURE 5.5 The scanning electron micrographs (a) CuCrCoFeNi, (b) $CuCrCoFeNiSi_{0.3}$, (c) $CuCrCoFeNiSi_{0.6}$, and (d) $CuCrCoFeNiSi_{0.9}$ high entropy alloys prepared by the arc melting process [16].

silicon-nickel has the highest negative enthalpy of mixing among the constituent elements. Therefore, the formation of Ni_3Si phase with a high value of silicon content (x = 0.9) is highly feasible. The elemental mappings are shown in Figure 5.6. After careful analysis of the elemental mapping, it can be concluded that as the silicon content reaches to x = 0.9, the dendritic region shows a higher concentration of nickel

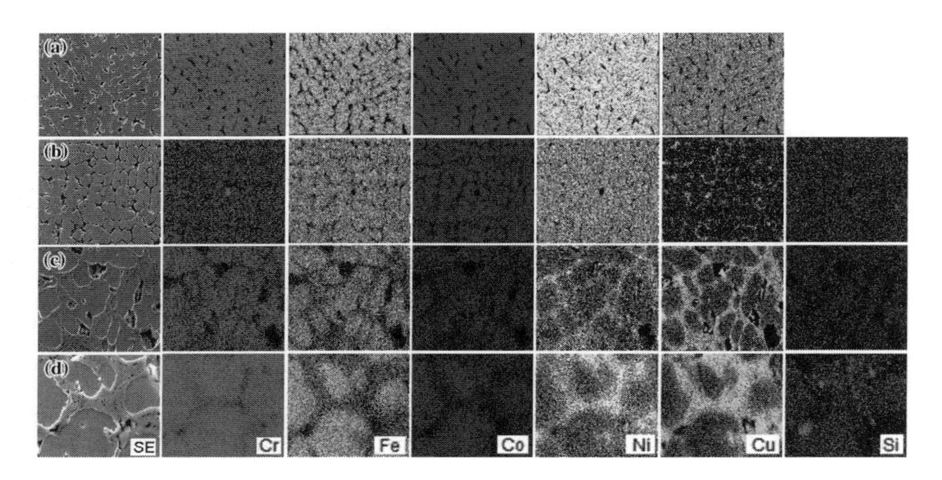

FIGURE 5.6 Elemental mapping of (a) CuCrCoFeNi, (b) $CuCrCoFeNiSi_{0.3}$, (c) $CuCrCoFeNiSi_{0.6}$, and (d) $CuCrCoFeNiSi_{0.9}$ high entropy alloys prepared by the arc melting process [16].

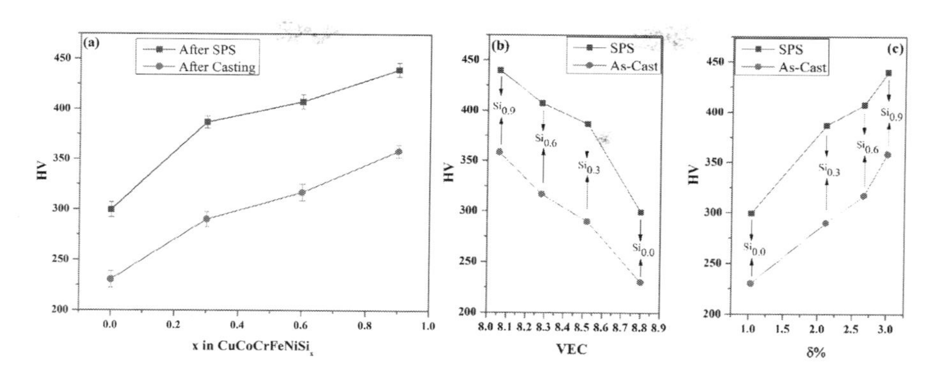

FIGURE 5.7 (a) Hardness values (HV) with respect to silicon content (at.%) in $CuCrCoFeNiSi_x$ (x = 0, 0.3, 0.6, 0.9) high entropy alloys, (b) hardness as a function of valence electron concentration, and (c) atomic size difference (δ %) [16].

and silicon elements which confirm the formation of Ni_3Si phase in $CoCrCuFeNiSi_{0.9}$ high entropy alloy.

5.2.3 Mechanical Properties

The hardness of spark plasma sintered and as-cast alloys are shown in Figure 5.7a. The addition of silicon has a beneficial effect on the hardness of the alloys. The hardness of the Si_0 alloy increases from 300 ± 6 hardness value to 440 ± 5 hardness value in $Si_{0.9}$ alloy after spark plasma sintering. The presence of silicon in the alloy enhances the hardness by a solid-solution strengthening mechanism as discussed previously. This strengthening may be attributed to hard sigma phase evolution after spark plasma sintering. Where in cast alloys, the increment in hardness values were from 230 ± 5 hardness value to 360 ± 7 hardness value for Si_0 and $Si_{0.9}$ high entropy alloys, respectively, therefore, the increase in hardness is observed with silicon addition in both i.e., spark plasma sintering and casting processed high entropy alloys.

The addition of silicon increasing the hardness of high entropy alloys was also reported previously by a few authors [46, 53]. Zuo and co-workers [53] observed that the hardness of the CoFeNi alloy enhanced with the amount of silicon addition. Liu and co-workers [46] also reported an increase in hardness and wear resistance of $Al_{0.5}CoCrCuFeNiSi_x$ (x = 0, 0.4, and 0.8) with silicon addition. The enhanced mechanical properties of the alloys investigated in the present work is in accordance with the aforementioned work done by Zuo and Liu [46, 53].

Additionally, Tian and co-workers [55] have worked on prediction of the structural properties, based on the simple parameters such as valence electron concentration (VEC) and atomic size difference (δ) on as-cast high entropy alloys and suggested a correlation that the hardness value, as a function of valence electron concentration, follows Gauss-type distribution, where an increase of 100 hardness value was observed with each percent increase of atomic size difference. These correlations are also checked for the alloys prepared in this study through spark plasma sintering and casting technique (Figure 5.7b and c). Increase in hardness was observed with a decrease in valence electron concentration and increase in δ%, similar to the theoretical model of Tian and co-workers. Therefore, this model seems valid for calculation of the physical properties of spark plasma sintered and as-cast high entropy alloys.

Further, a wear test has been carried out and it is observed that the alloys with the increasing amount of silicon content showed higher wear resistance compared to CoCrCuFeNi high entropy alloy (Figure 5.8). This strengthening may be attributed to hard sigma phase evolution after spark plasma sintering and due to Ni_3Si phase formation in cast high entropy alloys.

Scanning electron micrographs of worn surfaces of $CoCrCuFeNiSi_x$ high entropy alloys after dry sliding wear are shown in Figure 5.9. The captions used in the figure are as follows: Figure 5.9 (a, b, c, and d) are used for spark plasma sintering samples, whereas Figure 5.9 (e, f, g, and h) are used for samples prepared by casting route. The CoCrCuFeNi alloy without silicon addition (Figure 5.9a and e), clearly shows plastic deformation with some delamination wear on the worn surfaces, resulting in a high wear rate. The alloys show delamination and abrasive wear along with some debris on the worn surface (Figure 5.9b, c, and f, g). Figure 5.9d and h show that much less delamination wear and debris were observed in $CoCrCuFeNiSi_{0.9}$ alloy compared to the other three alloys. However, an increase in abrasive wear recorded with the increasing silicon content. Therefore, the wear resistance of the $CoCrCuFeNiSi_x$ alloys is sensitive to the hardness and presented a strong interrelationship. The increment in hardness can be attributed to the formation of hard and brittle sigma phase after spark plasma sintering and also Ni_3Si phase formation for as-cast high entropy alloys. The addition of silicon also increases the hardness of the face-centered cubic phase by solid solution strengthening mechanism. The silicon atom in the lattice acts

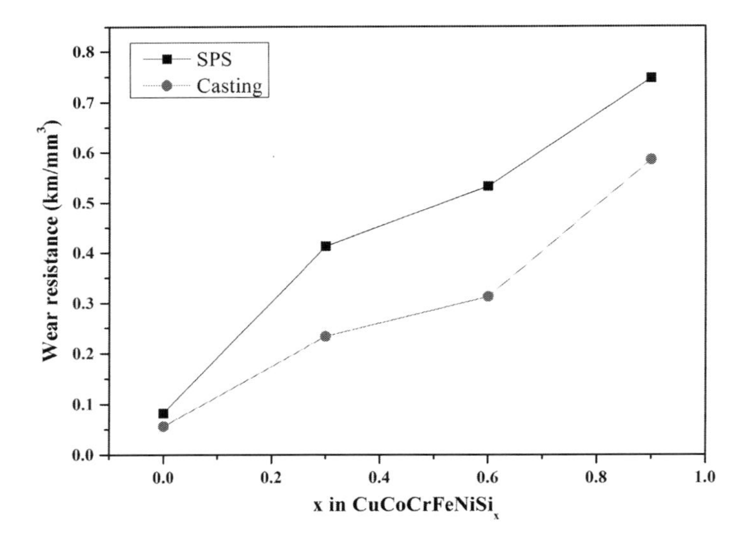

FIGURE 5.8 Wear resistance with respect to silicon content (at.%) in $CuCrCoFeNiSi_x$ (x = 0, 0.3, 0.6, 0.9) high entropy alloys [16].

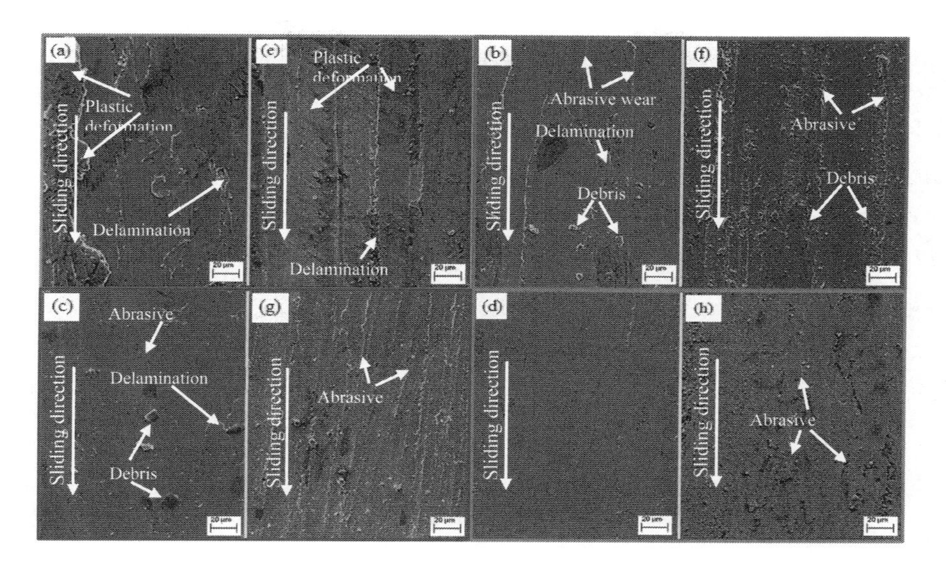

FIGURE 5.9 Microstructures of worn surfaces of spark plasma sintering (a, b, c, and d) and cast high entropy alloys (e, f, g, and h) [16].

as a barrier to the dislocation motion and prevents further deformation, resulting in improved hardness. Thus, both the sigma and face-centered cubic phases contribute to the increased hardness and wear resistance of the alloy.

5.3 SYNTHESIS THROUGH MECHANICAL ALLOYING AND SPARK PLASMA SINTERING

The elemental powders with more than 99.5% purity of cobalt, chromium, copper, iron, nickel, and silicon were mechanically alloyed for 20 h to prepare high entropy alloys. The high energy ball mill (Retsch: PM-400, Germany) was used for milling the powders. Tungsten carbide vials having a ball-to-powder ratio of 10:1 with 10 mm diameter balls were used for milling. The rotation speed of 300 rpm was set throughout the milling process and toluene was added as a process control agent to inhibit cold welding of the powders during milling. The crystal structure of mechanical alloyed powders and alloys after spark plasma sintering was observed by X-ray diffractometer with Cu-Kα radiation.

5.3.1 Phase Evolution during Mechanical Alloying

X-ray diffraction analysis of $CoCuCrFeNiSi_x$ alloys with respect to the milling time is shown in Figure 5.10. After 10 min of milling, the X-ray diffraction peak reveals the existence of all the elements of the alloys. As the milling time increased from

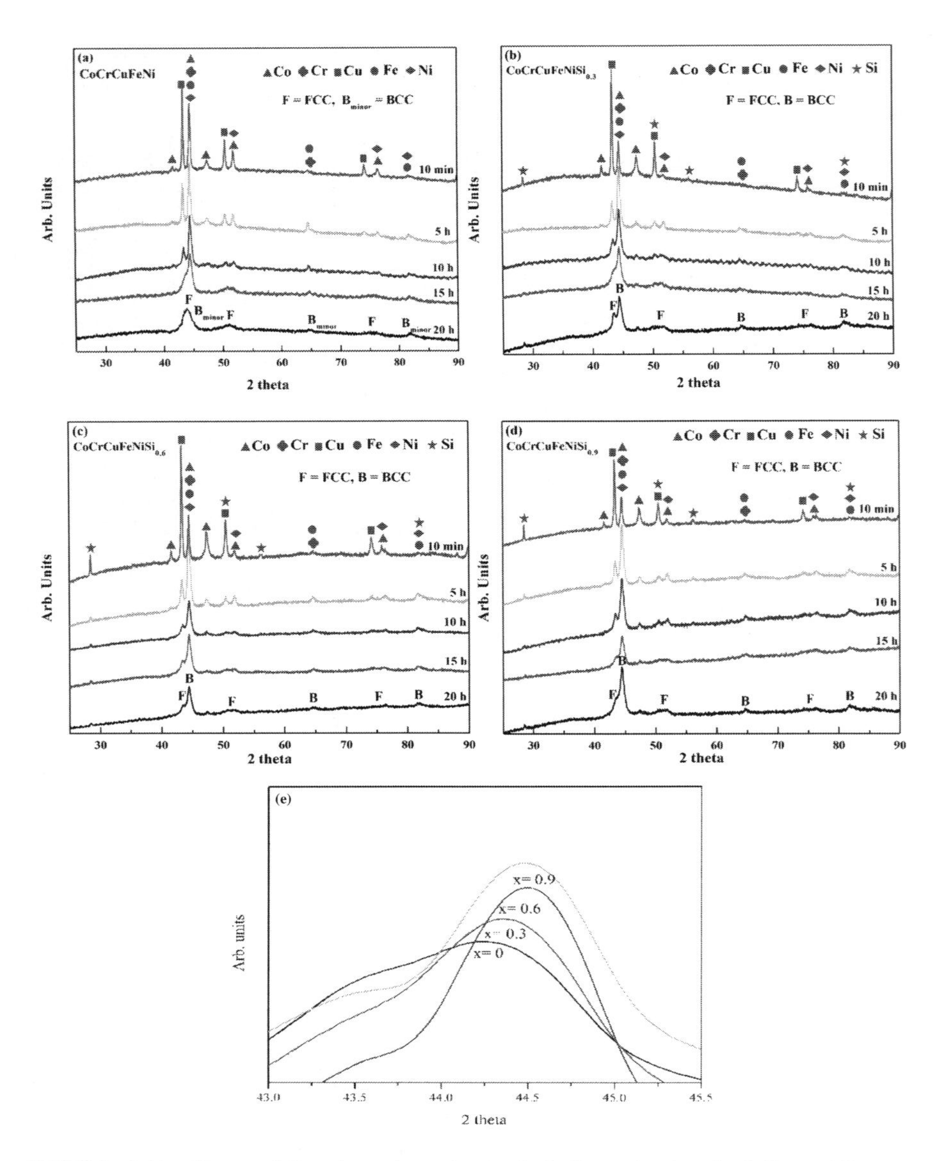

FIGURE 5.10 X-ray diffraction plot of (a) CoCrCuFeNi, (b) $CoCrCuFeNiSi_{0.3}$, (c) $CoCrCuFeNiSi_{0.6}$, (d) $CoCrCuFeNiSi_{0.9}$ with different milling time, and (e) shift in highest intensity peak of $CoCrCuFeNiSi_x$ (x = 0, 0.3, 0.6, and 0.9) toward right after 20 h milling [13].

15 h to 20 h, the slight decrement in the peak intensity was observed, which specifies the evolution of solid-solution phases in the high entropy alloys.

The X-ray diffraction result indicates that after 20 h of milling, the peak position of face-centered cubic phases is close to the peak of copper in $CoCrCuFeNiSi_x$ high entropy alloys. Therefore, all the other constituent elements seem to be dissolving in the copper element can be suggested. Similar phase evolution tendency was also investigated by Praveen and co-workers and Yu and co-workers [17, 56]. In Figure 5.10a, dissolution of copper in CoCrCuFeNi high entropy alloy was detected and as the milling time reached to 20 h, face-centered cubic and minor body-centered cubic (B_{minor}) structured phases evolved. Figure 5.10b illustrates that the addition of silicon in the $CoCrCuFeNiSi_x$ (x = 0.3) high entropy alloy favors the change of B_{minor} to major body-centered cubic phase. As silicon content increased (x = 0.6 and 0.9), the major body-centered cubic phase evolution over the face-centered cubic phase occurred (Figure 5.10c and d). Table 5.3 listed the phase fractions (PF) of $CoCrCuFeNiSi_x$ high entropy alloys after mechanical alloying and spark plasma sintering.

As the silicon content increases, the phase fraction of the body-centered cubic phase increased (Table 5.3). The increase in body-centered cubic phase after 20 h of milling may be attributed to the lower atomic radius of silicon compared to the other constituent elements. The evolution of body-centered cubic phase is more favorable as it has a lower packing efficiency and can accommodate elements with higher atomic volume without much distortion [57].

5.3.2 Microstructural Properties

Crystallite size and lattice strain of powders in different time intervals during milling were calculated using the Debye–Scherrer equation (Equation 5.2) and the Williamson–Hall method (Equation 5.3), respectively, after eliminating the instrumental broadening (Table 5.4a).

TABLE 5.3
Phase Fractions (PF) of $CoCrCuFeNiSi_x$ after 20 h of Mechanical Alloying and Spark Plasma Sintering and Crystallite Size (CS) after Spark Plasma Sintering [13]

	After 20 h of mechanical alloying		After spark plasma sintering		
Alloys	**Phases**	**PF (%)**	**Phases**	**PF (%)**	**CS (nm)**
CoCrCuFeNi	FCC	0.86	F1	0.63	91
	BCC	0.14	F2	0.32	79
			σ	0.05	58
$CoCrCuFeNiSi_{0.3}$	FCC	0.37	F1	0.59	69
	BCC	0.63	F2	0.30	55
			σ	0.11	49
$CoCrCuFeNiSi_{0.6}$	FCC	0.28	F1	0.55	61
	BCC	0.72	F2	0.25	49
			σ	0.20	45
$CoCrCuFeNiSi_{0.9}$	FCC	0.13	F1	0.52	56
	BCC	0.87	F2	0.23	41
			σ	0.25	41

TABLE 5.4

Microstructural Parameters Obtained by Scherrer Equation and Williamson–Hall Method: (a) Crystallite Size (CS) and Lattice Strain (LS) of the High Entropy Alloy Powders under Different Milling Times and (b) Lattice Strain after Spark Plasma Sintering [13]

(a)	CoCrCuFeNi		$CoCrCuFeNiSi_{0.3}$		$CoCrCuFeNiSi_{0.6}$		$CoCrCuFeNiSi_{0.9}$	
Milling time (h)	CS (nm)	LS (%)	CS (nm)	LS (%)	CS (nm)	LS (%)	CS (nm)	LS (%)
5	FCC-73	0.199	42	0.299	36	0.338	33	0.356
	BCC-37	0.323	31	0.575	28	0.592	26	0.623
10	FCC-46	0.278	37	0.328	32	0.373	26	0.427
	BCC-20	0.537	17	0.599	16	0.655	15	0.683
15	FCC-33	0.361	26	0.430	22	0.500	21	0.515
	BCC-16	0.645	14	0.642	12	0.737	11	0.787
20	FCC-32	0.364	24	0.435	21	0.508	19	0.519
	BCC-15	0.748	12	0.747	11	0.841	11	0.885

(b)	CoCrCuFeNi	$CoCrCuFeNiSi_{0.3}$	$CoCrCuFeNiSi_{0.6}$	$CoCrCuFeNiSi_{0.9}$
	LS (%)	LS (%)	LS (%)	LS (%)
	F2-0.061	0.069	0.075	0.081
	F1-0.094	0.101	0.127	0.131

$$\frac{0.89\gamma}{Bcos\theta} \tag{5.2}$$

$$B\cos\theta = \frac{0.89\gamma}{d} + 4\eta\sin\theta \tag{5.3}$$

where d is crystallite size, B is peak width at half the maximum intensity, θ is Bragg angle, λ is the wavelength of the radiation, and η is lattice strain [13].

Table 5.4a illustrates that after 15 h of mechanical alloying, the crystallite size reached about 14 nm and 26 nm of body-centered cubic and face-centered cubic phases, respectively. A further increase in milling time from 15 h to 20 h, significant variation is not observed in crystallite size which indicates that the equilibrium of the mechanical alloying process has been achieved, whereas the continuing increase in the lattice strain observed with milling time has a constructive influence on the solid solution strengthening. The rightward shifting of the body-centered cubic phase with silicon addition specifies that the increase in lattice strain is due to lattice distortion (Figure 5.10e). The enhancement in lattice strain is also attributed to (1) atomic mismatch among elements, as the atomic radius of silicon is smallest among all the elements in the compositions and (2) mechanical deformation [14, 58]. Higher lattice strain can also be attributed to the increase in dislocation density [59]. Therefore, the variance method being used in the present work to confirm the increase in dislocation densities during mechanical alloying and after spark plasma sintering. The calculation details are given below:

The variance method is an unintended method called variance range method [60]. Here, the variance (W) of a region $\pm\sigma$ can be expressed through Equation (5.4)

$$W(\sigma) = \frac{\int_{2\theta_0-\sigma}^{2\theta_0+\sigma} (2\theta - 2\theta_0)^2 I(2\theta)d(2\theta)}{\int_{-\infty}^{+\infty} I(2\theta)d(2\theta)} \tag{5.4}$$

where $2\theta_0$ is the centroid position and I is the intensity. The variance range function adopts the linear form for asymptotic function [60].

$$W(\sigma) = W_{o,\,\text{sample}} + \sigma k_{\text{sample}} \tag{5.5}$$

Similar calculation of $W_{o,\,\text{standard}}$ and k_{standard} has to be done on a standard silicon sample to subtract the instrumental influence. The instrument corrected parameters are:

$$W_{o,\,\text{Net}} = W_{o,\,\text{sample}} - W_{o,\,\text{standard}} \tag{5.6}$$

$$k_{\text{Net}} = k_{\text{sample}} - k_{\text{standard}} \tag{5.7}$$

The root mean square (RMS) strain and crystallite size can be estimated from Equations (5.8) and (5.9), respectively:

$$W_{o,\,\mathrm{Net}} = 4\tan^2\theta_0 \left\langle \varepsilon^2 \right\rangle \tag{5.8}$$

$$k_{\mathrm{Net}} = \frac{0.89\lambda}{\pi^2 \cos\theta_0\, d} \tag{5.9}$$

Finally, the dislocation density (ρ) can be calculated (Equation 5.10) by using the root mean square strain ($\langle \varepsilon^2 \rangle$) and crystallite size ($d$).

$$\rho = \frac{3.464\sqrt{\left\langle \varepsilon^2 \right\rangle}}{d^* b} \tag{5.10}$$

The dislocation density of the body-centered cubic phases at different milling time intervals calculated by the variance method are summarized in Table 5.5. The analysis of Tables 5.4 and 5.5 indicate that, with milling time and silicon addition, dislocation density increases, which contributes to an increase in lattice strain.

5.3.3 Phase Evolution after Spark Plasma Sintering of Mechanically Alloyed Powder

The X-ray diffraction patterns of sintered high entropy alloys are shown in Figure 5.11. The X-ray diffraction peak indicates that after spark plasma sintering, alloys consist of face-centered cubic (F1), (F2), and sigma phases unlike face-centered cubic and body-centered cubic phases in 20 h milled powder. New phase evolution after spark plasma sintering might be due to the transformation of metastable solid-solution phases to equilibrium phases [61, 62]. A significant amount of energy

TABLE 5.5
Dislocation Density Obtained by Variance Method: (a) Body-Centered Cubic Phase under Different Milling Times and (b) F1 Phase after Spark Plasma Sintering [13]

(a)	Dislocation density (m^{-2})			
Milling time (h)	CoCrCuFeNi	$CoCrCuFeNiSi_{0.3}$	$CoCrCuFeNiSi_{0.6}$	$CoCrCuFeNiSi_{0.9}$
5	2.0E+16	5.8E+16	6.8E+16	8.0E+16
10	3.5E+16	6.5E+16	7.8E+16	8.8E+16
15	4.2E+16	7.6E+16	8.0E+16	9.2E+16
20	5.8E+16	8.5E+16	8.8E+16	9.5E+16
(b)	**Dislocation density (m^{-2})**			
SPS	**CoCrCuFeNi**	**$CoCrCuFeNiSi_{0.3}$**	**$CoCrCuFeNiSi_{0.6}$**	**$CoCrCuFeNiSi_{0.9}$**

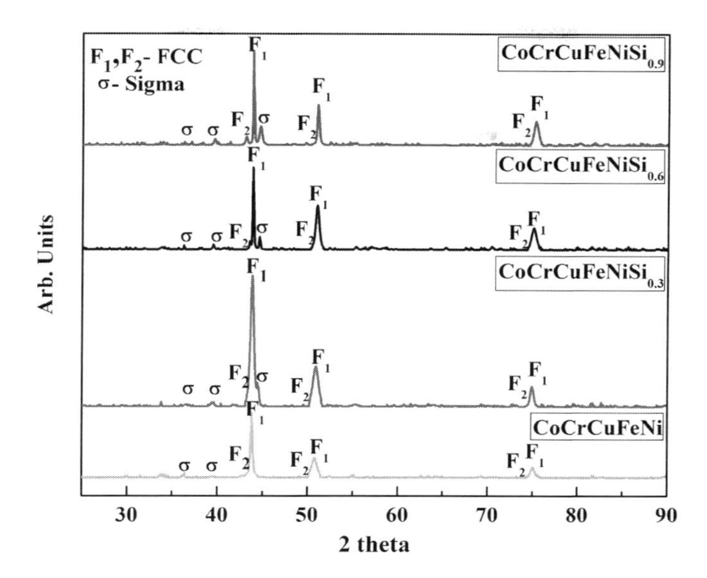

FIGURE 5.11 X-ray diffraction pattern of $CoCrCuFeNiSi_x$ high entropy alloys after spark plasma sintering [13].

is stored during milling due to higher dislocation density and volume fraction of grain boundaries [63, 64]. Therefore, the excess energy stored during milling may decrease the activation energy for phase evolution, whereas this phenomenon is in favor of the phase evolution after spark plasma sintering [61, 65]. As a result, a metastable supersaturated solid-solution of body-centered cubic phases evolved during milling transforms into the equilibrium of face-centered cubic and sigma phases under the influence of temperature and pressure during spark plasma sintering. A similar tendency of phase evolution was reported and discussed elsewhere for the same composition prepared by spark plasma sintering without mechanical alloying of powders [16]. Nickel is a well-known face-centered cubic austenite stabilizer in steels. Thus, the F1 phase should be the face-centered cubic phase of nickel-cobalt-iron with some amount of copper and chromium, whereas the F2 (copper) phase evolved after spark plasma sintering is from the initial face-centered cubic-nickel (copper) phase of 20 h milled powder. Table 5.6 indicates that copper has a high enthalpy of mixing (positive) with other elements of the alloy. This results in the low miscibility of copper with other constituent elements of the alloy [47]. Therefore, segregation of copper occurs at the grain boundaries.

The evolution of F2 phase after spark plasma sintering can be attributed to the strong affinity of copper and nickel due to positive mixing enthalpy which hinders the whole segregation of copper into the face-centered cubic phase [15, 39]. Evidently, the elimination of the body-centered cubic phase (evolved after mechanical alloying) after the consolidation is due to the migration of cobalt and iron above 422°C and 912°C, respectively. Consequently, the phase diagram of chromium-iron suggests that evolution of the sigma phase may be from the residual body-centered cubic phase above 770°C during consolidation and F1 phase could be nickel (cobalt,

TABLE 5.6
Hardness Value of CoCrCuFeNi of Spark Plasma Sintering $CoCrCuFeNiSi_x$ Alloy Prepared in This Work [13]

Alloy	Process	Hardness value (HV)	Ref.
CoCrCuFeNi	Arc melting	133 ± 4	[27]
CoCrCuFeNi	Mechanical alloying + spark plasma sintering	392 ± 6	Present work
$CoCrCuFeNiSi_{0.3}$	Mechanical alloying + spark plasma sintering	450 ± 8	Present work
$CoCrCuFeNiSi_{0.6}$	Mechanical alloying + spark plasma sintering	525 ± 5	Present work
$CoCrCuFeNiSi_{0.9}$	Mechanical alloying + spark plasma sintering	610 ± 6	Present work

iron) [39]. The evolution of the sigma phase is also observed in cobalt, chromium, and iron having alloys such as CoCrFeMnNi [49] and $AlCoCr_xFeMo_{0.5}Ni$ alloys [50]. These transformations of phases can be attributed to rearrangements of atoms due to lattice strain abatement after consolidation and decrement in grain boundaries.

Furthermore, Ji and co-workers [28] observed that the body-centered cubic phase is unstable above 773 K and heating the mechanical alloyed powder at 873 K promotes the transformation of the body-centered cubic phase. The present study also reveals a similar effect for the $CoCrCuFeNiSi_x$ high entropy alloys. Therefore, the body-centered cubic solid solution phase evolved after mechanical alloying undergoes a phase transformation following spark plasma sintering at 1,273 K which promotes the evolution of the two face-centered cubic (F1 and F2) and sigma phases.

The lattice strain of high entropy alloys after spark plasma sintering drastically decreased compared to the lattice strain of 20 h milled powders are illustrated in Table 5.4b. This validates that defects presented after mechanical alloying by severe plastic deformation are almost inhibited after spark plasma sintering. The decrease in the lattice strain can be attributed to the slight decrease in dislocation density after spark plasma sintering (Table 5.5b). The lower atomic radius of silicon could also be a reason for the decrease in lattice strain. Consequently, a small-sized atom in the lattice favors the solid-solution strengthening. The mechanical alloying process promotes higher strain and defects which could result in the delay of solubility; therefore, the powders after mechanical alloying are usually in a metastable state. Thus, the stable phase is certain in the present study, due to the rearrangement of atoms after spark plasma sintering.

Scanning electron micrographs, along with the compositional details by energy-dispersive X-ray spectroscopy of $CoCrCuFeNiSi_x$ high entropy alloys prepared by spark plasma sintering, are shown in Figure 5.12. Two different contrasts, i.e., light gray and gray regions can be recognized on the CoCrCuFeNi alloy (Figure 5.12a). The energy dispersive X-ray spectroscopy results (Figure 5.12a) indicates that the light gray regions have larger amounts of nickel, copper, cobalt, and iron, while the gray region is chromium-iron-rich. Combined analysis of X-ray diffraction patterns

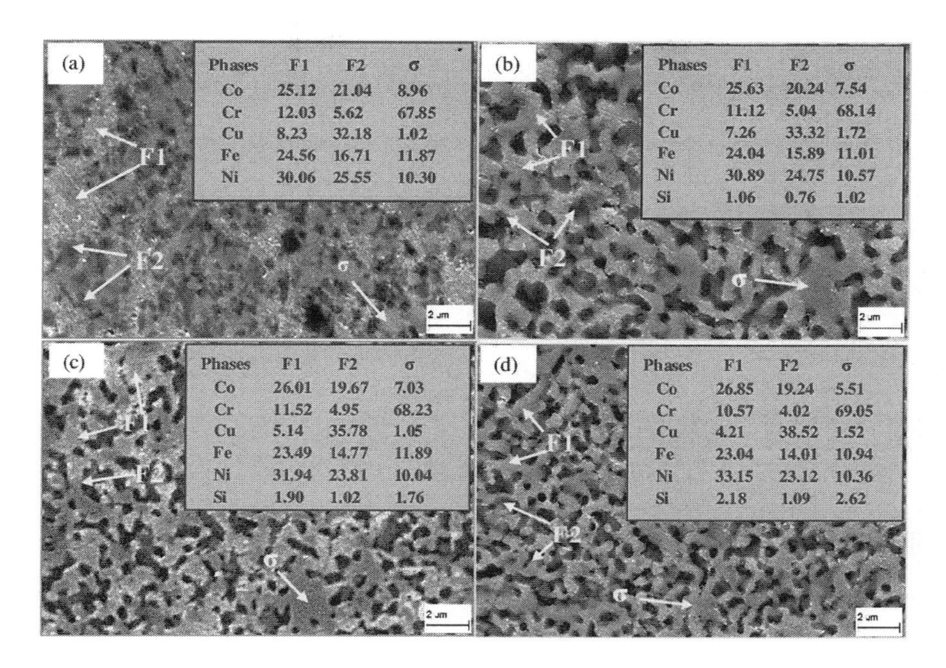

FIGURE 5.12 Scanning electron micrographs along with the energy dispersive X-ray spectroscopy analysis in inset (at.%) of (a) CoCrCuFeNi, (b) $CoCrCuFeNiSi_{0.3}$, (c) $CoCrCuFeNiSi_{0.6}$, and (d) $CoCrCuFeNiSi_{0.9}$ high entropy alloys after spark plasma sintering [13].

(Figure 5.11), scanning electron micrographs (Figure 5.12), and energy dispersive X-ray spectroscopy results indicate that the light gray region includes two face-centered cubic phases. The outer region contains more copper, while the inner region contains more nickel, cobalt, and iron. Therefore, one of the face-centered cubic phases is iron (cobalt, nickel), i.e., the F1 phase, and another might be copper (nickel) i.e., the F2 phase. The gray region is the chromium-iron-rich σ phase which might be a conversion of the body-centered cubic phase above 770°C temperature evolved after 20 h of mechanical alloying [31, 39]. The grain boundaries are rich in the copper element while chromium is gathered inside the matrix. These could be attributed to the higher binary enthalpy (positive) of chromium with copper, and negative binary enthalpies with other elements of the alloy. Accordingly, inside the matrix, chromium is well adopted, whereas copper is segregated in the grain boundaries. This promotes the separation of copper which forms the copper-rich face-centered cubic (F2) phase. Therefore, the chromium-rich gray region may correspond to the sigma phase, and the light gray region comprises the F1 and F2 phases in $CoCuCrFeNiSi_x$ high entropy alloys.

5.3.4 Mechanical Behavior after Spark Plasma Sintering

Microhardness and wear resistance after spark plasma sintering of the $CoCrCuFeNiSi_x$ high entropy alloys with silicon content is illustrated in Figure 5.13. Enhancement

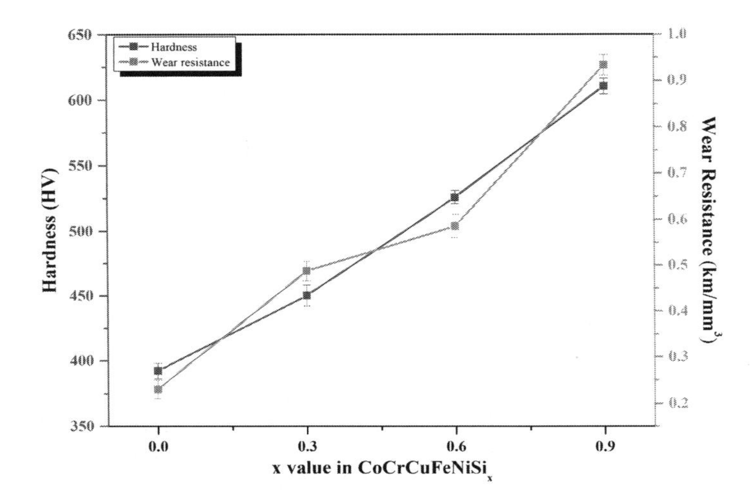

FIGURE 5.13 Variation in microhardness and wear resistance with silicon addition in the $CoCrCuFeNiSi_x$ (x = 0, 0.3, 0.6, and 0.9 in atomic ratio) high entropy alloys after spark plasma sintering [13].

in microhardness is observed with silicon addition from 0 to 0.9 in CoCrCuFeNi alloy. The hardness value increased from 392 ± 6 HV to 610 ± 6 HV with x = 0 to 0.9, respectively. Therefore, the alloys studied in this work show higher hardness values compared to the previously reported CoCrCuFeNi alloy prepared by arc melting (Table 5.6). The improved hardness of the alloys under consideration are ascribed by a solid solution of silicon element which results in the chromium-rich sigma phase. The enhancement in lattice distortion energy owing to the increase in lattice strain with silicon content also contributes to the enhancement of the solid solution strengthening. Thus, the microhardness of the high entropy alloys is enhanced greatly with the addition of the silicon element.

The strengthening mechanism after spark plasma sintering is investigated in detail by Sriharitha and co-workers [10] and Sathiyamoorthi and co-workers [67] based on the Hall–Petch and modified Hall–Petch analysis, respectively. Sriharitha and co-workers [10] observed that solid-solution strengthening together with order strengthening increased with Al content in $Al_xCoCrCuFeNi$ (x = 0.45, 1, 2.5, and 5 mol) high entropy alloys whereas Sathiyamoorthi and co-workers [67] concluded that the major contribution to strengthening comes from grain boundaries. The present study also claims an increase in hardness value with silicon content, hence it is worth incorporating the analysis based on the Hall–Petch equation as discussed by Sriharitha and co-workers [10]. The calculations made in the present study are carried out with the help of formulas and assumptions given in Sriharitha and co-workers' work [10]. The Hall–Petch coefficients (σ_0 and k) for all the elements were taken from this work [10], except silicon which is not available (Table 5.7). The approximate value of σ_0 for silicon is estimated to be 7.94 MPa. Similarly, the k value calculated for silicon is approximately 0.07 MPa $m^{1/2}$.

TABLE 5.7
σ_o and k (Hall–Petch Coefficients) Values for Various Metallic Elements Except for Silicon Present in the High Entropy Alloys of This Study Are Taken from Hsu and Co-Workers [42]

		k (MPa $m^{1/2}$)	
Element	**σ_o(MPa)**	**Microcrystalline**	**At 100 nm**
Co	10.3	0.08	–
Cr	454	0.95	–
Cu	25	0.11	0.05
Fe	100	0.60	0.35
Ni	22	0.16	0.12
Si	7.94	0.07	–

Employing the phase fractions of phases calculated from X-ray diffraction analysis (Table 5.3), the grain-size strengthening (σ_{gss}) in the high entropy alloys is calculated. Since the contributions of grain-size strengthening are calculated, the rest of the contribution to hardness can be entitled to a solid solution (HV_{ss}) together with order (HV_{os}) strengthening. Each of these contributions is presented in Table 5.8.

Hence, it is clear from Hall–Petch analysis proposed by Sriharitha and co-workers [10] that, apart from an increase in hardness of the alloys with an increase in silicon content, grain boundary strengthening also increases considerably with the silicon content. The enhancement in grain boundary strengthening can be correlated with the increase in dislocation density with the silicon content. Recently, a few studies reported the carbide formation after mechanical alloying and spark plasma sintering in high entropy alloys [68, 69]. Pohan and co-workers [68] have reported the high compressive yield strength for the high entropy alloys prepared by the spark plasma sintering process and concluded that major contributors to the alloy strengthening were grain boundary strengthening coupled with dispersion strengthening via metal carbides formation. Pohan and co-workers have confirmed the presence of carbide

TABLE 5.8
Various Strengthening Contributions to the Hardness of Spark Plasma Sintering $CoCrCuFeNiSi_x$ High Entropy Alloys [13]

Alloys	**σ_{gss} (HV)**	**HV_{ss+OS}**	**HV_{Total}**
CoCrCuFeNi	300	92	392
$CoCrCuFeNiSi_{0.3}$	377	73	450
$CoCrCuFeNiSi_{0.6}$	473	52	525
$CoCrCuFeNiSi_{0.9}$	529	81	610

in the alloy by utilizing X-ray diffraction and energy dispersive X-ray spectroscopy analysis. On the other hand, in the present study, the possibility of carbon contamination and carbide formation can be rejected, as none of the aforesaid features has been observed.

The wear coefficient of $CoCrCuFeNiSi_x$ high entropy alloys with respect to the variation in silicon content is illustrated in Figure 5.14. The values of the wear coefficient are calculated from the relation published by Chen and co-workers [32]. The result shows a decrease in the wear coefficient with increasing silicon content (Figure 5.14) Therefore, in other words, there is an enhancement of the wear properties of the $CoCrCuFeNiSi_x$ high entropy alloys with silicon. The higher wear resistance translates into higher energy required to remove the material from the surface as that of low wear resistance. There is also a direct relationship between the wear coefficients of $CoCrCuFeNiSi_x$ high entropy alloys to their hardness with an increase in sigma phase as previously discussed.

Scanning electron micrographs of worn surfaces after wear test of $CoCrCuFeNiSi_x$ high entropy alloys are shown in Figure 5.15. The worn surface (Figure 5.15a), shows immense plastic deformation with the grooves, causing lower wear resistance in the absence of silicon. The delamination and abrasive wear with some debris observed with silicon addition from x = 0.3 to 0.9 (Figure 5.15b, c, and d). With silicon content 0.3, 0.6, and 0.9 delamination wear and debris decreased, respectively. On the other hand, abrasive wear increased in $CoCrCuFeNiSi_{0.9}$. This observation can be explained on the basis of two factors. First, the presence of hard σ phase and second, the increased silicon content which is brittle in nature. The decrease in delamination wear confirms minor removal of materials during the wear test. Moreover, recently, Kumar and co-workers [13] studied the wear behavior of high entropy alloys with an

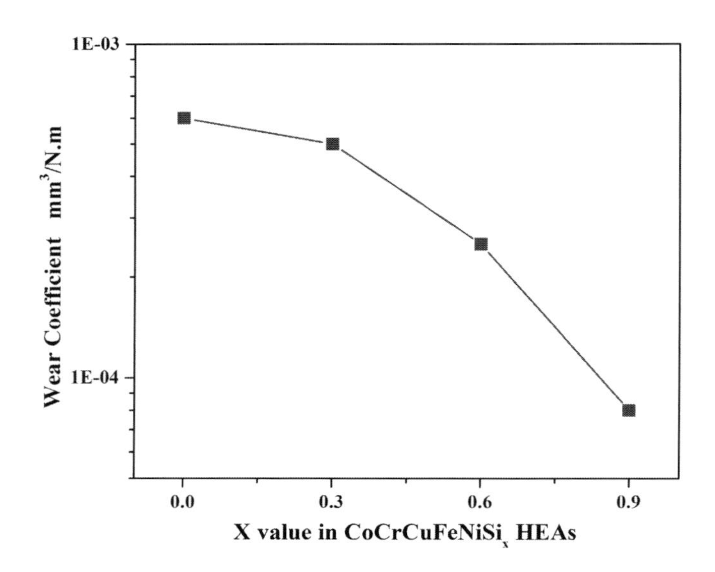

FIGURE 5.14 Variation in wear coefficient of $CoCrCuFeNiSi_x$ high entropy alloys with silicon addition [13].

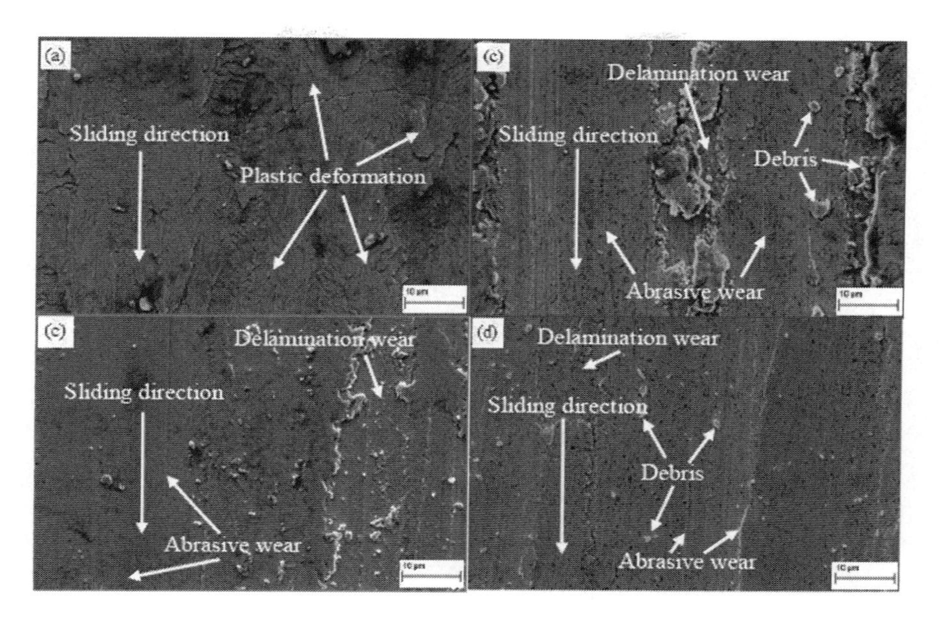

FIGURE 5.15 Scanning electron micrographs of spark plasma sintering samples after wear (a) CoCrCuFeNi, (b) $CoCrCuFeNiSi_{0.3}$, (c) $CoCrCuFeNiSi_{0.6}$, and (d) $CoCrCuFeNiSi_{0.9}$ [13].

increase in silicon content. They observed the increase in wear resistance with silicon content due to the formation of hard sigma phase. The above observation proposes that the wear resistance of the $CoCrCuFeNiSi_x$ high entropy alloys is sensitive to the microhardness, which in turn depends upon the silicon content.

5.4 CONCLUSION

1. A series of $CoCrCuFeNiSi_x$ high entropy alloys were prepared successfully through two different processing routes, i.e., spark plasma sintering and vacuum-arc melting.
2. The X-ray diffraction results show the formation of two face-centered cubic based phases F1 (nickel-cobalt-iron) and F2 (copper-rich) together with sigma (σ) phases (chromium-rich) in samples consolidated by spark plasma sintering. However, a single face-centered cubic phase was observed in $CoCrCuFeNiSi_x$ ($x = 0$, 0.3 and 0.6) high entropy alloy samples prepared by the vacuum-arc melting route.
3. The increase in silicon content from $x = 0.6$ to 0.9 leads to Ni_3Si phase formations due to the highest negative enthalpy of mixing value between silicon and nickel elements.
4. The empirical correlation of hardness (HV) with valence electron concentration and atomic size difference (δ) percentage for as-cast samples was also valid for samples prepared by spark plasma sintering in the present study.
5. The mechanical alloying route favors to a non-equilibrium microstructure while the stable microstructure evolved after spark plasma sintering.

6. After 20 h of mechanical alloying of CoCrCuFeNi high entropy alloy, major face-centered cubic and minor body-centered cubic phase were observed; however, after the addition of silicon element in CoCrCuFeNi, a major body-centered cubic phase compared to the face-centered cubic phase was observed.
7. Consolidation through spark plasma sintering leads to the transformation of metastable phase formed after 20h of mechanical alloying into the face-centered cubic based F1 (cobalt-iron-nickel) and copper-rich F2 phases along with σ phases (chromium-rich). Hall–Petch analysis carried out on sintered samples reveals that grain boundary strengthening increases with an increase in silicon content.
8. Improved wear resistance observed with silicon addition. The improvement in properties can be ascribed to the solid-solution strengthening effect due to the higher atomic mismatch between silicon and other elements of the alloy.
9. The excellent increase in corrosion resistance of CoCrCuFeNiSix high entropy alloys in 3.5% NaCl solution obtained with silicon addition.
10. The better mechanical properties for samples processed by spark plasma sintering and casting routes were associated with the hard σ phase and Ni3Si formation, respectively.

REFERENCES

1. B.J. Yeh, S. Chen, S. Lin, J. Gan, T. Chin, T. Shun, C. Tsau. 2004. Nanostructured high-entropy alloys with multiple principal elements : Novel alloy design concepts and outcomes. *Adv. Eng. Mat.* 6:299–303.
2. B.S. Li, Y.P. Wang, M.X. Ren, C. Yang, H.Z. Fu. 2008. Effects of Mn, Ti and V on the microstructure and properties of AlCrFeCoNiCu high entropy alloy. *Mater. Sci. Eng. A* 498(1–2):482–486.
3. B. Cantor. 2014. Multicomponent and high entropy alloys. *Entropy* 16(9):4749–4768.
4. T.M. Yue, H. Zhang. 2014. Laser cladding of FeCoNiCrAlCu xSi 0·5 high entropy alloys on AZ31 Mg alloy substrates. *Mater. Res. Innov.* 18(sup2):S2-624–S2-628.
5. S. Niu, H. Kou, T. Guo, Y. Zhang, J. Wang, J. Li. 2016. Materials science & engineering a strengthening of nanoprecipitations in an annealed Al 0.5 CoCrFeNi high entropy alloy. *Mater. Sci. Eng. A* 671:82–86.
6. S.T. Chen, W.Y. Tang, Y.F. Kuo, S.Y. Chen, C.H. Tsau, T.T. Shun, J.W. Yeh. 2010. Microstructure and properties of age-hardenable AlxCrFe1.5MnNi0.5 alloys. *Mater. Sci. Eng. A.* 527(21–22):5818–5825.
7. S.R.B.S. Murty. 1998. Novel materials synthesis by mechanical alloying/ milling. *Int. Mater.* 43:101–141.
8. M.-R. Chen, S.-J. Lin, J.-W. Yeh, S.-K. Chen, Y.-S. Huang, C.-P. Tu. 2006. Microstructure and properties of Al0.5CoCrCuFeNiTix (x=0–2.0) high-entropy alloys. *Mater. Trans.* 47(5):1395–1401.
9. M. Vaidya, S. Armugam, S. Kashyap, B.S. Murty. 2015. Amorphization in equiatomic high entropy alloys. *J. Non Cryst. Solids* 413:8–14.
10. R. Sriharitha, B.S. Murty, R.S. Kottada. 2014. Alloying, thermal stability and strengthening in spark plasma sintered AlxCoCrCuFeNi high entropy alloys. *J. Alloys Compd.* 583:419–426.
11. S. Mohanty, N.P. Gurao, P. Padaikathan, K. Biswas. 2017. Ageing behaviour of equiatomic consolidated Al 20 Co 20 Cu 20 Ni 20 Zn 20 high entropy alloy. *Mater. Charact.* 129:127–134.

12. A. Kumar, A.K. Swarnakar, M. Chopkar. 2018. Phase evolution and mechanical properties of AlCoCrFeNiSix high-entropy alloys synthesized by mechanical alloying and spark plasma sintering. *J. Mater. Eng. Perform.* 27(7):3304–3314.
13. A. Kumar, P. Dhekne, A.K. Swarnakar, M. Chopkar. 2019. Phase evolution of CoCrCuFeNiSix high-entropy alloys prepared by mechanical alloying and spark plasma sintering. *Mater. Res. Express* 6(2). http://www.ncbi.nlm.nih.gov/pubmed/026532.
14. Z. Fu, W. Chen, H. Xiao, L. Zhou, D. Zhu, S. Yang. 2013. Fabrication and properties of nanocrystalline Co0.5FeNiCrTi0.5 high entropy alloy by MA-SPS technique. *Mater. Des.* 44:535–539.
15. S. Praveen, B.S. Murty, R.S. Kottada. 2012. Alloying behavior in multi-component AlCoCrCuFe and NiCoCrCuFe high entropy alloys. *Mater. Sci. Eng. A* 534:83–89.
16. A. Kumar, A.K. Swarnakar, A. Basu, M. Chopkar. 2018. Effects of processing route on phase evolution and mechanical properties of CoCrCuFeNiSix high entropy alloys. *J. Alloys Compd.* 748:889–897.
17. S. Praveen, B.S. Murty, R.S. Kottada. 2013. Phase evolution and densification behavior of nanocrystalline multicomponent high entropy alloys during spark plasma sintering. *Jom* 65(12):1797–1804.
18. I. Moravcik, J. Cizek, P. Gavendova, S. Sheikh, S. Guo, I. Dlouhy. 2016. Effect of heat treatment on microstructure and mechanical properties of spark plasma sintered AlCoCrFeNiTi0.5 high entropy alloy. *Mater. Lett.* 174:53–56.
19. C. Zhang, G.F. Wu, P.Q. Dai. 2015. Phase transformation and aging behavior of Al0.5CoCrFeNiSi0.2 high-entropy alloy. *J. Mater. Eng. Perform.* 24(5):1918–1925.
20. P. Jinhong, P. Ye, Z. Hui, Z. Lu. 2012. Microstructure and properties of AlCrFeCuNix ($0.6 \leq x \leq 1.4$) high-entropy alloys. *Mater. Sci. Eng. A* 534:228–233.
21. A.K. Mishra, S. Samal, K. Biswas. 2012. Solidification behaviour of Ti-Cu-Fe-Co-Ni high entropy alloys. *Trans. Indian Inst. Met.* 65(6):725–730.
22. Y. Dong, Y. Lu, J. Kong, J. Zhang, T. Li. 2013. Microstructure and mechanical properties of multi-component AlCrFeNiMo x high-entropy alloys. *J. Alloys Compd.* 573:96–101.
23. W. Huo, H. Zhou, F. Fang, Z. Xie, J. Jiang. 2017. Microstructure and mechanical properties of CoCrFeNiZrx eutectic high-entropy alloys. *Mater. Des.* 134:226–233.
24. J. Dąbrowa, G. Cieślak, M. Stygar, K. Mroczka, K. Berent, T. Kulik, M. Danielewski. 2017. Influence of Cu content on high temperature oxidation behavior of AlCoCrCuxFeNi high entropy alloys (x = 0; 0.5; 1). *Intermetallics* 84:52–61.
25. R. Razuan, N.A. Jani, M.K. Harun, M.K. Talari. 2013. Microstructure and hardness properties investigation of Ti and Nb added FeNiAlCuCrTi x Nb y high entropy alloys. *Trans. Indian Inst. Met.* 66(4):309–312.
26. F.J. Baldenebro-Lopez, J.M. Herrera-Ramírez, S.P. Arredondo-Rea, C.D. Gómez-Esparza, R. Martínez-Sánchez. 2014. Simultaneous effect of mechanical alloying and arc-melting processes in the microstructure and hardness of an AlCoFeMoNiTi high-entropy alloy. *J. Alloys Compd.* 643:10–15.
27. C.-J. Tong, Y.-L. Chen, J.-W. Yeh, S.-J. Lin, S.-K. Chen, T.-T. Shun, C.-H. Tsau, S.-Y. Chang. 2005. Microstructure characterization of Al x CoCrCuFeNi high-entropy alloy system with multiprincipal elements. *Metall. Mater. Trans. A* 36(4):881–893.
28. W. Ji, Z. Fu, W. Wang, H. Wang, J. Zhang, Y. Wang, F. Zhang. 2014. Mechanical alloying synthesis and spark plasma sintering consolidation of CoCrFeNiAl high-entropy alloy. *J. Alloys Compd.* 589:61–66.
29. Y.P. Wang, B.S. Li, M.X. Ren, C. Yang, H.Z. Fu. 2008. Microstructure and compressive properties of AlCrFeCoNi high entropy alloy. *Mater. Sci. Eng. A* 491(1–2):154–158.
30. X.-W. Qiu. 2013. Microstructure and properties of AlCrFeNiCoCu high entropy alloy prepared by powder metallurgy. *J. Alloys Compd.* 555:246–249.
31. D.G. Shaysultanov, N.D. Stepanov, A.V. Kuznetsov, G.A. Salishchev, O.N. Senkov. 2013. Phase composition and superplastic behavior of a wrought AlCoCrCuFeNi high-entropy alloy. *Jom* 65(12):1815–1828.

32. J.M. Wu, S.J. Lin, J.W. Yeh, S.K. Chen, Y.S. Huang, H.C. Chen. 2006. Adhesive wear behavior of AlxCoCrCuFeNi high-entropy alloys as a function of aluminum content. *Wear* 261(5–6):513–519.
33. C.-H. Tsau, Y.-H. Chang. 2013. Microstructures and mechanical properties of TiCrZrNbNx alloy nitride thin films. *Entropy* 15(12):5012–5021.
34. H.R. Sistla, J.W. Newkirk, F. Frank Liou. 2015. Effect of Al/Ni ratio, heat treatment on phase transformations and microstructure of AlxFeCoCrNi2–x (x=0.3, 1) high entropy alloys. *Mater. Des.* 81:113–121.
35. Z. Hu, Y. Zhan, G. Zhang, J. She, C. Li. 2010. Effect of rare earth Y addition on the microstructure and mechanical properties of high entropy AlCoCrCuNiTi alloys. *Mater. Des.* 31(3):1599–1602.
36. Y.L. Chou, J.W. Yeh, H.C. Shih. 2010. The effect of molybdenum on the corrosion behaviour of the high-entropy alloys Co1.5CrFeNi1.5Ti0.5Mox in aqueous environments. *Corros. Sci.* 52(8):2571–2581.
37. M.-R. Chen, S.-J. Lin, J.-W. Yeh, M.-H. Chuang, S.-K. Chen, Y.-S. Huang. 2006. Effect of vanadium addition on the microstructure, hardness, and wear resistance of Al0.5CoCrCuFeNi high-entropy alloy. *Metall. Mater. Trans. A* 37(5):1363–1369.
38. L. Jiang, Y. Lu, Y. Dong, T. Wang, Z. Cao, T. Li. 2015. Effects of Nb addition on structural evolution and properties of the CoFeNi2V0.5high-entropy alloy. *Appl. Phys. A Mater. Sci. Process.* 119(1):291–297.
39. P. Wang, H. Cai, X. Cheng. 2016. Effect of Ni/Cr ratio on phase, microstructure and mechanical properties of NixCoCuFeCr2-x (x = 1.0, 1.2, 1.5, 1.8 mol) high entropy alloys. *J. Alloys Compd.* 662:20–31.
40. Z. Fu, W. Chen, S. Fang, X. Li. 2014. Effect of Cr addition on the alloying behavior, microstructure and mechanical properties of twinned CoFeNiAl0.5Ti0.5 alloy. *Mater. Sci. Eng. A.* 597:204–211.
41. X.W. Qiu, Y.P. Zhang, C.G. Liu. 2014. Effect of Ti content on structure and properties of Al 2 CrFeNiCoCuTi x high-entropy alloy coatings. *J. Alloys Compd.* 585:282–286.
42. C.Y. Hsu, T.S. Sheu, J.W. Yeh, S.K. Chen. 2010. Effect of iron content on wear behavior of AlCoCrFexMo0.5Ni high-entropy alloys. *Wear* 268(5–6):653–659.
43. N.H. Tariq, M. Naeem, B.A. Hasan, J.I. Akhter, M. Siddique. 2013. Effect of W and Zr on structural, thermal and magnetic properties of AlCoCrCuFeNi high entropy alloy. *J. Alloys Compd.* 556:79–85.
44. O. Maulik, D. Kumar, S. Kumar, D.M. Fabijanic, V. Kumar. 2016. Structural evolution of spark plasma sintered AlFeCuCrMgx (x= 0, 0.5, 1, 1.7) high entropy alloys. *Intermetallics* 77:46–56.
45. A. Kumar, P. Dhekne, A. Kumar, M. Kumar. 2017. Analysis of Si addition on phase formation in AlCoCrCuFeNiSix high entropy alloys. *Mater. Lett.* 188:73–76.
46. X. Liu, W. Lei, L. Ma, J. Liu, J. Liu, J. Cui. 2012. On the microstructures, phase assemblages and properties of Al0.5CoCrCuFeNiSix high-entropy alloys. *J. Alloys Compd.* 630:151–157.
47. X. Yang, Y. Zhang. 2012. Prediction of high-entropy stabilized solid-solution in multicomponent alloys. *Mater. Chem. Phys.* 132(2–3):233–238.
48. C. Allibert, C. Bernard, N. Valignat, M. Dombre. 1978. CoCr binary system: Experimental re-determination of the phase diagram and comparison with the diagram calculated from the thermodynamic data. *J. Less Common Met.* 59(2):211–228.
49. N. Park, B.-J. Lee, N. Tsuji. 2017. The phase stability of equiatomic CoCrFeMnNi high-entropy alloy: Comparison between experiment and calculation results. *J. Alloys Compd.* 719:189–193.
50. C.-Y. Hsu, C.-C. Juan, W.-R. Wang, T.-S. Sheu, J.-W. Yeh, S.-K. Chen. 2011. On the superior hot hardness and softening resistance of AlCoCrxFeMo0.5Ni high-entropy alloys. *Mater. Sci. Eng. A* 528(10–11):3581–3588.

51. S. Praveen, B.S. Murty, R.S. Kottada. 2014. Effect of molybdenum and niobium on the phase formation and hardness of nanocrystalline CoCrFeNi high entropy alloys. *J. Nanosci. Nanotechnol.* 14(10):8106–8109.
52. K.Y. Tsai, M.H. Tsai, J.W. Yeh. 2013. Sluggish diffusion in Co-Cr-Fe-Mn-Ni high-entropy alloys. *Acta Mater.* 61(13):4887–4897.
53. T.T. Zuo, R.B. Li, X.J. Ren, Y. Zhang. 2014. Effects of Al and Si addition on the structure and properties of CoFeNi equal atomic ratio alloy. *J. Magn. Magn. Mater.* 371:60–68.
54. A. Takeuchi, A. Inoue. 2005. Classification of bulk metallic glasses by atomic size difference, heat of mixing and period of constituent elements and its application to characterization of the main alloying element. *Mater. Trans.* 46(12):2817–2829.
55. F. Tian, L.K. Varga, N. Chen, J. Shen, L. Vitos. 2015. Empirical design of single phase high-entropy alloys with high hardness. *Intermetallics* 58:1–6.
56. P.F. Yu, L.J. Zhang, H. Cheng, H. Zhang, M.Z. Ma, Y.C. Li, G. Li, P.K. Liaw, R.P. Liu. 2016. The high-entropy alloys with high hardness and soft magnetic property prepared by mechanical alloying and high-pressure sintering. *Intermetallics* 70:82–87.
57. F.J. Wang, Y. Zhang, G.L. Chen. 2009. Atomic packing efficiency and phase transition in a high entropy alloy. *J. Alloys Compd.* 478(1–2):321–324.
58. H.X. Sui, M. Zhu, M. Qi, G.B. Li, D.Z. Yang. 1992. The enhancement of solid solubility limits of AlCo intermetallic compound by high-energy ball milling. *J. Appl. Phys.* 71(6):2945–2949.
59. F. Graner, J.A. Glazier. 1992. Mechanically driven alloying of immiscible elements. *Phys. Rev. Lett.* 69:2013–2016.
60. D.D.E. Electromecnica, E. De Industriales, U. De, U. De Extremadura, C. De Elvas. 1997. The use of the pseudo-voigt function in the variance method of X-ray line-broadening analysis. *J. Appl. Cryst.* 30:427–430.
61. C. Suryanarayana. 2001. Mechanical alloying and milling. *Prog. Mater. Sci.* 46(1–2):1–184.
62. Z. Fu, W. Chen, Z. Chen, H. Wen, E.J. Lavernia. 2014. Influence of Ti addition and sintering method on microstructure and mechanical behavior of a medium-entropy Al0.6CoNiFe alloy. *Mater. Sci. Eng. A* 619:137–145.
63. H. Wen, Y. Zhao, Y. Li, O. Ertorer, K.M. Nesterov, R.K. Islamgaliev, R.Z. Valiev, E.J. Lavernia. 2010. High-pressure torsion-induced grain growth and detwinning in cryomilled Cu powders. *Philos. Mag.* 90(34):4541–4550.
64. H. Wen, R.K. Islamgaliev, K.M. Nesterov, R.Z. Valiev, E.J. Lavernia. 2013. Dynamic balance between grain refinement and grain growth during high-pressure torsion of Cu powders. *Philos. Mag. Lett.* 93(8):481–489.
65. W. Chen, Z. Fu, S. Fang, H. Xiao, D. Zhu. 2013. Alloying behavior, microstructure and mechanical properties in a FeNiCrCo0.3Al0.7 high entropy alloy. *Mater. Des.* 51:854–860.
66. R.S. Kottada. 2017. Thermal stability and grain boundary strengthening in ultrafine-grained CoCrFeNi high entropy alloy composite. *Mater. Des.* 134:426–433.
67. R.M. Pohan, B. Gwalani, J. Lee, T. Alam, J.Y. Hwang, H.J. Ryu, R. Banerjee, S.H. Hong. 2018. Microstructures and mechanical properties of mechanically alloyed and spark plasma sintered Al0.3CoCrFeMnNi high entropy alloy. *Mater. Chem. Phys.* 210:62–70.
68. B.S.M.M. Vaidya, Anirudha Karati, A. Marshal, K.G. Pradeep. 2019. Phase evolution and thermal stability of nanocrystalline CoCrFeNi and CoCrFeMnNi high entropy alloys. *J. Alloys Compd.* 770:1004–1015.

6 The Corrosion and Thermal Behavior of $AlCr_{1.5}CuNi_2FeCo_x$ High Entropy Alloys

Vikas Kukshal, Amar Patnaik, and I.K. Bhat

CONTENTS

6.1 INTRODUCTION

Materials play a vital role in increasing the efficiency of the machine elements used in the areas of power generation and transportation. The key area of concern for the prevailing metals and alloys are the extreme environmental conditions that lead to the failure of the component. The aluminum alloys possess remarkable properties due to which they are extensively utilized in most areas in the past few decades [1]. The properties of the aluminum alloys include low density, comparatively high specific strength, improved corrosion in a normal environment, and better electric and thermal conductivity [2]. On the other hand, aluminum alloys can be used only within the specified range of values for all the properties, and hence new design strategies and processing routes are required for enhanced properties as compared to existing ones [3]. In addition, the economics and physical viability of the alloys are other important factors for developing novel alloys.

It is challenging to decide the requirements and finalize the number of elements to be included in the alloys. Exhaustive affirmation of limiting properties such as ductility, hardness, fatigue, creep, fracture, corrosion resistance, etc. is very much needed for its proper use. Controlled experimental techniques are required for the enhancement of the property. The use of superalloys in extreme service conditions such as high temperature and pressure enhances the chances of the failure of the alloys. There is a continuous need to develop alloys with low cost and improved properties for aeronautical, oil and gas, and marine sectors.

Currently, the superalloys are the backbone of various industries, including energy generation, oil and gas extraction, and the metallurgical and aerospace industries. Superalloys provide the benefits of a low weight material along with an increase in the efficiency of the machine. Despite all these advantages, there is a considerable challenge in the field of materials science to overcome the shortages of the existing superalloys. Compared to the predictable metallic alloys based on the combination of specific elements, high entropy alloys (HEAs) generally contain five or more principal elements with a concentration range of 5 and 35 at.% [4]. High entropy alloys exhibit exceptional properties such as improved hardness, high tensile strength, high compressive strength, excellent wear, and corrosion resistance, explaining their potential applications in different fields [5].

The alloy design for a particular application is based on the appropriate selection of elements for the development of high entropy alloys wherein each element plays a vital role in altering the property. Many high entropy alloys (HEAs) have been investigated extensively in recent years, focusing on the effect of elements such as aluminum, chromium, copper, molybdenum, and titanium [6–9]. The dominant factor controlling the properties of the high entropy alloys is the configuration of its crystal structure during the fabrication and homogenization process. The high entropy alloys may contain solid solution or multiple phases comprising of secondary solid-solution phases and/or intermetallic phases such as β, Laves, and σ [10].

Although numerous equiatomic high entropy alloys have been investigated in the past, there are very few studies examining the near-equiatomic multi-principal alloy system. Cobalt is selected as the alloying element, as it falls into the category of transition elements which are in close proximity to those of the majority of elements in the base alloy and is thus likely to result in the formation of a solid solution, desirable for high-temperature applications. The present work aims to study the effect of cobalt addition on the corrosion and thermal behavior of $AlCr_{1.5}CuNi_2FeCo_x$ (where $x = 0, 0.25, 0.5, 0.75$, and 1 in molar ratio) high entropy alloys.

6.1.1 Corrosion Behavior of High Entropy Alloys

The composition of the alloys and the microstructure are two significant factors that determine the corrosion behavior. Moreover, the addition of other elements also decides the nature of the alloy being subjected to the different environmental conditions. The degree of heterogeneity may accelerate pitting corrosion. Therefore, the solid solution provides more resistance to corrosion compared to the multiphase alloys. The corrosion behavior of various high entropy alloys as available in the literature is discussed in subsequènt sections.

Corrosion behavior of quinary high entropy alloys in 3.5% sodium chloride (NaCl) solution was reported by Hsu et al. [11]. It was shown that the increase in copper in high entropy alloys increased the tendency of these alloys to develop localized corrosion as revealed by immersion (for 30 days) and polarization tests. It may be attributed to the formation of interdendrites, which deplete against active cell segregation in the alloy and galvanic action attack. In another study, Nene and co-workers [12] developed $Fe_{38.5}Mn_{20}Co_{20}Cr_{15}Si_5Cu_{1.5}$ (Cu-HEA) by a friction stir process (FSP) and found that the minimization of copper partitioning within the γ matrix by FSP improved the corrosion-resistance of the alloy thermodynamically and kinetically in 3.5 wt.% sodium chloride.

The comparison of high entropy alloys (HEAs) for electrochemical kinetics with stainless steel in an aqueous environment has shown a higher degree of magnitude of the disorder. The investigation revealed that high entropy alloys have better corrosion resistance than the 304 steel in both the acidic and sodium chloride (NaCl) medium, but the pitting action was more predominant in high entropy alloys as studied by Chen and co-workers [13]. It was observed that the high entropy alloys and the steel possess energy values of 94.06, 310.43 KJ/mole and 219.97, 343.18 KJ/mole in sodium chloride and the acidic solution, respectively. It was concluded that the corrosion resistance of high entropy alloys was significantly higher than its counterpart. The investigation of Cu-rich high entropy alloys was carried out in 288°C high purity water that resulted in a bulk glassy-like structures. The weight loss was as low as 4.5 $\mu g/mm^2$ in a sodium sulphate solution for $Cu_{0.5}$ immersed for 12 weeks [14].

The effect of aluminum addition in the multicomponent high entropy alloys in both sulfuric acid and NaCl (base) solution was studied by Lee and co-workers [15, 16]. It was observed that aluminum-free high entropy alloys exhibited tremendous resistance to general corrosion than the high entropy alloys with aluminum in acidic solution. Chou and co-workers [17] studied the effect of molybdenum in three aqueous environments such as basic, acidic, and marine to study the electrochemical properties in an ambient temperature (25°C). The corrosion resistance of the molybdenum-free high entropy alloys was significantly higher than that of the high entropy alloys with a molybdenum contribution. It resulted due to the Mo-rich alloy being susceptible to the pitting action in the sodium chloride solution. Chou and co-workers [18] studied the behavior of high entropy alloys for pitting potential and pitting temperature in chloride solution. The critical pitting temperature obtained for 0.1, 0.5, and 1 mol/L NaCl solution was 70°C and 60°C, respectively. The positive effect was witnessed for both pitting potentials, as well as the pitting temperature as the ratio exceeds 0.5. The passivity and transpassive corrosion were overcoming pitting corrosion as shown by the potentio-dynamic polarization curve. Pang and co-workers [19] observed from the polarization curves that the CrMnFeCoNi without the oxide represents the passive region, whereas the CrMnFeCoNi with the oxide does not present passive region.

Lin and co-workers [20] examined the effect of heat treatment on microstructure and corrosion behavior of $FeCoNiCrCu_{0.5}$ high entropy alloys. The X-ray diffraction (XRD) revealed that the alloy consists of a face-centered cubic structure; however, on heat treatment at 1,250°C, the microstructure study revealed a Cu-rich phase with spinodal decomposition. The corrosion resistance of the alloy degraded in the presence of the copper-rich matrix in the alloy. Hsueh and co-workers [21] investigated the effect of nitride-rich $(AlCrSiTiZr)_{100-x}N_x$ high entropy alloys films on aluminum alloy and mild steel substrates deposited by magnetron sputtering technique. The difference in atomic size resulted in a low crystalline structure. The film deposited with 30% nitrogen showed better corrosion resistance among all the deposited films. It was also revealed that corrosion resistance of the film depends on the interface bonding, film structure, and nitrogen content. The increase in substrate bias from 0 to –100 V changes hardness from 16.9 to 19.6 GPa and Young's modulus from 231.5 to 227.5 GPa.

Qiu and co-workers [22] developed quinary high entropy alloys by the laser cladding technique to study the microstructure and corrosion resistance behavior. The solid-solution strengthening resulted in a combined face-centered cubic and body-centered cubic structure owing to the rapid solidification. This combined face-centered cubic and body-centered cubic structure minimizes the chances of the formation of brittle compounds. Better corrosion resistance was witnessed for 1 mol/L in sodium chloride than 0.5 mol/L sulfuric acid (H_2SO_4).

The migration and segregation of species in the CrCoFeNi high entropy alloys system creates a negative vacancy for chromium, whereas it is positive for the others in the alloy. Low activation energy during the migration of each element may have caused the negative vacancy for chromium in the alloy system. However, cobalt, iron, and nickel migrated with an activation energy of 1.07, 1.32, and 1.36, respectively, compared to a value of 0.68 for chromium. In addition, chromium is not thermodynamically stable in the quandary alloy system and thereby segregates out to form a chromium metal by creating a negative vacancy, as revealed by Middleburgh and co-workers [23]. Zhang and co-workers [24] investigated the synthesized laser surface alloying technique to improve the corrosion and cavitation resistance of the premixed high purity elemental powders. The synthesis method results in a good metallurgical bonding with the substrate. A single-phase solid solution of the alloy results in a body-centered cubic structure due to tremendous mixing entropy and the combined effect of the atomic size difference and valence electron concentration. The microhardness of the high entropy alloys was three times harder than the steel and improvement in corrosion, as well as the cavitation resistance being 7.6 times than steel. In another study, Shang and co-workers [25] studied the effect of CoCrFeNi (W1-xMox) high entropy alloys coating on Q235 steel substrate and concluded that the Mo addition also improves the corrosion resistance. Kukshal and co-workers [26, 27] studied the effect of manganese and titanium on the corrosion behavior of $AlCr_{1.5}CuFeNi_2$ and concluded that both the elements enhance the corrosion resistance of the alloy with the increase in the molar ratio.

The equimolar quinary Ni-rich alloy system with promising mechanical properties at cryogenic temperatures was achieved by Yea and co-workers [28]. The X-ray diffraction analysis revealed that the coating was formed with identical Fe-Co-Ni rich and Mn-Ni-rich dendrites. The nobler corrosion resistance of the coating in both 3.5% sodium chloride (NaCl) and 0.5 M sulfuric acid solution was witnessed as compared to steel. Energy dispersive X-ray spectroscopy (EDS) analysis on the corroded sample revealed that the corrosion initiated with the chromium dendrites. The Cr-Ti-rich equimolar high entropy alloys were synthesized to study the effect of chromium and titanium by non-consumable arc melting. Multiphase microstructures with body-centered cubic lattices, face-centered cubic lattices, and intermetallics were identified after the solid-solution strengthening. The septenary alloy system exhibited a short-range pair of Ti-Co, Cr-Fe, Al-Ni, and Co-Cr in liquid structure as found by *ab initio* molecular dynamics (AIMD). The addition of chromium and titanium facilitates the formation of body-centered cubic and face-centered cubic dendrites in the alloy system, as investigated by Xiao and co-workers [29]. Tian and co-workers [30] studied the potential polarization results of CrMnFeCoNi high entropy alloy coated Q235 steel and compared it with the uncoated samples. It was found that the coated samples depict more positive Ecorr and lower Icorr compared to the uncoated samples in 3.5 wt.% sodium chloride solutions, thus concluding that the coating possesses excellent corrosion resistance in comparison to the Q235 substrate.

Wang and co-workers [31] synthesized the high entropy alloys with a direct laser fabrication technique for high-temperature applications. The aged samples at a temperature in a range between 600 to 1,200°C were examined for microstructure, corrosion, and mechanical properties. The formation of a face-centered cubic phase for the synthesized high entropy alloys was the result of the fast rate of cooling during deposition, whereas the as-deposited alloy B2 solid structure was evident. The microstructure of the high entropy alloys aging between 800–1,200°C resulted in needle-like, plate-like, and wall-shaped precipitates across the boundaries. Microstructures obtained in the Al-Ni-rich alloy were more susceptible to the galvanic corrosion which corrodes preferentially than those in the Fe-Cr-rich alloy. High entropy alloy coatings have evoked the interest of researchers worldwide owing to their tremendous mechanical and corrosive properties. Depositions of Ti-rich high entropy alloys synthesized on steel substrate were studied for the determination of the microhardness and the mechanical and corrosive properties. Phase identification revealed that formation of FCC + BCC for $x = 0.5$ to 1, FCC + BCC + Ti_2Ni for $x = 1.5$, and FCC + BCC + Ti_2Ni + ordered BCC dendrites for $x = 2$. The fabricated alloy exhibited excellent cavitation resistance in distilled water but was weak in the sodium chloride solution [32].

The chemical segregation upon the homogenization by the addition of aluminum content resulted in an inevitable multiphase structure, as investigated by Shi and co-workers [33]. It was found that the heat treatment to a temperature of 1,250°C for 1,000 h to homogenize the microstructure with chemical segregation leading to improved corrosion resistance. The author also depicts further increase in corrosion resistance by adopting a suitable annealing methodology. Li and co-workers [34] analyzed nickel-rich high entropy alloys; both the as-cast and homogenized alloys represent strong passivity in 0.6 M NaCl and 1 M and 6 M HCl solutions. On exposure of alloys for longer periods, active dissolution corrosion was only observed in the 12 M HCl solution. Both as-cast and homogenized high entropy alloys exhibited strong passivity in 0.6 M NaCl, 1 M and 6 M HCl solutions, and transpassive breakdown was the only corrosion form in these solutions under the test conditions. Active dissolution corrosion was only observed in 12 M HCl after a long exposure time in the solution. In all these cases, no pits were observed on the HEA. Qiu [35] investigated the corrosion behavior of $Al_2CrFeCo_xCuNiTi$ high entropy alloy in two mediums i.e., alkaline and salt solution. The high entropy alloy showed excellent corrosion resistance in 1 mol/L NaOH and 3.5 wt.% NaCl solutions. The corrosion current density of the alloy was reduced in both the media.

A short communication by Luo and co-workers [36] on corrosion behavior showed that depleted chromium and enriched iron and manganese were evident during the solid-solution strengthening of CoCrFeMnNi high entropy alloy. In addition, the formation of the metal hydroxide in the fabricated alloy showed a weak corrosion resistance. It was also revealed that the corrosion resistance of the steel is not greatly affected by the grain size in the micrometer range. Microstructure studies depicted that pits were evident on the surface when tested in the H_2So_4 solution. Shi and co-workers [37] investigated the corrosion behavior of quinary $Al_xCoCrFeNi$ via the electrochemical-AFM technique. Micro and sub-micro surface studies were carried out to depict the changes in the surface affected by the corrosion in 3.5% sodium chloride (NaCl) solution. The microstructure of the Al-rich ($x = 0.7$) alloy showed the presence of face-centered cubic phases with the small traces of body-centered dendrites. The study revealed that the height difference of 40.33 ± 2.87 was evident between face-centered cubic and body-centered cubic dendrites. The formation of pits at the intersection between face-centered cubic and body-centered cubic and the inside body-centered cubic phase can be observed. Zhang and co-workers [38] investigated the effect of re-melting and annealing on corrosion resistance of AlFeNiCoCuCr high entropy alloy. Both the treatment process improved the corrosion resistance of the alloy. It was observed that the annealed samples show better corrosion resistance compared to the re-melting samples.

6.1.2 Thermal Behavior of High Entropy Alloys

The detailed structural evolution of hexanary high entropy alloys involving the variation of aluminum in forming amorphous alloys exhibited an equiatomic crystalline icosahedral phase as studied by Kim and co-workers [39]. The Al_{10} rich alloy exhibited a Zr_2Cu-type icosahedral phase at 970 K. The formation of a Zr_2Cu-type phase retards decomposition after the melting move beyond 970 K, thereby enhancing the thermal stability. In contrast, the other two alloys formed with Ti_2Ni-type and $MgZn_2$-type decrease thermal stability by accelerating the decomposition. It was also reported that the formation of atomic complexity dwells in the equiatomic substitution and alloy composition.

Studies on aluminum-rich quinary high entropy alloys by Chou and co-workers [40] on thermophysical and electrical properties by the arc melting method depicted that the change in aluminum from 0 to 2% resulted in phase transformation from a single face-centered cubic (FCC) to a single body-centered cubic (BCC) with duplex FCC/BCC dendrites. XRD analysis revealed a weak lattice distortion due to the variation of aluminum, thereby resulting in lower electrical and thermal conductivity as compared to pure metals. The transport property of the electron and the photon were similar and comparable to those of semi-metals. Dolique and co-workers [41] successfully deposited the thin film on AlCoCrCuFeNi high entropy alloys by magnetron sputtering on mosaic targets. The developed thin films depicted thermal stability at a temperature of 510°C of heat treatment (annealing), but this resulted in distorted thin films beyond 510°C. The phase transition took place at 310°C, involving the disappearance of body-centered cubic (BCC) dendrites to form the AlCr binary phase. Tariq and co-workers [42] studied AlCoCrCuFeNi HEA and concluded that the endothermic peaks occurring after the exothermic peaks is due to the phase change and disintegration of the crystal structure at the higher temperature.

Lu and co-workers [43] studied the heat transfer and thermal expansion capability of the quinary high entropy alloys. The increment in thermal diffusivity was 20% and 50% at 423 K and 573 K. The increase in thermal diffusivity was achieved by the lattice dilation for a different temperature range. The increase in thermal diffusivity is a result of slight shift in reflection peaks to lower angle forming body-centered cubic dendrites at a temperature range of 450 K to 600 K. The effect of the addition of aluminum in mole fraction to hexanary high entropy alloys fabricated by spark plasma sintering resulted in dense alloys as investigated by Sriharitha and co-workers [44]. The studies on thermal stability in the argon atmosphere for 2–10 h in a 400–600°C temperature range resulted in a fine crystal structure and excellent thermal stability. Laplanche and co-workers [45] studied the CoCrFeMnNi HEAs and correlated the elastic modulus with the different properties. It was concluded that there is no linear dependence of thermal coefficient on temperature.

Investigation of the multicomponent high entropy alloys by Caro and co-workers [46] revealed that HEAs are not only thermally stable, but also possess the potential to resist radiation. This is due to the equilibrium thermodynamic state resulting in the extreme disorder of the dendrites with the stabilized entropy. It is observed that disorder in the dendrites keeps the damage energy longer in such a way as to reduce the photon and electron conductivity. The results also suggested guidelines in the selection of HEAs possessing photon and electron conductivity. A classical theory on molecular dynamics based on the Lennard–Jones approach was adopted in the investigation of the effect of disorder in thermal conductivity. The addition of more impure atoms in the solid solution once the thermal conductivity has reached the critical point does not reduce the thermal conductivity. Therefore, to reduce thermal conductivity below the critical point (ultralow), a combined effect of local strain field and mass defect scattering should occur in the solid solution, surpassing the minimum limits suggested in the theoretical aspects studied by Giri and co-workers [47].

6.2 MATERIALS AND FABRICATION TECHNIQUE

Selection of elements while designing an alloy is the essential criterion. Each element selected for the developments of the high entropy alloy has a distinct property and plays a vital role in the microstructure of alloys. The elements used in the present study include aluminum, chromium, copper, iron, nickel, and cobalt. The characteristics of the elements used in the production of alloys are presented in Table 6.1.

6.2.1 Alloy Production

The $AlCr_{1.5}CuFeNi_2Co_x$ HEAs samples were cast using a high-temperature vacuum induction furnace (Make: Jet Technocrats, India), as shown in Figure 6.1 under an

TABLE 6.1
Characteristics of Elements Used for the Development of HEAs [48, 49]

Element	Atomic number	Atom radius (pm)	Pauling electronegativity	Valence electron concentration (VEC)	Structure at RT_a	Density (g/cm^3)	Melting temperature (K)
Al	13	143.17	1.61	3	FCC	2.70	933
Cr	24	124.91	1.66	6	BCC	7.19	2180
Cu	29	127.8	1.90	11	FCC	8.94	1358
Fe	26	124.12	1.83	8	BCC	7.88	1811
Ni	28	124.59	1.91	10	FCC	8.91	1728
Co	22	146	1.54	4	HCP	8.90	1768

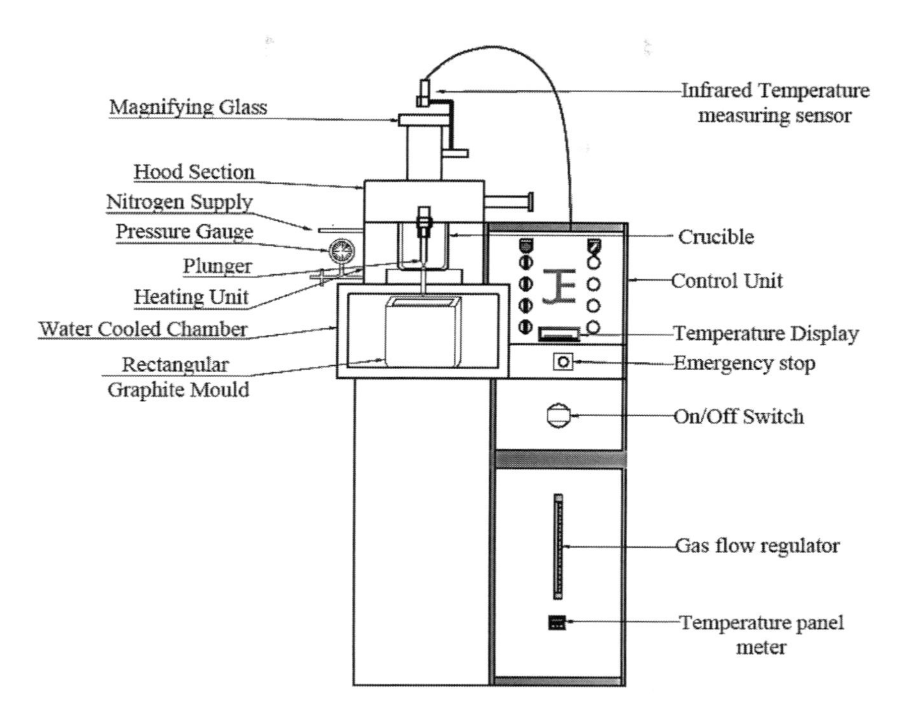

FIGURE 6.1 Schematic diagram of high-temperature vacuum induction furnace.

inert gas environment. The elements aluminum, chromium, copper, iron, nickel, and cobalt with high purity (above 99.9%) in granular form were used for the casting of the alloys.

The $AlCr_{1.5}CuFeNi_2Co_x$ alloys are abbreviated as Co_0, $Co_{0.25}$, $Co_{0.5}$, $Co_{0.75}$, and Co where x = 0, 0.25, 0.5, 0.75, and 1.0, respectively. The elements weighing approximately 800 gm were melted at 200°C in the ceramic crucible surrounded by a water-cooled chamber under the inert gas atmosphere. Each alloy was melted five times, with electromagnetic stirring to improve the chemical uniformity. The cast samples were homogenized at a temperature of 1,150°C for 48 h and then furnace cooled before the corrosion and thermal conductivity test were conducted.

6.2.2 Electrochemical Test

The electrochemical corrosion test was performed at room temperature on sample size 10 mm × 10 mm ×4 mm by potentio-dynamic-polarization measurement using a 3.5 wt.% sodium chloride (NaCl) solution. The specimens were finished using various grits of silicon carbide paper and then polished with 1 μm diamond suspension. The technique (GAMRY Reference 600TM) utilized for the corrosion test had measurements recorded at a scan rate of 0.5 mVs^{-1} in the range of −1.5 V to

0.5 V. The experiment is performed with three electrodes i.e., specimen, saturated calomel, and platinum electrodes. The electrochemical impedance spectra were attained at the open circuit potential in the frequency range of 10^5–10^{-2} Hz with an amplitude of 5–mV. All the samples were dipped in sodium chloride (NaCl) solution (pH ≈ 6) for one and half hours before the test, allowing the system to reach equilibrium with the electrolyte exposing the sample of size 1 cm^2 to the electrolyte. The schematic diagram of the corrosion testing machine is shown in Figure 6.2.

6.2.3 Thermal Conductivity Test

Thermal conductivity of the high entropy alloys sample was determined by the principle of transient plane source (TPS) using a Hot Disk Thermal Constants Analyzer (Model: TPS 500, Make: Gothenburg, Sweden). Rapid measurement of thermal conductivity within a span of 2.5 s is offered by the instrument. The samples of size of 25 mm × 25 mm × 2 mm were used for the measurement. A spiral-shaped hot disk, being coated with Kapton serves as heat source having a radius r, and the temperature response sensor was placed between two samples and data acquisition using Hot Disk analysis software. The thermal conductivity is measured as a function of time by the flow of electrical power through the high entropy alloys. The schematic of the thermal conductivity analyzer is shown in Figure 6.3.

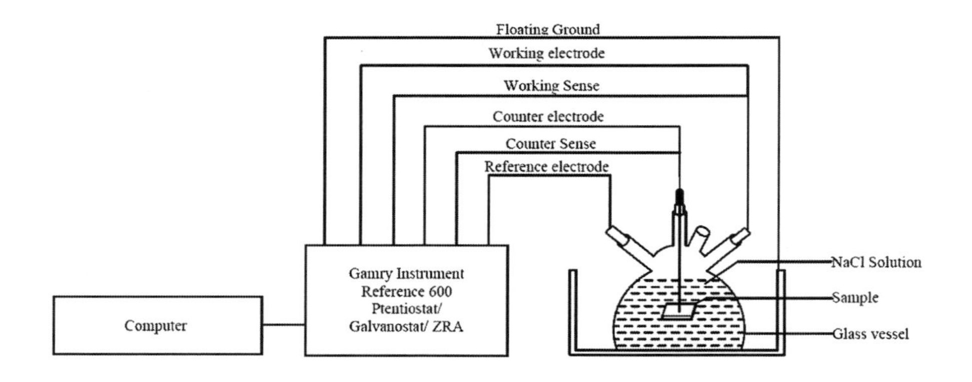

FIGURE 6.2 Schematic diagram of corrosion testing setup.

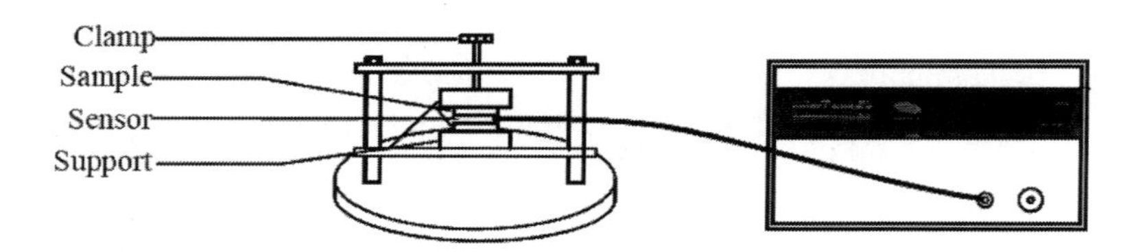

FIGURE 6.3 Schematic diagram of thermal conductivity analyzer.

6.3 RESULTS AND DISCUSSION

6.3.1 Corrosion Behavior of $AlCr_{1.5}CuFeNi_2Co_x$ High Entropy Alloys

The ability to form a stable passive film on the surface of the alloy governs the resistance of the passive film to the external environment. The potential polarization curve of $AlCr_{1.5}CuFeNi_2Co_x$ in HEAs in 3.5 wt.% sodium chloride (NaCl) at ambient temperatures is shown in Figure 6.4. Table 6.2 presents the electrochemical parameters of high entropy alloys. It is observed that the corrosion current density is decreasing as the cobalt content is increasing in the alloy. This is attributed due to the presence of highly corrosion-resistant elements such as nickel, chromium, and cobalt. These elements form a passive film over the surface, resulting in corrosion reduction.

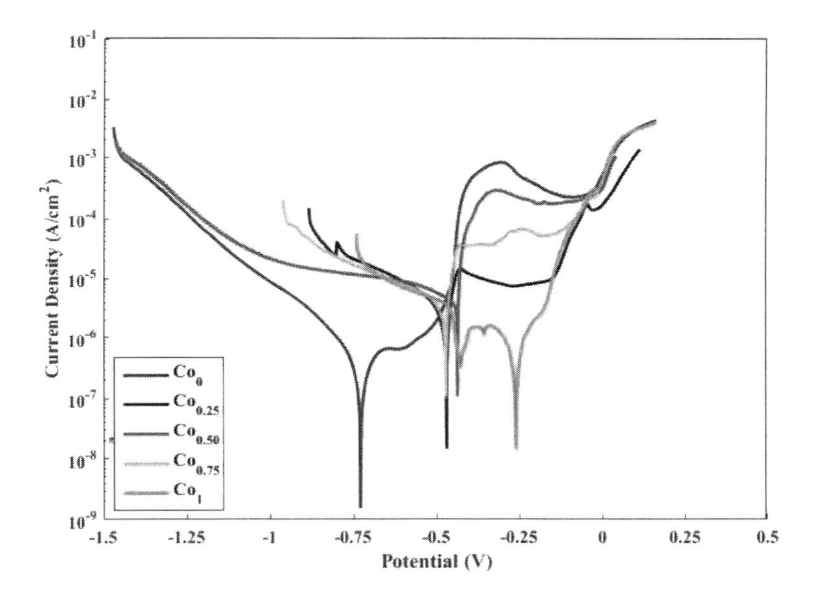

FIGURE 6.4 Potentiodynamic-polarization curves of $AlCr_{1.5}CuFeNi_2Co_x$ HEAs in 3.5 wt. % NaCl at ambient temperature.

TABLE 6.2
Electrochemical Parameters of $AlCr_{1.5}CuFeNi_2Co_x$ High Entropy Alloys

Alloy	E_{corr}(V)	I_{corr}(mA/cm^2)
Co_0	−0.728	0.089
$Co_{0.25}$	−0.440	0.071
$Co_{0.5}$	−0.436	0.056
$Co_{0.75}$	−0.354	0.045
Co_1	−0.259	0.033

The results indicate that the corrosion resistance of the HEAs is improved by the addition of cobalt content. A positive shift in the corrosion potential is observed for the alloys varying from Co_0 to Co_1 in the NaCl solution. A similar trend is also found in $Al_2CrFeCo_xCuNiTi$ high entropy alloy coatings in 3.5 wt.% NaCl solution. It is analyzed that the $Al_2CrFeCo_xCuNiTi$ high entropy alloy depicts better corrosion resistance in the NaCl solution compared to the Q235 steel [35]. The polarization curves of Co_1 and Co_0 alloys show the minimum and maximum corrosion rate compared to the other alloys. The HEAs with larger corrosion potential (E_{corr}) and smaller corrosion current density (i_{corr}) possess better corrosion resistance.

6.3.2 Thermal Conductivity of $AlCr_{1.5}CuNi_2FeCo_x$ High Entropy Alloys

Hung and co-workers [50] studied the thermal stability of V-Nb-Mo-Ta-W and V-Nb-Mo-Ta-W-Cr-B refractory high entropy alloy at 500°C for one hour in an air environment. It was found that the V-Nb-Mo-Ta-W-Cr-B possesses better thermal stability compared to V-Nb-Mo-Ta-W under the given condition. The lattice thermal conductivity of $Nb_{1-x}M_xFeSb$ (M = Hf, Zr, Mo, V, Ti) alloy was studied by Yan and co-workers [51]. It was concluded that the lattice thermal conductivity decreased as the M contents increased, which was attributed due to the high entropy effect. The thermal conductivity of 2.5 W/mK was found at 873, K thus showing a large potential for reduction in lattice thermal conductivity and enhancement of the thermoelectric performance.

The variation of thermal conductivity of $AlCr_{1.5}CuFeNi_2Co_x$ high entropy alloys as a function of x is shown in Figure 6.5. It is observed that the thermal conductivity of $AlCr_{1.5}CuFeNi_2Co_x$ alloys decreases with an increase in the cobalt content. This

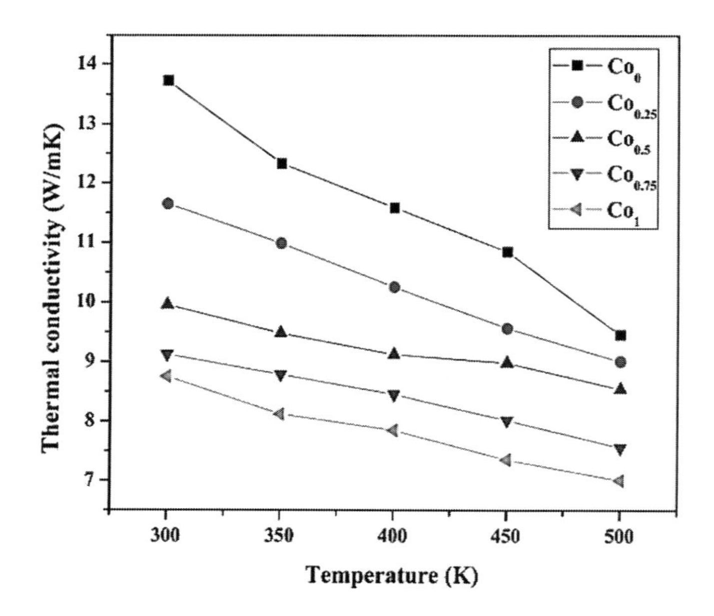

FIGURE 6.5 Variation of thermal conductivity of $AlCr_{1.5}CuFeNi_2Co_x$ alloy as a function of temperature.

is due to the presence of dual phases, i.e, face-centered and body-centered cubic phases in the high entropy alloys. The phenomenon can be explained with respect to the larger scattering effect due to the addition of the cobalt content and hence a reduction in the thermal conductivity. Chou and co-workers [40] found a similar observation while studying the thermal conductivity of $Al_xCoCrFeNi$. The duplex phase is characterized by the more interface boundary as compared to the single-phase resulting in the lowering of the heat transfer in the HEAs. The decrease in the thermal conductivity of the $AlCr_{1.5}CuFeNi_2Co_x$ alloys may also be due to the very low thermal conductivity of the cobalt. As the cobalt is added to the $AlCr_{1.5}CuFeNi_2$ alloy, there is a decrease in the mean free path of electrons, resulting in the decrease in the thermal conductivity of the alloys.

6.4 CONCLUSION

The $AlCr_{1.5}CuNi_2FeCo_x$ (where x = 0, 0.25, 0.5, 0.75, and 1) high entropy alloys were successfully cast by a high-temperature vacuum induction furnace in a controlled environment. The addition of Co improves the corrosion resistance of HEAs at room temperature in 3.5 wt.% sodium chloride (NaCl) solution. The elements nickel, chromium, and cobalt present in the high entropy alloys form a passive film over the surface resulting in the corrosion reduction. The formation of the protective layer and the body-centered cubic structure improves the corrosion resistance of the alloys. A positive shift in the corrosion potential is observed for the alloys varying from Co_0 to Co_1 in an NaCl solution. The $AlCr_{1.5}CuFeNi_2Co$ alloy depicts the highest corrosion resistance. The thermal conductivity of the alloys decreases with the increase in the molar ratio and increases with the increase in the temperature. The addition of cobalt decreases the mean free path of electron, and hence decreases the thermal conductivity.

REFERENCES

1. Huber, G., Djurdjevic, M.B. and Manasijevic, S., 2019. Determination some thermo-physical and metallurgical properties of aluminum alloys using their known chemical composition. *International Journal of Heat and Mass Transfer*, 139, pp.548–553.
2. Yang, L., Wei, S. and Zhang, Q., 2013. Aluminum reticulated spatial structures: State of the art and key issues. *Journal of Building Structures*, 34(2), pp.1–19.
3. Liu, Y., Liu, H. and Chen, Z., 2019. Post-fire mechanical properties of aluminum alloy 6082-T6. *Construction and Building Materials*, 196, pp.256–266.
4. Yeh, J.W., Chen, S.K., Lin, S.J., et al., 2004. Nanostructured high-entropy alloys with multiple principal elements: Novel alloy design concepts and outcomes. *Advanced Engineering Materials*, 6(5), pp.299–303.
5. Senkov, O.N., Senkova, S.V. and Woodward, C., 2014. Effect of aluminum on the microstructure and properties of two refractory high-entropy alloys. *Actamaterialia*, 68, pp.214–228.
6. Wang, Y.P., Li, B.S., Ren, M.X., et al., 2008. Microstructure and compressive properties of AlCrFeCoNi high entropy alloy. Materials Science and Engineering: *Part* A, 491, pp.154–158.
7. Poletti, M.G., Branz, S., Fiore, G., et al., 2016. Equilibrium high entropy phases in X-NbTaTiZr (X= Al, V, Cr and Sn) multiprincipal component alloys. *Journal of Alloys and Compounds*, 655, pp.138–146.

8. Wu, P.H., Liu, N., Yang, W., et al., 2015. Microstructure and solidification behavior of multicomponent CoCrCuxFeMoNi high-entropy alloys. *Materials Science and Engineering: Part A*, 642, pp.142–149.
9. Shun, T.T., Hung, C.H. and Lee, C.F., 2010. The effects of secondary elemental Mo or Ti addition in Al0. 3CoCrFeNi high-entropy alloy on age hardening at 700 C. *Journal of Alloys and Compounds*, 495(1), pp.55–58.
10. Kukshal, V., Patnaik, A. and Bhat, I.K., 2018. Effect of cobalt on microstructure and properties of AlCr1. 5CuFeNi2Cox high-entropy alloys. *Materials Research Express*, 5(4), p.046514.
11. Hsu, Y.J., Chiang, W.C. and Wu, J.K., 2005. Corrosion behavior of FeCoNiCrCux high-entropy alloys in 3.5% sodium chloride solution. *Materials Chemistry and Physics*, 92(1), pp.112–117.
12. Nene, S.S., Frank, M., Liu, K., et al., 2019. Corrosion-resistant high entropy alloy with high strength and ductility. *Scripta Materialia*, 166, pp.168–172.
13. Chen, Y.Y., Hong, U.T., Shih, H.C., et al., 2005. Electrochemical kinetics of the high entropy alloys in aqueous environments—A comparison with type 304 stainless steel. *Corrosion Science*, 47(11), pp.2679–2699.
14. Chen, Y.Y., Hong, U.T., Yeh, J.W., et al., 2006. Selected corrosion behaviors of a Cu0. 5NiAlCoCrFeSi bulk glassy alloy in 288 C high-purity water. *Scriptamaterialia*, 54(12), pp.1997–2001.
15. Lee, C.P., Chang, C.C., Chen, Y.Y., et al., 2008. Effect of the aluminium content of AlxCrFe1. 5MnNi0. 5 High-entropy alloys on the corrosion behaviour in aqueous environments. *Corrosion Science*, 50(7), pp.2053–2060.
16. Lee, C.P., Chen, Y.Y., Hsu, C.Y. et al., 2008. Enhancing pitting corrosion resistance of AlxCrFe1.5MnNi0.5 high-entropy alloys by anodic treatment in sulfuric acid. *Thin Solid Films*, 517(3), pp.1301–1305.
17. Chou, Y.L., Yeh, J.W. and Shih, H.C., 2010. The effect of molybdenum on the corrosion behaviour of the high-entropy alloys Co1. 5CrFeNi1. 5Ti0. 5Mox in aqueous environments. *Corrosion Science*, 52(8), pp.2571–2581.
18. Chou, Y.L., Wang, Y.C., Yeh, J.W. et al., 2010. Pitting corrosion of the high-entropy alloy Co1. 5CrFeNi1. 5Ti0. 5Mo0. 1 in chloride-containing sulphate solutions. *Corrosion Science*, 52(10), pp.3481–3491.
19. Pang, J., Xiong, T., Wei, X., et al., 2019. Oxide MnCr2O4 induced pitting corrosion in high entropy alloy CrMnFeCoNi. *Materialia*, 6, p.100275.
20. Lin, C.M. and Tsai, H.L., 2011. Effect of annealing treatment on microstructure and properties of high-entropy FeCoNiCrCu0. 5 alloy. *Materials Chemistry and Physics*, 128(1–2), pp.50–56.
21. Hsueh, H.T., Shen, W.J., Tsai, M.H. and Yeh, J.W, 2012. Effect of nitrogen content and substrate bias on mechanical and corrosion properties of high-entropy films (AlCrSiTiZr) 100– xNx. *Surface and Coatings Technology*, 206(19–20), pp.4106–4112.
22. Qiu, X.W., Zhang, Y.P., He, L. and Liu, C., 2013. Microstructure and corrosion resistance of AlCrFeCuCo high entropy alloy. *Journal of Alloys and Compounds*, 549, pp.195–199.
23. Middleburgh, S.C., King, D.M., Lumpkin, G.R., et al., 2014. Segregation and migration of species in the CrCoFeNi high entropy alloy. *Journal of Alloys and Compounds*, 599, pp.179–182.
24. Zhang, S., Wu, C.L., Zhang, C.H., et al., 2016. Laser surface alloying of FeCoCrAlNi high-entropy alloy on 304 stainless steel to enhance corrosion and cavitation erosion resistance. *Optics and Laser Technology*, 84, pp.23–31.
25. Shang, C., Axinte, E., Sun, J., et al., 2017. CoCrFeNi (W1– xMox) high-entropy alloy coatings with excellent mechanical properties and corrosion resistance prepared by mechanical alloying and hot pressing sintering. *Materials and Design*, 117, pp.193–202.

26. Kukshal, V., Patnaik, A. and Bhat, I.K., 2018, June. Effect of Mn on corrosion and thermal behaviour of AlCr1. 5CuFeNi2Mnx high-entropy alloys. In: IOP Conference Series: Materials Science and Engineering (Vol. 377, No. 1, p. 012023). IOP Publishing.
27. Kukshal, V., Patnaik, A. and Bhat, I.K., 2018. Corrosion and thermal behaviour of AlCr1. 5CuFeNi2Tix high-entropy alloys. *Materials Today: Proceedings*, 5(9), pp.17073–17079.
28. Ye, Q., Feng, K., Li, Z., et al., 2017. Microstructure and corrosion properties of CrMnFeCoNi high entropy alloy coating. *Applied Surface Science*, 396, pp.1420–1426.
29. Xiao, D.H., Zhou, P.F., Wu, W.Q., et al., 2017. Microstructure, mechanical and corrosion behaviors of AlCoCuFeNi-(Cr, Ti) high entropy alloys. *Materials and Design*, 116, pp.438–447.
30. Tian, Y., Lu, C., Shen, Y. and Feng, X., 2019. Microstructure and corrosion property of CrMnFeCoNi high entropy alloy coating on Q235 substrate via mechanical alloying method. *Surfaces and Interfaces*, 15, pp.135–140.
31. Wang, R., Zhang, K., Davies, C. and Wu, X., 2017. Evolution of microstructure, mechanical and corrosion properties of AlCoCrFeNi high-entropy alloy prepared by direct laser fabrication. *Journal of Alloys and Compounds*, 694, pp.971–981.
32. Wu, C.L., Zhang, S., Zhang, C.H., et al., 2017. Phase evolution and cavitation erosion-corrosion behavior of FeCoCrAlNiTix high entropy alloy coatings on 304 stainless steel by laser surface alloying. *Journal of Alloys and Compounds*, 698, pp.761–770.
33. Shi, Y., Collins, L., Feng, R., et al., 2018. Homogenization of AlxCoCrFeNi high-entropy alloys with improved corrosion resistance. *Corrosion Science*, 133, pp.120–131.
34. Li, T., Swanson, O.J., Frankel, G.S., et al., 2019. Localized corrosion behavior of a single-phase non-equimolar high entropy alloy. *Electrochimica Acta*, 306, pp.71–84.
35. Qiu, X.W., 2019. Corrosion behavior of Al2CrFeCoxCuNiTi high-entropy alloy coating in alkaline solution and salt solution. *Results in Physics*, 12, pp.1737–1741.
36. Luo, H., Li, Z., Mingers, A.M. and Raabe, D., 2018. Corrosion behavior of an equiatomic CoCrFeMnNi high-entropy alloy compared with 304 stainless steel in sulfuric acid solution. *Corrosion Science*, 134, pp.131–139.
37. Shi, Y., Collins, L., Balke, N., et al., 2018. In-situ electrochemical-AFM study of localized corrosion of AlxCoCrFeNi high-entropy alloys in chloride solution. *Applied Surface Science*, 439, pp.533–544.
38. Zhang, X., Guo, J., Zhang, X., et al., 2019. Influence of remelting and annealing treatment on corrosion resistance of AlFeNiCoCuCr high entropy alloy in 3.5% NaCl solution. *Journal of Alloys and Compounds*, 775, pp.565–570.
39. Kim, K.B., Warren, P.J., Cantor, B. and Eckert, J., 2006. Enhanced thermal stability of the devitrified nanoscale icosahedral phase in novel multicomponent amorphous alloys. *Journal of Materials Research*, 21(4), pp.823–831.
40. Chou, H.P., Chang, Y.S., Chen, S.K. and Yeh, J., 2009. Microstructure, thermophysical and electrical properties in AlxCoCrFeNi ($0 \leq x \leq 2$) high-entropy alloys. *Materials Science and Engineering: Part B*, 163(3), pp.184–189.
41. Dolique, V., Thomann, A.L., Brault, P., et al., 2010. Thermal stability of AlCoCrCuFeNi high entropy alloy thin films studied by in-situ XRD analysis. *Surface and Coatings Technology*, 204(12–13), pp.1989–1992.
42. Tariq, N.H., Naeem, M., Hasan, B.A., et al., 2013. Effect of W and Zr on structural, thermal and magnetic properties of AlCoCrCuFeNi high entropy alloy. *Journal of Alloys and Compounds*, 556, pp.79–85.
43. Lu, C.L., Lu, S.Y., Yeh, J.W. and Hsu, W., 2013. Thermal expansion and enhanced heat transfer in high-entropy alloys. *Journal of Applied Crystallography*, 46(3), pp.736–739.
44. Sriharitha, R., Murty, B.S. and Kottada, R.S., 2014. Alloying, thermal stability and strengthening in spark plasma sintered AlxCoCrCuFeNi high entropy alloys. *Journal of Alloys and Compounds*, 583, pp.419–426.

45. Laplanche, G., Gadaud, P., Horst, O., et al., 2015. Temperature dependencies of the elastic moduli and thermal expansion coefficient of an equiatomic, single-phase CoCrFeMnNi high-entropy alloy. *Journal of Alloys and Compounds*, 623, pp.348–353.
46. Caro, M., Beland, L.K., Samolyuk, G.D., et al., 2015. Lattice thermal conductivity of multi-component alloys. *Journal of Alloys and Compounds*, 648, pp.408–413.
47. Giri, A., Braun, J.L., Rost, C.M. and Hopkins, P.E., 2017. On the minimum limit to thermal conductivity of multi-atom component crystalline solid solutions based on impurity mass scattering. *Scripta Materialia*, 138, pp.134–138.
48. Guo, S., Hu, Q., Ng, C. and Liu, C.T., 2013. More than entropy in high-entropy alloys: Forming solid solutions or amorphous phase. *Intermetallics*, 41, pp.96–103.
49. Callister Jr, W.D. and Rethwisch, D.G., 2012. *Fundamentals of Materials Science and Engineering: An Integrated Approach*. John Wiley & Sons.
50. Hung, S.B., Wang, C.J., Chen, Y.Y., et al., 2019. Thermal and corrosion properties of V-Nb-Mo-Ta-W and V-Nb-Mo-Ta-W-Cr-B high entropy alloy coatings. *Surface and Coatings Technology*, 375, pp.802–809.
51. Yan, J., Liu, F., Ma, G., et al., 2018.Suppression of the lattice thermal conductivity in NbFeSb-based half-Heusler thermoelectric materials through high entropy effects. *Scriptamaterialia*, 157, pp.129–134.

7 Examining, Analyzing, Interpreting, and Understanding the Fracture Resistance of High Entropy Alloys

Weidong Li

CONTENTS

7.1 INTRODUCTION

Fracture is a problem that our society faces daily in engineered structural or even non-structural components and materials. Historically, human beings have paid huge costs for disasters caused by unpredicted fracture events [1, 2]. Today, the same stories are being repeated, or becoming even worse, with evolving technological complexity, despite our knowledge in this area being much richer. Materials can fracture in many different ways, depending on the microstructure, stress state, temperature, loading rate, etc. Broadly, they can be classified into brittle and ductile fractures on the basis of the strain leading to failure [3]. Brittle fractures usually happen in a catastrophic manner, with no evidence of plastic deformation, and therefore are more of a concern in practice. Ductile fractures, on the other hand, are more favorable, as they trigger appreciable plastic deformation to blunt crack tips and thus postpone the failure. Nevertheless, the strain is not the only factor determining the fracture resistance of a material. It is also closely related to strengths. For instance, a material with a high failure strain but extremely low strength may be more prone to a fracture than a material with a low failure strain but extremely high strength. By considering the effects from both the strain and strength, the fracture resistance of a material is more meaningfully quantified by the fracture toughness, e.g., K_{IC} or J_{IC} [2]. Materials that fail in a "brittle" manner usually have a fracture toughness of below 10 $\text{MPa}\sqrt{\text{m}}$, whereas the value for those "ductile fractured" materials could range from a few tens to hundreds [4–6]. All materials, prior to engineering applications, require extensive characterizations of their fracture resistance, with examples being found in traditional metals and alloys [2, 7], superalloys [8–11], and metallic glasses [12–14], among others.

As materials with superior performance are continuously being searched for, a novel alloy concept, termed high entropy alloys (HEAs), is emerging as a new class of metallic materials and is regarded as promising for structural applications [15–18]. Very distinct from the conventional physical metallurgy principle in which one or two principal metal elements are utilized for dictating the primary phase and properties, and minor alloying elements are added for modifying the properties toward the desired ones [19], HEAs are formulated by mixing five or more principal elements in equal or near-equal molar ratios [17, 20]. This new alloy design strategy essentially shifts the exploration of new metallic materials from the corners and edges of phase diagrams to the less exploited centers [20–23]. Interestingly, instead of forming multiple phases or intermetallic compounds as one may expect from the established wisdom, the metallurgy of multi-principal metal elements, in fact, could result in simple solid solution phases, such as face-centered cubic (FCC), body-centered cubic (BCC), or hexagonal close-packed (HCP) structures [24–28]. The formed solid-solution phases may be present as the sole phase in alloys or coexist with other phases to form multiphase alloys [29–34]. Thermodynamically, the formation of these solid solutions is attributed to the stabilization effect posed by the high configurational entropy (thus the low Gibbs free energy) and sluggish cooperative diffusion [31, 35]. Such a revolutionary alloy design strategy opens up enormous possibilities in terms of mechanical and functional properties of alloys by permitting extensive and flexible tuning of alloy compositions. Over a decade of dedicated research has revealed that many HEAs possess unparalleled properties in comparison with traditionally used alloys; for instance, great thermal and microstructural stability [36, 37], high hardness [38, 39], high strengths at elevated temperatures [40, 41], and excellent resistance to wear, corrosion, fatigue, creep, and high-temperature softening [42–49]. Given these merits, the applications of HEAs in different fields are being actively explored [48, 50].

In addition to the above-listed performance indices, it is also vital to gain a complete picture of the fracture resistance of HEAs in order to promote their engineering applications. Since the 2010s, a number of studies have been conducted to characterize the fracture resistance of various HEA systems (e.g., $FeCoNiCrCuTiMoAlSiBe_{0.5}$, CrMnFeCoNi, and AlCoCrCuFeNi) [1, 51–59]. The fracture toughness, K_{IC}, of the investigated alloys is found to vary from less than unity [58], to as high as exceeding 200 MPa$\sqrt{m}$ [53], depending not only on the composition but also on the phase constitution, temperature, and processing history [51, 53, 58]. Their fracture modes and underlying failure mechanisms also change from one type to another [52, 53, 55, 60–62]. There is no doubt that more fracture resistance characterizations need to be done to pinpoint definite correlations between the fracture toughness and the factors mentioned above. As such, the advances in understanding the fracture behavior of HEAs in the past decade are summarized in this chapter. The writing is on the basis of the author's published review article on the same subject [1], with some recent advances incorporated.

The rest of the chapter is arranged as follows: Section 7.2 begins by introducing the fracture resistance characterizations of HEAs by varied means, followed by discussing the effects of the temperature and phase constitution on their fracture resistance, and then by comparing the fracture resistance of HEAs with other commonly encountered materials. In Section 7.3, typical fractographic features of HEAs are surveyed, followed by unveiling the associated fracture mechanisms in different types of HEAs in Section 7.4. The chapter concludes in Section 7.5.

7.2 FRACTURE RESISTANCE

Fracture resistance represents the resistance of a material to the development of a fracture under loads. For the most frequently encountered Mode I fracture [63], the fracture resistance of a material may be quantified rigorously by fracture toughness or loosely by impact energy absorption or crack tip opening angle. The measurements of the fracture toughness of HEAs had been carried out based on the linear elastic fracture mechanics, elastic-plastic fracture mechanics, and nano- and micromechanics. The characterization of the impact energy absorption of HEAs primarily utilized the Charpy or other pendulum impact tests. Besides, the crack tip opening angle was also used to quantify the fracture resistance of some HEAs. The characterization of the fracture resistance of HEAs with various methods is first discussed in Section 7.2.1, following which the effects of temperature and phase constitution on the fracture resistance of HEAs are covered in Sections 7.2.2 and 7.2.3, respectively. Finally, the fracture resistance of the HEAs reported in the literature is compared with other materials in an Ashby fracture toughness–yield strength map.

7.2.1 Characterizations

7.2.1.1 Fracture Toughness

7.2.1.1.1 Based on Linear Elastic Fracture Mechanics

When a material behaves in an approximately linear manner with minimal plasticity prior to failure, from the fracture mechanics standpoint, the plastic zone around the crack tip is small compared to the specimen dimensions. In such a scenario, the fracture toughness of the material can be satisfactorily characterized by K_{IC}, the critical stress intensity factor at which the cracks start growing [2]. K_{IC} can be determined by failing a pre-cracked specimen, following the ASTM (American Society for Testing and Materials) standard E399 [64]. Simply put, it is derived by first calculating the provisional fracture toughness, K_Q, followed by checking against the validity criteria. For the most widely used single-edge notched bend (SENB) specimen in three-point bending tests, K_Q is calculated as

$$K_Q = \frac{P_Q S}{BW^{1.5}} f\left(\frac{a}{W}\right) \tag{7.1}$$

where B, W, and S are the sample thickness, width, and span, respectively, a is the average crack length, which can be measured from the fracture surface, P_Q is the critical load determined from the recorded load-displacement curve [64], and $f\left(\frac{a}{W}\right)$ is a geometry dependent polynomial function that can be looked up from the standard [64].

In order to make the computed K_Q from Equation (7.1) a valid K_{IC} and reflect the plane-strain fracture toughness of measured materials, it needs to pass a few validity checks, which include the following:

$$0.45 \le \frac{a}{W} \le 0.5 \tag{7.2}$$

$$B, a \ge 2.5\left(\frac{K_Q}{\sigma_y}\right)^2 \tag{7.3}$$

$$P_{\max} \le 1.1 P_Q \tag{7.4}$$

where σ_y is the yield strength of materials, and $P_{\max}$ is the maximum load on a load-displacement curve.

With the SENB specimens, the average K_{IC} of the as-cast $Al_{23}Co_{15}Cr_{23}Cu_8Fe_{15}Ni_{15}$ HEAs with a BCC structure was determined to be 5.8 $\text{MPa}\sqrt{\text{m}}$ [52]. Following the ASTM standard E1304, a verification measurement with a chevron-notched rectangular bar (CVNRB) specimens of the same HEA revealed a marginally small but overall consistent K_{IC} of 5.4 $\text{MPa}\sqrt{\text{m}}$ [52]. The marginal difference between the two measured K_{IC} values was ascribed to the fact that the SENB specimens were not pre-cracked by fatigue. Again with SENB specimens, Chen et al. [57] measured the fracture toughness of three as-cast HEAs, i.e., $Al_{18}Cr_{21}Fe_{20}Co_{20}Ni_{21}$ with a BCC structure, $Al_{15.5}Cr_{22.25}Fe_{20}Co_{20}Ni_{22.25}$ with BCC and FCC phases, and $Al_{13}Cr_{23.5}Fe_{20}Co_{20}Ni_{23.5}$ with an FCC structure, and reported 9 $MPa\sqrt{m}$, 11 $MPa\sqrt{m}$, and 53 $MPa\sqrt{m}$, respectively. Given that validity checks were not performed in the original work, here we attempt to check if the measured K_Q values for the three alloys are actually K_{IC}. First, we note that all samples used have $W = 6$ mm, $B = 3$ mm and $a\sim = 3$ mm. These dimensions lead to $\frac{a}{W} = 0.5$, meeting the requirement posted in Equation (7.2). Then, we substitute K_Q and σ_y listed in Table 7.1 into Equation (7.3), and the calculations suggest that B and a for the $Al_{18}Cr_{21}Fe_{20}Co_{20}Ni_{21}$, $Al_{15.5}Cr_{22.25}Fe_{20}Co_{20}Ni_{22.25}$, and $Al_{13}Cr_{23.5}Fe_{20}Co_{20}Ni_{23.5}$ HEAs need to be greater than 0.58, 2.21, and 12.16 mm, respectively. From these checks, it is clear that 9 $MPa\sqrt{m}$ and 11 $MPa\sqrt{m}$ for the first two alloys are valid K_{IC} values whereas 53 $MPa\sqrt{m}$ for the last alloy should be regarded as K_Q. Seifi et al. [55] utilized the same method to determine K_Q of the as-cast $Al_{0.2}CrFeNiTi_{0.2}$ to be 32–35 $MPa\sqrt{m}$, and K_Q of the as-cast $AlCrFeNi_2Cu$ HEAs at room temperature and 473 K to be 40–45 $MPa\sqrt{m}$ and 46–47 $MPa\sqrt{m}$, respectively. Following the same validity check procedure outlined above, 32–35 $MPa\sqrt{m}$ for the first alloy is decided to be a valid K_{IC} whilst the values reported for the second alloy at both temperatures are not. Zhang et al. [32] and Mohanty et al. [65] both measured the fracture toughness of the equiatomic AlCoCrFeNi alloy, but reported vastly different values. Zhang et al. [32] synthesized the alloy using spark plasma sintering at 1,473 K, which consists of a disordered FCC phase and a duplex BCC phase comprising spinodally decomposed ordered and disordered BCC structures. The fracture toughness measurements adopted SENB samples according to the ASTM standard E399 [64] and reported $K_{IC} = 25.2$ $MPa\sqrt{m}$. On the other hand, the same alloy was made by Mohanty et al. [65], also with spark plasma sintering, but at sintering temperatures of 973–1,273 K. The alloy consists of a disordered FCC phase, an Al-Ni rich $L1_2$ phase, and a tetragonal Cr-Fe-Co-based σ phase. The fracture toughness measurements utilized single-edge-V-notched beam (SEVNB) samples based on the European standard (EN) 843-1 [65]. And a minimum fracture toughness of 1 $MPa\sqrt{m}$ and a maximum value of 3.9 $MPa\sqrt{m}$ were reported for the alloys sintered at 1,273 K and 973 K, respectively. From the above comparison, we can reason that the discrepancy in the fracture toughness of the AlCoCrFeNi alloys from the two separate works [32, 65] has primarily originated from their dissimilar phase constitutions inherited from their distinct spark plasma sintering parameters. The difference in the testing standards on which their fracture toughness measurements were based may also have a contribution, but it should be minimal if both groups strictly followed their respective standard.

7.2.1.1.2 Based on Elastic-Plastic Fracture Mechanics

The K_{IC} measurements with the ASTM standard E399 [64] are based on the linear elastic fracture mechanics and pose very strict requirements on the sample preparation. Materials with even a small amount of plasticity exceeding the small-scale yielding assumption may require an unreasonably huge sample size to warrant a

TABLE 7.1
Fracture Toughness and Associated Information for a Variety of High Entropy Alloys

Composition	Phase	Processing	T (K)	σ_y (MPa)	ε_f	K_{IC} or K_Q (MPa m$^{0.5}$)	Test method	Ref.
$FeCoNiCrCuTiMoAlSiBe_{0.5}$	BCC + martensite	Laser-solidified coating (1.5 mm thick)	RT	3770^H	n/a	50.9	Nanoindentation	[51]
$FeCoNiCrAl_3$	BCC	Laser-solidified coating (1.5 mm thick)	RT	2500^H	n/a	7.6	Nanoindentation	[51]
$Al_{23}Co_{15}Cr_{23}Cu_8Fe_{15}Ni_{15}$	BCC	Vacuum induction melted, cast	RT	1260^C	0.21^C	5.4–5.8	K testing on SENB (ASTM E399) and CVNRB specimens (ASTM E1304)	[52]
AlCoCrFeNi(I)	FCC + duplex BCC	SPS at 1,473 K	RT	1262^T	$\sim 0.29^T$	25.2	K testing on SENB specimens (ASTM E399)	[32]
AlCoCrFeNi(II)	FCC + $L1_2$ + tetragonal σ	SPS at 973–1,273 K	RT	$333\text{-}2667^H$	n/a	1–3.9	K testing on SEVNB specimens (EN 843-1)	[65]
CrMnFeCoNi	FCC	Arc-melted, cast, cold forged and rolled, recrystallized	77 200 293	759^T 518^T 410^T	0.71^T 0.6^T 0.57^T	219 221 217	J-R curve testing on CT specimens (ASTM E1820)	[53]
$AlCrFeNi_2Cu$	FCC + BCC	Vacuum levitation melted, cast	RT 473	1049^H n/a	n/a n/a	40–45 46–47	K testing on SENB specimens (ASTM E399)	[55]
$Al_{0.2}CrFeNiTi_{0.2}$	FCC + BCC	Vacuum levitation melted, cast	RT	1666^H	n/a	32 - 35	K testing on SENB specimens (ASTM E399)	[55]
$Al_{18}Cr_{21}Fe_{20}Co_{20}Ni_{21}$	two BCC	Arc melted, cast	RT	$\sim 370^T$	0^T	9	K testing on SENB specimens (ASTM E399)	[57]

(Continued)

TABLE 7.1 (CONTINUED)
Fracture Toughness and Associated Information for a Variety of High Entropy Alloys

Composition	Phase	Processing	T (K)	σ_y (MPa)	ε_f	K_{IC} or K_Q (MPa m$^{0.5}$)	Test method	Ref.
$Al_{15.5}Cr_{22.25}Fe_{20}Co_{20}Ni_{22.25}$	FCC + BCC	Arc melted, cast	RT	~ 590^{T}	0^{T}	11	K testing on SENB specimens (ASTM E399)	[57]
$Al_{13}Cr_{23.5}Fe_{20}Co_{20}Ni_{23.5}$	FCC	Arc melted, cast	RT	760^{T}	0.06^{T}	53		[57]
Single-crystal $Nb_{25}Mo_{25}Ta_{25}W_{25}$	BCC	Arc melted, cast, homogenized	RT	$2640^{H, MC}$	0.026^{C}	1.6	Bending of notched micro-cantilever specimens (ASTM E399 and FEM)	[41, 58]
Bi-crystal $Nb_{25}Mo_{25}Ta_{25}W_{25}$	BCC	Arc melted, cast, homogenized	RT	$1058^{H, MC}$	0.026^{C}	0.2	Bending of notched micro-cantilever specimens (ASTM E399 and FEM)	[41, 58]
$CoCrFeNiNb_{0.5}$	eutectic, FCC + Laves	Arc melted, cast	RT	1700^{C}	~ 0.7^{C}	11.4-14.8	J-R curve testing on SENB specimens (ASTM E1820)	[33]
TiZrNbTa	BCC	Arc melted, cast	RT	1020^{C}	> 0.5^{C}	28.5	K testing on SENB specimens (ASTM E399)	[125, 126]
$(TiZrNbTa)_{95}Mo_5$	BCC	Arc melted, cast	RT	1180^{C}	~ 0.51^{C}	22.5	K testing on SENB specimens (ASTM E399)	[125, 126]
$(TiZrNbTa)_{90}Mo_{10}$	BCC	Arc melted, cast	RT	1300^{C}	~ 0.26^{C}	20	K testing on SENB specimens (ASTM E399)	[125, 126]
$(TiZrNbTa)_{80}Mo_{20}$	BCC	Arc melted, cast	RT	1460^{C}	~ 0.07^{C}	18.7	K testing on SENB specimens (ASTM E399)	[125, 126]

Notes: FCC and BCC denote face-centered cubic and body-centered cubic, respectively; RT denotes room temperature; SPS denotes spark plasma sintering; SENB denotes single-edge notched bend; CVNRB denotes chevron-notched rectangular bar; CT denotes compact tension; the superscript ^{T}, ^{C}, ^{H}, and ^{MC} indicate that the data are sourced from tensile tests, compressive tests, hardness (the yield strength is estimated from $\sigma_y = \frac{H}{3}$), and micro-compression tests, respectively; the superscript $^{(I)}$ and $^{(II)}$ indicate the same alloy from different works; K_Q is used for the shaded numbers in the fracture-toughness column; $\varepsilon_f = 0.026$ for the $Nb_{25}Mo_{25}Ta_{25}W_{25}$ HEAs are from their polycrystalline counterparts [41].

valid K_{IC} [2]. This makes it fairly inconvenient to use in practice for determining the fracture resistance of materials exhibiting appreciable plasticity. Thus, the fracture toughness of these materials should be measured by the *K-R* curve, the linear elastic fracture mechanics approach with the corrected plasticity effect, or J_{IC} or *J-R* curve on the basis of the elastic-plastic fracture mechanics [2]. In characterizing HEAs with significant plasticity, the J_{IC} and *J-R* curve approaches were employed. The detailed procedure for the determination of the critical *J*-integral near the onset of ductile crack extension, J_{IC}, is outlined in the ASTM standard E1820 [66]. Specifically, a *J-R* curve, i.e., *J*-integral as a function of crack extension, need to be constructed first. To do so, the crack growth in a test needs to be monitored using either a multiple specimen loading-unloading technique or unloading compliance method [2]. In the case of using the unloading compliance method [2, 53, 60], the crack length is periodically determined by unloading the sample and recording the elastic-unloading compliance. For side-grooved compact-tension samples, the instantaneous crack length, a_i, is calculated by

$$\frac{a_i}{W} = 1.000196 - 4.06319u + 11.242u^2 - 106.043u^3 + 464.335u^4 - 650.677u^5 \tag{7.5}$$

where W is the sample width, $u = \frac{1}{\left(B_e E C_{c(i)}\right)^{0.5} + 1}$ with B_e is the effective specimen thickness of side-grooved samples calculated from the sample thickness at the non-groove and groove locations, B and B_N, using $B - (B - B_N)^2/B$, and $C_{c(i)}$ is the corresponding instantaneous elastic unloading compliance. Once a_i is calculated, the crack extension at the ith instant, Δa_i, can be computed from its difference with the initial crack length, a_0, using $\Delta a_i = a_i - a_0$.

For each calculated instantaneous crack length, a_i, the corresponding *J*-integral, J_i, is calculated from the sum of the elastic and plastic component

$$J_i = J_{el(i)} + J_{pl(i)} \tag{7.6}$$

$$J_{el(i)} = \frac{K_i^2\left(1 - \nu^2\right)}{E} \tag{7.7}$$

$$J_{pl(i)} = \left[J_{pl(i-1)} + \frac{\eta_{i-1}\left(A_{pl(i)} - A_{pl(i-1)}\right)}{B_N b_{(i-1)}}\right]\left[1 - \gamma_{(i-1)}\left(\frac{a_{(i)} - a_{(i-1)}}{b_{(i-1)}}\right)\right] \tag{7.8}$$

where E is Young's modulus, ν is Poisson's ratio, K_i is the instantaneous linear elastic stress intensity factor, calculated as $K_i = \frac{P_i}{\sqrt{BB_N W}} f\left(a_i / W\right)$, A_{pl} is the plastic area under the recorded load-displacement curve, $b_i = W - a_i$ is the instantaneous ligament length, $\eta_i = 2 + 0.522 b_i / W$, and $\gamma_i = 1 + 0.76 b_i / W$.

Using Equations (7.5)–(7.8), the desired number of (Δa_i, J_i) pair can be generated to construct the *J-R* curve. The provisional fracture toughness, J_Q, is determined at the intersection of the *J-R* curve with the 0.2 mm offset line. J_Q is a valid, size-independent J_{IC} if $B, b_0 \geq \frac{10J_Q}{\sigma_y}$, where b_0 is the initial ligament length [53, 60]. J_{IC} can be further converted to K_{IC} by using $K_{IC} = \sqrt{\frac{EJ_{IC}}{1-\nu^2}}$, which is deduced from the fact the energy release rate, *G*, equals J_{IC} [2].

With SENB specimens, Chung et al. [33] used this approach to measure the room-temperature K_{IC} of the as-cast $CoCrFeNiNb_{0.5}$ HEA with an eutectic structure, concluding with 11.4–14.8 $MPa\sqrt{m}$. The FCC CrMnFeCoNi HEAs exhibit extensive plasticity (the tensile fracture elongation $\varepsilon_f > 0.55$) after undergoing casting, homogenization, cold forging, cross rolling, and annealing for recrystallization [53], and are also suited for the *J*-integral method. By applying this method for the compact-tension samples, Gludovatz et al. [53] determined that the K_{IC} for the CrMnFeCoNi HEAs at 77 K, 200 K, and 293 K were 219, 221, and 217 $MPa\sqrt{m}$, respectively.

7.2.1.1.3 Based on the Nano- and Micro-Mechanics

In addition to the classic fracture mechanics methods heretofore, fracture toughness can also be measured with different nano- or micro-mechanics approaches [45, 67, 68]. The approaches that have been applied to HEAs include the nanoindentation and micro-cantilever bending tests [51, 58]. The rationale behind using nanoindentation to probe the fracture toughness has its foundation in the well-established fracture toughness–crack length relation in the Vickers indentation [69–71]

$$K_c = \alpha \left(\frac{E}{H}\right)^{0.5} \frac{P}{c^{1.5}} \tag{7.9}$$

where *E* is the Young's modulus, *H* is the hardness, *P* is the indentation load, *c* is the radial crack length from the indentation center to the crack tip, and α is a proportional constant depending on the indenter geometry.

Equation (7.9) can be used in nanoindentation as well, but a shaper cube corner indenter with a centerline-to-face angle of 35.3° is recommended, as it can significantly bring down the material cracking threshold value and permit the occurrence of radial cracks for most tested brittle materials [67, 68, 72]. In a nanoindentation test of fracture toughness, *E* and *H* can be readily determined during the test, *c* is usually measured with the aid of a scanning electron microscope (SEM), and $\alpha \approx 0.036$ [67, 68]. The fracture toughness of the BCC $FeCoNiCrAl_3$ and $FeCoNiCrCuTiMoAlSiBe_{0.5}$ HEAs measured by this approach is 7.6 $MPa\sqrt{m}$ and 50.9 $MPa\sqrt{m}$, respectively [51]. The values obtained from this method, however, typically have 30–40% uncertainty, and one should keep this in mind when interpreting the data [67, 68, 73].

The *in situ* micro-cantilever bending is applied to measure the fracture toughness of the refractory $Nb_{25}Mo_{25}Ta_{25}W_{25}$ HEA with a BCC phase [58]. Since this alloy manifests brittle characteristics, its fracture toughness can be gauged with K_{IC} as in bulk brittle materials, and its value can be measured by following the ASTM standard E399 (Equations 7.1–7.4) [64]. But as a result of using the cantilever sample geometry, a different $f\left(\frac{a}{W}\right)$ needs to be used, which can be found from the reference [74]. One of the greatest merits of using micro-sized samples to conduct tests is that it allows for the deliberate inspection of impact of microstructural features on the fracture toughness. In the $Nb_{25}Mo_{25}Ta_{25}W_{25}$ HEA, the effect of grain boundaries on fracture toughness was examined through comparatively studying the single crystal and bi-crystal specimens [58]. The determined provisional fracture toughness, K_Q, were 1.6 $\text{MPa}\sqrt{\text{m}}$ and 0.2 $\text{MPa}\sqrt{\text{m}}$, respectively [58]. With Equation (7.3) and the reported strength data [41], the calculated critical dimensions for the single crystal and bi-crystal specimens are ~1.0 μm and 0.09 μm, respectively. Given that for both cases, the used specimens have a thickness (*B*) of 1.5–2.0 μm and crack length (*a*) of 0.3–0.5 μm, it is seen that 0.2 $\text{MPa}\sqrt{\text{m}}$ for the single crystal is strictly a valid K_{IC}. The validity check for the single crystal passes for $B \geq 2.5\left(\frac{K_Q}{\sigma_y}\right)^2$ but fails for $a \geq 2.5\left(\frac{K_Q}{\sigma_y}\right)^2$. However, since the confirmative computation with the finite element method under a plane strain condition gave very comparable fracture toughness values as the measured [58], 1.6 $\text{MPa}\sqrt{\text{m}}$ for the single crystal can still be viewed as a valid K_{IC}.

7.2.1.2 Impact Toughness

An alternative method to characterize the fracture resistance of materials is to use the old fashioned, easy-to-use, but less rigorous pendulum impact tests [2]. A Charpy-impact test is one such method, and some other variants also exist and are used [2]. In all these tests, a hammer at the end of the pendulum is swung to a given height, h_1, and then released to impact the notched sample located at the lowest position. After failure, the hammer swings to the other side with a height of h_2. The potential energy loss in the pendulum system, i.e., $h_1 - h_2$ times the weight of the pendulum, is the energy required to separate the sample. This energy is indicative of the resistance of the testing material to brittle fracture. These methods are usually referred to as a qualitative measure of the fracture toughness, because they lack the mathematical rigor of the fracture mechanics methods, and are not as reliable for predicting the structural integrity.

The impact toughness of a number of HEAs was measured with pendulum-impact tests, and their resulting impact energies are compiled in Table 7.2. With such a pendulum system, Chen et al. [57] determined that the impact-toughness density (per unit cross-sectional area) in the $Al_{18}Cr_{21}Fe_{20}Co_{20}Ni_{21}$, $Al_{15.5}Cr_{22.25}Fe_{20}Co_{20}Ni_{22.25}$, $Al_{13}Cr_{23.5}Fe_{20}Co_{20}Ni_{23.5}$, and $Al_{10.5}Cr_{24.75}Fe_{20}Co_{20}Ni_{24.75}$ HEAs are 12 kJ/m^2, 15 kJ/m^2, 106 kJ/m^2, and 477 kJ/m^2, respectively. Since they used $6 \times 6 \times 55$ mm^3 samples with a 2 mm U-notch, the cross-sectional area is calculated to be 24 mm^2, and the total impact toughness of the four alloys are thus 0.29 J, 0.36 J, 2.54 J, and 11.45 J, respectively. Utilizing the Charpy-impact tests standardized in the ASTM standard E-23 [75], Li et al. [56] measured impact toughness values of the FCC $Al_{0.1}CoCrFeNi$ and $Al_{0.3}CoCrFeNi$ HEAs hot forged at 1,323 K. The measurements were made on

TABLE 7.2
Impact Toughness and Associated Information for a Variety of High Entropy Alloys

Composition	Phase	Processing	T (K)	σ_y (MPa)	ε_f	Impact energy (J)	Test method	Ref.
$Al_{18}Cr_{21}Fe_{20}Co_{20}Ni_{21}$	BCC	Arc melted, cast	RT	$\sim 370^{T}$	0^{T}	0.29	Impact test on U-notched specimens	[57]
$Al_{15.5}Cr_{22.25}Fe_{20}Co_{20}Ni_{22.25}$	FCC + BCC	Arc melted, cast	RT	$\sim 590^{T}$	0^{T}	0.36	Impact test on U-notched specimens	[57]
$Al_{13}Cr_{23.5}Fe_{20}Co_{20}Ni_{23.5}$	FCC	Arc melted, cast	RT	760^{T}	0.06^{T}	2.54	Impact test on U-notched specimens	[57]
$Al_{10.5}Cr_{24.75}Fe_{20}Co_{20}Ni_{24.75}$	FCC + BCC	Arc melted, cast	RT	370^{T}	0.25^{T}	11.45	Impact test on U-notched specimens	[57]
$Al_{0.1}CoCrFeNi^{(I)}$	FCC	Vacuum levitation melted, cast, hot forged at 1,323 K	77	412^{T}	$\sim 0.9^{T}$	289	Charpy-impact test on V-notched specimens (ASTM E-23)	[56]
			200	295^{T}	$\sim 0.65^{T}$	318		
			298	250^{T}	$\sim 0.54^{T}$	420		
$Al_{0.3}CoCrFeNi$	FCC	Vacuum levitation melted, cast, hot forged at 1,323 K	77	515^{T}	$\sim 0.74^{T}$	328	Charpy-impact test on V-notched specimens (ASTM E-23)	[56]
			200	310^{T}	$\sim 0.66^{T}$	409		
			298	220^{T}	$\sim 0.65^{T}$	413		
CoCrFeNi	FCC	Vacuum levitation melted, cast	77	n/a	n/a	398	Charpy-impact test on V-notched specimens (ASTM E-23)	[76, 127]
			200	n/a	n/a	322		
			298	155^{T}	0.6^{T}	287		
$Al_{0.1}CoCrFeNi^{(II)}$	FCC	Vacuum levitation melted, cast	77	n/a	n/a	371	Charpy-impact test on V-notched specimens (ASTM E-23)	[76]
			200	n/a	n/a	327		
			298	n/a	n/a	294		

(Continued)

TABLE 7.2 (CONTINUED)

Impact Toughness and Associated Information for a Variety of High Entropy Alloys

Composition	Phase	Processing	T (K)	σ_y (MPa)	ε_f	Impact energy (J)	Test method	Ref.
$Al_{0.75}CoCrFeNi$	FCC + B2	Vacuum levitation melted, cast	77	n/a	n/a	1.82	Charpy-impact test on V-notched specimens (ASTM E-23)	[76, 128, 129]
			200	n/a	n/a	3.02		
			298	800–1,400^{C}	0.35–0.55^{C}	3.58		
$Al_{1.5}CoCrFeNi$	BCC + B2	Vacuum levitation melted, cast	77	n/a	n/a	0.64	Charpy-impact test on V-notched specimens (ASTM E-23)	[76, 130, 131]
			200	n/a	n/a	0.64		
			298	1,453–1,867^{H}	n/a	1.28		

Notes: FCC and BCC denote face-centered cubic and body-centered cubit, respectively; RT indicates room temperature; the superscript ^{T}, ^{C}, and ^{H} represent that the data are sourced from tensile tests, compressive tests, and hardness (the yield strength is estimated from $\sigma_y = \frac{H}{3}$); the superscript $^{(I)}$ and $^{(II)}$ denote the same alloy from different works.

$10 \times 10 \times 55$ mm^3 samples with a 2 mm deep V-notch in the middle, and impact toughness values of $Al_{0.1}CoCrFeNi$ at 77 K, 200 K, and 298 K were reported to be 292 J, 318 J, and 419 J, respectively; $Al_{0.3}CoCrFeNi$ had impact toughness values of 329 J, 409 J, and 415 J at 77 K, 200 K, and 298 K, respectively. Also using the $10 \times 10 \times 55$ mm^3 samples with a 2 mm deep V-notch in the middle and Charpy-impact tests, Xia et al. [76] measured the impact toughness of the FCC CoCrFeNi, FCC $Al_{0.1}CoCrFeNi$, $Al_{0.75}CoCrFeNi$ with FCC and B2 phases, and $Al_{1.5}CoCrFeNi$ with BCC and B2 phases at 77–298 K, and their corresponding values are listed in Table 7.1. What is worth noting is that the impact toughness of the as-cast FCC $Al_{0.1}CoCrFeNi$ HEAs measured by Xia et al. [76] (371 J, 327 J, and 294 J for the respective testing temperatures of 77 K, 200 K, and 298 K) are different than the values reported by Li et al. [56] By comparing their samples and testing conditions, it is inferred that one possible source of the difference is their distinct processing histories, namely, the as-cast alloys by Li et al. [56] underwent hot forging at 1,323 K, whereas the same processing was not applied to the as-cast alloys by Xia et al. [76].

7.2.1.3 Crack Tip Opening Angle

The crack tip opening angle (CTOA), defined as the angle made by two cracking flanks, is another measure characterizing the resistance of materials to plastic fracture. It is primarily applicable to thin-sheet structures subjected to a plane-stress condition. The rationale behind using CTOA as a fracture-resistance criterion is that the attainment of a steady-state CTOA normally implies that the stable crack growth prevails in materials. Usually, a high steady-state CTOA value is indicative of great crack propagation resistance and vice versa.

To determine the steady-state CTOA in a $CoCrFeNiMn_{0.2}$ HEA made by the powder metallurgy and composed of a FCC matrix and Cr-rich precipitates, Li et al. [34] conducted the *in situ* SEM fracture tests on single-edge-notched-tension samples ($40 \times 10 \times 1$ mm^3 with a 1 mm long notch) and measured the CTOA from the interrupted SEM images as a function of crack extension Δa. Ultimately, the steady-state CTOA was determined to be 18°. This value is believed by the authors to be a relatively high value, indicating that the alloy under investigation possesses high fracture resistance. Nevertheless, as this criterion is not used as frequently as others introduced before, its comparison with other criteria to rank the fracture resistance of different HEAs turns out to be unlikely.

7.2.2 Temperature Effect

The fracture toughness of the CrMnFeCoNi HEA was characterized at the temperature of 77 K, 200 K, and 298 K [53], corresponding to the K_{IC} values of 219 $\text{MPa}\sqrt{\text{m}}$, 221 $\text{MPa}\sqrt{\text{m}}$, and 217 $\text{MPa}\sqrt{\text{m}}$, respectively. Essentially, its fracture toughness remains almost constant as the temperature varies from 77 K to 298 K, indicating its insensitivity to temperature change.

In regard to the change of the impact toughness with temperature in HEAs, two opposing trends were reported. For the CoCrFeNi and $Al_{0.1}CoCrFeNi$ HEAs by Xia et al. [76], their impact toughness increases with the descending temperature. In contrast, reducing the testing temperature in the $Al_{0.75}CoCrFeNi$ [76], $Al_{1.5}CoCrFeNi$

[76], and $Al_{0.1}CoCrFeNi$ HEAs by Li et al. [56] leads to a gradual decrease in their impact toughness values. Both trends are evident from the data in Table 7.2. What is of interest is that in two separate works by Xia et al. [76] and Li et al. [56], the temperature affects the impact toughness of the $Al_{0.1}CoCrFeNi$ HEAs in completely opposite directions. The discrepancy might be due to the distinct processing histories undergone by the respective alloys.

In these HEAs investigated, no temperature-induced ductile-to-brittle transition is noted as in many conventional alloys such as steel, amorphous alloys, Mg alloys, porous metals, and nanocrystalline metals [12, 77–84]. This finding conveys an important implication that HEAs could be excellent candidate materials for applications under extremely cold conditions, for example, materials for the body of ships, aircrafts, and tanks for low-temperature storage. The ductile-to-brittle transition is detrimental to cold-condition applications and once caused the well-known World War II ship *Liberty*'s failure, due to the steel that made up the ship's body transitioning to brittle failure as the temperature drops below a critical value [3]. While material scientists are still striving to defeat the catastrophic ductile-to-brittle transition by bringing down the transition temperature of engineering materials, those HEAs without a ductile-to-brittle transition may offer a more easily accessible alternative to practical applications under extremely cold conditions.

7.2.3 Phase Effect

Those HEAs that have been studied for their fracture resistance exist in the form of a single FCC phase, a single BCC phase, or multiple phases. The correlation between fracture toughness and phase constitution is investigated by plotting K_{IC} as a function of the yield strength, σ_y, and elongation, ε_f, for HEAs composed of different phases. As seen in Figure 7.1, in terms of fracture toughness, the plots clearly separate the FCC, BCC, and multiphase HEAs, with the FCC HEAs floating on the top, BCC ones on the bottom, and multiphase ones located in between, but next to the BCC ones. Obviously, the FCC HEAs possess the most superior fracture resistance. In comparison, the fracture resistances of the BCC and multiphase HEAs are inferior. The strength, as expected from the classical mechanical metallurgy knowledge, increases in sequence from the FCC to the multiphase, and to the BCC HEAs. And the ductility, measured by the elongation, varies in the reverse order. In Figure 7.1a, the fracture toughness of HEAs decreases with the increasing yield strength, as indicated by the bounding band. On the other hand, in Figure 7.1b, the fracture toughness increases with increasing elongation.

7.2.4 Comparison with Other Materials

The Ashby plot of the fracture toughness against the yield strength is a useful tool to evaluate the overall damage tolerance of materials. With the valid K_{IC} data (K_Q values excluded) and corresponding σ_y in Table 7.1, in Figure 7.2 we outline the damage tolerance space for HEAs alongside many traditional metals and alloys and other material classes. The damage tolerance of HEAs covers a wide space, with the yield strength running from ~400 MPa to over 2 GPa and the fracture toughness from

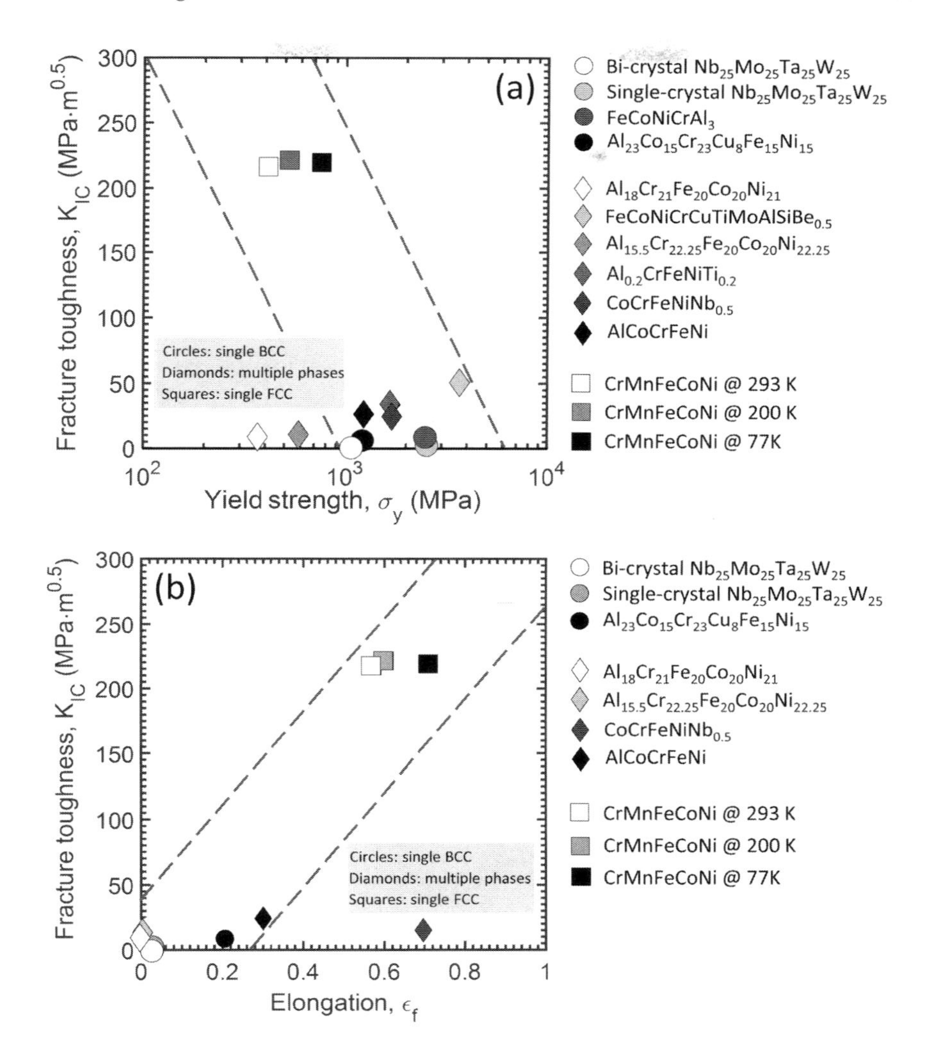

FIGURE 7.1 Fracture toughness plotted against the (a) yield strength and (b) elongation for high entropy alloys with a single face-centered cubic (FCC) phase, a single body-centered cubic (BCC) phase, and multiple phases. Dashed lines denote the trend bands.

0.2 to 221 MPa$\sqrt{\text{m}}$. Note that some unrealistically high or low yield strengths (e.g., 3,770 MPa for the $FeCoNiCrCuTiMoAlSiBe_{0.5}$ [51] and 333 MPa for the AlCoCrFeNi [65]) estimated from the hardness data with $\sigma_y = \frac{H}{3}$ are excluded from plotting considering the crudeness of this relation. Overall, the HEA domain manifests a slight backward tilting, with FCC HEAs (e.g., CrMnFeCoNi [53]) having high fracture toughness (217–211 MPa$\sqrt{\text{m}}$) but low strengths (410–760 MPa), BCC HEAs (e.g., $Al_{18}Cr_{21}Fe_{20}Co_{20}Ni_{21}$ [57]) having low fracture toughness (0.2–9 MPa$\sqrt{\text{m}}$) but high strengths (1,000–~2,640 MPa), and those with multiple phases in between. According to the phase constitution, the entire HEA damage-tolerance domain can be roughly subdivided into three subdomains, that is, the single-FCC-phase, single-BCC-phase, and multiphase subdomains, which are located at the top, bottom, and middle of the HEA domain, as demonstrated in Figure 7.2. In this subdivision, one exception is the spark-plasma-sintered AlCoCrFeNi HEA with FCC, $L1_2$, and tetragonal Cr-Fe-Co-based σ phases, which comprises multiple phases, but its damage tolerance falls in the BCC subdomain [65]. This is probably because the spark plasma sintering itself introduced certain fluctuations in the phase constitution and mechanical properties of the alloys compared to those made from traditional casting. However, the splitting of the damage-tolerance domain by the phase constitution is in general in line with the classical physical metallurgy wisdom that FCC phases are soft, yet ductile, while BCC phases are hard, yet brittle.

In comparison with other materials, a large portion of HEAs are located at the top-right corner of the Ashby plot, a direction in which material scientists have strived to search for extraordinary damage-tolerant materials. HEAs with a single FCC phase are usually limited by their low strength, while those having a single BCC phase are often brittle. With proper compositional and microstructural designs,

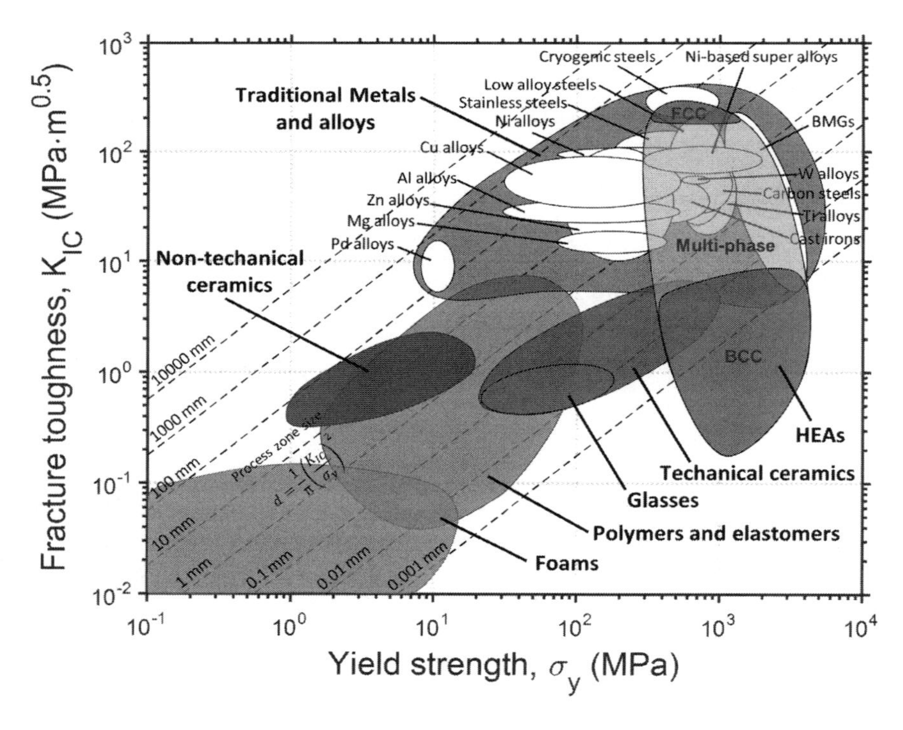

FIGURE 7.2 Ashby plot of fracture toughness versus the yield strength of high entropy alloys along with other materials, with the dashed lines roughly representing the process zone size, $\frac{1}{\pi}\left(\frac{K_{IC}}{\sigma_f}\right)^2$, at the crack tip [1].

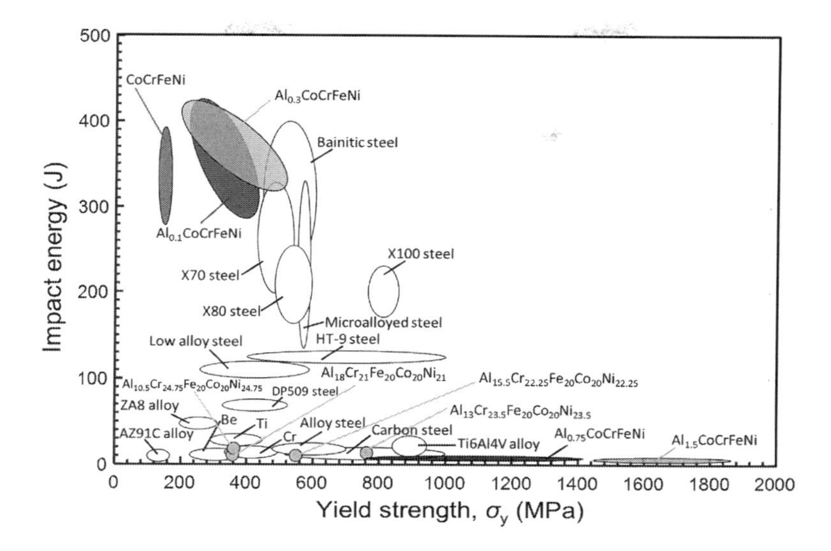

FIGURE 7.3 Impact energy plotted against the yield strength of some selected high entropy alloys in comparison with other metallic materials [56, 79, 132–145].

HEAs with multiple phases, however, are likely to overcome both the strength and ductility constraints, standing out as remarkable fracture-resistant materials.

Likewise, with the data at 77–298 K in Table 7.2, an impact energy–yield strength plot is generated for HEAs whose fracture resistances are quantified by impact toughness. Such a plot is given in Figure 7.3, together with some selected conventional metals and alloys. Based on the phase constitution, HEAs in Figure 7.3 are separated into two groups. The first group is located at the top left of the plot with high impact toughness (289–420 J) but low strengths (155–515 MPa), corresponding to those HEAs with a single FCC phase, such as CoCrFeNi [76], $Al_{0.1}CoCrFeNi$ [56, 76], and $Al_{0.3}CoCrFeNi$ [56, 76]. The second group sinks down at the bottom of the plot with the very low impact energy (<12 J), but a wide span of strengths. The majority of HEAs falling in this group comprise either a single BCC phase or combined phases, with one exception being $Al_{13}Cr_{23.5}Fe_{20}Co_{20}Ni_{23.5}$, which has a single FCC phase [57]. This trend, again, is roughly aligned with the long-established understanding of the mechanical behaviors of FCC and BCC metals. Compared with other metal and alloys, the FCC CoCrFeNi, $Al_{0.1}CoCrFeNi$, and $Al_{0.3}CoCrFeNi$ alloys indeed show superior impact toughness, but their strengths are relatively low, disqualifying them as materials of high damage tolerance. All other HEAs located on the bottom, on the other hand, possess extremely low impact toughness (0.29–11.45 J), and are not suitable for high damage-tolerance applications. Like the Ashby plot in Figure 7.2, the top-left corner in Figure 7.2 is a desirable direction for HEAs to be more damage-tolerant.

7.3 FRACTOGRAPHY

Visually examining the fracture surfaces of failed HEAs is a straightforward means to study their fracture modes, thereby gaining insights on microscopic deformation mechanisms responsible for high or low fracture toughness values measured. Based on the crystallographic mode, the fracture may be crudely classified as cleavage and shear fractures. The cleavage fracture is governed by the tensile stress normal to crystallographic cleavage planes, and in appearance, it is characteristic of bright facets or granular features, particularly at low magnifications. The shear fracture, on the other hand, is induced by slips driven by shear stresses, and usually finds gray or dark fibrous or dimple-like features. In ideal situations, the cleavage fracture is associated with low fracture toughness whereas the shear fracture is associated with high fracture toughness. But, in practice, things are much more complex, and one would often find a mixture of two fracture modes or different variants of each. An attempt at correlating these fractographic characteristics with the fracture toughness values will be made in this section.

7.3.1 Cleavage Fracture

Under uniaxial tension or compression, cleavage or quasi-cleavage fracture was observed on the fracture surfaces of the BCC $AlCoCrFeNb_{0.1}Ni$ [85], BCC $Cu_{0.5}AlCoCrFeNiTi_{0.5}$ [86], BCC AlCoCrFeNi [87], BCC CoCrFeNiTiAl [88], BCC MoNbTaV [89], BCC NbMoTaW and VNbMoTaW [41], BCC HfMoTaTiZr and HfMoNbTaTiZr [54], FCC + BCC $AlCrCuNiZr_x$ (x = 0, 1) [90], FCC + BCC FeCoNiCuAl [91], FCC + BCC $FeNiCrCo_{0.3}Al_{0.7}$ [92], and FCC + BCC AlCoCrCuMnFe [93], etc. When failed in fracture tests, the cleavage fracture characteristics are closely related to the measured fracture toughness values. The BCC $Al_{23}Co_{15}Cr_{23}Cu_8Fe_{15}Ni_{15}$ HEA has a fracture toughness value of 5.4–5.8 $MPa\sqrt{m}$, and the corresponding fractographs are shown in Figure 7.4a and b [52]. From Figure 7.4a, it is readily seen that island-like facets (B) surrounded by river lines (A) are formed on the fracture surface. Facets are typical of the cleavage fracture, and river lines are known to result from the joining of cleavage steps in a single grain [52]. Close inspection of the river lines in Figure 7.4b discloses that they resemble shear bands observed on the fracture surfaces of nano-crystalline materials and amorphous alloys [6, 94–97], which are also a typical brittle fracture feature that forms by the localization of the plastic flow and usually leads to catastrophic cracks [98–100]. Likewise, a fan-like (facets plus radial river lines) cleavage fracture is also seen in the BCC $Al_{18}Cr_{21}Fe_{20}Co_{20}Ni_{21}$ HEA, which has $K_{IC} = 9$ $MPa\sqrt{m}$, as shown in Figure 7.5a [57]. The fracture surface of the $Al_{15.5}Cr_{22.25}Fe_{20}Co_{20}Ni_{22.25}$ HEA (FCC + BCC, $K_{IC} = 11$ $MPa\sqrt{m}$) is also predominantly composed of river-like patterns, as revealed in Figure 7.5b [57].

In general, HEAs with a fracture toughness of ~10 $MPa\sqrt{m}$ or less would find facets and river-like patterns on their fracture surfaces. The percentage of each fracture feature appearing varies from material to material and depends on the composition and phase constitution. However, it is not yet clear as of now if the percentage of each is correlated with increasing or decreasing fracture toughness. As the fracture toughness of HEAs improves, some small-scale dimple-like zones isolated by facets or river lines may come into view.

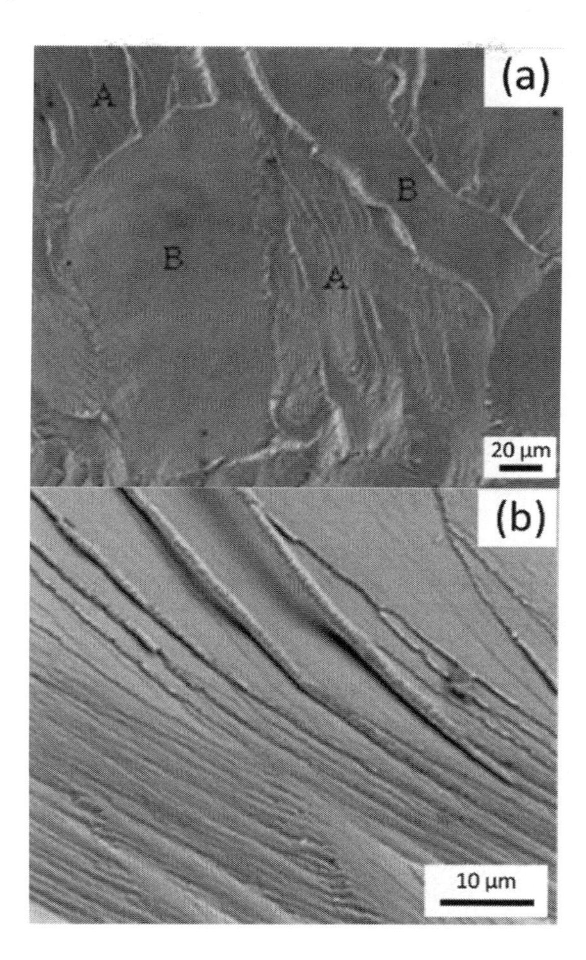

FIGURE 7.4 (a) Fracture surface of the body-centered cubic (BCC) $Al_{23}Co_{15}Cr_{23}Cu_8Fe_{15}Ni_{15}$ high entropy alloy from the three-point bending fracture test, showing island-like facets (location B) and surrounding river lines (location A), indicative of the cleavage fracture. (b) Close inspection of the river lines in (a), showing shear band-like textures [52].

7.3.2 Shear Fracture

In contrast to the cleavage fracture in the low fracture toughness HEAs, alloys with very high fracture toughness, for instance, the FCC CrMnFeCoNi HEA, exhibit tremendously distinct fractographic characteristics [53, 101]. As exemplified by the CrMnFeCoNi HEA tested at room temperature in Figure 7.6, the fracture, in fact, occurs as a result of the coalescence of microvoids, as evidenced by numerous equiaxed dimples [53, 101]. The close observation of the formed dimples reveals particles located at the bottom of depressions, which are thought to serve as the initiation sites of microvoids. Some particles are observed to fracture at the plane

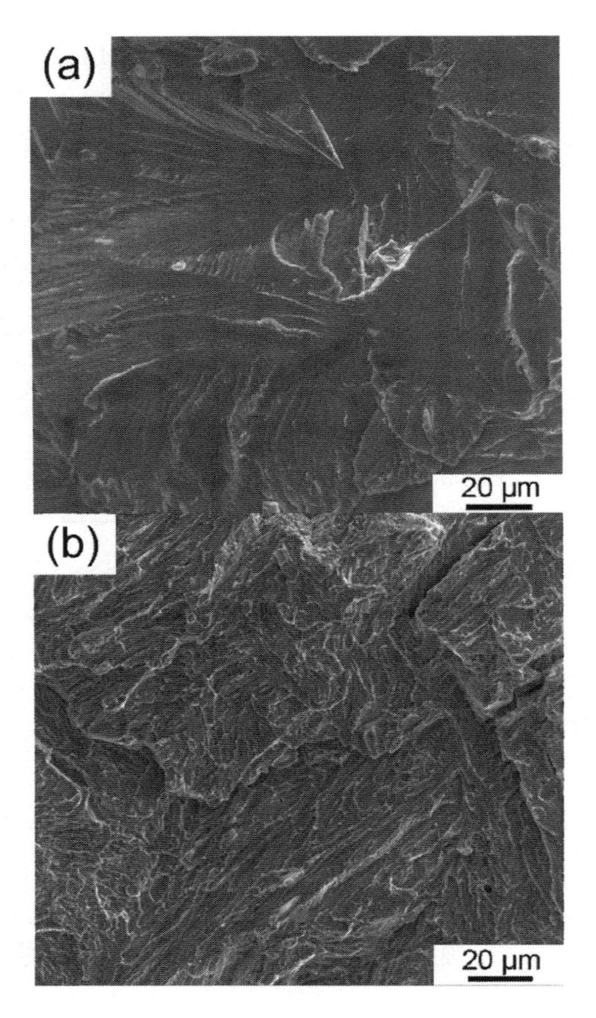

FIGURE 7.5 (a) Fracture surface of the body-centered cubic (BCC) $Al_{18}Cr_{21}Fe_{20}Co_{20}Ni_{21}$ high entropy alloy (HEA) revealing fan-like (facets plus radial river lines) cleavage fracture. (b) Fracture surface of the $Al_{15.5}Cr_{22.25}Fe_{20}Co_{20}Ni_{22.25}$ HEA with FCC and BCC phases, predominantly composed of river-like patterns [57].

normal to the tensile stress (i.e., the paper plane) or to de-bond the matrix, implying that the nucleation of microvoids is likely through particle cracking or interfacial debonding with the matrix. Following the nucleation, microvoids grow as the plastic flow evolves; the grown microvoids then meet each other and coalesce; the microvoids' coalescence creates thin walls, which subsequently suffer from necking at the action of the shear stress and lead to failure ultimately. With the aid of the energy-dispersive X-ray (EDX) spectroscopy, these particles are found to be either Cr-rich or Mn-rich oxide inclusions, presumably formed in the melting process. The size of the

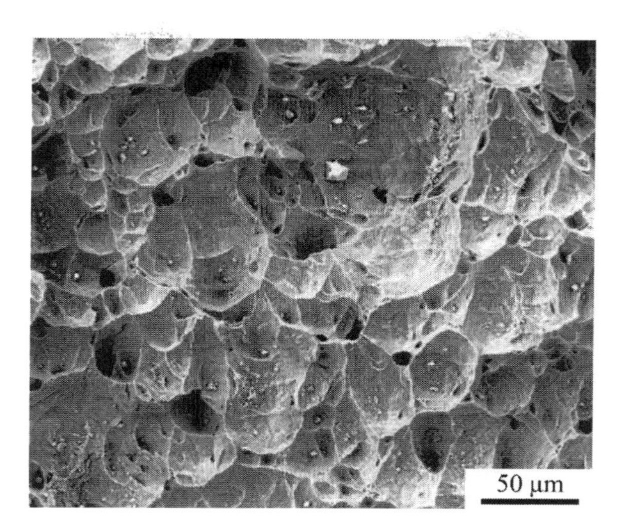

FIGURE 7.6 Ductile fracture characteristic of numerous dimples in the face-centered cubic (FCC) CrMnFeCoNi high entropy alloy [101], failed at room temperature.

microvoids and particle inclusions are noticed to vary from ~1 to tens of micrometers and <1 μm to ~5 μm, respectively.

In a similar fashion, the FCC $Al_{0.1}CoCrFeNi$, FCC $Al_{0.3}CoCrFeNi$, and FCC + BCC $Al_{10.5}Cr_{24.75}Fe_{20}Co_{20}Ni_{24.75}$ HEAs, measured with high impact energies, present dimples and tear edges on their fracture surfaces, as evidenced in Figure 7.7a–c [56, 57]. Overall, the fracture of HEAs with either high fracture toughness or impact energy is primarily progressed via the nucleation, growth, and coalescence of microvoids, manifesting in fractographs as dimples and tear edges. These ductile fracture morphologies are also observed in some HEAs failed by uniaxial tension or compression e.g., the FCC CoCrFeNi [102], FCC CoCrFeNiTi [88], FCC $Al_{0.25}CoCrFe_{1.25}Ni_{1.25}$ [103], FCC $Al_{0.5}CrCuFeNi_2$ [104], FCC CoCrFeMnNi [105], FCC $FeNiMnCr_{18}$ [106],

FIGURE 7.7 Fractographs of the (a) face-centered cubic (FCC) $Al_{0.1}CoCrFeNi$ [56], (b) FCC $Al_{0.3}CoCrFeNi$ [56], and (c) FCC + body-centered cubic (BCC) $Al_{10.5}Cr_{24.75}Fe_{20}Co_{20}Ni_{24.75}$ high entropy alloys [57], all characteristic of dimples and tear edges.

FCC $CoCrFeNiAl_{0.3}$ [107], FCC FeMnNiCuCo and $FeMnNiCuCoSn_{0.03}$ [108], and FCC $Fe_{0.4}NiCr_{0.4}Cu$ and $Fe_{0.4}MnNiCr_{0.4}Cu$ [109].

7.4 FRACTURE MECHANISMS

For a good understanding of why some HEAs possess superior fracture resistance (e.g., FCC CrMnFeCoNi [53]) while others suffer from catastrophic brittle failures (e.g., BCC $Al_{18}Cr_{21}Fe_{20}Co_{20}Ni_{21}$ [57]), examining their respective fracture mechanisms, from both macro and micro perspectives, is imperative. Besides, revealing their fracture mechanisms is also instructive to design future fracture-resistant HEAs with deliberate compositional, microstructural, and/or processing controls. We herein summarize the fracture mechanisms in a number of HEA systems and group them by the phase constitution. In general, the portrayal of the fracture mechanisms involves the initiation and propagation of cracks and their interactions with such microstructural features as phases, dislocations, twins, precipitates, etc.

7.4.1 In Face-Centered Cubic High Entropy Alloys

7.4.1.1 Intrinsic Toughening by Dislocation Activities

As in many other FCC metals, plastic deformation in HEAs with a FCC phase is fulfilled predominantly by dislocation motion [3, 53, 60]. Extensive dislocation activities at the vicinity of the crack tip can considerably blunt the crack tip, thereby offering a formidable toughening mechanism. This trend is exemplified by the CrMnFeCoNi HEA, whose toughening mechanisms at different stages of deformation in the vicinity of the crack tip are revealed by virtue of *in situ* transmission electron microscopy (TEM) straining of a crack containing thin foil sample [61]. At the early stage of deformation, it is observed that extensive and fast-moving Shockley partial dislocations of $\frac{1}{6}\langle 112\rangle$ type prevail near the crack tip, continuously forming numerous stacking faults on the {111} slip planes owing to the low stacking fault energy in this alloy [29, 110, 111], as indicated by the arrows in Figure 7.8a and b. The stacking faults formed are also found to be continuously annihilated by the rapid motion of trailing partial dislocations. The atomic scale dynamic process of stacking fault formation is shown in Figure 7.8c and d. At $t = 0$ in Figure 7.8c, no stacking fault is observed ahead of the crack tip. After applying several displacement pulses, at $t = 1.2$ s in Figure 7.8d, it is clearly noticed that stacking faults are formed near the crack tip as a consequence of the nucleation and propagation of multiple partial dislocations on the {111} slip planes. As shown by the magnified inverse fast Fourier transform image in Figure 7.8e, the stacking fault is confirmed to transit from an FCC to a local HCP stacking sequence. As individual partial dislocations move easily and promote the ductility of the HEA, three pairs of parallel {111} stacking faults may interact with each other, forming "stacking fault parallelepipeds." The formed stacking fault parallelepipeds can be viewed as three-dimensional defects, which serve as powerful obstacles to dislocation motions. In stark contrast to the motion of individual dislocations, the stacking fault parallelepipeds advocate strain hardening.

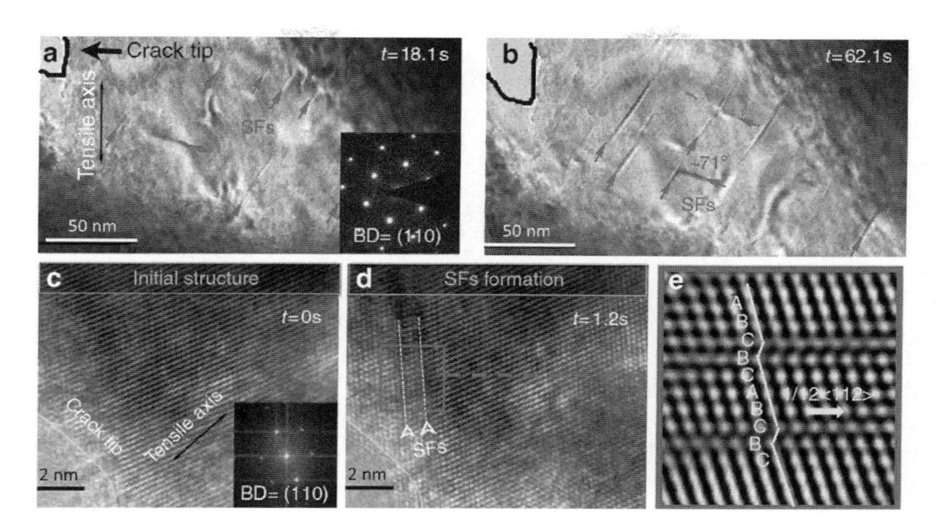

FIGURE 7.8 Fracture mechanism in the face-centered cubic (FCC) CrMnFeCoNi high entropy alloy [61]. (a, b) Bright-field images of the crack containing, thin foil sample stretched under *in situ* transmission electron microscopy (TEM), disclosing the formation of stacking faults, as indicated by arrows. High-resolution TEM images at (c) t = 0 with no stacking fault observed, and (d) t = 1.2 s showing the clear formation of stacking faults ahead of the crack tip. (e) The magnified inverse fast Fourier transform image, confirming the transition from the FCC to local hexagonal close-packed (HCP) stacking sequence with the closure failure being $\sim \frac{1}{12} <112>$. The beam direction is [110].

At the later stage of deformation, $\frac{1}{2} <110>$ type perfect dislocations start to move, but in the form of small segments with extremely low speeds. Due to the difficulty in motion, perfect dislocations tend to form close-packed dislocation arrays and move in localized bands, ultimately causing planar slip. The localized band of the planer slip and the slow-moving perfect dislocation segments inside act as impediments to the fast-moving partial dislocations. Partial often cease to move and start to pile up once they run into a perfect dislocation band. The hindrance to the motion of partial dislocations is deleterious to ductility but it boosts the strong interaction of the perfect and partial dislocations, yielding important strain hardening to plastic deformation. In a nutshell, during the entire process of plastic deformation, the fast motion of partial dislocations, the formation of stacking fault parallelepipeds, and partial dislocation/perfect dislocation interactions work jointly and coordinately to engender the high fracture toughness value of over 200 $\mathrm{MPa}\sqrt{\mathrm{m}}$ in the CrMnFeCoNi HEAs [53].

7.4.1.2 Intrinsic Toughening by Twinning

Apart from extensive dislocation activities, it is found that deformation induced nano-twinning supplies a supplemental deformation mechanism in certain FCC HEAs and medium entropy alloys (MEAs) to facilitate their high fracture toughness [53, 60]. Take the CrCoNi MEA as an example: TEM revealed the formation of multiple twinning systems in the majority of grains deformed uniaxially [60, 62]. As demonstrated in the bright-field TEM image of a grain in Figure 7.9a, three twinning systems are activated and interact with each other to produce a three-dimensional twin architecture. The twins have a typical thickness from ~100 nm to a few micrometers, and their appearances are further affirmed by the set of diffraction spots in addition to the matrix in the selected area diffraction (SAED) patterns in Figure 7.9c. As for the twin boundary, two categories are observed, namely, the $\Sigma_3\{111\}$ coherent twin boundaries and $\Sigma_3\{112\}$ incoherent twin boundaries. These twin boundaries act as barriers to dislocation motions. As dislocations glide on a {111} plane not parallel to the twin boundary, they are stopped and pile up once meeting twin boundaries, as shown representatively near a coherent twin boundary in Figure 7.9a. Besides, dislocation arrays are also found to form and gather at coherent twin boundaries or the intersections of coherent twin boundaries and incoherent twin boundaries, as displayed in Figure 7.9b. The high angle annular dark field scanning transmission electron microscopy (STEM) image in Figure 7.9d divulges that a 9R structure (a set of Shockley partial dislocations with a repetitive order of b2:b1:b3) forms in a {112} incoherent twin boundary intersecting with a {111} coherent twin boundary.

Twinning itself is an effective means to intensify the ductility of metals and alloys [112–119]. The formation of hierarchical twin architectures in HEAs by the

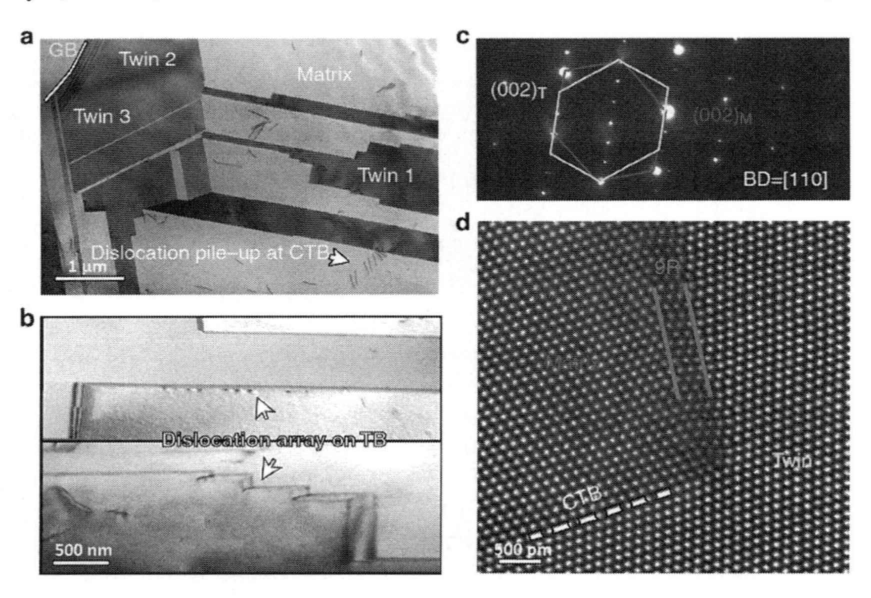

FIGURE 7.9 Bright-field transmission electron microscopy (TEM) images of the CrCoNi medium entropy alloy following uniaxial tension, showing (a) the three-dimensional twin architecture in a grain formed through the interaction of three different twinning systems, and (b) dislocation arrays gathered at twin boundaries. (c) Selected area diffraction pattern along the 110 beam direction showing the set of diffraction spots in addition to the matrix, which correspond to the twin structure. (d) The high angle annular dark field scanning transmission electron microscopy (STEM) image reveals the formation of a 9R structure (a set of Shockley partial dislocations with a repetitive order of b2:b1:b3) in a {112} incoherent twin boundary intersecting with a {111} coherent twin boundary [62].

inter-junctions of different twin systems further enhances the ductility through creating three-dimensional pathways and permitting facile cross-slips of dislocations among twin boundaries [62]. In the meantime, the twin–twin interactions and obstruction of dislocation motions by twins compromise ductility, but encourage strengthening. The synergistic operations of these micro-deformation processes are the attributed cause of the high fracture resistance of the CrCoNi MEA.

7.4.1.3 Extrinsic Toughening by Crack Bridging

Extrinsic toughening mechanisms are processes that are activated and act behind the crack tip as it propagates. Fiber or lamella bridging is one of the most important such toughening mechanisms, particularly useful in resisting the catastrophic crack propagation and retaining the desired fracture resistance in brittle materials [120–123]. Through *in situ* straining in an aberration-corrected transmission electron microscope, the crack propagation in a micro-sized CrMnFeCoNi HEA with an FCC phase is found to be impeded in a similar fashion at the final stages of deformation [61]. To be specific, the bright-field TEM image of the cracking region ~500 nm away from the crack tip is shown in Figure 7.10a, in which can be easily seen that nanoscale fibers form [61]. These nanoscale fibers bridge the two fracture surfaces at the wake of the crack tip and effectively resist crack opening, thereby offering a steady source of toughening the alloy. The emergence of these fibers is a direct result of the formed nanovoids at alternate locations through the intersection of two {111} slip planes, and is believed to be mechanistically related to the low stacking fault energy in the CrMnFeCoNi HEA [61]. Further opening of the crack extends and

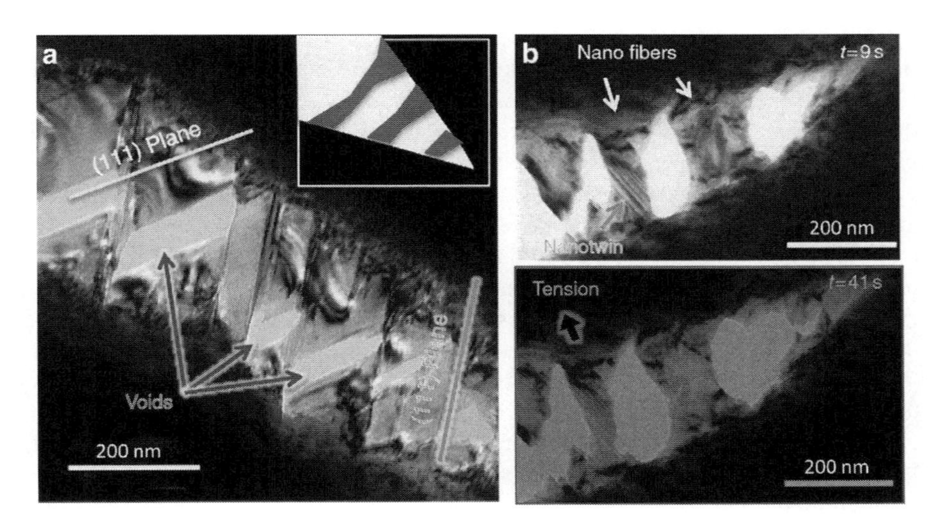

FIGURE 7.10 (a) Bright-field transmission electron microscopy (TEM) image of the CrMnFeCoNi high entropy alloy sample taken near the crack tip during the *in situ* straining, indicating the formation of nanovoids and nanoscale fiber-like bridges at alternate locations, which grow along two {111} slip bands. The inset schematically shows the bridged structure at the crack tip. (b) TEM images of the crack tip region at t = 9 and 41 s, showing that nanotwins appear inside the bridges to accommodate deformation [61].

tears the nano-bridges and activates the nanoscale twinning as the primary deformation mechanism to accommodate strain, as shown in Figure 7.10b [61]. As in some other FCC metals, the formed twins have a coherent twin boundary [117].

The twinned nanoscale bridges, by holding the two crack faces together and delaying the crack propagation, provide an additional source of toughening other than the intrinsic toughening mechanisms (e.g., partial dislocation motion and formation of stacking faults). This extrinsic toughening mechanism, in part, contributes to the high fracture toughness (over 200 $MPa\sqrt{m}$) of the CrMnFeCoNi HEAs [53].

7.4.2 In Body-Centered Cubic High Entropy Alloys

The HEAs with a single BCC phase can fail by either transgranular cleavage or intergranular separation. Since both processes involve very limited plasticity, the associated fracture mechanisms in BCC HEAs are not expected to be as complex as in FCC HEAs. Like all other materials, the cleavage in BCC HEAs also results from the breakage of weaker bonds between atoms on certain crystallographic planes. On the other hand, the intergranular fracture may be caused by the segregation of impurities at grain boundaries.

The refractory $Nb_{25}Mo_{25}Ta_{25}W_{25}$ HEA with a single BCC phase is brittle [41, 58]. By carrying out cantilever bending tests on the micro-sized single crystal and bi-crystal samples, it was determined that the bi-crystal samples possessed a mean fracture toughness (0.2 $MPa\sqrt{m}$) one order of magnitude smaller than the single crystal samples (1.6 $MPa\sqrt{m}$) [58]. Examination of the fractographs shows that the single crystal sample experiences a transgranular fracture with obvious river markings while the bi-crystal sample fails in an intergranular manner along the grain boundary [58]. This is a clear implication that significant embrittlement in the bi-crystal samples is associated with the presence of the grain boundary. To confirm this trend, the chemical compositions of the bi-crystal samples were analyzed with the atom probe tomography (APT) in a region enclosing the grain boundary. The reconstructed atom map of such a region showing the distribution of niobium, molybdenum, tantalum, and tungsten elements is given in Figure 7.11a [58]. It is seen that all these elements are homogeneously distributed over the domain. The one-dimensional concentration profiles across the boundary condition in Figure 7.11b do not see an appreciable congregation of any elements, further confirming their uniform distribution. On average, niobium, molybdenum, tantalum, and tungsten have a concentration of 22.3 at.%, 22.5 at.%, 26.3 at.%, and 25.4 at.%, close to the nominal composition of the alloy. Nevertheless, in the reconstructed atom map of the nitrogen, oxygen, and carbon elements in Figure 7.11c, a band of element clustering is noticed. This band is indicative of the position of the grain boundary and element segregation herein. A quantitative scanning of the elemental concentration across the grain boundary shows that the nitrogen and carbon elements are remarkably enriched at the grain boundary, whilst oxygen is slightly segregated, as shown in Figure 7.11e. It is inferred that the impurity

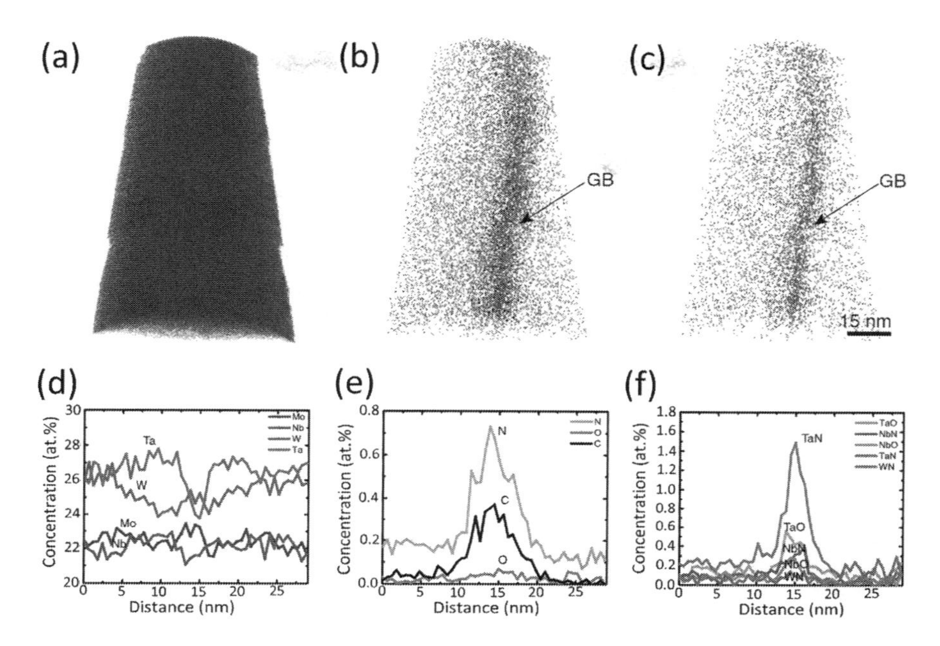

FIGURE 7.11 Reconstructed atom map from the atom-probe-tomography (APT) tip of a bi-crystal $Nb_{25}Mo_{25}Ta_{25}W_{25}$ high entropy alloy showing the elemental distribution of (a) niobium, molybdenum, tantalum, and tungsten; (b) nitrogen, oxygen, and carbon; and (c) tantalum nitride (TaN), tantalum oxide (TaO), niobium nitride (NbN), niobium oxide (NbO), and tungsten nitride (WN). Alongside shown in (d), (e), and (f) are the corresponding concentration profiles across the grain boundary [58].

elements, nitrogen and oxygen, are introduced during the sample preparation and annealing processes, but the carbon is inherently from raw materials. The high-concentration presence of these foreign elements, particularly nitrogen and oxygen, at the boundary promotes the formation and clustering of oxides and nitrides herein. The atom map of the tantalum nitride (TaN), tantalum oxide (TaO), niobium nitride (NbN), niobium oxide (NbO), and tungsten nitride (WN) compounds provided in Figure 7.11c, and the corresponding concentration profiles across the grain boundary in Figure 7.11f, testify to this reasoning. These oxides and nitrides are favorably formed due to relatively low formation enthalpies [58] or higher oxygen/nitrogen solubility [124].

The segregation of brittle intermetallic phases (e.g., tantalum oxide [TaO] and tantalum nitride [TaN]) at grain boundaries is detrimental to the fracture resistance of HEAs. In conjunction with the segregated nitrogen, oxygen, and carbon elements, these intermetallic compounds can easily embrittle the alloys, leading to a significantly reduced fracture toughness value and intergranular fracture in the bi-crystal samples. Despite the findings being based on the micro-sized HEA samples with a single grain boundary, the knowledge gained may also be extended to the polycrystalline HEAs of the same or different type.

7.4.3 In High Entropy Alloys with Face-Centered Cubic and Body-Centered Cubic Phases

The AlCoCrFeNi HEA constitutes an FCC phase and a duplex BCC phase accounting for approximately 17.2% and 82.8% volume fractions, respectively. The duplex BCC phases are enclosed by a net-like FCC phase, as illustrated in Figure 7.12a [32]. The duplex BCC phase is further composed of a spinodally decomposed ordered matrix rich in Al and Ni and disordered precipitates rich in Fe and Cr, as shown in the magnified microstructure in Figure 7.12b. The crack in this alloy is found to propagate inside the FCC phase or along the FCC-BCC interfaces. Given that the FCC phase has a net-like distribution in the alloy, the crack propagation is forced to be curvy, which means it retains a greater amount of the surface energy than the straight one. In addition, the crack could be blunted and arrested by the ductile FCC phase, bridged by the FCC phase or BCC-phase fragments, and deflected by the FCC-BCC interfaces. All these mechanisms seen in Figure 7.12c work synergistically to toughen the AlCoCrFeNi HEA, leading to intermediate fracture toughness of 25.2 MPa$\sqrt{m}$ [32].

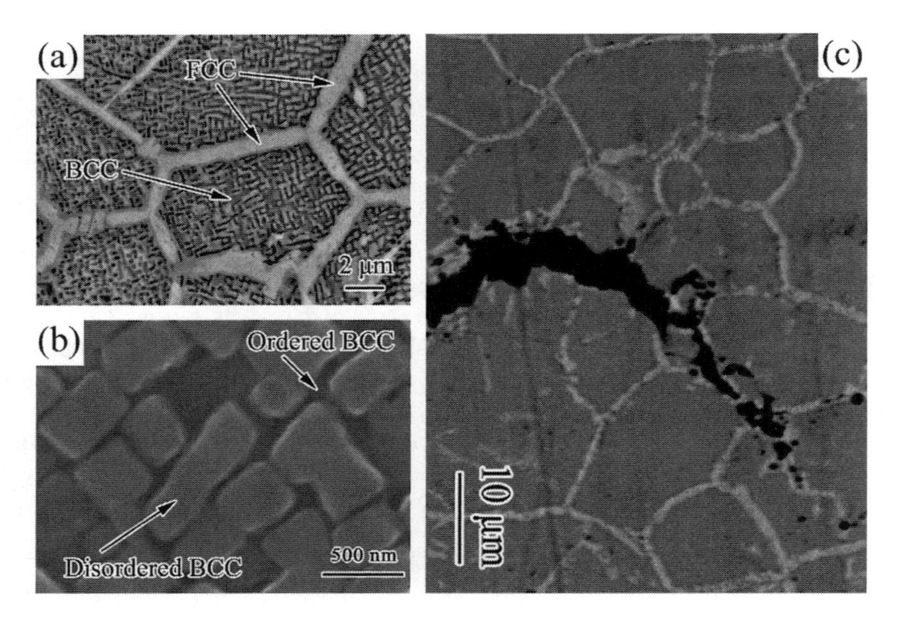

FIGURE 7.12 (a) Back-scattered image of the overall microstructure of the spark-plasma-sintered AlCoCrFeNi high entropy alloy consisting of spinodally modulated body-centered cubic (BCC) regions enclosed by face-centered cubic (FCC) "nets." (b) Secondary electron image of the magnified BCC region revealing a spinodally modulated structure composed of ordered and disordered BCC phases. (c) Crack propagation trajectory observed in a fracture toughness test [32].

7.4.4 In Eutectic High Entropy Alloys

Some carefully tailored eutectic microstructures may benefit the fracture resistance of alloys by positively affecting crack propagation. Such a demonstration is found in the $CoCrFeNiNb_{0.5}$ HEA with a hierarchical lamellar eutectic structure comprising Laves and FCC phases in approximately equal volume fractions [33]. The hierarchical eutectic microstructure overall has an appearance of a grain-like colonial network, as shown by the optical microscopy image in Figure 7.13a. Each colony is further composed of fine lamella structures and coarse lamella structures, as revealed by the top and bottom insets, respectively, in the example SEM image in Figure 7.13b. This particular hierarchical lamellar eutectic structure intrinsically toughens the alloy. In the coarse lamellar structure (spacing ≈250 nm), the crack tends to initiate and propagate inside the brittle Laves phase, but is arrested and blunted by the ductile FCC phase, as exemplified by the crack path rectangle in Figure 7.13c. As the crack

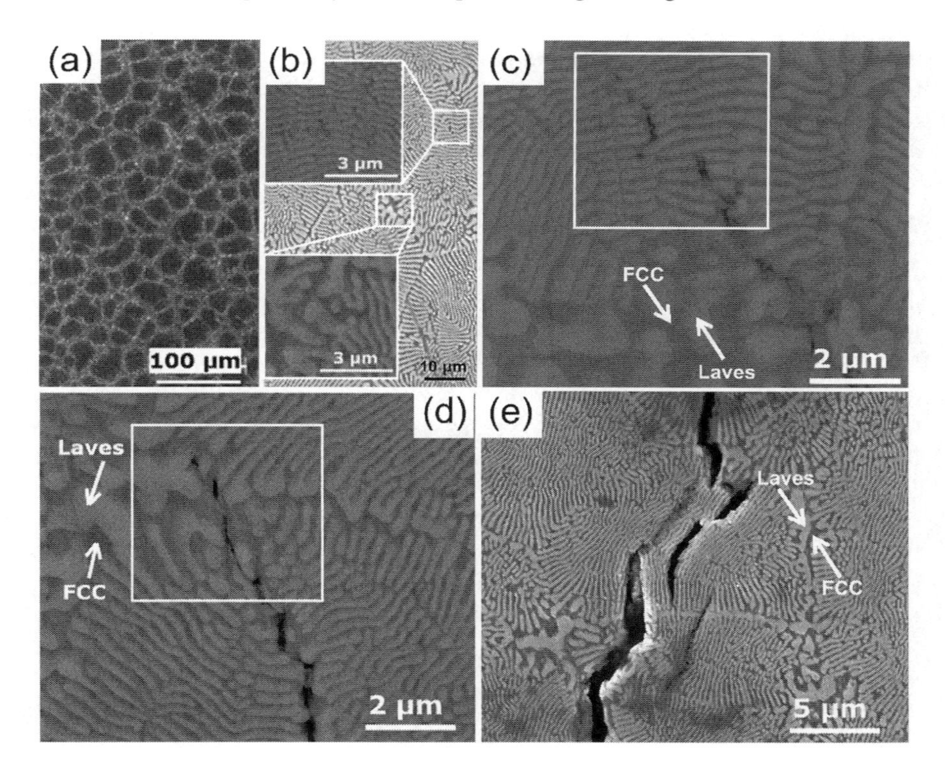

FIGURE 7.13 Microstructures and crack propagation in the $CoCrFeNiNb_{0.5}$ high entropy alloy with hierarchical, lamellar eutectic Laves, and face-centered cubic (FCC) phases [33]. (a) The dark field optical microscopy image of the eutectic structure. (b) Scanning electron microscopy (SEM) image of the eutectic structure with the top and bottom insets showing the fine and coarse lamella regions, respectively. The crack propagation paths (SEM images) in the (c) fine lamella region and (d) coarse lamella region, which contribute to intrinsic toughening. (e) Extrinsic toughening by crack deflection and bridging.

advances into the fine lamellar structure (spacing ≈70 nm), the roles of the Laves and FCC phases are reversed. Specifically, the crack propagation takes place within the embrittled FCC phase but is detoured by the ductilized Laves phase, as representatively illustrated by the rectangle region in Figure 7.13d. Cleavage fracture favors the FCC phase in this structure because the crack tip plasticity is suppressed in ductile lamellas whose thicknesses are smaller than two times the crack tip plastic zone, r_p [33]. On the other hand, the nano-sized Laves phase in the fine lamellar structure experiences a size-controlled brittle-to-ductile transition, becoming ductile [33].

Furthermore, as the crack prefers to grow in the direction parallel to the local lamella, randomly oriented lamellas in both the coarse and fine lamellar structures continuously alter the crack growth path. Simply put, the lamellar structures cause crack deflection and detouring, as seen in Figure 7.13e. Also, due to the vast difference in the toughness of the Laves and FCC phases, the crack propagation can easily become discontinuous, forming intermittent ligaments bridging crack faces, as seen in Figure 7.13e [33]. With SEM imaging, the bridging ligaments are found to form in the FCC phase in the coarse lamellar structure, but in the Laves phase in the fine lamellar structure [33]. The crack deflection and bridging triggered by the nature of the lamellar structure serve as extrinsic toughening mechanisms, additionally improving the fracture resistance of the alloy.

7.4.5 In Precipitation-Hardened High Entropy Alloys

In the precipitation-hardened HEAs, for example, the $CoCrFeNiMo_{0.2}$ alloy with an FCC phase and Cr-rich intermetallic particles of 1–3 μm in diameter [34], the high fracture resistance, as indicated by a 18° CTOA value [34], is suggested to be fulfilled by (1) the excellent ductility and work hardening of the FCC matrix; (2) the nucleation of microvoids by the debonding of intermetallic particles from the FCC matrix (Figure 7.14a), followed by their growth and coalescence; (3) the remarkable crack tip blunting through massive plastic deformation in the FCC matrix (Figure 7.14b), mediated by both dislocation slips and twinning; (4) the crack branching (Figure 7.14c); (5) the formation of the tortuous crack propagation path by merging microcracks (evolved from microvoids) into the propagating main crack (Figure 7.14d). The synergy of these mechanisms eventually leads to ductile fracture of the alloy, featuring a dimpled fracture morphology [34].

7.5 CONCLUSION

HEAs represent a juvenile class of metallic materials, and the investigations into their fracture resistance remain nascent. As the fracture resistance is being characterized, exceptional and fascinating fracture toughness, as high as 200 $MPa\sqrt{m}$, have been discovered for some compositions (e.g., CrMnFeCoNi) [53]. Such findings are encouraging for the development the toughest structural materials ever for demanding engineering applications.

However, this is merely one side of the story. The underlying microstructural reasons, particularly when associated with the peculiar lattice structure of high entropy solid solutions, remain elusive, although intricate dislocation activities and

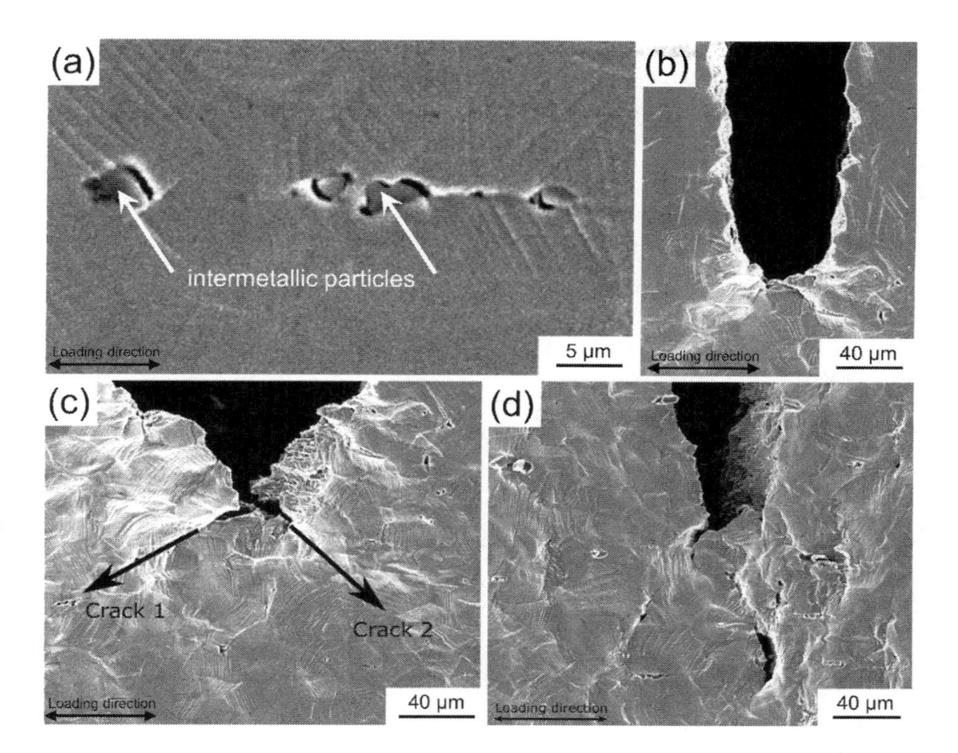

FIGURE 7.14 Crack initiation and propagation in the $CoCrFeNiMo_{0.2}$ high entropy alloy with a face-centered cubic (FCC) matrix and Cr-rich precipitates [34]. (a) Scanning electron microscopy (SEM) observation of microvoid nucleation by the debonding of the Cr-rich intermetallic particles. (b) Crack tip blunting by plastic deformation. (c) Crack branching. (d) Formation of the tortuous crack propagation path by merging microcracks (evolved from microvoids) into the main crack.

nano-twinning are observed as the predominant deformation micro-mechanisms in general. In other words, the micro-mechanisms unveiled essentially do not distinguish high entropy alloys from conventional alloys. Overall, extensive and rigorous fracture resistance characterizations are still needed for more high entropy alloys in order to search for candidate alloys of high fracture toughness as well as to select representative ones for fundamental understandings. Connecting the fracture resistance of alloys with their microstructures and lattice structures, in the background of multi-principal element solid solutions, remains the most pressing and challenging mission.

REFERENCES

1. W. Li, P.K. Liaw, Y. Gao, *Intermetallics* 99: (2018), 69–83.
2. T.L. Anderson, *Fracture Mechanics: Fundamentals and Applications*, 3rd ed., CRC Press, Boca Raton, FL, 2005.
3. G.E. Dieter, D.J. Bacon, *Mechanical Metallurgy*, McGraw-Hill, New York, 1986.

4. J.J. Lewandowski, W.H. Wang, A.L. Greer, *Philos. Mag. Lett.* 85(2): (2005), 77–87.
5. J.J. Lewandowski, M. Shazly, A. Shamimi Nouri, *Scr. Mater.* 54(3): (2006), 337–341.
6. W. Li, H. Bei, Y. Gao, *Intermetallics* 79: (2016), 12–19.
7. H. Liebowitz, *Fracture of Metals: An Advanced Treatise*, Academic Press, New York, 1969.
8. U. Krupp, W.M. Kane, C. Laird, C.J. McMahon, *Mater. Sci. Eng. A* 387: (2004), 409–413.
9. Y.C. Lin, J. Deng, Y.-Q. Jiang, D.-X. Wen, G. Liu, *Mater. Sci. Eng. A* 598: (2014), 251–262.
10. J.P. Dennison, P.D. Holmes, B. Wilshire, *Mater. Sci. Eng.* 33(1): (1978), 35–47.
11. R.C. Reed, *The Superalloys: Fundamentals and Applications*, Cambridge University Press, Cambridge, 2006.
12. W. Li, H. Bei, Y. Tong, W. Dmowski, Y.F. Gao, *Appl. Phys. Lett.* 103(17): (2013). http://www.ncbi.nlm.nih.gov/pubmed/171910.
13. W. Li, Y. Gao, H. Bei, *Sci. Rep.* 5: (2015), 14786.
14. H. Jia, X. Xie, L. Zhao, J. Wang, Y. Gao, K.A. Dahmen, W. Li, P.K. Liaw, C. Ma, J. Mater. *Res. Technol.* (2017).
15. B. Cantor, I.T.H. Chang, P. Knight, A.J.B. Vincent, *Mater. Sci. Eng. A* 375: (2004), 213–218.
16. J.W. Yeh, S.K. Chen, S.J. Lin, J.Y. Gan, T.S. Chin, T.T. Shun, C.H. Tsau, S.Y. Chang, *Adv. Eng. Mater.* 6(5): (2004), 299–303.
17. Y. Zhang, T.T. Zuo, Z. Tang, M.C. Gao, K.A. Dahmen, P.K. Liaw, Z.P. Lu, *Prog. Mater. Sci.* 61: (2014), 1–93.
18. S. Chen, X. Xie, W. Li, R. Feng, B. Chen, J. Qiao, Y. Ren, Y. Zhang, K.A. Dahmen, P.K. Liaw, *Mater. Chem. Phys.* 210: (2017), 20–28.
19. R. Abbaschian, R.E. Reed-Hill, *Physical Metallurgy Principles*, 4th ed., Cengage Learning, Stamford, CT, 2008.
20. O.N. Senkov, J.D. Miller, D.B. Miracle, C. Woodward, *Nat. Commun.* 6: (2015), 6529.
21. D.A. Porter, K.E. Easterling, M. Sherif, *Phase Transformations in Metals and Alloys*, 3rd ed., CRC Press, Boca Raton, FL, 2009.
22. D.B. Miracle, O.N. Senkov, *Acta Mater.* 122: (2017), 448–511.
23. Y.F. Ye, Q. Wang, J. Lu, C.T. Liu, Y. Yang, *Mater. Today* 19(6): (2016), 349–362.
24. Z. Wu, H. Bei, G.M. Pharr, E.P. George, *Acta Mater.* 81: (2014), 428–441.
25. L. Lilensten, J.-P. Couzinié, J. Bourgon, L. Perrière, G. Dirras, F. Prima, I. Guillot, *Mater. Res. Lett.* 5(2): (2017), 110–116.
26. C.L. Tracy, S. Park, D.R. Rittman, S.J. Zinkle, H. Bei, M. Lang, R.C. Ewing, W.L. Mao, *Nat. Commun.* 8: (2017), 15634.
27. F. Zhang, Y. Wu, H. Lou, Z. Zeng, V.B. Prakapenka, E. Greenberg, Y. Ren, J. Yan, J.S. Okasinski, X. Liu, Y. Liu, Q. Zeng, Z. Lu, *Nat. Commun.* 8: (2017), 15687.
28. S. Chen, K.-K. Tseng, Y. Tong, W. Li, C.-W. Tsai, J.-W. Yeh, P.K. Liaw, *J. Alloys Compd.* 795: (2019), 19–26.
29. F. Otto, A. Dlouhý, C. Somsen, H. Bei, G. Eggeler, E.P. George, *Acta Mater.* 61(15): (2013), 5743–5755.
30. Z. Li, K.G. Pradeep, Y. Deng, D. Raabe, C.C. Tasan, *Nature* 534(7606): (2016), 227–230.
31. X. Yang, Y. Zhang, *Mater. Chem. Phys.* 132(2): (2012), 233–238.
32. A. Zhang, J. Han, J. Meng, B. Su, P. Li, *Mater. Lett.* 181: (2016), 82–85.
33. D. Chung, Z. Ding, Y. Yang, *Adv. Eng. Mater.* (2018). http://www.ncbi.nlm.nih.gov/pubmed/1801060.
34. W.P. Li, X.G. Wang, B. Liu, Q.H. Fang, C. Jiang, *Mater. Sci. Eng. A* 723: (2018) 79–88.
35. K.Y. Tsai, M.H. Tsai, J.W. Yeh, *Acta Mater.* 61(13): (2013), 4887–4897.
36. Z. Wu, M.C. Troparevsky, Y.F. Gao, J.R. Morris, G.M. Stocks, H. Bei, *Curr. Opin. Solid State Mater. Sci.* 21(5): (2017), 267–284.

37. S.Y. Chen, Y. Tong, K.K. Tseng, J.W. Yeh, J.D. Poplawsky, J.G. Wen, M.C. Gao, G. Kim, W. Chen, Y. Ren, R. Feng, W.D. Li, P.K. Liaw, *Scr. Mater.* 158: (2019), 50–56.
38. K.M. Youssef, A.J. Zaddach, C. Niu, D.L. Irving, C.C. Koch, *Mater. Res. Lett.* 3(2): (2015), 95–99.
39. P.F. Yu, L.J. Zhang, H. Cheng, H. Zhang, M.Z. Ma, Y.C. Li, G. Li, P.K. Liaw, R.P. Liu, *Intermetallics* 70: (2016), 82–87.
40. O.N. Senkov, G.B. Wilks, D.B. Miracle, C.P. Chuang, P.K. Liaw, *Intermetallics* 18(9): (2010), 1758–1765.
41. O.N. Senkov, G.B. Wilks, J.M. Scott, D.B. Miracle, *Intermetallics* 19(5): (2011), 698–706.
42. O.N. Senkov, J.M. Scott, S.V. Senkova, F. Meisenkothen, D.B. Miracle, C.F. Woodward, *J. Mater. Sci.* 47(9): (2012), 4062–4074.
43. Y. Qiu, S. Thomas, M.A. Gibson, H.L. Fraser, N. Birbilis, *NPJ Mater. Degrad.* 1(1): (2017), 1–15.
44. A. Poulia, E. Georgatis, A. Lekatou, A.E. Karantzalis, *Int. J. Refract. Met. Hard Mater.* 57: (2016), 50–63.
45. M.G. Poletti, G. Fiore, F. Gili, D. Mangherini, L. Battezzati, *Mater. Des.* 115: (2017), 247–254.
46. M.A. Hemphill, T. Yuan, G.Y. Wang, J.W. Yeh, C.W. Tsai, A. Chuang, P.K. Liaw, *Acta Mater.* 60(16): (2012), 5723–5734.
47. Z. Tang, T. Yuan, C.-W. Tsai, J.-W. Yeh, C.D. Lundin, P.K. Liaw, *Acta Mater.* 99: (2015), 247–258.
48. D.B. Miracle, J.D. Miller, O.N. Senkov, C. Woodward, M.D. Uchic, J. Tiley, *Entropy* 16(1): (2014), 494–525.
49. S. Chen, W. Li, X. Xie, J. Brechtl, B. Chen, P. Li, G. Zhao, F. Yang, J. Qiao, P.K. Liaw, *J. Alloys Compd.* 752: (2018), 464–475.
50. B.S. Murty, J.-W. Yeh, S. Ranganathan, *High-Entropy Alloys*, Butterworth-Heinemann, London, 2014.
51. H. Zhang, Y. He, Y. Pan, *Scr. Mater.* 69(4): (2013), 342–345.
52. U. Roy, H. Roy, H. Daoud, U. Glatzel, K.K. Ray, *Mater. Lett.* 132: (2014), 186–189.
53. B. Gludovatz, A. Hohenwarter, D. Catoor, E.H. Chang, E.P. George, R.O. Ritchie, *Science* 345(6201): (2014), 1153–1158.
54. C.-C. Juan, M.-H. Tsai, C.-W. Tsai, C.-M. Lin, W.-R. Wang, C.-C. Yang, S.-K. Chen, S.-J. Lin, J.-W. Yeh, *Intermetallics* 62: (2015), 76–83.
55. M. Seifi, D. Li, Z. Yong, P.K. Liaw, J.J. Lewandowski, *JOM* 67(10): (2015), 2288–2295.
56. D. Li, Y. Zhang, *Intermetallics* 70: (2016), 24–28.
57. C. Chen, S. Pang, Y. Cheng, T. Zhang, *J. Alloys Compd.* 659: (2016), 279–287.
58. Y. Zou, P. Okle, H. Yu, T. Sumigawa, T. Kitamura, S. Maiti, W. Steurer, R. Spolenak, *Scr. Mater.* 128: (2017), 95–99.
59. W. Li, G. Wang, S. Wu, P.K. Liaw, *J. Mater. Res.* 33(19): (2018), 3011–3034.
60. B. Gludovatz, A. Hohenwarter, K.V.S. Thurston, H. Bei, Z. Wu, E.P. George, R.O. Ritchie, *Nat. Commun.* 7: (2016), 10602.
61. Z. Zhang, M.M. Mao, J. Wang, B. Gludovatz, Z. Zhang, S.X. Mao, E.P. George, Q. Yu, R.O. Ritchie, *Nat. Commun.* 6: (2015), 10143.
62. Z. Zhang, H. Sheng, Z. Wang, B. Gludovatz, Z. Zhang, E.P. George, Q. Yu, S.X. Mao, R.O. Ritchie, *Nat. Commun.* 8: (2017), 14390.
63. A.F. Bower, *Applied Mechanics of Solids*, CRC Press, Boca Raton, FL, 2009.
64. ASTM E399-17, *Standard Test Method for Linear-Elastic Plane-Strain Fracture Toughness K_{Ic} of Metallic Materials*, ASTM International, West Conshohocken, PA, 2017.
65. S. Mohanty, T.N. Maity, S. Mukhopadhyay, S. Sarkar, N.P. Gurao, S. Bhowmick, K. Biswas, *Mater. Sci. Eng. A* 679: (2017), 299–313.

66. ASTM E1820-17, *Standard Test Method for Measurement of Fracture Toughness*, ASTM International, West Conshohocken, PA, 2017.
67. D.S. Harding, W.C. Oliver, G.M. Pharr, *M.R.S. Proc.* 356: (2011), 663–668.
68. G.M. Pharr, D.S. Harding, W.C. Oliver. In: M. Nastasi, D.M. Parkin, H. Gleiter (Eds.), *Mechanical Properties and Deformation Behavior of Materials Having Ultra-Fine Microstructures*, Springer Netherlands, Dordrecht, 1993, 449–461.
69. G.R. Anstis, P. Chantikul, B.R. Lawn, D.B. Marshall, *J. Am. Ceram. Soc.* 64(9): (1981), 533–538.
70. P. Chantikul, G.R. Anstis, B.R. Lawn, D.B. Marshall, *J. Am. Ceram. Soc.* 64(9): (1981), 539–543.
71. G.D. Quinn, R.C. Bradt, *J. Am. Ceram. Soc.* 90(3): (2007), 673–680.
72. J.-i. Jang, G.M. Pharr, *Acta Mater.* 56(16): (2008), 4458–4469.
73. J.H. Lee, Y.F. Gao, K.E. Johanns, G.M. Pharr, *Acta Mater.* 60(15): (2012), 5448–5467.
74. F. Iqbal, J. Ast, M. Göken, K. Durst, *Acta Mater.* 60(3): (2012), 1193–1200.
75. ASTM E23-16b, *Standard Test Methods for Notched Bar Impact Testing of Metallic Materials*, ASTM International, West Conshohocken, PA, 2016.
76. S.Q. Xia, M.C. Gao, Y. Zhang, *Mater. Chem. Phys.* 210: (2018), 213–221.
77. M.A. Sokolov, H. Tanigawa, G.R. Odette, K. Shiba, R.L. Klueh, *J. Nucl. Mater.* 367(Part A): (2007), 68–73.
78. M. Tanaka, A.J. Wilkinson, S.G. Roberts, *J. Nucl. Mater.* 378(3): (2008), 305–311.
79. B. Hwang, C.G. Lee, *Mater. Sci. Eng. A* 527(16): (2010), 4341–4346.
80. H. Li, F. Ebrahimi, *Adv. Mater.* 17(16): (2005), 1969–1972.
81. R. Li, K. Sieradzki, *Phys. Rev. Lett.* 68(8): (1992), 1168–1171.
82. R. Raghavan, P. Murali, U. Ramamurty, *Acta Mater.* 57(11): (2009), 3332–3340.
83. C.-W. Yang, T.-S. Lui, L.-H. Chen, H.-E. Hung, *Scr. Mater.* 61(12): (2009), 1141–1144.
84. H.L. Jia, L.L. Zheng, W.D. Li, N. Li, J.W. Qiao, G.Y. Wang, Y. Ren, P.K. Liaw, Y. Gao, *Metall. Mater. Trans. A* 46(6): (2015), 2431–2442.
85. S.G. Ma, Y. Zhang, *Mater. Sci. Eng. A* 532: (2012), 480–486.
86. Y.J. Zhou, Y. Zhang, F.J. Wang, Y.L. Wang, G.L. Chen, *J. Alloys Compd.* 466(1): (2008), 201–204.
87. J. Chen, P. Niu, Y. Liu, Y. Lu, X. Wang, Y. Peng, J. Liu, *Mater. Des.* 94: (2016), 39–44.
88. K.B. Zhang, Z.Y. Fu, J.Y. Zhang, W.M. Wang, H. Wang, Y.C. Wang, Q.J. Zhang, J. Shi, *Mater. Sci. Eng. A* 508(1): (2009), 214–219.
89. H. Yao, J.-W. Qiao, M. Gao, J. Hawk, S.-G. Ma, H. Zhou, *Entropy* 18(5): (2016), 189.
90. X.R. Wang, P. He, T.S. Lin, Z.Q. Wang, *Mater. Sci. Technol.* 31(15): (2015), 1842–1849.
91. Y.X. Zhuang, H.D. Xue, Z.Y. Chen, Z.Y. Hu, J.C. He, *Mater. Sci. Eng. A* 572: (2013), 30–35.
92. W. Chen, Z. Fu, S. Fang, H. Xiao, D. Zhu, *Mater. Des.* 51: (2013), 854–860.
93. Z. Wang, X. Wang, H. Yue, G. Shi, S. Wang, *Mater. Sci. Eng. A* 627: (2015), 391–398.
94. Y. Ivanisenko, L. Kurmanaeva, J. Weissmueller, K. Yang, J. Markmann, H. Rösner, T. Scherer, H.J. Fecht, *Acta Mater.* 57(11): (2009), 3391–3401.
95. W. Li, Y. Gao, H. Bei, *Sci. Rep.* 6: (2016), 34878.
96. W. Li, H. Jia, C. Pu, X. Liu, J. Xie, *J. Mater. Sci. Technol.* 31(10): (2015), 1018–1026.
97. W. Li, K. Xu, H. Li, H. Jia, X. Liu, J. Xie, *J. Mater. Sci. Technol.* 33(11): (2017), 1353–1361.
98. H. Jia, F. Liu, Z. An, W. Li, G. Wang, J.P. Chu, J.S.C. Jang, Y. Gao, P.K. Liaw, *Thin Solid Films* 561(Supplement C): (2014), 2–27.
99. M.C. Liu, J.C. Huang, K.W. Chen, J.F. Lin, W.D. Li, Y.F. Gao, T.G. Nieh, *Scr. Mater.* 66(10): (2012), 817–820.
100. Z.N. An, W.D. Li, F.X. Liu, P.K. Liaw, Y.F. Gao, *Metall. Mater. Trans. A* 43(8): (2012), 2729–2741.
101. B. Gludovatz, E.P. George, R.O. Ritchie, *JOM* 67(10): (2015), 2262–2270.

102. W.H. Liu, J.Y. He, H.L. Huang, H. Wang, Z.P. Lu, C.T. Liu, *Intermetallics* 60: (2015), 1–8.
103. Z. Wang, M.C. Gao, S.G. Ma, H.J. Yang, Z.H. Wang, M. Ziomek-Moroz, J.W. Qiao, *Mater. Sci. Eng. A* 645: (2015), 163–169.
104. C. Ng, S. Guo, J. Luan, Q. Wang, J. Lu, S. Shi, C.T. Liu, *J. Alloys Compd.* 584: (2014), 530–537.
105. B. Schuh, F. Mendez-Martin, B. Völker, E.P. George, H. Clemens, R. Pippan, A. Hohenwarter, *Acta Mater.* 96: (2015), 258–268.
106. Z. Wu, H. Bei, *Mater. Sci. Eng. A* 640: (2015), 217–224.
107. S.G. Ma, S.F. Zhang, J.W. Qiao, Z.H. Wang, M.C. Gao, Z.M. Jiao, H.J. Yang, Y. Zhang, *Intermetallics* 54: (2014), 104–109.
108. L. Liu, J.B. Zhu, L. Li, J.C. Li, Q. Jiang, *Mater. Des.* 44: (2013), 223–227.
109. Z.Y. Rao, X. Wang, J. Zhu, X.H. Chen, L. Wang, J.J. Si, Y.D. Wu, X.D. Hui, *Intermetallics* 77: (2016), 23–33.
110. A. Gali, E.P. George, *Intermetallics* 39(Supplement C): (2013), 74–78.
111. A.J. Zaddach, C. Niu, C.C. Koch, D.L. Irving, *JOM* 65(12): (2013), 1780–1789.
112. M.R. Barnett, *Mater. Sci. Eng. A* 464(1): (2007), 1–7.
113. M.R. Barnett, *Mater. Sci. Eng. A* 464(1): (2007), 8–16.
114. G. Frommeyer, U. Brüx, P. Neumann, *ISIJ Int.* 43(3): (2003), 438–446.
115. O. Grässel, L. Krüger, G. Frommeyer, L.W. Meyer, *Int. J. Plast.* 16(10): (2000), 1391–1409.
116. H. Kou, J. Lu, Y. Li, *Adv. Mater.* 26(31): (2014), 5518–5524.
117. K. Lu, L. Lu, S. Suresh, *Science* 324(5925): (2009), 349–352.
118. L. Lu, X. Chen, X. Huang, K. Lu, *Science* 323(5914): (2009), 607–610.
119. Y. Wei, Y. Li, L. Zhu, Y. Liu, X. Lei, G. Wang, Y. Wu, Z. Mi, J. Liu, H. Wang, H. Gao, *Nat. Commun.* 5: (2014), 3580.
120. G. Bao, Z. Suo, *Appl. Mech. Rev.* 45(8): (1992), 355–366.
121. A.G. Evans, D.B. Marshall, *Acta Metall.* 37(10): (1989), 2567–2583.
122. D.B. Marshall, A.G. Evans, *J. Am. Ceram. Soc.* 68(5): (1985), 225–231.
123. S. Nemat-Nasser, M. Hori, *Mech. Mater.* 6(3): (1987), 245–269.
124. J.R. DiStefano, B.A. Pint, J.H. DeVan, *Int. J. Refract. Met. Hard Mater.* 18(4): (2000), 237–243.
125. S.-P. Wang, E. Ma, J. Xu, *Intermetallics* 103: (2018), 78–87.
126. S.-P. Wang, J. Xu, *Intermetallics* 95: (2018), 59–72.
127. W.H. Liu, T. Yang, C.T. Liu, *Mater. Chem. Phys.* 210: (2018), 2–11.
128. R.X. Li, P.K. Liaw, Y. Zhang, *Mater. Sci. Eng. A* 707: (2017), 668–673.
129. Y. Lv, R. Hu, Z. Yao, J. Chen, D. Xu, Y. Liu, X. Fan, *Mater. Des.* 132: (2017), 392–399.
130. W.-R. Wang, W.-L. Wang, J.-W. Yeh, *J. Alloys Compd.* 589: (2014), 143–152.
131. T. Yang, S. Xia, S. Liu, C. Wang, S. Liu, Y. Zhang, J. Xue, S. Yan, Y. Wang, *Mater. Sci. Eng. A* 648: (2015), 15–22.
132. W. Bolton, *Newnes Engineering Materials Pocket Book*, Heinemann Newnes, Oxford, 1989.
133. C. Buirette, J. Huez, N. Gey, A. Vassel, E. Andrieu, *Mater. Sci. Eng. A* 618: (2014), 546–557.
134. Y.J. Chao, J. Ward, R.G. Sands, *Mater. Des.* 28(2): (2007), 551–557.
135. S.Y. Han, S.Y. Shin, S. Lee, N.J. Kim, J.-H. Bae, K. Kim, *Metall. Mater. Trans. A* 41(2): (2010), 329.
136. B. Hwang, Y.G. Kim, S. Lee, Y.M. Kim, N.J. Kim, J.Y. Yoo, *Metall. Mater. Trans. A* 36(8): (2005), 2107–2114.
137. Handbook Committee, *Properties and Selection: Nonferrous Alloys and Special-Purpose Materials,* ASM International, 1990.
138. A. Karimpoor, K. Aust, U. Erb, *Scr. Mater.* 56(3): (2007), 201–204.

139. B.S. Louden, A. Kumar, F. Garner, M. Hamilton, W. Hu, *J. Nucl. Mater.* 155: (1988), 662–667.
140. P. Maziasz, D. Alexander, J. Wright, *Intermetallics* 5(7): (1997), 547–562.
141. Z. Oksiuta, N. Baluc, *J. Nucl. Mater.* 374(1): (2008), 178–184.
142. A. Rossoll, C. Berdin, C. Prioul, *Int. J. Fract.* 115(3): (2002), 205–226.
143. S.Y. Shin, B. Hwang, S. Lee, N.J. Kim, S.S. Ahn, *Mater. Sci. Eng. A* 458(1): (2007), 281–289.
144. W. Yan, Y. Shan, K. Yang, *Metall. Mater. Trans. A* 37(7): (2006), 2147–2158.
145. X.-L. Yang, Y.-B. Xu, X.-D. Tan, D. Wu, *Mater. Sci. Eng. A* 641: (2015), 96–106.

8 Thermodynamics of High Entropy Alloys

K. Guruvidyathri and B.S. Murty

CONTENTS

8.1 INTRODUCTION

The science of thermodynamics is the bedrock of the design philosophy of HEAs. Consequently, it provides the basis for the name HEA [1]. These are the first and only class of alloys named after a thermodynamic property so far.

Thermodynamic parameters that are not significant in conventional alloys become profoundly influential in the concentrated multi-element environment of HEAs. One such property is the configurational entropy arising from the inherent compositional nature of HEAs. Evaluation of its dominance has been one of the most active aspects of HEA investigations. There are ambiguities in its fundamental nature. In metallic glasses research and in HEA research, this quantity represents different physical phenomena. No discussion is available yet on this difference, which is attempted in the present chapter. It is also important to revisit the formulation of configurational entropy and the correctness of its use in the literature. For example, the assumption of ideal mixing to calculate this quantity may be too misleading in certain cases.

Focus on other entropy contributions is essential. For a given phase, the vibrational entropy, especially is much higher in magnitude. Unlike configurational entropy, vibrational entropy does not increase in the HEA regime as a mixing quantity [2]. However, there is an experimental claim of increasing vibrational contribution in the HEA regime [3]. This chapter attempts to review it in the light of other experimental reports.

A significant number of studies in HEA literature are dedicated to finding effective alloy design strategies. Thermodynamics is the basis for most of these strategies. Various parametric approaches provide phase formation guidelines. The success rate of these parameters and the reasons for exceptions are discussed in detail in other sources [2, 4, 5]. A brief explanation of other simple methods, such as phase diagram inspection and G-x plot methods is given. Application of the Calphad (CALculation of PHAse Diagram) method—which is the most promising and robust alloy design tool—its success, and its scope are discussed in detail.

8.2 THERMODYNAMICS AND THE DESIGN PHILOSOPHY OF HEAs

In thermodynamics, competition between Gibbs energy (G) of various phases in an alloy results in a certain phase or combination of phases becoming stable. The criterion of this phase selection is that the Gibbs energy of the alloy is minimized. Gibbs energy has enthalpy (H) and entropy (S) contributions. For a solution phase,

$$G = \sum_i x_i G_i^{\mathrm{o}} + \Delta G_{\mathrm{mix}} \tag{8.1}$$

where x_i is the mole fraction of ith element, G_i^{o} is the Gibbs energy contribution of the pure component i, ΔG_{mix} is the Gibbs energy of mixing, which can be written in terms of enthalpy (ΔH_{mix}) and entropy of mixing (ΔS_{mix}) as

$$\Delta G_{\mathrm{mix}} = \Delta H_{\mathrm{mix}} - T\Delta S_{\mathrm{mix}} \tag{8.2}$$

where T is the temperature. In ideal mixing, the enthalpy change, $\Delta H_{\mathrm{mix}}^{\mathrm{ideal}}$, is zero, indicating no interaction effects between components (see Section 8.4). On the other hand, the entropy change in ideal mixing, $\Delta S_{\mathrm{mix}}^{\mathrm{ideal}}$, is non-zero. In general, ΔS_{mix}, has configurational, vibrational, magnetic, and electronic contributions. However, in ideal mixing, only the configurational entropy contribution, ΔS_C, exists. It is a measure of the randomness in the mixing of the components. In the ideal case, the mixing takes up atomic configurations that are totally random. It is worth noting that, even in regular solutions, entropy of mixing is assumed to be equal to that of an ideal one [6], although the mixing is expected to deviate from total random configuration due to non-ideal interactions.

Boltzmann's entropy formula is the statistical mechanics basis for the calculation of entropy.

$$S = k_B \ln w \tag{8.3}$$

where k_B is Boltzmann's constant, and w is the number of microstates that belong to the observed macrostate. This equation has far-reaching implications, since it is the basis for any entropic contribution in statistical mechanics. The nature of microstates varies according to the type of entropy. For calculating the configurational entropy discussed here, microstates are the atomic configurations. The ideal mixing entropy, $\Delta S_c^{\text{ideal}}$ is derived from Equation (8.3), using Stirling's approximation. For one mole of a phase, it is represented as

$$\Delta S_{\text{mix}}^{\text{ideal}} = \Delta S_c^{\text{ideal}} = -R \sum_i x_i \ln x_i \tag{8.4}$$

where R is the universal gas constant. The nature of this equation is the key to HEAs. This quantity depends only on the number and amount of components. The nature of this quantity is that it reaches the maximum when the mole fraction of elements approaches the equimolar ratio. As the number of elements increases this quantity increases.

Therefore, compositions with a larger number of elements in equimolar or near-equimolar ratios will help in maximizing the configurational entropy of mixing. This entropy increase is expected to minimize the Gibbs energy of the solution phase according to Equations (8.1) and (8.2). Therefore, the chances of stabilizing the solid solution (SS) over intermetallic (IM) phase formation and/or unmixing is enhanced (see Section 8.4). Thus, the HEA design hypothesis targets SS. This expected behavior is termed as "high entropy effect," which is one of the four core effects expected in HEAs [4].

When configurational entropy, ΔS_C, is more than 1.5R, it is expected to compete effectively against enthalpy effects. Hence, such alloys are considered to be HEAs. A broad guideline in terms of composition is to have five or more elements, each in 5–35 at.%, which will mostly result in $\Delta S_C > 1.5R$. This is suggested as the composition-based definition of HEAs. The uniqueness of this alloy design is in stark difference to the conventional alloys, in the sense that HEAs have multiple principle elements.

The emphasis on configurational entropy of mixing is understandably greater in the HEA literature [1, 4]. However, the complex interplay of configurational and other contributions of entropy, along with enthalpy in the form of Gibbs energy, is understood to be the basis for phase formation, as discussed in the rest of the sections in this chapter. Achieving only the SS phase is also realized to be a more restrictive target for the field since several successful alloys consist of more than one phase, including IM phases. The biggest breakthrough of HEAs is the opening up of concentrated multicomponent composition space for consideration. The scope of the field has thus broadened to be the exploration of such a vast composition space. In order to represent this broadened scope, various other suitable names are suggested [2, 7]. However, the name HEAs is used widely as a legacy of the founding idea.

8.3 ENTROPY

The entropy contribution from the pure components (Equation 8.1) is substantial. This involves thermal (vibrational, magnetic, and electronic) contributions. As discussed in the previous section, ΔS_{mix} is treated to be equivalent to $\Delta S_c^{\text{ideal}}$, which is the case in several reports on HEAs. Non-deal mixing contributions, however, are often significant. There is another entropy proposed to be arising from atomic packing and mismatch in bulk metallic glasses. It is also considered to be present in HEA literature.

8.3.1 Difference between Configurational Entropy in Metallic Glasses and Crystalline Solids

Metallic glasses are configurationally frozen supercooled liquids with atoms in dense random packing arrangements [8]. Since the crystalline structure is absent, various positional possibilities exist for the atoms to arrange themselves. These structural configurations can result in different minima points in the potential energy landscape (also called inherent structures). Each minimum is enclosed in its own "basin," in which all locations go downhill motion to connect to the minimum. The lowest potential energy other than crystalline state occurs at "ideal glass" configuration. Because of the complexity of such many-body landscapes, a statistical description becomes unavoidable. Thus, exploration of different basins gives rise to a configurational entropy [9]. In this way, in glasses, dynamics connect to thermodynamics. This entropy shall be referred as structural configurational entropy, ΔS_C^{σ}. In principle, this entropy shall be present even in a glass of pure element. It is different from the configurational entropy arising due to arrangement of different species, ΔS_C, which can be referred to as chemical configurational entropy (Equation 8.4). As long as there are different chemical species mixing into an ideal solution, ΔS_C (Equation 8.4) arises, irrespective of the positions in the space. It is worth noting that Equation (8.4) applies to SS phases of different crystal structures as well as to gases and liquids in which there is no notion of lattices. It arises in a pure element when vacancies are treated as a second chemical specie. This distinction becomes relevant in the context of HEAs, which is not discussed in the HEA literature.

Equation (8.3) is applicable for the calculation of positional configurational entropy as well. The quantity of interest here (w in Equation 8.3) is the number of potential energy minima corresponding to a given depth (microstates corresponding to a macrostate) [10, 11].

$$\frac{dw}{d\phi} = Ce^{N\sigma(\phi)} \tag{8.5}$$

dw denotes the number of potential energy minima with a depth between $\phi \pm d\phi/2$ per particle. N is the number of particles, and C is an N-independent factor with units of inverse energy. A comparison of this equation with Equation (8.3), after taking the logarithm, reveals that $\sigma(\phi)$ is the configurational entropy per particle, which is also called the basin enumeration number. The Helmholtz energy (A) is given by

$$\frac{A}{Nk_BT} = \frac{\overline{\phi}}{k_BT} - \sigma(\phi) + \frac{a^V}{k_BT} \tag{8.6}$$

where $\overline{\phi}$ is the depth of the basins at temperature T. a^V is vibrational energy per particle. The minimization of A can be used as the criteria for finding the entropy contributions. Equation (8.6) indicates the temperature-dependent nature of ΔS_C^{σ}, indicating that it is measurable from calorimetric studies. It is estimated through residual entropy at 0 K, by extrapolation of finite temperature calorimetric data [12]. Residual entropy has both vibrational and configurational contributions (Equation 8.6). Recently, an attempt to separate these contributions was reported to be successful [13]. The residual entropy is virtually equal to only the structural configurational entropy. Vibrational entropy contribution is negligible.

Thus, the separation of ΔS_C^{σ} and vibrational parts in a metallic glass was possible through calorimetry [13], but no method is reported yet for the separation of ΔS_C and vibrational part in a HEA. ΔS_C does not have temperature dependency (T is not involved in Equation 8.4). As a consequence, it cannot be measured using calorimetric studies. Therefore, it is calculated and added as a separate contribution (as shown in Figure 8.1) to thermal entropy, discussed in the following section. It is correct to add such a contribution only when the temperature is finite. Equilibrium atomic configurations are achieved by atomic rearrangements at any temperature above 0 K, hence, the corresponding ΔS_C exists. At low temperatures, since atomic mobility exponentially decreases with temperature, atomic rearrangements and achievement of internal equilibrium become difficult. At 0 K, however, according to the third law of thermodynamics, there cannot be any entropy when the system is in internal equilibrium.

The ΔS_C^{σ} discussed above is based on the potential energy model. Another simpler formulation based on the solid sphere model uses packing fraction (ρ) and atomic

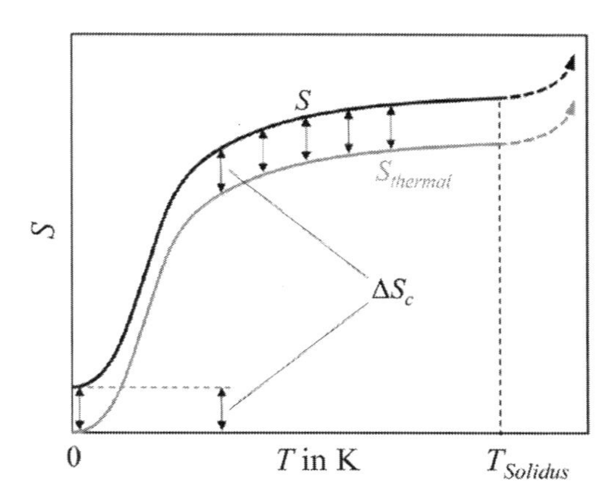

FIGURE 8.1 Thermal (S_{thermal}) and configurational entropy (ΔS_C) contributions to the total entropy of a phase (S). S_{thermal} is measured using calorimetric studies. ΔS_C is calculated and added [14].

sizes (d_i) [15–17]. Takeuchi [15] referred to it as mismatch entropy. This formulation, however, does not take into account the temperature dependency.

$$\frac{\Delta S_C^\sigma}{k_B} = \left\{ \frac{3}{2}(\zeta - 1) y_1 + \frac{3}{2}(\zeta - 1)^2 y_2 - \left[\frac{1}{2}(\zeta - 1)(\zeta - 3) + \ln \zeta \right](1 - y_3) \right\} \tag{8.7}$$

y_1, y_2, and y_3 in Equation (8.7) are dependent on atomic size and concentration as given in Equations (8.8)–(8.12) below.

$$y_1 = \frac{1}{\tau^3} \sum_{j>i=1}^{N} (d_i + d_j)(d_i - d_j)^2 C_i C_j \tag{8.8}$$

$$y_2 = \frac{\tau^2}{(\tau^3)^2} \sum_{j>i=1}^{N} d_i d_j (d_i - d_j)^2 C_i C_j \tag{8.9}$$

$$y_3 = \frac{(\tau^2)^3}{(\tau^3)^2} \tag{8.10}$$

$$\tau^k = \sum_{i=1}^{N} C_i d_i^k \tag{8.11}$$

$$\zeta = \frac{1}{(1 - \rho)} \tag{8.12}$$

The basis for ΔS_C^σ in the solid sphere model is that the atoms with different sizes can bring some uncertainty in atom location, thus giving a configurational entropy term. The uncertainty in atom location increases with increasing size differences and concentrations. Therefore, such entropy contribution is thought to be significant in HEAs and applied to understand phase formation [15, 18, 19].

8.3.2 Changes in Configurational Entropy Due to Interaction Effects

In reality, SS phases have non-ideal mixing behavior. Sub-regular solution behavior is by far the most common. Both positive and negative interaction effects deviate the arrangement of atoms from the most random configuration. Therefore, Equation (8.4) can no longer be used to calculate the value of configurational entropy. In these cases,

$$\Delta S_C \neq \Delta S_c^{\text{ideal}} \tag{8.13}$$

Treatment of ΔS_C for such non-random mixing were proposed for binary systems [20]. The short-range order parameter is used to quantify the extent of short-range ordering or clustering. It is used to find the ΔS_C [6]. Such treatments are not regularly applied for HEAs.

Monte Carlo studies revealed that about 20% deviation is expected due to short-range ordering in certain HEAs [21]. This indicates a significant degree of error is possible in applications of Equation (8.4) for calculating configurational entropy for certain systems. The higher the interaction effects, the higher the errors associated with application of it.

8.3.3 Calculation of Configurational Entropy

Similar to the enthalpy values discussed in Section 8.4, experimentally reported values of entropy for several lower-order phases are available in standard handbooks. For higher-order systems, similar to Gibbs energy (Section 8.5.2), thermal entropy is also extrapolated.

There are different ways of modeling the Gibbs energy of phases in Calphad method (Section 8.5.1). One noteworthy model is sublattice formalism, in which atoms are treated to be arranged in sublattices. The formalism is versatile and can be applied to SS, IM phases, salts, etc. In each sublattice, the components are assumed to be mixed in totally random configurations (ideal mixing). SS phases are usually considered to have one sublattice. Therefore, Equation (8.4) is applied without any changes for SS phases.

For chemically ordered phases, more than one sublattice is used in the model. Consider a hypothetical ordered phase of type AC_3 in A-C and A-B-C-D-E systems. In both systems, the phase is modeled to have two sublattices (Figure 8.2). In A-C, consider that the phase is completely ordered. Its sublattice formalism will then be $(A)_1(B)_3$. The configurational entropy will be zero for this case since there is no mixing of the elements in any sublattice. In A-B-C-D-E system, consider that the model is $(A,B)_1(C,D,E)_3$. The configurational entropy is calculated using the

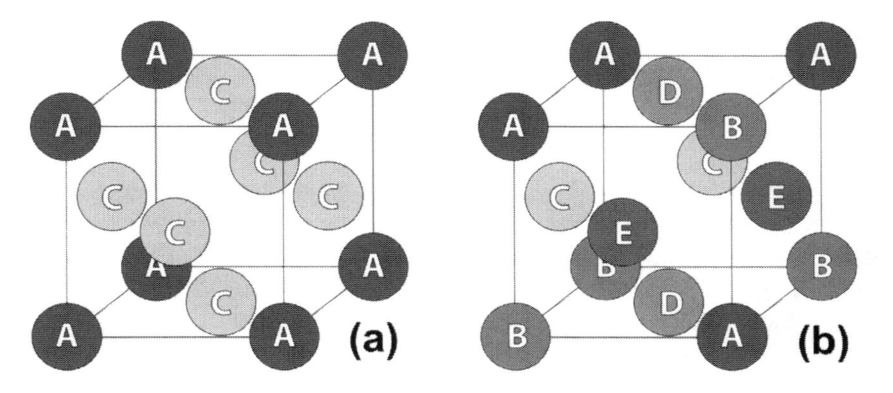

FIGURE 8.2 Ordered phases in (a) A-C system with A and C sublattices containing A and C atoms, respectively, and in (b) A-B-C-D-E system with A sublattice containing A and B atoms, and C sublattice containing C, D, and E atoms [2].

following equation, which is given for a similar hypothetical two sublattices model, $(A,B,\ldots)_{a1}(C,D,E,\ldots)_{a2}$:

$$\Delta S_C = -R\left\{\left(\frac{a_1}{a_1+a_2}\left(y_A^1 \ln y_A^1 + y_B^1 \ln y_B^1 + \ldots\right)\right) + \left(\frac{a_2}{a_1+a_2}\left(y_C^2 \ln y_C^2 + y_D^2 \ln y_D^2 + y_E^2 \ln y_E^2 + \ldots\right)\right)\right\} \tag{8.14}$$

where $a_1$and a_2 are the occupancies of sublattices 1 and 2, respectively, which will be 1 and 3 for $(A,B)_1(C,D,E)_3$, respectively. y_A^1 and $y_B^1$are the site fractions of elements A and B in sublattice 1. Similarly, y_C^2, y_D^2, and y_E^2 are site fractions of elements C, D, and E in sublattice 2. Site fractions are essentially mole fractions within a sublattice. Equation (8.14) is, in principle, the same as Equation (8.4) applied to each sublattice with the weightage of sublattice occupancies. The possible error in this method can be the overestimation of the configurational entropy since random mixing is assumed. More rigorous ways of treating configurational entropy are needed for more accurate calculations.

The calculated entropy values of the ordered phases using Equation (8.14) will be, in general, much smaller than 1.5*R*, to call "high entropy." However, it will be higher than the value of zero, which is the configurational entropy of a completely ordered binary phase. In other words, the Gibbs energy not just of an SS phase, but even an ordered phase can be reduced due to configurational entropy. When such an ordered phase is stabilized, it can be claimed to have "entropy stabilized." The treatment is extended to include any compound, e.g., entropy-stabilized oxides [22].

8.3.4 Thermal Entropy

Thermal entropy arises mostly from lattice vibrations, which are temperature-dependent. Electronic and magnetic contributions are also temperature-dependent, but are usually much lower than the vibrational part. Together vibrational, magnetic, and electronic contributions are called thermal entropy, although other contributions (Section 8.3.1) may be part of the calorimetric data.

Calorimetric measurements of thermal entropy were reported for HEAs [14, 23, 3, 24]. The contribution of thermal entropy to the Gibbs energy of a phase is about a few folds higher than the configurational entropy [14]. As shown in Figure 8.3, at 600°C, for Cantor alloy (CoCrFeMnNi), the total entropy contribution is about 64 $kJmol^{-1}$ in which the configurational entropy contribution is about 12 $kJmol^{-1}$. The rest (the thermal part) is about 52 $kJmol^{-1}$. The report also indicates that pure Ni and the Cantor alloy has virtually the same amount of thermal entropy as if the HEA had merely the averaged value of thermal entropy of its constituent elements. Another report claims that thermal entropy is not merely an average of the constituent elements, but has significantly deviated from it [3]. In other words, there is an excess

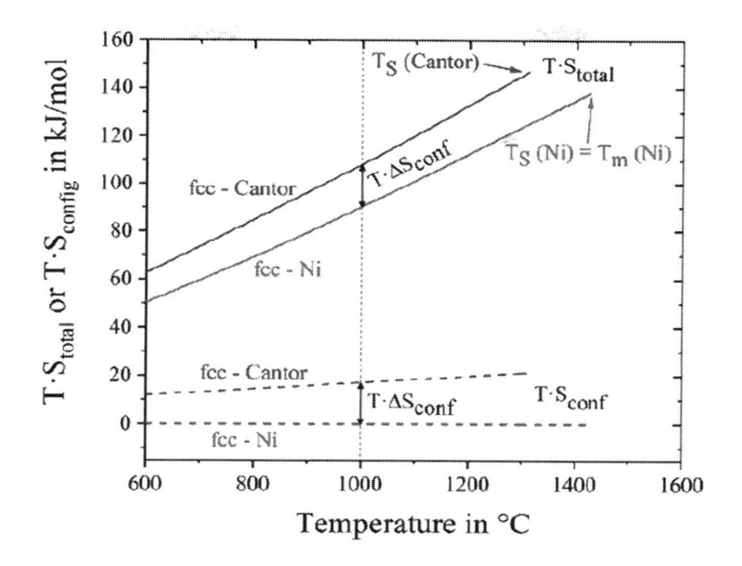

FIGURE 8.3 Contribution of total and configurational entropy of FCC-nickel and FCC-Cantor to Gibbs energy [14].

thermal entropy. The averaged heat capacity value is subtracted from the measured value (C_P) to obtain excess heat capacity as given in Equation (8.15).

$$C_P^{ex} = C_P - \sum_{i=1}^{n} x_i C_{P,i} \tag{8.15}$$

This excess heat capacity is then used to estimate the excess thermal entropy indicated as S^{ex} in Figure 8.4. Additionally, the report claims that the excess thermal entropy increases with the increase in the number of elements. In CoCrFeNi alloy, it is as high as 8 $kJmol^{-1}K^{-1}$ (Figure 8.4). However, the conclusions are not corroborated well by the other calorimetric reports on different HEAs [14, 23, 24]. It can be concluded that the observation of significant excess thermal entropy in CoCrFeNi by [3] is specific to the alloy studied, rather than a general thermodynamic trend in HEAs.

8.4 ENTHALPY

When the interaction between components is repulsive, then the enthalpy effects are considered positive. In this case, the number of bonds between like atoms increases, leading to clustering in single-phase SS. On the other hand, a negative enthalpy of mixing leads to short-range ordering in single-phase SS, due to attractive interaction between components. When the extent of positive interaction increases, unmixing happens, resulting in more than one phase. Alternatively, IM phases form with the increased negative interaction [25].

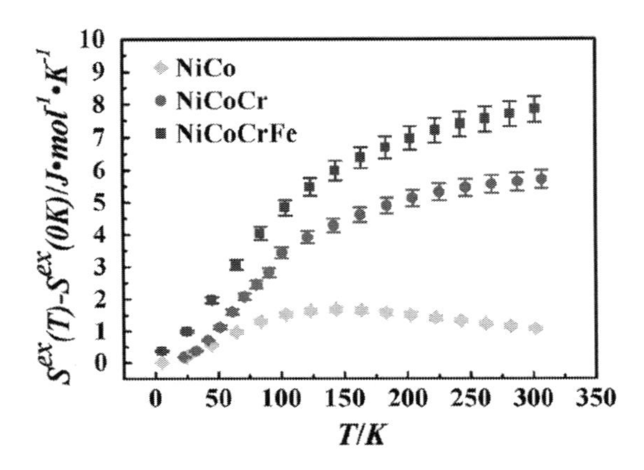

FIGURE 8.4 Excess thermal entropy calculated from excess heat capacity [44].

The enthalpy of mixing is calculated for HEAs in various ways. A simple method is based on Miedema's model (Miedema, de Châtel, and de Boer 1980). For HEAs, it is calculated as

$$\Delta H_{\text{mix}} = \sum_{i=1, i \neq j}^{n} \Omega_{ij} C_i C_j \tag{8.16}$$

where C_i is the atomic percentage of the ith component and $\Omega_{ij}\left(= 4\Delta H_{ij}^{\text{mix}}\right)$ is the regular melt interaction parameter between the ith and j^{th} element [26].

When certain constituent binary systems of an HEA have positive and the rest have negative mixing enthalpies, in the weighted-average method (Equation 8.16), "cancellation effect" occurs [27]. It is the case with the modulus of this value as used in the Ω parameter (Section 8.6.1) e.g., in $Al_{0.5}CrCuFeNi_2$ alloy, the $|\Delta H_{\text{mix}}|$ = 2.51 kJmol^{-1}. Such small value of $|\Delta H_{\text{mix}}|$ is expected to result in SS, but unmixing was experimentally observed [28]. The reason for such unmixing is that the positive enthalpy effects of the Cu-containing binaries Cu-Cr and Cu-Fe are dominant. During the averaging, the negative enthalpy effects in Al-Ni and Al-Fe binaries cancel the positive enthalpy effects to a misleadingly higher extent and result in small net value.

A different formulation is suggested in the form of $|\Delta H|_{\text{sum}}$ [27]. It is to avoid all the cancellation effects.

$$|\Delta H|_{\text{sum}} = \sum_{i=1, i<j}^{n} 4 C_i C_j \left|\Delta H_{ij}^{\text{mix}}\right| \tag{8.17}$$

Since the individual binary enthalpy effects are taken in modulus form, $|\Delta H|_{\text{sum}}$ is a measure of *accumulated* deviation from ΔH_{mix} value from the ideal case. The idea is that the higher the deviation, the lower the chance of achieving SS. Both IM phase formation and unmixing are considered to be the same effects in this treatment, destabilizing SS.

In reality, however, the expected behavior is that the cancellation effect would occur, but its extent can be significantly different from the mere averaged binary effects. Interactions can be more complex in a higher-order system. Besides, various entropy effects will also influence the phase formation. In the above discussed example of $Al_{0.5}CrCuFeNi_2$ alloy, the cancellation effect is presumably not high enough to suppress unmixing.

Therefore, the treatment using $|\Delta H|_{sum}$ can result in overestimation of the actual deviation from ideal mixing, since the negative enthalpy effects in this modulus form simply add up rather than causing a certain amount of cancellation effect.

The difference between the enthalpy of mixing (ΔH_{mix}) of an SS and the enthalpy of formation of an IM phase (ΔH_{IM}) was elucidated [2]. In the HEA literature, a large negative ΔH_{mix} value is treated to be a destabilizing factor for SS formation [26, 29, 30]. In fact, disordered SSs are expected to become more stable as their ΔH_{mix} values become more negative, according to Equation (8.2). However, this thermodynamic trend utilized in these studies is useful in the sense that systems with large, negative ΔH_{mix} values are also likely to have ΔH_{IM} values for IM phases, although the latter may be a slightly more negative. Therefore, ΔH_{mix} is a reasonable proxy for ΔH_{IM}.

Apart from Miedema's approach, *ab initio* calculations were also employed for the calculation of binary enthalpy values. Attempts to do these calculations with high-throughput possibilities were reported [31–33]. The experimentally reported values of enthalpy for such lower-order systems are available for several systems [34–36]. However, enthalpy for higher-order systems is barely available.

In the Calphad approach, enthalpy values from experimental sources as well as *ab initio* calculations and other simpler models were used as part of the input to generate Gibbs energy descriptions. The available Calphad Gibbs energy databases are thus a robust source for calculating enthalpy values using the relation

$$H = \frac{\partial(G/T)}{\partial(1/T)} \tag{8.18}$$

For HEAs, the enthalpy values of lower order are extrapolated to higher-order systems, as discussed in Section 8.5.2 for Gibbs energy.

8.5 GIBBS ENERGY

Gibbs energy captures the complex interplay of enthalpy and various entropy effects and ultimately decides the phase formation in HEAs. A systematic study on evaluating the high-entropy effect is based on the substitution of similar elements in the same concentration in a CoCrFeMnNi alloy [37]. It was observed that only CoCrFeMnNi alloy has single-phase FCC SS in the microstructure and none of the other alloys have a single phase. The conclusion is that the relative effects of ΔS_{mix} and ΔH_{mix} in the form of Gibbs energy dictate the phase formation in a HEA. Neither ΔS_{mix} nor ΔH_{mix} can decide it independently. Thus, attempts to maximize configurational entropy alone do not lead to the formation of single-phase SS, and one needs to consider the interactions between the elements, attractive or repulsive, which can lead to IM phases or unmixing.

The ideal disordered SS will have a ΔH_{mix} value of zero. If configurational entropy can counterbalance the effect of non-zero ΔH_{mix}, then the high entropy effect can be considered to have resulted in an SS. On the other hand, if the configurational entropy is unable to suppress compound formation or unmixing, then the high entropy effect can be considered to be not enough to stabilize the SS. Therefore, a few researchers suggest that the ability to form a SS should be the criteria for calling an alloy an HEA, rather than the value of ΔS_C or composition [5, 2].

The high entropy effect can be better analyzed by considering *G*-*x* plots of a hypothetical binary system A-B at a temperature as shown in Figure 8.5 [38]. The black solid curve represents the Gibbs energy of disordered SS, which has a regular solution behavior. Gibbs energy of two IM phases is represented with solid square and circle. The black solid curve is symmetric about the equimolar composition, which is the typical behavior of regular solution. Recall that the configurational entropy also has a symmetric shape (Equation 8.4). In this scenario, the tangent (solid gray line) is below both the IM phases, therefore, only SS is stable at equimolar ratio. In such cases, configurational entropy can be considered to have significantly assisted stabilization of SS.

Consider a different scenario where the SS has a sub-regular behavior. Its *G*-*x* plot is indicated using a dotted line in Figure 8.5. It is not symmetric about the equimolar composition. In this case, a common tangent construction (gray dotted line) indicates that at the equiatomic composition, the SS is in equilibrium with an IM phase. The maximization of configurational entropy at the equimolar ratio is not enough in this case to stabilize the SS. Only up to the composition of x_1 single-phase SS can be stable. Sub-regular solutions are by far the most common. Thus, the analysis indicates the possibility of vast number of non-equiatomic HEA compositions that can form SS. Thus, the enthalpy effects are also important in stabilizing such SS.

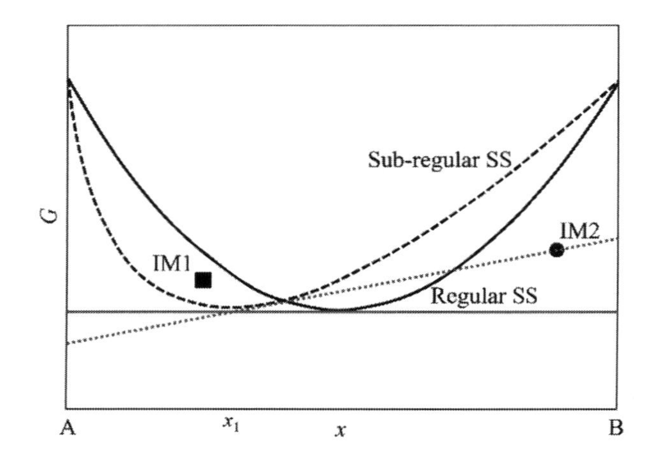

FIGURE 8.5 Gibbs energies of phases in a hypothetical A–B system (adopted from Zhang et al. [38]).

Moreover, the pure elements are considered in the reference state of the same crystal structure and at the same temperature. Therefore, the G-x curve is at the same G values at both terminals. In reality, the terminal points will shift, making the curve more skewed, which will again reduce the effect of maximum configurational entropy at the equimolar ratio.

8.5.1 Gibbs Energy in the Calphad Approach

Apart from standard handbooks for Gibbs energy data of the lower-order system, an extensive and robust source for Gibbs energy is available in the Calphad approach. Calphad is regarded as the direct method for designing HEAs [39]. The phase diagrams calculation from Gibbs energy models started more than a century ago [40]. Kaufman generalized it with the idea of lattice stability [41]. The approach grew over the last few decades and is now considered to have reached maturity [42]. The Calphad approach involves two major sets of activities: one is generation of Gibbs energy databases and the other is calculation of phase diagrams and thermodynamic properties using those databases.

Database generation involves various set of tasks. These are generation and critical evaluation of experimental and theoretical input (thermochemical and phase equilibria) data, selection of Gibbs energy models for phases, statistical fitting (optimization) of parameters in the Gibbs energy models to the input data, and finally, combining the assessed models (Gibbs energy descriptions) to make an internally consistent multicomponent database for a set of components.

Gibbs energy for a multicomponent phase:

$$G = \sum_i x_i G_i^{\mathrm{o}} + \Delta G_{\mathrm{mix}}^{\mathrm{Ex}} + \Delta G_{\mathrm{mix}}^{\mathrm{Ideal}} + \Delta G^{\mathrm{Phy}} \tag{8.19}$$

Configurational Gibbs energy ($\Delta G_{\mathrm{mix}}^{\mathrm{Ideal}}$) is the ideal mixing contribution, which is essentially the treatment of configurational entropy in the Calphad approach for any phase (Section 8.3.3). Gibbs energy due to physical phenomena such as magnetic ordering is ΔG^{Phy}. Selection of Gibbs energy model parameters is done for the excess Gibbs energy ($\Delta G_{\mathrm{mix}}^{\mathrm{Ex}}$), which is the mixing contribution from chemical interaction. It has temperature and composition dependence. The Calphad approach uses Redlich–Kister polynomials (L^{ν}) for temperature dependence terms.

$$L^{\mathrm{v}} = a_{\mathrm{v}} + b_{\mathrm{v}} T \tag{8.19}$$

where a_{ν} and b_{ν} are model parameters. The expression may become more complex depending on the system. The composition dependence is calculated using the expression for A-B binary system,

$$\Delta G_{\mathrm{mix}}^{\mathrm{Ex}} = x_A x_B \sum_{\mathrm{v}=0}^{k} L_{\mathrm{v}} (x_{\mathrm{A}} - x_{\mathrm{B}})^{\mathrm{v}} \tag{8.20}$$

8.5.2 Gibbs Energy for Ternary and Higher-Order Systems

Usually, Gibbs energy descriptions are available only up to binary or ternary cases. For higher-order systems, geometrical extrapolation is adopted [43]. Muggianu extrapolation is more common. It is described for obtaining ternary descriptions from binary descriptions as:

$$\Delta G_{\text{mix}}^{\text{Ex}} = x_A x_B \left\{ L_{AB}^0 + L_{AB}^1 \left(x_A - x_B \right) \right\} + x_B x_C \left\{ L_{BC}^0 + L_{BC}^1 \left(x_B - x_C \right) \right\} + x_B x_C \left\{ L_{AC}^0 + L_{AC}^1 \left(x_A - x_C \right) \right\} \tag{8.21}$$

8.6 APPLICATION OF THERMODYNAMICS FOR HEA COMPOSITIONAL DESIGN

The number of possible HEA compositions are huge. The choice of potential compositions for detailed experimental investigation is a vital step in effective alloy discovery. Phase diagrams are the roadmaps for such a process, which are scarcely available for the higher dimensional compositional space. The experimental efforts required to make it are immense [44]. Therefore, in the HEA literature, two major types of methods are used for compositional design. One type is simple methods, such as empirical rules, phase diagram inspection, and Gibbs energy composition (G-x) plots method. The other type is the robust Calphad method. Calculations using quantum physics are quite useful in getting insights into the HEAs' behavior, but these are not used on a routine basis for HEA compositional design.

8.6.1 Empirical Rules Based on Thermodynamic Parameters

Empirical rules give broadly indicative guidelines for predicting certain stable phases in the HEA microstructure. Since the original motivation of the HEA design idea is to achieve random SS, most of the empirical rules are developed to delineate SS-forming compositions from others. These rules rely on certain thermodynamic parameters and/or Hume–Rothery rules of SS formation. Experience from metallic glass research has also contributed significantly to developing a few of these rules.

Configurational entropy is proposed as a parameter. Though a broad guideline for HEAs is to have $\Delta S_C > 1.5R$, Guo and Liu [30] suggested that $11 \leq \Delta S_C \leq 19.5$ kJmol^{-1}K^{-1} favors SS formation.

According to Hume–Rothery rules, extended SSs are favored in alloys depending on the closeness of atomic size, crystal structure, electronegativity, and valence electron concentration of constituent components. The simple empirical rules developed for phase prediction in HEA literature rely on a *composition-weighted average* of these properties [45, 26, 29, 30]. Ranges for these values are suggested to delineate SS-forming compositions. Similarly thermodynamic properties were formulated.

The suggested range for enthalpy of mixing (Equation 8.16) for SS formation is $-15 \leq \Delta H_{\text{mix}} \leq 5$ kJmol^{-1} [26]. The range suggested in later reports were $-5 \leq \Delta H_{\text{mix}} \leq 5$ kJmol^{-1} [45] and $-22 \leq \Delta H_{\text{mix}} \leq 5$ kJmol^{-1} [30].

A parameter that compares the effect of enthalpy and entropy effects together is Ω. It is defined as

$$\Omega = \frac{T_m \Delta S_{\text{mix}}}{|\Delta H_{\text{mix}}|} \tag{8.23}$$

where T_m is the melting temperature, which is again averaged based on composition. ΔS_{mix} is assumed to be equal to $\Delta S_C^{\text{ideal}}$. The value of $\Omega > 1.1$ is suggested to favor SS [29].

These parameters can be calculated easily and are very useful to get a broad indication of the microstructures. Several exceptions to these rules, however, were reported [5, 2], indicating the need for better phase prediction tools.

8.6.2 Phase Diagram Inspection

Another simple method is inspecting phase diagrams of binary and ternary subsystems [38, 46]. The method by Zhang et al. [38] identifies matching elements for SS formation, e.g., to get a single-phase FCC in a HEA, the elements must be chosen in such a way that their respective binary phase diagrams should have a broad FCC phase region.

8.6.3 *G-x* Plot Methods

When applied for alloy design, a phase diagram inspection method like the matching element method can miss a beneficial component, e.g., Cr in the case of CoCrFeMnNi. Phase diagrams of Cr-Mn and Cr-Fe have only narrow FCC fields, whereas Cr-Co, Cr-Ni, Cr-Fe, and Cr-Mn systems have large BCC regions [47]. Therefore, according to the matching elements method, Cr would promote BCC phase rather than FCC phase. However, CoCrFeMnNi is a well-studied HEA with single-phase FCC microstructure [48–50]. Zhang et al. attribute the occurrence of FCC phase in this system to the mutual solubility of elements in the quinary FCC phase, since mere binary phase diagram inspection was not adequate to explain the behavior.

This is mainly because the phase diagram inspection method does not provide certain crucial information. For example, the stable phases can be seen from a binary phase diagram, but it is not straightforward to deduce how "far" a metastable phase is in terms of stability competition. Such pieces of information become important in higher-order systems. The *G-x* plots method is proposed to deduce such information [51]. Binary *G-x* plots of the subsystems are analyzed at a temperature. Parameters like the Gibbs energy of the single-phase (G_{sys}), the driving force (D), and the extent of unmixing tendency at equimolar composition (ΔG_{50}), can be obtained from these plots as shown in Figure 8.6.

The *G-x* plots method for designing an HEA with a single target phase can be summarized as follows. The lowest Gibbs energy envelope in most of the constituent binary systems consisting of the target phase as the single most stable one (Figure 8.6a) is advantageous. In such cases, a lower G_{sys} is better. If, in a constituent

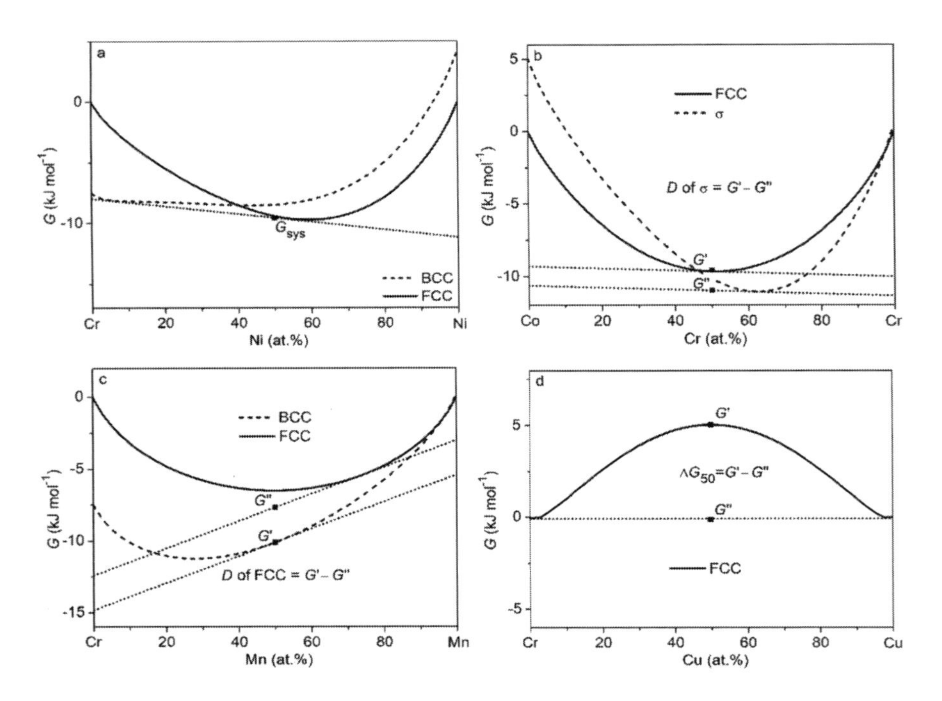

FIGURE 8.6 Quantities obtained from G-x plots corresponding to equimolar composition (a) G_{sys}, (b) D of the undesired phase, when equilibrium is between FCC and undesired phase, (c) D for FCC, when FCC is not an equilibrium phase, and (d) ΔG_{50} for FCC phase with unmixing tendency [51].

binary system, an undesired phase appears in equilibrium with the target phase (Figure 8.6b), it is better to have its D value with lower magnitude for the undesired phase. In a case where the target phase is metastable in a binary system (Figure 8.6c), it is better to have a smaller magnitude of its D. If there is a change in sign of curvature of G-x plot (Figure 8.6d), having a lower magnitude for unmixing tendency is advantageous.

The parameters G_{sys} and D are useful to deduce the phase competition behavior. D indicates how easy or difficult it is to stabilize a metastable phase or to destabilize an equilibrium phase. It is, in a way, a measure of the distance between the G-x curves of competing phases. G_{sys} indicates how deeply stable an equilibrium phase is. Consider two systems, where D is the same for a phase α to appear from an existing phase β. Though D is the same, if G_{sys} for β is larger in one system than the other, it makes the appearance of α more difficult in this system than in the other. Therefore, it is important to consider G_{sys} along with the D value. ΔG_{50} is to deduce the extent of immiscibility, which is not obtainable easily from D or G_{sys}. Thus, these three parameters deduced from G-x plots are complementary to each other in revealing phase formation characteristics of a system. Calculation of these parameters does not necessitate Calphad software and requires only binary Gibbs energy functions available in open literature.

8.6.4 Application of Calphad Method to HEAs

The field of HEAs has brought immense scope for Calphad studies. Commercial Calphad databases such as TCHEA [52] and PANHEA [53] were developed focusing on HEAs. The latest versions of databases include Gibbs energy descriptions of 26 and 13 elements, respectively. Besides, individual research groups have reported Calphad studies using in-house databases [54, 55].

In the case of HEAs, the phase diagram calculation involves newer challenges. The representation of the hyperdimensional composition space in a two-dimensional form severely limits the perception. However, isoplethal sections and phase fraction plots are widely used for HEAs. Isoplethal sections are similar to a binary phase diagram. These diagrams, however, involve restrictions on the composition. One example is that in a five-component system amount of three components is fixed and two components vary. Another example is that as the content of one element varies other elements vary in the equimolar ratio. Both these types were reported for HEAs [56, 54, 38, 49]. When it comes to fixed compositions, the widely considered representation is a phase fraction plot. In phase fraction plot, the amount of equilibrium phases is plotted against temperature [57, 58–60].

A notable application of the Calphad method is the rapid filtering of HEA compositions for targeted microstructures [61]. These are high-throughput calculations. The useful compositions are filtered based on application-driven criteria e.g., maximum use temperature (T_{use}) was used as a criterion for structural applications. The lowest of two values, $0.8T_m$ (T_m: melting temperature) and the temperature of the last first-order phase transformation below T_m, is taken as T_{use}. One of the conclusions from the study is that the maximum use temperature decreases considerably with an increase in the number of alloying elements. Another conclusion is that as the number of elements increases, the greater are the chances for multiphase microstructures, which is expected from the Gibbs phase rule. Since the last first-order transformation temperature below T_m is critical, accurate calculation of this point is important for reliable screening.

8.6.5 Evaluation of the Success of the Calphad Approach

From attempts like high-throughput screening, it is clear that the Calphad method has the capability of extensive HEA compositional exploration. Most of the available commercial multicomponent Gibbs energy databases are usually developed for specific compositional ranges. Evaluating the accuracies associated with such databases beyond the specified ranges is necessary for efficient HEA discovery.

In experimental validation, discrepancies can result from problems in either calculations or experiments. In calculations, the issues are mostly to do with missing or inaccurate Gibbs energy descriptions in a database. Credibility criteria were proposed [61] to quantify the reliability of a database for a specific alloy system. It is the number of included binary and ternary thermodynamic descriptions, called the fraction of fully assessed binaries (F_{AB}) and the fraction of fully assessed ternaries (F_{AT}), respectively. Higher values of these quantities imply that the database is more reliable.

Apart from these criteria, more rigorous reliability evaluations pertaining to specific scenarios were reported. In one report, ternary phase equilibria in the concentrated composition regime are focused [62], where ternary phase equilibria were calculated through extrapolation from mere binary data. Here, the "distance" of the point of interest in the compositional space from assessed binary subsystems is a measure of reliability. The conclusion suggests a guideline to apply such databases for compositions of distance within 25 at.% of an assessed binary. Another scenario is the usage of databases designed for a principal-element (PE) alloy system in predicting ternary phase equilibria. These calculations are accurate near the PE-rich region, but not in the composition domain containing the PE in minor quantity. The guideline is to use such databases when PE is more than 25 at.% in the composition. For higher-order systems, such guidelines are not yet explored. The study reveals certain insights. When different databases with $F_{AT} = 1$ are used for ternary calculations, there are variations in the results. The reasons are (i) inaccurate descriptions, and/or (ii) errors due to selective optimization of the models for a PE. Besides, significant ternary systems (14% in 72 systems) have ternary phases that are not part of binary subsystems. The influence of data for such a new ternary phase on higher-order systems was not explored in the report.

The effect of missing ternary data on the accuracy of the calculations for quaternary systems was reported later [63]. In this study, comparison of calculations using extrapolated data from mere binary descriptions and those resulting from complete binary and ternary descriptions was done. It was concluded that the quaternary alloy phase diagrams can be correctly predicted by mere binary data extrapolation only when (i) the binary miscibility gaps are not present, (ii) binary IM phases are not present or present in a few quantities, and (iii) ternary IM phases are not present. Both these studies [63, 62] signify the importance of having complete ternary descriptions of good quality.

Experimental difficulties can result in misleading discrepancies between calculations and experiments. These are errors due to the extent of equilibration. The reported HEA microstructures are mostly in as-synthesized conditions [2], which can be far from the calculated equilibrium state. In the HEA literature, the number of phases significantly changes from the as-cast to the heat-treated condition. After heat treatment, the number of alloys with only one phase or two phases decreases and the number of alloys with three or more phases increases from as-cast condition (Figure 8.7). Therefore, comparing equilibrium calculation results against the as-cast condition is often incorrect.

In order to have confidence in whether or not equilibrium is achieved, diffusivity data are required, on which there are only a few reports [64–70]. Comparing the calculations to microstructures after a certain amount of heat treatment is a reasonably better way. Though there is no guarantee that the equilibrium is achieved, at least a significant departure from the as-synthesized state toward equilibrium state is expected. There are not many long-term heat treatments. Such phase stability studies are thus necessary for better evaluation of calculations and the elevated temperature applicability. In order to achieve a near-equilibrium state, 100 h is suggested as a practical starting point; however, no standard thermal treatment is available presently to ensure attainment of equilibrium [2].

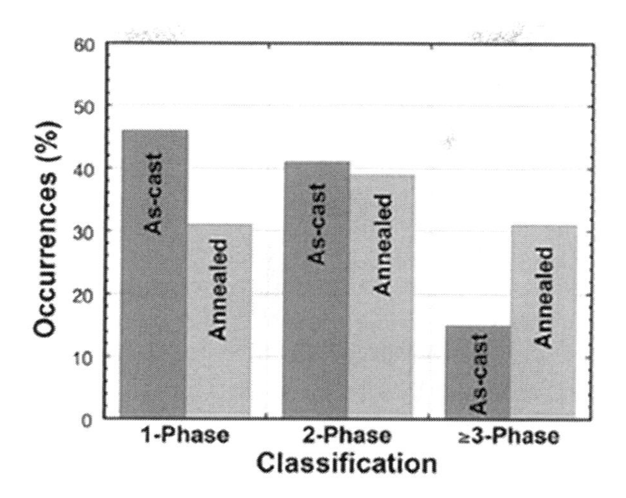

FIGURE 8.7 Comparison of as-cast and annealed HEAs by the number of phases, indicating a larger number of phases in a substantial number of cases in annealed condition [51].

Comparing Calphad and experimental results with regard to certain phases will be useful in understanding the ability of Gibbs energy databases in predicting those phases. In HEA literature, FCC and BCC structured SS phases are the most observed, followed by B2 and Topologically Close-Packed (TCP) phases [2]. Single-phase FCC HEAs were well-predicted using TCHEA1 and TCFE7 databases [48, 49]. The BCC phase was predicted correctly using Pandat databases, TTTI3, TCNI8, and in-house databases, especially in refractory HEAs [59, 71, 72, 54, 73].

The B2 phase is reported in HEAs that involve Al with Ni, Co, and Fe. One of the first Calphad studies for HEAs was an isoplethal section for the Al-Cr-Fe-Mn-Ni system [56]. In the Al-Co-Cr-Fe-Ni system, the fact that a minimum level of Al is required for B2 to appear is well-captured by Calphad studies [54]. In certain compositions of this system, only BCC and FCC phases were reported by a few experimentally [74], however, it was reported to be B2 and FCC phases by others, validating the calculations [75]. In AlCoFeMnNi alloy, B2 phase is predicted, however, experimentally BCC was reported. The compositions of the predicted B2 and the experimentally observed BCC are close. Therefore, the B2 phase had possibly undergone an order–disorder transformation [76]. The study revealed the scope for a better Gibbs energy data to calculate such order–disorder transformation in HEA.

A subset of IM phases is TCP phases [77]. σ-phase, Laves phases, and μ phase are the most reported TCP phases in HEA literature so far. These phases often appear after heat treatment. Evaluation of the success of the Calphad approach against 123 post-heat treatment microstructures of 52 HEAs, focusing on TCP phases, was reported [78]. The conclusion was that the calculations match in 64% of the cases with the experiments. The discrepancies are due to calculations as well as experiments. These are inaccurate/unavailable Gibbs energy databases and insufficient equilibration treatments, respectively.

8.7 AVERAGING AND MAXIMIZING OF THERMODYNAMIC PROPERTIES

It is worth emphasizing that Equations (8.16) and (8.21) for enthalpy and excess Gibbs energy of mixing are basically attempts of *averaging*. The treatment of these properties in HEA literature is such that when the number of elements is increased, properties only get averaged. However, the configurational entropy alone increases (Equation 8.4). One of the four core effects, the cocktail effect, is also contrary to this averaging since it conveys that HEAs may have synergetic effects. However, this maximizing of configurational entropy and averaging of other thermodynamic factors strengthen the design philosophy of HEAs.

8.8 SUMMARY

Thermodynamic trends in HEAs are attractive for fundamental exploration and are important for effective applications. The high configurational entropy and the other thermodynamic quantities that contribute to Gibbs energy eventually control the phase formation in HEAs. The term configurational entropy used in metallic glasses literature is essentially structural configurational entropy, dealing with the structural configurations resulting in energy variations. The one used as the basis for HEA design is chemical configurational entropy. The structural configurational entropy may be significant in HEAs due to the uncertainties in the atomic positions of different sized atoms. It is dependent on temperature, whereas chemical configurational entropy is not temperature dependent. The recent studies on the trends in thermal entropy and its excess mixing quantities in HEAs show that this excess contribution can be significant in certain systems. The application of thermodynamic concepts for compositional design is effective. Simple parametric guidelines based on entropy or enthalpy or both are in use to get an indicative idea about the microstructure. Phase diagram inspection and Gibbs energy-composition (G-x) plots methods give much better insights for compositional design. The robust Calphad method is far more effective since it can directly give the phase diagrams. Its accuracy depends on the availability of reliable multicomponent Gibbs energy databases. The databases available presently require improvement, especially with respect to certain IM phases. Developing such thermodynamic databases using experiments, *ab initio* calculations, and suitable modeling is the key to effective alloy design. Besides, long-term heat treatment and diffusivity studies will help us understand the equilibrium microstructure better.

REFERENCES

1. Yeh, J. W., S. K. Chen, S. J. Lin, J. Y. Gan, T. S. Chin, T. T. Shun, C. H. Tsau, and S. Y. Chang. 2004. "Nanostructured High-Entropy Alloys with Multiple Principal Elements: Novel Alloy Design Concepts and Outcomes." *Advanced Engineering Materials* 6(5): 299–303. doi:10.1002/adem.200300567.
2. Miracle, D. B., and O. N. Senkov. 2017. "A Critical Review of High Entropy Alloys and Related Concepts." *Acta Materialia* 122(January). Elsevier Ltd: 448–511. doi:10.1016/j.actamat.2016.08.081.

3. Wang, Jianbin, Junjie Li, Qing Wang, Jincheng Wang, Zhijun Wang, and C. T. Liu. 2019. "The Incredible Excess Entropy in High Entropy Alloys." *Scripta Materialia* 168(July). Elsevier Ltd: 19–22. doi:10.1016/j.scriptamat.2019.04.013.
4. Murty, B. S., J. W. Yeh, S. Ranganathan, and P. P. Bhattacharjee. 2019. *High-Entropy Alloys*, 2nd ed. London: Butterworth-Heinemann. doi:10.1016/B978-0-12-816067-1.09990-2.
5. Singh, Anil Kumar, Nitesh Kumar, Akanksha Dwivedi, and Anandh Subramaniam. 2014b. "A Geometrical Parameter for the Formation of Disordered Solid Solutions in Multi-Component Alloys." *Intermetallics* 53(February). Elsevier B.V.: 112–9. doi:10.1016/j.intermet.2014.04.019.
6. Swalin, Richard A. 1962. *Thermodynamics of Solids*.
7. Pickering, E. J., and N. G. Jones. 2016. "High-Entropy Alloys - A Critical Assessment of Their Founding Principles and Future Prospects." *International Materials Reviews* 61(3): 183–202. doi:10.1080/09506608.2016.1180020.
8. Suryanarayana, C., and A. Inoue. 2011. *Bulk Metallic Glasses*. Boca Raton, FL: CRC Press.
9. Debenedetti, Pablo G., and Frank H. Stillinger. 2001. "Supercooled Liquids and the Glass Transition." *Nature* 410(6825): 259–67. doi:10.1038/35065704.
10. Stillinger, Frank H. 1999. "Exponential Multiplicity of Inherent Structures." *Physical Review. Part E - Statistical Physics, Plasmas, Fluids, and Related Interdisciplinary Topics* 59(1): 48–51. doi:10.1103/PhysRevE.59.48.
11. Stillinger, Frank H., and Pablo G. Debenedetti. 1999. "Distinguishing Vibrational and Structural Equilibration Contributions to Thermal Expansion." *The Journal of Physical Chemistry. Part B* 103(20): 4052–9. doi:10.1021/jp983831o.
12. Kauzmann, Walter. 1948. "The Nature of the Glassy State and the Behavior of Liquids at Low Temperatures." *Chemical Reviews* 43(2): 219–56. doi:10.1021/cr60135a002.
13. Smith, Hillary L., Chen W. Li, Andrew Hoff, Glenn R. Garrett, Dennis S. Kim, Fred C. Yang, Matthew S. Lucas, et al. 2017. "Separating the Configurational and Vibrational Entropy Contributions in Metallic Glasses." *Nature Physics* 13(9): 900–5. doi:10.1038/nphys4142.
14. Haas, Sebastian, Mike Mosbachar, Olag N. Senkov, Michael Feuerbacher, Jens Freudenberger, Senol Gezgin, Rainer Völkl, and Uwe Glatzel. 2018. "Entropy Determination of Single-Phase High Entropy Alloys with Different Crystal Structures over a Wide Temperature Range." *Entropy* 20(9): 1–12. doi:10.3390/e20090654.
15. Takeuchi, Akira, Kenji Amiya, Takeshi Wada, Kunio Yubuta, Wei Zhang, and Akihiro Makino. 2013. "Entropies in Alloy Design for High-Entropy and Bulk Glassy Alloys." *Entropy* 15(9): 3810–21. doi:10.3390/e15093810.
16. Takeuchi, Akira, and Akihisa Inoue. 2000. "Calculations of Mixing Enthalpy and Mismatch Entropy for Ternary Amorphous Alloys." *Materials Transactions, JIM* 41(11): 1372–8. doi:10.2320/matertrans1989.41.1372.
17. Mansoori, G. A., N. F. Carnahan, K. E. Starling, and T. W. Leland. 1971. "Equilibrium Thermodynamic Properties of the Mixture of Hard Spheres." *The Journal of Chemical Physics* 54(4): 1523–5. doi:10.1063/1.1675048.
18. Ye, Y. F., Q. Wang, J. Lu, C. T. Liu, and Y. Yang. 2015a. "The Generalized Thermodynamic Rule for Phase Selection in Multicomponent Alloys." *Intermetallics* 59: 75–80. doi:10.1016/j.intermet.2014.12.011.
19. Ye, Y. F., Q. Wang, J. Lu, C. T. Liu, and Y. Yang. 2015b. "Design of High Entropy Alloys: A Single-Parameter Thermodynamic Rule." *Scripta Materialia* 104(July): 53–5. doi:10.1016/j.scriptamat.2015.03.023.
20. Guggenheim, E. A. 1952. *Mixtures*. London: Clarendon Press.
21. Tamm, Artur, Alvo Aabloo, Mattias Klintenberg, Malcolm Stocks, and Alfredo Caro. 2015. "Atomic-Scale Properties of Ni-Based FCC Ternary, and Quaternary Alloys." *Acta Materialia* 99. Acta Materialia Inc.: 307–12. doi:10.1016/j.actamat.2015.08.015.

22. Rost, Christina M., Edward Sachet, Trent Borman, Ali Moballegh, Elizabeth C. Dickey, Dong Hou, Jacob L. Jones, Stefano Curtarolo, and Jon Paul Maria. 2015. "Entropy-Stabilized Oxides." *Nature Communications* 6. Nature Publishing Group: 1–8. doi:10.1038/ncomms9485.
23. Jin, K., S. Mu, K. An, W. D. Porter, G. D. Samolyuk, G. M. Stocks, and H. Bei. 2017. "Thermophysical Properties of Ni-Containing Single-Phase Concentrated Solid Solution Alloys." *Materials and Design* 117(March): 185–92. doi:10.1016/j.matdes.2016.12.079.
24. Khavala, Vedasri Bai. 2019. "Undergraduate Thesis—Studies on Measuring Nonconfigurational Entropy of High-Entropy Alloys." Indian Institute of Technology Madras.
25. Porter, D. A., and K. E. Easterling. 1992. *Phase Transformations In Metals and Alloys*, 2nd ed. London: Chapman & Hall.
26. Zhang, Yong, Yun Jun Zhou, Jun Pin Lin, Guo Liang Chen, and Peter K. Liaw. 2008. "Solid-Solution Phase Formation Rules for Multi-Component Alloys." *Advanced Engineering Materials* 10(6): 534–8. doi:10.1002/adem.200700240.
27. Zheng, Mingjie, Wenyi Ding, Weitao Cao, Shenyang Hu, and Qunying Huang. 2019. "A Quick Screening Approach for Design of Multi-Principal Element Alloy with Solid Solution Phase." *Materials and Design* 179. The Authors: 107882. doi:10.1016/j.matdes.2019.107882.
28. Guo, Sheng, Chun Ng, and C. T. Liu. 2013. "Anomalous Solidification Microstructures in Co-Free Al XCrCuFeNi2 High-Entropy Alloys." *Journal of Alloys and Compounds* 557(April): 77–81. doi:10.1016/j.jallcom.2013.01.007.
29. Yang, X., and Y. Zhang. 2012. "Prediction of High-Entropy Stabilized Solid-Solution in Multi-Component Alloys." *Materials Chemistry and Physics* 132(2–3). Elsevier B.V.: 233–8. doi:10.1016/j.matchemphys.2011.11.021.
30. Guo, Sheng, and C. T. Liu. 2011. "Phase Stability in High Entropy Alloys: Formation of Solid-Solution Phase or Amorphous Phase." *Progress in Natural Science: Materials International* 21(6). Chinese Materials Research Society: 433–46. doi:10.1016/S1002-0071(12)60080-X.
31. Curtarolo, Stefano, Dane Morgan, Kristin Persson, John Rodgers, and Gerbrand Ceder. 2003. "Predicting Crystal Structures with Data Mining of Quantum Calculations." *Physical Review Letters* 91(13): 1–4. doi:10.1103/PhysRevLett.91.135503.
32. Hart, Gus L. W., Stefano Curtarolo, Thaddeus B. Massalski, and Ohad Levy. 2014. "Comprehensive Search for New Phases and Compounds in Binary Alloy Systems Based on Platinum-Group Metals, Using a Computational First-Principles Approach." *Physical Review X* 3(4): 1–33. doi:10.1103/PhysRevX.3.041035.
33. Troparevsky, M. Claudia, James R. Morris, Paul R. C. Kent, Andrew R. Lupini, and G. Malcolm Stocks. 2015. "Criteria for Predicting the Formation of Single-Phase High-Entropy Alloys." *Physical Review X* 5(1): 011041. doi:10.1103/PhysRevX.5.011041.
34. Hultgren, R., R. L. Orr, P. D. Anderson, and K. K. Kelley. 1963. *Selected Values of Thermodynamic Properties of Metals and Alloys*. New York: John Wiley & Sons.
35. Kubaschewski, Oswald, C. B. Alcock, and P. J. Spencer. 1993. *Materials Thermochemistry. Revised*. Oxford: Pergamon Press.
36. Brandes, E. A., and G. B. Brook. 2013. *Smithells Metals Reference Book*. Elsevier.
37. Otto, F., Y. Yang, H. Bei, and E. P. George. 2013. "Relative Effects of Enthalpy and Entropy on the Phase Stability of Equiatomic High-Entropy Alloys." *Acta Materialia* 61(7): 2628–38. doi:10.1016/j.actamat.2013.01.042.
38. Zhang, C. Zhang, S. L. Chen, J. Zhu, W. S. Cao, and U. R. Kattner. 2014. "An Understanding of High Entropy Alloys from Phase Diagram Calculations." *Calphad: Computer Coupling of Phase Diagrams and Thermochemistry* 45. Elsevier: 1–10. doi:10.1016/j.calphad.2013.10.006.

39. Gao, Michael C., Jien Wei Yeh, Peter K. Liaw, and Yong Zhang. 2016. *High-Entropy Alloys: Fundamentals and Applications*. Edited by Michael C. Gao, Jien-Wei Yeh, Peter K. Liaw, and Yong Zhang. Cham: Springer International Publishing. doi:10.1007/978-3-319-27013-5.
40. van Laar, J. J. . 1908. "Melting or Solidification Curves in Binary System." *Zeitschrift für Physikalische Chemie* 63: 216.
41. Kaufman, Larry. 1959. "The Lattice Stability of Titanium and Zirconium." *Acta Metallurgica* 7(August): 575–87.
42. Lukas, H. L., S. G. Fries, and Bo Sundman. 2007. *Computational Thermodynamics*. Cambridge: Cambridge University Press.
43. Saunders, N., and A. P. Miodownik. 1998. *CALPHAD (Calculation of Phase Diagrams): A Comprehensive Guide*. Pergamon. http://www.sciencedirect.com/science/bookseries/14701804/1.
44. Ranganathan, S. 2003. "Alloyed Pleasures: Multimetallic Cocktails." *Current Science* 85(10): 1404–6.
45. Guo, Sheng, Chun Ng, Jian Lu, and C. T. Liu. 2011a. "Effect of Valence Electron Concentration on Stability of FCC or BCC Phase in High Entropy Alloys." *Journal of Applied Physics* 109(10): 103505. doi:10.1063/1.3587228.
46. Gao, Michael, and David Alman. 2013. "Searching for Next Single-Phase High-Entropy Alloy Compositions." *Entropy* 15(10): 4504–19. doi:10.3390/e15104504.
47. Massalski, T. B., and H. Okamoto. 1990. *Binary Alloy Phase Diagrams*, 2nd ed. Materials Park, OH: ASM International.
48. Ma, Duancheng, Mengji Yao, K. G. Pradeep, Cemal C. Tasan, Hauke Springer, and Dierk Raabe. 2015. "Phase Stability of Non-Equiatomic CoCrFeMnNi High Entropy Alloys." *Acta Materialia* 98. Acta Materialia Inc.: 288–96. doi:10.1016/j.actamat.2015.07.030.
49. Bracq, Guillaume, Mathilde Laurent-Brocq, Loïc Perrière, Rémy Pirès, Jean Marc Joubert, and Ivan Guillot. 2017. "The FCC Solid Solution Stability in the Co-Cr-Fe-Mn-Ni Multi-Component System." *Acta Materialia* 128: 327–36. doi:10.1016/j.actamat.2017.02.017.
50. Vaidya, M., K. Guruvidyathri, and B. S. Murty. 2019. "Phase Formation and Thermal Stability of CoCrFeNi and CoCrFeMnNi Equiatomic High Entropy Alloys." *Journal of Alloys and Compounds* 774. Elsevier B.V: 856–64. doi:10.1016/j.jallcom.2018.09.342.
51. Guruvidyathri, K., B. S. Murty, J. W. Yeh, and K. C. Hari Kumar. 2018. "Gibbs Energy-Composition Plots as a Tool for High-Entropy Alloy Design." *Journal of Alloys and Compounds* 768. Elsevier B.V: 358–67. doi:10.1016/j.jallcom.2018.07.264.
52. Chen, Hai Lin, Huahai Mao, and Qing Chen. 2018. "Database Development and Calphad Calculations for High Entropy Alloys: Challenges, Strategies, and Tips." *Materials Chemistry and Physics* 210(May). Elsevier B.V: 279–90. doi:10.1016/j.matchemphys.2017.07.082.
53. Pandat. 2019. "CompuTherm LLC, Madison, WI 53719, USA." http://www.computherm.com.
54. Zhang, Chuan, Fan Zhang, Shuanglin Chen, and Weisheng Cao. 2012. "Computational Thermodynamics Aided High-Entropy Alloy Design." *Jom* 64(7): 839–45. doi:10.1007/s11837-012-0365-6.
55. Tapia, Antonio João Seco Ferreira, Dami Yim, Hyoung Seop Kim, and Byeong Joo Lee. 2018. "An Approach for Screening Single Phase High-Entropy Alloys Using an In-House Thermodynamic Database." *Intermetallics* 101(April). Elsevier: 56–63. doi:10.1016/j.intermet.2018.07.009.
56. Hsieh, Ker Chang, Cheng Fu Yu, Wen Tai Hsieh, Wei Ren Chiang, Jin Son Ku, Jiun Hui Lai, Chin Pan Tu, and Chih Chao Yang. 2009. "The Microstructure and Phase Equilibrium of New High Performance High-Entropy Alloys." *Journal of Alloys and Compounds* 483(1–2): 209–12. doi:10.1016/j.jallcom.2008.08.118.

57. Huang, Can, Yongzhong Zhang, Jianyun Shen, and Rui Vilar. 2011. "Thermal Stability and Oxidation Resistance of Laser Clad TiVCrAlSi High Entropy Alloy Coatings on Ti-6Al-4V Alloy." *Surface and Coatings Technology* 206(6). Elsevier B.V.: 1389–95. doi:10.1016/j.surfcoat.2011.08.063.
58. Ng, Chun, Sheng Guo, Junhua Luan, Sanqiang Shi, and C. T. Liu. 2012. "Entropy-Driven Phase Stability and Slow Diffusion Kinetics in an Al0.5CoCrCuFeNi High Entropy Alloy." *Intermetallics* 31(December). Elsevier Ltd: 165–72. doi:10.1016/j.intermet.2012.07.001.
59. Senkov, O. N., S. V. Senkova, C. Woodward, and D. B. Miracle. 2013. "Low-Density, Refractory Multi-Principal Element Alloys of the Cr-Nb-Ti-V-Zr System: Microstructure and Phase Analysis." *Acta Materialia* 61(5): 1545–57. doi:10.1016/j.actamat.2012.11.032.
60. Manzoni, A., H. Daoud, S. Mondal, S. Van Smaalen, R. Völkl, U. Glatzel, and N. Wanderka. 2013. "Investigation of Phases in Al23Co15Cr23Cu8Fe15Ni16 and Al8Co17Cr17Cu8Fe17Ni33 High Entropy Alloys and Comparison with Equilibrium phases predicted by Thermo-Calc." *Journal of Alloys and Compounds* 552. Elsevier B.V.: 430–36. doi:10.1016/j.jallcom.2012.11.074.
61. Senkov, O. N., J. D. Miller, D. B. Miracle, and C. Woodward. 2015. "Accelerated Exploration of Multi-Principal Element Alloys with Solid Solution Phases." *Nature Communications* 6(1): 6529. doi:10.1038/ncomms7529.
62. Wertz, Katelun N., Jonathan D. Miller, and Oleg N. Senkov. 2018. "Toward Multi-Principal Component Alloy Discovery: Assessment of CALPHAD Thermodynamic Databases for Prediction of Novel Ternary Alloy Systems." *Journal of Materials Research* 33(19): 3204–17. doi:10.1557/jmr.2018.61.
63. Gorsse, Stéphane, and Oleg Senkov. 2018. "About the Reliability of CALPHAD Predictions in Multicomponent Systems." *Entropy* 20(12): 1–9. doi:10.3390/e20120899.
64. Vaidya, M., K. G. Pradeep, B. S. Murty, G. Wilde, and S. V. Divinski. 2017. "Radioactive Isotopes Reveal a Non Sluggish Kinetics of Grain Boundary Diffusion in High Entropy Alloys." *Scientific Reports* 7(1). Springer US: 1–11. doi:10.1038/s41598-017-12551-9.
65. Vaidya, M., S. Trubel, B. S. Murty, G. Wilde, and S. V. Divinski. 2016. "Ni Tracer Diffusion in CoCrFeNi and CoCrFeMnNi High Entropy Alloys." *Journal of Alloys and Compounds* 688(December). Elsevier Ltd: 994–1001. doi:10.1016/j.jallcom.2016.07.239.
66. Gaertner, Daniel, Josua Kottke, Gerhard Wilde, Sergiy V. Divinski, and Yury Chumlyakov. 2018. "Tracer Diffusion in Single Crystalline CoCrFeNi and CoCrFeMnNi High Entropy Alloys." *Journal of Materials Research* 33(19): 3184–91. doi:10.1557/jmr.2018.162.
67. Vaidya, M., K. G. Pradeep, B. S. Murty, G. Wilde, and S. V. Divinski. 2018. "Bulk Tracer Diffusion in CoCrFeNi and CoCrFeMnNi High Entropy Alloys." *Acta Materialia* 146. Acta Materialia Inc.: 211–24. http://linkinghub.elsevier.com/retrieve/pii/S1359645418300089.
68. Nadutov, V. M., and V. F. Mazanko. 2017. "Tracer Diffusion of Cobalt in High-Entropy Alloys AlxFeNiCoCuCr." 348(3): 337–48. doi:10.15407/mfint.39.03.0337.
69. Verma, Vivek, Aparna Tripathi, and Kaustubh N. Kulkarni. 2017. "On Interdiffusion in FeNiCoCrMn High Entropy Alloy." *Journal of Phase Equilibria and Diffusion* 38(4). Springer US: 445–56. doi:10.1007/s11669-017-0579-y.
70. Kulkarni, Kaustubh, and Gyanendra Pratap Singh Chauhan. 2015. "Investigations of Quaternary Interdiffusion in a Constituent System of High Entropy Alloys." *AIP Advances* 5(9): 097162. doi:10.1063/1.4931806.
71. Senkov, O. N., S. V. Senkova, and C. Woodward. 2014. "Effect of Aluminum on the Microstructure and Properties of Two Refractory High-Entropy Alloys." *Acta Materialia* 68. Acta Materialia Inc.: 214–28. doi:10.1016/j.actamat.2014.01.029.

72. Stepanov, N. D., N. Yu. Yurchenko, E. S. Panina, M. A. Tikhonovsky, and S. V. Zherebtsov. 2017. "Precipitation-Strengthened Refractory Al0.5CrNbTi2V0.5 High Entropy Alloy." *Materials Letters* 188(October 2016). Elsevier: 162–4. doi:10.1016/j.matlet.2016.11.030.
73. Yao, H. W., J. W. Qiao, J. A. Hawk, H. F. Zhou, M. W. Chen, and M. C. Gao. 2017. "Mechanical Properties of Refractory High-Entropy Alloys: Experiments and Modeling." *Journal of Alloys and Compounds* 696. Elsevier B.V: 1139–50. doi:10.1016/j.jallcom.2016.11.188.
74. Kao, Yih Farn, Ting Jie Chen, Swe Kai Chen, and Jien Wei Yeh. 2009. "Microstructure and Mechanical Property of As-Cast, -Homogenized, and -Deformed AlxCoCrFeNi ($0 \leq x \leq 2$) High-Entropy Alloys." *Journal of Alloys and Compounds* 488(1): 57–64. doi:10.1016/j.jallcom.2009.08.090.
75. Chou, Hsuan Ping, Yee Shyi Chang, Swe Kai Chen, and Jien Wei Yeh. 2009. "Microstructure, Thermophysical and Electrical Properties in AlxCoCrFeNi ($0 \leq x \leq 2$) High-Entropy Alloys." *Materials Science and Engineering B: Solid-State Materials for Advanced Technology* 163(3): 184–9. doi:10.1016/j.mseb.2009.05.024.
76. Karati, Anirudha, K. Guruvidyathri, V. S. Hariharan, and B. S. Murty. 2019. "Thermal Stability of AlCoFeMnNi High-Entropy Alloy." *Scripta Materialia* 162. Elsevier Ltd: 465–7. doi:10.1016/j.scriptamat.2018.12.017.
77. Westbrook, J. H., and R. L. Fleischer. 2000. *Intermetallic Compounds-Crystal Structures of Intermetallic Compounds*. West Sussex: Wiley.
78. Guruvidyathri, K., K. C. Hari Kumar, J. W. Yeh, and B. S. Murty. 2017. "Topologically Close-Packed Phase Formation in High Entropy Alloys: A Review of Calphad and Experimental Results." *Jom* 69(11). Springer US: 2113–24. doi:10.1007/s11837-017-2566-5.

9 Electrodeposition of High Entropy Alloy Coatings

Microstructure and Corrosion Properties

Ahmed Aliyu, M.Y. Rekha, and Chandan Srivastava

CONTENTS

9.1 INTRODUCTION

The traditional alloy design approach involves mixing of a principal element with other minor additives in order to produce complex microstructures containing solid solutions and intermediate phases/compounds. A radically different approach in alloy design was developed by Cantor and co-workers [1] and Yeh and co-workers [2] which involved mixing of five or more elements in nearly equiatomic proportions to form alloys with a mixture of phases with simple cubic structures. Yeh and co-workers [2] proposed that the absence of intermetallic phases in high entropy alloys is due to increase in the system's configurational entropy because of the formation of solid solution phases. Over the years, it has been reported that high entropy alloys exhibit enhanced properties such as high corrosion resistance [3–9], high fatigue resistance and fracture toughness [10–14], unique electrical and magnetic responses [7, 10, 11, 15], resistance to heat softening [16–19], etc. These attributes are highly desirable in structural and functional materials.

Recently, research interests have been aroused in the application of high entropy alloy thin films and coatings for corrosion and oxidation resistance applications. Techniques that have been employed for producing thin films or coatings of high entropy alloys are magnetron sputtering [16–22], plasma-transferred arc cladding [23–26], laser cladding [27, 28], spraying [29, 30], and electrodeposition [3, 31, 32]. Among these techniques, electrodeposition of high entropy alloys on metallic substrates has been reported relatively rarely, primarily due to the complexity of the electrolyte bath with multiple metallic precursors and precise control of the deposition parameters for co-depositing multiple components with widely different reduction potentials. Use of the electrodeposition technique for producing high entropy alloy coatings, nonetheless, should be encouraged due to the fact that electrodeposition is a non-equipment-intensive technique which can be performed under ambient conditions, provides opportunities for easy tuning of the deposition parameters to perform microstructural engineering, and is suitable for large-scale coating.

This chapter describes the work done by the authors on the microstructural evolution and corrosion properties of CrFeCoNiCu, AlCrFeCoNiCu, and MnCrFeCoNi high entropy alloy systems electrodeposited on mild steel substrate.

9.2 ELECTRODEPOSITION OF HIGH ENTROPY ALLOY COATINGS: BATH CHEMISTRY AND DEPOSITION PARAMETERS

In the application of the electrodeposition technique to produce high entropy alloy coatings, electrolyte preparation is very crucial, and requires dissolving the desired metal precursors and other suitable constituents in the "right proportion" in the aqueous medium. The "right proportion" does not necessarily mean in equal amounts, but it refers to one that produces a uniform equiatomic coating. At the start, the substrate (cathode) is immersed in the electrolyte contained in a bath (cell) along with the counter electrode (anode) and connected to a direct current (DC) power source. The working electrode is then connected to the negative terminal, while the counter electrode is connected to the positive terminal of the power source, as illustrated in Figure 9.1. The substrate can be prepared by polishing with emery papers of different grit sizes to obtain a mirror finish, sonicating in ethanol for degreasing, and lastly,

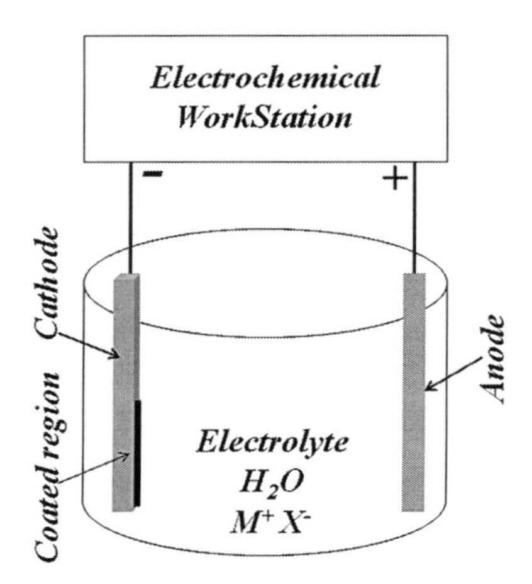

FIGURE 9.1 Illustration of electrodeposition technique.

treating with 10% v/v solution of hydrochloric acid to activate the surface before electroplating.

Electrodeposition of most common metallic alloys, over the years, are from sulphate [33–37], chloride [38, 39], fluorborate [38, 40], sulphamate [41, 42], or non-aqueous baths [43–46]. From the economic perspective, sulphate and chloride baths are the best choices for the electrodeposition process. Recently, electrodeposition of high entropy alloy coatings from chloride bath containing different additives have been reported by the authors [3, 31, 32]. Electrolyte preparation for the electrodeposition of the CuFeNiCoCr high entropy alloy coating system [31] involved mixing of hydrated chloride salts of copper, iron, nickel, cobalt, and chromium in water, followed by the addition of gelatine ($C_6H_{12}O_6$), L-ascobic acid ($C_6H_8O_6$), ammonium chloride (NH_4Cl), sulfanilic acid ($C_6H_7NO_3S$), sodium dodecyl sulphate ($C_{12}H_{25}O_4SNa$), boric acid (H_3BO_3), potassium chloride (KCl), and formic acid (HCO_2H) under vigorous stirring over a long period of time (~18–24 h) till the pH of the solution becomes stable. The chloride ion ion in the electrolyte from the hydrated chloride salts increases the conductivity of the solution and causes dissolution of the metals anode. Gelatine in the electrolyte functions as a bonding agent and also controls the deposition rate. The cohesion between the CuFeNiCoCr elements is achieved by the addition of sodium dodecyl sulphate, while the presence of L-ascobic acid in the electrolyte bath hinders the formation of some of the metallic hydroxide films during the coating process. During the electrodeposition process, the hydroxide layer forms in cathodic sites of the coatings, and the presence of ascorbic acid in the electrolyte then promotes discharge of metal from the hydroxide films in the coatings. Sulfanilic acid improves the brightness and uniformity of the high entropy alloy coatings. Boric acid in the electrolyte enables the application of a large range of current densities, enhances the coating appearance, acts as a catalyst for the reduction reaction of metallic ions, and prevents pH increment on the cathode surface during electrodeposition, as also reported by many other researchers for the electrodeposition of metallic alloys [47–51]. Formic acid is used for modification of the surface morphology.

For the electrodeposition of CrFeCoNiCu coatings as conducted by the authors [31] using a chloride bath containing chromium (III) ion (Cr^{3+}), iron (II) ion (Fe^{2+}), cobalt (II) ion (Co^{2+}), nickel (II) ion (Ni^{2+}), and copper (II) ion (Cu^{2+}), it was assumed that the reduction reaction of metals ions occurred in a two-step process as follows:

$$M(II) + e^{-} = M(I)_{ads} \tag{9.1}$$

$$M(I)_{ads} + e^{-} = M_{atom} \tag{9.2}$$

where M represent the desired metals (M = iron (Fe), cobalt (Co), nickel (Ni) or copper (Cu)). In the first step, reduction of iron (II) ion (Fe^{2+}), cobalt (II) ion (Co^{2+}), nickel (II) ion (Ni^{2+}), and copper (II) ion (Cu^{2+}), respectively, to iron ion (Fe^{+}), cobalt ion (Co^{+}), nickel ion (Ni^{+}), and copper ion (Cu^{+}) occurs, and these intermediate species then get adsorbed on the electrode surface. This is followed by the reduction of the intermediate ions iron ion (Fe^{+}), cobalt ion (Co^{+}), nickel ion (Ni^{+}), and copper ion (Cu^{+}) respectively to iron (Fe), cobalt (Co), nickel (Ni), and copper (Cu) atoms. For chromium (III) ion (Cr^{3+}), the above scheme involves conversion of chromium (III) ion (Cr^{3+}) to chromium (II) ion (Cr^{2+}) intermediate and then chromium (II) ion (Cr^{2+}) intermediate to chromium (Cr) atom. It has been observed by authors [3, 31, 32] that altering the concentration of the metal salts in electrolytes not only affects the coating composition, but also affects the microstructure (elemental distribution), morphology, and thickness of the high entropy alloy coatings. The possible mechanism of CrFeCoNiCu high entropy alloy electrodeposition can be visualized by taking into consideration the half cell reactions for the cathodic deposition of the individual elements:

$$Cr^{3+} + 3e^- \rightarrow Cr_{(s)},\ E^O = -0.74 \text{ V vs Standard Hydrogen Electrode} \quad (9.3)$$

$$Fe^{2+} + 2e^- \rightarrow Fe_{(s)},\ E^O = -0.44 \text{ V vs Standard Hydrogen Electrode} \quad (9.4)$$

$$Co^{2+} + 2e^- \rightarrow CO_{(s)},\ E^O = -0.28 \text{ V vs Standard Hydrogen Electrode} \quad (9.5)$$

$$Ni^{2+} + 2e^- \rightarrow Ni_{(s)},\ E^O = -0.25 \text{ V vs Standard Hydrogen Electrode} \quad (9.6)$$

$$Cu^{2+} + 2e^- \rightarrow Cu_{(s)},\ E^O = +0.337 \text{ V vs Standard Hydrogen Electrode} \quad (9.7)$$

From the half-reactions above, it is obvious that the deposition of copper with the highest reduction potential will be thermodynamically favored over that of nickel, cobalt, iron, and chromium in the respective sequence. As the reduction potentials of the individual elements vary (as shown in the half cell reactions above), and given that there is an attempt to deposit multicomponent coating with equiatomic composition, the relative amounts of the metal precursor salts in the electrolyte should be taken in such a ratio that the metal which has the lowest reduction potential is present in the largest amount. The relative amounts of the metallic precursor salts should qualitatively be guided by the relative difference in the reduction potential values.

The choice of deposition current density value should also be made such that a uniform coating microstructure and composition throughout the coating thickness is achieved. Current density value can affect the chemical composition and microstructure, and therefore the coating properties. The effect of current density can be explained by considering the half-reactions for the cathodic deposition of the individual elements in the CuFeNiCoCr high entropy alloy system illustrated above. The deposition of copper with highest reduction potential is expected to be activation controlled, while the deposition of chromium with the lowest reduction potential is considered to be controlled by the diffusion process. Therefore, an increase in the current density can probably lead to an increase in the cathodic overpotential thereby increasing the activation of reactions on the electrode surface, which in turn can cause an increase in the copper content and a decrease in the chromium content of the CuFeNiCoCr high entropy alloy deposit.

The electrolyte bath with pH values in the range of 1.5 to 2 were mostly used by the authors for the electrodeposition of high entropy alloy systems [3, 31, 32]. Formation of the metal hydroxides ions is an intermediate reaction which is highly pH dependent. Researchers have shown that a decrease in pH value reduces both current efficiency and content of elements with low reduction potential during coating [52–55]. This is because of the fact that the proton content at low pH values is very high and a high fraction of current is consumed through the reduction process of these protons, which in turn decreases the current efficiency. Therefore, in the electrodeposition of CuFeNiCoCr high entropy alloy, an increase of pH value on the cathodic surface favors or leads to increase in nickel, cobalt, iron, and chromium content in the respective sequence in the coating. Considering the complex nature of the high entropy alloy electrolyte, sufficient stirring of metal salts and the additive is very important because it enables the chemical reagents to becomes intimate and react with each other to increase the concentration of metals in the electrolyte, since it quickly compensates for the loss of the metal ions through discharge at the cathode surface. Sufficient stirring of the electrolyte during the electrodeposition process can reducing gas bubbles during coatings which can lead to pit formation. It is important to note that the stirring speed needs to be controlled properly because it can cause the formation of coarse-grained deposits due to the mechanical inclusion of sludge in the coating. The depositions' conditions and electrolyte bath constitution used by the authors for the electrodeposition of CrFeCoNiCu, AlCrFeCoNiCu, and MnCrFeCoNi high entropy alloy coatings on mild steel substrate are provided in Tables 9.1, 9.2, and 9.3, respectively.

TABLE 9.1
Deposition Conditions and Electrolyte Bath Constitution for the Electrodeposition of CrFeCoNiCu High Entropy Alloy (HEA) Coatings [31]

Bath composition	Concentration (gL^{-1})	Condition
$FeCl_2.4H_2O$	7.95	Current density (160 mA)
$NiCl_2.6H_2O$	16.64	Temperature (30°C)
$CoCl_2.6H_2O$	9.52	Deposition time (15 mins)
$CrCl_3.4H_2O$	37.30	pH 1.5
$CuCl_2.2H_2O$	6.82	Agitation speed (850 rpm)
Additives	30.64	

TABLE 9.2
Deposition Conditions and Electrolyte Bath Constitution for the Electrodeposition of AlCrFeCoNiCu High Entropy Alloy (HEA) Coatings [32]

Bath composition	Concentration (gL^{-1})	Condition
$AlCl_3.6H_2O$	24.14	Current density (200 mA)
$CrCl_3.4H_2O$	37.30	Temperature (30°C)
$FeCl_2.4H_2O$	7.95	Deposition time (15 mins)
$CoCl_2.6H_2O$	9.52	pH 1.5
$NiCl_2.6H_2O$	16.64	Agitation speed (850 rpm)
$CuCl_2.2H_2O$	6.82	
Additives	30.64	

TABLE 9.3
Deposition Conditions and Electrolyte Bath Constitution for the Electrodeposition of MnCrFeCoNi High Entropy Alloy (HEA) Coatings [3]

Bath composition	Concentration (gL^{-1})	Condition
$MnCl_2.4H_2O$	29.69	Current density (400 mA)
$CrCl_3.4H_2O$	33.30	Temperature (30°C)
$FeCl_2.4H_2O$	7.95	Deposition time (15 mins)
$CoCl_2.6H_2O$	9.517	pH 1.5
$NiCl_2.6H_2O$	19.02	Agitation speed (850 rpm)
Additives	30.64	

9.3 PHASE, MORPHOLOGY, AND MICROSTRUCTURE OF ELECTRODEPOSITED HIGH ENTROPY ALLOY COATINGS

The most studied high entropy alloy systems by researchers are the quinary CrFeCoNiCu and the senary AlCrFeCoNiCu systems [31, 32]. Authors have produced CrFeCoNiCu high entropy alloy (HEA) coatings on a mild steel substrate by the electrodeposition method [31]. This coating contained a mixture of face-centered cubic (FCC) and body-centered cubic (BCC) phases as revealed by the X-ray diffraction (XRD) pattern in Figure 9.2. In a separate system of electrodeposited AlCrFeCoNiCu high entropy alloy coating, it was observed that with the addition of aluminium in CrFeCoNiCu high entropy alloy, the volume fraction of the body-centered cubic phase increased. This enhancement in the BCC phase volume fraction can be attributed to aluminum-induced lattice distortion which destabilizes the close-packed face-centered cubic structure in the electrodeposited CrFeCoNiCu high entropy alloy. Also, the presence of the face-centered cubic phase was attributed to copper segregation, while chromium segregation led to the formation of the body-centered phase in both the electrodeposited CrFeCoNiCu and AlCrFeCoNiCu high entropy alloy systems. These observations by the authors [31] are similar to the ones shown in the case of high entropy alloys (of similar composition) produced by arc melting or mechanical alloying techniques as reported by Wang and co-workers [56], Guo and co-workers [57], and Singh and Subramaniam [58]. In another study by the authors [3], which replaced the copper in the electrodeposited CrFeCoNiCu system by manganese, formation of a single body-centered cubic phase was observed as shown in the X-ray diffraction (XRD) profile in Figure 9.4. This observation is in contrast to the results separately reported by Ye and co-workers [59], Otto and co-workers [60], and many others. They show formation of a single-phase face-centered cubic structure in MnCrFeCoNi high entropy alloys processed by magnetron sputtering. This discrepancy illustrates the influence of the synthesis conditions on the evolution of phases in high entropy alloy systems. The scanning electron microscopy (SEM) micrograph of the as-coated CrFeCoNiCu, AlCrFeCoNiCu, MCrFeCoNi high entropy alloys coatings shown in Figure 9.3 reveal that all the coatings possess

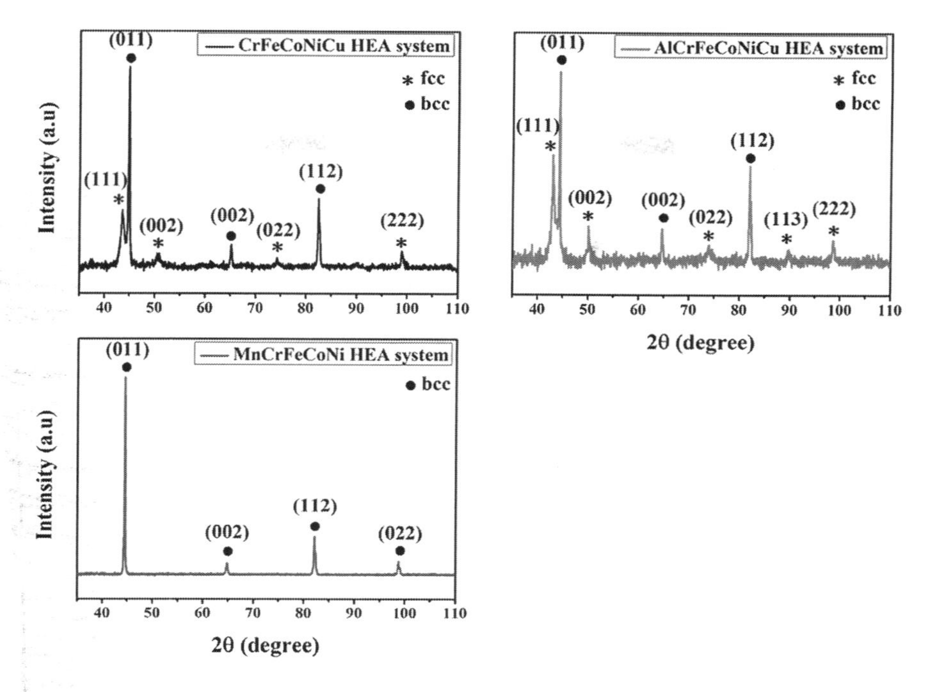

FIGURE 9.2 XRD pattern of as-coated high entropy alloy coatings of different systems.

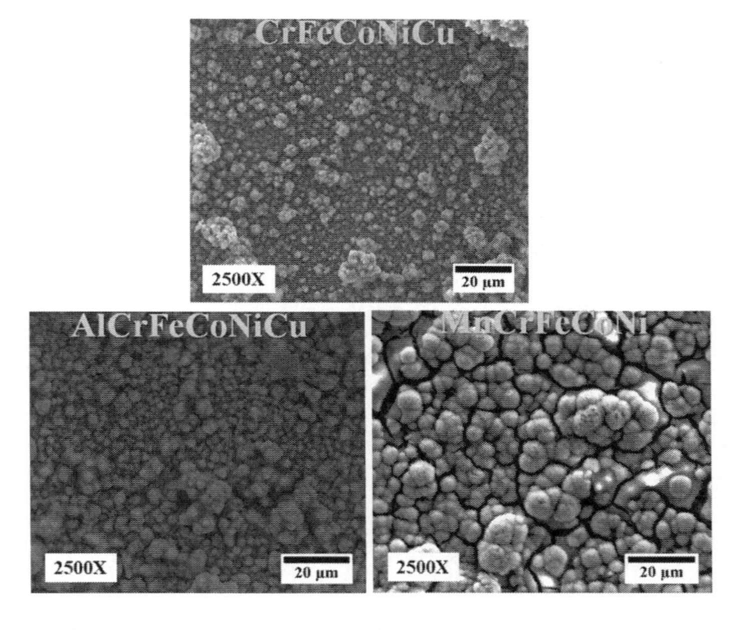

FIGURE 9.3 SEM micrograph of as-coated high entropy alloys (HEA) of different systems. Images from [3], [31], and [32].

crack-free granular morphology with distinctly different compactness and grain sizes. The average composition values of the electrodeposited high entropy alloy coatings obtained using the SEM-energy dispersive spectroscopy (SEM-EDS) method is provided in Table 9.4.

The authors have studied the microstructure of electrodeposited high entropy alloy coatings in detail. The microstructural examination was conducted on coating cross-section samples prepared using the scanning electron microscope-focused ion beam (SEM-FIB) instrument. The transmission electron microscopy (TEM) bright-field image of the CrFeCoNiCu high entropy alloy coating cross-section sample showed two distinct phases. Figure 9.4a: the chromium-rich relatively brighter contrast region and the copper-rich relatively darker contrast region. Figure 9.5a shows a high-angle annular dark-field (HAADF) image along with the scanning transmission electron microscopy-energy dispersive spectroscopy (STEM-EDS) compositional maps for as-coated CrFeCoNiCu coating. Copper, cobalt, iron, and nickel appear to be fairly uniformly distributed, while chromium is present in extremely low amount with these elements. On the other hand, the chromium-rich part tends to have all other four elements in trace amounts. It should also be noted that grain boundaries in the copper-rich regions are enriched with chromium. In contrast to the observations in the CrFeCoNiCu high entropy alloy coating, with the addition of aluminum, the microstructure of AlCrFeCoNiCu coating (cross-section in Figure 9.4b) showed copper-rich dendrites in an aluminium-rich matrix. Furthermore, the HAADF image along with the STEM-EDS compositional maps provided in Figure 9.5b confirmed that iron, cobalt, and nickel are populated more in the copper-rich dendrite phase whereas the aluminium-rich phase contained more chromium. Therefore, it can be deduced from the STEM-EDS compositional mapping that complete/partial solubility of the Co-Ni, Cu-Ni, and significant solubility of Fe in the Ni alleviate the face-centered cubic phase in these high entropy alloy systems, while chromium segregation tends to be responsible for the body-centered cubic phase. It is important to also observe that the presence of aluminium in the coating results in a dendritic

TABLE 9.4
SEM-EDS Average Atomic Percent Compositional Data of High Entropy Alloy (HEA) Coating Systems

Elements/System	CrFeCoNiCu HEA system (at.%)	AlCrFeCoNiCu HEA system (at.%)	MnCrFeCoNi HEA system (at.%)
Al	–	12.85 ± 3.67	–
Mn	–	–	25.92 ± 4.27
Cr	13.65 ± 1.40	17.81 ± 2.60	27.49 ± 2.55
Fe	20.19 ± 1.95	27.09 ± 2.79	21.00 ± 1.20
Co	18.60 ± 1.45	13.33 ± 1.70	12.06 ± 1.74
Ni	16.29 ± 3.18	13.80 ± 3.57	13.53 ± 0.98
Cu	31.27 ± 4.74	15.13 ± 3.14	–

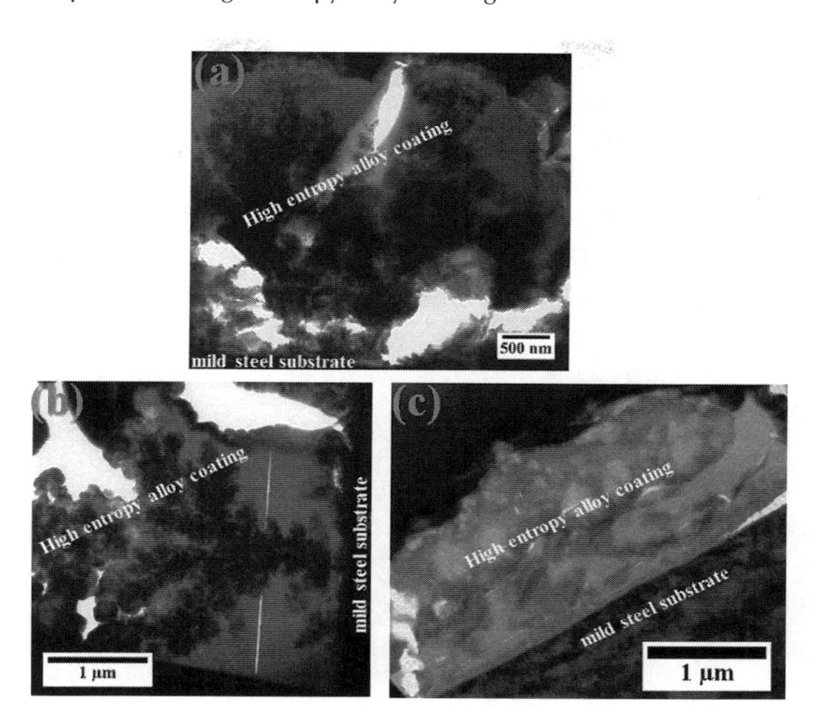

FIGURE 9.4 Bright-field TEM images of as-coated equiatomic (a) CrFeCoNiCu (b) AlCrFeCoNiCu and (c) MnCrFeCoNi high entropy alloy system. Images from [3], [31], and [32].

microstructure. In the case of an electrodeposited MnCrFeCoNi high entropy alloy system, a more homogenized microstructure was observed, as shown in the TEM bright-field image (Figure 9.4c). Also, the STEM-HAADF compositional map in Figure 9.5c shows that all the five elements (manganese, chromium, iron, cobalt, and nickel) are fairly uniformly distributed in the coating microstructure.

9.4 CORROSION BEHAVIOR OF HIGH ENTROPY ALLOY COATINGS

The overall elemental composition and microstructure of high entropy alloy systems tend to affect its corrosion properties. Over the years, corrosion properties of materials have been widely measured using the polarization curve. Shift of the potentiodynamic polarization curves toward lower corrosion current densities signifies decrease in the corrosion rate [61–67]. Authors have reported on the corrosion resistance of electrodeposited high entropy alloy coatings. Figure 9.6 shows the corrosion potential (E_{corr}) of the high entropy alloy coatings studied by the authors. It can be deduced that the CrFeCoNiCu coating (−0.881 V vs Ag/AgCl) is less noble than that of the MnCrFeCoNi coating (−0.845 V vs Ag/AgCl), followed by AlCrFeCoNiCu coating (−0.686 V vs Ag/AgCl) and the corrosion current density (i_{corr}) of the AlCrFeCoNiCu (24.66 $\mu A/cm^2$) is lower than that of the MnCrFeCoNi

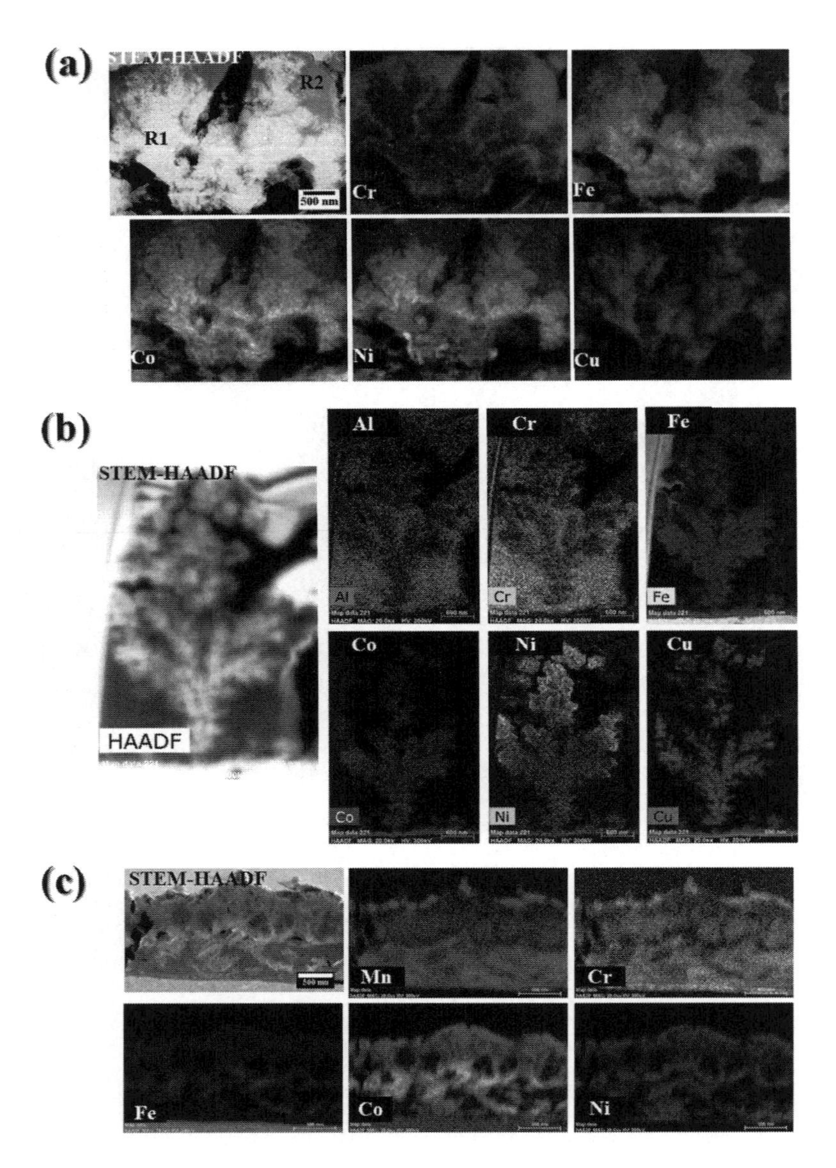

FIGURE 9.5 STEM-EDS compositional mapping of as-coated equiatomic (a) CrFeCoNiCu (b) AlCrFeCoNiCu, and (c) MnCrFeCoNi high entropy alloy system. Images from [3], [31], and [32].

(43.44 μA/cm^2) and CrFeCoNiCu coating (101.8 μA/cm^2) by an order of magnitude in a neutral 3.5 wt.% NaCl solution. It is important to note that all the polarization curves (Figure 9.6) show no passive or trans-passive region, which implies that only active dissolution of metals occurred in the applied potential range and that there was no pitting corrosion in the chloride ion environment for all the electrodeposited high entropy alloy presented.

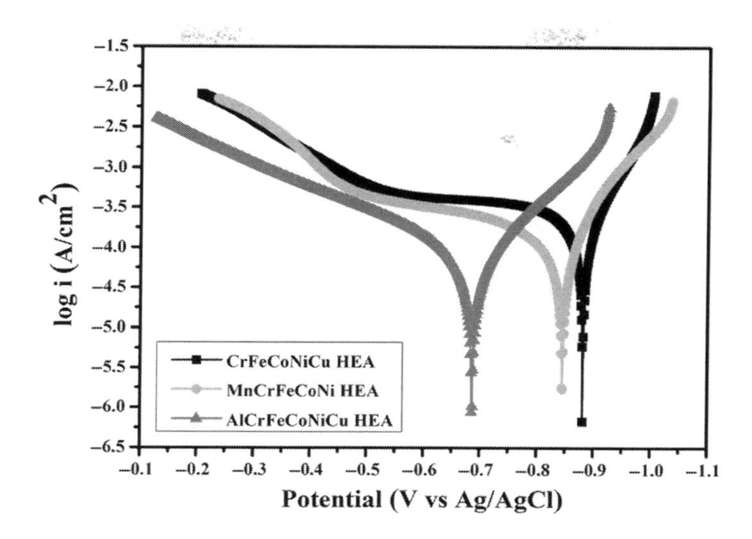

FIGURE 9.6 The potentiodynamic polarization curves of electrodeposited high entropy alloys (HEA) system.

9.5 SUMMARY AND CONCLUSIONS

High entropy alloy coatings have shown attractive and exclusive properties relative to the conventional metallic coatings. This chapter provides information on the electrodeposition of high entropy alloy coating (bath chemistry and deposition parameters), coating microstructure, and corrosion behavior. The contents of this chapter can be summarized as follows:

1. Electrolyte preparation is very crucial and requires dissolving the desired metal precursors in the right proportions guided primarily by their reduction potential.
2. Evolution of coating microstructure is very sensitive to the choice of the elements and their relative amounts in the electrolyte bath, deposition current density, electrolyte pH, deposition time, and stirring time/speed. It is emphasized here that coating microstructure should be examined in detail, as elemental partitioning and segregation in multicomponent systems are very much possible.
3. The CrFeCoNiCu and AlCrFeCoNiCu high entropy alloy coatings contained a mixture of face-centered cubic and body-centered cubic phases, with AlCrFeCoNiCu high entropy alloy coatings showing an increase in the volume fraction of the body-centered cubic phase primarily due to the presence of aluminium which causes lattice distortion, while the MnCrFeNiCo high entropy alloy coatings showed a single body-centered cubic phase structure.
4. The CrFeCoNiCu high entropy alloy microstructure consisted of a chromium-rich region with low amounts of iron, cobalt, and nickel, while the copper-rich region contained more of these elements. Addition of aluminum

in the AlCrFeCoNiCu high entropy alloy system tends to promote the formation of a dendritic copper-rich region in aluminium-rich matrix phases, while the MnCrFeNiCo high entropy alloy system showed a more homogenized microstructure compared to the other two systems.
5. Corrosion resistance of the coatings tend to increase in the order of CrFeCoNiCu → MnCrFeNiCo → AlCrFeCoNiCu, respectively, without any observable passive or trans-passive region on the anodic slope in a neutral 3.5 wt.% NaCl solution.

Finally, despite the complexity of the electrolyte baths and electrodeposition parameters which are needed for controlled co-deposition of several elements with varying deposition potentials, the authors believe that the electrodeposition of more high entropy alloy systems is possible and must be explored. Also, more efforts should be made to study the tribological and mechanical properties of the electrodeposited high entropy alloy coatings for their wide range of applications.

ACKNOWLEDGMENT

The authors acknowledge the research funding received from the Science and Engineering Research Board (SERB), Government of India.

REFERENCES

1. Cantor, B., Chang, I., Knight, P., Vincent, A. 2004. Microstructural development in equiatomic multicomponent alloys. *Mater. Sci. Eng. A* 375: 213–218.
2. Yeh, J.W., Chen, S.K., Lin, S.J., Gan, J.Y., Chin, T.S., Shun, T.T., Tsau, C.H., Chang, S.Y. 2004. Nanostructured high-entropy alloys with multiple principal elements: Novel alloy design concepts and outcomes. *Adv. Eng. Mater.* 6(5): 299–303.
3. Aliyu, A., Srivastava, C. 2019. Microstructure and corrosion properties of MnCrFeCoNi high entropy alloy-graphene oxide composite coatings. *Materilia* 5. http://www.ncbi.nlm.nih.gov/pubmed/100249.
4. Chen, Y.Y., Duval, T., Hung, U.D., Yeh, J.W., Shih, H.C. 2005. Microstructure and electrochemical properties of high entropy alloys – A comparison with type-304 stainless steel. *Corros. Sci.* 47(9): 2257–2279.
5. Chen, Y.Y., Hong, U.T., Shih, H.C., Yeh, J.W., Duval, T. 2005. Electrochemical kinetics of the high entropy alloys in aqueous environments - A comparison with type 304 stainless steel. *Corros. Sci.* 47(11): 2679–2699.
6. Hsu, Y.J., Chiang, W.C., Wu, J.K. 2005. Corrosion behavior of FeCoNiCrCux high-entropy alloys in 3.5% sodium chloride solution. *Mater. Chem. Phys.* 92(1): 112–117.
7. Kao, Y.F., Chen, S.K., Chen, T.J., Chu, P.C., Yeh, J.W., Lin, S.J. 2011. Electrical, magnetic, and hall properties of Al*x*CoCrFeNi high-entropy alloys. *J. Alloy Compd.* 509(5): 1607–1614.
8. Kozelj, P., Vrtnik, S., Jelen, A., Jazbec, S., Jaglicic, Z., Maiti, S., Feuerbacher, M., Steurer, W., Dolinsek, J. 2014. Discovery of a superconducting high-entropy alloy. *Phys. Rev. Lett.* 113(10): 107001.
9. Mishra, R.K., Sahay, P.P., Shahi, R.R. 2019. Alloying, magnetic and corrosion behavior of AlCrFeMnNiTi high entropy alloy. *J. Mater. Sci.* 54(5): 4433–4443.
10. Sekhar, R.A., Samal, S., Nayan, N., Bakshi, S.R. 2019. Microstructure and mechanical properties of Ti-Al-Ni-Co-Fe based high entropy alloys prepared by powder metallurgy route. *J. Alloys Compd.* 787: 123–132.

11. Zhang, Y., Zuo, T.T., Cheng,Y., Liaw, P.K. 2013. High-entropy alloys with high saturation magnetization, electrical resistivity, and malleability. *Sci. Rep.* 3: 1455.
12. Hemphill, M.A., Yuan, T., Wang, G.Y., Yeh, J.W., Tsai, C.W., Chuang, A., Liaw, P. 2012. Fatigue behavior of Al0.5CoCrCuFeNi high entropy alloys. *Acta Mater.* 60(16): 5723–5734.
13. Tang, Z., Yuan, T., Tsai, C.W., Yeh, J.W., Lundin, C.D., Liaw, P.K. 2015. Fatigue behavior of a wrought Al0.5CoCrCuFeNi two-phase high-entropy alloy. *Acta Mater.* 99: 247–258.
14. Gludovatz, B., Hohenwarter, A., Catoor, D., Chang, E.H., George, E.P., Ritchie, R.O. 2014. A fracture resistant high-entropy alloy for cryogenic applications. *Science* 345(6201): 1153–1158.
15. Karati, A., Nagini, M., Ghosh, S., Shabadi, R., Pradeep, K.G., Mallik, R.C., Murty, B.S., Varadaraju, U.V. 2019. Ti2NiCoSnSb - A new half-Heusler type high-entropy alloy showing simultaneous increase in Seebeck coefficient and electrical conductivity for thermoelectric applications. *Sci. Rep.* 9(1): 5331.
16. Eleti, R.R., Bhattacharjee, T., Shibata, A., Tsuji, N. 2019. Unique deformation behavior and microstructure evolution in high temperature processing of HfNbTaTiZr refractory high entropy alloy. *Acta Mater.* 171: 132–145.
17. Chen, S.T., Tang, W.Y., Kuo, Y.F., Chen, S.Y., Tsau, C.H., Shun, T.T., Yeh, J.W. 2010. Microstructure and properties of age-hardenable AlxCrFe1.5MnNi0.5 alloys. *Mat Sci Eng A-Struct.* 527(21–22): 5818–5825.
18. Liu, C.W., Wang, H.M., Zhang, S.Q., Tang, H.B., Zhang, A.L. 2014. Microstructure and oxidation behavior of new refractory high entropy alloys. *J. Alloy Compd.* 583: 162–169.
19. Kumar, D., Sharma, V.K., Prasad, Y.V.S.S., Kumar, V. 2019. Materials-structure-property correlation study of spark plasma sintered AlCuCrFeMnWx(x = 0, 0.05, 0.1, 0.5) high-entropy alloys. *J. Mater. Res.* 34(5): 767–776.
20. Wu, Z.F., Wang, X.D., Cao, Q.P., Zhao, G.H., Li, J.X., Zhang, D.X., Zhu, J.J., Jiang, J.Z. 2014. Microstructure characterization of AlxCo1Cr1Cu1Fe1Ni1 (x = 0 and 2.5) high entropy alloy films. *J. Alloys Compd.* 609: 137–142.
21. Dolique, V., Thomann, A.L., Brault, P., Tessier, Y., Gillon, P. 2009. Complex structure/composition relationship in thin films of AlCoCrCuFeNi high entropy alloy. *Mater. Chem. Phys.* 117(1): 142–147.
22. An, Z., Jia, H., Wu, Y., Rack, P.D., Patchen, A.D., Liu, Y., Ren, Y., Li, N., Liaw, P.K. 2015. Solid-solution CrCoCuFeNi high-entropy alloy thin films synthesized by sputter deposition. *Mater. Res. Lett.* 3(4): 203–209.
23. Liu, D., Cheng, J.B., Ling, H. 2015. Electrochemical behaviours of (NiCoFeCrCu)95 B5 high entropy alloy coatings. *Mater. Sci. Tech.* 31(10): 1159–1164.
24. Kumar, A., Dhekne, P., Swarnakar, A.K., Chopkar, M. 2019. Phase evolution of CoCrCuFeNiSix high-entropy alloys prepared by mechanical alloying and spark plasma sintering. *Mater. Res. Express* 6(2): 26532.
25. Vaidya, M., Anupam, A., Bharadwaj, J.V., Srivastava, C., Murty, B.S. 2019. Grain growth kinetics in CoCrFeNi and CoCrFeMnNi high entropy alloys processed by spark plasma sintering. *J. Alloys Compd.* 791: 1114–11121.
26. Sarkar, A., Wang, Q., Schiele, A., Chellali, M.R., Bhattacharya, S.S., Wang, D., Brezesinski, T., Hahn, H., Velasco, L., Breitung, B. 2019. High-entropy oxides: Fundamental aspects and electrochemical properties. *Adv. Mater.* http://www.ncbi.nlm.nih.gov/pubmed/1806236.
27. Ji, X., Duan, H., Zhang, H., Ma, J. 2015. Slurry erosion resistance of laser clad NiCoCrFeAl3 high-entropy alloy coatings. *Tribol. T.* 58(6): 1119–1123.
28. Zhang, H., Wu, W.F., He, Y.Z., Li, M.X., Guo, S. 2016. Formation of core–shell structure in high entropy alloy coating by laser cladding. *Appl. Surf. Sci.* 363: 543–547.

29. Yue, T., Xie, H., Lin, X., Yang, H., Meng, G. 2013. Microstructure of laser Re-melted AlCoCrCuFeNi high entropy alloy coatings produced by plasma spraying. *Entropy* 15(12): 2833–2845.
30. Anupam, A., Kumar, S., Chavan, N.M., Murty, B.S., Kottada, R.S. 2019. First report on cold-sprayed AlCoCrFeNi high-entropy alloy and its isothermal oxidation. *J. Mater. Res.* 34(5): 796–806.
31. Aliyu, A., Rekha, M.Y., Srivastava, C. 2019. Microstructure electrochemical property correlation in electrodeposited CuFeNiCoCr high-entropy alloy-graphene oxide composite coatings. *Phil. Mag.* 99(6): 718–735.
32. Aliyu, A., Srivastava, C. 2019. Microstructure-corrosion property correlation in electrodeposited AlCrFeCoNiCu high entropy alloys-graphene oxide composite coatings. *Thin Solid Films*. doi:10.1016/j.tsf.2019.137434.
33. Gupta, A., Srivastava, C. 2019. Enhanced corrosion resistance by SnCu-graphene oxide composite coatings. *Thin Solid Films* 669: 85–95.
34. Di Bari, G.A. 2000. Electrodeposition of nickel. *Mod. Electroplat* 5: 79–114.
35. Gupta, A., Srivastava, C. 2019. Correlation between microstructure and corrosion behaviour of SnBi-graphene oxide composite coatings. *Surf. Coat. Technol.* 375: 573–588.
36. Gezerman, A.O., Corbacioglu, B.D. 2010. Analysis of the characteristics of nickel plating baths. *Int. J. Chem.* 2(2): 124.
37. Kumar, P.M.K., Rekha, M.Y., Agarwal, J., Agarwal, T.M., Srivastava, C. 2019. Microstructure, morphology and electrochemical properties of ZnFe-graphene composite coatings. *J. Alloys Compd.* 783: 820–827.
38. Yusrini, M., Idris, Y.I. 2013. Dispersion of strengthening particles on the nickel-iron silicon nitride nanocomposite coating. *Adv. Mater. Res.* 647: 705–710.
39. Jyotheender, K.S., Srivastava, C. 2019. Ni-Graphene Oxide composite coatings: Optimum graphene oxide for enhanced corrosion resistance. *Compos. B Eng.* 175. http://www.ncbi.nlm.nih.gov/pubmed/107145.
40. Tabakovic, I., Inturi, V., Thurn, J., Kief, M. 2010. Properties of Ni1_xFex (0.1<x<0.9) and Invar (x¼0.64) alloys obtained by electrodeposition. *Electrochim. Acta* 55(22): 6749–6754.
41. Seo, M.H., Kim, D.J., Kim, J.S. 2005. The effects of pH and temperature on Ni-Fe-P alloy electrodeposition from a sulfamate bath and the material properties of the deposits. *Thin Solid Films* 489(1–2): 122–129.
42. Kim, D.J., Roh, Y.M., Seo, M.H., Kim, J.S. 2005. Effects of the peak current density and duty cycle on material properties of pulse-plated Ni-P-Fe electrodeposits. *Surf. Coat. Technol.* 192(1): 88–93.
43. Chaudhari, A.K., Singh, V.B. 2015. Studies on electrodeposition, microstructure and physical properties of Ni-Fe/In2O3 nanocomposite. *J. Electrochem. Soc.* 162(8): D341–D349.
44. Chaudhari, A.K., Singh, V. 2014. Structure and properties of electro Co-Deposited Ni-Fe/ZrO2 nanocomposites from ethylene glycol bath. *Int. J. Electrochem. Sci.* 9: 7021–7037.
45. Tripathi, M.K., Singh, D.K., Singh, V. 2015. Microstructure and properties of electrochemically deposited Ni-Fe/Si3N4 nanocomposites from a DMF bath. *J. Electrochem. Soc.* 162(3): D87–D95.
46. Dolati, A., Ghorbani, M., Afshar, A. 2003. The electrodeposition of quaternary Fe-Cr-Ni-Mo alloys from the chloride-complexing agents electrolyte. Part I. Processing. *Surf. Coat. Technol.* 166(2–3): 105–110.
47. Rekha, M.Y., Kamboj, A., Srivastava, C. 2017. Electrochemical behaviour of SnZn-graphene oxide composite coatings. *Thin Solid Films* 636: 593–601.

48. Saedi, A., Ghorbani, M. 2005. Electrodeposition of Ni-Fe-Co alloy nanowire in modified AAO template. *Mater. Chem. Phys.* 91(2–3): 417–423.
49. Orinakova, R., Turonova, A., Kladekova, D., Galova, M., Smith, R.M. 2006. Recent developments in the electrodeposition of nickel and some nickel-based alloys. *J. Appl. Electrochem.* 36(9): 957–972.
50. Tsuru, Y., Nomura, M., Foulkes, F.R. 2002. Effects of boric acid on hydrogen evolution and internal stress in films deposited from a nickel sulfamate bath. *J. Appl. Electrochem.* 32(6): 629–634.
51. Wu, Y., Chang, D., Kim, D., Kwon, S.C. 2003. Influence of boric acid on the electrodepositing process and structures of NieW alloy coating. *Surf. Coat. Technol.* 173(2–3): 259–264.
52. Vaes, J., Fransaer, J., Celis, J., 2000. The role of metal hydroxides in NiFe deposition. *J. Electrochem. Soc.* 147(10): 3718–3724.
53. Matlosz, M. 1993. Competitive adsorption effects in the electrodeposition of iron nickel alloys. *J. Electrochem. Soc.* 140(8): 2272–2279.
54. Rekha, M.Y., Kamboj, A., Srivastava, C. 2018. Electrochemical behaviour of SnNi-graphene oxide composite coatings. *Thin Solid Films* 657: 82–92.
55. Arora, S., Kumari, N., Srivastava, C. Microstructure and corrosion behaviour of NiCo-carbon nanotube composite coatings. *J. Alloys Compd.* 801: 449–459.
56. Wang, F.J., Zhang, Y., Chen, G.L. 2009. Atomic packing efficiency and phase transition in a high entropy alloy. *J. Alloys Compd.* 478(1–2): 321–324.
57. Guo, S., Ng, C., Wang, Z., Liu, C.T. 2014. Solid solutioning in equiatomic alloys: Limit set by topological instability. *J. Alloys Compd.* 583: 410–413.
58. Singh, A.K., Subramaniam, A. 2014. On the formation of disordered solid solutions in multicomponent alloys. *J. Alloys Compd.* 587: 113–119.
59. Ye, G.X., Wua, B., Zhang, C.H., Chen, T., Lin, M.H., Xie, Y.J., Xiao, Y.X., Zhang, W.J., Zhang, L.K., Zheng, Z.H., Wang, C. 2012. Study of solidification microstructures of multi-principal high-entropy alloy FeCoNiCrMn by using experiments and simulation. *Adv. Mater. Res.* 399–401: 1746–1749.
60. Otto, F., Dlouhy, A., Somsen, C., Bei, H., Eggeler, G., George, E.P. 2013. The influences of temperature and microstructure on the tensile properties of a CoCrFeMnNi high-entropy alloy. *Acta Mater.* 61(15): 5743–5755.
61. Mishra, A.K., Balasubramaniam, R. 2007. Corrosion inhibition of aluminum alloy 6061 by rare earth chlorides. *Corrosion* 63(3): 240–248.
62. Rekha, M.Y., Mallik, N., Srivastava, C. 2018. First report on high entropy alloy nanoparticle decorated draphene. *Sci. Rep.* 8(1): 8737.
63. Jin, W., Wang, G., Lin, Z., Feng, H., Li, W., Peng, X., Qasim, A.M., Chu, P.K. 2017. Corrosion resistance and cytocompatibility of tantalum-surface-functionalized biomedical ZK60 Mg alloy. *Corros. Sci.* 114: 45–56.
64. Rekha, M.Y., Srivastava, C. 2019. Microstructure and corrosion properties of Zinc-graphene oxide composite coatings. *Corros. Sci.* 152: 234–248.
65. Liu, Y., Li, S., Zhang, J., Liu, J., Han, Z., Ren, L. 2015. Corrosion inhibition of biomimetic super-hydrophobic electrodeposition coatings on copper substrate. *Corros. Sci.* 94: 190–196.
66. Lazar, A.M., Yespica, W.P., Marcelin, S., Pebere, N., Samelor, D., Tendero, C., Vahlas, C. 2014. Corrosion protection of 304L stainless steel by chemical vapor deposited alumina coatings. *Corros. Sci.* 81: 125–131.
67. Arora, S., Srivastava, C. 2019. Microstructure and corrosion properties of NiCo-graphene oxide composite coatings. *Thin Solid Films* 677: 45–54.

10 Understanding Disordered Multicomponent Solid Solutions or High Entropy Alloys Using X-Ray Diffraction

Anandh Subramaniam, Rameshwari Naorem, Anshul Gupta, Sukriti Mantri, K.V. Mani Krishna, and Kantesh Balani

CONTENTS

10.1 DEFECTS IN CRYSTALS

Under ideal Fraunhofer diffraction geometry conditions, the diffraction pattern from a perfect crystal consists of δ-peaks of intensity in reciprocal space. These δ-peaks (Bragg peaks) reside in a dark (zero intensity) background. A diffraction pattern obtained under "real" conditions has the following features: (i) the peaks are broadened and (ii) the background has a finite intensity (and is noisy). "Instrumental reasons" and defects in the crystal contribute to these two effects. Hence, the sample may contribute to the intensity under the Bragg peak and between the peaks. The intensity between the peaks is usually "diffuse" (referred to as diffuse scattering) and is much weaker than the Bragg peak intensity. Under certain conditions diffuse intensity is also observed close to the Bragg peaks.

X-ray diffraction has proved to be a powerful tool for the study of defects in crystals [1–4]. Defects in crystals have been classified into two types. In type-I defects the strain field decays as r^{-2} or faster, while in type-II defects the decay is $r^{-3/2}$ or slower. The effect of these two types of defects on the X-ray diffraction pattern is summarized in Table 10.1. The above implies that type-I defects typically have a "localized" effect, while type-II defects have a "long-range" distortion field. Needless to say, the actual defect structure in a solid will be much more complex, thus lending to a mixed character to some of the defects e.g., a straight dislocation is a type-II defect, while a "small" dislocation loop has type-I character. Large loops and dislocation networks tend to have a mixed character. It is to be noted that both in type-I and type-II defects, the defect can be visualized as a "local" deviation from a

TABLE 10.1
The Effect of Type-I and Type-II Defects on the X-ray Diffraction Pattern

Properties	**Type-I**	**Type-II**	**Distinct class**
Examples	**Thermal vibrations, point defects (and dilute solid solutions)**	**Dislocations, stacking faults**	**Concentrated solid solutions**
Variation in strain field (decay)	$\sim \frac{1}{r^2}$ or faster	$\sim \frac{1}{(r)^{3/2}}$ or slower	Permeates the whole crystal
Peak			
type	δ-function	No δ-function[a]	No δ-function
height	Decreases (↓)	Decreases (↓)	Decreases (↓)
width	No Peak broadening	Broadened "Peak"[b]	Broadening
Displacement	Yes[c]	No	No (only due to composition variation)
Integrated intensity	Decreases (↓)	No Change	Changes
Diffuse scattering	Yes	No	Yes
Structure	Reference lattice intact	Reference lattice intact	"Defected" crystal[d]

Note: The characteristics of a concentrated disordered solid solution is also listed—the details of which are considered later in the chapter.

[a] Functions used typically to fit X-ray diffraction peaks: Gaussian/Lorentzian/Pseudo-Voigt/Pearson-7.
[b] Continuous transition from background to peak. This also true for type-I defects.
[c] Peak shifts due to change in average lattice parameter.
[d] Disorder permeates the entire material and is not infinitesimal anywhere.

perfect lattice. Table 10.1 summarizes the effect of the formation of a concentrated disordered solid solution on the X-ray diffraction pattern—aspects which form the rest of the chapter.

The effect of temperature, atomic disorder, lattice strain, crystallite size, and crystalline defects on the X-ray diffraction pattern has been comprehensively described in the literature [5–8]. Considerable attention has been given to the effect of these factors on the diffuse scattering present between the Bragg peaks [9–12]. In the current chapter we will focus on X-ray powder diffraction patterns on polycrystalline samples, consisting of a disordered solid solution.

10.1.1 Effect of Thermal Vibrations

As an illustrative example let us consider the effect of temperature on an X-ray diffraction pattern. This is especially important, as often the effect of atomic disorder (on the X-ray diffraction pattern), which arises due to the formation of a disordered solid solution, is modeled akin to the effect of temperature [3, 13]. We will see later that this is technically incorrect. The key difference between thermal disorder and that arising from the formation of disordered solid solution is with regard to the dynamic nature of the former. "Temperature" results in atomic vibrations of atoms about their mean positions (i.e., the reference lattice is still intact and a time-averaged picture will reveal "smeared out" atoms in lattice sites), while in a solid solution the "lattice" is permanently distorted. This implies that in the timescale of the interaction of the X-rays, the waves will "see" a "enlarged atom" with a lower scattering power, decorating an enlarged average lattice. Needless to say, atoms in a solid solution also will experience thermal disorder in addition to the static displacement arising from atomic size difference.

"Temperature" has the following effects on the X-ray diffraction pattern (these aspects are schematically illustrated in Figure 10.1): (i) a shift in the Bragg peak due to an increase in the lattice parameter, (ii) a decrease in the intensity of the Bragg peaks (given by the Debye–Waller temperature factor), (iii) the presence of a diffuse background which increases with ($\sin\theta/\lambda$), and (iv) the presence of a diffuse scattering "tail" around the Bragg peaks. Standard formulations consider the effect of thermal vibrations as a decrease in the intensity of the Bragg peaks, via a decrease in peak height without an increase in peak width [14]. Advanced theoretical treatments on this subject matter also exist [3, 15]. With reference to point (iv) above, the plot of the diffuse intensity pattern around the Bragg peak is considerably different for a disordered solid solution compared to that arising from thermal disorder (Figure 10.2).

Note: the intensity level of the diffuse scattering is much lower than that of the Bragg peak (shown as "δ' peaks). The length of the reciprocal lattice vector is "q."

Let I_0 be the intensity of a Bragg peak in the absence of thermal disorder. The intensity of the peak in the presence of thermal disorder (I_T) is reduced by the Debye–Waller temperature factor $e^{(-2M^{\mathrm{T}})}$ i.e.,

$$I_T = I_0\left[e^{(-2M^{\mathrm{T}})}\right] \tag{10.1}$$

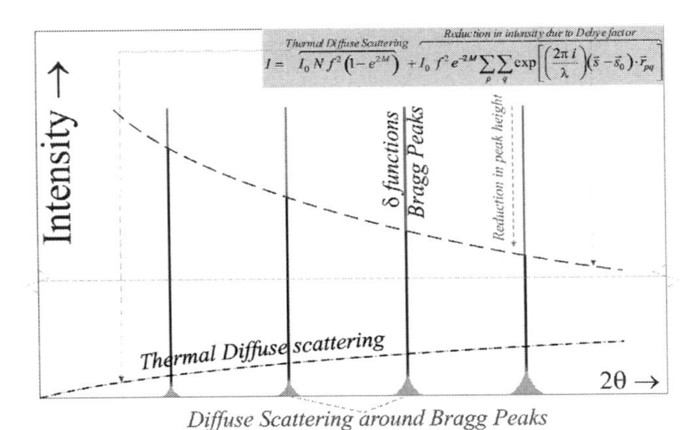

FIGURE 10.1 Schematic illustration of the effect of thermal vibrations on an X-ray diffraction pattern.

where I_0 is the intensity in the absence of thermal disorder. The parameter M^T is given by the following equation [3]:

$$M^T = 8\pi^2 \langle u^2 \rangle^T \left(\frac{\sin\theta}{\lambda}\right)^2 = \frac{6h^2 T}{mk\Theta_M^2}\left[\varphi(x) + \frac{x}{4}\right]\left(\frac{\sin\theta}{\lambda}\right)^2 \tag{10.2}$$

where $\langle u^2 \rangle^T$ is mean square displacement of atoms, h is Planck's constant, T is the temperature of the material, m is the atomic mass, k is the Boltzmann constant, φ is the Debye function, Θ_M is the characteristic temperature, and x is Θ_M/T (θ & λ have their usual meaning according to Bragg's equation). The parameter Θ_M is technically different from the Debye temperature (Θ_D), which is used in the context of the Debye theory of specific heat. However, the difference between the values of Θ_M and Θ_D is not usually large.

10.1.2 Effect of Dislocations

Dislocations, as has been noted in the previous section are primarily classified as type-II defects in the context of their influence on the X-ray diffraction peak profiles. This is due to the long-range stress fields associated with them. Prediction or modeling of the diffraction intensity variation due to dislocations gets complicated due to (a) a complex arrangement of individual dislocations in terms of spatial arrangement, presence of dislocation junctions, etc. and (b) the presence of different types of dislocations. Here, "type" describes the character of dislocation (edge or screw) and Burgers vector. For instance, deformed polycrystalline samples of the hexagonal close packed materials are known to exhibit dislocations of <a>, <c> and <a+c> type of Burgers vectors [16, 17]. Further, dislocations usually exist (especially in deformed materials) in the form of curved lines (in deformed materials) or loops (in irradiated materials). The inherent anisotropy of the crystal further

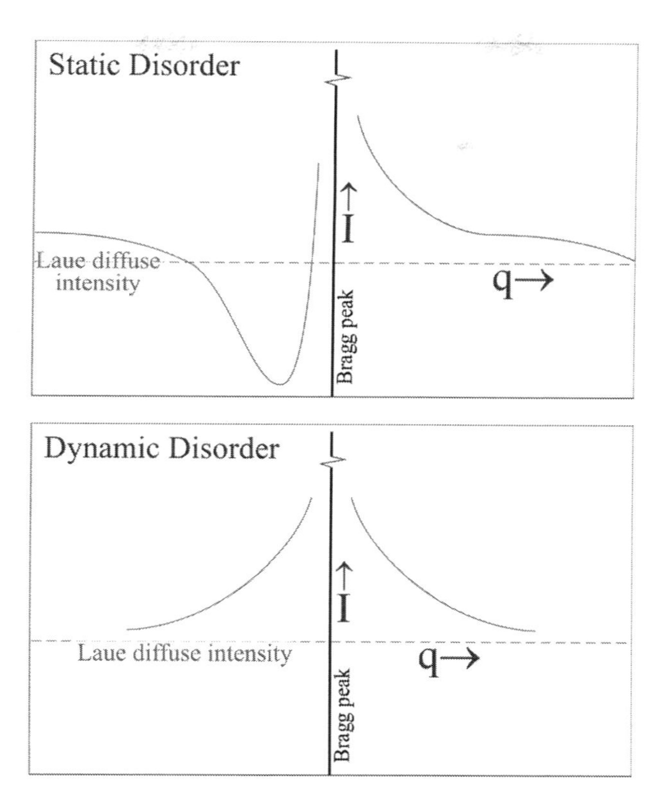

FIGURE 10.2 Schematic illustration of the nature of the diffuse scattering around the Bragg peak: (a) that arising from static disorder (e.g., in a disordered solid solution) and (b) that due to dynamic disorder (thermal vibrations).

adds to the complexity of the problem, as it leads to anisotropy in the displacement fields of the dislocations. Further, the efforts for developing analytical models for the estimation of scattering intensity from the dislocations have to deal with the singularities in displacement fields associated with analytical expressions of dislocations cores. These difficulties necessitate an incorporation of certain simplistic assumptions on the form of dislocation distribution (for example, completely random or uncorrelated distribution of dislocations), restricted random distribution [18, 19], and the presence of only straight dislocation segments in cylindrical imaginary blocks of crystals inside the crystal under consideration. In spite of these simplistic assumptions, the models were able to bring out certain specific characteristics of intensity variation of diffraction line profiles due to dislocation, as summarized below.

- In the case of homogenous dislocations (i.e., dislocations of specific Burgers vector family, say <110> as in face-centered cubic materials) with random distribution, the scattered intensity follows, particularly in the close vicinity of Bragg maxima Gaussian distribution. However, in regions away from

Bragg minima (i.e., in tail region of the Bragg peaks), the fall in intensity is much slower than that dictated by Gaussian distribution [2].

- For large scattering vector values (i.e., in regions sufficiently away from the Bragg maxima tail regions) the asymptotic fall in intensity is proportional to Q^{-3}, where Q is scattering vector [2, 18].
- The presence of dislocations with perfect random distribution and homogenous character leads to symmetric diffraction line profiles (i.e., the fall in the scattering intensity around the Bragg maxima is independent of direction). However, for realistic samples with highly correlated dislocation distributions (for instance, dislocation cell structures), significant asymmetry arises in the diffraction line profiles [20].
- The extent of line broadening caused by the dislocations is a function of the length of the reciprocal lattice vector and its orientation with respect to the Burgers vector of the dislocations. In the case of screw dislocations, the extent of broadening is expected to be higher for (h00) reflections than for the (hhh) reflections [2]. Exactly the opposite effect is seen in case of edge dislocations, where the broadening is more significant for (hhh) reflection in comparison to the (h00) reflections. These effects become even more significant when the underlying crystal undergoing diffraction is elastically highly anisotropic (such as Cu).
- The presence of a large density of dislocations in the form of loops can lead to significant asymmetry in the diffraction line profiles. A recent study, employing state-of-the-art high resolution measurements, reported extremely asymmetric peak profiles in case of irradiated zirconium (Zr) samples containing a large density of prismatic dislocation loops [21]. Further, the presence of very small dislocation loops were interpreted to cause significant rise in diffuse scattering, the intensity of which increases non-monotonically with increasing scattering angle [2].
- The second order moment of diffracted intensity in the tail region (an asymptotic part of the diffraction peak profile) varies linearly with mean dislocation density of the sample, and the third order moment of the intensity profile correlates with fluctuation in the dislocation density (polarization) [18].

10.2 THE FORMATION OF DISORDERED SOLID SOLUTIONS

A clarification of the terminology used in the literature and the meaning of the same in this chapter, with regard to multicomponent alloys may be useful here. The term "alloy" refers to a mixture of two or more elements (typically metals). An alloy may be single-phase, two-phase, or multiphase and the structure of the phase(s) is not kept in focus in the use of the term. A multicomponent alloy has many elements as its components. This phase/these phases may be solid solution(s) or compound(s). A solid solution implies a disordered solid solution, which has to be contrasted with its antithesis—an intermediate (or intermetallic) compound. In a disordered solid solution, the elements are positioned randomly in an "average" lattice, while in a compound, elements occupy distinct sublattices. A multicomponent alloy may have one or more principal elements (i.e., elements present in major fractions). If the elements are present in equal proportion, then we have an equimolar multicomponent alloy wherein the entropy is maximized. If this enhanced entropy can stabilize a disordered solid solution (a multicomponent disordered solid solution), in preference to phase separation or compound formation, the alloy is referred to as a high entropy alloy [22–28]. Often the term high entropy alloy is used in a "relaxed" sense, wherein the presence of multiple disordered solid solutions is "tolerated." Alloys obtained by metastable processing routes are also often included within the ambit of high entropy alloys. A disordered solid solution can be diluted or concentrated. A term related to the above discussion, often used in the literature, is "concentrated multicomponent alloy." In a concentrated multicomponent alloy, multiple elements are present in high concentration, with the equimolar alloy being a special case. *Prima facie*, there is no reference to the nature of phase(s) which form in a concentrated multicomponent alloy.

In a disordered solid solution, atoms are randomly positioned on a reference lattice. The formation of binary solid solutions is guided by the Hume–Rothery rules, which may be generalized to multicomponent alloys [16]. This implies that size difference between the atoms forming the alloys is expected to be small, and so will be the resultant distortion in the lattice. In the strictest sense, disordered solid solutions are "glasses/amorphous," and the tenets of crystallography break down at the "unit cell" level [29]. They are treated as crystals only in the "average probabilistic occupational order" sense, and give rise to sharp Bragg peaks, which is in sharp contrast to truly amorphous structures (Figure 10.3).

A disordered solid solution can be created by two modes: (i) start with a pure element and increase the concentration of a second element, and (ii) start with a pure element and increase the number of elements (keeping the composition equimolar). In both these modes we face a severe limitation with respect to the number of systems which form a disordered solid solution across compositions i.e., very few binary isomorphous systems exist and with an increase in the number of elements (in equimolar proportion), very few systems give rise to multicomponent disordered solid solution [30]. In the current chapter, we restrict ourselves to those systems which form disordered solid solution as a stable structure. Systems which are in a metastable state (i.e., quenched-in as a disordered solid solution) are not considered.

10.2.1 The Issue of Atomic Radius

The difference in the atomic radii of the elements participating in the disordered solid solution gives rise to lattice strain. This aspect may further restrict the compositions which form a disordered solid solution. Hence, it is important to know the atomic radii of the elements in the alloy. The answer to the question: "What is the diameter of an atom?" is a tricky one. This question is difficult to answer even for an isolated atom, but becomes especially tricky when the atom is present in a compound or an alloy. In a solid, the local environment will determine the diameter "assumed" by the atom, and hence there is no unique answer. An allied question of importance is: "What diameter should I use for an atom which forms a body-centered cubic crystal, but now is part of a face-centered cubic disordered solid solution?"

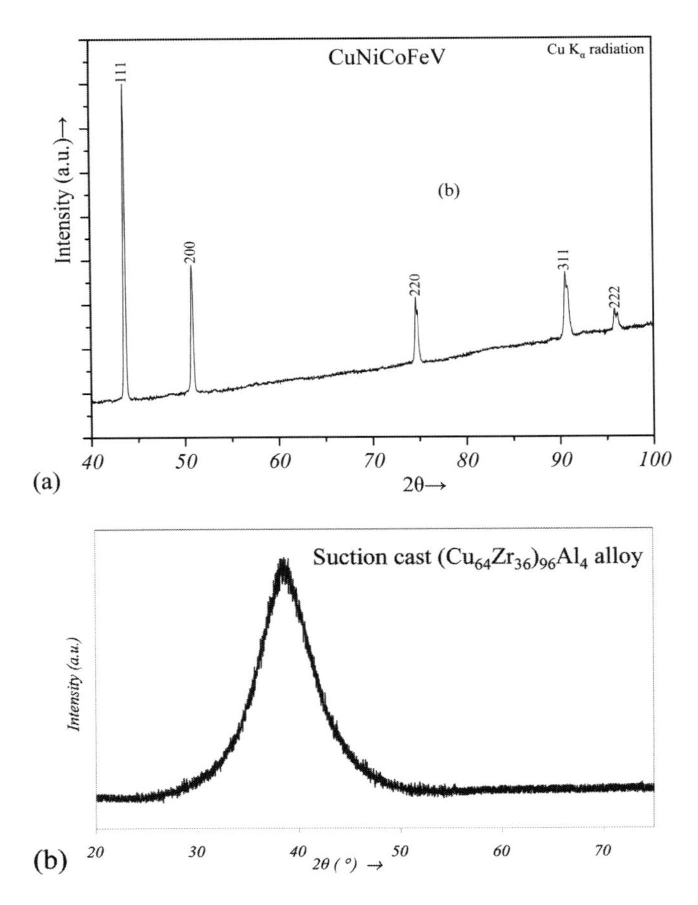

FIGURE 10.3 A comparison of the X-ray diffraction pattern obtained from a disordered solid solution (a), with that from a truly amorphous alloy (b). Note the "sharp" peaks in pattern from the disordered solid solution (obtained from an equimolar CuNiCoFeV single phase alloy) and the broad diffuse peak in the case of the amorphous alloy ($(Cu_{64}Zr_{36})_{96}Al_4$). Note the increasing background (with 2θ)—we will discuss more about this later.

In the context of the formation of concentrated disordered solid solution, the atomic radii listed in the literature are often used for the computation of a priori measures of "lattice strain" (Section 10.2.2). In this context, the following important points are to be noted: (1) The atomic radius of an element in an alloy is expected to be dependent on the local environment, although this effect may not be significant [31, 32], (2) Multiple listings of atomic radii exist in the literature (e.g., the Goldschmidt radius [33, 34] and radii based on empirical formulations [35, 36]). The Goldschmidt radii [37] seem to work well for metallic systems. In case of elements with body-centered cubic structure, which are part of a disordered solid solution with face-centered cubic structure, a correction factor for 12-coordination has to be employed (correction factor for 12-coordination: 1/0.97 = 1.03).

10.2.2 Measures of Lattice Strain and Bond Distortion

We have three options to get a handle on the "atomic level" strain in a multicomponent alloy: (S1) Use atomic sizes and composition of the components and utilize *a priori* measures of lattice strain; (S2) Computationally create the alloy and determine the lattice distortion; (S3) Experimentally determine the strain in the lattice. The advantage of method S1 is that it provides us an easy way to compute the lattice strain. On the other hand, approach S1 has some significant lacunae. As pointed out before, the radius of the atom in the alloy is expected to be different from the one used from a standard tabulation. The actual strain in the alloy may be lower as the system may reduce the strain by choosing an appropriate local configuration. The second aspect is especially true in multicomponent disordered solid solution (Section 10.3.3). Method S2 typically involves creating a disordered solid solution in a density functional theory (DFT) or molecular dynamics (MD) framework. Capturing true disorder requires a large system size (a large supercell), which can prove to be computationally "expensive." The computation may involve interatomic potentials, which are not suited for multicomponent alloys and hence may introduce considerable error. As we shall see later, the experimental method (S3) involves careful analysis of X-ray diffraction data. In practice, it may prove to be impossible to obtain the statistics related to bond distortion from experiments. Hence, a combination of experiments (S3) and computations (S2) may be the best option to determine the lattice strain, the starting point for which can be *a priori* measures (S1). It is important to note that bond distortion is not the same as lattice strain.

The strain in a lattice arising from atomic size differences between the elements is usually characterized by $\Delta R_{\max}$ or δ (or δ_r) parameters [38]:

$$\Delta R_{\max} = \frac{\mathrm{Max}(r_i - \bar{r})}{\bar{r}} \tag{10.3}$$

$$\delta = \left[\sum_{i=1}^{n} C_i \left(1 - \frac{r_i}{\bar{r}}\right)^2\right]^{1/2} \tag{10.4}$$

where $\bar{r} = \sum_{i=1}^{n} C_i r_i$, r_i is the radius of ith element in the alloy, n is the total number of elements in the alloy, and C_i is the atomic fraction of the ith element in the alloy. A product of δ and $\bar{r}$ (i.e., $\delta \times \bar{r} = \bar{u}^{\mathrm{D}}_{\mathrm{Owen}}$) can also be used as a measure of the lattice distortion [39]. The "intrinsic lattice distortion" ($\bar{u}_I^{\mathrm{D}}$) is another parameter, which has been used in conjunction with the Debye–Waller factor, for the computation of the intensity of the Bragg peaks [40]:

$$\bar{u}_I^{\mathrm{D}} = \left[\sum_{i=1}^{n} \left(d_i^{\mathrm{eff}} - \bar{d}\right)^2\right]^{1/2} \tag{10.5}$$

where $\bar{d}$ is the average lattice parameter and $d_i^{\mathrm{eff}} = \sum_i^n C_j \left(1 + \frac{\Delta V_{ij}}{V_i}\right)^{1/3} d_i$. The parameters C_j, d_i, V_i, and ΔV_{ij} are: the atomic fraction, the lattice constant of the ith

element, the unit cell volume of the ith element, and the difference in the unit cell volumes between the ith and the jth elements, respectively. It is to be noted that the intensity referred to above is the area under the Bragg peak (integrated intensity) above the baseline [41].

Variance as defined below has also been used as a measure of lattice distortion [42].

$$\sigma_{\mathrm{D}} = \sqrt{\frac{1}{n}\sum_i (r_i^2) - \left(\frac{1}{n}\sum_i r_i\right)^2} \tag{10.6}$$

An inspection shows that this parameter (σ_{D}) is identical to $\bar{u}^{\mathrm{D}}_{\mathrm{Owen}}$.

The average displacement of the atoms on the formation of the disordered solid solution (designated as $\bar{u}_{AC}$) can be computed from the position vectors of the atoms in the perfect lattice ($\vec{r}_o$), and that in an actual alloy ($\vec{r}_i$) using $\bar{u}_{AC} = \frac{1}{N}\sum_{i=1}^{N}(\vec{r}_i - \vec{r}_o)$. The values of $\vec{r}_i$ may be obtained from a MD or a DFT computation. The parameter $\bar{u}_{AC}$ is an *a posteriori* measure of the strain, arising from the displacement of the atoms from a perfect lattice.

A true measure of lattice strain, which is responsible for the strain energy in the crystal, should be computed from the actual bond length distortion. Bond length distortion in a face-centered cubic crystal can be defined as:

$$\overline{\Delta D} = \tfrac{1}{2N}\sum_{i=1}^{N}\sum_{j}^{12}\left(\left|D_{ij} - D_{ij}^{0}\right|\right) \tag{10.7}$$

where D_{ij} is the bond length between the ith atom and its nearest neighbor the jth atom, D_{ij}^0 is the equilibrium bond length $(r_i + r_j)$ between the atoms, and N is the number of atoms. The corresponding strain can be computed as: $\varepsilon_{\overline{\Delta D}} = \frac{1}{2N}\sum_{i=1}^{N}\sum_{j}^{12}\left(\left|\frac{D_{ij} - D_{ij}^{0}}{D_{ij}^{0}}\right|\right)$. It is to be noted that an accurate computation of bond distortion should involve higher order neighbors and distortions to the bond angle.

The terms "lattice strain" and "lattice distortion" have been used interchangeably to essentially depict atomic displacements from a perfect lattice. It is important to note that the strain computed using *a priori* parameters (e.g., $\bar{u}^{\mathrm{D}}_{\mathrm{Owen}}$ & $DR_{\max}$), may not be a good measure of the bond length distortion in a concentrated disordered solid solution. The lattice strain as measured by parameters such as δ & $\Delta R_{\max}$ is dimensionless, while lattice distortion as measured by parameters such as $\bar{u}^{\mathrm{D}}_{\mathrm{Owen}}$ & $\bar{r}$ have dimensions of length.

10.3 X-RAY DIFFRACTION FROM SOLID SOLUTIONS

As we shall see, solid solutions, based on their signature on the X-ray diffraction pattern, can be categorized as: (a) dilute, (b) transition, and (c) concentrated. This is important, as X-ray diffraction allows us to demarcate these regimes (terms), which were hitherto loosely defined.

In X-ray diffraction studies from these alloys, the pure element, based on which the binary and multicomponent alloys have been developed, serves as a good reference (standard) to study the variation of Bragg intensity, full width at half-maximum and diffuse scattering intensity. In the X-ray diffraction pattern, we would like to focus on two aspects: (i) the Bragg peaks and (ii) diffuse scattering between the peaks.

In the section on the effect of thermal vibrations on the X-ray diffraction pattern (Section 10.1.1), we had considered the parameter $\langle u^2 \rangle^{\mathrm{T}}$, which is the mean square displacement of atoms from their equilibrium lattice positions. A method to account for the contributions to lattice distortion arising from thermal and non-thermal effects is as follows [43].

$$8\pi^2 U_{\text{iso}} = \frac{6h^2 T}{mk\Theta_{\text{M}}^2}\left[\varphi(x) + \frac{x}{4}\right] + d^2 \tag{10.8}$$

where $8\pi^2 U_{\text{iso}}$ is the isotropic atomic displacement parameter (ADP). The first term in Equation (10.7) is due to thermal effects, and the second term can be considered to arise from static displacements. This parameter (U_{iso}) is to be used along with Equation (10.1) and the equation for I_T, to compute the reduction in intensity. However, as we pointed out, physically, thermal disorder is distinct from positional disorder.

10.3.1 Background and Fundamentals

In a classic work, Huang [44] studied the effect of the formation of dilute solid solutions on the X-ray diffraction intensities. The assertions made in the work of Huang [37], which include the "weakening of the maxima" and the "presence of diffuse maxima," have subsequently been experimentally confirmed [45–47]. At the other end of the spectrum is the study of Yeh et al. [33], who investigated the effect of an increase in the number of elements (in the Cu–Ni–Al–Co–Cr–Fe–Si system) on the X-ray diffraction powder pattern intensities. These compositions are equimolar and hence concentrated. Unfortunately, not all alloys in this system form disordered solid solutions—thus making it practically intractable to isolate the effect of lattice strain or to make comparison of parameters across alloys.

The questions which we would like to address regarding the formation of disordered solid solutions are as follows. (1) How are dilute alloys different from concentrated ones? (2) How are binary concentrated alloys different from concentrated disordered solid solution? (3) Are the lattices in concentrated disordered solid solution severely distorted and how can we get a handle on the true bond distortion? (4) Which parameter(s) give(s) a good measure of the bond distortion and how to experimentally access the bond distortion? (5) Does the lattice distortion in multicomponent alloys lead to severe dampening of the Bragg peak intensity? (6) What is the effect on the full width at half maximum and is it correct to model the effect of formation of a disordered solid solution as a thermal effect? (7) How to compare X-ray diffraction peaks across alloys and to isolate the effects of lattice strain.

The addition of alloying element(s), leading to the formation of a disordered solid solution, has the following major effects [12]. (1) Change in lattice parameter leading to a shift in peak position. (2) Alteration in the average atomic scattering factor leading to a change in the intensity of peaks. (3) Change in the solidus temperature. (4) Modification in the level of absorption of X-rays. (5) The elastic constants will change on the formation of a disordered solid solution. Point (4) is of importance, as for a given radiation some elements may cause fluorescence, thus leading to a reduced Bragg peak intensity along with a concomitant increase in background intensity. This aspect may be misconstrued as that arising from the formation of a multicomponent alloy (i.e., due to the disorder created by the formation of a multicomponent disordered solid solution).

The atomic "disorder" in disordered solid solutions arises because of three reasons: (i) probabilistic occupation of lattice sites according to the stoichiometry (referred to as "elemental disorder"); (ii) displacement of atoms from lattice sites due to atomic size differences (positional disorder); and (iii) thermal atomic vibrations. Point (i) increases the configurational entropy of the system and hence is expected to play a role in the stabilization of a disordered solid solution. The static (points i and ii) and dynamic (point iii) aspects of disorder are best understood by considering two model scenarios: the heating of a pure element (A) from 0 K (case-a) versus alloying it with another element (B) at 0 K (case-b).

As is evident by now, our goal here is to comprehend the effect of the formation of a disordered solid solution on the X-ray diffraction pattern. However, the task is complicated by the presence of the following "non-idealities" which real experimental conditions may suffer from: (1) the presence of other defects in the sample like dislocations, grain boundaries, stacking faults and thermal disorder; (2) the existence of short-range order (SRO) and/or local clustering in the sample; and (3) fluorescence. The kinematic approximation, which is used in much of the analysis in X-ray diffraction, simplifies matters. On the other hand, the presence of multiple scattering effects can further complicate the analysis.

If the atomic positions are known, then the intensity-2θ plots can be computed using the equation [3]:

$$I = \sum_p \sum_q f_p f_q \exp\left[\left(\frac{2\pi i}{\lambda}\right)\left(\vec{s} - \vec{s}_0\right) \cdot \vec{r}_{pq}\right] \tag{10.9}$$

where f_p & f_q are the atomic scattering factors of the pth and q^{th} atoms; $\vec{s}$ & $\vec{s}_0$ are unit vectors in the direction of diffracted and primary beams, respectively; and $\vec{r}_{pq}$ is the displacement vector between the pth and q^{th} atoms. It is worthwhile to note the following points: (1) for crystallites in random orientations this equation simplifies to the Debye scattering equation [48]; (2) this is the fundamental equation used for the derivation of other formulae, wherein the intensity is split into those arising from fundamental reflections, short-range order (diffuse intensity), size effect modulation, etc. (3) The splitting process typically involves approximations and is necessitated by the computational effort involved for large crystallites (using Equation 10.8).

In order to compare the intensity across the alloys of a series (i.e., as we increase the number of elements to form a multicomponent disordered solid solution), we need to know the factors which affect the intensity of the Bragg peaks and then go on to isolate the effects arising from atomic disorder. The integrated intensity per unit length of the diffraction line (J/s.m) in powder diffraction patterns obtained from a single-phase sample can be written as [49]:

$$I_{\text{net}} = I_1 \times I_2 = \left[\frac{I_0 A \lambda^3}{32\pi r}\left(\frac{\mu_0}{4\pi}\right)^2 \frac{e^4}{m^2}\right] \times I_2 \tag{10.10}$$

where A is the cross-sectional area of the incident beam, λ is the wavelength of the incident radiation, r is the radius of the diffractometer circle, μ_0 is the permeability constant ($4\pi \times 10^{-7}$ m.kg/C^2), m is the mass of the electron (kg), e is the charge on the electron (C), and I_0 ($= Bi(V-V_K)^n$) is the intensity of the incident beam (J/m^2.s). The term I_l can be written as [42]:

$$I_1 = \text{Constants}\,(I_0)(\lambda^3) = \text{Constants}\left[Bi\,(V - V_K)^n\right](\lambda^3) \tag{10.11}$$

where B is a proportionality constant, i and V are the applied current and voltage, respectively, V_K is the excitation voltage for the K-shell electron, and n is an exponent. The value of V_K can be calculated using the relation: $V_K = \frac{hc}{e\lambda_{K\alpha}}$, where h is the Planck's constant, c is the velocity of light, e is the charge of the electron, and $\lambda_{K\alpha}$ is the wavelength of Kα line, which depends on the type of radiation used. This implies that, in order to make a comparison of the intensities observed in an X-ray diffraction experiments, the wavelength of the radiation and the excitation voltage have to be taken into account.

The factors to be considered in the determination of I_2 are: (i) atomic scattering factor (f), (ii) mass absorption coefficient (μ/ρ), (iii) temperature factor (e^{-2M^T}), (iv) Lorentz-Polarization (LP) factor, (v) multiplicity factor (p), and (vi) the unit cell volume (v_{UC}).

The term I_2 in equation (10.9) is given by:

$$I_2 = [\text{Structure}][\text{Absorption}][\text{Temperature}][\text{Multiplicity}][LP]\left(\frac{1}{v_{UC}^2}\right)$$

$$= I_S \times I_A \times I_T \times (\text{three terms}) = I_{SAT} \times (\text{three terms}) \tag{10.12}$$

$$= \left[F^2\right]\left[\frac{1}{2\mu_{\text{alloy}}}\right]\left[e^{-2M^T}\right]\left[p\right]\overbrace{\left[\left(\frac{1+\cos^2 2\theta}{\sin^2\theta\cos\theta}\right)^2\right]}^{\text{Lorentz-Polarization Factor}}\left(\frac{1}{v_{UC}^2}\right)$$

where, F is the structure factor, μ is the linear absorption coefficient, p is the multiplicity factor, M^T is the Debye–Waller factor (which was encountered before), and v_{UC} is the unit cell volume. The value of μ_{alloy} can be computed using: $\mu_{\text{alloy}} = \rho_{\text{alloy}} \sum_{i=1}^{n} \left(\frac{\mu}{\rho}\right)_i w_i$, where ρ_{alloy} is the density of the alloy, w_i is the weight fraction of the ith element, and $(\mu/\rho)_i$ is the mass absorption coefficient of the ith element. The labels in the square brackets correspond to the various factors.

In addition to the Bragg peak intensity, we would like to compute the intensity of the diffuse scattering (I_{Diffuse}). I_{Diffuse} for a binary disordered solid solution is given by [3]:

$$I_{\text{Diffuse}} = Nx_A x_B \left(f_B - f_A\right)^2 \left[\sum_i \alpha(i) e^{i\vec{k}\cdot\vec{r}_i} + \sum_i \beta(i)\, i\vec{k}\cdot\vec{r}_i e^{i\vec{k}\cdot\vec{r}_i}\right] \tag{10.13}$$

where, I_{Diffuse} is the intensity of the diffuse scattering, x_A and x_B are the atomic fractions of A and B elements, f_A and f_B the corresponding atomic scattering factors, and $\alpha(i)$ and $\beta(i)$ are given by the expressions as follows: $\alpha(i) = 1 - \frac{p_A(i)}{x_A} = 1 - \frac{p_B(i)}{x_B}$.

$\beta(i) = \left(\frac{1}{f_B/f_A - 1}\right)\left[-\left(\frac{x_A}{x_B + \alpha_i}\right)\varepsilon_{AA}^i + \left(\frac{x_B}{x_A} + \alpha_i\right)\left(\frac{f_B}{f_A}\right)\varepsilon_{BB}^i\right]$, where $p_A(i)$ and $p_B(i)$ are the probabilities of finding "A" and "B" atoms. The quantities ε_{AA}^i & ε_{BB}^i are given by:

$\varepsilon_{AA}^i = \frac{r_{AA}^i - r_i}{r_i}$ and $\varepsilon_{BB}^i = \frac{r_{BB}^i - r_i}{r_i}$.

10.3.2 The Choice of Radiation and Experimental Procedure

The consideration of various factors and aspects in the previous subsection, leads us to the following "recipe" for a fruitful and meaningful comparison of peak intensities across alloys of a series. (1) An alloy system which forms a single phase disordered solid solution should be chosen. The tendency for short-range order and clustering should be minimum. (2) A radiation that shows minimum fluorescence (absorption) for the elements in the alloy system should be chosen and if this is not possible, a radiation must be chosen wherein the variation in absorption factor is small across the alloys. (3) The results must be corrected/normalized for variation of: (a) average atomic scattering factor, (b) absorption factor, and (c) solidus temperature across alloys. Once the above-mentioned factors are taken into account, any remaining variation in the intensity of the Bragg peaks across the alloys is purely due to the effect of the strain arising from atomic disorder. Needless to point out, the $\sin\theta/\lambda$ dependence of the atomic scattering factor, temperature factor, and the Lorentz-polarization factor have to be considered.

10.3.3 Intensity and Full Width at Half Maximum

Let us systematically investigate each aspect involved in the formation of disordered solid solutions. We would like to isolate each variable and its effect on the X-ray diffraction pattern, taking recourse to model computations and experiments. In the end we would like to have the answers for the following questions:

1. What is the effect of pure positional disorder (i.e., static disorder without elemental disorder)?

2. How does pure elemental disorder affect the X-ray diffraction pattern (i.e., without positional disorder)?
3. What is the effect of combined positional and elemental disorder (which is present in actual alloys)?
4. How does the full width at half maximum vary across a binary alloy and across an alloy series?
5. On the formation of a multicomponent alloy, what are the primary factors affecting the change in intensity and how to isolate the effect of lattice strain?

Table 10.2 summarizes the answers to these questions. Figures which illustrate the effect are also cited in the table. The gist of the results is considered in here and the reader may consult the works of Naorem et al. [50] to get an insight into the details.

Figure 10.4 shows the effect of pure positional disorder (in the absence of elemental disorder, model M1). The plot is generated using a computation model [51], in which atoms of Ni are displaced by a vector of fixed length from its lattice position. The orientation of the vector is randomly chosen and the distortion increases with an increase in the length of the displacement vector. A plot is created between the (111) Bragg peak intensity (or full width at half maximum) and distortion ($\overline{u}_{AC}$) using Equation (10.8). It is seen that such a disorder decreases the intensity of the peak and further leads to an increase in the full width at half maximum (broadening). Thus, the following important conclusions can be drawn from the figure: (i) pure positional disorder which permeates the entire crystal, leads to a broadening of the Bragg peaks; (ii) the magnitude of broadening is approximately linear with the distortion ($\overline{u}_{AC}$); (iii) the intensity monotonically decreases with distortion.

Next, the effect of pure elemental disorder (without positional disorder, model M2) is computationally investigated (results in Figure 10.5). A series of alloys are generated starting with Ni and going up to NiCoFeCrMn (all alloys are equimolar). The alloys of the series are labeled as follows: (A1) Ni, (A2) NiCo, (A3) NiCoFe, (A4) NiCoFeCr, and (A5) NiCoFeCrMn. The element(s) are placed on a perfect lattice and the intensity and the full width at half maximum are tracked as a function of the number of alloying elements. It is seen that elemental disorder does not lead to peak broadening; however, there is an intensity change due to change in the average atomic scattering factor (f_{avg}).

The intensity decreases due to a decrease in the average atomic scattering factor, while the full width at half maximum is unaltered.

The source of the X-rays (i.e., the interaction of the X-rays from a certain source with the elements of the sample) can be an important factor which determines the intensity of the Bragg peaks. This effect is clearly visible, when we analyze Figure 10.6 and Figure 10.7a in conjunction. In both cases, we study the X-ray diffraction patterns for alloys A1 to A5. In alloys A2–A5 the Bragg peak intensity is severely dampened compared to that from A1 (Figure 10.6), if Cu-K_α radiation is used. On the other hand, with Cr-K_α radiation this effect is absent (Figure 10.7a). This highlights the importance of absorption effects and the proper choice of radiation.

TABLE 10.2
A Compilation of the Effect of Various Variables on the X-ray Diffraction Pattern: Model Computations and Experiments Form the Basis of the Tabulation

Model (system) and composition (effect)	Experiment/ Computation	Configuration	Effect on peak intensity	Effect on the full width at half maximum	Effect on diffuse intensity	Figure
(M1) Pure Ni (Positional disorder)	Computation	Atoms displaced from lattice positions	Decreases	Increases with distortion *[N1]*	{Increases}	Figure 10.4
(M2) alloys A1 to A5 (Elemental disorder)	Computation	Different atoms on a perfect lattice	Depends on f_{avg}	No broadening	{Increases}	Figure 10.5
(M3) NiCoFeCrMn (Effect of radiation)	Experiment	Increasing number of elements from 1 to 5. Cu-K_α radiation	Fluorescence increases the background *[N2]*	(Increases with distortion)	*{[N3]}*	Figure 10.6
(M4) NiCoFeCrMn (Effect of radiation and increase in bond distortion)	Experiment	Increasing number of elements from 1 to 5. Cr-K_α radiation	Depends on various factors	Broadening increases with bond distortion	*{[N3]}*	Figure 10.7
(M5) Ag-Au [N4] (Effect of concentration)	Experiment and computation	Increasing the concentration of the second element in binary alloy	{Depends on f_{avg}}	No broadening for dilute alloys. Increases in concentrated alloys	Increases with strain	Figure 10.9a,c
(M6) Ag-Vacancy (Effect of severe lattice strain)	Computation	Vacancies in the Ag crystal	{Decreases due to decrease in f_{avg}}	No broadening for dilute alloys. Increases in concentrated alloys	Increases with strain	Figure 10.9b,d

Notes: f_{avg}: average atomic scattering factor. Text in curly brackets { } indicates that the effect will be observed, but not discussed in the current text.
[N1] Distortion implies bond distortion.
[N2] The increased background intensity effectively reduces the peak intensity.
[N3] Diffuse intensity is expected to increase with bond distortion. Refer model (M5-M7).
[N4] Isomorphous system.

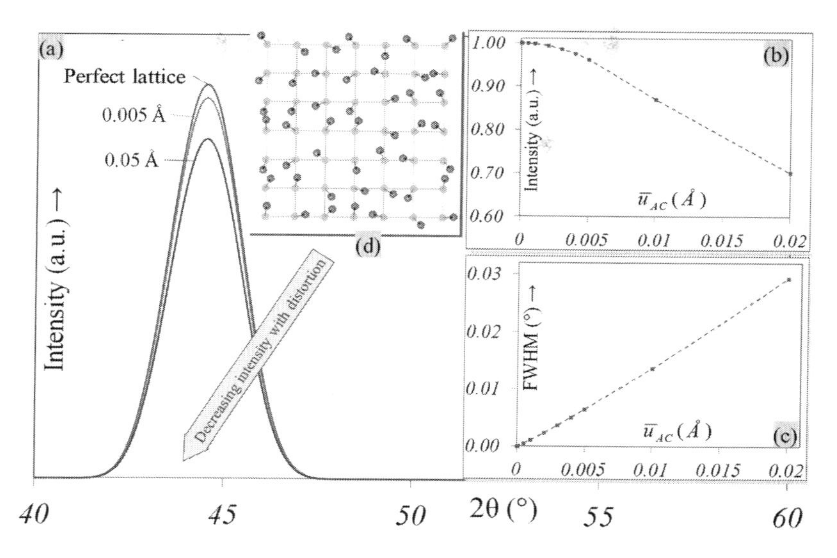

FIGURE 10.4 The case of pure positional disorder in Ni. (a) The effect of atomic displacement in Ni on the shape of the (111) Bragg peak for selected values of $\bar{u}_{AC}$. (b) The change in intensity on increasing the value of the distortion ($\bar{u}_{AC}$). (c) Plot of the full width at half maximum (FWHM) as a function of $\bar{u}_{AC}$. (d) A schematic of the model M1, wherein randomly oriented displacement vectors shift the atoms from lattice positions. The length of the vectors is varied to obtain the plots.

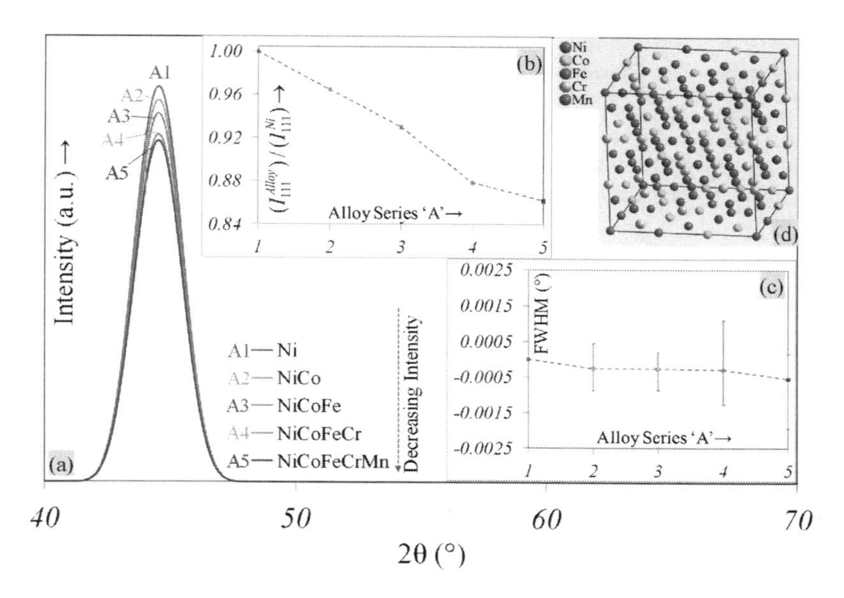

FIGURE 10.5 The effect of pure elemental disorder on the (111) Bragg peak (computational results). Elements are placed randomly on a perfect lattice. (a) The profile of the Bragg peaks. (b) A plot of the intensity of the (111) peak with the alloy number. (c) A plot of the full width at half maximum (FWHM) of the (111) peak with the alloy number. (d) A patch of a crystal (A5) with purely elemental disorder.

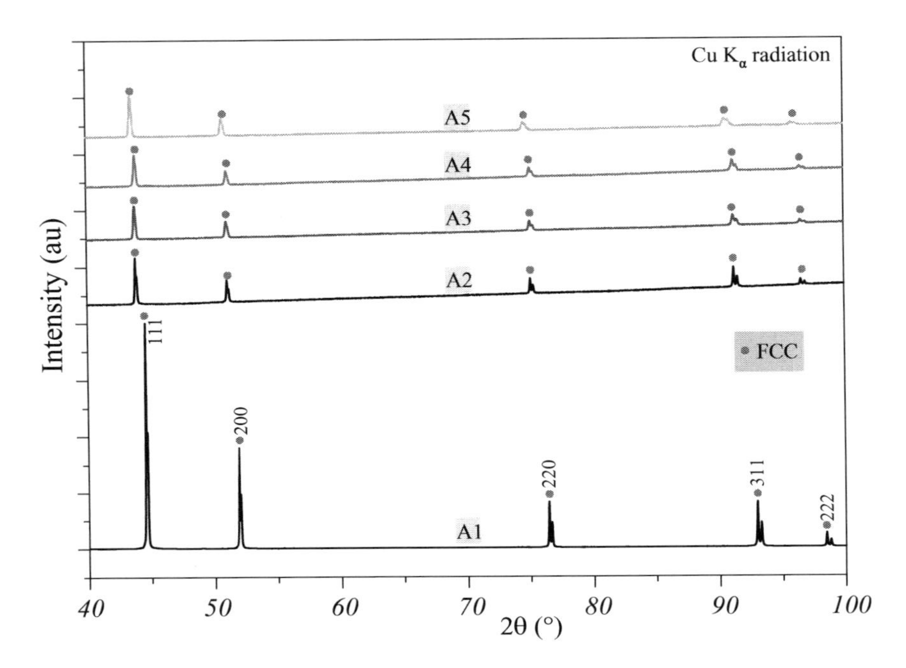

FIGURE 10.6 X-ray diffraction patterns obtained from the alloys A1 to A5 system using Cu-K$_\alpha$ radiation.

Next, we would like to address the important question: "Do Bragg peaks broaden in concentrated disordered solid solution?" Figure 10.7b shows a plot of the full width at half maximum with $u\frac{\text{D}}{\text{Owen}}$, while Figure 10.7c shows a plot of the full width at half maximum with bond distortion ($\overline{\Delta D}$). The variation in the intensity of the (111) peak with the number of elements (two to five elements), using Cu-K$_\alpha$ radiation, is shown in Figure 10.7d. The computational methodology is briefly described here and further details can be found in the work of Naorem et al. [43]. Based on the stoichiometry, elements are randomly populated on a supercell. The structure is allowed to relax to a local equilibrium configuration in a molecular dynamics framework. Five random configurations are used for each composition. The X-ray diffraction pattern is computed from the atomic coordinates using Equation (10.8).

The following observations can be made from Figure 10.7b–d. (1) The experimental results match well with the computed results. (2) Peaks are broadened in concentrated disordered solid solution. (2) The full width at half-maximum scales approximately linearly with $\overline{\Delta D}$ (but not with $u\frac{\text{D}}{\text{Owen}}$). (3) A steep decrease in the normalized intensity $((I_{111}^{\text{Alloy}})/(I_{111}^{\text{Ni}}))$ with alloy number (A2 to A5) is seen in the case where raw data is used. If the data is corrected for various "extraneous" factors affecting the intensity (using Equation 10.11), only a mild decrease in the normalized intensity is observed. (4) The intensity decrease due to the formation of concentrated disordered solid solution, which is about 10–20%, is not insignificant. However, this cannot be considered as anomalous either. Hence, we can conclude that: (i) atomic

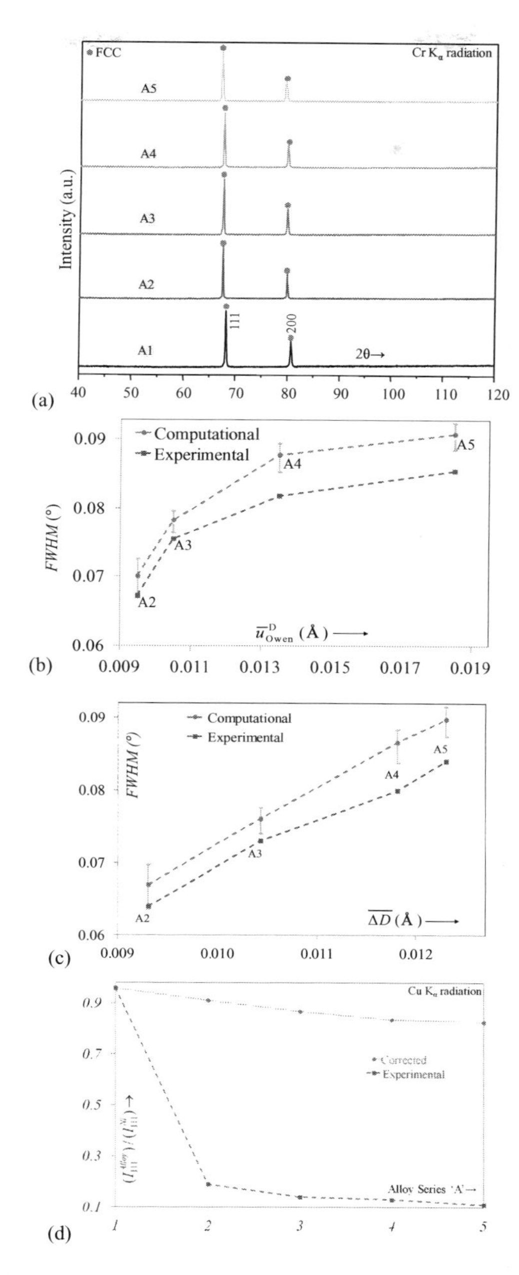

FIGURE 10.7 (a) Experimental X-ray diffraction patterns obtained from the alloys A1 to A5 using Cr-K$_\alpha$ radiation ($\lambda = 2.29$ Å). (b) A plot of the full width at half maximum (FWHM) with $\bar{u}^D_{\text{Owen}}$. (c) A plot of the full width at half maximum with bond distortion ($\overline{\Delta D}$). (d) Plot of the normalized intensity of the (111) peak $((I^{\text{Alloy}}_{111})/(I^{\text{Ni}}_{111}))$ with alloy numbers (A2–A5). Plot of the raw data and that corrected for various factors using Equation (10.11). The error bar in the computed values corresponds to five random configurations for each composition.

disorder leads to peak broadening, (ii) *a priori* measures like $\overline{u}^{D}_{Owen}$ do not accurately represent the bond distortion in a concentrated disordered solid solution, (iii) the parameter $\overline{\Delta D}$ is a good measure of bond distortion, and (iv) the intensity decrease depends on bond distortion and not on the number of elements. Thus, we can measure the full width at a half maximum of Bragg peaks to get a handle on the bond distortion, which arises from atomic disorder.

A reasoning based on the Bragg's viewpoint for the broadening of peaks in concentrated disordered solid solution is illustrated in Figure 10.8. This has been contrasted with the effect of thermal vibrations in the figure. A physical description of the two scenarios was considered in Section 10.1.1. Further details regarding the use of an appropriate measure of lattice distortion can be found in the work of Naorem et al. [44].

The next question we would like to address is "Are multicomponent alloys special?" The question can also be phrased as: "Is the key issue related to 'concentrated versus dilute' or is it an issue of 'binary versus multicomponent'?" To answer the above, we consider two binary alloys: Ag-Au and Ag-vacancy. It is to be noted that the Ag-vacancy system is a model system, which can be accessed only via computations. In fact, binary alloys are better from the viewpoint of a "control study." Figure 10.9a and b shows the plot of the full width at half maximum versus $\overline{\Delta D}$ for the alloys from the two systems. Figure 10.9c and d show the plot of the intensity of diffuse

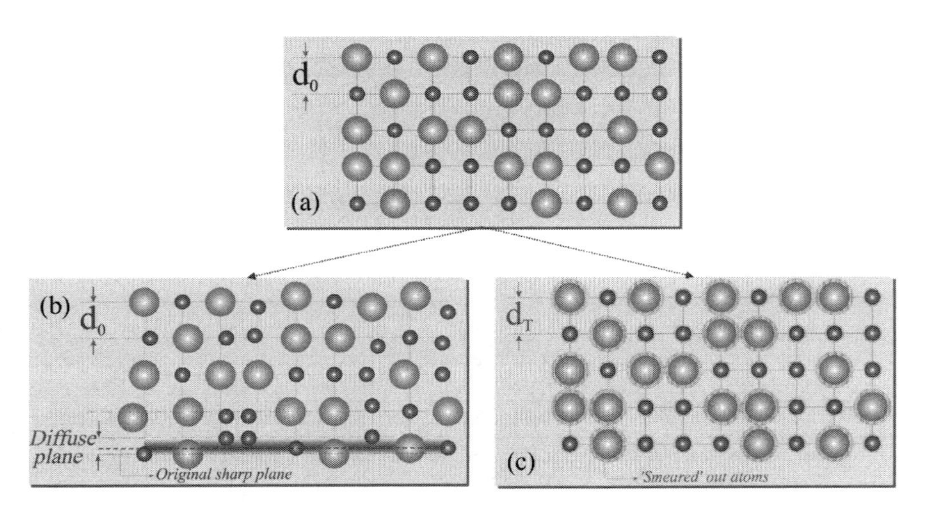

FIGURE 10.8 (a) An idealized crystal at 0 K, wherein atoms of two types are placed on a perfect lattice. The spacing of the Bragg planes was d_0. (b) A schematic illustration of the reason behind the peak broadening in a concentrated disordered solid solution, using the Bragg's viewpoint. This has to be contrasted with the effect of thermal vibrations (c). In (b) the atomic planes become "diffuse," with a range of d-spacings. Due to thermal vibrations, the X-rays "see" a "smeared" out atom occupying an average lattice with a slightly larger lattice parameter (due to thermal expansion). In (c), since we are considering the time-averaged picture, the atomic planes can be considered as sharp.

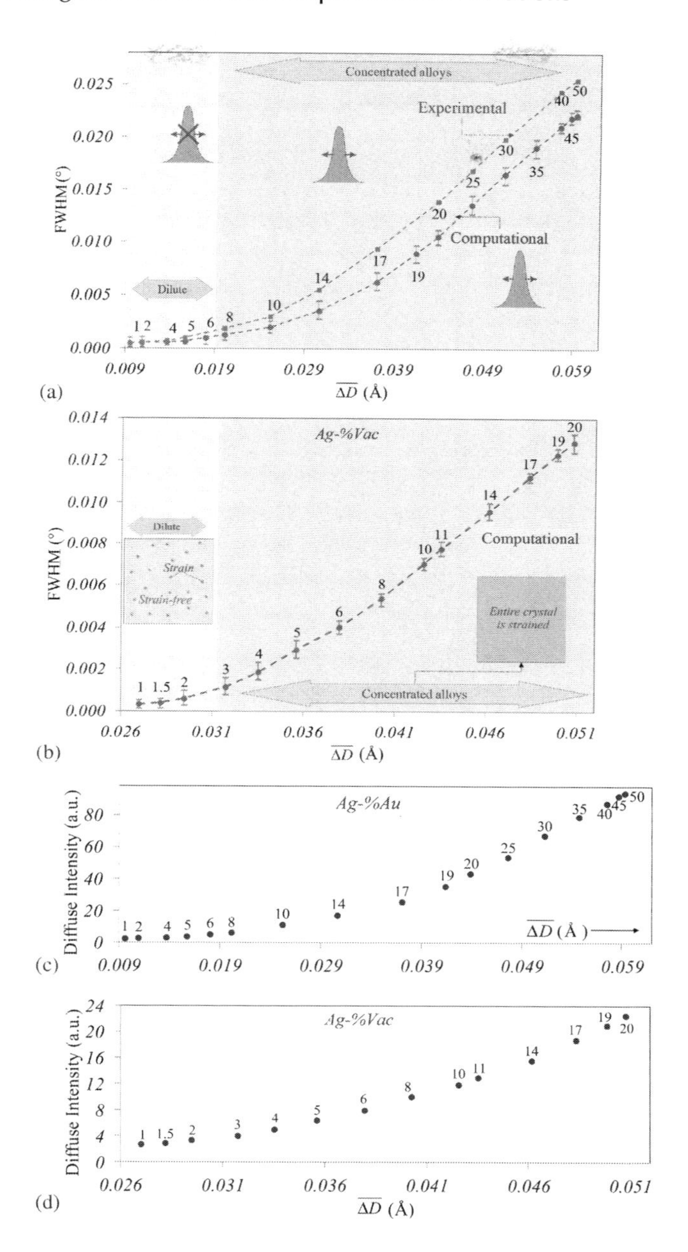

FIGURE 10.9 A plot of the full width at half maximum of the (111) peak versus bond distortion ($\overline{\Delta D}$) for the: (a) Ag-Au system and (b) the Ag-vacancy system. The error bar corresponds to five computational configurations. The numerical values above the data points correspond to the atomic percentage of the second component. Insets schematically illustrate the presence or absence of peak broadening [in (a)] and the presence of strained regions [in (b)]. A plot of diffuse scattering intensity (computational) versus bond distortion ($\overline{\Delta D}$) for the systems: (c) Ag-Au and (d) Ag-vacancy.

scattering versus $\overline{\Delta D}$. The diffuse scattering intensity is computed using Equation (10.4). Akin to the multicomponent alloys system we considered before, extreme care is taken to ensure that other contributions to the full width at half maximum are minimized. The insets to Figure 10.9b schematically illustrate the issue of strain field overlap in the lattice due to the point defects. In dilute alloys the strain fields arising from the point defects are isolated, and this implies that defects can be visualized as perturbations to a reference lattice. At the other end of the spectrum are concentrated disordered solid solutions, wherein the strain pervades the entire solid and is infinitesimal nowhere. This implies that in concentrated alloys the reference lattice is reduced to a mere "academic consideration."

The results obtained from the plots can be summarized as follows: (i) peak broadening does not occur in dilute alloys; (ii) highly concentrated alloys show a linear variation of the full width at half maximum with $\overline{\Delta D}$; (iii) in spite of the high strain in the Ag-Vac system, the dilute alloys show no broadening, and; (iv) the intensity of diffuse scattering increases with bond distortion. This implies that the key issue is of "dilute" versus "concentrated" and not binary versus multicomponent. Concentrated binary disordered solid solutions are akin to multicomponent disordered solid solution, in that both show peak broadening.

The increase in lattice strain on progressing across an alloy series (from a pure element to a concentrated disordered solid solution) is typically large using *a priori* measures of strain (e.g., the increase in strain using $\bar{u}^{D}_{Owen}$ is 100%); however, the increase in bond distortion (as measured by $\overline{\Delta D}$) is not very large (33%). This implies that the bond distortion, which is a true measure of strain in the alloy and which contributes to the strain energy, is not as large as anticipated from *a priori* measures of strain. This further implies that the strain energy penalty to the crystal is lower and explains the relative ease of the formation of concentrated disordered solid solution in the presence of multiple elements. Additionally, in the case of multicomponent disordered solid solution, entropy has to overcome the effects of a lower magnitude of "true strain" in the crystal and hence, it is not surprising that we obtain a disordered solid solution.

It is worthwhile to note that Figure 10.9a and b represents a simplified demarcation of regions. In practice we can subdivide the regions (dilute and concentrated) further. This can be arrived at by observing the curvature of the full width at half maximum versus $\overline{\Delta D}$ plot and the overlap of the strain fields [44]. The highly concentrated disordered solid solution has an approximately linear plot of the full width at half maximum with $\overline{\Delta D}$.

10.4 SUMMARY

Isomorphous binary compositions are rare and hence multicomponent disordered solid solutions are even rarer. Disordered solid solutions "suffer" from atomic disorder, which consists of elemental and positional disorder. X-ray diffraction is a powerful tool to study the atomic disorder in solid solutions. Important points arising out of X-ray diffraction study of disordered solid solutions are as follows.

1. X-ray diffraction data from single-phase disordered solid solutions can be analyzed to isolate the effect of lattice strain arising from atomic disorder.
2. Pure positional disorder (in the absence of elemental disorder) leads to an increase in the peak width (*full width at half maximum*), along with a decrease in the intensity.
3. Pure elemental disorder (without positional disorder) does not lead to peak broadening; however, there is an intensity change due to a change in the average atomic scattering factor (f_{avg}).
4. The *full width at half maximum* of a given X-ray diffraction peak increases with the strain in the lattice, which arises due to the formation of a multicomponent disordered solid solution. Among the many measures of lattice strain, the lattice distortion parameter ($\overline{u}^{D}_{Owen}$) seems to correlate best with the values of the *full width at half maximum*. However, the *a priori* computed measures of strain do not correspond to bond distortion.
5. The bond length distortion, which can be directly related to the strain energy cost to the crystal, can be perceived as a "true measure" of lattice strain. It is established that FWHM, which is an experimentally and computationally accessible quantity, is a measure of true lattice strain arising from bond length distortion. This can thus be used to separate the static and dynamic components of disorder.
6. Given that the peak broadening observed is an important signature of concentrated disordered solid solution, it is incorrect to model the effect of lattice distortion as a "temperature factor."
7. On the formation of a multicomponent disordered solid solution, the intensity of the Bragg peaks decreases with lattice strain, but this aspect cannot be correlated with the number of elements. The decrease in intensity is not insignificant; however, this cannot be termed as anomalous either.
8. The bond length distortion in multicomponent disordered solid solution is of significantly lower magnitude than that given by *a priori* measures of strain.
9. Dilute alloys are inherently different from concentrated ones and fall into the Huang scattering regime. Concentrated multicomponent disordered solid solutions are akin to concentrated binary alloys, with respect to their signature on the X-ray diffraction pattern.
10. In a concentrated disordered solid solution, the entire crystal can be construed as a defected solid, rather than a defect in a solid (i.e., the atomic disorder permeates the whole crystal and is of non-local nature).

ACKNOWLEDGMENTS

Prof. S. Anantha Ramakrishna (Department of Physics, IIT Kanpur), Prof. Madhav Ranganathan (Department of Chemistry, IIT Kanpur), and Prof. Raj Pala (Chemical Engineering, IIT Kanpur) are kindly acknowledged for their helpful discussions.

REFERENCES

1. Guinier, A. 2000. *X-Ray Diffraction: In Crystals, Imperfect Crystals, and Amorphous Bodies*. New York: Dover Publications, Incorporated.
2. Kirvoglas, M. A. 1969. *Theory of X-Ray and Thermal-Neutron Scattering by Real Crystals: Tr. from Russian*. New York: Plenum Press.
3. Warren, B. E. 1969. *X-Ray Diffraction*. New York: Addison-Wesley Pub. Co.
4. Guinebretière, R. 2007. *X-Ray Diffraction by Polycrystalline Materials*. London: ISTE Ltd.
5. James, R. W., W. H. Bragg, and W. L. Bragg. 1967. *The Optical Principles of the Diffraction of X-Rays*. London: Bell.
6. Waseda, Y., E. Matsubara, and K. Shinoda. 2014. *X-Ray Diffraction Crystallography Introduction, Examples and Solved Problems*. Berlin: Springer.
7. Seeck, O. H., and B. M. Murphy. 2015. *X-Ray Diffraction: Modern Experimental Techniques*. Singapore: Pan Stanford Publishing.
8. Chung, F. H., and D. K. Smith. 2005. *Industrial Applications of X-Ray Diffraction*. Boca Raton, FL: CRC/Taylor & Francis.
9. Dietrich, S., and W. Fenzl. 1989. "Correlations in Disordered Crystals and Diffuse Scattering of X-Rays or Neutrons." *Physical Review. Part B* 39(13): 8873–8899. doi:10.1103/physrevb.39.8873.
10. Welberry, T. R., and B. D. Butler. 1995. "Diffuse X-Ray Scattering from Disordered Crystals." *Chemical Reviews* 95(7): 2369–2403. doi:10.1021/cr00039a005.
11. Schweika, W. 1998. *Disordered Alloys: Diffuse Scattering and Monte Carlo Simulations*. Berlin: Springer.
12. Zachariasen, W. H. 2004. *Theory of X-Ray Diffraction in Crystals*. Mineola, NY: Dover Publications.
13. Cullity, B. D., and S. R. Stock. 2015. *Elements of X-Ray Diffraction*. Chennai: Pearson India Education Services.
14. Egami, T., and S. J. L. Billinge. 2003. *Underneath the Bragg Peaks: Structural Analysis of Complex Materials*. New York: Pergamon.
15. Welberry, T. R., and T. Weber. 2015. "One Hundred Years of Diffuse Scattering." *Crystallography Reviews* 22(1): 2–78. doi:10.1080/0889311x.2015.1046853.
16. Reed-Hill, R. E., and R. Abbaschian. 1992. *Physical Metallurgy Principles*. Boston, MA: PWS-Kent Pub.
17. Callister, W. D., and D. G. Rethwisch. 2018. *Materials Science and Engineering: An Introduction*. Hoboken, NJ: Wiley.
18. Wilkens, M. 1987. "X-Ray Line Broadening and Mean Square Strains of Straight Dislocations in Elastically Anisotropic Crystals of Cubic Symmetry." *Physica Status Solidi (A)* 104(1): K1–K6. doi:10.1002/pssa.2211040137.
19. Groma, I. 1998. "X-Ray Line Broadening due to an Inhomogeneous Dislocation Distribution." *Physical Review. Part B* 57(13): 7535–7542. doi:10.1103/physrevb.57.7535.
20. Groma, I., D. Tüzes, and P. D. Ispánovity. 2013. "Asymmetric X-Ray Line Broadening Caused by Dislocation Polarization Induced by External Load." *Scripta Materialia* 68(9): 755–758. doi:10.1016/j.scriptamat.2013.01.002.
21. Seymour, T., P. Frankel, I. Balogh, T. Ungár, S. P. Thompson, D. Jädernäs, J. Romero, L. Hallstadius, M. R. Daymond, G. Ribárik, and M. Preuss. 2017. "Evolution of Dislocation Structure in Neutron Irradiated Zircaloy-2 Studied by Synchrotron x-Ray Diffraction Peak Profile Analysis." *Acta Materialia* 126: 102–113. doi:10.1016/j.actamat.2016.12.031.
22. Murty, B. S., J. W. Yeh, S. Ranganathan, and P. P. Bhattacherjee. 2019. *High Entropy Alloys*. Amsterdam: Elsevier.

23. Gao, M. C., J. W. Yeh, P. K. Liaw, and Y. Zhang. 2018. *High-Entropy Alloys: Fundamentals and Applications*. Chalmers: Springer International Publishing.
24. Zhang, Y. 2019. *High-Entropy Materials: A Brief Introduction*. Singapore: Springer Verlag.
25. Zhang, Y., T. T. Zuo, Z. Tang, M. C. Gao, K. A. Dahmen, P. K. Liaw, and Z. P. Lu. 2014. "Microstructures and Properties of High-Entropy Alloys." *Progress in Materials Science* 61: 1–93. doi:10.1016/j.pmatsci.2013.10.001.
26. Miracle, D. B., and O. N. Senkov. 2017. "A Critical Review of High Entropy Alloys and Related Concepts." *Acta Materialia* 122: 448–511. doi:10.1016/j.actamat.2016.08.081.
27. Yeh, J.-W., S.-K. Chen, S.-J. Lin, J.-Y. Gan, T.-S. Chin, T.-T. Shun, C.-H. Tsau, and S.-Y. Chang. 2004. "Nanostructured High-Entropy Alloys with Multiple Principal Elements: Novel Alloy Design Concepts and Outcomes." *Advanced Engineering Materials* 6(5): 299–303. doi:10.1002/adem.200300567.
28. Tsai, M.-H., and J.-W. Yeh. 2014. "High-Entropy Alloys: A Critical Review." *Materials Research Letters* 2(3): 107–123. doi:10.1080/21663831.2014.912690.
29. Kirkwood, J. G. 1938. "Order and Disorder in Binary Solid Solutions." *The Journal of Chemical Physics* 6(2): 70–75. doi:10.1063/1.1750205.
30. Hukins, D. W. L. 1981. *X-Ray Diffraction by Disordered and Ordered Systems*. New York: Pergamon Press.
31. Okamoto, N. L., K. Yuge, K. Tanaka, H. Inui, and E. P. George. 2016. "Atomic Displacement in the CrMnFeCoNi High-Entropy Alloy – A Scaling Factor to Predict Solid Solution Strengthening." *AIP Advances* 6(12). doi:10.1063/1.4971371. http://www.ncbi.nlm.nih.gov/pubmed/1250081-1250088.
32. Ye, Y. F., C. T. Liu, and Y. Yang. 2015. "A Geometric Model for Intrinsic Residual Strain and Phase Stability in High Entropy Alloys." *Acta Materialia* 94: 152–161. doi:10.1016/j.actamat.2015.04.051.
33. Brandes, E. A. 1983. *Smithells Metals Reference Book*. London: Butterworths.
34. Goldschmidt, V. M. 1928. "Über Atomabstände In Metallen." *Zeitschrift für Physikalische Chemie* 133(1): 397–419.
35. Winter, M. "The Periodic Table of the Elements." *The Periodic Table of the Elements by WebElements*. Accessed June 7, 2019. http://www.webelements.com./.
36. Slater, J. C. 1964. "Atomic Radii in Crystals." *The Journal of Chemical Physics* 41(10): 3199–3204. doi:10.1063/1.1725697.
37. Pollock, D. D. 2018. *Physical Properties of Materials for Engineers: Volume 2*. Boca Raton, FL: CRC Press.
38. Singh, A. K., N. Kumar, A. Dwivedi, and A. Subramaniam. 2014. "A Geometrical Parameter for the Formation of Disordered Solid Solutions in Multi-Component Alloys." *Intermetallics* 53: 112–119. doi:10.1016/j.intermet.2014.04.019.
39. Owen, L. R., E. J. Pickering, H. Y. Playford, H. J. Stone, M. G. Tucker, and N. G. Jones. 2017. "An Assessment of the Lattice Strain in the CrMnFeCoNi High-Entropy Alloy." *Acta Materialia* 122: 11–18. doi:10.1016/j.actamat.2016.09.032.
40. Yeh, J.-W., S.-Y. Chang, Y.-D. Hong, S.-K. Chen, and S.-J. Lin. 2007. "Anomalous Decrease in X-Ray Diffraction Intensities of Cu–Ni–Al–Co–Cr–Fe–Si Alloy Systems with Multi-Principal Elements." *Materials Chemistry and Physics* 103(1): 41–46. doi:10.1016/j.matchemphys.2007.01.003.
41. Mittemeijer, E. J., and U. Welzel. 2013. *Modern Diffraction Methods*. Weinheim: Wiley-VCH.
42. Feng, B., and M. Widom. 2018. "Elastic Stability and Lattice Distortion of Refractory High Entropy Alloys." *Materials Chemistry and Physics* 210: 309–314. doi:10.1016/j.matchemphys.2017.06.038.

43. Calamiotou, M., D. Lampakis, N. D. Zhigadlo, S. Katrych, J. Karpinski, A. Fitch, P. Tsiaklagkanos, and E. Liarokapis. 2016. "Local Lattice Distortions vs. Structural Phase Transition in $NdFeAsO_{1-x}F_x$." *Physica C: Superconductivity and its Applications* 527: 55–62. doi:10.1016/j.physc.2016.05.019.
44. Huang, K. 1947. "X-Ray Reflexions from Dilute Solid Solutions." *Proceedings of the Royal Society of London Series A – Mathematical and Physical Sciences* 190(1020): 102–117. doi:10.1098/rspa.1947.0064.
45. Webb, W. W. 1962. "Atomic Displacements in Metallic Solid Solutions." *Journal of Applied Physics* 33(12): 3546–3552. doi:10.1063/1.1702444.
46. Herbstein, F. H., B. S. Borie, and B. L. Averbach. 1956. "Local Atomic Displacements in Solid Solutions." *Acta Crystallographica* 9(5): 466–471. doi:10.1107/s0365110x56001261.
47. Coyle, R. A., and B. Gale. 1955. "Integrated X-Ray Intensity Measurements from a Solid Solution of Copper–Gold." *Acta Crystallographica* 8(2): 105–111. doi:10.1107/s0365110x55000406.
48. Debye, P. 1915. "Zerstreuung Von Röntgenstrahlen." *Annalen der Physik* 351(6): 809–823. doi:10.1002/andp.19153510606.
49. Suryanarayana, C., and M. G. Norton. 1998. *X-Ray Diffraction A Practical Approach*. New York: Plenum Press.
50. Naorem, R., A. Gupta, S. Mantri, G. Sethi, K. V. Manikrishna, R. Pala, K. Balani, and A. Subramaniam. 2019. "A Critical Analysis of the X-Ray Diffraction Intensities in Concentrated Multicomponent Alloys." *International Journal of Materials Research* 110(5): 393–405. doi:10.3139/146.111762.
51. Naorem, R., A. Gupta, S. Mantri, K. Balani, and A. Subramaniam. *X-Ray Scattering from Disordered Alloys with Strain Field Correlations* (manuscript to be communicated).

11 Microstructure and Cracking Noise in High Entropy Alloys

Rui Xuan Li and Yong Zhang

CONTENTS

11.1 INTRODUCTION

The concept of high entropy is raised in the process of exploring bulk metallic glasses with large critical size. It is believed by Greer in his "Confusion Rule" that the more components contained in an alloy system, the higher the entropy is, and therefore the better the glass formation ability is. It is believed that when the configurational entropy is high enough, the complex atomic configuration will be easier to maintain after solidification, and thus the critical size will be larger. Later, some scientists found that high entropy isn't equal to high glass forming ability, while the single-phase solid solution can be found in some high mixing entropy alloys instead. Professor Yeh from the National Tsing Hua University believed that it is the high mixing entropy that stabilizes the solid solution, and thus the concept of a new king of alloys, which was called the high entropy alloy, was put forward.

At first, high entropy alloys are defined as a kind of typical multicomponent alloy which are made up of five or more elements in equal atomic ratio by Professor Yeh in

2004 [1]. As the field has evolved, high entropy alloys with unequal atomic ratio have been developed, which are even found to have better performance than the traditional equiatomic ratio high entropy alloys. With more and more research available, the concept of the high entropy alloys has been widely extended. The structure of high entropy alloys is not simply limited to single-phase solid solutions, and even a new definition of high entropy amorphous alloys has appeared [2, 3]. Nowadays, it is widely accepted that a molten alloy system where the effect of entropy plays the dominant role ($\Omega \geq 1.1$) and which can form a simple solid solution can be called the high entropy alloys.

11.2 THE MICROSTRUCTURE OF HIGH ENTROPY ALLOYS

11.2.1 Solidification Principles of High Entropy Alloys

Different from the pure metals, the solidification process of single-phase alloys is generally completed in a temperature range with solid–liquid phases coexisting. During equilibrium solidification, this temperature range begins at the liquidus temperature in the equilibrium phase diagram and ends at the solidus temperature. As the temperature changes, the composition of solid phase and liquid phase changes according to the solidus line and liquidus line, respectively. A mass transfer process will also occur during the solidification process, that is, the solute redistribution will continue on both sides of the solid–liquid interface. In fact, it is extremely difficult to achieve equilibrium solidification, especially in the solid phase where the uniformity of the components is achieved by the diffusion of atoms, while in the liquid phase, there are two mechanisms for the uniform mixing of atoms: diffusion and the flow of liquid. Under a situation of no agitation, it is also difficult to achieve uniformity in the liquid phase. Therefore, the enrichment of the solute is generated at the solidification interface, which in turn causes the change of the theoretical solidification temperature due to the difference in the composition of the liquid phase. When there is no constituent supercooling or a negative temperature gradient, the metal is the same as the pure metal, and the liquid–solid interface is dendritic. While when there is a positive temperature gradient, the crystal grows in a variety of ways: when there is a slightly supercooling, it is cell growth; as the temperature gradient decreases or as the degree of supercooling increases, the crystal changes from cellular to columnar, columnar dendrites and equiaxed dendrites. Among them, the conditions for forming a single-phase alloy to form columnar dendrites are:

$$G_L \rightarrow m_L C^i \left(1 - k_0\right)$$

$$R \rightarrow D_L k^\varphi$$

where G_L is the temperature gradient in the liquid phase beside the liquid–solid interface; R is the solidification rate; m_L is the slope of the liquidus; k_0 is the solute equilibrium partition coefficient; D_L is the diffusion coefficient of the solute in the liquid phase; and C_0 is the alloy composition.

The solidification process of disorder solid solution in the high entropy alloy is similar to that of traditional single-phase alloy, except that there are many kinds of elements in high entropy alloys, and the concentration of each component is equivalent. When forming a single-phase solid solution, it is generally considered that the alloy component is randomly occupied in the lattice, and all the elements can be regarded as solute, so the solute concentration of multicomponent high entropy alloy is very high. It can be seen from the formula (1-1) that, other things being equal, as the solute concentration C_0 of the alloy increases, the tendency of the supercooling in the alloy also increases. In addition, the high entropy alloy has a slow diffusion effect, and even in the case of a liquid state, the atomic diffusion rate is lowered due to the high mixing entropy, that is, the D_L is decreased, which also increases the tendency of the supercooling. Thus, during the solidification of the multicomponent alloys, the supercooling is obviously inevitable, and in many cases, it is even large. Therefore, the planar growth of the liquid–solid interface is destroyed, and the dendrites are formed along the growth direction of preferential crystallization. Usually, the growth direction of the primary dendrite arm is parallel to the direction of the heat flow. If the supercooled zone is sufficiently wide, the secondary dendrites will change into third dendrites at the front end during subsequent growth, and finally, a dendritic skeleton is rapidly formed in the supercooled zone. On the other hand, as the dendrite grows and branches, the solute of the remaining liquid phase is continuously enriched, and the melting point is continuously lowered, so that the supercooling of the melt quickly disappears. Since there is no supercooling at last, the side of the branch often completes its solidification process in a plane growth manner.

11.2.2 The Microstructure of High Entropy Alloys

Although high entropy alloys have many kinds of elements, they often form a simple solid solution phase structure. For example, a solid solution of simple face-centered cubic (FCC), body-centered cubic (BCC), and close-packed hexagonal (HCP) structures have all been found in the high entropy alloys. In addition, other phases such as amorphous phase and intermetallic compounds may also appear. Therefore, the phase structures and microstructures of the high entropy alloys are mainly affected by the composition, and the influence of the preparation process should also be considered. For high entropy alloys prepared by conventional casting processes such as arc melting, the alloy structure generally exhibits dendritic morphology due to elemental segregation. In addition, second phase particles of different shapes, amplitude decomposed structures, eutectic structures, and the like are also frequently present. The alloy can be grown into columnar crystal or single crystal by the special process such as directional solidification, and an amorphous high entropy alloy can be obtained by mechanical alloying. This chapter analyzes and summarizes the phase structure and microstructure of high entropy alloys that have appeared in the research.

11.2.2.1 Single Crystal Structure

The so-called single crystal is a kind of crystal where the particles inside are regularly and periodically arranged in a three-dimensional space. It can also be concluded that the single crystal does not contain grain boundaries in the macroscopic scale. The preparation and research of single crystals have been accompanied by the development of superalloys. Since the grain boundary is a weak link during high temperature deformation, the grain boundary is softened first during the deformation process, so the elimination or reduction of the grain boundary can effectively improve the mechanical properties of the alloy at high temperatures. In addition, the magnetic properties of materials are also closely related to the grain size and the density of the grain boundaries. The grain boundaries can seriously impede the movement of the magnetic domain walls, thereby increasing the coercivity of the alloy. Therefore, the preparation of a single crystal can not only improve the high-temperature mechanical properties of the alloy, but also contribute to the improvement of magnetic properties of soft magnetic materials.

At present, Zhang Yong's research group first used the Bridgeman directional solidification technology to prepare the CoCrFeNiAl0.3 high entropy alloy with single crystal FCC structure [1]. A schematic diagram of the CoCrFeNiAl0.3 sample after directional solidification of Bridgman, and the corresponding optical microstructure of each region are shown in Figure 11.1. The entire sample consists of an unmelted zone, a transition zone, and a directional solidification zone. In the bottom unmelted zone, the alloy maintains the dendritic morphology after casting, and in the transition zone, the dendrites are transformed into equiaxed crystals with grain sizes ranging from 50 μm to 300 μm. This change in morphology can be attributed to the re-melted interdendritic regions, large G/V, and residual thermal stress inside the sample. After the transition zone, it is the complete melting zone, where the alloy transforms from equiaxed crystals into columnar crystals and finally develops into single crystals. This is due to the preferential orientation of grain growth. In order to make the entire sample single crystal, the sample was inverted 180° after the first drawing to do the secondary drawing, leaving the single crystal region at the bottom of the sample in an unmelted state, which is similar to the seed crystal method. With the structural similarity, the crystal grows from the unmelted region to the molten metal, and the liquid atoms form a complete coherent interface with the unmelted atoms, so the sample tends to be singular after a second preferred orientation. Figure 11.2 shows the EBSD pattern of a CoCrFeNiAl0.3 alloy with a pronounced <001> orientation, which is the same as a conventional Ni-based superalloy. The grain boundaries are all small angular grain boundaries, and the grain

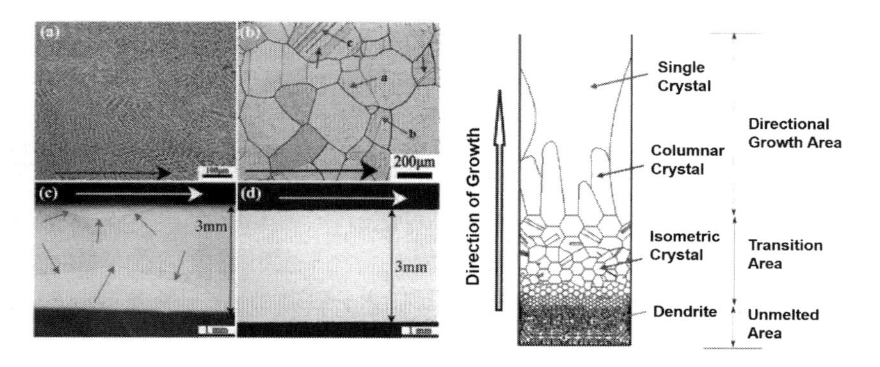

FIGURE 11.1 The microstructure of $CoCrFeNiAl_{0.3}$ after Bridgman directional solidification and the schematic of crystal grows after that [1].

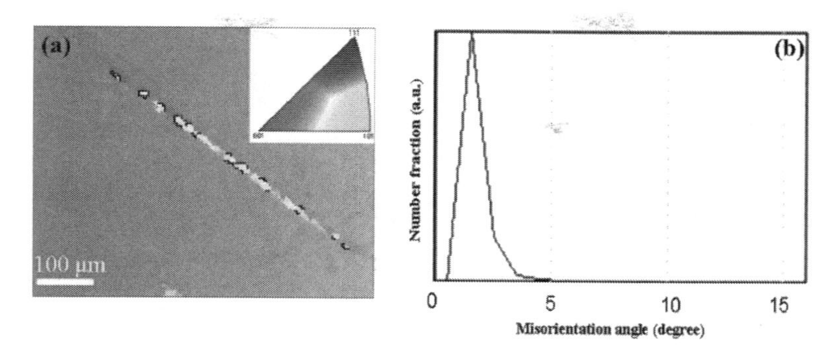

FIGURE 11.2 The EBSD map of $CoCrFeNiAl_{0.3}$ single crystal high entropy alloys.

boundary angle is mostly within 2°, which better proves that the crystal obtained by secondary drawing is a single crystal [1].

Patriarca et al. [3] present an experimental and theoretical study of slip nucleation in high entropy FeNiCoCrMn alloy which possesses superb mechanical properties, as shown in Figure 11.3. They conducted the uniaxial compression experiments on the oriented single crystals. The critical resolved shear stress (CRSS) to initiate slip for the new FeNiCoCrMn high entropy alloy was determined using DIC strain measurements to be τ77 K = 175 MPa and τ293 K = 70 MPa at 77 K and 273 K, respectively, which is in excellent agreement with the value obtained utilizing an advanced atomistic modified Peierls–Nabarro modeling formalism. This close agreement

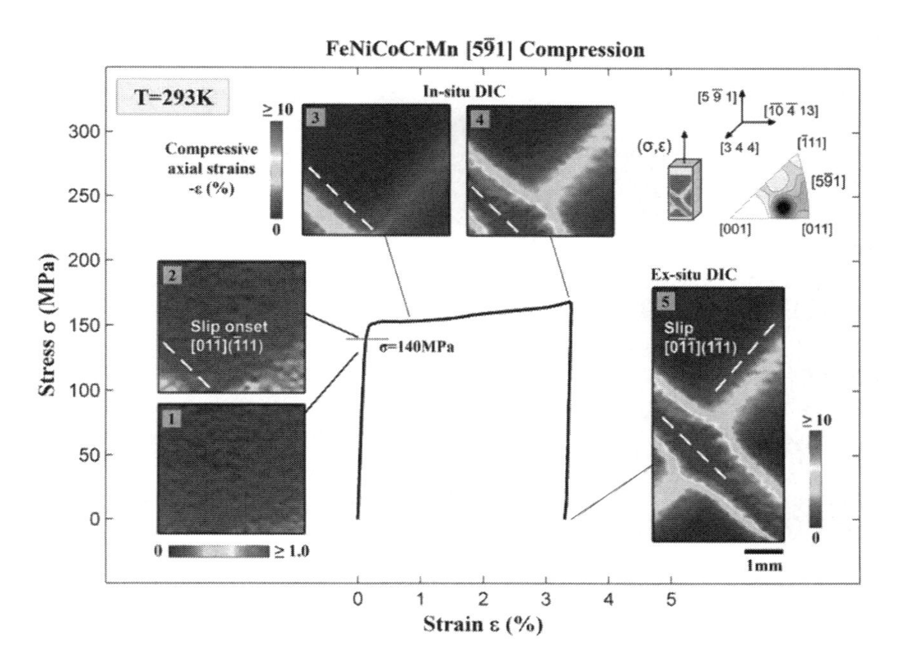

FIGURE 11.3 Stress-strain behavior in compression at room temperature [3].

demonstrates the efficacy of our methodology and has implications in design of new high entropy alloys.

11.2.2.2 Polycrystal Structure with Single Phase

High entropy alloys prepared for conventional casting processes such as arc melting tend to consist of multiple grains forming an equiaxed or columnar crystal structure. When the solidification process approaches equilibrium solidification, it will obtain a uniform single-phase solid solution. Such solidification process needs to achieve uniform diffusion in the liquid and solid phase at each temperature, as well as the uniform growth of the solid phase. However, in the actual solidification process, due to the fast cooling rate, sufficient diffusion time cannot be ensured at each temperature, so that the solidification process deviates from the equilibrium condition, which in turn results in the compositional segregation between interdendritic and dendrite arm regions.

CoCrFeMnNi is currently the most widely studied single-phase FCC solid solution alloy. In the as-cast state, the alloy is equiaxed, and since the five alloying elements are similar, no obvious compositional segregation occurs during solidification. Figure 11.4 shows the back-scattered structures of CoCrFeMnNi alloy during heat treatment at 800°, 900°, and 1,000°C for 1 h after cold rolling in different degrees [4]. It can be seen that the alloy after heat treatment is an equiaxed crystal structure, and the crystal grains are clearly visible. It was found that after heat treatment at 800°C for 1 h, only the sample with a large amount of deformation (61%, 84%, 92%, 96%) was completely recrystallized. When the temperature was raised, the amount of deformation required for crystallization is continuously decreasing. When the deformation is ≥80%, and complete recrystallization occurs at 800°C, the grain size is kept at 4–5 μm and almost unaffected by the amount of deformation. After complete recrystallization at 1,000°C, the grain size varies from 44 μm to 109 μm. In summary, in the cases where complete recrystallization occurs, the size of the crystal grains increases as the heat treatment temperature increases, and decreases as the amount of deformation increases. After recrystallization, the growth of the grains conforms to the law of grain growth. The grain growth index is $n = 3$, and the activation energy $Q = 325$ kJ/mol. As the heat treatment temperature increases (except for 1,000°C) and the amount of deformation increases, the percentage of the twin boundaries increases. However, when the grains after rolling are completely recrystallized, and when they grow at different temperatures, the density of twins in each grain and the percentage of twin boundaries in the grain boundaries are only related to the grain size, having nothing to do with the heat treatment process. This is consistent with the formation of Fullman–Fisher type twins during grain growth. Furthermore, Liu et al. [4] found that the relationship between hardness and grain size of CoCrFeMnNi alloy satisfies the Hall–Petch relationship.

Since high entropy alloys with BCC structure tend to have high yield strength and fracture strength, the development and study of them have become the focus of high entropy alloys. Senkov et al. [5] developed a variety of heat-resistant high entropy alloys, such as WNbMoTa and WNbMoTaV, through composition design, and their X-ray diffraction patterns are shown in Figure 11.5. Since the atomic radii and valence electron concentrations of these elements are very close, both alloys form a single-phase BCC structure. The back-scattered image (Figure 11.6) shows that both

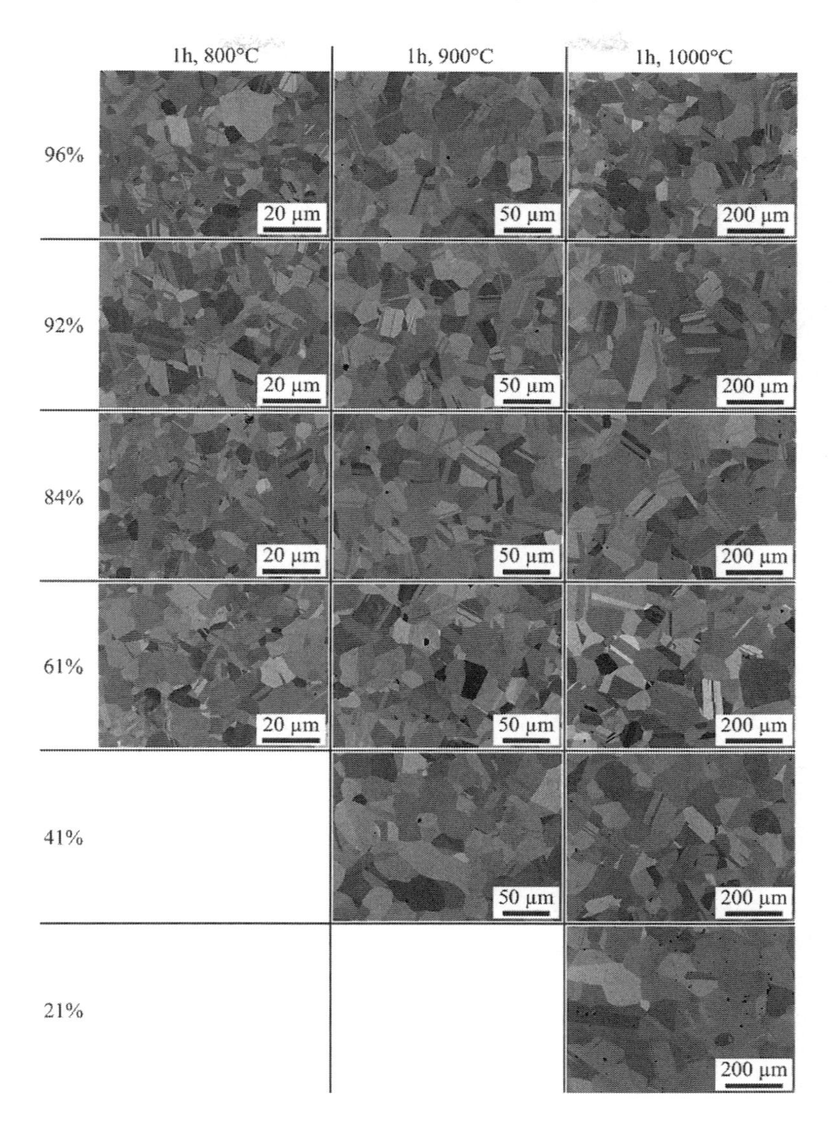

FIGURE 11.4 The back-scattered structures of CoCrFeMnNi alloy during heat treatment at 800, 900, and 1,000 °C for 1 h after cold rolling in different degrees [4].

WNbMoTa and WNbMoTaV are polycrystalline. The grain size of WNbMoTa is large, and the grain diameter is about 200 μm; the grain size of WNbMoTaV is relatively small, and the grain diameter is about 80 μm. A slight compositional contrast, which is also called the dendritic morphology, can be seen inside the two grains. It is found by compositional analysis that the distribution of Ta in the two alloys is relatively uniform, W is more likely to distribute in the dendrite region, and the other three elements of Mo, V, and Nb are tend to segregate between the dendrites. This indicates that the alloy has undergone non-equilibrium solidification. The degree

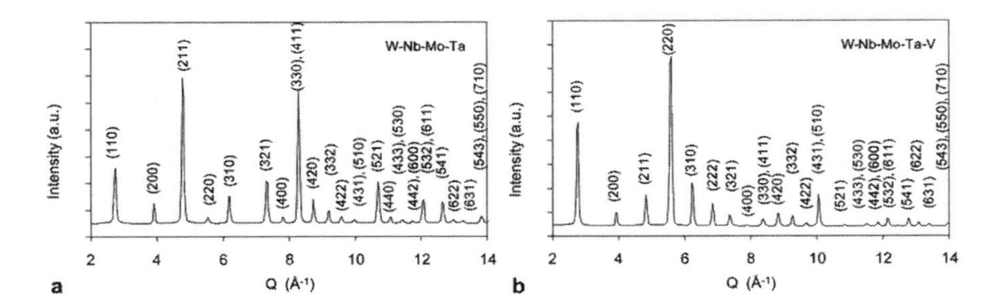

FIGURE 11.5 The XRD pattern of (a) WNbMoTa and (b) WNbMoTaV [5].

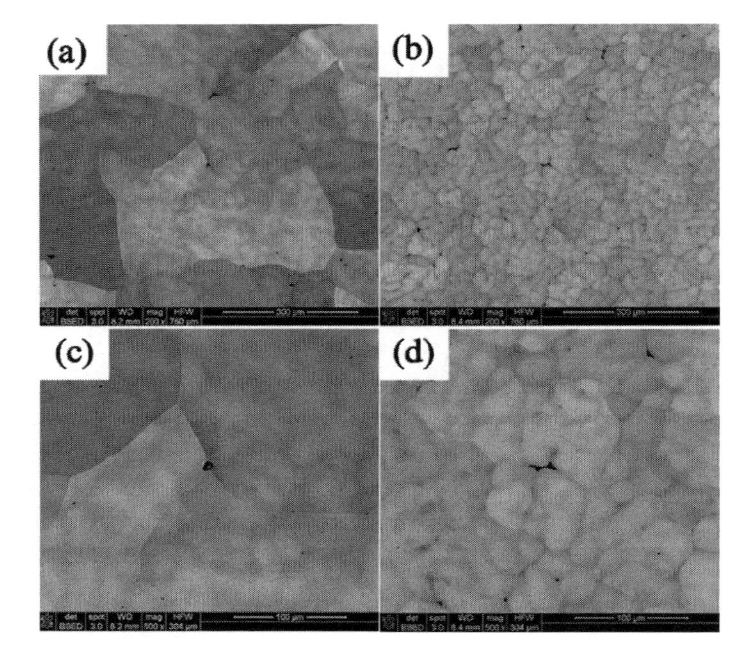

FIGURE 11.6 The back-scattered image of (a) (c) WNbMoTa and (b) (d)WNbMoTaV [5].

of element segregation increases as the temperature interval between the solid–liquid lines increases. However, dendrite segregation is a type of thermodynamically unstable intragranular segregation, so it can be eliminated after a long period of heat below the solidus line.

Figure 11.7 shows the microstructure of NbTiVMoAlx (x = 0, 0.25, 0.5, 0.75, 1.0, 1.5) multicomponent alloys with BCC structure. It can clearly be seen from the figure that all alloys exhibit a typical cast columnar dendritic microstructure and the secondary dendrite arms are approximately perpendicular to the primary dendrite arms. Such columnar dendrites and equiaxed dendrites are ubiquitous in multicomponent high entropy alloys. From the alloy solidification theory described above, it can be explained by the criterion of "supercooling" of the single-phase solid solution in the alloy.

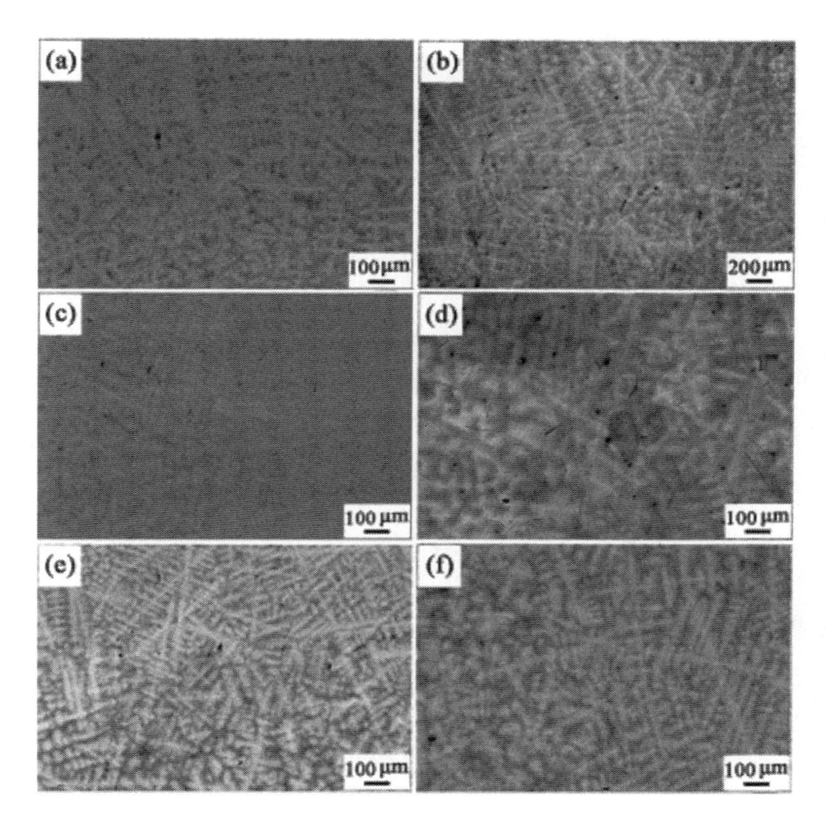

FIGURE 11.7 SEM image of NbTiVMoAl (a) x = 0; (b) x = 0.25; (c) x = 0.5; (d) x = 0.75; (e) x = 1; (f) x = 1.5.

The phase of the HCP structure is rarely observed in high entropy alloys. At present, there are a few reports on the single-phase HCP structure, mainly including precious metals and rare earth elements. Yusenko et al. [6] first prepared the Ir0.19Os0.22Re0.21Rh0.20Ru0.19 high entropy alloys containing five and six platinum group metals by thermal decomposition of single-source precursors not requiring high-temperature or mechanical alloying, as shown in Figure 11.8. They found that heat treatment up to 1,500 K and compression up to 45 GPa do not result in phase changes, which is a record temperature and pressure stability for a single-phase high entropy alloy.

11.2.2.3 Single-Phase Amorphous Structure

The high entropy amorphous alloy is an amorphous alloy designed based on the concept of a high entropy alloy. The phase formed is amorphous, but its composition characteristics satisfy the concept of high entropy alloy. This type of alloy not only has very good glass-forming ability, but also has a high mixing entropy. Therefore, it can be considered that the high entropy amorphous alloy has strong topological disorder and chemical disorder. The initial discovery of high entropy

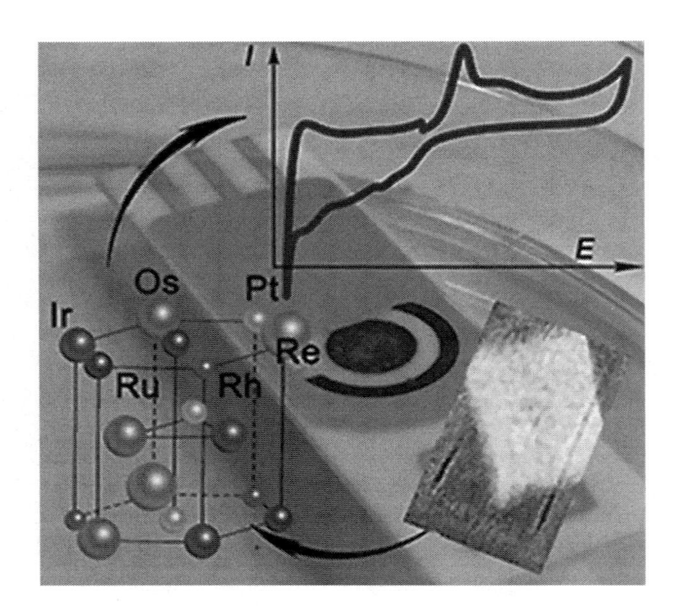

FIGURE 11.8 Schematic of the preparation method [6].

alloys was based through the development of bulk amorphous alloys in the 1990s, and efforts were made to find alloys with ultrahigh glass-forming abilities. In 1993, Greer in the University of Cambridge put forward the principle of the "Confusion Rule." He believed that the more components of the alloy, the higher the degree of chaos, the less likely it is to form a crystalline phase, and thus the more easily the amorphous phase is formed. However, the formation of amorphous alloys is also related to other factors such as atomic size difference. A variety of amorphous high entropy alloys have been discovered. Based on the Pd40Ni20Cu20P20 with great glass-forming ability and element substitution method, Takeuchi et al. [8] designed the Pd20Pt20Ni20Cu20P20 alloy and obtained the bulk amorphous alloys of 10 mm diameter by water quenching. The morphology and XRD pattern are shown in Figure 11.9. It can be seen that the surface of the rod sample exhibits a good metallic luster and is neither oxidized nor markedly defective. Its XRD pattern indicates that

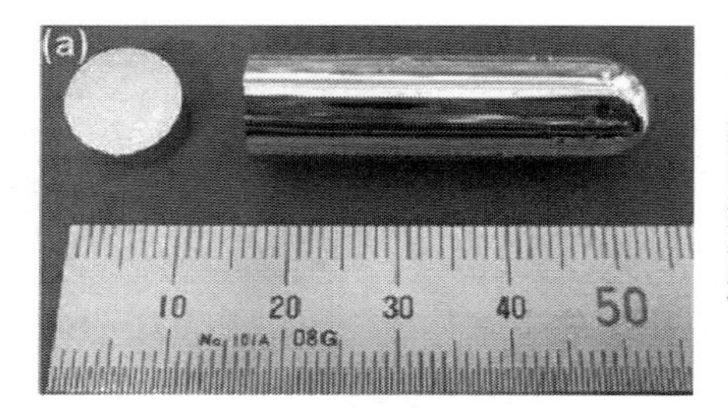

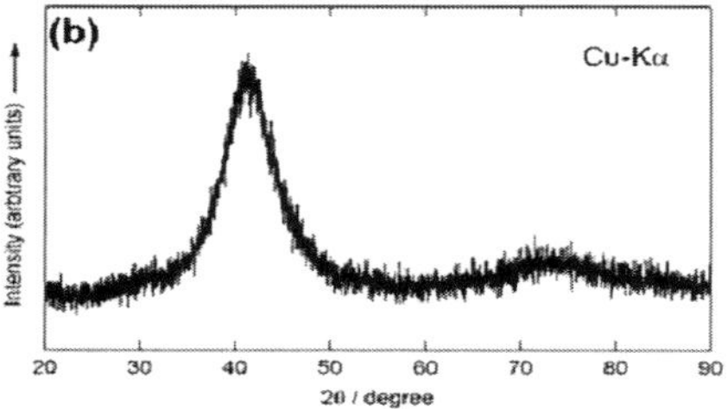

FIGURE 11.9 The morphology and XRD pattern of $Pd_{20}Pt_{20}Ni_{20}Cu_{20}P_{20}$ [8].

the alloy forms a single amorphous phase. In addition, through a similar composition substitution method, Takeuchi prepared CuNiPdTiZr amorphous alloy on the basis of Cu60Zr40, and Ding and Yao [7] prepared BeCuNiTiZr and ZrTiCuNiBe amorphous alloy on the basis of (ZrTi)40(CuNi)40Be20.

Zhao et al. [9] prepared the $Zn_{20}Ca_{20}Sr_{20}Yb_{20}(Li_{0.55}Mg_{0.45})_{20}$ alloy by induction melting and rapid suction casting to copper mold. The broadened diffraction peaks indicate that the alloy consists of a single amorphous phase in which no crystalline phase is present. In addition, significant glass transitions (Tg) and crystallization peaks can be observed on the DSC curve, which is consistent with the amorphous nature of the alloy. The alloy has a very low Tg point (323 K), high specific strength, good electrical conductivity, and an ultralow modulus of elasticity. When it is compressed and deformed at room temperature, it exhibits uniform steady-state rheological properties without any occurrence of shear bands. Furthermore, the *in vivo* animal tests showed that the Ca20Mg20Zn20Sr20Yb20 high entropy bulk metallic glass did not show any obvious degradation after four weeks of implantation, and they can promote osteogenesis and new bone formation after two weeks of implantation. The improved mechanical properties and corrosion behavior can be attributed to the different chemical composition, as well as the formation of a unique high entropy atomic structure with a maximum degree of disorder [10].

However, it is not clear whether the amorphization is due to the high mixing entropy effect. The high entropy amorphous phase can also be prepared by coating or mechanical alloying [11].

11.2.2.4 Multiphase Eutectic Structure

Although the high mixing entropy usually promotes the formation of a simple solid-solution structure, the complex interactions between various principal elements tend to make the alloy consist of multiple phases, including simple solid solutions, ordered solid solutions, intermetallic compounds, etc. The eutectic structure is a kind of very typical and special microstructure, and its formation is often associated with ordered phases and compounds. Eutectic alloys are very important in the foundry industry, and the reasons are that: (1) they have lower melting point than pure components, simplifying the operation of melting and casting; (2) they have better fluidity than pure metals, preventing the formation of dendrites that hinder liquid flow during solidification, thereby improving casting performance; (3) the constant temperature transition (no solidification temperature range) reduces casting defects, such as segregation and shrinkage; and (4) microstructures with different morphology can be obtained, especially the regularly arranged sheet or rod eutectic structure, which may become an *in situ* composite of superior performance. In addition, the eutectic structure has unique advantages in high-temperature use: (1) approximately balanced microstructure; (2) low phase boundary energy; (3) controllability of microstructure; (4) high breaking strength; (5) stable defect structure; and (6) strong resistance to high-temperature creep. Therefore, the design and preparation of eutectic high entropy alloys have important significance.

Lu et al. [12] used the vacuum induction furnace to prepare AlCoCrFeNi2.1 eutectic high entropy alloys. The microstructure is shown in Figure 11.10, which

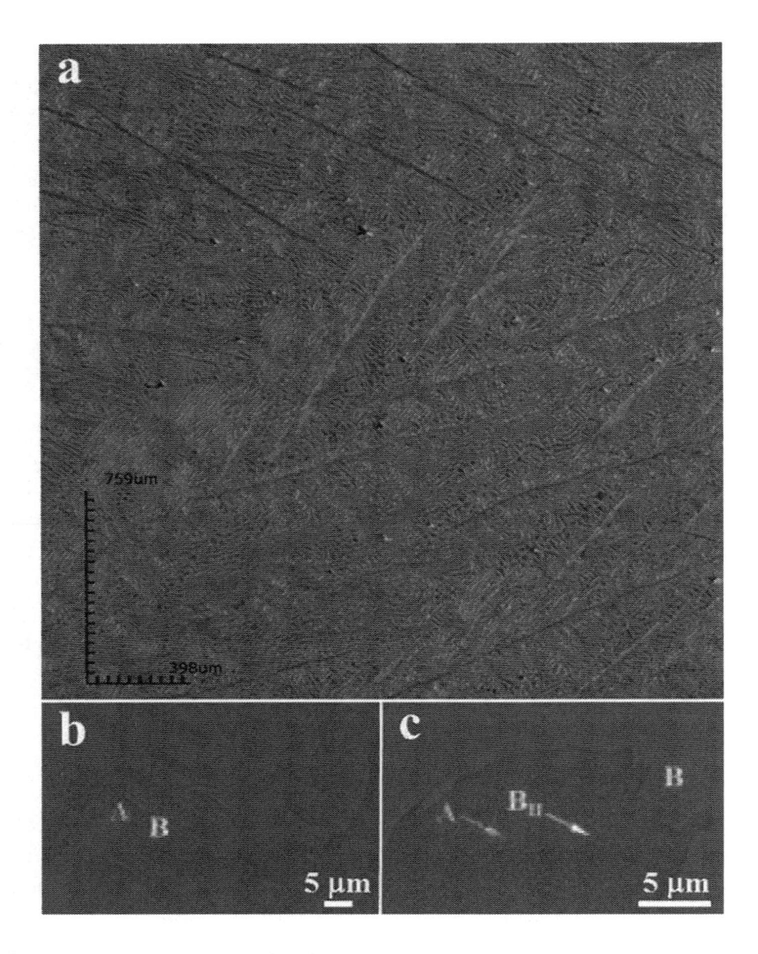

FIGURE 11.10 Microstructure of AlCoCrFeNi2.1 alloy: (a) photograph of lamellar eutectic structure of confocal laser scanning; (b) SEM image; (c) enlarged photo of (b), A, B is the phase corresponding to different slices, BII is the phase precipitated from A, and its composition is the same as B [12].

is a typical fine lamellar eutectic structure with a distance between the layers of approximately 2 μm.

Through compositional analysis, it is found that the layer B region is enriched with Al and Ni, and the interlayer A region is enriched with Fe, Co, and Cr. At the same time, a large number of nanoscale NiAl-based precipitates are dispersed in the interlayer A region. The XRD pattern indicated that the alloy exhibits FCC and B2 double phase (Figure 11.11), and then combined with the composition analysis, it can be inferred that the A region is the FCC phase, and the B region and the nano-precipitate phase BII are both the B2 phase. It can be seen from the DSC curve that the alloy has only one endothermic peak or exothermic peak during heating and cooling, which is a good proof that the alloy only undergoes eutectic reaction during solidification.

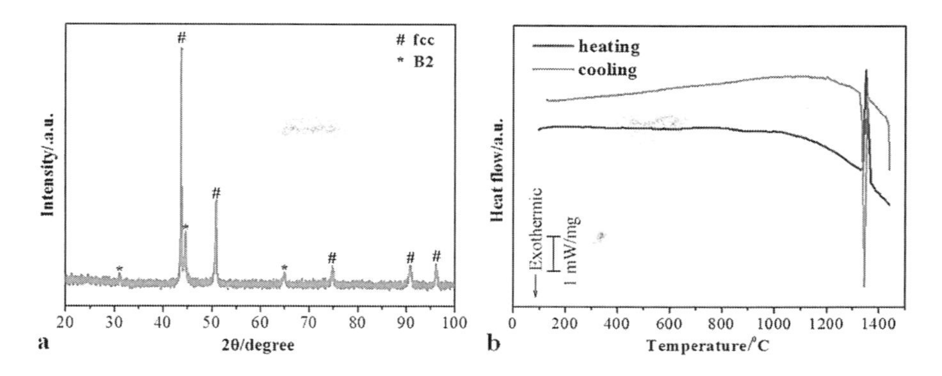

FIGURE 11.11 (a) The XRD pattern and (b) DSC curve of $AlCoCrFeNi_{2.1}$ [12].

Based on the mechanical properties, Guo et al. [13] designed an AlxCrCuFeNi2 high entropy alloy without a Co element. It is expected that the balance of plasticity and strength can be achieved by replacing Co with Ni, which has a high valence electron concentration. Figure 11.12 shows the corresponding microstructure of the alloy system.

There is a large number of uniformly distributed rod-shaped phases in the Al1.2CrCuFeNi2 alloy, and the average length is about 180 nm. This microstructure indicates that the alloy is close to the eutectic composition or is in the range of eutectic composition under non-equilibrium solidification. Combined with XRD analysis, the rod-shaped second phase is a B2 structure, and the matrix is an FCC structure. The microstructures of Al2.0CrCuFeNi2, Al2.2CrCuFeNi2, and Al2.5CrCuFeNi2 are similar, mainly composed of eutectic groups similar to a sunflower shape: the petal-like layer radiates outward from the central area, and the sunflower-like

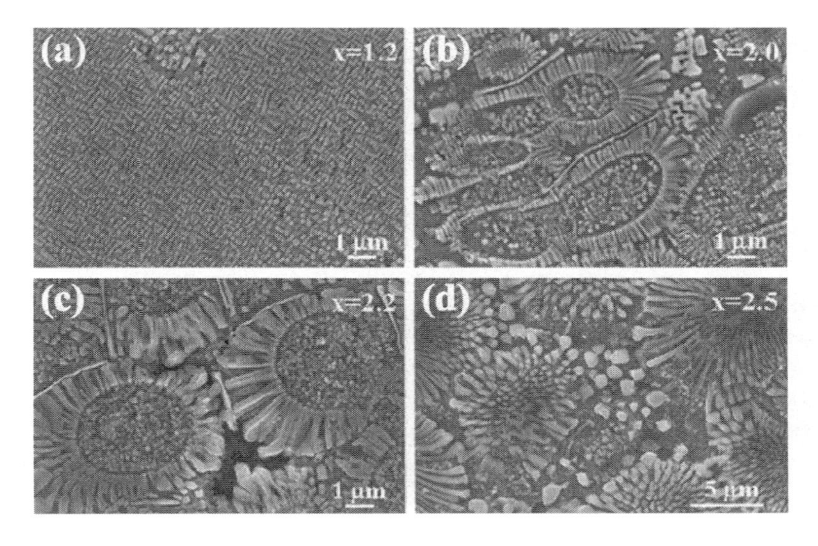

FIGURE 11.12 Microstructure of $Al_xCrCuFeNi_2$ high entropy alloys [13].

granular precipitate is homogeneously distributed in the central region. This morphology is rarely seen in conventional alloys, which can also be found in the as-cast pseudo-binary eutectic alloy NiAl (Ti)-Cr(Mo), and its schematic diagram is shown in Figure 11.13 [14].

In addition, eutectic structures are often found in alloys containing Nb and Mo.

Ma et al. [15] studied the effect of the addition of Nb on the microstructure and properties of AlCoCrFeNi alloy. It can be seen from the XRD pattern (Figure 11.14a) that the AlCoCrFeNi and AlCoCrFeNb0.1Ni alloys are single-phase BCC structures, while the AlCoCrFeNb0.25Ni, AlCoCrFeNb0.5Ni, and AlCoCrFeNb0.75Ni alloys are composed of a BCC phase and a Laves phase. The microstructures of AlCoCrFeNb0.25Ni and AlCoCrFeNb0.5Ni alloys are similar, and they all exhibit a hypo-eutectic structure. It can also be seen that the addition of Nb not only reduces the dendrite size, but also prepares the eutectic structure to be refined. Since the size of the interlayer spacing is inversely proportional to the degree of supercooling during solidification ($\lambda \propto (1/\Delta T)$), it is presumed that the addition of Nb decreases the degree of supercooling. Unlike the two alloys, the primary phase of the AlCoCrFeNb0.75Ni alloy is the Laves phase. Therefore, the alloy composition should be a hyper-eutectic. Combined with the phase formation and microstructure transformation of the alloy, a pseudo-binary phase diagram as a function of Nb content is plotted, as shown in Figures 11.14(b).

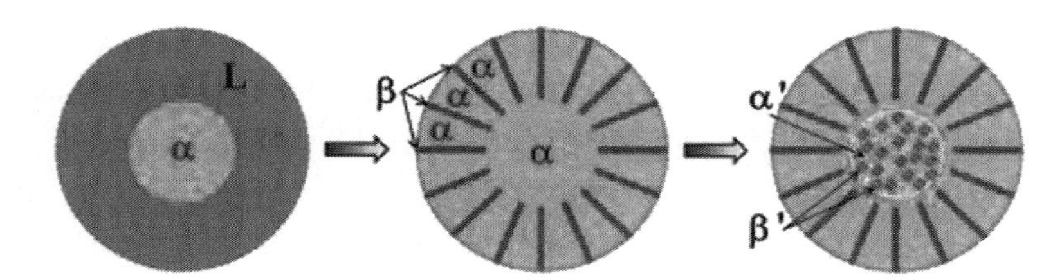

FIGURE 11.13 Schematic diagram of the formation mechanism of sunflower-like tissue: L is the liquid phase, α is the preliminary B2 phase, β is the A2 phase of eutectic layer, and α'β' is the B2 and A2 phase from α after spinodal decomposition [14].

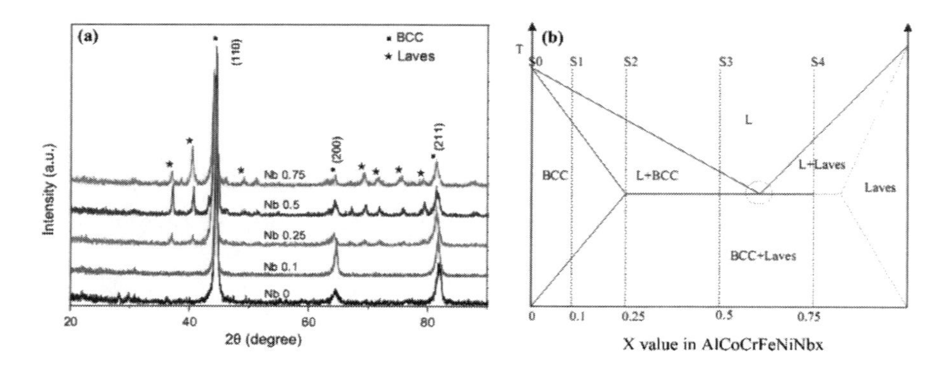

FIGURE 11.14 (a) XRD analysis of $AlCoCrFeNiNb_x$; (b) phase diagram of $AlCoCrFeNiNb_x$, S0–S4 corresponding to x = 0, 0.1, 0.25, 0.5, 0.75, respectively [15].

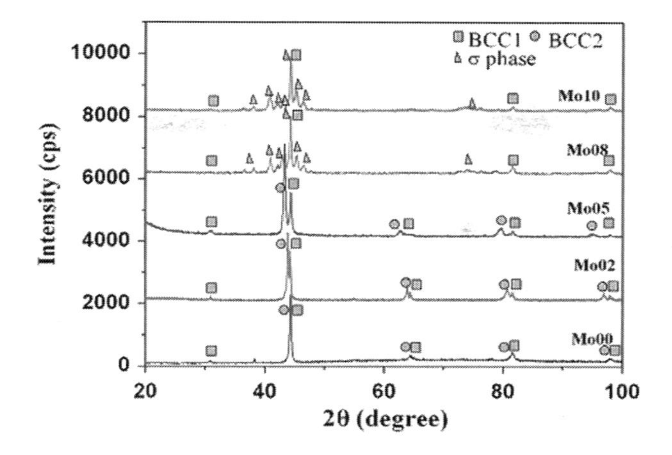

FIGURE 11.15 The XRD pattern of AlCrFeNiMox [16].

Dong et al. [16] studied the evolution of microstructure in AlCrFeNiMox (x = 0, 0.2, 0.5, 0.8, 1.0) alloys with Mo content. It can be seen from Figure 11.15 that Mo00, Mo02, and Mo05 alloys are two-phase BCC structures. When Mo is added to the alloy, the lattice constants of the two BCC phases are changed, and therefore, there are two sets of diffraction peaks in XRD. And as the Mo content increases, the diffraction peak shifts to the left, indicating that the crystal lattice is expanding and the lattice constant is gradually increasing. When the Mo content is higher, the solubility of Mo in the BCC2 phase reaches the limit, and BCC2 completely transforms into the FeCrMo-σ phase. It can be seen from the secondary electron scanning photograph (Figure 11.16) that the Mo00 alloy is a typical regular lamellar eutectic structure and the thickness of the sheet gradually increases. According to previous studies, the eutectic structure is a composite of an AlNi-type intermetallic compound and a FeCr-type solid solution. Compared with the Mo00 alloy, the layer in the Mo02 alloy is thickened, and the NiAl phase is rectangular or block-shaped at the edge of the grain boundary. When the Mo content is increased, the Mo05 alloy exhibits a hypoeutectic structure, and when it further increases to over 0.5, the Mo08 and Mo10 alloys exhibit a hypereutectic structure.

11.2.2.5 Other Structures with Multiphase

In addition to the eutectic structure, when the high entropy alloy contains multiphases, it will also exhibit different microstructures, for example, dendritic structures, amplitude-decomposed structures, and the like. In addition, high entropy alloy in the as-cast state is often obtained in a non-equilibrium solidified state, and its phase structure is in an unstable state. When the heat treatment is carried out at different temperatures, the phases and microstructures of the alloy change significantly.

Figure 11.17 shows the XRD pattern of the AlxCoCrFeNi high entropy alloy. [17] It can be seen from the figure that when $x \leq 0.3$, the alloy is single phase structure of FCC, and as the Al content increases, the proportion of BCC phase increases continuously. When $0.5 \leq x \leq 0.75$, the ordered BCC phase appears and there are

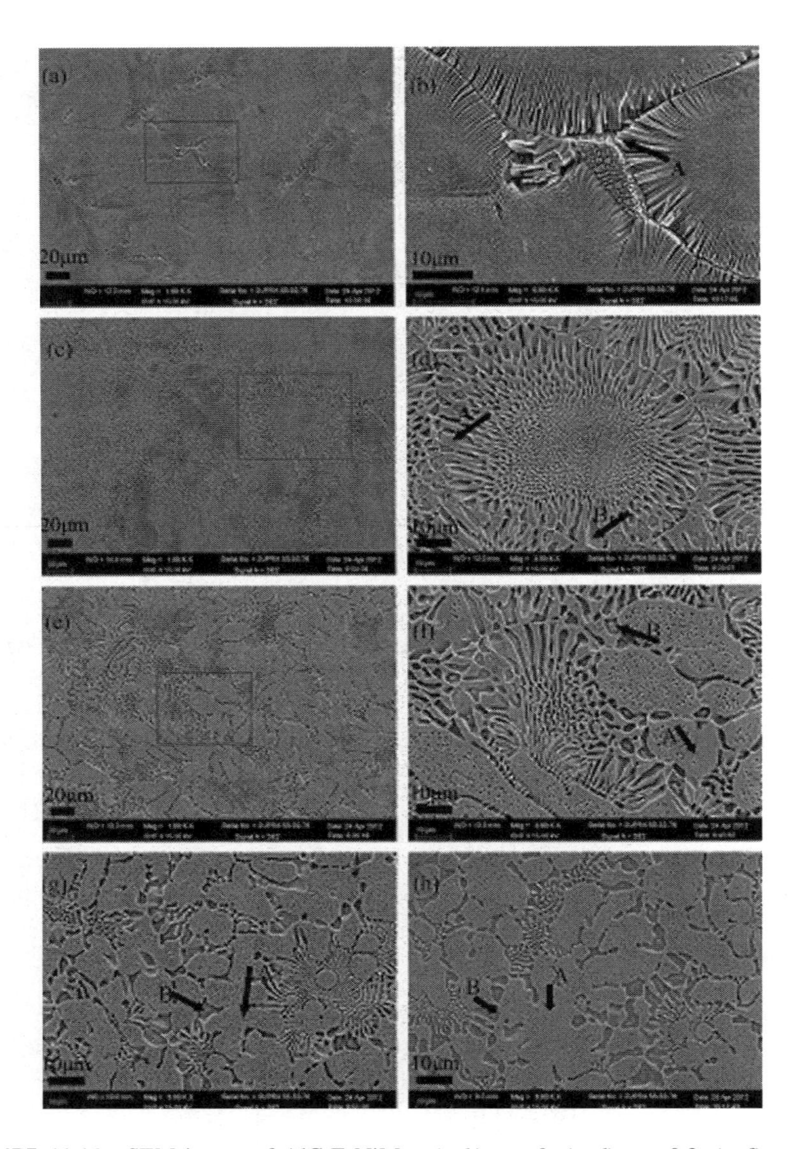

FIGURE 11.16 SEM image of AlCrFeNiMox (a, b) x = 0; (c, d) x = 0.2; (e, f) x = 0.5; (g) x = 0.8; (h) x = 1.0 [16].

BCC, FCC, and B2 structures coexisting. In addition, when x ≥ 1, the FCC phase disappears. Since the diffraction peak of the ordered B2 phase overlaps with that of the disordered BCC phase when the Al content is high, the alloy may be a disordered BCC phase and an ordered B2 phase double-phase structure or a B2 single-phase structure.

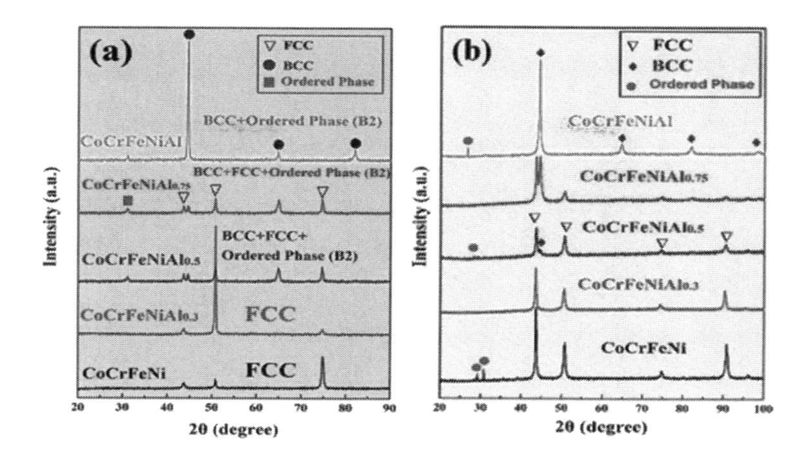

FIGURE 11.17 The XRD pattern of AlxCoCrFeNi prepared by (a) supergravity and (b) casting with increasing Al content [17].

In terms of a single-phase HCP structure, its appearance is often accompanied by the formation of other phases. Tsau et al. [18] added Ti in FeCoNi to prepare an alloy of HCP and FCC double phase structure, in which the ordered HCP phase is the dendritic region and the interdendritic region is the eutectic structure. That is to say, granular HCP phase distribution on the ordered FCC matrix. In addition, Shun et al. [19] found that the interdendritic region of the as-cast Al0.3CoCrFeNiTi0.1 alloy is an HCP structure. In 2016, Wu et al. [20] and Tracy et al. [21] simultaneously found that a CoCrFeMnNi high entropy alloy can be transformed from FCC to HCP structure under high pressure, and the alloy can maintain HCP structure after reducing pressure (as shown in Figure 11.18).

Wu believes the irreversible phase transition results from the similar free energy of the HCP and FCC, as well as the large energy barrier between them.

In high entropy alloys, in addition to the disordered solid-solution phase, ordered phases and intermetallic compounds often occur, especially when the atomic size difference in the alloy is relatively large, and the interaction between some elements is very strong, such as the B2 phase, σ phase, and Laves phase. In conventional steel, the σ phase often appears in the Cr-containing alloy, and its composition is close to the equiatomic ratio of FeCr with a tetragonal structure. When a high entropy alloy contains Fe, Co, Cr, or Mo elements, the σ phase is also often formed. However, this σ phase is a solid-solution phase with multiple elements. The larger-sized atom occupies one type of lattice, and a smaller-sized atom occupies another set of lattices, thereby forming a closely arranged crystal structure.

In short, the concept of a high entropy alloy breaks the design concept of traditional alloys and provides new ideas for the design of new alloys. High entropy alloys have become a new research hotspot in the field of materials due to their unique phase structure and microstructure.

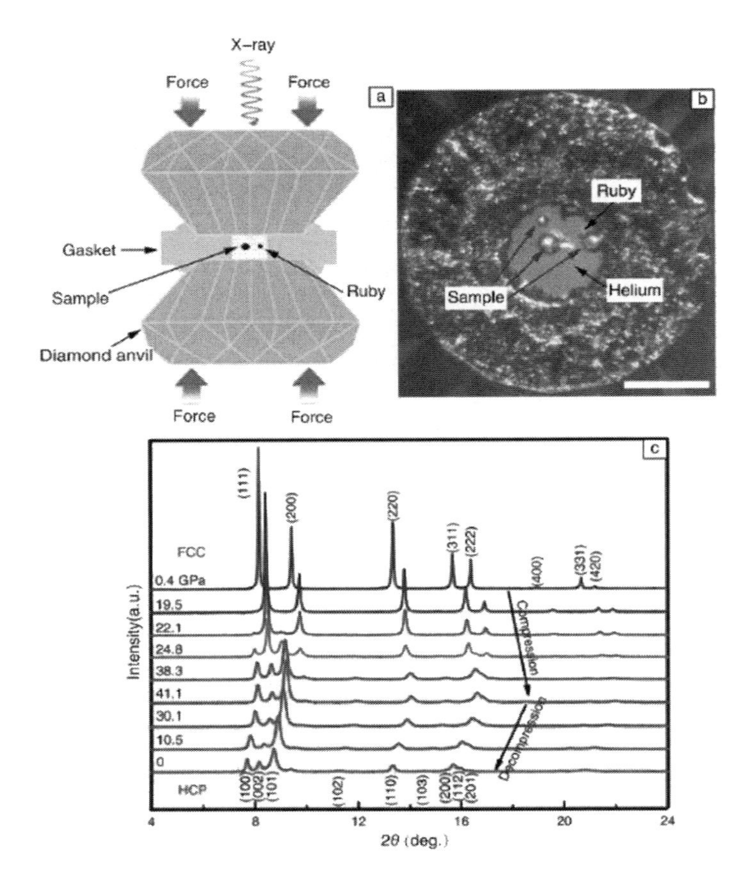

FIGURE 11.18 Experimental setup and the *in situ* high-pressure XRD patterns of the CoCrFeMnNi HEA in a DAC during compression and decompression at room temperature [20].

11.3 SERRATED FLOW IN HIGH ENTROPY ALLOYS

11.3.1 The Concept and Examples of Serrated Flow

Serrated flow refers to the stress-strain collapse phenomenon generated during the plastic deformation of the material, which is the external expression of the local instability of the material during the deformation. It exhibits disordered distribution in time and space, and its formation has a close relationship with the flow unit of materials. As a typical feature of the disordered response in the space-time correlation field, this phenomenon is widely present in several kinds of materials such as low carbon steel, aluminum-magnesium alloy, amorphous alloy, and nano-material, such as the yielding platform in low carbon steel. The serrated flow during plastic deformation can objectively reflect some characteristics of its deformation mechanism, such as the interaction of interstitial solute atoms or replacement solute atoms with dislocations, local shear instability, grain boundary migration, twinning etc. At the same time, the characteristics of the serrated flow are affected by many external factors such as temperature, strain rate, and the heat treatment process, and internal factors such as composition, grain morphology, size, and phase composition.

11.3.2 Serrated Flow in Steel

The study of serrated flow in steel materials was first proposed by Cottrell in the study of low carbon steel [22]. They found that in the tensile process of carbon steel, there is a phenomenon of a "yield platform," which is actually a typical performance of serrated flow in carbon steel materials. This occurs because of the interaction of solid solution or interstitial atoms with dislocations during deformation. Studies have shown that at different strain rates, the upper yield stress value of carbon steel changes significantly, while the lower yield stress does not change much. This is because the loading rate can affect the effect of interstitial atoms and dislocations in the material, while the lower yield stress reflects the intrinsic properties of the material and therefore has little effect on the loading rate.

Subsequently, in the study of Fe-Mn steel, due to the addition of 17–20% Mn, the steel remained in the austenite state at room temperature, and a large number of twins were generated during the deformation process. This allows the material to maintain significant strain strengthening during plastic deformation, delaying the occurrence of necking and improving the plasticity of the material. With the large number of deformation twins, the stress-strain curve of high Mn steel also shows obvious serrated flow. Figure 11.19 is a stress-strain curve of Fe-20 wt.% Mn-1.2 wt.% C TWIP steel at different temperatures [23]. It can be seen from the figure that even at the same strain rate, the difference in service temperature can lead to significantly different flow characteristics.

In 2013, Choudhary et al. [24] studied the serrated flow of 9Cr-1Mo steel and summarized the effects of temperature and strain rate on it. They believe that it is the repetitive and systematic fluctuation of the load on the stress-strain curve during plastic deformation. Different physical processes can cause different changes in the serrated flow characteristics. It is divided into seven categories: (1) sudden and instantaneous increase of movable dislocation density or velocity; (2) interaction of dislocation movement and solute atom diffusion; (3) order–disorder transition caused by dislocation movement; (4) continuous twinning deformation; (5) phase transition caused by stress; 6) local temperature rise resulting from adiabatic shear; (7) the yielding phenomenon of the fracture surface of the brittle material under specific stress state. The most widely recognized is the dynamic strain aging effect (DSA effect), that is, the dynamic interaction in the plastic deformation of the aggregation of solute atoms near dislocations, and the dislocation.

11.3.3 Serrated Flow in the Amorphous Alloys

Because there are no defects such as dislocations, the bulk metallic glass exhibits some excellent properties and also has a completely different plastic deformation mechanism than the conventional crystalline material. There are two main types

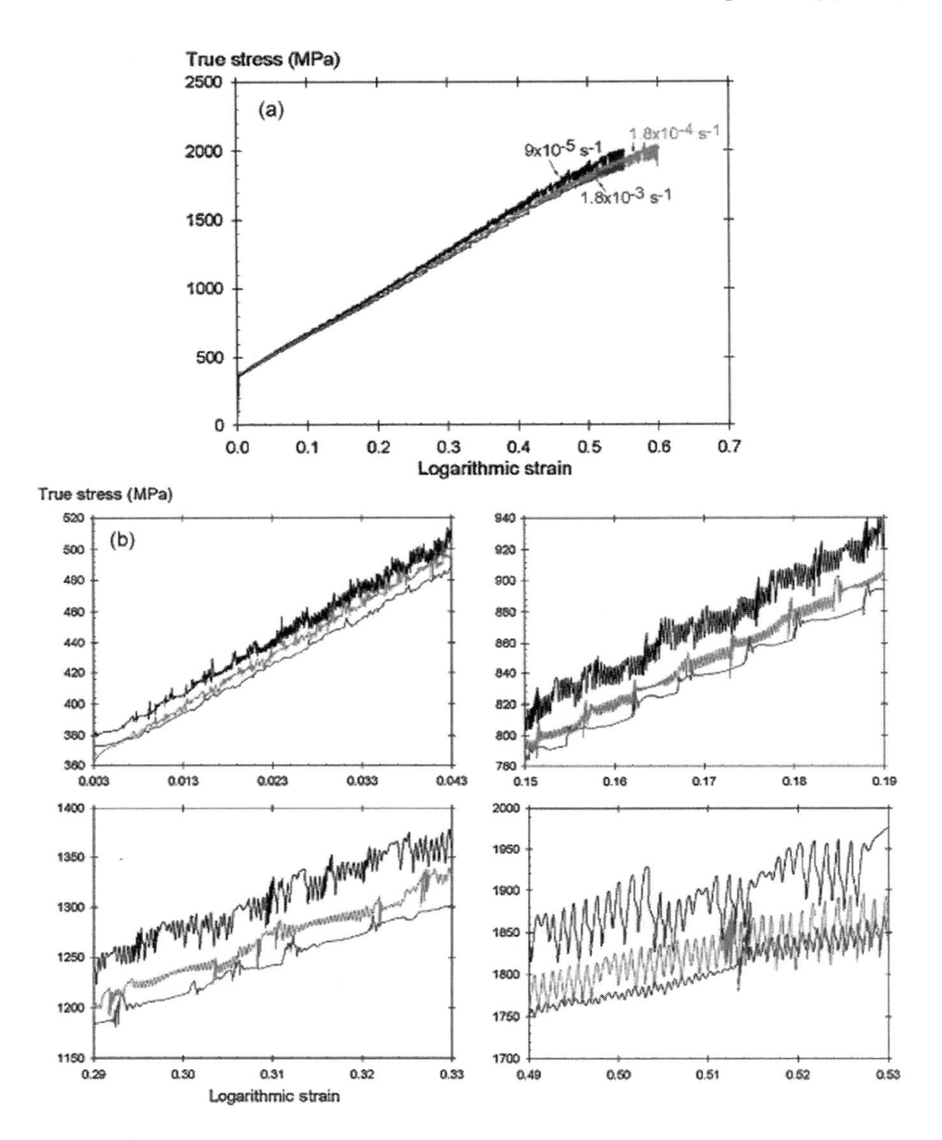

FIGURE 11.19 (a) True stress logarithmic strain curves at the different tested strain rates at room temperature, and (b) insets showing in more detail the evolution of the serrations at these strain rates [23].

of plastic deformation of amorphous alloys: one is homogeneous deformation below the glass transition temperature and below the crystallization temperature; the other is inhomogeneous plastic deformation below the glass transition temperature [25]. The free volume and shear transition regions are considered to be weak regions or flow units of amorphous alloy deformation [26], and recently the weak regions have been studied as soft points, approximate liquid regions, or

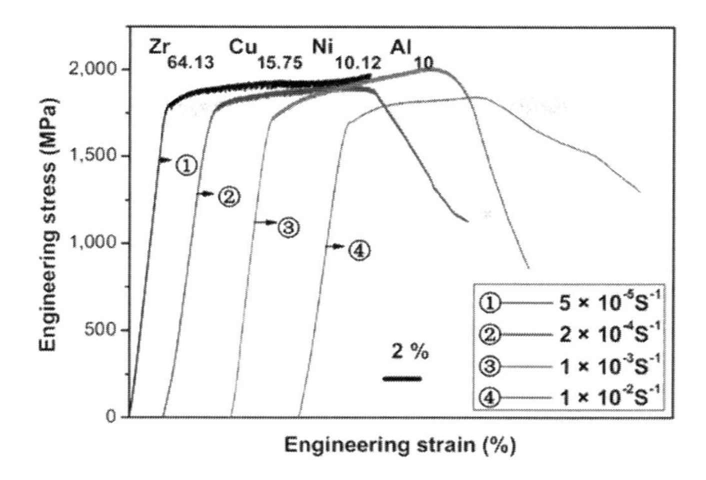

FIGURE 11.20 The compressive engineering stress-strain curves at different strain rates at room temperature [27].

β relaxation. In amorphous alloys, the serrated flow behavior is often observed. These features are primarily dependent on the loading rate. Figure 11.20 shows the serrated flow characteristics of a typical amorphous alloy [27]. It can be seen that as the strain rate increases, the flow becomes less and less noticeable. Figure 11.21 is a stress time curve of ZrCuNiAl bulk metallic glass at different

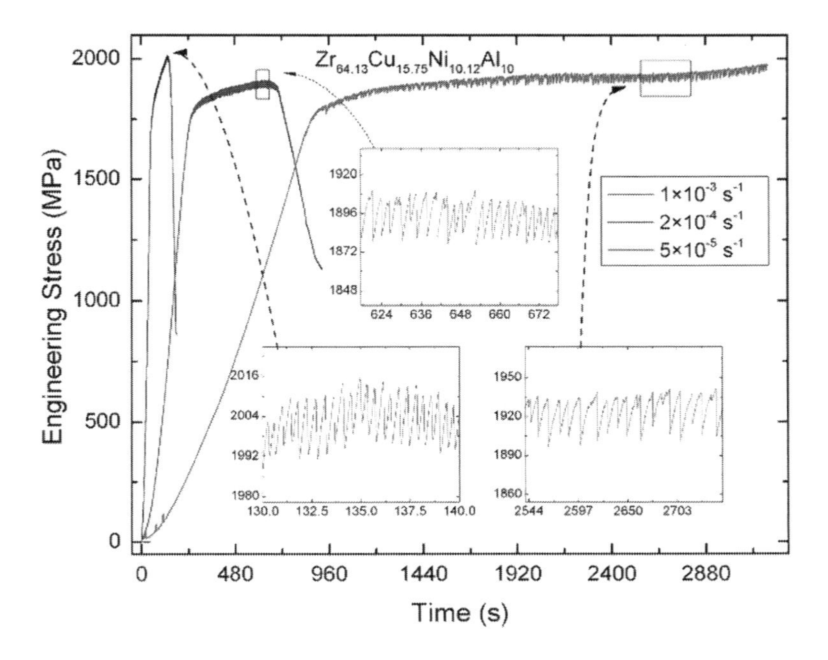

FIGURE 11.21 Compression stress time profiles of $Zr_{64.13}Cu_{15.75}Ni_{10.12}Al_{10}$ ingots [28].

strain rates [28]. It can be seen that as the strain rate increases, the waiting time between stress collapses increases, that is, the flow features are denser. The serrated flow deformation that occurs in amorphous materials is generally considered to be closely related to the generation and expansion of multiple shear bands.

11.3.4 Serrated Flow in High Entropy Alloys

High entropy alloys are usually solid-solution structure formed of four elements and above with high mixing entropy. There are no major elements in high entropy alloys, so it is difficult to distinguish between solute or solvent elements. The

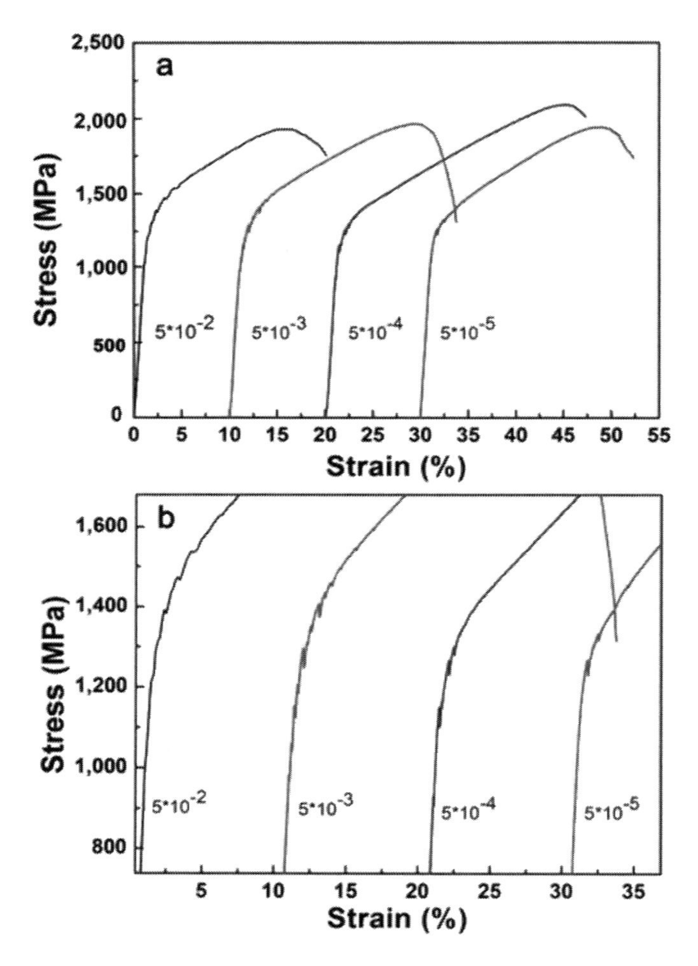

FIGURE 11.22 (a) Stress-strain curve of NbTiMoV alloy; (b) local magnification around the yield points [29].

unique solid-solution structure makes the plastic deformation mechanism of high entropy alloys different from that of conventional alloys. First, the dislocation structure of a high entropy alloy may be separated by solid solution atoms and is not as straight as a dislocation line in a pure metal. Chen et al. found that the serrated flow of high entropy alloys is sensitive to composition, and this represents the first such reported phenomenon at room temperature in HEAs. Figure 11.22 shows the serrated flow in NbTaMoV high entropy alloys at different strain rates [29]. These flows are concentrated in the transition region between elastic and plastic deformation.

High entropy alloys also exhibit serrated flow under extremely low-temperature conditions [30]. Liu et al. [31] established a self-consistent model for the serrated flow behavior of high entropy alloys under ultralow temperature environment. The fractal dimension and the largest Lyapunov exponent for low-temperature deformation of high entropy alloys are obtained by suitable time delay and phase space dimensions. The positive value of the largest Lyapunov exponent also indicates that the serrated flow of CoCrFeNi alloy during tensile deformation in an ultralow temperature environment is unstable, which is closely related to FCC-HCP phase transition. Figure 11.23 shows the tensile properties of the CoCrFeNi alloy at different temperatures.

In addition, the serrated flow behaviors are also observed in nano-sized high entropy alloys [32], and are closely related to crystal orientation and sample size, as shown in Figure 11.24.

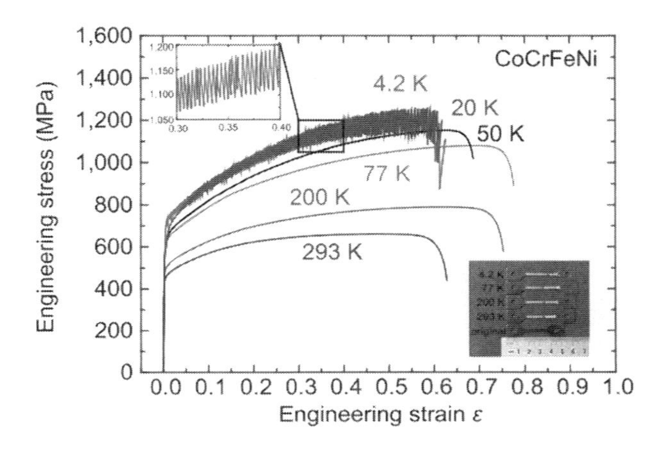

FIGURE 11.23 Tensile properties of the CoCrFeNi alloy at different temperatures [31].

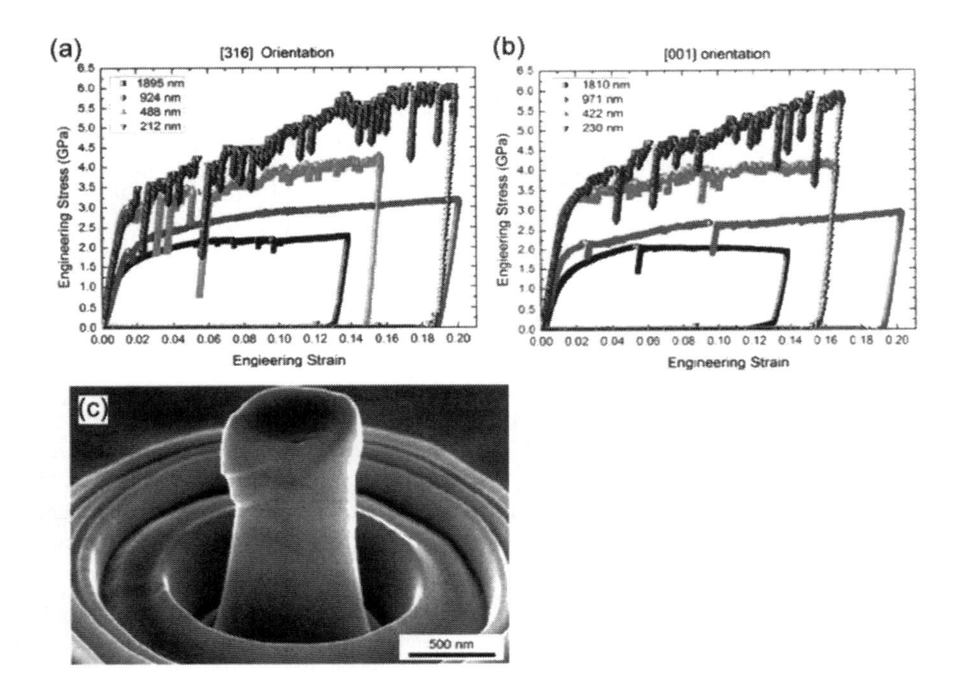

FIGURE 11.24 Compression properties of samples of different sizes: (a) [316]-oriented; (b) [001]-oriented single; (c) sample after compression [32].

REFERENCES

1. Ma, S., S. Zhang, M. Gao, P. Liaw, and Y. Zhang, A successful synthesis of the CoCrFeNiAl 0.3 single-crystal, high-entropy alloy by Bridgman solidification. *Jom*, 2013. 65(12): p. 17511758.
2. Ma, S.G., S.F. Zhang, J.W. Qiao, Z.H. Wang, M.C. Gao, Z.M. Jiao, H.J. Yang, and Y. Zhang, Superior high tensile elongation of a single-crystal CoCrFeNiAl0.3 high-entropy alloy by Bridgman solidification. *Intermetallics*, 2014. 54: p. 104–109.
3. Patriarca, L., A. Ojha, H. Sehitoglu, and Y.I. Chumlyakov, Slip nucleation in single crystal FeNiCoCrMn high entropy alloy. *Scripta Materialia*, 2016. 112: p. 54–57.
4. Otto, F., N.L. Hanold, and E.P. George, Microstructural evolution after thermomechanical processing in an equiatomic, single-phase CoCrFeMnNi high-entropy alloy with special focus on twin boundaries. *Intermetallics*, 2014. 54: p. 39–48.
5. Senkov, O.N., G.B. Wilks, D.B. Miracle, C.P. Chuang, and P.K. Liaw, Refractory high-entropy alloys. *Intermetallics*, 2010. 18(9): p. 1758–1765.
6. Yusenko, K.V., S. Riva, P.A. Carvalho, M.V. Yusenko, S. Arnaboldi, A.S. Sukhikh, M. Hanfland, and S.A. Gromilov, First hexagonal close packed high-entropy alloy with outstanding stability under extreme conditions and electrocatalytic activity for methanol oxidation. *Scripta Materialia*, 2017. 138: p. 22–27.
7. Ding, H.Y. and K.F. Yao, High entropy Ti20Zr20Cu20Ni20Be20 bulk metallic glass. *Journal of Non-Crystalline Solids*, 2013. 364: p. 9–12.
8. Takeuchi, A., N. Chen, T. Wada, W. Zhang, Y. Yokoyama, A. Inoue, and J.W. Yeh, Alloy design for high-entropy bulk glassy alloys. *Procedia Engineering*, 2012. 36: p. 226–234.

9. Zhao, K., X.X. Xia, H.Y. Bai, D.Q. Zhao, and W.H. Wang, Room temperature homogeneous flow in a bulk metallic glass with low glass transition temperature. *Applied Physics Letters*, 2011. 98(14): p. 141913.
10. Li, H.F., X.H. Xie, K. Zhao, Y.B. Wang, Y.F. Zheng, W.H. Wang, and L. Qin, In vitro and in vivo studies on biodegradable CaMgZnSrYb high-entropy bulk metallic glass. *Acta Biomaterialia*, 2013. 9(10): p. 8561–8573.
11. Zhang, Y., X. Yan, J. Ma, Z. Lu, and Y. Zhao, Compositional gradient films constructed by sputtering in a multicomponent Ti–Al–(Cr, Fe, Ni) system. *Journal of Materials Research*, 2018: p. 1–9.
12. Lu, Y., Y. Dong, S. Guo, L. Jiang, H. Kang, T. Wang, B. Wen, Z. Wang, J. Jie, Z. Cao, H. Ruan, and T. Li, A promising new class of high-temperature alloys: Eutectic high-entropy alloys. *Scientific Reports*, 2014. 4: p. 6200.
13. Guo, S., C. Ng, and C.T. Liu, Anomalous solidification microstructures in Co-free AlxCrCuFeNi2 high-entropy alloys. *Journal of Alloys and Compounds*, 2013. 557: p. 77–81.
14. Guo, S., C. Ng, and C.T. Liu, Sunflower-like solidification microstructure in a Near-eutectic high-entropy alloy. *Materials Research Letters*, 2013. 1(4): p. 228–232.
15. Ma, S.G. and Y. Zhang, Effect of Nb addition on the microstructure and properties of AlCoCrFeNi high-entropy alloy. *Materials Science and Engineering: Part A*, 2012. 532: p. 480486.
16. Dong, Y., Y. Lu, J. Kong, J. Zhang, and T. Li, Microstructure and mechanical properties of multicomponent AlCrFeNiMox high-entropy alloys. *Journal of Alloys and Compounds*, 2013. 573: p. 96–101.
17. Li, R.X., P.K. Liaw, and Y. Zhang, Synthesis of Al x CoCrFeNi high-entropy alloys by high gravity combustion from oxides. *Materials Science and Engineering: Part A*, 2017. 707: p. 668673.
18. Tsau, C.-H., Phase transformation and mechanical behavior of TiFeCoNi alloy during annealing. *Materials Science and Engineering: Part A*, 2009. 501(1): p. 81–86.
19. Shun, T.-T., C.-H. Hung, and C.-F. Lee, The effects of secondary elemental Mo or Ti addition in Al0. 3CoCrFeNi high-entropy alloy on age hardening at 700 C. *Journal of Alloys and Compounds*, 2010. 495(1): p. 55–58.
20. Zhang, F., Y. Wu, H. Lou, et al, *Nature Communications*, 2017. 8: p. 15687-1-7.
21. Tracy, C.L., S. Park, D.R. Rittman, et al, *Nature Communications*, 2017. 8: p. 15634-1-6.
22. Cottrell, A.H. LXXXVI. A note on the Portevin-Le Chatelier effect. *The London, Edinburgh, and Dublin Philosophical Magazine and Journal of Science*, 1953. 44(355): p. 829–832.
23. Renard, K., S. Ryelandt, and P.J. Jacques, Characterisation of the Portevin-LeChatelier effect affecting an austenitic TWIP steel based on digital image correlation. *Materials Science and Engineering: Part A*, 2010. 527(12): p. 2969–2977.
24. Choudhary, B.K., Influence of strain rate and temperature on serrated flow in 9Cr-1Mo ferritic steel. *Materials Science and Engineering: Part A*, 2013. 564: p. 303–309.
25. Spaepen, F., A microscopic mechanism for steady state inhomogeneous flow in metallic glasses. *Acta Metallurgica*, 1977. 25(4): p. 407–415.
26. Schuh, C.A., T.C. Hufnagel, and U. Ramamurty, Mechanical behavior of amorphous alloys. *Acta Materialia*, 2007. 55(12): p. 4067–4109.
27. Qiao, J.W., Y. Zhang, and P.K. Liaw, Serrated flow kinetics in a Zr-based bulk metallic glass. *Intermetallics*, 2010. 18(11): p. 2057–2064.
28. Antonaglia, J., X. Xie, G. Schwarz, M. Wraith, J. Qiao, Y. Zhang, P.K. Liaw, J.T. Uhl, and K.A. Dahmen, Tuned critical avalanche scaling in bulk metallic glasses. *Scientific Reports*, 2014. 4(3): p. 4382.
29. Chen, S.Y., X. Yang, K.A. Dahmen, P.K. Liaw, and Y. Zhang, Microstructures and crackling noise of AlXNbTiMoV high entropy alloys. *Entropy*, 2014. 16(12): p. 870–884.

30. Antonaglia, J., X. Xie, Z. Tang, C.W. Tsai, J.W. Qiao, Y. Zhang, M.O. Laktionova, E.D. Tabachnikova, J.W. Yeh, O.N. Senkov, M.C. Gao, J.T. Uhl, P.K. Liaw, and K.A. Dahmen, Temperature effects on deformation and serration behavior of high-entropy alloys (HEAs). *Jom*, 2014. 66(10): p. 2002–2008.
31. Liu, J., X. Guo, Q. Lin, Z. He, X. An, L. Li, P.K. Liaw, X. Liao, L. Yu, J. Lin, L. Xie, J. Ren, and Y. Zhang, Excellent ductility and serration feature of metastable CoCrFeNi high entropy alloy at extremely low temperatures. *Science China Materials*, 2018.
32. Zou, Y., S. Maiti, W. Steurer, and R. Spolenak, Size-dependent plasticity in an $Nb_{25}Mo_{25}Ta_{25}W_{25}$ refractory high-entropy alloy. *Acta Materialia*, 2014. 65(6): p. 85–97.

Section C

Applications

12 Benefits of the Selection and Use of High Entropy Alloys for High-Temperature Thermoelectric Applications

Samrand Shafeie and Sheng Guo

CONTENTS

12.1 INTRODUCTION

12.1.1 Thermoelectric Applications

12.1.1.1 Broad Background

Efficiently converting waste heat to electrical energy to meet the needs of a sustainable clean energy society is a great challenge for the scientific community. Thermoelectric materials are solid state materials that create a significant chemical potential difference, $\Delta\mu$, between the two regions of different temperatures. This chemical potential difference caused by the Seebeck effect creates a potential difference (i.e., Seebeck voltage, ΔV) which is expressed as the Seebeck coefficient, $|S|$ (i.e., $\left|\frac{\Delta V}{\Delta T}\right|$). The general expression is shown in Equation (12.1).

$$|S| = \left|\frac{\Delta V}{\Delta T}\right| = \left|\frac{\Delta\mu}{e}\right| \tag{12.1}$$

where ΔV, ΔT, and $\Delta\mu$ are the voltage difference, the temperature difference, and the chemical potential difference between the hot and the cold side, respectively, and e is the electronic charge (Figure 12.1).

The heat on one side will cause some of the energy being transferred through lattice vibrations (phonons) and electrons (or holes) to the cold side and eventually cause the cold side to heat up. This process will keep going until the difference in chemical potential has vanished, and it will eventually cause the Seebeck effect to vanish unless the temperature difference can be maintained. This is achieved by fine-tuning the material properties so that the thermoelectric effect can be maintained with maximum heat-to-electricity conversion. Different materials are usually compared through their thermoelectric figure of merit (zT), defined as:

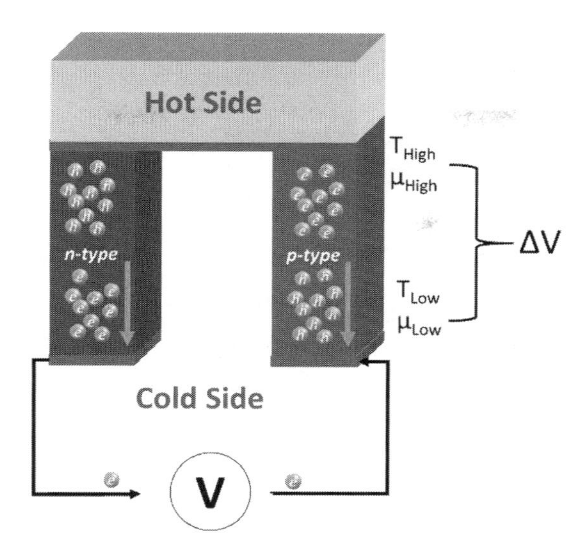

FIGURE 12.1 Illustration of the thermoelectric effect.

$$zT = \frac{S^2\sigma}{(\kappa_e + \kappa_{\text{Latt}})} * T \tag{12.2}$$

where σ is the electronic conductivity, T is the temperature in Kelvin, and κ_e and κ_{Latt} are the electronic and lattice vibration contributions to the total thermal conductivity (κ_{tot}), respectively. It becomes clear from Equation (12.2) that finding a highly efficient thermoelectric material will require optimization of interrelated material properties, and this is often illustrated through a simplified figure (see Figure 12.2).

The conversion efficiency for different temperature differences for a material can be calculated through Equation (12.3):

$$\eta(\%) = \frac{T_H - T_C}{T_H} * \frac{\sqrt{1 + zT} - 1}{\sqrt{1 + zT} + T_C / T_H} \tag{12.3}$$

where T_H and T_C are the temperatures at the hot and the cold side of the thermoelectric material, zT is the thermoelectric figure of merit from (2), and η the conversion efficiency (%) [1, 2]. Using Equation (12.3), we find that two different materials with $zT \approx 1$ and $zT \approx 3$, using 800°C, will have conversion efficiencies of ~18% and ~32%, respectively. It is therefore desirable to achieve a $zT > 3$ to be able to compete with e.g., heat engines ($\eta \approx$ 30–50%). However, the zT is assumed to be temperature independent when zT is calculated, and will only give an estimated efficiency at the peak zT temperature. In most materials, the thermoelectric properties need optimization and require the thermoelectric properties to be tuned to achieve the maximum efficiency. As shown in Figure 12.2, a good thermoelectric material should, according to the common understanding, have properties in between an insulator and a metal.

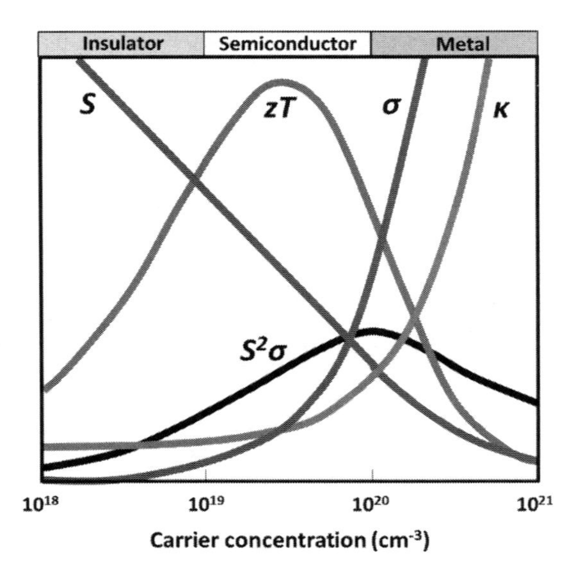

FIGURE 12.2 Interdependence of zT from S, σ, $S^2\sigma$, and κ as a function of the number of total charge carriers [2].

Since the k_{tot} is large for metallic materials, due to the large contribution from the k_e, which results in very low Seebeck coefficients, metal alloys have mostly been avoided in the search for highly efficient thermoelectric materials. Some of the more well-known metal alloy compounds are half-Heusler compounds and quasi-crystals, where their inherent properties make them suitable as potential thermoelectric materials [3–7].

From Table 12.1 it can be seen that the properties required from a thermoelectric material require that the materials have properties similar to a semi-metallic material to balance most of the interconnected properties.

12.1.1.2 Thermoelectric–Figure of Merit

As the general definition of zT suggests, σ, S, and κ are interrelated material properties that need to be separately adjusted in order to increase zT beyond the current state-of-the-art levels ~2 @ 727°C, in e.g., $Cu_2Se_{1-x}I_x$ [8]. More specifically, since

TABLE 12.1
Approximate Range of Values for *S*, and σ for Metals, Semiconductors, and Insulators [2]

Property	Metals	Semiconductor	Insulators
S (μVK^{-1})	~ 5	~ 200	~ 1,000
σ (Ω^{-1}cm^{-1})	~ 10^6	~ 10^3	~ 10^{-12}

the electrical conductivity is related to the charge carrier density(n) and the charge carrier mobility (μ) through:

$$\sigma = ne\mu \tag{12.4}$$

while the Seebeck coefficient is inversely related through a $n^{2/3}$ relationship, which renders highly conductive materials like metals and metal alloys, but also highly insulating materials like polymers and simple high band-gap oxides not suitable as starting points for highly efficient thermoelectric materials. Typically, n for good thermoelectric materials should be in the range of 10^{19}–10^{21} cm^{-3} [2]. These materials are usually highly doped semiconductors [2]. For metals and degenerate semiconductors, the Seebeck coefficient can be estimated as:

$$S = \frac{8\pi^2\kappa_B^2}{3eh^2} m^* T \left(\frac{\pi}{3n}\right)^{2/3} \tag{12.5}$$

where m^* is the effective mass of the carrier and κ_B is Boltzmann's constant [2]. Typically, m^* is high for compounds with flat and narrow bands such as in transition metal oxides. Flat bands decrease the mobility of the charge carriers, and thus the electrical conductivity (see Equation 12.4) [2]. In addition to the effect on the Seebeck coefficient, the number of charge carriers, n, also affects the electronic contribution to the total thermal conductivity (i.e., κ_e). The total thermal conductivity ($\kappa_{tot} = \kappa_e + \kappa_{Latt}$) can be related to the κ_e and the κ_{Latt} through the Wiedemann–Franz law [2]:

$$\kappa_e = L\sigma T = ne\mu LT \tag{12.6}$$

Where L is the Lorentz factor ($2.4{\cdot}10^{-8} J^2K^{-2}C^{-2}$) for free electrons, and e the electronic charge [2]. From the Wiedemann–Franz law, it is therefore evident that on the one hand, the κ_e will be difficult to minimize efficiently since it is coupled directly to S and σ, while on the other hand, the κ_{Latt}, which is directly related to the phonon transport of heat, can be separately minimized. This becomes true since the majority of phonons exist at longer wavelengths (lower frequencies) than the electrons, and can therefore be affected by defects that are not affecting the electrons to the same extent; hence, more directed phonon scattering can be achieved while maintaining a high electron mobility through the material. It is therefore important to find materials that are able to decouple these terms.

In terms of thermal conductivity, mostly κ_{tot} is reported for materials. The κ_{tot} is mainly composed of two parts according to:

$$\kappa_{tot} = \kappa_e + \kappa_{Latt} \tag{12.7}$$

where the electronic contribution, κ_e can be found using the Wiedemann–Franz law from Equation (12.6).

The k_{Latt} mediated through phonons has been one of the main targets to decrease in the pursuit for improvements of zT. Mainly, the k_{Latt} can be described according to:

$$\kappa_{Latt} = \frac{C_p \bar{v}_q \Lambda_{ph}}{3} \tag{12.8}$$

where $\bar{v}_q$ is the average phonon velocity (commonly the speed of sound in the material), C_p is the heat capacity, and Λ_{ph} the phonon mean free path.

The k_{tot} also depends on the density of the material, and will decrease for materials with low true density, ρ_{true} (i.e., fraction of $\rho_{theoretical}$). On the scale of the unit cell, however, the number of atoms in the unit cell should be maximized to increase the number of optic phonon branches. In practice, it means more ways for the heatwave to dissipate, and thus retard the movement of the heat transport. In Figure 12.3, the accumulated percentage of phonons with different wavelengths is reproduced from Biswas et al. [9] where it clearly illustrates the need to further focus efforts on defects on the length scale larger than 10 nm.

12.1.1.3 Requirements for High-Temperature Applications *T* > 700°C

In contrast to low-temperature thermoelectric materials, where degradation issues are not a major problem, at higher temperatures >700°C, the material stability will also play a major role in the selection of the materials since a material with good thermoelectric properties that will degrade quickly is of no use in practical applications. At high temperatures, most materials will experience a decrease in performance, since ambipolar diffusion (excitation of the opposite type of charge carriers) will become significant, and will contribute to lowering of the Seebeck coefficient, while at the same time increasing the κ_e. It is therefore necessary to find suitable materials with ideally, a temperature independent electronic conductivity and Seebeck coefficient. Some of the more successful *p*- and *n*-type materials such

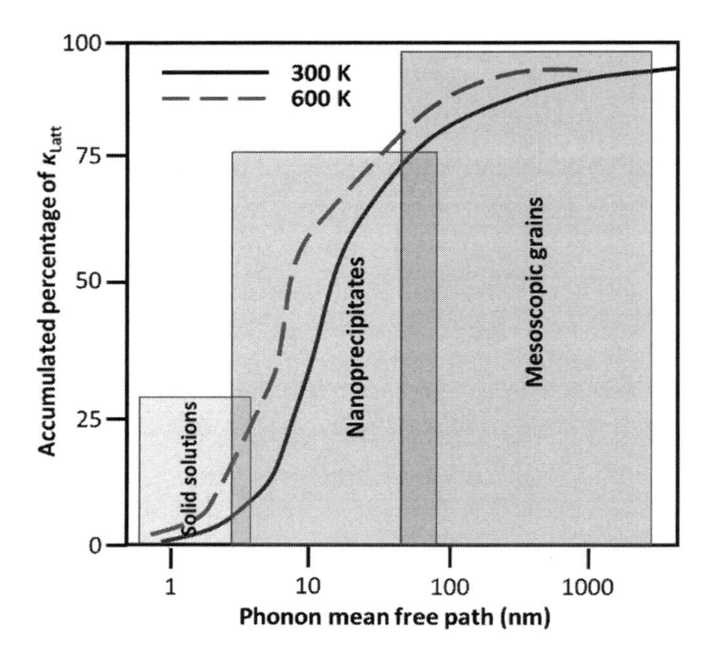

FIGURE 12.3 The accumulated percentage of contributions to κ_{Latt} from phonons with different wavelengths in PbTe redrawn from Biswas et al. [9].

as GeTe, SiGe, $Yb_{14}MnSb_{11}$, PbTe, $Cu_2Se_{1-x}I_x$, and $CuS_{1-x}Te_x$ reach a very low k_{tot} at high temperatures and are thus able to reach a peak in their zT between 700–900°C [2, 8, 10]. However, to obtain materials that are viable for higher temperatures and for long term use in high power applications they need to reach several basic requirements: (1) maintain a high zT and $S^2\sigma$ (power factor) [11]; (2) have a high melting point; (3) have a low reactivity with other components; and (4) stay stable against decomposition at the working temperature.

The second requirement is the one that is the main focus in the search for new high-temperature thermoelectric materials with a high zT. However, a high melting point often means a high bulk modulus, i.e., stiff bonds and thus higher k_{Latt} due to faster propagation of vibrations (phonons) through the material. It is therefore necessary to find a material that contains a large number of defects that will be able to scatter the phonons efficiently, while at the same time being stable against decomposition. A material with a large number of defects can also be more reactive. These three latter requirements, if fulfilled, will also pose problems to the maintenance of a high power factor and a high zT, since the high melting point will increase the k_{Latt}, large number of defects will increase the reactivity, and stability against decomposition will be low since the material will require some type of defects to scatter phonons with. If the working temperature of the thermoelectric material is too close to its melting point, the "useful" defects will be unstable and tempered away. To achieve high stability ultrahigh temperatures (>>1,000°C), refractory materials with high melting points and the capacity to form defects on all length scales are necessary.

Few materials are available with exceptionally high melting points; however, recently refractory high entropy alloys [12, 13] have become increasingly more popular for high-temperature mechanical properties due to their high intrinsic ultrahigh temperature (<2,000°C) stability in inert atmospheres. These alloys have not been explored for thermoelectric applications yet, but may provide the right matrix necessary for future ultrahigh thermoelectric materials, since high entropy alloys can easily be tuned to introduce secondary phases which may act to lower the k_{tot} further.

12.1.1.4 Common Thermoelectric Materials

Many popular thermoelectric materials like Zintl, Heusler, SiGe, MgSi, clathrates, and $CoSb_3$-Ba, La, Yb-based skutterudites have all been well studied over the years and many in-depth reviews are available [2, 10, 14, 15]. Popular p-type materials for high temperatures are Pb(Te, Se, S)-SeTe, SiGe, and skutterudites with peak zT temperatures of around 800–1,000°C. In skutterudites and clathrates, the structure contains a void which decreases the density of the unit cell, and thus decreases thermal transport. Clathrates e.g., $Ba_8Ga_{16}Ge_{30}$, can reach k_{tot} values ~1 $Wm^{-1}K^{-1}$, and are able to reach a $zT \approx 1$–1.7 in the temperature range ~427°C–627°C [16]. To further decrease the k_{Latt}, this empty void can be filled with a rattler atom of different size and mass that can dissipate some of the energy transported through phonons. For the Pb-chalcogenides, it was recently demonstrated that different types of defects can be introduced using a A-B system that exhibits a miscibility gap, where the phase separation occurs thermodynamically due to metastability of the solid solution. To make this separation possible, it is important to have two phases that are completely soluble in the liquid state, to ensure complete mixture and homogeneous precipitation. The phase separation will lead to nucleation and growth of metastable precipitates, but could also form the basis for a spinodal decomposition [14]. By adjusting the composition, and adding certain elements to the mixture, it becomes possible to attain nano-precipitates, grain boundaries, and point defects on a nm–μm scale while maintaining the electronic properties. In Pb-chalcogenides, it is also possible to adjust the types of phases that are precipitated, and thus select secondary phases that are slightly different in electronic structure to facilitate electron or hole filtering to improve the Seebeck coefficient. In addition, a material that can effectively achieve inclusions and defects on different length scales comparable to the wavelengths of the phonons will be able to achieve efficient scattering of phonons and thus minimize the k_{Latt}. For example, it has been demonstrated that ~25%, ~55%, and ~20% of the k_{Latt} comes from phonon modes with mean free paths < 5 nm, 5–100 nm, and 0.1–1.0 μm, respectively (see illustration for PbTe in Figure 12.3). Below 5 nm, atomic scale solid solution point defects will be the main scatterers. In the 5–100 nm range, nanoscale defects are the most efficient scatterers where interfaces between the matrix phase and the precipitate and/or mass contrast between the phases are believed to play important roles. In the 0.1–1 μm range, it is mainly the grain boundaries and larger precipitates that will effectively scatter the phonons [14]. Another state-of-the-art *p*-type thermoelectric material based on the Cu-chalcogenides, such as Cu_2(Se, Te, S, I), uses an additional scattering process that arises during its continuous phase transition to a superionic conductor of Cu ions [17]. By analogy, the Ag chalcogenides behave the same way, whereas the Ag/Cu ions behave as a liquid in a framework of Se/Te/S/I ions [8, 18, 19]. Since the metal moves freely as in the liquid state, it will also be able to scatter phonons and electrons similarly to an atom in the liquid state, and thus will lead to a greatly enhanced scattering of phonons and electrons. The improvements in the thermoelectric properties are seen as a sharp increase in the $|S|$ value, while a sharp drop in the κ_{tot} is observed [8]. In the Cu and the Ag systems, this leads to a remarkable increase in the zT by 3–7 times, thus reaching zT values ~2.3 @ 400 K [8]. This shows that novel approaches to improve the thermoelectric properties are still needed and can be the way to finally reach zT values > 3. For a more in-depth introduction to different thermoelectric materials systems the readers are referred to Ren et al. [10] and references within, where many comprehensive reviews have discussed them recently.

12.1.1.5 Advantages and Limitations in Alloys for Thermoelectric Applications

Metal alloys with simple face-centered cubic (FCC), base-centered cubic (BCC), and hexagonal cubic packing are in general highly conductive materials, with a large number of mobile charge carriers (> 21 cm^{-3}). However, the lack of control over the number of charge carriers is one of the main obstacles for using metal alloys. Since both S, σ, and κ_e are highly dependent on the number of charge carriers, the large number of charge carriers will most often lead to very bad thermoelectric properties. Overall, several different parameters are often used to control the properties in thermoelectric materials: (1) number of charge carriers; (2) charge carrier mobility; (3) band structure; (4) crystal structure; and (5) microstructure. In many semiconducting compounds, the number of charge carriers can be fine-tuned by small amounts of doping with different dopants.

12.1.1.6 Phonon Glass Electron Crystal Concept

During the last couple of decades, a new concept proposed by Slack [20] has been pursued. The concept proposed by Slack is a pathway toward maximizing the zT. The idea focuses on the fact that phonons are severely affected by defects on different lengths scales, while electrons are transported easily as long as the material maintains long-range order, as in a crystal [1, 2, 20]. This suggests that a material with defects that efficiently scatter phonons (similar to a glass), while at the same time providing a high level of electrical conductivity and mobility for the charge carriers, as in a crystalline ordered material, is a reliable way of improving thermoelectric materials. Such materials commonly involve the use of a wide variety of defects from the nm scale up to the mm scale [9, 14]. Commonly, materials systems where defects can be generated through, for example, precipitations, are used, or by selecting one or more materials systems that are combined through some method of mixing, such as ball milling. Both approaches have advantages and disadvantages. Essentially, all types of defects that affect the phonon transport with minimum implications on the electron mobility are desirable. The following parameters can be thought to be controllable parameters:

- Solid solution (e.g., different atoms on the same crystallographic position)
- Interstitial atoms/ions and vacancies
- Strain (i.e., similar bond strength but locally contracted or expanded bonds)
- Homo-, and hetero-interfaces (i.e., incoherent interfaces with high distortion due to mismatch between crystallites of the same phase or different phases)
- Atomic weight differences
- Complex crystal structures with 1-D and 2-D blocks (e.g., layered structures)
- Incommensurability and modulated structures (irrational repetition of unit cell periodicity)
- Quasi-crystallinity (crystalline materials with non-crystallographic point symmetries e.g., diffraction pattern symmetry of 5-, 7-, 10-fold; no translational symmetry in the crystal)
- Elastic modulus and bulk modulus
- Cage structures (e.g., rattling atoms that can be exchanged based on weight and size)
- Phase transitions (e.g., superionic transport properties, as in the (Cu, $Ag)_2$(Se, S, I, Te) compounds)

A phonon glass electron crystal can therefore be achieved if all these parameters can be combined to affect the electron mobility minimally while maintaining scattering and disrupting the phonon movement maximally in the material on all scales. By including as many of the different parameters illustrated in Figure 12.4 as possible, phonon scattering can be maximized. The phonon glass electron crystal concept can be thought of as a way to maximize the "entropy" of scattering pathways for the phonons, while decreasing the "entropy" of the scattering pathways for the electrons.

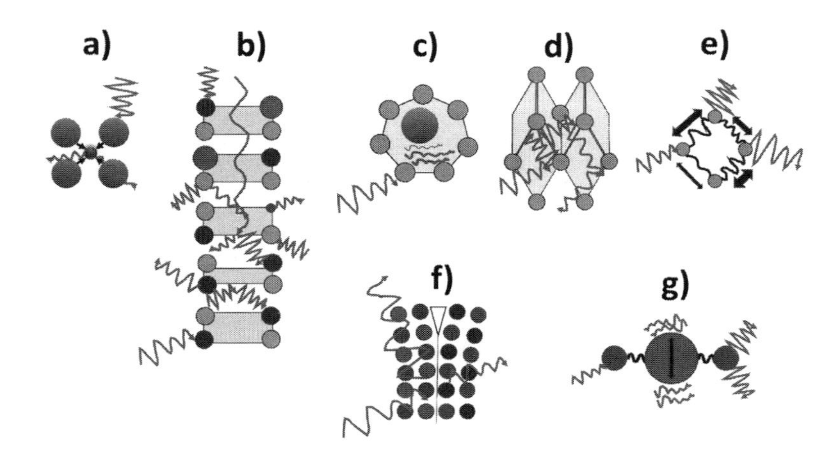

FIGURE 12.4 A cartoon illustration of (a) point defects e.g., interstitial, smaller atom; (b) layered structure with solid solution containing different atoms; (c) cage structure with rattler atoms that dissipates the incoming energy from the phonons; (d) quasi-crystalline materials without translational symmetry in 3D; (e) atoms with different bond properties combined to maximize the bond strength differences; (f) hetero-structured interfaces; and (g) heavy atoms that absorb phonon energy.

Some of the more successful materials recently reported where the phonon glass electron crystal concept is more easily applied are found among chalcogenides [1] and half-Heusler compounds [6] that can reach zT values > 1.

12.1.1.7 Minority Carriers

At high temperatures the excitation of a significant number of minority charge carriers will start to decrease the thermoelectric properties through a decrease in S and an increase in σ and κ_e. This effect is most prevalent in semiconductors where the minority charge carriers (i.e., holes or electrons) are excited to a larger degree and start impacting the total charge carrier content and the Seebeck coefficient negatively. In metals and alloys, this effect is not very dominant, since the number of charge carriers is large ($>10^{20}$ cm^{-3}), and therefore the effect of minority carriers is not as pronounced. For semiconductors and wide band-gap materials, this effect may be more significant and will affect the absolute value of S and the κ_e negatively. This effect is to some extent responsible for the rapid decrease of S after it reaches a maximum at higher temperatures. It is therefore important for semiconductors to have a means of decreasing the number of minority charge carriers, such as electron filtering through secondary phases or by selecting materials with larger band gaps [21].

12.1.1.8 Charge Carriers

Adjustment of charge carriers in metals and metal alloys is relatively limited, and usually very difficult to control. The charge carrier density in metals and alloys is usually found to be in the range of 10^{20}–10^{21} cm^{-3} and will therefore be in the region far away from where zT is maximized. Metals and metal alloys are therefore a very challenging class of materials that has been avoided for high zT thermoelectric applications. One approach that may be viable, but has not been utilized much is the use of the valence electrons, i.e., the valence electron count (VEC). The VEC is the total number of valence electrons in an element [22]. The authors have previously shown that the electrical resistivity can be decreased by decreasing the average VEC value for the $Al_xCoCrFeNi$ high entropy alloy system [23]; however, although this effect is very pronounced in this system, more studies are required to elucidate the VEC effects on charge carrier concentration and charge carrier mobility changes.

12.1.1.9 Charge Carrier Mobility

Since the charge carrier mobility is related to S through Equation (12.4) and Equation (12.5), the large charge carrier mobility (μ) of a metal would be desirable in thermoelectric materials. However, as described previously, metals and metal alloys have an excessive number of charge carriers that are fundamentally difficult to vary, and thus have left most of the research on thermoelectric materials to exclude metal alloys as prospective thermoelectric materials. In alloys with large differences in atomic number (Z), the charge carrier mobility, μ, will suffer due to large differences in the potential fields from the positively charged core or the atoms, where heavy atoms will influence the electrons strongly and induce further scattering. However, the number of charge carriers exists in such a large number that the scattering effect will work as a way to reduce the number of charge carriers. Another possible effect that will decrease the mobility of the charge carriers is the interface between different phases or crystallites in the materials. In thermoelectric materials, having a significant number of interfaces is an important ingredient in order to scatter phonons efficiently and thus decrease the κ_{Latt}. However, these interfaces will also to some extent decrease the mobility of the charge carriers, if the difference in band energy is not small enough. It is therefore also important to keep in mind the band alignment of the secondary phases when designing materials where precipitation of secondary phases for phonon scattering is of importance. In metal alloys, however, this effect can be assumed to be less significant, since the s-electrons move freely according to the free electron model, and will scatter to some extent with *d*-electrons. In addition, the mobility and band alignment between metallic alloy phases of simple BCC, FCC, and HCP metals are close enough to have electrons affected by minimum amount of scattering when moving between one phase to another in metals.

12.1.1.10 Band Structure

Periodic crystal structures form a band structure that fundamentally depends on the crystal structure symmetry. The band structure for the same crystal structure can to some extent be fine tuned by careful selection of dopants [24]. In compounds where solid solution exists for dopants, the dopants will, in addition to having an effect on the number of charge carriers, and to some extent the mobility, also affect the band structure by varying bond lengths in a certain direction so that the bands can be tuned to converge. In metal alloys, this option is more limited since the atoms are mainly packed randomly without directional bonding in simple cubic packing (BCC, FCC, HCP), thus limiting the option to specifically target certain crystallographic sites that may affect certain Brillouin zones in the band structure. However, such metal structures already have a highly symmetrical crystal structure and already have the potential to start with a high convergence of bands; that is advantageous for an improved Seebeck coefficient [24, 25]. The fine-tuning of the valleys and anisotropy of the charge carrier mobility in different directions is thus not easily achieved in metal alloys. If anisotropic band structures are desirable, the metal compounds composed of anisotropic crystal structures with lower symmetries should be selected from start, which in turn also will produce anisotropic charge carrier transport, related to the lower symmetry [25, 26].

12.1.1.11 Crystal Structure

In addition to the mentioned importance of the band structure for obtaining improved thermoelectric properties through dopants, the dimensionless thermoelectric quality factor defined in Pei et al. [24] shall also be considered:

$$B = \frac{2\kappa_B^2\hbar}{3\pi}\frac{N_v C_l}{m_l^* \Xi^2 \kappa_{\mathrm{Latt}}} T \tag{12.9}$$

where B is the thermoelectric quality factor, N_v the number of degenerate conducting bands, C_l the average elastic moduli tensor, m_l^* the inertial effective mass, κ_{Latt} the lattice thermal conductivity, and Ξ the deformation potential coefficient as described in Pei et al. [24]. Since N_v is already large for a metal with a simple cubic structure, and the effective mass m_l^* is small, the way to maximize B in a metal alloy would be to use light elements with a small Ξ, since this term is in the denominator, and is squared. However, a balance is needed with κ_{Latt}, since the use of only small elements would result in very little phonon scattering by mass fluctuations and hence, a large κ_{Latt}. In addition, the modification of the C_l, by using elements with stronger or softer bonds will also be one approach. It is very clear that strong bonds will conduct heat very efficiently like in, for example, boron nitride (BN), where a k_{tot} of > 1,500 $\mathrm{Wm^{-1}K^{-1}}$ is reached. This clearly indicates that soft bonds are highly advantageous to decrease phonon transport in thermoelectric materials, and thus compounds with soft bonds, i.e., high elastic modulus.

12.1.1.12 Changing Resistivity Using Differences in Atomic Numbers

Changing the electrical resistivity using VEC is only one tool available in simple metal alloys, while another tool more suitable to alloys with different elements is the use of the Linde's rule [27], which describes the empirical relationship between differences in atomic numbers and their effects on the resistivity, as described by:

$$\Delta\rho \propto \Delta Z^2 \tag{12.10}$$

where the resistivity difference arises from differences in atomic number (ΔZ) between the host matrix and the dopant. This effect is an effective way to increase the electrical resistivity in an alloy, while at the same time also increasing the phonon scattering to decrease κ_{Latt}. However, the negative effect of decreasing the mobility of the charge carriers will have to be considered to find an optimum in scattering. For example, carbon in Fe is found to increase the resistivity by ~7 times than in stainless steel. Although this method is useful to decrease the overall electrical conductivity in a metal alloy and the corresponding κ_{Latt}, it also comes with the downside of a lower mobility.

12.2 HIGH ENTROPY ALLOYS FOR HIGH-TEMPERATURE THERMOELECTRIC APPLICATIONS

During the last decade, the concept of high entropy stabilization of solid solution has been explored intensely in metal alloys [28–30]. The concept has now spread to other materials such as oxides, carbides, and nitrides [31] and is now intensely studied to achieve improved properties in current and new materials. The high entropy alloys (known as HEAs) have suitable properties for strong phonon scattering on all length scales due to their diverse microstructure [23, 32]; however, the difficulty in controlling the electronic transport properties in metal alloys is the main problem. However, the backside of a large carrier concentration in high entropy alloys may instead be useful in other applications where a very low κ_{Latt} is required, such as for infrared detectors [33].

12.2.1 CURRENT PROGRESS IN HIGH ENTROPY ALLOYS FOR THERMOELECTRIC APPLICATIONS

12.2.1.1 Thermoelectric Properties

Currently, high entropy alloys have not been widely explored for their thermoelectric properties. However, several reports indicate that they have a potential for reaching unusually low $\kappa_{latt} < 3–4$ Wm^{-1}K^{-1}, which is comparable to some of the best half-Heusler compounds [3, 23]. As has been reported in Shafeie et al. [23] and Zhang et al. and Kao et al. [34, 35], the k_{tot} decreases significantly in the reported $Al_xCoCrFeNi$ system with increasing x for $0 \leq x \leq 3$. Also, it appears that a significant enhancement in S is observed, while σ decreases to ~0.3 MSm^{-1}, which is also the upper limit of σ reported for "dirty" metals [36, 37] (i.e., metals or alloys containing a high degree of impurities or substitutional elements). However, the low $|S|$ values of metals and alloys ~1–5 μV/K is usually not of interest in high-performance thermoelectric applications where a $|S|$ value ~200 μVK^{-1} is aimed for to achieve a high zT. New approaches are therefore necessary if metal alloys, and, more specifically, high entropy alloys are to be seriously considered for thermoelectric applications, some of which will be described in the following sections.

12.2.1.2 Thermal

In pure metals phonon transport is mediated through the large number of charge carriers (free electrons) and lattice vibrations. Relatively few defects exist that will interfere significantly to lower κ_{Latt} from the theoretical values of the metal. The k_{tot} for pure metals is therefore usually also extremely high (i.e., > 50–100 Wm^{-1}K^{-1}). The thermal conductivity in pure metals is therefore not ideal for thermoelectric, since the total heat conduction κ_{tot} is too high >> 10–20 Wm^{-1}K^{-1}, which puts them in a category of bad thermoelectric materials. However, when metals alloyed with different elements (e.g., Fe alloyed with C, or Cu alloyed with Fe), the k_{tot} drops significantly, as in steels where κ_{tot} for Fe is ~80 $Wm^{-1}K^{-1}$, while it is ~52 $Wm^{-1}K^{-1}$ for low carbon steel and ~15 $Wm^{-1}K^{-1}$ for 304 stainless steel [38]. Other examples include Ti-6Al-4V with a κ_{tot} ~6 $Wm^{-1}K^{-1}$ where the pure elements Ti and Al have κ_{tot} ~22 $Wm^{-1}K^{-1}$, and ~237 $Wm^{-1}K^{-1}$, respectively [38]. The drop in κ_{tot} is mainly from the formation of novel microstructures [39], mass fluctuations [40], strain, and spin scattering from unpaired electrons. It is therefore interesting to note that for high-entropy alloys, the electrical resistivity appears to be higher despite similar κ_{tot} (e.g., for a κ_{tot} ~11 $Wm^{-1}K^{-1}$ AlCoCrFeNi has ρ ~221 $\mu\Omega$cm, and for a κ_{tot} ~15 $Wm^{-1}K^{-1}$ 304 stainless steels has ρ ~69 $\mu\Omega$cm), thus indicating that the resistivity increase in high entropy alloys might come from scattering of electrons in addition to phonons, which the current authors have recently also found to lead to a remarkably low-temperature coefficient of resistance (TCR) [32].

12.2.1.3 Composites

Attempts have been made with conventional composites [41, 42] in common thermoelectric materials. However, the reactivity between the components in the composite is usually critical in addition to the compatibility of their thermoelectric properties to avoid local eddy currents that may affect the zT negatively [43]. From a composite perspective, high entropy alloys have tremendous potential, since synergic effects [32] related to microstructure, crystal structure, and composition can be combined by forming different phases *in situ*. Much work remains in this area which has not yet been explored, especially combining high entropy alloys together with different semiconductors to achieve high zT values [42–44]. Theoretical studies have indicated that the composite of a thermoelectric material should be designed to avoid the formation of high quality thermoelements inside the main composites. Recently Riva et al. reported high entropy alloy composites with additions of nano-diamonds, SiC, Sc_2O_3, h-BN, c-BN, and CN [45], and realized the incompatibility of the additives due to low affinity to the high entropy alloy matrix. As a result, only 2 wt.% nano-diamonds was possible to study. The machinability for the composite was reported to be good, but with presence of porosity, thus requiring the use of spark plasma sintering (SPS) to achieve dense samples. This study indicates the difficulty to select non-metallic additives to high entropy alloys, and thus also points at a problem that needs to be further studied for other additives to exclude that incompatibility comes from the: (1) amount of additives; (2) particle size of additives; (3) sintering method; or (4) mixing process (Figure 12.5).

12.2.2 Electrical

12.2.2.1 Atomic Number Difference

As the electrons move through the periodic arrangement of atoms in a crystal, these electrons will be subjected to an attractive potential from the nucleus of the surrounding atoms that will to some extent exert an attractive force that will force the electrons to deviate from their original path, and to a higher probability scatter with neighboring atoms. The effect is described by the empirical Linde's rule [27], where the resistivity, ρ, increases with the square of the atomic number

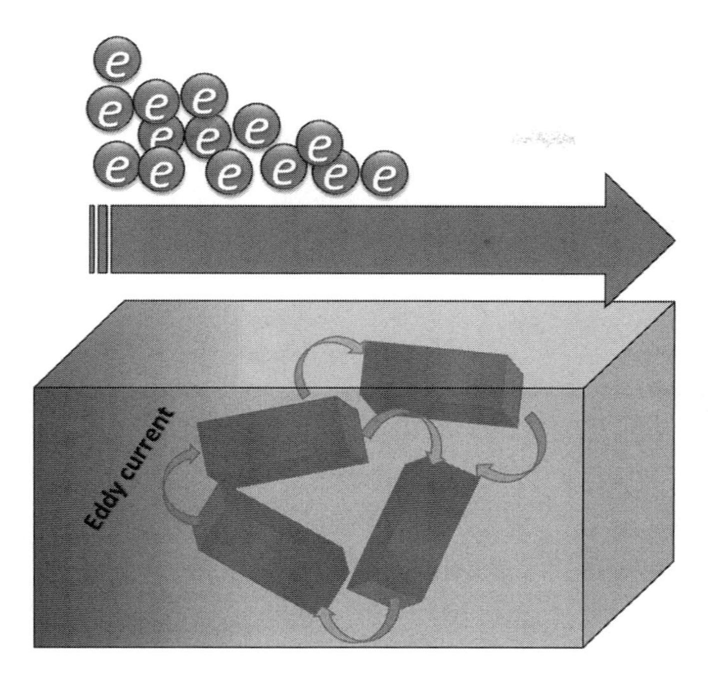

FIGURE 12.5 Heating a composite with components that form internal thermoelements with high zT, short-circuit locally, and decrease the overall efficiency of the full composite material [43].

difference, $(\Delta Z)^2$. The effect will be that a larger number of electrons will scatter and experience a lower μ, which will decrease σ and therefore also the κ_e. This effect, also affects the κ_{Latt} simultaneously, as described in the previous section, where atomic mass differences can be used to scatter phonons. However, this approach, although only helping in a couple of aspects, will not change the number of charge carriers, which significantly impacts the $|S|$ value mixing process. Figure 12.6.

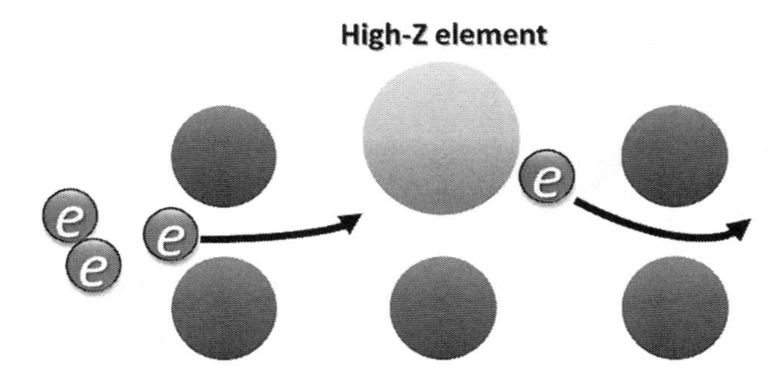

FIGURE 12.6 Illustration of the electrons being scattered by the larger atoms, thus decreasing the mean free path of the conduction electrons.

12.2.2.2 Electron Filtering

Low-energy electrons that only contribute very little to the Seebeck potential but still add to the number of charge carriers and heat conduction, can hypothetically be filtered to only let high energy electrons pass to maintain a high $|S|$ value. In certain materials such as superlattices of ErAs:InGaAs/InGaAlAs, and in, e.g., bulk PbTe-SrTe, it has been reported that nano-precipitates are responsible for the enhanced zT through "electron filtering" effects [14, 46]. For high entropy alloys, this may require very careful selection of elements in order to precipitate phases that have the right level of conduction band edge energy to efficiently filter electrons in the high entropy alloy and only let high energy electrons pass through. This approach will most likely require accurate prediction of phases through phase diagrams, and an accurate prediction of the required band structure of the high entropy alloy and the secondary phases to achieve this effect. For high entropy alloys, this remains an avenue to explore, since the endless possibilities of the *in situ* formation of secondary phases may open a path toward higher $|S|$ values Figure 12.7.

12.2.2.3 Phase Transitions (Structural, Magnetic, Microstructural) Seebeck-Charge Mobility Engineering

Recently, a novel approach was proposed to enhance the Seebeck coefficient in materials by charge mobility engineering [47]. The current authors suggested a new approach toward the enhancement of the Seebeck coefficient through charge mobility engineering [32]. By using a large variation of the mobility as a function of temperature, $d\mu/dT$, an additional component to the Seebeck can be added to the conventional Seebeck coefficient. This effect is expected to be observed in systems where an abrupt change in the charge mobility occurs as in systems with structural phase

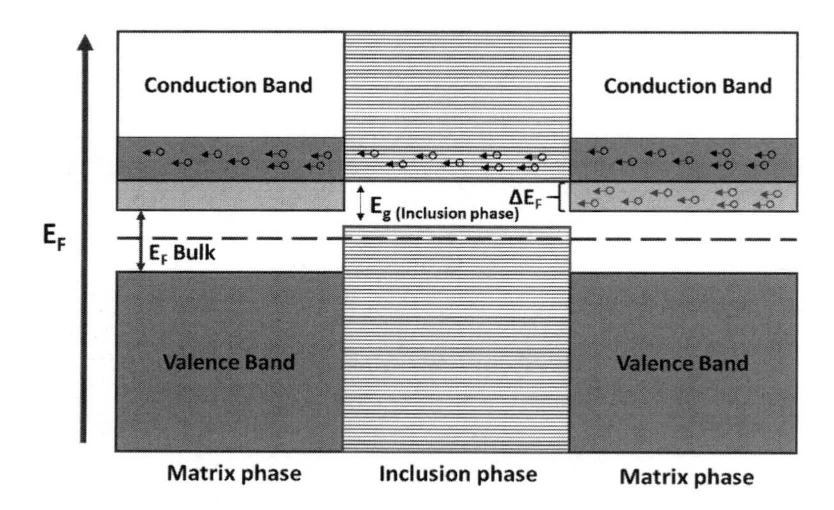

FIGURE 12.7 Conceptual illustration of the electron filtering, with the inclusion phase having a different band gap relative the host matrix phase, thus only allowing certain high energy level electrons or holes to be transported through. (Illustration redrawn based on Bos and Downey [6].)

transitions (e.g., Cu_2Se system where scattering of electrons is found to enhance the Seebeck coefficient significantly ~127°C), or in Kondo- and heavy-fermion systems where anomalous scattering may dominate. The effect can be thought of as electrons moving from a highly mobile state with broad bands, into a less mobile state with flat bands. On average, the electron will therefore experience a variation in the electronic states as if there was a resonant state as illustrated in Figure 12.8.

The Seebeck coefficient can therefore be written to include a term that is related to the mobility variation as shown in Equation (12.8):

$$S = \frac{\pi^2 \kappa_B^2 T}{3e}\left[\frac{\partial \ln \tau}{\partial \varepsilon} + \frac{\partial \ln N}{\partial \varepsilon}\right]_{\varepsilon_F} = S_\tau + S_N \tag{12.11}$$

where τ is the charge carrier mobility, ε the energy at the Fermi level, N the density of states, T the temperature, S_τ the Seebeck coefficient from charge relaxation, and S_N the usual Seebeck coefficient. It is therefore interesting to observe that the mobility-related term, which has not been utilized previously, may add significantly to the Seebeck coefficient if thermoelectric materials are selected wisely. Although, conceptually different, the Kondo effect where localized *d*- or *f*-electrons interact with conduction electrons in e.g., metals gives rise to a similar resonance state (see Figure 12.9 and Figure 12.10 redrawn from Kouwenhoven and Glazman [48]). The unpaired

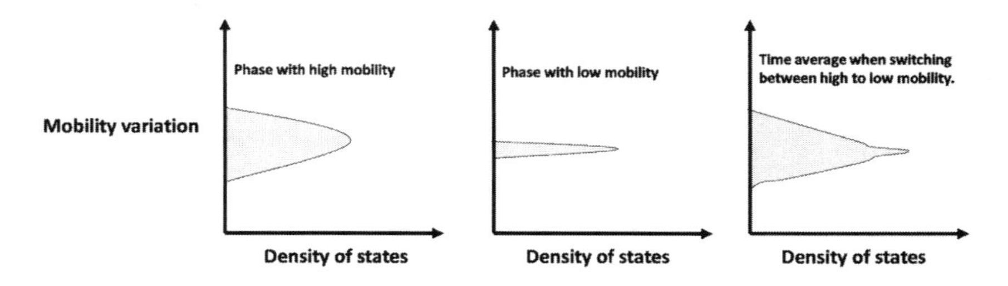

FIGURE 12.8 Simplified resulting time average between a high and low mobility phase resulting in a state similar to Kondo resonance that can enhance the Seebeck coefficient.

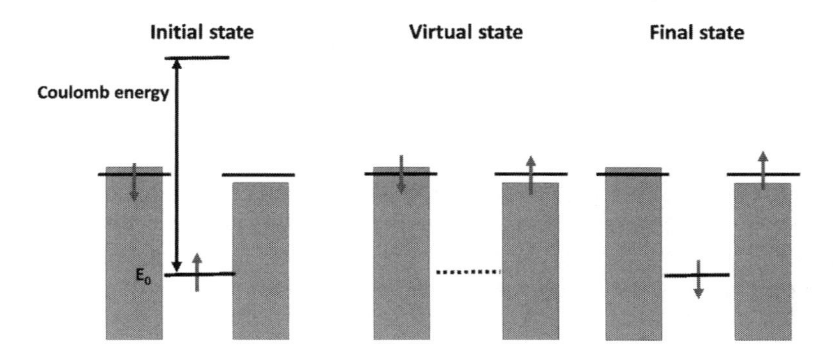

FIGURE 12.9 Simplified Anderson model of Kondo scattering redrawn according to Kouwenhoven and Glazman [48].

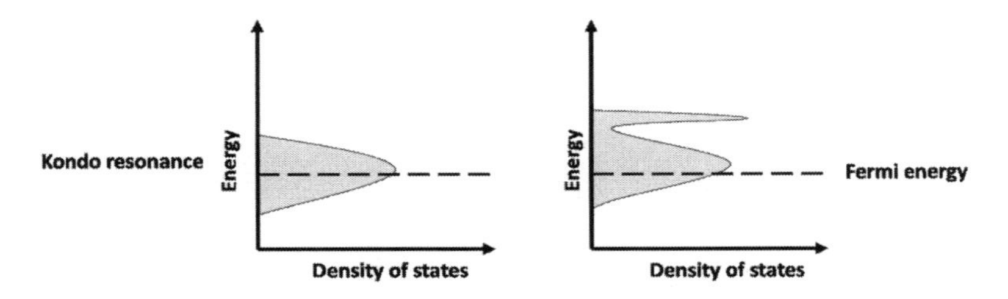

FIGURE 12.10 The resulting resonant state formed due to the Kondo scattering processes [48].

electron in a magnetic impurity surrounded by a non-magnetic conducting matrix will interact with the conduction electrons and after a brief period, give rise to a virtual state when the electron spin is flipped and excited to a higher energy at the Fermi level (see Figure 12.9). This process effectively creates a new resonant state which will be added to the DOS of the material, and hence enhance the Seebeck coefficient (see Figure 12.10) [48]. In, e.g., high entropy alloy systems like the $Al_xCoCrFeNi$ with $x > 2$ we believe effects from changes in microstructure in combination with magnetic transitions and Kondo-like scattering can lead to similar enhancements of the Seebeck coefficient. In the high entropy alloy system studied by the current authors [32], magnetic atoms in high entropy alloys can form additional phases where magnetic ordering or magnetic clusters may exist. Some of these phases are related to the Fe-Co-Cr or Fe-Cr system which contains a miscibility gap. The maximum in the Seebeck coefficient appears to coincide with the expected temperature range where the spinodal decomposition reaches its maximum temperature in combination with the Curie temperature for the magnetic atoms [49–51]. Overall, this novel way of achieving an enhanced Seebeck value presents the multifunctionality that is possible to achieve in high entropy alloys if the complex interplay between different phases, microstructures, and atoms can be combined by design.

12.2.2.4 Composite Formations *In Situ* in High Entropy Alloys

As discussed previously in the general case for thermoelectric materials, composites have been a very popular way of creating new grain boundaries for additional phonon scattering to help lower κ_{Latt}; however, the drawbacks may be thermodynamic instability of the different phases that are mixed, thus leading to secondary reactions and new undesirable phases forming. Recently, it was demonstrated that when yttrium is added to the CoCrFeNi [52] high entropy alloy, a microstructure with yttrium located in the interdendritic regions is found. This is similar to what the current authors have found (research unpublished), and it indicates that it is clearly possible to design the microstructure of a high entropy alloy for thermoelectric applications, but also for mechanical applications by adding certain elements that will phase segregate and improve the overall properties. When 0.05 yttrium is added to CoCrFeNi, the κ_{tot} decreases for the high entropy alloy from ~12.25 $Wm^{-1}K^{-1}$ to ~9.5 $Wm^{-1}K^{-1}$ at 27°C [52]. This decrease can be related to both the new boundaries that are formed between the dendritic and interdendritic regions, but also from the large difference in mass and volume of the yttrium relative the other elements. This avenue still remains an unexplored territory among high entropy alloys for thermoelectric applications, but will most probably lead to a larger decrease in the κ_{Latt}, but also the κ_e, due to large differences in the deformation potential, which leads to stronger electron scattering. Apart from adding specific elements that will phase segregate due to a large mismatch in atomic radii or $\Delta H_{A\text{-}B}$, other types of high entropy alloy composites that are still not explored may be composed of the precipitation of specific compounds with matching thermoelectric properties, as described in the previous section on composites.

The formation of a high entropy alloy quasi-crystals or high entropy alloy lanthanide composites, where the use of the significantly lower κ_{Latt} (~1–3 $Wm^{-1}K^{-1}$), and large S for the quasi-crystals [7] or the lower κ_{tot} values (lower electronic conductivity [53] and f-electrons in lanthanide atoms to form potential resonant levels, or Kondo systems for significantly enhanced S values [54, 55]). High entropy alloys should have significant potential in the use of different types of resonant scattering mechanisms using magnetic atoms, and the use of specific nanostructures to stabilize magnetic phases which would strongly affect the carrier mobility and thus enhance the Seebeck as described in the previous section [32, 47]. We therefore believe that the use of a matrix with inherently low electronic conductivity (e.g., lanthanides or low VEC elements) can be used to host the precipitation of novel phases and microstructures to enhance the otherwise low Seebeck values in metallic alloys.

12.3 OTHER APPLICATIONS

12.3.1 Alloys with Switchable Sign of the Temperature Coefficient of Resistance (TCR)

Other applications that require accurate control of the temperature dependence of a conductor usually require materials with a low-temperature coefficient of resistance (TCR). The TCR is defined as the change in resistivity of a material over a temperature range as described by [32]:

$$\text{TCR} = \frac{\Delta\rho}{\rho_0 \Delta T} = \frac{\rho_T - \rho_0}{\rho_0 \Delta T} \tag{12.12}$$

where ρ_0 is the resistivity at the lowest measured temperature, or ideally at 0 K, and ρ_T is the resistivity value at the highest measured temperature, or as commonly defined at 27°C. Depending on the application, or temperature range of interest, these parameters can be selected for the specific material in order to evaluate the suitability for the specific application. Commonly, alloys with very low TCR values are desirable in applications related to airplanes, sensors, and resistors, where a constant resistivity over a wide temperature range is desirable in combination with a significantly higher resistivity to minimize the noise signal. In addition, the material should be easy to shape or mold in different forms and show long-term stability. Historically, very few options have been available as low TCR materials, and over a century has passed since both constantan and manganin were discovered in the Cu-Ni alloy system (see Shafeie et al. [32] and references within). Recently, the current authors presented the study of a high entropy alloy system composed of $Al_xCoCrFeNiCu_y$ with $x = 1.0$, 2.0,

2.25, 2.5, 2.75, 3.0 and x, y=2.0, 1.0; 3.2855, 1.0, where they found that the electrical resistivity is strongly dependent on the microstructure of the material, but also on Kondo-type scattering that are stabilized up to very high temperatures > −23°C. They found from detailed microstructural studies that the microstructure, which consists of a dendritic and an interdendritic phase, consists of additional phases within these parts [32]. The different regions appear to be constituted of different types of atoms, and hence different phases with different properties are "encapsulated" within the matrix [32]. The decomposition can be described as a decomposition within the decomposed phases, which is illustrated in Figure 12.11.

Interestingly, the dendritic phase is highly concentrated in Al, Co, Ni, where the possibility of forming phases with Kondo-like scattering is increased, and where clusters of non-magnetic atoms (Al) surrounding magnetic atoms (Fe, Co) may form. The Kondo-like scattering was confirmed by low-temperature resistivity measurements, and through fitting [32], it was determined that the Kondo temperature, T_K, likely reaches high temperatures > −23°C for x ≈ 2.75 which is remarkable in a high-entropy alloy that is considered a dirty metal. Additionally, the current authors also

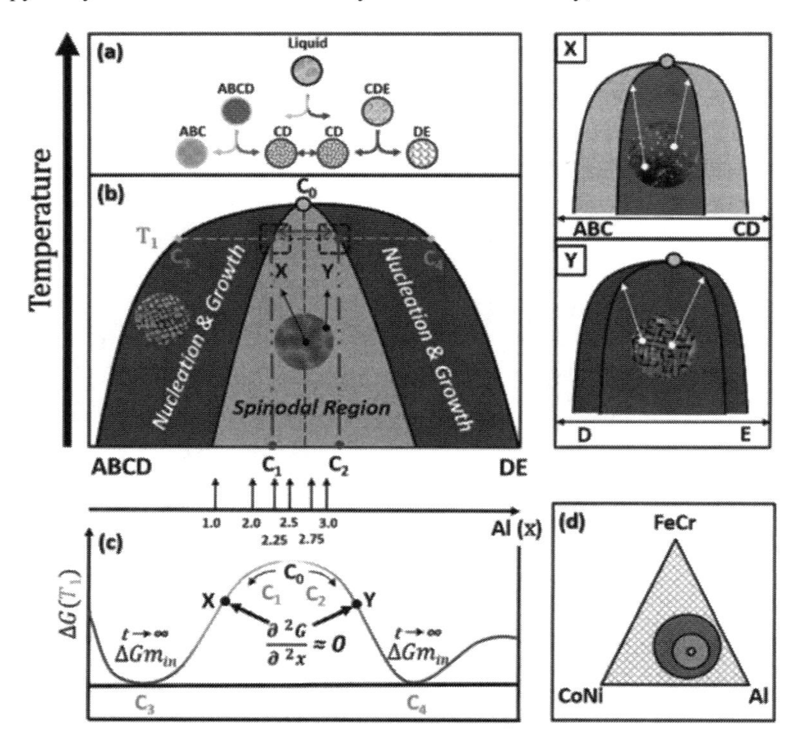

FIGURE 12.11 Decomposition of a multicomponent phase e.g., a high entropy alloy into separate phases from liquid state into subsystems (a) X, Y, (b) involving a miscibility gap (i.e., spinodal decomposition region), where (c) the phase separation is thermodynamically driven, (d) the pseudo ternary phase diagram [32]. (Copyright (2019) Wiley. Used with permission from S. Shafeie, S. Guo, P. Erhart, Q. Hu, A. Palmqvist, *Balancing Scattering Channels: A Panoscopic Approach toward Zero Temperature Coefficient of Resistance Using High-Entropy Alloys*. Publisher: John Wiley & Sons.)

found that the Seebeck coefficient, which reaches a maximum around ~427°C [23], coincides with the temperature range of the spinodal decomposition and magnetic transitions of the Cr-Fe phase [50, 56], which was found from EDS mapping to reside in the interdendritic phase of the high entropy alloy. It is therefore believed that the enhanced Seebeck coefficient in this high entropy alloy system may actually come not only from a combination of resonant scattering from Kondo-like clusters, but also from a large change in electron mobility as a function of temperature, $d\mu/dT$, as described above [47]. The practical implications of using high entropy alloys as materials that can host several different "sub-materials" are huge, and would imply that an accurate prediction of the temperature-dependent phase diagram for different high entropy alloys and their magnetic phase diagrams would open the possibility to design high entropy alloys for demanding high-temperature thermoelectric applications in power plants, airplanes, and scientific instrumentation.

12.4 DESIGNING HIGH ENTROPY ALLOYS FOR THERMOELECTRIC APPLICATIONS

It is obvious that the design of new highly efficient thermoelectric materials requires further understanding of unexpected synergic effects or the combination of novel phenomena to reach the goal of a $zT > 3$. Also new materials are needed that can combine multiple zT enhancing effects at the same time without loss of any other important property. One of these effects that may be promising in high entropy alloys is the above-mentioned charge mobility engineering approach [32, 47]. However, such approaches will only be readily available for designing novel materials once complex phase diagrams of high entropy alloys can be accurately calculated. This will open the door toward a hierarchical design using decomposition of different phases into desired microstructures and compositions with specific properties as described in Shafeie et al. [32].

12.4.1 3-D Printing High Entropy Alloys to Lower the Overall Thermal Conductivity

Currently the research on 3-D printing is in an intensive stage, both academically and industrially. 3-D printing for thermoelectric materials has not been explored significantly, and may potentially be an avenue to decrease the k_{tot} of metallic alloys beyond current levels using the following approaches:

1. Different patterns (artificial superlattices) on the μm–mm sale to further scatter lattice phonons.
2. Introduce voids that will significantly lower the k_{Latt} on average, while the highly mobile electrons can pass by freely.
3. Combine multiple layers of different types of high entropy alloys e.g., lanthanid-based (LHEAs), refractory-based (RHEAs), and usual HEAs (mainly 3-D elements)

As illustrated in Figure 12.12, 3-D printing will make it possible to introduce several new aspects that are at the moment not possible to control through the bulk

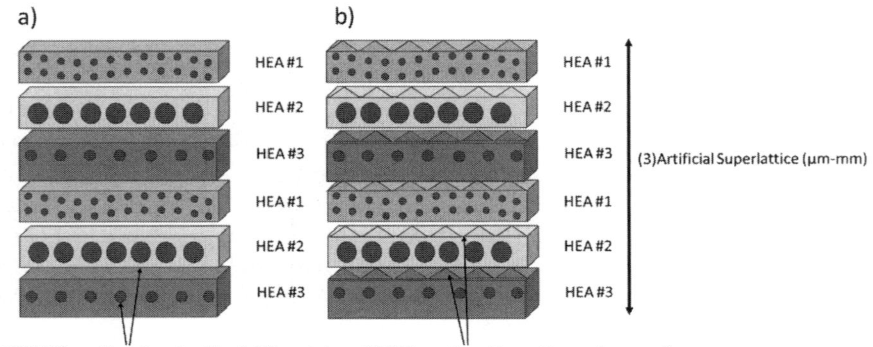

FIGURE 12.12 Illustration of different ways of utilizing high entropy alloys to reduce the thermal conductivity.

preparation methods used in the thermoelectric society, e.g., precipitations, thin films, ball milling, etc., 3-D printing allows for high precision control of the bulk 3-D structure and phase composition on a μm–mm scale, which is on the upper part of the mesoscale mostly dominated by phonons of long wavelength, as illustrated by Figure 12.3 [9]. However, this method will, despite the ease of using 3-D printing, be technically difficult to apply, since the different high entropy alloy materials have to match thermoelectrically, and in addition, it also requires appropriate phase diagrams for the heat treatments needed to introduce the right mixture of phases in the temperature range where it is to be used. Another level of complexity may be applied by using high entropy alloy layers or other compatible phases to induce electron-filtering properties and magnetic scattering. This approach will mostly be feasible in the future when phase diagrams of high entropy alloys can be predicted accurately together with their bulk electronic properties.

12.4.2 Proposed Flowchart to Design Novel Thermoelectric High Entropy Alloy Materials

To facilitate the discovery of new high entropy alloys with thermoelectric properties, a multifaceted approach is necessary since metal alloys possess certain challenges regarding control of charge carrier density. The κ_e, and the S especially attain too high and too low values, respectively. Based on the experience of the current authors, the following flowchart may serve as a guideline for selecting new compositions.

In Figure 12.13, we show that to design a high entropy alloy for thermoelectric applications or related fields, it is necessary to first select a composition that will be the matrix composition, where all elements exist in a solid solution. This can be ensured to a large extent by selecting elements with $\Delta H_{A\text{-}B}$ and $\Delta R_{A\text{-}B} \approx 0$ between all the pairs. This step is primarily important to ensure solid solution through the well-known Hume–Rothery rules. Once the matrix elements are selected, decisions about the secondary phases has to be made, in order to precipitate the desired phases. This is most easily done by varying the $\Delta H_{A\text{-}B}$ to be larger than ± 15 kJ/mol^{-1}, and if

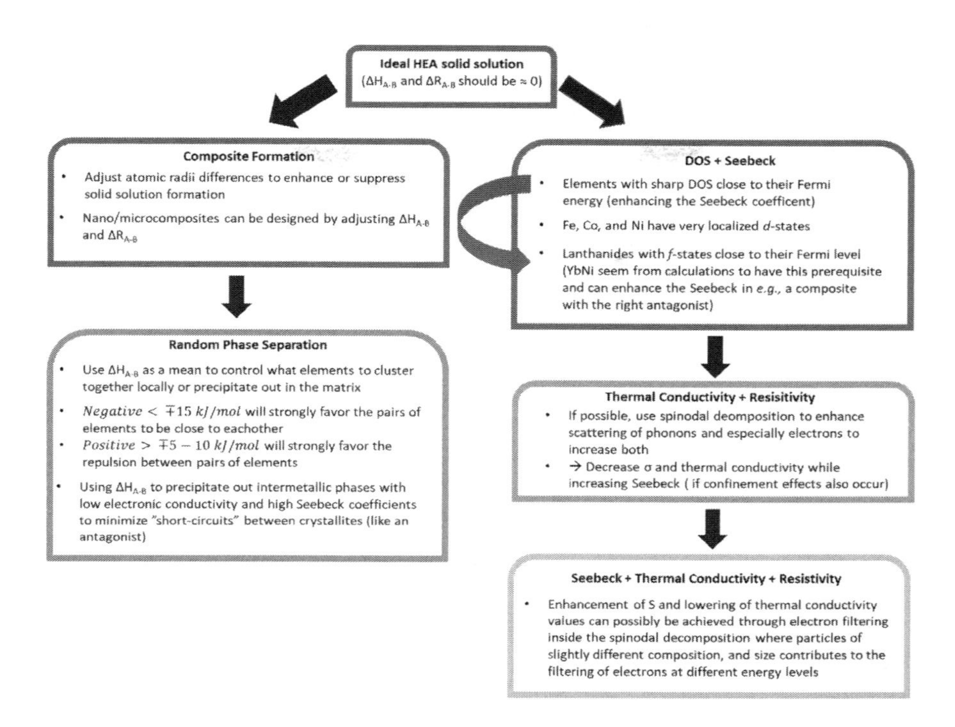

FIGURE 12.13 Overall flowchart on designing high entropy alloys with different basic properties and their extension to thermoelectric material and related fields where complex systems with phases of different electronic properties are important.

necessary, adjusting the atomic size difference between the matrix elements, and the elements in the desired secondary phase, ΔR_{A-B} to >10–15% (in accordance with the well-known Hume–Rothery rules) to further facilitate a separation into the desired phases. The secondary phases and the matrix elements also have to be selected based on their suitability for thermoelectric property enhancement. The Seebeck coefficient can be affected through the creation of resonant states [54] using elements that are well-known to form Kondo scattering clusters in non-magnetic matrices e.g., Al, Cu, Au. Thermal conductivity decrease can also be obtained by selecting elements that will form immiscibility regions so that stable nano-precipitates can be formed through spinodal decomposition. As described more in detail in Shafeie et al. [32], these phases and precipitates are also able to significantly affect the electrons through electron-scattering. In general, the multifunctionality of high entropy alloys has not been explored yet, but a bright future is expected for these types of alloys in novel applications that are not yet envisioned.

The conclusions of the topics in this chapter can be listed as points:

- High entropy alloys can be a viable option for wide temperature thermoelectric applications, where a high power factor is more important than a high zT such as in waste heat recovery in power plants and steel factories.

- Complex microstructures can be achieved by high entropy alloys, to introduce complexity on all length scales, to scatter both phonons and electrons, for low thermal conductivity and high electrical resistivity properties.
- Novel magnetic scattering and high-temperature Kondo scattering > 200 K can be stabilized in conjunction with spinodally decomposed phases, to achieve very low values of temperature-dependent resistivity and wide temperature ranges between 5–20°C up to 600°C.
- Novel complex composites between high entropy alloys, and for example quasi-crystals and semiconducting phases can be achieved *in situ* if the composition is correctly chosen to potentially improve thermoelectric properties significantly.
- A significantly enhanced Seebeck coefficient is possible in high entropy alloys through abrupt charge carrier mobility changes as a function of temperature, to use in, e.g., significantly enhanced high-temperature thermocouples and thermopile dependent applications.
- 3-D printing of complex high entropy alloys and microstructures can be achieved to further improve the thermoelectric properties.
- Adding porosity to high entropy alloys may work as an efficient method to decrease the thermal conductivity of a high entropy alloy and to increase its electrical resistivity.
- A flowchart is presented to design high entropy alloys by selecting elements depending on what type of phases and microstructural properties are necessary, as a starting point, for example, in thermoelectric materials.

REFERENCES

1. C. Xiao, Z. Li, K. Li, P. Huang, and Y. Xie, "Decoupling interrelated parameters for designing high performance thermoelectric materials," *Acc. Chem. Res.*, 47(4), pp. 1287–1295, Apr. 2014.
2. G. J. Snyder, and E. S. Toberer, "Complex thermoelectric materials," 7(February), pp. 105–114, 2008.
3. L. Chen, S. Gao, X. Zeng, A. Mehdizadeh Dehkordi, T. M. Tritt, and S. J. Poon, "Uncovering high thermoelectric figure of merit in (Hf,Zr)NiSn half-Heusler alloys," *Appl. Phys. Lett.*, 107(4), p. 041902, Jul. 2015.
4. S. Sportouch, M. A. Rocci-Lanet, P. Brazis, C. R., Kannewud, M. A. Rocci-Lane, J. Ireland, P. Brazis, C. R. Kannewurt, and M. G. G. Kanatzidis, "Thermoelectric properties of half-heusler phases : ErNi1–x,CuxSb, YNi1–xCuxSb and ZrxHfyTizNiSn," *2000 IEEE*, 18(1999), pp. 344–347, 1999.
5. K. Miyamoto, A. Kimura, K. Sakamoto, M. Ye, Y. Cui, K. Shimada, H. Namatame, M. Taniguchi, S.-I. Fujimori, Y. Saitoh, E. Ikenaga, K. Kobayashi, J. Tadano, and T. Kanomata, "In-gap electronic states responsible for the excellent thermoelectric properties of Ni-based half-Heusler alloys," *Appl. Phys. Express*, 1, p. 081901, Jul. 2008.
6. J. G. Bos, and R. A. Downie, "Half-Heusler thermoelectrics: A complex class of materials," *J. Phys. Condens. Matter*, 26(43), p. 433201, Oct. 2014.
7. I. T. Materials, T. M. Tritt, M. L. Wilson, A. L. Johnson, S. Legault, and R. Stroud, "Potential of quasicrystals and quasicrystal approximants for new and improved thermoelectric materials," in *16th international Conference on Thermoelectrics*, 1997, no. 1997, pp. 454–458.

8. H. Liu, X. Yuan, P. Lu, X. Shi, F. Xu, Y. He, Y. Tang, S. Bai, W. Zhang, L. Chen, Y. Lin, L. Shi, H. Lin, X. Gao, X. Zhang, H. Chi, and C. Uher, "Ultrahigh Thermoelectric Performance by Electron and Phonon Critical Scattering in Cu 2 Se 1-x I x," *Adv. Mater.*, 25(45), pp. 6607–6612, Dec. 2013.
9. K. Biswas, J. He, I. D. Blum, C.-I. Wu, T. P. Hogan, D. N. Seidman, V. P. Dravid, and M. G. Kanatzidis, "High-performance bulk thermoelectrics with all-scale hierarchical architectures," *Nature*, 489(7416), pp. 414–418, Sep. 2012.
10. P. Ren, Y. Liu, J. He, T. Lv, J. Gao, and G. Xu, "Recent advances in inorganic material thermoelectrics," *Inorg. Chem. Front.*, 5(10), pp. 2380–2398, 2018.
11. W. Liu, H. S. Kim, Q. Jie, and Z. Ren, "Importance of high power factor in thermoelectric materials for power generation application: A perspective," *Scr. Mater.*, 111, pp. 3–9, Jan. 2016.
12. S. Sheikh, S. Shafeie, Q. Hu, J. Ahlström, C. Persson, J. Veselý, J. Zýka, U. Klement, and S. Guo, "Alloy design for intrinsically ductile refractory high-entropy alloys," *J. Appl. Phys.*, 120(16), p. 164902, Oct. 2016.
13. O. N. Senkov, G. B. Wilks, D. B. Miracle, C. P. Chuang, and P. K. Liaw, "Refractory high-entropy alloys," *Intermetallics*, 18(9), pp. 1758–1765, Sep. 2010.
14. L.-D. Zhao, V. P. Dravid, and M. G. Kanatzidis, "The panoscopic approach to high performance thermoelectrics," *Energy Environ. Sci.*, 7(1), pp. 251–268, 2014.
15. M. Christensen, S. Johnsen, and B. B. Iversen, "Thermoelectric clathrates of type I," *Dalt. Trans*, 39(4), pp. 978–992, 2010.
16. L. Su, and Y. X. Y. Gan, "Advances in thermoelectric energy conversion nanocomposites," *Adv. Compos. Mater. Med. Nanotechnol.*, 2, pp. 119–131, 2011.
17. Y. He, P. Lu, X. Shi, F. Xu, T. Zhang, G. J. Snyder, C. Uher, and L. Chen, "Ultrahigh thermoelectric performance in mosaic crystals," *Adv. Mater.*, 27(24), pp. 3639–3644, Jun. 2015.
18. Y. Pei, A. F. May, and G. J. Snyder, "Self-tuning the carrier concentration of PbTe/Ag2Te composites with excess Ag for high thermoelectric performance," *Adv. Energy Mater.*, 1(2), pp. 291–296, Mar. 2011.
19. W. Mi, P. Qiu, T. Zhang, Y. Lv, X. Shi, and L. Chen, "Thermoelectric transport of Se-rich Ag 2 Se in normal phases and phase transitions," *Appl. Phys. Lett.*, 104(13), p. 133903, Mar. 2014.
20. G. A. Slack, *CRC Handbook of Thermoelectrics*. Boca Raton, FL: CRC, 1995.
21. J. Bahk, Z. Bian, and A. Shakouri, "Electron transport modeling and energy filtering for efficient thermoelectric Mg2Si1-xSnx solid solution," *Phys. Rev. B*, 89(7), p. 075204, Feb. 2014.
22. S. Guo, C. Ng, J. Lu, and C. T. Liu, "Effect of valence electron concentration on stability of FCC or BCC phase in high entropy alloys," *J. Appl. Phys.*, 109(10), p. 103505, May 2011.
23. S. Shafeie, S. Guo, Q. Hu, H. Fahlquist, P. Erhart, and A. Palmqvist, "High-entropy alloys as high-temperature thermoelectric materials," *J. Appl. Phys.*, 118(18), p. 184905, Nov. 2015.
24. Y. Pei, H. Wang, and G. J. Snyder, "Band engineering of thermoelectric materials," *Adv. Mater.*, 24(46), pp. 6125–6135, Dec. 2012.
25. Y. Pei, X. Shi, A. LaLonde, H. Wang, L. Chen, and G. J. Snyder, "Convergence of electronic bands for high performance bulk thermoelectrics," *Nature*, 473(7345), pp. 66–69, May 2011.
26. L.-D. Zhao, S. H. Lo, Y. Zhang, H. Sun, G. Tan, C. Uher, C. Wolverton, V. P. Dravid, M. G. Kanatzidis, "Ultralow thermal conductivity and high thermoelectric figure of merit in SnSe crystals," *Nature*, 508(7496), pp. 373–377, Apr. 2014.
27. F. Seitz, *The Modern Theory of Solids*, 1st ed. New York: McGraw-Hill, 1940.

28. Y. F. Ye, Q. Wang, J. Lu, C. T. Liu, and Y. Yang, "High-entropy alloy: Challenges and prospects," *Mater. Today*, 19(6), pp. 349–362, Jul. 2016.
29. D. B. Miracle, and O. N. Senkov, "A critical review of high entropy alloys and related concepts," *Acta Mater.*, 122, pp. 448–511, Jan. 2017.
30. M. Widom, "Modeling the structure and thermodynamics of high-entropy alloys," *J. Mater. Res.*, 33(19), pp. 2881–2898, Oct. 2018.
31. E. Castle, T. Csanádi, S. Grasso, J. Dusza, and M. Reece, "Processing and properties of high-entropy ultra-high temperature carbides," *Sci. Rep.*, 8(1), p. 8609, Dec. 2018.
32. S. Shafeie, S. Guo, P. Erhart, Q. Hu, and A. Palmqvist, "Balancing scattering channels: A Panoscopic approach toward zero temperature coefficient of resistance using high-entropy alloys," *Adv. Mater.*, 31(2), p. 1805392, Jan. 2019.
33. M.-A. Park, K. Savran, and Y.-J. Kim, "Weak localization and the Mooij rule in disordered metals," *Phys. Status Solidi*, 237(2), pp. 500–512, Jun. 2003.
34. Y. Zhang, T. Zuo, Y. Cheng, and P. K. Liaw, "High-entropy alloys with high saturation magnetization, electrical resistivity and malleability," *Sci. Rep.*, 3(1), p. 1455, Dec. 2013.
35. Y. Kao, S. Chen, T. Chen, P. Chu, J. Yeh, and S. Lin, "Electrical, magnetic, and Hall properties of Al x CoCrFeNi high-entropy alloys," *J. Alloys Compd.*, 509(5), pp. 1607–1614, 2011.
36. J. H. Mooij, "Electrical conduction in concentrated disordered transition metal alloys," *Phys. Status Solidi*, 17(2), pp. 521–530, Jun. 1973.
37. Y. Kim, "Weak localization effects on the electron-phonon interaction in disordered metals," May 2006.
38. M. Tsai, "Physical properties of high entropy alloys," *Entropy*, 15(12), pp. 5338–5345, Dec. 2013.
39. P. Norouzzadeh, and D. Vashaee, "The effect of grain size and volume fraction on charge transport in thermoelectric nanocomposite of Bi2Te3-Sb2Te3," in *2012 IEEE Green Technologies Conference*, 2012, pp. 1–3.
40. T. Shiga, T. Hori, and J. Shiomi, "Influence of mass contrast in alloy phonon scattering," *Jpn. J. Appl. Phys.*, 53(2), p. 021802, Feb. 2014.
41. T. Mori, and T. Hara, "Hybrid effect to possibly overcome the trade-off between Seebeck coefficient and electrical conductivity," *Scr. Mater.*, 111, pp. 44–48, Jan. 2016.
42. J. P. Heremans, and C. M. Jaworski, "Experimental study of the thermoelectric power factor enhancement in composites," *Appl. Phys. Lett.*, 93(12), p. 122107, Sep. 2008.
43. A. A. Snarskii, M. I. Zhenirovskii, and I. V. Bezsudnov, "Limiting values of the quality factor of thermoelectric composites," *Semiconductors*, 42(1), pp. 80–85, Jan. 2008.
44. J.-J. Gu, D. Zhang, and Q. X. Guo, "Giant Seebeck coefficient decrease in polycrystalline materials with highly anisotropic band structures: Implications in seeking high-quality thermoelectric materials," *Solid State Commun.*, 148(1–2), pp. 10–13, Oct. 2008.
45. S. Riva, A. Tudball, S. Mehraban, N. P. Lavery, S. G. R. Brown, and K. V. Yusenko, "A novel High-Entropy Alloy-based composite material," *J. Alloys Compd.*, 730, pp. 544–551, Jan. 2018.
46. J. M. O. Zide, D. Vashaee, G. Zeng, J. E. Bowers, A. Shakouri, and A. C. Gossard, "Demonstration of electron filtering to increase the Seebeck coefficient in ErAs:InGaAs/InGaAlAs superlattices," 39(5), pp. 561–563, Oct. 2005.
47. P. Sun, B. Wei, J. Zhang, J. M. Tomczak, A. M. Strydom, M. Søndergaard, B. B. Iversen, F. Steglich, "Large Seebeck effect by charge-mobility engineering," *Nat. Commun.*, 6(1), p. 7475, Dec. 2015.
48. L. Kouwenhoven, and L. Glazman, "Revival of the Kondo effect," pp. 33–38, 2001.
49. R. Tahara, Y. Nakamura, M. Inagaki, and Y. Iwama, "Mössbauer study of spinodal decomposition in Fe–Cr–Co alloy," *Phys. Status Solidi*, 41(2), pp. 451–458, Jun. 1977.

50. M. Okada, G. Thomas, M. Homma, and H. Kaneko, "Microstructure and magnetic properties of Fe-Cr-Co alloys," *IEEE Trans. Magn.*, 14(4), pp. 245–252, Jul. 1978.
51. V. Raghavan, "Co-Cr-Fe (cobalt-chromium-iron)," *J. Phase Equilib.*, 15(5), pp. 524–525, Oct. 1994.
52. W. Dong, Z. Zhou, L. Zhang, M. Zhang, P. Liaw, G. Li, R. Liu, "Effects of Y, GdCu, and al addition on the thermoelectric behavior of CoCrFeNi high entropy alloys," *Metals (Basel)*, 8(10), p. 781, Sep. 2018.
53. C. W. Chen, "Spin dependence of the electrical resistivities of gadolinium alloys," *Solid State Commun.*, 3(9), pp. 231–233, 1965.
54. J. P. Heremans, B. Wiendlocha, and A. M. Chamoire, "Resonant levels in bulk thermoelectric semiconductors," *Energy Environ. Sci.*, 5(2), pp. 5510–5530, 2012.
55. H. J. Born, *Low Temperature Thermoelectric Power of Rare Earth Metals*. Ames, IA: Iowa State University, Digital Repository, 1960.
56. M. Eibschütz, G. Y. Chin, S. Jin, and D. Brasen, "Observation of phase separation in a Cr-Co-Fe alloy (chromindur) by Mössbauer effect," *Appl. Phys. Lett.*, 33(4), pp. 362–363, Aug. 1978.

13 Fatigue Behavior of High Entropy Alloys

A Review

K. Liu, S.S. Nene, Shivakant Shukla, and R.S. Mishra

CONTENTS

13.1 INTRODUCTION

Fatigue is a damage accumulation process that results in catastrophic failure of structural materials, and thus is a threat to human safety. Therefore, designing fatigue-resistant materials to counteract or at least to mitigate compromises to human safety is imperative. Classically, total fatigue life has been improved by increasing fatigue crack initiation life and by reducing the crack propagation rate [1]. From a materials science point of view, the key to improving fatigue resistance is governed by engineering the initial microstructural condition, alloy chemistry, and the subsequent deformation mechanisms. In this report, certain specific alloy systems with unique intrinsic microstructural features were chosen to prove their effect on the fatigue behavior of the material. Refinement of grain size is known to result in a higher resistance to crack initiation, thereby improving fatigue initiation life [2–6]. However, ultrafine-grained materials exhibit a higher crack growth rate, which results in a lower fatigue propagation life, and ultimately in poor response at higher stress amplitude [3, 7]. To overcome this significant deficiency, additional mechanisms such as transformation induced plasticity (TRIP) or twinning induced plasticity (TWIP) can be used to arrest or retard crack propagation by localized work hardening (WH) activity within the crack tip plastic zone [8–11].

Conventional alloys exhibit certain responses to thermomechanical processing in terms of microstructural evolution and subsequent deformation behavior. A new class of alloy system known as high entropy alloys (HEAs) was developed independently in 2004 by Yeh et al. [12] and Cantor et al. [13]. The literal meaning of high entropy is that these alloys have maximized configurational entropy due to having more than five equiatomic or non-equiatomic principal elements to form an ordered or disordered solid solution. HEAs have attracted enormous attention recently due to their interesting mechanical properties, which are directly related to the ease in engineering the microstructural features to attain structural hierarchy. The concept of HEAs was established with the design of equiatomic CrMnFeCoNi (also known as Cantor alloy), which showed massive solid-solution hardening due to a stabilized single-phase FCC (face-centered cubic) microstructure. Thus, an excellent strength–ductility synergy was attained in the Cantor alloy using the HEA concept [14–16]. However, the work hardening ability was reduced due to the standalone contribution of solid-solution hardening in the material, which hardening is further enhanced by the development of multiphase HEAs that also contribute to multiple deformation modes during deformation. Multiphase HEAs in Fe-Mn-Co-Cr-Ni system were classified into two main subcategories; namely, twinning induced plasticity (TWIP) and transformation induced plasticity (TRIP) HEAs. A primary feature of TWIP HEAs is minor additions of Al Fe-Co-Cr-Ni matrix that lead to novel $Al_xFeCoCrNi$ HEAs (where x ranges from 0.1–0.7). These HEAs have mainly a single-phase γ-FCC structure that subsequently evolves to a dual-phase FCC + BCC (body-centered cubic) microstructure with increase in Al. The strengthening coherent FCC L12 (Ni_3Al)-type phase can also be obtained in the FCC matrix upon low-temperature aging, depending on the Ni/Al ratio, which further promotes the precipitation-hardening effect of these HEAs [17]. Near-equiatomic initial compositions limit the formation of L12 phases and favor the NiAl type B2 or complex σ phase formation in the microstructure [18]. Other classes of HEAs were designed recently with the notion that transformation during straining or heating can lead to enhanced work hardening ability of the material. This transformation can be triggered during straining by making the stable phase metastable, which was done by engineering the alloy Fe-Mn-Co-Cr system chemistry with the addition of non-transition Si. Increased metastability of the parent γ matrix triggered $\gamma \longrightarrow \varepsilon$ transformation under straining and resulted in a pronounced TRIP effect in these HEAs. Further minor additions of either Al or Cu in a Fe-Mn-Co-Cr-Si matrix altered the γ metastability drastically, thereby providing it with extremely good work hardening ability and yield strength values. Many results on these TWIP/TRIP HEAs agree that varying alloy composition could contribute significantly to overcoming strength–ductility trade-offs [14,18–23]. Indeed, the alloy properties could be tuned by varying second-phase type, size, shape, volume fraction, and their distribution [17, 22, 24–26] such that a structural hierarchy can be created in the microstructure. With tunable multiphase, multicomponent hierarchical microstructures of varying alloy chemistry, processing parameters, and heat treatment times have led to HEAs/CCAs often exhibiting exceptional mechanical properties.

Using the literature results on the fatigue behavior of HEAs/CCAs, here we review micro-mechanisms often observed during fatigue deformation of HEAs/CCAs, and focus on highlights of major accomplishments in high entropy alloys in terms of fatigue behavior as compared to conventional alloys. Our review also includes, albeit more briefly, expected future directions.

13.2 FATIGUE OF CONVENTIONAL MATERIALS

13.2.1 Theory

Fatigue is one of the most common damage modes of in-service components and accounts for ~90% of sudden failures [1, 11, 27]. At the beginning of the 20th century, researchers investigating material fatigue behavior noted that stress amplitude below the material's yield strength controls the material's response under cyclic loading. Material undergoes fatigue failure in two steps, namely, crack initiation and crack propagation. The time spent by the material in both these steps can be controlled by engineering its microstructure. Microstructurally, fatigue strength of the material is reduced by the presence of inclusions, brittle second-phase particles/phases, as these act as potential sites for microcrack initiation depending on the nature of the inclusions/matrix interface [4]. Moreover, the crack propagating rate can be reduced by attaining hierarchically microstructural features or by activating localized micro-mechanisms like TWIP or TRIP within the crack tip plastic zone [28, 29].

13.2.2 Effect of Inclusions/Second Phases on Fatigue Behavior

Chemical inhomogeneity is a potential stress concentration raiser and can act as a preferential site for crack initiation and propagation. Lankford et al. [30] identified the series of stages during initiation where the initial stage is interfaced de-bonding at the inclusion/matrix interface that leads to crack initiation. Morrow et al. [31] investigated the influence of inclusion size on the fatigue limit of high-strength steel by applying the concept of a notch fatigue strength reduction factor. Morris et al. [32] studied crack initiation at inclusions extensively on Al alloys, where they claim the de-bonding between matrix and inclusion interface is due mainly to a stress concentration raiser. Tanaka et al. [33] analyzed this de-bonding theoretically by dislocation pile-ups at the inclusion during cyclic loading. At the crack propagation stage, the earlier initiated microcracks contribute to localized inhomogeneous stress distribution that leads to crack tip stress concentration. In general, the microcracks grow perpendicular to the cyclic loading direction. The crack propagating rate could be reduced by minimizing the mean free path for crack propagation. Figure 13.1 summarizes the fatigue behavior of various alloys. The figure provides information on

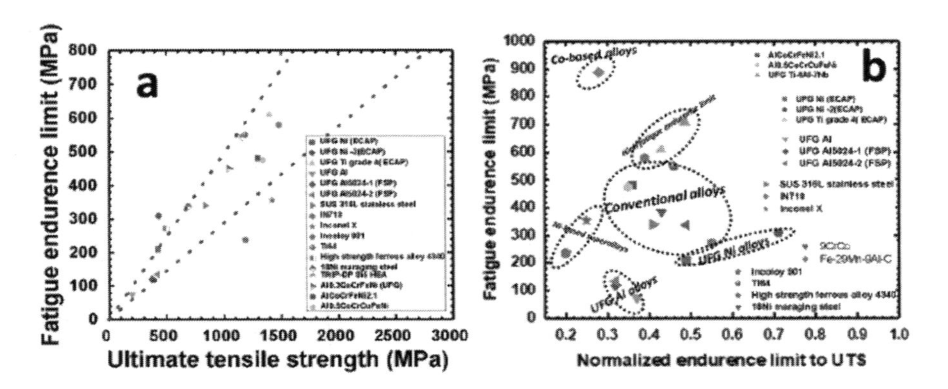

FIGURE 13.1 (a) Comparison plot of a fatigue endurance limit vs UTS for various alloys, and (b) comparison plot of a fatigue endurance limit vs UTS normalized endurance limit for various alloys.

the fatigue endurance limit of majority fatigue-resistant alloys; for example, steels, Ni-based and Co-based super alloys, and UFG materials. Also, Figure 13.1b highlights and compares fatigue strength to alloy ultimate tensile strength (UTS). The ultimate goal of enhancing the fatigue property of an alloy is to push its property to the upper right corner of Figure 13.1b, where the material exhibits high fatigue strength as well as high tensile strength.

13.2.3 Effect of Grain Size on Fatigue Behavior

The fatigue property of an alloy is often affected by the alloy grain size. Similar to the grain size effect on tensile property, an increase in the area fraction of grain boundaries enhances the ability of a material to store dislocations, and essentially improves material tolerance to damage, thus the importance of ultrafine- and fine-grained materials.

Ultrafine-grained (UFG) and fine-grained (FG) materials have high innovation potential for use in commercial applications [34]. UFG and FG materials by definition have a grain size in the range of 100 nm to 1 μm and 1–10 μm, respectively, which is demarcated by the dislocation mechanisms [35]. These materials have become well-known because of outstanding tensile strength properties [35, 36], based on the Hall–Petch relationship or grain boundary strengthening. Moreover, if the grains are reasonably stable, UFG materials could exhibit outstanding superplastic properties at elevated temperatures [34]. Further, low stress amplitude or high cycle fatigue property of a UFG material was proven to be superior compared to conventional grain size. The coarser grain materials yield at lower stress, and hence crack initiation becomes easy, whereas UFG materials could delay fatigue crack initiation and thus exhibit higher fatigue resistance than CG materials [37].

13.2.4 Effect of Deformation Mechanisms on Fatigue Behavior

Polycrystalline materials deform mainly with a dislocation motion along closed packed planes and directions through slip in uniaxial loading. However, the propensity of slip is controlled by material crystal structure [38, 39]. Extensive study on deformation accommodation through slip under cyclic loading has led to claims that formation of fatigue intrusions and extrusions is the result of preferential slipping tendency of each grain under alternative tensile and compressive loading. However, the thinking has been that activation of additional deformation mechanisms such as mechanical twinning (TWIP effect) and strain induced martensitic transformation (TRIP effect) can effectively tune the fatigue properties. Stacking fault energy (SFE) is a key factor that governs the activation of these additional deformation modes in the material [40–42]. Engineering SFE of a polycrystalline material was considered to be difficult in conventional alloy design; however, the abundant chemical space available in HEAs/CCAs design has enabled very easy tuning of alloy SFE [18]. An interesting point to note in HEAs is the slope of the Stage III work hardening curve, which was rather steep and was due to reduction in dynamic recovery processes via TWIP or TWIP effects facilitated by low SFE [38, 43].

The change in slope at the end of Stage III is generally attributed to the activation of those additional deformation mechanisms. In general, martensitic transformation is observed in very low SFE steels (below 20 mJ/m2), while twinning is observed in medium SFE steels (20–40 mJ/m2). When the SFE exceeds 45 mJ/m2, dislocation glide becomes the predominant mode of plastic deformation [44]. Liu et al. [45] further proved that lower SFE promoted thinner deformation twins or hexagonal close-packed (HCP)–ε phase under loading. The interaction between and among the newly formed deformation twins and/or HCP–ε phase and dislocations increased the defect storage rate via the dynamic Hall–Petch effect and led to slope change. The dislocation-mediated plasticity is attributed to the formation of nano-sized twins or Ɛ plates, which resulted in boundaries. This can hinder a dislocation motion followed by a pile-up at these boundaries, thereby promoting work hardenability [46]. According to the Considère criterion, the onset of necking starts when the work hardening rate quates with flow stress [27, 47]. The onset of necking was postponed due to a sustained work hardening rate above the flow stress through twinning or transformation, which ultimately prolongs plastic deformation. Based on all the information stated, the work hardening behavior of an alloy changes along with the deformation mechanism. This is useful in deciding fatigue crack propagation since it involves plastic deformation within the crack tip plastic zone and leads to crack blunting. More crack blunting at the tip results in energy dissipation to form crack branches, which in turn increase the mean free path for crack growth.

13.3 FATIGUE PROPERTIES OF HIGH ENTROPY ALLOYS

13.3.1 Effect of Impurities on Fatigue Behavior of CG HEAs

The impurities introduced during manufacturing or processing lead to stress inhomogeneity in the material, which initiates fatigue cracks [4]. Hemphill et al. [48] examined the fatigue behavior of cold rolled $Al_{0.5}CoCrCuFeNi$, where fatigue failure was initiated from Al_2O_3 inclusions and pre-existing microcracks. The initial microstructure consists of dark α-FCC dendrite phase and a light β-FCC Cu #-rich phase, with the β phase elongated along the rolling direction (Figure 13.2a). X-ray

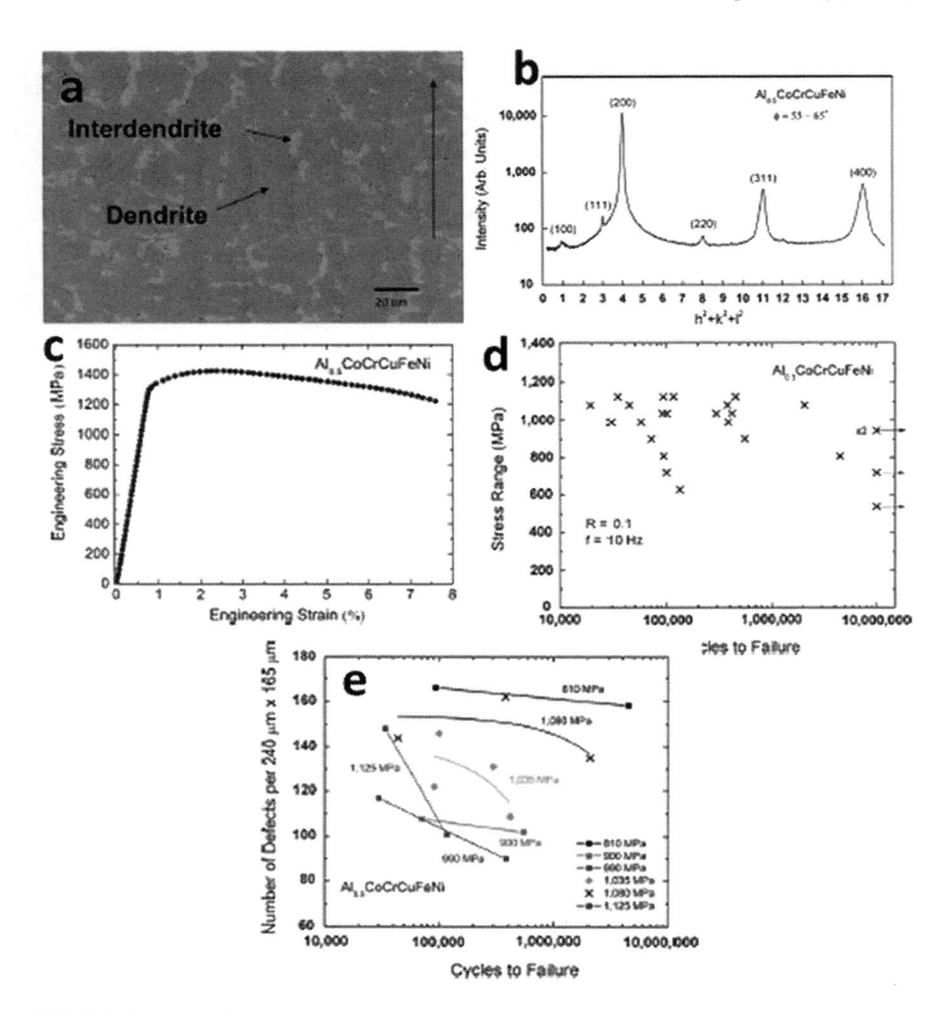

FIGURE 13.2 (a) SEM micrograph showing two phases with rolling direction directed by the arrow, (b) synchrotron X-ray diffraction pattern, (c) stress-strain curve, (d) fatigue S-N curve, (e) the effect of surface defects on the cycles to failure [48].

synchrotron analysis confirms the presence of ordered L12 phase with (100) peak. No clear BCC peaks in the X-ray diffraction pattern further confirm the observed structure is a two-phase FCC structure. Uniaxial tensile testing shows that the alloy exhibited ~1,284 MPa yield strength and 1,344 MPa of ultimate tensile strength, which is superior to most conventional alloys. The as-rolled alloy shows a uniform elongation of 7.6%. The experimental results confirm the effect of defects on the fatigue life of HEAs (Figure 13.2e). Similar to the conventional alloys, the fatigue property of an HEA is also related to the number of defects presented in the as-received material.

With the same composition ($Al_{0.5}CoCrCuFeNi$) as Hemphill et al. [48], Tang et al. [49] investigated the effect of raw material purity on fatigue behavior. Alloys made with high-purity elements exhibited better properties, due mainly to less fraction of the pre-existing microcracks, pores, and oxides (Figure 13.3a–c). Fractography confirms the presence of voids and/or inclusions at the fatigue crack initiation site in the wrought $Al_{0.5}CoCrCuFeNi$ made with less impure materials (Figure 13.3d). Similarly, Liu et al. [50, 51] have observed cracks along the secondary σ and b2 phases, which confirms that the presence of chemical inhomogeneity causes a decrease in the fatigue property of an alloy. In both $Al_{0.7}CoCrFeNi$ and UFG $Al_{0.3}CoCrFeNi$ HEAs, cracks propagating along the phase boundaries are evident. Detailed geometrically necessary dislocation (GND) analysis confirms that the accumulation of GNDs around the phase boundary triggers nucleation of the microcracks, and is due mainly to the raised strain/modulus mismatch between the soft matrix and harder phases, which resulted in non-uniform strain partitioning. Continued exposure to cyclic loading led to formation of the microcracks link into the major crack, which eventually causes material failure.

13.3.2 Effect of PSB Formation on Fatigue Behavior of CG HEAs

During cyclic deformation, formation of persistent slip bands (PSBs) in coarse-grained microstructures has always been observed in conventional alloys. The slip in cyclically deformed material is not uniform, unlike material deformed under uniaxial loading. Thus, the formed dislocation structure accommodates further

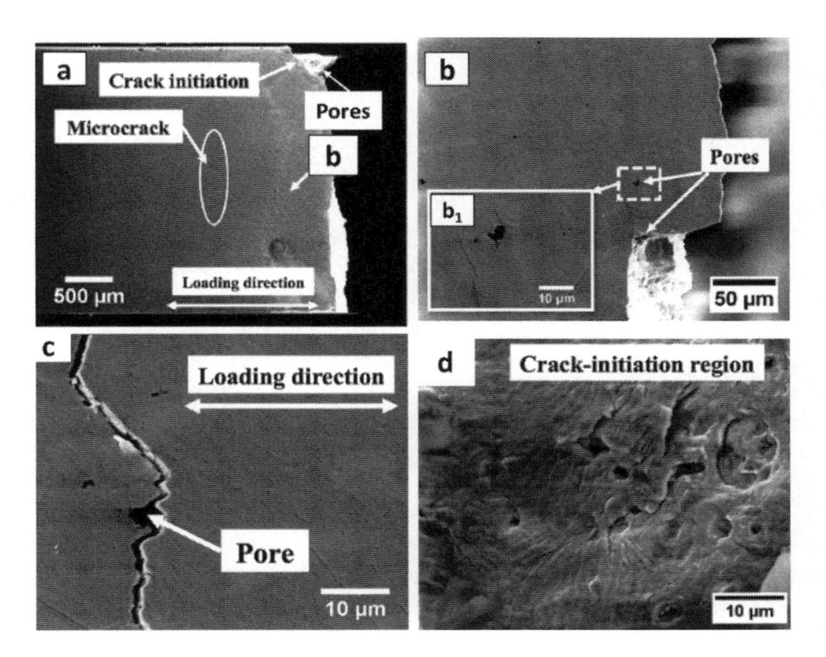

FIGURE 13.3 (a) Overview of surface of failed $Al_{0.5}CoCrCuFeNi$ specimen, (b), (c) high magnification showing shrinkage pore, (d) fractography of crack initiation region fatigue striations and pores.

plastic deformation, locally by formation of PSBs where slip activities higher than the matrix on preferential grains were observed [52–54]. Surface extrusions and intrusions formed due to the PSBs, which always act as stress concentration sites. Therefore, PSBs will always lead to crack initiation, and provide a preferred path for cracks to propagate during cyclic loading [15, 25, 4].

Tang et al. [49], Liu et al. [50], and Shukla et al. [55] observed the formation of PSBs after a significant number of loading cycles, which initiate fatigue crack followed by its propagation. Liu et al. carried out stress-controlled bending fatigue testing on a coarse-grained $Al_{0.7}CoCrFeNi$ HEA. With featured lamellar microstructure consisting of FCC+B2 phases, they reported that the PSBs are forming in the FCC matrix, and act as crack initiation sites. The PRIAS map from EBSD confirms the presence of intrusion and extrusion formed on the sample surface due to PSBs (Figure 13.4a). Angles among various PSBs are consistent and confirmed that various {111} <110> slip systems had been activated (Figure 13.4a_2, a_3). Thereafter, crack propagation followed by a zigzag crack-propagating pattern (Figure 13.4) was due to the activation of various slip systems [50]. Similar observations near the major crack were made by Tang et al. [49] in a wrought $Al_{0.5}CoCrCuFeNi$ (Figure 13.4b, b_1). Noticeably, no slip activities occurred in the harder B2 phase for both dual-phase eutectic HEAs [50, 55].

13.3.3 Effect of Precipitates on Fatigue Behavior of CG HEAs

In conventional alloys, fatigue crack initiates and propagates along with chemical inhomogeneity. With the unique microstructural flexibility characteristic of HEAs, researchers tried to engineer the microstructure and study the effect of precipitates on the fatigue behavior of an HEA. With the aid of CALPHAD, Liu et al. [50] predicted the phase diagram for $Al_{0.7}CoCrFeNi$. They chose two different heat treatment routes to vary the presence of L12 in the matrix. With this study, they were able to compare the effect of nano-sized L12 precipitates on the fatigue behavior of a lamellar structured HEA. Nano-sized (~2 nm) L12 precipitates were dispersed into the FCC matrix to strengthen the matrix. The study is important, as the effect of L12 precipitates on the uniaxial tensile property is tremendous—~300 MPa improvement in both yield and ultimate tensile strength, with a slight decrease in uniform elongation. However, the fatigue results indicate no obvious differences in terms of fatigue property with and without L12 precipitate. The detailed post-deformation TEM analysis reveals that L12 precipitates were sheared by PSBs during cyclic loading. Therefore, the effect of nano-precipitates on fatigue property is not pre-eminent Figure 13.5.

On the other hand, Shukla et al. [55] compared the fatigue behavior of as-cast conditions and thermomechanically treated (EHEAw) conditions. One of the distinct microstructural features between the two conditions is the presence of various morphology B2/BCC in the FCC lamella introduced by rolling and annealing. Of the unique microstructural features, one major observation was that cracks were arrested at relatively larger B2 precipitates, and subsequently delayed crack initiation. The prolonged crack initiation period therefore led to enhanced fatigue property after rolling and annealing Figure 13.6.

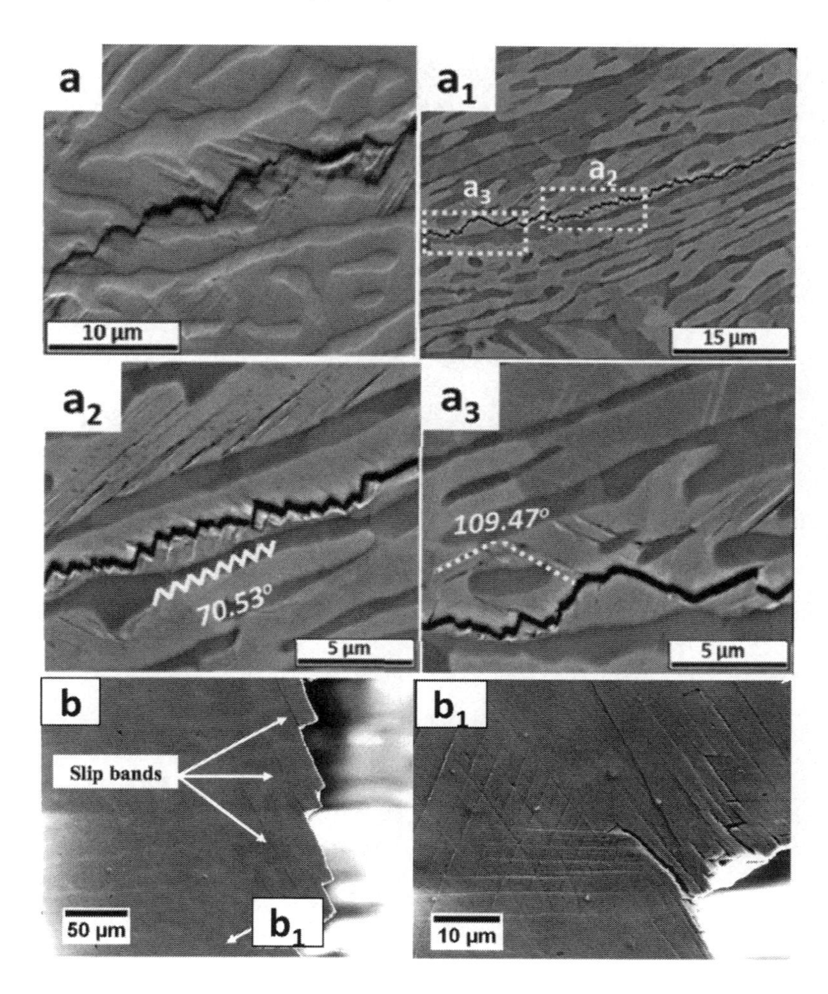

FIGURE 13.4 Presence of PSBs near the major fatigue crack region. (a) PRIAS map confirms PSBs within the FCC phase of $Al_{0.7}CoCrFeNi$, (a_{1-3}) high magnification images show the morphology of the PSBs, (b and b_1) the PSBs observed in wrought $Al_{0.5}CoCrCuFeNi$.

13.3.4 Fatigue Behavior of TWIP HEAs

In uniaxial tensile testing, formation of mechanical twins along with deformation could alter the work hardening behavior of the alloy locally and establish a superior combination of strength and ductility in conventional TWIP steels [56, 57]. Mechanical twinning introduced from pre-strained high manganese TWIP steels shows enhanced fatigue behavior due to its high work hardenability [58]. Therefore, researchers have studied the effect of twinning on the fatigue behavior of HEAs.

Initially, Tang et al. [49] reported more deformation twins near the crack initiation sites compared to the adjacent region and away from the crack initiation sites. Based on detailed TEM analysis, two sets of nano-twins with orientation identified

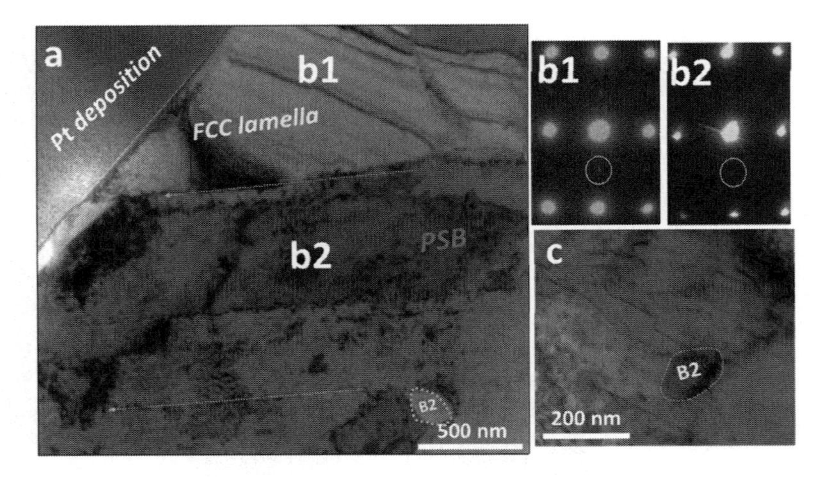

FIGURE 13.5 (a) BFTEM results for AH+LTA condition, (b1) SADPs obtained from FCC lamella.

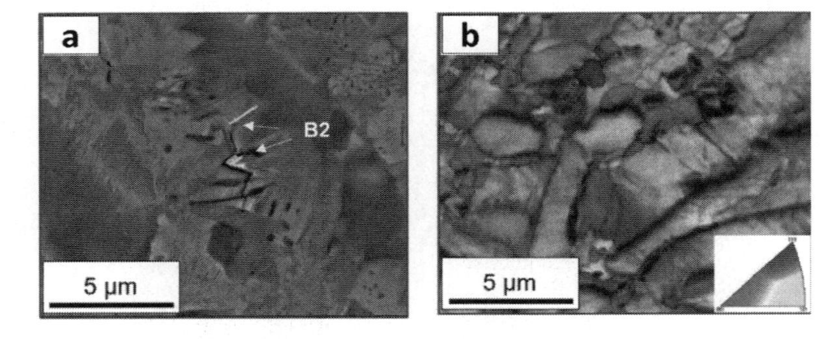

FIGURE 13.6 (a) BSE image of w condition showing PSBs and crack deviation by B2 precipitate (pointed with arrows), (b) EBSD image of fatigued EHEAw condition.

as $[2\bar{1}1]$ and $[21\bar{1}]$ were observed near the crack initiation sites within the FCC phase (Figure 13.7a–f). Large amounts of tangled dislocations have also been observed at the same region. Due to low stacking fault energy, the formation of the nano-twins and tangled dislocations strengthened the FCC matrix, thereby enhancing the alloy fatigue limit due to the reduced dislocation mean free path.

Liu et al. [51] proved that during cyclic loading, deformation twins formed in the low SFE FCC matrix. The microstructure prior to cyclic deformation consists of three phases (FCC, B2, σ). Minimal casting defects were observed in the initial microstructure after thermomechanical processing. The researchers believe fatigue crack initiated at the phase boundaries, and this is further proved by post-deformation GNDs analysis (Figure 13.8a–c). However, the dislocation accumulation capacity of the alloy, as improved with ultrafine-grained microstructure, delayed crack initiation and prolonged total fatigue life. Further, the formation of deformation

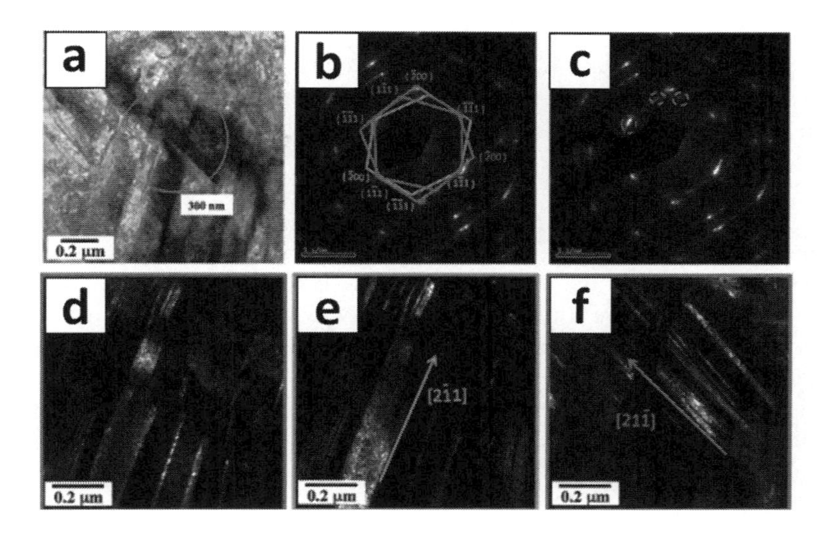

FIGURE 13.7 (a) Two different orientation sets of dense nano-twins with a high density of tangled dislocations, with zone axis of [0 1 1]; (b) the matrix and two orientations of twins; (c) three chosen spots for dark field images. Three dark-field images are in the matrix in combination with the first set of twins. (d,e) the first set of twins; (f) and the second set of twins.

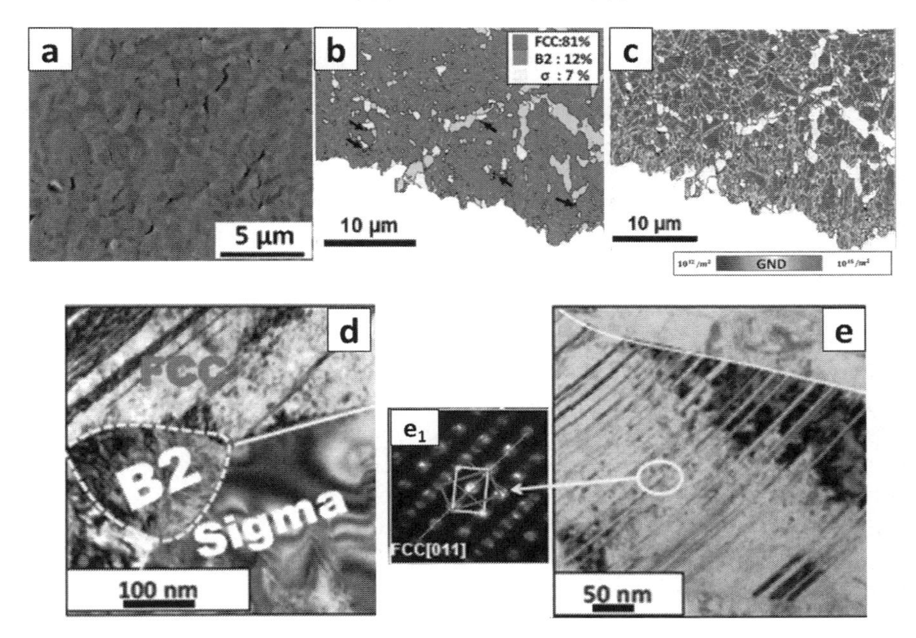

FIGURE 13.8 Region along the major crack. (a) Microcracks along the second phase boundaries, (b) phase map at the fracture surface with black phase boundaries, white CSL boundaries, and dark arrows pointed at microcracks, and (c) GND analysis corresponding to (b), (d) undeformed B2 and σ phase, and deformation twins in FCC matrix, (e) micro-twinning within FCC further confirmed with micro-diffraction (e_1).

twins in the FCC matrix within the major crack plastic zone was a primary contributor to the excellent fatigue property (Figure 13.8d,e). The work hardening behavior of the material within the crack plastic zone changing during cyclic loading therefore enables greater damage tolerance of the material.

13.3.5 Fatigue Behavior of TRIP HEAs

Another often-mentioned deformation phenomenon is altered work hardening behavior due to activation of deformation-induced transformation. Consequently, work hardenability of the alloy has been altered with newly formed phase interfaces. Further, transformation observed at the crack tip arrests the propagating crack due to strain mismatch from the phase interfaces [32, 8, 59–62].

Liu et al. [63] studied the effect of TRIP on the fatigue behavior of a fine-grained metastable high entropy alloy. With friction stir processing, average grain size could be maintained below 2 μm. The extremely metastable γ phase provided an opportunity for the material to accommodate increased deformation. Clear strain-induced transformation has been observed within the plastic zone near the major crack (Figure 13.9$a_{1\text{-}4}$). The work hardening behavior within the plastic zone, having been altered, apparently accommodates more energy from the major crack and thus slows the crack propagation rate. The fatigue behavior of this alloy system is superior to other conventional fatigue-resistant TRIP steels, and is due mainly to improved γ metastability and reduced grain size. Fractography features with Stage II crack propagation (Figure 13.9b) and crack branching on the specimen surface (Figure 13.9c) confirm the presence of crack blunting, and ultimately reduce the crack propagation rate.

13.3.6 HEA Design for Improved Fatigue Properties

Based on the observations cited above, beneficial microstructural features for HEAs fatigue properties are ultrafine-grained matrix enabled with TWIP, and/or TRIP effects. Therefore, an alloy design approach is based on the earlier findings. Nene et al. [64] designed and Liu et al. [65] confirmed improved mechanical and fatigue properties in a $Fe_{38.5}Mn_{20}Co_{20}Cr_{15}Si_5Cu_{1.5}$ HEA compared to conventional fatigue-resistant alloys. They altered the stacking fault energy of the alloy by interchanging the constituent element in the alloy, which enables change in subsequent deformation mechanisms. With the addition of Cu in the solid solution (Figure 13.10a), the γ matrix becomes stronger and tougher by increased TRIP stress. Moreover, they reduced the heat input from friction stir processing to further reduce grain size in the ultrafine-grained regime. Unsurprisingly, the alloy demonstrated fatigue properties that were superior to other fatigue-resistant alloys. Figure 13.10b presents a comparison plot of the fatigue property of this alloy system with other conventional fatigue-resistant alloys.

A detailed deformation mechanism study included the observation of deformation twinning and strain-induced transformation within the crack plastic zone. The significant crack branching near the major crack is an indicator of energy dissipation during crack propagation. The energy needed for propagating the major crack was

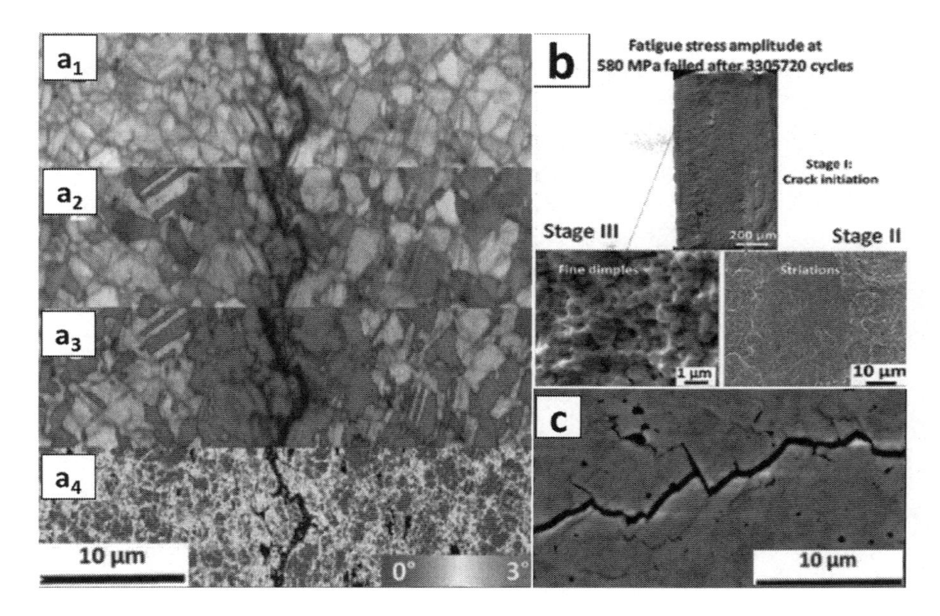

FIGURE 13.9 (a_{1-4}) higher magnification scan around the crack with image quality map, IPF map, phase map, and KAM, respectively, (b) fractographs of a sample tested at the cyclic stress of 580 MPa after failure at 3,305,720 cycles, (c) BSE image at the crack reveals the fatigue branching cracks.

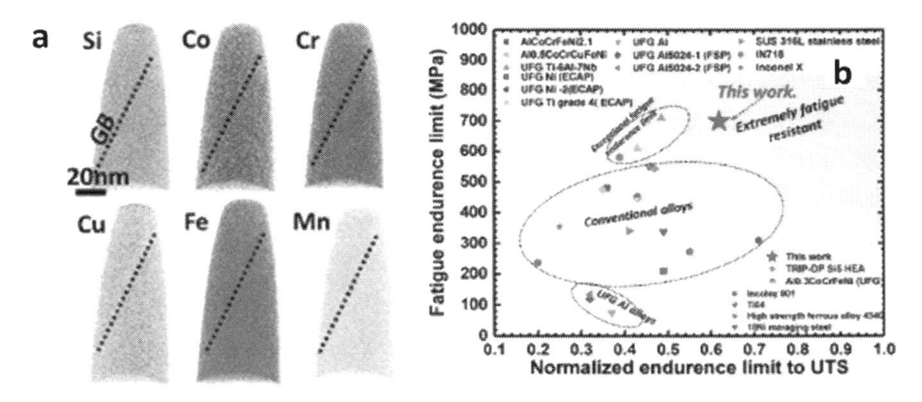

FIGURE 13.10 (a) Atom probe analysis confirms the composition of the alloy with Cu distributed in the grain interior. (b) Fatigue property comparison plot confirms the superior fatigue property of Cu-HEA as compared to other conventionally fatigue-resistant alloys.

shared during formation of the microcracks, as well as during transformation and twinning formation. On the other hand, due to the nature of ultrafine grain, the material's ability to store dislocations prior to crack initiation is enhanced (Figure 13.11). Overall, the fatigue property of the alloy was enhanced significantly by delayed crack initiation of the UFG microstructure, whereas the crack propagation rate was reduced by local plasticity via TRIP in the crack tip plastic zone.

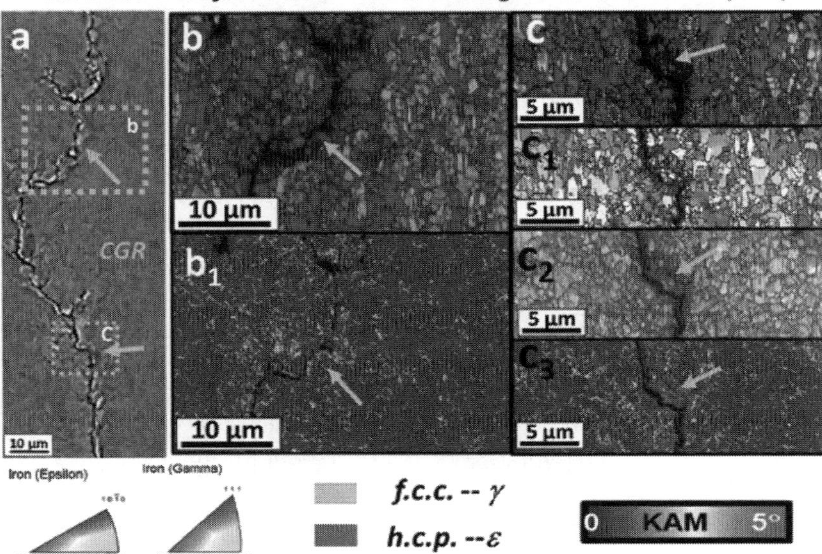

FIGURE 13.11 Evidence of crack growth retardation (CGR). (a) BSE image of a crack region, (b) region b EBSD phase map shows phase distribution, (b_1) KAM analysis corresponds to (b), (c, $c_{1\text{-}3}$) higher magnification scan around the crack with image phase map, IPF map, image quality map, and KAM, respectively.

13.4 MAJOR ACCOMPLISHMENTS

Figure 13.12 summarizes the HEA fatigue properties available to date in terms of fatigue endurance limit and ultimate tensile strength. Figure 13.12 makes it clear that conventional materials like steels and Al-Mg alloys have limited fatigue endurance when compared with Ni-based superalloys and Ti-based alloys. Moreover, dual-phase Al_xFeCrCoNi HEAs showed similar fatigue properties at ambient temperature when compared with their steel counterpart. However, the design of metastable HEAs includes the feasibility of attaining an exceptional endurance limit more than the traditionally fatigue-resistant materials in high-cycle fatigue (star in Figure 13.12). Thus, the abundant compositional space available in the hyperdimensional phase diagram for HEA design opens a pathway to realizing the extremely fatigue-resistant materials with exceptional strength–ductility in uniaxial loading. Figure 13.12 also confirms that the results included in this report contribute significantly to this relatively new field of research and present a new alloy design strategy with a focus on fatigue properties.

13.5 PLANNED WORK

The following planned research is designed to deepen the understanding of deformation mechanisms. (1) The effect of grain size on the fatigue property of a high entropy alloy will be studied in depth. (2) For metastable HEAs, detailed nano-deformation

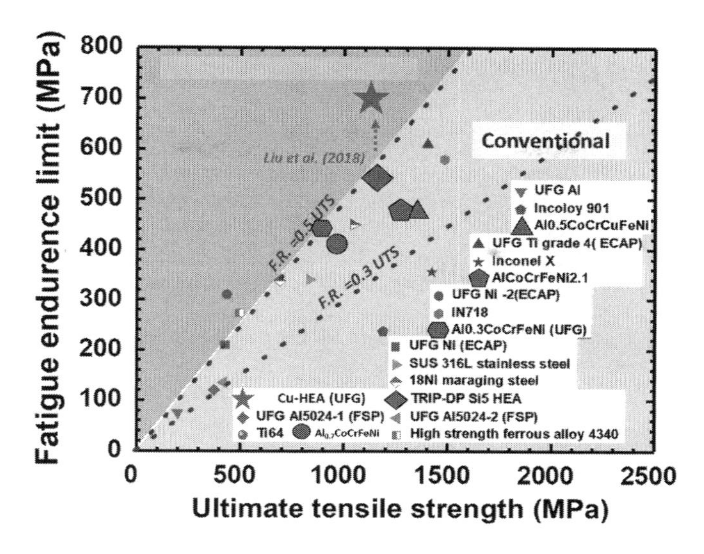

FIGURE 13.12 Comparison plot shows the fatigue property of high entropy alloys compared to the other conventional fatigue-resistant alloys.

analysis should be examined. (3) Secondary phases, i.e., B2 and L12, can be introduced in the metastable HEAs through appropriate alloying additions and annealing treatment to investigate the effect of strengthening phases in the metastable high entropy alloy system.

13.6 CONCLUSION

Compared to conventional fatigue-resistant alloys, the microstructure of HEAs is more flexible. The classical fatigue-metallurgy combined concepts could be proved easily with various thermomechanical processing paths. To date, some major highlights of high entropy alloy fatigue property are: (i) crack initiation delays in UFG materials; (ii) deformation mechanism is strongly dependent on the variation of stacking fault energy of the alloy system, where the SFE is heavily related to the constituent alloying element; (iii) localized work hardening behavior at the crack tip alters with activated deformation mechanisms; (iv) crack retardation due to altered crack tip work hardening behavior reduces crack propagating rate.

ACKNOWLEDGMENT

The authors gratefully acknowledge the support of the National Science Foundation through Grant No. 1435810.

REFERENCES

1. Suresh, S. *Fatigue of Materials*. Cambridge University Press, 1998.
2. Lerch, B.A.; Jayaraman, N.; Antolovich, S.D. A study of fatigue damage mechanisms in Waspaloy from 25 to 800 C. *Mater. Sci. Eng.* 1984, *66*(2), 151–166.

3. Mughrabi, H. On the grain-size dependence of metal fatigue: Outlook on the fatigue of ultrafine-grained metals. In: *Investigations and Applications of Severe Plastic Deformation*; Springer, 2000; pp. 241–253.
4. Sangid, M.D. The physics of fatigue crack initiation. *Int. J. Fatigue* 2013, *57*, 58–72.
5. Vinogradov, A. Mechanical properties of ultrafine-grained metals: New challenges and perspectives. *Adv. Eng. Mater.* 2015, *17*(12), 1710–1722.
6. Lu, K.; Lu, L.; Suresh, S. Strengthening materials by engineering coherent internal boundaries at the nanoscale. *Science* 2009, *324*(5925), 349–352.
7. Mughrabi, H.; Höppel, H.W. Cyclic deformation and fatigue properties of ultrafine grain size materials: Current status and some criteria for improvement of the fatigue resistance. *MRS Online Proc. Libr. Arch.* 2000, *634*.
8. Cheng, X.; Petrov, R.; Zhao, L.; Janssen, M. Fatigue crack growth in TRIP steel under positive R-ratios. *Eng. Fract. Mech.* 2008, *75*(3–4), 739–749.
9. De, P.S.; Obermark, C.M.; Mishra, R.S. Development of a reversible bending fatigue test bed to evaluate bulk properties using sub-size specimens. *J. Test. Eval.* 2008, *36*, 402–405.
10. Nene, S.S.; Liu, K.; Frank, M.; Mishra, R.S.; Brennan, R.E.; Cho, K.C.; Li, Z.; Raabe, D. Enhanced strength and ductility in a friction stir processing engineered dual phase high entropy alloy. *Sci. Rep.* 2017.
11. Reed-Hill, R.E.; Abbaschian, R.; Abbaschian, R. *Physical Metallurgy Principles*, 1973.
12. Yeh, J.; Chen, S.; Lin, S.; Gan, J.; Chin, T.; Shun, T.; Tsau, C.; Chang, S. Nanostructured high-entropy alloys with multiple principal elements: Novel alloy design concepts and outcomes. *Adv. Eng. Mater.* 2004, *6*(5), 299–303.
13. Cantor, B.; Chang, I.T.H.; Knight, P.; Vincent, A.J.B. Microstructural development in equiatomic multicomponent alloys. *Mater. Sci. Eng. A* 2004, *375*, 213–218.
14. Gludovatz, B.; Hohenwarter, A.; Catoor, D.; Chang, E.H.; George, E.P.; Ritchie, R.O. A fracture-resistant high-entropy alloy for cryogenic applications. *Science* 2014, *345*(6201), 1153–1158.
15. Otto, F.; Dlouhý, A.; Somsen, C.; Bei, H.; Eggeler, G.; George, E.P. The influences of temperature and microstructure on the tensile properties of a CoCrFeMnNi high-entropy alloy. *Acta Mater.* 2013.
16. Gali, A.; George, E.P. Tensile properties of high- and medium-entropy alloys. *Intermetallics* 2013.
17. Wang, Z.G.; Zhou, W.; Fu, L.M.; Wang, J.F.; Luo, R.C.; Han, X.C.; Chen, B.; Wang, X.D. Effect of coherent L12 nanoprecipitates on the tensile behavior of a FCC-based high-entropy alloy. *Mater. Sci. Eng. A* 2017, *696*, 503–510.
18. Miracle, D.B.; Senkov, O.N.; Ye, Y.F.; Wang, Q.; Lu, J.; Liu, C.T.; Yang, Y.; Li, Z.; Pradeep, K.G.; Deng, Y. et al. A critical review of high entropy alloys and related concepts. *Acta Mater.* 2013.
19. Huang, H.; Wu, Y.; He, J.; Wang, H.; Liu, X.; An, K.; Wu, W.; Lu, Z. Phase-transformation ductilization of brittle high-entropy alloys via metastability engineering. *Adv. Mater. Weinheim* 2017, *29*(30). http://www.ncbi.nlm.nih.gov/pubmed/1701678.
20. Lu, Y.; Gao, X.; Jiang, L.; Chen, Z.; Wang, T.; Jie, J.; Kang, H.; Zhang, Y.; Guo, S.; Ruan, H. et al. Directly cast bulk eutectic and near-eutectic high entropy alloys with balanced strength and ductility in a wide temperature range. *Acta Mater.* 2017, *124*, 143–150.
21. Wani, I.S.; Bhattacharjee, T.; Sheikh, S.; Bhattacharjee, P.P.; Guo, S.; Tsuji, N. Tailoring nanostructures and mechanical properties of AlCoCrFeNi2. 1 Eutectic high entropy alloy using thermo-mechanical processing. *Mater. Sci. Eng. A* 2016, *675*, 99–109.
22. Ming, K.; Bi, X.; Wang, J. Precipitation strengthening of ductile Cr15Fe20Co35Ni20Mo10 alloys. *Scr. Mater.* 2017, *137*, 88–93.

23. Liu, W.H.; Yang, T.; Liu, C.T. Precipitation hardening in CoCrFeNi-based high entropy alloys. *Mater. Chem. Phys.* 2018, *210*, 2–11.
24. He, J.Y.; Wang, H.; Huang, H.L.; Xu, X.D.; Chen, M.W.; Wu, Y.; Liu, X.J.; Nieh, T.G.; An, K.; Lu, Z.P. A precipitation-hardened high-entropy alloy with outstanding tensile properties. *Acta Mater.* 2016, *102*, 187–196.
25. Zhao, Y.L.; Yang, T.; Tong, Y.; Wang, J.; Luan, J.H.; Jiao, Z.B.; Chen, D.; Yang, Y.; Hu, A.; Liu, C.T. Heterogeneous precipitation behavior and stacking-fault-mediated deformation in a CoCrNi-based medium-entropy alloy. *Acta Mater.* 2017, *138*, 72–82.
26. Li, D.; Li, C.; Feng, T.; Zhang, Y.; Sha, G.; Lewandowski, J.J.; Liaw, P.K.; Zhang, Y. High-entropy Al 0.3 CoCrFeNi alloy fibers with high tensile strength and ductility at ambient and cryogenic temperatures. *Acta Mater.* 2017, *123*, 285–294.
27. Dieter, G.E.; Bacon, D.J. *Mechanical Metallurgy*. McGraw-hill New York, 1986; Vol. 3.
28. Koyama, M.; Zhang, Z.; Wang, M.; Ponge, D.; Raabe, D.; Tsuzaki, K.; Noguchi, H.; Tasan, C.C. Bone-like crack resistance in hierarchical metastable nanolaminate steels. *Science* 2017, *355*(6329), 1055–1057.
29. Hamada, A.S.; Karjalainen, L.P.; Puustinen, J. Fatigue behavior of high-Mn TWIP steels. *Mater. Sci. Eng. A* 2009.
30. Lankford, J. The growth of small fatigue cracks in 7075–T6 aluminum. *Fatigue Fract. Eng. Mater. Struct.* 1982, *5*(3), 233–248.
31. Morrow, J. Cyclic plastic strain energy and fatigue of metals. In: *Internal Friction, Damping, and Cyclic Plasticity*; ASTM International, 1965.
32. Fine, M.E. Fatigue resistance of metals. *Metall. Trans. A* 1980, *11*(3), 365–379.
33. Tanaka, K.; Mura, T. A theory of fatigue crack initiation at inclusions. *Metall. Trans. A* 1982, *13*(1), 117–123.
34. Huang, Y.; Langdon, T.G. Advances in ultrafine-grained materials. *Mater. Today* 2013, *16*(3), 85–93.
35. Gleiter, H. Nanocrystalline materials. In: *Advanced Structural and Functional Materials*; Springer, 1991, 1–37.
36. Valiev, R.Z. Structure and mechanical properties of ultrafine-grained metals. *Mater. Sci. Eng. A* 1997, *234*, 59–66.
37. Liu, R.; Tian, Y.; Zhang, Z.; An, X.; Zhang, P.; Zhang, Z. Exceptional high fatigue strength in Cu-15at.%Al alloy with moderate grain size. *Sci. Rep.* 2016.
38. Hull, D.; Bacon, D.J. *Introduction to Dislocations*. Butterworth-Heinemann, 2001.
39. Hull, D.; Bacon, D.J. Movement of dislocations. In: *Introduction to Dislocations*, 2007.
40. Piercea, D.T.; Jiménezc, J.A.; Bentley, J.; Raabe, D.; Oskaya, C.; Wittiga, J.E. The influence of manganese content on the stacking fault and austenite/e-martensite interfacial energies in Fe–Mn–(Al–Si) steels investigated by experiment and theory. *Acta Mater.* 2014, *68*, 238–253.
41. Frommeyer, G.; Brüx, U.; Neumann, P. Supra-ductile and high-strength manganese-TRIP/TWIP steels for high energy absorption purposes. *ISIJ Int.* 2003, *43*(3), 438–446.
42. Grässel, O.; Krüger, L.; Frommeyer, G.; Meyer, L.W. High strength Fe–Mn–(Al, Si) TRIP/TWIP steels development—Properties—Application. *Int. J. Plast.* 2000, *16*(10–11), 1391–1409.
43. Bouaziz, O.; Allain, S.; Scott, C.P.; Cugy, P.; Barbier, D. High manganese austenitic twinning induced plasticity steels: A review of the microstructure properties relationships. *Curr. Opin. Solid State Mater. Sci.* 2011, *15*(4), 141–168.
44. Allain, S.; Chateau, J.-P.; Bouaziz, O.; Migot, S.; Guelton, N. Correlations between the calculated stacking fault energy and the plasticity mechanisms in Fe–Mn–C alloys. *Mater. Sci. Eng. A* 2004, *387*, 158–162.
45. Liu, S.F.; Wu, Y.; Wang, H.T.; He, J.Y.; Liu, J.B.; Chen, C.X.; Liu, X.J.; Wang, H.; Lu, Z.P. Stacking fault energy of face-centered-cubic high entropy alloys. *Intermetallics* 2017.

46. Dao, M.; Lu, L.; Asaro, R.J.; De Hosson, J.T.M.; Ma, E. Toward a quantitative understanding of mechanical behavior of nanocrystalline metals. *Acta Mater.* 2007, *55*(12), 4041–4065.
47. Courtney, T.H. *Mechanical Behavior of Materials*. Waveland Press, 2005.
48. Hemphill, M.A.; Yuan, T.; Wang, G.Y.; Yeh, J.W.; Tsai, C.W.; Chuang, A.; Liaw, P.K. Fatigue behavior of Al 0.5CoCrCuFeNi high entropy alloys. *Acta Mater.* 2012.
49. Tang, Z.; Yuan, T.; Tsai, C.-W.; Yeh, J.-W.; Lundin, C.D.; Liaw, P.K. Fatigue behavior of a wrought Al 0.5 CoCrCuFeNi two-phase high-entropy alloy. *Acta Mater.* 2015, *99*, 247–258.
50. Liu, K.; Gwalani, B.; Komarasamy, M.; Shukla, S.; Wang, T.; Mishra, R.S. Effect of nano-sized precipitates on the fatigue property of a lamellar structured high entropy alloy. *Mater. Sci. Eng. A* 2019, *760*.
51. Liu, K.; Komarasamy, M.; Gwalani, B.; Shukla, S.; Mishra, R.S. Fatigue behavior of ultrafine grained triplex Al0. 3CoCrFeNi high entropy alloy. *Scr. Mater.* 2019, *158*, 116–120.
52. Lukáš, P.; Kunz, L. Role of persistent slip bands in fatigue. *Philos. Mag.* 2004, *84*(3–5), 317–330.
53. Repetto, E.A.; Ortiz, M. A micromechanical model of cyclic deformation and fatigue-crack nucleation in f.c.c. single crystals. *Acta Mater.* 1997.
54. Hunsche, A.; Neumann, P. Quantitative measurement of persistent slip band profiles and crack initiation. *Acta Metall.* 1986.
55. Shukla, S.; Wang, T.; Cotton, S.; Mishra, R.S. Hierarchical microstructure for improved fatigue properties in a eutectic high entropy alloy. *Scr. Mater.* 2018, *156*, 105–109.
56. De Cooman, B.C. ; Estrin, Y.; Kim, S.K. Twinning-induced plasticity (TWIP) steels. *Acta Mater.* 2018, *142*, 283–362.
57. Niendorf, T.; Lotze, C.; Canadinc, D.; Frehn, A.; Maier, H.J. The role of monotonic pre-deformation on the fatigue performance of a high-manganese austenitic TWIP steel. *Mater. Sci. Eng. A* 2009, *499*(1–2), 518–524.
58. Hamada, A.S.; Järvenpää, A.; Honkanen, M.; Jaskari, M.; Porter, D.A.; Karjalainen, L.P. Effects of cyclic pre-straining on mechanical properties of an austenitic micro-alloyed high-Mn twinning-induced plasticity steel. In: *Proceedings of the Procedia Engineering* 2014.
59. Pineau, A.G.; Pelloux, R.M. Influence of strain-induced martensitic transformations on fatigue crack growth rates in stainless steels. *Met. Trans.* 1974.
60. Abareshi, M.; Emadoddin, E. Effect of retained austenite characteristics on fatigue behavior and tensile properties of transformation induced plasticity steel. *Mater. Des.* 2011, *32*(10), 5099–5105.
61. Sugimoto, K.; Fiji, D.; Yoshikawa, N. Fatigue strength of newly developed high-strength low alloy TRIP-aided steels with good hardenability. *Procedia Eng.* 2010, *2*(1), 359–362.
62. Sugimouto, K.-I.; Kobayashi, M.; Yasuki, S.-I. Cyclic deformation behavior of a transformation-induced plasticity-aided dual-phase steel. *Metall. Mater. Trans. A* 1997, *28*(12), 2637–2644.
63. Liu, K.; Nene, S.S.; Frank, M.; Sinha, S.; Mishra, R.S. Metastability-assisted fatigue behavior in a friction stir processed dual-phase high entropy alloy. *Mater. Res. Lett.* 2018.
64. Nene, S.S.; Frank, M.; Liu, K.; Sinha, S.; Mishra, R.S.; McWilliams, B.A.; Cho, K.C. Corrosion-resistant high entropy alloy with high strength and ductility. *Scr. Mater.* 2019, *166*.
65. Liu, K.; Nene, S.S.; Frank, M.; Sinha, S.; Mishra, R.S. Extremely high fatigue resistance in an ultrafine grained high entropy alloy. *Appl. Mater. Today* 2019, *15*.

14 Functional Properties of High Entropy Alloys

Anirudha Karati, Joydev Manna, Soumyaranajan Mishra, and B.S. Murty

CONTENTS

14.1 INTRODUCTION

High entropy alloys are made up of five, or more than five, elements in equiatomic or near-equiatomic compositions. Although major work on these alloys has focused on mechanical properties, several functional properties have also come into prominence for high entropy alloys. Most of these reports relate the properties to the microstructure. Additionally, close attention has been also paid to varying the compositions to generate non-equiatomic high entropy alloys that have improved properties in comparison to their equiatomic counterparts. The increased lattice strain, the complicated assembly of atoms in a simple unit cell, and the possibility of exploring a multitude of phase spaces are the common attractive points that have attracted attention to high entropy alloys coming out richer over the last decade.

14.2 MAGNETIC PROPERTIES IN HIGH ENTROPY ALLOYS

Iron, cobalt, and nickel are the ferromagnetic elements that have largely been used as alloys for magnetic applications. The discovery of high entropy alloys has the paved way for the exploration of a vast number of alloys in equiatomic or near equiatomic compositions. Intrinsic magnetization of these high entropy alloys can be predicted using first principle or density functional theory calculations. The magnetic exchange in these alloys can be given as:

$$J = \sum_{ij} J_{ij} z_p S_i . S_j \tag{14.1}$$

where J_{ij} is the strength of exchange interaction constant, z_p is the coordination number of pth coordination shell, and S_i, S_j are the magnetic moments of the elements i and j.

CoCrFeMnNi is predicted to have a magnetic ordering tendency and lower critical temperature for face-centered cubic than body-centered cubic structure. Magnetic exchange interactions in body-centered cubic structures are due to iron–iron, iron–cobalt, cobalt–cobalt, and cobalt–manganese pairs and the antiferromagnetic interactions arises due to chromium–chromium, chromium–manganese, and manganese–manganese pairs. In hindsight, for face-centered cubic, magnetism is due to the antiferromagnetic coupling between iron–manganese and manganese–manganese pairs. Curie temperature (T_c) of these high entropy alloys can be predicted using density functional theory and Monte Carlo simulation or in a simple approach such as mean field approximation. It has been observed that alloys with a relatively higher magnetic fraction have a higher T_c as observed in the case of CoFeNi (804 K) and CoFeNiPt (837 K). However, with an increasing fraction of non-magnetic elements, the calculated T_c drops to as low a temperature as 10 K for CoCrFeMnNiV and 21 K for CrFeMoV. Therefore, high entropy alloys can typically be expected to exhibit significantly lower T_c than traditional low entropy magnetic alloys [1].

Lucas and co-workers [2] reported CoCrFeNi to have a M_s value of 3 emu/g. The introduction of aluminum to generate the quinary equiatomic alloy (AlCoCrFeNi) converts the structure from face-centered to B2 (cesium chloride structure). The M_s value rose to 65 emu/g [3]. The introduction of further aluminum to generate $Al_2CoCrFeNi$, however, drops the M_s to 13 emu/g [2]. The senary as-cast alloy AlCoCrCuFeNi exhibited a M_s value of 41 emu/g [2] which decreased to 16 emu/g upon annealing the cast alloy [4].

Zuo and co-workers [5] reported that CoFeMnNi crystallizes as a single phase face-centered cubic and exhibits a M_s value of 18 emu/g. Subsequently, they substituted gallium (80 emu/g), tin (80 emu/g), and chromium (1 emu/g), as shown in Figure 14.1. The quinary equiatomic alloy, AlCoFeMnNi, shows the maximum M_s value of 148 emu/g reported to date [5]. It has also been observed that this alloy exhibits soft magnetic properties which are magnetized to saturation with a relatively weak magnetic field. Also, they have low coercivity which makes the switching of spin states easy. Lucas and co-workers [2] have shown that CoCrFeMnNiV synthesized by mechanical alloying exhibits a M_s of 100 emu/g. The addition of germanium to form CoCrFeNiGe has been reported by induction melting to form a dual phase (body-centered cubic + face-centered cubic) that exhibits a M_s of 50 emu/g [6]. Additionally, the as-cast $CoCrFeNiPd_2$ alloy was subjected to rolling. Both the a-cast and rolled alloys exhibited a M_s of 34 emu/g [2]. Although most of the reported alloys synthesized by the casting route have low H_c, high entropy alloys reported by mechanical alloying route have exhibited higher coercivity values of 515 Oe [7].

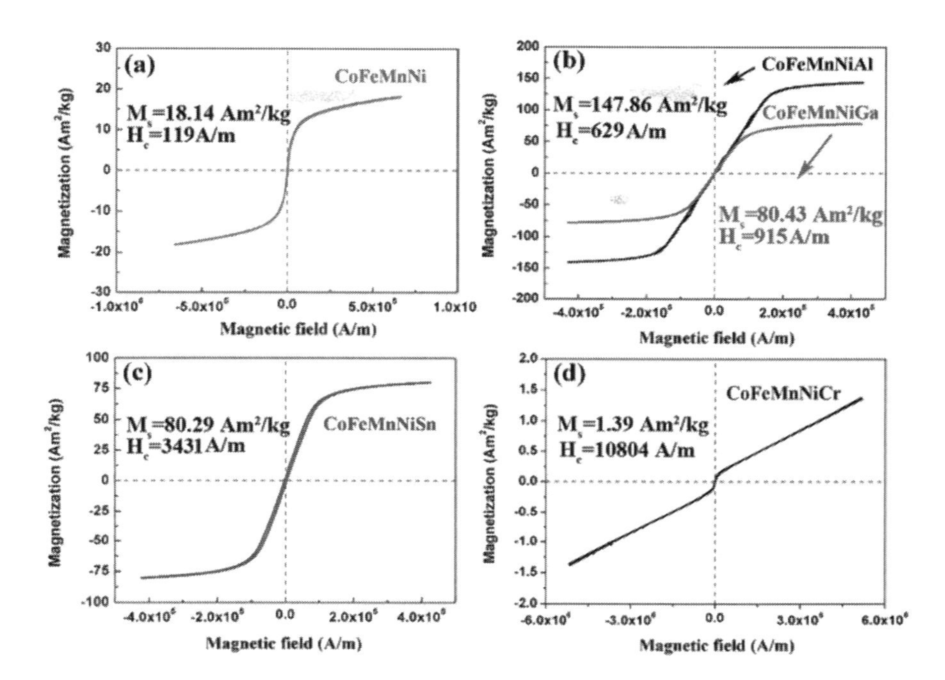

FIGURE 14.1 Hysteresis loops of (a) CoFeMnNiMn, (b) CoFeMnNiAl and CoFeMnNiGa, (c) CoFeMnNiSn, and (d) CoFeMnNiCr alloys at room temperature [5].

In our group, we have shown that the addition of copper to AlNiCo increases the hard-magnetic character. The addition of iron to AlNiCo on the other hand increases the soft magnetic property and saturation magnetization as shown in Figure 14.2. The maximum M_s of 84 emu/g and coercivity (H_c) of 162 emu/g have been reported for AlCoCuFeNi [8]. Table 14.1 gives a detailed analysis of the magnetic high entropy alloys reported in the literature to date.

To conclude, high entropy alloy magnets exhibit lower M_s due to lower ferromagnetic elements content. Intrinsic coercivity and electrical resistivity are higher due to lattice structure. Lattice distortion either pins the domain walls, leading to increase in coercivity, or causes more scattering of electrons due to its distorted structure giving rise to high electrical resistivity. Thus, designing of a magnetic material constitutes some set of agreement including cost and properties.

14.3 MAGNETOCALORIC APPLICATIONS

Magnetocaloric effect or adiabatic magnetism or reverse thermomagnetism refers to heat generation or liberation under the effect of a magnetic field. In recent times, the magnetocaloric effect has come to replace regular gas refrigerants owing to greater efficiency [38]. We have rare earth-containing magnetocaloric materials, and those without rare earths, the so-called eco-friendly magnetocaloric materials. Nowadays, the major focus is on getting magnetocaloric materials with magnetic transition temperature or Curie temperature (T_C) near room temperature and with a high change

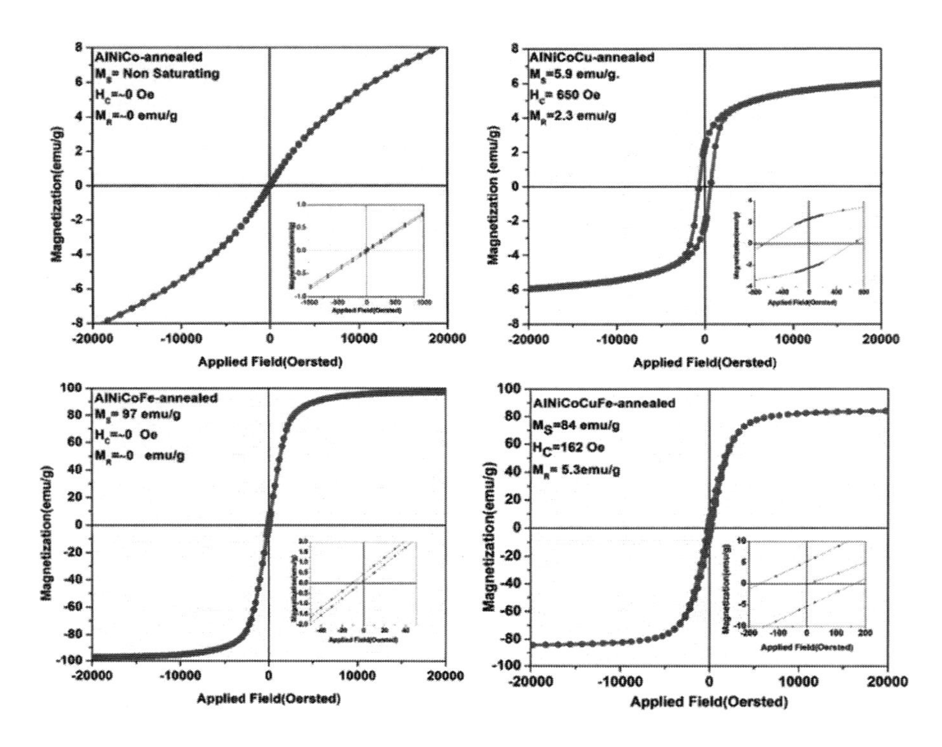

FIGURE 14.2 M–H loops for annealed alloys at room temperature [8].

in magnetic entropy (ΔS_M) and refrigeration capacity (R_C). ΔS_M is due to change in magnetic spin alignment under external field. It is generally measured at 5T for standard comparison. In magnetocaloric materials, it is seen that with increasing enthalpy under magnetic field, there is a rise in lattice entropy which compensates for the drop-in spin entropy of the subsystems. This causes the thermal effect. Materials possessing high magnetic entropy hence often possess high refrigeration capacitance too.

ΔS_M and R_C can be calculated using the following equations [1]:

$$\Delta S_M = \int_0^H \left(\frac{\partial M}{\partial T} \right)_{Hdh} \tag{14.2}$$

$$R_C = \left| \Delta S_M \Delta T_{FWHM} \right| \tag{14.3}$$

In high entropy alloys, entropy stabilization causes various elements to be arranged in a single crystalline phase. The exchange interaction in these systems and consequent changes in magnetic spin alignment has drawn us recently toward high entropy alloys as suitable magnetocaloric materials. At the same time, the chemical chaos in these systems aids in lowering the phase transition rate and thus enhances the magnetocaloric materials [39]. We generally see equiatomic 5 or 6 element systems with maximum configurational entropy and we generally have one or two dia- or

TABLE 14.1
Magnetic Properties of HEAs Reported to Date

Composition	Processing condition	SS or IM	Phases (major)	Ms (emu/g)	Hc (Oe)	Ref
CoFeNi	AC	SS	FCC	150	2	[9]
$Al_{0.25}CoFeNi$	AC	SS	FCC	134	3	
$Al_{0.5}CoFeNi$	AC	SS+IM	FCC+B2	105	5	
$Al_{0.75}CoFeNi$	AC	SS+IM	FCC+B2	110	5	
AlCoFeNi	AC	IM	B2	105	4	
CoCrFeNi	AC+CR	SS	FCC	3	–	[2]
CoFeMnNi	AC	SS	FCC	18	1.5	[5]
$CoFeNiSi_{0.25}$	AC	SS	FCC	133	3	[9]
$CoFeNiSi_{0.5}$	AC	SS	FCC	100	5	
$sCoFeNiSi_{0.75}$	AC	SS	FCC	80	58	
$CoFeNi_2V_{0.5}$	AC	SS	FCC	70	2	[10]
$Al_{0.1}(CoCr_{0.4}Fe_{1.6}Mn)_{0.9}$	AC	SS+IM	BCC+B2	135	9	[11]
$Al_{0.25}CoFe\ Mn_{0.25}Ni$	AC+Ann	SS	FCC	104	3	[12]
$Al_{0.25}CoFe\ Mn_{0.25}Ni$	AC	SS	FCC	101	3	
$Al_{0.25}CoFeMn_{0.25}Ni$	AC+CR	SS	FCC	104	8	
$Al_{0.2}CoFeNiSi_{0.2}$	AC	SS	FCC	117	18	[13]
$Al_{0.2}CoFeNiSi_{0.2}$	AC	SS	FCC	77	–	[14]
$Al_{0.2}CoFeNiSi_{0.2}$	BS	SS	FCC	120	12	[15]
$Al_{0.375}CoFeNiSi_{0.375}$	30h BM	SS	FCC	91	64	[16]
$Al_{0.5}CoFeMn_{0.5}Ni$	AC	SS	FCC	101	3	[12]
$Al_{0.5}CoFeMn_{0.5}Ni$	AC+CR	SS	FCC	104	8	
$Al_{0.5}CoFeMn_{0.5}Ni$	AC+Ann	SS	FCC	104	3	
$Al_{0.5}CoFeNiSi_{0.5}$	AC	SS	FCC	108	–	[14]
$Al_2CoCrFeNi$	AC	IM	B2	13	–	[2]
AlBFeNiSi	MA	SS+SS	FCC+BCC	50	200	[17]
AlCoCrCuNi	SG	SS+SS	BCC+FCC	13	11	[18]
AlCoCrFeNi	AC	SS	BCC	65	53	[3]
$AlCoCrFeNi_{2.1}$	GA	SS+SS	FCC+BCC	9	19	[19]
AlCoCuFeNi	AC	SS+SS	BCC+FCC	84	162	[8]
AlCoFeMnNi	AC	SS	B2	148	8	[5]
$AlCrFeNiMo_{0.1}$	AC	SS+SS	BCC+BCC	31	75	[20]
$CoCr_{0.2}FeNiSi_{0.2}$	AC	SS	FCC	70	2	[21]
$Co_{0.5}\ Fe_{0.5}GaMnNi$	AC	SS	BCC	78	30	[22]
$Co_{28}Cu_7Fe_{29}Ni_{29}Ti_7$	AC	SS	FCC	112	4	[23]
$Co_{35}Cr_5Fe_{20}Ni_{20}Ti_{20}$	20h BM	SS+SS	BCC+FCC	46	15	[24]
CoCrCuFeNi	AC	SS	2FCC	2	0	[25]
CoCrCuFeNi	MA–HPS	SS+SS	BCC+FCC	53	166	[26]
CoCrFeGaNi	AC	SS+SS	BCC+FCC	38	–	[27]
CoCrFeMnNi	AC	SS	FCC	1	135	[5]

(Continued)

TABLE 14.1 (CONTINUED)
Magnetic Properties of HEAs Reported to Date

Composition	Processing condition	SS or IM	Phases (major)	Ms (emu/g)	Hc (Oe)	Ref
CoCrFeMnNi	MA–HPS	SS+SS	BCC+FCC	1	0	[26]
CoCrFeNiGe	IM	SS+SS	BCC+FCC	50	–	[6]
CoCrFeNiPd	AC+CR	SS	FCC	33	–	[2]
CoCrFeNiPd	AC	SS	FCC	33	–	
$CoCrFeNiPd_2$	AC+CR	SS	FCC	34	–	
$CoCrFeNiPd_2$	AC	SS	FCC	34	–	
CoFeGaMnNi	AC	SS+SS	FCC+BCC	80	11	[5]
CoFeMnNiSn	AC	SS	BCC	80	43	
CoFeMnNiV	48h MA	–	FCC+BCC+BCC+ IM	100	150	[28]
CrCuFeMnNi	50h BM	SS+SS	BCC+FCC	16	56.2	[29]
$Al_{0.8}CoCu_{0.8}Fe\ Ga_{0.08}Ni$	AC	SS+SS	BCC+FCC	83	9	[30]
$Al_{1.3}CrCuFeNi_2$	LENS	SS+SS	B2+ IM	21	74	[31]
$Al_5Co_{35}Cr_5Fe_{40}Ni_5Si_{10}$	MS	SS	BCC	10	1	[32]
AlBCeFeNiSi	MA	SS	BCC	1	240	[33]
AlBCFeNiSi	MA	SS+SS	BCC+FCC	25	70	
AlCoCrCuFeNi	AC+Ann	SS+IM	2FCC+B2	10	10	[4]
AlCoCrCuFeNi	SQ	IM	B2	41	11	[34]
AlCoCrCuFeNi	AC	SS+SS	BCC+FCC	41	44	[2]
AlCoCrCuFeNi	AC+Ann	SS+SS	BCC+FCC	16	15	[4]
AlCoCrCuFeNi	MS	–	Amorphous	60	80	[35]
$AlCoCrFeNb_{0.1}Ni$	AC	SS	BCC	65	53	[3]
$AlCoCrFeNb_{0.1}Ni$	AC	SS	BCC	50	140	
$AlCoCrFeNb_{0.25}Ni$	AC	SS	BCC	35	95	
$AlCoCrFeNb_{0.25}Ni$	AC	SS	BCC	40	95	
$AlCoCrFeNb_{0.5}Ni$	AC	SS	BCC	20	45	
$AlCoCrFeNb_{0.5}Ni$	AC	SS	BCC	20	50	
$AlCoCrFeNb_{0.75}Ni$	AC	SS	BCC	10	95	
$AlCoCrFeNb_{0.75}Ni$	AC	SS	BCC	10	15	
AlCrFeMnNiTi	25h BM	SS+SS	FCC+BCC	18	154	[36]
$Co_2CrCuFeMnNi$	50h BM	SS	FCC	52	14	[37]
CoCrCuFeNiTi	AC	SS	FCC	2	0	[25]
$CoCrCuFeNiTi_{0.5}$	AC	SS+SS	FCC+FCC	0.3	0	
$CoCrCuFeNiTi_{0.8}$	AC	SS	FCC	1	0	
CoCrFeMnNiZn	MA	SS	FCC	50	515	[7]

*AC – As Cast, Ann – Annealed, BS – Bridgman Solidification, BM – Ball Milling, CR – Cold Rolled, GA – Gas Atomisation, HIP – Hot Isostatic Pressing, IM – Induction Melting, LENS – Laser Engineered Net Shaping, MA – Mechanical Alloying, MS – Magnetron Sputtering, SG – Sol Gel, SPS – Spark Plasma Synthesis, SQ – Splat Quenching, CR – Cold Rolled.

non-magnetic elements to bring down T_C to near room temperature. To tune the T_C and get spin glass-like behavior, the focus has been mainly on bulk metallic glass type high entropy alloys. As such, most of these materials are synthesized by arc melting followed by melt spinning routes. Various such materials investigated to date is listed in Table 14.2.

As can be concluded from the ensuing discussion, the rare earth-containing systems have high ΔS_M and R_C, while ferrous-type non-rare earth-containing systems have high T_C. Kurniawan and co-workers [48] have tuned CoCuFeMnNi to $CoCu_{0.95}FeMnNi_{1.05}$ to lower T_C from 400 K to 280 K, while Na and co-workers [46] have tuned AlCoCFeNi to $AlCoCr_{0.5}FeNi_{1.5}$ to reduce T_C from 380 K to 275 K. Similarly, the highest reported R_C is for AlDyFeGdTb with 691 J/kg [42], while ΔS_M as high as 15 J/kgK is reported in AlCoErHoTm [41] under a 5T external field. Huo and co-workers [45] have also designed denary alloy AgAlCoDyErGdHoNiTbY with ΔS_M up to 10.5 J/kgK and R_C at 532 J/kg. They have a low T_C of 24 K only. Considering that, R_C as high as 1,346 J/kg and ΔS_M up to 29 J/kgK [49] has been

TABLE 14.2
Reported High Entropy Alloys for Magnetocaloric Applications

Composition	Phases	ΔSM (J/kgK) @ 5T	RC (J/Kg)	TC (K)	Ref
AlCoDyGd	MG	8.7	567	60	[40]
AlCoGdHo	MG	9.8	626	50	
AlCoGdTb	MG	8.9	577	73	
AlCoDyErGd	MG	9.1	619	43	[39]
AlCoDyErTb	MG	8.6	525	29	
AlCoDyErTm	MG	11.9	405	13	
AlCoDyErHo	MG	12.6	468	36	[41]
AlCoDyGdTb	MG	9.4	632	58	[42]
AlCoErGdHo	MG	11.2	627	40	[41]
AlCoErHoTm	MG	15.0	375	32	
AlDyFeGdTb	MG	5.9	691	112	[42]
AlDyGdNiTb	MG	7.3	507	45	
AlErFeGdHo	MG	5.1	446	55	[43]
AlErGdHoNi	MG	9.5	511	25	
DyErGdHoTb	HCP	8.6	627	186*	[44]
AgAlCoDyEr GdHoNiTbY	MG	10.6	532	24	[45]
AlCoCrFeNi	FCC	0.51 @ 2T	242.6	380	[46]
$AlCoCr_{0.5}FeNi_{1.5}$	FCC+BCC	0.277 @ 2T	–	275	
CoCuFeMnNi	FCC	0.115 @0.55T	127	395	[47]
CoCuFeMnNi	FCC	0.8	–	400	[48]
$CoCu_{0.95}FeMnNi_{1.05}$	FCC	0.46	~120	280	
$CoCu_{0.9}FeMnNi_{1.1}$	FCC	0.39	–	065	

*MG – Metallic Glass.

reached, current research aims at exploring high entropy alloys as possible materials with R_C and ΔS_M as high as that, and also with T_C above 300 K.

14.4 HYDROGEN STORAGE PROPERTIES OF HIGH ENTROPY ALLOYS

Sufficient and benign supply of energy is one of the most important challenges for the future of human society. Hydrogen has appeared as one of the most likely potential candidates for the replacement of the existing fossil fuel-based economy, due to several promising features such as great abundance, high energy density, and environmental benignity of the oxidation product. However, the biggest challenge toward the use of hydrogen as energy carrier for fuel cell applications is its very low density at standard temperature and pressure, highly energy intensive processes of liquefaction, and safety issues in handling of pressurized hydrogen canisters. The conventional storage options of compressed gaseous storage, liquid storage, etc. are relatively less attractive due to either high pressure or very low temperatures involved, which in turn raises concerns related to safety and involvement of energy intensive process for storage. Physical storage of molecular hydrogen inside the porous cavities of different high surface area materials viz., carbon-based materials, metal-oxide frameworks, zeolites, etc., is a conceivable option, but is not suitable for practical applications due to low gravimetric hydrogen storage capacity of these materials.

Metal hydrides are regarded as an excellent option for hydrogen storage application over the last few decades because of their high storage capacity per volume and reversibility. Lanthanum nickel metal hydrides has been commercialized and used in metal hydride batteries. However, the low gravimetric storage capacity is one of the major drawbacks of these hydrides being used as hydrogen storage materials in fuel cell vehicular applications. The multicomponent high entropy alloys provide an opportunity to manipulate their structure and properties as per our demands. Recently, it has been reported that the body-centered cubic or Laves phase high entropy alloys tend to absorb hydrogen at ambient temperature [50]. The ability of hydrogen absorption by high entropy alloys shows great promise to develop numerous metal-hydrogen systems for practical application, such as hydrogen storage materials for fuel cell vehicles and metal hydride batteries, etc.

Kao and co-workers [51] first reported the use of high entropy alloys in hydrogen storage applications. They designed an equimolar CoFeMnTiVZr HEA system and analyzed the hydrogen absorption–desorption properties. Hydrogen storage properties can be improved by changing the amount of titanium, vanadium, and zirconium contents in the alloy. Maximum hydrogen absorption of 1.8 wt.% is reported for this alloy system with $CoFeMnTi_2VZr$ composition at ambient temperature. Among the reported high entropy alloys, HfNbTiVZr shows the highest hydrogen absorption (around 2.7 wt.%). Pressure-composition isotherms for hydrogen absorption and van 't Hoff plot for HfNbTiVZr high entropy alloy are shown in Figure 14.3. It has been suggested that this high entropy alloy has a large difference in the atomic radii of the elements (56.82%), and accordingly there is a large lattice distortion. This lattice distortion is believed to favor the hydrogen absorption in both the tetrahedral and octahedral interstitial sites of the lattice. The studied high entropy alloys for hydrogen storage applications are listed in Table 14.3.

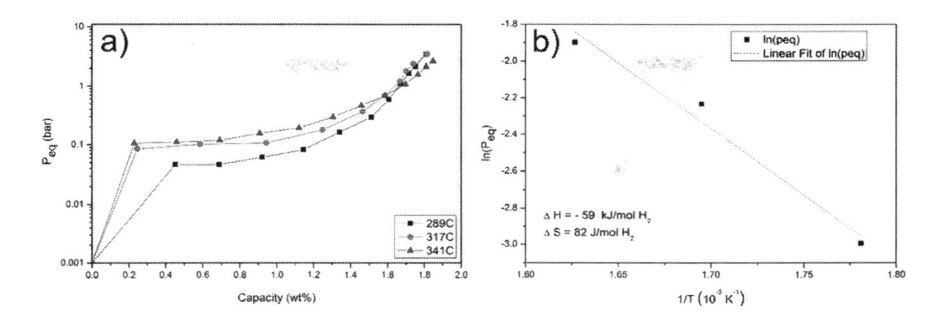

FIGURE 14.3 (a) Hydrogen absorption pressure composition isotherms (PCI) for HfNbTiVZr HEA at different temperature and corresponding (b) van 't Hoff plot [52].

TABLE 14.3
Hydrogen Storage Properties of HEAs with Their Synthesis Procedure and Phase Composition

Composition	Processing condition	Phases	Max H2 storage capacity (wt.%)	Ref
$Ti_xV_yZr_zCoFeMn$	AM	C14 Laves phase	1.8 wt.% at 100 bar and 25°C	[51]
TiVZrCrFeNi	LENS	C14 Laves phase + α–Ti (minor)	1.81 wt.% at 100 bar and 50°C	[53]
TiVZrMoNb	LENS	Two phase dendric matrix (BCC + orthorhombic)	2.3 wt.% at 85 bar and 50°C	[50]
TiVZrHfNb	AM	BCC (W–type)	2.7 wt.% at 53 bar and 300°C	[54]
La–Ni–Fe–V–Mn	LENS	Hexagonal ($CaCu_5$ type)	0.83 wt.% at 50 bar and 35°C	[55]
TiVZrHfNb	AM	BCC	2.5 wt.% at 53 bar and 300°C	[52]
$MgZrTiFe_{0.5}Co_{0.5}Ni_{0.5}$	HEBM	BCC	1.2 wt.% at 20 bar and 350°C	[56]
TiZrNbHfTa	AM	BCC	2.5 wt.% at 50 bar and 300°C	[57]

AM – Arc Melting, LENS – Laser Engineered Net Shaping, HEBM – High Energy Ball Milling.

14.5 CORROSION BEHAVIOR OF HIGH ENTROPY ALLOYS

High entropy alloys tend to show unique corrosion-resistant properties because of their random arrangement of multiple elements in a locally disordered chemical environment [58]. High entropy alloys could also be coated on metal surfaces by various techniques such as laser cladding, electro-spark deposition, magnetron sputtering, etc. to prevent corrosion of certain metals and alloys. The corrosion behavior of alloys is dependent on the interactions between materials and the environments.

In general, aqueous environments containing salts, acids, bases, etc. affect the alloys. High temperature and pressure in these environments have also an influential effect on the corrosion of alloys. The constituent element(s) of a high entropy alloy system highly affect its corrosion properties due to the change in the phase structure, chances of elemental segregation, and formation of a different or new passive film during corrosion.

A standard electrochemical technique known as potentiodynamic polarization test is generally performed to investigate the corrosion properties of materials. In this process, the potential of the electrode is varied at a constant rate by applying current through the electrolyte (salts or acid solutions). A wide variety of functions could be obtained from the curve, such as: the corrosion potential (E_{corr}), corrosion-current density (i_{corr}), pitting potential (E_{pit}), or breakdown potential (E_b), and passive current density (i_{pass}), as shown in Figure 14.4. Average corrosion rates (r_{corr}) of the materials can be obtained using the following equation:

$$\text{corrosion rate}\left(\text{mm}/\text{year}\right) = 3.27\times10^{-3}\times\frac{i_{corr}}{d}\times EW \tag{14.4}$$

where, d = density (g/cc), i_{corr} = corrosion current density ($\mu A/cm^2$), and EW (equivalent weight) = $\left(\sum\frac{n_i f_i}{W_i}\right)^{-1}$, where n_i, f_i, and W_i are valence, mass fraction, and atomic weight of the ith element, respectively. Another way to investigate corrosion is by immersion test where the corrosion rate is calculated by measuring the mass loss of the materials. Electrochemical-impedance spectroscopy is also used to provide more information on the electrochemical processes occurring during the corrosion of the alloy.

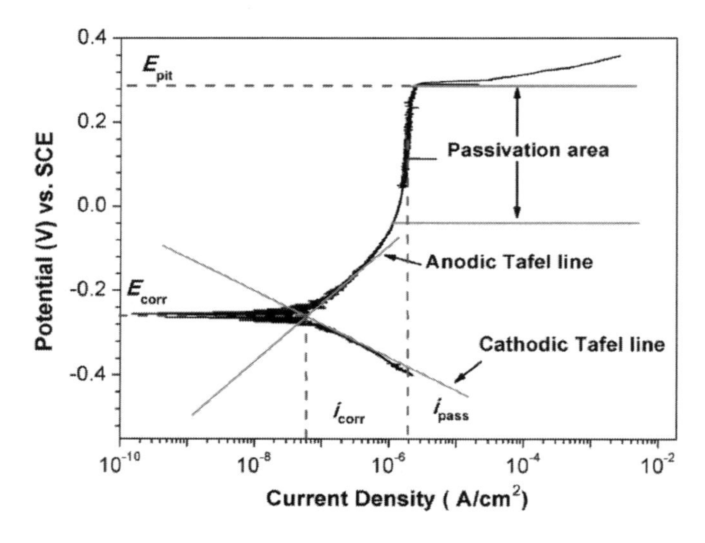

FIGURE 14.4 A typical potentiodynamic-polarization curve for a material showing passivation area, and other corrosion parameters, such as, i_{corr}, E_{corr}, E_{pit}, i_{pass}, etc. [59].

Among the first few investigations, Chen and co-workers [60] studied the electrochemical properties of $Cu_{0.5}NiAlCoCrFeSi$ high entropy alloy in various concentration (0.1 to 1 M) of sodium chloride and sulfuric acid media. They compared the result of $Cu_{0.5}NiAlCoCrFeSi$ with 304SS and reported that the general corrosion resistance of the high entropy alloy was better than that of 304SS while pitting corrosion resistance of the high entropy alloy was reported to be poor than that of 304SS. The corrosion behavior of high entropy alloys in salt (NaCl) and acid solution (H_2SO_4, HCl, and HNO_3) is discussed in the next section of this chapter.

14.5.1 Electrochemical Properties of High Entropy Alloys in Salt–Aqueous Medium

The presence of copper in high entropy alloy is highly detrimental for its corrosion properties. Corrosion properties of high entropy alloys composed of CoCrFeNi, $CoCrFeNiCu_{0.5}$ and CoCrFeNiCu alloys were investigated by Hsu and co-workers [61]. The order of weight loss after immersion tests in an aerated 3.5% NaCl solution after 30 days was reported to be FeCoNiCrCu > FeCoNiCrCu0.5 > FeCoNiCr. No uniform corrosion has been observed on the alloy surfaces. Localized and pitting corrosion was observed on the copper-rich phase and with increase in copper content, corrosion resistance of the alloy was observed to deteriorate. A similar observation has been reported for the $AlCoCrNiTiCu_x$ high entropy alloy [62]. For this alloy, corrosion increases with the increase in copper concentration. Weight-loss in $AlCoCrNiTiCu_{1.5}$ alloy was increased two-fold in comparison to that of the $AlCoCrNiTiCu_{0.5}$ alloy. Addition of manganese in $Al_{0.3}CoCrFeNiMn_x$ high entropy alloys also worsens the corrosion resistance of the high entropy alloy [63]. Similarly, with increase in iron content in the $CoCrFe_xNi$ high entropy alloy, corrosion resistance decreased [64].

The presence of aluminum in the salt solution diminishes the corrosion resistance properties of high entropy alloys. It has been reported that a secondary precipitation effect will accelerate the corrosion of aluminum in presence of other higher electropositive metal ions (e.g., Fe^{2+}, Ni^{2+}, and Cu^{2+}, etc.) [65]. Shi and co-workers [59] have studied the effect of aluminum content in $Al_xCoCrFeNi$ alloys ($x = 0.3, 0.5, 0.7$). They observed that with an increase in the aluminum content in the high entropy alloys, the volume of the (Al,Ni)-rich phase and the chromium-depleted body-centered cubic phase increased. These phases are prone to attack by chloride ions and accordingly, the amount of localized corrosion was observed to increase with the increase in aluminum content. The thickness of the passive films also increased with the increasing aluminum content due to the formation of aluminum oxide on the films. However, a higher amount of aluminum oxide weakened the protection ability of the passive films. After homogenization at 1,250°C, improvement in local corrosion resistance was observed for these high entropy alloys [66]. Galvanic corrosion was also noted for aluminum containing high entropy alloys [e.g., $Al_xCoCrFeNi(B)$ ($x = 0.5, 1, 1.5, 2$)]. The corrosion potential (E_{corr}) was increased for these high entropy alloys while increasing the aluminum content and varied between −0.353 to −0.482 V_{SCE} [67].

In case of $Al_xCoCrFeNiCu_{0.5}$ (x = 0.5, 1.0, 1.5) high entropy alloys, the $Al_{1.0}$ high entropy alloy showed the best corrosion resistance while the corrosion resistance of the $Al_{1.5}$ alloy was the poorest [65]. Heat treatment of the $Al_{1.5}$ alloy transforms its phase from a single-phase body-centered cubic to face-centered cubic + body-centered cubic duplex phase, and accordingly corrosion resistance of the alloy increased after the heat treatment. However, corrosion resistances of other two ($Al_{0.5}$ and $Al_{1.0}$) alloys decreased after heat treatment. Sun and co-workers [68] studied the effect of aluminum in $Al_xCoCrCuFeNiTi$ alloys (x = 0, 0.5, 1.0, 1.5, 2.0) in 3.5% NaCl solution. The $Al_{1.5}$ high entropy alloy showed maximum corrosion resistance properties. The passivation phenomenon was observed for all the high entropy alloys due to the presence of chromium. However, the passivation effect was reduced with the addition of aluminum. Similar observations were also noted by Qiu and co-workers [69]. The passivity current was maintained as lowest for the $Al_{1.5}$ high entropy alloy, and consequently corrosion speed was the slowest for this high entropy alloy. The presence of higher amounts of aluminum and chromium with self-passivation properties, along with the relatively fine microstructure of the $Al_{1.5}CoCrCuFeNiTi$ alloy has improved the corrosion resistance property of this high entropy alloy. Qiu and co-workers [70] has reported the corrosion behavior of AlCoCrCuFeNi in 1 M NaCl solution and compared it with 304 SS. Small micro-etching batteries were formed due to the elemental segregation. Free corrosion potential (E_{corr}) and corrosion current density (I_{corr}) for this alloy were observed to be −0.012 V and 0.003 μA/cm^2. Corrosion current density of this high entropy alloy was two orders of magnitude smaller than 304 SS. The high corrosion resistance of this high entropy alloy was mainly due to the presence of a chromium-rich phase.

It is well known that the pitting corrosion resistance of stainless steels in chloride containing solution can be improved by addition of molybdenum [71–73]. According to the Pourbaix diagram, molybdenum would transform into molybdate anion (MoO_2^{-4}) layers in sodium chloride solutions. This ionic oxide layer is a very effective pitting inhibitor and resists other anions (e.g., Cl^-, Br^-, etc.). Chou and co-workers [74] studied the effect of molybdenum content in $Co_{1.5}CrFeNi_{0.5}Ti_{1.5}Mo_x$ (x = 0, 0.1, 0.5, 0.8) high entropy alloys and its effect on corrosion properties in 1 M sodium chloride solution. The molybdenum-free high entropy alloys tend to experience pitting corrosion whereas the molybdenum-containing high entropy alloys were not susceptible to pitting corrosion in same environment. As expected, the addition of molybdenum in the high entropy alloys improved the pitting resistance, and the breakdown potential (E_b) for the molybdenum-containing alloys was comparably higher than the molybdenum-free alloy. It has also been reported that the addition of 0.5 M sodium sulphate as inhibitors to a 1 M sodium chloride solution increased the pitting corrosion resistance of $Co_{1.5}CrFeNi_{0.5}Ti_{1.5}Mo_{0.1}$ alloy [75]. The effects of other inhibitors were also studied in sodium chloride solution and it has been observed that nitrates and benzoates have the best inhibition effect on this alloy [76].

The effects of tin content in $CoCuFeNiSn_x$ alloys and its effect on corrosion properties were studied by Zheng and co-workers [77]. They compared corrosion properties of the tin-containing high entropy alloys with 304 SS and reported that the E_{corr} and the i_{corr} values of the $CoCuFeNiSn_x$ high entropy alloys in sodium chloride solution are less than those of the 304 SS. Corrosion properties of the $CoCuFeNiSn_{0.04}$

high entropy alloy is the best among the synthesized high entropy alloys. In this high entropy alloy ($Sn_{0.04}$), mainly galvanic action between the interdendrite and dendrite phase is observed which resulted in localized corrosion. The superior corrosion resistance of the $CoCuFeNiSn_{0.04}$ high entropy alloy is attributed to the fine grain and absence of segregation in the microstructure of the alloy.

Han and co-workers [78] reported the change in corrosion properties due to the vanadium content in $AlCoCrFeNiTi_{0.5}V_x$ (x = 0, 0.5, 1, 1.5, 2) high entropy alloys in sodium chloride solution. Corrosion resistance of these alloys decreased with the increase in vanadium content up to 1. Afterward, corrosion rate increased with increase in the vanadium content. It has been suggested that passivation of the alloy is mainly depended on the distribution of aluminum, chromium, nickel, and titanium elements [79]. In the case of $AlCoCrFeNiTi_{0.5}V_x$ high entropy alloys, the relative amount of vanadium in the dendritic region was increased with the addition of higher vanadium content, and accordingly the interdendritic region became rich with titanium and chromium elements. Therefore, the corrosion resistance of the interdendritic region was increased as the alloy gained a protective effect from titanium and chromium elements. The presence of pores could also accelerate the corrosion rate, as observed for $AlCoCrFeNiTi_{0.5}V_{1.5}$ high entropy alloy.

The effect of other elements (e.g., titanium, silicon, carbon, nitrogen, etc.) on corrosion properties of various high entropy alloys were also reported. For example, Ren and co-workers [80] studied the effect of titanium and silicon content in $Al_{0.3}CrFe_{1.5}MnNi_{0.5}Ti_x$ and $Al_{0.3}CrFe_{1.5}MnNi_{0.5}Si_x$ high entropy alloys in 3.5 wt.% sodium chloride solution. The corrosion resistance of the $Al_{0.3}CrFe_{1.5}MnNi_{0.5}$ high entropy alloy decreased after the addition of titanium and silicon elements. Addition of silicon carbide by hot isostatic pressing improved the corrosion properties of CoCrFeMnNi high entropy alloy for the first 16 hours, and later pitting corrosion was noted [81]. Nitriding of CuCrFeMnNi high entropy alloy was also reported to be effective in improving the corrosion resistance due to the formation of chromium nitride [82]. Zhang and co-workers [83] have synthesized a series of high entropy alloys (CuHfTiYZr, AlCuHfYZr, and AlCuTiYZr) having hexagonal close-packed (HCP) dendrites with body-centered cubic (CuY-type) interdendrite phases. The CuHfTiYZr and AlCuHfYZr high entropy alloys showed better corrosion resistance properties, while AlCuTiYZr showed better pitting resistance. The interdendritic structures (body-centered cubic phases) were mainly corroded for all the high entropy alloys. As can be seen from Figure 14.5, coral-like hexagonal close-packed dendrites are revealed after the corrosion tests and some small body-centered cubic lamella structures remained in the interdendritic region. The hexagonal close-packed phase was not affected by the corrosion and only a small part of the dendrite region was corroded (circled in Figure 14.5d).

Synthesis process, heat treatments, and processing routes highly affect the corrosion resistance of high entropy alloys. Corrosion resistance of directionally solidified CoCrCuFeNi high entropy alloy was reported to be higher than the as-cast alloy in a 3.5% sodium chloride solution [84]. In this report, the as-cast alloy samples were heated at 1,600°C in argon atmosphere and then pulled downward at a uniform rate to form the directionally solidified high entropy alloys. A wider passive region was also noted for the directionally solidified high entropy alloy. Microstructure of the high entropy alloys reportedly transformed after annealing or aging heat treatment

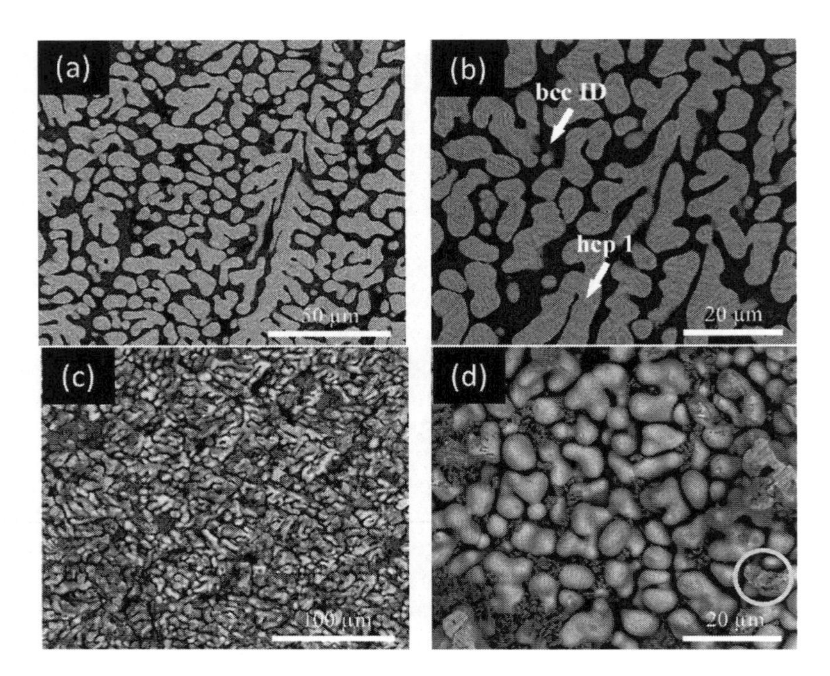

FIGURE 14.5 (a, b) BSE image of as-cast CuHfTiYZr HEA. (c, d) corroded surface of the alloy in seawater solution [83].

processes. Several reports have been published where annealing improved the corrosion properties of the as-cast high entropy alloys. Lin and co-workers [85] observed the effect of aging heat treatment (350–1,350°C) on $Cu_{0.5}CoCrFeNi$ high entropy alloys. Serious corrosion was observed in 3.5 wt.% NaCl solution for the as-cast and the high entropy alloys that had been heat treated at 350–950°C. However, corrosion properties were improved when the high entropy alloys were heat treated at 1,100–1,350°C. It has been suggested that at lower temperature (< 950°C), the copper-rich phase precipitated in the face-centered cubic matrix, while at elevated temperature this phase dissolved into the face-centered cubic matrix.

With increase in the aging temperature from 350 to 950°C, the E_{corr} value decreased from −0.29 to −0.51 V_{SCE}. At a higher aging temperature (1,100–1,350°C), the E_{corr} value was observed to be in the range of −0.27 to −0.41 V_{SCE}. However, the pitting corrosion potential (E_p) was increased with the aging temperature which is due to the breakdown of the copper-rich or chromium-rich phases by chloride ions. However, for $Al_{0.5}CoCrFeNi$ HEA, the aging treatment was observed to be detrimental [86], although, this high entropy alloy was subjected to comparatively lower temperature than the previous case (350–950°C for 24 h). The corrosion resistance of these high entropy alloys aged at higher temperatures was significantly lower than the alloys aged at lower temperatures. With an increase in the aging temperature, the amount of the aluminum–nickel-rich phase decreases, leading to an increase in the amount of the aluminum (nickel, cobalt, chromium, iron) phase and accordingly, corrosion resistance degraded. The as-cast and aged specimens have E_{corr} value in the range of 0.04 to −0.64 V_{SCE} and E_p in the range from 0.13 to 0.71 V_{SCE}. Tsao and co-workers [87] investigated the effect of annealing at 650–750°C for 8 h on corrosion properties of the $Al_{0.3}CrFe_{1.5}MnNi_{0.5}$ high entropy alloy.

The as-cast high entropy alloy had a chromium-rich dendritic body-centered cubic phase, nickel-rich first face-centered cubic interdendrite phases, and a small fraction of a nickel-rich second face-centered cubic phase. The E_{corr} and i_{corr} value of the as-cast HEA were 0.845 V_{SCE} and 6.98 μA cm^{-2}, respectively. During the annealing treatment, a σ phase ($Cr_5Fe_6Mn_8$) was formed along with precipitation of an aluminum–nickel phase. The precipitated aluminum–nickel phase coarsens with an increase in the annealing temperature and the σ phase transformed into a chromium-rich body-centered cubic phase during the slow-cooling process. The presence of a chromium-rich phase, an aluminum–nickel phase, and a σ–phase exhibited superior corrosion behavior. Therefore, the corrosion potential (E_{corr}) value of the as-cast high entropy alloy shifted toward noble values after the annealing process. The E_{corr} values were reported to be −0.697 V_{SCE} for 650°C and −0.677 V_{SCE} for 750°C annealed high entropy alloys. Soare and co-workers [88] studied the effect of re-melting on corrosion properties of AlCrCuFeMnNi alloy. The as-cast high entropy alloy was re-melted to homogenize the elemental distribution in the high entropy alloy. It has been reported that in consecutive re-melting of the high entropy alloy, the copper-rich phases were partially dissolved and consequently the galvanic corrosion diminished. Negligible copper segregation was found in the re-melted alloy, and accordingly, the corrosion resistance of the re-melted high entropy alloys improved in comparison to the as-cast alloy. Similar observations have been reported after the re-melting of the AlCoCrFeNiTi high entropy alloy [89]. Spark plasma sintering could also improve the corrosion resistance properties of high entropy alloys. The AlBFeNiSi high entropy alloy showed passivation after sintering at 520°C due to the formation of an oxide film on the alloy surface, and accordingly it exhibited better pitting resistance [90]. Improvement of overall corrosion resistance is reported after sintering this high entropy alloy at 1,080°C.

Various techniques such laser cladding, sputtering, pressing, etc., have been used to prepare high entropy alloy coatings on different substrates. The high entropy alloy coatings were reported to be highly corrosion-resistant and could be applied for practical applications. The boron content in $CoCrFeNiB_x$ (x = 0.5, 0.75, 1.0, and 1.25) coatings prepared by the laser cladding process is found to be beneficial for corrosion resistance [91]. The corrosion resistance of the high entropy alloy coatings increased with increasing boron content. The synthesized high entropy alloy coatings were found to be in face-centered cubic solid solution phase with minor boride phase ($(Cr,Fe)_2B$). The face-centered cubic phase showed lower galvanic corrosion potential than the boride phase which improved the corrosion resistance of the coating. However, when x = 1.25 for $CoCrFeNiB_{1.25}$ coatings, the orthorhombic borides phase transformed to a tetragonal boride phase, which deteriorated the corrosion resistance of the high entropy alloy coatings. This was probably occurred due to the observed stacking faults found in tetragonal boride phase.

Qiu and co-workers [92] investigated the corrosion properties of AlCoCrCuFe high entropy alloy coatings on Q235 steel prepared by laser cladding process at various scanning speed (ν = 3, 4, 5, and 6 mm/s). The high entropy alloy coatings showed excellent corrosion resistance in 1 M sodium chloride solution due to the finer microstructure with reduced grain boundaries formed during the cladding process. With an increase in the scanning speed during the cladding process, enhancement in the corrosion resistance was first noted, and then at higher speed, corrosion resistance was found to be low. It has been suggested that at fast scanning speed, the cladding layer surface became rough due to increase in the convection process during cooling of the coatings, and accordingly corrosion resistance of the coatings decreased. Li and co-workers [93] synthesized AlCoCrFeNi high entropy alloy coatings on AISI 405 carbon steel by electrospark deposition (ESD) techniques and reported the effect of coating pass on corrosion properties of the high entropy alloy coatings. Corrosion properties were reported to be more noble than those of the substrate material due to the presence of relatively higher chromium and aluminum content at the coating surface. Passivation was observed for the coatings and the passive zone increased with the number of coating pass. Coating pass also improved the overall corrosion resistance of the high entropy alloy coatings. As the number of coating pass increased, the amount of transition phases was reduced, and the coating level was diluted. Transition phases generally enhance the chance of formation of a micro-galvanic cell and hence weaken the corrosion resistance of the coatings. An $Al_2CoCrCuFeTiNi_{1.0}$ high entropy alloy coating on Q235 steel prepared by laser cladding process also showed improved corrosion resistance properties with $E_{corr} = -0.65$ V and $i_{corr} = 0.026$ μA/cm^2 [94]. Similarly, improved corrosion resistance properties were noted for $AlCoCrCuFeNi_{1.0}$ coating on copper substrate prepared by a laser surface alloying process [95].

The addition of molybdenum in high entropy alloy coatings was also found to be effective at improving their corrosion resistance. Superior corrosion resistance was noted for $CoCrFeNiW_{0.5}Mo_{0.5}$ high entropy alloy coatings prepared by mechanical alloying followed by vacuum hot-pressing sintering on Q235 steel substrate [96]. A wider passive region was noted for AlCoCuFeNiV high entropy alloy coatings on Quartz glass wafers prepared by direct current magnetron [97]. In comparison with 201 SS, the synthesized high entropy alloy coatings showed better corrosion resistivity. Presence of high corrosion resistance elements (e.g., cobalt and nickel) along with no observable segregation of elements are reported to be the reasons behind the high corrosion resistance of this high entropy alloy coatings. Wu and co-workers [98] investigated the effect of titanium content on corrosion properties of $AlCoCrFeNiTi_x$ (x = 0.5, 1.0, 1.5, and 2.0) high entropy alloy coatings. The high entropy alloy coatings were prepared on 304 SS by laser surface alloying process. The $AlCoCrFeNiTi_{1.0}$ showed the best corrosion resistance in 3.5% NaCl solution and the descending order of resistance properties of the synthesized coatings can be arranged as follows: $Ti_{1.0} > Ti_{0.5} > Ti_{1.5} > Ti_{2.0}$. A spontaneous passivation phenomenon was also observable for $Ti_{1.0}$ coating. This passive film formation acted as a resistive layer for this high entropy alloy coatings and improved its corrosion resistivity. For the $Ti_{1.5}$ and $Ti_{2.0}$ high entropy alloy coatings, the presence of Ti_2Ni and NiAl intermetallic compounds minimized the passivation tendency. Also, these intermetallic compounds could act as micro-cells during the corrosion process, which may have reduced the corrosion resistance of these high entropy alloy coatings. Table 14.4 gives details of all the high entropy alloys studied in salt-aqueous media.

TABLE 14.4

Corrosion Properties of Reported HEAs in Salt Solution with Their Synthesis Processes and Phase Composition

Composition	Processing condition	Phases	Test solution	E_{corr} (VSHE)	i_{corr} (μA/cm²)	E_{pit} (VSHE)	Ref
$Al_{0.1}CoCrFeNi$	IM + HIP	FCC	3.5 wt.% NaCl + Sand	−0.225	0.638	0.49	[99]
$Al_{0.3}CoCrFeNi$	MS	FCC	3.5 wt.% NaCl	−0.451	0.103	1.07	[100]
$Al_{0.3}CoCrFeNi$	AM	FCC	3.5 wt.% NaCl	−0.195	0.0835	0.460	[59]
$Al_{0.3}CrFe_{1.5}MnNi_{0.5}$	AM	FCC + BCC	3.5 wt.% NaCl	−0.845	6.98	−0.344	[87]
$Al_{0.3}CoCrFeNiMn_{0.3}$	IM	FCC	3.5 wt.% NaCl	−0.215	0.324	0.613	[63]
$Al_{0.3}CrFe_{1.5}MnNi_{0.5}$	AM	FCC + BCC	3.5 wt.% NaCl	−0.677	6.03	−0.227	[87]
$Al_{0.3}CrFe_{1.5}MnNi_{0.5}Ti_{0.2}$	AM	BCC + FCC	3.5 wt.% NaCl	−0.31	0.734	–	[80]
$Al_{0.3}CrFe_{1.5}MnNi_{0.5}Si_{0.2}$	AM	BCC + FCC	3.5 wt.% NaCl	−0.30	0.0471	–	
$Al_{0.5}CoCrFeNi$	AM	FCC + BCC	3.5 wt.% NaCl	−0.225	0.252	0.385	[59]
$Al_{0.5}CoCrFeNi$	AM	FCC	3.5 wt.% NaCl	−0.57	0.17	0.13	[86]
$Al_{0.5}CoCrFeNi(B)$	IM	BCC + FCC	3.5 wt.% NaCl	−0.482	0.701	–	[67]
$Al_{0.5}CoCrCu_{0.5}FeNi$	AM	FCC	0.5 M NaCl	−0.286	1.306	–	[65]
$Al_{0.7}CoCrFeNi$	AM	FCC + BCC	3.5 wt.% NaCl	−0.275	0.429	0.052	[59]
$Al_{0.9}CoCrFeNi$	AM	BCC + FCC + B2	0.6 M NaCl	−0.216	0.093	0.184	[69]
AlCoCrFe	LDC	BCC	0.6 M NaCl	−0.732	0.036	–	[101]
AlCrFeNi	MA + VHPS	BCC	3.5 wt.% NaCl	−1.08	20.77	0.07	[102]
AlCoCrCuFe	LC	FCC + BCC	1 M NaCl	−0.117	60.2	–	[92]
AlCoCrFeNi	IM	BCC + FCC	3.5 wt.% NaCl	−0.185	5.70	–	[103]
AlCoCrFeNi	LSA	BCC	3.5 wt.% NaCl	−0.167	0.073	–	[104]
AlCoCrFeNi	AM	BCC	3.5 wt.% NaCl	−0.294	0.093	–	[93]
AlCoCrFeNi	AM + ESD	BCC	3.5 wt.% NaCl	−0.349	0.016	–	
AlCoCrFeNi	LSA	BCC+BCC	3.5 wt.% NaCl	−0.73	0.3	–	[105]

(Continued)

TABLE 14.4 (CONTINUED)
Corrosion Properties of Reported HEAs in Salt Solution with Their Synthesis Processes and Phase Composition

Composition	Processing condition	Phases	Test solution	E_{corr} (VSHE)	i_{corr} (μA/cm^2)	E_{pit} (VSHE)	Ref
AlCoCrFeNi	DLF	BCC + FCC	0.6 M NaCl	−0.240	0.089	0.035	[106]
AlCoCrFeTi	MA + SPS	BCC + FCC	3.5 wt.% NaCl	−0.217	1.475	–	[107]
AlCoCuFeNiV	MS	FCC + BCC	3.5 wt.% NaCl	−0.747	10.609	–	[97]
AlCoCuCrFeMn	MS	FCC	3.5 wt.% NaCl	−0.702	2.6651	–	[108]
AlCrCuFeMnNi	IM	BCC + FCC	3.5 wt.% NaCl	−0.282	3.9	–	[88]
AlCoCrFeNiTi	IM	BCC + B2	3.5 wt.% NaCl	−0.479	5.47	–	[89]
AlCoCrCu$_{0.5}$FeNi	AM	FCC + BCC	0.5 M NaCl	−0.288	0.8214	–	[65]
AlCoCrCuFeNi	PM	FCC + BCC	1 M NaCl	−0.012	0.003	–	[70]
AlCoCrFeMoNi	IM	BCC + FCC	3.5 wt.% NaCl	−0.257	5.50	–	[103]
AlCoCrFeMoNiSi	IM	BCC + FCC	3.5 wt.% NaCl	−0.387	6.60	–	
AlCoCrFeMoNiSiB$_{0.2}$	IM	BCC + FCC	3.5 wt.% NaCl	−0.285	1.20	–	
AlCoCrFeNiTi	LSA	BCC + FCC	3.5 wt.% NaCl	−0.203	0.073	–	[98]

(Continued)

TABLE 14.4 (CONTINUED)
Corrosion Properties of Reported HEAs in Salt Solution with Their Synthesis Processes and Phase Composition

Composition	Processing condition	Phases	Test solution	E_{corr} (VSHE)	i_{corr} (μA/cm²)	E_{pit} (VSHE)	Ref
$AlCoCrCu_{0.9}FeNi$	LC	BCC	3.5 wt.% NaCl	−1.40	0.025	–	[109]
$AlCoCrCu_{0.5}FeNiSi_{0.5}$	LC	BCC + FCC	3.5 wt.% NaCl	−0.09	4.5	–	[110]
$AlCoCrCu_{0.5}FeNiSi$	AM	Amorphous + BCC	1 M NaCl	−0.53	3.16	−0.25	[60]
AlCuNiTiZr	MA + VHPS	FCC + BCC	Seawater	−1.03	89.0	–	[111]
AlCuHfYZr	AM	HCP + BCC	Na,Mg, Ca salt*	−1.201	146	−0.907	[20]
AlCuTiYZr	AM	HCP + BCC	Na,Mg, Ca salt*	−1.290	235	−0.808	
AlCuNiTiZr	MA + SPS	BCC + FCC	Artificial seawater*	−1.19	0.748	−0.03	[112]
AlCuMgMnNiPbSnZn	IM	FCC	3.5 wt.% NaCl	−1.33	0.293	–	[113]
$Al_{1.5}CoCrCu_{0.5}FeNi$	AM	BCC	0.5 M NaCl	−0.443	18.68	–	[65]
$Al_{1.5}CrCoCuFeNiTi$	AM	BCC + FCC	3.5 wt.% NaCl	−0.720	–	–	[68]
$Al_{2.0}CoCr$ FeNi(B)	IM	BCC + FCC	3.5 wt.% NaCl	−0.465	0.968	–	[67]
$Al_{2.0}CoCrCuFeTi$	LC	FCC + BCC	3.5 wt.% NaCl	−0.57	710	–	[94]
$Al_{2.0}CoCrCuFeTiNi$	LC	FCC + BCC	3.5 wt.% NaCl	−0.22	13	–	
BCoCrFeNi	LC	FCC + borides	3.5 wt.% NaCl	−0.084	0.0198	–	[91]
$CoCrFe_{0.2}Ni$	LC	FCC	3.5 wt.% NaCl	−0.433	0.056	0.464	[64]
$CoCrFe_{0.6}Ni$	LC	FCC	3.5 wt.% NaCl	−0.454	0.102	0.262	
CoCrFeNi	AM	FCC	3.5 wt.% NaCl	−0.26	0.031	0.31	[61]
CoCrFeNi	MA + VHPS	FCC	3.5 wt.% NaCl	−1.08	9.44	0.12	[114]
CoCrNiV	LC	BCC + HCP + FCC	3.5 wt.% NaCl	−0.280	0.575	–	[115]
$CoCrCu_{0.5}FeNi$	AM	FCC	3.5 wt.% NaCl	−0.29	0.723	0.090	[61]
$CoCrCu_{0.5}FeNi$	AM	FCC	3.5 wt.% NaCl	−0.41	0.045	0.05	[85]

(*Continued*)

TABLE 14.4 (CONTINUED)
Corrosion Properties of Reported HEAs in Salt Solution with Their Synthesis Processes and Phase Composition

Composition	Processing condition	Phases	Test solution	E_{corr} (VSHE)	i_{corr} (μA/cm²)	E_{pit} (VSHE)	Ref
CoCrCuFeNi	AM	FCC	3.5 wt.% NaCl	−0.33	1.32	0.08	[61]
CoCrCuFeNi	IM	FCC	3.5 wt.% NaCl	−0.235	31.35	–	[84]
CoCrCuFeNi	DS	FCC	3.5 wt.% NaCl	−0.384	3.473	–	
CoCrFeMnNi	MA + LSA	FCC	3.5 wt.% NaCl	−0.099	0.105	0.218	[116]
CoCrFeMnNi	Electrolysis	FCC	3.5 wt.% NaCl	−0.390	0.879	–	[117]
CoCrFeMnNi	MA + Coating	FCC	3.5 wt.% NaCl	−0.06	9.81	–	[118]
$CoCuFeNiSn_{0.04}$	AM	FCC	3.5 wt.% NaCl	−0.722	0.969	–	[119]
CuHfTiYZr	AM	HCP + BCC	Na, Mg, Ca–salt*	−1.183	159	−0.903	[83]
CoCrFeNiW	MA + VHPS	FCC+ FCC	3.5 wt.% NaCl	−0.99	1.42	−0.37	[96]
$CoCrFeNiW_{0.5}Mo_{0.5}$	MA + VHPS	FCC1+ FCC2	3.5 wt.% NaCl	−1.02	8.90	−0.26	
$(CoCrFeNi)_{80}B_{20}$	AM	Amorphous	3.5 wt.% NaCl	−0.153	0.366	–	Ding et al. (2017b)
CoCrCuFeNi	MA + VHPS	FCC	3.5 wt.% NaCl	−1.05	17.7	0.09	[114]

IM – Induction Melting, AM – Arc Melting, HIP – Hot Isotactic Pressing, DLF – Direct Laser Fabrication, PM – Powder Metallurgy, DS – Directional Solidification, LAC – Laser Additive Coating, ESD – Electro Spark Deposition, MA – Mechanical Alloying, LC – Laser Cladding, SPS – Spark Plasma Sintering, VHPS – Vacuum Hot Press Sintering, LSA – Laser Surface Alloying, MS – Magnetron Sputtering.

* – NaCl, $MgCl_2$, $MgSO_4$, $CaCl_2$ with distilled water (salt and water mass ratio, 1:30).

14.5.2 Electrochemical Properties of High Entropy Alloys in Acid-Aqueous Medium

In acid solution (H_2SO_4, HNO_3), the chance of pitting corrosion reduces due to the absence of chloride ions. However, the properties of passivation film are highly controlled by the H^+ ions of the acid and accordingly, the general corrosion behavior of the alloy changes. Elemental composition, segregations, and phase structure of the high entropy alloys mainly influence the corrosion resistance. For example, the presence of copper and segregation was noted to be detrimental for CuCrFeNiMn high entropy alloys in a 1 M sulfuric acid solution [120].

The presence of boron in $Al_{0.5}CoCrCuFeNiB_x$ (x = 0, 0.2, 0.6, 1.0) high entropy alloys diminished the corrosion resistance of the high entropy alloys in a 1 N H_2SO_4 solution [121, 122]. With an increase in boron content in the high entropy alloys, the general corrosion resistance of the alloy decreased. The $Al_{0.5}CoCrCuFeNiB_x$ high entropy alloys are not susceptible to localized corrosion. Stringy precipitates of the chromium, iron, and cobalt borides were noticed after the corrosion process. On the other hand, the boron content in $(CoCrCuFeNi)_{95}B_5$ synthesized by Liu and co-workers [123] were effective for improvement of the corrosion resistance in 1N HCl solution. The potentiodynamic polarization curves exhibited excellent corrosion resistance for this high entropy alloy with a passive region and lower i_{corr} value. The corrosion properties of $(CoCrCuFeNi)_{95}B_5$ are reported to be more noble than those of 304 SS in similar environment.

Aluminum content in high entropy alloys enhances the chance of passivation film formation in acidic media e.g., $Al_xCrFe_{1.5}MnNi_{0.5}$ (x = 0, 0.3, 0.5) high entropy alloys exhibited a wide passive region ($\Delta E > 1,000$ mV) in 0.5 M H_2SO_4 solution (Lee 2008a). However, the aluminum-free HEA was more resistant to general corrosion than the aluminum-containing high entropy alloys. X-ray photoelectron spectroscopy analysis confirmed the formation of aluminum oxide and chromium oxide after corrosion [124]. Similarly, $Al_xCoCrCu_{0.5}FeNi$ (x = 0.5, 1.0, 1.5) high entropy alloys also exhibited a passive region in a 0.5 M H_2SO_4 solution [65]. The lowest passivation potential and widest passive region was observed for $Al_{1.0}$ HEA. The general corrosion resistances of as-cast $Al_{1.0}$ HEA was also stronger than that of the other two high entropy alloys and 321 SS. The formation of a single body-centered cubic phase in $Al_{1.0}$ is comparatively more corrosion-resistant than that of the face-centered cubic phase which was observed in the $Al_{0.5}$ and $Al_{1.5}$ high entropy alloys. It has been suggested that the aluminum content in the body-centered cubic phase was more corrosion-resistant than that of the face-centered cubic phase. Therefore, the proportion of the body-centered cubic phase with aluminum content plays an important role to improve the corrosion resistance of aluminum containing high entropy alloys. Similar observations are also reported by Zhang and co-workers [119] for $Al_xCoCrFeNiTi_{0.5}$ (x = 0, 1.0, 1.5) high entropy alloys in a 0.5 M sulfuric acid solution. In the case of $Al_xCoCrFeNiTi_{0.5}$ high entropy alloys, with an increase in aluminum content from 0 to 1.0, the proportion of the body-centered cubic phase was increased more than the face-centered cubic phase and accordingly, corrosion resistance of the $Al_{1.0}$ alloy was observed to be highest among the reported high entropy alloys. The $Al_{7.5}Cr_{22.5}Fe_{35}Mn_{20}Ni_{15}$ high entropy alloy showed almost equivalent or even better overall corrosion resistance than 304 SS in 1 M de-aerated H_2SO_4, HNO_3, and HCl solutions [125]. Almost complete surface corrosion was noted for the high entropy alloy in acidic media. For AlCrCuMnNi high entropy alloy, the presence of aluminum and manganese showed a negative effect on the corrosion properties, whereas chromium and nickel increased the corrosion resistance of the high entropy alloy [126]. Around 4%, 0.4%, and 0.7% weight loss have been observed for this high entropy alloy in 3% HF+10% HNO_3, 10% H_2SO_4, and 5% HCl, respectively, after 120 hours.

Corrosion resistance of an AlCoCuFeNi high entropy alloy was improved after chromium addition, whereas it decreased after titanium addition in a 0.5 M H_2SO_4 solution [127]. Formation of stress corrosion cells was noted due to the synthesis and casting process. Accordingly, the high entropy alloys with two-phase composition were corroded spontaneously. A passivation layer of Cr_2O_3 was formed for the chromium-added high entropy alloys. Higher chromium content (31.2 atomic %) also improved the corrosion resistance of the $Al_5Cr_{32}Fe_{35}Ni_{22}Ti_6$ high entropy alloy [128]. Corrosion resistances of this high entropy alloy was reported to be better than that of 316L SS. On the other hand, an improvement in corrosion properties after an increase in titanium content was observed for $Al_2CoCrCuFeNiTi_x$ (x = 0, 0.5, 1.0, 1.5, 2.0) high entropy alloys in 0.5 M nitric acid solution due to the formation of surface passivation layers [129]. Addition of suitable amounts of titanium content could resist the corrosion rate whereas immoderate titanium addition can increase the corrosion [130, 131].

Chou and co-workers [74] have investigated the effect of molybdenum content in corrosion properties of $Co_{1.5}CrFeNi_{1.5}Ti_{0.5}Mo_x$ (x = 0, 0.1, 0.5, 0.8) high entropy alloys in 0.5 M H_2SO_4. It has been reported that the corrosion current density of molybdenum-containing alloys increased with the molybdenum addition. The general corrosion resistance of the molybdenum-free high entropy alloy ($Co_{1.5}CrFeNi_{1.5}Ti_{0.5}$) was better than that of the molybdenum-containing high entropy alloys in acidic solution.

The presence of niobium has a positive effect on corrosion properties of the CoCrCuFeNbNi high entropy alloy in 6 N HCl solution [132]. The general corrosion rates improved from 1.14 to 0.74 mm/yr after addition of niobium. A spontaneous passivation and a pseudo-passive behavior were also noted for HfNbTaTiZr HEA in 11.5 M HNO_3 and 11.5 M HNO_3 + 0.05 M NaF solutions [133]. The corrosion rates increased in fluorinated nitric acid more than HNO_3. Figure 14.6 shows the micrographs of the alloy after potentiodynamic polarization tests in nitric acid and fluorinated nitric acid. X-ray photoelectron spectroscopic studies revealed that the passive film was enriched with Ta_2O_5, Nb_2O_5, HfO_2, ZrO_2, and TiO_2. Formation of fluoride (i.e., ZrF_4, $ZrOF_2$, and HfF_4) was also observed in fluorinated HNO_3 solution.

The effect of nitrogen content and substrate bias voltage during synthesis of $(AlCrSiTiZr)_{100-x}N_x$ high entropy alloy coatings on various substrates (silicon wafers, 6,061 aluminum alloy, and mild steel) by a direct current magnetron sputtering system on corrosion properties of the coatings were studied in 0.1 M H_2SO_4 by Hsueh and co-workers [134]. The elemental composition of the studied alloy was $Al_{21.6}Cr_{17.8}Si_{21.6}Ti_{17.2}Zr_{21.8}$. The added nitrogen content was dependent on the nitrogen flowrate during the coating process. It has been reported that the high entropy alloy

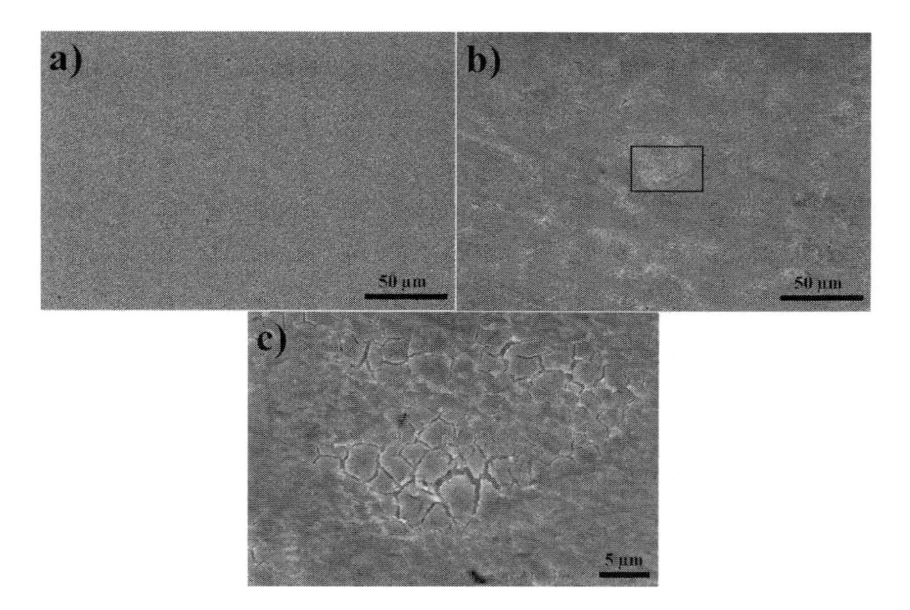

FIGURE 14.6 SEM micrographs of HfNbTaTiZr HEA after potentiodynamic polarization tests in (a) nitric acid, (b) fluorinated nitric acid, (c) magnified region marked in image (b) [133].

coating deposited at nitrogen flowrate of 30% (where x = 53.2 in alloy composition) showed the best corrosion resistance. The high corrosion resistiveness of this alloy coating is reported to be due to the strong metal-nitrogen bonding and the dense columnar structure of the coatings. The AlCrSiTiZr film without nitrogen doping also showed good corrosion properties, which was mostly due to the amorphous nature of the film. The substrate bias improved the film densification and consequently corrosion resistance was improved. Multicomponent carbide thin films (CrNbTaTiW)C with 30–40 at.% C showed better corrosion resistance than hyper-duplex stainless steel in 0.1 M hydrochloric acid [135]. Tantalum- and tungsten-rich films showed better corrosion resistance, whereas niobium-rich film exhibited a passive region. Table 14.5 gives details of all the high entropy alloys studied in acid-aqueous media.

14.6 THERMOELECTRIC APPLICATIONS

With the prime focus on renewable, clean, and safe technologies, a large amount of investigation has been dedicated to searching for new and alternative sources of energy. Presently, in the running of industries, automobiles, and other heavy machineries, a large amount of heat is lost irreversibly. If this heat can be trapped and converted into useful energy, it could well be a solution, at least partly, to the current energy crisis faced worldwide. The thermoelectric efficiency of any material used in devices can be expressed in terms of its thermoelectric figure of merit (Z).

$$ZT = (S^2\sigma / k)T \tag{14.5}$$

TABLE 14.5
Corrosion Properties of Reported HEAs in Acidic Solution with Their Synthesis Processes and Phase Composition

Composition	Processing condition	Phases	Solution	E_{corr} (VSHE)	i_{corr} (μA/cm²)	ΔEb (VSHE)	Ref
$AlCoCrCu_{0.5}FeNiSi$	AM	Amorphous + BCC	1 M H_2SO_4	−0.170	251	–	[60]
$Al_{0.5}CoCrCuFeNc$	IM	FCC	1 N H_2SO_4	−0.115	780	273	[122]
$Al_{0.5}CoCrCuFeNiB_{0.2}$	IM	FCC + borides	1 N H_2SO_4	−0.121	1,025	256	
$Al_{0.5}CoCrCuFeNiB_{0.6}$	IM	FCC + borides	1 N H_2SO_4	−0.148	2,626	–	
$Al_{0.5}CoCrCuFeNiB_{1.0}$	IM	FCC + borides	1 N H_2SO_4	−0.159	2,848	–	
$CrFe_{1.5}MnNi_{0.5}$	AM	FCC + α–FeCr	0.5 M H_2SO_4	−0.229	686	1.2	[124]
$Al_{0.3}CrFe_{1.5}MnNi_{0.5}$	AM	BCC + FCC	0.5 M H_2SO_4	−0.194	2,390	1.2	
$Al_{0.5}CrFe_{1.5}MnNi_{0.5}$	AM	BCC	0.5 M H_2SO_4	−0.206	5,080	1.1	
$Co_{1.5}CrFeNi_{1.5}Ti_{0.5}$	IM	FCC	0.5 M H_2SO_4	−0.092	30	1.1	[74]
$Co_{1.5}CrFeNi_{1.5}Ti_{0.5}Mo_{0.1}$	IM	FCC	0.5 M H_2SO_4	−0.071	78	1.1	
$Co_{1.5}CrFeNi_{1.5}Ti_{0.5}Mo_{0.8}$	IM	FCC + σ (Mo, Cr)	0.5 M H_2SO_4	−0.070	69	1.1	
AlCoCrFeNi	LC	FCC + BCC	0.05 M HCl	0.55	29.2	–	[136]
$Al_{1.8}CoCrFeNi$	LC	FCC + BCC	0.05 M HCl	0.673	7.6	–	
$CuCr_2Fe_2Ni_2Mn_2$	AM	FCC	1 M H_2SO_4	−0.73	2.1	–	[120]
$Cu_2CrFe_2Ni_2Mn_2$	AM	FCC + BCC	1 M H_2SO_4	−0.90	40.2	–	
$Al_{21.6}Cr_{17.8}Si_{21.6}Ti_{17.2}Zr_{21.8}$	AM + SC (on 6061 Al)	Amorphous	0.1 M H_2SO_4	−0.304	7.3	–	[134]
$(Al_{21.6}Cr_{17.8}Si_{21.6}Ti_{17.2}Zr_{21.8})_{46.8}N_{53.2}$	AM + SC (on 6061 Al)	Amorphous + FCC	0.1 M H_2SO_4	−0.442	3.1	–	
AlCoCrCuFe	LC	FCC + BCC	0.5 M H_2SO_4	−0.109	10.4	–	[92]
CoCrCuFeNi	PTAC	FCC	1 N HCl	−0.235	9.0	–	[137]
CoCrCuFeNiNb	PTAC	FCC + laves phase	1 N HCl	−0.305	10	–	[138]

(Continued)

TABLE 14.5 (CONTINUED)
Corrosion Properties of Reported HEAs in Acidic Solution with Their Synthesis Processes and Phase Composition

Composition	Processing condition	Phases	Solution	E_{corr} (VSHE)	i_{corr} (μA/cm²)	ΔEb (VSHE)	Ref
$Al_{0.5}CoCrCu_{0.5}FeNi$	AM	FCC	0.5 M H_2SO_4	–0.112	4.2	–	[65]
$Al_1CoCrCu_{0.5}FeNi$	AM	FCC	0.5 M H_2SO_4	–0.220	50.9	–	
$Al_{1.5}CoCrCu_{0.5}FeNi$	AM	FCC	0.5 M H_2SO_4	–0.172	0.2	–	
$Al_2CrFeNiCoCuTi_{2.0}$	LC	BCC	0.5 M HNO_3	–0.15	2,700	–	[129]
AlCoCrFeNiTi0·5	AM	BCC + FCC	0.5 M H_2SO_4	–0.1	20.6	–	[119]
$(CoCrCuFeNi)_{95}B_5$	PTAC	FCC + σ	1 N HCl	–0.219	13.7	–	[123]
$Al_5Cr_{32}Fe_{35}Ni_{22}Ti_6$	AM	BCC + FCC + Heusler	0.5 M H_2SO_4	–0.13	35.2	–	[128]
AlCoCuFeNiV	MS	FCC + BCC	10% H_2SO_4	–0.281	580	–	[97]
AlCoCrCuFeMn	AM+ MS	FCC	10% H_2SO_4	–0.312	359	–	[20]
AlCoCuFeNi	AM	FCC + A2/B2	0.5 M H_2SO_4	–0.058	8	–	[127]
AlCoCrCuFeNi	AM	A2/B2 + FCC	0.5 M H_2SO_4	–0.075	5.1	–	
AlCoCuFeNiTi	AM	FCC + A2/B2 + Laves	0.5 M H_2SO_4	–0.253	44.8	–	
AlCoCrCuFeNiTi	AM	FCC + A2/B2 + Laves	0.5 M H_2SO_4	–0.256	40	–	
$(CoCrFeNi)_{80}B_{20}$	AM	Amorphous	0.1 M H_2SO_4	–0.226	0.2	–	[139]
HfNbTaTiZr	AM	BCC	11.5 M HNO_3	0.0682	0.2	–	[133]
HfNbTaTiZr	AM	BCC	11.5 M HNO_3 + 0.05 M NaF	0.115	2.1	–	
$Al_2NbTi_3V_2Zr$	HEBM + Sintering	BCC + Zr_3Al + Ti_2ZrAl	10 wt.% HNO_3	0.015	5.6	–	[140]
CoCrFeMnNi	EL	FCC	1 M HCl	–0.350	0.3	–	[117]
$CoCr_2FeNiTi$	LC	FCC + laves phase + TiN	35 wt.% H_3PO_4 + 40 wt.% H_2SO_4	–0.231	0.03	–	[130]

*AM – Arc Melting, IM – Induction Melting, LC – Laser Cladding, PTAC – Plasma Transferred Arc Cladding, MS – Magnetron Sputtering, HEBM – High Energy Ball Milling, EL – Electrolysis

Here, S = Seebeck coefficient, σ = Electrical conductivity, T = Temperature, and K = Thermal conductivity. Furthermore, K is a combination of $_{Ke}$ (electronic thermal conductivity) and $_{Kl}$ (lattice thermal conductivity).

Since, the three physical properties (S, σ, and κ) are closely interrelated, the ZT value cannot be related so easily. Alteration of any of these properties often results in negative influence on the other properties. The conversion efficiency of the thermoelectric device is determined by both the Carnot efficiency (it can be expressed as $\mu_C = \frac{(T_h - T_c)}{T_h}$) and the ZT values of the thermoelectric materials. For example, the maximum conversion efficiency of thermoelectric power generation, η, is given by:

$$\eta_{max} = \frac{T_h - T_c}{T_h} \frac{\sqrt{1+ZT} - 1}{\sqrt{1+ZT} + \frac{T_c}{T_h}} \quad (14.6)$$

where T_h and T_c are the hot side and cold side temperatures, respectively. For a given working temperature, the efficiency of conversion depends on ZT values that can be obtained from Equation (14.6). The maximum η is about 10%, which is much less than the maximum attainable Carnot efficiency (62.5%). The low efficiency obtained from current thermoelectric materials is one of the greatest hurdles to be overcome in the present scenario [141].

There exists a plethora of thermoelectric materials operating at a wide range of temperatures. Bismuth–Telluride systems are suitable for relatively low-temperature operation (room temperature to 200°C), whilst silicon–germanium alloys work best for high-temperature applications (>800°C). For moderate temperatures (T = 500–800°C), heat sources such as automobile exhaust and industrial waste heat, lead chalcogenides (PbTe and PbSe), skutterudites, and half-Heuslers are mostly studied. Among them, PbTe/PbSe (maximum ZT = 1.4 at 527°C for n-type and 1.7 at 600°C for p-type) and skutterudites (maximum ZT = 1.7 at 577°C for n-type and 1.1 at 527°C for p-type) have relatively high ZT values. For large-scale device applications, PbTe/PbSe are infamous for the great toxicity of lead and their mechanical properties, while skutterudites suffer from low thermal stability and limited supply of rare earth elements. In comparison, half-Heusler (maximum ZT = 1.0 for both n- and p-type at 500–800°C) is environmentally friendly and mechanically and thermally robust, albeit there are cost issues if a lot of Hf is used. Therefore, they have attracted intensive research interest over the past decade [141].

Major research over the last decade on the improvement of ZT has been focused on the reduction of $_{Kl}$ by a series of methods like (1) heavy atom substitution in the lattice to generate point defects, (2) introduction of rattling centers to initiate phonon dispersion, (3) *in situ* and *ex situ* addition of a second phase, (4) grain boundary and interfacial scattering by nanostructuring, to name a few. Furthermore, it has also been observed lately that tetrahedrites being highly symmetric (cubic), yet with a complex crystal structure are useful materials for reducing the thermal conductivity [142].

High entropy alloys typically crystallize in body-centered cubic and face-centered cubic crystal structures. Substitution of multiple elements in the unit cell leads to a unit cell that has an increased lattice strain. Thus, high entropy alloys were proposed to be materials that can exhibit reduced $_{Kl}$ due to the complex nature that these materials possess. Shafeie and co-workers [143] have shown that increased addition of aluminum in $Al_xCoCrFeNi$ can lead to an increase in the thermopower value. Additionally, low Ke leads to a net low K which further helped in improving thermoelectric properties. The maximum reported ZT of 0.013 at 873 K was observed in $Al_2CoCrFeNi$.

Yan and co-workers [144] applied the high entropy methodology by doping equimolar concentration of five elements in the niobium center in $Nb_{1-x}M_xFeSb$ (M = hafnium, zirconium, molybdenum, vanadium, titanium). The x = 0.4 alloy showed $K = 4.2\ Wm^{-1}K^{-2}$, which is much lower than other NbFeSb systems. Due to combined effects of a high power factor and low K, x = 0.2 alloy showed a maximum ZT of 0.88 at 873 K. An increase by 1,367% was observed in comparison to the pristine alloy as shown in Figure 14.7 [144].

Karati and co-workers [145] reported the synthesis of a single-phase half-Heusler high entropy alloy $Ti_2NiCoSnSb$. A combination of scanning electron microscopy and 3-D-atom probe tomography studies confirmed the absence of any secondary phase in the HH phase. They reported a simultaneous increase in S and σ in the nanocrystalline high entropy alloy. A maximum ZT of 0.144 at 860 K is shown in Figure 14.8. The thermoelectric properties of high entropy alloys are tabulated as shown in Table 14.6 [145].

14.7 SUPERCONDUCTIVITY APPLICATIONS

Superconductivity refers to infinite conductivity and expelling of magnetic flux at below some temperatures called the critical temperature (T_C). Depending upon the presence of singular or dual critical fields (H_{C1} and H_{C2}), they are broadly classified as type I and type II superconductors [153]. Even though it has been

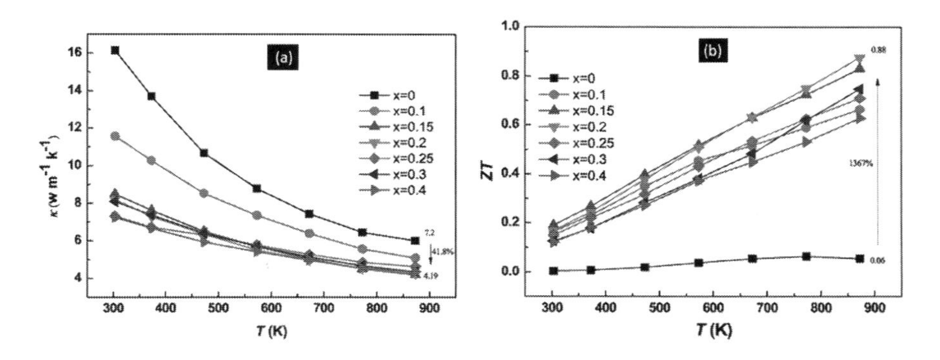

FIGURE 14.7 (a) Total thermal conductivity and (b) ZT of $Nb_{1-x}M_xFeSb$ where x = 0, 0.1, 0.15, 0.2, 0.25, 0.3, 0.4 and M = Ti, Zr, Hf, V, Mo [144].

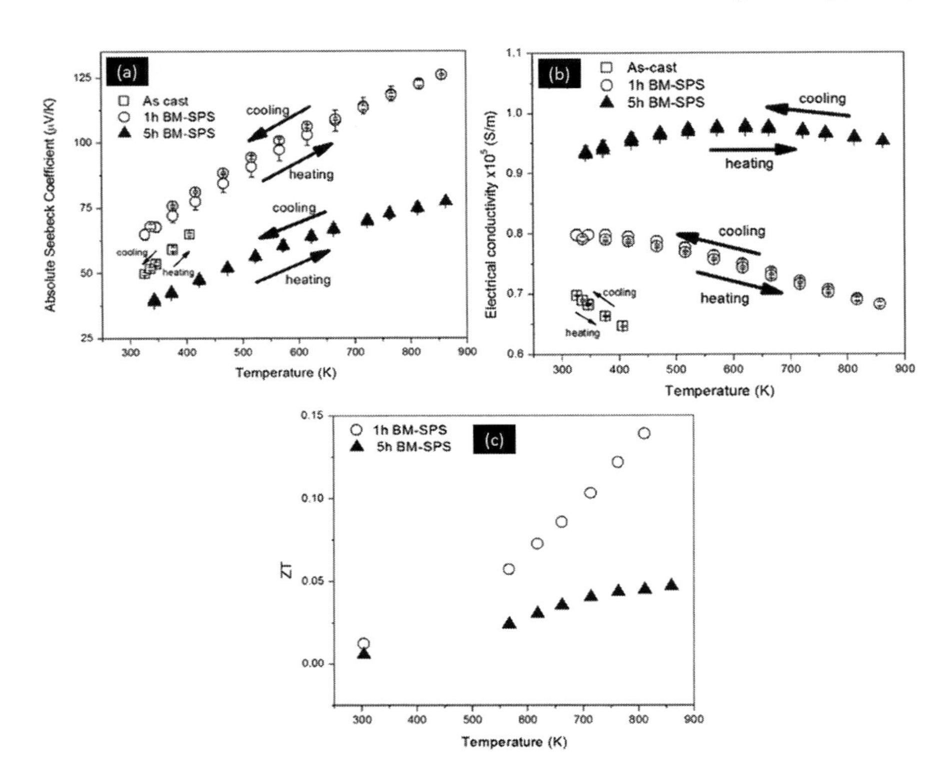

FIGURE 14.8 Variation of (a) Seebeck coefficient, (b) electrical conductivity as a function of temperature for as-cast, 1 h BM–SPS and 5 h BM–SPS samples (heating and cooling cycles). Each data point is an average of three measurements. The error bars are indicated. (c) ZT as a function of temperature [145].

TABLE 14.6
Thermoelectric Properties of HEAs Reported in the Literature

Composition	Processing condition	Phases	$S^2\sigma$ ($mWm^{-1}K^{-2}$)	ZT	Ref
$Al_{2.25}CoCrFeNi$	AM	BCC	0.23 at 873 K	0.0125 at 873 K	[143]
$Pb_{0.94}La_{0.06}SeSnTe$	AM+SPS	$Fm\bar{3}m$	1.30 at 873 K	0.8 at 873 K	[146]
$(BiSbTe_{1.5}Se_{1.5})_{0.091}Ag_{0.009}$	AM+SPS	$R\bar{3}m$	0.65 at 423 K	0.63 at 450 K	[147]
$Al_{0.3}CoCrFeNi$	AM+Ann	FCC	–	0.018 at 750 K	[148]
$Sn_{0.555}Ge_{0.15}Pb_{0.075}Mn_{0.275}Te$	AM+SPS	$Fm\bar{3}m$	1.75 at 900 K	1.42 at 900 K	[149]
$Al_2CoCrFeNi$+3 wt.% Sc	IM+Ann	BCC	0.25 at 923 K	0.014 at 923 K	[150]
$Nb_{0.6}(HfMoTiVZr)_{0.4}FeSb$	AC+Ann+SPS	$F\bar{4}3m$	3.75 at 873 K	0.88 at 873 K	[144]
$Cu_5Sn_{1.2}MgGeZnS_9$	MA–SPS	$I\bar{4}2m$	–	0.6 at 800 K	[151]
$Ti_2NiCoSnSb$	AM+BM+SPS	$F\bar{4}3m$	1.02 at 860 K	0.144 at 860 K	[145]
$(Sn_{0.5}Ge_{0.4875})_{0.5}Pb_{0.5}Te$	AM+SPS	$Fm\bar{3}m$	–	1.61 at 773 K	[152]

more than a century since its discovery by Onnes [154], it still remains one of the most fascinating field of research for physicists, with the major aim being to reach near-room temperature superconductivity. Efforts in the BaCuLaO and BaCuOY system [155, 156] from three decades ago has been the most promising revolution in this regard, with recent works in H_2S system [157] and last year the controversial silver–gold nanocomposites have brought great hopes to the superconducting community. The other challenge in these systems deals with operating superconductors under extreme conditions for applications like the Large Hadron Collider [158].

It has been observed in multi-principal element systems that they crystallize as simple single or dual-phase crystals owing to high configurational entropy stabilizing the random solid solutions at their lattice sites [159]. In these high entropy alloys, the mixing entropy, valence electronic configuration, and their correlation with phase selection, and ultimately their resulting properties have been extensively studied over the past decade. As such, Kozeli and co-workers (2014) have investigated the well-commercialized niobium–titanium superconductors and have extended it to high entropy domain by designing a mix of ordered body-centered cubic and glassy phase compound $Ta_{34}Nb_{33}Hf_8Zr_{14}Ti_{11}$, which even though it has a very low T_C of 7.3 K, has displayed stable zero resistance to as high as 190.6 GPa pressure. This opened the possibilities of using high entropy alloys for applications such as superconductors under extreme conditions. Since then many other systems have been investigated and are drafted under Table 14.7.

Stolze and co-workers [167] have found correlations between lattice parameter (Figure 14.9a), valence electron count (Figure 14.9b), and T_C in the case of these high entropy alloy superconductor, while Rohr and co-workers [165] and Ishizu and Kitawaga [166] have deduced the impact of both valence electron count and mixing entropy (Figure 14.10a and b) on the "cocktail effect" of the critical temperature in these superconductors. Rohr and Cava [168] have also focused on isoelectronic substitution and aluminum doping in the $(TaNb)_{67}(ZrHfTi)_{33}$ system and have concluded that isoelectronic substitution and the resulting change in T_C depends upon the group the element are in, that overall valence electron count dominates over disorder in the lattice, and that crystallinity of above high entropy alloys depends upon the extent of aluminum doping and the overall T_C of these systems. And Vrtnik and co-workers [160] have investigated the effect of annealing time on overall microstructure, how it was being influenced by the overall minimization of mixing enthalpy and maximization of mixing entropy, and how that has an overall influence on the superconductivity in these samples.

Just like magnetization, specific heat is also affected by superconducting transition indicated by sharp jump in specific heat near T_C. Specific heat, as we know, has an electronic as well as phononic component. And it has been proved that the change is due to change in the electronic component and it obeys the BCS theory of superconductors. As such, the difference between superconducting and normal states in terms of specific heat (ΔC) is given by following equation:

$$\Delta C = C_{es} - C_{en} \tag{14.7}$$

TABLE 14.7
List of HEAs Showing Superconducting Properties

Composition	Space group	TC (K)	HC (T)	ρ* (μΩcm)	γ (mJ/molK²)	θD (K)	Ref
$Hf_{26}Nb_{25}Ta_{25}Zr_{24}$	$Im\bar{3}m$	7.6	13.4	57.5	6.6	224	[160]
$Nb_{35}Ta_{35}Ti_{15}Zr_{15}$	$Im\bar{3}m$	8.0	11.6	16.5	9.3	191	[161]
$Hf_{10}Ir_{10}Re_{60}Ta_{10}W_{10}$	$I\bar{4}3m$	4.0	4.64	~530	–	–	[162]
$Hf_{23}Nb_{21}Re_{16}Ti_{20}Zr_{20}$	$Im\bar{3}m$	5.3	8.9	120	5.8	–	[163]
$Hf_8Nb_{33}Ta_{34}Ti_{11}Zr_{14}$	$Im\bar{3}m$	7.3	8.2	36.5	8.3	238	[164]
$Hf_8Nb_{33}Ta_{34}Ti_{11}Zr_{14}$	$Im\bar{3}m$	7.3	8.1	36.5	8.2	238	[160]
$Hf_{20}Nb_{21}Ta_{20}Ti_{19}Zr_{20}$	$Im\bar{3}m$	6.0	10.5	77.0	7.9	212	
$Hf_{21}Nb_{24}Ta_{22}Ti_{10}Zr_{23}$	$Im\bar{3}m$	5.0	14.2	93.4	6.8	238	
$Hf_{11}Nb_{33}Ta_{34}Ti_{11}Zr_{11}$	$Im\bar{3}m$	7.7	7.7	107	7.9	225	[165]
$Hf_{28}Nb_8Ta_8Ti_{28}Zr_{28}$	$Im\bar{3}m$	4.5	9.0	–	6.5	184	
$Hf_{21}Nb_{25}Ti_{15}V_{15}Zr_{24}$	$Im\bar{3}m$	5.3	–	121	–	–	[166]
$Hf_{10}Pt_{10}Re_{60}Ta_{10}W_{10}$	$I\bar{4}3m$	4.4	5.9	585	–	–	[162]
$Mo_{30}Nb_5Re_{30}Ru_{30}Zr_5$	$I\bar{4}3m$	5.3	7.9	103	–	–	
$Nb_{22}Pd_{18}Rh_{17}Sc_{21}Zr_{22}$	$Pm\bar{3}m$	9.7	10.7	26.5	–	–	[167]
$Al_{30}Hf_8Nb_{23}Ta_{23}Ti_8Zr_8$	$Im\bar{3}m$	7.0	–	–	–	–	[168]
$Cr_{22}Hf_{11}Nb_{34}Sc_{11}Ti_{11}Zr_{11}$	$Im\bar{3}m$	5.6	–	–	–	–	
$Cr_{22}Hf_{11}Sc_{11}Ta_{34}Ti_{11}Zr_{11}$	$Im\bar{3}m$	4.4	–	–	–	–	
$Hf_{11}Mo_4Nb_{33}Ta_{34}Ti_{11}Y_7$	$Im\bar{3}m$	7.6	–	–	–	–	
$Hf_{11}Mo_{22}Nb_{34}Sc_{11}Ti_{11}Zr_{11}$	$Im\bar{3}m$	4.4	–	–	–	–	
$Hf_{11}Nb_{22}Ta_{23}Ti_{11}V_{22}Zr_{11}$	$Im\bar{3}m$	4.1	–	–	7.9	236	[165]
$Mo_4Nb_{33}Ti_{11}Ta_{34}Zr_{11}Y_7$	$Im\bar{3}m$	6.7	–	–	–	–	[168]
$Mo_4Nb_{33}Sc_7Ta_{34}Ti_{11}Zr_{11}$	$Im\bar{3}m$	7.5	–	–	–	–	
$Nb_{17}Pd_{16}Rh_{16}Sc_{17}Ta_{17}Zr_{17}$	$Pm\bar{3}m$	6.4	8.8	–	–	–	[167]

(Continued)

TABLE 14.7 (CONTINUED)
List of HEAs Showing Superconducting Properties

Composition	Space group	TC (K)	HC (T)	ρ* (μΩcm)	γ (mJ/molK²)	θD (K)	Ref
$BiCe_{0.1}F_{0.5}La_{0.1}Nd_{0.3}Pr_{0.2}S_2Sm_{0.3}O_{0.5}$	*P4/nmm*	4.5	–	~22.8	–	–	[169]
$BiCe_{0.1}F_{0.5}La_{0.1}Nd_{0.3}Pr_{0.3}S_2Sm_{0.2}O_{0.5}$	*P4/nmm*	4.2	–	~24.0	–	–	
$BiCe_{0.2}F_{0.5}La_{0.2}Nd_{0.2}Pr_{0.2}S_2Sm_{0.2}O_{0.5}$	*P4/nmm*	3.8	–	~13.8	–	–	
$BiCe_{0.3}F_{0.5}La_{0.3}Nd_{0.1}Pr_{0.2}S_2Sm_{0.1}O_{0.5}$	*P4/nmm*	3.0	–	~13.9	–	–	

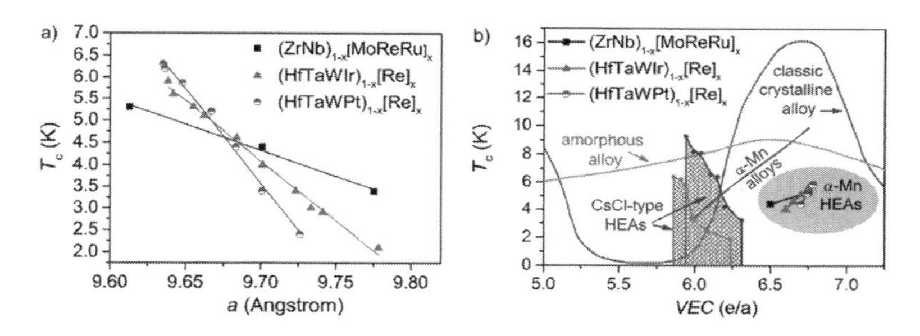

FIGURE 14.9 Dependence of T_C on (a) lattice parameter and (b) VEC [167].

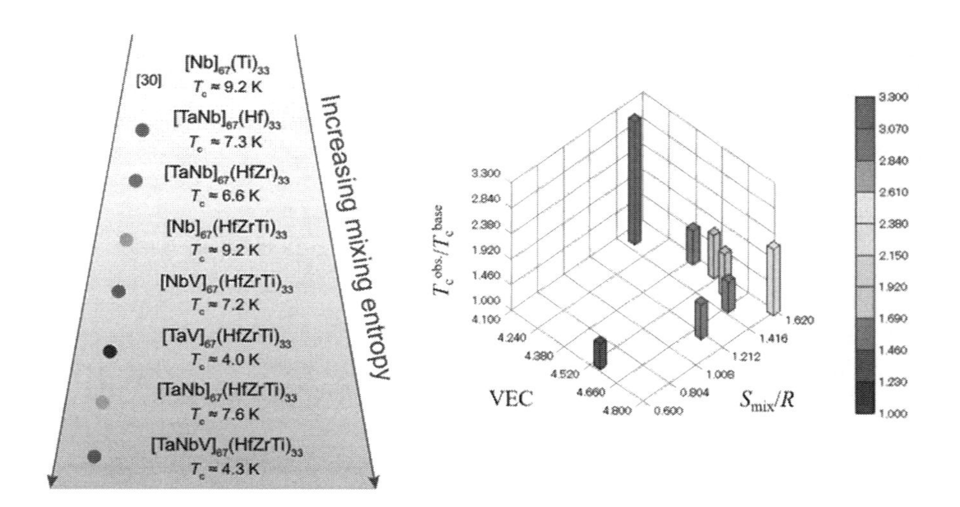

FIGURE 14.10 Dependence of T_C on (a) mixing entropy [165] and (b) combination of VEC and mixing entropy [166].

where for normal state, specific heat (C_{en}) is given as:

$$\frac{C_{en}}{T} = \gamma + \alpha T^2 \tag{14.8}$$

with γ as the electronic coefficient and α as the phononic coefficient. As such, the ratio of $\Delta C/\ \gamma T_C$ gives the Bardeen–Cooper–Schrieffer (BCS) prediction for the superconducting states [160]. The phononic component term directly correlates to the Debye temperature (θ_D). Hence, apart from T_C, H_C, and resistivity in vicinity of superconducting transition, γ and θ_D are the parameters that determine the BCS ratio used to study the superconductors. The BCS ratio was found to be 1.63 for $Hf_8Nb_{33}Ta_{34}Ti_{11}Zr_{14}$, 1.62 for $Hf_8Nb_{33}Ta_{34}Ti_{11}Zr_{14}$, 1.81 for $Hf_{23}Nb_{21}Re_{16}Ti_{20}Zr_{20}$, and 1.98 for $Hf_{28}Nb_8Ta_8Ti_{28}Zr_{28}$ which is higher than the lower limit of weak coupling as per BCS theory, which is 1.43, hence justifying the presence of superconductivity in high entropy alloys. Even though with high entropy alloys, the stability of superconductors at high pressures has been addressed, they still have extremely low T_C. Considering that the cuprites, iron chalcogenides, and hydrides have T_C between 80 and 200 K [170], high entropy alloys have comparatively very low T_C and hence, future research should also focus on increasing the T_C in high entropy alloys.

14.8 CONCLUSION

In summary, functional properties of high entropy alloys such as magnetic properties, magnetocaloric applications, hydrogen storage properties, corrosion resistance properties, thermoelectric applications, and superconducting behavior were discussed. It has been observed that increasing entropy is not always favorable for the improvement of functional properties. Functional properties of high entropy alloys are not just dependent on core high entropy alloy structural properties, but also can be related to the elemental composition and the electronic structure of the alloy.

1. In the case of magnetic materials, it can be observed that lattice distortion leads to pining of domains that enhances the coercivity though at the expense of electrical conductivity. But due to lower fractions of ferromagnetic content, high entropy alloys actually have lower saturation magnetization and lower Curie point temperature.
2. For the magnetocaloric applications, it is observed that rare earth-free high entropy alloys have a higher Curie point temperature, while rare earth-containing high entropy alloys have higher magnetic entropy and refrigeration capacity. Future research should aim at finding a balance between them.
3. The effects of environments, elemental additions, and synthesis procedures, etc. on the corrosion-resistance properties of high entropy alloys are discussed. Corrosion properties of high entropy alloys are comparable to the conventional alloys and could be used in various environments. It has been noted that surface passive films play an important role in improving the corrosion resistance behavior of high entropy alloys. Addition of molybdenum and inclusion of inhibitors could be useful to improve the

corrosion resistance. However, elemental addition is not always beneficial for improvement of the corrosion resistance. Addition of elements such as aluminum, copper, etc. leads to the formation of non-uniform passive films due to elemental segregation, and consequently galvanic corrosion increases and breakdown of the films are reported. Homogeneous elemental distribution and uniform microstructure formation improves the corrosion resistance properties of high entropy alloys. Synthesis procedure such as laser cladding, electrospark deposition, and magnetron sputtering are useful to improve the corrosion resistance due to the rapid cooling during these processes. Heat treatment at elevated temperature is also found to be effective as it could remove the elemental segregations and homogeneous microstructure could be attained.

4. Only a few reports are available on hydrogen storage property analysis of high entropy alloys. Mainly, titanium–vanadium–zirconium (Ti–V–Zr)-based BCC and Laves phase high entropy alloys are analyzed and around 1.8–2 wt.% hydrogen absorption is noted. Highest storage capacity (around 2.7 wt.%) is observed for the TiVZrHfNb alloy.
5. High crystal symmetry and single-phase formation in high entropy alloys can enhance the power factor, while lattice distortion and large crystal size can lower lattice thermal conductivity, leading to more efficient thermoelectric materials. Hence, entropy engineering of thermoelectric materials has been trending recently. In polycrystalline SnTe systems, unlike their single crystalline counterpart, a figure of merit > 1.2 was difficult to obtain, but recently through entropy engineering, a figure of merit of 1.61 was achieved in $(Sn_{0.5}Ge_{0.4875})_{0.5}Pb_{0.5}Te$.
6. It seems that high entropy alloys can answer the high-pressure stability of superconductors, but their Curie point temperature is still quite low and a considerable amount of research is being done in that direction. One approach can be to move toward high entropy chalcogenides and hydrides.

REFERENCES

1. Gao, M. C., D. B. Miracle, D. Maurice, X. Yan, Y. Zhang, and J. H. Hawk. 2018. High-Entropy Functional Materials. *Journal of Materials Research* 33(19): 3138–55.
2. Lucas, M. S., L. Mauger, J. A. Muoz. 2011. Magnetic and Vibrational Properties of High-Entropy Alloys. *Journal of Applied Physics* 109(7): 07E307.
3. Ma, S. G., and Y. Zhang. 2012. Effect of Nb Addition on the Microstructure and Properties of AlCoCrFeNi High-Entropy Alloy. *Materials Science and Engineering A* 532: 480–86.
4. Zhang, K. B., Z. Y. Fu, J. Y. Zhang, J. Shi, W. M. Wang, H. Wang, Y. C. Wang, and Q. J. Zhang. 2010. Annealing on the Structure and Properties Evolution of the CoCrFeNiCuAl High-Entropy Alloy. *Journal of Alloys and Compounds* 502(2): 295–99.
5. Zuo, T., M. C. Gao, L. Ouyang, X. Yang, Y. Cheng, R. Feng, S. Chen, P. K. Liaw, J. A. Hawk, and Y. Zhang. 2017. Tailoring Magnetic Behavior of CoFeMnNiX (X = Al, Cr, Ga, and Sn) High Entropy Alloys by Metal Doping. *Acta Materialia* 130: 10–8.
6. Huang, S., Á. Vida, D. Molnár, K. Kádas, L. K. Varga, E. Holmström, and L. Vitos. 2015. Phase Stability and Magnetic Behavior of FeCrCoNiGe High-Entropy Alloy. *Applied Physics Letters* 107(25): 1–5.

7. Zaddach, A. J., C. Niu, A. A. Oni, M. Fan, J. M. L. Beau, D. L. Irving, and C. C. Koch. 2016. Structure and Magnetic Properties of a Multi-Principal Element Ni-Fe-Cr-Co-Zn-Mn Alloy. *Intermetallics* 68: 107–12.
8. Kulkarni, R., B. S. Murty, and V. Srinivas. 2018. Study of Microstructure and Magnetic Properties of AlNiCo(CuFe) High Entropy Alloy. *Journal of Alloys and Compounds* 746: 194–99.
9. Zuo, T. T., R. B. Li, X. J. Ren, and Y. Zhang. 2014. Effects of Al and Si Addition on the Structure and Properties of CoFeNi Equal Atomic Ratio Alloy. *Journal of Magnetism and Magnetic Materials* 371: 60–8.
10. Jiang, L., Y. Lu, Y. Dong, T. Wang, Z. Cao, and T. Li. 2015. Effects of Nb Addition on Structural Evolution and Properties of the $CoFeNi_2V_{0.5}$ High-Entropy Alloy. *Applied Physics. Part A: Materials Science and Processing* 119(1): 291–97.
11. Jung, C., K. Kang, A. Marshal, K. G. Pradeep, J. B. Seol, H. M. Lee, and P. K. Choi. 2019. Effects of Phase Composition and Elemental Partitioning on Soft Magnetic Properties of AlFeCoCrMn High Entropy Alloys. *Acta Materialia* 171: 31–9.
12. Li, P., A. Wang, and C. T. Liu. 2017. Composition Dependence of Structure, Physical and Mechanical Properties of $FeCoNi(MnAl)_x$ High Entropy Alloys. *Intermetallics* 87: 21–6.
13. Zhang, Y., T. Zuo, Y. Cheng, and P. K. Liaw. 2013. High-Entropy Alloys with High Saturation Magnetization, Electrical Resistivity, and Malleability. *Scientific Reports* 3: 1–7.
14. Zhang, Y., and W. J. Peng. 2012. Microstructural Control and Properties Optimization of High-Entrop Alloys. *Procedia Engineering* 27: 1169–78.
15. Zuo, T., Y. Xiao, P. K. Liaw, and Y. Zhang. 2015. Influence of Bridgman Solidification on Microstructures and Magnetic Behaviors of a Non-Equiatomic FeCoNiAlSi High-Entropy Alloy. *Intermetallics* 67: 171–76.
16. Bazzi, K., A. Rathi, V. M. Meka, R. Goswami, and T. V. Jayaraman. 2019. Significant Reduction in Intrinsic Coercivity of High-Entropy Alloy $FeCoNiAl_{0.375}Si_{0.375}$ Comprised of Supersaturated f.c.c Phase. *Materialia* 6. 10002 93.
17. Wang, J., Z. Zheng, J. Xu, and Y. Wang. 2014. Microstructure and Magnetic Properties of Mechanically Alloyed FeSiBAlNi(Nb) High Entropy Alloys. *Journal of Magnetism and Magnetic Materials* 355: 58–64.
18. Niu, B., F. Zhang, H. Ping, N. Li, J. Zhou, L. Lei, J. Xie, J. Zhang, W. Wang, and Z. Fu. 2017. Sol-Gel Autocombustion Synthesis of Nanocrystalline High-Entropy Alloys. *Scientific Reports* 7(1): 1–7.
19. Ding, P., A. Mao, X. Zhang, X. Jin, B. Wang, M. Liu, and X. Gu. 2017a. Preparation, Characterization and Properties of Multicomponent $AlCoCrFeNi_{2.1}$ Powder by Gas Atomization Method. *Journal of Alloys and Compounds* 721: 609–14.
20. Li, X. C., D. Dou, Z. Y. Zheng, and J. C. Li. 2016b. Microstructure and Properties of $FeAlCrNiMo_x$ High-Entropy Alloys. *Journal of Materials Engineering and Performance* 25(6): 2164–69.
21. Zhang, H., Y. Yang, L. Liu, C. Chen, T. Wang, R. Wei, T. Zhang, Y. Dong, and F. Lia. 2019b. A Novel $FeCoNiCr_{0.2}Si_{0.2}$ High Entropy Alloy with an Excellent Balance of Mechanical and Soft Magnetic Properties. *Journal of Magnetism and Magnetic Materials* 478: 116–21.
22. Zuo, T., M. Zhang, P. K. Liaw, and Y. Zhang. 2018. Novel High Entropy Alloys of $Fe_xCo_{1-x}NiMnGa$ with Excellent Soft Magnetic Properties. *Intermetallics* 100: 1–8.
23. Fu, Z., B. E. MacDonald, T. C. Monson, B. Zheng, W. Chen, and E. J. Lavernia. 2018. Influence of Heat Treatment on Microstructure, Mechanical Behavior, and Soft Magnetic Properties in an FCC-Based $Fe_{29}Co_{28}Ni_{29}Cu_7Ti_7$ High-Entropy Alloy. *Journal of Materials Research* 33(15): 2214–22.
24. Mishra, R. K., and R. R. Shahi. 2019b. Novel $Co_{35}Cr_5Fe_{20}Ni_{20}Ti_{20}$ High Entropy Alloy for High Magnetization and Low Coercivity. *Journal of Magnetism and Magnetic Materials* 484: 83–7.

25. Wang, X. F., Y. Zhang, Y. Qiao, and G. L. Chen. 2007. Novel Microstructure and Properties of Multicomponent $CoCrCuFeNiTi_x$ Alloys. *Intermetallics* 15(3): 357–62.
26. Yu, P. F., L. J. Zhang, H. Cheng, H. Zhang, M. Z. Ma, Y. C. Li, G. Lia, P. K. Liaw, and R. P. Liu. 2016. The High-Entropy Alloys with High Hardness and Soft Magnetic Property Prepared by Mechanical Alloying and High-Pressure Sintering. *Intermetallics* 70: 82–7.
27. Na, S. M., J. H. Yoo, P. K. Lambert, and N. J. Jones. 2018. Room-Temperature Ferromagnetic Transitions and the Temperature Dependence of Magnetic Behaviors in FeCoNiCr-Based High-Entropy Alloys. *AIP Advances* 8(5).
28. Alijani, F., M. Reihanian, and K. Gheisari. 2019. Study on Phase Formation in Magnetic FeCoNiMnV High Entropy Alloy Produced by Mechanical Alloying. *Journal of Alloys and Compounds* 773: 623–30.
29. Zhao, R. F., B. Ren, G. P. Zhang, Z. X. Liu, and J. J. Zhang. 2018b. Phase Transition of as-Milled and Annealed CrCuFeMnNi High-Entropy Alloy Powder. *Nano* 13(9) 1850100.
30. Li, Z., H. Xu, Y. Gu, M. Pan, L. Yu, X. Tan, and X. Hou. 2018a. Correlation between the Magnetic Properties and Phase Constitution of $FeCoNi(CuAl)_{0.8}Ga_x$ ($0 \leq x \leq 0.08$) High-Entropy Alloys. *Journal of Alloys and Compounds* 746: 285–91.
31. Borkar, T., B. Gwalani, D. Choudhuri, C. V. Mikler, C. J. Yannetta, X. Chen, R. V. Ramanujan, M. J. Styles, M. A. Gibson, and R. Banerjee. 2016. A Combinatorial Assessment of $Al_xCrCuFeNi_2$ ($0 < x < 1.5$) Complex Concentrated Alloys: Microstructure, Microhardness, and Magnetic Properties. *Acta Materialia* 116: 63–76.
32. Lin, P. C., C. Y. Cheng, J. W. Yeh, and T. S. Chin. 2016. Soft Magnetic Properties of High-Entropy Fe-Co-Ni-Cr-Al-Si Thin Films. *Entropy* 18(8): 1–9.
33. Xu, J., E. Axinte, Z. Zhao, and Y. Wang. 2016. Effect of C and Ce Addition on the Microstructure and Magnetic Property of the Mechanically Alloyed FeSiBAlNi High Entropy Alloys. *Journal of Magnetism and Magnetic Materials* 414: 59–68.
34. Kao, Y. F., S. K. Chen, T. J. Chen, P. C. Chou, J. W. Yeh, and S. J. Lin. 2011. Electrical, Magnetic, and Hall Properties of $Al_xCoCrFeNi$ High-Entropy Alloys. *Journal of Alloys and Compounds* 509(5): 1607–14.
35. Kourov, N. I., V. G. Pushin, A. V. Korolev, Y. V. Knyazev, M. V. Ivchenko, and Y. M. Ustyugov. 2015. Peculiar Features of Physical Properties of the Rapid Quenched AlCrFeCoNiCu High-Entropy Alloy. *Journal of Alloys and Compounds* 636: 304–9.
36. Mishra, R. K., P. P. Sahay, and R. R. Shahi. 2019a. Alloying, Magnetic and Corrosion Behavior of AlCrFeMnNiTi High Entropy Alloy. *Journal of Materials Science* 54(5): 4433–43.
37. Zhao, R. F., B. Ren, G. P. Zhang, Z. X. Liu, and J. J. Zhang. 2018a. Effect of Co Content on the Phase Transition and Magnetic Properties of $Co_xCrCuFeMnNi$ High-Entropy Alloy Powders. *Journal of Magnetism and Magnetic Materials* 468: 14–24.
38. Gschneidner, K. A., and V. K. Pecharsky. 2008. Thirty Years of near Room Temperature Magnetic Cooling: Where We Are Today and Future Prospects. *International Journal of Refrigeration* 31(6): 945–61.
39. Li, J., L. Xue, W. Yang, C. Yuan, J. Huo, and B. Shen. 2018b. Distinct Spin Glass Behavior and Excellent Magnetocaloric Effect in $Er_{20}Dy_{20}Co_{20}Al_{20}RE_{20}$ (RE = Gd, Tb and Tm) High-Entropy Bulk Metallic Glasses. *Intermetallics* 96: 90–3.
40. Xue, L., L. Shao, Q. Luo, and B. Shen. 2019. $Gd_{25}RE_{25}Co_{25}Al_{25}$ (RE = Tb, Dy and Ho) High-Entropy Glassy Alloys with Distinct Spin-Glass Behavior and Good Magnetocaloric Effect. *Journal of Alloys and Compounds* 25: 633–39.
41. Huo, J., L. Huo, J. Li, H. Men, X. Wang, A. Inoue, C. Chang, J. Q. Wang, and R. W. Li. 2015b. High-Entropy Bulk Metallic Glasses as Promising Magnetic Refrigerants. *Journal of Applied Physics* 117(7) 073902.

42. Huo, J., L. Huo, H. Men, X. Wang, A. Inoue, J. Wang, C. Chang, and R. W. Li. 2015a. The Magnetocaloric Effect of Gd-Tb-Dy-Al-M (M = Fe, Co and Ni) High-Entropy Bulk Metallic Glasses. *Intermetallics* 58: 31–5.
43. Sheng, W., J. Q. Wang, G. Wang, J. Huo, X. Wang, and R. W. Li. 2018. Amorphous Microwires of High Entropy Alloys with Large Magnetocaloric Effect. *Intermetallics* 96: 79–83.
44. Yuan, Y., Y. Wu, X. Tong, H. Zhang, H. Wang, X. J. Liu, L. Ma, H. L. Suo, and Z. P. Lu. 2017. Rare-Earth High-Entropy Alloys with Giant Magnetocaloric Effect. *Acta Materialia* 125: 481–89.
45. Huo, J., J. Q. Wang, and W. H. Wang. 2019. Denary High Entropy Metallic Glass with Large Magnetocaloric Effect. *Journal of Alloys and Compounds* 776: 202–6.
46. Na, S. M., P. K. Lambert, H. Kim, J. Paglione, and N. J. Jones. 2019. Thermomagnetic Properties and Magnetocaloric Effect of FeCoNiCrAl-Type High-Entropy Alloys. *AIP Advances* 9(3).
47. Perrin, A., M. Sorescu, M. T. Burton, D. E. Laughlin, and M. McHenry. 2017. The Role of Compositional Tuning of the Distributed Exchange on Magnetocaloric Properties of High-Entropy Alloys. *The Journal of the Minerals, Metals and Materials Society* 69(11): 2125–29.
48. Kurniawan, M., A. Perrin, P. Xu, V. Keylin, and M. McHenry. 2016. Curie Temperature Engineering in High Entropy Alloys for Magnetocaloric Applications. *IEEE Magnetics Letters* 7: 1–5.
49. Li, L. W. 2016a. Review of Magnetic Properties and Magnetocaloric Effect in the Intermetallic Compounds of Rare-Earth with Low Boiling Point Metal (s). *Chinese Physics. Part B* 25(3): 1–15.
50. Kunce, I., M. Polanski, and J. Bystrzycki. 2014. Microstructure and Hydrogen Storage Properties of a TiZrNbMoV High Entropy Alloy Synthesized Using Laser Engineered Net Shaping (LENS). *International Journal of Hydrogen Energy* 39(18): 9904–10.
51. Kao, Y. F., S. K. Chen, J. H. Sheu, J. T. Lin, W. E. Lin, J. W. Yeh, S. J. Lin, T. H. Liou, and C. W. Wang. 2010a. Hydrogen Storage Properties of Multi-Principal-Component $CoFeMnTi_xV_yZr_z$ Alloys. *International Journal of Hydrogen Energy* 35(17): 9046–59.
52. Karlsson, D., G. Ek, J. Cedervall, C. Zlotea, K. T. Møller, T. C. Hansen, J. Bendnarcik, M. Paskevicius, M. H. Sorby, T. R. Jensen, U. Jansson, and M. Sahlberg. 2018. Structure and Hydrogenation Properties of a HfNbTiVZr High-Entropy Alloy. *Inorganic Chemistry* 57(4): 2103–10.
53. Kunce, I., M. Polanski, and J. Bystrzycki. 2013. Structure and Hydrogen Storage Properties of a High Entropy ZrTiVCrFeNi Alloy Synthesized Using Laser Engineered Net Shaping (LENS). *International Journal of Hydrogen Energy* 38(27): 12180–89.
54. Sahlberg, M., D. Karlsson, C. Zlotea, and U. Jansson. 2016. Superior Hydrogen Storage in High Entropy Alloys. *Scientific Reports* 6: 1–6.
55. Kunce, I., M. Polański, and T. Czujko. 2017. Microstructures and Hydrogen Storage Properties of La–Ni–Fe–V–Mn Alloys. *International Journal of Hydrogen Energy* 42(44): 27154–64.
56. Zepon, G., D. R. Leiva, R. B. Strozi, A. Bedocha, S. J. A. Figueroa, T. T. Ishikawa, and W. J. Bottaa. 2018. Hydrogen-Induced Phase Transition of $MgZrTiFe_{0.5}Co_{0.5}Ni_{0.5}$ High Entropy Alloy. *International Journal of Hydrogen Energy* 43(3): 1702–8.
57. Zlotea, C., M. A. Sow, G. Ek, J. P. Couzinie, L. Perriere, I. Guillot, J. Bourgon, K. T. Moller, T. R. Jensen, E. Akiba, and M. Sahlberg. 2019. Hydrogen Sorption in TiZrNbHfTa High Entropy Alloy. *Journal of Alloys and Compounds* 775: 667–74.
58. Diao, H., L. J. Santodonato, Z. Tang, T. Egami, and P. K. Liaw. 2015. Local Structures of High-Entropy Alloys (HEAs) on Atomic Scales: An Overview. *The Journal of the Minerals, Metals and Materials Society* 67(10): 2321–25.

59. Shi, Y., B. Yang, X. Xie, J. Brechtl, K. A. Dahmen, and P. K. Liaw. 2017. Corrosion of $Al_xCoCrFeNi$ High-Entropy Alloys: Al-Content and Potential Scan-Rate Dependent Pitting Behavior. *Corrosion Science* 119: 33–45.
60. Chen, Y. Y., T. Duval, U. D. Hung, J. W. Yeh, and H. C. Shih. 2005. Microstructure and Electrochemical Properties of High Entropy Alloys-A Comparison with Type-304 Stainless Steel. *Corrosion Science* 47(9): 2257–79.
61. Hsu, Y. J., W. C. Chiang, and J. K. Wu. 2005. Corrosion Behavior of FeCoNiCrCux High-Entropy Alloys in 3.5% Sodium Chloride Solution. *Materials Chemistry and Physics* 92(1): 112–17.
62. Wang, C. W., Y. Mo, Z. Mo, J. Tang, and H. Wei. 2013. The Study about Property of Corrosion Resistance of $AlCoCrTiNiCu_x$ High-Entropy Alloys. *Applied Mechanics and Materials* 395–396: 214–17.
63. Wong, S. K., T. T. Shun, C. H. Chang, and C. F. Lee. 2018. Microstructures and Properties of $Al_{0.3}CoCrFeNiMn_x$ High-Entropy Alloys. *Materials Chemistry and Physics* 210: 146–51.
64. Cai, Y., Y. Chen, S. M. Manladan, Z. Luo, F. Gao, and L. Li. 2018. Influence of Dilution Rate on the Microstructure and Properties of FeCrCoNi High-Entropy Alloy Coating. *Materials and Design* 142(31): 124–37.
65. Li, B. Y., K. Peng, A. P. Hu, L. P. Zhou, J. J. Zhu, and D. Y. Li. 2013a. Structure and Properties of $FeCoNiCrCu_{0.5}Al_x$ High-Entropy Alloy. *Transactions of Nonferrous Metals Society of China* 23(3): 735–41.
66. Shi, Y., L. Collins, R. Feng, C. Zhang, N. Balke, P. K. Liaw, and B. Yang. 2018. Homogenization of $Al_xCoCrFeNi$ High-Entropy Alloys with Improved Corrosion Resistance. *Corrosion Science* 133: 120–31.
67. Zhu, S., W. Du, X. Wang, and G. Han. 2013b. Microstructure and Electrochemical Performances of $Al_xFeCoNiCr(B)$ High Entropy Alloys. *Materials Science Forum* 745–746: 728–33.
68. Sun, H. F., C. M. Wang, X. Zhang, R. Z. Li, and L. Y. Ruan. 2015. Study of the Microstructure and Performance of High-Entropy Alloys $Al_xFeCuCoNiCrTi$. *Materials Research Innovations* 19(S8): 89–93.
69. Qiu, Y., S. Thomas, D. Fabijanic, A. J. Barlow, H. L. Fraser, and N. Birbilis. 2019. Microstructural Evolution, Electrochemical and Corrosion Properties of $Al_xCoCrFeNiTi_y$ High Entropy Alloys. *Materials and Design* 170.107698.
70. Qiu, X. W. 2013a. Microstructure and Properties of AlCrFeNiCoCu High Entropy Alloy Prepared by Powder Metallurgy. *Journal of Alloys and Compounds* 555: 246–49.
71. Brigham, R. J. 1972. Pitting of Molybdenum Bearing Austenitic Stainless Steel. *Corrosion* 28(5): 177–79.
72. Sedriks, A. J. 1986. Effects of Alloy Composition and Microstructure on the Passivity of Stainless Steels. *Corrosion* 42(7): 376–89.
73. Ameer, M. A., A. M. Fekry, and F. E. T. Heakal. 2004. Electrochemical Behaviour of Passive Films on Molybdenum-Containing Austenitic Stainless Steels in Aqueous Solutions. *Electrochimica Acta* 50(1): 43–9.
74. Chou, Y. L., J. W. Yeh, and H. C. Shih. 2010a. The Effect of Molybdenum on the Corrosion Behaviour of the High-Entropy Alloys $Co_{1.5}CrFeNi_{1.5}Ti_{0.5}Mo_x$ in Aqueous Environments. *Corrosion Science* 52(8): 2571–81.
75. Chou, Y. L., Y. C. Wang, J. W. Yeh, and H. C. Shih. 2010b. Pitting Corrosion of the High-Entropy Alloy $Co_{1.5}CrFeNi_{1.5}Ti_{0.5}Mo_{0.1}$ in Chloride-Containing Sulphate Solutions. *Corrosion Science* 52(10): 3481–91.
76. Chou, Y. L., J. W. Yeh, and H. C. Shih. 2011. Effect of Inhibitors on the Critical Pitting Temperature of the High-Entropy Alloy $Co_{1.5}CrFeNi_{1.5}Ti_{0.5}Mo_{0.1}$. *Journal of the Electrochemical Society* 158(8): C246.

77. Zheng, Z. Y., X. C. Li, C. Zhang, and J. C. Li. 2014. Microstructure and Corrosion Behaviour of $FeCoNiCuSn_x$ High Entropy Alloys. *Materials Science and Technology* 31(10): 1148–52.
78. Han, L. X., C. M. Wang, and H. F. Sun. 2016. Microstructure and Anticorrosion Property of High-Entropy Alloy $AlFeNiCrCoTi_{0.5}V_x$. *Materials Transactions* 57(7): 1134–37.
79. Kao, Y. F., T. D. Lee, S. K. Chen, and Y. S. Chang. 2010b. Electrochemical Passive Properties of $Al_xCoCrFeNi$ (x = 0, 0.25, 0.50, 1.00) Alloys in Sulfuric Acids. *Corrosion Science* 52(3): 1026–34.
80. Ren, B., R. F. Zhao, Z. X. Liu, S. K. Guan, and H. S. Zhang. 2014. Microstructure and Properties of $Al_{0.3}CrFe_{1.5}MnNi_{0.5}Ti_x$ and $Al_{0.3}CrFe_{1.5}MnNi_{0.5}Si_x$ High-Entropy Alloys. *Rare Metals* 33(2): 149–54.
81. Szklarz, Z., J. Lekki, P. Bobrowski, M. B. Szklarz, and Ł. Rogal. 2018. The Effect of SiC Nanoparticles Addition on the Electrochemical Response of Mechanically Alloyed CoCrFeMnNi High Entropy Alloy. *Materials Chemistry and Physics* 215: 385–92.
82. Nishimoto, Akio, T. Fukube, and T. Maruyama. 2018. Microstructural, Mechanical, and Corrosion Properties of Plasma-Nitrided CoCrFeMnNi High-Entropy Alloys. *Surface and Coatings Technology*. In press.
83. Zhang, Z., E. Axinte, W. Ge, C. Shang, and Y. Wang. 2016b. Microstructure, Mechanical Properties and Corrosion Resistance of CuZrY/Al, Ti, Hf Series High-Entropy Alloys. *Materials and Design* 108: 106–13.
84. Cui, H., L. Zheng, and J. Wang. 2011. Microstructural Evolution and Corrosion Behavior of Directionally Solidified FeCoNiCrAl High Entropy Alloy. *Applied Mechanics and Materials* 66–68: 146–49.
85. Lin, C. M., H. L. Tsai, and H. Y. Bor. 2010. Effect of Aging Treatment on Microstructure and Properties of High-Entropy $Cu_{0.5}CoCrFeNi$ Alloy. *Intermetallics* 18(6): 1244–50.
86. Lin, C. M., and H. L. Tsai. 2011. Evolution of Microstructure, Hardness, and Corrosion Properties of High-Entropy $Al_{0.5}CoCrFeNi$ Alloy. *Intermetallics* 19(3): 288–94.
87. Tsao, L. C., C. S. Chen, K. H. Fan, and Y. T. Huang. 2013. Effect of the Annealing Treatment on the Microstructure, Microhardness and Corrosion Behaviour of $Al_{0.3}CrFe_{1.5}MnNi_{0.5}$ High-Entropy Alloys. *Advanced Materials Research* 748: 79–85.
88. Soare, V., D. Mitrica, I. Constantin, G. Popescu, I. Csaki, M. Tarcolea, and I. Carcea. 2015b. The Mechanical and Corrosion Behaviors of As-Cast and Re-Melted AlCrCuFeMnNi Multi-Component High-Entropy Alloy. *Metallurgical and Materials Transactions. Part A: Physical Metallurgy and Materials Science* 46(4): 1468–73.
89. Soare, V., D. Mitrica, I. Constantin, V. Badilita, F. Stoiciu, A. M. J. Popescu, and I. Carcea. 2015a. Influence of Remelting on Microstructure, Hardness and Corrosion Behaviour of AlCoCrFeNiTi High Entropy Alloy. *Materials Science and Technology* 31(10): 1194–200.
90. Wang, H. L., T. X. Gao, J. Z. Niu, P. J. Shi, J. Xu, and Y. Wang. 2016. Microstructure, Thermal Properties, and Corrosion Behaviors of FeSiBAlNi Alloy Fabricated by Mechanical Alloying and Spark Plasma Sintering. *International Journal of Minerals, Metallurgy and Materials* 23(1): 77–82.
91. Zhang, C., G. J. Chen, and P. Q. Dai. 2016c. Evolution of the Microstructure and Properties of Laser-Clad $FeCrNiCoB_x$ High-Entropy Alloy Coatings. *Materials Science and Technology* 32(16): 1666–72.
92. Qiu, X. W., Y. P. Zhang, L. He, and C. G. Liu. 2013c. Microstructure and Corrosion Resistance of AlCrFeCuCo High Entropy Alloy. *Journal of Alloys and Compounds* 549: 195–99.
93. Li, Q. H., T. M. Yue, Z. N. Guo, and X. Lin. 2013b. Microstructure and Corrosion Properties of AlCoCrFeNi High Entropy Alloy Coatings Deposited on AISI 1045 Steel by the Electrospark Process. *Metallurgical and Materials Transactions A* 44A: 1767–78.

94. Qiu, X. W., and C. G. Liu. 2013b. Microstructure and Properties of $Al_2CrFeCoCuTiNi_x$ High-Entropy Alloys Prepared by Laser Cladding. *Journal of Alloys and Compounds* 553: 216–20.
95. Wu, C. L., S. Zhang, C. H. Zhang, H. Zhang, and S. Y. Dong. 2017b. Phase Evolution and Properties in Laser Surface Alloying of $FeCoCrAlCuNi_x$ High-Entropy Alloy on Copper Substrate. *Surface and Coatings Technology* 315: 368–76.
96. Shang, C., E. Axinte, J. Sun, X. Li, P. Li, J. Du, P. Qiao, and Y. Wang. 2017a. $CoCrFeNi(W_{1-x}Mo_x)$ High-Entropy Alloy Coatings with Excellent Mechanical Properties and Corrosion Resistance Prepared by Mechanical Alloying and Hot Pressing Sintering. *Materials and Design* 117: 193–202.
97. Dou, D., X. C. Li, Z. Y. Zheng, and J. C. Li. 2016. Coatings of FeAlCoCuNiV High Entropy Alloy. *Surface Engineering* 32(10): 766–70.
98. Wu, C. L., S. Zhang, C. H. Zhang, H. Zhang, and S. Y. Dong. 2017a. Phase Evolution and Cavitation Erosion-Corrosion Behavior of $FeCoCrAlNiTi_x$ High Entropy Alloy Coatings on 304 Stainless Steel by Laser Surface Alloying. *Journal of Alloys and Compounds* 698: 761–70.
99. Nair, R. B., H. S. Arora, A. Ayyagari, S. Mukherjee, and H. S. Grewal. 2018. High Entropy Alloys: Prospective Materials for Tribo-Corrosion Applications. *Advanced Engineering Materials* 20(6): 1–9.
100. Gao, L., W. Liao, H. Zhang, J. U. Surjadi, D. Sun, and Y. Lu. 2017. Microstructure, Mechanical and Corrosion Behaviors of $CoCrFeNiAl_{0.3}$ High Entropy Alloy (HEA) Films. *Coatings* 7(10): 156.
101. Argade, G. R., S. S. Joshi, A. V. Ayyagari, S. Mukherjee, R. S. Mishra, and N. B. Dahotre. 2019. Tribocorrosion Performance of Laser Additively Processed High-Entropy Alloy Coatings on Aluminum. *Applied Physics. Part A: Materials Science and Processing* 125(4): 1–9.
102. Shang, C. Y., and Y. Wang. 2017c. AlCrFeNi High-Entropy Coating Fabricated by Mechanical Alloying and Hot Pressing Sintering. *Materials Science Forum* 898: 628–37.
103. Zhu, S., W. B. Du, X. M. Wang, and G. F. Han. 2013a. High Mixing Entropy Alloys Design with High Anticorrosion and Wear-Resistance Properties. *Advanced in Materials Research* 815: 19–24.
104. Zhang, S., C. L. Wu, C. H. Zhang, M. Guan, and J. Z. Tan. 2016a. Laser Surface Alloying of FeCoCrAlNi High-Entropy Alloy on 304 Stainless Steel to Enhance Corrosion and Cavitation Erosion Resistance. *Optics and Laser Technology* 84: 23–31.
105. Shon, Y., S. S. Joshi, S. Katakam, R. S. Rajamure, and N. B. Dahotre. 2015. Laser Additive Synthesis of High Entropy Alloy Coating on Aluminum: Corrosion Behavior. *Materials Letters* 142: 122–25.
106. Wang, R., K. Zhang, C. Davies, and X. Wu. 2017b. Evolution of Microstructure, Mechanical and Corrosion Properties of AlCoCrFeNi High-Entropy Alloy Prepared by Direct Laser Fabrication. *Journal of Alloys and Compounds* 694: 971–81.
107. Raphel, A., S. Kumaran, K. V. Kumar, and L. Varghese. 2017. Oxidation and Corrosion Resistance of AlCoCrFeTi High Entropy Alloy. *Materials Today* 4(2): 195–202.
108. Li, X., Z. Zheng, D. Dou, and J. Li. 2016c. Microstructure and Properties of Coating of FeAlCuCrCoMn High Entropy Alloy Deposited by Direct Current Magnetron Sputtering. *Materials Research* 19(4): 802–6.
109. Huang, K. J., X. Lin, Y. Y. Wang, C. S. Xie, and T. M. Yue. 2014. Microstructure and Corrosion Resistance of $Cu_{0.9}NiAlCoCrFe$ High Entropy Alloy Coating on AZ91D Magnesium Alloys by Laser Cladding. *Materials Research Innovations* S2: 1008–11.
110. Yue, T. M., and H. Zhang. 2014. Laser Cladding of $FeCoNiCrAlCu_xSi_{0.5}$ High Entropy Alloys on AZ31 Mg Alloy Substrates. *Materials Research Innovations* 18: S2-624-628.

111. Ge, W., B. Wu, S. Wang, S. Xu, C. Shang, Z. Zhang, and Y. Wang. 2017a. Characterization and Properties of CuZrAlTiNi High Entropy Alloy Coating Obtained by Mechanical Alloying and Vacuum Hot Pressing Sintering. *Advanced Powder Technology* 28(10): 2556–63.
112. Ge, W., Y. Wang, C. Shang, Z. Zhang, and Y. Wang. 2017b. Microstructures and Properties of Equiatomic CuZr and CuZrAlTiNi Bulk Alloys Fabricated by Mechanical Alloying and Spark Plasma Sintering. *Journal of Materials Science* 52(10): 5726–37.
113. Gan, Z. H., L. M. Xu, Z. H. Lu, H. H. Zhou, C. H. Song, and F. Huang. 2013. A Novel High-Entropy Alloy AlMgZnSnPbCuMnNi with Low Free Corrosion Potential. *Applied Mechanics and Materials* 327: 103–7.
114. Shang, C., E. Axinte, W. Ge, Z. Zhang, and Y. Wang. 2017b. High-Entropy Alloy Coatings with Excellent Mechanical, Corrosion Resistance and Magnetic Properties Prepared by Mechanical Alloying and Hot Pressing Sintering. *Surfaces and Interfaces* 9: 36–43.
115. Cai, Z. B., X. J. Pang, X. F. Cui, X. Wen, Z. Liu, M. L. Dong, Y. Li, and G. Jin. 2017. *In Situ* Laser Synthesis of High Entropy Alloy Coating on Ti-6Al-4V Alloy: Characterization of Microstructure and Properties. *Materials Science Forum* 898: 643–50.
116. Ye, Q., K. Feng, Z. Li, F. Lu, R. Li, J. Huang, and Y. Wu. 2017. Microstructure and Corrosion Properties of CrMnFeCoNi High Entropy Alloy Coating. *Applied Surface Science* 396: 1420–26.
117. Wang, B., J. Huang, J. Fan, Y. Dou, H. Zhu, and D. Wang. 2017a. Preparation of FeCoNiCrMn High Entropy Alloy by Electrochemical Reduction of Solid Oxides in Molten Salt and Its Corrosion Behavior in Aqueous Solution. *Journal of the Electrochemical Society* 164(14): E575–79.
118. Tian, Y., C. Lu, Y. Shen, and X. Feng. 2019. Microstructure and Corrosion Property of CrMnFeCoNi High Entropy Alloy Coating on Q235 Substrate via Mechanical Alloying Method. *Surfaces and Interfaces* 15: 135–40.
119. Zhang, J. J., X. L. Yin, Y. Dong, Y. P. Lu, L. Jiang, T. M. Wang, and T. J. Li. 2014. Corrosion Properties of $Al_xCoCrFeNiTi_{0.5}$ High Entropy Alloys in 0·5M H_2SO_4 Aqueous Solution. *Materials Research Innovations* 18: S4-756-760.
120. Ren, B., Z. X. Liu, D. M. Li, L. Shi, B. Cai, and M. X. Wang. 2012. Corrosion Behavior of CuCrFeNiMn High Entropy Alloy System in 1 M Sulfuric Acid Solution. *Materials and Corrosion* 63(9): 828–34.
121. Lee, C. P., Y. Y. Chen, C. H. Wu, C. Y. Hsu, J. W. Yeh, and H. C. Shih. 2007a. Effect of Boron on the Corrosion Properties of $Al_{0.5}CoCrCuFeNiB_x$ High Entropy Alloys in 1N Sulphuric Acid. *ECS Transactions* 2(26): 15–33.
122. Lee, C. P., Y. Y. Chen, C. Y. Hsu, J. W. Yeh, and H. C. Shih. 2007b. The Effect of Boron on the Corrosion Resistance of the High Entropy Alloys $Al_{0.5}CoCrCuFeNiB_x$. *Journal of the Electrochemical Society* 154(8): C424.
123. Liu, D., J. B. Cheng, and H. Ling. 2015. Electrochemical Behaviours of $(NiCoFeCrCu)_{95}B_5$ High Entropy Alloy Coatings. *Materials Science and Technology* 31(10): 1159–64.
124. Lee, C. P., C. C. Chang, Y. Y. Chen, J. W. Yeh, and H. C. Shih. 2008b. Effect of the Aluminium Content of $Al_xCrFe_{1.5}MnNi_{0.5}$. High-Entropy Alloys on the Corrosion Behaviour in Aqueous Environments. *Corrosion Science* 50(7): 2053–60.
125. Tsau, C. H., and P. Y. Lee. 2016. Microstructures of $Al_{7.5}Cr_{22.5}Fe_{35}Mn_{20}Ni_{15}$ High-Entropy Alloy and Its Polarization Behaviors in Sulfuric Acid, Nitric Acid and Hydrochloric Acid Solutions. *Entropy* 18(8): 288.
126. Florea, I., G. Buluc, R. M. Florea, V. Soare, and I. Carcea. 2015. Study on Corrosion Resistance of High - Entropy Alloy in Medium Acid Liquid and Chemical Properties. *IOP Conference Series: Materials Science and Engineering* 95(1).

127. Xiao, D. H., P. F. Zhou, W. Q. Wu, H. Y. Diao, M. C. Gao, M. Song, and P. K. Liaw. 2017. Microstructure, Mechanical and Corrosion Behaviors of AlCoCuFeNi-(Cr,Ti) High Entropy Alloys. *Materials and Design* 116: 438–47.
128. Lin, C. W., M. H. Tsai, C. W. Tsai, J. W. Yeh, and S. K. Chen. 2015. Microstructure and Aging Behaviour of $Al_5Cr_{32}Fe_{35}Ni_{22}Ti_6$ High Entropy Alloy. *Materials Science and Technology* 31(10): 1165–70.
129. Qiu, X. W., Y. P. Zhang, and C. G. Liu. 2014. Effect of Ti Content on Structure and Properties of $Al_2CrFeNiCoCuTi_x$ High-Entropy Alloy Coatings. *Journal of Alloys and Compounds* 585: 282–86.
130. Guo, Y. X., X. J. Shang, and Qibin Liu. 2018. Microstructure and Properties of in-Situ TiN Reinforced Laser Cladding $CoCr_2FeNiTi_x$ High-Entropy Alloy Composite Coatings. *Surface and Coatings Technology*: 353–58.
131. Guo, Y. X., Q. B. Liu, and X. J. Shang. 2019. *In Situ* TiN-Reinforced $CoCr_2FeNiTi_{0.5}$ High-Entropy Alloy Composite Coating Fabricated by Laser Cladding. *Rare Metals: Nonferrous Metals Society of China*: 1–5.
132. Cheng, J. B., D. Liu, X. B. Liang, and B. Xu. 2014a. Microstructure and Electrochemical Properties of CoCrCuFeNiNb High-Entropy Alloys Coatings. *Acta Metallurgica Sinica* 27(6): 1031–37.
133. Jayaraj, J., C. Thinaharan, S. Ningshen, C. Mallika, and U. K. Mudali. 2017. Corrosion Behavior and Surface Film Characterization of TaNbHfZrTi High Entropy Alloy in Aggressive Nitric Acid Medium. *Intermetallics* 89: 123–32.
134. Hsueh, H. T., W. J. Shen, M. H. Tsai, and J. W. Yeh. 2012. Effect of Nitrogen Content and Substrate Bias on Mechanical and Corrosion Properties of High-Entropy Films $(AlCrSiTiZr)_{100-x}N_x$. *Surface and Coatings Technology* 206(19–20): 4106–12.
135. Malinovskis, P., S. Fritze, L. Riekehr, L. V. Fieandt, J. Cedervall, D. Rehnlund, L. Nyholm, E. Lewin, and U. Jansson. 2018. Synthesis and Characterization of Multicomponent (CrNbTaTiW)C Films for Increased Hardness and Corrosion Resistance. *Materials and Design* 149: 51–62.
136. Ye, X., M. Ma, Y. Cao, W. Liu, X. Ye, and Y. Gu. 2011. The Property Research on High-Entropy Alloy $Al_xFeCoNiCuCr$ Coating by Laser Cladding. *Physics Procedia* 12: 303–12.
137. Cheng, J. B., X. B. Liang, Z. H. Wang, and B. S. Xu. 2013. Formation and Mechanical Properties of CoNiCuFeCr High-Entropy Alloys Coatings Prepared by Plasma Transferred Arc Cladding Process. *Plasma Chemistry and Plasma Processing* 33(5): 979–92.
138. Cheng, J. B., X. B. Liang, and B. S. Xu. 2014b. Effect of Nb Addition on the Structure and Mechanical Behaviors of CoCrCuFeNi High-Entropy Alloy Coatings. *Surface and Coatings Technology* 240: 184–90.
139. Ding, J., A. Inoue, Y. Han, F. L. Kong, S. L. Zhu, Z. Wang, E. Shalaan, and F. A. Marzouki. 2017b. High Entropy Effect on Structure and Properties of (Fe,Co,Ni,Cr)-B Amorphous Alloys. *Journal of Alloys and Compounds* 696: 345–52.
140. Tan, X. R., Q. Zhi, R. B. Yang, F. Z. Wang, J. R. Yang, and Z. X. Liu. 2017. Effects of Milling on the Corrosion Behavior of $Al_2NbTi_3V_2Zr$ High-Entropy Alloy System in 10% Nitric Acid Solution. *Materials and Corrosion* 68(10): 1080–89.
141. Fitriani, R. Ovik, B. D. Long, M. C. Barma, M. Riaz, M. F. M. Sabri, S. M. Said, and R. Saidur. 2016. A Review on Nanostructures of High-Temperature Thermoelectric Materials for Waste Heat Recovery. *Renewable and Sustainable Energy Reviews* 64: 635–59.
142. Xie, W., A. Weidenkaff, X. Tang, Q. Zhang, J. Poon, and T. Tritt. 2012. Recent Advances in Nanostructured Thermoelectric Half-Heusler Compounds. *Nanomaterials* 2(4): 379–412.

143. Shafeie, S., S. Guo, Q. Hu, H. Fahlquist, P. Erhart, and A. Palmqvist. 2015. High-Entropy Alloys as High-Temperature Thermoelectric Materials. *Journal of Applied Physics* 118(18).
144. Yan, J., F. Liu, G. Ma, B. Gong, J. Zhu, X. Wang, W. Ao, C. Zhang, Y. Li, and J. Li. 2018. Suppression of the Lattice Thermal Conductivity in NbFeSb-Based Half-Heusler Thermoelectric Materials through High Entropy Effects. *Scripta Materialia* 157: 129–34.
145. Karati, A., M. Nagini, S. Ghosh, R. Shabadi, K. G. Pradeep, R. C. Mallik, B. S. Murty, and U. V. Varadaraju. 2019. $Ti_2NiCoSnSb$—A New Half-Heusler Type High-Entropy Alloy Showing Simultaneous Increase in Seebeck Coefficient and Electrical Conductivity for Thermoelectric Applications. *Scientific Reports* 9(1): 1–12.
146. Fan, Z., H.Wang, Y. Wu, X. Liu, and Z. Lu. 2017. Thermoelectric Performance of PbSnTeSe High-Entropy Alloys. *Materials Research Letters* 5(3): 187–94.
147. Fan, Z., H. Wang, Y. Wu, X. J. Liu, and Z. P. Lu. 2016. Thermoelectric High-Entropy Alloys with Low Lattice Thermal Conductivity. *RSC Advances* 6(57): 52164–70.
148. Dong, W., Z. Zhou, L. Zhang, M. Zhang, P. K. Liaw, G. Li, and R. Liu. 2018. Effects of Y, GdCu, and Al Addition on the Thermoelectric Behavior of CoCrFeNi High Entropy Alloys. *Metals* 8(10): 781.
149. Hu, L., Y. Zhang, H. Wu, J. Li, Y. Li, M. Mckenna, J. He, F. Liu, S. J. Pennycock, and X. Zeng. 2018. Entropy Engineering of SnTe: Multi-Principal-Element Alloying Leading to Ultralow Lattice Thermal Conductivity and State-Of-the-Art Thermoelectric Performance. *Advanced Energy Materials* 8(29): 1–14.
150. Riva, S., S. Mehraban, N. P. Lavery, S. Schwarzmuller, O. Oeckler, S. G. R. Brown, and K. Yusenko. 2018. The Effect of Scandium Ternary Intergrain Precipitates in Al-Containing High-Entropy Alloys. *Entropy* 20(7): 1–16.
151. Zhang, R. Z., F. Gucci, H. Zhu, K. Chen, and M. J. Reece. 2018. Data-Driven Design of Ecofriendly Thermoelectric High Entropy Sulphides. *Inorganic Chemistry* 57(20): 13027–33.
152. Zhao, S. Y., R. Chen, J. Q. Li, L. Yang, C. H. Zhang, Y. Li, F. S. Liu, and W. Q. Ao. 2019. Synergistic Effects on Thermoelectric Properties of $Sn_{0.5}Ge_{0.4875}Te$ with Pb Alloying. *Journal of Alloys and Compounds* 777: 1334–39.
153. Kleiner, R., and W. Buckel. 2016. *Superconductivity an Introduction*. 3rd edition. Weinheim, Germany: Wiley-VCH Verlag GmbH & Co.
154. Onnes, H. K. 1911. Further Experiments with Liquid Helium. D. On the Change of the Electrical Resistance of Pure Metals at Very Low Temperatures, Etc. V. The Disappearance of the Resistance of Mercury. *Comm. Phys. Lab. Univ. Leiden* 124c.
155. Bednorz, J. G., and K. A. Muller. 1986. Possible High Tc Superconductivity in the Ba-La-Cu-O System. *Zeitschrift für Physik B: Condensed Matter* 64(2): 189–93.
156. Katano, S., S. Funahashi, T. Hatano, A. Matsushita, K. Nakamura, T. Matsumoto, and K. Ogawa. 1987. Structure of High-Tc Superconductor $YBa_2Cu_3O_{6.6}$ at Low Temperatures. *Japanese Journal of Applied Physics* 26(6): 1046–48.
157. Drozdov, A. P., M. I. Eremets, I. A. Troyan, V. Ksenofontov, and S. I. Shylin. 2015. Conventional Superconductivity at 203 Kelvin at High Pressures in the Sulfur Hydride System. *Nature* 525(7567): 73–6.
158. Rossi, L. 2010. Superconductivity: Its Role, Its Success and Its Setbacks in the Large Hadron Collider of CERN. *Superconductor Science and Technology* 23(3).
159. Murty, B. S., J. W. Yeh, S. Ranganathan, and P. P. Bhattacharjee. 2019. *High-Entropy Alloys*. Elsevier.
160. Vrtnik, S., P. Koželj, A. Meden, S. Maiti, W. Steurer, M. Feuerbacher, and J. Dolinšeka. 2017. Superconductivity in Thermally Annealed Ta-Nb-Hf-Zr-Ti High-Entropy Alloys. *Journal of Alloys and Compounds* 695: 3530–40.

161. Yuan, Y., Y. Wu, H. Luo, Z. Wang, X. Liang, Z. Yang, H. Wang, X. Liu, and Z. Lu. 2018. Superconducting $Ti_{15}Zr_{15}Nb_{35}Ta_{35}$ High-Entropy Alloy with Intermediate Electron-Phonon Coupling. *Frontiers in Materials* 5: 1–6.
162. Stolze, K., F. A. Cevallos, T. Kong, and R. J. Cava. 2018a. High-Entropy Alloy Superconductors on an α-Mn Lattice. *Journal of Materials Chemistry C* 6(39): 10441–49.
163. Marik, S., M. Varghese, K. P. Sajilesh, D. Singh, and R. P. Singh. 2018. Superconductivity in Equimolar Nb-Re-Hf-Zr-Ti High Entropy Alloy. *Journal of Alloys and Compounds* 769: 1059–63.
164. Koželj, P., S. Vrtnik, A. Jelen, S. Jazbec, Z. Jaglicic, S. Maiti, M. Feuerbacher, W. Steurer, and J. Dolinšek. 2014. Discovery of a Superconducting High-Entropy Alloy. *Physical Review Letters* 113(10): 1–5.
165. Rohr, F. O. V., M. J. Winiarski, J. Tao, T. Klimczuk, and R. J. Cava. 2016. Effect of Electron Count and Chemical Complexity in the Ta-Nb-Hf-Zr-Ti High-Entropy Alloy Superconductor. *Proceedings of the National Academy of Sciences* 113(46): 7144–50.
166. Ishizu, N., and J. Kitagawa. 2019. New High-Entropy Alloy Superconductor $Hf_{21}Nb_{25}Ti_{15}V_{15}Zr_{24}$. *Results in Physics* 13. 102275.
167. Stolze, K., J. Tao, F. O. V. Rohr, T. Kong, and R. J. Cava. 2018b. Sc-Zr-Nb-Rh-Pd and Sc-Zr-Nb-Ta-Rh-Pd High-Entropy Alloy Superconductors on a Cscl-Type Lattice. *Chemistry of Materials* 30(3): 906–14.
168. Rohr, F. O. V., and R. J. Cava. 2018. Isoelectronic Substitutions and Aluminium Alloying in the Ta-Nb-Hf-Zr-Ti High-Entropy Alloy Superconductor. *Physical Review Materials* 2(3): 1–7.
169. Sogabe, R., Y. Goto, and Y. Mizuguchi. 2018. Superconductivity in REO0.5F0.5BiS2 with High-Entropy-Alloy-Type Blocking Layers. *Applied Physics Express* 11(5). 053102.
170. Yao, Y., and J. S. Tse. 2018. Superconducting Hydrogen Sulfide. *Chemistry - a European Journal* 24(8): 1769–78.
171. Lee, C. P. 2008. Enhancing Pitting Corrosion Resistance of $Al_xCrFe_{1.5}MnNi_{0.5}$ High-Entropy Alloys by Anodic Treatment in Sulfuric Acid. *Thin Solid Films* 517(3): 1301–5.
172. Zhang, J., Y. Zhu, L. Yao, C. Xu, Y. Liu, and L. Li. 2019a. State of the Art Multi-Strategy Improvement of Mg-Based Hydrides for Hydrogen Storage. *Journal of Alloys and Compounds* 782: 796–823.

15 High Entropy Alloys

Challenges in Commercialization and the Road Ahead

P. Neelima, S.V.S. Narayana Murty, P. Chakravarthy, and T.S. Srivatsan

CONTENTS

15.1 SUMMARY

The ultimate goal of any alloy development effort is to put the newly developed alloys to use in one or more practical applications. Every alloy designer's dream is to see the proposed new material gets selected for use in applications spanning the domains of both performance-critical and non-performance-critical. To achieve this objective, the newly developed alloys must demonstrate an overall superiority over the existing alloys for selection and use in applications in which they are currently being used and under identical conditions. Replacement of an existing alloy with a high entropy alloy (HEA) does necessitate the need for a detailed understanding of the system in which the material has to perform, namely: loading conditions, environmental conditions to include temperature range, fluids the material will come in contact with in the case of materials used in structures, and a host of related and relevant physical-chemical parameters that the currently used conventional counterpart offers. A few of the other considerations include the pros and cons of replacement, overall ease of availability of the proposed materials, repeatability of quality and reproducibility of properties, an extensive database on properties of these materials, aspects specific to the ease of processability of the alloys at a commercial level, and finally, cost of the material coupled with the lead time essential for procurement.

The first and foremost problem for commercialization of high entropy alloys (HEAs) comes from a scaling-up of alloy quantity from laboratory scale tablet-sized specimens to large-sized ingots melted in an industrial environment setting. It is easy to melt in an arc-melter any number of alloying elements any number of times for the purpose of obtaining homogeneity and report the property data without much emphasis on the losses incurred during alloy preparation. The results are reported as the composition of the engineered alloy based on targeted chemistry rather than the actual chemistry. Also, mechanical properties of small test specimens deformed in compression are usually reported (since a tensile test specimen takes a larger quantity of material, even for a non-standard test specimen) and the database generated will be of limited use to facilitate further expansion of the system. The problem does get complicated for the case of HEAs processed using the technique of powder metallurgical processing. This is essentially because impurities play a key role in contaminating the alloy with irreproducible results coupled with the inherent limitations of the powder metallurgy (PM) processing technique. Researchers should make the proposed alloys look attractive for purpose of commercial development. This essentially necessitates the researchers studying the range of composition for which a guaranteed minimum physical-chemical-mechanical properties combination can be obtained for comparison with the conventional alloys. The selected alloy systems for which some breakthrough properties have been reported can be taken as systems which can be seriously pursued for additional research with the intention of generating an extensive test database, under standard test conditions and/or procedures, on large-scale ingots of a single (heat) melt composition. The problem further gets complicated once these alloys are fusion welded since the microstructural modification that is imparted to the as-cast microstructure through thermomechanical processing gets nullified and reverts back to the as-cast structure subsequent to welding. Therefore, a primary challenge for alloy designers lies in obtaining a balanced combination of strength and toughness under a stable microstructural condition for the chosen high entropy alloys processed by ingot metallurgy to get large-sized ingots, on a repeatable basis, in a manufacturing environment.

A mere improvement in any one of the properties is no valid reason, even though strong claims have been made by several researchers in their scientific publications, for a gradual replacement of the conventional alloys with the HEAs. The road ahead for HEAs is challenging and the progress that can be made depends on how researchers leave the visible path and find alternative pathways to enable them to get to their logical conclusion for the purpose of selection and use of these alloys for critical functional applications while the momentum continues in terms of scientific interest and funding opportunities.

The general purpose of this chapter is to focus on the issue of the road ahead for these exotic alloys. Its main purpose is to make the "concerned" researchers aware of how alloy development efforts progress from laboratory-scale development to an industrial-scale output and the few to many challenges that can arise during processing. The problem of scaling-up is noticeably more complex than for the regular conventional alloy systems. This is because HEAs are line compositions with no fixed composition ranges. The overall sensitivity of the composition ranges obtained based on both physical and mechanical properties is an area that needs to be examined. In order to get an idea about the conventional alloys and to concurrently focus on specific HEA development for a given application, commercial alloys and their properties are presented in the tables for selected grades of stainless steels, maraging steels, aluminum alloys, titanium alloys, superalloys, and refractory materials along with some typical applications of these materials, such as bearings and fasteners. The tables include materials from low strength to medium strength steels having excellent ductility, maraging steels with excellent strength–toughness combinations, bearing steels with high compressive strength and excellent wear resistance, superalloys with good corrosion resistance and high-temperature strength, and commercial refractory alloys with adequate high-temperature resistance, which can be taken as reference data for purpose of comparison. The data published in the "open" literature on HEA systems (Al-Co-Cr-Fe-Mn-Ni, Al-Co-Cr-Cu-Fe-Mn-Ni, and the refractory high entropy alloys) has been tabulated and presented for the purpose of comparison with the conventional alloys. It is hoped that the limited data neatly compiled and presented in this chapter will be of use to designers that opt to select and put to use the high entropy alloys (HEAs) to solve the challenges ahead.

15.2 INTRODUCTION AND EARLY STUDIES

Through the years, alloy design, alloy development, and alloy selection efforts have been confined to one-element, one compound, and with continued research, to the development and use of the two-element, three-element, and even the four-element concept. Alloy development efforts even up to three elements did impose a limit on composition and the concomitant development of microstructure and properties. All of this in synergism with numerous other research studies did succeed in promoting and consolidating the area of materials science and engineering, which up until now has largely been confined to the domain of pure metals and their alloy counterparts, and subsequently to the development and emergence of the compositionally complex alloys (CCAs). These alloys have attracted increased attention and resultant action by way of highly focused and/or directed research and development efforts due essentially to their high entropy. This led this alloy family to be categorized as high entropy alloys (HEAs). The high entropy alloys (HEAs) normally consist of five or more number of elements in the range of 5–35 atomic percentage. The key driving force for studying these alloys is to stabilize the strong and ductile solid-solution phases over the brittle intermetallic phases by increasing the configurational entropy [$\Delta S_{conf.}$]. These alloys have been shown to provide the much-needed impetus for a gradual search into new and improved alloys that offer a combination of exceptional properties to essentially include both physical properties and mechanical properties. Commensurate with observable technological advances and needs, the arduous journey in search of the new and improved alloys will no longer be through the well laid-out, age-old highways that fall both within and along the boundaries of phase diagrams. This has necessitated the need for unknown paths to be traversed prior to identifying the fruit-bearing tree, or trees, in the vast realm of the multi-alloy kingdom. The search has enabled identifying a few to many exotic compositions that offered an attractive combination of physical properties, mechanical properties, and functional properties that would enable in their selection for use in specific applications. It is always the dream of a physical metallurgist to be involved in such an exotic journey. The absence of a well-defined map to precisely locate the intricacies governing selection and use of a HEA often makes the journey both challenging and thrilling.

In an endeavor to identify, index, and involve the unknown in the vast unmapped territory spanning the family of materials, researchers have attempted to use the existing knowledge to locate one or more of the hidden treasures. Some have taken the traditional approach of conducting extensive experimentation on both existing alloys and the emerging alloys, be it one or more. In either of the cases, both resources of a significant amount and precious time must be allotted for accomplishing the needful tasks. The end result may or may not be striking or useful, but the knowledge gained will certainly be unparalleled and importantly beneficial.

A growing recognition of this need and the much-valued tradition behind it, a brand-new alloy concept termed HEAs emerged with the primary intent of breaking up the confinements imposed by the family of classical alloys or the traditional engineering alloys. This thought was put forth and soon initiated into exploration as early as 1995 [1–5]. Sustained progress, at an incremental level, was being made in few selected nations scattered through the globe. The preponderance of the research and development efforts was mostly being undertaken at educational institutions, national research laboratories, and even selected industries. These studies did provide a significant breakthrough and interest in the development and emergence of several new and improved alloys belonging to the high entropy family of alloys [2, 5–14]. Through the years following their initial development and emergence coupled with the potential for continued study, since as early as 1995 [2], the HEAs have been the subject of several key research studies with the primary objective of synergizing an alloy that can offer a good combination of properties, to essentially include both physical and mechanical. The mechanical properties of interest to the engineer were the following: (i) high hardness [2], (ii) capacity to work harden or strain harden during deformation [14, 2], (iii) good wear resistance [14], (iv) strength at elevated temperatures through precipitation-hardening, (v) good anti-oxidation properties, and (vi) an overall good resistance to corrosion [14–16]. The newly developed and put forth HEAs are fundamentally different from the vast majority of the traditionally developed and used alloys. The key difference is that the HEAs have a significantly higher amount of mixing energy, which is conducive for enhancing overall stability of the solid solution when compared to the intermetallic compounds. This is especially true at the higher temperatures. It is by now well documented in the published literature that the newly developed and emerging HEAs must contain at least five principal elements, with each element having an atomic percentage between 5 pct. and 35 pct. [4, 14, 17]. During the last two decades, i.e., essentially since 1995, several independent research studies have clearly shown that the HEAs possess a relatively simplified structure while concurrently offering a combination of acceptable to good properties, to include both physical and mechanical [17–29]. In a few of these independent research studies, the alloys that were engineered, developed, and put forth, and subsequently both examined and studied from a scientific perspective consisted mainly of solid-solution phases in combination with a small number of intermetallic compounds [18].

An observable combination of good to exceptional properties of a few of the developed alloys has provided the much-desired objective of enabling them to be considered for use in a spectrum of industrial applications. One of the alloys that was engineered and thoroughly studied, i.e., CoCrFeMnNi, was found to be appropriate for use in cryogenic applications, such as liquefied gas storage tanks. This is primarily because this alloy offered the capability of retaining the desired mechanical properties at temperatures as low as −196°C [19]. As research efforts gradually progressed through the months and years, even coatings of an HEA were found to be suitable and appropriate for use in: (i) food preparation dishes, (ii) cooking ware, and (iii) utensils [20]. This was essentially because the specific high entropy alloys (HEAs) developed for this purpose offered a combination of anti-corrosion, anti-oxidation, and wear resistant properties. With continued focus and development efforts, these alloys were also found to be suitable for use in electronics, primarily because they offered the capability of suppressing electromagnetic interference. A few of the HEAs that were developed and studied, such as (i) AlCoFeNiTi, and (ii) CoCrFeNiTi, exhibited good binding properties making them suitable for use as hard metal alloy binders [21]. Sustained progress in the domain specific to research being made at an observable pace eventually led to the development of powders of a HEA chosen for use as a coating on tools, dies, nozzles, and molds of both thermal spray equipment and plasma spray equipment [22].

Several of these studies conducted during the last 25 years [1995–present] have found that these alloys can offer a combination of high hardness, wear resistance, resistance to softening at elevated temperatures, and an overall good resistance to environment-induced degradation, or corrosion [18, 22, 23]. A combination of these properties made the family of HEAs both appropriate and suitable for the protection of surfaces of machine components and even for tools used for both machining and general manufacturing.

Early development of the HEAs was essentially confined to high-temperature applications to replace the superalloys [24–26]. This led to the development of HEAs with compositions that often resulted in a high value for the density. For purposes of using these alloys in applications specific to: (i) transportation, (ii) energy-related matters, and (iii) bio-medical-related matters, there was a need for alloys that offered a combination of low density and high strength. This led to an interest in the development and emergence of the high entropy alloys with low density referred to as lightweight high entropy alloys (LWHEAs). Senkov and co-workers [27, 28] developed a CuNbTiVZr refractory alloy that had a density of 6.49 g/cm^3 but offered high strength even at temperatures as high as 1,000°C. However, the continuing need for even lower density alloys for selection and use in lightweight or weight-critical applications led to continuing research studies [29]. This need resulted in Stepanov and co-workers [29] replacing the heavier element chromium (Cr) with the lighter element aluminum (Al) to bring forth their refractory alloy $Al_XNbTiVZr$. The density of this alloy was 5.55 g/cm^3 [29]. This value was still much higher than the value of 3.0 g/cm^3 that both designers and materials selectors were seeking for in a new and improved alloy for use in weight-critical applications [30]. Juan and co-workers [31] did synthesize and put forth an $Al_{20}Be_{20}Fe_{10}Si_{15}Ti_{35}$ alloy using the technique of casting, which had a density of 3.91 g/cm^3. The microstructure of this alloy, in the as-cast condition, comprised of one major phase and two minor phases. Around the same time, using the technique of induction melting and casting, Li and co-workers [32, 33] prepared a high entropy alloy containing magnesium and with the composition Mg_X $(AlCuMnZn)_{100-X}$. The developed alloy had a multiphase structure, composed of hexagonal close-packed (HCP) phases and Al-Mn icosahedral quasi-crystalline phases. Depending on the percentage of magnesium (Mg) in the alloy, the resultant alloy had a density in the range 2.20–4.29 g/cm^3. However, the magnesium-containing alloys did exhibit good compressive strength [400 to 500 MPa] at room temperature and with ductility values in the range 3–5% in compression. Near-identical microstructure and mechanical properties were also obtained for the alloys that had an equiatomic composition and were initially cast and subsequently cooled in laboratory air or by immersion in water or a brine solution [33]. Chen and co-workers [31] used the technique of mechanical alloying (MA) to prepare the BeCoMgTi and BeCoMgTiZn alloys. In both of these alloys the presence and distribution of amorphous phases in the microstructure was evident [31, 32]. In a subsequent study, Youssef and co-workers [34] developed and studied an alloy having a composition of $Al_{20}Li_{20}Mg_{10}Sc_{20}Ti_{30}$. This alloy was developed using the mechanical alloying (MA) technique. The resultant alloy had a density of 2.67 g/cm^3, a single-phase nanocrystalline structure, and an overall high hardness. Overall, of the few alloys that were developed and thoroughly studied, only the alloys having compositions of Mg_x $(AlCuMnZn)_{100-X}$ and $Al_{20}Li_{20}Mg_{20}Sc_{20}Ti_{20}$ had a density that was less than 3.0 g/cm^3 and could be classified as a lightweight high entropy alloy (LWHEA) that made them suitable for selection and use in weight-critical applications.

Through the years, it has been shown by many researchers that the alloys that were explored are capable of being synthesized, processed, and even analyzed, contrary to the misconceptions that prevailed based on traditional experiences. Emergence of active research and development efforts on the high entropy alloy system did provide the much-needed impetus for a spectrum of opportunities in this field for the purpose of scientific study and use in industry-related applications. The Ashby map of fracture toughness as a function of yield strength for high-entropy alloys in relation to a wide range of material systems is shown in Figure 15.1 [19]. The excellent damage tolerance (toughness combined with strength) of the CrMnFeCoNi alloy is evident, in that this high entropy alloy exceeds the toughness of most pure metals and most metallic alloys and has a strength comparable to that of structural ceramics and close to that of a few bulk-metallic glasses. Despite the claims of a new class of alloys, they are at the boundary of conventional metals and alloys (circle in the Figure 15.1), which is just an extension of them without any appreciable difference or surprisingly improved properties for the HEAs.

As the field of HEAs enters its teenage years, a proper thrust in synergism with directions for future research is both desired and essential to ensure its success as a preferential material for selection and use in structures and in a spectrum of

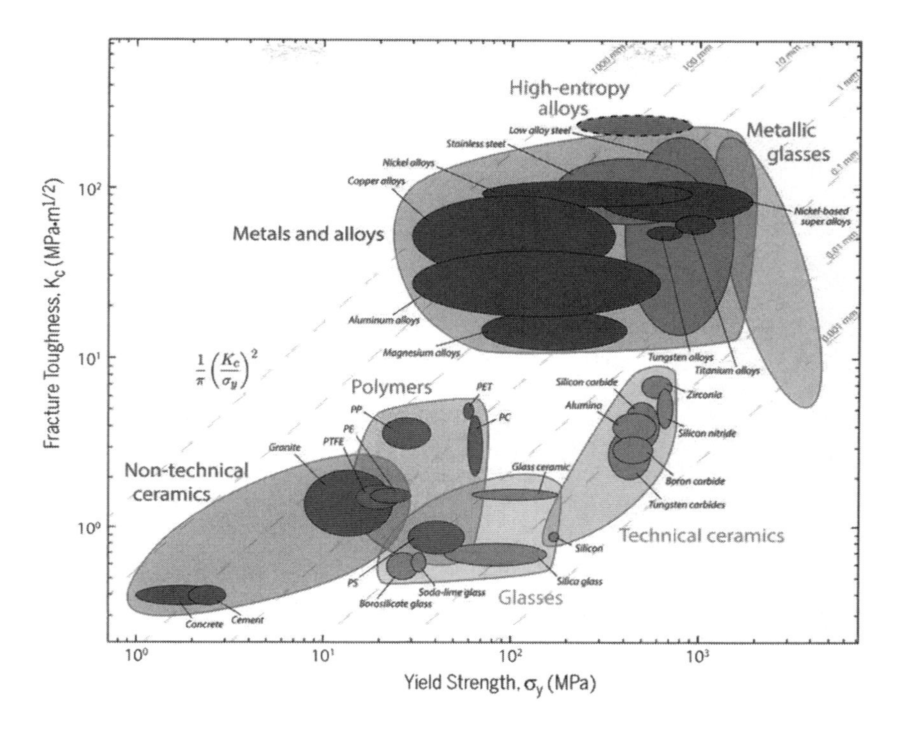

FIGURE 15.1 Ashby map showing fracture toughness as a function of yield strength for high entropy alloys in relation to a wide range of material systems. The excellent damage tolerance (toughness combined with strength) of the CrMnFeCoNi alloy is evident in that the high entropy alloy exceeds the toughness of most pure metals and most metallic alloys and has a strength comparable to that of structural ceramics and close to that of some bulk-metallic glasses [20].

function-critical applications. The exotic alloys can merely end up being of scientific curiosity with little to no commercial success. This field will be alive, active, and continue to attract attention only if a few commercially successful alloys can be engineered and put forth for purpose of use. This specific task can safely be considered to be as complex as finding new forms of life outside Earth and is a field as vast as the universe itself. An attempt is made in this chapter to identify, present, and discuss the challenges that arise during industrial processing of these alloys. It is often the dream of any alloy designer to see his compositions are effective and can be easily produced for purpose of selection and use.

The elements from the periodic table, their alloying characteristics, and remarks on their suitability for consideration for an HEA or compositionally correct alloy (CCA) are presented in Table 15.1. It can be noted from the table that the elements iron (Fe), nickel (Ni), manganese (Mn), chromium (Cr), vanadium (V), aluminum (Al), and copper (Cu) are inexpensive, while the elements cobalt (Co), titanium (Ti), and zirconium (Zr), are slightly more expensive, and all these elements are suitable as additions to the HEA. The elements hafnium (Hf), niobium (Nb), and rhenium (Re) are very expensive and their additions should justify the advances that can be

TABLE 15.1
Suitability of Elements in the Periodic Table as Constitutents in High Entropy Alloys

S. No	Group	Elements in the group	Remarks about the elements	Remarks about the possible elements in HEAs
1	1A	H, Li, Na, K, Rb, Cs, Fr	Except lithium as a minor alloying element (for example, in Al-Li or Mg-Li alloys), no other element is used structural applications.	Lithium is highly reactive and has very low density. Not suitable for PM processing. Gets oxidized easily and processing needs inert gas covering. **Not suitable for making HEA.**
2	IIA	Be, Mg, Ca, Sr, Ba, Ra	Mg is a base forming Mg alloys. Be is harmful and is used in mirrors and as a minor addition in Al/Cu alloys for improving stiffness. Ca is used as a minor alloying element in Mg alloys.	Be is processed by PM route. Highly toxic. Should be handled only in glove boxes. Mg is highly reactive and has low density (1.7 g/cc) and when added in significant proportions with other liquid elements, it floats to the top. **Not suitable for making HEA.**
3	IIIB	Sc, Y	Sc is used in Al alloys as a trace addition and a minor quantity of Y is used in superalloys.	Sc is very expensive and is rarely available in large quantities (mines are mostly in Russia). **Not suitable for making HEA.**
4	IVB	Ti,Zr,Hf	Ti and Zr are base materials and Hf is an addition in superalloys and niobium alloys (viz., C103).	Ti and Zr are widely used and have their own system of alloys commercially available catering to aerospace and nuclear industries. The low density and high specific strength/specific stiffness and corrosion resistance are key points for using Ti alloys. This advantage cannot be compromised if added with other elements and returns should be guaranteed for penalty. Zr alloys are used in nuclear applications and sensitive to hydrogen. Hf is very expensive and alloys containing significant additions of Hf should show commensurate benefits for applications. **All three elements can be used with caution.**

(*Continued*)

TABLE 15.1 (CONTINUED)
Suitability of Elements in the Periodic Table as Constitutents in High Entropy Alloys

S. No	Group	Elements in the group	Remarks about the elements	Remarks about the possible elements in HEAs
5	VB	V, Nb, Ta,	Nb is a base alloy system. V and Ta are used only as minor alloying additions.	Nb is expensive element and has high affinity to Oxygen. Difficult to process. V and Ta can be used as additions for ingot processing. **V and Ta can be used in HEA and Nb with caution.**
6	VIB	Cr, Mo, W	Mo and W are powder metallurgy materials with some alloys (TZM or W-Co) and Cr is an alloying element (in corrosion resistant steels).	Mo and W are high melting point elements and are produced by PM route. Added as ferro-alloys (Iron containing; ferro-tungsten/ferro-molybdenum). Not suitable for ingot metallurgical processing, if Fe is not present in the HEA system. **Cr can be used as additions for ingot processing.**
7	VIIB	Mn, Tc, Re	Mn is an addition in steels and Re is a high-temperature material (used as alloying element in superalloy). No alloys with Tc.	Re is a very expensive material and significant additions should significantly improve the properties. **Mn can be used as additions for ingot processing.**
8	VIII	Fe, Co, Ni, Ru, Rh, Pd, Os, Ir, Pt	Fe is the base for all steels, low cost and widely available. Co is the base for cobalt-based superalloys (KC20WN) and as an addition to a large number of alloys, and is expensive. Ni is a base for nickel-based superalloys and as an alloying addition in steels. Rh is used in thermocouples. Pd, Ir and Pt are noble elements.	Ru, Rh are costly. Pd, Ir and Pt are prohibitively expensive elements. Co is very expensive and large quantities should justify its addition. **Fe, Ni can be used as additions for ingot processing.**

(Continued)

TABLE 15.1 (CONTINUED)
Suitability of Elements in the Periodic Table as Constitutents in High Entropy Alloys

S. No	Group	Elements in the group	Remarks about the elements	Remarks about the possible elements in HEAs
9	IB	Cu, Ag, Au	Cu is a base for a number of copper-based alloys for high heat flux applications. Ag and Au are noble elements.	Ag and Au are prohibitively expensive alloying elements. Cu has high affinity for oxygen. Low cost is an added advantage. **Cu is the base material and can be used for additions in HEAs.**
10	IIB	Zn, Cd, Hg	Zn is the alloying addition for brasses. Low vapor pressure and low melting point. Cd and Hg are not suitable for making alloys.	Zn is a low melting element and has a high vapor pressure. Control of chemistry is difficult in vacuum melting. **Not suitable for structural applications.**
11	IIIA	B, Al, Ga, In, Tl	Al is a base for a host of aluminum alloys. B is a grain refiner and additive in a number of alloys. Other elements are not used for making structural materials.	Al is base systems alloys with varying strength properties. Most important aerospace material and gives alloys in different temper conditions. B, Ga, In, and Tl are not suitable for structural applications. **Al can be used for additions in HEAs.**
12	IVA	C, Si, Ge, Sn, Pb	Minor amount of C is the principal strengthener in steels. Minor quantity of Si is an alloying element in several systems.	Minor C and Si can be added to improve the properties of HEAs. **None of the elements are useful for structural applications.**
13	VA	N, P, As, Sb, Bi	N is a strengthener for steels. Other elements are not used for making structural materials.	N can be useful for strengthening some steels. **None of the elements are useful for structural applications.**

(*Continued*)

TABLE 15.1 (CONTINUED)
Suitability of Elements in the Periodic Table as Constitutents in High Entropy Alloys

S. No	Group	Elements in the group	Remarks about the elements	Remarks about the possible elements in HEAs
14	VIA	O, S, Se, Te, Po	O is an interstitial element in Ti alloys. Other elements are not used for making structural materials.	**None of the elements are useful for structural applications.**
15	VIIA	F, Cl, Br, I, At	Gases.	Not useful
16	VIIIA	He, Ne, Ar, Kr, Xr,Rn	Nobel gases.	Not useful.
17	Lanthanides	La-Lu	Rare earths used in a number of electronic and magnetic materials—yet to be explored in HEAs.	Very strategic and expensive. Can be useful. Yet to be initiated.
18	Actinides	Ac-Lr	Radioactive, difficult to handle.	Not suitable.

Among the elements from the periodic table, Fe, Ni, Mn, Cr, V, Co, Ti, Zr, Al, and Cu are inexpensive and suitable for additions in HEA. Hf, Nb, and Re are very expensive and their additions should justify the improvements in properties. Mo and W are high melting point elements and are added in the form of ferro-alloys (in fact, in steels, Ni, Mn, Cr and V are also added as ferro-alloys as ferro-nickel, ferro-manganese, ferro-chrome, and ferro-vanadium) during ingot metallurgy, and this will place limitations on the possible alloy compositions. Even though it appears as though a large number of elements are available for developing HEAs (this is true in powder metallurgical processing as there are no limitations), when it comes to ingot metallurgical processing, the elemental options seem to be few, when considered in tandem with the cost of elements. Segregation of alloying additions during solidification, homogenization of the alloys and obtaining the desired quality ingot for subsequent thermomechanical processing are challenging.

achieved through an improvement in properties. Molybdenum (Mo) and tungsten (W) are high melting point elements and are often added in the form of ferroalloys. For the case of steels, nickel (Ni), manganese (Mn), chromium (Cr), and vanadium (V) are added as ferro-alloys, i.e., ferro-nickel, ferro-manganese, ferro-chrome, and ferro-vanadium, respectively, during ingot metallurgy, and this will have limitations on the possible alloy compositions that can be synthesized or engineered. Even though it appears that a large number of elements are available for developing HEAs (true in the case of PM processing, as there are no limitations), when it comes to ingot metallurgy (IM) processing, the elemental options seem to be few, when considered in tandem with cost of the elements. Also, segregation of the alloying additions during solidification, homogenization of the alloys, and obtaining the desired quality ingot for purpose of subsequent thermomechanical processing are challenging issues.

In this chapter, an attempt is made to present the findings and inferences of some of the early studies aimed at developing and putting forth HEAs having essentially a low density and to make them available for use in a spectrum of weight-critical applications. The basic principles behind designing a HEA by appropriate choice or combination of elements, the key parameters governing the formation and stability of solid solution, the presence of impurities, and the role of gas content in influencing the behavioral kinetics or response kinetics of the engineered HEAs are also presented and briefly discussed. Having fixed the compositions and set the specifications and the presence and role of both impurity content and gas content, the ability to obtain reproducibility in properties of interest falls to the engineer assigned with the task of choosing or selecting an alloy for a particular end product of interest to the concerned end user. Key aspects, issues, and variables governing and influencing the importance and role of alloy selection on the cost of the end product are presented and briefly discussed.

15.3 STANDARDS FOR INDUSTRIAL PROCESSING AND QUALIFICATION OF MATERIALS

Usually all metals and their alloy counterparts are processed in conformance with some specified standard. Once an alloy has been developed and carefully studied through a comprehensive characterization of its various physical properties, mechanical properties, and functional properties in a laboratory setting, and if it was found to be suitable for a particular application, several scale-up melts of the same alloy can be processed to systematically determine and concurrently establish the variation in properties within an allowable band of permissible chemistry. Based on experiences gained from both the scale-up melts and resultant properties being established, the next step is to produce the materials in the desired forms using thermomechanical processing (TMP). Based on this, a broad list of specifications will be generated and put forth. Industrially, the commercial alloys will be produced based on the specifications put forth by the following:

(i) American Society for Testing and Materials (ASTM)
(ii) Aerospace Material Specification (AMS)
(iii) Military Specification/Standard (MIL)

(iv) German Specification: Deutsches Institut für Normung (DIN)
(v) French Specification: Association Française de Normalisation (AFNOR)
(vi) Russian Specification: ГОСТ (GOST)
(vii) Japanese Industrial Standards (JIS)

The specifications essentially include the following:

(a) Method of manufacture
(b) Chemical composition of alloys
(c) Minimum guaranteed mechanical properties
(d) Applicable standards
(e) Testing and analysis
(f) Qualification methodologies for a given alloy

Any industrial application requires a standard against which the developed alloy has to be qualified and certified. Therefore, it is important that the standards are best understood in the context of commercialization and to ensure both the suitability and ease of selection of the HEAs for one or more applications.

15.4 COMPOSITIONS AND COMPOSITION RANGE

15.4.1 Alloying Elements: Selection, Amount, and Compatibility

The permissible range for the composition of a HEA that ensures reproducibility in properties is often a challenge. Usually, for any alloy, the addition of alloying elements is permitted within a specific range. Similarly, the presence and distribution of impurities should not exceed a certain limit. The permitted alloying elements, their composition range, and ease of mixing with the other elements in the alloy will provide the desired physical and mechanical properties to the final product. A given alloy having a certain chemical composition range will tend to exhibit a given set of minimum guaranteed physical and mechanical properties. The ranges are often arrived at following extensive experimentation with the intention of ensuring that both microstructure and mechanical properties are reproducible. This is crucial not only to the designer, but also the manufacturer, since they are now in a position to choose the alloy with a complete understanding of what the desired HEA has to offer. To the metallurgist on the shop floor, once the desired HEA is well within the permitted composition range, he can safely tap the liquid metal. During melting, he can safely incorporate corrections to chemistry of the HEA with the prime objective of obtaining the targeted chemistry, which is often well within the center of the range permitted. When chemistry of the specific HEA alloy is within the range, it is identified by a grade and must exhibit all of the characteristics of that grade. Jiang and co-workers [35] have attempted direct solidification of bulk ultrafine-microstructure eutectic HEAs having outstanding thermal stability. Figure 15.2 shows the macro-profile of the bulk CoCrFeNiNb0.45 EHEA ingot along with the microstructure observed in the upper and bottom part of the ingots. This is one of the few works reporting on the ingot metallurgical processing of high entropy alloys.

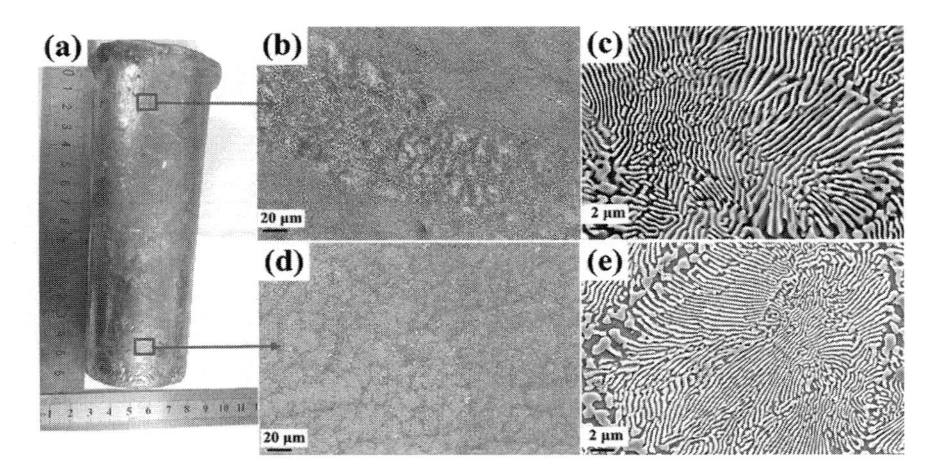

FIGURE 15.2 (a) Macro-profile of the bulk $CoCrFeNiNb_{0.45}$ EHEA ingot. (b, c) The microstructure observed in the upper part of the ingot. (d, e) The microstructure observed in the bottom part of the ingots [36].

Having appreciated the important role of composition range for the emerging family of HEAs, one must look into their role in contributing to both physical properties and mechanical properties. Line compositions are often difficult to control during the melt processing of alloys. For example, the nickel-titanium (Ni-Ti) shape memory alloy having an exact line composition of 49.5 and 50.5 atomic percentage does pose a challenge during processing. In fact, small deviations in alloy chemistry, i.e., selection of elements and the resultant composition can exert a large deviation in the transformation temperature of the alloy. Therefore, there is a need to both study and concurrently understand the role and/or influence of deviation in alloy composition on various properties of the engineered alloy. For the HEAs having five or more elements in them, a deviation in each of the elements from a mean value and its associated influence on properties is an important challenge that must be addressed.

The challenges are often more complex than expected. This is primarily because many of the alloy compositions have to be melted through vacuum induction melting (VIM) to obtain the desired composition, while concurrently controlling the presence of impurities and gases. During VIM, the alloying elements are added either as a virgin material or as a "master" alloy. The sequence of addition of the elements and the time allowed prior to pouring are important parameters, since element loss can often occur even in an inert environment, such as a vacuum, depending on the vapor pressure of the specific element. Elements like cadmium (Cd), zinc (Zn), and manganese (Mn), which have a high vapor pressure are often prone to experience significant loss during melting and immediately prior to pouring. On the other hand, melting can be also be conducted in an environment of argon (Ar) gas pressure to minimize loss of the specific element. To obtain the required composition and to concurrently take in to account an apparent loss of the high vapor pressure elements, an additional amount of the specific elements must be added to compensate for the loss.

The parameters that exert an influence on the loss must be studied to provide an estimate of the loss experienced by a specific element for purposes of compensation.

Several of the HEAs are processed by arc melting coupled with melting multiple times with the intent of achieving better homogeneity. A good balance must be struck between the time subsequent to melting and prior to pouring to minimize loss experienced by the alloying elements with the desired objective of obtaining the right chemistry and the resultant microstructure. In industrial practice, melting multiple times for the purpose of achieving improved compositional homogeneity is not economically viable.

Another important method for processing of the HEAs is using the technique of powder metallurgy. High purity alloy powders, targeting a specified composition, are milled for a given time period under a given set of processing conditions to form the desired solid solution phase. The resultant powders are compacted and subsequently sintered using one of the following techniques: (i) cold press, (ii) hot press, (iii) hot isostatic pressing (HIP), and (iv) spark plasma sintering (SPS). Each of these techniques has their own advantages and disadvantages to offer when it comes to processing bulk materials. Here, the most important challenge is a contamination that could occur from both the balls and jar used for milling the powders. The sequence of milling, time for milling, and environment or medium chosen does play an important role in realizing the final product. Since many of the HEAs often require a prolonged milling time for the formation of a stable solid solution phase, the contamination levels can be much higher than what occurs in conventional powder metallurgy (PM) products. A few authors have reported contamination from carbon that is present in the milling medium. Oxidation of the products is another problem, which does contribute to the formation and presence of unwanted brittle phases in the microstructure that are often detrimental to overall characteristics of the specific HEA of interest. A few studies have successfully synthesized refractory high entropy alloys through powder metallurgy processing followed by HIP. A prudent use of the consolidation processes is essential for obtaining near-theoretical density in the final product. However, an adequate amount of care must be taken to use them only for those products that are intended for "niche" applications where cost is not the primary driving factor. Furthermore, utilization of the compacting equipment is often expensive when: (i) the products must be batch processed, and (ii) quality of raw material chosen, processing sequence used, and the relevant steps in processing need to be monitored for obtaining repeatability or reproducibility in properties. Powder metallurgy processing is suitable for many small-sized "niche" products on a commercial scale, provided it offers a few to several distinct advantages over the conventional processing (CP) techniques used for the same product.

15.4.2 Role of Impurities and Gas Content

The presence of impurities is often undesirable, but unavoidable, in an alloy. The detrimental influence of impurities and gases is many times more influential than the beneficial effects achieved through alloying additions on microstructural development and resultant properties, to include both physical and mechanical, of both metals and their alloy counterparts. The presence of impurities above a certain minimum will exert a serious influence on properties and must therefore be kept to an absolute minimum. The specifications for an alloy containing impurities must often be restricted to a minimum. Care should be exercised to restrict the amount of impurities by using raw materials having a low level of impurity content or by selectively removing the impurities during melting. For the HEAs, the impurities can be contributed by each of the five or more alloying elements chosen. For the purpose of laboratory-related research studies, high purity elements (>99.9+ purity) are often chosen and used. However, for industrial melts, elements having a high level of purity are seldom used, primarily because the cost of a high purity element, at the bulk level, is prohibitively high. As five or more elements are often used in near equal atomic percentages, a higher total impurity content will be present in the alloy than in the individual elements. Furthermore, there is a need to study the influence of the presence and role of impurities on mechanical performance of the engineered HEAs, should they result in the formation and presence of undesirable phases in the microstructure.

Gas contents are often specified for each of the commercial alloys either as a range or the maximum permitted. Titanium alloys are available as extra-low interstitial (ELI) or as regular grades depending on the oxygen content. For titanium alloys, an oxygen content greater than 1,300 ppm is the regular grade and below is the ELI grade. On the other hand, there are few other gases that must be well-controlled. For example, for high strength steels, the maximum permitted level of hydrogen is 2 ppm. Beyond this level the specific steel can be considered to be susceptible to hydrogen-induced embrittlement. Therefore, a careful control of the gaseous environment is important in the engineering of HEAs.

15.5 MECHANICAL PROPERTIES AND THEIR REPRODUCIBILITY

A key aspect to remember in enabling engineering applications of the HEAs is an ability to reproduce the physical properties, mechanical properties, and functional properties. Several points exert an influence on the reproducibility of properties and include the following: (i) chemical composition, (ii) homogeneity of the alloy, and (iii) local variations in microstructure. Should the density of the chosen alloying elements be varying by an observable amount, then there is every possibility that the lighter element will tend to float (like lithium (Li) additions to an aluminum (Al) melt) and the heavier element would settle at the bottom. It is here that induction melting (IM) would help in obtaining a thorough mixing of the chosen elements. In the case of vacuum-arc re-melting (VAR) of the chosen elements, obtaining the right microstructure is important. A healthy combination of VIM and VAR would be an appropriate choice for performance-critical applications. For melting compositions that contain the reactive elements like titanium (Ti), zirconium (Zr), niobium (Nb), and hafnium (Hf), suitable melt practices aimed at reducing unavoidable contamination are essential to obtain the desired properties.

Adequate homogenization times are also both essential and needed. At times, this can exceed a few hundreds of hours. This will aid in bringing about uniformity in chemical composition. In industrial practices, implementing such steps is usually not encouraged, since the furnaces are rated for operating at a given maximum temperature for a specified maximum time.

Extensive studies have been reported on a wide range of materials to include the following systems: (i) the Al-Co-Cr-Fe-Mn-Ni alloy, and (ii) the Al-Co-Cr-Cu-Fe-Mn-Ni alloy. This is because of an overall ease in the availability of raw materials coupled with ease of processing and their low cost. Except for the presence of aluminum in these compositions, these multimetallic alloys form the basis for conventional stainless steels comprising ferritic stainless steels, austenitic stainless steels, and martensitic stainless steels used for wide range of applications at both cryogenic temperatures and room temperature.

Tong and co-workers [36] have studied and reported outstanding tensile properties of the precipitation-strengthened $FeCoNiCrTi_{0.2}$ high entropy alloy at both room temperature and cryogenic temperature (77 K/–196°C) and highlighted an overall superiority of the alloy over other conventional solid-solution medium entropy/high entropy alloys and the data is as shown in Figure 15.3. Jiang and co-workers [35] have reported on mechanical properties of directly solidified bulk ultrafine-microstructure eutectic high entropy alloys having an outstanding thermal stability. The variation of Vickers hardness with temperature of the $CoCrFeNiNb_{0.45}$ alloy as well as traditional nano-grained (NG)/ultrafine-grained (UFG) alloys at both room temperature and high temperatures, along with a comparison of yield strength variation with temperature for this alloy and a few other typical traditional NG/UFG alloys are shown in Figure 15.4. A comparison of properties of the HEAs with other alloys strengthened through grain refinement revealed higher hardness/yield strength. However, the comparison is not relevant since the other chosen materials are the

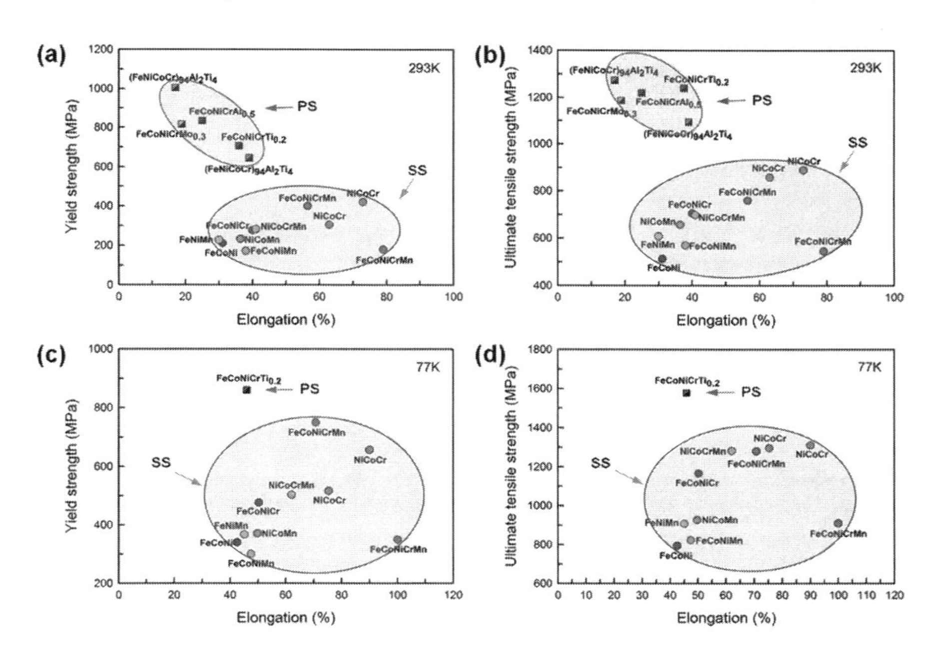

FIGURE 15.3 Tensile properties of the aged FeCoNiCrTi $_{0.2}$ high entropy alloy at 293 K (a, b) and 77 K (c, d) compared with solid-solution (SS) MEAs/HEAs and precipitation-strengthened (PS) FeCoNiCr HEAs [36].

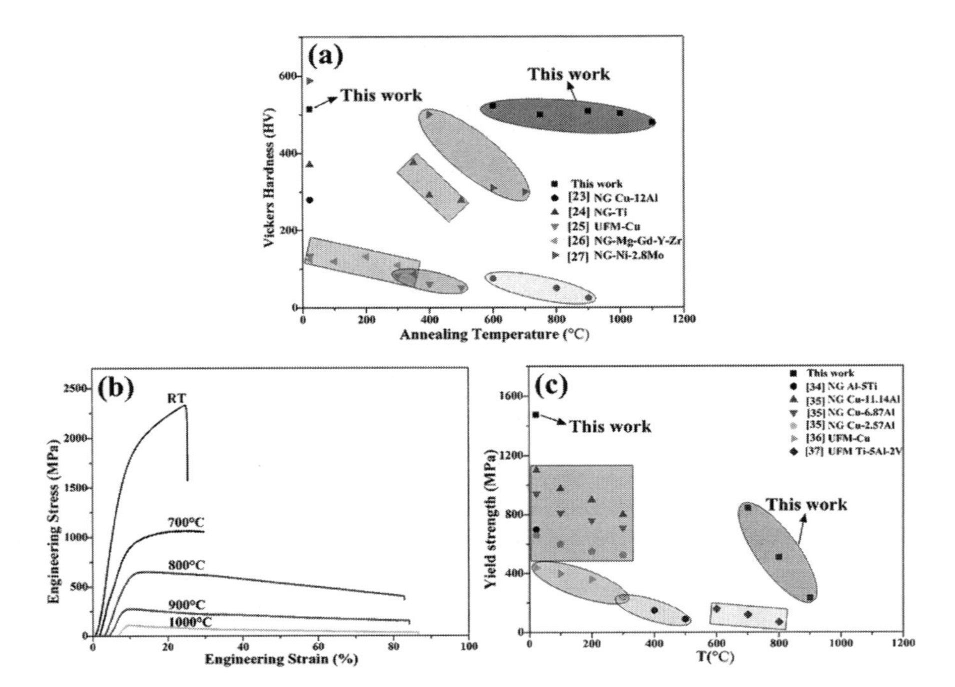

FIGURE 15.4 (a) Vickers hardness versus temperature (T) of CoCrFeNiNb $_{0.45}$ EHEA as well as traditional nano-grain/ultrafine microstructure alloys. (b) Room temperature and high-temperature compressive curves of CoCrFeNiNb $_{0.45}$EHEA. (c) Yield strength versus temperature of CoCrFeNiNb $_{0.45}$ EHEA and some typical traditional nano-grain/ultrafine grain materials [35].

aluminum/copper alloys having a lower inherent strength and a noticeably lower melting point.

Yang and co-workers [37] studied the nanoparticle-strengthened high entropy alloys for use in cryogenic applications that revealed an exceptional combination of strength and ductility. The engineering tensile curves of the HEA at both ambient temperature and cryogenic temperature are shown in Figure 15.5. Deformation of the high entropy alloy at 77 K (–196°C) revealed a ductile dimpled fracture with no evidence of macroscopic necking. A comparison was made of the developed high entropy alloys (HEA) with the mechanical properties of other selected cryogenic alloys.

Jiang and co-workers [38] developed and reported a $CoFeNi_2V_{0.5}Mo_{0.2}$ HEA having exceptional ductility. The tensile properties of the $CoFeNi_2V_{0.5}Mo_{0.2}$ HEA are shown in Figure 15.6. The engineering stress versus engineering strain curves tested at the six temperatures reveal the alloy to have excellent ductility at both room temperature (300 K/27°C) and high-temperature ductility, especially at 873 K (600°C), to a maximum value of 189.2 percentage. The strength–ductility relationships of the different alloys and the HEAs at room temperature (300 K/27°C) indicate the current HEA to be superior to the other HEAs and most conventional alloys. The strength–ductility relationships of both the conventional alloys and the HEAs at

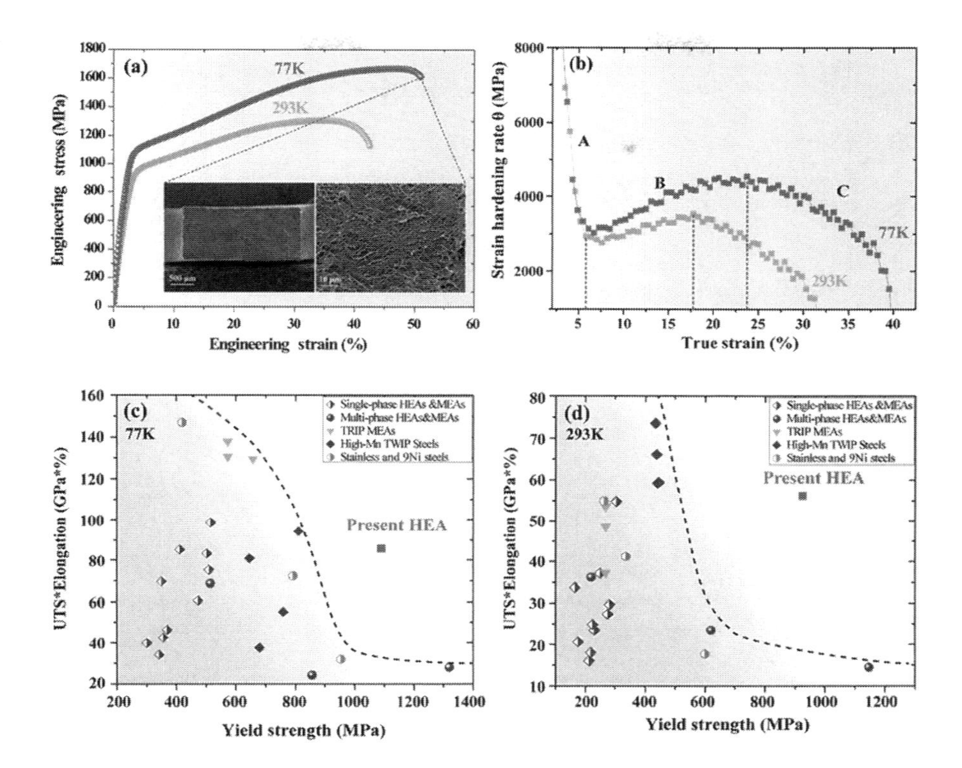

FIGURE 15.5 (a) Engineering tensile curves of HEA at ambient and cryogenic temperatures. Insets show the typical fracture surfaces of the present HEA deformed at 77 K, exhibiting ductile dimpled microstructures with almost no macroscopic necking. (b) Strain-hardening responses of HEA. Comparison of the mechanical properties of HEA with other selected cryogenic alloys obtained at (c) 77 K and (d) 293 K, respectively [37].

600°C reveal the ductility of the current alloy to be three times that of the other alloys at a strain rate of 1×10^{-3} s^{-1}.

Gwalani and co-workers [39] did report an interesting development on HEA where the tensile yield strength of a single bulk $Al_{0.3}CoCrFeNi$ high entropy alloy can be tuned from 160 MPa to 1,800 MPa. The mechanical behavior at room temperature (27°C) of the $Al_{0.3}CoCrFeNi$ CCA is shown in Figure 15.7. This figure also shows the ultimate tensile strength (σ_{UTS}) and ductility (ε_f) achieved by Al0.3CoCrFeNi in comparison with the main classes of structural alloys to include the present generation of alloys used in both automotive applications and aerospace applications. Further, in this figure, the materials property space for room temperature (27°C) elongation versus tensile strength of the steels and the Al0.3CoCrFeNi CCA along with the conventional steels, conventional high-strength steels, and the ferritic-bainitic (FB) steels is also shown. The area suggested by Gwalani and co-workers overlaps with the conventional steels currently being used.

Raabe and co-workers [40] have summarized the variation of ultimate tensile strength versus ductility in terms of total elongation to fracture plot for a number

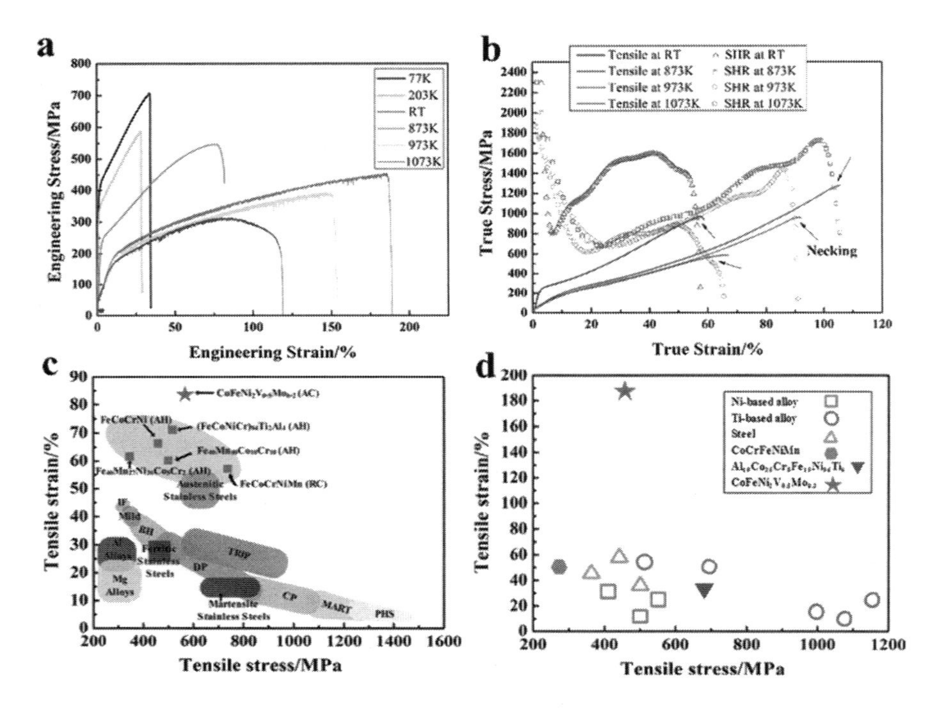

FIGURE 15.6 Tensile properties of $CoFeNi_2V_{0.5}Mo_{0.2}$ high entropy alloy. (a) Engineering stress-strain curves tested at six temperatures show that the alloy has excellent room temperature and high-temperature ductility, especially at 873 K, to a maximum value of 189.2%. (b) True stress-strain curves and SHR curves, indicates the on-site necking points of the alloy. The SHR curves display a first decreasing, then increasing, and finally decreasing tendency with increasing strain. (c) Strength–ductility relationships of different alloys and HEAs at room temperature indicate that the current HEA is superior to the other HEAs and most conventional alloys. The data of the other HEAs are all at their as-homogenized (AH) or recrystallization (RC) state, with few casting defects. While the results of the current $CoFeNi_2V_{0.5}Mo_{0.2}$ high entropy alloy are directly based on the as-cast (AC) ingot. (d) Strength–ductility relationships of conventional alloys and HEAs at 873 K show that the ductility of the current alloy is three times that of the other alloys at the same strain rate of 1×10^{-3} s^{-1} [38].

of different classes of steels, showing high entropy (HE) steel with superior performance, as shown in Figure 15.8. The figure reveals that HE steels are capable of covering a wide range of mechanical properties which is enabled by composition and temperature-dependent tuning of phase stability for activating transformation induced plasticity (TRIP), twinning induced plasticity (TWIP), and precipitation effects.

Gludovatz and co-workers [19] have reported the extraordinary property of fracture-resistance of a high entropy alloy (HEA) for use in cryogenic applications. A fully recrystallized microstructure having a near-equiaxed grain structure with a grain size of ~6 μm along with the mechanical properties of the CrMnFeCoNi high-entropy alloy is shown in Figure 15.9. The yield strength (σ_{YS}), ultimate

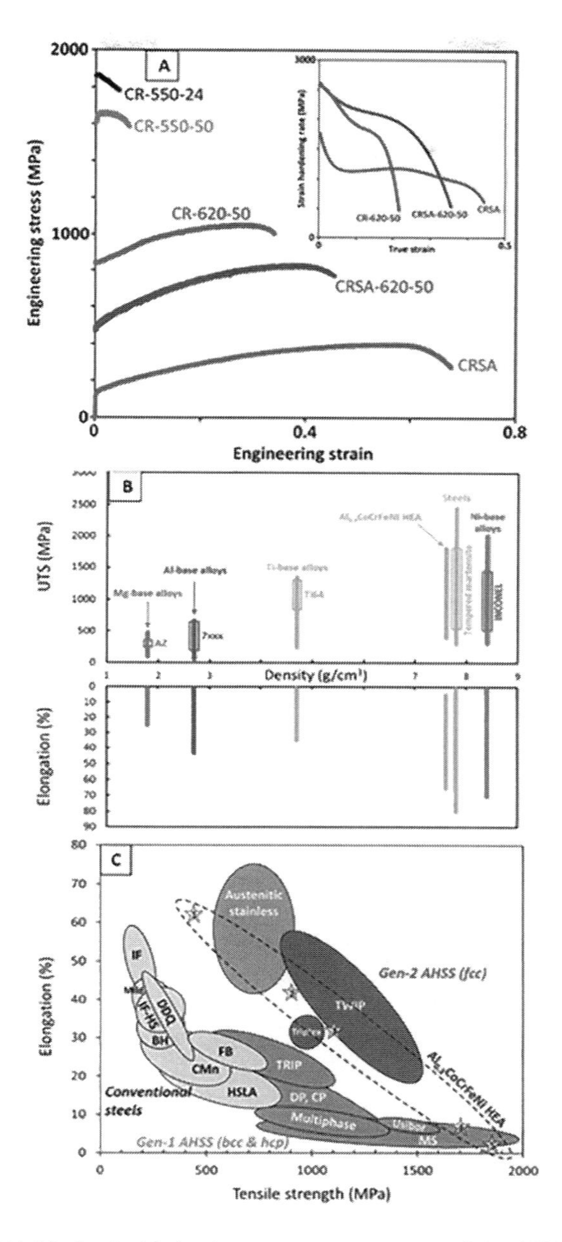

FIGURE 15.7 (A) Mechanical behavior at room temperature of the Al0.3CoCrFeNi CCA. Engineering stress-strain and strain hardening vs true strain curves (inset). (A) Ranges of ultimate tensile strength and ductility achieved by Al0.3CoCrFeNi in comparison with the main classes of structural alloys including present generation automotive and aerospace engineering alloys. (C) Materials property space for room temperature elongation vs tensile strength of steels and Al0.3CoCrFeNi CCA. This strength–ductility diagram shows the range of properties achieved by Al0.3CoCrFeNi CCA (represented by stars 1–4) in comparison with all steel grades which are classified in three groups in addition to stainless steels [39].

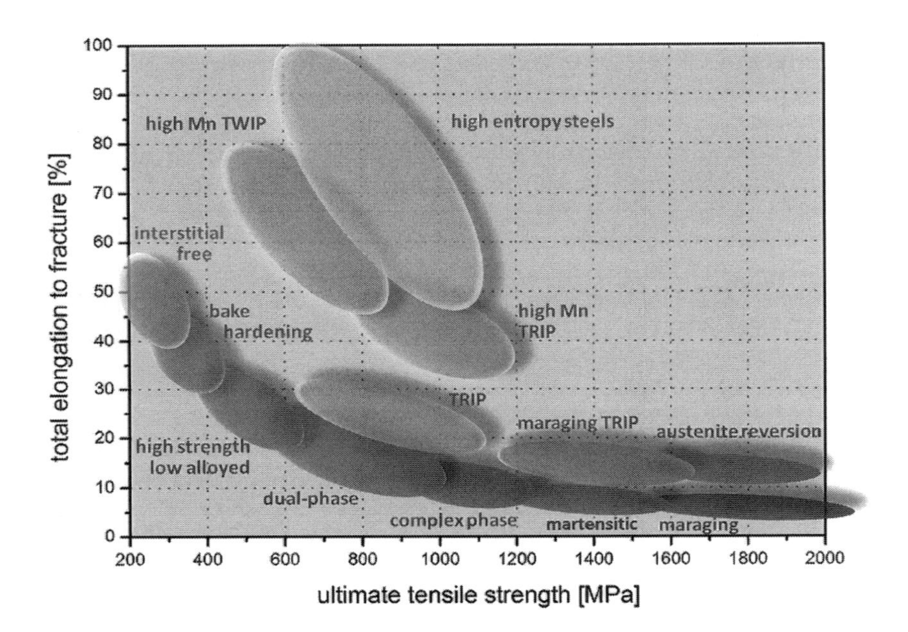

FIGURE 15.8 Ultimate tensile strength versus ductility in terms of total elongation to fracture plot for a number of different classes of steels showing high entropy (HE) steel with superior performance Additionally, the diagram includes data for high entropy steels with quaternary FeMnAlC and quinary FeMnAlSiC composition, which in part exceed the mechanical properties known from many other types of alloys. The diagram reveals that the high entropy steels are capable of covering a wide range of mechanical properties which is enabled by composition- and temperature-dependent tuning of phase stability for activating TRIP, TWIP, and precipitation effects [40].

tensile strength (σ_{UTS}), and ductility (strain to failure (ε_f)) of the alloy all increase with a decrease in test temperature. The fracture toughness measurements reveal KJ_{Ic} values of 217 MPa·m$^{1/2}$ at 293 K (20°C), 221 MPa·m$^{1/2}$ at 200 K (–73°C), and 219 MPa·m$^{1/2}$ at 77 K (–196°C), coupled with an increase in fracture resistance in terms of the J integral as a function of crack extension. Like austenitic stainless steels (e.g., Type 304, Type 316, or cryogenic nickel-containing steels), the strength of the CrMnFeCoNi HEA increases with a decrease in test temperature. Although the toughness of other alloys decreases with a decrease in temperature, the toughness of the HEA remains essentially unchanged, and by some measure it marginally increases at the lower test temperatures.

Xu and co-workers [41] have designed novel low-density refractory high entropy alloys (RHEAs) for selection and use in high-temperature applications. A comparison between reported RHEAs and other alloys is shown in Figure 15.10. A comparison of RHEAs with that of conventional nickel-based superalloys (IN718/MAR M247) is not appropriate, because the nature and application of the superalloys is completely different when compared one-on-one with that of the refractory alloys. The RHEAs should be compared with the conventional refractory alloys or vice versa. The overall superiority of the HEAs should be compared with that of the RHEAs.

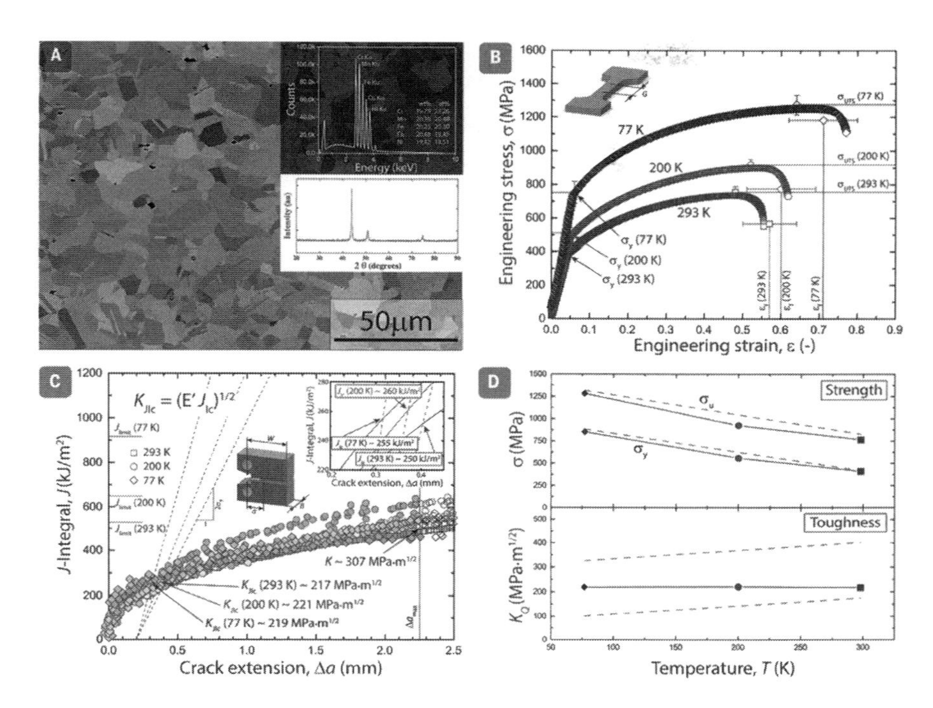

FIGURE 15.9 Microstructure and mechanical properties of the CrMnFeCoNi high-entropy alloy. (A) Fully recrystallized microstructure with an equiaxed grain structure and grain size of ~6 μm; the composition is approximately equiatomic, and the alloy is single-phase, as shown from the EDX spectroscopy and XRD insets. (B) Yield strength, ultimate tensile strength, and ductility (strain to failure) all increase with decreasing temperature. (C) Fracture toughness measurements show KJ_{Ic} values of 217 MPa·m$^{1/2}$, 221 MPa·m$^{1/2}$, and 219 MPa·m$^{1/2}$ at 293 K, 200 K, and 77 K, respectively, and an increasing fracture resistance in terms of the J integral as a function of crack extension. (D) Similar to austenitic stainless steels (e.g., 304, 316, or cryogenic Ni steels), the strength of the high entropy alloy (solid lines) increases with decreasing temperature [19].

15.6 COST

On many occasions, cost of the final product often dictates both its selection and its use for practical applications. Barring a few defense-related and space-related applications, for most commercial alloys, the bottom line is cost of the product. The cost of an element increases with its purity level and often a portion of the high-purity elements are thoroughly mixed with small quantities of scraps of the alloy during melting while concurrently ensuring effective utilization of alloy scrap as a viable method for recycling. In the case of HEAs, the number of alloying elements is basically five or more. The elements magnesium (Mg), aluminum (Al), iron (Fe), manganese (Mn), chromium (Cr), copper (Cu), zinc (Zn), and lead (Pb) are relatively inexpensive when compared to the other transition elements, and even the refractory elements. This has often raised the question of whether these alloys must be designed based on their cost or the possible combination of properties they have to offer. The

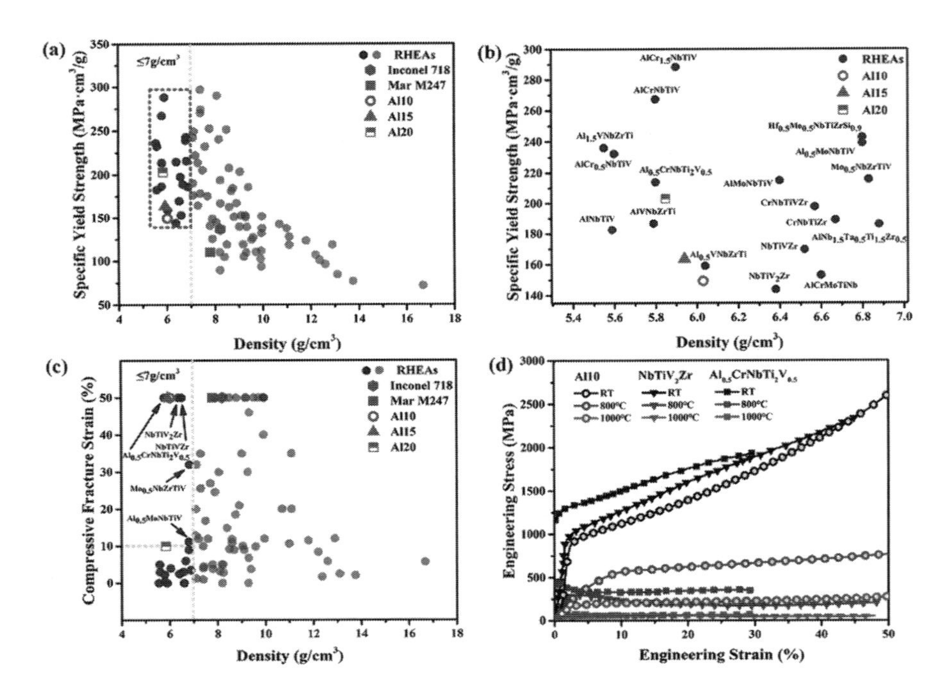

FIGURE 15.10 Comparisons between reported RHEAs and alloys. (a) Specific yield strength versus density. (b) The alloys in the dashed square in (a). (c) Compressive engineering strain versus density. (d) Compressive stress-strain curves of Al10, NbTiV2Zr, and Al0·5CrNbTi2V0.5 [41].

latter will be an option for engineering novel materials that have a unique combination of properties to offer that are much better when compared to the existing alloys.

Extensive data on structural HEA has to date not shown any remarkable properties that they have to offer that surpass the conventional alloys. For the structural alloys and with specific reference to strength: (i) the conventional carbon steels have the capability to offer strength as high as 400 MPa, (ii) the high strength low alloy steels can offer strengths in the range of 800–1,200 MPa, and (iii) the ultrahigh-strength steels have strengths in the range of 1,200 MPa to 2,400 MPa. These ultrahighstrength steels are based on Fe-Ni-Co-Mo and have been successfully chosen for use in several performance-critical applications spanning the aerospace, nuclear, and defense industries. The total alloying additions are less than 35% and through a healthy synergism of heat treatment, they can offer an excellent combination of properties including good strength and fracture toughness.

The researchers working on engineering the development of HEAs should take a hint from these highly alloyed, high-strength steels to understand the philosophy behind alloy design principles and to design new and improved alloys that can offer improved performance. A significant improvement in the performance characteristics over the presently chosen and used alloys in terms of properties, including endurance, is essential for these alloys to be considered for selection and use in both performance-critical and non-performance-critical applications, with all other properties being unaltered. When the desired properties can be obtained from the inexpensive plain carbon steel and the family of high strength low alloy (HSLA) steels, why should HEAs be considered for selection and use for an application? Further, the bottom line with specific reference to applications would be the cost of the alloy under comparable conditions for the chosen application.

Questions that often arise in engineering the development of HEAs are the following:

(a) Does interstitial solid solution (ISS) give higher strength than a substitutional solid solution (SSS); and
(b) Whether precipitation-hardening (PH) can overtake the other two, i.e., ISS and SSS.

Alloy designers should seriously consider incorporating elements that can enhance precipitation-hardening with the primary objective of enabling high strength in the HEA rather than adhere to the conventional concept of solid-solution elements. This is the fundamental basis on which the family of superalloys was designed.

Alloys should be designed for specific applications while concurrently offering more than a few of the desired properties. If the alloy is for high strength, the design should consist of using elements that are focused on enhancing strength. On the other hand, if the newly developed alloy is for a combination of strength and toughness that is desired for the purpose of damage-tolerant applications, the alloy should exhibit a balance of properties. For the purpose of functional applications, the designed HEA should exhibit superior performance characteristics.

15.7 MANUFACTURING AND WELDABILITY

When two HEAs are welded, or a dissimilar weld between a HEA and a conventional alloy is made, the weld bead will tend to have a cast microstructure with the possible occurrence of extensive coring. The weld bead is most likely to be the weakest region, and failure of the joint often tends to occur both at and around the weld bead when deformed in tension. Post-weld heat treatment of the welds with the primary intent of homogenizing the cast microstructure would tend to deteriorate properties of the base metal, thereby necessitating the need for a suitable weld filler wire to be used to alleviate this problem. The filler metal often contains grain refiners that will contribute in a noticeable way to minimizing size of the as-cast grains in the microstructure while concurrently reducing the degree of coring and contributing to improving the strength of the weld. Alternatively, non-arc-based techniques such as friction stir welding (FSW) can also be used to alleviate any and all of the problems that arise with specific reference to welding of these alloys. Similarly, the 3-D printing of components made from HEAs and vacuum plasma spraying (VPS) of HEA powders onto a shaped object or pattern that can be gradually dissolved to obtain the desired component are also being tried. HEA coatings on substrates that offer case properties that are different from properties of the core are short-term options that researchers are currently working on in an attempt to facilitate ease in both the selection and implementation of these novel alloys.

Holmstrom and co-workers [42] have attempted substituting for cobalt (Co) in cutting-edge technology for tools and have successfully demonstrated an overall superiority of HEA cutting tools over conventional tools. An overview of the microstructure at the cutting edge along with a depletion of cubic carbides in the gradient being visible and cubic carbides at the edge is shown in Figure 15.11. This figure also shows the gradient zone following an interrupted sintering as well as the gradient and coatings following full sintering. A schematic view of the test machine along with results of the HEA bonded product compared to a state-of-the-art reference with cobalt (Co) as the binder phase is shown in Figure 15.12. The actual shape of the tested inserts is also shown in this figure.

Gorssea and co-workers [43] have consolidated the mechanical properties of high entropy alloys and presented them in the form of materials property spaces for HEAs and CCAs and this is shown in Figure 15.13. This figure shows the variation of remnant induction with coercivity for both conventional magnets and magnetic HEAs and CCAs. Further, a variation of room temperature yield strength (σ_{YS}) with density (ρ) of the conventional metal alloys, HEAs, and CCAs is also shown. The variation of yield strength with density at 1,000°C for the refractory metal HEAs and CCAs and commercial nickel (Ni)-based superalloys is also shown in this figure. The mechanical properties of the RHEAs are compared with the nickel-based superalloys to claim an overall superiority of the HEAs over the conventional alloys. However, this can be safely considered to be fully illogical since the purposes of superalloys and their intrinsic characteristics are completely different to those of the refractory materials. Superalloys serve completely different kinds of engineering

FIGURE 15.11 (a) Overview of the microstructure at the cutting edge. The depletion of cubic carbides in the gradient is visible, as well as the remaining cone of cubic carbides at the edge. (b) The gradient zone after an interrupted sintering. (c)Gradient and coatings after the full sintering. Light gray grains are WC, dark gray grains are cubic carbides of (Ti,Ta,Nb)(C,N), and the binder phase is white [42].

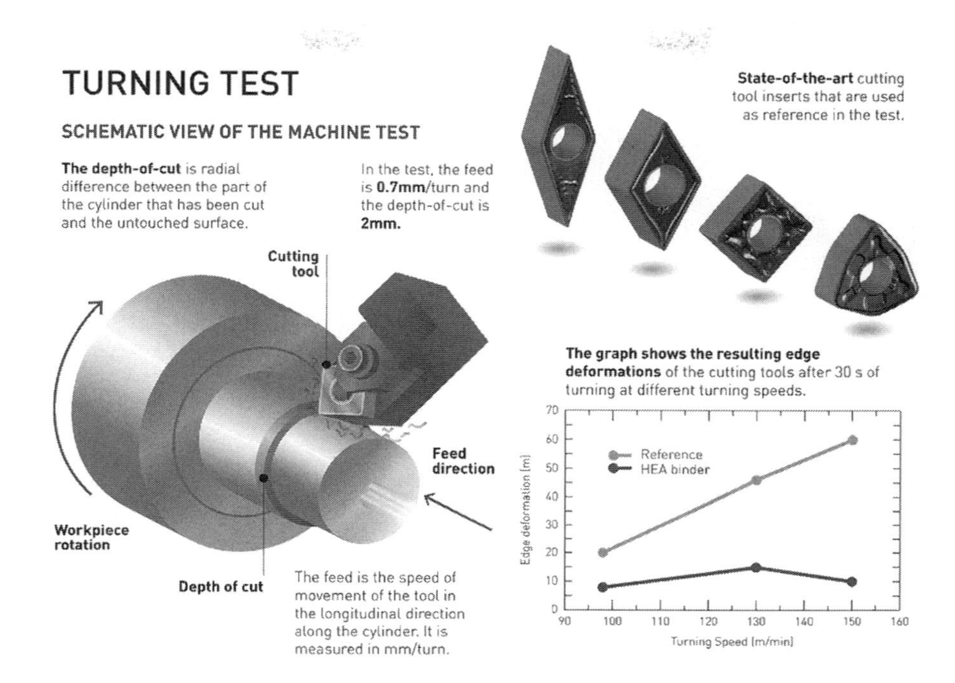

FIGURE 15.12 A schematic view of the machine test, the results of the HEA bonded product compared to a state-of-the-art reference with Co as binder phase. The actual shape of the tested inserts is the third from the top [42].

applications and the RHEAs are not a substitute. Further, for the applications where the refractory alloys are currently being used, the HEAs can be suggested as a viable alternative. A comparison of the materials should be on "equal terms," and there are several instances where this is not done in order to claim/demonstrate an overall superiority of the developed HEA compositions. This unnecessarily jeopardizes the chances for improvements through conducting thorough research to further study the possibilities of making the HEAs better.

15.8 THE FUTURE: STRUCTURAL OR FUNCTIONAL MATERIALS

It is interesting to note which of the two important areas, namely (i) structural materials, and (ii) functional materials, will opt for HEA as a viable substitute to the existing conventional alloys. Conventional structural alloys have a long history coupled with adequate usage and an extensive test database that is easily available to designers for the purpose of considering these alloys for use in a spectrum of applications. It is important for HEAs to have an equivalent comparative database that can be of valuable use to the designers, manufacturers, and even end-users of finished products to understand the overall superiority these alloys have to offer while concurrently providing an insight into the numerous advantages these alloys have to offer over their conventional alloy counterpart. It will be interesting to note whether a traditional understanding of the principles of metallurgy will be followed by the HEA,

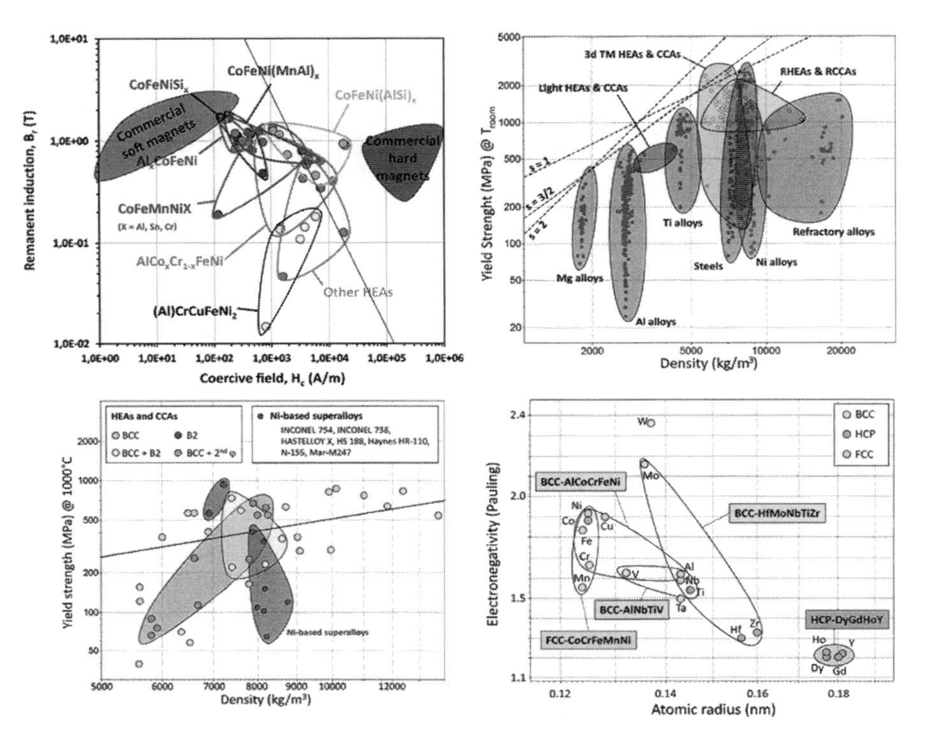

FIGURE 15.13 Materials property spaces for HEAs and CCAs. (a) Remnant induction versus coercivity for conventional magnets and magnetic HEAs and CCAs. (b) Room temperature yield strength versus density of conventional metal alloys, HEAs and CCAs. (c) Yield strength versus density at 1,000°C for refractory metal HEAs and CCAs (RHEAs and RCCAs) and commercial nickel-based superalloys. (d) Electronegativity against atomic radius [43].

namely, an increase in strength is balanced by a decrease in ductility and toughness. Similarly, it would be interesting to know how the addition of selective minor alloying elements would tend to exert an influence on mechanical properties and how an ability of these elements to form a compound could be easily affected and/or influenced in a multi-element-containing alloy.

The mechanical properties of precipitation hardenable hard aluminum alloys is summarized in Table 15.2. As can be seen from this table, the 2XXX-series, 6XXX-series, and 7XXX-series alloys can be obtained in different tempers with varying mechanical properties and corrosion/stress corrosion resistance, thereby serving the needs of the different engineering requirements of both the automotive (ground transportation) industry and the aerospace industry [44]. This is apart from the non-precipitation hardenable aluminum alloys that are chosen for use in domestic applications and are made available in different strain hardenable tempers. Further, the latest generation (third generation) Al-Cu-Li alloys have been chosen for use in lightweight, high-strength applications both in the wrought condition (i.e., structural applications in commercial and military aircraft) and in welded constructions (e.g., propellant storage tanks in satellite launch vehicles). The field of aluminum alloys

TABLE 15.2
Mechanical Properties of Precipitation Hardenable Wrought Aluminum Alloys [44]

		Nominal composition								Mechanical properties		
S. No	Type	Al	Cu	Si	Mg	Cr	Mn	Others	Form	UTS (MPa)	0.2% YS (MPa)	Hardness (BHN)
1	AA2014	93.5	4.4	0.8	0.5	0.1	0.6	–	O	185	97	45
									T4	425	290	105
									T6	485	415	135
2	AA2024	93.5	4.4	–	1.5	–	0.6	–	O	185	76	47
									T3	485	345	120
									T4, T351	470	325	120
3	AA2219	93	6.3	–	–	–	0.3	0.18Zr,0.1V,0.06Ti	O	170	76	–
									T81, T851	455	350	–
									T87	475	395	–
4	AA6061	97.9	0.28	0.6	1	0.2	–	–	O	125	55	30
									T4, T451	240	145	65
									T6, T651	310	275	95
5	AA6351	97.8	–	1	0.6	–	0.6	–	T4	250	150	–
									T6	310	285	95
6	AA7075	90	1.6	–	2.5	0.23	–	5.6Zn,	O	230	105	60
									T6, T651	570	505	150
									T73	505	435	–
7	AA7175	90	1.6	–	2.5	0.23	–	–	T66	595	525	150
8	AA7475								T61	525	460	–

is very wide and up until now unsurpassed with respect to alloy systems and temper designations tailored for a specific application. These alloys are inexpensive in this category for strength–toughness–damage tolerance for the high specific strength (strength to density ratio) applications. Little room exists to provide viable alternatives to the family of aluminum alloys, as any alloying addition to this system will increase the density, which is unattractive for aerospace-related applications.

The mechanical properties of ferritic stainless steels characterized by low to medium strength, good elongation and reduction in area, and containing chromium in the range of typically 13–20 weight percentage is provided in Table 15.3 [45]. The strength–ductility balance varies depending on the heat treatment condition. The mechanical properties of austenitic stainless steels with chromium in the range of 16–20 weight percentage and nickel in the range of 6–14 weight percentage is provided in Table 15.4 [46]. These steels can be strengthened by strain hardening and are typically characterized by an excellent combination of yield strength (σ_{YS}) and ductility (ε_f) and are widely used for a number of applications even including some low-temperature applications. The mechanical properties of high nitrogen austenitic stainless steels with the addition of the elements nickel (Ni), chromium (Cr), and manganese (Mn) is provided in Table 15.5 [47]. These alloys have yield strength–ultimate strength spread and are characterized by both good ductility and reduction in area. Some of these alloy steels exhibit excellent weldability and are being chosen for use in performance-critical applications at temperatures as low as –253°C, such as rotating components of an impeller in a cryogenic engine. The mechanical properties of martensitic stainless steels containing chromium in the range of 11–20%, along with a high percentage of carbon, are provided in Table 15.6 [48]. The microstructure of these steels essentially consists of a dispersion of carbides of chromium that provide high hardness. These steels can be heat-treated to very high hardness levels and are chosen for use in applications where the stress is primarily compressive in nature. The research efforts on HEAs have primarily been focused on alloy systems that are in the group of ferritic, austenitic, martensitic, and high nitrogen steels. In a majority of the cases, good mechanical properties have been reported under compression, and this was attributed to be due to the presence of impurities and brittle phases. On the other hand, the room temperature tensile properties are quite similar to those of the conventional alloys. The low cost of elements coupled with their easy availability is the prime reason for researchers to take up alloy design efforts on this class of alloys. Barring a few interesting results on low-temperature mechanical properties, the properties are quite similar to those of the conventional alloys presented in Table 15.3 to Table 15.6. Those compositions with breakthrough results have been pursued for repeatability and for possible additional improvements in mechanical properties.

The mechanical properties of prominent martensitic [49] and semi-austenitic precipitation hardenable stainless steels are summarized in Table 15.7 and Table 15.8 [50]. These steels are often chosen for use in performance-critical applications and the properties can be varied by the heat treatment adopted during processing. These steels are strengthened by the different types of precipitates that form as a result of heat treatment and an exceptional combination of properties can be obtained for this class of steels. Some of the latest commercial steels in these categories can easily

TABLE 15.3
Mechanical Properties of Ferritic Stainless Steel [45]

		Composition										Mechanical properties			
S.No	Type	Fe	Cr	Mn	Si	C	P	S	Others	Form	Condition	UTS (MPa)	0.2%YS (MPa)	%elongation	%RA
1	AISI405	85	13	1	1	0.08	0.04	0.03	0.2Al	Wire	Annealed	480	275	20	45
											Annealed, cold finished	480	275	16	45
2	AISI430	Balance	16	<1	<1	0.12	<0.04	<0.03	<0.75Ni	Bar	Annealed, hot finished	480	275	20	45
											Annealed, cold finished	480	275	16	45
3	AISI430Ti	76–80	16–19.5	1	1	0.1	0.04	0.03	0.8Ni, 0.5Ti	Bar	Annealed	515	310	30	65
4	AISI430F	81	16	<1.25	<1	<0.12		>0.15	0.6Mo,0.06K	Wire	Annealed	585–860	–	–	–
5	AISI446	73	23	1.5	1	0.2	0.04	0.03	0.25Ni	Bar	Annealed, hot finished	480	275	20	45
											Annealed, cold finished	480	275	16	45

TABLE 15.4
Mechanical Properties of Austenitic Stainless Steel [46]

		Composition								Mechanical properties			
S.No	Type	C	Mn	Cr	Mo	Ni	Others	Form	Condition	UTS (MPa)	0.2%YS (MPa)	%e	%RA
1	AISI301	0.15	2	16	–	6–8	–	Bar	Annealed	515	205	–	–
2	AISI302	0.01	0.6	17.7	–	9.7	3Cu	Bar	Hot finished and annealed	515	205	–	50
									Cold finished and annealed[a]	620	310	–	40
									Cold finished and annealed[b]	515	205	–	40
3	AISI304	0.08	2	18–20	–	8–12	–	Bar	Hot finished and annealed	515	205	–	50
									Cold finished and annealed[a]	620	310	–	40
									Cold finished and annealed[b]	515	205	–	40
4	AISI304L	0.03	2	18–20	–	8–12	–	Bar	Hot finished and annealed	480	170	–	50
									Cold finished and annealed[a]	620	310	–	40
									Cold finished and annealed[b]	480	170	–	40
5	AISI304LN							Bar	Annealed	515	205	–	–
6	AISI316L	0.03	2	16–18	2–3	10–14		Bar	Hot finished and annealed	480	170	–	50
									Cold finished and annealed[a]	620	310	–	40
									Cold finished and annealed[b]	480	170	–	40
7	AISI316LN							Bar	Annealed	515	205	–	70

[a] Up to 13 mm thick, [b] over 13 mm thick.

TABLE 15.5
Mechanical Properties of High Nitrogen Austenitic Stainless Steel [47]

		Composition										Mechanical properties			
S.No	Type	Fe	Cr	Mn	Ni	Si	N	C	Others	Form	Condition	UTS (MPa)	0.2%YS (MPa)	% elongation	%RA
1	AISI201	72	16	5.5	3.5	1	0.25	0.15	0.06P, 0.03S	Bar	Annealed	515	275	40	45
2	AISI202	68	17	7.5	4	<1	<0.25	<0.15	<0.06P, <0.03S	Bar	Annealed	515	275	40	–
3	AISI205	Balance	16.5	14	1	1	0.32	0.12	0.06P, 0.03S	Plate	Annealed	830	475	58	62
4	AISI304N	66.4	18	<2	8	<0.75	0.1	<0.08	<0.045P, <0.03S	Bar	Annealed	550	240	30	–
5	AISI304HN	66	18	<2	8	<1	0.16	<0.08	<0.045P, <0.03S	Bar	Annealed	620	345	30	50
6	AISI316N	61.9	16	<2	10	<0.75	0.1	<0.08	2Mo, <0.045P, <0.03S	Bar	Annealed	550	240	30	–

TABLE 15.6
Mechanical Properties of Martensitic Stainless Steel [45]

		Composition										Mechanical Properties				
S.No	Type	Fe	Cr	Mn	Ni	Si	Mo	C	Others	Form	Condition	UTS (MPa)	0.2% YS (MPa)	% elongation	% RA	Hardness
1	AISI410	Balance	11.5	<1	0.75	<1	–	<0.15	<0.04P, <0.03S	Bar	Annealed, hot finished	485	275	20	45	–
											Annealed, cold finished	485	275	16	45	–
											Intermediate temper, hot finished	690	550	15	45	–
											Intermediate temper, cold finished	690	550	12	40	–
											Hard temper, hot finished	825	620	12	40	–
											Hard temper, cold finished	825	620	12	40	–
2	AISI420	Balance	14	<1	–	–	–	0.15	<0.04P, <0.03S	Bar	Tempered 205°C	1720	1480	8	25	52 HRC
										Wire	Annealed, cold finished	860	–	–	–	
3	AISI440A	Balance	16	<1	–	<1	<0.75	0.6	<0.04P, <0.03S	Bar	Annealed	725	415	20	–	95 HRB
											Tempered 315°C	1790	1650	5	–	51 HRC
4	AISI440B	Balance	16	<1	–	<1	<0.75	0.75	<0.04P, <0.03S	Bar	Annealed	740	425	18	–	96 HRB
											Tempered 315°C	1930	1860	3	–	55 HRC
5	AISI440C	Balance	16	<1	–	<1	<0.75	0.95	<0.04P, <0.03S	Bar	Annealed	760	450	14	–	97 HRB
											Tempered 315°C	1970	1900	2	–	57 HRC

TABLE 15.7
Chemical Composition, Properties, and Types of Precipitates for Prominent Commercially Available Martensitic PH Stainless Steel [49]

Sl. No.	Alloy designation	Chemical composition	Type of precipitate	Mechanical properties
1.	17-4 PH	C<0.05, Cr-17, Ni-4.3, Cu-3.2, Mn-0.6, & Si-0.6	Copper nano-clusters	UTS- 1310 MPa, YS-1210 MPa, %El.-14, & %RA-41 (H925 condition)
2.	15-5 PH	C<0.05, Cr-15, Ni-4.3, Cu-3.2, Mn-0.6, & Si-0.6	Copper nano-clusters	UTS- 1300 MPa, YS-1210 MPa, %El.-15, & %RA-41 (H950 condition)
3.	13-8Mo PH	C<0.05, Cr-13, Ni-8.5, Mo-2,Al-1.1, Mn-0.1, & Si-0.1	NiAl	UTS- 1550 MPa, YS-1450 MPa, %El.-12, & %RA-47 (H950 condition)
4.	Custom 455	C<0.05, Cr-12, Ni-8.5, Cu-2.5,Ti-1, Mn-0.3, Si-0.3 & Nb0.3	Ni_3Ti & copper-nano clusters	UTS- 1585 MPa, YS-1515 MPa, %El.-10, & %RA-48 (H950 condition)
5.	Custom 465	C<0.02, Cr-12, Ni-11, Ti-2, Mo-1, Mn-0.2, & Si-0.2	Ni_3Ti	UTS- 1765 MPa, YS-1650 MPa, %El.-11, & %RA-49 (H950 condition)
6	Custom 475	C<0.02, Cr-11, Ni-8, Co-8, Mo-5, Al-1.2, Mn-0.4 &Si-0.4	Ni3Mo & Ni3Ti	UTS- 2005 MPa, YS1855 MPa, %El.-5, & %RA-54. (H975 condition)
7	Ferrium S53	C<0.2, Cr-10, Ni-5.5, Co14, W-1, Mo-2, V-0.3, Mn-0.1, & Si-0.1	Mo2C	UTS- 1985 MPa, YS1565 MPa & %El.-15.

TABLE 15.8
Chemical Composition, Properties, and Types of Precipitates for Prominent Commercially Available Semi-Austenitic PH Stainless Steel [50]

Sl. No.	Alloy designation	Chemical composition	Type of precipitate	Mechanical properties	
1.	A-286	C-0.05, Cr-15, Ni-25.5, Ti2.15, Mn-1.5, Mo-1.3, Si0.5, V-0.3 & Al-0.15	Ni3Ti	UTS- 1080 MPa, YS-760 MPa, %El.-28 & %RA-48. (Aged at 720°C/16 hours)	
2.	17-7 PH	C-0.1, Cr-17, Ni7.1, Al-1, Mn-0.5 & Si-0.3.	Ni3Al	TH1050	UTS- 1310 MPa, YS1100 MPa & %El.-10.
				RH950	UTS- 1520 MPa, YS1380 MPa & %El.-9.
				CH900	UTS- 1450 MPa, YS1585 MPa & %El.-2.
3.	15-7 PH	C-0.1, Cr-15, Ni7.1, Al-1, Mo-2.2, Mn-0.5 & Si-0.3.	Ni3Al	TH1050	UTS-1450 MPa, YS- 1380 MPa & %El.-7.
				RH950	UTS-1650 MPa, YS1550 MPa & %El.-6.
				CH900	UTS-1790 MPa, YS1720 MPa & %El.-2.
4.	AM-350	C-0.1, Cr-17, Ni4.3, Mo-2.8, Mn0.8, Si-0.4 & N0.1.	(Cr,Fe)2N	SCT850	UTS-1420 MPa, YS1210 MPa & %El.-12.
				SCT1000	UTS-1165 MPa, YS1020 MPa & %El.-15.
5.	AM-355	C-0.1, Cr-16, Ni4.3, Mo-2.8, Mn0.9, Si-0.4 & N0.1.	(Cr,Fe)2N	SCT850	UTS-1510 MPa, YS1250 MPa & %El.-13.
				SCT1000	UTS-1124 MPa, YS1035 MPa & %El.-22.

reach 2 GPa in tension coupled with excellent ductility, as can be seen from the data summarized in Table 15.7 and Table 15.8. The high-performance steels are tailored to obtain high strength by the precipitation of fine intermetallic phases and are often tempered to obtain good strength and resistance to stress corrosion cracking. The properties obtained for these steels have not as yet been surpassed by the high entropy alloys in tension up until now, even with the addition of expensive alloying elements like cobalt (Co). Here lies the challenge for alloy developers to combine precipitating elements in the compositions to facilitate an overall improvement in the properties to enhance the selection and use of these alloys.

The commercial tool steels that have been traditionally used in industry and their properties are summarized in Table 15.9 [51]. The microstructure of these steels permits them to reach a high hardness and strength levels, while they are relatively low cost. Tensile yield strength of approximately 2 GPa is obtained for the simple compositions and should serve as a guiding principle for alloy designers. The ultra-high-strength martensitic steels [52] based on Fe-Ni-Co-Mo system and having an exceptional strength-fracture toughness combination for damage tolerant applications is summarized in Table 15.10. Tensile yield strength levels of nearly 2.5 GPa can be obtained for these steels along with good ductility. These conventional steels developed over 50 years ago are being chosen for use in number of applications including solid rocket motor casings and aircraft landing gears. They are candidates that should be challenged by high entropy alloys for use in structural applications.

The functional applications for possible high entropy alloy alternatives to the conventional alloys are summarized in Tables 15.11 and 15.12. These two tables present the properties of materials for aerospace fasteners [53] and bearings [54], respectively. Depending on alloy composition, fasteners of different grades, ranging from 1 GPa to 1.8 GPa tensile strength, can be obtained from conventional alloys. Similarly, for bearing-relevant applications, steels with a tensile strength of 2.5 GPa are available on a commercial basis for critical applications as bearing elements for reaction wheels in spacecraft that work continuously for more than 10–15 years without the occurrence of failure.

The details specific to the commercially available superalloys strengthened by different phases and overall mechanical properties at both room temperature (21°C) and at 760°C or 870°C along with their ductility is summarized in Table 15.13 [55]. Some of the alloys retain more than 80% of their room temperature (21°C) strength even at 650°C along with good ductility. The mechanical properties of commercially available cobalt-based superalloys are summarized in Table 15.14 [56]. Many of the alloys listed in Table 15.13 and Table 15.14 are used in a welded condition and in different forms for a wide spectrum of applications. Their tensile properties can be used as the guiding values for equivalent HEA development.

The mechanical properties of titanium alloys at both room temperature (27°C) and elevated temperature for the alpha/near-alpha, alpha+beta, and beta alloys [57] are summarized in Table 15.15. These alloys have high specific strength (σ/ρ) and are often chosen for use in aerospace-related applications. Controlled processing of these alloys is essential to obtain both the desired microstructure and repeatable properties. The mechanical properties of commercial refractory alloys [58], namely niobium (Nb), molybdenum (Mo), tantalum (Ta), and tungsten (W) at elevated

TABLE 15.9
Typical Materials for Tool Steel [51]

S. No	Type of tool steel	Composition (wt.%)	Condition	Mechanical properties UTS (MPa)	0.2% YS (MPa)	% Elongation	% RA	Hardness (HRC)	Impact strength (J)
1	L2	0.7Cr-0.5Si-0.45C-0.25Mo-0.1 Mn-0.1V-0.03P-0.03S- Fe Balance	Annealed	710	510	25	50	96 HRB	–
			Oil quenched from 855°C + aged at						
			205°C	2000	1790	5	15	54	28[a]
			650°C	930	760	25	55	30	125[a]
2	L6	0.6Cr-0.5Si-0.65C-0.5Mo-0.25 Mn-0.2V-0.03P-0.03S-1.25Ni-0.25Cu-Fe Balance	Annealed	655	380	25	55	93 HRB	–
			Oil quenched from 855°C + aged at						
			315°C	2000	1790	4	9	54	12[a]
			650°C	965	830	20	48	32	81[a]
3	S1	1Cr-0.15Si-0.4C-0.5Mo-0.1 Mn-0.15V-0.03P-0.03S-0.3 Ni-0.25Cu-1.5W-Fe Balance	Annealed	690	415	24	52	96 HRB	–
			Oil quenched from 855°C + aged at						
			205°C	2070	1895	–	–	57.5	249[b]
			540°C	1680	1525	9	23	–	230[b]
			650°C	1345	1240	12	37	42	–
4	S5	0.5C-0.6 Mn-1.75Si-0.2Mo-0.25Cu-Fe Balance with <0.35Cr, <0.35V, <0.03P, and <0.03S	Annealed	725	440	25	50	96 HRB	–
			Oil quenched from 855°C + aged at						
			205°C	2345	1930	5	20	59	206[b]
			540°C	1520	1380	10	30	–	188[b]
			650°C	1035	1170	15	40	37	–
5	S7	0.45C-0.2 Mn-0.2Si-3Cr-1.3Mo-0.2V-0.25Cu-0.03 P-0.03S-Fe Balance	Annealed	640	380	25	55	95 HRB	–
			Oil quenched from 855°C + aged at						
			205°C	1820	1450	7	20	58	244[b]
			650°C	1240	1035	14	45	39	358[b]

[a] Charpy V-notch and [b] Charpy unnotched.

TABLE 15.10
Ultrahigh Strength Maraging Steels Based on Fe-Ni-Mo for Performance Critical Aerospace Applications [52]

Element (all in weight percentage)	18Ni 1400	18Ni 1700	18Ni 1900	18Ni 2400
Fe	Balance	Balance	Balance	Balance
Ni	17–19	17–19	18–19	18–19
Co	8–9	7.0–8.5	8.5–9.5	11.5–12.5
Mo	3.0–3.5	4.6–5.2	4.6–5.2	4.6–5.2
Ti	0.15–0.25	0.3–0.5	0.5–0.8	1.3–1.6
Al	0.05–0.15	0.05–0.15	0.05–0.15	0.05–0.15
Mechanical properties in solution treated condition (820°C) for 1 h				
0.2% Yield strength (MPa)	800	800	790	830
Ultimate tensile strength (MPa)	1300	1010	1010	1150
% Elongation	17	19	17	18
% Reduction in area	79	72	76	70
Hardness (Rc)	27	29	32	35
Mechanical properties in solution treated condition (820°C) for 1 h followed by aging at 480°C for 3 h				
0.2% Yield strength (MPa)	1310–1550	1650–1830	1790–2070	2460
Ultimate tensile strength (MPa)	1340–1590	1690–1860	1830–2100	2460
% Elongation	6–12	6–10	5–10	8
% Reduction in area	35–67	35–60	30–50	36
Hardness (Rc)	44–48	48–50	51–55	56–59
Young modulus (E)	181	186	190	195
Impact (Charpy) Toughness J	35–68	24–45	16–26	11
Fracture toughness (MPa-$m^{1/2}$	–	100	66	33

TABLE 15.11
Typical Aerospace Fastener Materials with Their Mechanical Properties [53]

				Typical mechanical properties				
S. No	Material	Tensile strength (MPa)	Nominal chemical composition	UTS (MPa)	0.2 YS (MPa)	% elongation ($GL = 5.65\sqrt{S}$)	Impact energy (J/cm²)	Hardness
1	A-286	1080	0.05C–15Cr-25Ni-2-1.5 Mn-1.3 Mo-0.5Si-0.3V-0.15Al	1080	760	28	–	–
2	30 NCD 16	1100	0.3C-0.25Si-0.4 Mn-1.25Cr-3.8Ni-0.5Mo-Fe Balance with S < 0.020 and P < 0.025	1080–1230	≥ 880	≥10	70	321–360 HB 35–39 HRC
3	35 NCD 16	1250	0.35C-0.3Si-0.4 Mn-1.8Cr-3.8Ni-0.5Mo-Fe Balance with S<0.020 and P<0.025	1230–1380	≥1030	≥8	50	363–401 HB 39–43 HRC
4	E 40 CDV 20	1550	0.4C-0.9Si-0.4 Mn-1.8Cr-3.8Ni-0.5Mo-Fe Balance with S<0.020 and P<0.025	1500–1700	≥1300	≥9	40	45–48 HRC
5	MP35N	1760	0.02C-0.15Mn-20Cr-35Ni-10M0-1Ti-1Fe-Balance Co with S<0.010 and P.0.015	1760	1550	12	–	50 HRC

TABLE 15.12
Materials for Bearing Applications

			Mechanical properties					
S. No	Grade	Composition	Yield strength (MPa)	Ultimate strength (MPa)	% elongation	% reduction in area	Fracture toughness (KIC) MPa m$^{1/2}$	Charpy (J)
1	AISI 52100	1.0C-0.35Mn-1.45Cr-0.3Si-Fe Balance	2034	2240	–	–	–	–
2	AISI440C	1C-0.5Mn-0.4Si-17Cr-0.5Mo-Fe-Balance	1273 (Tensile) 2027 (Compressive)	1748 (Tensile) 2516 (Compressive)	2	–	–	19
3	Cronidur30	0.33C-0.4Mn-0.8Si-15.7Cr-0.25Ni-0.9Mo-0.35N-Fe Balance	1862 (Tensile) 1864 (Compressive)	2169 (Tensile) 2610 (Compressive)	4	–	–	120
4	M50	0.8C-0.3Mn-0.2Si-4Cr-0.15Ni-4.25Mo-1V-0.2W-0.2C0-Fe Balance	2110	2600	5	10	17–20	–
5	M50NiL	0.13C-0.25Mn-0.2Si-4Cr-3.4Ni-4.25Mo-1.2V-0.1W0.2C0-Fe Balance	1175	1387	15	74	55	–
6	RBD	0.18C-0.3Mn-0.2Si-3Cr-0.75Ni-0.4V-10W-Fe Balance	1125	1410	16	52	25	–
7	T1(18-4-1)	0.75C-0.28Mn-0.28Si-4.3Cr-0.06Ni-0.07Mo-1.1V-18.2W-Fe Balance	2100	2500	<1	–	20–22	–

Note: All these bearing steels are commercially available with AISI52100 being used widely in a number of domestic appliances. Note the strength levels achieved by simple compositions with high carbon content with no expensive alloying additions. The Cr containing bearing are for niche applications that even include the momentum and reaction wheels of the satellites, yet are simple binary alloys [54].

TABLE 15.13
Mechanical Properties of Nickel-Based Superalloys [55]

S. No	Type	Composition	Form	Condition, temperature (°C)	Mechanical properties		
					UTS (MPa)	0.2% YS (MPa)	% elongation
1	Astroloy	55Ni-15Cr-17Co-5.3Mo-4Al-3.5Ti-0.06C–0.03B	bar	21	1410	1050	16
				870	770	690	25
2	Inconel 600	76Ni-15.5Cr-8Fe-0.5Mn-0.2Si-0.08C	bar	21	620	250	47
				870	105	62	80
3	Inconel 625	61Ni-21.5Cr-9Mn-3.6Nb-0.2Al-0.2Ti-2.5Fe-0.2Mn -0.2Si-0.05C	bar	21	855	490	50
				870	285	475	125
4	Inconel 718	52.5Ni-19Cr-3Mo-5.1Nb-0.5Al-0.9Ti-18.5Fe-0.2M n-0.2Si-0.04C	bar	21	1430	1190	21
				870	340	330	88
5	Nimonic 90	59Ni-19.5Cr-16.5Co-1.5Al-2.5Ti-0.3Mn-0.3Si-0.07C-0 .003B-0.06Zr	bar	21	1240	805	23
				870	430	260	16
6	Nimonic 115	60Ni-14.3Cr-13.2Co-4.9Al-3.7Ti-0.15C-0.16B-0.04Zr	bar	21	1240	860	25
				870	825	550	18
7	Pyromet 860	43Ni-12.6Cr-4Co-6Mo-1.25Al-3Ti-30Fe-0.05Mn-0. 05Si-0.05C-0.01B	bar	21	1300	835	22
				870	910	835	18
8	René 95	61Ni-14Cr-8Co-3.5Mo-3.5W–3.5Nb-3.5Al-2.5Ti-0. 15C-0.01B-0.05Zr	bar	21	1620	1310	15
				870	1170	1100	15
9	Udimet 700	55Ni-15Cr-17Co-5Mo-4Al-3.5Ti-0.06C-0.03B	bar	21	1410	965	17
				870	690	635	27
10	Waspaloy	58Ni-19.5Cr-13.5Co-4.3Mo-1.3Al-3Ti-0.08C-0.006Zr	bar	21	1280	795	25
				870	525	515	35

TABLE 15.14
Mechanical Properties of Cobalt-Based Superalloys [56]

S. No	Type	Composition	Form	Condition, temperature (°C)	Mechanical properties		
					UTS (MPa)	0.2% YS (MPa)	% elongation
1	Haynes 25 (L–605)	20Cr-15W-0.1C-3Fe-10Ni-1Si-1.5 Mn	sheet	21	1010	460	64
				870	325	240	30
2	Haynes 188	22Cr-14W-0.1C-3Fe-22Ni-0.35Si-1.25Mn-0.05La	sheet	21	960	485	56
				870	420	260	73
3	S-816	20Cr-4W-0.38C-4Fe-20Ni-4Mo-4Nb/Cb	bar	21	965	385	30
				870	360	240	16

TABLE 15.15
Mechanical Properties of Titanium Alloys at Room Temperature [57]

S. No	Class	Material	Composition	Condition	Mechanical properties					
					UTS (MPa)	0.2% YS (MPa)	% elongation	% RA	Hardness (HRC)	Impact strength (J)
1	Commercially pure	99.5% Ti (Grade 1)	0.1C-0.18O-0.03N-0.2Fe	Annealed	RT: 331 315°: 152	RT: 241 315°C: 97	RT: 30 315°C: 25	RT: 55 315°C: 80	RT: 120 (HB) 315°C:	RT: 43 (99.2%) 315°C:
2	Alpha alloys	Ti–5Al–2.5Sn	5Al-2.5Sn-0.4Fe-0.2O-0.1C-0.05N-0.0125H	Annealed	RT: 862 315°C: 565	RT: 807 315°C: 458	RT: 16 315°C: 18	RT: 40 315°C: 45	RT: 36	RT: 26
		Ti–5Al–2.5Sn (ELI)	5Al-2.5Sn-0.25Fe-0.12O-0.05C-0.035N-0.0125H	Annealed	RT: 807 –195°C: 1241	RT: 745 –195°C: 1158	RT: 16 195°C: 16	RT: – –	RT: 35	RT: 27
3	Near alpha alloys	Ti–8Al–1Mo–1V	8Al-1Mo-1V-0.015N-0.035C-0.005H-0.2Fe-0.12O	Duplex annealed	RT: 1000 540°C: 621	RT: 951 540°C: 517	RT: 15 540°C: 25	RT: 28 540°C: 55	RT: 35	RT: 32
		Ti–6Al–2Sn–4Zr–2Mo	6Al-2Sn-4Zr-2Mo-0.05N-0.05C-0.0125H-0.25Fe-0.15O	Duplex annealed	RT: 979 540°C: 648	RT: 896 540°C: 489	RT: 15 540°C: 26	RT: 35 540°C: 60	RT: 32	RT: 31

(Continued)

TABLE 15.15 (CONTINUED)
Mechanical Properties of Titanium Alloys at Room Temperature [57]

S. No	Class	Material	Composition	Condition	Mechanical pruoperties					
					UTS (MPa)	0.2% YS (MPa)	% elongation	% RA	Hardness (HRC)	Impact strength (J)
4	Alpha–beta Alloys	Ti–3Al–2.5V	3Al-2.5V-0.2Fe-0.15O-0.05C-0.03N-0.015H	Annealed	RT: 689 315°C: 483	RT: 586 315°C: 345	RT: 20 315°C: 25	RT: – –	RT: –	RT: –
		Ti–6Al–4V	6Al-4V-0.03N-0.05C-0.0125H-0.25Fe-0.12O-0.005Y	Annealed	RT: 993 540°C: 531	RT: 924 540°C: 427	RT: 14 540°C: 35	RT: 30 540°C: 50	RT: 36	RT: 19
				Solution + age	RT: 1172 540°C: 655	RT: 1103 540°C: 483	RT: 10 540°C: 22	RT: 25 540°C: 45	RT: 41	RT: –
		Ti–6Al–4V(ELI)	6Al-4V-0.1Sn-0.1Zr-0.1Mo-0.04N-0.1C-0.0125H-0.3Fe-0.16O	Annealed	RT: 896 160°C: 1517	RT: 827 160°C: 1413	RT: 15 160°C: 14	RT: 35 –	RT: 35	RT: 24
		Ti–10V–2Fe–3Al	10V-3Al-2Fe-0.13O-0.05C-0.05N-0.015H	Solution + age	RT: 1276 315°C: 1103	RT: 1200 315°C: 979	RT: 10 315°C: 13	RT: 19 315°C: 42	RT: –	RT: –
5	Beta alloys	Ti–13V–11Cr–3Al	13V-11Cr-3Al-0.05N-0.05C-0.025H-0.35Fe-0.17O	Solution + age	RT: 1220 425°C: 1103	RT: 1172 425°C: 827	RT: 8 425°C: 12	–	RT: 40	RT: 11
		Ti-15V-3Cr-3Al-3Sn*	76Ti-15V-3Al-3Sn-3Cr-0.3Fe-0.03N-0.03C-0.015H-0.13O	Annealed	RT: 785	RT: 773	RT: 22	–	RT: 95 (Rockwell)	–

*ASM Handbook, Volume 2.

temperatures are summarized in Table 15.16. Some of these alloys are used for very high-temperature applications. However, oxidation is a problem at high temperatures and a suitable coating (like the silicide coating on a niobium alloy C-103) is often given to overcome the problem.

The tables on the mechanical properties of commercial alloys spanning steels, aluminum alloys, titanium alloys, superalloys, and even the refractory alloys can be used as guiding principles for engineering the targeted properties for a HEA. The mechanical properties in tension of the developed compositions should be compared in a stable microstructural condition with the conventional alloys.

The mechanical properties of the Al-Co-Cr-Fe-Mn-Ni and the Al-Co-Cr-Cu-Fe-Mn-Ni HEAs are provided in Table 15.17 and Table 15.18 respectively. These two families form the bulk of the HEAs due essentially to an easy availability of the raw materials and affordability. Table 15.17 and Table 15.18 present both the tensile properties and compressive properties of these two alloy systems. Further, it is observed that the properties reported under compression loading are much higher than those reported under tension loading.

The mechanical properties of RHEAs at both room temperature and elevated temperatures are summarized in Table 15.19 and Table 15.20. The mechanical properties at room temperature (27°C) are predominantly reported under compression loading, whereas data is available for both compression loading and tension loading at the elevated temperatures.

The variation of yield strength (σ_{YS}) as a function of density (ρ) and specific yield strength (σ/ρ) as a function of density (ρ) for both compression test data and tensile test data for the Al-Co-Cr-Fe-Mn-Ni alloy is shown in Figure 15.14. As can be seen from Figure 15.14, the compressive test data shows higher properties than the tensile test data. Similar data for the Al-Co-Cr-Cu-Fe-Mn-Ni HEA is shown in Figure 15.15(a–d). The trend for this family is like that seen for the Al-Co-Cr-Fe-Mn-Ni alloy system.

The strength versus ductility of three commercial steels, namely a highly ductile high nitrogen steel, a highly brittle bearing steel, and a maraging steel having high strength and good fracture toughness is shown in Figure 15.16. Even though the concept of HEAs is to produce solid solution phases over intermetallic, strengthening from solid solutioning is certainly not the most efficient method, if one looks at the mechanical properties of this class of alloys. The solid solution alloys offer a good combination of strength and ductility. On the other hand, alloys strengthened by both dispersion hardening and/or precipitation-hardening by the presence of fine precipitating phases from the supersaturated solid solution upon aging supersede the solid solution alloys in terms of strength–ductility balance and form the basis for the development of a family of a large number of commercial alloys for viable engineering applications. Therefore, for structural applications, the HEAs by definition may not be an ideal choice or can challenge the existing conventional alloy systems, unless they are further strengthened by other mechanisms. These three alloy systems can be taken as references by the designers of high entropy alloys for the purpose of setting targets for the structural materials.

The high-temperature compressive flow stress data (at a strain of 0.1) of several commercial superalloys (Figure 15.17a) and commercial titanium alloys

TABLE 15.16
Mechanical Properties of Refractory Alloys [58]

S. No	Type	Material	Composition	Form	Condition	Test temperature (°C)	Mechanical property UTS (MPa)
1	Niobium and its alloys	Unalloyed Nb	–	All	Recrystallized	1093	67
		Nb–1Zr	1Zr	All		1093	158
		C103	10Hf-1Ti-0.7Zr	All		1093	186
		C129	10W-10Hf-0.1Y	Sheet		982	179
2	Molybdenum and its alloys	Unalloyed Mo	–	All	Stress-relieved annealed	1315	358
		TZM	0.5Ti-0.08Zr-0.015C	All		1315	310
		TZC	1Ti-0.14Zr-0.05C	All		1315	379
		Mo–30 W	30 W	All		1093	344
3	Tantalum alloys	Unalloyed Ta	–	All	Recrystallized	1315	58
		TA–10W	10W	All		1315	344
		FS61	7.5W	Wire, sheet	Cold worked	24	1137
4	Tungsten alloys	Unalloyed W	–	Bar, sheet, wire	Stress-relieved annealed	1648	172
		W–15 Mo	15Mo	Bar, wire		1648	248
		W–1 ThO2	1ThO2	Bar, sheet, wire		1648	255
		W–25 Re	25Re	Bar, sheet, wire		1648	227

TABLE 15.17
Mechanical Properties of Al-Co-Cr-Fe-Mn-Ni System

S. No	Material	Prior condition	Density (ρ)	Type of test	Strain rate (s^{-1})	Mechanical properties			Reference
						UTS (MPa)	0.2% YS (MPa)	% elongation	
1	CoFeNi	As cast	8.5	C	2×10^{-4}	–	204	–	[59]
2	CoFeNiSi0.75	As cast	6.6	C	2×10^{-4}	–	1301	–	[59]
3	Al0.75CoFeNi	As cast	7	C	2×10^{-4}	–	794	–	[59]
4	CoCrFeNiTi	As cast	7.2	C	1 mm/min	–	2020	9	[60]
5	Al0.7Co0.3CrFeNi	Processed by mechanical alloying and spark plasma sintering (SPS)	6.8	C	1×10^{-3}	2635	2033	8	[61]
6	AlCoCrFeNi	As cast	6.7	C	2×10^{-4}		1051		[62]
7	AlCoCrFeNi	As cast	6.7	C	2×10^{-4}	3531	1373	25	[63]
8	AlC1.5CoCrFeNi	As cast – water cooled	7	C	1×10^{-4}	2083	1255	6	[64]
9	AlCoCrFeNb0.5Ni	As cast	7	C	2×10^{-4}	3170	2473	4	[63]
10	AlCoCrFeNiSi0.4	As cast	6.2	C	1×10^{-4}	2444	1481	13	[65]
11	AlCoCrFeNiTi0.5	As cast	6.4	C	1×10^{-4}	3135	2040	24	[66]
12	AlCoCrFeNiTi0.5	As cast	6.4	C	1×10^{-4}	3140	2260	23	[67]
13	AlCoCrFeNiTi	As cast	6.2	C	1 mm/min	–	2280	6	[60]
14	CoCrFeMnNiV1.0	As cast, annealed 1000°C 24 h	7.7	C	1×10^{-3}	1845	1660	–	[68]
15	AlCrFeNi	As cast, water cooled	6.3	C	3 mm/min	2927	1406	29	[69]
16	Al1.3CrFeNi	As cast	5.9	C	2×10^{-3}	3705.3	1122.8	24.7	[70]
17	AlCrFeNiMo0.5	As cast	6.8	C	–	2644	1749	13	[69]

(Continued)

TABLE 15.17 (CONTINUED)
Mechanical Properties of Al-Co-Cr-Fe-Mn-Ni System

S. No	Material	Prior condition	Density (ρ)	Type of test	Strain rate (s^{-1})	Mechanical properties: UTS (MPa)	0.2% YS (MPa)	% elongation	Reference
18	Al0.62CoCrFeMnNi	Drop cast	7.2	T	1×10^{-3}	–	833	5	[71]
19	CoFeNi	1200°C 24 h—annealed 900°C 1 h	8.5	T	1×10^{-3}	513	211	31	[72]
20	CoCrFeNi	Drop cast	8.2	T	1×10^{-3}	413	148	48	[73]
21	CoCrFeMnNi	Drop cast	8	T	1×10^{-3}	–	208	62	[71]
22	CoCrNi	1200°C 24 h—annealed 900°C 1 h	8.3	T	1×10^{-3}	860	300	60	[72]
23	CoCrFeNi	1200°C 4 h	8.1	T	1×10^{-3}	453	190	70	[74]
24	Co35Cr15Fe20Ni20Mo10	1200°C 48 h—hot rolled 1100°C—cold rolled—800°C 1 h, air cooled	8.6	T	1×10^{-3}	1410	1311	12.1	[75]
25	Co23Cr23Fe23Ni23Mo7	Powder metallurgy, cold rolling—850°C 1 h	8.4	T	1×10^{-3}	1186.5	815.5	18.9	[76]
26	Co10Cr10Fe50Mn30	Hot rolling 900°C, homogenized 1200°C 2 h—cold rolling—900°C 3 min	7.8	T	1×10^{-3}	1380	315	50	[77]

(Continued)

TABLE 15.17 (CONTINUED)
Mechanical Properties of Al-Co-Cr-Fe-Mn-Ni System

S. No	Material	Prior condition	Density (ρ)	Type of test	Strain rate (s^{-1})	Mechanical properties UTS (MPa)	0.2% YS (MPa)	% elongation	Reference
27	Co10Cr10Fe50Mn30	Hot rolling—1200°C 5 h—friction stirred 350 rpm	7.8	T	1×10^{-3}	1400	298	45	[78]
28	AlCoCrFeNi	As cast	6.7	T	1×10^{-4}	@700°C: 400	@700°C: 395	@700°C: 1	[79]
29	AlCoCrFeNi	1000°C 6 h	6.7	T	3×10^{-4}	935	900	0.8	[80]
30	Al4Co24Cr24Fe24Ni24Ti2	1200°C 4 h—cold rolling—700°C 4 h– water quenching (P2)	7.99		1×10^{-3}	1273	1005	17	[74]
31	Al7Co23Cr23Fe23Ni23	1100°C 48 h—cold rolling—800°C 1 h	7.64	T	1.7×10^{-3}	1050	860	48	[81]
32	Al0.3CoCrFeNi	Cold rolling—1150°C 2 min—620°C 50 h	7.6	T	1×10^{-3}	840	490	45	[82]
33	Al0.3CoCrFeNi	900°C 72 h	7.6	T	4×10^{-4}	570	280	44	[83]
34	Al10Co25Cr8Fe15Ni36Ti6	1220°C 20 h—900°C 50 h	7.9	T	3.3×10^{-3}	1039	596	20	[84,85]
35	Al0.7CoCrFe2Ni	As cast	7.2	T	2×10^{-4}	@RT: 1223 @700°C: 340	@RT: 866 @700°C: 220	@RT: 7.9 @700°C: 5.4	[86]
36	Al17Co17Cr17Fe17Ni33.3	960°C 50h	7.04	–	1×10^{-3}	@RT: 1030 @−70°C: 1034 @−192°C: 1050	@RT: 545 @−70°C: 580 −192°C: 700	@RT: 16.5 @−70°C: 10.5 −192°C: 4	[85]

(Continued)

TABLE 15.17 (CONTINUED)
Mechanical Properties of Al-Co-Cr-Fe-Mn-Ni System

S. No	Material	Prior condition	Density (ρ)	Type of test	Strain rate (s^{-1})	Mechanical properties: UTS (MPa)	Mechanical properties: 0.2% YS (MPa)	Mechanical properties: % elongation	Reference
37	Al0.3CoCrFeNi	As cast, 900°C 72 h, water quenched	7.6	T	4×10^{-4}	570	240	45	[83]
38	CoCrFeMnNi	1200°C 48 h, cold rolling—800°C 1 h	8	T	1×10^{-3}	@RT: 651 @800°C: 145	@RT: 362 @800°C: 127	@RT: 51 @800°C: 51	[87]
39	CoCrFeMnNi (Cantor+2 at.% C)	Forged 700°C—1000°C 6 h—cold rolling—900°C 6 h	8	T	1×10^{-3}	857	581	28	[88]
40	CoCrFeMnNi	As cast, cold-forged and cross-rolled 60%, 800°C 1 h	8	T	–	@−196°C: 1280 @RT: 763	@−196°C: 759 @RT: 410	@−196°C: 71 @RT: 57	[19]
41	CoCrFeNi	As cast, 1000°C 24 h, Hot rolled to 92% @ 1000°C, 900°C 1 h	8.2	T	1×10^{-3}	@−196°C: 1109 @RT: 671 @800°C: 320	@−196°C: 480 @RT: 300 @800°C: 80	@−196°C: 66 @RT: 42 @800°C: 28	[89]
42	CoCrMnNi	As cast, 1100°C 24 h, cold rolled 90%, 1000°C 1 h	8.1	T	1×10^{-3}	@−196°C: 1150 @RT: 699 @400°C: 550	@−196°C: 499 @RT: 280 @400°C: 186	@−196°C: 62 @RT: 43 @400°C: 28	[72]
43	CoFeMnNi	As cast, 1100°C 24 h, cold rolled 90%, 1000°C 1 h	8.2	T	1×10^{-3}	@−196°C: 835 @RT: 551 @400°C: 465	@−196°C: 300 @RT: 175 @400°C: 116	@−196°C: 48 @RT: 41 @400°C: 37	[72]

TABLE 15.18
Mechanical Properties of Al-Co-Cr-Cu-Fe-Mn-Ni System

						Mechanical properties			
S. No	Material	Prior condition	Density (ρ)	Type of test	Strain rate (s^{-1})	UTS (MPa)	0.2% YS (MPa)	% elongation	References
1	AlCrCuFeNi0.8	650°C 4h	6.7	C	3×10^{-4}	2000	1000	45	[90]
2	CoCrCuFeNiTi0.5	As cast	7.8	C	1×10^{-4}	1650	700	29	[66]
3	CoCrCuFeNiTi	As cast	7.4	C	1×10^{-4}	1272	1272	2	[91]
4	Al0.25CoCrCu0.75FeNiTi0.5	As cast	7.5	C	1×10^{-4}	1970	750	39	[66]
5	AlCoCrCuFeNi	As cast	7.1	C	–	–	1303	24	[92]
6	AlCoCrCuFeNiV	As cast	6.9	C	–	–	1469	16	[92]
7	Al0.75CoCrCu0.25FeNiTi0.5	As cast	6.8	C	1×10^{-4}	2697	1900	12	[66]
8	AlCoCrCuNiTi	As cast	6.4	C	–	1495		8	[93]
9	AlCoCrCuNiTiY0.8		5.9	C	–	1325	–	5	[93]
10	AlCuFeNiTi	500°C and 900°C 30 min	6.1	C	5×10^{-4}	1617	1074	8	[94]
11	Al4Co19Cr19Cu4Fe19Ni37	1200°C 24 h – cold-rolling – 700°C 20 h, 800°C 1 h	7.4	T	1.7×10^{-3}	1048	719	30	[95]
12	Al0.5CoCrCuFeNi	1000°C 6 h, quenched, cold rolled	7.6	T	–	1344	1284	7.6	[96]
13	AlCoCrCuFeNi	As cast	7.1	T	1×10^{-3}	@700°C: 360	@700°C: 350	@700°C: 5	[97]

(*Continued*)

TABLE 15.18 (CONTINUED)
Mechanical Properties of Al-Co-Cr-Cu-Fe-Mn-Ni System

S. No	Material	Prior condition	Density (ρ)	Type of test	Strain rate (s^{-1})	Mechanical properties			References
						UTS (MPa)	0.2% YS (MPa)	% elongation	
14	Al0.5CoCrCuFeNi (9.1 at.% Al)	As cast, 1000°C 6 h, cold rolling 84%	7.6	T	1×10^{-3}	1344	1284	23	[98]
15	AlCoCrCuFeNi (16.7 at.% Al)	As cast	7.1	T	1×10^{-3}	@1000°C: 44	@1000°C: 37	@1000°C: 77	[97]
16	Al0.5CrCuFeNi2 (9.1 at.% Al)	As cast, cold rolling 43%, 900°C 24 h	7.6	T	1×10^{-3}	1088 ± 20	704 ± 180	5.6 ± 3.2	[99]
17	CoCuFeNi	As cast	8.6	T	0.2 mm/min	480	–	15	[100]
18	CoCuFeNiSn0.07	As cast	8.6	T	0.2 mm/min	632	–	19	[100]
19	CoCuFeMnNi	As cast	8.4	T	0.1 mm/min 1×10^{-4}	478	–	14	[101]

TABLE 15.19
Mechanical Properties of High Entropy Alloys at Room Temperature

S. No	Material	Prior condition	Density (ρ)	Type of test	Strain rate (s^{-1})	Mechanical properties: 0.2% YS (MPa)	Mechanical properties: σy/ρ (MPa/g-cm^{-3})	References
1	AlMoNbTiV	As cast	6.4	C	5×10^{-4}	1375	214.9	[102]
2	Al0.25MoNbTiV	As cast	7.1	C	5×10^{-4}	1250	176.9	[102]
3	Al0.5MoNbTiV	As cast	6.8	C	5×10^{-4}	1625	238.3	[102]
4	Al0.75MoNbTiV	As cast	6.6	C	5×10^{-4}	1260	191	[102]
5	AlNbTaTiV	As cast	7.9	C	2×10^{-4}	991	125.6	[24]
6	Al0.25NbTaTiV	As cast	8.8	C	2×10^{-4}	1330	151.2	[24]
7	Al0.5NbTaTiV	As cast	8.5	C	2×10^{-4}	1012	119.6	[24]
8	AlNbTa0.5TiZr0.5	HIP @ 1400°C and 207 MPa 2 h, 1400°C 6 h	6.9	C	1×10^{-3}	1352	195.3	[103]
9	Al0.25NbTaTiZr	HIP @ 1400°C and 207 MPa 2h, 1400°C 6 h	8.6	C	1×10^{-3}	1745	203.1	[103]
10	AlMo0.5NbTa0.5TiZr	HIP @ 1400°C and 207 MPa 2 h, 1400°C 6 h	7.1	C	1×10^{-3}	2197	307.4	[103]
11	AlMo0.5NbTa0.5TiZr0.5	HIP @ 1400°C and 207 MPa 2 h, 1400°C 6 h	7.2	C	1×10^{-3}	1320	183.3	[103]
12	Al0.5Mo0.5NbTa0.5TiZr	HIP @ 1400°C and 207 MPa 2 h, 1400°C 6 h	7.6	C	1×10^{-3}	2350	309.7	[103]
13	AlMoTaTiV	As cast	8.2	C	1×10^{-3}	735	89.6	[104]
14	Al0.2MoTaTiV	As cast	9.3	C	1×10^{-3}	1021	110.3	[104]
15	Al0.6MoTaTiV	As cast	8.7	C	1×10^{-3}	962	110.9	[104]

(Continued)

TABLE 15.19 (CONTINUED)
Mechanical Properties of High Entropy Alloys at Room Temperature

						Mechanical properties		
S. No	Material	Prior condition	Density (ρ)	Type of test	Strain rate (s^{-1})	0.2% YS (MPa)	σy/ρ (MPa/ g-cm^{-3})	References
16	AlHfNbTaTiZr	As cast	8.9	C	1×10^{-3}	1489	168	[105]
17	Al0.3HfNbTaTiZr	As cast	9.6	C	1×10^{-3}	1188	124.4	[105]
18	Al0.5HfNbTaTiZr	As cast	9.3	C	1×10^{-3}	1302	139.4	[105]
19	Al0.75HfNbTaTiZr	As cast	9.1	C	1×10^{-3}	1415	155.6	[105]
20	Al0.4Hf0.6NbTaTiZr	HIP @ 1200°C and 207 MPa 2 h, 1200°C 24 h	9.1	C	1×10^{-3}	1841	202.5	[106,107]
21	AlNb1.5Ta0.5Ti1.5Zr0.5	HIP @ 1400°C and 207 MPa 2 h, 1400°C 24 h	6.8	C	1×10^{-3}	1280	186.9	[106]
22	Al0.3NbTa0.8Ti1.4V0.2Zr1.3	HIP @ 1200°C and 207 MPa 2 h, 1200°C 24 h	7.7	C	1×10^{-3}	1965	255	[106]
23	Al0.3NbTaTi1.4Zr1.3	HIP @ 1200°C and 207 MPa 2 h, 1200°C 24 h	8.1	C	1×10^{-3}	1965	242.9	[106]
24	Al0.5NbTa0.8Ti1.5V0.2Zr	HIP @ 1200°C and 207 MPa 2 h, 1200°C 24 h	7.6	C	1×10^{-3}	2035	269.2	[106]
25	Al0.5CrNbTi2V0.5	1200°C 24 h	5.8	C	1×10^{-4}	1340	232.4	[108]
26	AlCrNbTiV	1200°C 24 h	5.8	C	1×10^{-4}	1550	269.2	[109]
27	AlCr0.5NbTiV	1200°C 24 h	5.6	C	1×10^{-4}	1300	230.6	[109]
28	AlCr1.5NbTiV	1200°C 24 h	5.9	C	1×10^{-4}	1700	290.1	[109]
29	AlMo0.5NbTa0.5TiZr	HIP @ 1400°C and 207 MPa 2 h, 1400°C 24h	7.1	C	1×10^{-3}	2000	279.8	[106,107,110]

(Continued)

TABLE 15.19 (CONTINUED)
Mechanical Properties of High Entropy Alloys at Room Temperature

S. No	Material	Prior condition	Density (ρ)	Type of test	Strain rate (s^{-1})	Mechanical properties 0.2% YS (MPa)	Mechanical properties σy/ρ (MPa/g-cm^{-3})	References
30	AlNbTiV	1200°C 24 h	5.5	C	1×10^{-3}	1020	185.6	[111]
31	AlNbTiV	1200°C 24 h	5.5	C	1×10^{-4}	1000	181.9	[109,112]
32	AlNbTiVZr1.5	1200°C 24 h	5.8	C	1×10^{-4}	1535	262.6	[112]
33	C0.3Hf0.5Mo0.5NbTiZr	As cast	7.7	C	1×10^{-3}	1201	156.2	[113]
34	CoCrMoNbTi0.2	As cast	8.5	C	1×10^{-3}	1905.6	224.2	[114]
35	CrHfNbTiZr	500°C 600 s	8.2	C	5×10^{-4}	1457	176.9	[115]
36	CrMo0.5NbTa0.5TiZr	HIP @1450°C and 207 MPa 3 h	8	C	1×10^{-3}	1595	199.5	[116]
37	CrNbTiVZr	HIP @1450°C and 207 MPa 3 h, 1200°C 24h	6.6	C	1×10^{-3}	1298	197.8	[28]
38	CrNbTiZr	As cast, HIP @ 1200°C 207 MPa 2 h, 1200°C 24 h	6.67	C	1×10^{-3}	1260	188	[28]
39	CrTaVW	Sintered at 1500°C and 50 MPa	13	C	1×10^{-3}	2327	178.5	[117]
40	Hf0.4Nb1.54Ta1.54Ti0.89Zr0.64	As cast	10.4	C	1×10^{-4}	822	79.1	[118]
41	Hf0.5Mo0.5NbSi0.9TiZr	As cast	6.8	C	1×10^{-3}	1650	241.5	[119]
42	Hf0.5Mo0.5NbTiZr	As cast	7.9	C	1×10^{-3}	1176	149.4	[113]
43	Hf0.5Nb0.5Ta0.5Ti1.5Zr	As cast	8.2	C	1×10^{-3}	903	110.3	[120]
44	Hf15Ti30Zr25Nb20Ta10	As cast	8.4	C	3×10^{-4}	1150	136.7	[121]
45	HfMo0.5NbSi0.7TiV0.5	As cast	7.9	C	1×10^{-3}	2134	270.1	[122]

(Continued)

TABLE 15.19 (CONTINUED)
Mechanical Properties of High Entropy Alloys at Room Temperature

S. No	Material	Prior condition	Density (ρ)	Type of test	Strain rate (s^{-1})	Mechanical properties 0.2% YS (MPa)	σy/ρ (MPa/ g-cm^{-3})	References
46	HfMoNbTaTiZr	As cast	9.9	C	1×10^{-3}	1512	152.1	[123,124]
47	HfMoNbTiZr	As cast	8.7	C	1×10^{-3}	1719	197.9	[125]
48	HfMoNbTiZr	As cast, 1100°C 10 h, Slow cooled	8.7	C	1×10^{-3}	1575	181	[125]
49	HfMoTaTiZr	As cast	10.2	C	1×10^{-3}	1600	157	[124]
50	HfNbSi0.5TiV	As cast	7.8	C	1×10^{-3}	1399	179.3	[126]
51	HfNbSi0.5TiVZr	As cast	7.5	C	1×10^{-3}	1540	204.9	[127]
52	HfNbTaZr	1800°C 24 h	11.1	C	1×10^{-3}	2310	208.8	[128]
53	HfNbTiVZr	As cast	11.1	C	5×10^{-4}	1170	105.4	[115]
54	HfNbTiVZr	500°C 600 s	8.1	C	5×10^{-4}	1253	155.5	[115]
55	Mo2NbTiVZr	1000°C 72 h	7.6	C	2×10^{-4}	1765	232.6	[129]
56	MoNbTaTi0.75W	As cast	12.2	C	5.6×10^{-4}	1304	107.3	[130]
57	MoNbTaTiV	As cast	9.4	C	5×10^{-4}	1400	149.4	[131]
58	MoNbTaTiVW	As cast	11	C	1×10^{-3}	1515	138.1	[132]
59	MoNbTaTiW	As cast	11.8	C	5.6×10^{-4}	1455	123.8	[130]
60	MoNbTaTiW	As cast	11.8	C	1×10^{-3}	1343	114.2	[132]
61	MoNbTaTiZr	As cast	9.1	C	1×10^{-3}	1390	152.2	[133]
62	MoNbTaTiZr	As cast	9.1	C	1.67×10^{-4}	1375	150.5	[134]
63	MoNbTaV	As cast	10.7	C	5×10^{-4}	1525	142.7	[135]
64	MoNbTaVW	SPS @ 1500°C,	12.4	C	1×10^{-3}	2612	211	[136]

(*Continued*)

TABLE 15.19 (CONTINUED)
Mechanical Properties of High Entropy Alloys at Room Temperature

S. No	Material	Prior condition	Density (ρ)	Type of test	Strain rate (s^{-1})	Mechanical properties 0.2% YS (MPa)	σy/ρ (MPa/ g-cm^{-3})	References
65	MoNbTaVW	1400°C 19 h,	12.4	C	1×10^{-3}	1246	100.7	[137]
66	MoNbTaW	As cast	13.7	C	5.6×10^{-4}	996	72.9	[130]
67	MoNbTaW	1400°C 19 h,	13.7	C	1×10^{-3}	1058	77.5	[137]
68	MoNbTiV	As cast	7.3	C	5×10^{-4}	1200	163.4	[102]
69	MoNbTiVZr	1000°C 72 h	7.1	C	2×10^{-4}	1779	249.7	[129]
70	MoNbTiVZr	As cast	7.1	C	2×10^{-4}	1770	248.5	[138]
71	MoTaTiV	As cast	9.6	C	1×10^{-3}	1221	127.4	[104]
72	NbTaTiV	As cast	9.2	C	5×10^{-4}	965	105.2	[139]
73	NbTaVW	As cast	12.9	C	5×10^{-4}	1530	118.7	[139]
74	NbTiVZr	HIP @1450°C and 207 MPa 3h, 1200°C 24 h	6.5	C	1×10^{-3}	1105	171.1	[28]
75	NbTiVZr	1000°C 72 h	6.5	C	2×10^{-4}	1104	170.9	[129]
76	NbTiV2Zr	As cast, HIP @ 1200°C/ 207 MPa 2 h, 1200°C/24 h	6.34	C	1×10^{-3}	918	145.7	[28]
77	HfNbTaTiZr	As cast	9.9	T	5×10^{-3}	828	83.7	[140]
78	HfNbTiZr	1300°C 6 h	8.4	T	1×10^{-3}	879	104.8	[141]
79	HfTaTiZr	As cast	10.2	T	1×10^{-3}	1500	147.3	[142]

TABLE 15.20
Mechanical Properties of High Entropy Alloys at High Temperatures

S. No	Material	Prior condition	Density (ρ)	Type of test	Temperature, °C	Strain rate, s^{-1}	Mechanical properties: 0.2% YS (MPa)	Mechanical properties: σy/ρ (MPa/g-cm^{-3})	References
1	Al0.25NbTaTiZr	HIP @ 1400°C and 207 MPa 2 h, 1400°C 6 h	8.6	C	1000	1×10^{-3}	366	42.6	[103]
2	AlNbTa0.5TiZr0.5	HIP @ 1400°C and 207 MPa 2 h, 1400°C 6 h	6.9	C	1000	1×10^{-3}	535	77.3	[103]
3	Al0.3NbTa0.8Ti1.4V0.2Zr1.3	HIP @ 1200°C and 207 MPa 2 h, 1200°C 24 h	7.7	C	1000	1×10^{-3}	166	21.5	[106]
4	Al0.3NbTaTi1.4Zr1.3	HIP @ 1200°C and 207 MPa 2 h, 1200°C 24 h	8.1	C	1000	1×10^{-3}	236	29.2	[106]
5	AlNb1.5Ta0.5Ti1.5Zr0.5	HIP @ 1400°C and 207 MPa 2 h, 1400°C 6 h	6.8	C	1000	1×10^{-3}	403	58.8	[106]
6	Al0.4Hf0.6NbTaTiZr	HIP @ 1200°C and 207 MPa 2 h, 1200°C 24 h	9.1	C	1000	1×10^{-3}	298	32.8	[106]
7	Al0.4Hf0.6NbTaTiZr	HIP @ 1200°C and 207 MPa 2 h, 1200°C 24 h	9.1	C	1200	1×10^{-3}	89	9.8	[107]
8	Al0.5CrNbTi2V0.5	1200°C 24 h	5.8	C	1000	1×10^{-4}	90	15.6	[108]
9	Al0.5Mo0.5NbTa0.5TiZr	HIP @ 1400°C and 207 MPa 2 h, 1400°C 6 h	7.6	C	1000	1×10^{-3}	579	76.3	[103]
10	AlMo0.5NbTa0.5TiZr	HIP @ 1400°C and 207 MPa 2 h, 1400°C 6 h	7.1	C	1000	1×10^{-3}	745	104.2	[103,106]
11	AlMo0.5NbTa0.5TiZr	HIP @ 1400°C and 207 MPa 2 h, 1400°C 6 h	7.1	C	1200	1×10^{-3}	250	35	[107,110]

(Continued)

TABLE 15.20 (CONTINUED)
Mechanical Properties of High Entropy Alloys at High Temperatures

S. No	Material	Prior condition	Density (ρ)	Type of test	Temperature, °C	Strain rate, s^{-1}	Mechanical properties 0.2% YS (MPa)	Mechanical properties σy/ρ (MPa/g-cm^{-3})	References
12	AlMo0.5NbTa0.5TiZr0.5	HIP @ 1400°C and 207 MPa 2 h, 1400°C 6 h	7.2	C	1000	1×10^{-3}	935	129.1	[103]
13	AlMo0.5NbTa0.5TiZr	As cast, HIP @ 1400°C/ 207 MPa/2 h, 1400°C/24 h	7.4	C	1200	1×10^{-3}	255	34.4	[107]
14	Al0.5NbTa0.8Ti1.5V0.2Zr	HIP @ 1200°C and 207 MPa 2 h, 1200°C 24 h	7.6	C	1000	1×10^{-3}	220	29.1	[106]
15	AlCr0.5NbTiV	1200°C 24 h	5.6	C	1000	1×10^{-4}	40	7.1	[109]
16	AlCr1.5NbTiV	1200°C 24 h	5.9	C	1000	1×10^{-4}	75	12.8	[109]
17	AlCrMoNbTi	1300°C 20 h	6.6	C	400	1×10^{-3}	1080	0.02	[143]
18	AlCrMoNbTi	1300°C 20 h	6.6	C	1200	1×10^{-3}	105	16	[143]
19	AlCrMoNbTi	1300°C 20 h	6.6	C	1200	1×10^{-3}	150	22.8	[144]
20	AlCrMoTi	1200°C 20 h	6	C	1000	1×10^{-3}	375	62.7	[144]
21	AlCrNbTiV	1200°C 24 h	5.8	C	1000	1×10^{-4}	65	11.3	[109]
22	AlMoNbTi	1500°C 20 h	6.5	C	1200	1×10^{-3}	200	31	[144]
23	AlNbTiV	1200°C 24 h	5.5	C	800	1×10^{-4}	560	101.9	[112]
24	AlNbTiV	1200°C 24 h	5.5	C	1000	1×10^{-4}	110	20	[109]
25	AlNbTiV	1200°C 24 h	5.5	C	1000	1×10^{-3}	158	28.7	[111]
26	AlNbTiVZr0.1	1200°C 24 h	5.5	C	800	1×10^{-4}	865	156.4	[112]
27	CrMo0.5NbTa0.5TiZr	HIP @1450°C and 207 MPa 3 h	8	C	1200	1×10^{-3}	170	21.3	[116]
28	CrNbTiVZr	HIP @1200°C and 207 MPa 2 h, 1200°C 24 h	6.6	C	1000	1×10^{-3}	259	39.5	[28]

(Continued)

TABLE 15.20 (CONTINUED)
Mechanical Properties of High Entropy Alloys at High Temperatures

S. No	Material	Prior condition	Density (ρ)	Type of test	Temperature, °C	Strain rate, s^{-1}	Mechanical properties: 0.2% YS (MPa)	Mechanical properties: σy/ρ ($MPa/g\text{-}cm^{-3}$)	References
29	CrNbTiZr	As cast, HIP @ 1200°C/ 207 MPa/2 h, 1200°C/24 h	6.7	C	1000	1×10^{-3}	115	17.2	[28]
30	CrTaVW	Sintered at 1500°C and 50 MPa	13	C	1200	1×10^{-3}	979	75.1	[117]
31	Hf0.4Nb1.54Ta1.54Ti0.89Zr0.64	As cast	10.4	C	300	1×10^{-4}	590	56.8	[118]
32	Hf15Ti30Zr25Nb20Ta10	Cold rolled 40%, 997°C 1 h,	8.4	C	72	3×10^{-4}	1040	123.7	[121]
33	HfMo0.5NbSi0.7TiV0.5	As cast	7.9	C	1200	1×10^{-3}	235	29.7	[122]
34	HfMoNbTaTiZr	As cast	9.9	C	1200	1×10^{-3}	556	55.9	[124]
35	HfMoNbTiZr	1100°C 10 h	8.7	C	1200	1×10^{-3}	187	21.5	[125]
36	HfMoTaTiZr	As cast	10.2	C	1200	1×10^{-3}	404	39.6	[124]
37	HfNbSi0.5TiV	As cast	7.8	C	1000	1×10^{-3}	240	30.8	[126]
38	HfNbSi0.5TiVZr	As cast	7.5	C	800	1×10^{-3}	371	49.4	[127]
39	HfNbTaTiZr	HIP @ 1200°C and 207 MPa 2 h, 1200°C 24 h	9.9	C	1200	1×10^{-3}	92	9.3	[145]
40	MoNbTaTiVW	As cast	11	C	1200	1×10^{-3}	659	60.1	[132]
41	MoNbTaTiW	As cast	11.8	C	1200	1×10^{-3}	586	49.8	[132]
42	MoNbTaVW	1400°C 19 h,	12.4	C	1600	1×10^{-3}	477	38.5	[137]
43	MoNbTaW	1400°C 19 h,	13.7	C	1600	1×10^{-3}	405	29.6	[137]
44	NbTiV2Zr	HIP @1200°C and 207 MPa 2 h, 1200°C 24 h	6.4	C	1000	1×10^{-3}	72	11.2	[28]
45	NbTiVZr	As cast, HIP @ 1200°C 207 MPa 2 h, 1200°C 24 h	6.5	C	1000	1×10^{-3}	58	8.9	[28]

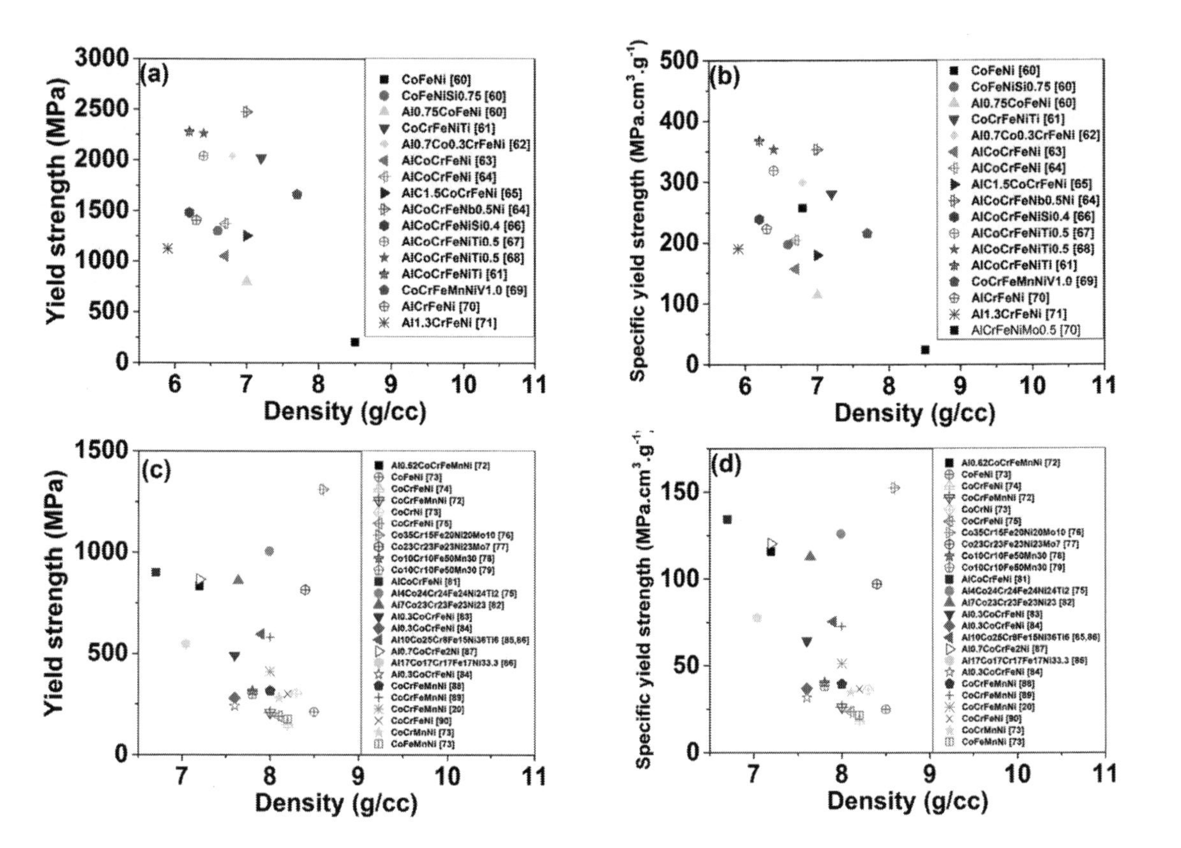

FIGURE 15.14 The variation of yield strength and specific yield strength as a function of density in compression (a and b) and tension (c and d) for Al-Co-Cr-Fe-Mn-Ni high entropy alloys. Note the significant difference in the compressive and tensile properties for these materials. The data has been obtained from the literature.

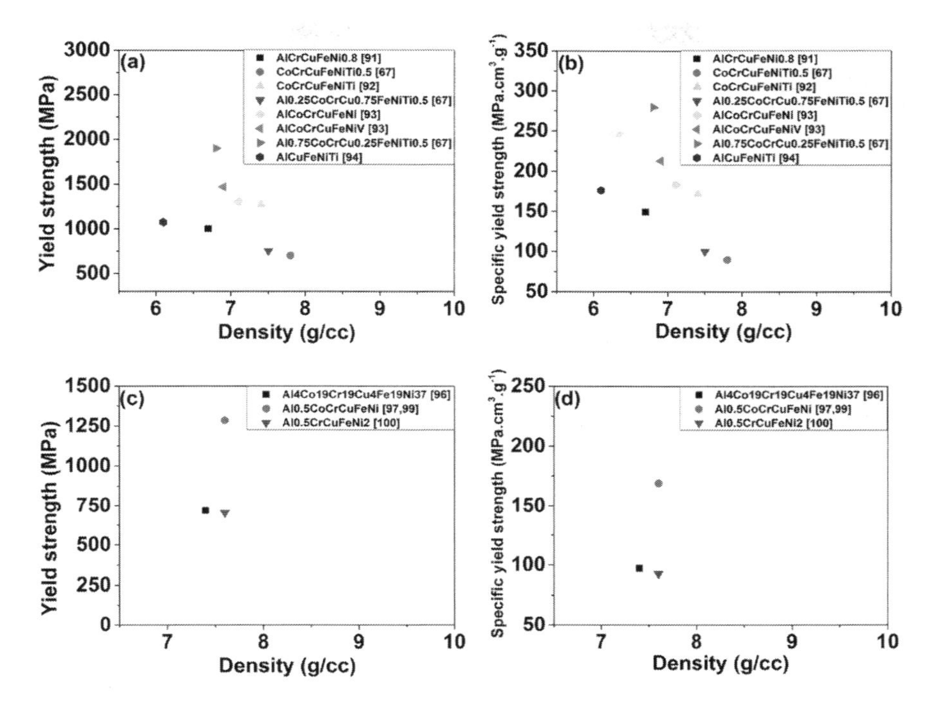

FIGURE 15.15 The variation of yield strength and specific yield strength as a function of density in compression (a and b) and tension (c and d) for Al-Co-Cr-Cu-Fe-Mn-Ni high entropy alloys. Note the significant difference in the compressive and tensile properties for these materials. The data has been obtained from the literature.

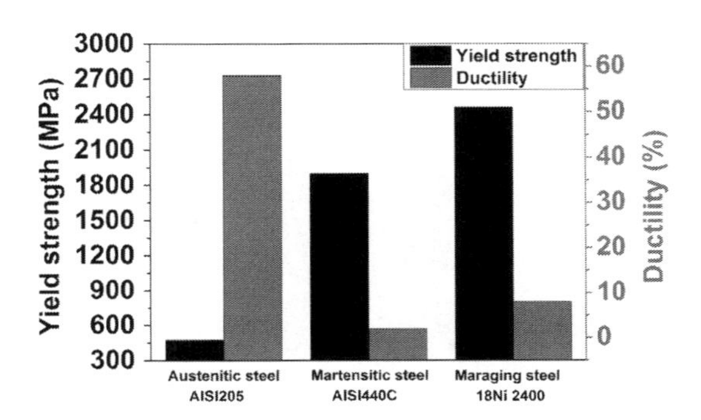

FIGURE 15.16 Data of tensile yield strength and ductility for three different types of materials viz., highly ductile nitrogen-bearing austenitic stainless steel AISI205, high strength martensitic steel AISI440C used in stainless bearings and ultrahigh strength maraging steels of 18Ni2400.

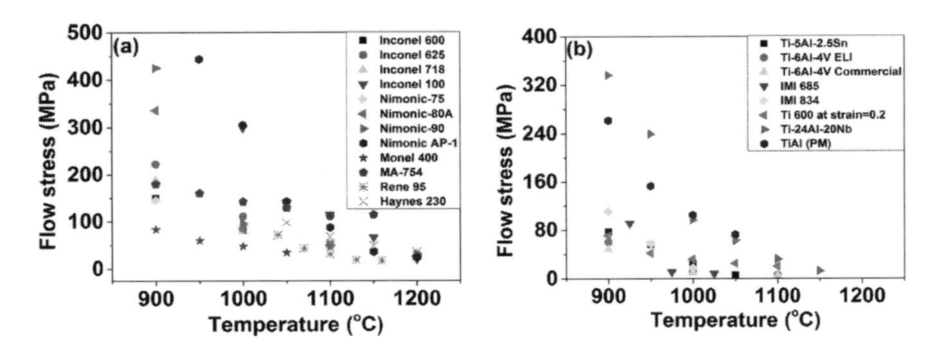

FIGURE 15.17 Variation of flow stress of commercial alloys as a function of test temperature for: (a) superalloys, and (b) titanium alloys [146].

(Figure 15.17b) at four different temperatures [146] is as shown in Figure 15.17. The test data reveals a decrease in flow stress with an increase in test temperature, as expected for both materials. The compressive yield strength of RHEAs as a function of density at room temperature is shown in Figure 15.18a, and yield strength as a function of temperature is shown in Figure 15.18b.

The key and observable advantages of alloying are that the mechanical properties of an alloy are often superior to that of the individual elements that are chosen to form the alloy. The alloys are often designed by selecting the base metal and looking at how additions of substitutional elements and/or interstitial elements could bring about an

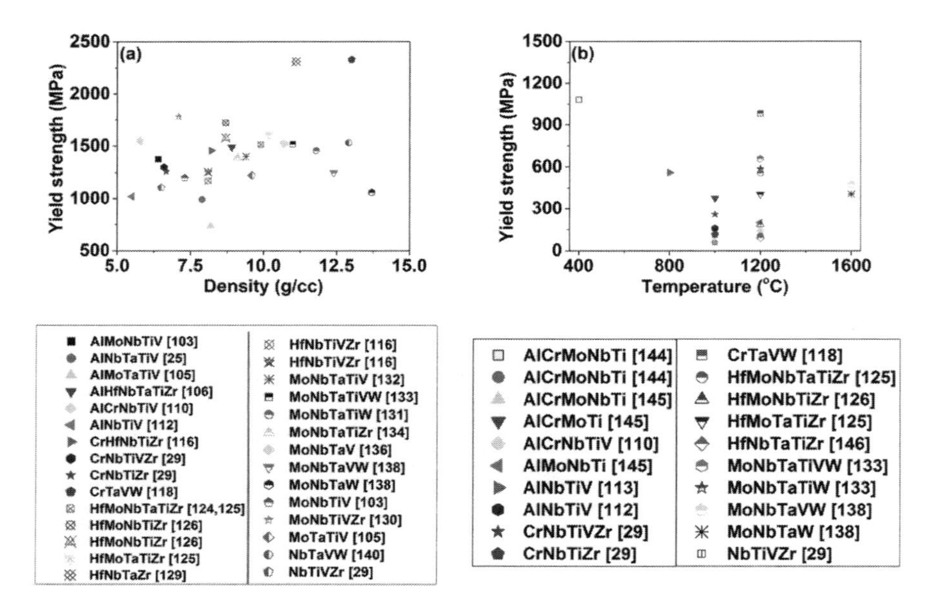

FIGURE 15.18 (a) The variation of compressive yield strength as a function of density, and (b) the variation of compressive yield strength with temperature, for the refractory high entropy alloys (RHEAs).

improvement in the desired properties. Further, the addition of major and/or minor alloying elements to the alloy is often made with a specific purpose as the end objective. Based on both chemistry and heat treatment conditions, they are often tuned to obtain a combination of mechanical properties that is both essential and desired for operating under the applied service conditions. Alloys exhibiting high strength are often subjected to aging and tempering treatments to provide a good balance of strength, ductility, and toughness so as to enable in their selection and use for structural applications. The strength of a metal can be significantly increased by either alloying additions or through thermomechanical processing, but it may not be useful for structural applications involving tensile loads, as it may lose its two other important design properties, namely: (i) toughness, and (ii) ductility. Also, the HEAs are highly concentrated compositions and can be a nightmare for environmentalists. A separation of the elements from the alloy is often as difficult as their addition, thereby making them least suitable for re-use. Unless there are strong reasons for their re-use, their production and future at the bulk/mass level is often questionable. Thus, it is important that research be aimed at finding the exotic properties these alloys have to offer, which otherwise cannot be obtained through the conventional compositions. Overall, considerable effort in the domains spanning research and development is being made with the prime objective of replacing the time-tested existing metals and their alloy counterparts. This is possible by creating a confidence level in the minds of designers that the new and improved HEAs are superior in all aspects while concurrently offering better properties.

For the purpose of functional applications, many avenues exist for the emerging family of HEAs as a viable substitute for the conventional alloys. The list is unending and can include the following:

1. A much improved and an overall better alloy having superior electrical conductivity than the presently used aluminum alloys for use as overhead electric transmission lines.
2. An alloy having a higher thermal conductivity than pure copper for use in high heat flux applications.
3. A better alloy for use as bearings that offer improved wear and corrosion resistance.
4. A non-rare earth-containing permanent magnetic alloy as a superior alternative to the existing counterparts.
5. An alternative to copper-beryllium (Cu-Be) alloys, without toxic beryllium (Be) for the purpose of applications requiring high stiffness, such as springs.
6. Non-nickel (Ni)- and cobalt (Co)-containing low expansion alternative to both Invar and Kovar.
7. A stainless steel without 18 weight percentage chromium (Cr) and offering good stress corrosion resistance.
8. A high strength steel (>1,250 MPa strength) without being easily susceptible to stress corrosion cracking.

Unlimited opportunities exist for the physical metallurgists to design and develop new and improved HEAs and the sky is the limit for their ideas.

Until now, a sizeable majority of the efforts in the domain specific to research on HEAs has been directed toward demonstrating the few to many attractive properties and advantages the family of HEAs have to offer. However, up until now there has been little emphasis with specific reference to working on superior alternate HEAs both as a viable challenge to the existing alloys and a potentially viable replacement. A few claims have been made in the published research on the possible areas where the family of HEAs could be put to effective use. Now it is time to move ahead and give serious thought to those areas where these alloys could be put to use while concurrently identifying the compositions that show much promise for continued interest and studies. The interested researcher should not be lost in the overall vastness of the field of HEAs. Also, due consideration and importance must be given to cost, ease of manufacture, and the assurance of recyclability to have a final say on the future of the family of HEAs, be it one or more. Now that the idea behind developing, characterizing, testing, and analyzing the high entropy alloys have been put forth by several independent research studies conducted during the last two decades i.e., essentially after 1995, the material's community is now looking forward to the commercial production and emergence of these alloys for use in a spectrum of applications ranging from structures to functional products.

15.9 CONCLUDING HIGHLIGHTS

Ever since the introduction of multicomponent alloys (with no specific matrix base in which alloying additions are made) against conventional alloys (having a matrix base to which alloying additions are made for specific reasons), there has been significant curiosity about developing materials that outsmart the conventional alloys. The foundations for conventional alloys are based on certain principles (e.g., Hume–Rothery rules for both interstitial solid solution and substitutional solid solution) and invoke the principles of precipitation-hardening, dispersion hardening, grain-size strengthening, and even thermomechanical processing for obtaining the desired strength–toughness combination for use in engineering applications. On the other hand, no such specific rules exist for the multicomponent alloys, even though it was originally restricted to the alloys having five or more number of alloying elements and named as high entropy alloys. These alloys were subsequently rechristened as complex concentrated alloys (CCA) to include the commercial alloys with a high concentration of the alloying elements, like the families of (i) superalloys, (ii) austenitic stainless steel, and (iii) maraging steels. This leaves one to wonder whether these alloys have been in commercial use for several decades, and attempts are only being made to reinvent the wheel. Developers of the HEA choose the multicomponent material systems predominantly based on the availability of the raw materials, ending up with systems having multiple phases including the intermetallic phases. With a significant number of researchers adopting the powder metallurgical processing route to realize the compositions, the inherent problems of contamination during ball milling lead to unwanted or undesirable additional phases, which impairs the properties spanning both physical and mechanical. Another popular route adopted is the arc melting route to prepare the compositions using multiple melting steps (often as high as 7–8 times) with the prime objective of achieving homogeneity of the material. Usually the raw materials are the high purity virgin elements. While powder metallurgy processing is limited by size of the component that can be obtained, the arc melted compositions cannot be directly scaled-up to industrial melts and it is rare that pure raw materials are used for large scale melting. Several times, virgin elements are mixed with recycled scrap of the same composition in industrial operations. Further, multiple meltings are often not used in the industry and melts have to be cast before there are noticeable losses of the low vapor pressure elements and/or low melting point elements in the composition. The scientific community should take note of the challenges that can often arise in the scaling-up of experimental compositions to the industrial level. The repeatability of composition and reproducibility of properties obtained should be demonstrated.

Even in the domain specific to materials testing and characterization, researchers tend to use compression testing (due to an overall ease in making test specimens and the small size of the test specimen, thereby consuming little material) and report the test data, which does not truly represent the mechanical properties, as the defects present/generated during loading will not open up during compression testing. Except in select cases where the loading conditions are purely compressive in nature, the developed compositions should demonstrate mechanical properties to which the "actual" components are subjected i.e., tensile properties to include 0.2% proof strength, ultimate tensile strength, percentage elongation, reduction in cross-section area, hardness, impact toughness (Charpy), and fracture toughness. All conventional commercial alloys have these properties documented, and equivalent multicomponent alloys should be characterized for these properties to demonstrate their overall superiority over the conventional alloys. A comparison of the compression test data of multicomponent alloys with tensile data of conventional alloys is misleading and the comparisons should be on a one-to-one basis under identical test conditions. In fact, the ranges of compositions for which the test data is available should be compared and the lowest guaranteed mechanical properties for a given composition should be the one to compare for the same heat-treated condition to enable an overall good comparison.

It is a misnomer to say that "focus was lost," as focus was never given to develop these HEAs as viable alternatives to the existing conventional alloys. Based on the experience gained during the last 15 years, it is time to give proper focus to the field of multi-component alloys. This is possible by selecting compositions that are promising to study in detail for reproducibility of mechanical properties and to demonstrate their overall superiority over the conventional alloys. In the absence of this, the field remains similar to a kitchen that prepares food but has no eaters. Before the funding dries up for this field, researchers should convincingly demonstrate the applications of multicomponent alloys to ensure their survival and growth in the years ahead, without joining the scrapheap of overhyped materials/processes that have been discarded without useful applications.

From the mechanical property data that has been reported until now on several alloy systems, it does not seem that the multicomponent alloys are ready to replace the low-cost conventional steels, aluminum alloys, and other materials chosen for use in structural applications. A primary reason for this is the low cost of the conventional materials and the well-established processes for producing them. It is impossible that in the near future these multicomponent alloys can replace the materials chosen and used for bridges, in the aerospace industry, and even in the civil construction industry. A reason for this is essentially the low cost. Simple carbon steel or low-alloy steel can give strength as high as 1,000 MPa along with good elongation, and near-identical properties have been reported for the HEAs after the addition of expensive alloying elements, such as nickel (Ni), chromium (Cr), and cobalt (Co). For example, the strength–toughness balance of 18Ni–2,400 MPa grade maraging steel has not been challenged until now by the multimetallic alloys with an expensive combination of alloying additions. Similarly, with the high-temperature materials like superalloys, no significant improvement in properties has been reported. Near-zero recyclability of the HEAs is a major handicap for recommending their selection for large-scale utilization of these materials, making them more relevant as functional materials of high value for small weight and low volume components. It is time that multimetallic alloys are identified for use in specific applications, preferably for a specific commercial application as a viable substitute for conventional alloys.

REFERENCES

1. J. W. Yeh, *Jom*, *65*(12), 1759–1771, (2013). doi:10.1007/s11837-013-0761-6
2. C. Y. Hsu, J. W. Yeh, S. K. Chen, and T. T. Shun, *Metall. Mater. Trans. A*, *35*(5), 1465–1469, (2004).
3. S. Ranganathan, *Curr. Sci.*, *85*(10), 1404–1406, (2003).
4. J. Yeh, *Ann. Chim. Sci. des Mater.*, *31*(6), 633–648, (2006). doi:10.3166/acsm.31.633-648
5. B. J. Yeh, S. Chen, S. Lin, J. Gan, T. Chin, T. Shun, and C. Tsau, *Adv. Eng. Mater.*, *6*(5), 299–303, (2004). doi:10.1002/adem.200300567
6. J. Wei Yeh, S. Kai Chen, J. Yiew Gan, S. Jien Lin, T. Shune Chin, T. T. Shun, C. H. Tsau, and S. Y. Chang, *Metall. Mater. Trans. A*, *35*(8), 2533–2536, (2004).
7. C. J. Tong, Y. L. Chen, S. K. Chen, J. W. Yeh, T. T. Shun, C. H. Tsau, S. J. Lin, and S. Y. Chang, *Metall. Mater. Trans. A*, *36*(A), 881–893, (2005). doi:10.1007/s11661-005-0283-0
8. T. K. Chen, T. T. Shun, J. W. Yeh, and M. S. Wong, *Surf. Coat. Technol.*, *188–189*, 193–200, (2004). doi:10.1016/j.surfcoat.2004.08.023
9. C. J. Tong, M. R. Chen, S. K. Chen, J. W. Yeh, T. T. Shun, S. J. Lin, and S. Y. Chang, *Metall. Mater. Trans. A*, *36*(5), 1263–1271, (2005).
10. Y. Zhang, *Mater. Sci. Forum*, *654–656*, 1058–1061, (2010). doi:10.4028/www.scientific.net/MSF.654-656.1058
11. Y. Zhang, and Y. J. Zhou, *Mater. Sci. Forum*, *561–565*, 1337–1339, (2007). doi:10.4028/www.scientific.net/MSF.561-565.1337
12. H. Zhang, Y. Pan, Y. He, and H. Jiao, *Appl. Surf. Sci.*, *257*(6), 2259–2263, (2011). doi:10.1016/j.apsusc.2010.09.084
13. C. M. Lin, H. L. Tsai, and H. Y. Bor, *Intermetallics*, *18*(6), 1244–1250, (2010). doi:10.1016/j.intermet.2010.03.030
14. B. P. Huang, and J. Yeh, 6, 74–78, *Adv. Eng. Mater.* (2004). doi:10.1002/adem.200300507
15. Y. Y. Chen, T. Duval, U. D. Hung, J. W. Yeh, and H. C. Shih, *Corros. Sci.*, *47*(9), 2257–2279, (2005). doi:10.1016/j.corsci.2004.11.008
16. Y. Y. Chen, U. T. Hong, H. C. Shih, J. W. Yeh, and T. Duval, *Corros. Sci.*, *47*(11), 2679–2699, (2005). doi:10.1016/j.corsci.2004.09.026
17. M. C. Gao, J. W. Yeh, P. K. Liaw, and Y. Zhang, *On High-Entropy Alloys: Fundamentals and Applications*, Springer International Publishing AG Switzerland, (2016). doi:10.1007/978-3-319-27013-5

18. Industrial Development of High Entropy Alloys, Masson, Available online at https://wwwscribd.com/document269966894.
19. B. Gludovatz, A. Hohenwarter, D. Catoor, E. H. Chang, E. P. George, and R. O. Ritchie, *Science (80-.).*, *345*(6201), 1153–1158, (2014). doi:10.1126/science.1254581
20. M. E. Glicksman, *Principles of Solidification*, On Springer, Springer US, (2011). doi:10.1007/978-1-4419-7344-3
21. Y. Zhang, T. T. Zuo, Z. Tang, M. C. Gao, K. A. Dahmen, P. K. Liaw, and Z. P. Lu, *Prog. Mater. Sci.*, *61*, 1–93, (2013). doi:10.1016/j.pmatsci.2013.10.001
22. Y. S. Huang, *Recent Parent Mater., Sci.*, *2*(2), 154–157, (2009). doi:10.2174/18744648 10902020154
23. W. E. Frazier, E. W. Lee, M. E. Donnellan, and J. J. Thompson, *Jom*, *41*(5), 22–26, (1989).
24. X. Yang, Y. Zhang, and P. K. Liaw, *Procedia Eng.*, *36*, 292–298, (2012). doi:10.1016/j.proeng.2012.03.043
25. O. N. Senkov, J. M. Scott, S. V. Senkova, D. B. Miracle, and C. F. Woodward, *J. Alloys Compd.*, *509*(20), 1–17, (2011).
26. L. Lilensten, J. P. Couzinié, L. Perrière, J. Bourgon, N. Emery, and I. Guillot, *Mater. Lett.*, *132*, 123–125, (2014). doi:10.1016/j.matlet.2014.06.064
27. O. N. Senkov, S. V. Senkova, C. Woodward, and D. B. Miracle, *Acta Mater.*, *61*(5), 1545–1557, (2013). doi:10.1016/j.actamat.2012.11.032
28. O. N. Senkov, S. V. Senkova, D. B. Miracle, and C. Woodward, *Mater. Sci. Eng. A*, *565*, 51–62, (2013). doi:10.1016/j.msea.2012.12.018
29. N. D. Stepanov, N. Y. Yurchenko, D. G. Shaysultanov, G. A. Salishchev, and M. A. Tikhonovsky, *Mater. Sci. Technol.*, *31*(10), 1184–1193, (2015). doi:10.1179/17432847 15Y.0000000032
30. A Novel Light High Entropy Alloy Al20Be20Fe10Si15Ti35, Available online at https://www.science24.com.
31. Y. Chen, C. Tsai, C. Juan, M. Chuang, J. Yeh, T. Chin, and S. Chen, *J. Alloys Compd.*, *506*(1), 210–215, (2010). doi:10.1016/j.jallcom.2010.06.179
32. R. Li, J. C. Gao, and K. Fan, *Mater. Sci. Forum*, *650*, 265–271, (2010). doi:10.4028/www.scientific.net/MSF.650.265
33. R. Li, J. C. Gao, and K. Fan, *Mater. Sci. Forum*, *686*, 235–241, (2011). doi:10.4028/www.scientific.net/MSF.686.235
34. K. M. Youssef, A. J. Zaddach, C. Niu, D. L. Irving, and C. C. Koch, *Mater. Res. Lett.*, *3*(2), 95–99, (2015). doi:10.1080/21663831.2014.985855
35. H. Jiang, D. Qiao, Y. Lu, Z. Ren, Z. Cao, T. Wang, and T. Li, *Scr. Mater.*, *165*, 145–149, (2019). doi:10.1016/j.scriptamat.2019.02.035
36. Y. Tong, D. Chen, B. Han, J. Wang, R. Feng, T. Yang, C. Zhao, Y. L. Zhao, Y. Shimizu, C. T. Liu, et al., *Acta Mater.*, *165*, 228–240, (2019). doi:10.1016/j.actamat.2018.11.049
37. T. Yang, Y. L. Zhao, J. H. Luan, B. Han, J. Wei, J. J. Kai, and C. T. Liu, *Scr. Mater.*, *164*, 30–35, (2019). doi:10.1016/j.scriptamat.2019.01.034
38. L. Jiang, Y. P. Lu, M. Song, C. Lu, K. Sun, Z. Q. Cao, T. M. Wang, F. Gao, and L. M. Wang, *Scr. Mater.*, *165*, 128–133, (2019). doi:10.1016/j.scriptamat.2019.02.038
39. B. Gwalani, S. Gorsse, D. Choudhuri, and Y. Zheng, *Scr. Mater.*, *162*, 18–23, (2019). doi:10.1016/j.scriptamat.2018.10.023
40. D. Raabe, C. C. Tasan, H. Springer, and M. Bausch, *Steel. Res. Int.*, *86*(10), 1127–1138, (2015). doi:10.1002/srin.201500133
41. Z. Q. Xu, Z. L. Ma, M. Wang, Y. W. Chen, Y. D. Tan, and X. W. Cheng, *Mater. Sci. Eng. A*, *755*, 318–322, (2019). doi:10.1016/j.msea.2019.03.054
42. E. Holmström, R. Lizárraga, D. Linder, A. Salmasi, W. Wang, B. Kaplan, H. Mao, H. Larsson, and L. Vitos, *Appl. Mater. Today*, *12*, 322–329, (2018). doi:10.1016/j.apmt.2018.07.001

43. S. Gorsse, J. Couzinié, and D. B. Miracle, *C. R. Phys.*, *19*(8), 721–736, (2018). doi:10.1016/j.crhy.2018.09.004
44. *ASM Metals Reference Book*, Second Edn., American Society for Metals, Metals Park, OH, p 299–302, (1984).
45. J. F. Shackelford, Y. H. Han, S. Kim, and S. H. Kwon, CRC Press, Boca Raton, FL, New York, p 67,90,136,153, (2016).
46. J. F. Shackelford, Y. H. Han, S. Kim, and S. H. Kwon, CRC Press, Boca Raton, FL, New York, p 65,88,152, (2016).
47. J. F. Shackelford, Y. H. Han, S. Kim, and S. H. Kwon, CRC Press, Boca Raton, FL, New York, p 68,91,137,154, (2016).
48. J. F. Shackelford, Y. H. Han, S. Kim, and S. H. Kwon, CRC Press, Boca Raton, FL, New York, p 68,90,115,136,154, (2016).
49. G. E. Dieter, H. A. Kuhn, and S. L. Semiatin, ASM international, Materials Park, OH, (2003).
50. M. Pietrzyk, J. G. Lenard, and G. M. Dalton, *Ann. CIRP*, *42*(1), 331–334, (1993).
51. *ASM Metals Reference Book*, Second Edn., American Society for Metals, Metals Park, OH, p 241, (1984).
52. 18 Percentage Nickel Maraging Steels Engineering Properties No. 4419, INCO Data Books, 1976, INCO Europe Limited.
53. S. K. Manwatkar, S. V. S. Narayana Murty, P. Ramesh Narayanan, S. C. Sharma, and P. V. Venkitakrishnan, *Pract. Metallogr.*, *54*(1), 19–38, (2017).
54. H. K. D. H. Bhadeshia, *Prog. Mater. Sci.*, *57*(2), 268–435, (2012). doi:10.1016/j.pmatsci.2011.06.002
55. Michael Bauccio (Ed.), *ASM Metals Reference Book*, Third Edn., ASM International, Materials Park, OH, p 387–389, (1993).
56. Michael Bauccio (Ed.), *ASM Metals Reference Book*, Third Edn., ASM International, Materials Park, OH, p 387, (1993).
57. Michael Bauccio (Ed.), *ASM Metals Reference Book*, Third Edn., ASM International, Materials Park, OH, p 512, (1993).
58. Michael Bauccio (Ed.), *ASM Metals Reference Book*, Third Edn., ASM International, Materials Park, OH, p 390, (1993).
59. T. T. Zuo, R. B. Li, X. J. Ren, and Y. Zhang, *J. Magn. Magn. Mater.*, *371*, 60–68, (2014). doi:10.1016/j.jmmm.2014.07.023
60. K. B. Zhang, Z. Y. Fu, J. Y. Zhang, W. M. Wang, H. Wang, Y. C. Wang, Q. J. Zhang, and J. Shi, *Mater. Sci. Eng. A*, *508*(1–2), 214–219, (2009). doi:10.1016/j.msea.2008.12.053
61. W. Chen, Z. Fu, S. Fang, H. Xiao, and D. Zhu, *Mater. Des.*, *51*, 854–860, (2013). doi:10.1016/j.matdes.2013.04.061
62. J. M. Zhu, H. M. Fu, H. F. Zhang, A. M. Wang, H. Li, and Z. Q. Hu, *527*(26), 6975–6979, (2010). doi:10.1016/j.msea.2010.07.028
63. S. G. Ma, and Y. Zhang, *Mater. Sci. Eng. A*, *532*, 480–486, (2012). doi:10.1016/j.msea.2011.10.110
64. J. M. Zhu, H. M. Fu, H. F. Zhang, A. M. Wang, H. Li, and Z. Q. Hu, *J. Alloys Compd.*, *509*(8), 3476–3480, (2011). doi:10.1016/j.jallcom.2010.10.047
65. J. M. Zhu, H. M. Fu, H. F. Zhang, A. M. Wang, H. Li, and Z. Q. Hu, *Mater. Sci. Eng. A*, *527*(27–28), 7210–7214, (2010). doi:10.1016/j.msea.2010.07.049
66. F. J. Wang, Y. Zhang, and G. L. Chen, *J. Alloys Compd.*, *478*, 321–324, (2009). doi:10.1016/j.jallcom.2008.11.059
67. Y. J. Zhou, Y. Zhang, Y. L. Wang, and G. L. Chen, *Appl. Phys. Lett.*, *90*, 181904, *1–3*, (2007). doi:10.1063/1.2734517
68. N. D. Stepanov, D. G. Shaysultanov, G. A. Salishchev, M. A. Tikhonovsky, E. E. Oleynik, A. S. Tortika, and O. N. Senkov, *J. Alloys Compd.*, *628*, 170–185, (2015). doi:10.1016/j.jallcom.2014.12.157

69. Y. Dong, Y. Lu, J. Kong, J. Zhang, and T. Li, *J. Alloys Compd.*, *573*, 96–101, (2013). doi:10.1016/j.jallcom.2013.03.253
70. X. Chen, J. Q. Qi, Y. W. Sui, Y. Z. He, F. X. Wei, Q. K. Meng, and Z. Sun, *Mater. Sci. Eng. A*, *681*, 25–31, (2017). doi:10.1016/j.msea.2016.11.019
71. J. Y. He, W. H. Liu, H. Wang, Y. Wu, X. J. Liu, T. G. Nieh, and Z. P. Lu, *Acta Mater.*, *62*, 105–113, (2014). doi:10.1016/j.actamat.2013.09.037
72. Z. Wu, H. Bei, G. M. Pharr, and E. P. George, *Acta Mater.*, *81*, 428–441, (2014). doi:10.1016/j.actamat.2014.08.026
73. W. H. Liu, J. Y. He, H. L. Huang, H. Wang, Z. P. Lu, and C. T. Liu, *Intermetallics*, *60*, 1–8, (2015). doi:10.1016/j.intermet.2015.01.004
74. J. Y. He, H. Wang, H. L. Huang, X. D. Xu, M. W. Chen, Y. Wu, X. J. Liu, T. G. Nieh, K. An, and Z. P. Lu, *Acta Mater.*, *102*, 187–196, (2016). doi:10.1016/j.actamat.2015.08.076
75. K. Ming, X. Bi, and J. Wang, *Scr. Mater.*, *137*, 88–93, (2017). doi:10.1016/j.scriptamat.2017.05.019
76. W. H. Liu, Z. P. Lu, J. Y. He, J. H. Luan, Z. J. Wang, B. Liu, Y. Liu, M. W. Chen, and C. T. Liu, *Acta Mater.*, *116*, 332–342, (2016). doi:10.1016/j.actamat.2016.06.063
77. Z. Li, C. C. Tasan, K. G. Pradeep, and D. Raabe, *Acta Mater.*, *131*, 323–335, (2017). doi:10.1016/j.actamat.2017.03.069
78. S. S. Nene, K. Liu, M. Frank, R. S. Mishra, R. E. Brennan, K. C. Cho, Z. Li, and D. Raabe, *Nature*, *7*, 16167, 1–7, (2017). doi:10.1038/s41598-017-16509-9
79. Z. Tang, O. N. Senkov, C. M. Parish, C. Zhang, F. Zhang, L. J. Santodonato, G. Wang, G. Zhao, F. Yang, and P. K. Liaw, *Mater. Sci. Eng. A*, *647*, 229–240, (2015). doi:10.1016/j.msea.2015.08.078
80. E. Ghassemali, R. Sonkusare, K. Biswas, and N. P. Gurao, *J. Alloys Compd.*, *710*, 539–546, (2017). doi:10.1016/j.jallcom.2017.03.307
81. H. Y. Yasuda, H. Miyamoto, K. Cho, and T. Nagase, *Mater. Lett.*, *199*, 120–123, (2017). doi:10.1016/j.matlet.2017.04.072
82. B. Gwalani, V. Soni, M. Lee, S. A. Mantri, Y. Ren, and R. Banerjee, *Mater. Des.*, *121*, 254–260, (2017). doi:10.1016/j.matdes.2017.02.072
83. T. T. Shun, and Y. C. Du, *J. Alloys Compd.*, *479*(1–2), 157–160, (2009). doi:10.1016/j.jallcom.2008.12.088
84. H. M. Daoud, A. M. Manzoni, N. Wanderka, and U. Glatzel, *Jom*, *67*(10), 2271–2277, (2015). doi:10.1007/s11837-015-1484-7
85. A. M. Manzoni, and U. Glatzel, *Mater. Charact.*, *147*, 512–532, (2018). doi:10.1016/j.matchar.2018.06.036
86. Q. Wang, Y. Ma, B. Jiang, X. Li, Y. Shi, C. Dong, and P. K. Liaw, *Scr. Mater.*, *120*, 85–89, (2016). doi:10.1016/j.scriptamat.2016.04.014
87. F. Otto, A. Dlouhý, C. Somsen, H. Bei, G. Eggeler, and E. P. George, *Acta Mater.*, *61*(5), 5743–5755, (2013). doi:10.1016/j.actamat.2013.06.018
88. H. Cheng, H. Y. Wang, Y. C. Xie, Q. H. Tang, and P. Q. Dai, *Mater. Sci. Technol.*, *33*(17), 2032–2039, (2017). doi:10.1080/02670836.2017.1342367
89. A. Gali, and E. P. George, *Intermetallics*, *39*, 74–78, (2013). doi:10.1016/j.intermet.2013.03.018
90. P. Jinhong, P. Ye, Z. Hui, and Z. Lu, *Mater. Sci. Eng. A*, *534*, 228–233, (2012). doi:10.1016/j.msea.2011.11.063
91. X. F. Wang, Y. Zhang, Y. Qiao, and G. L. Chen, *Intermetallics*, *15*(3), 357–362, (2007). doi:10.1016/j.intermet.2006.08.005
92. B. S. Li, Y. P. Wang, M. X. Ren, C. Yang, and H. Z. Fu, *Mater. Sci. Eng. A*, *498*(1–2), 482–486, (2008). doi:10.1016/j.msea.2008.08.025
93. Z. Hu, Y. Zhan, G. Zhang, J. She, and C. Li, *Mater. Des.*, *31*(3), 1599–1602, (2010). doi:10.1016/j.matdes.2009.09.016

94. É. Fazakas, V. Zadorozhnyy, and D. V. Louzguine-Luzgin, *Appl. Surf. Sci.*, *358*(B), 549–555, (2015). doi:10.1016/j.apsusc.2015.07.207
95. Z. G. Wang, W. Zhou, L. M. Fu, J. F. Wang, R. C. Luo, X. C. Han, B. Chen, and X. D. Wang, *Mater. Sci. Eng. A*, *696*, 503–510, (2017). doi:10.1016/j.msea.2017.04.111
96. Z. Tang, T. Yuan, C. Tsai, J. Yeh, C. D. Lundin, and P. K. Liaw, *Acta Mater.*, *99*, 247–258, (2015). doi:10.1016/j.actamat.2015.07.004
97. A. V. Kuznetsov, D. G. Shaysultanov, N. D. Stepanov, G. A. Salishchev, and O. N. Senkov, *Mater. Sci. Eng. A*, *533*, 107–118, (2012). doi:10.1016/j.msea.2011.11.045
98. M. A. Hemphill, T. Yuan, G. Y. Wang, J. W. Yeh, C. W. Tsai, A. Chuang, and P. K. Liaw, *Acta Mater.*, *60*(16), 5723–5734, (2012). doi:10.1016/j.actamat.2012.06.046
99. C. Ng, S. Guo, J. Luan, Q. Wang, J. Lu, S. Shi, and C. T. Liu, *J. Alloys Compd.*, *584*, 530–537, (2014). doi:10.1016/j.jallcom.2013.09.105
100. L. Liu, J. B. Zhu, C. Zhang, J. C. Li, and Q. Jiang, *Mater. Sci. Eng., A.*, *548*, 64–68, (2012). doi:10.1016/j.msea.2012.03.080
101. L. Liu, J. B. Zhu, L. Li, J. C. Li, and Q. Jiang, *Mater. Des.*, *44*, 223–227, (2013). doi:10.1016/j.matdes.2012.08.019
102. S. Y. Chen, X. Yang, K. A. Dahmen, P. K. Liaw, and Y. Zhang, *Entropy*, *16*(2), 870–884, (2014). doi:10.3390/e16020870
103. O. N. Senkov, J. K. Jensen, A. L. Pilchak, D. B. Miracle, and H. L. Fraser, *Mater. Des.*, *139*, 498–511, (2018). doi:10.1016/j.matdes.2017.11.033
104. D. Qiao, H. Jiang, X. Chang, Y. Lu, and T. Li, *Mater. Sci. Forum*, *898*, 638–642, (2017). doi:10.4028/www.scientific.net/MSF.898.638
105. C. M. Lin, C. C. Juan, C. H. Chang, C. W. Tsai, and J. W. Yeh, *J. Alloys Compd.*, *624*, 100–107, (2014). doi:10.1016/j.jallcom.2014.11.064
106. O. N. Senkov, C. Woodward, and D. B. Miracle, *Jom*, *66*(10), 2030–2042, (2014). doi:10.1007/s11837-014-1066-0
107. O. N. Senkov, S. V. Senkova, and C. Woodward, *ACTA Mater.*, *68*, 214–228, (2014). doi:10.1016/j.actamat.2014.01.029
108. N. D. Stepanov, N. Y. Yurchenko, E. S. Panina, M. A. Tikhonovsky, and S. V. Zherebtsov, *Mater. Lett.*, *188*, 162–164, (2017). doi:10.1016/j.matlet.2016.11.030
109. N. D. Stepanov, N. Y. Yurchenko, D. V. Skibin, M. A. Tikhonovsky, and G. A. Salishchev, *J. Alloys Compd.*, *652*, 266–280, (2015). doi:10.1016/j.jallcom.2015.08.224
110. O. N. Senkov, D. Isheim, D. N. Seidman, and A. L. Pilchak, *Entropy*, *18*(3), 1–13, (2016). doi:10.3390/e18030102
111. N. D. Stepanov, D. G. Shaysultanov, G. A. Salishchev, and M. A. Tikhonovsky, *Mater. Lett.*, *142*, 153–155, (2015). doi:10.1016/j.matlet.2014.11.162
112. N. Y. Yurchenko, N. D. Stepanov, S. V. Zherebtsov, M. A. Tikhonovsky, and G. A. Salishchev, *Mater. Sci. Eng. A*, *704*, 82–90, (2017). doi:10.1016/j.msea.2017.08.019
113. N. N. Guo, L. Wang, L. S. Luo, X. Z. Li, R. R. Chen, Y. Q. Su, J. J. Guo, and H. Z. Fu, *Intermetallics*, *69*, 74–77, (2016). doi:10.1016/j.intermet.2015.09.011
114. M. Zhang, X. Zhou, and J. Li, *J. Mater. Eng. Perform.*, *26*(8), 3657–3665, (2017). doi:10.1007/s11665-017-2799-z
115. É. Fazakas, V. Zadorozhnyy, L. K. Varga, A. Inoue, D. V. Louzguine-Luzgin, F. Tian, and L. Vitos, *Int. J. Refract. Met. Hard Mater.*, *47*, 131–138, (2014). doi:10.1016/j.ijrmhm.2014.07.009
116. O. N. Senkov, and C. F. Woodward, *Mater. Sci. Eng. A*, *529*, 311–320, (2011). doi:10.1016/j.msea.2011.09.033
117. O. Ahmed, J. Lee, H. Mo, and H. Jin, *J. Mater. Chem. Phys.*, *210*, 87–94, (2017). doi:10.1016/j.matchemphys.2017.06.054
118. M. Feuerbacher, M. Heidelmann, and C. Thomas, *Philos. Mag.*, *95*(11), 1221–1232, (2015). doi:10.1080/14786435.2015.1028506

119. N. N. Guo, L. Wang, L. S. Luo, X. Z. Li, R. R. Chen, Y. Q. Su, J. J. Guo, and H. Z. Fu, *J. Alloys Compd.*, *660*, 197–203, (2016). doi:10.1016/j.jallcom.2015.11.091
120. S. Sheikh, S. Shafeie, Q. Hu, J. Ahlström, C. Persson, J. Veselý, J. Zýka, U. Klement, and S. Guo, *J. Appl. Phys.*, *120*, 16, (2016). doi:10.1063/1.4966659
121. V. F. A. V. Podolskiy, E. D. Tabachnikova, V. V. Voloschuk, and S. A. F. Gorban, *Mater. Sci. Eng. A*, *710*, 136–141, (2018). doi:10.1016/j.msea.2017.10.073
122. Y. Liu, Y. Zhang, H. Zhang, N. Wang, X. Chen, H. Zhang, and Y. Li, *J. Alloys Compd.*, *694*, 869–876, (2016). doi:10.1016/j.jallcom.2016.10.014
123. C. C. Juan, K. Kai Tseng, W. I. Hsu, M. Hung Tsai, C. Wei Tsai, C. M. Lin, S. K. Chen, S. jien Lin, and J. W. Yeh, *Mater. Lett.*, *175*, 284–287, (2016). doi:10.1016/j.matlet.2016.03.133
124. C. Juan, M. Tsai, C. Tsai, C. Lin, W. Wang, C. Yang, S. Chen, S. Lin, and J. Yeh, *Intermetallics*, *62*, 76–83, (2015). doi:10.1016/j.intermet.2015.03.013
125. N. N. Guo, L. Wang, L. S. Luo, X. Z. Li, Y. Q. Su, J. J. Guo, and H. Z. Fu, *Mater. Des.*, *81*, 87–94, (2015). doi:10.1016/j.matdes.2015.05.019
126. Y. Zhang, Y. Liu, Y. Li, X. Chen, and H. Zhang, *Mater. Lett.*, *174*, 82–85, (2016). doi:10.1016/j.matlet.2016.03.092
127. Y. Zhang, Y. Liu, Y. Li, X. Chen, and H. Zhang, *Mater. Sci. Forum*, *849*, 76–84, (2016). doi:10.4028/www.scientific.net/MSF.849.76
128. S. Maiti, and W. Steurer, *Acta Mater.*, *106*, 87–97, (2016). doi:10.1016/j.actamat.2016.01.018
129. Y. D. Wu, Y. H. Cai, X. H. Chen, T. Wang, J. J. Si, L. Wang, Y. D. Wang, and X. D. Hui, *Mater. Des.*, *83*, 651–660, (2015). doi:10.1016/j.matdes.2015.06.072
130. Z. D. Han, H. W. Luan, X. Liu, N. Chen, X. Y. Li, Y. Shao, and K. F. Yao, *Mater. Sci. Eng. A*, *712*, 380–385, (2017). doi:10.1016/j.msea.2017.12.004
131. H. W. Yao, J. W. Qiao, J. A. Hawk, H. F. Zhou, M. W. Chen, and M. C. Gao, *J. Alloys Compd.*, *696*, 1139–1150, (2016). doi:10.1016/j.jallcom.2016.11.188
132. Z. D. Han, N. Chen, S. F. Zhao, L. W. Fan, G. N. Yang, Y. Shao, and K. F. Yao, *Intermetallics*, *84*, 153–157, (2017). doi:10.1016/j.intermet.2017.01.007
133. S. Ping Wang, and J. Xu, *Mater. Sci. Eng. C*, *73*, 80–89, (2016). doi:10.1016/j.msec.2016.12.057
134. M. Todai, T. Nagase, T. Hori, A. Matsugaki, A. Sekita, and T. Nakano, *Scr. Mater.*, *129*, 65–68, (2017). doi:10.1016/j.scriptamat.2016.10.028
135. H. Yao, J. W. Qiao, M. C. Gao, J. A. Hawk, S. G. Ma, and H. Zhou, *Entropy*, *18*(5), 1–15, (2016). doi:10.3390/e18050189
136. B. Kang, J. Lee, H. J. Ryu, and S. H. Hong, *Mater. Sci. Eng. A*, *712*, 616–624, (2018). doi:10.1016/j.msea.2017.12.021
137. O. N. Senkov, G. B. Wilks, J. M. Scott, and D. B. Miracle, *Intermetallics*, *19*(5), 698–706, (2011). doi:10.1016/j.intermet.2011.01.004
138. Y. Zhang, X. Yang, and P. K. Liaw, *Jom*, *64*(7), 830–838, (2012). doi:10.1007/s11837-012-0366-5
139. H. W. Yao, J. W. Qiao, M. C. Gao, J. A. Hawk, S. G. Ma, H. F. Zhou, and Y. Zhang, *Mater. Sci. Eng. A*, *674*, 203–211, (2016). doi:10.1016/j.msea.2016.07.102
140. G. Dirras, L. Lilensten, P. Djemia, D. Tingaud, J. Couzinié, L. Perrière, and I. Guillot, *Mater. Sci. Eng. A*, *654*, 30–38, (2015). doi:10.1016/j.msea.2015.12.017
141. Y. D. Wu, Y. H. Cai, T. Wang, J. J. Si, J. Zhu, Y. D. Wang, and X. D. Hui, *Mater. Lett.*, *130*, 277–280, (2014). doi:10.1016/j.matlet.2014.05.134
142. H. Huang, Y. Wu, J. He, H. Wang, X. Liu, K. An, W. Wu, and Z. Lu, *Adv. Mater.*, *29*(30), 1–7, (2017). doi:10.1002/adma.201701678
143. H. Chen, A. Kauffmann, B. Gorr, D. Schliephake, C. Seemüller, J. N. Wagner, H. J. Christ, and M. Heilmaier, *J. Alloys Compd.*, *661*, 206–215, (2016). doi:10.1016/j.jallcom.2015.11.050

144. H. Chen, A. Kauffmann, S. Laube, I.-C. Choi, R. Schwaiger, Y. Huang, K. Lichtenberg, F. Muller, B. Gorr, H.-J. Christ, et al., *Metall. Mater. Trans. A*, *49*(3), 772–781, (2017). doi:10.1007/s11661-017-4386-1
145. O. N. Senkov, J. M. Scott, S. V. Senkova, F. Meisenkothen, D. B. Miracle, and C. F. Woodward, *J. Mater. Sci.*, *47*(9), 4062–4074, (2012). doi:10.1007/s10853-012-6260-2
146. Y. V. R. K. Prasad, K. P. Rao, and S. Sasidhara, *On Hot Working Guide: A Compendium of Processing Maps*, ASM international, Materials Park, OH, p 440–486, 510–579, (2015). doi:10.1016/B978-0-08-033454-7.50019-X

16 Fracture and Fatigue Behavior of High Entropy Alloys

A Comprehensive Review

Kalyan Kumar Ray

CONTENTS

16.1 INTRODUCTION

Understanding related to fracture and fatigue apart from deformation behavior of materials is considered to be essential information for assessing the potential of an existing or an emerging material for its engineering applications through structural integrity philosophy. The term "structural integrity" refers to the ability of a structure or a component to withstand its intended service conditions, specifically loading without failure due to fracture, fatigue, or deformation. In an engineering sense, production of structures and components with the adoption of conventional or emerging materials should serve the design purpose for the desired service life. The human race has painfully suffered numerous engineering failures associated with loss of life and impediment to societal growth toward advancement for improved life quality. To avoid engineering failures, one needs to gain continued information and knowledge about several mechanical properties of material-like strength, ductility, toughness, fatigue, creep, wear, and degradation resistance to find the isolated or combined mechanical and chemical environments. This task also requires continued development of materials possessing superior combinations of the mechanical properties. High entropy alloys (HEAs) are one such group of emerging potential materials. Interestingly, the development of high entropy alloys in its first decade has shown the material to exhibit high hardness, good strength even at elevated temperature, and excellent resistance to wear and corrosion. With the background knowledge of these material properties, researchers have started exploration of their applications in structural engineering. This has naturally led to gaining knowledge on the fracture and fatigue behavior of these materials to assess their structural integrity issues. This search only started about six years ago, and the information gained so far is limited. This chapter aims to make a comprehensive projection of the current status of the fracture and fatigue behavior of high entropy alloys on the background of our existing knowledge and information on these disciplines.

16.1.1 Conventional Advanced Structural Materials vis-à-vis High Entropy Alloys

The emergence of the term "high entropy alloy" (HEA) is due to Yeh et al. [1], but a similar material philosophy with the term "equiatomic multicomponent alloy" was introduced by Cantor et al. [2] in the same year. Currently, high entropy and allied group of alloys are referred by several other terminologies like "multi-principal element alloys (MPEA)," "complex concentrated alloys (CCA)," "multicomponent alloys (MCA)," or "baseless alloys (BA)." But the terminology "high entropy alloy" appears to be most popular and will be used throughout this report, though there exists some debate and distinct variations on the magnitude of entropy of the allied alloys and use of the terminologies CCA, MPEA, MCA, or BA [3, 4]. The later terminologies may refer to "medium entropy alloys" (MEA) also. The term "high entropy" originates from the consideration that these alloys have "configurational entropy" > 1.61 R compared to classical alloys showing entropy values of < 0.69 R, while the medium entropy alloys possess entropy between these two boundary values. High entropy alloys are usually defined as materials consisting of "at least five principal elements;" the concentration of the elements could be equiatomic or non-equiatomic ranging between 5 at.% and 35 at.% [3–9]. The basic principle of the alloy design of HEAs is different from that of the conventional structural materials: HEAs could possess a superior combination of the desired mechanical properties for applications in challenging service conditions.

Conventional structural materials (like iron-, aluminum-, titanium-, or nickel-based alloys) are designed and fabricated to achieve a superior combination of mechanical properties (like strength, ductility, fracture toughness, fatigue, and creep resistance) with fewer manufacturing defects either using one or two principal elements or composites having the matrix of a single principal element based on metallic or ceramic material, or on a polymer. The physical metallurgy principles to develop the advanced alloys are based on familiar concepts of strengthening mechanisms by solid solution, precipitation, dispersion, transformation, and grain-size reduction [10–19], while the development of the composites is with the consideration of suitable reinforcement, continuous or discontinuous, on the metallic, ceramic, or polymeric base [20, 21]. The characteristics of the precipitates, dispersions, and reinforcing materials with those of the matrix in the microstructural constituents govern the overall mechanical response of the materials in neutral or aggressive environments. Usually a single (occasionally two) principal element governs the formation of the primary crystalline phase and the properties of the material, while minor additions of secondary elements induce some of the desired properties. In addition to the above-mentioned crystalline metallic and ceramic materials, bulk metallic glasses are also occasionally considered in the category of advanced structural materials; but these are distinct in having their amorphous constituent.

In contrast, the design philosophy of HEAs considers mixing five or more principal elements in equal or near-equal molar ratio [3, 22, 23] to achieve one or two solid solutions in the material instead of resulting in several intermetallics or precipitates of the concerned elements; the difference originates primarily from the higher configurational entropy of the chosen system. The mixing of five or more elements in near-equal molar ratio can result in single solid solution phases with face-centered cubic (FCC) [6, 24–27], body-centered cubic (BCC) [3, 6, 22, 28], or hexagonal close-packed (hcp) [6, 22, 29–32] crystal structures, or dual-phase materials [18, 33–40]. The formation of one or two solid solutions is attributed to the stabilization effect due to the lowering of the Gibbs free energy due to increase in the configurational entropy and sluggish diffusivity of the elements [4, 41]. Even with sluggish diffusivity, thermomechanical treatment of the alloys can result in a few precipitates of finer sizes [9, 42–44]. Miracle and Senkov [3] have provided some classification of the nature of high entropy alloys. These can be listed as:

(i) Transition metal-based HEAs [2, 24, 45, 46], such as CrMnFeCoNi;
(ii) Refractory HEAs [11, 47], such as NbMoTaW;
(iii) Low-density HEAs [48, 49], such as $Li_{20}Mg_{10}Al_{20}Sc_{20}Ti_{30}$ HEA;
(iv) Ceramic HEAs [17, 50–53], such as "equimolar mixture of MgO, NiO, ZnO, CuO, and CoO," $(Hf_{0.2}Zr_{0.2}Ta_{0.2}Nb_{0.2}Ti_{0.2})B_2$, and $Sr(Zr_{0.2}\ Sn_{0.2}\ Ti_{0.2}\ Hf_{0.2}\ Mn_{0.2})O_3$;
(v) Other types of HEAs [32, 54, 55], such as lanthanide: YGdTbDyLu; brasses and bronzes: $CuZnMnNiSn_{0.2}$; and noble metal: PtPdRhIrCuNi.

It is worth mentioning at this stage that studies on fracture and fatigue behavior of HEAs are primarily confined to the first two groups of materials to date.

16.1.2 Conventional Mechanical Properties of High Entropy Alloys

Mechanical properties of HEAs are governed by chemistry, microstructure, and intrinsic properties, as well as processing defects. These parameters decide the nature of the phases, their characteristics, amounts of the phases, their size, morphology, and distributions, intrinsic defects like vacancies, dislocations, grain boundaries, or twins, and extrinsic defects like pores, microcracks, segregation, or residual stresses. In general, as-cast materials possess most of the extrinsic defects which degrade their mechanical properties, while thermal or thermomechanical treatments considerably reduce them. The possible number of HEAs is several billion, and the number of reports emerging over the last five years is a few thousand, in which preparation and fabrications of the materials also have wide variations. Thus, any categorization of the "microstructural state of a HEA" with respect to its composition, microstructure, and defect content is a difficult task. This naturally leads to increased complexity in understanding the mechanical behavior of these materials with respect to their microstructural states. Initial interest in the assessment of mechanical properties of HEAs are confined to achieving information on the conventional mechanical properties like hardness, compressive, and tensile strengths and ductility. Interestingly Grosse et al. [56] and Couizine et al. [57] have compiled the information related to these properties generated in the period 2004–2017. A brief background on this information is presented here prior to bringing forward the fracture and fatigue behavior of these alloys.

One of the first reports by Yeh et al. [1] indicated that HEAs can exhibit high hardness values. These authors reported that, depending on the crystal structure of the solid solutions $CuCoNiCrAl_xFe$ alloys (x varying in the range 0–3 at.%), the hardness can be modulated between > 100 HV (0.98 GPa) and > 650 HV (6.37 GPa), as shown in Figure 16.1a [1, 4]. A further nine HEAs of TiVFeNiZr system with additions of one or two elements of Cu, Al, Mo, and Cr showed hardness in the range 590 (5.78 GPa)–890 HV (8.73 GPa) in as-cast or annealed conditions. Cantor et al. (2004) reported that hardness of $Fe_{20}Cr_{20}Mn_{20}Ni_{20}Co_{20}$ alloy is ~300 kg/mm^2 (2.94 GPa) and with the addition of Nb, V, and Ti it can go up to 1,000 kg/mm^2 (9.8 GPa)

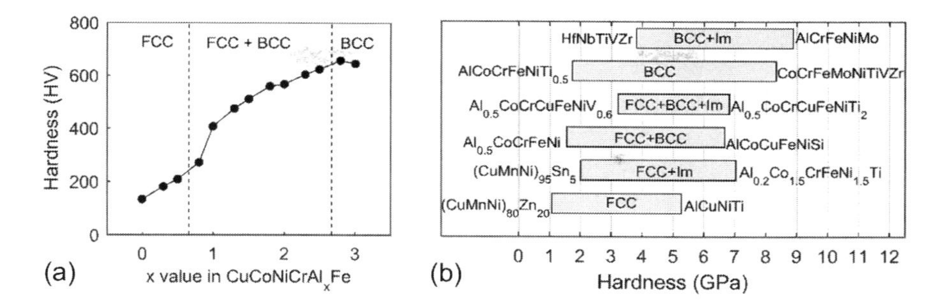

FIGURE 16.1 (a) Hardness of the AlxCoCrCuFeNi alloys as a function of Al content [1], and (b) range of hardness values for different crystal structures of various high entropy alloys [1, 25, 61–69] following the data bank of Gorsse et al. [56] (Im = Intermetallics).

when microhardness measurements are made at 10 gf load. Zhang et al. [58] have reported that hardness of FeCoNiCrCuTiMoAlSiB$_{0.5}$ alloy prepared by laser cladding can exhibit nano-hardness of 11.6 GPa (~1183 HV). Absolute comparisons of the hardness values from different reports are difficult because the estimated hardness is dependent on the magnitude of load employed and the technique of measurement [59, 60]. The range of hardness values for different HEAs is presented in Figure 16.1(b) [1, 25, 61–69] considering the data compiled by Grosse et al. [56] to illustrate the dependence of this property on the microstructure and the crystal structure of the phases.

The stress response of HEAs in tension and compression is dependent on the composition and structure of the concerned material. The HEAs are constituted of wide variety of elements, and their phases could be FCC, BCC, hcp, and dual phase with or without embedment of precipitates of different crystal structures. The reported yield strength of HEAs could vary between 148 MPa for CoCrFeNi alloy [70] and 1,435 MPa for CoCrNi [71], in tension, while that in compression varies between 140 MPa for (CuMnNi)$_{80}$Zn$_{20}$ [25] and 2,757 MPa for AlCoCrFeMo$_{0.5}$Ni alloy [72]. A compilation of the strength values of various types of HEAs is shown as a function of the density of these alloys in Figure 16.2 following Grosse et al. [56]; the density of the alloys is estimated using the rule of mixture as $\rho = \sum x_i M_i / \sum x_i V_i$ where x_i, M_i, and V_i correspond to the atomic fraction, molar mass, and molar volume of the element i. The stress response of the HEAs is dictated by the imposed test conditions, like strain rate, temperature, and environment, as well as by the deformation behavior which encompasses dislocation slip, lattice friction, stacking fault energy, twinning, and phase transformation. These parameters characterize the strain hardening nature of the materials as well as the strength–ductility behavior. It is well known that increasing strength is associated with loss of ductility, and HEAs indicate the same trend like other materials, as demonstrated using some recent reports on the strength–ductility properties of a few FCC HEAs in Figure 16.3 [26, 71, 73–80].

Compared to the number of investigations on the conventional mechanical properties, the number of investigations on fracture and fatigue behavior on HEAs by the standard test techniques is quite limited in number (< 20) at the time of preparation

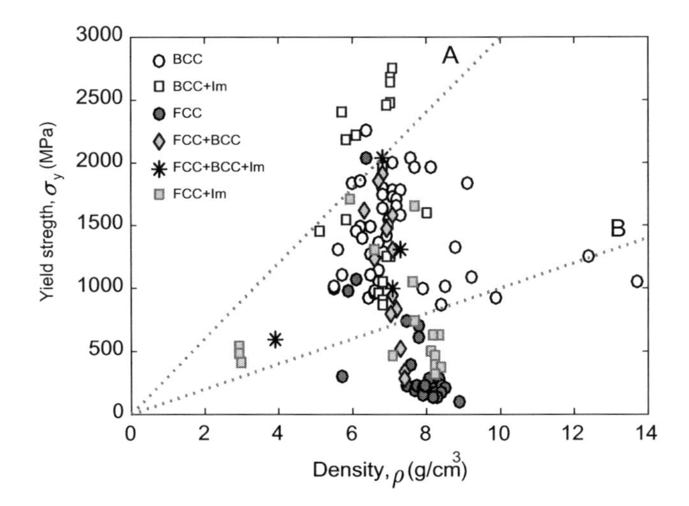

FIGURE 16.2 Yield strength of different high entropy and/or complex concentrated alloys as function of their density. The density of the alloys is estimated using the rule of mixture as $\rho=\sum xiMi/\sum xiVi$ where xi, Mi, and Vi correspond to atomic fraction, molar mass, and molar volume of the element i. The variations in crystal structures of the different HEAs are depicted using different symbols (following Grosse et al. [56]). A and B refers to materials with strength/density ratio as 300 and 100 MPa per unit density in g/cm^3, while Im indicates intermetallics.

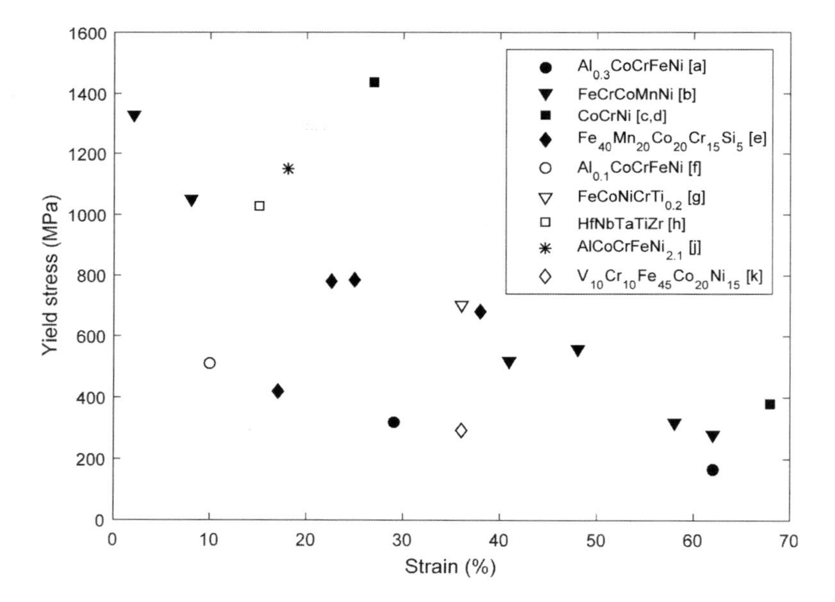

FIGURE 16.3 Typical variations of strength versus ductility of a few recently reported high entropy alloys with FCC matrix (a) [73], (b) [74], (c) [75], (d) [71], (e) [26], (f) [76], (g) [77], (h) [78], (j) [79], (k) [80]. The observed trend of variation between these properties follows that for common structural materials.

of this report. Interestingly, three reviews are already available on mechanical properties [17], fracture [81], and fatigue [82] behavior of HEAs apart from several scattered works documenting these properties in structure-property relations of HEAs. It is reiterated that the objective of this chapter is to specifically focus on the state of our understanding on fracture and fatigue behavior of HEAs, with reference to the existing knowledge on these disciplines.

16.2 FRACTURE OF SOLIDS

16.2.1 General

Fracture of engineering components at different scale lengths (usually micro- to mega-meters) has been encountered by the human race with the advent of science and technology for various material development programs. Fracture of a solid simply implies its breaking into two or more parts when the component may or may not contain defects at microscale at the start of its application. Preventing the occurrence of fracture and keeping it in one integrated piece is governed by the fracture behavior of materials. Fracture toughness is a quantitative measure of the resistance to cracking of materials and is one of the critical parameters to control the occurrence of failure of structural engineering components in monotonic loading, like transport vehicles in sea, land, air, or space; nevertheless, it is also of significance at microscale to protect against the breaking of an electronic chip in a circuit. More specifically, fracture toughness is a quantitative measure of the resistance to cracking at a critical point during loading of a notched/defect-containing specimen. The integrity of structural components, however, is dependent on two mutually exclusive properties: strength and fracture toughness; the first one has been dealt previously to some extent, while the property fracture toughness is the concern of this section. The magnitude of the latter property depends on the nature of the material, its sub-micro- and microstructural features, type of loading, state of stress, and loading conditions like loading rate, temperature, and environment, and the scale of yielding at the crack tip.

The initiation and growth of a crack or defect in a component is governed by the nature of the material, and the state of stress indicates the extent of yielding at the tip of the crack. In plane strain condition and/or for materials exhibiting low ductility, the initiation of cracking occurs almost in an unstable manner and the nature of fracture is termed as brittle. In contrast, the initiation of cracking for materials with considerable ability for plastic flow under a plane stress condition exhibits a stable crack growth regime between the initiation of cracking and unstable fracture, and this middle region, influenced by the plastic flow and strain hardening/softening of the material, is responsible for the ductile fracture. The critical value of the stress intensity factor (K), strain energy release rate (G), the J-integral value, or the crack tip opening displacement (CTOD or δ) at the onset of crack initiation is termed the fracture toughness criterion of the material in a plane strain or plane stress state of loading. The critical stress intensity factor under the plane strain state of loading is commonly known as the fracture toughness (K_{Ic}) of a material with the unit $MPam^{1/2}$ and the critical values of "G and J" and δ have energy units in joules (J) and length units in mm, respectively.

The investigations of the fracture resistance of high entropy alloys are primarily aimed at searching for their possible structural applications and to promote this category of materials for innovative structural applications, as they are often associated with good strength, improved wear [83–85], and corrosion resistance properties [26, 86, 87]. The investigations related to the fracture behavior of these alloys are few in number, but the existing results indicate that the material could be extremely brittle like glass, having fracture toughness (K_{IC}) less than the unity [88] for refractory high entropy alloys and sufficiently ductile like single phase metallic materials such as stainless steel [89], with a fracture toughness well over 200 MPam$^{1/2}$ for cantor alloy [24].The chemistry, microstructural constituents, processing history, and test conditions affect the fracture resistance of the reported HEAs. This section incorporates at the beginning: (i) the terminologies and popular test methods adopted for assessing fracture toughness of all types of materials by conventional standard techniques, prior to (ii) describing the reported results on fracture toughness values of HEAs by these methods. This is followed by (iii) alternate methods for estimating the fracture properties of materials and their adoption for examining fracture resistance of HEAs. The description of (iv) fracture-signatures and the fracture micro-mechanism and (v) the effect of the nature of the material on the fracture resistance of HEAs remain as some essential parts of this section.

16.2.2 Fracture Toughness: Preliminaries and Test Methods

The control of failure of structural materials due to fracture can be quantitatively done using the fracture toughness property. This property, in a quantitative manner, describes uniquely the propensity of a crack or a defect to extend in a material whether it is ductile or brittle in nature. As a generic term, fracture toughness is known as a material's resistance to fracture. The origin of fracture control is in the early report by Griffith [90] almost hundred years before now. Griffith's concept of strain energy release rate (G) or the crack driving force is still considered for fracture control in the discipline of fracture mechanics. This discipline incorporates the emergence of different fracture toughness criteria and provides the guideline for formulating tests for fracture toughness parameters. The fracture toughness property finds broad applications in damage tolerance design, assessment of residual strength, and structural integrity. As a final consequence, these assessments indicate the fitness for service of a structural material as a component in an assembly. The standardization of the fracture toughness property has been carried out by the American Society for Testing and Materials (ASTM), the British Standards (BS) Institution, the International Standards Organization (ISO), and by several other standards from different countries, but the ASTM standard is the most popular one from a global perspective. There are several standards for measurement of different fracture toughness-related parameters. The values of these parameters also serve as the basis for material characterization, their performance evaluation, and specifically for assessing the applicability of emerging materials like high entropy alloys (HEAs) or multi-principal element alloys (MPAs) for engineering structures.

The major factors that influence the fracture toughness properties of a material are its intrinsic fracture and deformation behavior, as well as the specimen size and geometry with respect to the constraint effect [91, 92]. The fracture behavior of a material relates to the micro-mechanism of cracking, and is popularly described as being "ductile" or "brittle," whereas the deformation behavior governs the fracture toughness parameters of concern, and their methods of measurement through the assessment of the extent of plastic deformation ahead of the crack tip, commonly termed the plastic zone. Considering the extent of plastic deformation, the fracture behavior of a material is characterized as being linear elastic, nonlinear elastic, elastic-plastic and gross yielding plastic. The regimes of fracture mechanics are broadly categorized on the basis of associated plasticity with the fracture phenomenon; these are: linear elastic fracture mechanics (LEFM), elastic-plastic fracture mechanics (EPFM), and gross yielding fracture mechanics (GYFM). In general, the fracture toughness property commonly bears an inverse qualitative relation with the strength of a material; higher the strength, the lower the fracture toughness. In LEFM, an insignificantly small zone of deformation occurs around the crack tip, and the initiation toughness governs the resistance to crack extension, causing the specimen to fail in an unstable manner; the phenomenon is termed "brittle fracture," and the toughness is measured as a unique point value. The emergence of the plane strain fracture toughness test and the corresponding ASTM E399 [93] standard to determine K_{Ic} is based on this unique point value. When the plastic deformation dominates at the crack tip for ductile fracture in EPFM, the material's resistance against fracture increases as the crack grows, and the fracture toughness property is commonly described as a fracture resistance curve. The property is then considered as a measure of the elastic-plastic initiation toughness, like J_c or J_{Ic} for ductile fracture. The role of plastic deformation on the measure of the fracture toughness property is schematically described in Figure 16.4a.

The fracture toughness parameters of interest in fracture mechanics are the stress intensity factor (K), elastic energy release rate (G), J-integral, crack tip opening

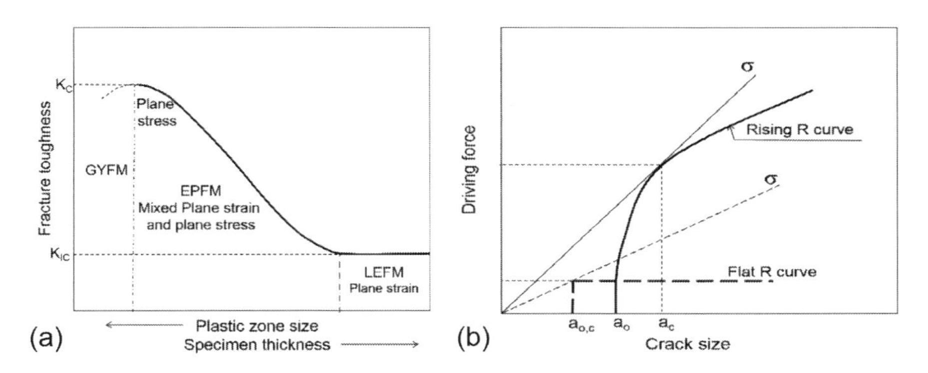

FIGURE 16.4 (a) Schematic variation of fracture toughness with thickness and the effect of plasticity in terms of plastic zone size ahead of the crack tip. The entire domain of the variation is marked with the regimes of linear elastic fracture mechanics (LEFM), elastic-plastic fracture mechanics (EPFM), and gross yielding fracture mechanics (GYFM); (b) schematic illustrations of flat and rising crack-growth resistance curves (R-curves) for brittle and ductile materials, showing the instability points at which the loading curve (σ) is at a tangent to the R-curve.

displacement (CTOD), and crack tip opening angle (CTOA). The parameter *K*, describes the crack tip stress field in LEFM, and the parameter *G* is related to *K*. In EPFM, the *J*-integral parameter illustrates the intensity of elastic-plastic crack tip fields. The parameter CTOD serves as an engineering fracture parameter with equivalence to *K* and *J* parameters. The CTOA parameter is confined to thin structural components with relatively higher toughness and usually depicts the stable crack extension. The introduction of the fracture toughness parameters *K*, *J*, and CTOD is due to Irwin [94], Rice [95], and Wells [96] and the detailed descriptions of these parameters and their applications are found in several text books like the ones by Anderson [91], Hertzberg [15], or Broek [97]. The descriptions of these terminologies and the test methods to determine these fracture toughness parameters are also well documented in several international standards [93, 98–101]. The *K*, *G*, *J*, and CTOD parameters are usually employed for relatively tough to ductile materials with tests in specimens with sharp pre-cracks. To characterize fracture toughness of brittle materials, one finds another *K*-based parameter, chevron-notched fracture toughness (K_{Icv}) in which the notch geometry in the specimen is different [102], but the intensity of the stress field is based on LEFM principles. All these fracture toughness parameters are conventional; but several non-conventional measurements of fracture toughness parameters also exist, like the ones based on hardness measurements in different scales [103, 104], through empirical correlations with other qualitative toughness parameters [15], or the GYFM parameters [105–107]. The fracture toughness of high entropy alloys so far has been assessed by *K*- or *J*-based parameters or by non-standard techniques using hardness measurements.

The development of the different techniques for measuring fracture toughness parameters is dependent on the degree of deformation and the extent of stable crack extension, the latter representing the overall fracture behavior. The constraint at the crack tip considerably dominates these phenomena. High constraint leads to lower yielding and consequently lower extent of plastic deformation and thus results in higher intensity of stresses at the crack tip, which promotes brittle fractures by reducing the fracture resistance curve. The opposite trend is encountered with lower constraint at the crack tip, which raises the fracture resistance curve toward ductile fracture. The degree of crack tip constraints is dictated by the size and geometry of fracture toughness specimens, as well as that of the inserted notch or the crack, apart from the type of loading. The existing fracture toughness standards usually demand higher specimen thickness and deep cracks to achieve higher constraints, and thus the conservative fracture resistance. The constraint due to specimen thickness is usually referred to as plane stress or plane strain state, the plane strain fracture toughness (K_{Ic}) determined using specimens having thickness over a critical value is considered as the conservative lower bound value.

In general, high constraint results in higher crack tip stresses with less crack tip yielding and either promotes brittle fracture or lowers the toughness by reducing the slope of the resistance curve toward ductile fracture, while low constraint results in lower crack tip stresses with more crack tip yielding, tends to reduce the possibility of brittle fracture and raises the resistance curve toughness for ductile fracture. The thickness constraint is commonly specified as being a plane stress and/or plane strain state of loading, with plane strain fracture toughness being regarded as the lower bound value. In order to obtain conservative fracture resistance, most of the fracture test standards suggest the adoption of deep-cracked bending specimens with high crack tip constraint conditions. Among these, compact tension (CT) specimen and single edge notched bends (SENB) specimen are the ones used most often in fracture toughness tests. However, many experiments showed that the crack tip constraint levels due to crack size, specimen geometry, and loading type have a substantial effect on fracture resistance of the material in terms of *J-R* or CTOD-*R* curves. In general, a high-constraint specimen leads to a lower *R*-curve, while a low-constraint specimen yields a higher *R*-curve; brittle materials show a flat R-curve, while ductile materials exhibit a rising *R*-curve, as illustrated in Figure 16.4b.

Attempts at understanding fracture behavior of HEAs have been made with limited efforts toward estimation of K_{Ic}, J_c or J_{Ic} and K_{Icv}. The magnitude of K_{Ic} is commonly determined with the help of a single edge notched bend (SENB) or compact tension C(T) specimen following ASTM standard E-399 [93]. In this method, simply a provisional value of plane strain fracture toughness K_Q is estimated, and it is then subjected to examinations through a few validity criteria in order to qualify it as K_{Ic}. The principal types of load-displacement curves encountered in this type of test are shown in Figure 16.5a. The magnitude of K_Q is expressed as [93]:

$$K_Q = \frac{P_Q.S}{BW^{3/2}} f(a/W)\left(\text{For SENB specimen}\right) \tag{16.1}$$

$$K_Q = \frac{P_Q.}{BW^{1/2}} f(a/W)\left(\text{For C}\left(\text{T}\right)\text{ specimen}\right) \tag{16.2}$$

where P_Q is the characteristic load determined from the recorded load (*P*)-crack tip opening displacement (δ) plot, *B* (thickness), *W*(width), and *S* (span) are the characteristic dimensions of the specimen, *a* is the average crack length determined on the fractured surface, and f(*a*/*W*) is a geometry factor for the finite fracture toughness specimen. The value of f(*a*/*W*) is calculated from polynomial functions as described in ASTM E-399 [93] or ASTM E-1820 [98]. The provisional fracture toughness

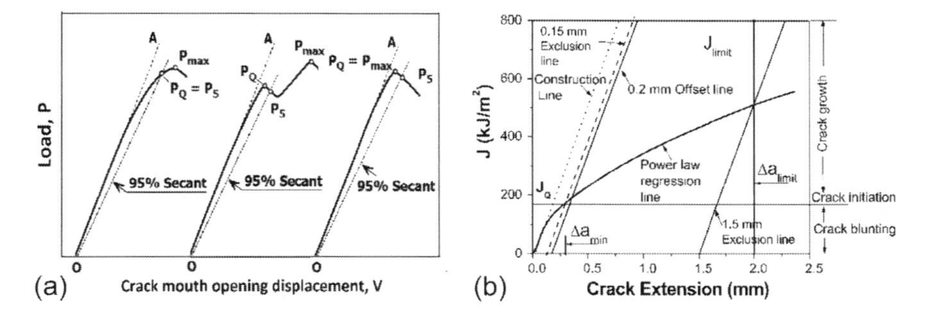

FIGURE 16.5 (a) Types of load-displacement plots encountered during a plane strain fracture toughness test following the ASTM E399 standard [93]; (b) typical plot of *J* vs crack extension (Δa) with the various construction lines to qualify the data and to estimate the critical value of *J* (J_C or J_{IC}) following the ASTM E1820 standard [98].

value K_Q is subjected to three major validity checks before it is referred to as plane strain fracture toughness, K_{Ic}. The validity criteria are:

(i) Thickness criterion

$$B \geq 2.5(K_Q / \sigma_{ys})^2 \tag{16.3}$$

(ii) Maximum load criterion

$$P_{\max} / P_Q < 1.10 \tag{16.4}$$

(iii) Displacement criterion

$$V_{\text{at } 0.8P_Q} < 0.25V_{\text{at } P_Q} \tag{16.5}$$

where σ_{ys} is the yield strength of the material, $P_{\max}$ is the maximum load in the P-δ record, and V refers to non-elastic displacement with suffixes representing the loads P_Q and $0.8P_Q$.

The detailed procedure for the measurement of the J-resistance (J-R) curve and subsequently for estimating J_c/J_{Ic} is described in ASTM standard E-1820 [98]. The magnitude of J is determined either using a multi-specimen or using a single specimen technique such as:

$$J = J_{el} + J_{pl} \tag{16.6}$$

$$J_{el} = K_i^2(1-\upsilon^2)/E \tag{16.7}$$

$$J_{pl(i)} = \left[J_{pl(i-1)} + \left(\frac{\eta_{(i-1)}}{(W-a_{(i-1)})}\right)\frac{\Delta A_{pl(i)}}{B_N}\right]\left[1-\gamma_{(i-1)}\frac{\alpha_{(i)}-a_{(i-1)}}{W-a_{(i-1)}}\right] \tag{16.8}$$

where J_{el} and J_{pl} are the elastic and plastic components of J, and the subscript (i) or ($i-1$) refers to the instantaneous components for crack length a_i. A desired number of (J_i, Δa_i) points are generated to construct the J-R curve; a schematic J-Δa plot is shown in Figure 16.5(b) (ASTM E 1820). The value of K_i is estimated from Equation (16.1) or Equation (16.2) depending on the type of specimen, and the magnitude of $\Delta A_{pl(i)}$ is estimated as the area under the P-δ plot up to the crack length a_i. E and ν are the elastic modulus and the Poisson's ratio of the material. η and γ are functions of instantaneous crack (a_i) or ligament length ($b = W - a_i$) as described in the standards [98].

One of the specific requirements to obtain the crack growth resistance curve is to estimate the crack length (a_i) and the crack extension (Δa_i) at the ith instant. The crack length during the fracture toughness testing can be determined using several techniques [108], but for a single specimen J-integral test, it is commonly done by compliance measurement by making restricted loading and unloading during the generation of the P-δ plot. The crack length (a_i) is estimated from the following expression [98] as:

$$a_i / W = \Big[0.999748 - 3.9504u + 2.9821u^2 - 3.21408u^3 + 51.51564u^4 - 113.031u^5\Big] \tag{16.9}$$

The term u in Equation (16.9) is described as:

$$u = \frac{1}{\left[(B_e WEC_i)/(S/4)\right]^{1/2} + 1} \tag{16.10}$$

where C_i (=$\Delta V_m/\Delta P$) is the elastic compliance (obtained during the loading/unloading sequence of the test) in which V_m is the crack mouth opening displacement at notch edge, and B_e is expressed as = B-$(B$-$B_N)^2/B$, in which B_N represents the thickness of the side-grooved specimen. The expression for C_i in terms of specimen dimensions and side groove is available in ASTM E-1820 [98]. The value of Δa is next calculated as $(a_i$-$a_o)$ where a_o is the initial crack length in the specimen at the start of the test. The construction of the J-Δa curve by the above procedure is next used to determine the provisional value of J_Q. It is estimated at the intersection of J-resistance curve and a 0.2 mm offset line. The validity check for J_Q to be J_{Ic} is done by the updated standard [98] given as B, $b > (10\ J_Q)/\sigma_{ys}$. If the validity criterion fails, J_Q is often referred to as J_c only.

The measurement of fracture toughness using a chevron-notched specimen (K_{Iv} or K_{IvM}) as per ASTM standard E-1304 [102] is commonly done using short bar or short rod specimens. Attempts exist in the literature to determine fracture toughness using chevron-notched rectangular or cylindrical bend bar specimens too [109, 110]. Evaluation of fracture toughness values for chevron-notched rectangular bend (CVNRB) bar specimens for some HEAs are reported in the literature [111], and a brief note on this measurement is thus incorporated. The estimation of K_{Icv} (the deviation from the notation K_{Iv} is incorporated to demark it from the standard practice) by this method is done using the Irwin–Kies relation [112]:

$$K_{Icv} = \frac{P_{\max}}{B\sqrt{W}} Y_c(\alpha_o, \alpha_1) \tag{16.11}$$

where $P_{\max}$ = maximum load applied on the specimen when a crack grows to a critical length a_c, B = specimen thickness, W = specimen width, $Y_c(\alpha_o, \alpha_1)$ = minimum value of the compliance function, $\alpha_o = a_o/W$, and $\alpha_1 = a_1/W$. The value of $Y_c(\alpha_o, \alpha_1)$ has been estimated by the expression suggested by Shang-Xian [112] as a function of W/B values and θ. For $W/B = 1.5$ and chevron-notched angle $\theta = 60°$, the expression converges to:

$$Y_c = 5.639 + 27.44\alpha_o + 18.93\alpha_o^2 - 43.42\alpha_o^3 + 338.9\alpha_o^4 \tag{16.12}$$

16.2.3 Fracture Toughness Assessment of High Entropy Alloys by Conventional Approaches

Limited attempts have been directed at assessing fracture toughness of high entropy alloys over the last five years using both conventional and alternate approaches.

Information and knowledge that are generated on the basis of the examinations of fracture toughness assessments of HEAs by the conventional approaches are the content of this section. The microstructure and the relevant mechanical properties of the concerned materials that have been subjected to fracture toughness examinations are also included for the sake of understanding the toughness of the investigated materials with regards to their structural features and tensile/hardness properties.

The first attempt to understand the order of fracture toughness of high entropy alloys using standard techniques was by Roy et al. [111]. The material used by the authors is $Al_{23}Co_{15}Cr_{23}Cu_8Fe_{15}Ni_{15}$ which comprised BCC Al-Ni plates with BCC Fe-Cr inter-plates as the matrix with a compressive yield strength of 1,230 MPa. These investigators reported the plane strain fracture toughness of the HEA, determined using single edge-notched bend (SENB) and chevron-notched rectangular bar (CVNRB) specimens. A typical load-displacement plot obtained during fracture toughness tests using an SENB specimen is shown in Figure 16.6. The principle of estimating K_{Ic} and K_{Icv} using Equation (16.1) and Equation (16.11) by these investigators are in accordance with ASTM E-399 [93] and are following ASTM E-1304 [102], respectively. The average magnitudes of K_{Ic} and K_{Icv} for the investigated alloy are reported to be 5.8 ± 0.2 MPam$^{1/2}$and 5.4 ± 0.2 MPam$^{1/2}$, respectively. Using the yield strength of the alloy as 1,208 MPa, the critical thickness for confirming the estimated values as plane strain fracture toughness are reported as 0.056 mm and 0.028 mm per Equation (16.3) and Equation (16.4), which are sufficiently below the specimen thickness used by the authors. The fracture toughness measured using the CVNRB specimens has been suggested by the authors as conservative, because the estimation of SENB specimens were done on sharp notched but non-fatigue pre-cracked specimens. The poor fracture toughness of the alloy is attributed to its BCC

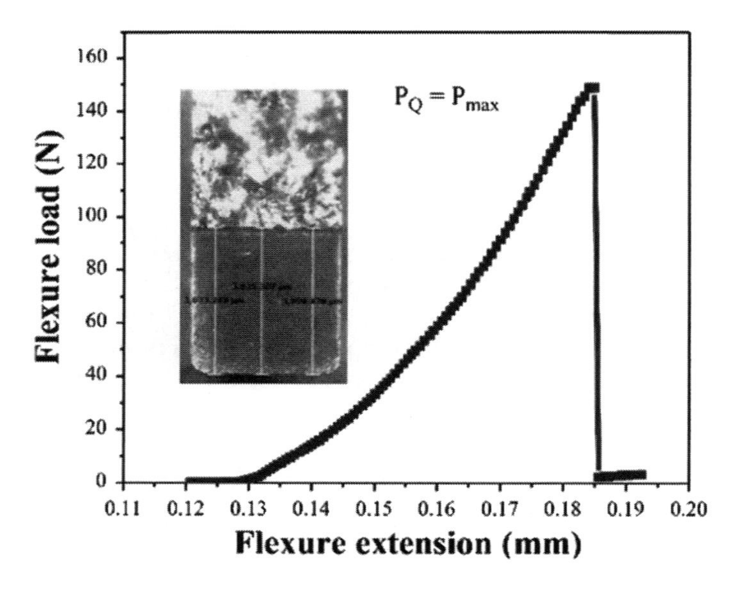

FIGURE 16.6 Typical load-displacement plot for a SENB specimen; photograph of a specimen is shown in the inset [111]. (Reprinted with permission from Elsevier.)

crystal structure having ordered Ni-Al as one of the major constituents and having relatively poor deformability, but the order of the estimated fracture toughness is found to be close to that of Ni-Al intermetallic [113, 114] having K_{Ic} as 5.08 MPam$^{1/2}$.

Chen et al. [115] examined the fracture toughness of four $Al_{20-x}Cr_{20+0.5x}Fe_{20}Co_{20}Ni_{20+0.5x}$ HEAs in which x was varied between 2, 4.5, 7, and 9.5 at.%. Depending on the amount of Al at.%, the alloys are referred to as AL18 (BCC), AL15.5 (BCC+FCC), AL13 (BCC+FCC), and AL10.5 (FCC), respectively; the crystal structures of the alloys are indicated in the parenthesis. The higher Al-containing alloys showed no plastic deformation compared to the lower Al-containing alloys, the AL13 and AL10.5 alloys. The authors have not compiled the detailed tensile properties for all the alloys. From their reported stress-strain curves, the values of yield strength appear to be ~390 MPa (AL18), ~600 MPa (AL15.5), 760 MPa (AL13), and 370 MPa (AL10.5) for the investigated materials. The investigators have examined the fracture toughness of the alloys using SENB specimens having dimension *B*, *W*, and *S* as 3 mm, 6 mm, and 24 mm, respectively, with U notches of root radius 150 μm. Typical load-displacement plots recorded by the authors during the fracture toughness tests for the different alloys are depicted in Figure 16.7. The fracture toughness of the first three HEAs, as estimated using Equation (16.1), are 9, 11, and 53 MPam$^{1/2}$, respectively, without any validation check for the estimated toughness values. The results indicate that increased content of the FCC phase results in a higher amount of lamellar structure which, in turn, increases the fracture toughness of the alloy.

Seifi et al. [116] carried out fracture toughness determination on $Al_{0.2}CrFeNiTi_{0.2}$ and $AlCrFeNi_2Cu$ HEAs using 3 mm×6 mm×20 mm bend bar specimens following ASTM E-399 [93] procedure. Both the cast alloys exhibit (BCC+FCC) dendritic structures with different Rockwell hardness values of 36 and 46 for $Al_{0.2}CrFeNiTi_{0.2}$

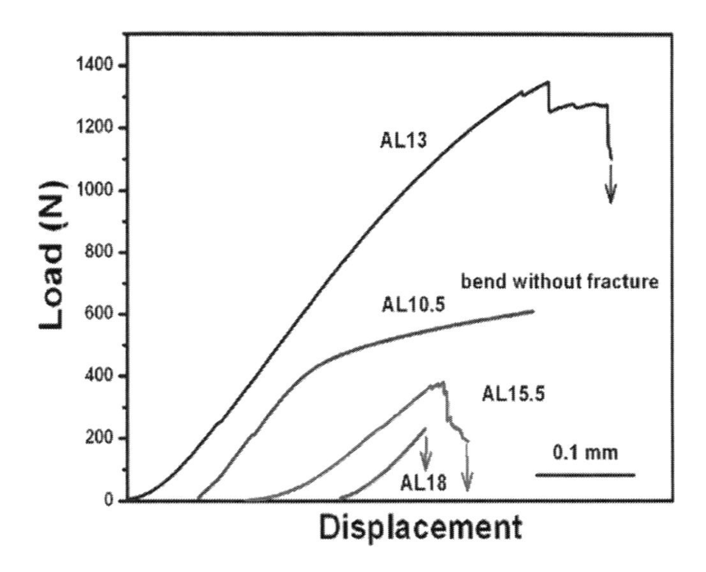

FIGURE 16.7 Typical load-displacement plots for notched SENB specimens of AlCrFeCoNi HEAs with atomic fraction of Al as 18,15.5,13, and 10.5 [115]. (Reprinted with permission from Elsevier.)

and $AlCrFeNi_2Cu$ HEA, respectively. The fracture toughness of the alloys are reported as 40–45 and 32–35 $MPam^{1/2}$, respectively. The fracture toughness of $Al_{0.2}CrFeNiTi_{0.2}$ HEA was found to be 46–47 $MPam^{1/2}$ at 200°C; without any validation check, the authors have indicated the obtained toughness values to be K_Q.

Higher fracture toughness values are reported for CrCoNi [117] and CrMnFeCoNi [24] alloys; the former alloy is usually called a medium entropy alloy (MEA), while the second one is currently a well-characterized HEA. Both the materials possess an FCC crystal structure, and their strength and ductility properties get enhanced at lower temperatures. These alloys were prepared by forging, rolling, and recrystallization treatment and exhibited extensive plasticity. The CrCoNi alloy is reported to exhibit nearly a 50% increase in strength with the lowering of the test temperature to $\sigma_y = 657$ MPa and $\sigma_{uts} = 1{,}311$ MPa at 77 K with associated increase in strain to failure (ductility) from ~25% (293 K) to over 90% (77 K). For the CrMnFeCoNi alloy also, similar increases in strength and ductility are also reported by the authors [24]. The CrMnFeCoNi alloy indicated $\sigma_y = 759$ MPa and $\sigma_{uts} = 1{,}280$ at 77 K with associated ductility of ~70%; the strength values increased by > 70% while the ductility increased by nearly 200% over the values obtained at 293 K. Due to their intrinsic nature, estimation of fracture toughness of these alloys could not be attempted with LEFM approach, and their toughness characterization was done by constructing *J-R* curve and thereby estimating their J_{IC} values. The procedure to construct *J-R* curve and to determine J_{IC} was in following the ASTM standard E-1820 [98] and with the outline described using Equation (16.6) to Equation (16.9) earlier. Gludovatz et al. [117] noted the value of J at crack initiation for the CrCoNi alloy at room temperature to be more than 200 kJm^{-2}, which increased to almost 400 kJm^{-2} at higher Δa values. At 77 K, this alloy indicated abnormal J value of 350 kJm^{-2} at crack initiation. The obtained J_{Ic} values were converted to K_{JIc} values using the relation

$$K_{JIc} = (J.E')^{1/2} \tag{16.13}$$

where E' is E in plane stress and $E/(1-\upsilon^2)$ in plane strain conditions. The values of K_{JIc} were reported to be as 208 $MPam^{1/2}$ and 273 $MPam^{1/2}$ at 293 K and 77 K, respectively. The fracture toughness values for the CrMnFeCoNi [24] alloys were obtained in an identical manner as 217 $MPam^{1/2}$, 221 $MPam^{1/2}$ and 219 $MPam^{1/2}$ at 77 K, 200 K and 273 K, respectively. The astonishingly high toughness values of these alloys are attributed to their FCC structure, mode of plastic deformation which induces a steady state of strain hardening, and micro-twinning at low temperatures in addition to planar slip.

Jo et al. [80] examined the fracture toughness of the non-equiatomic $V_{10}Cr_{10}Fe_{45}Co_{20}Ni_{15}$ alloy at 298 and 77 K in order to assess the potential of cracking susceptibility of CoCrFeMnNi and CrCoNi alloys. The tensile properties of the investigated non-equiatomic alloy at the two temperatures are reported as: $\sigma_y = 294 \pm 4$ (298 K) and 470 ± 6(77), $\sigma_{uts} = 626 \pm 5$ (298 K), and $1{,}000 \pm 1$(77 K), and %Elongation = 36 ± 1 (298 K) and 62 ± 3.9 (77 K). The fracture toughness tests of the alloys were conducted by the authors using *J*-integral technique following the ASTM standard E1820 [98] using 13 mm thick pre-cracked specimens. Typical *J*-Δ*a* plots for the alloy at 298 K and 77 K are shown in Figure 16.8; these are similar to the *J*-Δ*a* plots obtained for CoCrFeMnNi and CrCoNi alloys by Gludovatz et al. [24, 117]. The magnitudes of fracture toughness K_{JC} are found to be 219 $MPam^{1/2}$(at 298 K) and 232 $MPam^{1/2}$ (at 77 K) at the investigated temperatures. Comparison of the obtained fracture toughness values of the non-equiatomic alloy with those of equiatomic CoCrFeMnNi and CrCoNi alloys indicated that the V-containing alloy exhibits marginally higher fracture toughness than the equiatomic alloys. The improved properties of the alloy at lower temperatures are attributed to the predominance of twins over planar slip or dislocation cell structure.

A comparative assessment of the fracture toughness values of HEAs reported so far is illustrated using a bar diagram in Figure 16.9; in this diagram, the results on CrCoNi alloy are excluded as the alloy belongs to the allied HEA group and is commonly referred as a medium entropy alloy (MEA). The fracture toughness values in Figure 16.9 vividly bring forward the dependence of fracture toughness property on the crystalline structure of the concerned materials. The fracture toughness of HEAs is found to be highest for FCC alloys and lowest for BCC alloys while that for the dual phase structures (FCC+BCC) are between the two boundaries. The dependence of fracture toughness of the HEAs on different crystal structures is related to their plastic deformation ability as roughly expressed by their ductility. The associated plasticity of the investigated materials also necessitates different methodologies for fracture toughness assessments.

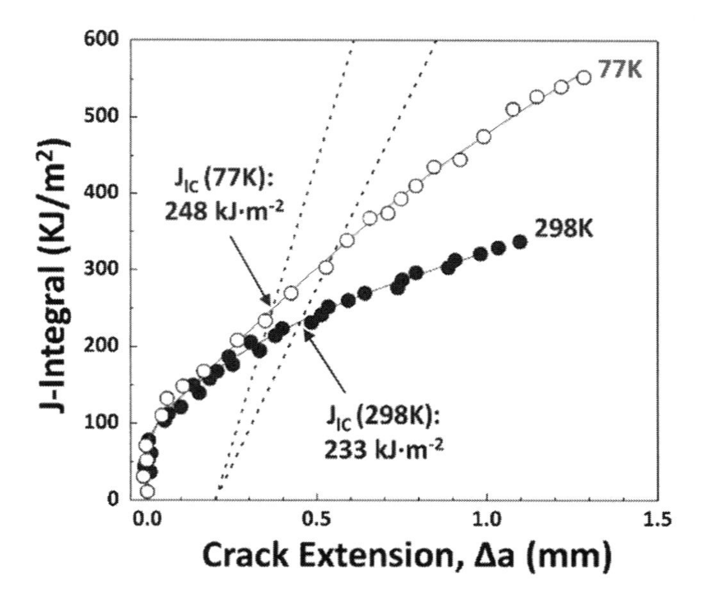

FIGURE 16.8 Typical crack resistance (*J*-Δ*a*) curves for $V_{10}Cr_{10}Fe_{45}Co_{20}Ni_{15}$ HEA [80] tested at 298 K and 77 K. The critical values of J_{IC} as determined by the 0.2 mm offset lines are also shown. (Reprinted with permission from Elsevier.)

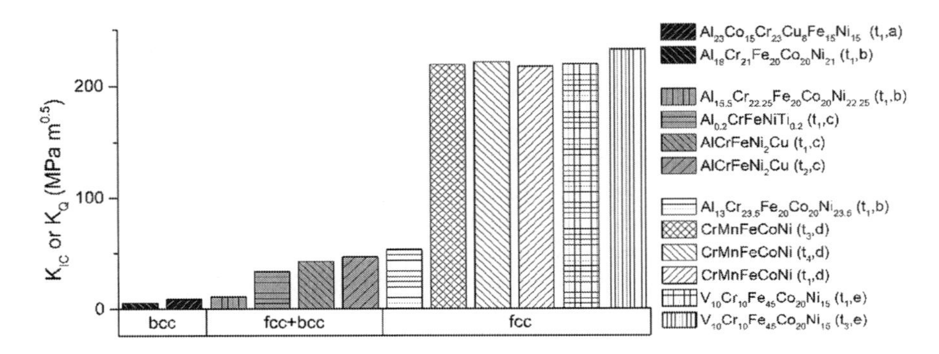

FIGURE 16.9 Comparative assessment of the fracture toughness values of high entropy alloys illustrating the dependence of the property on the structure of the constituent phases. The temperature and the references for the materials are given in the parenthesis as ti and a to e, respectively. $t_1 = 293$ K, $t_2 = 473$ K, $t_3 = 77$ K, $t_4 = 200$ K. a[111], b[115], c[116], d[117], e[80].

16.2.4 Toughness Measurement of High Entropy Alloys by Alternate Approaches

The toughness of materials has been assessed over a long time using Charpy (or Izod) impact energy values specifically to examine the temperature dependence of toughness and the possible working temperature range for structural materials through ductile-brittle transition temperature. These tests are based on the measurement of energy absorbed by a notched specimen when it receives a blow from a moving mass like a pendulum hammer. The estimated energy qualitatively indicates the propensity of material toward brittle fracture. The test procedure is described in ASTM standard E-23 [118]. This toughness measurement technique is considered useful for making a qualitative comparison of materials through their cracking resistance assessment in notched specimens, but it is not reliable for design and application for ensuring the structural integrity of components. Interestingly, this method of toughness assessment has also been employed for high entropy alloys.

Li and Zhang [119] examined the Charpy impact energy of $Al_{0.1}CoCrFeNi$ and $Al_{0.3}CoCrFeNi$ alloys using standard Charpy specimens of 10 mm × 10 mm × 55 mm at three different temperatures: 298 K, 200 K, and 77 K. The Charpy impact energy value of the 0.1 at.% Al alloy was found to increase from 289 J at 77 K to 420 J at 298 K. The corresponding values for the 0.3 at.% Al-containing HEA are 328 J (77 K) and 413 J (298 K). The impact energy values of these HEAs are considerably higher when compared to their strength values as above 620 MPa at room temperature and 1,020 MPa at 77 K. The authors noted that the strength and ductility of both the alloys increased with the decrease in test temperature in contrast to conventional metallic materials. They assumed that higher ductility at 77 K probably originates due to the predominance of plastic flow by mechanical nano-twinning rather than by planar slip of dislocations. The higher Charpy impact energy of the alloys at 77 K is also considered to emerge for the same reason.

Chen et al. [115] examined both fracture and impact toughness of four HEAs with the composition $Al_{20\_x}Cr_{20+0.5x}Fe_{20}Co_{20}Ni_{20+0.5x}$ (for x = 2, 4.5, 7, and 9.5 at.%), referred to as AL18, AL15.5, AL13, and AL10.5 at room temperature. The Charpy impact energy was determined by the authors using sub-size U-notched specimens with a cross-section of 6 mm× 6 mm. The estimated impact energy values are 12±2, 15±3, 106±23, and 477±15 kJ/m² for AL18, AL15.5, AL13, and AL10.5 alloys, respectively. With lower Al content, the increase in impact toughness bears the same trend of variation for the fracture toughness versus Al-content.

While impact toughness measurement provides us with qualitative toughness measures, fracture toughness values of materials are also estimated using macro-, micro-, or nanoindentations and micro-cantilever bending technique. These measurements provide quantitative fracture toughness values of the material by non-standard techniques, but are limited by their reliability and possible deviation from the bulk property. The measurement of fracture toughness by indentation (K_{IFT}) is simple and rapid, and it is popular for estimating toughness of brittle materials like ceramics, glasses, or intermetallics. The measurement of fracture toughness by indentations [103, 120] incorporates (i) recording the applied indentation load and (ii) measuring the length of the crack which emanates from the corners of the indentation, while estimation of K_{IFT} requires knowledge of the elastic modulus, Poisson's ratio, and true hardness of the material. The method of estimation of fracture toughness by indentation technique is based on empirical formulations [104] as the stress-field associated with indentation cracking is complex, and different types of cracks like Palmqvist or median, can originate due to indentation. One of the simplest formulations is:

$$K_{IFT} = \alpha(E / H)^{1/2}\left(P / C^{3/2}\right) \quad (16.14)$$

where α is a constant depending on the indenter geometry, the type of crack produced by indentation, and the stress-field associated with the indenter, E and H are elastic modulus and true hardness of the material, P is the applied load in indentation, and c is the crack length from the indentation center to the crack tip. The values of E and H are to be estimated by separate tests for the micro-indentation technique [104], while these can be determined simultaneously while performing the nano-indentation test [103].

Zhang et al. [58] estimated the indentation fracture toughness (K_{IFT}) values of two HEAs $FeCoNiCrAlSiCuTiMoB_{0.5}$ and $FeCoNiCrAl_3$ following the earlier reports of Zhang et al. [120] and Lee et al. [121] and indicated that the fracture toughness values of these HEAs are 50.9 MPa m$^{1/2}$ and 7.6 MPa m$^{1/2}$, respectively. The E and H values for the alloys with and without boron were estimated as "11.3 GPa and 187.1 GPa" and "7.5 GPa and 117.9 GPa," respectively. Typical load vs depth of indentation plots for the two alloys of concern are shown in Figure 16.10. The authors inferred that boron as solute interstitial, triggers the nucleation of martensitic phase in the boron-containing alloy and this, in turn, significantly contributes to the enhancement of the hardness and fracture toughness of this alloy compared to the $FeCoNiCrAl_3$ HEA.

Fracture toughness of HEAs has been measured using *in situ* micro-cantilever beam testing by Zou et al. [88]. In this procedure, the specimen and the notch are prepared by milling with a focused ion beam using coarse and fine milling by varying the current intensity, and then the beam is loaded *in situ* using a nano-indenter [122]. A typical cantilever beam specimen during loading by a micro-indenter is

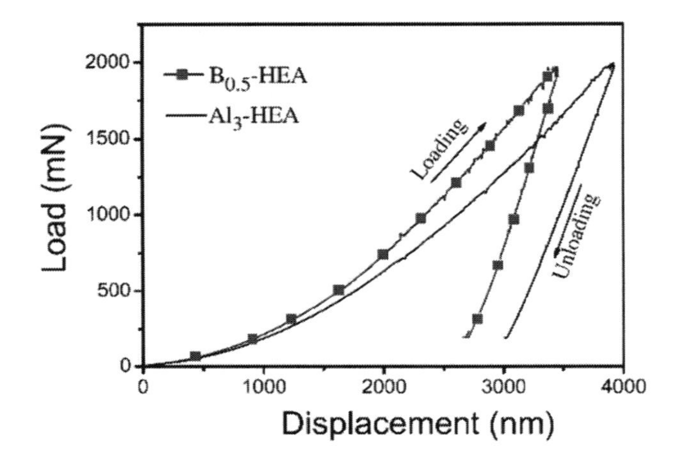

FIGURE 16.10 Load displacement plots obtained during fracture toughness measurement using nanoindentation for $FeCoNiCrAlSiCuTiMoB_{0.5}$ ($B_{0.5}$-HEA) and $FeCoNiCrAl_3$ (Al_3-HEA) high entropy alloys [58]. (Reprinted with permission from Elsevier.)

shown in Figure 16.11a and a typical load-displacement plot is given in Figure 16.11b. The fracture toughness of the micro-SENB-type specimen is estimated using the expression similar to Equation (16.1) as:

$$K_Q = \frac{F_{max}.L}{BW^{3/2}} f(a/W) \tag{16.15}$$

where F_{max} is the fracture force, and L is the distance between the notch and the loading point; other symbols are as defined in Equation (16.1). Using this methodology, Zou et al. [88] determined the fracture toughness of single and bi-crystals of refractory $Nb_{25}Mo_{25}Ta_{25}W_{25}$ HEA, and found extremely low values of fracture toughness.

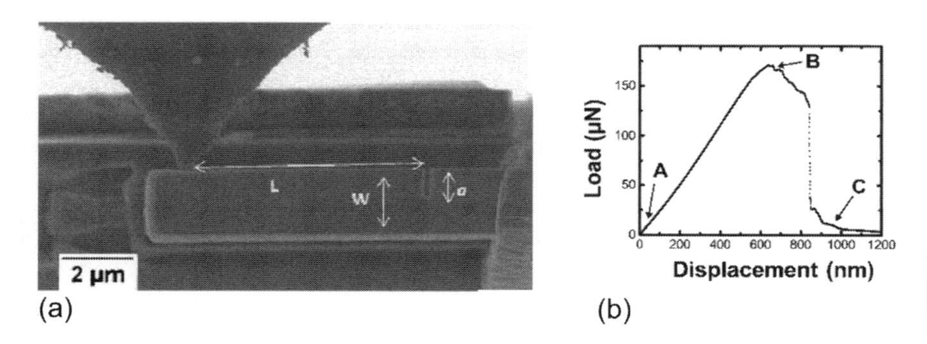

FIGURE 16.11 (a) Typical micro-cantilever beam loading setup [122] for measurement of fracture toughness for single crystals of $Nb_{25}Mo_{25}Ta_{25}W_{25}$, and (b) load-displacement plots recorded during tests on single crystal of the alloy. In subplot (a), L = half span length, W = width of specimen, and a = notch depth, in subplot (b) A = initial contacts, B = crack tip opening at the maximum load, C = fracture load. (Reprinted with permission from Elsevier.)

The investigators reported that single crystal exhibited fracture toughness of 1.3–2.1 $MPam^{1/2}$, while the bi-crystal indicated fracture toughness of only 0.2 $MPam^{1/2}$; the failure mechanism is quasi-cleavage for the single crystals, unlike intergranular fracture in bi-crystals.

16.2.5 Micro-Mechanism of Fracture in High Entropy Alloys

Examinations of the fracture surfaces are essential supplements to understand the resistance to fracture of any material, as these examinations provide the information about the initiation and the growth of cracks in a material. The features on the fracture surface bear the signatures of the (i) nature of the material and its microstructure, (ii) magnitude of applied stresses/strains, (iii) mode of fracture, (iv) rate of loading, (iv) temperature and environmental influences. The overall pattern and the isolated signatures on the fractured surface also act as an important tool for analyzing the failure of materials. Broadly, the fracture surface signatures in monotonic loading can be categorized into two groups [123]:

(a) Brittle-cleavage and intergranular fracture, in materials where the crack initiation is associated with limited plasticity: cleavage fracture is characterized by bright facets often marked with river lines, cleavage tongues and herringbone patterns [97, 124], or at times as granular features (at low magnification). Transgranular cleavage is triggered by tensile stresses acting normal to the crystallographic cleavage planes. Stresses acting normal to the interfaces and grain boundaries govern the formation of intergranular fracture. These are not as bright as the cleavage planes and do not bear the typical cleavage surface marks; and
(b) Ductile-dimple, "mixed dimple and cleavage" or quasi-cleavage fracture, where crack initiation and growth is associated with considerable plasticity: the common signature in the ductile fracture is dimples. This mode of the fracture involves nucleation of voids from second phase particles followed by their growth or the growth of existing voids like pores during deformation, and finally, their linking to create a rough fracture path. Alternatively, the fracture path can also be created by plastic instability, where void formation usually occurs, but is not essential [97, 124, 125]. Depending on the extent of plasticity and the microstructural constituents, the formation of "mixed dimple and cleavage" or "quasi-cleavage" fracture is noted on fracture surfaces.

Usually, the former (category-a) is locally stress-controlled, whereas the latter (category-b) is strain-controlled and is dominated by crack blunting exhibiting microvoid coalescence during the growth of the crack. Cleavage fracture is dominated by tensile stresses, whereas ductile fracture is primarily shear-stress governed. In category-b, the fracture surface is usually dull, rough, and fibrous, unlike the relatively flat and bright surfaces exhibited by category-a. The "mixed dimple and cleavage" is a combination of category-a and category-b at local scale. The fracture surface signatures are, on a broad scale, related to the fracture toughness of the concerned materials.

16.2.5.1 Brittle Fracture in High Entropy Alloys

Brittle fracture features in HEAs are in close agreement with the relatively low fracture toughness values of the concerned materials under investigation. Roy et al. [111] illustrated the brittle fracture characteristics of BCC $Al_{23}Co_{15}Cr_{23}Cu_8Fe_{15}Ni_{15}$ HEA (K_{Ic} = 5.4–5.8 MPam$^{1/2}$) as depicted in Figure 16.12. The fractographs in Figure 16.12 exhibit four distinct characteristics: (i) river line-type features, (ii) featureless facets, (iii) regions exhibiting mixed (a) and (b) type fracture features, and (iv) secondary cracks. Island-like facets (marked B) surrounded by river lines (marked A) are formed on the fracture surfaces. The facets are typical of the cleavage fracture, and river lines are known to result from the joining of cleavage steps in a single grain [111]. The river lines are termed "apparent river lines" by the investigators, as they appear to resemble shear bands on nano-crystalline materials [126]. Herringbone-type patterns are also noted by the investigators, which indicates the micro-mechanism of fracture being dictated by some nano-twins [97, 127]. Wang et al. [128] demonstrated that the notch fracture toughness of BCC $TiZrNbTa_{(100-x)}Mo_x$ HEA in as-cast condition decreases with increasing Mo concentration at room temperature from 28.5 MPa$\sqrt{m}$ (x=0) for the quaternary alloy, to 18.7 MPa$\sqrt{m}$ (x=20) for the quinary alloy. The features on the fractured surfaces under mode-I loading exhibit a transition from intergranular fracture for Mo-free TiZrNbTa to completely transgranular cleavage for the TiZrNbTaMo alloy. A set of well-depicted fracture surfaces, from this report, are shown in Figure 16.13 [128]. The gradual transition from

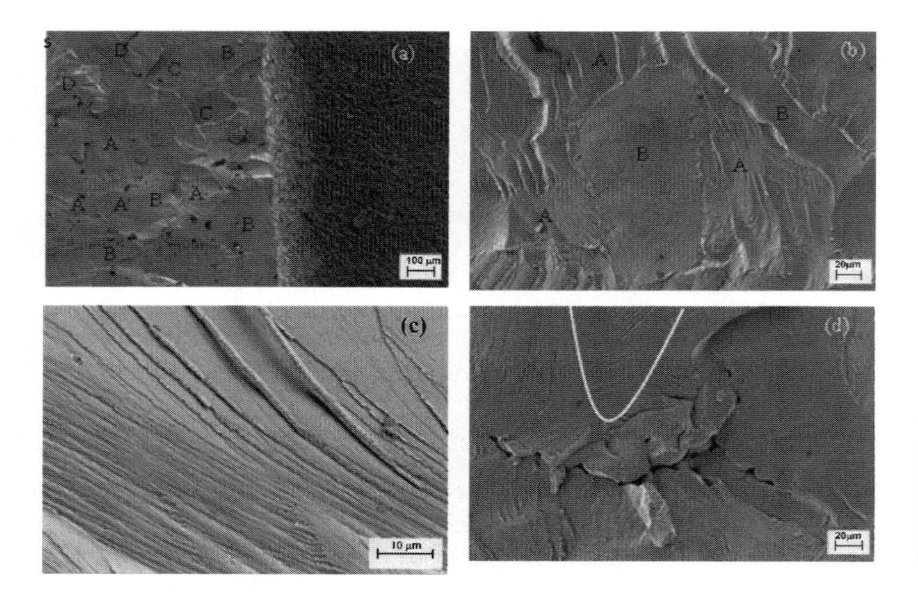

FIGURE 16.12 Fractographs (a) showing transition region between machined notch root and fractured surface of the alloy; (b) regions exhibiting apparent river lines adjacent to featureless facets; (c) enlarged view of apparent river lines which identifies these to be shear bands; (d) a region of herringbone-type pattern [111]. (Reprinted with permission from Elsevier.)

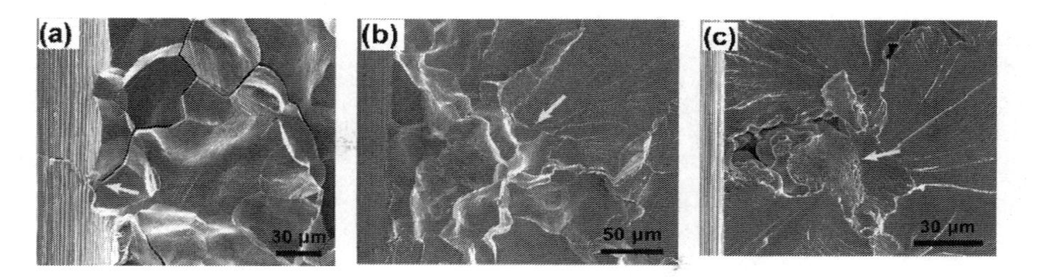

FIGURE 16.13 Fractured surfaces of $(TiZrNbTa)_{100-x}Mo_x$ HEA for (a) x=0, (b) x = 10, and (c) x=20, indicating gradual transition from intergranular to transgranular fracture [128]. (Reprinted with permission from Elsevier.)

intergranular to transgranular cleavage via mixed domains of "intergranular+transgranular" features is vivid in Figure 16.13.

16.2.5.2 Ductile Fracture in High Entropy Alloys

High or medium entropy alloys having an FCC structure and possessing high fracture toughness values exhibit distinctly different fracture surface features compared to BCC or "BCC+FCC" alloys with low or medium fracture toughness values, and show typical ductile fracture signatures, like dimples. Gludovatz et al. [24] first reported the high fracture toughness of CrMnFeCoNi HEA at ambient to liquid nitrogen temperature, and later, Gludovatz et al. [117] presented another report on fracture toughness of the medium entropy alloy, CrCoNi, from the ambient to the cryogenic temperature. Both the alloys exhibit distinct ductile fracture surfaces by initiation, growth, and coalescence of voids, and the voids are found to initiate from particles. Typical fracture surfaces of CrMnFeCoNi samples tested at room temperature are illustrated in Figure 16.14a [24] by ductile dimpled fracture, in which the voids are initiated from particles (Figure 16.14b). The latter were detected primarily as Mn-rich or Cr-rich particles, as evidenced by EDAX analysis (Figure 16.14c [24]). The coalescence of the voids is also illustrated by the authors using examinations of polished surfaces of the broken samples as shown in Figure 16.15 [24]. Similar observations were also evidenced by Gludovatz et al. [117] in the CrCoNi alloy as illustrated in Figure 16.16. The volume fraction of the inclusions which trigger void initiation is reported to be lower in the CrCoNi alloy than in the five-component CrMnFeCoNi HEA. The particles in the three-component alloy are reported as Cr-rich, whereas those in the five-component alloy are both Cr- and Mn- rich. The absence of Mn in the ternary alloy was considered by the authors to be responsible for lower volume fraction of particles.

Li and Zhang [119] also examined the fracture surfaces of FCC $Al_{0.1}CoCrFeNi$ and $Al_{0.3}CoCrFeNi$ HEAs broken by impact tests. These alloys showed increased strength, ductility, and impact toughness at lower temperatures (77 K) compared to room temperature (298 K). The investigators reported dimple-like features on the broken fracture surfaces of these alloys, and indicated that the observed higher mechanical properties of these alloys at lower temperatures are governed by nanotwins rather than planar slip. Chen et al. [115] examined the fracture surfaces of

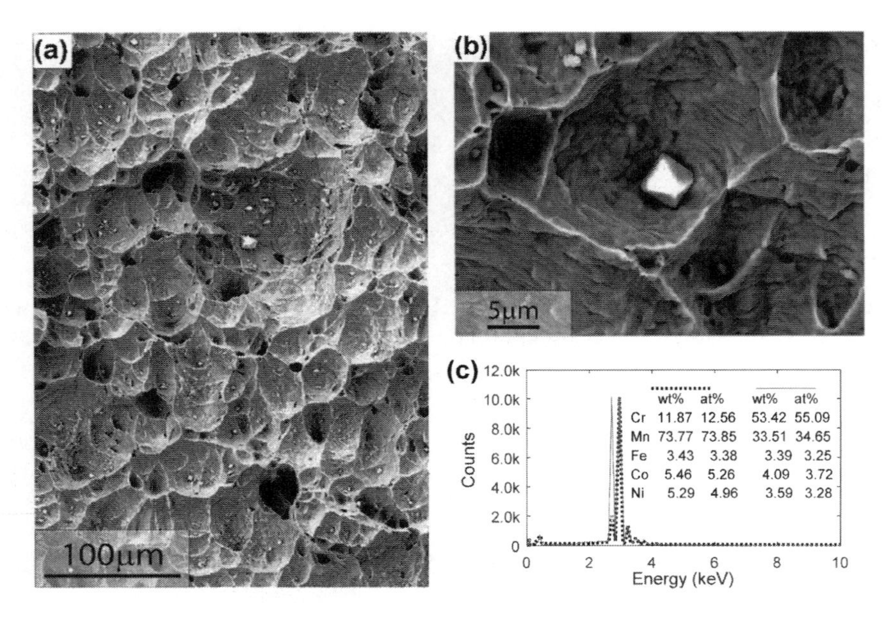

FIGURE 16.14 Fractured surface of a C(T) sample of CrMnFeCoNi HEA tested at room temperature [24]. Image (a) illustrates dimple formation, in which (b) the voids are initiated from Mn-rich or Cr-rich particles. (c) The nature of the particles is confirmed by EDAX analysis.

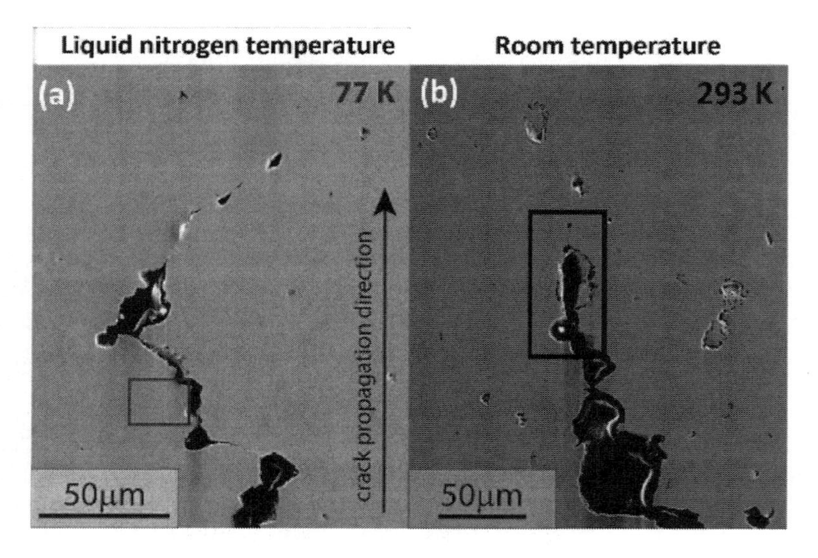

FIGURE 16.15 Examples of microvoid coalescence in CrMnFeCoNi HEA at (a) liquid nitrogen temperature 77 K, and (b) room temperature 293 K [24].

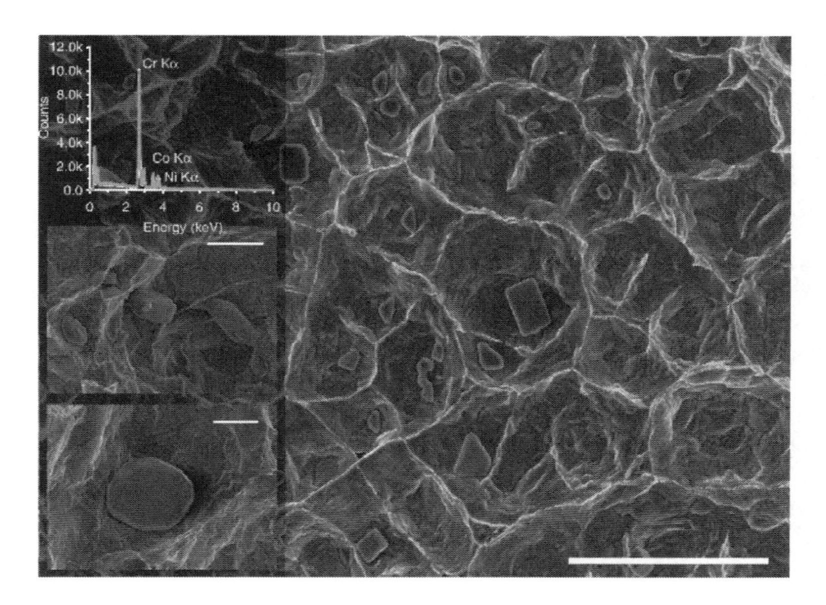

FIGURE 16.16 Fracture surface of CrCoNi medium entropy alloy exhibits ductile dimpled fracture, in which void initiation occurs from Cr-rich particles, as confirmed by the EDAX analysis (shown as an insert) [117]. The scale bar for the fractograph is 5 μm, and those for the insets are 2 μm.

$Al_{20-x}Cr_{20+0.5x}Fe_{20}Co_{20}Ni_{20+0.5x}$ HEAs with four compositions in which x varied between 2 and 9.5 at.%. These investigators noted that with decrease in Al content, the alloys exhibit FCC structure and high notch toughness as evidenced by tear ridges and dimples on the fracture surfaces. The fractured surfaces of HEA samples broken by compression or tension tests often indicate quasi-cleavage fractures, like in compression-tested $AlCoCrFeNb_{0.1}Ni$ and $AlCoCrFeNb_{0.25}Ni$ HEAs, and in tension-tested $Al_{0.6}$ CrMnFeCoNi HEA [129].

16.3 FATIGUE RESISTANCE OF STRUCTURAL ALLOYS

16.3.1 Fatigue and Its Like and Significance

Fatigue in engineering implies accumulation of damage and subsequent failure of structural components under repeated cyclic load. The damage due to fatigue occurs in three distinct major stages from nucleation and coalescence of microcracks into a macrocrack, followed by its stable growth, which ultimately leads to failure. Failure due to this mechanism occupies the majority of the mechanical failures on a global scale and causes enormous loss to society. An example of the magnitude of this loss was reported to constitute about 4% of GDP in the economy of the United States [130] almost four decades ago, and the quoted percentage of GDP is comparable or higher to the expenditure on the education budget in several developing countries. This socio-economic scenario has led to enormous research efforts to reduce fatigue failure by suitable design and development of structural materials. To harness the economic benefits of reduction in fatigue failure, significant attention is paid to engineering critical applications like in the nuclear industry, the construction of air and space vehicles, the ship-building industry, the automobile sector, and in several manufacturing processes and components. Since these industrial applications are primarily concerned with the bulk volume of structural materials like iron-base alloys and steel, aluminum alloys, titanium alloys, nickel-base alloys, and superalloys, the primary focus lies in the development of these materials with improved fatigue resistance. But with the emergence and applications of all new materials like various types of composites, bulk metallic glasses, or intermetallics, attention has also been directed to understanding the physics of fatigue for assessing their potential for applications [131–136]. The past background has emphasized the need to seek understanding of the fatigue behavior of high entropy alloys; the efforts directed so far are limited to these emergent materials, but are essential to assessing their potential for structural applications.

16.3.2 Brief Pertinent Background Related to Fatigue

A material component can withstand only a specific number of stress/strain cycles prior to its failure (N_f), depending on the characteristics of the stress or strain cycles, and the capability of the component to withstand N_f is the measure of fatigue life of the component. There are two broad ways for determining N_f experimentally, either by determining the value of the total number of cycles to cause fatigue failure (N_{ft}) or to estimate only the number of cycles to require a defect or a macrocrack to propagate through the specimen/component (N_{fp}), the second one being significantly dominant over the number of cycles to initiate a fatigue crack, N_{fi} (= N_{ft}-N_{fp}). The measurement of N_{ft} is commonly called a "total life approach," whereas the estimation of N_{fp} is termed a "defect tolerant approach" for fatigue life estimation. The phenomenology governing the magnitudes of N_{fi} and N_{fp} are different in nature. The associated phenomena toward failure of a component under cyclic loading can be considered to have four major steps: (i) damage of a single or multiple locations under stress/strain-controlled cycling by enhanced deformation at those locations due to the characteristics of the concerned material, (ii) formation of an initial flaw or microdamage zone where the micro-/macrocrack initiates, (iii) propagation of the flaw in micro- or macroscale, and (iv) final catastrophic failure of the component/specimen [137]. The number of cycles to initiate a fatigue crack (N_{fi}) accounts for step (i), whereas the number of cycles required for the crack propagation (N_{fp}) remains primarily in step (iii); the number of cycles required for crack propagation in step (ii) and step (iv) is insignificant compared to that in step (iii) and is thus usually not considered for estimating N_{fp} for engineering design considerations in a defect tolerant approach. The stages of damage in "defect tolerant" and "total life" approaches are illustrated in Figure 16.17.

The major quantitative assessments of fatigue damage are inherent in the works of Wohler [138], Basquin [139], Coffin [140], Manson [141], and Paris [142]. Wohler proposed the basic method of analyzing the high cycle fatigue (HCF) behavior using the *S-N* curve, where *S* represents characteristic stress, and *N* is the number of cycles to failure. The magnitude of *S* is usually either the maximum nominal stress (σ_{max}) or

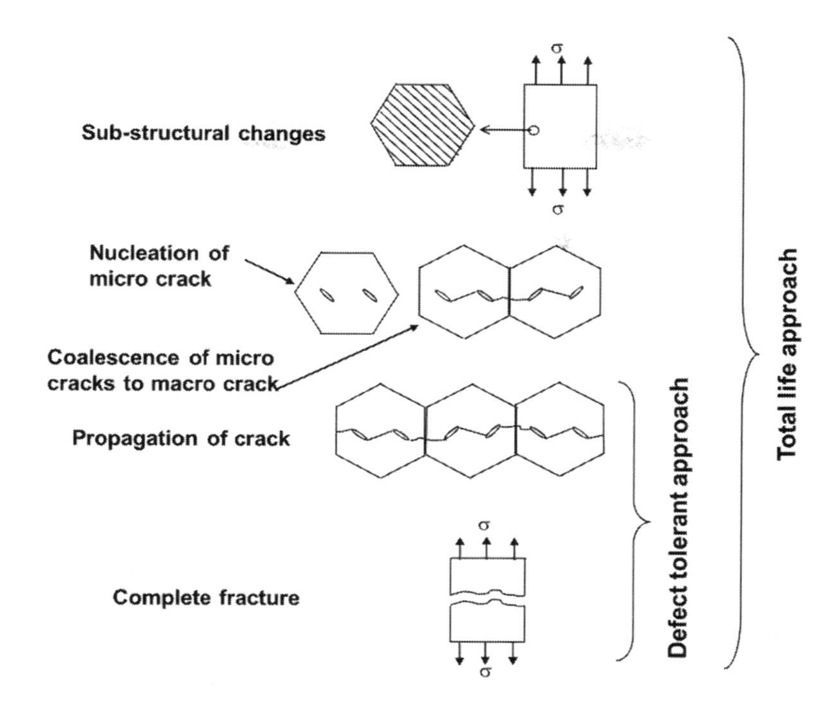

FIGURE 16.17 Schematic illustration of the major steps in fatigue failure in defect tolerant approach: (i) micro-damage, (ii) nucleation of crack, (iii) coalescence of macrocrack, (iv) propagation of crack, and (v) complete fracture.

the alternating stress (σ_a), given as $(\sigma_{max}$-$\sigma_{min})/2$ in which σ_{min} is the minimum nominal stress in a cycle. Typical *S-N* curves are schematically illustrated in Figure 16.18a, where any point on the curve represents the number of cycles (*N*) that a material can withstand at an applied stress (*S*) prior to failure. The figure depicts that the capability of a material to withstand higher *N* increases with decreasing *S*, and below a limiting value of *S*, the curve becomes horizontal for several materials, like steels. This limiting value of the stress is termed the endurance limit or fatigue limit for the concerned material, following Wohler [138]. A distinct fatigue limit is often not observed for several non-ferrous materials, like aluminum alloys, and in such cases, the fatigue strength is usually assigned arbitrarily at some specific number of cycles as 5×10^{-7}. Basquin [139] later described the high cycle region of the *S-N* curve with the expression:

$$N(\sigma_a)^p = C \tag{16.16}$$

where *p* and *C* are empirical material constants.

In HCF, the nature of the *S-N* curve is influenced by mean stress (σ_m), stress ratio (*R*), and the amplitude ratio (*A*) of the stress cycles, which are expressed as $(\sigma_{max}+\sigma_{min})/2$, $(\sigma_{min}/\sigma_{max})$ and (σ_a/σ_{max}), respectively. The *S-N* curves are usually determined by reversed bending using rotating-beam tests [143] or by tension–compression cycles following the ASTM standard E-466 [144].

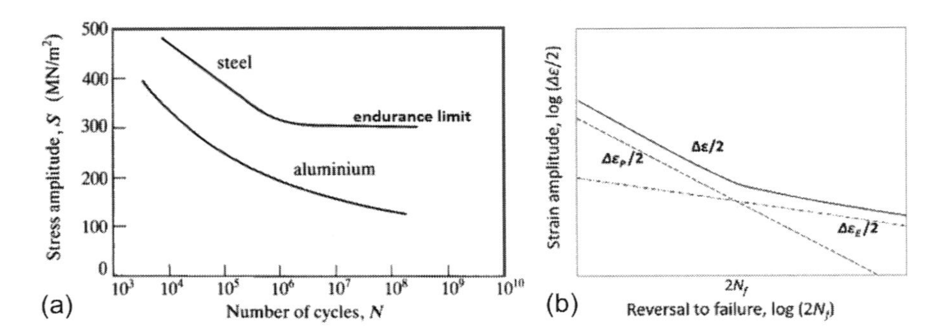

FIGURE 16.18 (a) Typical variation of stress (S) against number of cycles to failure (N) for ferrous (steel) and non-ferrous (aluminum) alloys with distinct endurance limit marked for steel. Values of N are in log-scale. (b) Fatigue strain life curve as obtained from superimposition of elastic and plastic strain-life equations. The linear plot between plastic strain range and cycles to failure is represented by the Coffin–Manson equation.

The stresses applied in the high cycle fatigue are considerably lower than the yield strength of the concerned material, and the estimated value of N_f is commonly greater than 10^4 to 10^5 cycles; a component having a fatigue life lower than this range of N_f value is usually subjected to higher stresses with some amount of plastic strain, and is considered to being subjected to low cycle fatigue (LCF). These tests are carried out using ASTM standard E-606 [145]. Typical incidences encountered in engineering applications for $N_f < 10^4$ to 10^5 cycles are in power-generating assemblies like steam turbines and nuclear pressure vessels. The fatigue life at relatively high stress and low number of cycles is proposed to be governed by the plastic strain, rather than stress, by Coffin and Manson [140, 141]. Following these investigators, the low cycle fatigue results are plotted as variation of plastic strain range ($\Delta\varepsilon_p/2$) versus the number of load reversal to failure ($2N$) in log–log scale as shown in Figure 16.18b considering their proposed relation between $\Delta\varepsilon_p$ and N as:

$$\left(\Delta\varepsilon_p/2\right) = (\varepsilon_f)'\left(2N\right)^c \tag{16.17}$$

where $(\varepsilon_f)'$ and c are referred to as fatigue ductility coefficient and fatigue ductility exponent, respectively; the magnitude of $(\varepsilon_f)'$ is approximately equal to fracture strain ε_f and that of c ranges between −0.5 and −0.7. Equation (16.17) is applicable specifically for a low cycle, high strain regime. But for a high cycle, low strain regime, this equation can be obtained by alteration of Basquin's equation. The overall view can be achieved by the superposition of elastic and plastic strains as function of N. This is illustrated diagrammatically in Figure 16.18b.

The measurement of the extent of propagation of a crack, its growth rate, and subsequently understanding the variation of the growth rate with the stress intensity factor is the concern of the fatigue-crack growth rate (FCGR) studies and is commonly done following the ASTM standard E-647 [146]. The test incorporates measurement of the crack length (a) at different number of cycles (N) in order to estimate the crack growth rate da/dN, the maximum (σ_{max}) and the minimum (σ_{min}) stresses operative in the cycle from which the range of cyclic stress ($\Delta\sigma$) is estimated, and subsequently the stress-intensity factor range (ΔK) is calculated from the knowledge of $\Delta\sigma$. The magnitude of ΔK is calculated as:

$$\Delta K = K_{max} - K_{min} = \alpha\sigma_{max}(\pi a)^{1/2} - \alpha\sigma_{min}(\pi a)^{1/2} = \alpha\Delta\sigma(\pi a)^{1/2} \quad (16.18)$$

where α, a dimensionless parameter, depends on the size and geometry of the specimen and is typically a polynomial function of a/W for a specimen having width W [93, 98].

The variation of crack growth rate (da/dN) in a wide spectrum of ΔK qualitatively exhibits three stages, commonly referred to as (i) near-threshold, stage-I, (ii) steady state, stage-II and (iii) fast fracture, stage-III, as shown in Figure 16.19. In stage-I, the rate of fatigue crack propagation is extremely slow and is of the order of nanometer per cycle. The linear variation of crack growth rate, (da/dN), with ΔK for stage-II is usually expressed by Paris' law as:

$$da/dN = c(\Delta K)^m \quad (16.19)$$

where c and m are material constants. However, the crack growth rate (da/dN) in a material during cyclic deformation depends on several factors, like the stress intensity factor differential (ΔK), stress or load ratio (R), frequency of cycling (ν), wave

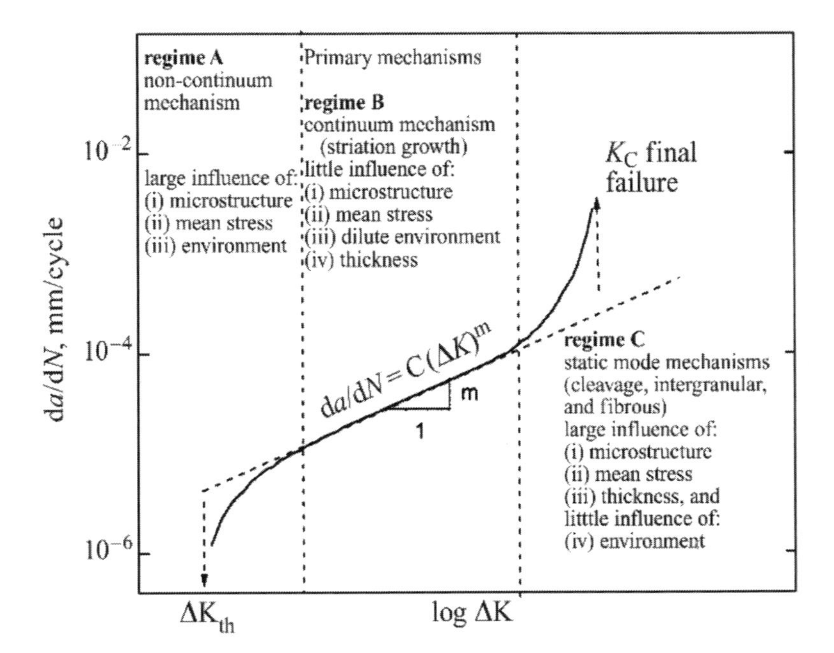

FIGURE 16.19 Schematic variation of fatigue crack growth rate (da/dN) with stress intensity factor range (ΔK) showing the three distinct regimes of fatigue crack propagation and indicating their dependent variables, like microstructure, environment, etc. [108].

form (*WF*), and the environment (*Env*). Thus d*a*/d*N* in a generalized form can be expressed as:

$$da/dN = f\left(\Delta K, R, \upsilon, WF, Env\right). \tag{16.20}$$

where ΔK can be denoted by Equation (16.18) or as:

$$\Delta K = K_{max} - K_{cl}\left(\text{when } K_{cl} > K_{min}\right) \tag{16.21}$$

in which K_{cl} is the crack closure stress intensity factor.

16.3.3 Fatigue Behavior of High Entropy Alloys

The characteristics of fatigue resistance of high entropy alloys have been examined by:

(i) high cycle fatigue for $Al_{0.5}CoCrCuFeNi$ HEA by Hemphill et al. [147] and Tang et al. [148], for HfNbTaTiZr by Guennec et al. [149], for $AlCoCrFeNi_{2.1}$ by Shukla et al. [150], and for $Al_{0.3}CoCrFeNi$ by Liu et al. [151];
(ii) low cycle fatigue for $Fe_{50}Mn_{30}Co_{10}Cr_{10}$ HEA alloy by Niendorf et al. [152]; and
(iii) fatigue crack growth rate analysis for $Al_{0.2}CrFeNiTi_{0.2}$ and $AlCr\text{-}FeNi_2Cu$ by Seifi et al. [116], for CrMnFeCoNi by Thurston et al. [153, 154].

16.3.3.1 High Cycle Fatigue

Until now, all investigations for high cycle fatigue tests to determine the *S-N* curves for HEAs have been done using only four-point bend (FPB) tests. Hemphill et al. [147] are the first to demonstrate the fatigue behavior of HEAs indicating their capability for high endurance limit. These investigators studied the HCF behavior of the $Al_{0.5}CoCrCuFeNi$ HEA using specimens of 3 mm× 3 mm cross-section with inner and outer span lengths of 10 mm and 20 mm only for the FPB tests at load ratio (*R*) of 0.1 and frequency of 10 Hz. The investigators observed significant scatter in the *S-N* plots and attributed the origin of the scatter to aluminum oxide particles as well as microcracks generated during the processing of the materials; the material was processed by repeated arc melting in vacuum, annealed, water quenched, and then cold rolled. The microstructure of the materials showed an α-FCC dendritic matrix with a β-Cu-rich FCC interdendritic phase; the latter was found to contain Ni-Al rich nanoparticles [148]. The authors [147] examined the scatter in the *S-N* plots by Weibull predictive model [155] and Weibull mixture predictive model [156]. The scatter in the *S-N* plot as shown in Figure 16.20 was characterized into a strong and weak group by the Weibull mixture predictive model. The strong group indicated the median time to failure to be greater than 10^7 cycles at 858 MPa and the fatigue ratio (ratio between endurance limit to tensile strength) was estimated to lie between 0.402 and 0.703. The endurance limit was found to compare well with several structural materials.

Tang et al. [148] further investigated the fatigue behavior of another two sets of newly prepared $Al_{0.5}CoCrCuFeNi$ HEAs following Hemphill et al. [147] in the

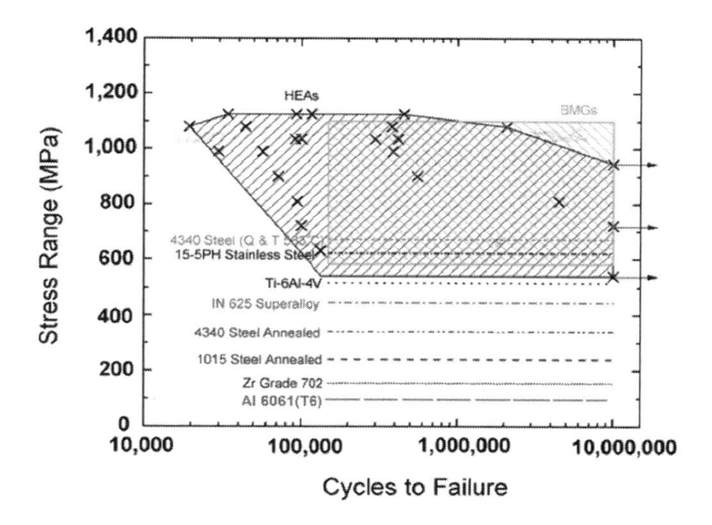

FIGURE 16.20 S–N curves showing the endurance limits of the $Al_{0.5}CoCrCuFeNi$ HEA [147] with its comparison with those for several conventional structural materials [157–161], like superalloys, steels, aluminum- and titanium-based alloys. (Reprinted with permission from Elsevier.)

same department. The additional preparation procedure for the alloys included removal of cast defects like shrinkage pores and macrosegregation by machining and cold rolling as well as preparation of the alloys using high purity elements (> 099.9% purity). The use of high purity elements reduced the inclusions and the microcracks. The investigators carried out tests on three conditions of the alloy (a) partially retaining the defects before rolling, (b) removing the defects before rolling, and (c) removal of defects in materials prepared using high purity materials. These investigators observed less scatter in the *S-N* plots as shown in Figure 16.21 compared to the results of Hemphill et al. [147], the minimum being for the materials with high purity, and this could achieve *S-N* curves with distinct endurance limit for the materials. The highest endurance limit was reported as 383 ± 71 MPa with associated fatigue ratio of 0.29. The scatter in the fatigue data was attributed to their inherent content of pores, microcracks, and oxide particles. An interesting observation by these investigators is the nano-twinning during initiation and growth of fatigue cracks which has been considered to increase the strength and fatigue endurance limit of the investigated HEAs.

Guennec et al. [149] are possibly the first to examine the HCF behavior of BCC HfNbTaTiZr HEA using FPB loading like the earlier investigators [147, 148]. The material selected by the authors was prepared by arc melting two master alloys of Nb-Ta and Hf-Ti-Zr followed by cold rolling and annealing. The thermomechanically treated material showed an equiaxed grain structure with some grain-size gradient from the surface to the center of the specimens. The *S-N* diagram determined by these researchers (Figure 16.22) indicated that the applied maximum stress level

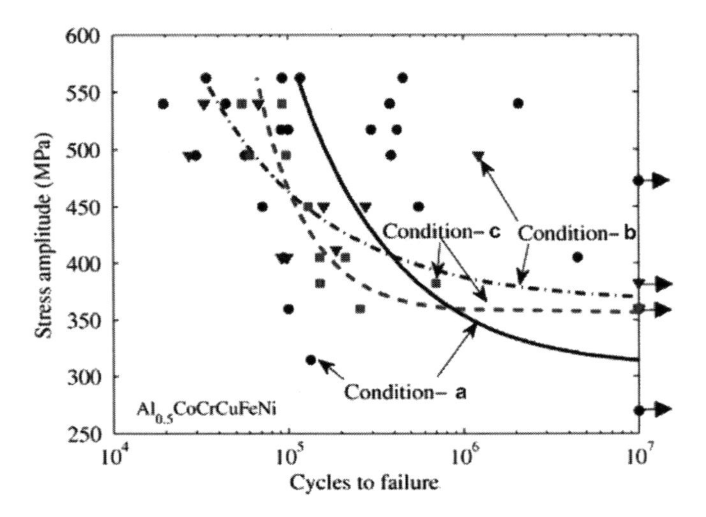

FIGURE 16.21 The S–N curves of the $Al_{0.5}CoCrCuFeNi$ HEA under four-point bending fatigue with an R ratio of 0.1 in air at room temperature. The predicted median fatigue life vs stress amplitude relationship for the three conditions of HEAs are shown by dashed lines. Arrows indicate run-out without failure [148]. (Reprinted with permission from Elsevier.)

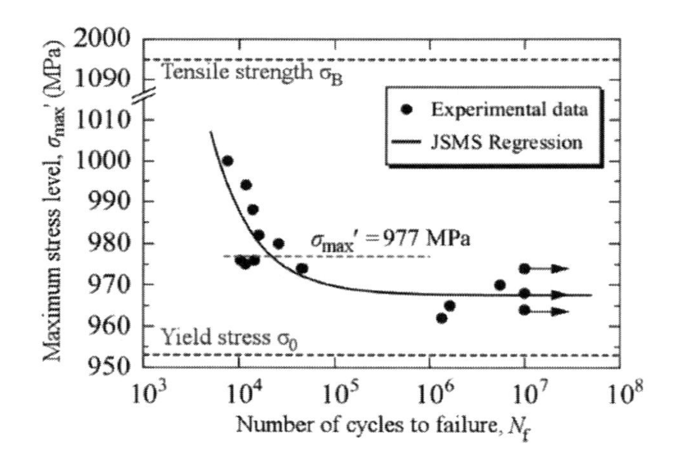

FIGURE 16.22 *S–N* curve for HfNbTaTiZr HEA indicating endurance limit to be higher than the yield strength. The ordinate is plotted here as the corrected maximum stress level during four-point bend fatigue test [149]. (Reprinted with permission from Elsevier.)

exceeds the yield strength of the material, and thus calculation of maximum stress using the following equation:

$$\sigma_b = 3PL/\left(bh^2\right) \tag{16.22}$$

where P = applied load and b, h and L are specimen width, thickness, and span-length, respectively, would not be appropriate considering only elastic analysis.

The authors made a correction for the maximum stress while plotting the *S-N* curve. With the consideration of elastic-plastic deformation and that fatigue damage in the specimens is induced by plastic deformation, the authors re-examined the data as plots of plastic width versus N_f, and found the plot to exhibit a linear relation. Guennec et al. [149] estimated the maximum stress endurance limit for the material as 966 MPa at 10^7 cycles as per the Japanese standard JSMS-SD-11-07 [162], but the elastic-plastic analyses indicate the fatigue endurance stress amplitude to be 431 MPa and the fatigue ratio for the material to be 0.43 at a stress ratio of 0.1, a high value compared to several structural materials. The authors reported that during HCF of the HfNbTaTiZr HEA, crack initiation occurs by intergranular cracking, while crack propagation is marked by transgranular fracture features.

Shukla et al. [150] investigated HCF by bending fatigue of miniature dual phase $AlCoCrFeNi_{2.1}$ HEAs in two processed conditions: as-cast and rolled followed by heat treating at 700°C for 12 h. The HEA materials exhibited eutectic morphology. The constituents in the processed materials are a hard BCC phase and a soft FCC phase in lamellar morphology and the investigators termed it "eutectic HEA" (EHEA). The fraction of the FCC ($L1_2$) to BCC (B2) constituents were in the approximate ratio of 70:30. The morphology of the FCC and the BCC phases were either lamellar or deformed and recrystallized with Cr-rich precipitates. The tensile properties of the cast alloy (YS = 746 MPa, UTS = 1057MPa, and Elong = 8%) was inferior to that (YS = 1110 MPa, UTS = 1340 MPa, and Elong = 10%) of the rolled and heat-treated alloy, and interestingly, the HCF response of the rolled alloy is also superior to that of the cast alloy. The stress at which a sample does not fail at 10^7 cycles for the cast alloy is ~390 MPa compared to that of the rolled alloy as ~500 MPa. The authors [150] also examined the number of cycles for the fatigue-crack initiation and found this to be higher for the rolled alloy; crack initiation in the cast alloy occurred around 30×10^3 cycles compared to 90×10^3 cycles in the rolled alloy. From the microstructural studies using EBSD, the authors inferred that PSB formation in FCC(L12) lamellae was the main reason for crack initiation in the cast alloy, while crack initiation in the rolled alloy was delayed due to (BCC)B2 precipitates which hinder the path of PSBs.

Liu et al. [151] examined the fully reversible bending fatigue behavior of $Al_{0.3}CoCrFeNi$ HEA having ultrafine grain size (0.71 ± 0.35 μm) with an emphasis on understanding the associated deformation mechanism in the alloy. The microstructure of the processed and heat-treated alloy exhibited FCC matrix with ordered BCC(B2) and σ precipitates the BCC precipitates were Ni-Al rich (50%:29%), whereas the σ precipitates are constituted of a Cr-Fe (78%:17.5%)-rich constituent. The material exhibited an excellent combination of strength and ductility (YS: 900 MPa, UTS: 1074 MPa, and total plastic strain of ~25%.); the higher strength is attributed to grain boundary strengthening and composite strengthening by hard precipitates, while higher ductility originated from the formation of nano-twins during deformation. The authors analyzed the fatigue run out stress amplitude to be 450 MPa, corresponding to an excellent fatigue strength ratio of 0.43. The enhanced fatigue resistance of the material is also attributed to nano-twin formation during fatigue damage of the HEA.

16.3.3.2 Low Cycle Fatigue

Niendorf et al. [152] are the only investigators so far to examine the low cycle fatigue behavior of high entropy alloy. They studied $Fe_{50}Mn_{30}Co_{10}Cr_{10}$ alloys that are subjected to two different thermomechanical routes to have dual-phase structures with coarse and fine grain sizes, pertinent details of the material preparation being reported by Li et al. [35]. The material consists of two phases and the FCC phase is metastable; on deformation the FCC phase transforms to an ε- martensite with a hexagonal closed packed (HCP) structure as one finds in materials exhibiting transformation induced plasticity (TRIP), and as a consequence, the investigators termed the alloy "TRIP-HEA." Subjected to plastic straining, the microstructure of the investigated material shows mechanical twins and ε-martensite along with the usual dislocations and stacking faults [35]. The material on LCF tests exhibited unusual cyclic plastic response compared to monotonic deformation, which indicates considerable strain hardening due to martensitic transformation. The authors noted that despite the evolution of considerable SIM, the strain hardening is negligible during cycling at strain amplitudes ($\Delta\varepsilon/2$) of 0.23% (up to 90,000 cycles) and 0.6% (up to 10,000 cycles) as illustrated in Figure 16.23. They noted the accumulation of plastic strains over 1,000% in all the studied cyclic tests and reasoned it was due to the absence of strain hardening. Interestingly, almost perfect Masing behavior was noted for the coarse-grained material while the fine-grained material showed non-perfect but near-Masing behavior. The authors inferred that partial reversibility of deformation occurs in the TRIP-HEA, possibly due to the planarity of slip. Results related to Coffin–Manson type behavior have yet to emerge for HEAs.

16.3.3.3 Fatigue Crack Growth Rate

Selfi et al. [116] studied the fracture and fatigue crack growth rate behavior of two cast high entropy alloys $Al_{0.2}CrFe\text{-}NiTi_{0.2}$, and $AlCrFeNi_2Cu$ using the standard test

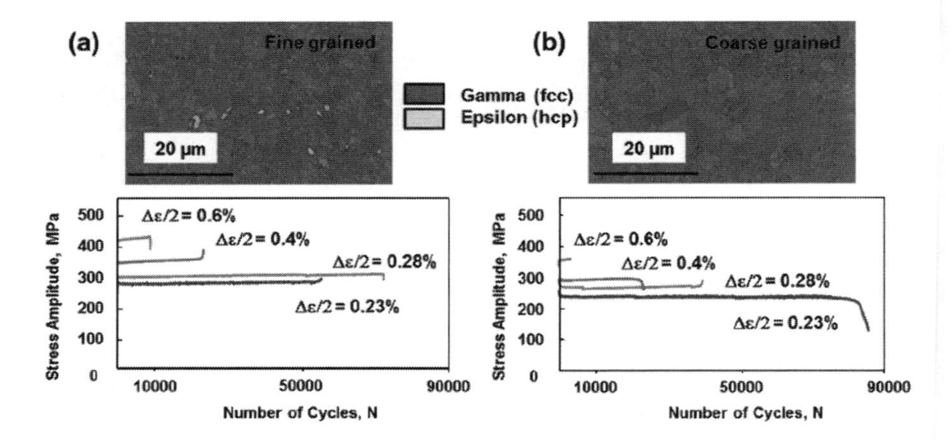

FIGURE 16.23 Cyclic stress response under various strain amplitudes of (a) fine-grained (~5 μm grain size) and (b) coarse-grained (~10 μm grain size) of $Fe_{50}Mn_{30}Co_{10}Cr_{10}$ TRIP-HEA tested at room temperature. The microstructural conditions are illustrated as insets (darker γ-phase, lighter ε-phase) taken from EBSD phase maps [152]. (Reprinted with permission from Elsevier.)

method [146]. The alloys were constituted of two dominant phases with FCC and BCC structures with some unidentified minor constituents, as revealed by the X-ray profiles. The matrix of the Ti-bearing first alloy is noted to be enriched with Cr and Fe, and the particles/precipitates are enriched with Al, Ni, and Ti. Almost the opposite chemistry nature is noted in the second alloy, $AlCrFeNi_2Cu$, in which the matrix is enriched with Al, Ni, and Cu, while the particles/precipitates primarily contain Fe and Cr. The Vickers microhardness and the fracture toughness of the $Al_{0.2}CrFe\text{-}NiTi_{0.2}$, and $AlCrFeNi_2Cu$ alloys are reported to be as "510 ± 7.7 HV and 32–35 $MPam^{1/2}$" and "320.9 ± 7.8 HV and 40–45 $MPam^{1/2}$," respectively. The investigators [116] constructed the fatigue crack growth rate curves (Figure 16.24) and analyzed the Paris' law exponents for the HEAs using Equation (16.19) apart from estimating the fatigue threshold values at three different load ratios. The Paris' law exponents (m) are found to increase with increasing R, and the fatigue threshold values are found to decrease with increasing R. The value of m for the da/dN vs ΔK plot at lower R ratios (0.1–0.3) is in agreement with that observed for metallic materials, but that at high R (0.7) is found to be unusually high. The increase in the Paris law exponent with higher R values was postulated due to the dominance of higher extent of brittle fracture. The high threshold values were considered to be the result of the crack closure effect due to surface roughness at lower R-values.

Subsequent to Seifi et al. [116], Thurston et al. [153] revealed the FCGR behavior of single phase FCC CrMnFeCoNi alloy. The material was prepared by vacuum induction melting, homogenized, rotary swaged, and recrystallized to achieve a fine grain size of 7 ± 3 μm. Interestingly evaluation of the yield strength, tensile strength, ductility, elastic modulus, and fracture toughness of the HEA showed considerable temperature dependence. At 293 K these values are approximately 760 MPa (UTS),

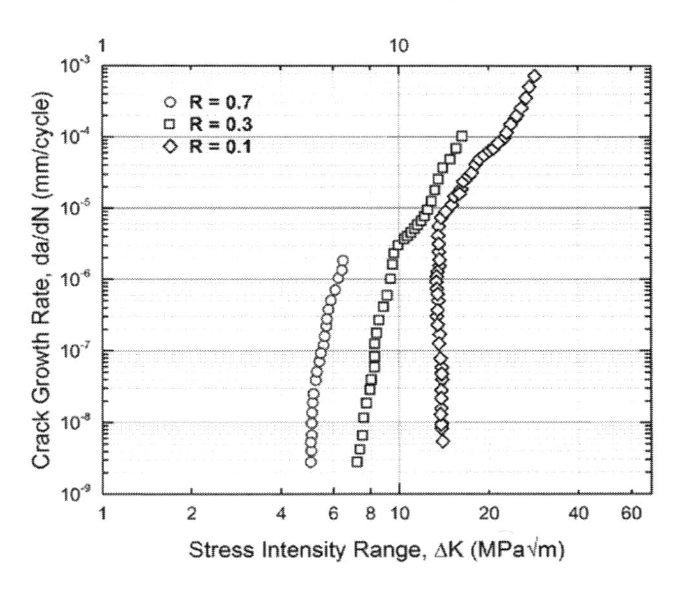

FIGURE 16.24 Typical fatigue-crack growth behavior of $Al_{0.2}CrFeNiTi_{0.2}$ HEA at different R ratios tested at room temperature [116]. (Reprinted with permission from Springer Nature.)

410 (YS) MPA, 0.6 (%Elong), 202 GPa (E), and 217 MPam$^{1/2}$ (K_{Ic}) in contrast to their values at 198 K as 925 MPa (UTS), 520 (YS) MPA, 0.7 (%Elong), 209 GPa (E), and 221 MPam$^{1/2}$ (K_{Ic}). The FCGR values of this material at temperatures of 293 K and 198 K were determined by the investigators using DC(T) specimens having the dimensions B = 6 mm, W=12.5 mm, a_o=3.6–5.1 mm, following the ASTM standard E-1820 [98]. The FCGR tests were made at f=25 Hz using a sinusoidal wave with a decreasing ΔK gradient, following ASTM standard E-647 [146]. The low-temperature tests were made in a dry ice–ethanol bath. Analyses of the FCGR results indicated that both the Paris' law exponent and the fatigue threshold increase when the test is done at a lower temperature. The m values were found to be 3.5 (293 K) and 4.5 (198 K), while the K_{th} values were estimated as 4.8 MPam$^{1/2}$ (293 K) and 6.3 MPam$^{1/2}$(198 K) for the investigated CrMnFeCoNi alloy. The increase in the fatigue threshold value was attributed to the dominance of the intergranular fracture at a lower temperature, which results into contact of neighboring crack flanks, leading to pronounced roughness induced crack closure.

Thurston et al. [153] later extended their work in order to reveal the FCGR characteristics of the same alloy at still lower temperatures up to 77 K and at higher load ratios of R=0.4 and R=0.7 with additional experiments to measure the crack closure by compliance method. For approximate closure measurement, a sample was uniaxially loaded to the maximum load value at the threshold, followed by unloading to the corresponding minimum load value of the cycle, and the process was iterated five times to record the point at which the slope of the unloading curve changes. The magnitude of the crack closure stress intensity factor was estimated following the earlier reports of one of the investigators [163, 164]. The results at different stress ratios and temperatures are summarized in Figure 16.25; the qualitative trend distinctly illustrated that as the temperature and the load ratio is decreased, the value of the fatigue threshold (ΔK_{th}) increases. With a decreasing load ratio, the Paris' law exponent also appears to increase, but its value for all the test conditions are in the characteristic range of 2–4, like that for metallic materials. The values of the reported ΔK_{th} ranged from 6.3 to 2.5 MPa$\sqrt{m}$, corresponding to load ratios between 0.1 and 0.7 and that of $\Delta K_{eff,th}$ (K_{max}-K_{cl} when K_{cl}> K_{min}) are found to range between 2.5–4.8 MPa$\sqrt{m}$; the values of ΔK_{th} and $\Delta K_{eff,th}$ are in the characteristic range for

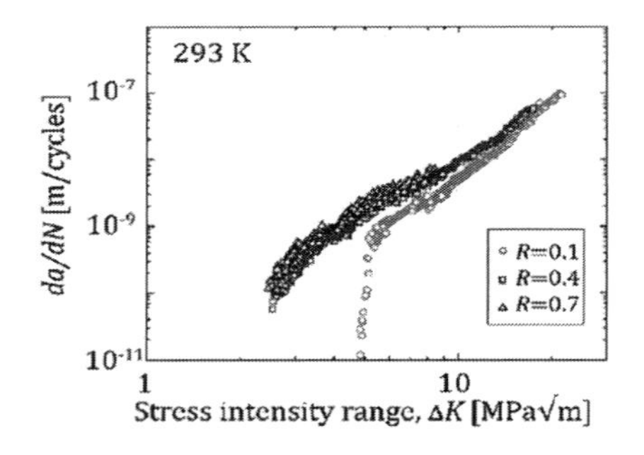

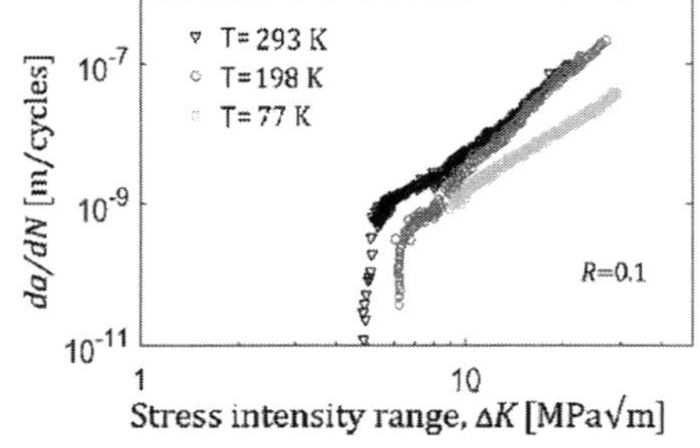

FIGURE 16.25 Fatigue-crack growth rates (FCGR) as a function of the stress-intensity range for the CrMnFeCoNi HEA under various testing conditions. (a) at various R ratios at 293 K, and (b) at various temperatures at constant R=0.1 [154]. (Reprinted with permission from Elsevier.)

metallic materials as can be seen from the summary of the FCGR parameters for CrMnFeCoNi alloy reported by Thurston et al. [154].

16.3.4 Fractography and Micro-Mechanism

Examination of the appearances of the fracture surfaces after the fatigue tests is the topic of fractography [123], a postmortem analysis to assess the fatigue failure mechanism and to understand the role of the microstructural constituents of a material to create such surface signatures under the applied state of cyclic stress/strain. Almost all the investigations related to the analysis of fatigue behavior of high entropy alloys as discussed in the previous section incorporate these postmortem analyses. Hemphill et al. [147] are one of the earliest workers to examine fatigue behavior of $Al_{0.5}CoCrCuFeNi$ (molar ratio) HEA. Later, Tang et al. [148] also worked on the same system in more details. These authors reported that crack initiation (during FPB tests on the material) occurs from the microcracks that are formed from defects on the surface or from the corner of the specimen, as these locations are sites for higher stress concentration. Examination of the fatigue crack propagation path showed distinct striation marks at different locations, while the final fracture path indicated dimples (Figure 16.26), the typical mark for ductile fractures. The characteristic striation marks on fatigue fracture surface and dimples on the final fracture path are in correspondence with the material microstructure (two-phase FCC structure) and tensile properties (YS: 1284 MPa, % elongation: 7.6%).

Seifi et al. [116] carried out fatigue crack growth studies in two as-cast $Al_{0.2}CrFeNiTi_{0.2}$ ($R_c = 46$, $K_{IC} = 32–35$ MPam$^{1/2}$) and $AlCrFeNi_2Cu$ HEAs ($R_c = 46$, $K_{IC} = 32–35$ MPam$^{1/2}$). These alloys showed a combination of BCC and FCC regions with considerable Rockwell hardness and fracture toughness values, as shown in the parenthesis of the concerned alloys. The FCGR studies indicated limited fatigue striations in the Paris' law regime and combinations of brittle and ductile/dimpled regions in the overload region. The fatigue striation marks, predominantly dimple-like features, brittle-like features and combinations of ductile and brittle features, as reported by the investigators, are depicted in Figure 16.27. The authors reported that the ductile and brittle features closely correspond to the volume fraction and

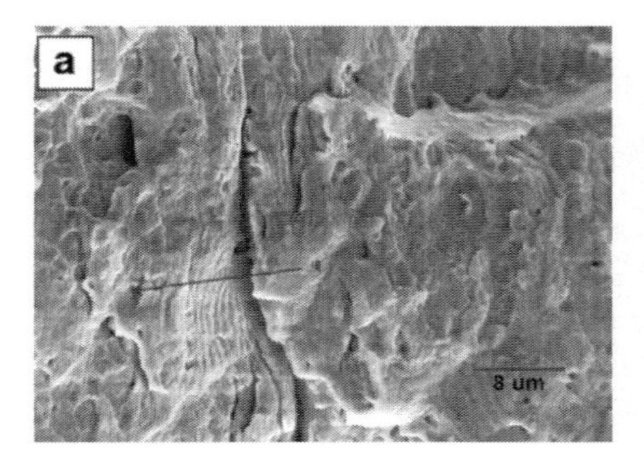

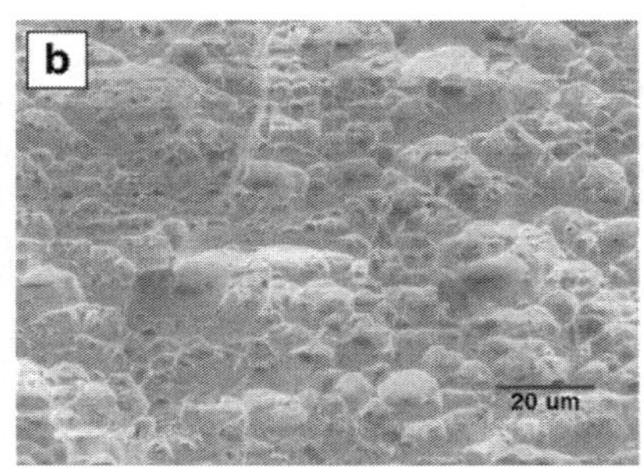

FIGURE 16.26 SEM factographs indicating (a) fatigue striations in the crack propagation regime, and (b) dimples in the final fracture region for $Al_{0.5}CoCrCuFeNi$ HEA [147]. (Reprinted with permission from Elsevier.)

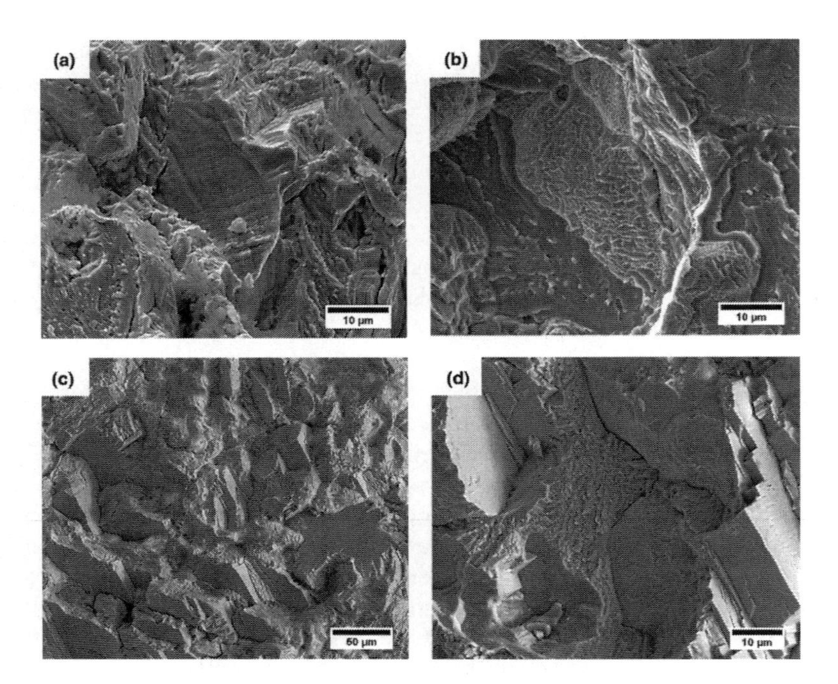

FIGURE 16.27 SEM images of AlCrFeNi2Cu HEA alloy tested in fatigue at 298 K (a) fatigue striation features, (b) ductile/dimple-like features, and (c and d) brittle and ductile/dimple features [116]. (Reprinted with permission from Springer Nature.)

size of the microstructural constituents having different crystal structures i.e., FCC and BCC. Thurston et al. [153, 154] investigated the FCGR of a single-phase cantor alloy, CrMnFeCoNi, with mechanical properties as shown in the parenthesis (yield strength: 760 MPa, ductility: 0.6, fracture toughness: 217 $MPam^{1/2}$). These investigators reported that fatigue fracture surfaces of CrMnFeCoNi exhibit primarily transgranular crack propagation, with some intergranular regions (Figure 16.28) at 293 K indicating fatigue striations at several locations. The amount of the intergranular regions increased for tests at lower temperatures. The striation spacing is found to increase for tests at increasing stress levels (Figure 16.29).

Guennec et al. [149] examined the fracture surfaces of BCC HfNbTaTiZr alloy by FPB tests and examined the fracture surfaces of this material. The authors indicated from fracture surface examinations that crack initiation in this material occurred by intergranular nucleation, while crack propagation was by transgranular crystallographic features with some unusual torn patterns (Figure 16.30). However, these torn patterns appear to resemble fatigue striation in brittle materials [15]. Shukla et al. [150] revealed the higher fatigue resistance of worked and annealed eutectic high entropy alloy AlCoCrFeNi2.1 compared to its cast condition by crack initiation sites using BSE images and EBSD phase maps. These investigators inferred that better fatigue resistance of EHEAw than that of EHEAc is due to BCC (B2) precipitates on the path of the persistent slip bands in EHEAw. Liu et al. [151] examined the fatigue behavior of rolled and annealed $Al_{0.3}CoCrFeNi$ alloy with BCC (B2) and σ phases in

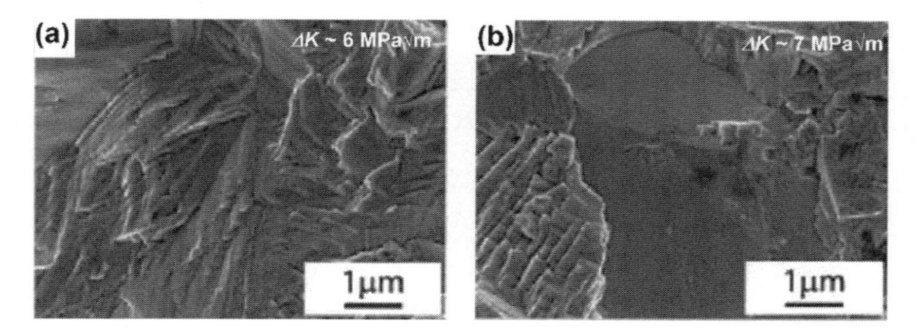

FIGURE 16.28 Fracture surfaces of CrMnFeCoNi HEA samples fatigue tested at 293 K with different ΔK values; (a) intergranular failure region, and (b) transgranular failure region. The latter is likely to be associated with cyclic slip steps resulting from dislocation motion by planar slip [153]. (Reprinted with permission from Elsevier.)

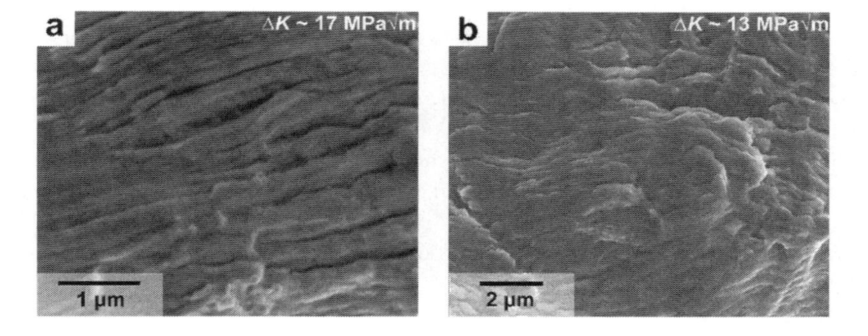

FIGURE 16.29 Different striation spacings in CrMnFeCoNi HEA subjected to fatigue crack growth rate testing at 198 K for (a) $\Delta K = 17$ MPam$^{0.5}$, (b) $\Delta K = 13$ MPam$^{0.5}$. Higher ΔK results in higher striation spacing [154]. (Reprinted with permission from Elsevier.)

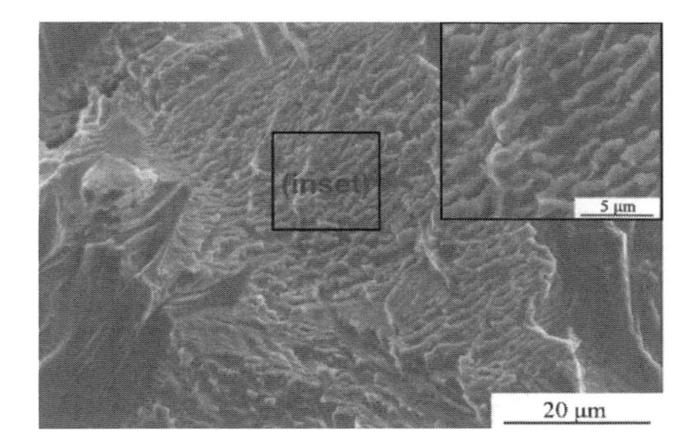

FIGURE 16.30 Typical saw-tooth edge features on HfNbTaTiZr HEA subjected to four-point bending fatigue test [149]. (Reprinted with permission from Elsevier.)

FCC matrix (YS of 900 MPa and total plastic strain of ~25%). These investigators reported that the alloy indicates fatigue-crack initiation due to the accumulation of geometrically necessary dislocations (GND) around harder phases on the basis of EBSD GND map near the fracture surface of the investigated samples. Niendorf et al. [152] are the only group who have carried out LCF of the $Fe_{50}Mn_{30}Co_{10}Cr_{10}$ alloy, and they have reported the high capability of the materials to accumulate large plastic strains in LCF, but have not reported its effect on the fracture surfaces.

16.4 DISCUSSIONS

The major emphasis in the discussions is to elucidate the current state of understanding on the fracture and fatigue behavior of high entropy alloys. Investigations of the conventional mechanical properties like strength, ductility, and hardness of high entropy alloys have generated a reasonable amount of information from the time of emergence of the concept of high entropy or complex concentrated or multi-principal element alloys. However, attempts to understand fracture and fatigue behavior of these alloys are few in number, and thus these disciplines can be referred to as being in an evolving embryonic state. The theoretically possible number of high entropy and allied alloys like MPEAs, MCAs, and CCAs is several million; out of these, only a few hundred have been synthesized so far, and their conventional mechanical properties have been habitually estimated. But the number of alloys examined to understand fracture and fatigue behavior has just passed single-digit numbers. Interestingly, the promising fracture toughness and resistance to fatigue of only a few HEAs have created enormous interest in the scientific community to search for their potential for possible engineering applications. The structure and the various operative deformation mechanisms dictate both the conventional mechanical properties, as well as the fracture and fatigue behavior of these alloys. The structure could consist of single or two-phase constituents with or without precipitates, while the deformation mechanisms in these alloys encompass slip, twinning, solid solution hardening, order-hardening, and occasionally, phase transformation. Currently it appears that the complexities that govern the mechanical response of established engineering materials are also operative for the signature of mechanical responses in HEAs, but possible generalization of the mechanical properties would be a distant task because of the nature of this group of materials. An overview of the current information and the gaps in the understanding of fracture and fatigue behavior of HEAs, and some possible directions in these disciplines for investigation in the immediate future are included below.

16.4.1 Problems in Material Availability

One of the simple queries that may emerge in the mind of researchers engaged with investigations related to HEAs is why only a few studies have surfaced to date on fracture toughness and resistance to fatigue behavior of these materials? The simple answer to this query is: the emergence of HEAs is not even two decades old, and the initial searches are more focused on playing with different compositions and microstructures with some simple assessments of conventional mechanical properties in order to justify their possible applicability, if any. But another critical issue is often not considered carefully is the relatively high cost of preparation of these alloys (usually from high purity elements) with a lower defect content. This has often restricted the volume of the material being prepared in sufficient amounts in laboratory-scale research; the amount is frequently not enough to characterize the advanced mechanical properties related to fracture and fatigue resistance. However, this aspect does not impede the growth of microstructural science of various HEAs by generating numerous structures using different thermal or thermomechanical treatments. From the technological viewpoint, economic and structural integrity issues grow synergistically toward the development of new materials for immediate applications. Currently, the potential of only a few HEAs (primarily based on an FCC matrix) appears to be getting established for encountering challenging applications.

16.4.2 Issues Related to Characterization Practices

Standard characterization practices for estimation of fracture toughness parameters demand the employment of the procedures laid down in the international standards, carrying out the minimum number of tests, assessing the validity of the measurements, and finally, the reliability of the assessments. An overview of the results on fracture toughness parameters indicate that majority of the tests on HEAs reported so far are on sub-size specimens following the standard methods, and often the investigators have indicated the measured values as K_Q rather than K_{Ic} or K_{Jc}. The magnitude of K_Q is indicative of the presence of an uncertain amount of plasticity being associated with the crack initiation in monotonic loading and is of little significance for comparative assessment of different types of materials. Estimation of fracture toughness by alternative approaches like indentation techniques through hardness tests (nano- to microscale) is coupled with high deviations; the scatter may even exceed 50% [165–167]. Attempts to convert Charpy impact energy values to fracture toughness are linked with considerable uncertainty because of the empiricism in conversion formulae and their applicability from one set of materials to another set [15]. The results obtained from micro-beam tests or others based on micro or miniature specimens are capable of providing the order of fracture toughness. But such results are usually confined to local microstructural states, and their equivalence with macro test results needs to be established first before using these results for comparative assessments. The future direction for research activities on HEAs can be considered to be progress based on small specimens specifically to evaluate the possible order of fracture toughness as a property index. But emphasis must be laid on estimating fracture toughness of HEAs using standard practices on larger specimens for examining their potential with respect to the existing engineering materials for structural applicability.

Assessment of the resistance to fatigue of structural materials is done using high cycle fatigue (HCF) and low cycle fatigue (LCF) in a total life design philosophy, whereas fatigue-crack growth rate (FCGR) studies are made in a defect-tolerant design philosophy. The reports generated so far related to fatigue studies on HEAs indicate that all the three approaches have been employed to understand fatigue resistance of HEAs. Careful examination of the investigated reports shows that most of the HCF studies have been carried out using specimens with smaller cross-sectional areas. This issue is possibly the natural consequence related to the lack of availability of larger amount of materials. A smaller cross-sectional area naturally increases the surface to volume ratio of the specimens, leading to higher probability of surface crack initiation and a lower crack growth regime. So probabilistic analysis may lead to an estimation of relatively lower fatigue limit; to the contrary, a lower cross-sectional area usually provides a plane-stress state of loading with an increased plastic regime which may result in a higher endurance limit. This issue can be overcome with the availability of a larger volume of materials. In spite of this difficulty, the reported results on HCF studies on HEAs indicate good promise for fatigue resistance, as discussed in Section 16.3.

The number of studies related to LCF and the FCGR behavior of HEAs is only about three in number to date. These tests were made on materials possessing either FCC or FCC + BCC microstructures. The materials investigated so far by these techniques had FCC as the primary matrix phase, and the materials were prepared by thermomechanical processing. The associated higher ductility of these materials assisted in the fabrication of relatively larger specimens for examination of their fatigue life. Fatigue-crack growth studies on HEAs with a major BCC phase and that have some technologically significant ductility need to be given attention, because these materials possess high strength and bear great potential for application as engineering structural materials.

16.4.3 Influence of Phase Constituents on Fracture and Fatigue Characteristics

High entropy alloys possessing major phases with crystal structures of FCC, FCC + BCC, or BCC, and with or without the embedment of precipitates, are the systems of investigations for fracture and fatigue studies to date. High fracture toughness has been obtained in FCC HEAs, while the other types are usually less fracture-resistant and can be referred to as almost brittle. The characteristics of the properties related to cracking sensitivity are governed by plasticity and, in turn, by deformation mechanisms associated with initiation of cracks as well as its stable and unstable propagation. The micro-deformation mechanisms in HEAs are similar to those of conventional metallic materials, but differ due to complex dislocation activities, considerable compositional dependence on stacking fault energy, and the nature of micro-twinning. The differences originate from sluggish diffusivity and different lattice resistance to dislocation movement in HEAs compared to metals and alloys. The intrinsic toughening mechanisms which dictate the strength–toughness relation of HEAs are dependent on the evolution of the materials with different compositions, microstructural features, refined constituent sizes of the microstructure and sensitive phase constituents, which can undergo stress- or strain-induced transformations, and with low fabrication defects.

Transition metal-based HEAs have already demonstrated high fracture toughness with suitable strength and ductility and are emerging as challenging competitors for several advanced structural materials. But the brittleness in BCC and BCC + FCC alloys with less fracture resistance could restrict their applicability. The strategy of phase transformation has been included in the development of some refractory BCC alloys to improve their ductility and natural toughness index. The strength–fracture toughness combination is possible to modulate in FCC+BCC alloys with an optimized amount of BCC phases or that for FCC alloys with suitable amount, size, distribution, and characteristics of the embedded precipitates. However, fracture and fatigue studies can only be carried out after the optimized structures are developed.

Fatigue-resistance characteristics have been evaluated for both transition element-based HEAs and refractory HEAs. Unlike fracture toughness, the fatigue-resistance of materials increases with increased strength and hardness with less defect content. The reported results on the characterization of fatigue are confined to a few systems with single or dual phases. Considerable fatigue lives at relatively high stresses have been encountered in a few HEAs; this is a promising possibility for developing these materials specifically for high-temperature applications. However, the analysis of defect content (fabrication), and the extent of the intrinsic defects like micro- or nano-twins are relevant for analyzing the role of the constituent phases on the fatigue resistance of HEAs.

16.4.4 Knowledge of Scientific Intricacies

The fracture toughness tests on HEAs have been estimated by both conventional and alternate approaches, and the evaluated results of these examinations are usually associated with studies on their micro-mechanisms. The intrinsic toughening mechanisms in this alloy group have been somewhat revealed, but the extrinsic toughening mechanisms need to be explored. Attempts to bring forward the deformation mechanisms around the point of crack initiation in the investigated systems are generally based on electron microscopic examinations. These examinations are necessary for the understanding of the mechanisms at microscale or sub-microscale, but are limited in the sense that these are localized in nature. Deformation mechanism at macroscale is better understood through thermal activation analysis. Some studies on thermal activation analyses on HEAs are available by now, but their correspondence with fracture behavior is yet to be explored. The effect of the external variables during fracture toughness tests like temperature and strain rate is another area which needs some emphasis in future research activities for assessing applicability of HEAs in different temperature ranges. Interestingly some of the 3-D transition element-based HEAs have shown excellent promise for low-temperature applications for their strength–ductility combinations, but elevated temperature fracture toughness, specifically that for refractory HEAs, needs careful investigations. The effect of strain rate on fracture toughness of HEAs is almost non-existent. Neither are there any investigations directed toward revealing the fracture toughness in aggressive environments, though these materials often indicate significant corrosion resistance.

Fatigue resistance of HEAs has been evaluated through HCF, LCF, or by FCGR tests and subsequent analyses. The number of investigations in LCF or FCGR is extremely limited. However, all the reported investigations indicate great promise for the potential applications of HEAs in cyclic loading. To establish the reliability of the forthcoming fatigue information, priority should be pre-fixed for screening these through systematic statistical analysis. Investigations related to fatigue resistance of HEAs at elevated temperatures and in corrosive environment are almost non-existent. In FCGR, the effect of *R*-ratio on the fatigue threshold has been evaluated for the 3-D-transition element-based alloys, but careful analysis is required to obtain information about the crack closure phenomenon. The dependence of ΔK_{th} on R-ratio also needs to be understood for the variation of the constituent phases in the path of evolution for several dual- or multiphase HEAs. Needless to mention, several HEAs in cast condition have shown promising mechanical properties, but the effect of the residual stresses in the fabricated specimen on their fatigue-resistance has not been carefully assessed. Information about ratcheting behavior of HEAs is an unexplored area. Finally, modeling studies related to fatigue and fracture characteristic of HEAs are yet to emerge.

16.4.5 Generalization

The fracture toughness and the fatigue resistance of HEAs have been compared frequently with the established advanced structural materials. Fascinating fracture toughness values in some FCC matrix HEAs with good strength and ductility have brought in encouragement for developing these materials for future challenging applications. The futuristic views are being represented by Ashby-type strength–fracture toughness diagrammatic relations. But the rationale behind such comparisons with limited results is questionable. The fatigue resistance of HEA and the allied group of materials is also equally promising, as discussed in Section 16.3. However, the overall information acquired so far on fracture and fatigue behavior of HEAs is not sufficient for rational comparisons. Some sparks for high promise for structural applications for a few HEA materials are just scintillating. Attempts to understand the in-depth scientific intricacies should be pursued by the material scientists, but the technological demands need continued research on all emerging HEAs for assessing their structural integrity potential for early applications in structural engineering.

16.5 SUMMARY

High entropy and the allied alloys like multiple principal element alloys or complex concentrated alloys are an emerging class of materials with almost an unbounded number, but the number of investigations on their fracture toughness and fatigue life are rather limited till the middle of 2019. Of the several existing categories of these alloys, only the transition metal-based and the refractory HEAs have been in the focus for characterizations of their fracture and fatigue behavior. In this perspective, it would not be rational to draw generalized inferences on fracture resistance or fatigue behavior of HEAs. But the trend of the reported results indicates that the factors which dictate the resistance to fracture and fatigue failure in conventional advanced structural materials are also operative in HEAs with minor variations, if any. The chemistry and the processing decide the amount, distribution and the nature of the constituents in the microstructure of the HEAs and the latter, in turn, governs their response to fracture and fatigue behavior almost in an identical way to that for advanced structural alloys. The most prominent trends of the fracture and fatigue behavior of HEAs are summarized in the following:

1. The resistance to fracture of transition metal-based HEAs is significantly superior to refractory HEAs. For example, fracture toughness (K_{Jc}) of CrMnFeCoNi or CrCoNi high entropy and allied alloys (which are

transition metal-based) are 219 $MPam^{1/2}$ at 273 K [24] and 208 $MPam^{1/2}$ at 293 K [117] respectively at the near ambient temperature, whereas fracture toughness of refractory high entropy alloys like $Nb_{25}Mo_{25}Ta_{25}W_{25}$ is only between 0.2–2.1 $MPam^{1/2}$ [88].

2. The fracture toughness of HEAs is considerably higher when the matrix constituent is FCC compared to that when the matrix phase possesses BCC crystal structure. If the matrix is of dual phase type with a mixture of FCC and BCC phases, the fracture toughness would be between the two extremities. A simple illustration of this trend is exhibited in Figure 16.9.
3. Higher fracture toughness in FCC high entropy or medium entropy alloys like CrMnFeCoNi and CrCoNi alloys are governed by the associated plastic deformation controlled by the movement of ½(110) dislocations on {111} planes [24]. On the other hand, lower fracture toughness in ordered B2 BCC structures like $Al_{23}Co_{15}Cr_{23}Cu_8Fe_{15}Ni_{15}$ alloy [111] is expected to be governed by the slip of <111> screw dislocations gliding on {112} planes, as reported by Feuerbacher [168] for $Al_{28}Co_{20}Cr_{11}Fe_{15}Ni_{26}$ HEA having a similar crystal structure. Interestingly, the associated dislocation participations in the plastic deformations processes are identical for metallic materials possessing these crystal structures.
4. Conventionally fracture toughness of metallic materials bears an inverse relation with the strength of the concerned material. A similar trend is also noticed for high entropy alloys if one compares the strength and fracture toughness values of FCC and BCC alloys in Figure 16.2 and Figure 16.9. However, simultaneous increase of strength and fracture toughness in some HEAs like CrMnFeCoNi and CrCoNi at low temperatures of 77 K [17, 24, 81, 117] remains a piece of puzzling but exciting information. It is already discerned that these phenomena occur due to additional features like micro-twinning and phase transformation, but in-depth studies of deformation and crack initiation studies at cryogenic temperatures can only reveal the intricacies.
5. The signatures on the fracture surfaces of specimens after fracture toughness tests have been found as typical dimple fracture for FCC materials [24, 117] and intergranular or transgranular cleavage [111, 128] in BCC materials. Mixed dimple with cleavage and quasi-cleavage fracture has also been recorded in some investigations [88]. These fractographic features are similar to what have been usually recorded on metallic materials.
6. Fatigue strength in terms of fatigue endurance limit for the investigated HEAs indicates a promising trend. The proportionality coefficient between endurance limit and strength of HEAs is higher than that for conventional metallic materials [82]. This information points toward better resistance to crack initiation and propagation in HEAs compared to that of established structural components made of metallic materials for similar strength level. Typically, $Al_{0.5}CoFeCrNiCu$ HEA exhibit excellent fatigue resistance even at relatively high stresses due to nano-twinning [148].
7. Careful manufacturing and processing of HEAs with less defect population are essential to achieve better fatigue resistance in these alloys like other materials. Statistical analyses of fatigue data of $Al_{0.5}CoFeCrNiCu$ HEA

with different levels of defect populations [147–148] have already supported this well-known phenomenon.

8. The variation of (da/dN) versus (DK) plots for HEAs exhibits the classical nature indicating the three distinct stages of crack growth, depending on the test conditions. Based on the few available reports, the Paris law exponents for HEAs are found to range between 3 and 5 [17] like metallic materials.
9. The effect of stress ratio on FCGR is similar to that of other structural materials. Higher stress ratio leads to higher crack growth rate and different fatigue threshold values. In the absence of any results on crack opening or crack closure stress intensity factors (K_{op} or K_{cl}), for the investigated materials, definite inferences on the magnitude of threshold stress intensity factor (DK_{th}) is difficult to draw.
10. Some limited work exists on the effect of temperature on the crack growth rate for transition metal-based HEAs specifically at lower temperatures. The results are highly encouraging concerning the fact that lower temperature enhances fatigue threshold value; this is an essential message for the potential application of these materials at lower temperatures.
11. The signatures on the fatigue crack propagation regime in high entropy alloys do exhibit well-known striation marks and herringbone pattern [111], and the fatigue-fractured regime indicates dimples or cleavage as observed on specimens of HEAs subjected to fracture toughness tests.
12. High entropy alloys in comparison to conventional structural materials can exhibit better resistance to fracture and fatigue endurance for similar strength levels; this information remains as the dazzling signal for their potential applications. Optimization of suitable chemistry and microstructural constituents of HEAs are expected to act as the driving force to evolve these materials for potential structural applications.

ACKNOWLEDGMENTS

The author takes this opportunity to gratefully acknowledge the kind invitation and the patient encouragement by the Editor, Professor T.S. Srivatsan, Professor (Emeritus) of the Department of Mechanical Engineering, University of Akron, Ohio, through a number of correspondences which is the primary driving force for the emergence of this chapter. The author thankfully acknowledges the excellent assistance provided by one of his research scholars, Mr. Atri Nath, Department of Civil Engineering, IIT Kharagpur, India, who has kindly helped in drafting the figures, arranging the references, and formatting the entire text, apart from providing some critical comments at times. The author also acknowledges the help rendered by Prof. Rahul Mitra and Prof. Debalay Chakrabarti of the Department of Metallurgical and Materials Engineering, IIT Kharagpur, India for providing computer facilities and working space. Finally, the author would like to thank Prof. B.S. Murty, currently Director of IIT Hyderabad, India, for introducing the author to this discipline through his enthusiastic work in high entropy and the allied group of alloys.

REFERENCES

1. J.-W. Yeh, S.-K. Chen, S.-J. Lin, J.-Y. Gan, T.-S. Chin, T.-T. Shun, C.-H. Tsau, and S.-Y. Chang, *Adv. Eng. Mater.* 6(5), 299–303, (2004).
2. B. Cantor, I. T. H. Chang, P. Knight, and A. J. B. Vincent, *Mater. Sci. Eng. A* 375–377, 213–218, (2004).
3. D. B. Miracle, and O. N. Senkov, *Acta Mater.* 122, 448–511, (2017).
4. M.-H. Tsai, and J.-W. Yeh, *Mater. Res. Lett.* 2(3), 107–123, (2014).
5. J. Chen, X. Zhou, W. Wang, B. Liu, Y. Lv, W. Yang, D. Xu, and Y. Liu, *J. Alloys Compd.* 760, 15–30, (2018).
6. M. C. Gao, *High-Entropy Alloys: Fundamentals and Applications*, Springer, Cham, (2016).
7. S. Gorsse, J.-P. Couzinié, and D. B. Miracle, *C.R. Phys.* 19(8), 721–736, (2018).
8. Y. Ikeda, B. Grabowski, and F. Körmann, *Mater. Charact.* 147, 464–511, (2019).
9. B. S. Murty, J. W. Yeh, S. Ranganathan, and P. P. Bhattacharjee, *High-Entropy Alloys*, Elsevier, 119–144, (2019).
10. I. Basu, V. Ocelík, and J. T. De Hosson, *Acta Mater.* 157, 83–95, (2018).
11. F. G. Coury, M. Kaufman, and A. J. Clarke, *Acta Mater.* 175, 66–81, (2019).
12. F. G. Coury, P. Wilson, K. D. Clarke, M. J. Kaufman, and A. J. Clarke, *Acta Mater.* 167, 1–11, (2019).
13. H. Y. Diao, R. Feng, K. A. Dahmen, and P. K. Liaw, *Curr. Opin. Solid State Mater. Sci.* 21(5), 252–266, (2017).
14. N. Gao, D. H. Lu, Y. Y. Zhao, X. W. Liu, G. H. Liu, Y. Wu, G. Liu, Z. T. Fan, Z. P. Lu, and E. P. George, *J. Alloys Compd.* 792, 1028–1035, (2019).
15. R. W. Hertzberg, R. P. Vinci, and J. L. Hertzberg, *Deformation and Fracture Mechanics of Engineering Materials*, John Wiley & Sons, Inc, Hoboken, NJ, (2012).
16. C. R. LaRosa, M. Shih, C. Varvenne, and M. Ghazisaeidi, *Mater. Charact.* 151, 310–317, (2019).
17. Z. Li, S. Zhao, R. O. Ritchie, and M. A. Meyers, *Prog. Mater. Sci.* 102, 296–345, (2019).
18. Y. Ma, J. Hao, J. Jie, Q. Wang, and C. Dong, *Mater. Sci. Eng. A* 764, (2019).
19. J. Yang, J. W. Qiao, S. G. Ma, G. Y. Wu, D. Zhao, and Z. H. Wang, *J. Alloys Compd.* 795, 269–274, (2019).
20. K. K. Chawla, *Composite Materials: Science and Engineering*, Springer, New York, NY, (2013).
21. F. C. Campbell, Structural Composite Materials, ASM Internat, Materials Park, OH, (2010).
22. Y. Zhang, T. T. Zuo, Z. Tang, M. C. Gao, K. A. Dahmen, P. K. Liaw, and Z. P. Lu, *Prog. Mater. Sci.* 61, 1–93, (2014).
23. O. N. Senkov, J. D. Miller, D. B. Miracle, and C. Woodward, *Nat. Commun.* 6, 6529, (2015).
24. B. Gludovatz, A. Hohenwarter, D. Catoor, E. H. Chang, E. P. George, and R. O. Ritchie, *Science* 345(6201), 1153–1158, (2014).
25. K. J. Laws, C. Crosby, A. Sridhar, P. Conway, L. S. Koloadin, M. Zhao, S. Aron-Dine, and L. C. Bassman, *J. Alloys Compd.* 650, 949–961, (2015).
26. S. S. Nene, M. Frank, K. Liu, S. Sinha, R. S. Mishra, B. A. McWilliams, and K. C. Cho, *Scr. Mater.* 166, 168–172, (2019).
27. G. Bracq, M. Laurent-Brocq, C. Varvenne, L. Perrière, W. A. Curtin, J.-M. Joubert, and I. Guillot, *Acta Mater.* 177, 266–279, (2019).
28. O. N. Senkov, S. Gorsse, and D. B. Miracle, *Acta Mater.* 175, 394–405, (2019).
29. A. Takeuchi, K. Amiya, T. Wada, and K. Yubuta, *Intermetallics* 69, 103–109, (2016).
30. S. Vrtnik, J. Lužnik, P. Koželj, A. Jelen, J. Luzar, Z. Jagličić, A. Meden, M. Feuerbacher, and J. Dolinšek, *J. Alloys Compd.* 742, 877–886, (2018).

31. M. C. Gao, B. Zhang, S. M. Guo, J. W. Qiao, and J. A. Hawk, *Metall. Mater. Trans. A* 47(7), 3322–3332, (2016).
32. A. Takeuchi, K. Amiya, T. Wada, K. Yubuta, and W. Zhang, *JOM* 66(10), 1984–1992, (2014).
33. W. Wang, Z. Hou, R. Lizárraga, Y. Tian, R. P. Babu, E. Holmström, H. Mao, and H. Larsson, *Acta Mater.* 176, 11–18, (2019).
34. R. Song, L. Wei, C. Yang, and S. Wu, *J. Alloys Compd.* 744, 552–560, (2018).
35. Z. Li, K. G. Pradeep, Y. Deng, D. Raabe, and C. C. Tasan, *Nature* 534(7606), 227–230, (2016).
36. Y. P. Cai, G. J. Wang, Y. J. Ma, Z. H. Cao, and X. K. Meng, *Scr. Mater.* 162, 281–285, (2019).
37. Z. Zhang, H. Sheng, Z. Wang, B. Gludovatz, Z. Zhang, E. P. George, Q. Yu, S. X. Mao, and R. O. Ritchie, *Nat. Commun.* 8, 14390, (2017).
38. C. L. Tracy, S. Park, D. R. Rittman, S. J. Zinkle, H. Bei, M. Lang, R. C. Ewing, and W. L. Mao, *Nat. Commun.* 8, 15634, (2017).
39. J. Hou, X. Shi, J. Qiao, Y. Zhang, P. K. Liaw, and Y. Wu, *Mater. Des.* 180, (2019).
40. F. Otto, Y. Yang, H. Bei, and E. P. George, *Acta Mater.* 61(7), 2628–2638, (2013).
41. M. Annasamy, N. Haghdadi, A. Taylor, P. Hodgson, and D. Fabijanic, *Mater. Sci. Eng. A* 754, 282–294, (2019).
42. B. Gwalani, S. Gorsse, D. Choudhuri, M. Styles, Y. Zheng, R. S. Mishra, and R. Banerjee, *Acta Mater.* 153, 169–185, (2018).
43. J. W. Bae, J. M. Park, J. Moon, W. M. Choi, B.-J. Lee, and H. S. Kim, *J. Alloys Compd.* 781, 75–83, (2019).
44. A. Shafiee, J. Moon, H. S. Kim, M. Jahazi, and M. Nili-Ahmadabadi, *Mater. Sci. Eng. A* 749, 271–280, (2019).
45. G. Laplanche, A. Kostka, O. M. Horst, G. Eggeler, and E. P. George, *Acta Mater.* 118, 152–163, (2016).
46. K. V. S. Thurston, A. Hohenwarter, G. Laplanche, E. P. George, B. Gludovatz, and R. O. Ritchie, *Intermetallics* 110, (2019).
47. H. Kim, S. Nam, A. Roh, M. Son, M.-H. Ham, J.-H. Kim, and H. Choi, *Int. J. Refract. Met. Hard Mater.* 80, 286–291, (2019).
48. W. Sun, X. Huang, and A. A. Luo, *Calphad* 56, 19–28, (2017).
49. R. Li, J. C. Gao, and K. Fan, *Mater. Sci. Forum* 650, 265–271, (2010).
50. J. Gild, Y. Zhang, T. Harrington, S. Jiang, T. Hu, M. C. Quinn, W. M. Mellor, N. Zhou, K. Vecchio, and J. Luo, *Sci. Rep.* 6, 37946, (2016).
51. S. Jiang, T. Hu, J. Gild, N. Zhou, J. Nie, M. Qin, T. Harrington, K. Vecchio, and J. Luo, *Scr. Mater.* 142, 116–120, (2018).
52. M. Balcerzak, K. Kawamura, R. Bobrowski, P. Rutkowski, and T. Brylewski, *J. Electron. Mater.*, (2019).
53. C. M. Rost, E. Sachet, T. Borman, A. Moballegh, E. C. Dickey, D. Hou, J. L. Jones, S. Curtarolo, and J.-P. Maria, *Nat. Commun.* 6, 8485, (2015).
54. T. Nagase, A. Shibata, M. Matsumuro, M. Takemura, and S. Semboshi, *Mater. Des.* 181, (2019).
55. S. Sohn, Y. Liu, J. Liu, P. Gong, S. Prades-Rodel, A. Blatter, B. E. Scanley, C. C. Broadbridge, and J. Schroers, *Scr. Mater.* 126, 29–32, (2017).
56. S. Gorsse, M. H. Nguyen, O. N. Senkov, and D. B. Miracle, *Data Brief* 21, 2664–2678, (2018).
57. J.-P. Couzinié, O. N. Senkov, D. B. Miracle, and G. Dirras, *Data Brief* 21, 1622–1641, (2018).
58. H. Zhang, Y. He, and Y. Pan, *Scr. Mater.* 69(4), 342–345, (2013).
59. A. Udalov, S. Parshin, and A. Udalov, *Mater. Today, Proc.* S2214785319321583, (2019).
60. W. D. Nix, and H. Gao, *J. Mèch. Phys. Solids* 46(3), 411–425, (1998).

61. M.-R. Chen, S.-J. Lin, J.-W. Yeh, M.-H. Chuang, S.-K. Chen, and Y.-S. Huang, *Metall. Mater. Trans. A* 37(5), 1363–1369, (2006).
62. M.-R. Chen, S.-J. Lin, J.-W. Yeh, S.-K. Chen, Y.-S. Huang, and C.-P. Tu, *Mater. Trans.* 47(5), 1395–1401, (2006).
63. M.-H. Chuang, M.-H. Tsai, W.-R. Wang, S.-J. Lin, and J.-W. Yeh, *Acta Mater.* 59(16), 6308–6317, (2011).
64. Y. Dong, Y. Lu, J. Kong, J. Zhang, and T. Li, *J. Alloys Compd.* 573, 96–101, (2013).
65. É. Fazakas, V. Zadorozhnyy, and D. V. Louzguine-Luzgin, *Appl. Surf. Sci.* 358, 549–555, (2015).
66. É. Fazakas, V. Zadorozhnyy, L. K. Varga, A. Inoue, D. V. Louzguine-Luzgin, F. Tian, and L. Vitos, *Int. J. Refract. Met. Hard Mater.* 47, 131–138, (2014).
67. C.-Y. Hsu, C.-C. Juan, T.-S. Sheu, S.-K. Chen, and J.-W. Yeh, *JOM* 65(12), 1840–1847, (2013).
68. Y. J. Zhou, Y. Zhang, Y. L. Wang, and G. L. Chen, *Appl. Phys. Lett.* 90, (2007).
69. Y. X. Zhuang, W. J. Liu, Z. Y. Chen, H. D. Xue, and J. C. He, *Mater. Sci. Eng. A* 556, 395–399, (2012).
70. W. H. Liu, J. Y. He, H. L. Huang, H. Wang, Z. P. Lu, and C. T. Liu, *Intermetallics* 60, 1–8, (2015).
71. P. Sathiyamoorthi, J. Moon, J. W. Bae, P. Asghari-Rad, and H. S. Kim, *Scr. Mater.* 163, 152–156, (2019).
72. J. M. Zhu, H. M. Fu, H. F. Zhang, A. M. Wang, H. Li, and Z. Q. Hu, *Mater. Sci. Eng. A* 527(26), 6975–6979, (2010).
73. H. Diao, D. Ma, R. Feng, T. Liu, C. Pu, C. Zhang, W. Guo, J. D. Poplawsky, Y. Gao, and P. K. Liaw, *Mater. Sci. Eng. A* 742, 636–647, (2019).
74. J. Gu, S. Ni, Y. Liu, and M. Song, *Mater. Sci. Eng. A* 755, 289–294, (2019).
75. S. J. Sun, Y. Z. Tian, H. R. Lin, S. Lu, H. J. Yang, and Z. F. Zhang, *Scr. Mater.* 163, 111–115, (2019).
76. G. Chen, J. W. Qiao, Z. M. Jiao, D. Zhao, T. W. Zhang, S. G. Ma, and Z. H. Wang, *Scr. Mater.* 167, 95–100, (2019).
77. Y. Tong, D. Chen, B. Han, J. Wang, R. Feng, T. Yang, C. Zhao, Y. L. Zhao, W. Guo, Y. Shimizu, C. T. Liu, P. K. Liaw, K. Inoue, Y. Nagai, A. Hu, and J. J. Kai, *Acta Mater.* 165, 228–240, (2019).
78. J. Čížek, P. Haušild, M. Cieslar, O. Melikhova, T. Vlasák, M. Janeček, R. Král, P. Harcuba, F. Lukáč, J. Zýka, J. Málek, J. Moon, and H. S. Kim, *J. Alloys Compd.* 768, 924–937, (2018).
79. Y. Zhang, J. Li, X. Wang, Y. Lu, Y. Zhou, and X. Sun, *J. Mater. Sci. Technol.* 35(5), 902–906, (2019).
80. Y. H. Jo, K.-Y. Doh, D. G. Kim, K. Lee, D. W. Kim, H. Sung, S. S. Sohn, D. Lee, H. S. Kim, B.-J. Lee, and S. Lee, *J. Alloys Compd.* 809, (2019).
81. W. Li, P. K. Liaw, and Y. Gao, *Intermetallics* 99, 69–83, (2018).
82. P. Chen, C. Lee, S.-Y. Wang, M. Seifi, J. J. Lewandowski, K. A. Dahmen, H. Jia, X. Xie, B. Chen, J.-W. Yeh, C.-W. Tsai, T. Yuan, and P. K. Liaw, *Sci. China Technol. Sci.* 61(2), 168–178, (2018).
83. S. Alvi, and F. Akhtar, *Wear* 426–427, 412–419, (2019).
84. Z.-S. Nong, Y.-N. Lei, and J.-C. Zhu, *Intermetallics* 101, 144–151, (2018).
85. A. Verma, P. Tarate, A. C. Abhyankar, M. R. Mohape, D. S. Gowtam, V. P. Deshmukh, and T. Shanmugasundaram, *Scr. Mater.* 161, 28–31, (2019).
86. Y. Qiu, S. Thomas, D. Fabijanic, A. J. Barlow, H. L. Fraser, and N. Birbilis, *Mater. Des.* 170, (2019).
87. C. Xiang, Z. M. Zhang, H. M. Fu, E.-H. Han, H. F. Zhang, and J. Q. Wang, *Intermetallics* 114, (2019).
88. Y. Zou, P. Okle, H. Yu, T. Sumigawa, T. Kitamura, S. Maiti, W. Steurer, and R. Spolenak, *Scr. Mater.* 128, 95–99, (2017).

89. A. J. Sedriks, and O. S. Zaroog, *Reference Module in Materials Science and Materials Engineering*, Elsevier, B9780128035818030000, (2017).
90. A. A. Griffith, *Philos. Trans. R. Soc. Math. Phys. Eng. Sci.* 221, 163–198, (1921).
91. T. L. Anderson, *Fracture Mechanics: Fundamentals and Applications*, 3rd Edition, CRC Press, (2005).
92. X.-K. Zhu, and J. A. Joyce, *Eng. Fract. Mech.* 85, 1–46, (2012).
93. ASTM E399-17, *Test Method for Linear-Elastic Plane-Strain Fracture Toughness KIc of Metallic Materials*, ASTM International, West Conshohocken, PA, (2017).
94. G. R. Irwin, *Trans ASME Ser E, J. Appl. Mech.* 24, 361–364, (1957).
95. J. R. Rice, *J. Appl. Mech.* 35(2), 379–386, (1968).
96. A. Wells, Br. *Weld. J.* 10, (1963).
97. D. Broek, *Elementary Engineering Fracture Mechanics*, Springer Netherlands, Dordrecht, (1986).
98. ASTM E1820-18ae1, *Test Method for Measurement of Fracture Toughness*, ASTM International, West Conshohocken, PA, (2018).
99. ASTM E2472-12R18, *Test Method for Determination of Resistance to Stable Crack Extension under Low-Constraint Conditions*, ASTM International, West Conshohocken, PA, (2018).
100. ASTM E1823-13, *Terminology Relating to Fatigue and Fracture Testing*, ASTM International, West Conshohocken, PA, (2013).
101. ASTM E1290-08e1, *Standard Test Method for Crack-Tip Opening Displacement (CTOD) Fracture Toughness Measurement (Withdrawn 2013)*, ASTM International, West Conshohocken, PA, (2008).
102. ASTM E1304-97, *Test Method for Plane-Strain (Chevron-Notch) Fracture Toughness of Metallic Materials*, ASTM International, West Conshohocken, PA, (2014).
103. M. Sebastiani, K. E. Johanns, E. G. Herbert, and G. M. Pharr, *Curr. Opin. Solid State Mater. Sci.* 19(6), 324–333, (2015).
104. K. K. Ray, and A. K. Dutta, *Br. Ceram. Trans.* 98(4), 165–171, (1999).
105. G. C. Sih, and E. Madenci, *Eng. Fract. Mech.* 18(3), 667–677, (1983).
106. S. K. Chandra, R. Sarkar, A. D. Bhowmick, P. S. De, P. C. Chakraborti, and S. K. Ray, *Eng. Fract. Mech.* 204, 29–45, (2018).
107. S. S. Javaid, W. R. Lanning, and C. L. Muhlstein, *Eng. Fract. Mech.* 218, (2019).
108. K. K. Ray, *Encyclopedia of Materials: Science and Technology*, Elsevier, 1741–1744, (2001).
109. K. K. Ray, and G. P. Poddar, *Fatigue Fract. Eng. Mater. Struct.* 27(3), 253–261, (2004).
110. K. K. Ray, D. Chakraborty, and S. Ray, *J. Mater. Sci.* 29(4), 921–928, (1994).
111. U. Roy, H. Roy, H. Daoud, U. Glatzel, and K. K. Ray, *Mater. Lett.* 132, 186–189, (2014).
112. W. Shang-Xian, in *Chevron-Notched Specim. Test. Stress Anal.* (Eds: J. Underwood, S. Freiman, and F. Baratta), ASTM International, West Conshohocken, PA, 176-176–17, (1984).
113. T. Kim, K. T. Hong, and K. S. Lee, *Intermetallics* 11(1), 33–39, (2003).
114. K. S. Kumar, S. K. Mannan, and R. K. Viswanadham, *Acta Metall. Mater.* 40(6), 1201–1222, (1992).
115. C. Chen, S. Pang, Y. Cheng, and T. Zhang, *J. Alloys Compd.* 659, 279–287, (2016).
116. M. Seifi, D. Li, Z. Yong, P. K. Liaw, and J. J. Lewandowski, *JOM* 67(10), 2288–2295, (2015).
117. B. Gludovatz, A. Hohenwarter, K. V. S. Thurston, H. Bei, Z. Wu, E. P. George, and R. O. Ritchie, *Nat. Commun.* 7, 10602, (2016).
118. ASTM E23-18, *Test Methods for Notched Bar Impact Testing of Metallic Materials*, ASTM International, West Conshohocken, PA, (2018).
119. D. Li, and Y. Zhang, *Intermetallics* 70, 24–28, (2016).
120. T. Zhang, Y. Feng, R. Yang, and P. Jiang, *Scr. Mater.* 62(4), 199–201, (2010).
121. J. H. Lee, Y. F. Gao, K. E. Johanns, and G. M. Pharr, *Acta Mater.* 60(15), 5448–5467, (2012).

122. F. Iqbal, J. Ast, M. Göken, and K. Durst, *Acta Mater.* 60, 1193–1200, (2012).
123. ASM International, and K. Mills, (Eds.), *Fractography*, ASM International, Materials Park, OH, (2009).
124. A. Pineau, A. A. Benzerga, and T. Pardoen, *Acta Mater.* 107, 424–483, (2016).
125. W. M. Garrison, and N. R. Moody, *J. Phys. Chem. Solids* 48(11), 1035–1074, (1987).
126. Yu. Ivanisenko, L. Kurmanaeva, J. Weissmueller, K. Yang, J. Markmann, H. Rösner, T. Scherer, and H.-J. Fecht, *Acta Mater.* 57(11), 3391–3401, (2009).
127. S. P. Lynch, *Scr. Metall.* 20(7), 1067–1072, (1986).
128. S.-P. Wang, E. Ma, and J. Xu, *Intermetallics* 103, 78–87, (2018).
129. C. M. Cao, W. Tong, S. H. Bukhari, J. Xu, Y. X. Hao, P. Gu, H. Hao, and L. M. Peng, *Mater. Sci. Eng. A* 759, 648–654, (2019).
130. R. P. Reed, *The Economic Effects of Fracture in the United States*, US Department of Commerce, National Bureau of Standards, (1983).
131. H. Jia, G. Wang, S. Chen, Y. Gao, W. Li, and P. K. Liaw, *Prog. Mater. Sci.* 98, 168–248, (2018).
132. M. D. Hayat, H. Singh, Z. He, and P. Cao, *Compos. Part Appl. Sci. Manuf.* 121, 418–438, (2019).
133. S. Teoh, *Int. J. Fatigue* 22(10), 825–837, (2000).
134. P. Alam, D. Mamalis, C. Robert, C. Floreani, and C. M. Ó. Brádaigh, *Compos. Part B Eng.* 166, 555–579, (2019).
135. S. Liang, P. B. Gning, and L. Guillaumat, *Compos. Sci. Technol.* 72(5), 535–543, (2012).
136. K. Shirvanimoghaddam, S. U. Hamim, M. Karbalaei Akbari, S. M. Fakhrhoseini, H. Khayyam, A. H. Pakseresht, E. Ghasali, M. Zabet, K. S. Munir, S. Jia, J. P. Davim, and M. Naebe, *Compos. Part Appl. Sci. Manuf.* 92, 70–96, (2017).
137. R. O. Ritchie, *Met. Sci.* 11(8–9), 368–381, (1977).
138. A. Wöhler, *Über Die Festigkeitsversuche Mit Eisen Und Stahl*, Ernst & Korn, (1870).
139. O. Basquin. *Proc. Am. Soc. Test. Mater.*, 625–630, (1910).
140. L. F. Coffin Jr, *Trans. Am. Soc. Mech. Eng. N. Y.* 76, 931–950, (1954).
141. S. S. Manson, NACA-TR-1170, Cleveland, OH, USA. (1954).
142. P. C. Paris, *Trend Eng.* 13, 9, (1961).
143. P. S. De, C. M. Obermark, and R. S. Mishra, *J. Test. Eval.* 36, (2008).
144. ASTM E0466-15, *Practice for Conducting Force Controlled Constant Amplitude Axial Fatigue Tests of Metallic Materials*, ASTM International, (2015).
145. ASTM E0606/E0606M-12, *Test Method for Strain-Controlled Fatigue Testing*, ASTM International, West Conshohocken, PA, (2012).
146. ASTM E647-15E01, *Test Method for Measurement of Fatigue Crack Growth Rates*, ASTM International, West Conshohocken, PA, (2015).
147. M. A. Hemphill, T. Yuan, G. Y. Wang, J. W. Yeh, C. W. Tsai, A. Chuang, and P. K. Liaw, *Acta Mater.* 60(16), 5723–5734, (2012).
148. Z. Tang, T. Yuan, C.-W. Tsai, J.-W. Yeh, C. D. Lundin, and P. K. Liaw, *Acta Mater.* 99, 247–258, (2015).
149. B. Guennec, V. Kentheswaran, L. Perrière, A. Ueno, I. Guillot, J.-Ph. Couzinié, and G. Dirras, *Materialia* 4, 348–360, (2018).
150. S. Shukla, T. Wang, S. Cotton, and R. S. Mishra, *Scr. Mater.* 156, 105–109, (2018).
151. K. Liu, M. Komarasamy, B. Gwalani, S. Shukla, and R. S. Mishra, *Scr. Mater.* 158, 116–120, (2019).
152. T. Niendorf, T. Wegener, Z. Li, and D. Raabe, *Scr. Mater.* 143, 63–67, (2018).
153. K. V. S. Thurston, B. Gludovatz, A. Hohenwarter, G. Laplanche, E. P. George, and R. O. Ritchie, *Intermetallics* 88, 65–72, (2017).
154. K. V. S. Thurston, B. Gludovatz, Q. Yu, G. Laplanche, E. P. George, and R. O. Ritchie, *J. Alloys Compd.* 794, 525–533, (2019).
155. F. G. Pascual, and W. Q. Meeker, *Technometrics* 41(4), 277–289, (1999).

156. D. G. Harlow, *Acta Mater.* 59(12), 5048–5053, (2011).
157. K. M. Flores, W. L. Johnson, and R. H. Dauskardt, *Scr. Mater.* 49(12), 1181–1187, (2003).
158. ASM International, and J. R. Davis, (Eds.), *Properties and Selection: Nonferrous Alloys and Special-Purpose Materials*, ASM International, Materials Park, OH, (2000).
159. J. Y. Mann, Camb. Univ. Press, London, 155 page (1967).
160. CINDAS/USAF, *Aerospace Structural Metals Handbook*, Purdue University, West Lafayette, IN, (2001).
161. G. Y. Wang, P. K. Liaw, Y. Yokoyama, A. Inoue, and C. T. Liu, *Mater. Sci. Eng. A* 494(1–2), 314–323, (2008).
162. JSMS SD-11-07, *Standard Evaluation Method of Fatigue Reliability for Metallic Materials, Standard Regression Method of S-N Curves*, JSMS Committee on Fatigue of Materials, Japan, (2007).
163. R. Ritchie, and W. Yu, *Small Fatigue Cracks*, 167–189, (1986).
164. K. T. Venkateswararao, W. Yu, and R. O. Ritchie, *Metall. Trans. A* 19(3), 549–561, (1988).
165. C. B. Ponton, and R. D. Rawlings, *Mater. Sci. Technol.* 5(9), 865–872, (1989).
166. G. D. Quinn, and R. C. Bradt, *J. Am. Ceram. Soc.* 90(3), 673–680, (2007).
167. J. J. Kruzic, D. K. Kim, K. J. Koester, and R. O. Ritchie, *J. Mech. Behav. Biomed. Mater.* 2(4), 384–395, (2009).
168. M. Feuerbacher, *Sci. Rep.* 6, 29700, (2016).

17 Welding of High Entropy Alloys—Techniques, Advantages, and Applications: A Review

R. Sokkalingam, K. Sivaprasad, and V. Muthupandi

CONTENTS

17.1 INTRODUCTION

Welding is a permanent joining process used for fabrication of complex structures from a number of simple structures in all kinds of structural applications from the fabrication of household utensils to aerospace structures [1–8]. As manufacturing of complex shaped structures or parts by other processes like casting, machining, and forming has size limitations and difficulties in manufacturing [9], welding is often considered a major fabrication method in construction and structural building works over many decades. Varieties of welding processes have emerged over the years, depending on the material and design, and the need for welding remains unchanged, as there are no other equivalent means to substitute welding [10]. Even though, additive manufacturing is envisaged as an alternative to welding in building complex shapes through layer by layer addition of powders [11–13], additive manufacturing of huge complex shapes is still difficult and time consuming. Moreover, the construction of huge structural parts with different materials can be accomplished only through welding [14–16].

In most of the cases, the structural materials used for various applications cannot be in the as-cast condition. Some post-processing (like heat treatment, mechanical processing, thermomechanical processing, etc.) must be carried out for tailoring mechanical properties and microstructures in order to meet the design needs of the structural parts [9]. However, the welding could deteriorate the mechanical as well as electrochemical properties of the actual material, as it could change the microstructure locally by rapid heating and cooling and lead to the formation of intermetallic compounds/low melting point elements segregation [17]. Hence, in order to withstand the design loads, the structures can be used with a higher thickness than the needed one to overcome the loss of properties. However, the increasing thickness would end up increasing the weight of the structure and causing a higher material cost. In some applications having rotating shafts, as in power generation and pump sectors, and in automotive and aerospace applications, the increase in weight could decrease the overall efficiency of the system and increase power consumption [18]. Considering these facts, many of the studies are focused on minimizing the deterioration of properties by employing proper welding processes, procedures, and methods [19]. Each welding method has its own advantages and limitations. The selection of the process should be made with consideration of the design needs and customer's needs.

17.2 A BRIEF OVERVIEW OF STRUCTURAL HIGH ENTROPY ALLOYS AND THEIR CLASSIFICATIONS

Miracle and co-workers have reviewed the high entropy alloys (HEAs) and their applications in the structural works of the transportation sectors (aerospace, automobiles) and energy sectors (power plants) [20]. They indicated the extended applications of the high entropy alloys in these sectors ranging from low-temperature to high-temperature requirements. For comparison, the properties of the high entropy alloys along with the present structural alloys [20–22] are given in Table 17.1. High entropy alloys first evolved in 2004 with the work of two independent research groups Cantor [23] and Yeh [24]. Cantor developed the equimolar CoCrFeNiMn high entropy alloy with simple microstructure. Further studies on Cantor alloys have revealed its exceptional damage tolerance property at room and cryogenic temperatures [25, 26]. Also, the Yeh alloy system (Al_xCoCrCuFeNi) provides superior mechanical properties. In this system, the hardness and the tensile properties increase with an increase in aluminum (Al) concentration due to the cock-tail effect [22, 27]. Alloys modified from the Yeh alloy system, with elimination of the copper (Cu) element have shown even better strength and ductility [27–38]. In this chapter, an alloy having no copper (Cu) or with very little copper (Cu) is designated as the modified Yeh alloy. Since the Cantor alloy and Yeh alloys were developed first, extensive studies have been conducted on the microstructure and the properties of these alloys and published. These studies have been carried out in the as-cast as well as deformed and recrystallized conditions to understand their mechanical properties such as strength, ductility, and fracture toughness. Similarly, the weldability studies of the high entropy alloys are also concentrated on the Cantor alloys and the modified Yeh alloy system.

17.3 COMMONLY USED WELDING TECHNIQUES

Welding is the process of establishing the metallurgical bond between two materials by the application of heat or pressure or a combination of both at the interfaces [39]. Over the years, depending on the available technologies at the time, various welding processes have been evolved and applied to the joining of materials. Though the evidence for the use of joining technologies by Iron Age people is available, the welding processes that are being used routinely in present-day life were found to evolve from the 18th century onward. The welding processes can be grouped into two main categories, i.e., fusion welding and solid-state welding processes. Fusion welding is a welding process that involves melting of the base materials to be joined to provide mobile atoms at the interface, that produces a permanent joint on the solidification by the contribution of atoms/molecules from the base materials [40]. Due to the complete melting and re-solidification, the weld that shows the as-cast microstructure, the temperature distribution is varying from the center of the heat source to the periphery. The center of the weld experiences the highest temperature and it decreases exponentially as we move from the center to the periphery. The boundary, where the temperature equals the solidus temperature of the material, is the fusion line. However, a region in the parent material adjacent to the fusion zone, experiencing temperatures above the recrystallization/transformation temperature, that undergoes either phase transformation or grain growth, or both is called the heat affected zone (HAZ). The base metal region, despite experiencing temperature above ambient temperature, does not undergo any microstructural variation and is called the unaffected base metal. The variation of microstructures across the weld adversely deteriorates the mechanical properties locally.

TABLE 17.1
The Comparison of the Properties of High Entropy Alloys and Present Structural Alloys

Alloys	Density [g/cc]	Working temperature [°C]	Young's modulus [GPa]	Yield strength [MPa]	Ultimate tensile strength [MPa]	Elongation [%]	K_{IC} [MPa-m$^{-1/2}$]	Ref.
Al alloys	2.6–2.9	≤ 150	~70	250–550	300–600	≥ 10	≥ 30	[20]
Ti alloys	4.4–4.6	≤ 450	100–120	800–1400	900–1600	3–15	20–110	[20]
Ni alloys	8–9	≤ 1100	210–220	400–1300	1000–1600	15–50	80–120	[20]
Co-Cr alloys	–	–	210–232	241–310	793–860	20–50	–	[21]
Stainless steel 316	–	–	193	190	490	40	–	[21]
Stainless steel 304	7.8	–	–	179	248	9.1	–	[21]
Duplex stainless steel	7.8	–	–	329	366	2.4	–	[22]
CoCrFeMnNi (77 K)	–	–	214.5	759	1280	71	219	[25]
CoCrFeMnNi (200 K)	–	–	209	518	925	60	221	[25]
CoCrFeMnNi (293 K)	–	–	202	410	763	57	217	[25]
CoCrFeMnNi (As-cast, Annealed and rolled, Ann)	–	–	189	200–280	450–590	44–51	–	[21]
$Al_{0.5}CoCuCrFeNi$	7.6	–	–	600–1300	700–1350	7–30	–	[27]
AlCoCuCrFeNi (As-cast)	7.1	–	–	350	360	4.7	–	[22]
$Al_{0.1}CoCrFeNi$ (As-cast)	–	–	–	140–199	370–531	60–65	–	[21, 28]
$Al_{0.1}CoCrFeNi$ (Cold rolled and annealed)	–	–	–	167–366	570–895	46–60	–	[21,28,29]
$Al_{0.1}CoCrFeNi$ (200°C)	–	–	–	140	798	49	–	[29]
$Al_{0.1}CoCrFeNi$ (300°C)	–	–	–	124	717	52	–	[29]
$Al_{0.1}CoCrFeNi$ (400°C)	–	–	–	129	686	49	–	[29]
$Al_{0.1}CoCrFeNi$ (Hot isostatic pressed—tested at 25°C)	–	–	–	300	420	27	–	[30]

(Continued)

TABLE 17.1 (CONTINUED)
The Comparison of the Properties of High Entropy Alloys and Present Structural Alloys

Alloys	Density [g/cc]	Working temperature [°C]	Young's modulus [GPa]	Yield strength [MPa]	Ultimate tensile strength [MPa]	Elongation [%]	K_{IC} [MPa-m$^{-1/2}$]	Ref.
$Al_{0.1}$CoCrFeNi (Hot isostatic pressed—tested at 500°C)	–	–	–	150	300	34	–	[30]
$Al_{0.1}$CoCrFeNi (Hot isostatic pressed—tested at 600°C)	–	–	–	110	210	24	–	[30]
$Al_{0.1}$CoCrFeNi (Hot isostatic pressed—tested at 700°C)	–	–	–	100	130	6	–	[30]
$Al_{0.25}$CoCrFeNi (Cold rolled + annealed)	–	–	–	118–150	700–807	46–55	–	[31,32]
$Al_{0.3}$CoCrFeNi (as-cast)	–	–	–	150	300	60	–	[33]
$Al_{0.3}$CoCrFeNi (annealed)	–	–	–	250–300	500–550	45	–	[33,34]
$Al_{0.3}$CoCrFeNi (as-cast)	–	–	–	275	528	37	–	[35]
$Al_{0.3}$CoCrFeNi (single crystal)	–	–	–	185	399	80	–	[35]
$Al_{0.3}$CoCrFeNi (Cold rolled + annealed in range of 973–1373 K)	–	–	201.4	200–1200	600–1400	10–95	–	[36]
$Al_{0.3}$CoCrFeNi (Fiber at 298 K)	–	–	–	1147	1207	12	–	[37]
$Al_{0.3}$CoCrFeNi (Fiber at 77 K)	–	–	–	1320	1600	17.5	–	[37]
$Al_{0.5}$CoCrFeNi (as-cast)	–	–	–	300	594	25	–	[38]
$Al_{0.5}$CoCrFeNi (annealed)	–	–	–	300–1000	600–1200	20–38	–	[38]
AlCoCrFeNi (as-cast)	7.0	–	–	395	400	1	–	[22]
AlCoCrFeNi (as-cast + hot isostatic pressed)	7.0	–	–	295	393	11.7	–	[22]

Solid-state welding involves joining of materials by the application of pressure on the materials that are heated below solidus temperature. Depending on the source of heat generation and heat input, many welding processes are available at present, such as the following [39–53].

1. Fusion welding
    a. Gas welding (exothermic combustion of acetylene and oxygen gases)
    b. Arc welding (electrical energy)
        i. Shielded metal arc welding (SMAW)
        ii. Gas-tungsten arc welding (GTAW)
        iii. Gas-metal arc welding (GMAW)
        iv. Flux cored arc welding (FCAW)
        v. Plasma arc welding (PAW)
        vi. Submerged arc welding (SAW)
    c. Resistance welding (electrical energy)
    d. High energy beam welding (conversion of the kinetic energy of beam to heat on collision with workpiece)
        i. Laser beam welding (LBW)
        ii. Electron beam welding (EBW)
2. Solid state welding
    a. Friction welding
        i. Rotary friction welding (RFW)
        ii. Linear friction welding (LFW)
        iii. Friction stir welding (FSW)
    b. Diffusion bonding
    c. Explosive welding

17.3.1 Selection of Welding Processes

As each welding process has its own advantages and disadvantages, the welding processes must be selected with careful evaluation of the welding procedures involved in welding an alloy. When considering the cost as the main criterion in the selection, the gas-metal arc welding (GMAW) and gas-tungsten arc welding (GTAW) are the best welding processes to select [55, 56]. Out of these two, gas-tungsten arc welding (GTAW) could give better quality weld, as the arc spattering is absent in GTAW compared to the similar cost GMAW process [10]. However, some components in critical applications (severe working conditions like aerospace components, space rocket components, etc.) could require high structural integrity. In such cases, despite the cost, the welding processes that provide concentrated heat input, like plasma arc welding and the high-energy beam welding processes (laser beam welding and electron beam welding) are preferred to reduce the width of the heat affected zone [18]. Considering the cost of the high entropy alloy, most applications will be toward critical applications like the aerospace sectors, power generation industries, and chemical plants. Hence for joining high entropy alloys, the welding process must be selected, based on its ability to provide a quality weld as well as good structural integrity. Mendez and Eagar have concluded that the most suitable welding processes in welding of the aeronautical structures are: GTAW, plasma arc welding (PAW), laser beam welding (LBW) and electron beam welding (EBW) from fusion welding, and the rotary friction welding (RFW), linear friction welding (LFW), and friction stir welding (FSW) from the solid-state welding categories [18]. Out of these welding processes, weldability studies on high entropy alloys have been carried out effectively with fusion welding processes like GTAW, laser beam welding (LBW), and electron beam welding (EBW), and friction stir welding (FSW), a solid-state welding process. Hence a brief introduction of these welding processes is given below.

17.3.2 Gas-Tungsten Arc Welding

Gas-tungsten arc welding (GTAW) is the arc welding process that utilizes the electric arc as the heating source. When the non-consumable tungsten electrode and workpiece are connected to the welding power source, with the reduction in the standoff distance, the gases present between them will get ionized and pave the way for electric discharge and closing the circuit. This process is illustrated in Figure 17.1.

When the electrode is moved close to the workpiece or touched and withdrawn from the workpiece, the available gas in between them ionizes and establishes the electric discharge, resulting in the formation of a high-temperature (~6,000°C) arc. This concentrated high temperature melts the base metal and the filler, and some electric and/magnetic fluxes produced in the circuit lead to intermixing in the molten pool, which on solidification provides the joint with a metallurgical bond. The introduction of the shielding gases provides perfect shielding for the solidifying weld metal from the external atmosphere. Thus, GTAW can provide a clean and quality weld, ensuring the weld quality even for welding reactive metals. However, GTAW can only effectively weld thin plates (~3 mm) due to its lower heat input and power density. Sometimes, the fluxes like SiO_2, TiO_2, and Cr_2O_3 are applied on the surface

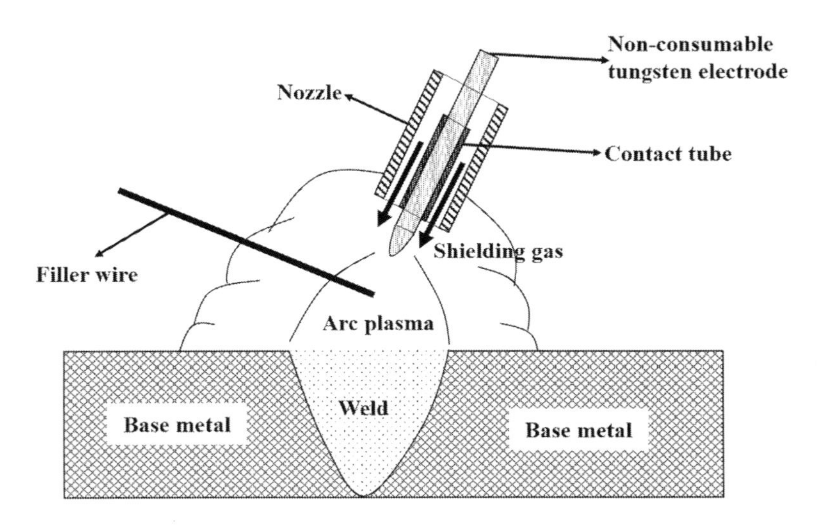

FIGURE 17.1 Schematic of the gas-tungsten arc welding process.

of the base plate to increase the depth of penetration (~5 mm) [57]. However, welding thicker plates can be completed by introducing multiple passes [58].

17.3.3 High-Energy Beam Welding

Laser beam welding (LBW) and electron beam welding (EBW) are high-energy beam welding processes [59–61]. These processes provide highly dense energy in a smaller area to melt the abutting base metal. When the accelerated electrons are focused on the surface of the metal, the heat is generated by conversion of its kinetic energy into the heat energy in EBW (Figure 17.2a), whereas in LBW, the extremely focused photons generate heat at narrow region (Figure 17.2b). Both electron EBW and LBW can provide a weld with a high depth to width ratio as they heat up a narrower region than the GTAW by keyhole welding phenomenon. Also, the production time of the weld joint is less in high energy density welding processes, as they can melt and/or vaporize the metals more rapidly than the arc welding processes. By these processes, the plates with higher thicknesses can also be welded in a single pass [61]. A shorter heating cycle results in a narrow heat affected zone and less distortion than the GTAW process. Since the EBW is carried out in vacuum conditions, highly reactive materials can be welded easily and effectively.

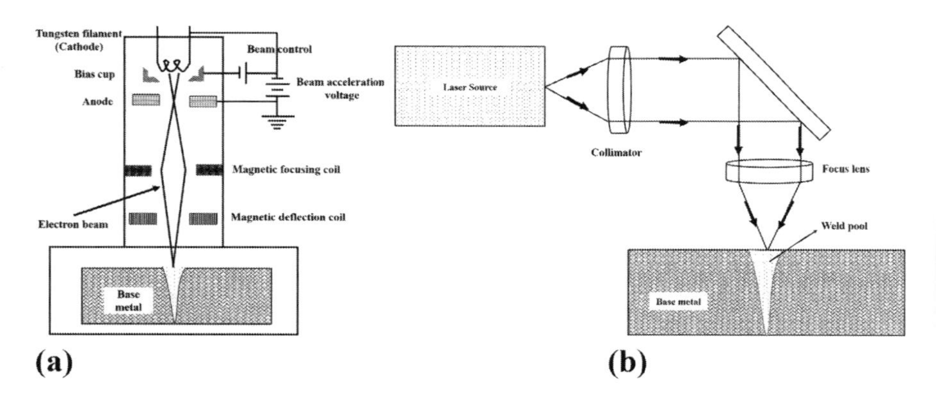

FIGURE 17.2 Schematic of (a) electron beam welding (EBW) and (b) laser beam welding (LBW) processes.

17.3.4 Friction Stir Welding

Friction stir welding (FSW) is one of the best solid-state welding processes, developed by the researchers in The Welding Institute (TWI), United Kingdom in 1991 [62]. A schematic diagram of the FSW is given in Figure 17.3 [63] in which a non-consumable tool is rotated along the locus, where the weld is to be made on the rigidly fixed plate.

The tool is specially designed with two parts: (a) shoulder and (b) pin. During welding, the pin is inserted into the plate and the shoulder is kept in touch with the surface of the plate around the pin. The pin region plays the main role in welding by

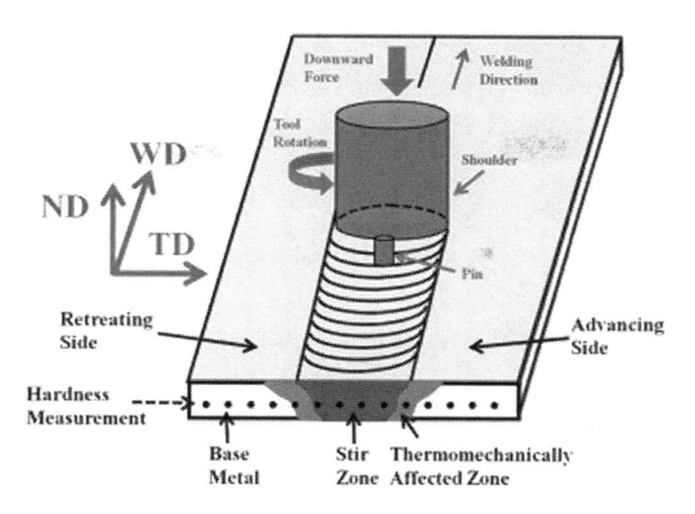

FIGURE 17.3 Schematic of friction stir welding (FSW) process [63]. (Reprinted from *Mater Sci Eng A*, 711, Zhu, Z.G., Sun, Y.F., Ng, F.L., et al., Friction-stir welding of a ductile high entropy alloy: microstructural evolution and weld strength, 524–32. Copyright (2018), with permission from Elsevier (License No. 4605310648768).)

producing about 75% of the heat input during welding from the friction between the rotating pin and the fixed base plate and exothermic heat during plastic deformation by stirring the softened material from one location to another. The shoulder part supports the welding by producing 25% of the heat input by friction between the surface of the shoulder and the plate as it is supported firmly by a fixed axial mechanical load. The shoulder also supports the welding by eliminating the pinhole formation in between the plates to ensure the soundness of the joint. This process can effectively produce the joint configurations of butt, lap, fillet, and "T," like the fusion welding processes. It is considered to be green technology as it does not require any shielding gas (like argon, helium, carbon dioxide, etc.) and consumes much less energy compared to conventional welding processes. In addition, FSW does not produce any harmful emissions as in arc welding and high-energy beam welding processes [64].

Like the fusion weldment, the friction stir welded sample also shows different microstructurally distinct regions: the stir zone (region near the pin), the thermomechanically affected zone (TMAZ), the heat affected zone (HAZ), and the unaffected base metal due to varying strain, temperature, and strain rates in the weldment from the pin portion to the base material. The metals near the rotating pin reach a temperature of around 0.8 Tm of the alloy and deform heavily; the locally softened material is extruded between the pin and the plate and forged in between the shoulder and the plate. The region where the pin is located before translation could produce fully recrystallized, fine, and equiaxed grains in the stir zone due to higher temperature and severe deformation, while the region next to the stir zone which experiences less heat input and deformation from the tool and will result in non-recrystallized, elongated, and thin grains is known as the thermomechanically affected zone (TMAZ).

17.4 WELDING OF CANTOR ALLOYS

So far, the equiatomic and non-equiatomic CoCrFeMnNi high entropy alloy (Cantor alloy) prepared in vacuum-arc melting [65, 66] and vacuum induction melting furnaces [67–70], and the non-equimolar Cantor type alloys [71] and carbon-doped Cantor alloy [72] prepared by self-propagating high-temperature synthesis have been studied for their weldability.

17.4.1 Base Materials Analysis

Wu and co-workers [65, 66] have prepared the Cantor alloy with a vacuum-arc melting (VAM) furnace with > 99.99% purity elemental granules. The as-cast alloy was drop casted and homogenized for 24 h at 1,200°C with subsequent air/water cooling. Then the billet was cold rolled with 86% thickness reduction and annealed at 900°C for 1 h (BM_{VAM}). Nam and co-workers [67, 68] and Jo and co-workers [69] have produced an equiatomic Cantor alloy, and Nene and co-workers [70] have produced a non-equiatomic ($Fe_{50}Mn_{30}Co_{10}Cr_{10}$ [atomic percent]) Cantor alloy using vacuum induction melting (VIM). The weldability of these alloys has been studied at the following initial conditions:

(a) Homogenized at 1,100°C for 24 h and subsequent air cooling (BM_H) (1.5 mm thick) [67];
(b) Homogenized at 1,100°C for 24 h and hot rolled with 98% thickness reduction from 145 mm to 3 mm with subsequent air cooling. Then the sample was cold rolled at 25°C with 50% thickness reduction (BM_R) [68];
(c) Homogenized at 1,100°C for 24 h and hot rolled with 87.5% thickness reduction from 16 mm to 2 mm with subsequent furnace cooling. Then the sample was cold rolled at 25°C with 25% thickness reduction and was annealed at 800°C for 1 h (BM_{RA}) (1.5 mm thick) [68]; and
(d) As-cast ingot was reduced 88% by hot rolling at 1,100°C and was annealed at 1,050°C for 1 h and quenched in water (BM_{HRA}) [69];
(e) As-cast non-equiatomic Cantor alloy was hot rolled at 900°C with 50% thickness reduction with subsequent homogenization at 1,200°C for 5 hours (BM_{NEC}) [70];
(f) Kashaev and co-workers have prepared the Cantor-type alloy (BM_{CT}) [71] and carbon doped Cantor alloy ($BM_{C\text{-}CT}$) [72] using self-propagating high-temperature synthesis (SHS) by using target element oxides with aluminum as the metal reducer.

17.4.1.1 Microstructure and Phase Analysis

The as-cast Cantor alloy prepared by the vacuum-arc melting (VAM) technique is composed of large, elongated grains extending from the edge to the center of the rectangular cross-section mold, parallel to the heat flow direction during solidification. Elongated grains are composed of the dendritic sub-structures, with the dendritic arm spacing (DAS) of ~15 μm [65]. The dendrite core is enriched with cobalt, chromium, and iron, while the interdendritic region is enriched with manganese and nickel. Manganese has undergone severe segregation during casting in the interdendritic region, where the manganese fraction in dendrite and interdendrite is reported as ~16 atomic percentage and ~26 atomic percentage, respectively. However, the segregation of nickel was not as high as manganese; it has shown only 4 at.% difference to manganese (10%). Some chromium- and manganese-rich oxides observed in the as-cast specimens could be due to the presence of oxygen in the raw materials themselves and/or contamination of melt pool due to insufficient shielding in vacuum-arc melting. Thermomechanical processing of the as-cast alloy has yielded the equiaxed grains in BM_{VAM} with an average grain size of ~30 μm. Many grains possess the annealing twins with micrometer range twin thickness. BM_{VAM} shows the homogeneous distribution of alloying elements [65].

The microstructure of the Cantor alloy prepared by vacuum induction melting and subsequent thermomechanical treatments is given in Figure 17.4a–d. The Cantor alloy has shown coarse equiaxed grains of size ~1 mm in BM_H condition [67]. While the BM_R condition has revealed a severely deformed microstructure with very fine grains of ~2 μm in size [68]. At BM_{HRA} condition, the alloy shows several annealing twins. The grain size is ~70 μm [38]. At BM_{RA} condition, the fine equiaxed grains with obvious annealing twins (~3.3 ± 0.3 μm) are produced due to heavy deformation and subsequent recrystallization during cold rolling and subsequent heat treatment [67]. Sometimes large non-metallic inclusions can be visible, as it is inevitable in a vacuum induction melting process [68]. In all four cases, the Cantor alloy exhibited a single FCC phase [67–69]. The BM_{NEC} alloy has shown a coarse microstructure of about 100 μm with two phases (87% hcp and 13% FCC) [70].

In BM_{CT}, columnar dendritic grains with too many manganese sulfide (MnS) inclusions are noticed, as seen in Figure 17.4e. The columnar grains are of major and minor axis length of ~285 μm and ~98 μm, respectively, with an aspect ratio ~0.3 with mean misorientation angle 35.6°. The actual composition of the alloy is measured to be ~18–24 % of cobalt, chromium, iron, and nickel, ~11.03% of manganese, ~2.64% of aluminum, and ~0.16–0.23% of sulfur and silicon. Transmission electron microscopy (TEM) analysis of the BM_{CT} reveals the continuous chain of rectangular $M_{23}C_6$ particles (most likely $Cr_{23}C_6$) with average length and width of 130 nm and 50 nm, respectively, along the grain boundaries [71]. $BM_{C\text{-}CT}$ is cold rolled to 75% thickness reduction with subsequent annealing (900°C for 1 h). It exhibits equiaxed grains with annealing twins of grain size ~ 9 μm (Figure 17.4f). Transmission electron microscopy (TEM) analysis indicates the presence of 2 volume percentage of $M_{23}C_6$ particles of average size ~138 nm in FCC matrix (inset of Figure 17.4f) [72].

17.4.1.2 Mechanical Properties

The tensile properties of the BM_{VAM} are given in Table 17.2. BM_{VAM} exhibits a good combination of strength and ductility. Its strength and ductility increase simultaneously at the cryogenic temperature (77 K). In the case of the vacuum induction melting processed alloy, the rolling has increased its tensile strength considerably (BM_R) at the expense of ductility. Recrystallization has recovered ductility with some drop in tensile strength in BM_{RA} and BM_{HRA}. Even though heat treatment after the rolling decreased the strength more than BM_R, it still retains higher strength compared

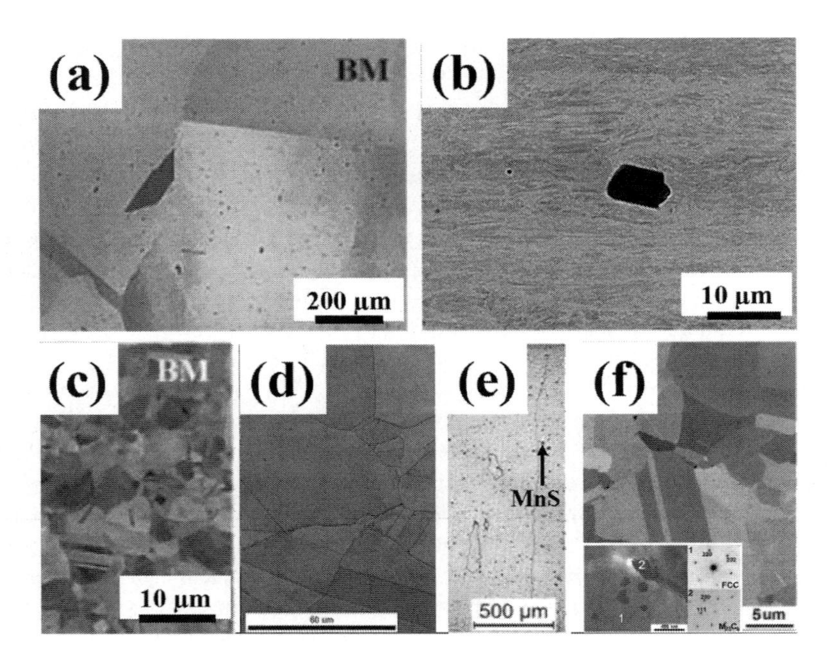

FIGURE 17.4 The base metal microstructure of the Cantor alloy at (a) BM_H [67], (b) BM_R [68], (c) BM_{RA} [67], (d) BM_{HRA} [69], (e) BM_{CT} [71], and (f) $BM_{C\text{-}CT}$ [72]. (Sources: [67] Reprinted from *Mater Sci Eng A, 742,* Nam, H., Park, C., Moon, J., Na, Y., Kim, H. and N. Kang, Laser weldability of cast and rolled high-entropy alloys for cryogenic applications, 224–30. Copyright (2019), with permission from Elsevier (License No. 4605300684732); [68] Reprinted from *Sci Technol Weld Join*, 23, Nam, H., Park, C., Kim, C., Kim, H. and N. Kang, Effect of post weld heat treatment on weldability of high entropy alloy welds, 420–27. Copyright (2018) with Permission from Taylor & Francis (License No. 4605880643136); [69] Reprinted from *Met Mater Int, 24,* Jo, M.G., Kim, H.J., Kang, M. et al., Microstructure and mechanical properties of friction stir welded and laser welded high entropy alloy CrMnFeCoNi, 73–83. Copyright (2018) with permission from Springer Nature (License No. 4594820609008); [71] Reprinted from *Intermetallics*, 96, Kashaev, N., Ventzke, V., Stepanov, N., Shaysultanov, D., Sanin, V. and S. Zherebtsov, Laser beam welding of a CoCrFeNiMn-type high entropy alloy produced by self-propagating high-temperature synthesis, 63–71. Copyright (2018) with permission from Elsevier (License No. 4594840879681); and [72] Reprinted from *Mater Charact*, 145, Shaysultanov, D., Stepanov, N., Malopheyev, S., et al., Friction stir welding of a carbon-doped CoCrFeNiMn high-entropy alloy, 353–61. Copyright (2018) with permission from Elsevier (License No. 4594811397480).)

to the coarser-grained BM_H with improved ductility. It could be attributed to the smaller grain size of BM_{RA} and BM_{HRA}. The alloy deformed at 77 K has shown the presence of Σ3 parallel twin bands [65, 66]. Such formation of twin bands reduces the effective grain size dynamically with time during straining and enhances both strength and ductility through the interaction between the twins and dislocation by the dynamic Hall–Petch effect. Twin bundles formed during deformation have significantly enhanced mechanical properties such as tensile strength and ductility by increasing the strain hardening capabilities and postponing necking activity.

TABLE 17.2
The Tensile Properties of Base Metals and Welds

S. No	Test sample conditions	Testing temperature [K]	YS [MPa]	UTS [MPa]	Elongation [%]	References
1	BM_{VAM}	298	273	633	38	[65]
2	BM_{VAM}	77	481	1,095	59	[65]
3	GTAWed BM_{VAM}	298	297	530	15	[65]
4	GTAWed BM_{VAM}	77	510	880	33	[65]
5	EBWed BM_{VAM}	298	320	617	27	[65]
6	EBWed BM_{VAM}	77	567	1,057	39	[65]
7	BM_{H}	298	188	363	54	[67]
8	BM_{R}	298	–	1,134	8.5	[65]
9	BM_{HRA}	298	345	570	54	[64]
10	BM_{RA}	298	442	693	51	[67]
11	BM_{H} LBWed at traverse speed 6 m/min	298	188	345	45.5	[67]
12	BM_{H} LBWed at traverse speed 8 m/min	298	188	378	36.5	[67]
13	BM_{H} LBWed at traverse speed 10 m/min	298	170	377	36.3	[67]
14	BM_{H} LBWed at traverse speed 10 m/min	77	334	607	65	[67]
15	BM_{R} LBWed at traverse speed 7 m/min	298	–	676	3	[68]
16	BM_{R} LBWed at traverse speed 8 m/min	298	–	638	2.8	[68]
17	BM_{R} LBWed at traverse speed 9 m/min	298	–	665	2.2	[68]
18	BM_{HRA} LBWed at traverse speed 9 m/min	298	337	539	38	[64]
19	BM_{RA} LBWed at traverse speed 6 m/min	298	412	585	18.3	[67]
20	BM_{RA} LBWed at traverse speed 8 m/min	298	442	605	14.7	[67]
21	BM_{RA} LBWed at traverse speed 10 m/min	298	432	616	26.5	[67]
22	BM_{RA} LBWed at traverse speed 10 m/min	77	646	1,013	33.5	[67]
23	BM_{R} heat treated at 1,000°C	298	118	421.77	57.75	[68]
24	BM_{R} heat treated at 900°C	298	213	547.90	53.72	[68]
25	BM_{R} heat treated at 800°C	298	314	605.58	42.7	[68]
26	BM_{R} LBWed at traverse speed 8 m/min and subsequent PWHT at 1000°C	298	79	403.47	65.62	[68]

(*Continued*)

TABLE 17.2 (CONTINUED)
The Tensile Properties of Base Metals and Welds

S. No	Test sample conditions	Testing temperature [K]	YS [MPa]	UTS [MPa]	Elongation [%]	References
27	BM_R LBWed at traverse speed 8 m/min and subsequent PWHT at 900°C	298	213	517.77	52.15	[68]
28	BM_R LBWed at traverse speed 8 m/min and subsequent PWHT at 800°C	298	301	554.12	38.5	[68]
29	FSWed BM_{HRA}	298	368.8	562	44	[69]
30	BM_{NEC} alloy	298	–	850*	37*	[70]
31	BM_{NEC} alloy FSPed at 350 rpm	298	–	1400*	45*	[70]
32	BM_{NEC} alloy FSPed at 650 rpm	298	–	1200*	42*	[70]
33	$BM_{C\text{-}CT}$ tested ‖ to RD	298	330	683	70	[72]
34	$BM_{C\text{-}CT}$ tested ⊥ to RD	298	290	630	65	[72]
35	FSWed $BM_{C\text{-}CT}$ tested along the seam (⊥ to RD)	298	490	713	52	[72]
36	FSWed $BM_{C\text{-}CT}$ tested across the seam (‖ to RD)	298	397	698	31	[72]

*Indicates true stress and true strain, while the others are engineering stress and engineering strain, respectively.

17.4.2 Fusion Welding of Cantor Alloys

17.4.2.1 Arc Welding of Cantor Alloys

The Cantor alloy of 1.6 mm thickness was gas-tungsten arc welded with welding parameters; 8.4 V, 75 A at the torch traverse speed 25.4 mm min^{-1} with full penetration [65].

17.4.2.1.1 Microstructure and Phase Analysis

The solidified weld has coarse columnar grains with dendritic sub-structures as revealed in the as-cast condition. The weldment is free from solidification cracks and heat affected zone (HAZ) cracks. The grains in the weld have grown epitaxially from the partially melted grains of the base material. Since the traverse speed is slow, the resulted weld pool shape is elliptical. The grains growing from the fusion line are curving toward the maximum thermal gradient and attain the crystal orientation like the associated grains in the base metal. As the preferred growth direction in the FCC, as well as in BCC alloy, is <100> direction, during initial grain growth, the grains oriented in an easy growth direction could grow faster and longer by competitive growth phenomenon. Even though the columnar grains are oriented in the preferred grain growth direction at the initial stage of solidification, as over the length scale, growth of some grains, that lagging optimal alignment with the thermal gradient of the weld will be stopped and the adjacent grain that optimally aligned will grow further. Hence, over the length from the fusion line to the center of the weld, the grain orientation is changing continuously. Like the as-cast alloy, the cobalt-chromium-iron (Co-Cr-Fe)-rich dendrites and manganese-nickel (Mn-Ni) rich interdendrites are also present in the weld. However, gas-tungsten arc welding (GTAW) has resulted in finer dendritic (less dendritic arm spacing) and homogeneous (less manganese-nickel [Mn-Ni] segregation) microstructure structure compared to the as-cast Cantor alloy due to reduced diffusion distances and increased nucleation rates during rapid cooling of the weld pool. The composition difference between the dendrites and interdendrites is observed to be very small (i.e., ~4% for manganese and ~3% for the rest of the elements) [65].

17.4.2.1.2 Mechanical Properties

Gas-tungsten arc welding has resulted in a 20% reduction in tensile strength and a 50% reduction in its ductility (Table 17.2). This reduction in tensile strength is due to the coarse-grained structure in the weld zone. The tensile fracture has occurred at the weld center. In the tensile deformed gas-tungsten arc welded sample, twin bands are observed. Even though the twin bands are formed in grains during deformation, the reduced fraction of the grain boundary for pinning the dislocation movement in the coarse structure lowers the strength by the Hall–Petch effect. Coarser dimples in the fractured weld than in the base metal suggests that the lower ductility is with the formation of voids filled up with some oxide inclusions formed due to the influence of the environment. However, like the BM_{VAM}, the gas-tungsten arc welded sample also shows an increase in tensile strength when tested at a cryogenic temperature [65].

17.4.2.2 High-Energy Beam Welding of Cantor Alloys

17.4.2.2.1 Electron Beam Welding

The electron beam welding was carried out at 125 kV at two different weld currents and a traverse speed of 2.2 mA and 0.63 mm s^{-1} [65], and 5 mA and 9.53 mm s^{-1} [66], respectively, on ~1.5 mm Cantor plate in BM_{VAM}.

17.4.2.2.1.1 Microstructure and Phase Analysis Back-scattered electron images of BM_{VAM}, top, and transverse microstructure of the electron beam weld are given in Figure 17.5a–c. Like in the gas-tungsten arc weld, coarse columnar grains

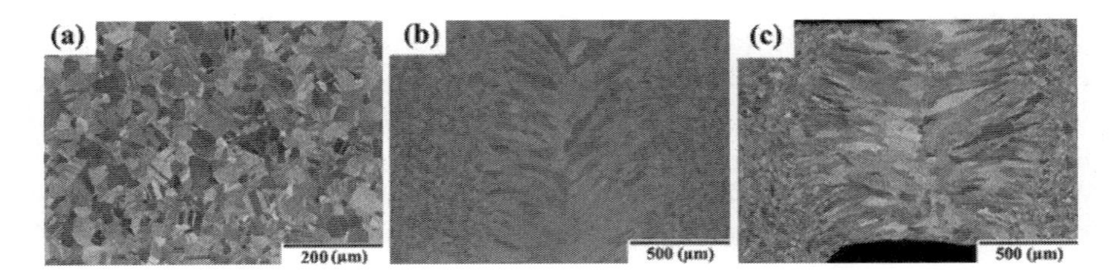

FIGURE 17.5 Back-scattered electron images of (a) BM_{VAM}, (b) top, and (c) transverse microstructure of the electron beam welded Cantor alloy [66]. (Reprinted from Scripta Mater, 124, Wu, Z., David, S. A., Feng, Z. and H. Bei, Weldability of a high entropy CrMnFeCoNi alloy, 81–85. Copyright (2016), with permission from Elsevier (License No. 4611230148044).)

epitaxially grown from partially melted equiaxed grains in the fusion line toward the center of the weld are also observed in electron beam welds. The electron beam weld is also free from solidification cracks and heat affected zone (HAZ) cracks. However, in the electron beam weld, the elongated grains have a random orientation. The average fraction of manganese in the electron beam welded alloy is far below than that in the base metal (20 at.%). The manganese segregated interdendrites are showing only about 16.8 at.%. Possibly, manganese with a lower melting point and higher vapor pressure is evaporated when it is scanned with a high energy electron beam. However, like an alloy in as-cast and gas-tungsten arc welded condition, the electron beam weld also shows segregation of manganese and nickel in interdendritic regions [65]. Compared to gas-tungsten arc weld, in electron beam weld, the variation in the composition between the dendrite and interdendrites is very small. The difference between the elements in different regions is about 3 at.% for manganese, while the rest of the elements varies only by ~2 at.%.

17.4.2.2.1.2 Mechanical Properties Electron beam weld has shown only meagre variation in strength, like BM_{VAM} with 30% reduction in ductility (Table 17.2). This reduction in tensile strength is due to the coarser grain size in the weld zone. Despite the presence of twin bands, grain coarsening has caused a reduction in strength, as in the gas-tungsten arc weld. Tensile specimens failed at the weld fusion zone. The finer grain size of the weld with limited microsegregation has resulted in higher strength in electron beam weld than in the gas-tungsten arc weld. Dimples in the fracture surface are coarser than the base metal and finer than the gas-tungsten arc weld, also very few microvoids having oxide inclusions has led to higher ductility than the gas-tungsten arc weld. However, like BM_{VAM} and gas-tungsten arc weld, the electron beam weld also revealed enhanced strength at cryogenic temperature compared to at room temperature [65, 66].

17.4.2.2.2 Laser Beam Welding

17.4.2.2.2.1 Microstructure and Phase Analysis The 1.5 mm thick Cantor alloy (BM_R, $BM_{H,}$ and BM_{RA}) welded with 3.5 kW laser power with varying traverse speed (5–10 m/min) have shown an "X"-shaped weld bead for full penetration [67]. Figure 17.6 shows that bead width decreases with an increase in travel speed in BM_R. In all the welds, the bead width at the top face is wider than at the root. The bead morphology changes from an "X" shape to an "I" shape and finally to a "U" shape with the increase in travel speed, and also the depth of penetration is reduced from full penetration to partial penetration. Weld bead formation in the BM_H and BM_{RA} have also followed a similar trend with the welding speed [68], suggesting that the change in bead morphology with travel speed in laser welding is irrespective of the base metal condition. Jo and co-workers have also welded a 1 mm thick BM_{HRA} plate with 3 kW laser power at welding speeds 8 and 9 m/min and 3.5 kW power and 9 m/min welding speed combinations with argon gas shielding at both sides of the plate [69]. The latter combination has resulted in full penetration and an "I2-shaped weld as shown in Figure 17.7a. Kashaev and co-workers [71] have welded a 2 mm thick plate of BM_{CT} alloy with 2 kW laser power at a welding speed of 3 and 6 mm/min. BM_{CT} also shows an "X"-shaped weld bead (Figure 17.7b).

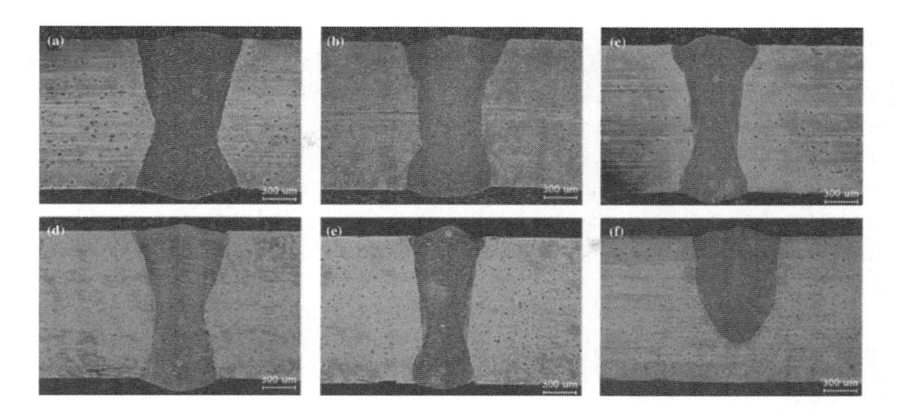

FIGURE 17.6 Laser weld produced in BM_R at traverse speed: (a) 5 m/min^{-1}, (b) 6 m/min^{-1}, (c) 7 m/min^{-1}, (d) 8 m/min^{-1}, (e) 9 m/min^{-1}, and (f) 10 m/min^{-1} [68]. (Reprinted from *Sci Technol Weld Join*, 23, Nam, H., Park, C., Kim, C., Kim, H. and N. Kang, Effect of post weld heat treatment on weldability of high entropy alloy welds, 420–27. Copyright (2018) with permission from Taylor & Francis (License No. 4605880643136).)

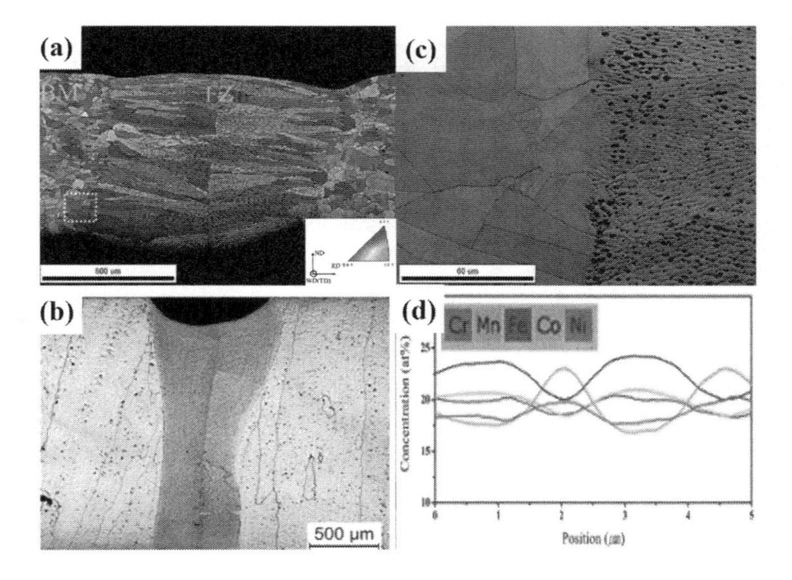

FIGURE 17.7 The weld cross-section of laser beam welded (a) BM_{HRA} [69], (b) BM_{CT} [71], (c) showing epitaxial growth of the columnar grains from partially melted base metal [69], and (d) microsegregation analysis across the dendrites near the fusion line [69]. (Sources: [69] Reprinted from *Met Mater Int, 24,* Jo, M.G., Kim, H.J., Kang, M. et al., Microstructure and mechanical properties of friction stir welded and laser welded high entropy alloy CrMnFeCoNi, 73–83. Copyright (2018) with permission from Springer Nature (License No. 4594820609008); and [71] Reprinted from *Intermetallics*, 96, Kashaev, N., Ventzke, V., Stepanov, N., Shaysultanov, D., Sanin, V. and S. Zherebtsov, Laser beam welding of a CoCrFeNiMn-type high entropy alloy produced by self-propagating high-temperature synthesis, 63–71. Copyright (2018) with permission from Elsevier (License No. 4594840879681).)

However, the weld is showing underfill due to the vaporization of some metallic materials, as observed in the welding plume during laser welding. The welding at a lower speed (3 m/min) has resulted in a wider weld bead, while at higher speed, the welding has resulted in partial penetration [68]. Due to columnar grain structure and its inclined orientation, the weld bead has shown an unsymmetrical shape as the weld grains are formed by epitaxial growth and the difference in heat input at the top (radiation exposure side), middle, and bottom (weld root) regions. The weld fabricated at 5 m/min has a nearly symmetrical and "X"-shaped weld with full penetration.

The welds prepared from the BMs, BM_H, BM_{RA}, and BM_{HRA} does not have any significant variation in grain size in the heat affected zone (HAZ) and base metal (BM). While in BM_R-based weld, the grains in the heat affected zone are coarsened significantly; about double the size of base metal grains. Also, equiaxed grains are formed in the heat affected zone, where the base metal is influenced by heat from laser melting, leading to recrystallization of heavily deformed grains. In all the conditions, the columnar grains with dendritic sub-structures have grown epitaxially from the heat affected zone at the weld fusion line. Examples of epitaxial growth in these alloys are presented in Figure 17.7a–c. The microsegregation analysis near the fusion line (Figure 17.7d) and at the weld center have shown that the interdendrites are rich in manganese and the dendrites are rich in iron, as observed in the as-cast alloy by Cantor [23]. The dendritic arm spacing (DAS), and the dendritic packet size (DPS) measured at the center of the weld are found to be refined with an increase in welding speed [67, 68]. The reduction in dendritic arm spacing and dendritic packet size with increased welding speed is compared to the faster cooling rate of the weld pool associated with higher speed. The dendritic arm spacings in welds made from BM_H, BM_{RA}, and BM_R are measured to be ~4.5 μm, 3.4 μm, and 4 μm, respectively. The variations in dendritic arm spacing and dendritic packet size values are decided by the size of the grain from where the columnar grains are formed at the fusion zone.

The weld defects such as internal pores, cracks, undercut, humping, and melt-through are insignificant at the macro level in all the welds. When the welds are evaluated at higher magnification, the presence of shrinkage voids [68] and intermetallic compounds inclusion [68–70] are reported. The shrinkage voids are observed at the interdendrites in the weld center, where the weld metal will solidify last. The intermetallic compound formed is identified as oxides of chromium–manganese [68] and manganese sulfides/oxides [71], which commonly occur in the as-cast and the thermomechanically processed Cantor alloy. The intermetallic compounds are very fine, with an average size of 1 μm, and are evenly distributed in the weld fusion zone. The volume fraction of these inclusions is estimated to be around 1%. Even though the base metal/heat affected zone have shown bigger inclusions in base metals and heat affected zones (~7/7 μm in BM_H, ~5/5 μm in BM_{RA}, and ~6/4 μm in BM_R), the laser beam melting has refined these inclusions and distributed evenly in the weld bead. A representative image showing the variation of the inclusion at different regions in BM_H and BM_{RA} is given in Figure 17.8. Other than shrinkage voids and inclusion, no major cracks or pores are found in the weld.

The laser welding of the high entropy alloys has not resulted in any phase transformation, like in stainless steels [73, 74] and in intermetallic compound formations,

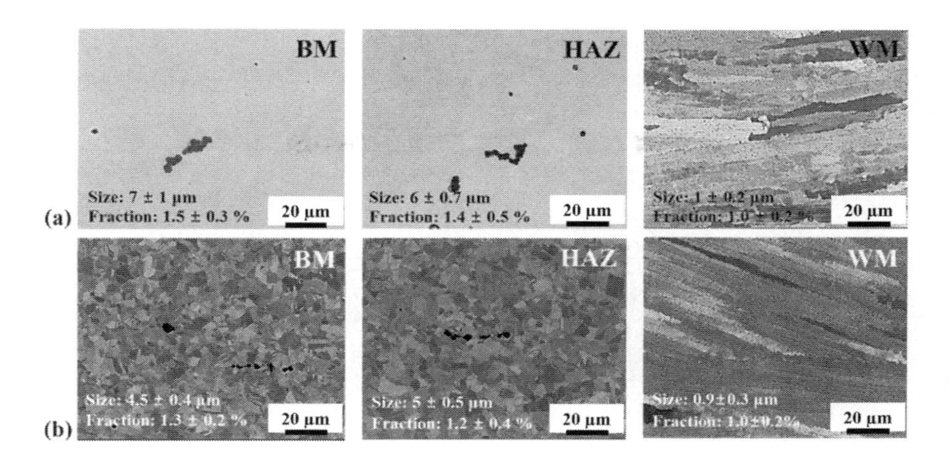

FIGURE 17.8 Size and volume fraction of the inclusions distributed in different regions of the laser beam welded (a) BM_H and (b) BM_{RA} [67]. (Reprinted from *Mater Sci Eng A, 742,* Nam, H., Park, C., Moon, J., Na, Y., Kim, H. and N. Kang, Laser weldability of cast and rolled high-entropy alloys for cryogenic applications, 224–30. Copyright (2019), with permission from Elsevier (License No. 4605300684732).)

like the Laves phase formation in Inconel 718 alloys [49,75] after welding. The solidified weld has shown a single FCC phase like the base metal [67]. Transmission electron microscopy (TEM) analysis has been carried out by Jo and co-workers [69] to study the compositional fluctuation-driven phase transformation in dendrites and interdendrites. Both regions have shown the diffraction patterns of FCC crystal structures, indicating that the laser beam welding has not created any phase transformation even at nano-level. However, the TEM analysis of the laser weld of BM_{CT} (Figure 17.9) has shown that the matrix of the solid solution is having two different precipitates. Selected-area electron diffraction (SAED) analysis of these particles has shown the particles as the coarse spherical shaped $M_{23}C_6$ carbides (diameter ~110 nm) and the very fine B2 structure (~5 nm). It also shows that the regions near the $M_{23}C_6$ particles are free from B2 structure distribution [72].

17.4.2.2.2.2 Mechanical Properties

17.4.2.2.2.2.1 Microhardness The fusion zone records a higher hardness than the corresponding base metals in BM_H, BM_{RA}, and BM_{HRA} welds (Table 17.3). However, the variation in hardness with respect to the base metal is more significant in the BM_H and BM_{HRA} than that of the BM_{RA}. The increase in hardness at fusion zone (FZ) is connected to the grain size of base metal and dendritic arm spacing of the weld metal. Since the difference in the dendritic arm spacing (2.9 µm) of weld and base metal (3.3 µm) is lower in BM_{RA} than that in BM_H (200 times), the weld made from BM_H shows huge variation in hardness compared to that from BM_{RA}. Also, since the weld made from BM_{RA} has shown narrower dendritic arm spacing than that from BM_H, the weld in the rolled sample is showing higher hardness than the BM_H. The local variation in dendrites and interdendrites also could lead to hardness

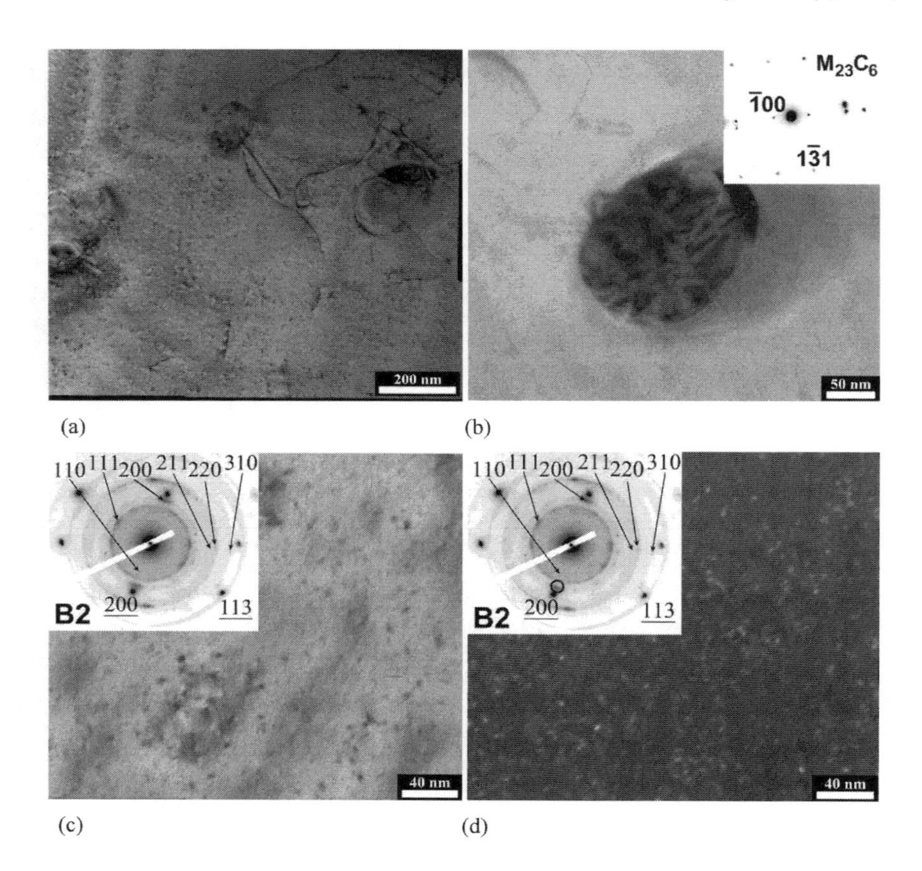

FIGURE 17.9 (a–c) Transmission electron microscopy (bright-field images) of fusion zone showing coarse $M_{23}C_6$ carbides and fine B2 structured precipitates, and (d) Transmission electron microscopy (dark-field images) of fusion zone with fine B2 structured precipitates. Insets in (b–d) indicate selected-area electron diffraction (SAED) patterns to carbides and B2 structures, respectively [71]. (Reprinted from *Intermetallics*, 96, Kashaev, N., Ventzke, V., Stepanov, N., Shaysultanov, D., Sanin, V. and S. Zherebtsov, Laser beam welding of a CoCrFeNiMn-type high entropy alloy produced by self-propagating high-temperature synthesis, 63–71. Copyright (2018) with permission from Elsevier (License No. 4594840879681).)

TABLE 17.3
Variation of Microhardness after Laser Welding

S. No	Sample conditions	Base metal hardness [HV]	Weld metal hardness [HV]	Reference
1	Laser beam welded BM_H	125	173	[67]
2	Laser beam welded BM_{RA}	177	185	[67]
3	Laser beam welded BM_R	321	189	[68]
4	Laser beam welded BM_{HRA}	140	190	[69]
5	Laser beam welded BM_{CT}	150	210	[70]

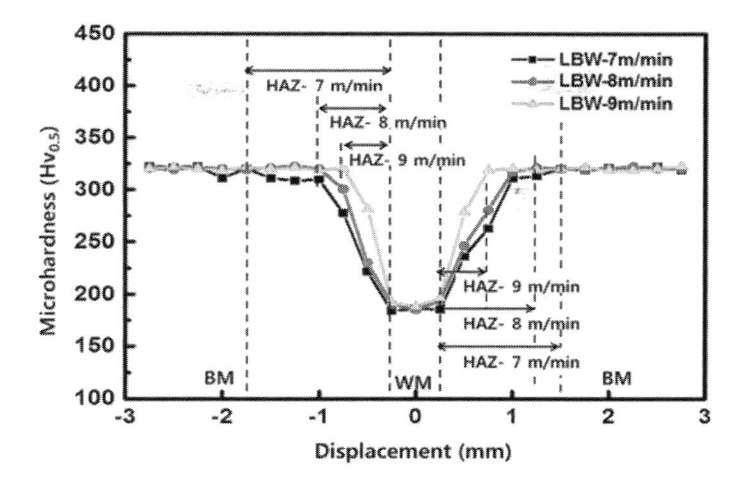

FIGURE 17.10 Microhardness survey across the laser welded BM_R weldment (at different welding speeds) [68]. (Reprinted from *Sci Technol Weld Join*, 23, Nam, H., Park, C., Kim, C., Kim, H. and N. Kang, Effect of post weld heat treatment on weldability of high entropy alloy welds, 420–27. Copyright (2018) with permission from Taylor & Francis (License No. 4605880643136).)

improvement by 3–4%. Also, the even distribution of chromium-manganese (Cr-Mn) oxide/manganese sulfide/oxide (Mn-S/O) inclusion in the weld could also enhance the hardness in the weld more than in base metal, while in case of the laser welded BM_R, the weld is showing lower hardness than the base metal. The hardness of the base metal is also found to be higher than that of other conditions. It could be due to the deformed fine grain size of the BM_R and the internal stress in the base metal.

In the case of BM_R weld, the base metal exhibits higher hardness than that of the fusion zone (Figure 17.10, Table 17.3). Since the welding has resulted in the recrystallization of the deformed grains and removal of the dislocations formed during the rolling by localized heating and re-solidification, the fusion zone is showing lower hardness. Also, the grain size of the base metal (~2 μm) is lower than the dendritic arm spacing of the weld (~4 μm). Hence, the hardness of the base metal is higher than the weld. The medium hardness in the heat affected zone is attributed to the partial recrystallization. The width of the heat affected zone is increased when the welding speed is lower and this is due to increased heat dissipation from the weld metal. The hardness is high in the heat affected zone and the fusion zone of the weld compared to the BM_{CT}. The average hardness in the base metal and the fusion zone are found to be ~153 $HV_{0.5}$ and 208 $HV_{0.5}$, respectively. The fine and even distribution of $M_{23}C_6$ particles and nanoscale B2 precipitates in fusion zone have improved the strength in the weld.

17.4.2.2.2.2.2 Tensile Properties The tensile properties of the laser welds made from different base metal conditions are given in Table 17.2. The welds made from the BM_H have shown similar strength, like the BM_H with 20–30% drop in ductility at all weld speeds. The fracture location in BM_H is found to be either in the heat affected zone or base metal because of their lower hardness, larger grain size, or presence of larger size inclusions. However, the welds made from BM_{RA} and BM_R have shown deterioration in both strength and ductility. Also, both tensile samples failed in the weld metal for all welding speeds due to large dendritic packet size in the weld than the grain size of the BM. Deterioration in strength and ductility is very high in BM_R sample where the residual stresses formed during rolling might have initiated and fastened the cracking tendency during tensile testing. The increase in tensile properties with increasing speed can be attributed to the reduced dendritic packet size and volume fraction of the shrinkage voids. The weld from BM_{HRA} has a strength equal to that of base metal. This is possibly due to a reduction in the inclusion fraction and residual stress by hot rolling and subsequent annealing treatment.

The tensile testing at 77 K has shown improved strength and ductility compared to that measured at the 298 K in both conditions. Though twins are observed at the crack tip in BM_H at 298 K, at 77 K they are present in larger amounts with tangled dislocations. In BM_{RA}, the fraction of deformation twins is less than that of BM_H at the room temperature, while at 77 K, a large fraction of deformation twins and tangled dislocations are formed. The larger grain size of BM_H has caused a reduction in the critical stress for twin formation, which has led to the greater reduction in strength than in BM_{RA} by forming twins and dislocation tangles.

Both BM_H and BM_{RA} fractured in ductile mode ascertained by the presence of dimples. When the fracture position is at the heat affected zone/base metal in BM_H, the dimple size is larger, which corresponds to the larger grain size at the fracture location. Similarly, the BM_{RA} exhibits the finer dimple size because of the smaller dendritic packet size at the weld as it fractured. However, at 77 K, the BM_H has revealed the combination of the finer and coarser dimples, and the BM_{RA} have also shown finer dimples than that tested at room temperature due to the formation of the deformation twins and tangled dislocations during the deformation.

17.4.2.2.2.3 Effect of the Post-Weld Heat Treatment Tensile strength and the ductility of the BM_R weld are lowered from the respective base metal. The post-weld heat treatment (PWHT) is one phenomenon which regains the strength of the weld. Since the Cantor alloy could exhibit a single FCC phase in the range of 670–1,280°C, the welds are post-weld heat treated at 800°, 900°, and 1,000°C [68].

17.4.2.2.2.3.1 Microstructure No cracks are observed in the heat affected zone/weld metal after the post-weld heat treatment. PWHT at 800°C has resulted in grain coarsening in the heat affected zone and base metal (Figure 17.11a), while the dendritic packet size (DPS) of the fusion zone remains unaltered. Fine fragmented elongated grains (~2 μm) in base metal along the rolling direction are transformed into coarser equiaxed grains with some annealing twins (~7 μm). Also, the heat affected zone is showing equiaxed grains with annealing twins; however, it has experienced a lower coarsening rate than the base metal and attained ~7 μm similar to that of base metal from ~4 μm. When post-weld heat treated at 1,000°C, the grains in base metal coarsened up to 20 μm, while the grains in the heat affected zone grew to 19 μm. Also, the dendritic packet size of weld metal has increased from 30 to 35 μm. The

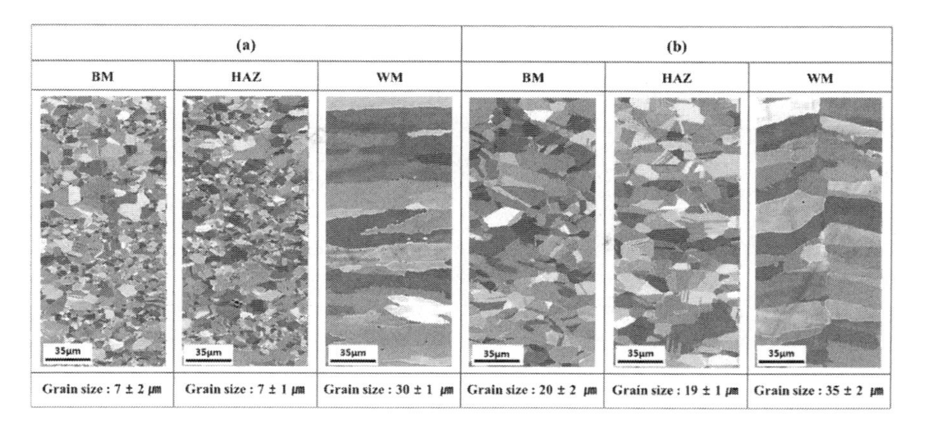

FIGURE 17.11 Electron back-scattered diffraction (EBSD) analysis of various regions of BM_R post-weld heat treated at (a) 800°C and (b) 1,000°C, respectively [68]. (Reprinted from *Sci Technol Weld Join*, 23, Nam, H., Park, C., Kim, C., Kim, H., and N. Kang, Effect of post weld heat treatment on weldability of high entropy alloy welds, 420–27. Copyright (2018) with permission from Taylor & Francis (License No. 4605880643136).)

rolled alloy microstructure could have more residual stresses and dislocation densities in it, hence the application of the heat treatment has stimulated the recrystallization and the grain growth even at a lower temperature like 800°C. However, since the weld metal has already undergone recrystallization, it does not respond to 800°C treatment. But with the increase in temperature to 1000°C, the dendritic packet size is coarsened to some extent, while in the base metal and heat affected zone, the effect is noticeable, which resulted in nearly the same grain sizes (Figure 17.11b). X-ray diffraction patterns of the post-weld heat-treated samples confirm the absence of the phase transformation by showing single FCC peaks (Figure 17.12a).

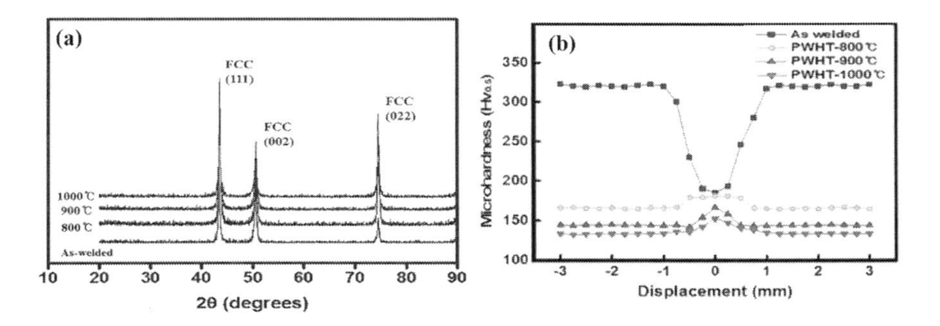

FIGURE 17.12 (a) X-ray diffraction (XRD) analysis of as-laser welded BM_R and post-weld heat treated samples and (b) microhardness survey across the welds at various post-weld heat treatments [68]. (Reprinted from *Sci Technol Weld Join*, 23, Nam, H., Park, C., Kim, C., Kim, H. and N. Kang, Effect of post weld heat treatment on weldability of high entropy alloy welds, 420–27. Copyright (2018) with permission from Taylor & Francis (License No. 4605880643136).)

17.4.2.2.2.3.2 Mechanical Properties Post-weld heat treatment at 800°C results in a reduction in hardness of the base metal and heat affected zone, while the weld metal is showing a similar hardness compared to the as-welded sample (Figure 17.12b). An increase in the post-weld heat treatment temperature resulted in a reduction in the hardness values of all regions—base metal, heat affected zone, and fusion zone—and is shown in Figure 17.12(b). This trend is attributed to the grain coarsening during the post-weld heat treatment. The net effect of post-weld heat treatment is an improvement in hardness than the BM_R like the welds made from BM_H and BM_{RA} conditions.

Post-weld heat treatment has significantly increased the ductility of the weld, while the strength remained the same as in the as-welded sample. The ductility increased from 3–4% in the as-welded samples to 40%, 55%, and 66%, respectively (Table 17.2). The ductility of the post-weld heat-treated BM_R weld sample is showing better ductility than the as-welded BM_{RA}.

17.4.3 Solid-State Welding of Cantor Alloys

17.4.3.1 Friction Stir Welding

The BM_{HRA}, BM_{NEC}, and $BM_{C\text{-}CT}$ of 2 mm thickness was friction stir welded with the process parameters as in Table 17.4. Figure 17.13a and Figure 17.14a illustrate the macrostructure of the friction stir welded BM_{HRA} and $BM_{C\text{-}CT}$ weldments, respectively. The single side welded BM_{HRA} exhibits a basin shaped weld, while a sand glass shape is formed in the double side welded $BM_{C\text{-}CT}$. Both welds have not shown any volumetric defects; however, in $BM_{C\text{-}CT}$, white irregular-shaped patterns at the stir zone in the advancing side on both sides of the plate are noticed.

17.4.3.1.1 Microstructure

Grain refinement has been noticed in friction stir welded BM_{HRA} in the region between the –2 mm and 2 mm from the weld center from about 70 μm in base metal to 5 μm due to dynamic recrystallization (Figure 17.13b–j). The recrystallization temperature of the Cantor alloy is reported to be $0.62T_m$ [76]. Generally, the materials with low stacking fault energy like stainless steel could experience more than $0.8T_m$ under the rotating tool, which can lead to dynamic recrystallization in that region [78]. Dwelling time of the tool being very short at a region during the friction stir welding process, the dynamically recrystallized deformed fine microstructure can never coarsen [72]. As the Cantor alloys are of low stacking fault energy, they exhibited higher grain coarsening resistance even up to 800°C [76, 77, 78]. Hence the friction stir welding of the Cantor alloy leaves a fine-grained structure.

Friction stir processing (FSP) has refined the microstructure of BM_{NEC} alloy from ~100 μm to 6.5 μm and 5.2 μm at 350 rpm and 650 rpm, respectively [70]. Also, with severe plastic deformation in friction stir processing, the fraction of the FCC phase increases to 90% and 92% at 350 rpm and 650 rpm respectively [70].

Grain size distribution at different regions of friction stir welded $BM_{C\text{-}CT}$ (as indicated in Figure 17.14a): (1) the mid-thickness of the weld, where grains are deformed two-fold, (2) stir zone at the top of the weld, and (3) weld near advancing

TABLE 17.4
The Friction Stir Welding Parameter for High Entropy Alloys

S. No	Material	Material thickness	Tool material	Tool specifications	Rotation speed (rpm)	Traverse speed (mm/min)	Force applied (kN)/Plunge depth (mm)	Tilt angle (degree)	Ref.
1	BM_{HRA}	2 mm	WC-12Co	Shoulder dia: 12 mm Tapered pin of height 1.85 mm.	600 and 700	150	–	3°	[69]
2	BM_{NEC}	5 mm	–	Shoulder dia: 12 mm Tapered pin: Root dia: 7.5 mm, pin dia: 6 mm and height 3.5 mm	350 and 650	50.8	3.65 mm	2.5°	[70]
3	$BM_{C\text{-}CT}$	2 mm	WC-Co based tool	Shoulder dia: 12.5 mm Hemispherical pin of height 1.5 mm	1000 rpm	30	11.1 kN	2.5°	[72]

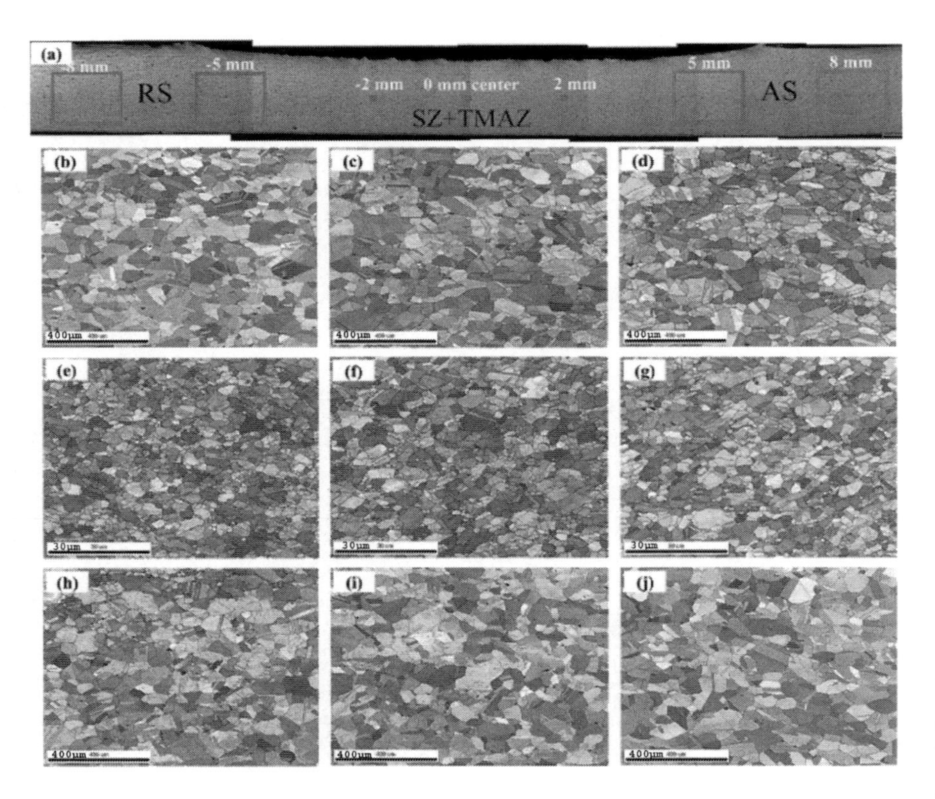

FIGURE 17.13 (a) Collage of optical micrographs from the cross-section of the friction stir welded specimen. The boxes indicate weld center and at 2 mm, 5 mm, 8 mm, 10 mm from the weld center where electron back-scattered diffraction (EBSD) analysis was carried out to obtain the micrographs below (b–j). Advancing Side (AS), Retreating (RS), Stir Zone (SZ), Thermomechanically Affected Zone (TMAZ) [69]. (Reprinted from *Met Mater Int*, 24, Jo, M.G., Kim, H.J., Kang, M. et al., Microstructure and mechanical properties of friction stir welded and laser welded high entropy alloy CrMnFeCoNi, 73–83. Copyright (2018) with permission from Springer Nature (License No. 4594820609008).)

side (AS) have been analyzed by Shaysultanov and co-workers [72]. The weld at mid-thickness has shown the distribution of higher fractions of fine grains (~2 μm) with few fractions of coarse grains (~7 μm). Fine grains are equiaxed with annealing twins, while curved and bulged-out grain boundaries are noted for coarser grains. Such refinement could be attributed to multiple deformations imposed on the grains when welded from both sides. Grains of the stir zone at the top of the weld are slightly coarse (4.5 μm), with a smaller number of annealed twins containing grains than that in the mid-thickness of the weld. The presence of polygonal shaped $M_{23}C_6$ particles with sharp straight boundaries in the stir zone (SZ) is noticed through the transmission electron microscopy (TEM) study (Figure 17.15). The microstructure of the stir zone possesses very low dislocation density. The size and volume fraction of the carbide particle are found to be 150 nm and 7%, respectively. The high fraction of carbides and lower dislocation density at the stir zone is due to the attainment

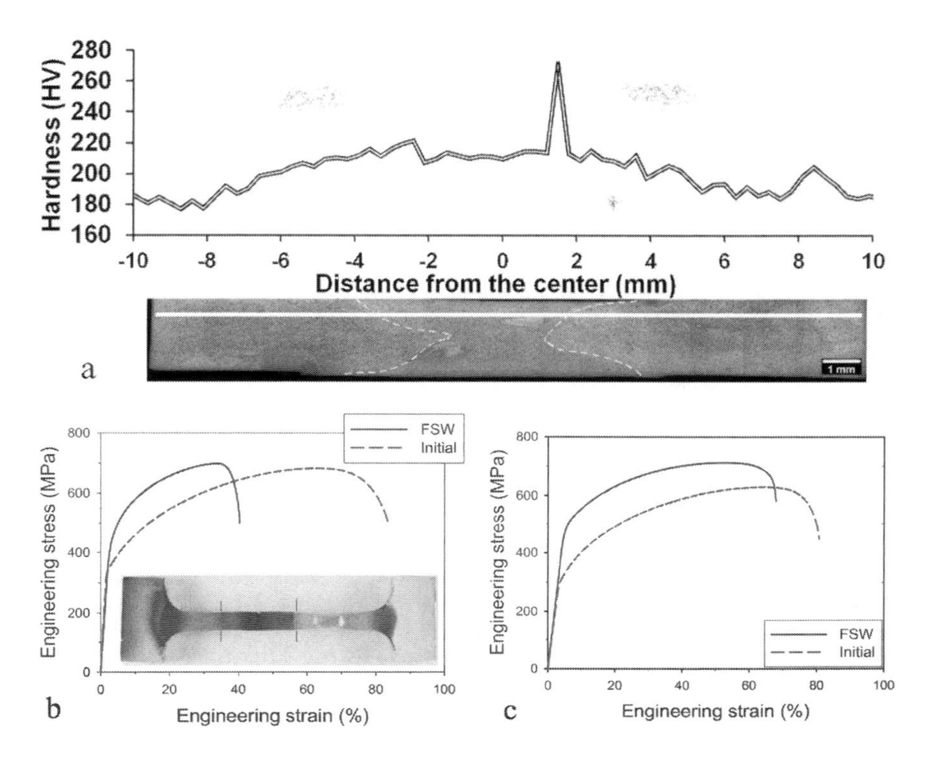

FIGURE 17.14 (a) Microhardness survey across the friction stir welded BM_{C-CT} weldment and its macrostructure (arrow indicating tungsten (W) rich white pattern) and engineering stress-strain curves of the friction stir welded BM_{C-CT} specimens (b) cut across the seam (along rolling direction [RD]) (c) or along the seam (across rolling direction [RD]) [72]. (Reprinted from *Mater Charact*, 145, Shaysultanov, D., Stepanov, N., Malopheyev, S., et al., Friction stir welding of a carbon-doped CoCrFeNiMn high-entropy alloy, 353–61. Copyright (2018) with permission from Elsevier (License No. 4594811397480).)

of a higher temperature (~1,200°C) at this region. The white pattern is visible at the advancing side (the arrow in Figure 17.14a) of the weld. The white patterns of stir zone have finer grains with W-particles (0.5–2µm) dispersion due to wear of the tungsten carbide-cobalt (WC-Co)-based tool during welding. However, in the weld region near the advancing side, a gradient microstructure varying from finer (2 µm) to coarser (4 µm) grains can be seen.

17.4.3.1.2 Mechanical Properties

The hardness of the weld nugget zone is higher than the hardness of the base metal, as shown in Figure 17.14a and Table 17.5. Also, the thermomechanically affected zone records lower hardness than the nugget zone, but higher hardness than the base metal. The hardness improvement is the result of the Hall–Petch relation and the presence of the residual stresses introduced during the heavy deformation by the rotating tool. In the nugget zone, the grains are very fine and heavily deformed by the pin, hence the residual stress is greater at this region compared to the thermomechanically

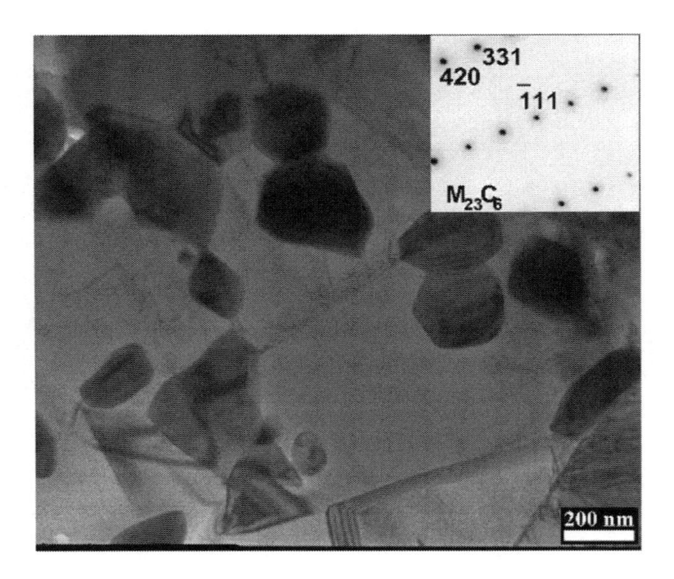

FIGURE 17.15 Transmission electron microscopy image of stir zone of the weld $BM_{C\text{-}CT}$ showing metal carbides [72]. (Reprinted from *Mater Charact*, 145, Shaysultanov, D., Stepanov, N., Malopheyev, S., et al., Friction stir welding of a carbon-doped CoCrFeNiMn high-entropy alloy, 353–61. Copyright (2018) with permission from Elsevier (License No. 4594811397480).)

TABLE 17.5
Hardness at Different Regions of Friction Stir Welded High Entropy Alloys

S. No.	Condition	Base metal hardness [HV]	Thermomechanically affected zone (TMAZ) hardness [HV]	Stir zone (SZ) hardness [HV]	White zone (WZ) hardness [HV]	Ref.
1	BM_{HRA}	140	160–190	210	–	[69]
2	$BM_{C\text{-}CT}$	180	200	220	260	[72]

affected zone at the shoulder region. Hence the nugget zone showing higher hardness than the thermomechanically affected zone and base metal.

Friction stir welded BM_{HRA} weldment has retained the tensile strength of the base metal with a 10% reduction in ductility (Table 17.2) [69]. As the weld possesses higher hardness and a finer microstructure, fracture occurs at the base metal. The relatively higher yield strength of the friction stir welded sample than the laser beam welded one is due to the finer grain size. The friction stir processed BM_{NEC} alloy has experienced an increase in tensile strength with improved ductility as shown in Table 17.2, and this is attributed to the pronounced transformation induced plasticity (TRIP) effect during deformation of more FCC-phase containing friction stir processed BM_{NEC} alloy, while the undeformed BM_{NEC} alloy has been found to experience a limited transformation induced plasticity effect as the phase transformation from FCC to HCP is limited in the alloy with a low fraction of the FCC phase [70]. In friction stir welded $BM_{C\text{-}CT}$, anisotropy is seen from the higher strength and ductility along the rolling direction, while having reduced strength across the rolling direction, as shown in Table 17.2 and Figure 17.14b [72]. As the weldment has increased yield strength, the failure takes place at the base metal and the ductility is reduced to 45%. However, the sample extracted along the weld has shown a higher tensile strength with improved ductility than the sample extracted across the weldment. The improvement in yield strength is contributed by Hall–Petch strengthening (55 MPa), by grain refinement and precipitation strengthening (~145 MPa) by precipitation of $M_{23}C_6$ carbide particles.

17.5 MODIFIED YEH ALLOY WELDING

Out of the modified Yeh high entropy alloys, AlxCoCrFeNi-high entropy alloys with "x" values 0 [63, 79], 0.1 [80, 81], 0.3 [82], 0.5 [83, 84], 0.6 [85], 0.8 [85], and 1 [86] were studied for their weldability by gas-tungsten arc welding (GTAW), laser beam welding (LBW), electron beam welding (EBW), resistance welding (RW), and friction stir welding (FSW) techniques. Further, the $Al_{0.5}$CoCrFeNi-high entropy alloy with 0.1 molar fraction of copper was also studied for its weldability by gas-tungsten arc welding (GTAW) and laser beam welding (LBW) processes and its hot cracking susceptibility by cast pin tear test (CPTT), a self-restrained weldability test [87]. For simplicity, the Al_xCoCrFeNi- high entropy alloys can be represented by their aluminum mole fraction, for example, $Al_{0.1}$CoCrFeNi-high entropy alloy with 0.1 mole fraction of aluminum can be expressed as $A_{0.1}$ alloy.

17.5.1 Alloys Used in Welding Studies

At present, studies on the weldability of high entropy alloys are very scarce. More gaps are available for weldability studies. Not all materials in Al_xCoCrFeNi high entropy alloys were studied, and not all the welding techniques were used to weld these alloys. Hence, a lot of research opportunities are available in the welding of high entropy alloys. Since the studies were carried out separately by various research groups, the base metals used for the welding were different in composition and condition, and the processing techniques were also different. Some alloys were welded in as-cast conditions, while some other alloys were welded after different thermomechanical processing conditions.

A_0, and its variants with a lower fraction of cobalt ($Co_{16}Cr_{28}Fe_{28}Ni_{28}$), $A_{0.3}$, $A_{0.5}$, and A_1 alloys that were prepared with vacuum=arc melting (VAM) facility [63, 79, 82, 83, 84, 86]. As-cast A_0 alloy variant and $A_{0.3}$ alloy were friction stir welded and the evolution of microstructure was studied along with microhardness distribution at different regions of the weldments. A_0 alloy after the two-step thermomechanical processing viz., (i) cold rolling and homogenization (1,100°C for 6 h) of as-cast alloy and (ii) subsequent cold rolling for 50% thickness reduction and annealing (900°C for 1 h) was welded by laser beam welding and their microstructure and mechanical properties were evaluated. As-cast $A_{0.1}$ alloy (produced by vacuum induction melting) was hot isostatic pressed (HIP) at 103 MPa and 1,204°C for 4 h with subsequent cooling to 343°C in 3 h and to 191°C in 1 h, was friction stir processed and the mechanical properties and associated microstructure were studied in detail [80, 81]. A vacuum-arc melted $A_{0.5}$ alloy in forged and homogenized (1,150°C for 24 h) condition was welded with gas-tungsten arc welding and laser welding processes [83, 84]. A variant of $Al_{0.5}$ alloy with 0.1 molar fraction of Cu ($A_{0.5+Cu}$) was used in three different conditions viz., (a) as-cast (by vacuum-arc melting), (b) heat-treated (650°C for 2.5 h) and (c) hot rolled and heat-treated (the as-cast $A_{0.5+Cu}$ alloy at 800°C were reduced 50% in thickness by rolling, and heat treated at 650°C for 2.5 h) and was welded with gas-tungsten arc welding [87]. Al $_{(x=0.6 \text{ and } 0.8)}$ CoCrFeNi-high entropy alloys [85] produced by vacuum-arc melting technique were studied for their weldability in the as-cast condition. Also, the A_1 alloy prepared by a vacuum-arc melting facility was used for a resistance welding study in as-cast condition by Cui and co-workers [86]. Hot cracking susceptibility of an $A_{0.5+Cu}$ alloy was assessed by the cast pin tear test [87].

17.5.1.1 Microstructure and Phase Analysis

A_0 alloy variant with lower Co fraction and $A_{0.3}$ alloy revealed elongated columnar grains with clear grain boundaries without any obvious precipitate formation (Figure 17.16b and d) [79, 82]. A thermomechanically treated A_0 alloy showed equiaxed grains of 6.6 μm in size along with many annealing twins [63]. Hot isostatic pressed $A_{0.1}$ alloy was observed to have coarser equiaxed grains in the range of millimeters (Figure 17.16c) [81]. All these alloys were found to have a single FCC phase in their initial conditions.

The forged and homogenized $A_{0.5}$ alloy developed equiaxed grains of 60 μm from the elongated dendritic grains due to recrystallization (Figure 17.16d) [83]. The energy dispersive X-ray spectroscopy analysis (Figure 17.17) of the $A_{0.5}$ alloy revealed that the grains were the iron-chromium rich FCC phase and the grain boundaries were the aluminum-nickel rich BCC phase. The microstructure of the as-cast $A_{0.6}$ alloy depicted columnar dendritic microstructure with clear interdendrites, while the $A_{0.8}$ alloy had a poorly aligned dendritic structure (Figure 17.16e and f) [85]. Both the alloys composed of two phases: FCC and BCC phase. With increasing aluminum content, the fraction of the BCC phase increased and led to the cock-tail effect [24]. The energy dispersive X-ray spectroscopy analysis of the dendrites and interdendrites showed that the dendrites were iron-chromium-rich FCC phases and the interdendrites were rich in aluminum-nickel. Further, the A_1 alloy was found to have the dendritic structure with a single BCC phase [86]. From the microstructural evaluation of AlxCoCrFeNi-high entropy alloys, it could be seen that the alloys with lower aluminum fraction (A_0, $A_{0.1}$, and $A_{0.3}$) are of simple microstructure with FCC crystal structure, while with an increase in aluminum fraction, the fraction of the phase with the BCC crystal structure was found to increase and was found to show phases with mixed crystal structures (FCC+BCC) in $A_{0.5}$, $A_{0.6}$, and $A_{0.8}$ alloys. And finally, A_1 showed only a BCC crystal structure with simple dendritic structure. The $A_{0.5+Cu}$ alloy prepared with the above-mentioned three conditions, was found to have the dendritic microstructure (as in Figure 17.16g, h and i) with BCC phase distribution in interdendrites in-between the FCC-phased dendrites. The interdendrites were aluminum-nickel rich and iron-cobalt-deficient with even distribution of

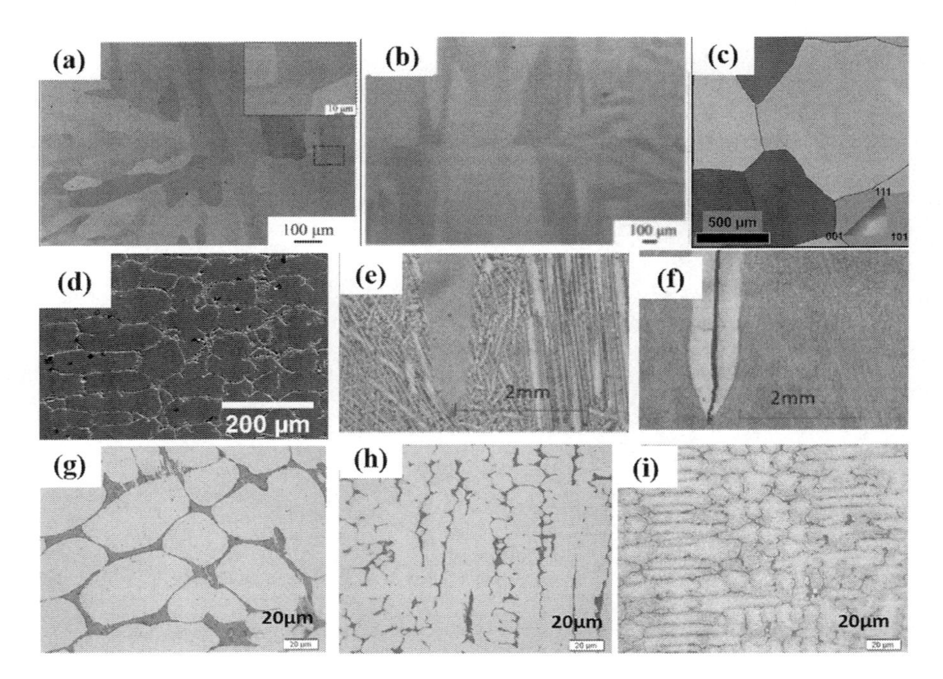

FIGURE 17.16 Microstructure of base metals (a) A_0 alloy variant with low Co fraction [63], (b) $A_{0.3}$ alloy [82], (c) $A_{0.1}$ alloy [81], (d) $A_{0.5}$ alloy [83], (e) $A_{0.6}$ alloy [85], (f) $A_{0.8}$ alloy [85], (g) as-cast $A_{0.5+Cu}$ alloy [87], (h) heat treated $A_{0.5+Cu}$ alloy [87], and (i) rolled and heat treated $A_{0.5+Cu}$ alloy [87]. (Sources: [63] Reprinted from *Mater Sci Eng A*, 711, Zhu, Z.G., Sun, Y.F., Ng, F.L., et al., Friction stir welding of a ductile high entropy alloy: microstructural evolution and weld strength, 524–32. (Copyright (2018), with permission from Elsevier (License No. 4605310648768); [81] Reprinted from *JOM*, 67, Kumar, N., Komarasamy, M., Nelaturu, P., Tang, Z., Liaw, P.K. and R.S. Mishra, Friction Stir Processing of a High Entropy Alloy $Al_{0.1}CoCrFeNi$, 1007–13. Copyright (2015) with permission from Springer Nature (License No. 4605311062272); [82] Reprinted from Mater Lett, 205, Zhu, Z.G., Sun, Y.F., Goh, M.H., et al., Friction stir welding of a $CoCrFeNiAl_{0.3}$ high entropy alloy, 142–44. Copyright (2017) with permission from Elsevier (License No. 4605311327429); [83] Reprinted from *Metall Mater Trans A*, 48, Sokkalingam, R., Mishra, S., Cheethirala, S.R., Muthupandi, V. and K. Sivaprasad, Enhanced Relative Slip Distance in Gas-Tungsten-Arc-Welded $Al_{0.5}CoCrFeNi$ High-Entropy Alloy, 3630–34. Copyright (2017) with permission from Springer Nature (License No. 4605320415540); [85] Source: Reprinted from *Metallogr Microstruct Anal*, 5, Nahmany, M., Hooper, Z., Stern, A., Geanta, V. and I. Voiculescu, Alx CrFeCoNi High-Entropy Alloys: Surface Modification by Electron Beam Bead-on-Plate Melting, 229–40. Copyright (2016) with permission from Springer Nature (License No. 4605320216581); and [87] Reprinted from *Weld World*, 63, Martin, A. C. and C. Fink, Initial weldability study on $Al_{0.5}CrCoCu_{0.1}FeNi$ high-entropy alloy, 739–50. Copyright (2019), with permission from Springer Nature (License No. 4604990792539).)

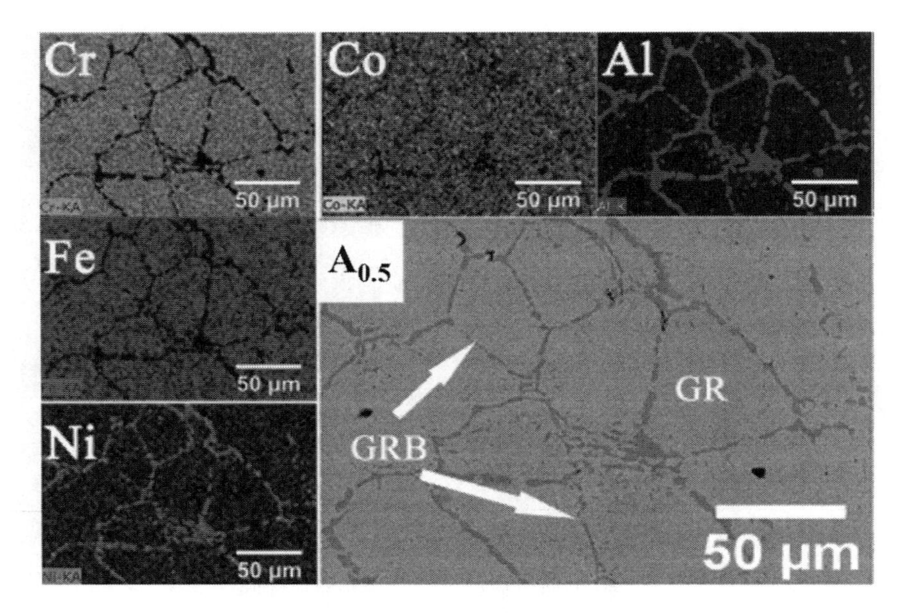

FIGURE 17.17 Scanning electron microscopy of $A_{0.5}$ alloy in back-scattered electron (BSE) mode and energy dispersive X-ray spectroscopy (EDS) composition distribution of alloying elements (Al, Co, Cr, Fe and Ni) [83]. (Reprinted from *Metall Mater Trans A*, 48, Sokkalingam, R., Mishra, S., Cheethirala, S.R., Muthupandi, V. and K. Sivaprasad, Enhanced Relative Slip Distance in Gas-Tungsten-Arc-Welded $Al_{0.5}$CoCrFeNi High-Entropy Alloy, 3630–34. Copyright (2017) with permission from Springer Nature (License No. 4605320415540).)

other elements. Heat treatment and hot rolling with subsequent heat treatment did not cause recrystallization in this alloy [87].

17.5.1.2 Mechanical Properties

The reported hardness and the tensile test results of the modified Yeh alloy are given in Table 17.6. It can be seen that the hardness and tensile properties of the alloys increase with an increase in aluminum molar fraction due to the evolution of the BCC phase and the resultant cocktail effect. The A_0 alloy has recorded higher tensile strength due to its initial heat-treated condition [63]. In the case of the $A_{0.5}$ alloy, even though the base material condition is same, the heavily deformed surface has been used for laser welding while the less deformed interior part has been gas-tungsten arc welded [83, 84]. Hence, the hardness of the $A_{0.5}$ alloy used for gas-tungsten arc welding was found to have a lower hardness than $A_{0.5}$ alloy used for laser welding. A finer microstructure in the base metal (that used for laser weld) extracted from the surface was found to have higher hardness. The hardness of the $A_{0.5+Cu}$ alloy is nearly the same as that of the $A_{0.5}$ alloy, as it was doped with a very small fraction of copper element [87]. However, the hardness of the $A_{0.5+Cu}$ alloy was increased from 200 HV to 287 HV on heat treatment due to the formation of B_2 precipitates in the dendritic and interdendritic region upon the heat treatment. Similarly, the heat-treated and hot rolled alloy experienced a huge increase in hardness (429 HV) because of increased

TABLE 17.6
Mechanical Properties of Modified Yeh Alloy and Its Welds

S/No	Materials	Condition	Microhardness [HV]	Yield strength [MPa]	Ultimate tensile strength [MPa]	Ductility [%]	Ref.
1	CoCrFeNi-HEA (A_0 alloy)	BM: Thermomechanically treated	160	242	635	53.4	[63]
		LBWed: tensile sample extracted across the weld	185	271	585	26.5	
		LBWed: tensile sample extracted across the weld	185	260	622	46.7	
2	$Co_{16}Cr_{28}Fe_{28}Ni_{28}$ HEA (A_0 alloy variant with lower Co fraction)	BM-(as-cast)	160	109	474	79	[79]
		$FSW_{50mm/min}$	240	–	–	–	
		$FSW_{30mm/min}$	220	–	–	–	
3	$Al_{0.1}$CoCrFeNi-HEA ($A_{0.1}$ alloy)	BM-HIPed	132	160	389	44.0	[80, 81]
		FSPed (Grain size – 0.95 μm)	181	544	730	27.5	
		FSPed (12 μm) (Grain size – 12 μm)	–	315	~600	75	
4	$Al_{0.3}$CoCrFeNi-HEA ($A_{0.3}$ alloy)	BM (as-cast)	180	–	–	–	[82]
		$FSW_{50mm/min}$	220	–	–	–	
		$FSW_{30mm/min}$	220	–	–	–	
5	$Al_{0.5}$CoCrFeNi-HEA ($A_{0.5}$ alloy)	BM – (Forged and homogenized) (Interior part)	280	380	810	67	[83]
		GTAW	232	257	689	56	
		BM – (Forged and homogenized) (Surface part)	325	–	–	–	[84]
		LBW	280	–	–	–	

(Continued)

TABLE 17.6 (CONTINUED)
Mechanical Properties of Modified Yeh Alloy and Its Welds

S/No	Materials	Condition	Microhardness [HV]	Yield strength [MPa]	Ultimate tensile strength [MPa]	Ductility [%]	Ref.
6	$Al_{0.6}$CoCrFeNi-HEA ($A_{0.6}$ alloy)	BM – as-cast	292	–	–	–	[85]
		EBW	400	–	–	–	
7	$Al_{0.8}$CoCrFeNi-HEA ($A_{0.8}$ alloy)	BM – as-cast	394	–	–	–	[85]
		EBW	600	–	–	–	
8	$Al_{0.5}Cu_{0.1}$CoCrFeNi-HEA ($A_{0.5+Cu}$ alloy)	BM – as-cast	200–220	–	–	–	[85]
		BM-HT	287	–	–	–	
		BM-HT-HR	429	–	–	–	
		Spot GTAW of as-cast alloy	200–220	–	–	–	
		Bead on plate GTAW of HT alloy	210	–	–	–	
		Butt welding of HT-HR alloy	200	–	–	–	
		Bead on plate laser welding of HT alloy	252	–	–	–	

dislocation density during the rolling, and precipitation of nano-sized B_2 precipitates in dendrites and interdendrites during heat treatment.

17.5.2 Fusion Welding

So far, laser welding was employed on A_0, $A_{0.5}$, and $A_{0.5+Cu}$ alloys, and $A_{0.5}$ and $A_{0.5+Cu}$ alloy were studied for their weldability under arc welding (GTAW).

17.5.2.1 Arc Welding

Sokkalingam and co-workers [83] have produced an autogenous GTA melt run on 2.5 mm thick $A_{0.5}$ alloy plate with the process parameters 2 mm – arc gap, 40 A – current, and 12 V – voltage at 80 mm/min – weld speed. In the case of $A_{0.5+Cu}$ alloy, the welding was carried out in three different ways [87]: (a) gas-tungsten arc spot welding on as-cast alloy by keeping the welding torch (operating at welding current and voltage of 120 A and 11 V, respectively) idle for 3 s at a standoff distance of 5 mm, (b) gas-tungsten arc autogenous melt run on the heat treated (650°C for 2.5 h) $A_{0.5+Cu}$ alloy with process parameters 95 A current, 11 V voltage, and 4.3 mm/s travel speed, and (c) gas-tungsten arc closed square butt welding of two rigidly clamped plates (hot rolled and heat treated condition) with welding parameters 100 A current, 10.5 V voltage, and 3.0 mm/s travel speed.

17.5.2.1.1 Microstructure

In the $A_{0.5}$ alloy weldment, three different zones were noticed as in other weld structures: (a) the fusion zone (FZ), (b) the heat affected zone (HAZ) next to the fusion zone (FZ), and (c) the unaffected base metal (BM) as indicated in Figure 17.18a. The heat affected zone in the gas-tungsten arc welded $A_{0.5}$ alloy was found to be very narrow and was confined within the partially melted grains of the base metal

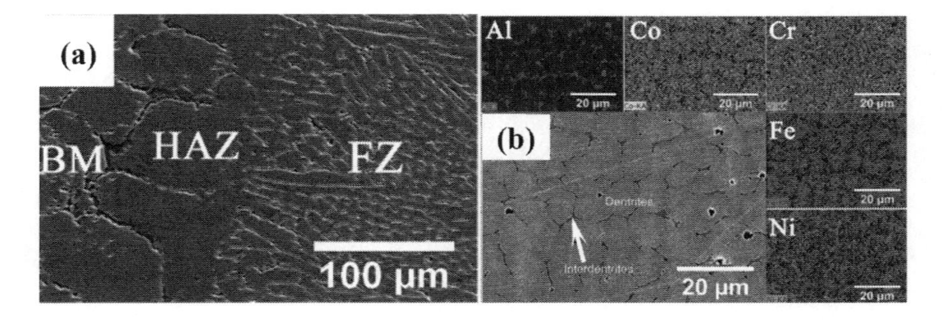

FIGURE 17.18 The microstructure of the (a) base metal/weld fusion zone interface and (b) scanning electron microscopy (SEM) of fusion zone of gas tungsten arc welded $A_{0.5}$ alloy in back scattered electron (BSE) mode and corresponding energy dispersive X-ray spectroscopy (EDS) maps for Al, Co, Cr, Fe, and Ni elements [83]. (Reprinted from *Metall Mater Trans A*, 48, Sokkalingam, R., Mishra, S., Cheethirala, S.R., Muthupandi, V. and K. Sivaprasad, Enhanced Relative Slip Distance in Gas-Tungsten-Arc-Welded $Al_{0.5}$CoCrFeNi High-Entropy Alloy, 3630–34. Copyright (2017) with permission from Springer Nature (License No. 4605320415540).)

adjacent to the fusion zone [83]. Even though the heat affected zone was narrow, the grains in the heat affected zone were found to be double the size of grains in the base metal. These partially melted grains in the heat affected zone provided the nucleation sites for the growth of the columnar dendritic grains from the fusion line. The energy dispersive X-ray spectroscopy (EDS) analysis of the dendrites and interdendrites showed that the dendrites were the iron-chromium (Fe-Cr)-rich FCC phase and the interdendrites were the aluminum-nickel (Al-Ni)-rich BCC phase (Figure 17.18 b). The dissolution of the aluminum-nickel (Al-Ni)-rich phases from the grain boundaries into the FCC phase during welding and subsequent rapid cooling of the weld have resulted in a reduction in the fraction of the BCC phase [83].

The gas-tungsten arc spot welding of the as-cast $A_{0.5+Cu}$ alloy (TW_S) showed crater cracks at the center of the weld due to sudden shutting off the arc (Figure 17.19a). Other than that, no solidification cracks or porosity are visible in the weld. Epitaxial growth of columnar dendrites across the fusion line is obvious (Figure 17.19b). These dendrites are finer than in the as-cast alloy due to the higher cooling rate experienced by the weld. The linear gas-tungsten arc welding of the heat treated $A_{0.5+Cu}$ alloy (TW_L) (Figure 17.19c) have shown a wide heat affected zone (1.5 mm width) due to the dissolution of the strengthening B2 precipitates. As in the spot weld, the epitaxial growth of the finer dendritic grains without any solidification cracks (Figure 17.19d) can be observed in the linear weld. The gas-tungsten arc butt weld made on the hot rolled and heat-treated $A_{0.5+Cu}$ alloy (TW_B) is of shallow depth of penetration [87]. Even though the weld was made from both sides, a 0.24 mm gap is left without melting. The heat affected zone in this weld is about 0.5 mm, narrower than that of the heat-treated samples as the abutting surface acts as the heat sink. The

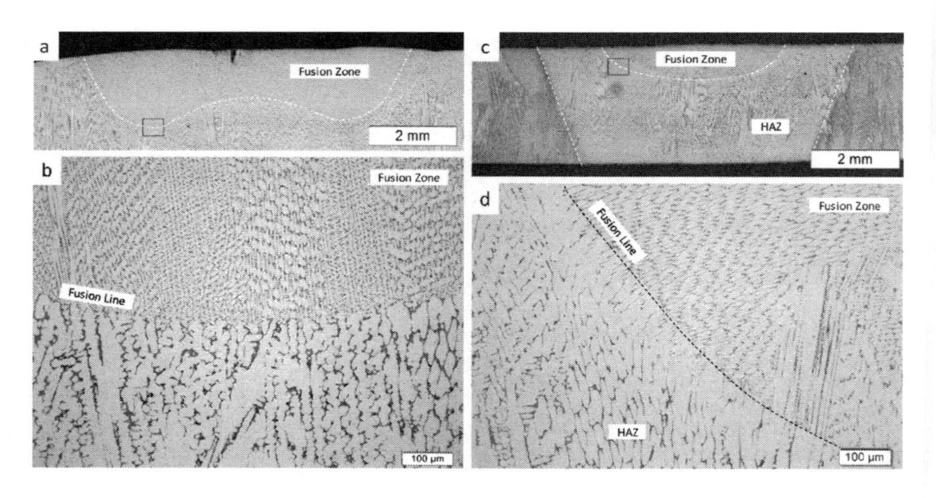

FIGURE 17.19 (a) Cross-section and (b) weld interface microstructure of autogenous gas tungsten arc spot welded as-cast $A_{0.5+Cu}$ alloy (TW_S) and (c) cross-section and (d) weld interface microstructure of gas-tungsten arc linear welded heat-treated $A_{0.5+Cu}$ alloy (TW_L) [87]. (Reprinted from *Weld World*, 63, Martin, A. C. and C. Fink, Initial weldability study on $Al_{0.5}CrCoCu_{0.1}FeNi$ high-entropy alloy, 739–50. Copyright (2019), with permission from Springer Nature (License No. 4604990792539).)

weld has produced a solidification crack at the starting point, which disappears on a further translation of the welding torch. Hence, in regular practice, hot cracking can be avoided by using a startup plate for arc ignition.

17.5.2.1.2 Mechanical Properties

Gas-tungsten arc welding has resulted in a reduction in strength (~17%) in $A_{0.5}$ alloy weld with enhancement of the relative slip distance and work hardening, as in Figure 17.20a–c. The weld metal of this alloy has revealed lower hardness due to the reduction in hardening precipitate/phases as they dissolved completely or partially during welding Figure 17.20d [83]. The TW_S weld metal has shown nearly the same hardness value (~200–220 HV) as the as-cast base alloy (Table 17.6), as the base alloy initial condition and weld condition are similar (i.e., as-cast) [87], whereas a slight drop in the hardness of the weld and the heat affected zone is observed in TW_L due to the dissolution of the strengthening phases/precipitates in the heat-treated base alloy [87]. The wide heat affected zone region can be seen from the microstructure (Figure 17.19). TW_B experiences a larger drop in hardness than others due to the dissolution of hardening precipitates in the weld. The narrow heat affected

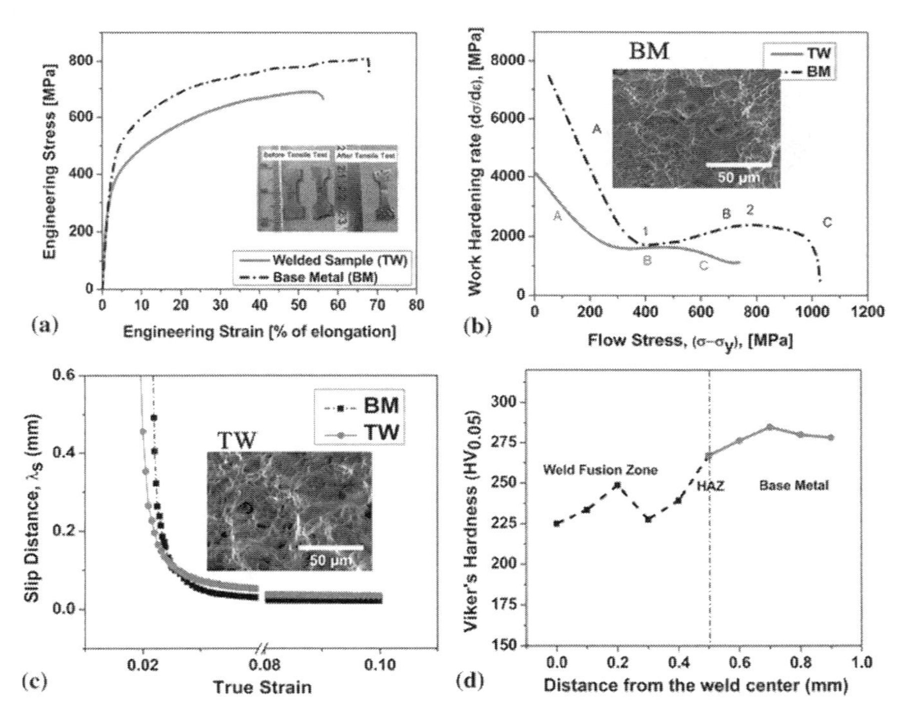

FIGURE 17.20 (a) Engineering stress-strain plot, (b) work-hardening plot, and (c) relative slip distance as a function of true strain for the $A_{0.5}$ alloy (BM) and its gas-tungsten arc weld (GTAW) and (d) microhardness survey across the weld [83]. (Reprinted from *Metall Mater Trans A*, 48, Sokkalingam, R., Mishra, S., Cheethirala, S.R., Muthupandi, V. and K. Sivaprasad, Enhanced Relative Slip Distance in Gas-Tungsten-Arc-Welded $Al_{0.5}CoCrFeNi$ High-Entropy Alloy, 3630–34. Copyright (2017) with permission from Springer Nature (License No. 4605320415540).)

zone observed in the weld is a result of larger heat sink available at the joining setup than in the bead on plate weld.

17.5.2.2 Laser Welding

A_0 and $A_{0.5}$ plates of 1 mm thickness were welded at laser power of 1.5 kW with a robotic fiber optic laser system [63, 84]. Welding speed of 40 mm/s and 10 mm/s were used to weld the A_0 and $A_{0.5}$ alloy, respectively. The heat treated $A_{0.5+Cu}$ alloy was linear welded with Nd: YAG pulsed laser with varying laser power and welding speeds: (1) 1 kW and 1.66 mm/s, (2) 5kW and 0.78 mm/s, and (3) 1 kW and 2.50 mm/s, respectively [87], and is designated here as LW_L.

17.5.2.2.1 Microstructure

Zhu and co-workers reported the anisotropic nature of the columnar grains with orientation along the <100> preferred growth direction [79]. The growth of the columnar grains was not disturbed until the center of the weld due to the lower temperature gradient and rapid solidification in the laser beam weld. The grains in the fusion zone of the laser weld of A_0 alloy showed local misorientation in the range of 0.1–0.6°, with geometrically necessary dislocations in it. The columnar grains in $A_{0.5}$ alloy laser weld having fine columnar dendritic sub-structures are shown in Figure 17.21(a and b). The interdendrites in the weld are aluminum-nickel (Al-Ni) rich, while dendrites are iron-chromium (Fe-Cr)-rich like in the gas-tungsten arc weld, but the degree of segregation in the laser weld is lower than in the gas-tungsten arc weld [83, 84]. Welds of both A_0 and $A_{0.5}$ alloys have shown epitaxially grown columnar grains from the partially melted base metal grains [79, 84].

During laser welding of the $A_{0.5+Cu}$ alloy with welding parameter 1, providing moderate power input was found to result in the keyhole mode of welding with a higher depth of penetration (Figure 17.22a), while welding parameters 2 and 3, as they provide very low and higher average input power, respectively, have resulted in a shallow and narrow bead. Due to the high energy density in the laser beam welding, the dendrites are finer, and the heat affected zone is very narrow. Porosity and solidification cracking are visible (Figure 17.22a–b) in the weld that was produced from parameter 1 due to the high strain rate imposed on the weld as a consequence

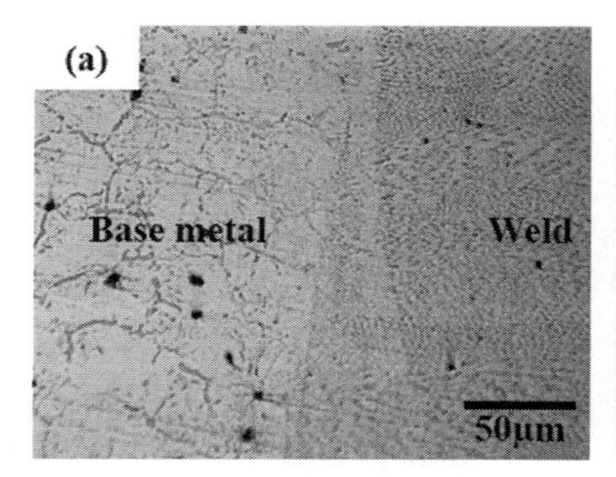

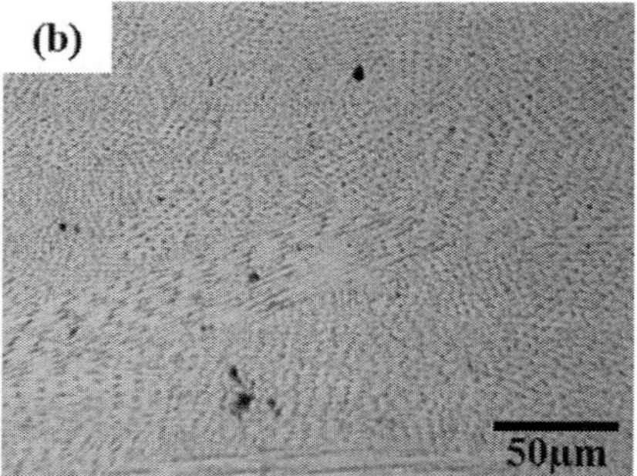

FIGURE 17.21 Optical micrograph of laser welded $A_{0.5}$ alloy (a) at the weld interface (showing base metal and weld fusion zone) and (b) weld fusion zone.

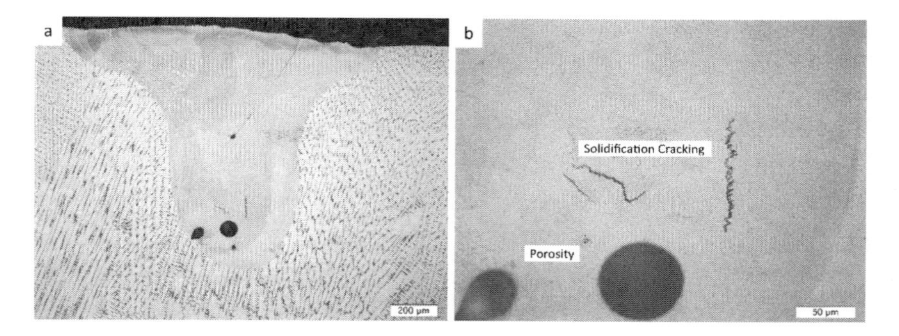

FIGURE 17.22 Optical micrograph of (a) cross-section of laser welded $A_{0.5+Cu}$ alloy using parameter 1 (LW_1) and (b) detailed view of solidification cracking and porosity in the fusion zone [87]. (Reprinted from *Weld World*, 63, Martin, A. C. and C. Fink, Initial weldability study on $Al_{0.5}CrCoCu_{0.1}FeNi$ high-entropy alloy, 739–50. Copyright (2019), with permission from Springer Nature (License No. 4604990792539).)

of the high cooling rate during welding. Since the welds made with the parameters 2 and 3 are of the conductive mode, they resulted in shallow and wide weld beads.

17.5.2.2.2 Mechanical Properties

Hardness of the laser beam weld made from the A_0 alloy is higher (185 HV) compared to the base metal (163 HV) [79], while the hardness of the welds of $A_{0.5}$ and $A_{0.5+Cu}$ alloys is lower than that of the corresponding base metal, which is similar to that noticed in the gas-tungsten arc weld [84, 87]. Laser beam welds attain a higher hardness than the gas-tungsten arc weld due to their finer microstructure resulting from the higher cooling rate of laser beam welding. Due to a very narrow heat affected zone, the hardness was found to drop suddenly when moving across the weldment.

The tensile properties of the A_0 alloy and its laser weld in transverse as well as in welding directions are given in Table 17.6. The weld is showing lower strength and ductility in the transverse direction while the sample extracted in the welding direction has retained its strength and ductility. Fracture surface analysis has concluded that the tensile deformation in A_0 alloy is by mechanical twinning as well as dislocation motion mechanisms. The former plays a dominant role in weld metal deformation while the latter plays a dominant role in base metal deformation [79].

17.5.2.2.3 Cast Pin Tear Test

The hot cracking susceptibility of the $A_{0.5+Cu}$ alloys is given as the circumferential cracking percentage as the function of the pin length in Figure 17.23. With the increase in pin length, the cracking tendency increases due to the high amount of restraint provided by the lengthy pin during solidification. The threshold value for $A_{0.5+Cu}$ alloy is the 0.625 inch pin length for cracking. As the alloys having threshold values of less than one inch are considered to be susceptible to cracking, it is taken that the laser beam welds are susceptible to cracking. It has been found that increasing the copper fraction in the alloy decreases the threshold value and hence increases the

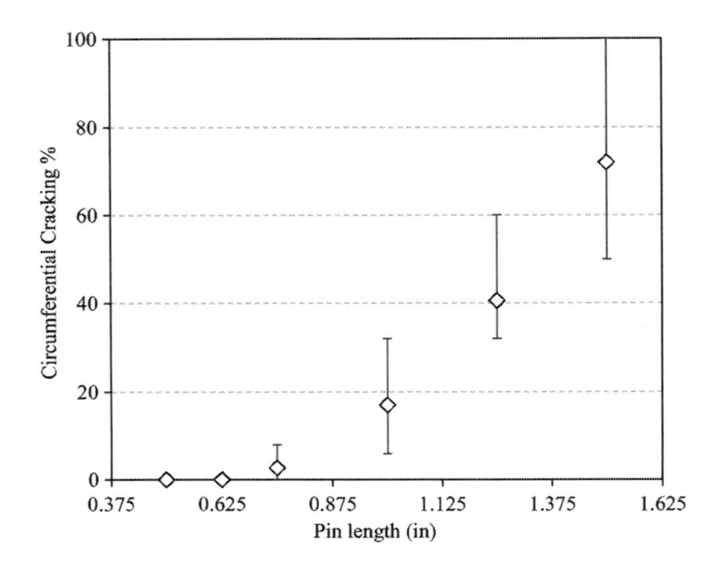

FIGURE 17.23 Cast pin tear testing (CPTT) of $A_{0.5+Cu}$ alloys shown as circumferential cracking (in %) measured as a function of cast pin length (in inches) [87]. (Reprinted from *Weld World*, 63, Martin, A. C. and C. Fink, Initial weldability study on $Al_{0.5}CrCoCu_{0.1}FeNi$ high-entropy alloy, 739–50. Copyright (2019), with permission from Springer Nature (License No. 4604990792539).)

solidification cracking tendency. Studies on the cracked surface have suggested that the copper-rich interdendrite that solidifies last is the pathway of the cracking [87].

17.5.2.3 Electron Beam Welding

$A_{0.6}$ and $A_{0.8}$ alloys were welded with electron beam welding technique using the beam current of 10–30 mA (heat input: 36–108 J/mm) and 10–35 mA (heat input: 36–126 J/mm), respectively [85].

17.5.2.3.1 Microstructure Analysis

Both melt runs show the wine glass shaped weld bead. $A_{0.6}$ has good weldability without any solidification cracking, while the solidification cracking is observed for the $A_{0.8}$ alloy at the center of the weld, especially when the depth to width ratio of the weld bead exceeded 5 (Figure 17.24). Also, the solidification cracking is found to be more pronounced in the weld that was made at higher heat input. But the single and multi-pass surface melting at lower heat input have yielded quality welds without any cracks [85]. The weld fusion zone in $A_{0.6}$ alloy is made of finer cellular dendritic structures and in the welds of $A_{0.8}$ alloy, an interconnected structure consisting of thin nanoscale wide alternating bright and dark phases is observed as in Figure 17.25a and b.

17.5.2.3.2 Mechanical Properties

The microhardness of the weld is higher than that of the base metal alloys ($A_{0.6}$ [374 HV] and $A_{0.8}$ alloys [530 HV]), as indicated in Table 17.6. The $A_{0.8}$ alloy records a

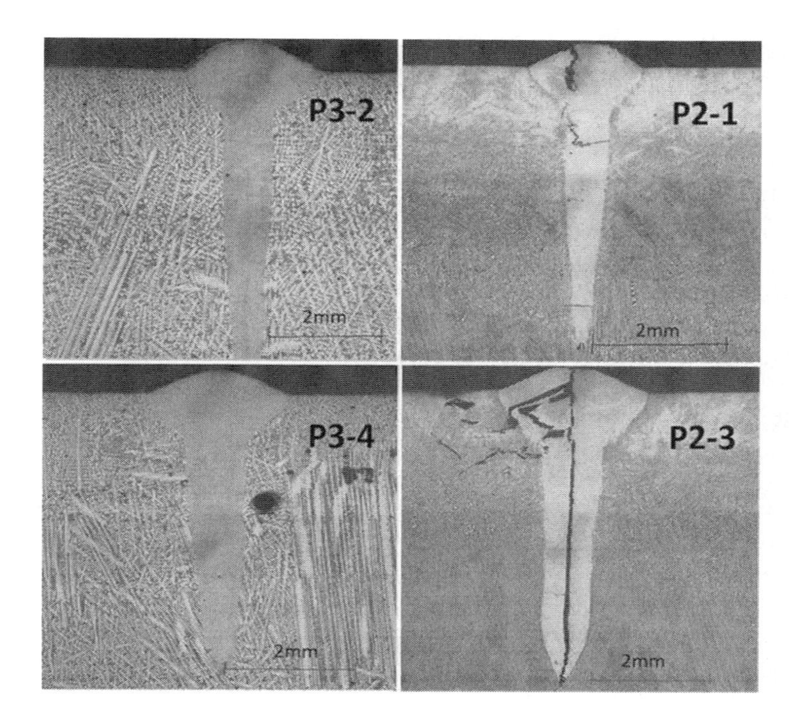

FIGURE 17.24 (a) and (b) Cross-section of bead on plate welded of $A_{0.6}$ and $A_{0.8}$ alloys at heat input 72 J/mm and (c) and (d) Cross-section of bead on plate welded of $A_{0.6}$ and $A_{0.8}$ alloys at heat input 108 J/mm [85]. (Reprinted from *Metallogr Microstruct Anal*, 5, Nahmany, M., Hooper, Z., Stern, A., Geanta, V. and I. Voiculescu, AlxCrFeCoNi High-Entropy Alloys: Surface Modification by Electron Beam Bead-on-Plate Melting, 229–40. Copyright (2016) with permission from Springer Nature (License No. 4605320216581).)

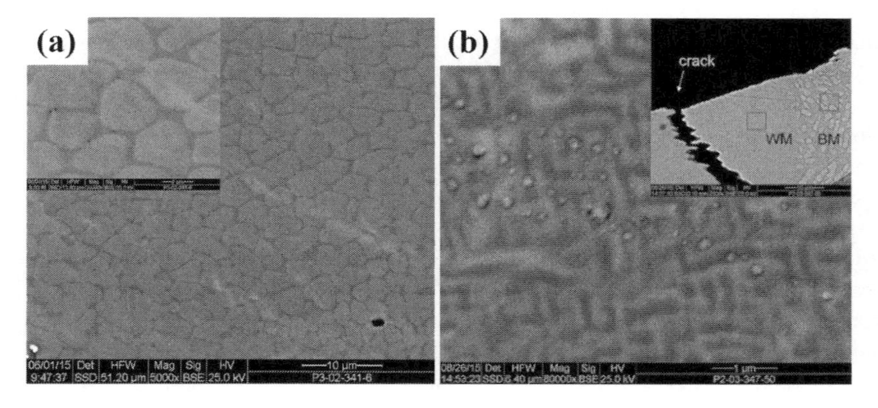

FIGURE 17.25 Scanning electron micrographs of weld fusion zone in (a) $A_{0.6}$ and (b) $A_{0.8}$ alloy [85]. (Reprinted from *Metallogr Microstruct Anal*, 5, Nahmany, M., Hooper, Z., Stern, A., Geanta, V. and I. Voiculescu, AlxCrFeCoNi High-Entropy Alloys: Surface Modification by Electron Beam Bead-on-Plate Melting, 229–40. Copyright (2016) with permission from Springer Nature. (License No. 4605320216581).)

higher hardness value than that of $A_{0.6}$ alloy due to the formation of the interconnected microstructures with nano-sized platelet-shaped precipitates [85].

17.5.3 Solid State Welding

17.5.3.1 Friction Stir Welding

Among the AlxCoCrFeNi-high entropy alloys, so far, the friction stir welding/processing has been carried out on the FCC alloys with Al=0, 0.1, and 0.3 molar ratio. The stacking fault energy (SFE) of the CoCrFeNi high entropy alloy is estimated to be 17 mJ/m^2 [78]. Generally, the microstructural evolution characteristics like the discontinuous dynamic recrystallization and the grain growth in materials with low to medium stacking fault energy depend on the strain, strain rate, and temperature at the particular region. The low stacking fault energy retards the dynamic recovery and hence increases the fraction of the dislocations in the processed alloy. This in turn, along with the already available grain boundary, acts as a nucleation site for recrystallization to occur. Also, the dislocation fraction is directly related to the intensity of the deformation received by the region. Hence, the nugget in the advancing side that experiences extensive deformation retains a higher fraction of dislocations compared to the retreating side. Even though the temperature at the stir zone reaches 0.85 Tm, the sluggish diffusion in high entropy alloys hinders the grain growth in the weld. Due to severe lattice distortion and sluggish diffusion, high entropy alloys exhibit rather a high recrystallization temperature (0.62 Tm). Grain coarsening in the CoCrFeMnNi high entropy alloy is observed only on annealing at 850°C [76, 78].

The scientific group of Mishra performed the friction stir processing (FSP) on the $A_{0.1}$CoCrFeNi high entropy alloy ($A_{0.1}$). Kumar and co-workers and Komarasamy and co-workers have carried out friction stir processing on the $A_{0.1}$ alloy with tools and processing parameters as in Table 17.7 [80, 81]. The friction stir welding has been carried out on 2 mm plates of A_0 in recrystallized condition and $A_{0.3}$ in as-cast condition at processing parameters and tools as given in Table 17.7 [82]. Generally, typical friction stir welded weldment exhibits different regions: unaffected base metal (BM), heat affected zone (HAZ), thermomechanically affected zone (TMAZ), and stir zone (SZ) based on the microstructure and the thermal conditions. The $A_{0.3}$ weld also shows all such regions; however, the heat affected zone in A_0 is insignificant, as the base metal used is already recrystallized at 900°C for 1 hour.

17.5.3.1.1 Microstructure

The top surface appearance of friction stir processed $A_{0.1}$ alloy and the respective tools used are given in Figure 17.26a and b. Figure 17.26c reveals a cross-sectional macrograph of the nugget. The nugget zone has experienced a heavy grain refinement from the several millimeters (2 mm) range of the grain size to the micrometer (~12 μm) range after friction stir processing (Figure 17.27). The filling defects were observed in nuggets at the advancing side of the weld [81]. The grains in the nugget are inhomogeneously distributed, as the grains at the advancing side of the nugget are four times finer than the grains at the retreating side.

In A_0 variant alloy with lower cobalt fraction, the weld produced at lower traverse speed has shown full penetration, while the weld made at a higher traverse speed has

TABLE 17.7
FSW Parameters for Modified Yeh Alloys

Material	Tool material	Tool shape	Shoulder diameter (mm)	Plunge depth (mm)/Axial load (kg)	Pin diameter and depth	Traverse speed	Tool rotation speed (rpm)	Tool tilt angle (degree)	Ref.
A_0	WC-Co alloy	–	12 mm	1,500 kg	Pin diameter and depth of 4 mm and 1.8 mm	30 and 50 mm/min	400 rpm	–	[63]
$A_{0.1}$	Polycrystalline cubic boron nitride (PCBN)	Convex, scrolled shoulder and conical pin with spiral feature	–	5.00 mm		25 mm/min	600 rpm	2.5°	[81]
	Tungsten carbide tool	Cylindrical tool	–	1.80 mm	1.5 mm	25.4 mm/min	1,000 rpm	0.5°	[80]
$A_{0.3}$	WC-Co alloy	–	12 mm	1,500 kg	Pin diameter and depth of 4 mm and 1.8 mm	30 and 50 mm/min	400 rpm	–	[82]

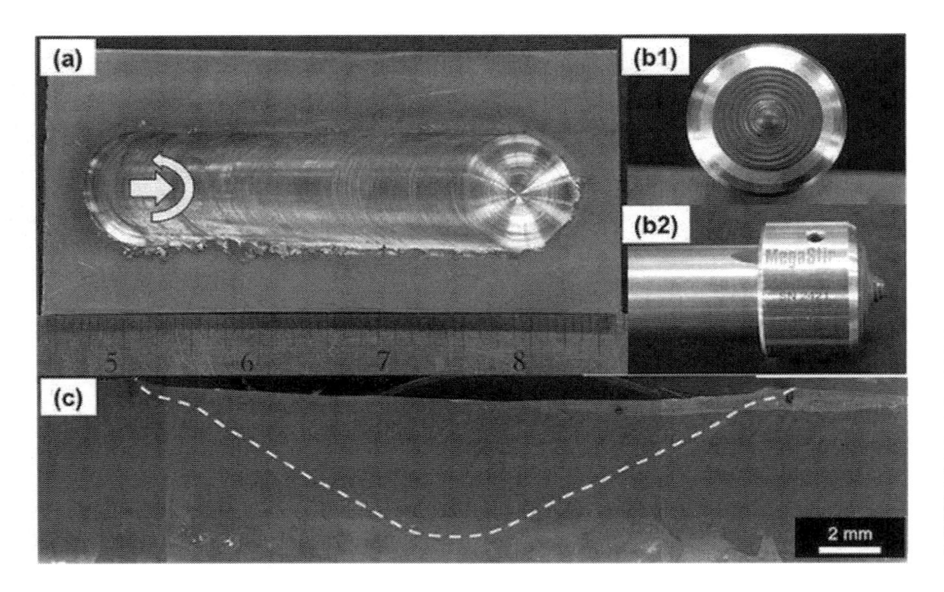

FIGURE 17.26 (a) The top surface of the friction stir processed $A_{0.1}$ alloy, (b) friction stir processing tool used, and (c) macrostructure of the weld nugget [81]. (Reprinted from *JOM*, 67, Kumar, N., Komarasamy, M., Nelaturu, P., Tang, Z., Liaw, P.K. and R.S. Mishra, Friction Stir Processing of a High Entropy Alloy $Al_{0.1}CoCrFeNi$, 1007–13. Copyright (2015) with permission from Springer Nature (License No. 4605311062272).)

FIGURE 17.27 Electron back-scattered diffraction (EBSD) images of the friction stir processed $A_{0.1}$ alloy [81]. (Reprinted from *JOM*, 67, Kumar, N., Komarasamy, M., Nelaturu, P., Tang, Z., Liaw, P.K. and R.S. Mishra, Friction Stir Processing of a High Entropy Alloy $Al_{0.1}CoCrFeNi$, 1007–13. Copyright (2015) with permission from Springer Nature (License No. 4605311062272).)

resulted in only 70% penetration, with an apparent kissing bond defect at the abutting line. Lower penetration is the result of insufficient metal flow due to low heat input at the higher traverse speed, whereas in the $A_{0.3}$ alloy, both the welding speeds have resulted in full depth of penetration; the only difference is the weld produced at the lower speed exhibiting a wider stir zone (Figure 17.28). Also, the presence of a white band is apparent in both the alloy welds and has been found to be composed of tungsten (W)-rich and chromium (Cr)-rich particles due to wearing of the tool during welding (Figure 17.28g). When the A_0 alloy at recrystallized condition is friction stir welded, the equiaxed grains are transformed to coarse elongated grains with a few finer grains in thermomechanically affected zone, while the stir zone is showing very fine grains (1.8 μm) [63]. The formation of fine grains in the thermomechanically affected zone in between coarse grains is due to discontinuous recrystallization mechanisms. Fine grains in the stir zone are formed by grain rotation and reorientation by the strain effect along with the transformation of low angle boundaries to high angle boundaries. The annealing twins are also found at the recrystallized grains.

The microstructure evolution in welds of $A_{0.3}$ alloy and A_0 alloy is similar. However, the grain coarsening is seen more in the $A_{0.3}$ alloy due to the as-cast nature of the base alloy. The grain size in the stir zone is 2.8 μm. The grains in the advancing side (higher temperature) are found to be coarser than the grains in the retreating side. Also, grains at the bottom of the retreating side are finer than those at the top (higher

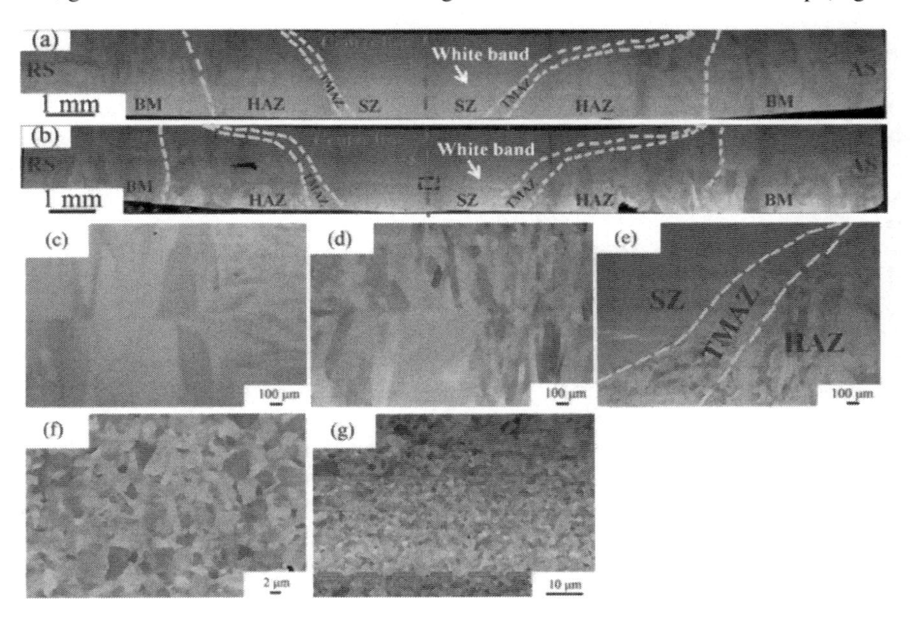

FIGURE 17.28 Macrostructure of the cross section of $A_{0.3}$ alloy friction stir welded at speed of (a) 50 mm/min and (b) 30 mm/min, respectively, and (c–g) Scanning electron microscopy of the base metal, heat affected zone, thermomechanically affected zone, stir zone and the region under the rectangle in (b), respectively, in friction stir welded with speed of 30 mm/min [82]. (Reprinted from *Mater Lett*, 205, Zhu, Z.G., Sun, Y.F., Goh, M.H., et al., Friction stir welding of a $CoCrFeNiAl_{0.3}$ high entropy alloy, 142–44. Copyright (2017) with permission from Elsevier. (License No. 4605311327429).)

temperature). The grain size in the stir zone of the weld made at a lower traverse speed is coarser than the weld produced at a higher speed. The variation in grain size could be attributed to the availability of the heat at the region, since the region experiencing a higher temperature could produce coarser grains as it has more time for grain growth than in the region with a lower temperature. The higher temperature (i) in the advancing side is about 100 K greater than in retreating side; (ii) at the top side of the retreating side compared to the bottom, as the heat is dissipated rapidly at the bottom by the backing plate; and (iii) in the weld with the lower traverse speed (longer dwelling time) leads to coarsening of the grains. Also, at a lower speed, the density of high angle grain boundaries and twin boundaries increases due to discontinuous recrystallization. Transmission electron microscopy (TEM) analysis of stir zone has shown dislocations in the equiaxed grains due to severe plastic deformation, without any second phase. Also, the white band region contains very much finer microstructure than in the other regions due to the particle-stimulated nucleation effect.

17.5.3.1.2 Mechanical Properties

After friction stir welding/processing (FSW/FSP), the nugget is showing higher hardness (Table 17.6) than the base metal due to grain refinement. The hardness of the weld increased from 160 HV to 240 HV in A_0, from 132 HV to 181 HV in $A_{0.1}$ and from 175 HV to 220 HV in $A_{0.3}$. The intrinsic hardness and Hall–Petch coefficient of A_0 alloy were found to be very high (74 HV and 90 HV $\mu m^{1/2}$) compared to conventional alloys due to its high hardening efficiency by severe lattice distortion in high entropy alloy and twin formation at the stir zone. The yield strength of $A_{0.1}$ in hot isostatic pressed and friction stir processed conditions are assessed to be ~150 MPa and ~315 MPa, respectively. The friction stress, critical resolved shear stress, and Hall–Petch constant in $A_{0.1}$ are estimated as 174 MPa, 56 MPa, and 371 MPa, respectively.

17.5.4 Resistance Welding

The as-cast vacuum arc melted A_1 alloy was resistant spot welded with varying heat input: 0.052 Q, 0.069 Q, 0.101 Q, and 153 Q, respectively, by keeping pressing time of (1 s), power on time (0.6 s), cooling time (0.4 s), maintenance time (0.4 s), and off time (1 s) constant [86]. As-cast A_1 alloy has shown dendritic morphology. With increasing heat input the grain size in the weld metal decreases gradually and the corresponding hardness of the weld has increased due to the grain boundary strengthening phenomenon. At higher heat input (0.153 Q), the weldment has shown defects such as cracks and splashes.

17.6 DISSIMILAR WELDING OF HIGH ENTROPY ALLOYS WITH CONVENTIONAL ALLOYS

Structural parts of power plants and the chemical, automobile, and aerospace industries require dissimilar joints to make the design simpler [88–89]. In the view of the applications of high entropy alloys in such industries, Sokkalingam and co-workers have produced dissimilar joints between $A_{0.1}$ alloy and AISI 304 stainless steel through the gas-tungsten arc welding technique [90].

17.6.1 Microstructure

Vacuum-arc melted $A_{0.1}$ alloy that was drop forged and homogenized at 1,050°C for 24 h was gas-tungsten arc welded autogenously with the as-received AISI 304 stainless steel with welding parameters: 10–11 V voltage, current of 50 A, and speed of 100 mm/min [90]. The macrograph of the dissimilar joint is given in Figure 17.29. It reveals different regions viz., unaffected $A_{0.1}$ alloy, interface between $A_{0.1}$ alloy and fusion zone (IF-1), the weld fusion zone (FZ), the interface between the fusion zone and the AISI 304 stainless steel, and the unaffected AISI 304 stainless steel. Scanning electron microscope images of each region are given in Figure 17.30. The $A_{0.1}$ alloy has shown the equiaxed grains with annealing twins. However, the weld metal shows a simple dendritic microstructure without any complex phase or

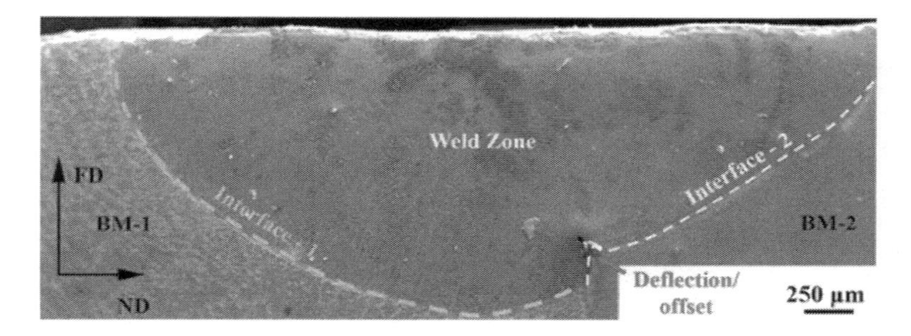

FIGURE 17.29 Macrostructure of the dissimilar joint between $A_{0.1}$ alloy (BM-1) and AISI-304 stainless steel (BM-2) [90]. (Reprinted from *J Mater Res*, Sokkalingam, R., Sivaprasad, K., Muthupandi, V. and K.G. Prashanth, Dissimilar welding of $Al_{0.1}CoCrFeNi$ high-entropy alloy and AISI304 stainless steel, https://doi.org/10.1557/jmr.2019.186. Copyright (2019), with permission from Cambridge University Press (License No. 4605380536561).)

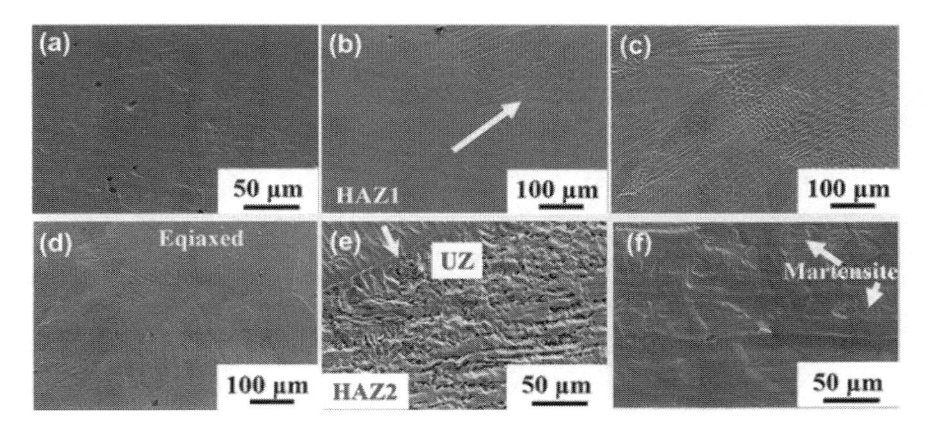

FIGURE 17.30 Scanning electron microscopy of unaffected $A_{0.1}$ alloy, interface between $A_{0.1}$ alloy and fusion zone (IF-1), weld fusion zone (FZ), interface between fusion zone and AISI 304 stainless steel and unaffected AISI 304 stainless steel [90]. (Reprinted from *J Mater Res*, Sokkalingam, R., Sivaprasad, K., Muthupandi, V. and K.G. Prashanth, Dissimilar welding of $Al_{0.1}CoCrFeNi$ high-entropy alloy and AISI304 stainless steel, https://doi.org/10.1557/jmr.2019.186. Copyright (2019), with permission from Cambridge University Press (License No. 4605380536561).)

compound formation. IF-1 shows epitaxial growth of the columnar elongated grains, while in IF-2 columnar grains are formed after the unmixed zone (UZ) [6]. When these columnar grains approached the top center of the weld, they transformed to equiaxed grains due to high degree of constitutional cooling during welding.

17.6.2 Mechanical Properties

The hardness distribution across a dissimilar weld is given in Figure 17.31. $A_{0.1}$ alloy shows hardness of ~178 $HV_{0.5}$. This is higher than that of the values observed by Kumar and co-workers [81] and is attributed to the grain refinement in the forged

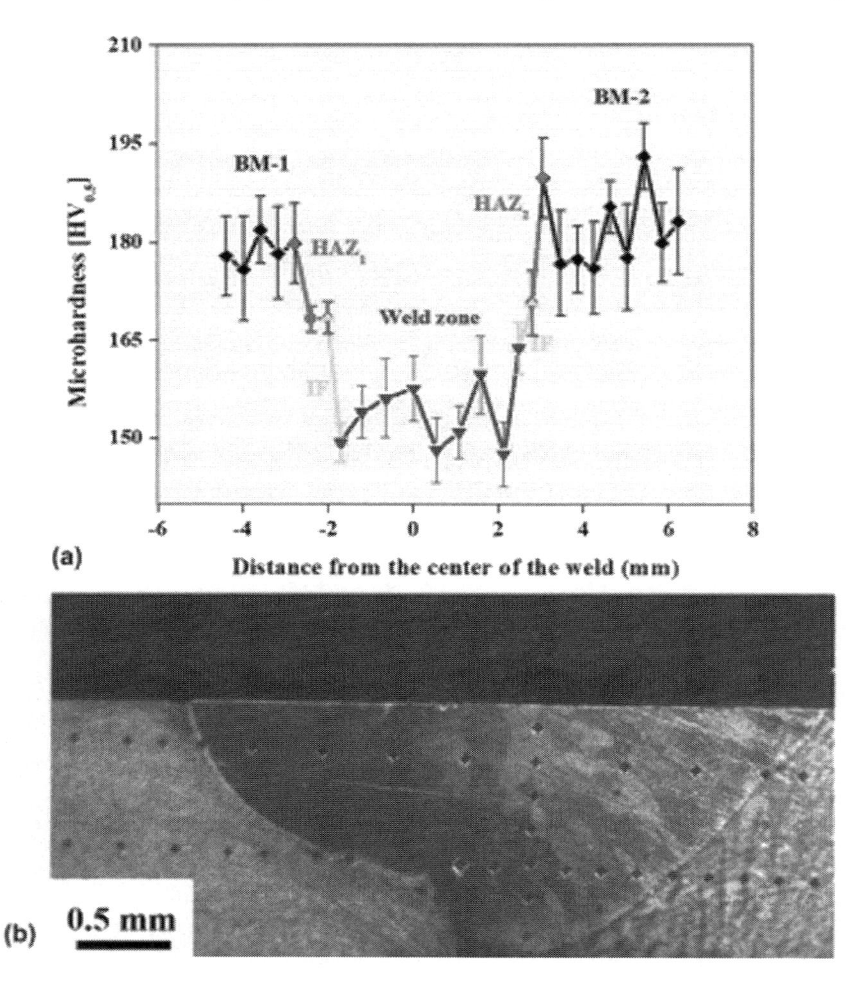

FIGURE 17.31 (a) Microhardness survey across the dissimilar joint and (b) macrograph with indents [90]. (Reprinted from *J Mater Res*, Sokkalingam, R., Sivaprasad, K., Muthupandi, V. and K.G. Prashanth, Dissimilar welding of $Al_{0.1}$CoCrFeNi high-entropy alloy and AISI304 stainless steel, https://doi.org/10.1557/jmr.2019.186. Copyright (2019), with permission from Cambridge University Press (License No. 4605380536561).)

and heat-treated condition. Also, AISI 304 stainless steel is showing ~180 $HV_{0.5}$. The hardness survey across the weld reveals the hardness drop at the heat-affected zone in both the base metals, where marginal grain coarsening occurred [90] and finally, the weld metal is showing lower hardness of ~150 $HV_{0.5}$ and could be due to coarsening of the grains and elimination of the twin boundaries in the weld fusion zone. The weld has a strength of ~540 MPa with the ductility of ~38% [90], which is higher than that of the coarse-grained $A_{0.1}$ alloy in a heat-treated condition [80, 81].

17.7 SUMMARY

From the available research articles on welding of high entropy alloys, the following conclusions can be made:

1. Cantor alloy (CoCrFeMnNi-high entropy alloy) with a single FCC crystal structure is weldable by fusion welding processes like gas-tungsten arc welding, laser beam welding, and electron beam welding, and by friction stir welding, a solid-state welding process without major welding defects or significant microsegregation.
2. Chromium-manganese (CrMn)-rich oxide inclusions were found to occur commonly in the weld metal of welds produced with fusion welding processes.
3. Welds of Cantor alloys retain the yield strength equivalent to that of the base metal at the expense of ductility. Compared to the base alloy, the welds exhibit mechanical properties in the order: gas-tungsten arc welding (GTAW) < high-energy beam welding (EBW/LBW) < solid-state welding (friction stir welding [FSW]).
4. Tensile properties of the Cantors alloys at the cryogenic temperature (77 K) are better than at room temperature and the difference in strengths is due to the formation of nano-twinning at cryogenic temperature.
5. Post-weld heat treatment (PWHT) has been observed to result in a reduction in strength along with an improvement in ductility of the laser welded Cantor alloy in the as-rolled condition. Reduction in dislocation density and associated stress relieving, and reduction in the size and fraction of CrMn-oxide inclusion by rolling are attributed to the change in mechanical properties.
6. The modified Yeh alloy (AlxCoCrFeNi-high entropy alloy) with lower molar fraction of aluminum (x = 0–0.6) can be welded by gas-tungsten arc welding (GTAW), laser beam welding (LBW), and electron beam welding (EBW) without any solidification cracking. However, alloys with higher aluminum mole fraction (x = 0.8 and 1) were found to show solidification cracking when welded with electron beam welding (EBW) and resistance welding.
7. As in Cantor alloy, fusion welding of modified Yeh alloy (AlxCoCrFeNi-high entropy alloy) results in a reduction in ductility due to the grain coarsening effect.
8. Modified Yeh alloys with a minor fraction of copper (Cu) are also weldable by gas-tungsten arc welding (GTAW) and laser beam welding (LBW) techniques without any solidification cracking.

9. In the weld metals of $Al_{(x=0.5,0.6)}$CoCrFeNi-HEA and $Al_{0.5}CoCu_{0.1}$CrFeNi-HEA microsegregation of aluminum-nickel (Al-Ni)-rich BCC phase at the interdendritic regions is noticed due to the non-equilibrium solidification.
10. The cast pin tear test (CPTT) shows that modified Yeh alloy with minor copper addition ($Al_{0.5}CoCrCu_{0.1}$FeNi alloy) is much more resistant to solidification cracking than the alloy with a higher molar fraction of copper (AlCoCrCuFeNi alloys).
11. Friction stir welding/processing of $Al_{(x=0,\ 0.1\text{and}\ 0.3)}$CoCrFeNi-alloy results in grain refinement assisted enhancement in mechanical properties (hardness and tensile properties).
12. $Al_{0.1}$CoCrFeNi high entropy alloy is weldable with AISI304 stainless steel by the conventional gas-tungsten arc welding technique. The dissimilar weld joint is found to have appreciable mechanical properties.

17.8 FUTURE OPPORTUNITIES

- Since the high entropy alloys are not well-established, the available knowledge base on the properties of these alloys is scanty. Also, the high entropy alloy concept can lead to the design and development of a large number of alloys with different compositions and crystal structures.
- As the study on weldability to date is mostly concentrated on the CoCrFeNi-based alloy system, other high entropy alloy systems also must be studied in the future, while considering the industrial application of these alloys.
- Also, available weldability studies are centered around gas-tungsten arc welding, electron beam welding, laser beam welding, resistance welding, and friction stir welding processes. In order to produce efficient weld joints through proper welding procedures, a lot of opportunities are available in weldability studies of plasma arc welding, gas metal arc welding, shielded metal arc welding, linear friction welding, friction welding, and other processes.
- Evaluation of the performance of the welded joints of high entropy alloys in the industrial applications and environments calls for studies on fracture toughness, fatigue, and corrosion behavior of these joints.

REFERENCES

1. Lumley, R. 2011. *Fundamentals of Aluminium Metallurgy: Production, Processing and Applications*. Woodhead Publishing Ltd.
2. Thomas, W.M. and Nicholas, E.D. 1997. Friction stir welding for the transportation industries. *Mater Des* 18(4–6):269–73.
3. Meco, S., Pardal, G., Ganguly, S., Williams, S. and McPherson, N. 2015. Application of laser in seam welding of dissimilar steel to aluminium joints for thick structural components. *Opt Laser Eng* 67:22–30.
4. Sivaprasad, K. and Raman, S.G.S. 2008. Influence of weld cooling rate on microstructure and mechanical properties of alloy 718 weldments. *Metall Mater Trans A* 39(9):2115–27.
5. Ramkumar, P., Prakash, F.G., Karthikeyan, M.K., Gupta, R.K. and Muthupandi, V. 2018. Development of copper coating technology on high strength low alloy steel filler wire for aerospace applications. *Mater Today Proc* 5(2):7296–302.

6. Ravikiran, K., Das, G., Kumar, S., Singh, P.K., Sivaprasad, K. and Ghosh, M. 2019. Evaluation of microstructure at interfaces of welded joint Between low alloy steel and stainless steel. *Metall Mater Trans A* 50(6):2784–97.
7. Anand, A. and Khajuria, A. 2013. Welding processes in marine applications: A review. *Int J Mech Eng & Rob Res* 2:215–25.
8. Ramakrishnan, M., Padmanaban, K. and Muthupandi, V. 2008. Effect of cold wire addition technique in tandem submerged arc welding for ring header fabrication in boiler manufacturing. Proc. Symposium on Joining of Materials (SOJOM), WRI, IIM Tiruchirappalli, India.
9. Youssef, H.A., El-Hofy, H.A. and Ahmed, M.H. 2011. *Manufacturing Technology Materials, Processes, and Equipment*. Boca Raton, FL: CRC Press.
10. Ravisankar, V., Balasubramanian, V. and Muralidharan, C. 2006. Selection of welding process to fabricate butt joints of high strength aluminium alloys using analytic hierarchic process. *Mater Des* 27(5):373–80.
11. Prashanth, K.G., Scudino, S. and Eckert, J. 2017. Defining the tensile properties of Al-12Si parts produced by selective laser melting. *Acta Mater* 126:25–35.
12. Zhang, S., Ma, P., Jia, Y. et al. 2019. Microstructure and mechanical properties of Al–(12–20)si bi-material fabricated by selective laser melting. *Materials* 12(13):1–11. doi:10.3390/ma12132126.
13. Suryawanshi, J., Prashanth, K.G. and Ramamurty, U. 2017. Tensile, fracture, and fatigue crack growth properties of a 3D printed maraging steel through selective laser melting. *J Alloy Compd* 725:355–64.
14. Eisazadeh, H., Bunn, J., Coules, H.E., Achuthan, A., Goldak, J. and Aidun, D.K. 2016. A residual stress study in similar and dissimilar welds. *Weld J* 95:111–19.
15. Nivas, R., Singh, P.K., Das, G., et al. 2017. A comparative study on microstructure and mechanical properties near interface for dissimilar materials during conventional V-groove and narrow gap welding. *J Manuf Process* 25:274–83.
16. Srinivas, B., Sivaprasad, K., Babu, N.K., Muthupandi, V. and Susila, P. 2013. Studies on dissimilar welding of AA5083 and AA6061 alloys by laser beam welding. *Adv Mater Res* 626:701–05.
17. Kou, S. 2003. *Welding Metallurgy*, 2nd ed. Hoboken, NJ: Wiley-Interscience.
18. Mendez, P.F. and Eagar, T. 2001. Welding processes for aeronautics. *Adv Mater Process* 159(5):39–43.
19. Darwish, S.M., Tamimi, A.A. and Al-Habdan, S. 1997. A knowledge base for metal welding process selection. *Int J Mach Tool Manufact* 37(7):1007–23.
20. Miracle, D.B., Miller, J.D., Senkov, O.N., Woodward, C., Uchic, M.D. and Tiley, J. 2014. Exploration and development of high entropy alloys for structural applications. *Entropy* 16(1):494–25.
21. Alagarsamy, K., Fortier, A., Komarasamy, M., et al. 2016. Mechanical properties of high entropy alloy Al0.1CoCrFeNi for peripheral vascular stent application. *Cardiovasc Eng Technol* 7(4):448–54.
22. Tang, Z., Senkov, O.N., Parish, C.M., et al. 2015. Tensile ductility of an AlCoCrFeNi multi-phase high-entropy alloy through hot isostatic pressing (HIP) and homogenization. *Mater Sci Eng A* 647:229–40.
23. Cantor, B., Chang, I.T.H., Knight, P. and Vincent, A.J.B. 2004. Microstructural development in equiatomic multicomponent alloys. *Mater Sci Eng A* 213:375–77.
24. Yeh, J.W., Chen, S.K., Lin, S.J., et al. 2004. Nanostructured high-entropy alloys with multiple principal elements: Novel alloy design concepts and outcomes. *Adv Eng Mater* 6(5):299–303.
25. Gludovatz, B., Hohenwarter, A., Catoor, D., Chang, E.H., George, E.P. and Ritchie, R.O. 2014. A fracture-resistant high-entropy alloy for cryogenic applications. *Science* 345(6201):1153–57.

26. Zhang, Z.J., Mao, M.M., Wang, J., et al. 2015. Nanoscale origins of the damage tolerance of the high-entropy alloy CrMnFeCoNi. *Nat Commun* 6:10143.
27. Tsai, C.W., Tsai, M.H., Tsai, K.Y., Chang, S.Y., Yeh, J.W. and Yeh, A.C. 2015. Microstructure and tensile properties of Al0.5CoCrCuFeNi alloys produced by simple rolling and annealing. *Mater Sci Technol* 31(10):1178–83.
28. Wu, S.W., Wang, G., Yi, J., et al. 2017. Strong grain-size effect on deformation twinning of an Al0.1CoCrFeNi high-entropy alloy. *Mater Res Lett* 5(4):276–83.
29. Komarasamy, M., Alagarsamy, K. and Mishra, R.S. 2017. Serration behavior and negative strain rate sensitivity of Al0. 1CoCrFeNi high entropy alloy. *Intermetallics* 84:20–4.
30. Yang, T., Tang, Z., Xie, X., et al. 2017. Deformation mechanisms of Al0.1CoCrFeNi at elevated temperatures. *Mater Sci Eng A* 684:552–58.
31. Hou, J., Zhang, M., Ma, S., Liaw, P.K., Zhang, Y. and Qiao, J. 2017. Strengthening in Al0.25CoCrFeNi high-entropy alloys by cold rolling. *Mater Sci Eng A* 707:593–601.
32. Hou, J., Zhang, M., Yang, H. and Qiao, J. 2017. Deformation behavior of Al0.25CoCrFeNi high-entropy alloy after recrystallization. *Metals* 7(4):111.
33. Shun, T.T. and Du, Y.C. 2009. Microstructure and tensile behaviors of FCC Al0.3CoCrFeNi high entropy alloy. *J Alloys Compd* 479(1–2):157–60.
34. Gwalani, B., Soni, V., Lee, M., Mantri, S.A., Ren, Y. and Banerjee, R. 2017. Optimizing the coupled effects of Hall-Petch and precipitation strengthening in a Al0.3CoCrFeNi high entropy alloy. *Mater Des* 121:254–60.
35. Ma, S.G., Zhang, S.F., Qiao, J.W., et al. 2014. Superior high tensile elongation of a single-crystal CoCrFeNiAl0.3 high-entropy alloy by Bridgman solidification. *Intermetallics* 54:104–09.
36. Yasuda, H.Y., Miyamoto, H., Cho, K. and Nagase, T. Formation of ultrafine-grained microstructure in Al0.3CoCrFeNi high entropy alloys with grain boundary precipitates. *Mater Lett* 199:120–23.
37. Li, D., Li, C., Feng, T., et al. 2017. High-entropy Al0. 3CoCrFeNi alloy fibers with high tensile strength and ductility at ambient and cryogenic temperatures. *Acta Mater* 123:285–94.
38. Niu, S.Z., Kou, H.C., Wang, J. and Li, J.S. 2017. Improved tensile properties of Al0.5CoCrFeNi high-entropy alloy by tailoring microstructures. *Rare Met*:1–6. doi:10.1007/s12598-016-0860-y.
39. Parmar, R.S. 2001. *Welding Processes and Technology*. Khanna Publishers.
40. Babu, N.K., Talari, M.K., Zheng, S., Dayou, P., Jerome, S. and Muthupandi, V. 2015. Arc welding. In: *Handbook of Manufacturing Engineering and Technology*, ed. A. Nee, 593–615. London: Springer.
41. Srinivasan, P.B., Muthupandi, V., Dietzel, W. and Sivan, V. 2006. An assessment of impact strength and corrosion behaviour of shielded metal arc welded dissimilar weldments between UNS 31803 and IS 2062 steels. *Mater Des* 27(3):182–91.
42. Lin, H.L. and Chou, C.P. 2009. Optimization of the GTA welding process using combination of the Taguchi method and a neural-genetic approach. *Mater Manuf Process* 25(7):631–36.
43. Srinivasan, K. and Balasubramanian, V. 2011. Effect of heat input on fume generation and joint properties of gas metal arc welded austenitic stainless steel. *J Iron Steel Res Int* 18(10):72–9.
44. Li, H.L., Liu, D., Yan, Y.T., Guo, N., Liu, Y.B. and Feng, J.C. Effects of heat input on arc stability and weld quality in underwater wet flux-cored arc welding of E40 steel. *J Manuf Process* 31:833–43.
45. Ramkumar, P., Karthikeyan, M.K., Gupta, R.K., Kumar, V.A., Magadum, C. and Muthupandi, V. 2017. Plasma arc welding of high strength 0.3% C–CrMoV (ESR). *Steel, Trans Indian Inst Met* 70(5):1317–22.

46. Li, K., Wu, Z., Zhu, Y. and Liu, C. 2017. Metal transfer in submerged arc welding. *J Mater Process Technol* 244:314–19.
47. Zhou, K. and Yao, P. 2019. Overview of recent advances of process analysis and quality control in resistance spot welding. *Mech Syst Signal Pr* 124:170–98.
48. Muthupandi, V., Srinivasan, P.B., Shankar, V., Seshadri, S.K. and Sundaresan, S. 2005. Effect of nickel and nitrogen addition on the microstructure and mechanical properties of power beam processed duplex stainless steel (UNS 31803) weld metals. *Mater Lett* 59(18):2305–09.
49. Sivaprasad, K., Raman, S.G.S., Murthy, C.V.S. and Reddy, G.M. 2006. Coupled effect of heat input and beam oscillation on mechanical properties of alloy 718 electron beam weldments. *Sci Technol Weld Join* 11(1):127–34.
50. Muralimohan, C.H., Ashfaq, M., Ashiri, R., Muthupandi, V. and Sivaprasad, K. 2016. Analysis and characterization of the role of Ni interlayer in the friction welding of titanium and 304 austenitic stainless steel. *Metall Mater Trans A* 47(1):347–59.
51. Geng, P., Qin, G., Li, T., Zhou, J., Zou, Z. and Yang, F. 2019. Microstructural characterization and mechanical property of GH4169 superalloy joints obtained by linear friction welding. *J Manuf Process* 45:100–14.
52. Sabari, S.S., Malarvizhi, S. and Balasubramanian, V. 2016. Characteristics of FSW and UWFSW joints of AA2519-T87 aluminium alloy: Effect of tool rotation speed. *J Manuf Process* 22:278–89.
53. Thiyaneshwaran, N., Sivaprasad, K. and Ravisankar, B. 2018. Nucleation and growth of TiAl3 intermetallic phase in diffusion bonded Ti/Al Metal Intermetallic Laminate. *Sci Rep* 8(1):16797.
54. Carvalho, G.H.S.F.L., Galvão, I., Mendes, R., Leala, R.M. and Loureiro, A. 2018. Explosive welding of aluminium to stainless steel. *J Mater Process Technol* 262:340–49.
55. Sokkalingam, R., Venkatesan, K., Sabari, S.S., Malarvizhi, S. and Balasubramanian, V. 2014. Effect of post-weld aging treatment on tensile properties of GTAW welded armor grade AA2519-T87 aluminum alloy joints. *Int J Res Sci Eng* 3(11):316–20.
56. Prasad, M.S., Ashfaq, M., Babu, N.K., Sreekanth, A., Sivaprasad, K. and Muthupandi, V. 2017. Improving the corrosion properties of magnesium AZ31 alloy GTA weld metal using microarc oxidation process. *Int J Min Met Mater* 24(5):566–73.
57. Venkatesan, G., Muthupandi, V. and Fathaha, A.B. 2017. Effect of oxide fluxes on depth of penetration in flux bounded tungsten inert gas welding of AISI 304L stainless steel. *Trans Indian Inst Met* 70(6):1455–62.
58. Ganguly, S., Sule, J. and Yakubu, M.Y. 2016. Stress engineering of multi-pass welds of structural steel to enhance structural integrity. *J Mater Eng Perform* 25(8):3238–44.
59. Phillips, D.H. 2016. *Welding Engineering: An Introduction*. John Wiley & Sons, Ltd.
60. Angella, G., Barbieri, G., Donnini, R., Montanari, R., Richetta, M. and Varone, A. 2017. Electron beam welding of IN792 DS: Effects of pass speed and PWHT on microstructure and hardness. *Materials* 10(9):1033. doi:10.3390/ma10091033.
61. Muthupandi, V., Srinivasan, P.B., Seshadri, S.K. and Sundaresan, S. 2003. Effect of weld metal chemistry and heat input on the structure and properties of duplex stainless steel welds. *Mater Sci Eng A* 358(1–2):9–16.
62. Mishra, R.S. and Ma, Z.Y. 2005. Friction stir welding and processing. *Mater Sci Eng R* 50(1–2):1–78.
63. Zhu, Z.G., Sun, Y.F., Ng, F.L., et al. 2018. Friction-stir welding of a ductile high entropy alloy: Microstructural evolution and weld strength. *Mater Sci Eng A* 711:524–32.
64. Dawood, H.I., Mohammed, K.S. and Rajab, M.Y. 2014. Advantages of the green solid state FSW over the conventional GMAW process. *Adv Mater Sci Eng* 2014(105713):1–10. doi:10.1155/2014/105713.

65. Wu, Z., David, S.A., Leonard, D.N., Feng, Z. and Bei, H. 2018. Microstructures and mechanical properties of a welded CoCrFeMnNi high-entropy alloy. *Sci Technol Weld Join* 23(7):585–95.
66. Wu, Z., David, S.A., Feng, Z. and Bei, H. 2016. Weldability of a high entropy CrMnFeCoNi alloy. *Scr Mater* 124:81–5.
67. Nam, H., Park, C., Moon, J., Na, Y., Kim, H. and Kang, N. 2019. Laser weldability of cast and rolled high-entropy alloys for cryogenic applications. *Mater Sci Eng A* 742:224–30.
68. Nam, H., Park, C., Kim, C., Kim, H. and Kang, N. 2018. Effect of post weld heat treatment on weldability of high entropy alloy welds. *Sci Technol Weld Join* 23(5):420–27.
69. Jo, M.G., Kim, H.J., Kang, M. et al. 2018. Microstructure and mechanical properties of friction stir welded and laser welded high entropy alloy CrMnFeCoNi. *Met Mater Int* 24(1):73–83.
70. Nene, S.S., Liu, K., Frank, M., et al. 2017. Enhanced strength and ductility in a friction stir processing engineered dual phase high entropy alloy. *Sci Rep* 7(1):16167.
71. Kashaev, N., Ventzke, V., Stepanov, N., Shaysultanov, D., Sanin, V. and Zherebtsov, S. 2018. Laser beam welding of a CoCrFeNiMn-type high entropy alloy produced by self-propagating high-temperature synthesis. *Intermetallics* 96:63–71.
72. Shaysultanov, D., Stepanov, N., Malopheyev, S., et al. 2018. Friction stir welding of a carbon-doped CoCrFeNiMn high-entropy alloy. *Mater Charact* 145:353–61.
73. Muthupandi, V., Srinivasan, P.B., Seshadri, S.K. and Sundaresan, S. 2013. Effect of nitrogen addition on formation of secondary austenite in duplex stainless steel weld metals and resultant properties. *Sci Technol Weld Join* 9:47–52.
74. Muthupandi, V., Srinivasan, P.B., Seshadri, S.K. and Sundaresan, S. 2013. Corrosion behaviour of duplex stainless steel weld metals with nitrogen additions. *Corros Eng Sci Techn* 38(4):303–08.
75. Reddy, G.M., Murthy, V.S., Rao, K.S. and Rao, K.P. 2009. Improvement of mechanical properties of Inconel 718 electron beam welds-influence of welding techniques and postweld heat treatment. *Int J Adv Manuf Technol* 43(7–8):671–80.
76. Bhattacharjee, P.P., Sathiaraj, G.D., Zaid, M., et al. 2014. Microstructure and texture evolution during annealing of equiatomic CoCrFeMnNi high-entropy alloy. *J Alloys Compd* 587:544–52.
77. Pal, S. and Phaniraj, M.P. 2015. Determination of heat partition between tool and workpiece during FSW of SS304 using 3D CFD modeling. *J Mater Process Tech* 222:280–86.
78. Zaddach, A.J., Niu, C., Koch, C.C. and Irving, D.L. 2013. Mechanical properties and stacking fault energies of NiFeCrCoMn high-entropy alloy. *Jom* 65(12):1780–89.
79. Zhu, Z.G., Ng, F.L., Qiao, J.W., et al. 2019. Interplay between microstructure and deformation behavior of a laser-welded CoCrFeNi high entropy alloy. *Mater Res Express* 6(4). http://www.ncbi.nlm.nih.gov/pubmed/046514.
80. Komarasamy, M., Kumar, N., Tang, Z., Mishra, R.S. and Liaw, P.K. 2015. Effect of microstructure on the deformation mechanism of friction stir-processed Al0.1CoCrFeNi high entropy alloy. *Mater Res Lett* 3(1):30–4.
81. Kumar, N., Komarasamy, M., Nelaturu, P., Tang, Z., Liaw, P.K. and Mishra, R.S. 2015. Friction stir processing of a high entropy alloy Al0.1CoCrFeNi. *Jom* 67(5):1007–13.
82. Zhu, Z.G., Sun, Y.F., Goh, M.H., et al. 2017. Friction stir welding of a CoCrFeNiAl0.3 high entropy alloy. *Mater Lett* 205:142–44.
83. Sokkalingam, R., Mishra, S., Cheethirala, S.R., Muthupandi, V. and Sivaprasad, K. 2017. Enhanced relative slip distance in gas-tungsten-arc-welded Al0.5CoCrFeNi high-entropy alloy. *Metall Mater Trans A* 48(8):3630–34.

84. Sokkalingam, R., Sivaprasad, K., Muthupandi, V. and Duraiselvam, M. 2018. Characterization of laser beam welded Al0.5CoCrFeNi high-entropy alloy. *Key Eng Mater* 775:448–53.
85. Nahmany, M., Hooper, Z., Stern, A., Geanta, V. and Voiculescu, I. 2016. Al x CrFeCoNi high-entropy alloys: Surface modification by electron beam bead-on-plate melting. *Metallogr Microstruct Anal* 5(3):229–40.
86. Cui, L., Ma, B., Feng, S. and Wang, X. 2014. Microstructure and mechanical properties of high-entropy alloys CoCrFeNiAl by welding. *Adv Mater Res* 936:1635–40.
87. Martin, A.C. and Fink, C. 2019. Initial weldability study on Al0.5CrCoCu0.1FeNi high-entropy alloy Al0.5CrCoCu0.1FeNi high-entropy alloy. *Weld World* 63(3):739–50.
88. Srinivasan, P.B., Muthupandi, V., Dietzel, W. and Sivan, V. 2006. Microstructure and corrosion behavior of shielded metal arc-welded dissimilar joints comprising duplex stainless steel and low alloy steel. *J Mater Eng Perform* 15(6):758–64.
89. Mortezaie, A. and Shamanian, M. 2014. An assessment of microstructure, mechanical properties and corrosion resistance of dissimilar welds between Inconel 718 and 310S austenitic stainless steel. *Int J Press Vessel Pip* 116:37–46.
90. Sokkalingam, R., Sivaprasad, K., Muthupandi, V. and Prashanth, K.G. 2019. Dissimilar welding of Al0.1CoCrFeNi high-entropy alloy and AISI304 stainless steel. *J Mater Res*. doi:10.1557/jmr.2019.186.

18 High Entropy Alloys
A Potentially Viable Magnetic Material

Rohit R. Shahi and Rajesh K. Mishra

CONTENTS

18.1 INTRODUCTION

Depending upon the requirements of mankind, alloys have been designed from simple to complex compositions. The resulting improvement in the performance of alloys empowered the advancement in civilization. In the past 100 years, the persistent work of researchers has led the significant evolution and progress in the area of metallic materials, which results in the invention of special alloys, such as stainless steel and superalloys. The conventional alloy design strategy relies on the mixing of one or two elements with the parent metal for the enhancement of the properties of individual elements. For example, the addition of carbon improved the strength of the steel, which showed higher strength than iron, and the addition of a small number of impurities to the ferromagnetic alloys changed their magnetic properties drastically. A shift in this conventional alloy design paradigm was encountered more than a decade ago when Cantor and co-workers [1] and Yeh and co-workers [2] independently announced the feasibility of multicomponent alloys later termed high entropy alloys (HEAs). The multi-principal element alloys (MPEAs) or high entropy alloys (HEAs) are based on multi-principal elements, which are defined in two different ways, based on composition and based on entropy [3,4]. Initially, HEAs were defined as alloys which contain at least five principal elements in equiatomic or near-equiatomic concentration (each having in the range of 5 to 35 at.%). If a minor element is added to the HEAs, their concentration should be less than 5 at.%. Later, the system was that those whose configurational entropy is greater than 1.5 R (R is the gas constant) are considered as HEA. Thus, based on the configurational entropy (ΔS_{con}), the alloys are classified as low, medium, and high entropy alloys (as shown in Figure 18.1) [3,4].

18.2 CORE EFFECTS OF HIGH ENTROPY ALLOYS

The presence of multi-principal elements in HEAs led to some interesting characteristics, which are very important for their technological applications. These characteristics are referred as "core effects," which includes high entropy effects, kinetic hysteresis diffusion effects, structural lattice distortion effects, and cocktail effects [3,4].

18.2.1 High Entropy Effect

Entropy is the key factor in stabilizing the random solid solution in HEAs. According to the high entropy effect, the higher mixing entropy (mainly configurational entropy) in HEAs lowers the free energy of the solid-solution phase at high temperatures and facilitates the formation of random solid-solution phases. It has been experimentally evidenced that due to the high entropy of mixing, the number of phases formed in HEAs is much lower, as predicted by the Gibbs phase rule. Thus, the entropy effect of HEAs minimized the Gibbs free energy and stabilized the solid-solution phase.

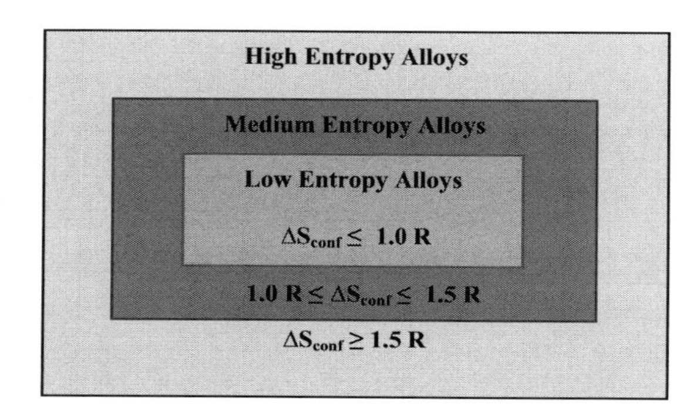

FIGURE 18.1 Classification of alloys based on configurational entropy.

TABLE 18.1
Configurational Entropy in Terms of R for Equiatomic Alloys

No. of elements (N)	2	3	4	5	6	7	8	9	10	11	12	13	14	15	16
ΔS_{conf} (in terms of R)	0.69	1.10	1.39	1.61	1.79	1.95	2.08	2.20	2.30	2.40	2.49	2.57	2.63	2.70	2.77

Moreover, the configurational entropy of the system is defined as $\Delta S_{conf} = R \ln N$ (N is the number of elements). As the number of alloying elements increases, the configurational entropy of the alloys increases. Table 18.1 represents the variation of configurational entropy with the variation of number of alloying elements. It can be observed that the enhancement in the value of entropy is negligible when the value of $N > 13$. Thus, the probability of formation of a solid-solution phase in the range of $N = 5$ to 13 is high, and for $N > 13$ the entropy effect becomes saturated.

18.2.2 Sluggish Diffusion Effect

The diffusion and phase transition kinetics in HEAs are slower than those in their conventional counterparts because in HEAs, the neighboring atoms of each lattice site is somewhat different than in the conventional alloys. This difference in local atomic configuration leads different atomic bonding and hence different local energy for each lattice site [5]. Usually, the phase transformation in HEAs requires interactive diffusion between the components to accomplish the partitioning of composition between phases. The slow kinetics in HEAs might provide several important advantages, such as nano-sized precipitates, slower grain growth, increased creep resistance, better high-temperature strength, and structural stability [6–9].

18.2.3 Severe Lattice Distortion Effect

Severe lattice distortion suggests that the crystalline structure in HEAs is deformed. The lattice in HEAs is composed of different sizes of atoms, and the difference in atomic size generates the distortion. This impedes the dislocation movements and leads to solid-solution strengthening. The severe lattice distortion effectively decreases the electrical and thermal conductivity because it markedly increases the scattering of propagating electrons and phonons [10].

18.2.4 Cocktail Effect

The HEAs are composed of different type of elements having different characteristics; the interaction between these different kinds of elements make the HEAs exhibit composite characteristics that are known as the "cocktail effect." It implies that the alloy properties can be effectively adjusted by changes in composition and alloying elements [2, 11–14]. For example, it has been reported that the addition of low-strength and low-density alloying element aluminum (Al) in $Al_xCoCrCuFeNi$ HEA

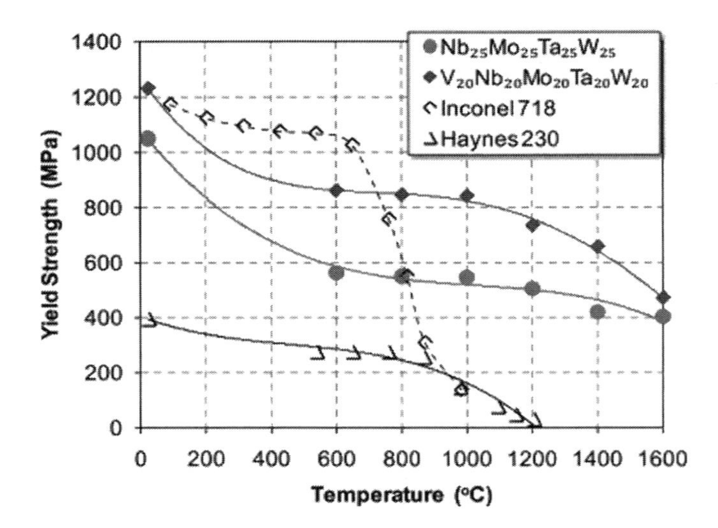

FIGURE 18.2 Temperature dependence yield strength of $Nb_{25}Mo_{25}Ta_{25}W_{25}$ and $V_{20}Nb_{20}Ta_{20}Mo_{20}W_{20}$ HEAs and two superalloys, Inconel 718 and Haynes 230. (Reprinted with permission from Senkov et al. [16].)

increased the strength and hardness of the resultant alloy [2]. The increased hardness of the HEA is associated with the formation of a body-centered cubic (BCC) phase and the stronger cohesive bonding between aluminum (Al) and other elements. It was also related to the larger atomic size of aluminum (Al) compared to the other elements present in HEA [2]. Moreover, a refractory HEA synthesized through arc melting has shown a very high melting point compared to those of nickel-based and cobalt-based superalloys [15, 16]. This is simply because of the selection of refractory elements as composing elements. They also reported that the synthesized alloys have extremely high yield strength of over 400 MPa at 1,600°C compared to the Inconel 718 and Haynes 230 conventional superalloys (as shown in Figure 18.2) [16]. Hence, the average properties of HEAs do not only exist due to the alloying elements, but exist due to their mutual interactions of alloying elements and severe lattice distortion. Therefore, for designing a suitable HEA, it is important to understand the related factors involved before selecting compositions based on the cocktail effect.

18.3 PREDICTION OF PHASE FORMATION BASED ON IMPORTANT THERMODYNAMIC PARAMETERS

From the classical Hume–Rothery rules, the formation of a solid-solution phase in conventional alloys is governed by atomic size difference, the structure of alloying elements, and the difference in electronegativity [17]. In addition to these parameters, enthalpy of mixing and entropy of mixing are the two prominent phase formation parameters for HEAs. Zhang and co-workers [18] and Guo and co-workers [19] studied the effect of these parameters on the phase formation for HEAs. They reported that the formation of a solid-solution phase in HEAs is governed mainly by three factors: these are enthalpy of mixing (ΔH_{mix}), entropy of mixing (ΔS_{mix}), and atomic size difference (δ) [18, 19]. It has been reported that the formation of simple solid-solution phases (i.e., face-centered cubic (FCC), body-centered cubic (BCC), or their mixture, including both ordered/disordered cases) will occur when the enthalpy of mixing, entropy of mixing, and atomic size difference satisfy simultaneously $-22 \leq \Delta H_{mix} \leq 7$ kJ/mol, $11 \leq \Delta S_{mix} \leq 19.5$ J/K.mol and $\delta \leq 8.5$ conditions [19]. It has also been reported that the large ΔH_{mix} leads to phase separation and large negative ΔH_{mix} favors the formation of intermetallic phases (such as μ, σ, Laves phase, etc.) [20–22]. Moreover, atomic size difference should be small enough, because large atomic size difference favors the formation of amorphous phase [19]. Apart from these thermodynamic parameters, Yang and co-workers proposed that the formation of simple solid-solution phase in multicomponent HEAs can be effectively predicted by the Ω parameter, where Ω represents the effect of entropy relative to that of enthalpy [23]. They concluded that when the value of $\Omega \geq 1.1$ and $\delta \leq 6.6\%$, the formation of a simple solid-solution phase will occur. The parameters Ω and δ are defined as follows [23]:

$$\Omega = \frac{T_m \Delta S_{mix}}{|\Delta H_{mix}|} \tag{18.1}$$

$$\delta = \sqrt{\sum_{i=1}^{n} C_i (1 - r_i/\bar{r})^2} \tag{18.2}$$

where $T_m = \sum_{i=1}^{n} C_i (T_m)_i$ is the average melting temperature of the nth element, C_i is the atomic concentration of the ith component, and $(T_m)_i$ is the melting point of the ith component of the alloy. $\Delta S_{mix} = -R\sum_{i=1}^{n} C_i \ln C_i$ is the configurational entropy of the alloy and R (=8.314 J/K.mol) is the gas constant. $\Delta H_{mix} = \sum_{i=1, i\neq j}^{n} \Omega_{ij} C_i C_j = \sum_{i=1, i\neq j}^{n} 4\Delta H_{AB}^{mix} C_i C_j$ is the enthalpy of mixing of the alloy, and ΔH_{AB}^{mix} is the enthalpy of mixing for the binary liquid in an A-B system at an equiatomic composition. $\bar{r} = \sum_{i=1}^{n} C_i r_i$ is the average atomic radius of the alloy and r_i is the radius of ith component of the alloy. In addition to this, the formed random solid-solution phase is face-centered cubic (FCC), body-centered cubic (BCC), or their mixture—it will depend on the value of valence electron count (VEC), where VEC is defined as $VEC = \sum_{i=1}^{n} C_i (VEC)_i$ and $(VEC)_i$ is the valence electron count of ith component of the alloy [24]. Guo and co-workers reported that for a large value of VEC (≥ 8.0), the face-centered cubic (FCC) phase is stabilized. For VEC ≤ 6.87, the body-centered cubic (BCC) structure is stabilized, and a mixture of these two (FCC and BCC) phases is observed when the value of the VEC lies in between 6.87 to 8.00 [24].

18.4 DIFFERENT SYNTHESIS ROUTES FOR HIGH ENTROPY ALLOYS

The traditional manufacturing process for the synthesis of conventional alloys can also be used for the synthesis of HEAs. Depending on how the composing elements

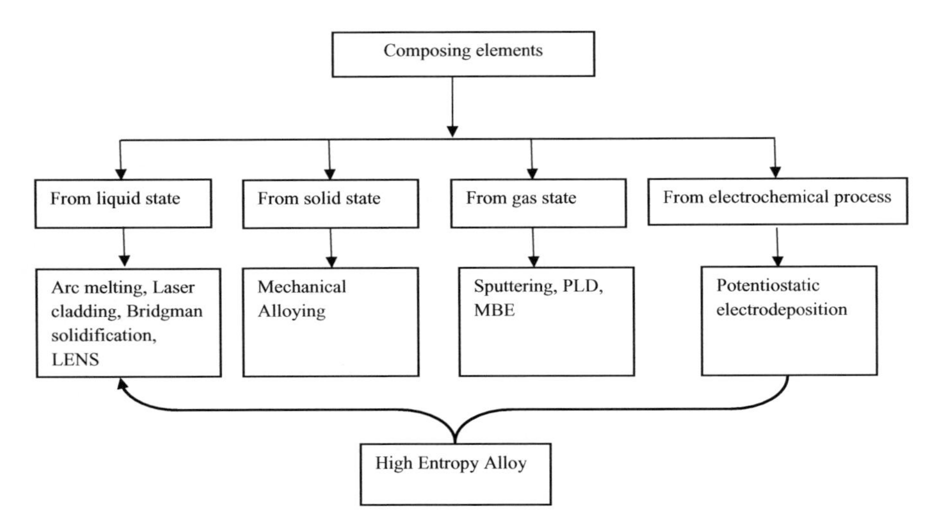

FIGURE 18.3 Different synthesis routes for high entropy alloys.

are mixed, the processes used for the synthesis of HEAs are classified into four categories [25] and shown in Figure 18.3. The first route includes arc melting, inductive melting, laser cladding, Bridgman solidification casting, and the laser-engineered net-shaping (LENS) process in which arc melting is a widely used processing route for the synthesis of HEAs [26,27]. The solid-state route is the second route, and is mainly based on mechanical alloying (MA), and subsequent consolidation processes such as spark plasma sintering (SPS) or hot-pressed sintering (HPS) [28]. The third route is the gas or vapor route, which includes sputter deposition and pulse laser deposition to prepare thin films on substrates [29,30]. The fourth electrodeposition process is also used to prepare thin films by the potentiostatic electrodeposition technique [31].

HEAs are generally synthesized through the physical method, mostly through liquid state synthesis (arc melting or induction melting) [25–27] and solid-state synthesis (mechanical alloying [MA]) [28]. The synthesis process has significant impacts on phase formation and microstructure, which affects the properties (such as mechanical, magnetic, corrosion, etc.) of the resultant HEAs [32–34]. Compared to the mechanical alloying process, the casting route has some drawbacks regarding phase segregation and inhomogeneous microstructure [35–37]. It has been reported that MA is a technique through which homogeneous nano-crystalline HEAs can be synthesized. The MA involves repeated cold welding, fracturing, and re-welding of powder particles in a high energy ball mill. Through MA one can synthesize a variety of equilibrium and non-equilibrium alloys starting from blended elemental powders, which are immiscible at liquid state [38–39].

18.5 MAGNETIC HIGH ENTROPY ALLOYS

Magnetic materials are essential components for energy applications (such as motors, generators, transformers, actuators, etc.) and improvement in magnetic materials will have a significant impact on their performance. It has been reported that about 9%

of generated electricity is lost during transmission and distribution [40]. For greater affordability in both military and commercial applications, the trend has provoked the electro-mechanical designer toward higher speed and better performance devices. Therefore, new magnetic materials are the requirement for the success of many of these power generation, distribution, and utilization systems [41]. On the other hand, the available conventional soft magnetic materials have their shortcomings, such as brittleness and low magnetization of soft ferrites, limited ductility, and low electrical resistivity of iron-nickel (Fe-Ni) alloys [42]. A better soft magnetic material is comprised of high saturation magnetization (Ms), low coercivity (Hc), high electrical resistivity (to suppress eddy current loss), and better mechanical properties (for high performance). Recently, some FeCoNi-based HEAs have shown appealing soft magnetic characteristics (high Ms and low Hc) with high electrical resistivity and excellent mechanical behavior [14, 34, 42, 43]. The reported $FeCoNiAl_{0.2}Si_{0.2}$ HEAs have shown optimal balance of better soft magnetic characteristics (Ms = 1.15 T, Hc = 1,400 A/m) with high electrical resistivity (69.5 μΩ.cm) and excellent mechanical properties (yield strength = 342 MPa and strain without fracture of 50%) [14]. Reported FeCoNiPdCu HEA has also exhibited better soft magnetic characteristics with high Curie temperature (Tc = 962 K) [43]. In addition, $FeCoNiAl_{0.25}Mn_{0.25}$ HEA synthesized trough arc melting has also evinced high Ms (101.0 emu/g), low Hc (268 A/m), high electrical resistivity (100 μΩ.cm), high Tc (1,078 K) as well as good tensile–ductility behaviors [42.] Thus, it is expected that the newly emerged soft magnetic HEAs have opened a new class of soft magnetic material which can meet all the requirements for a better soft magnetic material for high-performance applications.

18.5.1 Effect of Synthesis Routes on Structural/Microstructural Evolution and Magnetic Behavior of High Entropy Alloys

The study by Uporov and co-workers described the effect of processing of synthesized HEAs on phase evolution and magnetic behavior [44]. Equiatomic AlCoCrFeNi alloys were synthesized through arc melting and divided into three pieces. These pieces are further processed through vacuum suction casting (quenched), slow remelting in a resistance furnace (slowly re-melted), and longtime high-temperature annealing in the crystalline state (homogenized). A rod-shaped specimen (height 30 mm and diameter 3 mm) was obtained through suction casting. The re-melting process was performed in an electric vacuum furnace under a helium atmosphere at 1,600 K, and melting was done for 1 h, then cooled down to room temperature at a rate of 50 K/min. The third sample was obtained by solid-state annealing under vacuum at 1,400 K for 50 h and cooling was performed at a rate of 50 K/min. These samples were abbreviated as A, Q, R, and H for the as-cast, quenched, slowly re-melted, and homogenized samples, respectively. X-ray diffractometer (XRD) analysis confirmed that all the samples had a face-centered cubic (FCC) and a body-centered cubic (BCC) phase. The BCC solid-solution phase was dominant in as-cast, quenched and re-melted samples, while FCC phase was dominant in homogenized sample. Thus, the alloy structure is highly dependent on the synthesis/processing route, and the structure modification occurred with the processing route. As the crystal structure changed, the structural dependent properties must be changed after the processing route. Figure 18.4 represents the magnetization curves measured at two different temperatures, 300 K and 4 K for all four samples. For all the samples, the value of Hc lies in the range of 60–160 Oe. Thus, all these samples are semi-hard magnetic materials. Also, the value of Ms lies in the range 15 emu/g to 54 emu/g at room temperature. It is interesting to note that the value of Ms is lowest for homogenized sample (H) and highest for non-homogenized samples (A, Q, and R). This might be related to the formation of the FCC phase for the homogenized sample. Along with the magnetic properties, electrical and thermal conductivity also varied widely with the variation of processing route for the selected HEAs. The study also summarized that AlCoCrFeNi alloy is a promising magnetic material for functional applications and may become the alternative to amorphous ferromagnets.

The alloy ingot of $Fe_{26.7}Co_{28.5}Ni_{28.5}Si_{4.6}B_{8.7}P_3$ (atomic percent) was synthesized through induction melting [45]. Further, the ribbons of alloys were synthesized by the rapid solidification technique at two different wheel speed of 8 m/s and 32 m/s which provide different cooling rates. Figure 18.5a represents the XRD pattern of melt-spun ribbons at two different wheel speed. At high cooling rate 32 m/s only amorphous

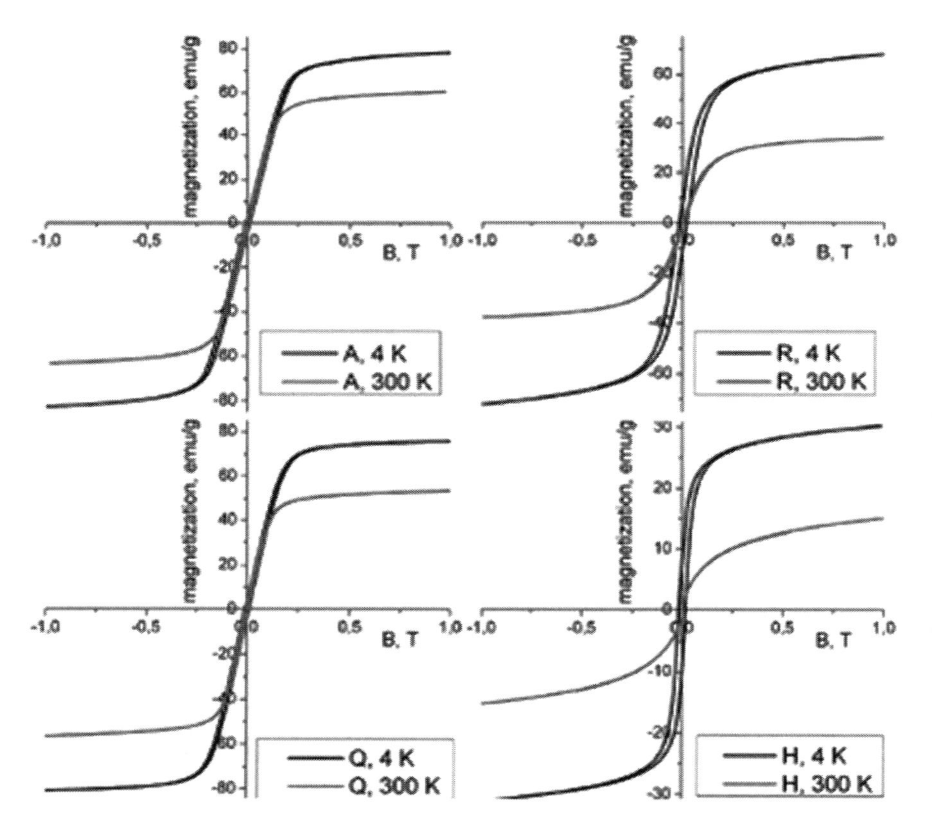

FIGURE 18.4 Magnetic-hysteresis (M-H) loop of AlCoCrFeNi high entropy alloy prepared by different routes; the symbol references the alloys synthesized through different ways mentioned in the text. (This figure is reproduced with permission from Uporov et al. [44].)

phase structure are formed for $Fe_{26.7}Co_{28.5}Ni_{28.5}Si_{4.6}B_{8.7}P_3$. While the face-centered cubic (FCC) phase along with the minor ordered phase has been observed for ribbons synthesized with a low cooling rate of 8 m/s, thus, an amorphous, as well as a crystalline solid-solution phase, can be synthesized for $Fe_{26.7}Co_{28.5}Ni_{28.5}Si_{4.6}B_{8.7}P_3$ alloy by controlling the cooling rate. Figure 18.5b represents the M-H loop of $Fe_{26.7}Co_{28.5}Ni_{28.5}Si_{4.6}B_{8.7}P_3$ HEA synthesized at different cooling rates. Both the samples exhibit typical soft magnetic characteristics. The amorphous phase has larger Ms (1.07 T) and low Hc (4 A/m) than the crystalline solid-solution phase (Ms = 1 T and Hc = 168 A/m). Ms is highly dependent upon the phase structure. The low Ms of solid-solution phase related to the fact that high symmetry of FCC phase neutralized atomic magnetic moment. However, amorphous alloys are without symmetry as compared to solid-solution phase, and hence neutralization is minimum. The high Hc of FCC phase is related to the larger grain size of HEAs.

The magnetic behavior of HEAs is sensitive to the content of ferromagnetic elements and phase formation. In the MA process, the formation of phases in an alloy can be affected by changing the MA parameters (such as the ball to powder ratio [BPR], RPM [revolutions per minute], etc.). Therefore, to improve the magnetic properties of CrFeMnNiTi HEA, we have studied the effect of the ball to powder ratio (BPR, 40:1) on the phase formation and their correlation with magnetic properties of CrFeMnNiTi HEA [46]. We also explored the effect of annealing temperatures and conditions on phase formation and magnetic behavior of CrFeMnNiTi HEA [46]. In the as-synthesized state, XRD analysis confirmed the formation of a mixture of face-centered cubic (FCC), body-centered cubic (BCC), and a minor fraction of σ-phase. After annealing at 500°C and 700°C under vacuum conditions, the relative volume fraction of synthesized phases has been changed (as shown in Figure 18.6a). Moreover, after annealing at the same temperatures under open air condition, the relative volume fraction of the synthesized FCC and BCC phases have also been changed, along with the formation of $(Fe,Ti)_3O_4$ type spinel and Mn_2O_3 type oxide phase (Figure 18.6b). The synthesized HEA exhibited ferromagnetic behavior with Ms (13.39 emu/g) and Hc (162.2 Oe) (Figure 18.6c). After annealing at 500°C under vacuum and open-air, the value of Ms increased more than twice compared to the synthesized HEA (Figure 18.6d,e). For 500°C vacuum and open-air annealed HEAs,

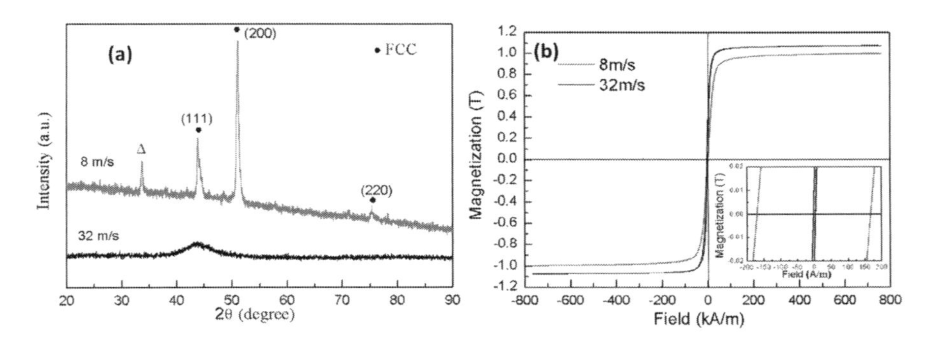

FIGURE 18.5 (a) XRD and (b) M-H loop of alloys synthesized at different cooling rates. (The figure is reproduced with permission from Wei et al. [45].)

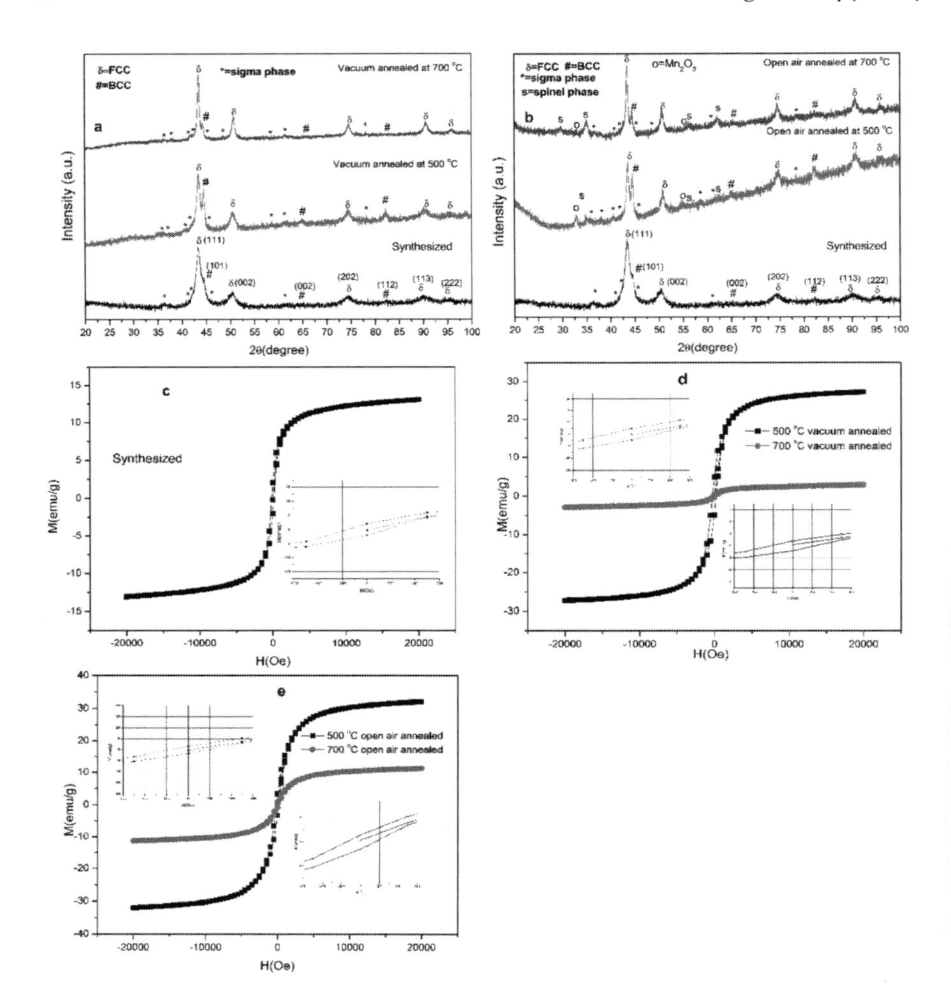

FIGURE 18.6 XRD patterns of (a) synthesized, vacuum annealed at 500°C and 700°C, (b) open-air annealed at 500°C and 700°C, M-H curve of (c) synthesized, (d) vacuum annealed at 500°C and 700°C, and (c) open-air annealed at 500°C and 700°C CrFeMnNiTi HEAs. (Reprinted with permission from Mishra and Shahi [46].)

the value of Ms was found to 27.96 emu/g and 32.85 emu/g, respectively. For 700°C annealed HEAs (under vacuum and open-air conditions), the value of Ms decreased substantially. Thus, it has been found that 500°C open-air annealed HEA exhibited better semi-hard magnetic behavior compared to the synthesized and annealed HEAs. The enhanced value of Ms for 500°C open-air annealed HEA was associated with the increase in the volume fraction of BCC phase and the formation of $(Fe,Ti)_3O_4$ type spinel and Mn_2O_3 type oxide phases. However, the decrement in the value of Ms for 700°C annealed HEAs was associated with the decrease in BCC phase content [46].

The study of Zuo and co-workers described that the directional solidification technique could also be applied to improve the soft magnetic properties of HEA [47]. Bridgman solidification has been utilized for the improvement of soft magnetic properties of the $FeCoNiAl_{0.2}Si_{0.2}$ alloy by controlling the grain morphology and crystallographic texture. The results indicated that for the sample produced by this technique, coercivity could be greatly reduced to 3.97 Oe (approx. 315 A/m) at the withdrawal velocity of 200 mm/s when the applied magnetic field direction was parallel to the crystal growth direction, which was much smaller than the value obtained for the as-cast sample (Hc approx. 1,400 A/m), although the saturation magnetization changed slightly. However, when the direction of the applied magnetic field was different, the alloy possessed high magnetic anisotropy. Thus, both the grain size and crystal orientation can influence the coercivity value significantly. At a withdrawal rate that was very slow (30 mm/s) or fast (200 mm/s), a comprehensive magnetic performance was obtained.

A simple solid solution of face-centered cubic (FCC) and body-centered cubic (BCC) phase of an equiatomic CoCrFeNiMn high entropy alloy with a refined microstructure of 10 nm has been synthesized through 60 h mechanical alloying at 250 rpm [48]. Since MA is a non-equilibrium technique for the synthesis of HEAs, an extensive amount of internal stress is stored in the lattice and defects such as lattice distortion and twins. The internal stress can be released by sintering at high temperatures, and the metastable state was transformed into a stable state. After spark plasma sintering (SPS) consolidation, the BCC phase disappeared, and only one FCC phase was preserved along with a peak shift toward a lower 2θ position. It was found that as-milled powder has agglomerated elliptical shape particle of size ~2.36 μm with thickness less than 1 μm (shown in Figure 18.7a). The nano-crystallinity of the synthesized CoCrFeNiMn HEA powder were also characterized by the transmission electron microscopy (TEM) bright-field image and the selected area

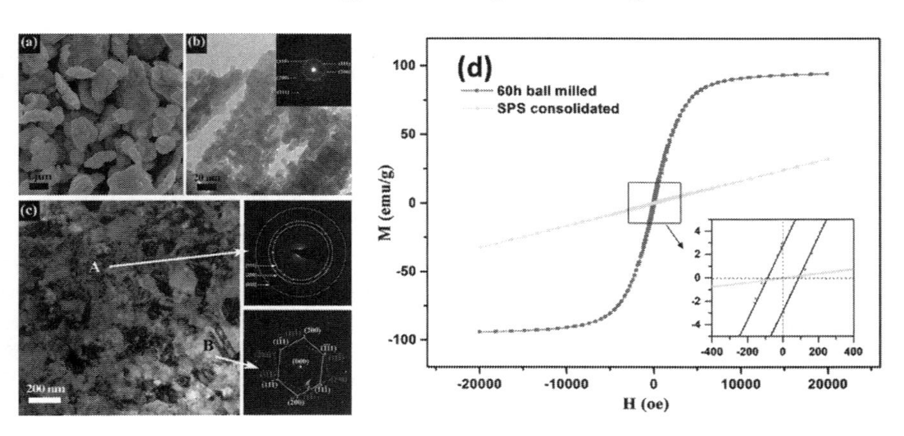

FIGURE 18.7 (a) Scanning electron microscopy (SEM), (b) transmission electron microscopy (TEM) micrographs of as prepared CoCrFeNiMn HEA, (c) TEM bright-field image of SSPed CoCrFeNiMn HEA, (d) M-H loop of as-prepared and SSPed CoCrFeNiMn HEA. (Reprinted with permission from Ji et al. [48].)

electron diffraction (SAED) patterns (shown in Figure 18.7b), and it was found that the average grain size of the particles is 10 nm in the bright-field TEM image with rings in the SAED pattern. These rings are indexed with BCC and FCC phase. After SPS, it has been found that sample has two different sizes of grains: one ranges from 100–200 nm and the other is approximately 50 nm (Figure 18.7c). Both have an FCC phase with a lattice parameter of 3.589 Å and 3.590 Å, respectively. The SPS sample exhibited high compressive strength of 1,987 MPa. Figure 18.7d represents the M-H loop of CoCrFeNiMn high entropy alloys. It exhibited better magnetic properties with saturated magnetizations (Ms), remanence ratio (Mr/Ms), and coercivity force (Hc) of 94.29 emu/g, 2.95%, and 175.68 Oe, respectively.

$FeCoNiMn_{0.25}Al_{0.25}$ HEA has been synthesized through the vacuum-arc melting [42]. The plates of the synthesized alloys were cold rolled and recrystallization was performed through 1 h annealing of the rolled sample at 900°C. It has been reported that the synthesized phase has face-centered cubic (FCC) structure, and this structure is stable after rolling and annealing treatment of synthesized HEA. The as-cast, cold rolled, and annealed $FeCoNiMn_{0.25}Al_{0.25}$ alloy is stable up to 1,000°C. Thus, the synthesized HEA was quite stable with variation of stress and temperature. The HEA at different states is easily magnetized and has Ms ~100 emu/g and coercivity lower than 1,000 A/m, suggested by the soft magnetic properties of $FeCoNiMn_{0.25}Al_{0.25}$ HEA. Moreover, these alloys showed high Curie temperature as well as good tensile ductility. Table 18.2 showed the variation of magnetic properties with the variation of shape deformation and annealing of $FeCoNiMn_{0.25}Al_{0.25}$ HEA. The Ms slightly increased, while the coercivity and permeability changed significantly because the Ms is affected by composition and crystal structure while Hc is highly sensitive to grain size, stress, presence of impure phase, deformation, and heat treatment. The reason behind the increase of Hc from 268 A/m (as-cast) to 625 A/m (cold rolled) was related to the fact that the internal stress and defect increased after cold rolling. Furthermore, Hc decreased to 230 A/m after annealing was related to fact that the internal stress released after heat treatment. Also, it is a well-established fact that the initial permeability and Hc behave in a contrary manner [49]. The high Curie temperature (Tc) 800°C has been found for annealed $FeCoNiMn_{0.25}Al_{0.25}$ alloy, which is suitable for its potential application in high-temperature area [50, 51].

Bazzi and co-workers successfully demonstrated the significant decrease in the intrinsic coercivity (Hc_i) for $FeCoNiAl_{0.375}Si_{0.375}$ HEA [52]. The X-ray diffractometer (XRD) and transmission electron microscopy (TEM) analysis of the sample confirmed the formation of a face-centered cubic (FCC) phase and the reduction in Hci obtained with the formation of this supersaturated FCC phase [52]. Further, the detailed study described the evolution of magnetic properties of nano-crystalline $FeCoNiAl_{0.375}Si_{0.375}$ HEA as a function of temperature. The Ms, Mr, and Hci were found to be ~90.8 ± 2.7 Am^2/kg and ~3.7 ± 0.3 Am^2/kg and ~5.1 ± 0.1 kA/m, respectively. The value of Ms is 1.5 times less than the Ms (~136 Am^2/kg) of the nano-crystalline FeCoNi alloy, while the reported value of Hci for FeCoNi alloy is ~2 kA/m [53]. The presence of aluminum (Al) and silicon (Si) in the FCC phase is responsible for the slight increase in the value of Hci. It can be observed that the value of Ms decreased as the content of Al and Si increased, this happens because the Ms is sensitive to composition and as the content of ferromagnetic elements

TABLE 18.2
Crystal Structures, Saturated Magnetization (Ms), Coercivity (Hc), Permeability (μ_e) at 1 kHz, Curie Temperature (Tc), Electrical Resistivity (ρ), Yield strength ($\sigma_{0.2}$), Tensile Strength (σ_{ul}), Elongation to Failure (δ_r), and Hardness of $FeCoNiMn_{0.25}Al_{0.25}$ Alloys [42]

Alloy state	Crystal structure	Ms (emu/g)	μ_e	Hc (A/m)	Tc (°C)	ρ ($\mu\Omega$.cm)	$\sigma_{0.2}$ (MPa)	σ_{ul} (MPa)	δ_r (%)	Hardness (HV)
As-cast	FCC	101.0	297.2	268	805	100	138.1	483.9	58.1	150.6
Cold rolled	FCC	104.1	71.5	625	810	110	662.7	1029.6	7.9	357.2
Annealed	FCC	104.2	479.6	230	812	91	331.4	651.0	47.9	175.8

decreased the value of Ms decreased. An abundance of body-centered cubic (BCC) phase has been obtained for arc-melted $FeCoNiAl_{0.4}Si_{0.4}$ alloy and the value Hci is ~18 kA/m [14]. The value of Hci for the same composition synthesized through the melt-spun/mechanically milled $FeCoNiAl_{0.4}Si_{0.4}$ alloy decreased from ~14 kA/m to as low as ~8 kA/m as the phase fraction of BCC/FCC decreased from ~9 to ~2. It was concluded by Bazzi and co-workers in the composition range $0.2 < x < 0.5$, the non-equilibrium fabrication technique like MA facilitates the formation of the FCC phase that exhibited a significantly low Hci [52]. At cryogenic temperatures, the Ms increased with the decrease in temperature, while Hci and Mr exhibited a negligible effect. At elevated temperatures, above ~800 K, the supersaturated FCC phase was unstable and dissociated into BCC and FCC phases. The heat treatment (400–800 K) of synthesized HEA resulted in a significant improvement in soft magnetic properties of both Hci and Mr (decreased by ~9% and ~24%, respectively), while Ms increased by ~3% [52].

18.5.2 Effect of the Content of Alloying Elements on Phase Evolution and Magnetic Properties of High Entropy Alloys

Qiushi and co-workers investigated the effect of content of boron (B) on microstructure and properties of AlCoCrFeNiBx HEA (x = 0, 0.1, 0.25, 0.50, 0.75, 1.0, denoted by B0, B0.1, B0.25, B0.5, B0.75 and B1.0, respectively) synthesized through vacuum arc melting [54]. It was reported that the mixture of BCC and B2 phase had been found for x = o and when the content of boron increased (x= 0.1 to 0.25) FCC phase started to appear. Moreover, with a further increase of x from 0.25, the intensity of both phases decreased, and the formation of (Cr, Co)-borides phase occurred. Face-centered cubic (FCC) phase increased with the addition of boron (B) element. Hence the plasticity and strength increased. Moreover, with the increase in the content of boron (B), value of Ms decreased because the content of ferromagnetic elements decreased and the coercivity are found to be 3.68×10^3, 2.48×10^3 and 1.76×10^3 A/m for x = 0, x = 0.25, x = 0.75, respectively.

$FeCoNi(CuAl)_{0.8}Ga_x$ ($0 \leq x \leq 0.08$) high entropy alloys (HEAs) have been synthesized by vacuum arc melting and studied for the phase formation and their correlation with magnetic properties [55]. The XRD analysis confirmed that all the samples contain a duplex phase of BCC and FCC. Moreover, the formed BCC phase is copper (Cu) and gallium (Ga) enriched and existed in the phase boundary region. The M-H loops for the alloys are shown in Figure 18.8a, the enlarged view of M-H loop shown in Figure 18.8b. The found values of maximum magnetic flux density (Bm), remanence (Br), coercivity (Hc), hysteresis loss (Pu), initial permeability (μ_i) and maximum permeability (μ_{max}) of the as-cast $FeCoNi(CuAl)_{0.8}Ga_x$ ($0 \leq x \leq 0.08$) HEAs shown in Figure 18.8c. How these parameters are varied with the intensity of FCC and BCC phases are also shown in the figure. The value of Ms for x = 0, i.e. $FeCoNi(CuAl)_{0.8}$ HEA is 78.6 Am^2/kg. The found value of Ms, in this case, is similar to the value of Ms reported for the cylindrical rod of $FeCoNi(CuAl)_{0.8}$ HEA (78.9 Am^2/kg) [56]. It is clear from Figure 18.8c. The value of Ms increased gradually as the content of Ga increased to 0 to 0.08. However, other magnetic parameters Bm, Br, Hc, and Pu increased monotonously as x increased while, value of μ_i and μ_{max}

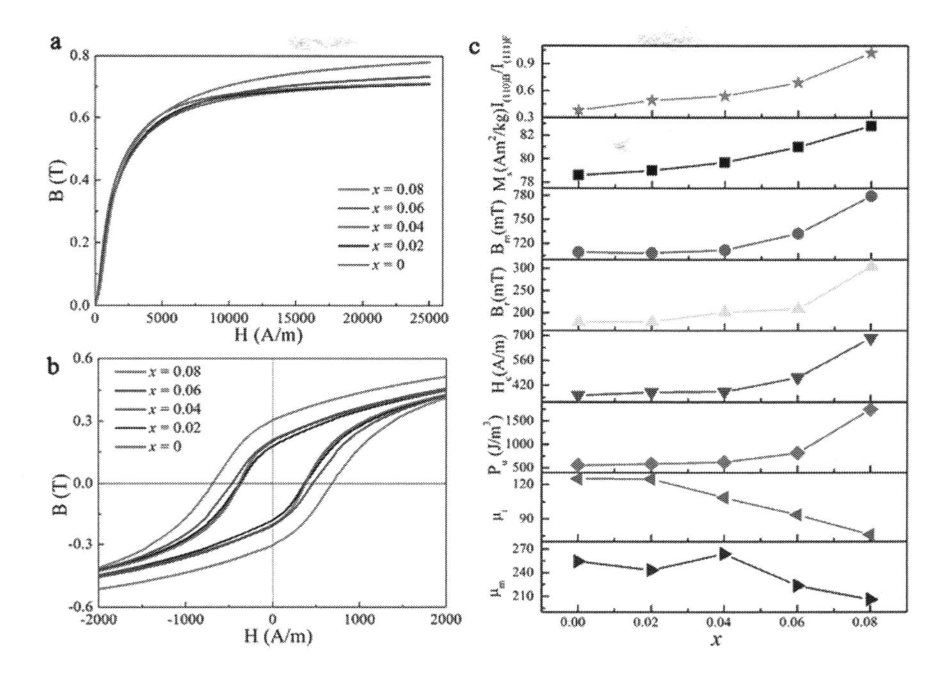

FIGURE 18.8 (a) Magnetization curve, (b) portion of hysteresis loop, (c) magnetic properties and $I_{(110)B}/I_{(111)F}$ as a function of x for the as-cast $FeCoNi(CuAl)_{0.8}Ga_x$ ($0 \leq x \leq 0.08$) HEAs. (The figure is reproduced with permission from Li et al. [55].)

decreased and have decreasing trend with increasing value of x. Because, the value of Ms, Bm, and Br are mainly dependent upon the crystal structure and composition while, Hc, μ_i, and μ_{max} dominantly affected by internal stress, grain size, impurity, heat treatment, etc. [42, 57]. For $FeCoNi(CuAl)_{0.8}$ HEA, the values of Bm, Br, Hc, Pu, μ_i, and μ_{max} was found to be 709.3 mT, 179.5 mT, 362.0 A/m, 558.6 J/m³, 124.8 and 254.3, respectively. Interestingly from Figure 18.8c, it can also be concluded that the values of Ms, Bm, Br, Hc, and Pu depend on the ratio of $I_{(110)B}/I_{(111)F}$ and consistent with it. On the other hand, the change of μ_i and μ_{max} is contrary to that of $I_{(110)B}/I_{(111)F}$. It clearly confirmed that the body-centered cubic (BCC) phase formation enhanced the value of Ms, Bm, Br, Hc, and Pu while the values of μ_i and μ_{max} increased by the formation of FCC phase [55].

A series of HEAs CoxCrCuFeMnNi (x=0.5, 1.0, 1.5, 2.0 mol) were synthesized by the variation of content of cobalt (Co) through mechanical alloying technique and studied for the effect of milling duration, Co content and annealing on phase structure stability and magnetic properties [58]. The alloys were synthesized by 45 h dry milling followed by 5 h wet milling in ethanol. Furthermore, synthesized alloys were heat-treated at three different temperatures 700°C, 800°C, and 900°C under the vacuum of $6{\times}10^{-3}$ Pa. After 30 h of MA, a primary FCC and secondary BCC phase were formed for $Co_{0.5}CrCuFeMnNi$ HEA. However, only one FCC phase was detected for all other HEA after 30h of MA. Further milling up to 45 and 50 h, the FCC peak gradually broadened and this is associated with crystallite size refinement.

The detailed analysis of 50 h milled XRD pattern of CoxCrCuFeMnNi HEA confirmed that for x = 0.5 mixed FCC and BCC phase was formed which was further converted to single FCC phase with the increase of the content of Co. Thus, the formation of higher atomic packing efficiency structure is facilitated with the addition of Co. Similar behavior was also observed for several other HEAs $CoxCrFeNiTi_{0.3}$, $Ti_{0.5}CrFeNiAlCox$, and FeCoxNiCuAl alloys [59–61]. The formation of nano-crystalline phases has also been confirmed through TEM analysis. The particle size analysis of $Co_{1.0}CrCuFeMnNi$ HEA powders with different milling duration and 50 h MA CoxCrCuFeMnNi HEA alloy powder is also shown in Figure 18.9a. It was found that the particle size of initial powder, 5 h MA, 10 h, 15 h, and 50 h MA powder sample of $Co_{1.0}CrCuFeMnNi$ was found to be 25 μm, 65 μm, 47 μm, 36 μm and 21 μm, respectively. Moreover, the particles of CoxCrCuFeMnNi alloys are distributed in size range of 10–40 μm (Figure 18.9b). The annealing experiment has also been performed for CoxCrCuFeMnNi HEA in the temperature range of 700–900°C. For annealed $Co_{0.5}CrCuFeMnNi$ HEA powder at 700°C, body-centered cubic (BCC) phase disappeared, and two FCC 1 and FCC 2 phases were formed along with some metastable ρ phase ($Cr_5Fe_6Mn_8$). This ρ phase ($Cr_5Fe_6Mn_8$) disappeared after annealing at 900°C. All other synthesized HEAs undergo a similar phase transformation after annealing at 700–900°C. The Figure 18.9c–f, represents the M-H loops for the 50 h ball milled CoxCrCuFeMnNi HEA powders and the value of Ms and Hc for x = 0.5, 1,1.5 and 2 are found to be 21, 32, 40, 52 emu/g and 63, 27, 19, and 14 Oe, respectively. The value of Ms increased, and Hc decreased gradually with the increase in the content of Co. For $Co_{2.0}CrCuFeMnNi$ HEA powder, the value of Hc is 14 Oe, which is lowest among other four HEAs. Figure 18.9g shows the relationship between Hc and crystallite size and found that the Hc decreased with the decrement of crystallite size. Further the effect of MA on the magnetic characteristics has also been investigated for $Co_{2.0}CrCuFeMnNi$ HEA. The value of Ms, Hc and remanence ratio of initial powder were found to be 105 emu/g, 56 Oe, and 4%, respectively. As the milling duration increased the value of Ms gradually decreased while the behavior in the variation of Hc is irregular (Figure 18.9 h) and unpredictable similar to the case of FeSiBAlNi(Nb) [62] and FeSiBAlNiC(Ce) [63] HEA powders. The value of Ms and Hc for 50 h MA $Co_{2.0}CrCuFeMnNi$ HEA are found to be 52 emu/g and 14 Oe, respectively. In comparison to other recently reported HEAs, 50 h MA $Co_{2.0}CrCuFeMnNi$ HEA showed smaller coercivity force and hence better soft magnetic properties. It is well known that saturation magnetization depends on the crystal structure, composition, and phase structure amount in the alloys. For CoxCrCuFeMnNi HEA, as the content of Co increased the mixed-phase structure of major FCC and minor BCC transformed into a single FCC phase. Therefore, the value of Ms for $Co_{2.0}CrCuFeMnNi$ HEA powder was found to be the highest as compared to other HEAs synthesized in this study. Furthermore, the value of Ms decreased as the milling duration increased. This is related to the fact that the number of boundaries and high dislocation density increases with increasing milling duration.

The effect of phase formation and magnetic and mechanical properties of FeCoNiCuAl has been studied with the variation of the content of copper-aluminum (CuAl) [56]. Different HEAs with the variation of CuAl content, i.e., FeCoNi(CuAl)*x*

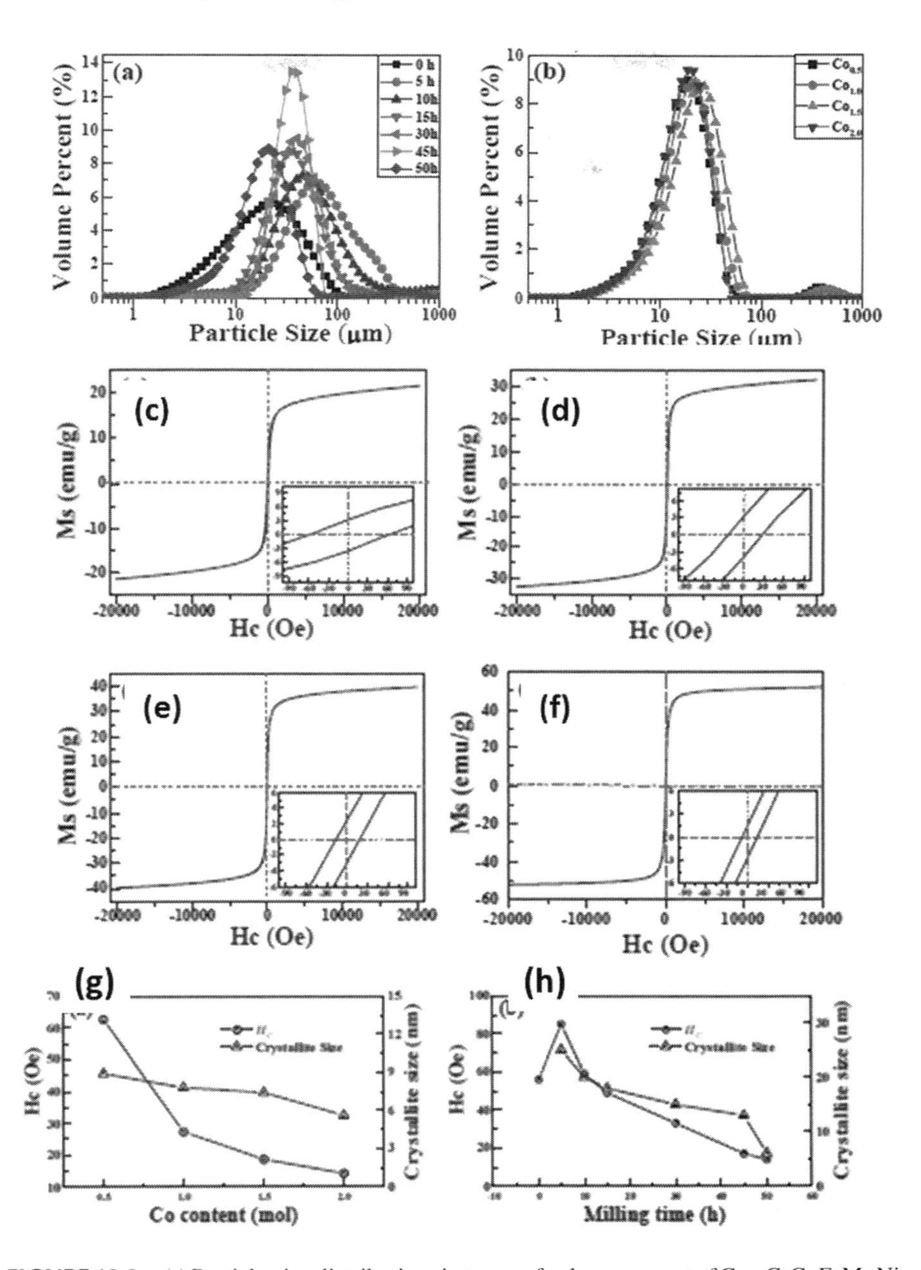

FIGURE 18.9 (a) Particle-size distributions in terms of volume percent of $Co_{1.0}$CrCuFeMnNi HEA powders under different milling times, (b) particle-size distributions of the 50 h ball milled CoxCrCuFeMnNi HEA powders, M-H loops of the 50 h ball milled CoxCrCuFeMnNi HEA powders (c) $Co_{0.5}$, (d) $Co_{1.0}$, (e) $Co_{1.5}$, (f) $Co_{2.0}$, relationship between Hc and crystallite size of (g) CoxCrCuFeMnNi HEA after 50 h of milling, and (h) $Co_{2.0}$CrCuFeMnNi powder after different milling times. (Reprinted with permission from Zhao et al. [58].)

($x = 0 - 1.2$) have been synthesized through the arc melting. XRD analysis confirmed that for x = 0–0.6 FCC phase was formed for the as-cast FeCoNi(CuAl) x HEAs. With increasing the content of x to 0.7 most prominent peak of BCC is started to appear and as x reached to 0.9 the X-ray diffraction peaks of FCC phase become nearly invisible. For x lies in the range of 0.9–1.2, BCC has appeared as a major phase and FCC phase as a minor content. Thus, for $x \leq 0.6$ single FCC phase, for $0.7 \leq x \geq 0.9$ duplex (FCC + BCC) phase and for x lies in range 0.9 to 1.2, BCC phase fraction was major. It has been found that Ms decreased as the content of CuAl increased with slight deviation in Ms when x lies in between 0.8–0.9. Generally, the Ms is highly dependent on the phase constitution (i.e., composition and atomic structure) other than the microstructure, e.g. morphology and grain size. It can also understand by the fact that at room temperature iron (Fe), cobalt (Co), and nickel (Ni) are ferromagnetic. However, copper (Cu) is diamagnetic, aluminum (Al) is paramagnetic [64]. Thus, as the content of non-magnetic atoms increased the Ms of the FeCoNi(CuAl)x (x = 0–1.2) HEAs decreased. However, an interesting phenomenon has been observed in the range of x = 0.8 to 0.9 similar behavior had also been reported for AlxCoCrFeNi [10] and AlxCoFeNi [65] HEAs. The effect of annealing (at 573–673 K) on magnetic behavior was also investigated by the author for FeCoNi(CuAl)x (x = 0-1.2) HEAs. Heat treatment has no effect on single-phase alloys. Only the duplexed phase alloys showed improvement in Ms after annealing. The value of Ms for $FeCoNi(CuAl)_{0.8}$ alloy increased from 78.9 Am^2/kg (as-cast) to 93.1 Am^2/kg after annealing at 673 K. Thus, the variation of particular element in the alloy system and heat treatment can tune the magnetic properties of HEAs.

An investigation has also been performed for the effect of content of manganese–aluminum (MnAl) on magnetic and other physical properties of FeCoNi(MnAl)x ($0 \leq x \leq 2$) high entropy alloys [57]. FeCoNi(MnAl)x ($0 \leq x \leq 2$) alloys with compositions FeCoNi (x = 0), $FeCoNiMn_{0.25}Al_{0.25}$ (x = 0.5), $FeCoNiMn_{0.5}Al_{0.5}$ (x = 1), $FeCoNiMn_{0.75}Al_{0.75}$ (x = 1.5), and FeCoNiMnAl (x = 2) were synthesized through vacuum-arc melting. Figure 18.10a represents the X-ray diffractometer (XRD) patterns of as-cast FeCoNi(MnAl)x HEAs for different values of x lies in between $0 \leq x \leq 2$. From this figure, it can be concluded that both FeCoNi and $FeCoNiMn_{0.25}Al_{0.25}$

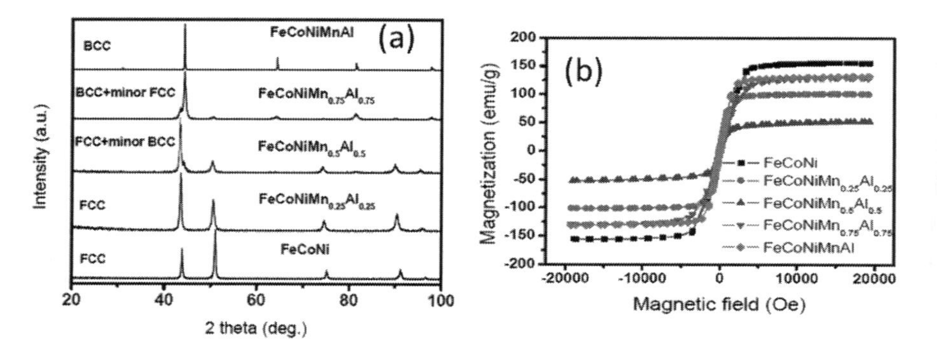

FIGURE 18.10 (a) XRD patterns (b) M-H loops at room temperature of as-cost FeCoNi(MnAl)x HEAs. (The figure is reproduced with permission from Li et al. [57].)

alloys have an FCC solid-solution phase with $FeCoNiMn_{0.25}Al_{0.25}$ alloy has larger lattice parameter than FeCoNi due to the large atomic radii of aluminum (Al) and manganese (Mn) as compared to iron (Fe), cobalt (Co), and nickel (Ni). The dual FCC and BCC phase are formed for x = 0.5 and x = 0.75. However, for $FeCoNiMn_{0.5}Al_{0.5}$ alloy FCC phase is dominated as compared to the BCC phase, while BCC is a major phase for $FeCoNiMn_{0.75}Al_{0.75}$ alloy. Furthermore, for equiatomic FeCoNiMnAl alloy, a single BCC phase is formed. Thus, it can be concluded that as the content of x increases, the phase structure gradually changed from FCC to BCC; this may have happened because the increase in the content of aluminum (Al), and it is well-known that aluminum (Al) is a BCC phase stabilizer [66]. The magnetization curves for the as-cast FeCoNi(MnAl)x ($0 \leq x \leq 2$) alloys at room temperature are as shown in Figure 18.10b. As can be seen, all these alloys are easily magnetized and have high magnetization in the rage of Ms = 51.9–155.7 emu/g with low Hc in the range Hc = 2.37–9.15 Oe (or 189–730 A/m) indicated that all these alloys have the potential to be used as soft magnetic materials. Moreover, the Ms first decreased (from FeCoNi to $FeCoNiMn_{0.5}Al_{0.5}$ alloy), then increased (from $FeCoNiMn_{0.5}Al_{0.5}$ to FeCoNiMnAl alloy) as the content of x increased, as we know the saturation magnetization highly dependent on the composition and crystal structure, which affect the magnetic exchange interaction. It is interesting to note that as the lattice parameter of the FCC phase increased, the value of Ms decreased, while the value of Ms increased as the lattice parameter of the BCC phase increased from $FeCoNiMn_{0.75}Al_{0.75}$ alloy (129.6 emu/g) to FeCoNiMnAl alloy (132.2 emu/g). This is the reason for the lowest value of Ms (59.1 emu/g) for $FeCoNiMn_{0.5}Al_{0.5}$ alloy. However, the coercivity (Hc) is affected by the ease of movement of domain walls, which depends on many factors like particle size, stress, defects, microstructural variables, etc. The enhancement in the content of AlMn increased the stress and may unavoidably affect the magnetic domain wall movement. This effect is pronounced for the case of $FeCoNiMn_{0.5}Al_{0.5}$ alloy due to the coexistence of both the BCC and FCC phase, which hindered the domain wall movement greatly. The study demonstrated that FeCoNi(MnAl)x HEAs possessed good mechanical and magnetic properties indicating their potential applications as soft magnetic materials.

Liu and co-workers synthesized AlCoCuFeNix HEAs through arc melting and investigated the effect of composition and phase constitution on the mechanical properties and magnetic performance for different concentrations of nickel (Ni) (x = 0.5, 0.8, 1.0, 1.5, 2.0, 3.0 in molar ratio) [67]. For the low content of Ni, a mixture of FCC, BCC, and ordered BCC phase has been formed. As the content of Ni increased, a single FCC phase was formed. The plasticity enhanced with reduction in hardness as the change in phase occurred. An interesting soft magnetic property has been observed for AlCoCuFeNix HEAs. As we know, the soft magnetic properties depend on composition and phase transformation. The optimal balance of mechanical and magnetic properties was found for $AlCoCuFeNi_{1.5}$ HEA. The Curie temperature for the $AlCoCuFeNi_{1.5}$ HEA has been found to be >900 K, and phase stability is below 1,350 K. The compressive strength, fracture strain, and Ms was found to be 1,725 MPa, 35.9%, and 63.58 emu/g, respectively.

The $Fe_xCo_{1-x}NiMnGa$ (FeNiMnGa, $Fe_{0.5}Co_{0.5}NiMnGa$, and CoNiMnGa) alloys were synthesized by arc melting for better soft magnetic and mechanical behavior [68].

The XRD analysis confirmed that as-cast alloys have a BCC structure along with some other peaks for FeNiMnGa and $Fe_{0.5}Co_{0.5}NiMnGa$ alloys. With the addition of cobalt (Co), the most intense (110) peaks of CoNiMnGa and $Fe_{0.5}Co_{0.5}NiMnGa$ alloys become stronger, it clearly indicated that these two alloys might have the preferred grain orientation. The typical hysteresis loops of FeNiMnGa, CoNiMnGa, and $Fe_{0.5}Co_{0.5}NiMnGa$ at T = 5 K and T = 300 K are shown in Figure 18.11, before the saturation. All three alloys having similar magnetization behavior which clearly indicates that they have similar magnetic susceptibility at these temperatures. The values of saturation magnetization for CoNiMnGa, $Fe_{0.5}Co_{0.5}NiMnGa$, and FeNiMnGa alloys are found to be 115.92 emu/g, 78.48 emu/g, and 37.79 emu/g, respectively at 300 K. The value of Ms increased to 124.73 emu/g, 92.24 emu/g, and 81.73 emu/g for CoNiMnGa, $Fe_{0.5}Co_{0.5}NiMnGa$, and FeNiMnGa alloys, respectively at T = 5 K. The value of Ms increased as the temperature decreased and for FeNiMnGa alloy at 5 K, the value of Ms is two times larger than that of Ms at 300 K. For FeNiMnGa, $Fe_{0.5}Co_{0.5}NiMnGa$, and CoNiMnGa alloys, the value of Hc is 28.4 Oe, 27.4 Oe, and 27.9 Oe, respectively, at 300 K. As the temperature decreased to 5 K, the value of Hc is about 26.6 Oe and 27.4 Oe for the $Fe_{0.5}Co_{0.5}NiMnGa$ and CoNiMnGa alloys, respectively. However, the value of Hc decreased for FeNiMnGa at 5 K to 20.8 Oe which is significantly smaller than its value at 300 K. Moreover, the yield strength for FeNiMnGa, $Fe_{0.5}Co_{0.5}NiMnGa$, and CoNiMnGa alloys was found to be 663.5 MPa, 801.8 MPa, and 567.2 MPa, respectively.

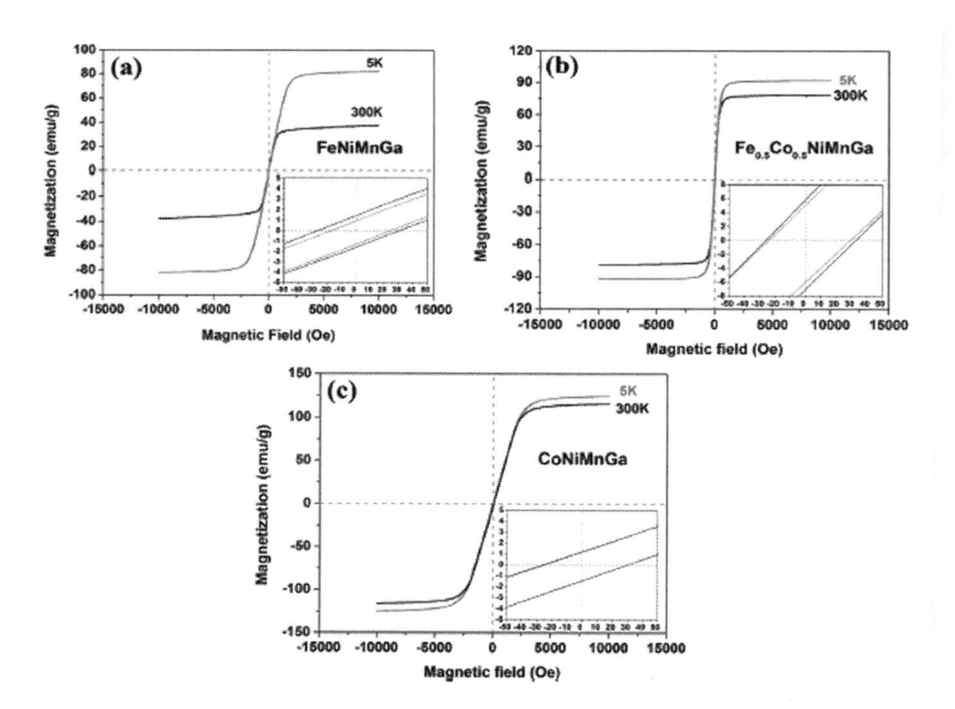

FIGURE 18.11 Magnetic-hysteresis (M-H) loop at 5 K and 300 K for $Fe_xCo_{1-x}NiMnGa$ HEA. (The figure is reprinted with permission from Zuo et al. [68].)

The modified arc discharge method was used for the synthesis of a series of alloys of composition AlxCoCrCuFeNi (x = 0, 0.3,1, 2 in molar ratio) [69]. These samples are abbreviated as Al-0, Al-0.3, Al-1, and Al-2, for x = 0, 0.3, 1, 2, respectively. The XRD analysis confirmed that for Al-0, the formation of FCC phase with a lattice parameter similar to the pure copper (Cu) element (3.615 Å) had been observed. As the content of aluminum (Al) increased to 0.3, lattice modulation occurred, and a new superlattice phase precipitated along with the FCC phase. With the increase in the content of aluminum (Al) up to x = 2, the BCC phase appeared along with FCC phase, and also the volume fraction of BCC increased with the aluminum (Al) content. The value of Ms decreased from 2.79 emu/g to 1.45 emu/g as the content of aluminum (Al) increased from x = 0 to x = 0.3. With increasing content of aluminum (Al) to a higher level, the Ms increased. The value of Ms for sample Al-2 is 5.84 emu/g. All the synthesized samples showed a low value of Ms, because the composition analysis of these alloys confirmed that the formed phases are Cu-Cr rich and the content of other ferromagnetic elements are less.

A systematic study on the effect of aluminum (Al) and silicon (Si) on the structure and properties of CoFeNi was performed by Zuo and co-workers [65]. They synthesized a series of AlxCoFeNi and CoFeNiSix high entropy alloys (HEAs) with different aluminum (Al) and silicon (Si) molar ratio (x = 0, 0.25, 0.5, 0.75 and 1) and studied the separate effect of Al and Si content on the phase, microstructure, mechanical behavior, electrical, and magnetic properties. It was found that the high molar ratio of Al element transformed the FCC phase to BCC phase. However, more Si addition enhanced the formation of new intermetallic compounds. The addition of both aluminum (Al) and silicon (Si) separately enhanced the yield strength and hardness with the compromise of plasticity. Also, the effect of addition of Si on mechanical properties is more significant than the addition of Al. Magnetization studies confirmed that all these alloys showed ferromagnetic behavior (Figure 18.12a,b). As the content of aluminum (Al) increased from x = 0 to 1, the saturation magnetization decreased from 151.3 emu/g to 101.8 emu/g. Similar behavior has also been observed for increasing the content of silicon (Si), Ms decreased from 151.3 emu/g (x = 0) to 80.5 emu/g (x = 0.75). Hence the Si content decreased the Ms more significantly than the Al content (Figure 18.12b). The opposite trend has been observed for the electrical resistivity: addition of Si increased electrical resistivity from 16.7 μΩ.cm to 82.89 μΩ.cm. It was also reported that both alloys undergo very small magnetostriction (the value lower than 35 ppm), which ensured that these materials are not stressed in the presence of external magnetic field. Conversely, it can conclude that external stress does not disrupt the magnetic properties of samples. Moreover, experimental data suggested that lower magnetostriction can be achieved by tuning the amount of Si content. This work described that the balance of mechanical and magnetic properties can be optimized by varying the ratios of aluminum (Al) or silicon (Si).

The equiatomic multi-principal CoCrFeCuNi and CoCrFeMnNi high entropy alloys (HEAs) were synthesized by mechanical alloying (MA) and consolidated through high-pressure sintering (HPS) [70]. MA was performed through a planetary ball miller with a ball-to-powder weight ratio of 10:1 with 450 rpm speed, and ethanol was used for process controlling agent. The milling was performed for 25–30 h. Both MAed (mechanically alloyed) powder samples were comprised of FCC and minor

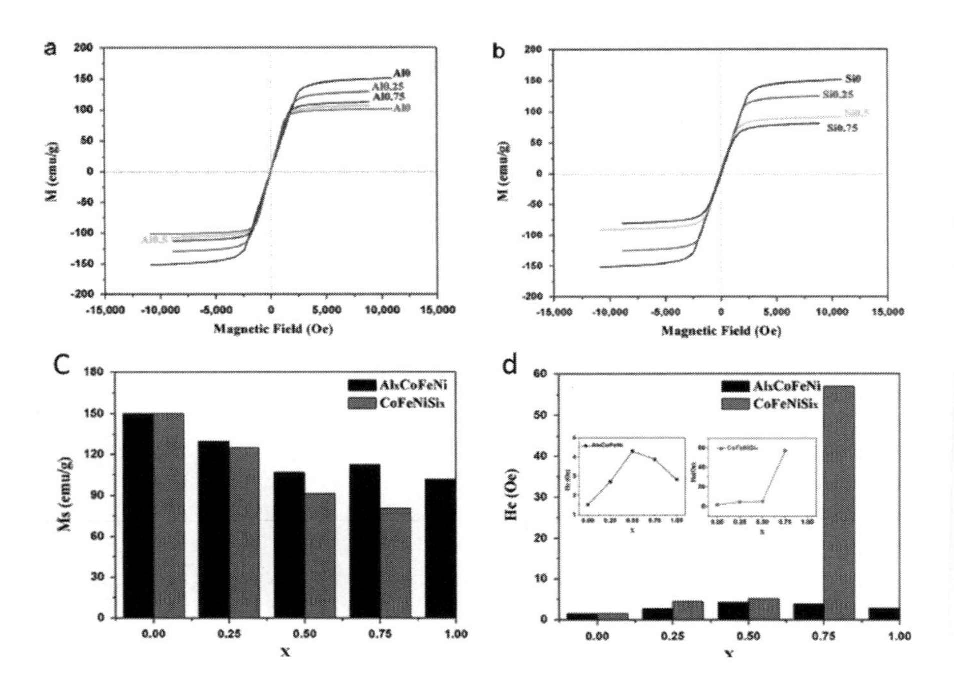

FIGURE 18.12 (a, b) M-H loops of AlxCoFeNi and CoFeNiSix alloy. (c, d) variation of Ms and Hc with the variation of x. Reproduced with permission from Zuo et al. [65].

BCC phase. After HPS, the BCC phase disappeared, and the pure FCC phase remains with lattice parameters of 3.54 Å and 3.56 Å for CoCrFeCuNi and CoCrFeMnNi HEAs, respectively. The BCC phase disappeared after HPS was attributed to the transformation of a metastable solid-solution phase into equilibrium phases. The decrease in the lattice parameter of the FCC phase after HPS is related to the fact that during mechanical alloying, a substantial amount of energy was stored due to high dislocation density and the large extent of grain boundaries. They reported that the particle size of MA-HPS CoCrFeCuNi and CoCrFeMnNi HEAs were about 100 nm. The CoCrFeCuNi and CoCrFeMnNi HEAs have hardnesses of 494 HV and 587 HV, respectively. The found value of saturation magnetization (Ms) for CoCrFeCuNi HEA is high (53.41 emu/g) as compared to the CoCrFeMnNi HEA (1.34 emu/g). Both synthesized alloys have the same phase and similar particle size, but their saturation magnetizations vary significantly. Thus, the study clearly confirmed that the selection of elements is crucial for the magnetic properties of HEAs.

18.5.3 Effect of Composition of Alloying Elements on Phase Evolution and Magnetic Behavior of High Entropy Alloys

The study by Na and co-workers described the effect of the addition of non-magnetic elements on ferromagnetic transitions and magnetic behavior of FeCoNiCrX (X = Al, Ga, Mn, and Sn) HEAs [71]. All these alloys were prepared by the melting technique and the ingot cut in a circular shape for magnetization characteristics. It has been found that the synthesized FeCoNiCr and FeCoNiCrMn alloys have a face-centered cubic (FCC) phase. However, FeCoNiCrSn alloy has Co-Sn intermetallic compounds within an FCC matrix, while FeCoNiCrAl and FeCoNiCrGa HEAs have a duplex phase of FCC and BCC. The magnetic behaviors of synthesized FeCoNiCrX HEAs are shown in Figure 18.13. From this figure, the FeCoNiCr alloy (labeled as FCNC in the figure) shows the paramagnetic behavior, and the nature is changed to ferromagnetic when Al and Ga are added as alloying elements in the FeCoNiCr, while with the addition of manganese (Mn) or tin (Sn) the paramagnetic behavior of the parent alloys does not change. Figure 18.13c,d represents the Curie temperature (Tc) obtained from the temperature dependence magnetization (M-T) in a temperature range of 2 K up to 1,123 K. The Tc of FeCoNiCr and FeCoNiCrMn HEA is found to be 104 K and 41 K, respectively. Thus, the magnetic transition temperature decreased with the addition of Mn in FeCoNiCr HEA. The Tc of FeCoNiCrAl and FeCoNiCrGa HEAs are significantly high and found to be 277 K and 703 K. Thus, a dramatic change has been observed in magnetic properties by the addition of non-magnetic element in FeCoNiCr alloy. This change in magnetic property was observed due to the change in the synthesized phases after adding aluminum (Al) and gallium (Ga).

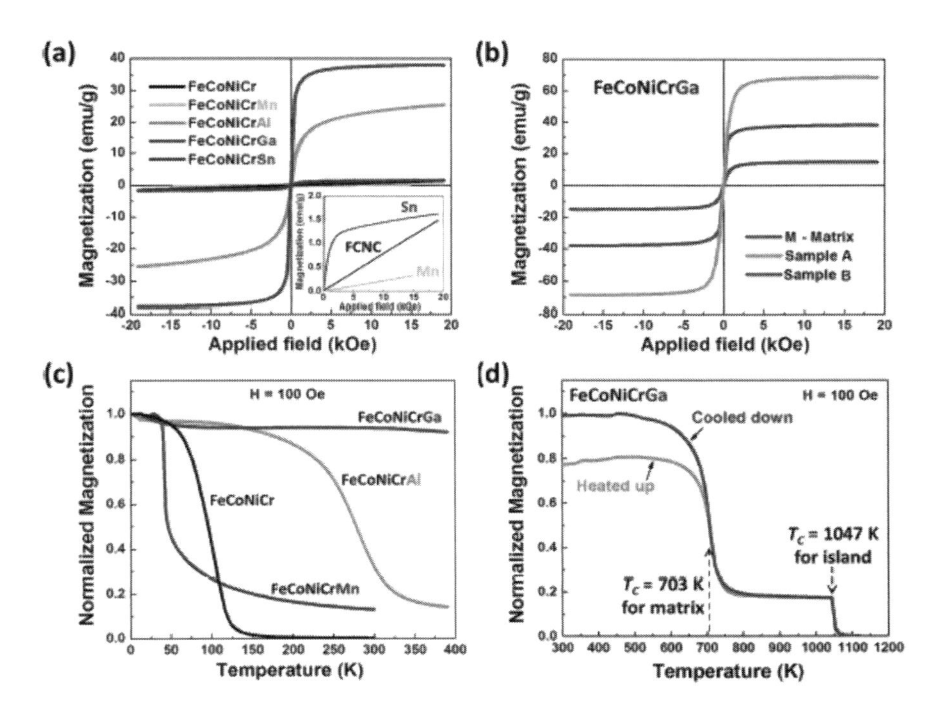

FIGURE 18.13 Magnetic properties of the FeCoNiCr and FeCoNiCrX (X= Al, Ga, Mn and Sn) (a) M-H loop at room temperature (RT), (b) M-H Curves measured at RT for different samples of FeCoNiCrGa HEA, (c) M-T curve at temperature range 2 K–380 K for FeCoNiCrX (X=Al, Ga, Mn and Sn), and M-T curve at temperature range of 300 K–1,123 K for FeCoNiCrGa HEA. (Figure reprinted with permission from Na et al. [71].)

Zuo and co-workers synthesized and investigated the magnetic properties of CoFeMnNiX (X = Al, Cr, Ga, and Sn) [72]. Alloy ingots of CoFeMnNi, CoFeMnNiAl, CoFeMnNiCr, CoFeMnNiGa, and CoFeMnNiSn HEAs synthesized through arc melting. For the CoFeMnNi and CoFeMnNiCr alloy possessed FCC phase, CoFeMnNiAl alloy have ordered a BCC structure, and the CoFeMnNiGa alloy possessed FCC and BCC phases with the volume fraction of the BCC phase is higher than that of the FCC phase. It has been found that the Co_2MnSn and BCC mixed-phase formed for CoFeMnNiSn alloy. Figure 18.14 represents the M-H loop for the different CoFeMnNiX (X = Al, Ga, and Sn) alloys. The characteristic M-H loop clearly represents typical ferromagnetic behavior for CoFeMnNiAl, CoFeMnNiGa, and CoFeMnNiSn alloy. However, the M-H loop for CoFeMnNiCr alloy is paramagnetic in nature. For CoFeMnNi alloy, the value of Ms and Hc are 18.14 Am^2/kg and 119 A/m, respectively. Both Ms and Hc increased with the addition of aluminum (Al), gallium (Ga), and tin (Sn) to the CoFeMnNi parent alloy separately. The value of Ms and Hc for the CoFeMnNiAl alloy was found to be 147.86 Am^2/kg and 629 A/m, respectively. Moreover, the value of Ms was nearly same and about 80 Am^2/kg for CoFeMnNiGa and CoFeMnNiSn alloys with Hc 915 A/m and 3,431 A/m, respectively. Moreover, CoFeMnNiCr alloy is not showing saturation magnetization, even with the field 5×10^6 A/m and have the largest Hc value of 10,804 A/m with paramagnetic nature. The phase change is responsible for the enhancement of Ms. Moreover, the theoretical calculations confirmed that antiferromagnetic nature of

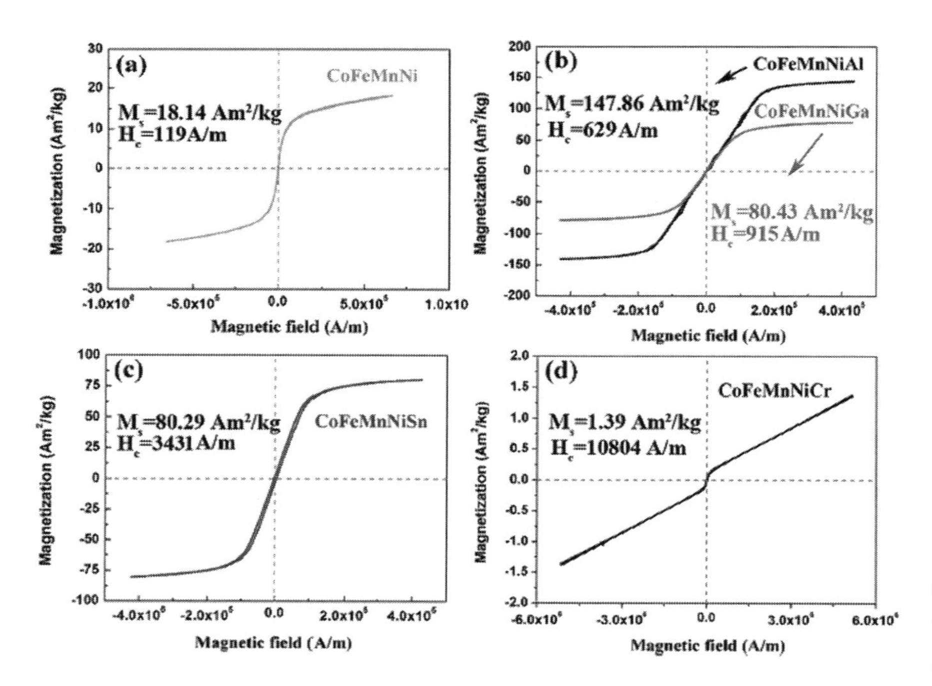

FIGURE 18.14 Magnetic-hysteresis loop of (a) CoFeMnNi, (b) CoFeMnNiAl and CoFeNiMnGa, (c) CoFeMnNiSn, and (d) CoFeMnNiCr HEAs. (Reprinted with permission from Zuo et al. [72].)

manganese (Mn) atoms in CoFeMnNi is compensated by the addition of Al, because the Al changed the Fermi level and itinerant electron-spin coupling which enhanced ferromagnetism. They also suggested that a potential soft magnetic material can be searched for by the composition variation, for example, CoFe(Cr,Mn)xNi(Al,Ga)y.

Alloy ingots of AlNiCo, AlNiCoCu, AlNiCoFe, and AlNiCo(CuFe) HEAs have been prepared by arc melting [73]. These alloys are also annealed at the temperature 1,000°C for 48 h in vacuum-sealed quartz ampules. The as-cast and annealed AlNiCo exhibited B2 phase while AlNiCoFe has a B2 phase which is converted to a disorder body-centered cubic (BCC) phase after annealing. A face-centered cubic (FCC) phase formed for the AlNiCoCu alloy. However, the BCC phase is stabilized with the addition of iron (Fe) in the AlNiCoCu alloy, which clearly indicates that phase separation occurs due to the presence of copper (Cu). The equiatomic AlNiCo alloy showed low magnetization (non-saturation, even at 20 kOe) and with approximately zero Hc, as shown in Figure 18.15. From the fitting through the Langevin function, the magnetic moment per particle was found to be 11,166 μ_B. The high value of the magnetic moment per particle and the absence of coercivity (Hc) and remanence (Mr) in the M-H curve clearly indicated the presence of a single domain and confirmed the presence of superparamagnetic behavior. Further with addition of copper (Cu), the value of magnetization is same as to AlNiCo, but the Hc increased. This happened due to a change in microstructure and pinning of the domain wall

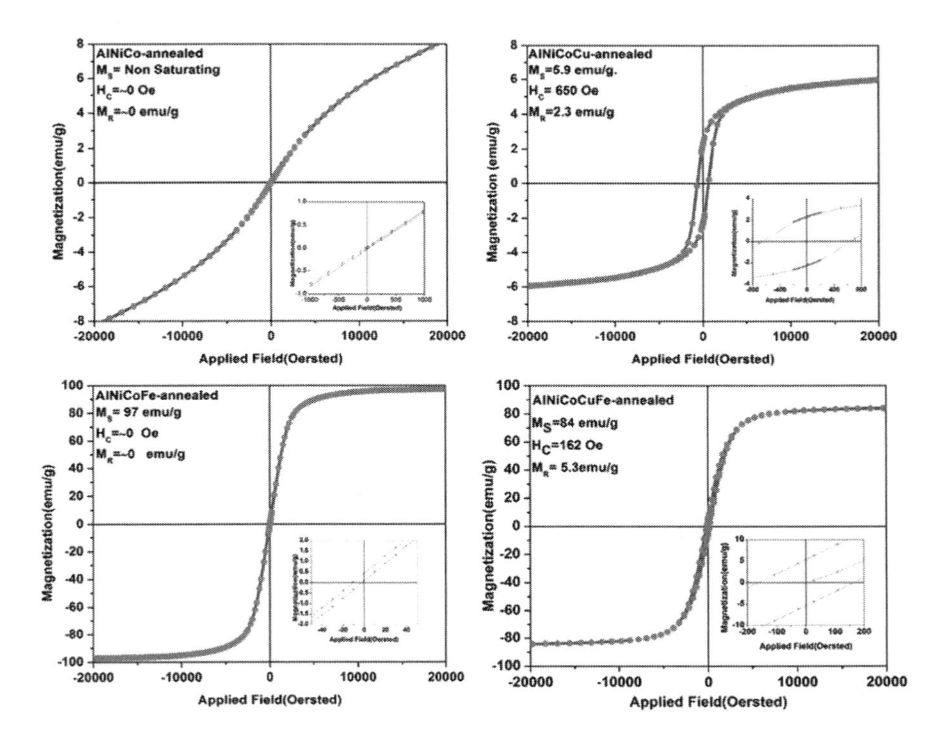

FIGURE 18.15 Magnetic-hysteresis (M-H) loops for annealed alloys at room temperature. (Reprinted with permission from Kulkarni et al. [73].)

or magnetic cluster. Further, with the addition of iron (Fe) to the AlNiCo alloy, the value of Ms increased significantly, and Hc decreased, and hence the magnetic properties of AlNiCo improved with addition of Fe. Moreover, the addition of Cu in AlNiCoFe alloy resulted the development of finite value of Hc and Mr. High value of Ms is in Fe-containing samples related to improvement in Fe-Co magnetic interaction due to the presence of both cobalt (Co) and Fe in alloys. The reasonably good magnetization and value of Hc suggested that the AlNiCoCuFe HEA belongs to the semi-hard magnetic materials.

Wang and co-workers studied the effect of mechanical milling duration and composition on the magnetic and mechanical properties of the FeSiBAlNi and FeSiBAlNiNb high entropy alloys [62]. An amorphous phase has been detected for FeSiBAlNi, and the results confirmed that the Nb addition enhances the milling time to produce a fully amorphous phase for the FeSiBAlNiNb high entropy alloy, and decreases the glass-forming ability. The amorphous phase of the FeSiBAlNiNb high entropy alloy possessed higher thermal stability and heat-resisting properties. The as-milled FeSiBAlNi(Nb) powders were soft magnets, confirmed by their low coercivity. As the milling duration increased, the value of Ms decreased and the value of Ms was lowest when the amorphous high entropy alloys were formed. The results confirmed that the as-synthesized solid-solution phase showed better soft magnetic properties than those with fully amorphous phases. Moreover, the study by Wang and co-workers confirmed that the addition of Nb is not beneficial for magnetic properties of FeSiBAlNi high entropy alloys.

We have also investigated the effect of the addition of another element (Mn and Co) on phase formation and magnetic behavior of CrFeNiTi-based HEAs synthesized through MA [74]. A simple solid solution (double FCC) along with a minor fraction of an intermetallic σ-phase was evolved for CrFeNiTi and CrFeMnNiTi HEAs, whereas a single FCC phase was formed for CoCrFeNiTi HEA at the as-synthesized state. The synthesized HEAs exhibited semi-hard ferromagnetic characteristics (shown in Figure 18.16). The value of Ms and Hc for CrFeNiTi alloy was found to be 13.38 emu/g and 166.9 Oe, respectively. With the addition of Mn and Co separately in CrFeNiTi alloy, the magnetic behavior of the base alloy changed drastically. For CrFeMnNiTi HEA, the value of Ms (2.28 emu/g) decreased, and Hc (225.8 Oe) increased compared to the CrFeNiTi alloy. However, for CoCrFeNiTi HEA, the value of Ms (24.44 emu/g) increased and Hc (149.5 Oe) decreased compared to the synthesized CrFeNiTi alloy. The change in magnetic behavior for CrFeNiTi alloy with the addition of Mn and Co was associated with the content of a ferromagnetic element present in the synthesized HEAs. Moreover, after annealing at 700°C, the value of Ms decreased dramatically for all three synthesized alloys (as shown in Figure 18.16d,e,f, due to the change in lattice parameter of annealed HEAs, and hence the separation between magnetic elements has been changed after annealing, which affects the magnetic exchange [74].

Further, to improve the magnetic behavior of CrFeMnNiTi HEA, we have studied the effect of the addition of aluminum (Al) element on structural and magnetic behavior AlCrFeMnNiTi HEA [75]. With the addition of Al in CrFeMnNiTi HEA, the formation of the σ-phase was suppressed and only simple solid-solution face-centered cubic (FCC) and body-centered cubic (BCC) phases were formed, because

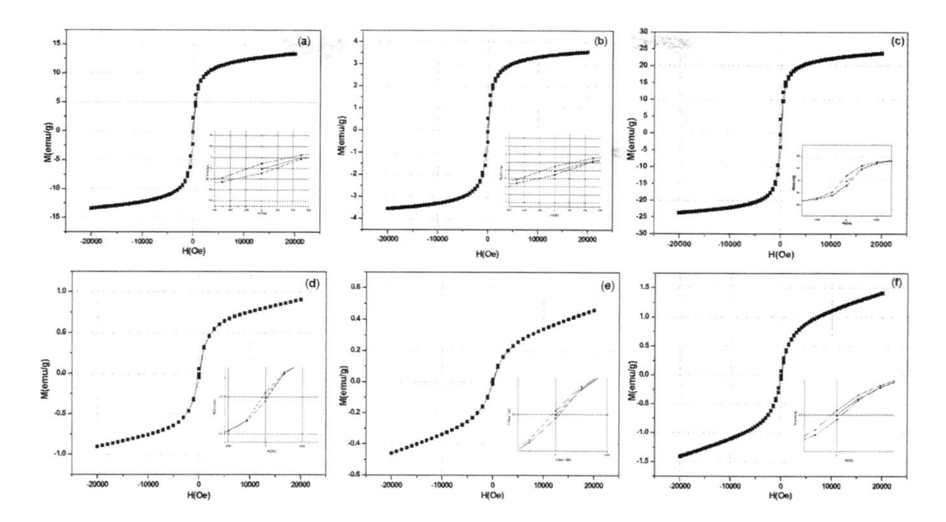

FIGURE 18.16 Magnetic-hysteresis curve of (a) CrFeNiTi, (b) CrFeMnNiTi, (c) CoCrFeNiTi as synthesized, and (d) CrFeNiTi, (e) CrFeMnNiTi, (f) CoCrFeNiTi annealed HEAs. The inset of the figure represents the enlarge view of selected region. (Reprinted with permission from Mishra and Shahi [74].)

the addition of one extra element enhanced the configurational entropy (ΔS_{con}) of the system and thus resulted in the formation of only a solid-solution phase, instead of the intermetallic compound (shown in Figure 18.17a). After annealing at 700°C, the volume fraction of the synthesized FCC and BCC phases changed. The relative volume fraction of the FCC phase increased, while the same for the BCC phase decreased after annealing. We have found that the synthesized and annealed HEAs exhibited ferromagnetic behavior (shown in Figure 18.17b). With the addition of Al in CrFeMnNiTi HEA, the value of Ms increased from 2.28 emu/g (for CrFeMnNiTi) to 17.55 emu/g, and Hc decreased to 153.8 Oe from 225.8 Oe (for CrFeMnNiTi HEA). The enhancement in the value of Ms for AlCrFeMnNiTi HEA was associated with the presence of Al.

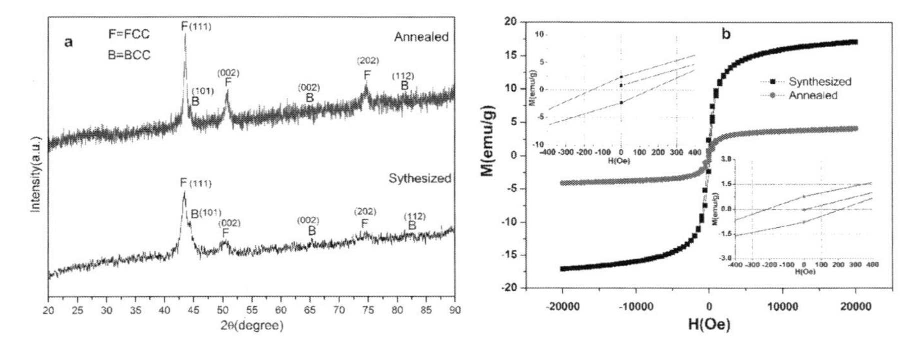

FIGURE 18.17 (a) XRD patterns, and (b) M-H loop of synthesized and annealed AlCrFeMnNiTi HEA. (Reprinted with permission from Mishra et al. [75].)

Zuo and co-workers reported that the addition of Al in FeCoNiMn alloy suppressed the antiferromagnetic ordering of manganese (Mn) and enhanced the value of Ms significantly [72]. Moreover, Zuo and co-workers also reported that the addition of Al changed the Femi level and the itinerant electron-spin coupling that leads to ferromagnetism in FeCoNiMnAl HEA [72]. After annealing at 700°C, the value of Ms and Hc for AlCrFeMnNiTi HEA changed appreciably. The value of Ms decreased to 4.22 emu/g and Hc increased to 215 Oe compared to the synthesized HEA. The drastic change in the value of Ms after annealing was associated with the change in the atomic ordering and/or the change in the volume fraction of BCC and FCC phases.

It is well known that soft magnetic materials required a high saturation magnetization, low coercivity, high electrical resistivity, and better mechanical properties. From the literature, it has been found that CoCrFeNiTi-based HEAs exhibited excellent mechanical properties. In order to develop a better CoCrFeNiTi-based soft magnetic material, we have changed the Co/Cr ratio and studied their structural and magnetic behavior [76]. The synthesized $Co_{35}Cr_5FeNiTi$ HEA was comprised of a mixture of face-centered cubic (FCC), body-centered cubic (BCC), and minor fraction of Ni_4Ti_3 type R-phase. The synthesized $Co_{35}Cr_5FeNiTi$ HEA exhibited better magnetic characteristics compared to the equiatomic composition. The value of Ms and Hc for $Co_{35}Cr_5FeNiTi$ HEA was found to be 46 emu/g and 15 Oe, respectively. In order to preserve the nano-crystalline effect and relieve the structural strain produced by MA, the synthesized HEA was annealed at 200°C for 10 h (as shown in Figure 18.18a). It was found that the XRD peaks were the same after annealing at 200°C for 10 h, and only the crystallinity of synthesized $Co_{35}Cr_5FeNiTi$ HEA improved. The value of Ms for 200°C annealed $Co_{35}Cr_5FeNiTi$ HEA increased approximately twice (Ms = 81 emu/g) compared to the synthesized HEA (Figure 18.18b). The value of Hc was found to be similar (Hc = 15 Oe) to the synthesized HEA. The enhancement in the value of Ms after annealing was associated with the improvement in the crystalline disorder, minimization of the surface/interface anisotropy and the change in the lattice parameter of the synthesized face-centered cubic (FCC) and body-centered cubic (BCC) phases, which affects the magnetic exchange coupling between neighboring elements [76].

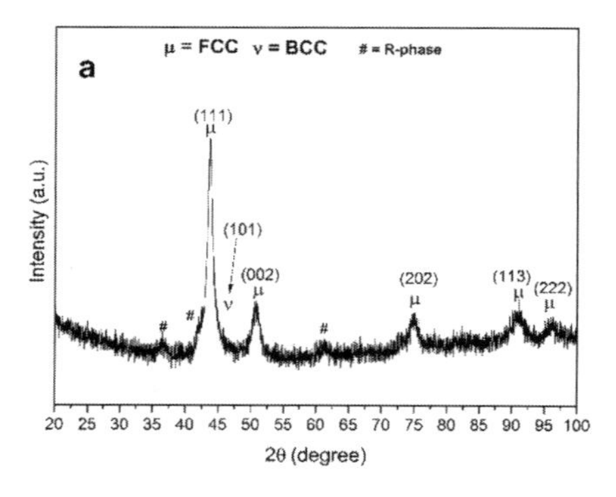

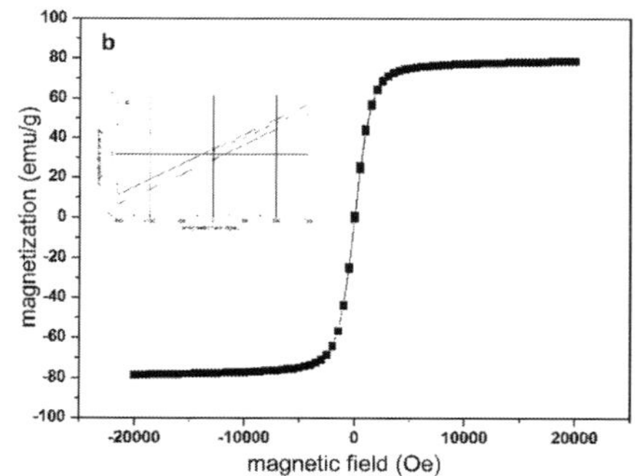

FIGURE 18.18 (a) XRD patterns and (b) M-H curve of 200°C annealed $Co_{35}Cr_5FeNiTi$ HEA. (Reprinted with permission from Mishra and Shahi [76].)

Recently, Alijani and co-workers synthesized and studied the phase formation and magnetic properties of FeCoNiMnV high entropy alloy synthesized through MA [77]. The structural analysis confirmed that the face-centered cubic (FCC) and body-centered cubic (BCC) phase coexist during the MA process. As the milling duration increased, the formation of a σ phase has also been observed due to the presence of vanadium (V). The obtained MA powder can be considered semi-hard, with the maximum Ms of about 100 emu/g for 48 h milled sample. As the milling duration increased, the value of Ms decreased due to the introduction of a high density of interfaces and defects at the grain boundaries, which pinned the domain wall movement and hence the value of Ms decreased.

18.6 CONCLUSIONS AND FUTURE PROSPECTS

From this discussion of magnetic properties of newly emerged HEAs, the following conclusions can be drawn:

1. HEAs can provide a wide range of magnetic properties in regard to soft, hard, and semi-hard magnetic materials. Among the reviewed HEAs, some have been proven to have comparable magnetic properties with respect to conventional materials. However, their mechanical behavior and electrical resistivity have been found to be superior compared to conventional materials.
2. Addition of alloying elements affects the microstructure, and thus the magnetic properties, of HEAs. The addition of paramagnetic or ferromagnetic element will effectively affect the magnetic properties of the HEA. As in the results discussed, addition of the aluminum (Al) element changed the FCC structure to BCC, while more addition of silicon (Si) leads to new compounds. Addition of Si enhanced the mechanical properties (strength and hardness) and electrical resistivity significantly compared to the Al element. The opposite trend was observed for saturation magnetization since the saturation magnetization prominently depends on composition and atomic-level structure.
3. Besides the concentrations of alloying elements, external parameters such as pressure and temperature also affect the phase transformation, and thus the properties of alloys. After the heat treatment process, a phase transition from amorphous to the BCC phase and then to the FCC phase was observed in $Fe_{30}Co_{29}Ni_{29}Zr_7B_4Cu_1$ HEA. The amorphous and FCC HEA exhibited a similar value of saturation magnetization (115 emu/g). However, after BCC phase precipitation in amorphous matrix, the value of Ms was enhanced considerably to 125 emu/g from the initial (115 emu/g).
4. For the case of CrFeMnNiTi HEA, annealing under vacuum and open-air conditions, the value of saturation magnetization also changed significantly compared to the synthesized HEA. Thus, the selection of synthesis route, alloying element and external parameters (such as temperature and pressure) are the crucial factors that influence the phase formation, microstructure and thus properties of HEAs.

5. The reported HEAs have shown similar or low value of Ms to the conventional magnetic materials. However, their electrical resistivity and mechanical behavior has been found outstanding. The magnetic properties of a system are sensitive to the composition and crystal structure. Therefore, from the reviewed results, it can be expected that the enhancement in the ferromagnetic content, mostly in Co (CoxCrCuFeMnNi, $Fe_xCo_{1-x}NiMnGa$) and Ni ($AlCoCuFeNi_x$) relative to the Fe element in a given system, will enhance their soft magnetic properties along with their mechanical properties.

Hence, more and more research is required to explore the magnetic properties of HEAs and HEA-related materials from both academic and applications viewpoints.

ACKNOWLEDGMENTS

The authors would like to acknowledge the research grant from the Innovation in Science Pursuit for Inspired Research (INSPIRE) project of the Department of Science and Technology (DST) India (IFA-12-PH-43) and the Science and Engineering Research Board–Research Scientists Scheme (SERB-SRS) Research Grant.

REFERENCES

1. Cantor, B., Chang, I. T. H., Knight, P., Vincent, A. J. B. 2004. Microstructural development in equiatomic multicomponent alloys. *Mater. Sci. Eng. A* 375–377:213–218.
2. Yeh, J. W., Chen, S. K., Lin, S. J., et al. 2004. Nanostructured high entropy alloys with multiple elements: Novel alloy design concept and outcomes. *Adv. Engin. Mater.* 6:299–303.
3. Murti, B. S., Yeh, J. W., Ranganathan, S. 2014. *High Entropy Alloys*. London: Elsevier.
4. Gao, M. C., Yeh, J. W., Liaw, P. K., Zhang, Y. 2016. *High Entropy Alloys: Fundamental and Application*. Switzerland: Springer.
5. Cheng, K. H., Lai, C. H., Lin, S. J., Yeh, J. W. 2006. Recent progress in multi-element alloy and nitride coatings sputtered from high-entropy alloy targets. *Ann. Chim. Sci. Mat.* 31(6):723–736.
6. Tsai, M. H., Wang, C. W., Tsai, C. W., et al. 2011. Thermal stability and performance of NbSiTaTiZr high-entropy alloy barrier for copper metallization. *J. Electrochem. Soc.* 158(11):H1161–H1165.
7. Tsai, M. H., Yeh, J. W., Gan, J. Y. 2008. Diffusion barrier properties of AlMoNbSiTaTiVZr high entropy alloy layer between copper and silicon. *Thin Solid Films* 516(16):5527–5530.
8. Shun, T. T., Hung, C. H., Lee, C. F. 2010. Formation of ordered/disordered nanoparticles in FCC high entropy alloys. *J. Alloys Compd.* 493(1–2):105–109.
9. Liu, W. H., Wu, Y., He, J. Y., Nieh, T. G., Lu, Z. P. 2013. Grain growth and the Hall–Petch relationship in a high-entropy FeCrNiCoMn alloy. *Scr. Mater.* 68(7):526–529.
10. Kao, Y. F., Chen, S. K., Chen, T. J., Chu, P. C., Yeh, J. W., Lin, S. J. 2011. Electrical, magnetic, and hall properties of AlxCoCrFeNi high-entropy alloys. *J. Alloys Compd.* 509(5):1607–1614.
11. Zhou, Y. J., Zhang, Y., Wang, Y. L., Chen, G. L. 2007. Solid solution alloys of AlCoCrFeNiTix with excellent room-temperature mechanical properties. *Appl. Phys. Lett.* 90(18). http://www.ncbi.nlm.nih.gov/pubmed/181904.

12. Li, C., Li, J. C., Zhao, M., Jiang, Q. 2009. Effect of alloying elements on microstructure and properties of multiprincipal elements high entropy alloys. *J. Alloy Compd.* 475(1–2):752–757.
13. Qin, G., Xue, W., Fan, C., et al. 2018. Effect of Co content on phase formation and mechanical properties of (AlCoCrFeNi)100-xCox high-entropy alloys. *Mater. Sci. Eng. A* 710:200–205.
14. Zhang, Y., Zuo, T. T., Cheng, Y. Q., Liaw, P. K. High entropy alloys with high saturation magnetization, electrical resistivity, and malleability. *Sci. Rep.* 3:1455.
15. Senkov, O. N., Wilks, G. B., Miracle, D. B., Chuang, C. P., Liaw, P. K. 2010. Refractory high-entropy alloys. *Intermetallics* 18(9):1758–1765.
16. Senkov, O. N., Wilks, G. B., Scott, J. M., Miracle, D. B. 2011. Mechanical properties of Nb25Mo25Ta25W25 and V20Nb20Ta20Mo20W20 refractory high entropy alloys. *Intermetallics* 19(5):698–706.
17. Cahn, R. W., Haasen, P. 1996. *Physical Metallurgy.* Amsterdam: Elsevier Science B.V.
18. Zhang, Y., Zhou, Y. J., Lin, J. P., Chen, G. L., Liaw, P. K. 2008. Solid solution phase formation rules for multi-component alloys. *Adv. Eng. Mater.* 10(6):534–538.
19. Guo, S., Liu, C. T. 2011. Phase stability in high entropy alloys: Formation of solid solution or amorphous phase. *Prog. Nat. Sci. Mater. Int.* 21:433–446.
20. Tsai, M. H., Tsai, R. C., Chang, T., Huang, W. F. 2019. Intermetallic phases in high-entropy alloys: Statistical analysis of their prevalence and structural inheritance. *Metals* 9(2):1–18.
21. Tsai, M. H., Yuan, H., Cheng, G., et al. 2013. Significant hardening due to the formation of a sigma phase matrix in a high entropy alloy. *Intermetallics* 33:81–86.
22. Ma, S. G., Zhang, Y. 2012. Effect of Nb addition on the microstructure and properties of AlCoCrFeNi high entropy alloy. *Mater. Sci. Eng. A* 532:480–486.
23. Yang, X., Zhang, Y. 2012. Prediction of high entropy stabilized solid solution in multi-component alloys. *Mater. Chem. Phy.* 132(2–3):233–238.
24. Guo, S., Ng, C., Lu, J., Liu, C. T. 2011. Effect of valance electron concentration on stability of FCC or BCC phase in high entropy alloys. *J. Appl. Phys.* 109. http://www.ncbi.nlm.nih.gov/pubmed/103505.
25. Zhang, Y., Zuo, T. T., Tang, Z., et al. 2014. Microstructures and properties of high-entropy alloys. *Prog. Mater. Sci.* 61:1–93.
26. Miracle, D. B., Senkov, O. N. 2017. A critical review of high entropy alloys and related concepts. *Acta Mater.* 122:448–511.
27. Chen, J., Zhou, X., Wang, W., et al. 2018. A review on fundamental of high entropy alloys with promising high–temperature properties. *J. Alloys Compd.* 760:15–30.
28. Vaidya, M., Muralikrishna, G. M., Murty, B. S. 2019. High-entropy alloys by mechanical alloying: A review. *J. Mater. Res.* 34(5):664–686.
29. Cheng, K. H., Lai, C. H., Lin, S. J., Yeh, J. W. 2011. Structural and mechanical properties of multi-element (AlCrMoTaTiZr)Nx coatings by reactive magnetron sputtering. *Thin Solid Films* 519(10):3185–3190.
30. Cropper, M. D. 2018. Thin films of AlCrFeCoNiCu high-entropy alloy by pulsed laser deposition. *App. Surf. Sci.* 455:153–159.
31. Yao, C. Z., Zhang, P., Liu, M., et al. 2008. Electrochemical preparation and magnetic study of Bi–Fe–Co–Ni–Mn high entropy alloy. *Electro. Acta.* 53(28):8359–8365.
32. Alaneme, K. K., Bodunrin, M. O., Oke, S. R. 2016. Processing, alloy composition and phase transition effect on the mechanical and corrosion properties of high entropy alloys: A review. *J. Mater. Res. Technol.* 5(4):384–393.
33. Ghazi, S. S., Ravi, K. R. 2016. Phase-evolution in high entropy alloys: Role of synthesis route. *Intermetallics* 73:40–42.
34. Mishra, R. K., Shahi, R. R. 2018. Magnetic characteristics of high entropy alloys. In: *Magnetism and Magnetic Materials*, ed. N. Panwar, 67–80. IntechOpen.

35. Ji, W., Fu, Z., Wang, W., et al. 2014. Mechanical alloying synthesis and spark plasma sintering consolidation of CoCrFeNiAl high-entropy alloy. *J. Alloys Compd.* 589:61–66.
36. Moravcik, I., Cizek, J., Gavendova, P., Sheikh, S., Guo, S., Dlouhy, I. 2016. Effect of heat treatment on microstructure and mechanical properties of spark plasma sintered AlCoCrFeNiTi0.5 high entropy alloy. *Mater. Lett.* 174:53–56.
37. Cai, Z., Jin, G., Cui, X., Li, Y., Fan, Y., Song, J. 2016. Experimental and simulated data about microstructure and phase composition of a NiCrCoTiV high-entropy alloy prepared by vacuum hot-pressing sintering. *Vacuum* 124:5–10.
38. Varalakshmi, S., Kamaraj, M., Murty, B. S. 2008. Synthesis and characterization of nanocrystalline AlFeTiCrZnCu high entropy solid solution by mechanical alloying. *J. Alloys Compd.* 460(1–2):253–257.
39. Chen, Y. L., Hu, Y. H., Hsieh, C. A., Yeh, J. W., Chen, S. K. 2009. Competition between elements during mechanical alloying in an octonary multi-principal-element alloy system. *J. Alloys Compd.* 481(1–2):768–775.
40. Gutfleisch, O., Willard, M. A., Brück, E., Chen, C. H., Sankar, S. G., Liu, J. P. 2011. Magnetic Materials and Devices for the 21st Century: Stronger, Lighter, and More Energy Efficient. *Adv. Mater. Weinheim* 23(7):821–842.
41. Fingersa, R. T., Rubertusa, C. S. 1999. Air force application of advanced magnetic materials. *MRC Proc.* 577:481–486.
42. Li, P., Wang, A., Liu, C. T. 2017. A ductile high entropy alloy with attractive magnetic properties. *J. Alloys Compd.* 694:55–60.
43. Kozelj, P., Vrtnik, S., Jelen, A., et al. 2019. Discovery of a FeCoNiPdCu high-entropy alloy with excellent magnetic softness. *Adv. Eng. Mater.* 21. http://www.ncbi.nlm.nih.gov/pubmed/1801055.
44. Uporov, S., Bykov, V., Pryanichnikov, S., Shubin, A., Uporova, N. 2017. Effect of synthesis route on structure and properties of AlCoCrFeNi high-entropy alloy. *Intermetallics* 83:1–8.
45. Wei, R., Sun, H., Chen, C., Han, Z., Li, F. 2017. Effect of cooling rate on the phase structure and magnetic properties of Fe26.7Co28.5Ni28.5Si4.6B8.7P3 high entropy alloy. *J. Magn. Magn. Mater.* 435:184–186.
46. Mishra, R. K., Shahi, R. R. 2018. Effect of annealing conditions and temperatures on phase formation and magnetic behavior of CrFeMnNiTi High Entropy Alloy. *J. Magn. Magn. Mater.* 465:169–175.
47. Zuo, T. T., Yang, X., Liaw, P. K., Zhang, Y. 2015. Influence of Bridgman solidification on microstructures and magnetic behaviors of a non-equiatomic FeCoNiAlSi high-entropy alloy. *Intermetallics* 67:171–176.
48. Ji, W., Wang, W., Wang, H., et al. 2015. Alloying behavior and novel properties of CoCrFeNiMn high-entropy alloy fabricated by mechanical alloying and spark plasma sintering. *Intermetallics* 56:24–27.
49. Boll, R. 2006. *Soft Magnetic Metals and Alloys*. Mater. Sci. Technol.
50. Willard, M. A., Johnson, F., Claassen, J. H., Stroud, R. M., McHenry, M. E., Harris, V. G. 2002. Soft magnetic nanocrystalline alloys for high temperature applications. *Mater. Trans.* 43(8):2000–2005.
51. Yu, R. H., Basu, S., Ren, L., et al. 2000. High temperature soft magnetic materials: FeCo alloys and composites. *IEEE Trans. Magn.* 36(5):3388–3393.
52. Bazzi, K., Rathi, A., Meka, V. M., Goswami, R., Jayaraman, T. V. 2019. Significant reduction in intrinsic coercivity of high-entropy alloy FeCoNiAl0.375Si0.375 comprised of supersaturated f.c.c. phase. *Materialia* 6. http://www.ncbi.nlm.nih.gov/pubmed/100293.
53. Rathi, A., Meka, V. M., Jayaraman, T. V. 2019. Synthesis of nanocrystalline equiatomic nickel-cobalt-iron powders by mechanical alloying and their structural and magnetic characterization. *J. Magn. Magn. Mater.* 469:67–482.

54. Qiushi, C., Yong, D., Junjia, Z., Yiping, Y. 2017. Microstructure and Properties of AlCoCrFeNiB*x* (*x*=0, 0.1, 0.25, 0.5, 0.75, 1.0) High Entropy Alloys. *Rare Met. Mater. Eng.* 46(3):0651–0656.
55. Li, Z., Xu, H., Gu, Y., et al. 2018. Correlation between the magnetic properties and phase constitution of FeCoNi(CuAl)0.8Gax ($0 \leq x \leq 0.08$) high-entropy alloys. *J. Alloys Compd.* 746:285–291.
56. Zhang, Q., Xu, H., Tan, X. H., et al. 2017.The effects of phase constitution on magnetic and mechanical properties of FeCoNi(CuAl)x (x = 0 - 1.2) high-entropy alloys. *J. Alloys Compd.* 693:1061–1067.
57. Li, P., Wang, A., Liu, C. T. 2017. Composition dependence of structure, physical and mechanical properties of FeCoNi(MnAl)x high entropy alloys. *Intermetallics* 87:21–26.
58. Zhao, R. F., Ren, B., Zhang, G. P., Liu, Z. X., Zhang, J. J. 2018. Effect of Co content on the phase transition and magnetic properties of CoxCrCuFeMnNi high-entropy alloy powders. *J. Magn. Magn. Mater.* 468:14–24.
59. Hung, W. J., Shun, T. T., Chiang, C. J. 2017. Effects of reducing Co content on microstructure and mechanical properties of CoxCrFeNiTi0.3 high-entropy alloys. *Mater. Chem. Phys.* 210:170–175.
60. Wang, F. J., Zhang, Y. 2008. Effect of Co addition on crystal structure and mechanical properties of Ti0.5CrFeNiAlCo high entropy alloy. *Mater. Sci. Eng. A* 496(1–2):214–216.
61. Zhuang, Y. X., Liu, W. J., Xing, P. F., Wang, F., He, J. C. 2012. Effect of Co element on microstructure and mechanical properties of FeCoxNiCuAl alloys. *Acta Metall. Sin.* 25:124–130.
62. Wang, J., Zhang, Z., Xu, J., Wang, Y. 2014. Microstructure and magnetic properties of mechanically alloyed FeSiBAlNi(Nb) high entropy alloys. *J. Magn. Magn. Mater.* 355:58–64.
63. Xu, J., Axinte, E., Zhao, Z. F., Wang, Y. 2016. Effect of C and Ce addition on the microstructure and magnetic property of the mechanically alloyed FeSiBAlNi high entropy alloys. *J. Magn. Magn. Mater.* 414:59–68.
64. Wan, D. F., Ma, X. L. 1994. *Magnetic Physics*. Beijing: Publishing House of Electronics Industry, pp. 38.
65. Zuo, T. T., Li, R. B., Ren, X. J., Zhang, Y. 2014. Effects of Al and Si addition on the structure and properties of CoFeNi equal atomic ratio alloy. *J. Magn. Magn. Mater.* 371:60–68.
66. Kao, Y. F., Chen, T. J., Chen, S. K., Yeh, J. W. 2009. Microstructure and mechanical property of as-cast, homogenized, and deformed AlxCoCrFeNi ($0 \leq x \leq 2$) high-entropy alloys. *J. Alloys Compd.* 488(1):57–64.
67. Liu, C., Peng, W., Jiang, C. S., et al. 2019. Composition and phase structure dependence of mechanical and magnetic properties of AlCoCuFeNix High Entropy Alloys. *J. Mater. Sci. Tech.* 35(6):1175–1183.
68. Zuo, T. T., Zhang, M., Liaw, P. K., Zhang, Y. 2018. Novel high entropy alloys of FexCo1–xNiMnGa with excellent soft magnetic properties. *Intermetallics* 100:1–8.
69. Mao, A., Ding, P., Quan, F., et al. 2018. Effect of aluminum element on microstructure evolution and properties of multicomponent Al-Co-Cr-Cu-Fe-Ni nanoparticles. *J. Alloys Compd.* 735:1167–1175.
70. Yu, P. F., Zhang, L. J., Cheng, H., et al. 2016. The high-entropy alloys with high hardness and soft magnetic property prepared by mechanical alloying and high-pressure sintering. *Intermetallics* 70:82–87.
71. Na, S. M., Yoo, J. H., Lambert, P. K., Jones, N. J. 2018. Room-temperature ferromagnetic transitions and the temperature dependence of magnetic behaviors in FeCoNiCr-based high entropy alloys. *AIP Adv.* 8(5). http://www.ncbi.nlm.nih.gov/pubmed/056412.

72. Zuo, T. T., Gao, M. C., Ouyang, L., et al. 2017. Tailoring magnetic behavior of CoFeMnNiX (X = Al, Cr, Ga, and Sn) high entropy alloys by metal doping. *Acta Mater.* 130:10–18.
73. Kulkarni, R., Murty, B. S., Srinivas, V. 2018. Study of microstructure and magnetic properties of AlNiCo(CuFe) high entropy alloy. *J. Alloys Compd.* 746:194–199.
74. Mishra, R. K., Shahi, R. R. 2017. Phase evolution and magnetic characteristics of TiFeNiCr and TiFeNiCrM (M=Mn, Co) high entropy alloys. *J. Magn. Magn. Mater.* 442:218–223.
75. Mishra, R. K., Sahay, P. P., Shahi, R. R. 2019. Alloying, Magnetic and Corrosion behavior of AlCrFeMnNiTi High Entropy Alloy. *J. Mater. Sci.* 54(5):4433–4443.
76. Mishra, R. K., Shahi, R. R. 2019. Novel Co35Cr5Fe20Ni20Ti20 High Entropy Alloy for high magnetization and low coercivity. *J. Magn. Magn. Mater.* 484:83–87.
77. Alijani, F., Reihanian, M., Gheisari, Kh. 2019. Study on phase formation in magnetic FeCoNiMnV high entropy alloy produced by mechanical alloying. *J. Alloys Compd.* 773:623–630.

19 High Entropy Alloy Fibers Having High Tensile Strength and Ductility

Dong Yue Li and Yong Zhang

CONTENTS

19.1 INTRODUCTION

Currently, high entropy alloy fiber is still in the beginning stage.

There are three basic types of high entropy fibers from now on i.e., $Al_{0.3}CoCrFeNi$ alloy fiber [1, 2], CoCrFeNi high entropy alloy fiber [3], and CoCrFeNiMn high entropy alloy fiber rod [4]. All of these alloys have mainly face-centered cubic (FCC) structure and low stacking fault energy, rendering them potential possessors of deformation ability over a wide temperature range.

19.2 HIGH ENTROPY FIBERS

19.2.1 $Al_{0.3}CoCrFeNi$ Alloy Fiber

$Al_{0.3}CoCrFeNi$ quinary alloy is a typical FCC high entropy alloy and widely studied [5–9]. The $Al_{0.3}CoCrFeNi$ fiber was fabricated into continuous fibers with diameters of 1.00–3.15 mm using traditional deformation processing techniques [1]. The macroscopic views of the fibers are shown in Figure 19.1. These fibers were cylindrically shaped and flawless implying that the hot drawing method is an effective technique for the production of high-quality high entropy fibers. The tensile strength and plasticity of the fibers were determined at both 298 K (1,207 MPa/7.8%) and 77 K (1,600 MPa/17.5%). The improvement of mechanical properties at cryogenic

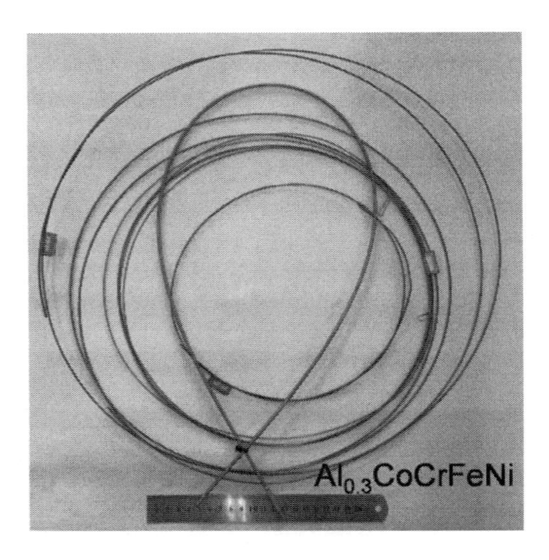

FIGURE 19.1 Macroscopic views of hot-drawn smooth $Al_{0.3}CoCrFeNi$ fibers. The diameters range from 1.00 to 3.15 mm [1].

temperatures can be attributed to a change in deformation mechanisms from the planar slip of dislocations to nano-twinning, which is beneficial for cryogenic applications [10–13].

The structure analysis revealed that the $Al_{0.3}CoCrFeNi$ high entropy fiber obtains a homogeneous face-centered cubic (FCC) structure in the as-cast material, while post-processing produced a nano-sized B2 precipitated phase in the FCC matrix. The TEM investigation (Figure 19.2) indicated that the fibers consisted of a (Co, Cr, and Fe)-rich matrix and a NiAl-type ordered BCC (e.g., B2) phase. The selected area electron diffraction (SAED) patterns confirmed the coexistence of the FCC matrix and B2 secondary phase. Table 19.1 listed the chemical composition in the matrix and particles of the fibers further confirm the phase structure of the fiber. The precipitates and fine grains both contribute to the sustained high strength and extended ductility observed during the tensile tests.

In terms of mechanical properties, Figure 19.3 revealed the measured uniaxial stress-strain curves of $Al_{0.3}CoCrFeNi$ fibers at room temperature (298 K) and in liquid nitrogen (77 K). The yield strength and ultimate tensile strength increased, from 1,136 to 1,320 MPa and from 1,207 to 1,600 MPa as the temperature drops from 298 K to 77 K. Similarly, the tensile ductility also increased from 7.8% to 17.5%. Compared to this alloy system before the tensile test, the dislocation density of the fibers after tensile testing was significantly enhanced. Figure 19.4 displays the TEM micrographs and deformation substructures of the 3.15 mm diameter fibers after tension testing at 298 and 77 K, respectively. The multiple interactions between the dislocations and the fine-scaled precipitates further increased the resistance of the dislocation motion with increasing deformation, thereby leading to the significant strengthening. However, deformation-induced nanoscaled twinning at 77 K also contributed to both the increases in strength and ductility.

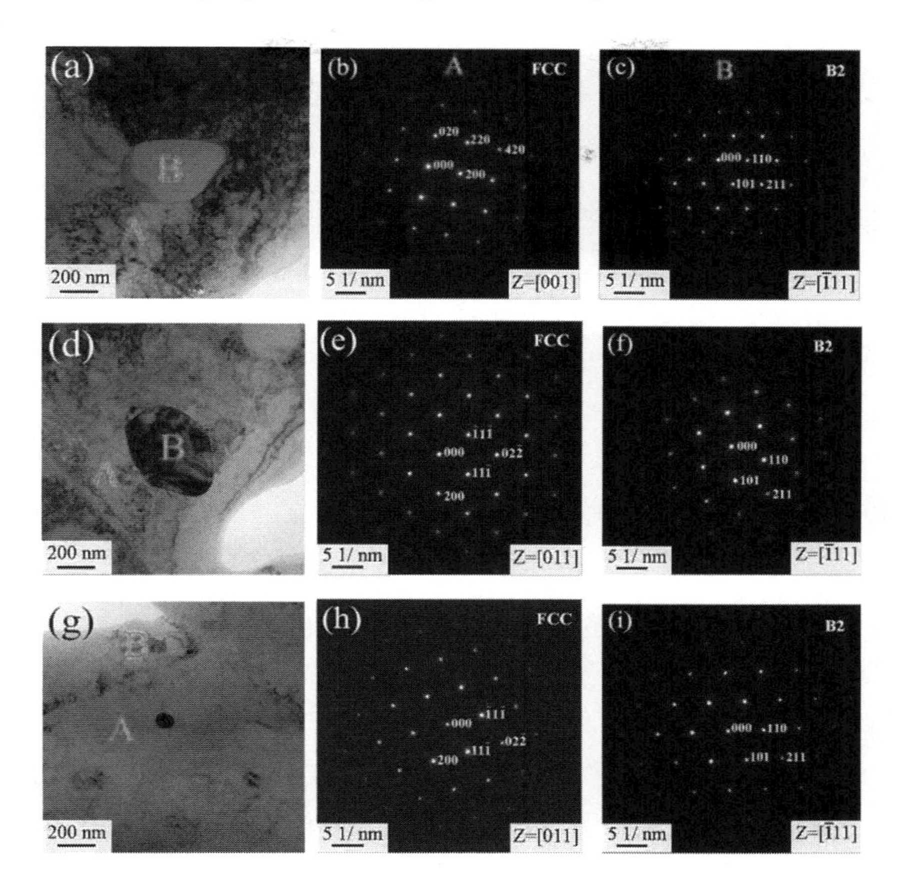

FIGURE 19.2 TEM of the (a) 3.15, (b) 1.60, and (c) 1.00 mm $Al_{0.3}CoCrFeNi$ fibers and the SADP for each (b, c), (e, f), and (h, i), respectively. The matrix (A) and precipitate particles (B) are identified for the TEM (a, d, g) and SADP in (b, e, h) and (c, f, i), respectively.

TABLE 19.1
Composition of the $Al_{0.3}CoCrFeNi$ HEA Wires [1]

Atomic %		Al	Co	Cr	Fe	Ni
Nominal		6.98	23.26	23.26	23.26	23.26
3.15 mm	A (matrix)	28.16	3.79	9.55	16.09	42.40
	B (particle)	5.18	22.86	25.02	24.03	22.87
		29.70		4.83	10.72	49.39
		3.22		24.90	23.64	26.37
1.00 mm	A (matrix)	22.99	16.43	4.45	11.13	44.97
	B (particle)	3.99	24.46	24.09	24.44	23.00
1.60 mm	A (matrix)	13.77				
	B (particle)	23.33				

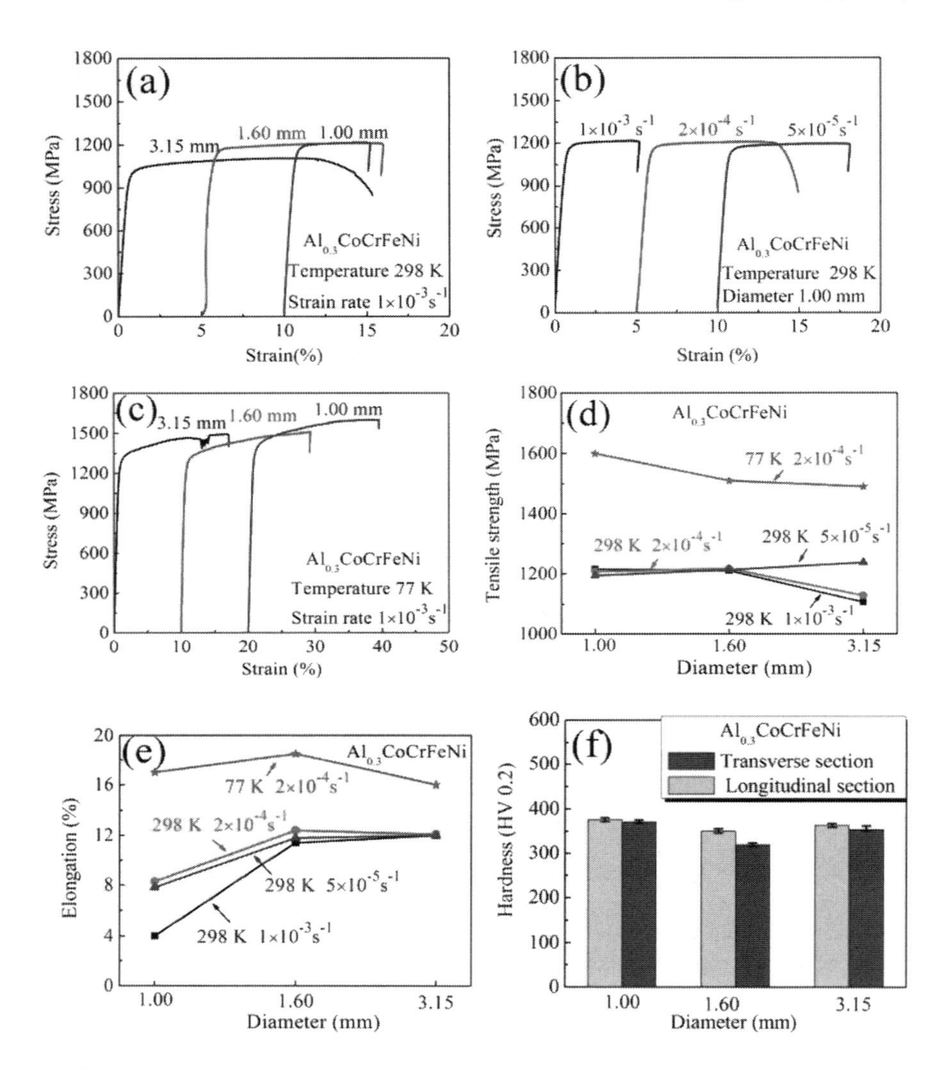

FIGURE 19.3 Mechanical properties of the $Al_{0.3}CoCrFeNi$ high entropy fibers: (a) and (b) are engineering stress-strain curves with different diameters and strain rates at room temperature, respectively; (c) is engineering stress-strain curves at 77 K; (d) and (e) are tensile strength and elongation as a function of diameter, respectively; and (f) is Vickers-hardness variation with change in diameter and section (i.e., longitudinal vs transverse directions) [1].

Although the $Al_{0.3}CoCrFeNi$ high entropy fiber exhibits considerable strength, the ductility is relatively limited. In order to balance ductility and strength in $Al_{0.3}CoCrFeNi$ fibers, annealing at an appropriate temperature is an effective approach [2]. Therefore, a systematic study on the microstructure and mechanical properties after annealing is carried out. Figure 19.5 shows microstructural features, including grain morphologies and texture evolution in the 1.0 mm diameter $Al_{0.3}CoCrFeNi$ fibers after annealing at 900°C. There are a large quantity of

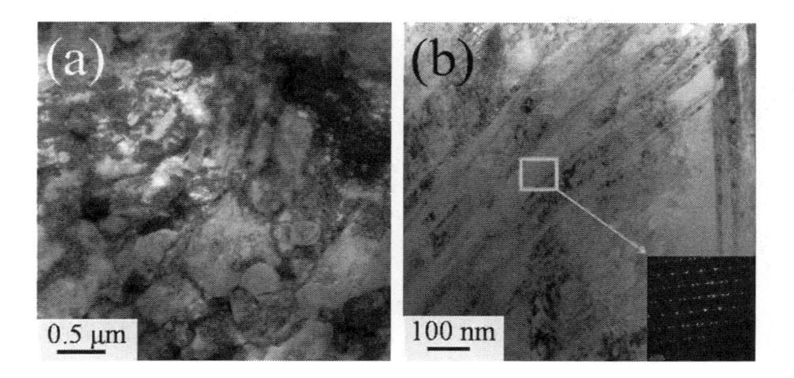

FIGURE 19.4 TEM micrographs of the 3.15 mm-diameter $Al_{0.3}CoCrFeNi$ fibers tested at: (a) 298 K, (b) 77 K [1].

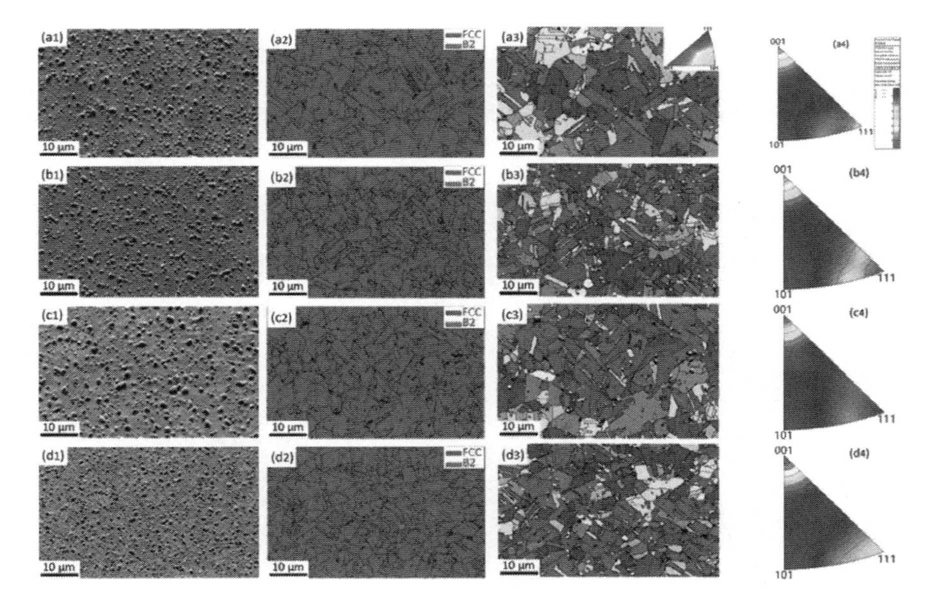

FIGURE 19.5 SEM images (a1–d1), phase map of the FCC phase (green color) and B2 phase (pink color) (a2–d2), EBSD images (a3–d3), and IPF patterns (a4–d4) of the 1.0 mm diameter $Al_{0.3}CoCrFeNi$ fibers annealed at 900°C for 10 min, 30 min, 300 min, and 720 min, respectively [2].

nano-sized spherical precipitates and annealing twins in most of the annealed-state grains. According to the statistics, the average size of random-oriented grains was around 2 μm. The volume fraction and size of the B2 phase increased slightly with when increasing the annealing time from 0 to 300 min. When continuing to increase the annealing time to 720 min, the fraction of B2 phase decreases. These precipitates are rich in the Ni element and the Al element, and are consistent with prior experimental and modeling studies [5, 14, 15].

Figure 19.5a3–d3 presents grain orientation maps, which show that annealing at 900°C resulted in partial recrystallization. The <111> fiber texture and <001> fiber texture were also observed in Figure 19.5a3–d3. On the right side of Figure 19.5 (i.e., Figure 19.5a4–d4), grain orientation maps indicate crystal orientations parallel to the longitudinal axis. The stereographic triangle is at the top right of the grain orientation map in Figure 19.5a3 for color decoding. The inverse pole figures (IPF) show the texture along the fiber axis. The pole density scale is on the right of Figure 19.5a4 where the orientation distribution function was normalized to express densities in multiples of random distribution.

It has been proved that precipitation strengthening primarily depend on the precipitate type, size, density, and distribution [16, 17]. Aiming to further investigate the role and the phase structures of the precipitate, TEM of annealed fibers was performed. Figure 19.6 demonstrates the TEM results for the 1.0 mm diameter fibers at 900°C for 720 min exposure. The microstructure consists of an FCC matrix with B2 precipitate particles. This was verified from selected area diffraction patterns (SADP)

The change in grain size has been shown as a function of annealing time. Grain growth was rapid up to 30 min exposure, and then decreased substantially with increasing exposure time. After annealing for 720 min, the average grain sizes, measured using linear intercept method, were smaller than 3 mm for both 1.0 mm and 1.6 mm diameter fibers. For many polycrystalline materials, the relationship between grain diameters d with time t can be described by the following Equation (19.1) [18, 19]:

$$d_f - d_0 = Kt^n \tag{19.1}$$

where d_f and d_0 are the final and initial grain diameter at time t and time t = 0, respectively, K is the grain growth rate coefficient, and n is the growth exponent Figure 19.7.

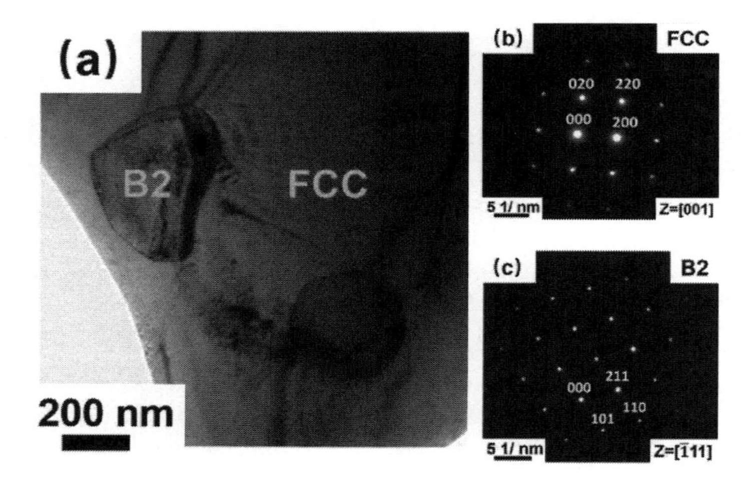

FIGURE 19.6 (a) TEM images of the 1.0 mm diameter $Al_{0.3}CoCrFeNi$ fibers after annealing at 900°C for 720 min, (b) SADP for FCC matrix, and (c) B2 precipitate particles [2].

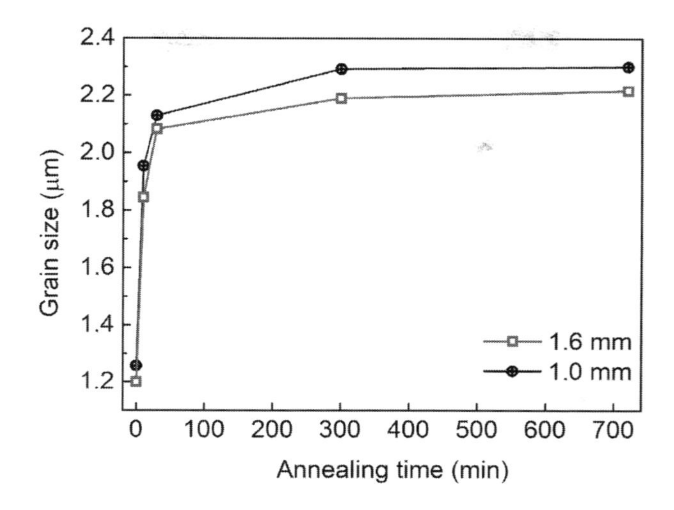

FIGURE 19.7 Grain sizes of the $Al_{0.3}$CoCFeNi HEA 1.0 mm diameter fibers and 1.6 mm diameter fibers after exposure at 900°C for different annealing times.

Figure 19.8 contrasts the ultimate tensile strength and yield strength against elongation-to-fracture for $Al_{0.3}$CoCrFeNi HEA fibers and other traditional alloy fibers (including Cu-Sn, Cu-Al-Ni, Cu-Al, Ni-Mn-Ga, ZnLi, Cu, Mg-Y-Zn, and 316 stainless) at room temperature. The mechanical properties of $Al_{0.3}$CoCrFeNi HEA fibers vary considerably, depending on the processing conditions. Obviously, both tensile strength and yield strength of $Al_{0.3}$CoCrFeNi fibers greatly exceed those of other traditional alloy fibers, and this can be attributed to the solid solution hardening characteristic of HEAs, grain boundary hardening, precipitation hardening, and dislocation hardening.

19.2.2 CoCrFeNi High Entropy Alloy Wires

CoCrFeNi high entropy alloy wires (9.3 mm in diameter) with nano-sized deformation twins were produced by the heavy cold-drawing process. The alloy wires exhibit a high tensile yield strength of 1.2 GPa, as well as a considerably high percentage elongation of 13.6% at 223 K. The dominant deformation mechanism shifted from secondary deformation twinning at cryogenic temperatures to dislocation slip and dynamic recovery at elevated temperatures. The wire was detected to contain a single FCC structure and a well-developed <111> texture aid at <100> texture, as shown in Figure 19.9.

The tensile strength of cold drawn CoCrFeNi HEAs wire was higher than that of undeformed HEAs at elevated temperatures. It should be caused by deformation-induced grain refinement, dislocation and nano-twinning.

According to the nanoindentation results (Figure 19.10a), the hardness of microcrystalline CoCrFeNi HEA wires was significantly improved compared to the recrystallized state. Furthermore, the HEA wires were proved to have a good combination of high strength and ductility over a wide range of temperatures, as demonstrated in

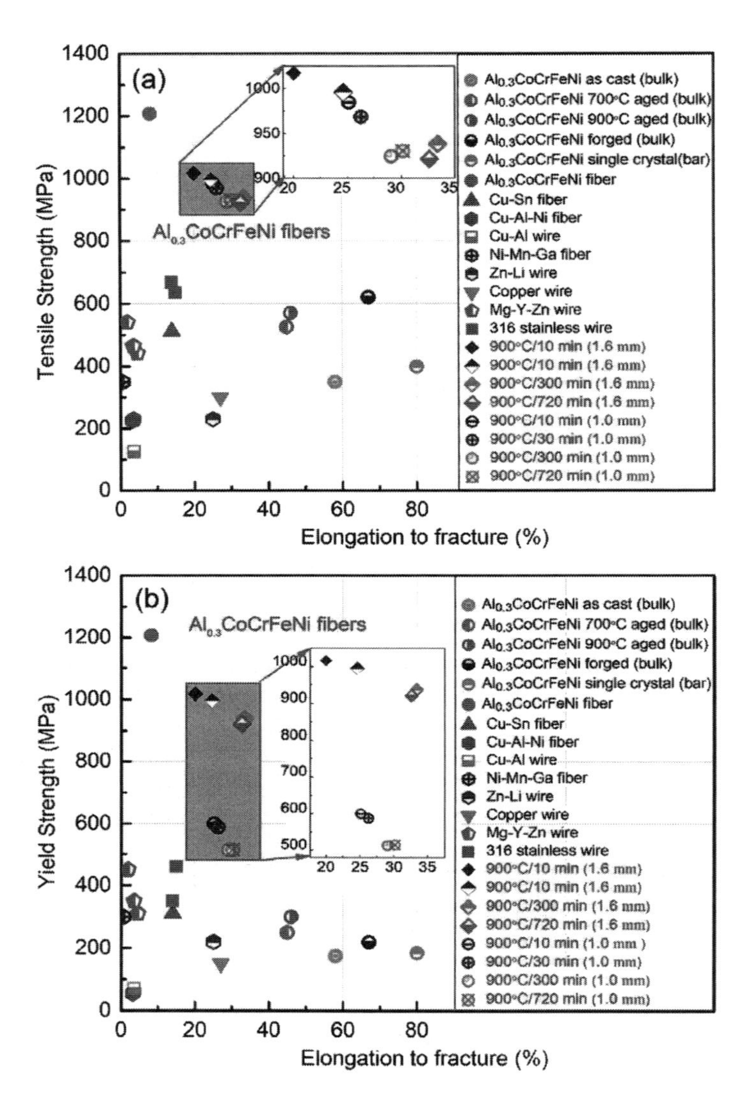

FIGURE 19.8 Comparison in mechanical properties of $Al_{0.3}$CoCrFeNi HEA in bulk and fiber forms and other conventional alloy fibers: (a) ultimate tensile strength and (b) yield stress versus elongation to fracture.

Figure 19.10b. Elongation increases significantly as the test temperature increases, accompanied by a modest drop of the yield strength. According to the Hall–Petch relation, the high density of the twin boundaries resulting in grain refinement is to a great extent responsible for enhanced the yield strength of alloy. However, it is worth noting that there is no strain hardening in the stress-strain curve of Figure 19.10c, which can be attributed to the suppression of dislocation glide by the high density of twin boundaries (Figure 19.10).

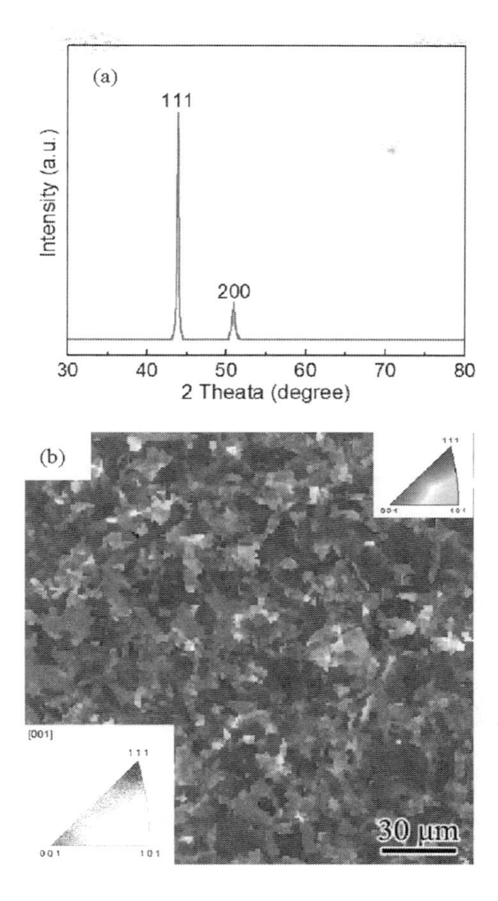

FIGURE 19.9 (a) XRD patterns, (b) grain orientation and [001] inverse pole figure map obtained from the longitudinal section of as-drawn CoCrFeNi HEA wires [3].

19.2.3 CoCrFeMnNi High Entropy Alloy Wire Rod

CoCrFeMnNi high entropy alloy wire rods (7.5 mm in diameter) were produced by cryogenic temperature caliber rolling (CTCR). On account of the highly increased twinning activity caused by lowering the temperature to 77 K, significant twinning-induced grain refinement occurred. Consequently, an ultrafine-grained structure could be achieved in the processed material. The processed material showed a tensile strength of about 1.7 GPa, and also had excellent resistance to hydrogen embrittlement (HE), in contrast to the typical trade-off relationship between these two properties. The exceptionally high resistance to HE was attributed to the combined effects of (1) difficulties in accumulating hydrogen owing to the sluggish hydrogen diffusion caused by the face-centered cubic crystal structure and the severe lattice distortion, (2) the high hydrogen threshold required for HE at the dominant cracking sites of twin boundaries, and (3) absence of martensite transformation.

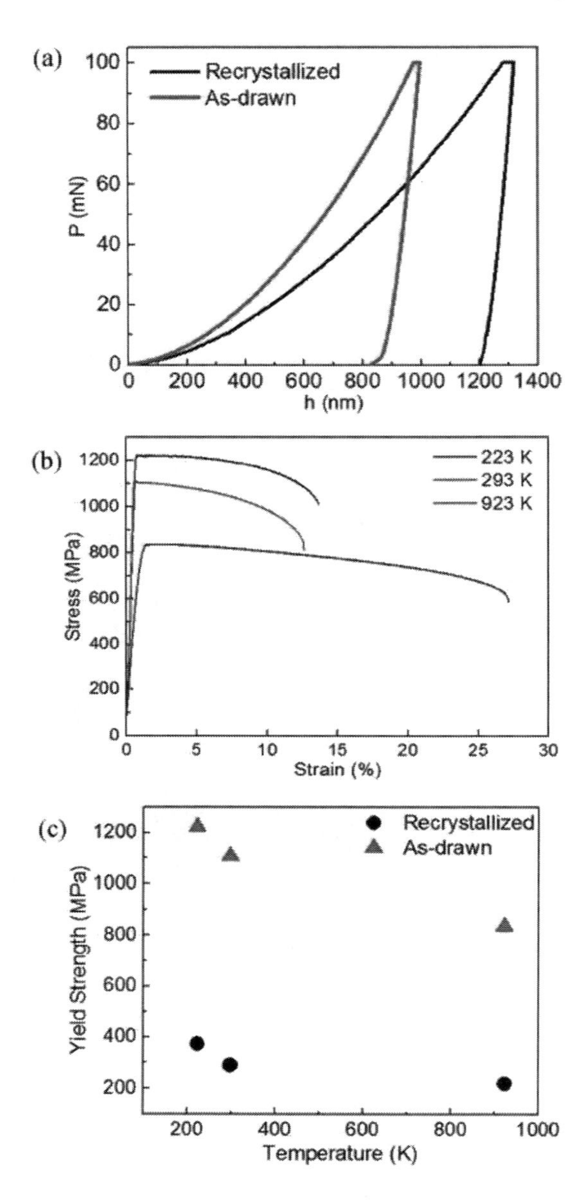

FIGURE 19.10 (a) Load-displacement curves of CoCrFeNi HEA wires, (b) stress-strain curves of as-drawn CoCrFeNi HEA wires, (c) yield strength of CoCrFeNi HEA wires with increasing temperature [3].

The EBSD band contrast map showed that the initial HEA had a fully recrystallized equiaxed grain structure with coarse annealing twins (Figure 19.11a). The XRD patterns indicate the initial HEA had a single FCC crystal structure (Figure 19.11b).

In the CTCR-ed HEA, abundant twins were introduced, which is not possible when employing room temperature processes. These twins were found to be extremely thin (5 to 20 nm) and intersected with each other. The intersection of twins was caused by the operation of different twin variants. As caliber rolling imposes stress in all directions along the circumference of the sample, the 12 possible twinning variants in the FCC crystal structure can operate. Notably, secondary twins were formed within a very narrow matrix bounded by primary twins, even though twinning activity significantly decreases as the matrix becomes narrow. Moreover, the intersected twin morphology led to effective refinement of the grain matrix as compared to that observed in the case of parallel twins (Figure 19.12).

Figure 19.13 shows that the tensile strength of the CTCR-ed HEA was remarkably higher than that of the initial HEA. The yield stress was more than four times higher than that of the initial HEA. Simultaneously, the ultimate tensile stress was over two times higher than that of the initial HEA. The improvement in strength can be attributed to the combined effect of dislocation accumulation strengthening and Hall–Petch strengthening. The fracture elongation of CTCR-ed HEA achieved around 10% in spite of it being significantly strengthened. The tensile properties of the CTCR-ed

HEA were in comparison with those of the tempered-martensitic steel and the pearlitic steel in Figure 19.13b. Both yield strength and ultimate tensile strength were higher in the CTCR-ed HEA than in the tempered-martensitic steel, although these two materials had a similar fracture elongation. These comparisons show that the CoCrFeMnNi HEA has better tensile properties than high-strength steels used for fasteners.

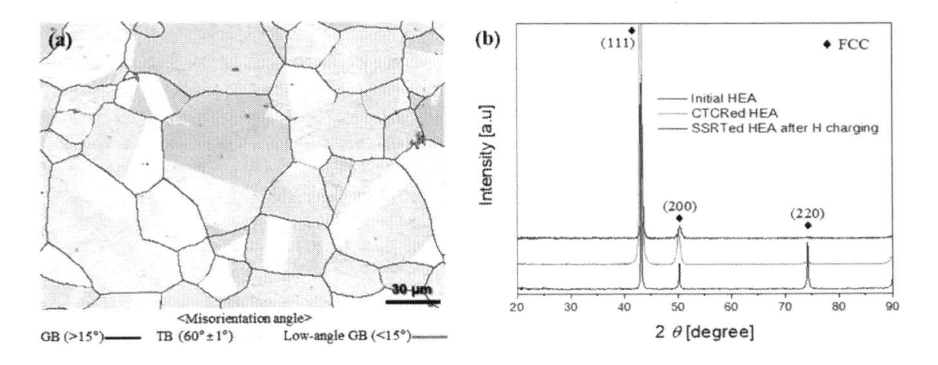

FIGURE 19.11 (a) EBSD band contrast map and (b) XRD patterns of the initial HEA [4].

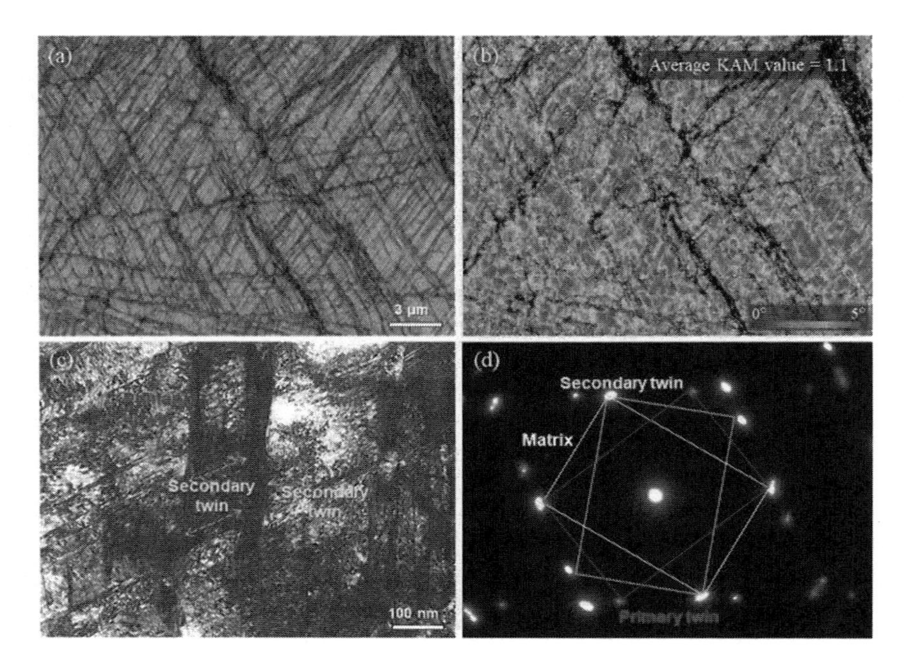

FIGURE 19.12 (a) EBSD band contrast map and (b) kernel average misorientation map of CTCR-ed HEA. (c) TEM bright-field image of the CTCR-ed HEA and (d) corresponding [011] diffraction patterns [4].

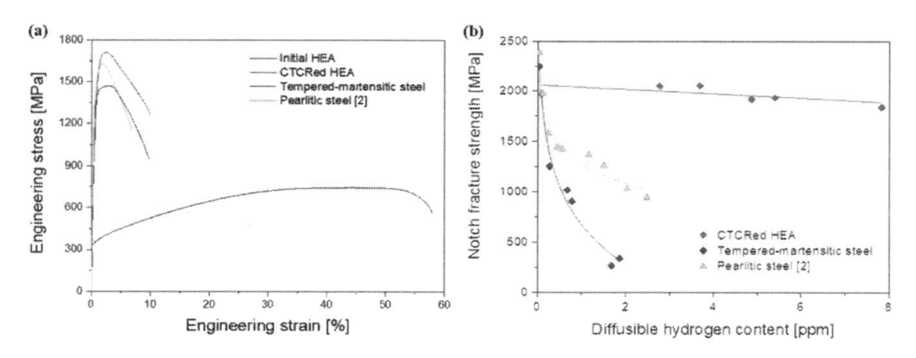

FIGURE 19.13 (a) Stress vs strain curves of the initial HEA, the CTCR-ed HEA, and the tempered-martensitic steel. (b) Slow strain rate test notch fracture stress with diffusible-hydrogen content for the CTCR-ed HEA and the tempered-martensitic steel. Curves for a 1.6 GP-grade pearlitic steel are also presented for comparison [4].

19.3 CONCLUSION

Fibers have a wide range of applications in functional and structural materials. High entropy alloy fibers are a new and potentially promising field. In the above, we have summed up the current understanding of the microstructure-mechanical property correlation of high entropy fibers, which contains the phase structure, microstructure, strengthening mechanisms, and deformation behavior. We can conclude that high entropy fibers have the following unique characteristics:

1. There are many methods to develop high entropy fibers, for instance, hot drawing, cold drawing, and cryogenic temperature caliber rolling. Accordingly, it is feasible and effective to produce high entropy fibers in significant quantities, which paves the way for the popularization and industrial applications.
2. Moreover, the fibers currently studied mainly maintain a simple FCC phase structure. The strength of the alloy has relatively low strength under as-cast conditions, whereas fibers of the same composition exhibit a quite high strength.
3. The microstructure of the high entropy fiber is dissimilar from the as-cast state ingots. Therefore, we have more space to manipulate the microstructure and thus affect the performance.

At the same time, the research into high entropy fiber still possesses multiple important and challenging issues. We summarized the following perspective for future research associated with the development of high entropy fibers.

1. Current studies of high entropy fiber are primarily limited to face-centered cubic structural alloys, which we can extend to body-centered cubic and hexagonal close-packed alloys.
2. For the performance of the high entropy fibers, we mainly focus on the mechanical properties. In fact, we can explore more potential valuable properties, including magnetic properties and electrical properties.
3. To take advantage of the fibers, we can also prepare them into composite materials to achieve their target performance.

REFERENCES

1. Li, D., Li, C., Feng, T., Zhang, Y., Sha, G., Lewandowski, J.J., et al. High-entropy Al 0.3 CoCrFeNi alloy fibers with high tensile strength and ductility at ambient and cryogenic temperatures. *Acta Materialia* 2017;123:285–94.
2. Li, D., Gao, M.C., Hawk, J.A., Zhang, Y. Annealing effect for the Al0.3CoCrFeNi high-entropy alloy fibers. *Journal of Alloys and Compounds* 2019;778:23–9.
3. Huo, W., Fang, F., Zhou, H., Xie, Z., Shang, J., Jiang, J. Remarkable strength of CoCrFeNi high entropy alloy wires at cryogenic and elevated temperatures. *Scripta Materialia* 2017;141:125–8.
4. Kwon, Y.J., Won, J.W., Park, S.H., Lee, J.H., Lim, K.R., Na, Y.S., Lee, C.S. Ultrahigh-strength CoCrFeMnNi high-entropy alloy wire rod with excellent resistance to hydrogen embrittlement. *Materials Science and Engineering: Part A* 2018;732:105–11.
5. Gwalani, B., Gorsse, S., Choudhuri, D., Zheng, Y., Mishra, R.S., Banerjee, R. Tensile yield strength of a single bulk Al0.3CoCrFeNi high entropy alloy can be tuned from 160 MPa to 1800 MPa. *Scripta Materialia* 2019;162:18–23.
6. Liao, W., Lan, S., Gao, L., Zhang, H, Xu S, Song J, et al. Nanocrystalline high-entropy alloy (CoCrFeNiAl 0.3) thin-film coating by magnetron sputtering. *Thin Solid Films* 2017;638:383–8.

7. Ma, S.G., Zhang, S.F., Qiao, J.W., Wang, Z.H., Gao, M.C., Jiao, Z.M., et al. Superior high tensile elongation of a single-crystal CoCrFeNiAl0.3 high-entropy alloy by Bridgman solidification. *Intermetallics* 2014;54:104–9.
8. Yasuda, H.Y., Shigeno, K., Nagase, T. Dynamic strain aging of Al 0.3 CoCrFeNi high entropy alloy single crystals. *Scripta Materialia* 2015;108:80–3.
9. Kireeva, I.V., Chumlyakov, Y.I., Pobedennaya, Z.V., Vyrodova, A.V., Kuksgauzen, I.V., Kuksgauzen, D.A. Orientation and temperature dependence of a planar slip and twinning in single crystals of Al0.3CoCrFeNi high-entropy alloy. *Materials Science and Engineering: Part A* 2018;737:47–60.
10. Gludovatz, B., Hohenwarter, A., Catoor, D., Chang, E.H., George, E.P., Ritchie, R.O. A fracture resistant high-entropy alloy for cryogenic applications. *Science* 2014;345(6201):1153–8.
11. Tong, Y., Chen, D., Han, B., Wang, J., Feng, R., Yang, T., et al. Outstanding tensile properties of a precipitation-strengthened FeCoNiCrTi0. 2 high-entropy alloy at room and cryogenic temperatures. *Acta Materialia* 2019;165:228–40.
12. Stepanov, N., Tikhonovsky, M., Yurchenko, N., Zyabkin, D., Klimova, M., Zherebtsov, S., et al. Effect of cryo-deformation on structure and properties of CoCrFeNiMn high-entropy alloy. *Intermetallics* 2015;59:8–17.
13. Qiu, Z., Yao, C., Feng, K,. Li, Z., Chu, P.K. Cryogenic deformation mechanism of CrMnFeCoNi high-entropy alloy fabricated by laser additive manufacturing process. *International Journal of Lightweight Materials and Manufacture* 2018;1(1):33–9.
14. Shun, T.-T, Du, Y.-C. Microstructure and tensile behaviors of FCC Al0.3CoCrFeNi high entropy alloy. *Journal of Alloys and Compounds* 2009;479(1–2):157–60.
15. Gwalani, B., Soni, V., Choudhuri, D., Lee, M., Hwang, J.Y., Nam, S.J., et al. Stability of ordered L1 2 and B 2 precipitates in face centered cubic based high entropy alloys - Al 0.3 CoFeCrNi and Al 0.3 CuFeCrNi 2. *Scripta Materialia* 2016;123:130–4.
16. Yang, T., Zhao, Y., Tong, Y., Jiao, Z., Wei, J., Cai, J., et al. Multicomponent intermetallic nanoparticles and superb mechanical behaviors of complex alloys. *Science* 2018;362(6417):933–7.
17. Liang, Y-J., Wang, L., Wen, Y., Cheng, B., Wu, Q., Cao, T, et al. High-content ductile coherent nanoprecipitates achieve ultrastrong high-entropy alloys. *Nature Communications* 2018;9(1):4063.
18. Hu, H., Rath, B. On the time exponent in isothermal grain growth. *Metallurgical Transactions* 1970;1:3181–4.
19. Beck, P.A. Effect of recrystallized grain size on grain growth. *Journal of Applied Physics* 1948;19(5):507–9. 01023493000

20 A Useful Review of High Entropy Films

Xue Hui Yan and Yong Zhang

CONTENTS

20.1 INTRODUCTION

Materials show a rising trend of alloy chemical complexity versus time. High entropy alloys (HEAs) have become another new research hotspot after the research system of bulk amorphous and intermetallic compounds due to their excellent performance [1]. The high mixing entropy allows alloys to have a lower free-energy and higher phase-stability. Experimental results indicate that the higher mixing entropy in these alloys enhances the formation of random solid-solution phases with simple structures, such as face-centered cubic (FCC), body-centered cubic (BCC) structures, or hexagonal close-packed (HCP) structures [2–4]. Due to severe lattice distortion and solid-solution strengthening attributed to the multi-component, the HEAs show many excellent mechanical properties, such as high strength, high hardness, high low-temperature fracture toughness, excellent oxidation resistance, anti-friction performance, and excellent soft magnetic properties [5, 6]. The superior properties also make it a potential structural and functional material.

High entropy films (HEFs) are a brand-new type of alloy films, which have been developed recently based on the design concept of high entropy alloys (HEAs). HEFs can be defined as multiple component films with high entropy of mixing. Based on a similar scientific concept, the HEFs have been designed and investigated gradually, and show a great potential application in coating industry. At present, binary or ternary film have been widely explored, such as AlN, TiN, TiZrN, etc. [7–9]. However, due to the limitations of alloy systems, the traditional metal nitride, carbide, and oxide film which are low entropy films, could not meet the higher requirements of modern industry. With the development of HEAs, research in the field of HEFs has gradually been carried out.

High entropy alloy films not only exhibit superior performance similar to high entropy alloys, but even some properties are superior to alloy bulks. In 2005, high entropy alloy nitride film was reported for the first time in the literature. Chen et al. [10] used FeCoNiCrCuAlMn and FeCoNiCrCuAl0.5 high entropy alloys as targets to prepare high entropy alloy thin films and two high entropies by magnetron sputtering the nitride film of the alloy and exploring the changes in its microstructure. Subsequently, Huang et al. [11] prepared AlCoCrCu0.5NiFe oxide film and analyzed its microstructure, hardness, and thermal stability. With the in-depth study of high entropy alloy films, research work on high entropy alloy films as diffusion barriers [12], hard coatings [13], and anti-corrosion coatings [14] has been carried out. High entropy alloy films have shown far-reaching prospects in the fields of tool coatings, diffusion barriers, and photothermal conversion.

20.2 COMPONENT DESIGN

Statistical results show that there are about 25 kinds of elements that can be used for the design of high entropy alloy films, which are mainly concentrated in the transition metal region. In general, the alloying elements used in the high entropy alloy films are consistent with those of bulk alloys.

The types of high entropy alloy films are similar to those of traditional alloy films, including alloy films, nitride films, carbonized films, and oxide films.

From the perspective of component design, the elements used in the design of HEFs can be loosely divided into "basic element" and "functional element" and "non-metal element," as shown in Figure 20.1. Metal elements having similar atomic sizes and tending to form a simple face-centered cubic (FCC) or body-centered cubic (BCC) solid-solution structure may be referred to as "basic elements," such as Co, Cr, Cu, and Fe. Similarly, elements having excellent properties such as wear resistance, heat resistance, and corrosion resistance may be referred to as a "functional elements," such as W, Mn, and V. Therefore, based on the "basic element," the "functional element" should be selected according to the required performances of alloy films. It should be noted that "non-metal elements" play an important role in engineering coatings, such as hard coatings. Thus, some small-sized non-metallic elements such as B, C, N, and O are usually added to the coating, which can fill the gap position of the film to improve the hardness characteristics of the film.

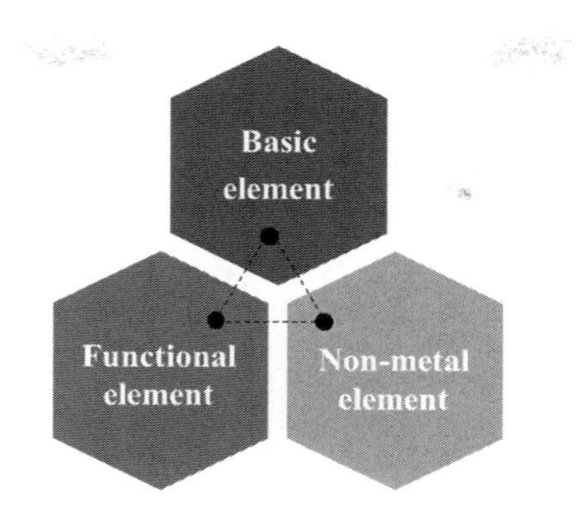

FIGURE 20.1 Composition design of HEFs.

20.3 PREPARATION TECHNOLOGY OF HIGH ENTROPY FILMS

The preparation method of the high entropy alloy film is consistent with the conventional alloy film. It is mainly divided into two categories: physical preparation and chemical preparation. Physical preparation mainly refers to physical vapor deposition (PVD), including vacuum sputtering, vacuum evaporation, and ion plating. Meanwhile, chemical preparation mainly includes chemical vapor deposition (CVD), and the liquid phase deposition (LPD). At present, the preparation methods of high entropy alloy films reported in the literature include magnetron sputtering [15–18], the laser cladding method [19–21], the electrochemical deposition method [22], the arc thermal spraying method [23], the cold spray method [24], the electron beam evaporation deposition method [25], the plasma cladding method [12], etc. Among them, magnetron sputtering and laser cladding techniques are the more common preparation methods for high entropy alloy films. The process characteristics of the two coating preparation techniques are analyzed and summarized in Table 20.1.

TABLE 20.1
Features of Magnetron Sputtering and Laser Cladding

Preparation methods	Characteristics
Magnetron sputtering	I. No special requirement for the conductivity of the target.
	II. Slow heating rate of the substrate, and high rate of deposition.
	III. Good compositional consistency and structural stability.
	IV. Good bonding of the film to the substrate.
	V. Flexible control of film properties and thickness.
	VI. Low target utilization.
Laser cladding	I. High bonding strength, coating and substrate are metallurgically bonded.
	II. Small thermal impact on the substrate.
	III. Coating thickness up to several millimeters.
	IV. High sensitivity of coating crack.

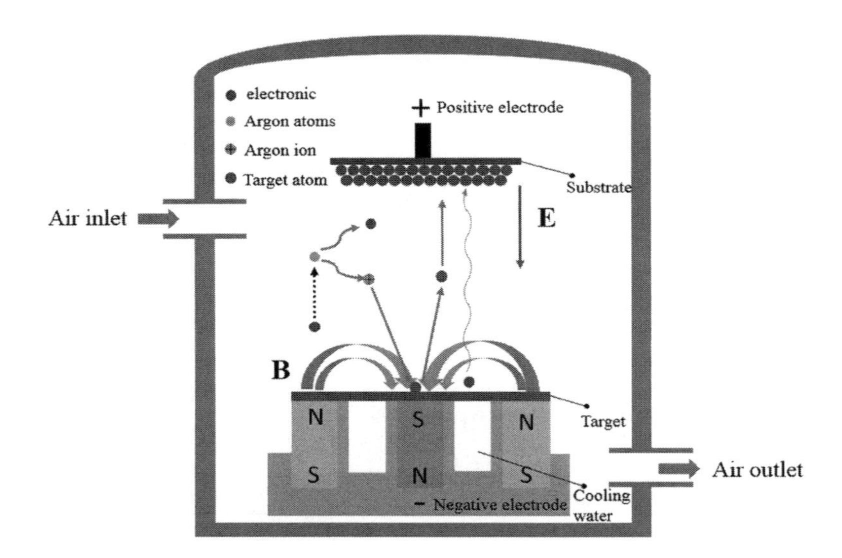

FIGURE 20.2 Schematic of the magnetron sputtering process.

The principle of magnetron sputtering deposition is a sputtering effect. The high energy particles bombard the surface of the target (Figure 20.2), causing the target atoms to escape and move in a certain direction, and finally form a thin film on the substrate. The dual functions of magnetic and electric fields increase the collision probabilities of the electron, charged particles, and gas molecules. In the usual cases, the magnetron sputtering target material is prepared by the arc melting and powder metallurgy methods. If the melting point of each principal component is relatively different, a powder metallurgy method is usually the priority choice. Also, it is difficult to obtain an equiatomic ratio thin film, since the different elements have different sputtering output capacities. Therefore, "multi-target sputtering" has been proposed [27]. According to the composition design of HEFs, many single element targets or alloy targets can be prepared, and the atomic ratio of elements can be controlled by adjusting the target sputtering power.

Laser cladding technology has indicated that melting metal powder by high-power and high-speed laser shows certain physical, chemical, or mechanical properties. A layer combining with the matrix in the way of metallurgy bonding can improve the mechanical properties between the layer and the matrix. Laser cladding technology is divided into two types of methods, referred to as pre-powder and synchronous feeding, as shown in the schematic diagram of Figure 20.3.

20.4 PHASE STRUCTURE

Due to the difference in the preparation process, the cooling rate of film is faster than that of bulk alloys. It means that the alloy films cannot reach the equilibrium state achieved by the bulk alloys. Thus, the phase structure between the high entropy alloy film and the bulk alloys has no correlation, even if they have the same composition.

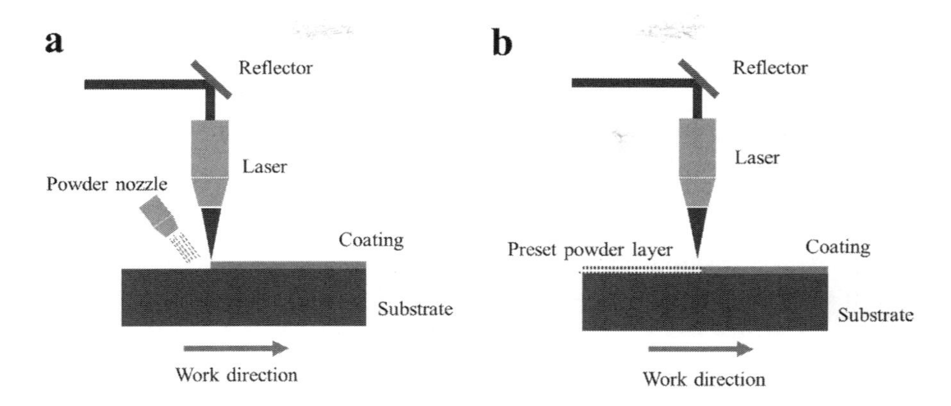

FIGURE 20.3 Schematic of laser cladding: (a) synchronous powder; (b) pre-powder layer.

In addition to the influence of component elements, the working atmosphere, substrate bias, and substrate temperature also have a significant effect on the phase structure of HEFs.

20.4.1 Nitrogen Flow Rate

Nitride films are a common class of engineering materials. For a high entropy alloy nitride film, the working atmosphere is a mixed gas of argon and nitrogen. It is found that the phase structure of the alloy film is usually amorphous under no nitrogen or low nitrogen conditions. With the increase of the nitrogen flow rate, the phase structure gradually changes from the initial amorphous state to the crystalline state, and tend to form a single solid-solution structure. Chang et al. [18] reported that the TiVCrAlZr films exhibited amorphous structures at low nitrogen gas flow rates (Figure 20.4). However, with increases to the N_2 flow rate, the HEFs phase structure transforms from an amorphous to a simple FCC solid-solution structure. Liang et al. [28] found a similar phenomenon in the preparation of TiVCrZrHf nitride films. When the N_2 flow rate was between 0 and 2 standard cubic cm per minute (SCCM), the XRD pattern of the film showed a broad diffraction peak. This phenomenon indicated that the phase structure of the TiVCrZrHf film was amorphous, as shown in Figure 20.5. Similar results have also been reported in other HEFs, such as AlCrTaTiZr [29], AlCrMoSiTi [30], and TiAlCrSiV [31].

The reasons for the formation of the amorphous phase structure of the high entropy alloy film can be summarized as follows: (1) The cooling rate is reason faster during the deposition process. Thus, the crystal grain cannot obtain sufficient energy to grow up. (2) High mixing entropy and large size differences are facilitated to the stability of amorphous structures [32]. With the increase of the N_2 flow rate, the surface migration energy of the alloy film increases, and the phase structure of the alloy film generally changes from the amorphous to the solid-solution phase. Taking the (TiVCrZrHf)N film as an example, the constituent elements should form several binary nitrides, such as TiN, VN, CrN, ZrN, and HfN. However, in the XRD diffraction pattern, the phase structure of the high entropy nitride film is a simple FCC

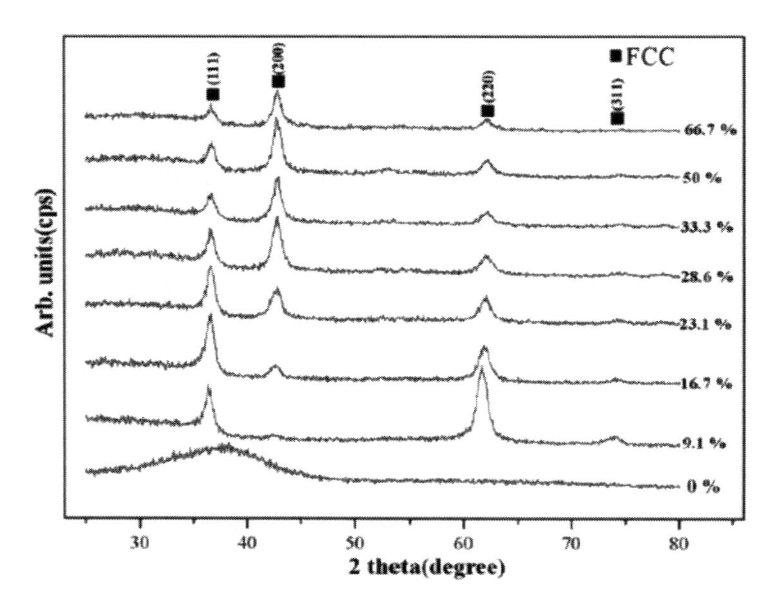

FIGURE 20.4 XRD patterns of TiVCrAlZr films with different N_2 flow rates.

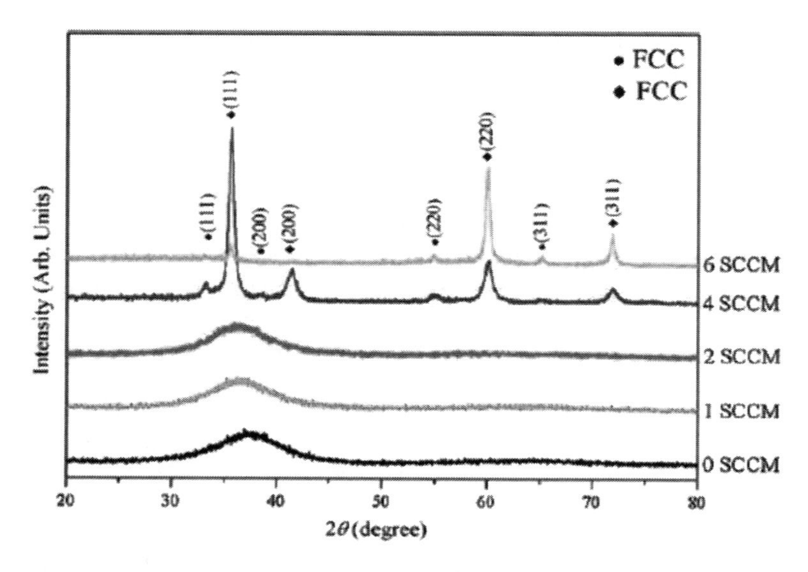

FIGURE 20.5 XRD patterns of the TiVCrZrHf films with different N2 flow rates

phase rather than multiple nitride phases. It indicates that a solid solution process has occurred between the nitrides in the alloy films. As the N_2 flow rate increases, the formation of nitride is promoted. The binary nitrides have similar sizes and structures, which makes it easier for the solid solution to occur between the nitrides so that the nitride alloy film finally obtains a single solid-solution phase.

In binary and ternary systems, a solid solution between nitrides and carbides has been demonstrated [33, 34]. Therefore, the crystal phase of the high entropy alloy nitride film is a multicomponent solid-solution phase. The solid solution unit is different from the metal atom of the alloy block, and the nitride is a solid solution unit. In addition, as the flow rate of N_2 increases, the migration energy of the surface of the film becomes larger, which is favorable for the diffusion of atoms and the growth of crystal grains.

20.4.2 Substrate Bias

The substrate bias is an effective method to improve the structure and properties of films. Shen et al. prepared $(Al_{1.5}CrNb_{0.5}Si_{0.5}Ti)N_x$ thin films using magnetron sputtering techniques [16] and investigated the effects of substrate bias (Vs) on the phase structures of films. It was found that the phase structures of the films tended to form simple FCC structures. With the change of Vs, the crystal diffraction peak became gradually weakened, and the structure slowly changed from a simple columnar crystal to a dense phase structure. At the same time, the grain was refined, and the grain size decreased from 70 nm to 5 nm. Huang et al. [17] also studied the effects of substrate bias on the phase structure of HEFs. The results show that when the Vs were weakened, the phase structure tended to be amorphous.

During film deposition, increasing the substrate bias increases the ability of atoms to diffuse across the surface of the film and participate in chemical reactions. Accordingly, the density and film-forming ability of the film are also improved. At the same time, various defects of the film are induced, the growth of the columnar crystals is suppressed, and the film grains are refined.

20.4.3 Substrate Temperature

Huang et al. prepared AlCrNbSiTiV nitride film with substrate temperature ranging from 100°C to 500°C, and the temperature gradient was 100°C, as shown in Figure 20.6. From the XRD diffraction pattern, the phase structure was a single-phase

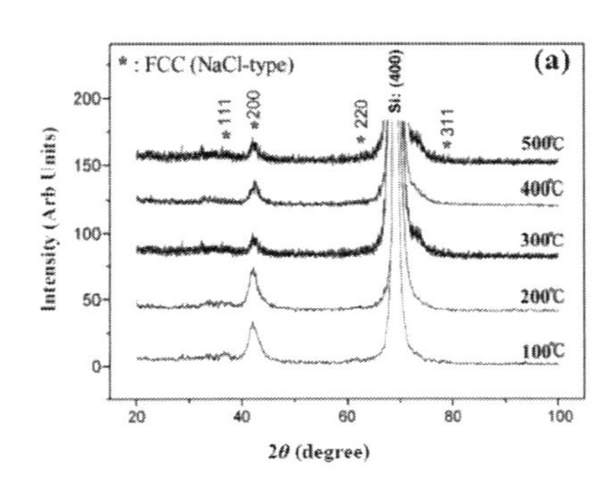

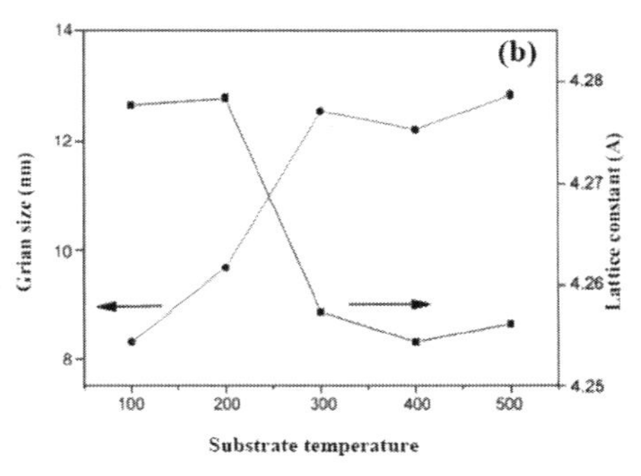

FIGURE 20.6 (a) XRD patterns of the AlCrNbSiTiV nitride films at different substrate temperatures; (b) Variations of the grain sizes and lattice constant of the AlCrNbSiTiV nitride film at different substrate temperatures.

FCC structure, and an increased tendency in grain size was been observed [35]. Liang et al. also reported similar conclusions when investigating the influence of substrate temperature on the phase structure of high-entropy films.

With the increase in substrate temperature, high entropy alloy nitride films still exhibit a simple FCC solid-solution phase structure, and the grain size shows an upward trend [36]. As the substrate temperature increases, atomic adsorption capacity and surface migration ability are enhanced, so that the grain size is increased subsequently.

20.5 MECHANICAL PROPERTIES

High entropy alloy films have excellent mechanical properties compared to conventional alloy films. Figure 20.7 compares the hardness and Young's modulus of high entropy alloy films with common alloys and amorphous materials. High entropy alloy films show significant advantages in hardness and modulus. In addition, the introduction of elements such as N, B, and C can further improve the mechanical properties of the high entropy alloy film. The hardness and modulus of nitride, boride, and oxide alloy films generally show an increasing trend compared to alloy films.

Generally, with the increase of the non-metallic elements, such as N, B, and C, the hardness of the alloy film shows a clear upward trend. It is mainly due to the fact that small-sized non-metal atoms such as N, C, and O are usually present in the lattice gap, that is, interstitial atoms. As the non-metallic elements increase, the content of solid solution interstitial atoms increases. According to Equation (20.1),

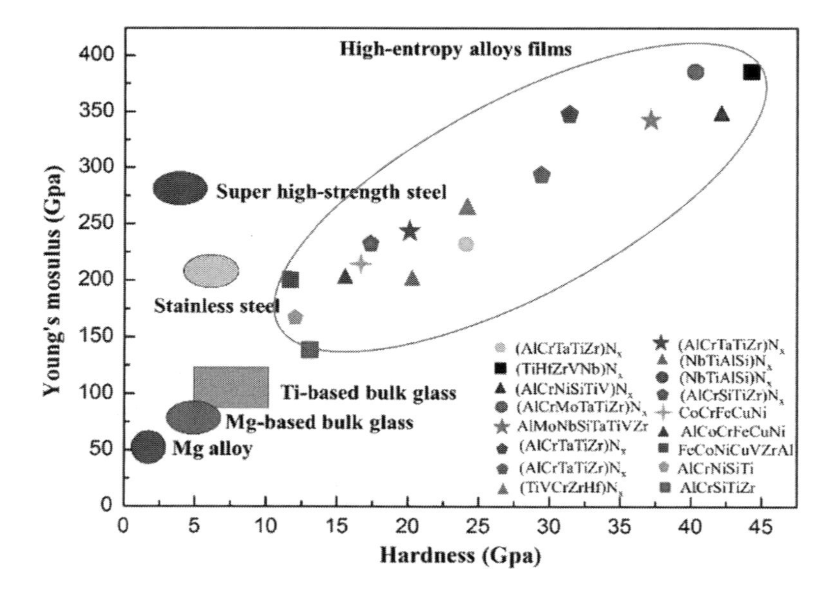

FIGURE 20.7 The relationship between hardness and Young's modulus for common alloys and high entropy thin films.

solid-solution strengthening increases the yield strength of the material by increasing the stress τ to move the dislocations [38]:

$$\Delta = \tau\varepsilon G b^{3/2} c \quad (20.1)$$

where c is the concentration of the solute atoms, G is the shear modulus, b is the size of the Burgers vector, and ε is the lattice strain caused by the solute. As the content of interstitial atoms in B, C, N increases, the concentration of interstitial atoms in the system increases, and the yield strength of the corresponding materials increases. However, the amount of solid solution of interstitial atoms is very low, and the increase in hardness value will decrease after reaching saturation.

20.6 HIGH-TEMPERATURE PERFORMANCE

The high entropy films have good high-temperature properties. It can maintain high strength and the stability of phase structure at high temperatures. Zhang et al. studied the phase structure, mechanical properties, and thermal stability of CoCrCuFeNi high entropy alloy coatings. The CoCrCuFeNi coating was annealed at a high temperature of 500° C for 5 h. It found that the hardness was almost unchanged. After treatment at a higher temperature, it was found that there was a hardness loss of 5.5% after annealing at 750° C for 5 h. In addition, the alloy film exhibited excellent phase structure stability at high temperatures. The phase structure of a high entropy alloy coating is a simple FCC solid-solution phase. After annealing at different temperatures, the phase structure did not change significantly [39]. Similarly, the NbTiAlSiWxNy high entropy alloy film also has excellent high-temperature stability. No change occurred in alloy films after annealing at 700° C for 24 h [40].

Table 20.2 summarizes the high-temperature resistance of high entropy alloy films reported in some literature. After the high entropy alloy film is kept at a high temperature for a period of time, its phase structure still has good stability. However, there are few reports on the changes in mechanical properties of alloy films after high-temperature treatment.

TABLE 20.2
Phase Structures of the HEFs after Annealing

Composition	Temperature (°C)	Time (h)	Phase structure Before/After annealing	Ref
AlCoCrCuFeNi	510	–	Amorphous/Amorphous	[43]
(AlCrTaTiZr)N	700	–	Amorphous/Amorphous	[44]
(NbTiAlSiW)N	700	24	Amorphous/Amorphous	[45]
TaNbTiW	700	1.5	BCC/BCC	[46]
FeNiCoCrAlTiSi	750	5	BCC/BCC	[47]
FeCrNiCoMn	900	2	FCC+BCC/FCC+BCC	[48]
CoCrCuFeNi	1000	5	FCC/FCC	[39]

20.7 CORROSION RESISTANCE AND SOFT MAGNETIC PROPERTIES

Corrosion resistant coatings are an effective measure to improve the life of engineering materials. Generally, a high entropy alloy coating containing elements such as Ni, Cr, Co, Ti, Cu, etc., has excellent mechanical properties and good corrosion resistance to high-concentration acids. Ren et al. studied the corrosion behavior of CuCrFeNiMn coating by immersion test and potential dynamic polarization measurement. The results show that FeCrNiMnCu is high after immersion in 1 mol/L sulfuric acid solution for 100 h at 25°C. The decay candle rate of the entropy alloy is only 0.074 mm/y, which is much lower than the 1.710 mm/y of 304 stainless steel [14]. The corrosion resistance of AlxCoCrFeNi high entropy alloy was compared with traditional materials such as stainless steel, Al-based, Ti-based, Cu-based, and Ni-based. Detection in a 3.5% NaCl solution at room temperature found that the high entropy alloy has a higher corrosion potential and a lower corrosion current [41]. In addition, with the development of high entropy alloys, its application in the field of soft magnetic materials has also been explored. Yao et al. reported the soft magnetic properties of NiFeCoMnBiTm high entropy alloy films, which exhibited soft magnetic properties after annealing. After holding at 873 K for 2 h, the coercive force (Hc) and saturation magnetization (Ms) of the alloy film were about 100 kA/m, and 0.11 $A{\cdot}m^2/kg^{-1}$ [42], respectively.

20.8 HEFs FOR HIGH-THROUGHPUT EXPERIMENTS

In the course of material development, a rising trend of chemical complexity versus time is exhibited. In other words, materials are gradually developed from the initial simple system to multiple components [1]. At present, in the field of new materials, multicomponent materials are the hotspot of new materials research. Common multicomponent materials are amorphous alloys, high-temperature alloys, and high entropy alloys. For multicomponent materials, the design, preparation, and optimization process are more complicated than traditional materials. In addition, there is no linear relationship between the performance and multicomponent material entropy, that is, the mixed entropy value cannot be used as an effective criterion for predicting material properties. This makes the screening of multicomponent materials more complicated and lengthier. At present, few theoretical calculations can accurately predict the structure and properties of materials, especially for multicomponent complex materials. Therefore, the preparation of component gradient materials to achieve parallel preparation of multicomponent materials is necessary and desirable for providing a component gradient platform for high-throughput screening of multicomponent materials.

Compared with traditional materials, the design and preparation of multicomponent materials are more complicated. Amorphous alloys are usually incorporated with three components, while HEAs possess more components. Initially, HEAs were loosely defined as solution alloys containing more than five principal elements in equal or near-equal atomic percentages (at.%) [5, 49]. To date, three and four principal element HEAs has been proposed. It was found that alloys containing three or four components have high mixing entropy which can suppress the precipitation of ordered phases and facilitate the generation of a solid solution, for instance, NbMoTaW and ZrNbHf alloys with BCC structure, FeCoCrNi alloys with BCC structure [2, 3, 50–52]. Obtaining better performance through adjusting the proportion of the components is a common approach during the research. There are approximately 25 kinds of elements used for designing multicomponent materials. Through the algorithm of combinatorial computing: Cn/m, where m≈25, and n is the number of components, we estimated the various systems of multicomponent material, as shown in Figure 20.8. For a certain system, each component varies 10 points (i.e., from 0.1 to 1), and there will be 10n different compositions. It is undeniable that the preparation and verification is a large workload.

The "trial and error" method relying on scientific intuition is still the main method of material development. Researchers preliminarily selected a composition of an alloy based on a theoretical model, and then confirmed it through a series of experimental approaches, such as alloy preparation, structural characterization, and performance testing. The process is time-consuming and inefficient. Few theoretical calculations can predict the structure and performance of materials, especially for such complex materials. Hence, high-throughput technologies are an urgently need for the multicomponent material. Therefore, a combinational synthesis of material library is the key step for achieving high-throughput screening. In 1970, Hanak first proposed the concept of a "multisample experiment" and applied it to superconducting materials in the form of films [53]. With the further development, the form of high-throughput has been extended from the film form to liquid, block, and powder. In this chapter, we concluded the high-throughput preparation of multicomponent materials from the form of film and block, and in the final section, some suggestions are put forward for the existing problems.

In contrast to other high-throughput preparation technologies, the film method is more widely used in multicomponent materials. It can mainly be divided into four categories: co-deposition, order deposition, physical masking, and additive manufacturing. More specific classification is shown in Figure 20.9. There are two steps for combinational sample synthesized by film method: 1) combination, mixing of multiple elements according to an expected system, and obtaining a combination of required components, 2) phase formation, homogenization of thin film material by post-treatment, and formation of an amorphous or crystal phase.

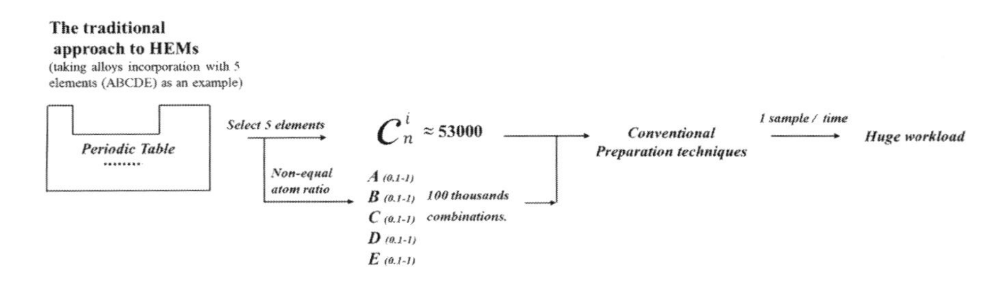

FIGURE 20.8 The conventional method for designing multicomponent alloys.

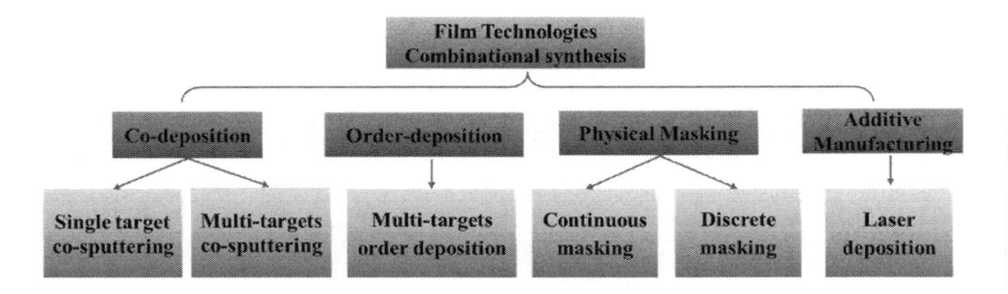

FIGURE 20.9 High-throughput preparation technology of thin film methods.

20.8.1 Co-Deposition Method

Utilizing the different relative angle between the substrate and various components, a composition gradient can be obtained in the horizontal direction. Co-deposition can be divided into multi-target co-sputtering (mc-sputtering) and single-target co-sputtering (sc-sputtering). Ternary co-deposition is being taken as an example, the schematic diagram of which was shown in Figure 20.10. For sc-sputtering, different materials were merged into one target and sputtering was carried out simultaneously. The area occupied by various components can be allocated according to the ratio of the expected system. In contrast to sc-sputtering, different materials were assigned to multiple targets, for the reason that a larger composition gradient was obtained in the space. Hence, the material library synthesized by mc-sputtering can cover more extensive range of phase diagrams. Moreover, the content of various components is more controllable, and it can be achieved by adjusting the tilt angle and working power of each deposition source, etc.

The material library of Mg-Cu-Y amorphous alloys has been synthesized by three targets co-sputtering; approximately 3,000 members with 500 um in diameter were fabricated simultaneously [54]. The composition gradient between the adjacent membranes was 0.53% on average. According to the relationship between the glass-forming ability and thermoplastic formability, the better the thermoplastic formability, the stronger the glass-forming ability; the optimum alloy was selected through the parallel blow forming. Some other alloy systems also achieve parallel preparation by co-deposition, such as Au-Cu-Si[55], Zr-Cu-Al-Ag[56].

In the multi-target co-sputtering, material distribution usually follows the principle: (1) atomic radius: elements with a larger difference in atomic radius should

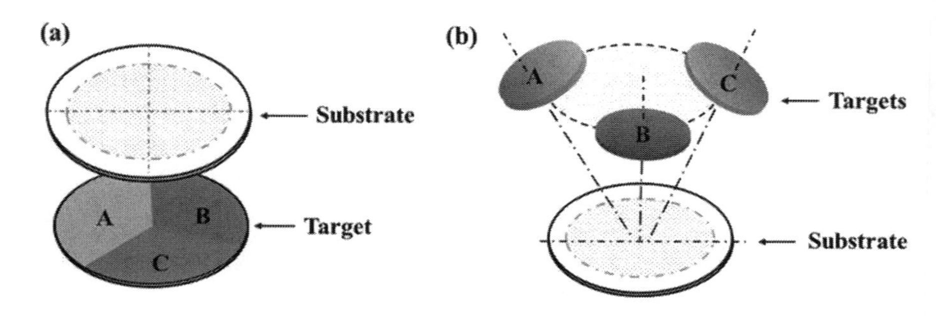

FIGURE 20.10 Schematic of multi-target co-sputtering.

be prepared as a single element target, while on the contrary, the alloy target can be considered; (2) sputtering yield: a large area of material with a low sputtering yield is necessary to reach the required ratio, and the sputtering yield of common materials was shown in Table 20.3 [57]; (3) physical properties: a pure ferromagnetic-element target should be avoided, as target built with more than one magnetic element can avoid the local perturbation of the permanent magnetic field.

20.8.2 Continuous Masking Method

The continuous masking method is a multilayer-film deposition technique that combines film technique with a physical template, using a mask that moves over time to form a multicomponent sample with a continuously graded gradient distribution in the direction of movement. Figure 20.11 is a schematic diagram of preparing a ternary sample by a continuous masking method. A template was installed on the

TABLE 20.3
Sputtering Yield of Common Materials

Factors	Order of common materials	Relationship with ionization rate
Melting point	Al < Cu < Ni < Co < Fe < Ti < Zr < Cr	Positive correlation
Saturated vapor pressure	Al > Cu > Cr > Fe > Ni > Co > Ti < Zr	Negative correlation

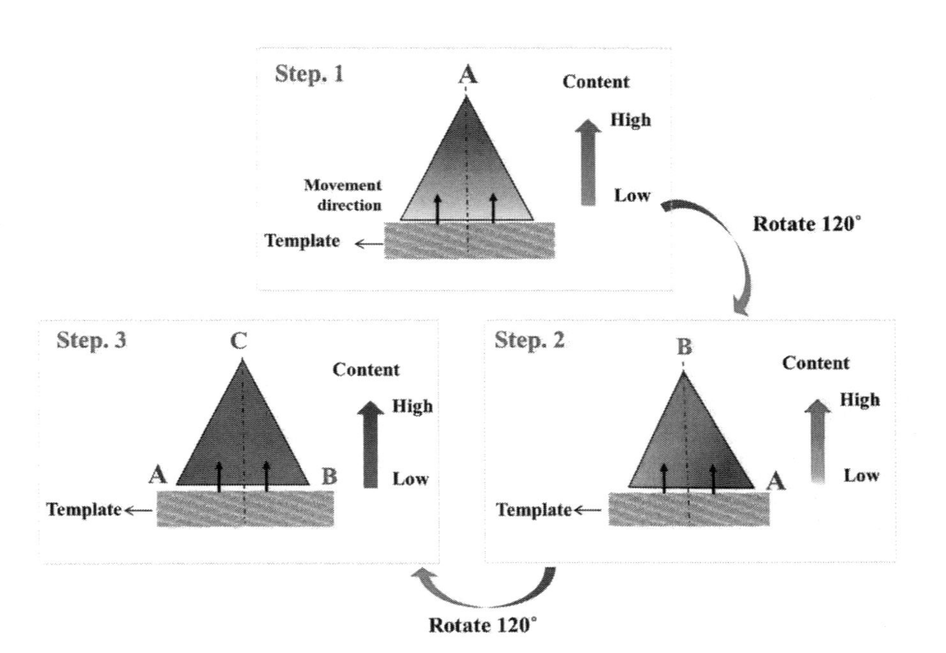

FIGURE 20.11 Schematic of continuous masking method.

substrates, and move over time during the deposition. The bottom of the triangle is masked first, while the top of the triangle is masked finally. Hence, the thickness increases from the bottom to top, and a composition gradient is formed in the direction of movement of the template. The substrate is rotated 120° after the deposition of material A, and material B is deposited in the same approach. The deposition of material C is the same as the deposition of material B. Finally, a multilayer film with a continuous composition gradient is obtained. In this way, Chang et al. investigated the microwave dielectric properties of (Ba-Sr-Ca)TiO_3 system alloys [58]. TiO_2 was first deposited uniformly on the entire substrate. The $SrCO_3$, $BaCO_3$, and $CaCO_3$ was deposited separately through the method described above. Yoo et al., using a continuous masking method, studied the ternary phase diagram of Fe-Co-Ni alloys, and identified the range of ingredients that can form metal glass [59].

In a difference from the co-deposition method, the composition gradient material obtained by the continuous making method is a multilayer gradient film. Afterward, the multilayer films were diffused by post-processing, and a composite material library with a gradient change in the horizontal direction of the film was obtained. The method can obtain a continuous linear gradient component distribution, and the controllability of component distribution is strong. In addition, the continuous masking method can apply to the study of material phase diagram.

20.8.3 Discrete Masking Method

The discrete masking method is a film deposition technology combined with a set of physical templates. In contrast to the continuous masking method, the templates used in discrete masking method are discrete, and need to be masked several times. To date, binary template [60] and quaternary template [61] are the most common forms of multicomponent materials combinatorial synthesis. The number of material library members increases exponentially by 2^n and 4^n for using binary template and quaternary template, respectively. Xiang et al. [61] synthesized 128 combinational members with size of 1 mm × 2 mm using seven discrete templates. Each template corresponds to a specific target: they are $BaCO_3$, $Bi2O_3$, CaO, CuO, PbO, $SrCO_3$, and YO_3, respectively. After successive deposition, a gradient material consisting of multilayers was obtained. Various systems of semiconductor materials were obtained by sintering the layered material library at 840°C. Chang et al. [62] investigated the ferroelectric materials of Ba-Sr-TiO_3 system using the quaternary template. A material library with 256 members of 650 um × 650 um was obtained. The layered material library was annealed at 400°C in flowing oxygen for 24 h to facilitate mixing of the precursors TiO_2, BaF_2, and SrF_2 as well as diffusion of the dopants. Afterward, a further annealing in flowing oxygen at 900°C for 1.5 h was carried out to facilitate phasing. Blue photoluminescent composite material was also synthesized by the discrete masking method, and 1,024 different compositions on substrates 2.5 cm square were obtained at a time [63].

The most significant advantage of the discrete masking method is that the method is not limited by the number of components and can obtain various systems with arbitrary composition distribution at a time. The discrete status between members reduces the proliferation of components during post-processing.

20.9 OUTLOOK

The high entropy alloy films feature many excellent properties, such as high stress, high-temperature resistance, wear resistance, and corrosion resistance. It has shown a huge potential in the fields of solar thermal conversion, surface engineering of workpieces, and diffusion barriers in the field of integrated circuits. However, the research on high entropy alloy thin films is relatively recent, and there is still a long way to go to achieve its industrial application. At present, the research on high entropy alloy thin films mainly focuses on mechanical properties and physical properties. There are relatively few studies on the phase formation of thin films. It is necessary to establish the phase formation law of the high entropy alloy film and realize the phase structure design, which will effectively promote industrial applications.

Recently, multicomponent materials have become one of the most promising materials in engineering and biomedical applications. Compared with traditional alloys, the composition design of multicomponent materials is more complicated, and lots of alloys with different compositions need to be prepared and tested. In addition, the relationship between mixing entropy and performance of multicomponent materials is nonlinear, therefore the structure and performance cannot be effectively predicted by mixing entropy values, which makes it more difficult to design the alloys efficiently. Taking biomedical materials as an example, the obtained low Young's modulus value is generally a relatively low value in a small composition region, rather than the lowest value of a global system. Therefore, the conventional "trial-and-error" method inevitably causes incompleteness and contingency in research results. In this case, high-throughput technology is an effective way to solve this issue.

The thin film method is a common method for obtaining a component gradient material. The material library prepared by the thin film method has high component coverage and large component gradient resolution. For high-throughput techniques, how to increase the composition gradient is the key issue in current technology.

REFERENCES

1. He QF, Ding ZY, Ye YF, Yang Y. Design of high-entropy alloy: A Perspective from nonideal mixing. *Journal of the Minerals* 2017, 69(11):2092–2098.
2. Senkov ON, Wilks GB, Miracle DB, Chuang CP, Liaw PK. Refractory high-entropy alloys. *Intermetallics* 2010, 18(9):1758–1765.
3. Senkov ON, Wilks GB, Scott JM, Miracle DB. Mechanical properties of $Nb_{25}Mo_{25}Ta_{25}W_{25}$ and $V_{20}Nb_{20}Mo_{20}Ta_{20}W_{20}$ refractory high entropy alloys. *Intermetallics* 2011, 19(5):698–706.
4. Liu Shi, Gao Michael C, Liaw Peter K, Zhany Yong. Microstructures and mechanical properties of $AlxCrFeNiTi_{0.25}$ alloys. *Journal of Alloys and Compounds* 2015, 619:610–615.
5. Zhang Yong, Zuo Ting-ting, Tang Zhi, Gao Michael C, Dahmen Karin A, Liaw Peter K, Lu Zhao-ping. Microstructures and properties of high-entropy alloys. *Progress in Materials Science* 2014, 61:1–93.
6. Zhang Wei-ran, Liaw Peter K, Zhang Yong. Science and technology in high-entropy alloys. *Science China Materials* 2018, 61(1):2–22.
7. Hoerling A, Sjolen J, Willmann H, Larsson T, Oden M, Hultman L. Thermal stability, microstructure and mechanical properties of $Ti_{1-x}Zr_xN$ thin films. *Thin Solid Films* 2008, 516(18):6421–6431.

8. Carvalho S, Rebouta L, Ribeiro E, Vaz F, Denannot MF, Pacaud J, Riviere JP, Paumier F, Gaboriaud RJ, Alves E. Microstructure of (Ti,Si,Al)N nanocomposite coatings. *Surface and Coatings Technology* 2004, 177–178:369–375.
9. Zhou M, Makino Y, Nose M, Nogi K. Phase transition and properties of Ti–Al–N thin films prepared by r.f.-plasma assisted magnetron sputtering. *Thin Solid Films* 1999, 339(1–2):203–208.
10. Chen TK, Shun TT, Yeh JW, Wong MS. Nanostructured nitride films of multi-element high-entropy alloys by reactive DC sputtering. *Surface & Coatings Technology* 2004, s188–189(1):193–200.
11. Huang Yuan-sheng, Chen Ling, Lui Hong-wei, Cai Ming-hong, Yeh Jien-wei. Microstructure, hardness, resistivity and thermal stability of sputtered oxide films of $AlCoCrCu_{0.5}NiFe$ high-entropy alloy. *Materials Science & Engineering A* 2007, 457(1–2):77–83.
12. Tsai Ming-hung, Yeh Jien-wei, Gan John-yiew. Diffusion barrier properties of AlMoNbSiTaTiVZr high-entropy alloy layer between copper and silicon. *Thin Solid Films* 2008, 516(16):5527–5530.
13. Lin Dan-yang, Zhang Nan-nan, He Bin, Zhang Guang-wei, Zhang Yue, Li De-yuan. Tribological properties of $FeCoCrNiAlB_x$ high-entropy alloys coating prepared by laser cladding. *Journal of Iron and Steel Research, International* 2017, 24:184–189.
14. Ren B, Liu Z-x, Li D-m, Shi L, Cai B, Wang M-X. Corrosion behavior of CuCrFeNiMn high entropy alloy system in 1 M sulfuric acid solution. *Materials & Corrosion* 2011, 63(9):828834.
15. Yu Ruei-sung, Huang Chueh-jung, Huang Rong-hsin, Sun Chung-hsing, Shieu Fuh-sheng. Structure and optoelectronic properties of multi-element oxide thin film. *Applied Surface Science* 2011, 257(14):6073–6078.
16. Shen Wan-jui, Tsai Ming-hung, Chang Yee-shyi, Yeh Jien-wei. Effects of substrate bias on the structure and mechanical properties of $(Al_{1.5}CrNb_{0.5}Si_{0.5}Ti)N_x$ coatings. *Thin Solid Films* 2012, 520(19):6183–6188.
17. Huang Ping-kang, Yeh Jien-wei. Effects of substrate bias on structure and mechanical properties of (AlCrNbSiTiV)N coatings. *Journal of Physics D: Applied Physics* 2009, 42(11):115401–115407.
18. Chang Zue-chin, Liang Shi-chang, Han Sheng, Chen Yi-kun, Shieu Fuh-sheng. Characteristics of TiVCrAlZr multi-element nitride films prepared by reactive sputtering. *Nuclear Instruments & Methods in Physics Research* 2010, 268(16):2504–2509.
19. Zhang Hui, He Yi-zhu, Pan Ye, He Yin-sheng, Shin Keesam. Synthesis and characterization of $NiCoFeCrAl_3$ high entropy alloy coating by laser cladding. *Advanced Materials Research* 2010, 97–101:1408–1411.
20. Zhang Hui, Pan Ye, He Yi-zhu. The preparation of $FeCoNiCrAl_2Si$ high-entropy alloy coating by Laser cladding.
21. Huang Can, Zhang Yong-zhong, Shen Jian-yun, Vilar R. Thermal stability and oxidation resistance of laser clad TiVCrAlSi high entropy alloy coatings on Ti–6Al–4V alloy. *Surface and Coatings Technology* 2011, 206(6):1389–1395.
22. Yao Chen-zhong, Wei Bo-hui, Zhang Peng, Lu Xi-hong, Liu Peng, Tong Ye-xiang. Facile preparation and magnetic study of amorphous Tm-Fe-Co-Ni-Mn multicomponent alloy nanofilm. *Journal of Rare Earths* 2011, 29(2):133–137.
23. Li Q, Yue T, Guo Z. Electro-spark deposition of multi-element high entropy alloy coating. *ASM International*, Member/Customer Service Center Materials Park OH 44073-0002 United States 2010.
24. Zhu Sheng, Du Wen-bo, Wang Xiao-ming, Yao Ju-kun. Surface protection technology of magnesium alloy based on high-entropy alloy. *Journal of Armored Force Engineering College* 2013, 27(6):78–84.

25. Yue Tai-M, Xie Hui, Lin Xin, Yang Hai-ou, Meng Guang-hui. Microstructure of laser remelted AlCoCrCuFeNi high entropy alloy coatings produced by plasma spraying. *Entropy* 2013, 15(7):2833–2845.
26. Yan Xue-hui, Li Jin-shan, Zhang Wei-ran, Zhang Yong. A brief review of high-entropy films. *Materials Chemistry and Physics* 2018, 210(1):12–19.
27. Feng Xing-guo, Tang Guang-ze, Sun Ming-ren, Ma Xin-xin, Wang Li-qing, Yukimura Ken. Structure and properties of multi-targets magnetron sputtered ZrNbTaTiW multi-elements alloy thin films. *Surface & Coatings Technology* 2013, 228(9):S424–S427.
28. Liang Shi-chang, Tsai Du-cheng, Chang Zue-chin, Sung Huan-shin, Lin Yi-chen, Yeh Yi-jung, Deng Min-jen, Shieu Fuh-sheng. Structural and mechanical properties of multi-element (TiVCrZrHf)N coatings by reactive magnetron sputtering. *Applied Surface Science* 2011, 258(1):399–403.
29. Lai Chia-han, Lin Su-jie, Yeh Jien-wei, Chang Shou-yi. Preparation and characterization of AlCrTaTiZr multi-element nitride coatings. *Surface & Coatings Technology* 2006, 201(6):3275–3280.
30. Chang Hui-wen, Huang Ping-kang, Davison Andrew, Yeh Jien-wei, Tsau Chun-huei, Yang Chih-chao. Nitride films deposited from an equimolar Al–Cr–Mo–Si–Ti alloy target by reactive direct current magnetron sputtering. *Thin Solid Films* 2008, 516(18):6402–6408.
31. Lin CH, Duh JG, Yeh JW. Multi-component nitride coatings derived from Ti–Al–Cr–Si–V target in RF magnetron sputter. *Surface & Coatings Technology* 2007, 201(14):6304–6308.
32. Cheng Keng-hao, Lai Chia-Han, Lin Su-jien, Yeh Jien-wei. Structural and mechanical properties of multi-element (AlCrMoTaTiZr)Nx coatings by reactive magnetron sputtering. *Thin Solid Films* 2011, 519(10):3185–3190.
33. Zhang Yong, Lin Jun-pin, Chen Guo-liang, Yang Ji-xian, Ren Yong-gang, Zhang Shizhong. Uranium carbide powder metallurgy. *Powder Metallurgy Technology* 1995, 4:303–306.
34. Vetter J, Scholl HJ, Knotek O. (TiCr)N coatings deposited by cathodic vacuum arc evaporation. *Surface & Coatings Technology* 1995, s 74–75:286–291.
35. Huang Ping-kang, Yeh Jien-wei. Effects of substrate temperature and post-annealing on microstructure and properties of (AlCrNbSiTiV)N coatings. *Thin Solid Films* 2009, 518(1):180184.
36. Liang Shih-chang, Chang Zue-chin, Tsai Du-cheng, Lin Yi-chen, Sung Huan-shin, Deng Min-jen, Shieu Fuh-sheng. Effects of substrate temperature on the structure and mechanical properties of (TiVCrZrHf)N coatings. *Surface & Coatings Technology* 2011, 257(17):7709–7713.
37. Yan Xue-hui, Zhang Yong. *High-entropy Films and Compositional Gradient Materials.*
38. Pelleg J. Mechanical properties of materials. *Solid Mechanics & Its Applications* 2013, 113–177.
39. Zhang Hui, He Yi-zhu, Pan Ye, Guo Sheng. Thermally stable laser cladded CoCrCuFeNi high-entropy alloy coating with low stacking fault energy. *Journal of Alloys & Compounds* 2014, 600(600):210–214.
40. Sheng Wen-jie, Yang Xiao, Zhu Jie, Wang Cong, Zhang Yong. Amorphous phase stability of NbTiAlSiNX high-entropy films. *Rare Metals* 2017, 5:1–18.
41. Shi Yun-zhu, Yang Bin, Xie Xie, Brechtl Jamieson, Dahmend Karin A., Liaw Peter K. Corrosion of Al_xCoCrFeNi high entropy alloys: Al-content and potential scan-rate dependent pitting behavior. *Corrosion Science* 2017, 119:33–45.
42. Yao Chen-zhong, Zhang Peng, Tong Ye-xiang, Xia Dao-cheng, Ma Hui-xuan. Electrochemical synthesis and magnetic studies of Ni-Fe-Co-Mn-Bi-Tm high entropy alloy film. *Chemical Research in Chinese Universities* 2010, 26(4):640–644.

43. Dolique V, Thomann AL, Brault P, Tessier Y, Gillon P. Thermal stability of AlCoCrCuFeNi high entropy alloy thin films studied by in-situ XRD analysis. *Surface and Coatings Technology* 2010, 204(12–13):1989–1992.
44. Chang Shou-yi, Chen Ming-ku. High thermal stability of AlCrTaTiZr nitride film as diffusion barrier for copper metallization. *Thin Solid Films* 2009, 517(17):4961–4965.
45. Sheng Wen-jie, Yang Xiao, Wang Cong, Zhang Yong. Nano-Crystallization of high entropy amorphous NbTiAlSiWxNy films prepared by magnetron sputtering. *Entropy* 2016, 18(6):226.
46. Feng Xing-guo, Tang Guang-ze, Gu Le, Ma Xin-xin, Sun Ming-ren, Wang Lin-qing. Preparation and characterization of TaNbTiW multi-element alloy films. *Applied Surface Science* 2012, 261(1):447–453.
47. Zhang Hui, Pan Ye, He Yi-zhu. Effects of annealing on the microstructure and Properties of 6FeNiCoCrAlTiSi high-entropy alloy coating prepared by laser cladding. *Journal of Thermal Spray Technology* 2011, 20(5):1049–1055.
48. Wong Zi-qing, Dong Gang, Zhang Qun-li, Guo Shi-rui, Yao Jian-hua. Effect of annealing on microstructure and properties of FeCrNiCoMn high-entropy alloy coating by laser cladding. *Laser Journal* 2014, 3:59–64.
49. Yeh JW, Chen SK, Lin SJ, Gan JY, Chin TS, Shun TT, Tsau CH, Chang SY. Nanostructured high-entropy alloys with multiple principal elements: novel alloy design concepts and outcomes. *Advanced Engineering Materials* 2004, 6(5):299–303.
50. Senkov ON, Senkova SV, Woodward C, Miracle DB. Low-density, refractory multi-principal element alloys of the Cr–Nb–Ti–V–Zr system: Microstructure and phase analysis. *Acta Materialia* 2013, 61(5):1545–1557.
51. Lucas MS, Wilks GB, Mauger L, Munoz JA, Senkov ON, Michel E, Horwath J, Semiatin SL, Stone MB, Abernathy DL, Karapetrova E. Absence of long-range chemical ordering in equimolar FeCoCrNi. *Applied Physics Letters* 2012, 100(25):299.
52. Guo W, Dmowski W, Noh JY, Rack P, Liaw PK, Egami T. Local Atomic Structure of a High Entropy Alloy: An X-Ray and Neutron Scattering Study. *Metallurgical & Materials Transactions Part A* 2013, 44(5):1994–1997.
53. Hanak JJ. The "multiple-sample concept" in materials research: Synthesis, compositional analysis and testing of entire multicomponent systems. *Journal of Materials Science* 1970, 5(11):964–971.
54. Ding S, Liu Y, Li Y, Liu Z, Sohn S, Walker FJ, Schroers J. Combinatorial development of bulk metallic glasses. *Nature Materials* 2014, 13(5):494–500.
55. Li Y, Jensen KE, Liu Y, Liu J, Gong P, Scanley E, Broadbridge CC, Schroers J. Combinatorial strategies for synthesis and characterization of alloy microstructures over large compositional ranges. *ACS Combinatorial Science* 2016, 18(10):630.
56. Liu Y, Padmanabhan J, Cheung B, Liu J, Chen Z, Scanley BE, Wesolowski D, Pressley M, Broadbridge CC, Altman S. Combinatorial development of antibacterial Zr-Cu-Al-Ag thin film metallic glasses. *Scientific Reports* 2016, 6:26950.
57. Kimblin CW, Lowke JJ. Decay and thermal reignition of low-current cylindrical arcs. *Journal of Applied Physics* 1973, 44(10):4545–4547.
58. Chang H, Takeuchi I, Xiang XD. A low-loss composition region identified from a thin-film composition spread of (Ba1–x–ySrxCay)TiO3. *Applied Physics Letters* 1999, 74(8):1165–1167.
59. Yoo YK, Xue Q, Chu YS, Xu S, Hangen U, Lee H-C, Stein W, Xiang X-D. Identification of amorphous phases in the Fe–Ni–Co ternary alloy system using continuous phase diagram material chips. *Intermetallics* 2006, 14(3):241–247.
60. Xiang XD, Sun X, Briceño G, Lou Y, Wang KA, Chang H, Wallace-Freedman WG, Chen SW, Schultz PG. A combinatorial approach to materials discovery. *Science* 1995, 268(5218):1738.

61. Xiang XD, Schultz PG. The combinatorial synthesis and evaluation of functional materials. *Physica C Superconductivity* 1997, s 282–287(Part 1):428–430.
62. Chang H, Gao C, Takeuchi I, Yoo Y, Wang J, Schultz PG, Xiang XD, Sharma RP, Downes M, Venkatesan T. Combinatorial synthesis and high throughput evaluation of ferroelectric/dielectric thin-film libraries for microwave applications. *Applied Physics Letters* 1998, 72(17):2185–2187.
63. Wang J, Yoo Y, Gao C, Takeuchi II, Sun X, Chang H, Xiang X, Schultz PG. Identification of a blue photoluminescent composite material from a combinatorial library. *Science* 1998, 279(5357):1712–1714.

Index